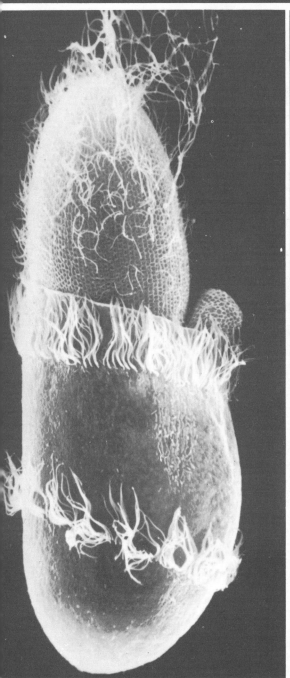

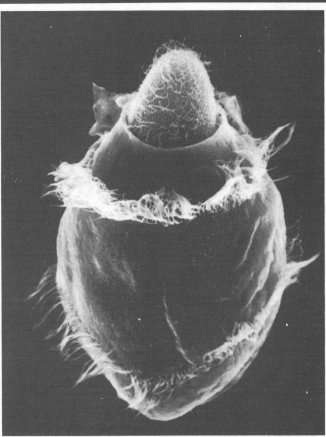

Randy Barton

Randy Barton

Biology

SEVENTH EDITION

CLAUDE A. VILLEE

Andelot Professor of Biological Chemistry
Harvard University Medical School

W. B. SAUNDERS COMPANY

Philadelphia • London • Toronto

W. B. Saunders Company: West Washington Square
                          Philadelphia, Pa. 19105

                          1 St. Anne's Road
                          Eastbourne, East Sussex BN21 3UN, England

                          1 Goldthorne Avenue
                          Toronto, Ontario M8Z 5T9, Canada

Library of Congress Cataloging in Publication Data

Villee, Claude Alvin,

Biology

First ed. published in 1950 under title: Biology, the human approach.

Bibliography: p.

Includes index.

1. Biology.

QH308.2.V56 1977   570   76–14694

ISBN 0–7216–9023–8

Listed here is the latest translated edition of this book together with the language of the translation and the publisher.

Spanish (*6th Edition*) — Nueva Editoria Interamericana,
                          Mexico 4 D.F., Mexico

Chinese (*6th Edition*) — National Book Company,
                          Taipei, Taiwan

Polish (*5th Edition*) — Panstwowe Wydawnictwo
                          Rolnicze i Lesne,
                          Warsaw, Poland

Russian (*4th Edition*) — Mir Publishers,
                          2 Pervy Rizhsky pereulok,
                          Moscow I–110, GSP, USSR

Front cover illustration: Fern spore. (Courtesy of Phillip A. Harrington.)

Biology   ISBN   0-7216-9023-8

Last digit is the print number: 9   8   7   6   5   4

PREFACE

The principles of biology can be learned using as a model the frog, dogfish, daisy, fern or even the colon bacillus. However, most students have a special interest in human biology—in the structure, function and development of the human body—generated perhaps by their plans for a career in medicine, dentistry, or another of the health sciences, or simply by a narcissistic interest in how their own body is put together and how it works. The earlier editions of this text have emphasized the human aspects of biology and the biology of human beings; this edition extends and strengthens that emphasis. Like the earlier ones, this edition is neither encyclopedic nor cursory but attempts to present the concepts of biology and their relevance to human beings in an interesting and understandable fashion.

The changes in the seventh edition reflect the continuing rapid development of the biological sciences. Our understanding of the events of life at the molecular, cellular, organismal and population levels of biological organization has been increasing quickly. Remarkable advances have been made in recent years in our understanding of many aspects of biological sciences. However, to be understood and fully appreciated, these newer discoveries must be viewed against the background of the more classical aspects of biology. These advances not only have reemphasized the basic unity of life but also have demonstrated the fundamental similarity of life processes that occur in all organisms.

The discussions of cell structure and organelles, cell cycle, intermediary metabolism, molecular function, protein synthesis, bioenergetics, endocrinology, immunology, and immunogenetics have been completely revised. The chapters dealing specifically with plant biology have been completely rewritten. I am especially indebted to Dr. Knut Norstog, who read the manuscript of these chapters, made many helpful suggestions for their improvement and permitted me to borrow several illustrations from his and the late Dr. Robert Long's *Plant Biology*. The chapter on human evolution was reviewed by Dr. Michael Charney and rewritten in the light of his helpful suggestions. Many illustrations have been replaced and new ones added. The new line drawings for this edition were made by John Hackmaster, Celeste Brennan, Linda Downham, and Susan O'Neill. The appendices contain a discussion of some of the basic physical and chemical principles on which modern biology is founded and a taxonomic summary of the many kinds of living organisms arranged according to the Five Kingdom System proposed by R. H. Whittaker.

A course in general biology attempts to provide the student with an understanding and appreciation of the vast diversity of living things, their special adaptations to their environment, and their evolutionary and ecologic relationships. It should also emphasize the basic unity of life and the fundamental similarities of the problems that have been faced and solved by all living things. However, there has never been general agreement among biologists as to the sequence in which the several topics in a general biology course should be taught. This is understandable, for reasonable arguments can be advanced for each of the many possible combinations and permutations. The various aspects of biology are intimately related and each could be grasped much more readily if all the other aspects had been learned previously. Since this cannot be done (except perhaps by a student repeating the course!), each instructor must choose the sequence that seems optimal to him. The various parts and chapters in this text can be taken up in any of a number of sequences.

An appreciation of science requires not only a grasp of the product of science but also an insight into the processes by which scientific knowledge is acquired. An introduction to the methods of science is given in Chapter 1, and throughout the text examples of experimental work are presented to illustrate modern methods in biology. The second introductory chapter presents the major generalizations of modern biology. Part 1 discusses the molecular and cellular basis of biology, the architecture of cells and tissues, and the properties and constituents of enzyme systems, ending with a description of the grand metabolic design by which living systems obtain and utilize free energy by photosynthesis and oxidative phosphorylation.

Part 2, concerned with single-celled organisms and plants, opens with a discussion of biological interrelationships, the cyclic use of matter and the interactions among species of plants and animals. It continues with a consideration of prokaryotes, viruses, protists and fungi. The special characteristics of each division of algae and fungi are described, together with their life cycles and economic importance to humans. Chapter 9 presents the primitive land plants and discusses the evolution of mosses and ferns. The gymnosperms and angiosperms are described in Chapter 10. These, the largest in size and in number of types, are of special importance to human life as sources of food, flowers, scents, seasonings and lumber. A discussion of the general physiologic and morphologic attributes of green plants, including plant hormones, follows, and Part 2 ends with a description of plant development and the functions of roots, soil, stems and leaves. A similar survey of the invertebrate and vertebrate animals living today and their structural and functional adaptations is presented in Part 3. The organ systems of the frog are described in some detail, since this animal is frequently used to demonstrate vertebrate characteristics.

In Part 4 the structure and functions of organ systems of the human body are presented and compared with those of other vertebrates. Part 5 is concerned with the biological basis of behavior, and in these chapters the essential features of the human nervous system are described and compared with the nervous systems of other organisms. The section continues with descriptions of receptors and effectors and ends with a chapter on integration by endocrines and their role in determining behavior.

Part 6 deals with reproductive and developmental biology. The subject is presented primarily from the viewpoint of human reproduction and development, but these are compared with similar processes in other vertebrates. Part 7 deals with heredity and evolution, beginning with the physical basis of inheritance and the principles of classic transmission genetics. A greatly revised and rewritten discussion of the genetic code and the transfer of biological information in DNA molecules from generation to generation is given in Chapter 29. The principles of population genetics and differential reproduction and the basis for our present concepts of the mechanisms of evolution are detailed in Chapter 30. This is followed by a description of the principles of evolution, the evidence for evolutionary changes, and some theories regarding the mechanisms of evolution and the role of differential reproduction as a driving force in evolution. It ends with a discussion of the possible mechanisms of the origin of life itself on this planet.

The evolutionary and ecologic relationships of living organisms are the two major unifying threads running through the text. Part 8 presents a discussion of the principles of ecology; those aspects concerned primarily with the ecologic relationships of individuals are considered in Chapter 34, and those concerned with populations and communities and their characteristics are described in Chapter 35, which also includes a discussion of adaptations, ecosystems and the various types of biomes found on land, in the sea, and in fresh water lakes. The final chapter on human ecology considers some of our present predicaments involving overpopulation, pollution, and the depletion of our natural resources.

A glossary giving the derivations and definitions of some important biological terms is also included. The student will find further definitions of biological terms in an unabridged dictionary. Special biological dictionaries have been published; an excellent and inexpensive one is *A Dictionary of Biology* by Abercrombie, Hickman and Johnson, published by Penguin Books.

The questions at the end of each chapter are designed to assist the student in reviewing the material presented and in testing his comprehension of the principles and facts discussed. Suggestions for further reading are given at the end of each chapter. Complete citations for these references will be found in the bibliography in the appendix.

I am greatly indebted to the many instructors who have made suggestions for revisions based upon their experience in using previous editions of the text. My special thanks are due to Edwin M. Banks, Arthur C. Borror, John Bauman, William H. Behle, Jeffrey M. Camhi, Herbert M. Clarke, Carl Hammen, Donald O. Hebb, Joseph Hichar, Bartholemew Kunny, Ralph W. Lewis, Irving McNulty, Fred R. Rickson, James Slater, Elizabeth Souter, Shirley Sparling, and Richard T. Wilfong.

I want to express my thanks to the artists, Julia Child, Margaret Croup, Frank Fancher, John Hackmaster, Betsy Holbert, Paul Larkin, Grant Lashbrook, Murial McLatchy and William Osburn, who transformed my rough sketches into finished line drawings. I am indebted to many scientists, museums and publishers who have furnished me with original photographs to use as illustrations; each of these is acknowledged individually in the legend to the photograph. I am especially grateful to Mr. Richard Lampert and the editorial staff of the W. B. Saunders Company for their many helpful suggestions for revisions. My thanks are also due to Kathleen Callinan, Janet Loring and Suzanne Villee for their assistance in reading proof and preparing the index. Finally, I want to express my deep appreciation to my wife, Dorothy, who, despite the demands of a busy career in medicine, found time to help me in many ways in the preparation of this volume.

CLAUDE A. VILLEE

CONTENTS

Chapter 11   GENERAL PROPERTIES OF GREEN PLANTS...................................... 232

Chapter 12   PROCURING AND DISTRIBUTING NUTRIENTS IN SEED PLANTS.......... 251

Part 3   THE WORLD OF LIFE: ANIMALS

Chapter 13   THE ANIMAL KINGDOM: LOWER INVERTEBRATES............................ 278

Chapter 25

CONTROL SYSTEMS: HORMONAL INTEGRATION 546

Part 6

REPRODUCTIVE PROCESSES

Chapter 26

REPRODUCTION 586

Chapter 27

EMBRYONIC DEVELOPMENT 613

Part 7

HEREDITY AND EVOLUTION

Chapter 28

TRANSMISSION GENETICS: THE CHROMOSOME THEORY
OF HEREDITY 653

Chapter 33

THE PRIMATES AND HUMAN EVOLUTION 783

Part 8

ECOLOGY

Chapter 34

PRINCIPLES OF ECOLOGY 811

Chapter 35

SYNECOLOGY: COMMUNITIES, BIOMES AND LIFE ZONES 826

CHAPTER 1

INTRODUCTION: BIOLOGY AND THE SCIENTIFIC METHOD

In one sense, biology is an ancient science, for men began many centuries ago to catalogue living things and to study their structure and function. A considerable body of knowledge and theories about living things existed in the time of Aristotle (384–322 B.C.). Even in the older civilizations of Egypt, Mesopotamia and China much was known about the practical uses of plants and animals. The cave men who lived 20,000 and more years ago drew on the walls of their caves accurate and artistic pictures of the deer, cattle and mammoths that lived around them. Their survival depended on a knowledge of such fundamental biologic facts as which animals were dangerous and which plants could be safely eaten.

Yet in another sense biology is a young science. The major generalizations which are the foundations of any science have been formulated only comparatively recently in biology; many of them are still being revised. The development of the electron microscope and the scanning electron microscope has revealed a whole new order of complexity in living matter. In recent years the application of new physical and chemical methods to problems in biology has revealed orders of complexity in the functions of living things that parallel the structural complexity seen in the electron microscope.

1–1 EARLY HISTORY OF BIOLOGY

Biology as an organized body of knowledge can be said to have begun with the Greeks. They and the Romans described the many kinds of plants and animals known at the time. Galen (131–200 A.D.) was the first experimental physiologist and performed many experiments to analyze the functions of nerves and blood vessels. His descriptions of human anatomy were the unchallenged authority for 1300 years. They were based, however, on dissections of apes and pigs and contained some remarkable errors. Men such as Pliny (23–79 A.D.) prepared encyclopedias which were strange mixtures of facts and fiction about living things. In the succeeding centuries of the Middle Ages men wrote "herbals" and "bestiaries," cataloguing and describing plants and animals respectively. With the Renaissance interest in natural history revived and more accurate studies of the structure, functions and life habits of countless plants and animals were made. Vesalius (1514–1564), Harvey (1578–1657) and John Hunter (1728–1793) studied the structure and functions of animals in general and man in particular and laid the foundations of anatomy and physiology. Vesalius carefully dissected human bodies and made clear drawings of his observations, revealing many of the inaccuracies in Galen's descriptions. He emphasized the importance of relying on careful, first-hand observation, rather than on Galenic authority, and laid, in this way, the groundwork for the modern approach to anatomy. With the invention of the microscope early in the seventeenth century, Malpighi (1628–1694), Swammerdam (1637–1680) and Leeuwenhoek (1632–1723) investigated the fine structure of a variety of plant and animal tissues. Leeuwenhoek was the first to describe bacteria, protozoa and sperm.

Biology expanded and altered greatly in the nineteenth century and has continued this trend at an accelerated pace in the twentieth. This is due in part to the broader scope and more detailed knowledge available today and in part to the new approaches made possible by the discoveries and techniques of physics and chemistry. These technical advances have led to quantitative studies of the molecular structures and events underlying biological processes, a facet of the science which has been termed *molecular biology.* This includes (1) analyses of gene structure and function and of the mechanisms by which genes are replicated and transcribed and control the synthesis of enzymes and other proteins; (2) studies of subcellular structures and their roles in adaptive and regulatory processes within the cell; (3) investigations of the mechanisms underlying cellular differentiation; and (4) analyses of the molecular basis of evolution by comparative studies of the molecular structure of specific proteins—hemoglobins, enzymes and hormones—in different species. These studies were made possible by the development of methods to determine the sequence of amino acids in a protein molecule and the sequence of nucleotides in RNA and DNA molecules. Many of these studies, for technical reasons, have used the simpler microorganisms, bacteria and viruses, but the principles discovered appear to apply to all living things. These investigations, indeed, by elucidating the nature of the chemical and energy transformations that characterize biological phenomena have reemphasized the fundamental unity of life.

1–2 THE BIOLOGICAL SCIENCES

The usual definition of biology as the "science of life" is meaningful only if we have some idea of what "life" and "science" mean. Life does not lend itself to a simple definition; the characteristics of living things—growth, movement, metabolism, reproduction and adaptation—will be discussed in Chapter 3. Biology is concerned with the myriad forms that living things may have, with their structure, function, evolution, development and relations to their environment. It has grown to be much too broad a science to be investigated by one man or to be treated thoroughly in a single textbook, and most biologists are specialists in some one of the biological sciences.

The *botanist* and *zoologist* respectively study types of plants and animals and their relationships. There are specialists who deal with one kind of living thing—ichthyologists, who study fish, mycologists, who study fungi, ornithologists, who study birds, and so on. The sciences of *anatomy*, *physiology* and *embryology* deal with the structure, function and development of an organism; these can be further subdivided according to the kind of organism investigated: e.g., animal physiology, mammalian physiology, human physiology. The *parasitologist* studies those forms of life that live in or on and at the expense of other forms, the *cytologist* investigates the structure, composition and function of cells, and the *histologist* inquires into the properties of tissues.

The science of *genetics* is concerned with the mode of transmission of the characteristics of one generation to another, and is closely related to the study of *evolution*, which attempts to discover how new species arise, as well as how the present forms evolved from previous ones. The study of the classification of plants and animals and their evolutionary relations is known as *taxonomy*. *Ecology* is the study of the relations of a group of organisms to its environment, including both the physical factors and other living organisms which provide food or shelter for it, or compete with or prey upon it. It deals with such questions as why natural communities are composed of certain organisms and not others, how the various organisms interact with each other and with the physical environment, and how we can control and maintain these natural communities.

1–3 SOURCES OF SCIENTIFIC INFORMATION

Where, you may ask, do all the facts about biology described in this book come from? And how do we know they are true? The ultimate source of each fact, of course, is in some carefully controlled observation or experiment made by a biologist. In earlier times, some scientists kept their discoveries to themselves, but now there is a strong tradition that scientific discoveries are public property and should be freely published. It is not enough for a scientist to report in a scientific publication the discovery of a certain fact; all the relevant details by which the fact was discovered must be given so that others can repeat the observation. It is this criterion of *repeatability* that makes us accept a certain observation or experiment as representing a true fact; observations that cannot be repeated by competent investigators are discarded.

When a biologist has made a discovery, he writes a report, called a "paper," in which he describes his methods in sufficient detail that another can repeat them, gives the results of his observations, discusses the conclusions to be drawn from them, perhaps formulates a theory to explain them, and indicates the place of these new facts in the present body of scientific knowledge. The knowledge that his discovery will be subjected to the keen scrutiny of his colleagues is a strong stimulus for carefully repeating the observations or experiments an adequate number of times before publishing them. He then submits his paper for publication in one of the professional journals in the particular field of his discovery (it is estimated that there are more than 9700 of them published over the world in the various fields of biology!) and it is read by one or more of the editors of the journal, all of whom are experts in the field. If it is approved, it is published and thus becomes part of "the literature" of the subject.

At one time, when there were fewer journals, it might have been possible for a scientist to read them each month as they appeared, but this is obviously impossible now. Journals such as *Biological Abstracts* assist the hard-pressed biologist by publishing, classified by fields, very short reports or *abstracts* of each paper published—giving the facts found, and a reference to the journal. A step beyond the publishing of abstracts is the journal *Current Contents*, which simply lists the titles and authors of the research papers appearing in each of several hundred journals, together with the name, volume and pages of the journal where the paper may be found.

Many journals are devoted solely to reviewing the newer developments in one particular field; some of these are *Physiological Reviews, The Botanical Review, Quarterly Review of Biology, Annual Review of Microbiology* and *Nutrition Reviews*. The new fact or theory thus becomes widely known through publication in a professional journal and by reference in abstract and review journals, and eventually may become a sentence or two in a textbook.

Other means for the dissemination of new knowledge are the annual meetings held by the professional societies of botanists, geneticists, physiologists and other specialists at which papers are read and discussed. Prominent among these are the meetings of the American Institute of Biological Sciences (AIBS) and the Federation of American Societies for Experimental Biology (FASEB). There are, from time to time, national and international gatherings, called *symposia* (Gr. *syn*, together, + *posis*, drinking), of specialists in a given field to discuss the newer findings and the present status of the knowledge in that field. The discussions of these symposia are usually published as books.

1–4 THE SCIENTIFIC METHOD

The goal of every science is to provide explanations for observed phenomena and to establish generalizations that can predict the relations between these and other phenomena. These explanations and generalizations are achieved by a kind of organized common sense called the *scientific method*, but it is difficult to reduce this method to a simple set of rules that apply to all the branches of science. One of the basic tenets of the scientific method is the *rejection of authority*—the refusal to accept a statement just because someone says it is so. The scientist is always skeptical and requires confirmation of the observation by an independent person. The essence of the scientific method is the posing of questions and the search for answers, but they must be "scientific" questions, arising from observations and experiments, and "scientific" answers, ones that are testable by further observation and experiment.

The bases of the scientific method and the ultimate sources of all the facts of science are careful, close observations and experiments, free of bias, with suitable controls, made as quantitatively as possible. The observations and experiments may then be analyzed, or simplified into their constituent parts, so that some sort of order can be brought into the observed phenomena. Then the parts can be synthesized or reassembled and their interactions discovered. On the basis of these observations, the scientist makes a generalization, or constructs a *hypothesis,* a trial idea about the nature of the observation, or possibly the connections between a chain of events, or even cause and effect relationships between different events. Predictions made on the basis of the hypothesis can then be tested by further controlled experiments. It is in the construction of hypotheses that scientists differ most and that true genius shows itself. The ability to see through a mass of data and suggest a reason for their interrelations is all too rare. It must be emphasized that science does not advance by the mere accumulation of facts, or by the mere postulation of hypotheses. The two go hand-in-hand in scientific investigations: observation, hypothesis, observation, revised hypothesis, further observation, and so on.

Any scientist embarking upon an investigation has the advantage of the relevant facts already known with which to build a "working hypothesis" to guide the design of his experiments. When a scientist makes an observation that does not agree with his hypothesis he may conclude that either his hypothesis or his observation is wrong. He then repeats his observation, perhaps altering the design of his experiment to get at the relationship in a new way, or perhaps using a different technique. If he can satisfy himself that his observation is valid, he either discards his hypothesis or amends it to account for the new observation. Ideally, each new observation should either agree or disagree clearly with the hypothesis but it is often difficult to design experiments that will give clearcut yes or no answers.

Hypotheses are constantly being refined and elaborated. There are few scientists who

consider any hypothesis, no matter how many times it may have been tested, to be a statement of absolute and universal truth. The hypothesis is simply regarded as the best available approximation to the truth for some finite range of circumstances. The Law of the Conservation of Energy (p. 18), for example, was widely accepted until the work of Einstein showed that it had to be modified to allow for the possible interconversion of matter and energy. Although at one time this might have seemed to be an inconsequential distinction, for it has no importance at all in ordinary chemical processes, it is the theoretical basis of atomic power.

Once a hypothesis has been set up to explain a certain body of facts, the rules of formal logic can be used to deduce certain consequences. In physics and chemistry, and to a lesser extent in biology, the hypotheses and deductions can be stated in mathematical terms and elaborate and far-reaching conclusions can be drawn. On the basis of these deductions the results of other observations and experiments can be predicted and the hypothesis can be tested by its ability to make valid predictions. If the hypothesis is a simple generalization, it may be enough simply to examine more examples and determine whether the generalization holds true. More complex hypotheses, that perhaps cannot be tested directly, can be tested by seeing whether certain logical deductions from the hypothesis hold true. A hypothesis must be subject to some sort of experimental test—it must make a prediction that can be verified in some way—or it is mere speculation.

A hypothesis supported by a large body of different types of observations and experiments becomes a *theory,* which is defined by Webster as "a scientifically acceptable general principle offered to explain phenomena; the analysis of a set of facts in their ideal relations to one another." A good theory relates, from one point of view, facts which previously appeared unrelated and which could not be explained on common ground. A good theory grows; it relates additional facts as they become known. Indeed, it predicts new facts and suggests new relationships between phenomena.

A good theory, by showing the relationships between classes of facts, simplifies and clarifies our understanding of natural phenomena. In the words of Einstein, "In the whole history of science from Greek philosophy to modern physics, there have been constant attempts to reduce the apparent complexity of natural phenomena to some simple, fundamental ideas and relations." Science is really the search for simplicity. William of Occam, a fourteenth century philosopher, made the dictum, *"Essentia non sunt multiplicanda praeter necessitatem,"* or "Entities should not be multiplied beyond necessity." This principle of parsimony (often called *Occam's razor* because it pares a theory to its bare essentials) means that no more forces or causes should be postulated than are necessary to account for the phenomena observed. In practice, this means that the simplest explanation which will account satisfactorily for all the known facts is to be preferred. A new theory in biology, by clearing away previous misconceptions and by pointing up new interrelations of phenomena, stimulates further research in theoretical biology and may provide the basis for a host of practical advances in medicine, agriculture and similar fields.

A poor theory, in contrast, when its consequences are followed, will sooner or later lead to absurdities and clear, irreconcilable contradictions. It frequently happens that at some stage in our knowledge two, or even more, alternative theories provide equally acceptable explanations for the data at hand. But as more observations or experiments are made, one or the other (or perhaps both!) are ruled out.

The scientific method, then, consists of making careful observations and arranging these observations so as to bring order into the observed phenomena. Then we try to find a hypothesis or a *conceptual scheme* which will explain not only the facts previously observed but also new facts as they are discovered. Sciences differ widely in the extent to which their phenomena can be predicted and there are some who claim that biology is not a science because it is not completely predictable. However, even physics, generally regarded as the most "scientific" of the sciences, is far from completely predictable. Although we can predict with high precision the occurrence of eclipses, we cannot make predictions in the field of quantum mechanics, nor can we predict an earthquake, or even tomorrow's weather.

In most scientific studies one of the ulti-mate goals is to explain the cause of some phenomenon, but hard and fast proof that a cause and effect relationship exists between two events is extremely difficult to obtain. If the circumstances leading to a certain event always have a certain factor in common in a variety of cases, that factor may be the cause of the event. The difficulty lies in making sure that the factor under consideration is the *only* one common to all the cases. Although Scotch and soda, bourbon and soda, and rye and soda all may produce intoxication, it would be incorrect to conclude that soda is the only factor in common and therefore the cause of the intoxication!

This method of discovering the common factor in a variety of cases that may be the cause of the event (known as the *method of agreement*) can seldom be used as a valid proof of cause-and-effect relationship because of this difficulty in being sure that it really is the *only* common factor. The finding that all people suffering from beriberi have diets that are low in thiamine is not proof that this deficiency causes the disease, for there may be many other factors in common.

Another method for unraveling cause and effect relations is the *method of difference:* If two sets of circumstances differ in only one factor, and the one containing the factor leads to an event and the other does not, the factor may be considered the cause of the event. For example, if two groups of rats are fed diets which are identical except that one contains all the vitamins and the second contains all but thiamine, and if the first group grows normally and the second group fails to grow and ultimately develops polyneuritis, this would be a strong suggestion, but not absolute proof, that polyneuritis or beriberi in rats is caused by a deficiency of thiamine. By using an inbred strain of rats that are as alike as possible in inherited traits, and by using litter mates (brothers and sisters) of this strain, one could make certain that there were no hereditary differences between the *controls* (the ones getting the complete diet) and the *experimentals* (the ones getting the thiamine-deficient diet).

If the diet without thiamine does not have as attractive a taste as the one with it, the experimental group might simply eat less food and fail to grow and develop the deficiency symptoms because they are partially starved. This source of error can be avoided by "pair-feeding," by pairing a control and an experimental animal, weighing the food eaten each day by each of the experimental animals and then giving only that much food to each control member of the pair.

A third way of detecting cause and effect relationships is the *method of concomitant variation:* If a variation in the amount of a given factor produces a parallel variation in the effect, the factor may be the cause. Thus if other groups of rats are given diets with varying amounts of thiamine and if the amount of protection against beriberi varies directly with the amount of thiamine in the diet, we could be reasonably sure that thiamine deficiency is the cause of beriberi.

It must be emphasized that it is seldom that we can be more than "reasonably sure" that X is the cause of Y. As more experiments and observations lead to the same result, the probability increases that X is the cause of Y. When experiments or observations can be made in a quantitative fashion—when their results can be counted or measured in some way—one can, by the methods of statistical analysis, determine the probability that X is the cause of Y, or the probability that Y follows X simply as a matter of chance. Scientists are usually satisfied that there is some sort of cause and effect relationship between X and Y if they can show that there is less than one chance in a hundred that the observed X-Y relationship could be due to chance alone. A statistical analysis of a set of data can never give a flat yes or no to a question—it can only state that something is very probable or very improbable. It can also tell an investigator approximately how many more experiments he must do to reach a given probability that Y is caused by X.

Each experiment must contain a control group—one treated exactly like the experimental group in all respects but one, the factor whose effect is being tested. The use of controls in medical experiments raises the difficult question of the ethical justification of withholding treatment from a patient who might be benefited by it. If there *is* sufficient evidence that one treatment is better than a second one, a physician would hardly be justified in further experimentation. However, the medical literature is full of treatments now known to be useless or even harmful, which were used for years but finally were abandoned as experience showed they were ineffective and that the evidence which had suggested their use originally was improperly controlled. There is a time in the development of any new treatment when the medical profession is not only morally justified but really morally required to do carefully controlled tests on human beings to be sure that the new treatment is better than the former one.

In such tests it is not sufficient simply to give a treatment to one group of patients and not to give it to another, for it is widely known that there is a strong psychologic effect in simply giving any treatment at all. For example, a group of students at a large western university served as subjects for a test of the hypothesis that daily doses of extra amounts of vitamin C might help prevent colds. This grew out of the observation that people who drank lots of fruit juice seemed to have fewer colds. The group receiving the vitamin C showed a 65 per cent reduction in the number of colds contracted during the winter when they were receiving treatment compared to the previous winter when they were not receiving treatment. There were enough students in the group (208) to make this result statistically significant. In the absence of controls, one would have been led to conclude that vitamin C does help prevent colds. But a second group was given "placebos," pills identical in size, shape, color and taste to the vitamin C pills but without any vitamin C. The students were not told who was getting vitamin C and who was not, they only knew they were getting pills that might help prevent colds. The group getting placebos showed a 63 per cent reduction in the number of colds; thus, vitamin C had nothing to do with the result and the reported reductions in both groups were probably psychological effects. In a "double-blind" study the physician doesn't know which subject is receiving the experimental compound and which one the placebo. Each patient is given pills with a different code number and only after the experiment has been completed is the code revealed.

In all experiments, the scientist must ever be on his guard against bias in himself, bias in the subject, bias in his instruments, and bias in the way the experiment is designed. The proper design of experiments is a science in itself, but one for which only general rules can be made.

A hypothesis that has been tested and found to fit the facts and capable of making valid predictions may then be called a *theory*, a *principle* or a *law*. Although there is some connotation of greater reliance in a statement called a "law" than in one called a "theory," the two words are used interchangeably.

1–5 BIOLOGICAL NOMENCLATURE AND UNITS

The student of biology is confronted with an extensive list of terms for the kinds of plants and animals, their structures and functional mechanisms, their chemical constituents and their interrelationships. To be as precise as possible, and to have an internationally accepted system, it is customary to use Latin or Greek words where possible or to manufacture new words using Greek or Latin roots, casting the new words in Latin form, when describing a newly discovered structure or process. The number of new words introduced in this text has been minimized as much as possible, but many terms are really intrinsic parts of the con-

cepts and principles under discussion and cannot be eliminated.

In dealing with cellular dimensions and the amounts of material present at the cellular level, units of an appropriately small size are necessary. Units of length include the micrometer (0.001 mm.), the nanometer (10^{-6} mm.) and the Ångström unit (10^{-7} mm.). Weights are expressed in milligrams (10^{-3} gm.), micrograms (10^{-6} gm.), nanograms (10^{-9} gm.), picograms (10^{-12} gm.) or in daltons. A *dalton,* the unit of molecular weight, is the weight of a hydrogen atom. One molecule of water (H_2O) weighs 18 daltons and a molecule of hemoglobin, a medium-sized protein, weighs 64,500 daltons. The range of sizes of biological structures is depicted in Figure 1–1, in which the sizes of organisms, cells, viruses and molecules are arranged on a logarithmic scale.

1–6 APPLICATIONS OF BIOLOGY

Some of the practical uses of a knowledge of biology will become apparent as the student

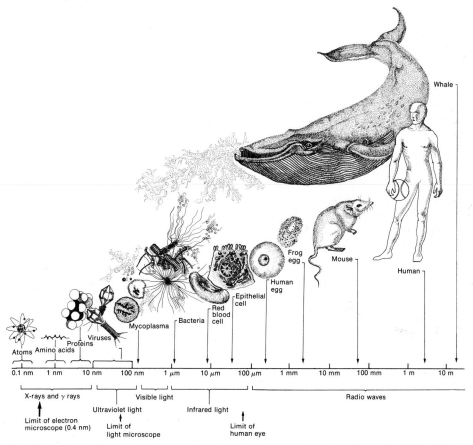

Figure 1–1 The dimensions of biology. To accommodate the wide range of sizes in the world of living things, the unit scale is drawn logarithmically.

reads on through this text—its applications in the fields of medicine and public health, in agriculture and conservation, its basic importance to the social studies, and its contributions to the formulation of a philosophy of life. There are esthetic values in a study of biology as well. A student cannot expect to learn all or even many of the names and characteristics of the vast variety of plants and animals, but a knowledge of the structure and functions of the major types will greatly increase the pleasure of a stroll in the woods or an excursion to the seashore. The average city-dweller gets only a small glimpse of the vast panorama of living things, for so many of them live in places where they are not easily seen—the sea, or parts of the earth that are not easily visited. Trips to botanical gardens, zoos, aquariums and museums will help give one an appreciation of the tremendous variety of living things.

It is impossible to describe the forms of life without reference to their habitats, the places in which they live. This brings us to one of the major unifying conceptual schemes of biology, that the living things of a given region are closely interrelated with each other and with the environment. The study of this is basic to sociology. The present forms of life are also related more or less closely by evolutionary descent. As we deal with each of the major life forms, the facts about them will be easier to understand and remember if we try to fit them into their place in the closely interwoven tapestry of life.

In our discussions of biologic principles we will focus our attention primarily on the human to gain an appreciation of our place in the biologic world. In numbers, size, strength, endurance and adaptability humans are inferior to many animals and in our adjustment to the environment—which, as we shall see, may be considered to be the most important biologic attribute of any living organism—we often fail. However, in a survey of general biology, both practical considerations and interest demand that our discussions focus on humans, for we are primarily concerned with such things as the human stomach ache, the human gestation period and the endurance of the human body.

QUESTIONS

1. How would you define "science"?
2. Contrast a hypothesis and a law.
3. How would you go about testing the hypothesis that beriberi is caused by a deficiency of thiamine?
4. What would you consider to be proof that beriberi is caused by thiamine deficiency?
5. To which of the biological sciences would you assign the following scientific papers?
   The Flora of Northern Michigan
   The Fate of the Aortic Arches in the Development of the Chick
   The Regulation of the Heart Rate
   The Geographical Distribution of the Species of Wheat
6. Describe in your own words the mode of operation of the scientific method.
7. Contrast the "method of agreement" and the "method of difference" as means of establishing cause and effect relationships.
8. What characteristics and attitudes do you think would be helpful for a career in science?
9. What is meant by a "controlled experiment"?
10. Devise a suitably controlled experiment to show:
    a. Whether a strain of mold found in your garden produces an effective antibiotic.
    b. Whether the rate of growth of a bean seedling is affected by temperature.
    c. Whether bees have color vision.
11. Design a suitably controlled "double-blind" experiment to test whether a new drug is effective in the treatment of some disease. What ethical, moral and legal problems might arise in the course of the experiment?

SUPPLEMENTARY READING

There are a number of fine books on the history of science: The development of the sciences in general is described in Sedgwick, Tyler and Bigelow's *A Short History of Science,* and *The Making of Modern Science* by Gerald Holton gives biographical sketches of several biologists. The histories of the biologic sciences by Nordenskiöld and by Singer are well written and informative. The *History of Medicine* written by Douglas Guthrie describes the beginnings of anatomy, physiology and bacteriology. The scientific method and its application to research problems are discussed in Conant's *Science and Common Sense,* Beveridge's *The Art of Scientific Investigation* and Baker and Allen's *Hypothesis, Prediction and Implication in Biology.* E. Bright Wilson's *An Introduction to Scientific Research* gives an excellent discussion in nontechnical terms of the methods of science and some of the problems involved in scientific investigation. W. B. Cannon's *The Way of an Investigator* gives some interesting examples of the scientific method in medical research. An insight into the motivations and thought processes of one of the classic molecular biologists is provided by James Watson's *The Double Helix. In the Name of Science,* by Martin Gardner, describes many pseudosciences and, in showing up their shortcomings, gives an appreciation for scientific evidence and standards. An excellent, brief discussion of scientific method is given in J. K. Feibleman, *Testing Hypotheses by Experiment.*

CHAPTER 2

SOME MAJOR GENERALIZATIONS OF THE BIOLOGICAL SCIENCES

Biology, like physics and chemistry, is a science composed of thousands of facts derived from a multitude of individual observations. Yet, to understand this science, the student need not memorize all of these facts, or even any considerable part of them. As in physics and chemistry, but perhaps not to the same extent, there are broad generalizations—theories, principles and laws—which have been inferred from careful study and evaluation of these individual observations. Since these generalizations are the foundation of present-day biology, it seems worthwhile to discuss them briefly at this point before continuing with the more detailed presentation of plant and animal form and function. The subsequent, more detailed discussions should be more meaningful if viewed against the background of these theories. These generalizations, in turn, should become clearer and more firmly fixed in your mind as you examine the relevant observations and experimental results which will be presented in the succeeding chapters. We cannot discuss these generalizations fully at this time, for that would obviously require a large book, but the present summarization should aid each student in getting a firm grasp of the broad picture of biology. It will probably be helpful to reread this chapter after reading the subsequent ones.

As with most generalizations, there are exceptions to many of the statements in this chapter, some of them unimportant, others of considerable theoretical and practical importance. These exceptions will be dealt with in the later, more detailed discussions.

2-1 PHYSICAL AND CHEMICAL PRINCIPLES GOVERN LIVING SYSTEMS

One of the basic tenets of modern biology is that all of the phenomena of life are governed by, and can be explained in terms of, chemical and physical principles. Until early in this century most people, biologists and laymen alike, believed that life processes differed in some fundamental way from those of nonliving systems. With the vast increase since then in our understanding of chemical and physical principles it has become clear that the myriad phenomena of life, although much more complex than nonliving systems, can be explained in chemical and physical terms without postulating some mysterious and unique vital force. The properties of living cells and organisms, which seemed at one time to be so inexplicable, now appear to be quite straightforward. Many of the complex phenomena of life can be reproduced in the test tube under appropriate conditions. One corollary of this is the belief that if we knew enough about the chemistry and physics of living systems, we could create life in the test tube.

2–2 BIOGENESIS

There appear to be no exceptions to the generalization that "all life comes only from living things." The idea that even large organisms, such as worms, frogs and rats, could arise by spontaneous generation was widely held until the seventeenth century, when it was disproved by the experiments of Redi and Spallanzani.

The experiments of Pasteur, Tyndall and others over a century ago finally provided convincing proof that microorganisms such as bacteria also cannot originate from nonliving material by spontaneous generation. Previously the scientists who believed that organisms could not arise by spontaneous generation had shown that if nutrient medium is placed in a stoppered flask and the contents are boiled no organisms subsequently appear when the flask is cooled. The proponents of spontaneous generation replied that the boiling process had destroyed some essential life-forming nutrient and for this reason no spontaneous generation was observed. *Pasteur's experiments* involved putting nutrient broth into flasks with long **S**-shaped necks (Fig. 2–1). He boiled the broth in the flasks to kill any bacteria that might be present. Flasks with straight necks permitted dust particles and their adhering bacteria to settle into the broth; bacterial colonies appeared very quickly. However, although the **S**-shaped flasks were equally open to the air, bacteria could not penetrate to the nutrient broth, because they were trapped in the film of moisture in the **S**-shaped curve of the neck. This

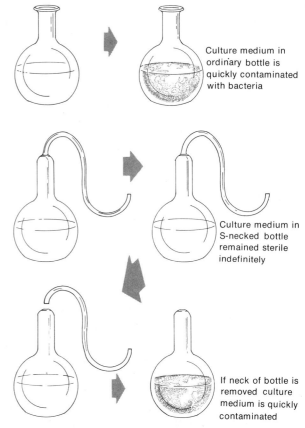

Culture medium in ordinary bottle is quickly contaminated with bacteria

Culture medium in S-necked bottle remained sterile indefinitely

If neck of bottle is removed culture medium is quickly contaminated

Figure 2–1 Pasteur's experiments disproving the spontaneous generation of microorganisms. Nutrient broth (sugar and yeast) was placed in flasks with long S-shaped necks and boiled to kill any bacteria present. Flasks with straight necks permitted bacteria to settle into the broth, and it was quickly teeming with bacteria. The S-shaped neck did not permit bacteria to enter, and although the flask was open to the air, its contents did not become contaminated unless the neck was removed.

acted as a filter. The broths could be left in such flasks for weeks or months without any bacteria appearing in them. Pasteur showed further that if the S-shaped neck were broken off, the broth rapidly developed an extensive population of bacteria. By a series of such experiments, Pasteur showed that the bacteria that appeared in the nutrient broth did not develop by spontaneous generation. Rather, these bacteria were present in the air and were carried to the broth attached to dust particles.

Whether the submicroscopic, filtrable viruses should be considered as living or not is perhaps arguable, but it seems clear that the multiplication of viruses requires the presence of preexisting viruses; viruses do not arise *de novo* from nonviral material.

In recent years biologists have come to realize that although spontaneous generation does not occur under the conditions prevailing on this planet at present, it must have occurred some billions of years ago when life initially appeared. At that time conditions in the atmosphere and in the sea were quite different from those that exist today, and not only was the spontaneous generation of living things possible, it was highly probable.

2-3 THE CELL THEORY

One of the broadest and most fundamental biological generalizations is the **cell theory.** The cell theory, as presently formulated, states that all living things, animals, plants and bacteria, are composed of cells and cell products, that new cells are formed by the division of preexisting cells, that there are fundamental similarities in the chemical constituents and metabolic activities of all cells, and that the activity of an organism as a whole is the sum of the activities and interactions of its independent cell units.

Cells were first described by Robert Hooke, who examined a piece of cork using one of the crude microscopes of the seventeenth century. What Hooke saw (Fig. 2-2) were the cell walls of the dead cork cells, and it was not until some two centuries later that it was realized that the important part of the cell was its contents, not its walls.

Like most broad theories, the cell theory is not the product of any single person's research and thought. The German botanist, Matthias Schleiden, and zoologist, Theodor Schwann, are usually credited with this theory, for in 1838 they pointed out that plants and animals are aggregates of cells arranged according to definite laws. However, the French biologist, Dutrochet, clearly stated (1824) that "all organic tissues are actually globular cells of exceeding smallness, which appear to be united only by simple adhesive forces; thus all tissues, all animal organs are actually only a cellular tissue variously modified." Even earlier, Lamarck (1809) stated that "no body can have life if its constituent parts are not cellular tissue or are not formed by cellular tissue." Dutrochet recognized that growth is the result of increases in the volumes of individual cells

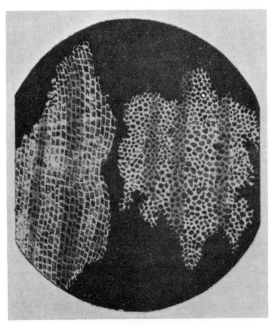

Figure 2-2 Drawing by Robert Hooke of the microscopic structure of a thin slice of cork. From the book *Micrographia*, published in 1665, in which Hooke described many of the objects he had viewed using the compound microscope he constructed.

and of the addition of new little cells. The presence of a nucleus within the cell, now recognized as an almost universal feature of cells, was first described by Robert Brown in 1831. As in many other fields of science, Schleiden and Schwann, although not the first to enunciate a principle, stated it with such force and clarity that the idea gained general acceptance with biologists of the time.

2-4 THE THEORY OF ORGANIC EVOLUTION

That all of the many kinds of plants and animals existing at the present time were not created *de novo* but have descended from previously existing, simpler organisms by gradual modifications which have accumulated in successive generations is one of the great unifying concepts of biology. Elements of this were implicit in the writings of certain Greek philosophers before the Christian era, from Thales to Aristotle. The theory of organic evolution was considered by a number of philosophers and naturalists from the fourteenth to the nineteenth century. However, not until the publication in 1859 of Charles Darwin's *On the Origin of Species by Means of Natural Selection* was the theory brought to general attention. In his book Darwin presented a wealth of detailed evidence and cogent argument to show that organic evolution had occurred. He also presented a theory, that of Natural Selection, to explain how evolution may occur.

According to Darwin's **Theory of Natural Selection**, every group of animals or plants tends to undergo variation. More organisms of each kind are produced than can possibly obtain food and survive. There is a struggle for survival among the many individuals born, and those individuals which possess characters that give them some advantage in the struggle for existence will be more likely to survive than those without them. The survivors will pass these advantageous characters on to their offspring, so that successful variations will be transmitted to future generations. The core of Darwin's theory is this concept of the struggle for existence, the "survival of the fittest," and the inheritance of the advantageous characters by the offspring of the surviving individuals. This concept has had a central role in biological theory for the past century, and, with suitable amendments to bring it into line with subsequent discoveries in genetics and evolution, is held by most present-day biologists.

Studies of the development of many kinds of animals and plants from fertilized egg to adult have led to the interesting generalization that organisms tend to repeat, in the course of their embryonic development, some of the corresponding stages of their evolutionary ancestors. The generalization, called the **Theory of Recapitulation**, was once interpreted as meaning that embryos have in succession the appearance of the *adult* forms of their ancestors. Most biologists would now prefer the statement that embryos recapitulate some of the *embryonic* forms of their ancestors. The human being, at successive stages in development, resembles in certain respects first a fish embryo, then an amphibian embryo, then a reptilian embryo, and so on.

2–5 THE GENE THEORY

Once it had been established that a new organism originates from the union of an egg and a sperm, the question arose of how the parental characters are transmitted to the offspring through these tenuous bits of living material. Among the many theories regarding this was the one stated by Charles Darwin, that each tissue or organ of the parent contributed some sort of models, or "pangenes," which were incorporated into the egg or sperm and thus transmitted to the offspring. There they guided development so as to produce in the offspring a duplicate of the organ from which they came.

The theory of the "continuity of the germ plasm" was formulated by August Weismann in 1889. His answer to the question of how a single germ cell, an egg or a sperm, can contain all the hereditary tendencies of the whole organism was that these germ cells are in turn derived from the parent germ cell and not from the body (somatic) cells of the individual. He suggested that from the very first division of the fertilized egg one line of cells, the **germ plasm**, was distinct from the body cells, or **somatoplasm.** The germ plasm was unaffected by the somatoplasm or by external influences. He realized, before chromosomes or genes were discovered, that heredity involves the transfer of particular molecular constitutions from one generation to the next. A little thought reveals the obvious corollary of Weismann's theory: acquired characteristics are not inherited. Only changes in the germ plasm, not in the somatoplasm, can be transmitted to successive generations.

In the development of certain invertebrate animals, the continuity of the germ plasm from generation to generation is clear. Early in cleavage one cell can be distinguished as the precursor of the germ cells; its descendants can be traced ultimately to their final location in the testis or ovary. In most animals the distinction between germ plasm and somatoplasm is not clearly evident, and germ cells appear to arise from unspecialized somatic cells. As more has been learned about chromosomes and genes, it has become clear that genetic continuity from generation to generation lies in the chromosomes, present in all cells, and not in some peculiar property of the germ line itself.

A dramatic demonstration that the germ cells are indeed uninfluenced by the somatic cells was provided by W. E. Castle and J. C. Philips in 1909. They removed the ovaries from a white (albino) guinea pig and implanted in her an ovary from a black guinea pig. The animal was later mated with a white male guinea pig and produced offspring all of which were black.

The generalizations about the mechanism of inheritance are among the most exact and quantitative of biological theories. They permit us to make predictions as to the probability that the offspring of two given parents will have a particular characteristic. These generalizations are called **Mendel's Laws,** for they were first enunciated by the Austrian abbot Gregor Mendel in 1866. He was led to infer these generalizations from his careful breeding experiments with peas. The importance of Mendel's findings was not recognized until 1900, when the principles were independently rediscovered by Correns, de Vries and von Tschermak.

Mendel's First Law, the **Law of Segregation,** states that genes, the units of heredity, exist in individuals as pairs. In the formation of gametes the two genes separate or segregate and pass into different gametes, so that each gamete has one and only one of each kind of gene. Mendel's Second Law, the **Law of Independent Segregation,** states that the segregation of each pair of genes in the process of gamete formation is independent of the segregation of the members of other pairs of genes. Thus the members of the pairs come to be assorted at random in the resulting gamete. Mendel's keen insight is truly remarkable, for he made these generalizations despite the fact that the details of chromosomes, meiosis and fertilization were unknown. Later, when chromosomes were discovered and genetic and cytologic evidence was available, the modern concept that the units of inheritance are arranged in linear order in the chromosomes was stated by W. S. Sutton in 1902 and T. H. Morgan in 1911.

2–6 GENETIC EQUILIBRIUM AND DIFFERENTIAL REPRODUCTION

The question that may puzzle a beginning biologist is why, if the genes for brown eyes are dominant to the genes for blue eyes, haven't all the blue-eyed genes and all the blue-eyed individuals disappeared? The answer lies partly in the fact that a recessive gene, such as

the one for blue eyes, is not changed by having existed for a generation in the same cell with a brown-eyed gene. The remainder of the explanation lies in the fact that as long as there is no selection for either eye color, that is, as long as people with blue eyes are just as likely to marry and have as many children as people with brown eyes, successive generations will have the same proportions of blue- and brown-eyed people as the initial one.

The principle that a population of a given species of animals or plants is in genetic equilibrium, in the absence of natural selection, and tends to have the same proportion of organisms with a given characteristic in successive generations, was arrived at independently by the English mathematician, G. H. Hardy, and the German physician, G. Weinberg, in 1908. They pointed out that the frequency of the possible combinations of a pair of genes in a population may be calculated from the expansion of the binomial equation $(pA + qa)^2$. When we consider all of the matings of all of the individuals in any given generation, a p number of **A**-containing eggs and a q number of **a**-containing eggs are fertilized by a p number of **A**-containing sperm and a q number of **a**-containing sperm: $(pA + qa) \times (pA + qa)$. The offspring of these matings and their relative frequency are described by the algebraic product: $p^2AA + 2\,pq\,Aa + q^2aa$. Any population in which the distribution of a pair of alleles **A** and **a** conforms to the relation $p^2AA + 2\,pq\,Aa + q^2aa$ is in **genetic equilibrium.** The proportions of these alleles in the members of successive generations will be the same, unless they are altered by selection, by mutation or by chance. This relationship, referred to as the **Hardy-Weinberg Law,** has been of great importance in genetics, particularly in human genetics, for it is the basis of the statistical methods used in determining the mode of inheritance of a given trait in the absence of control and test matings.

This Hardy-Weinberg principle is also fundamental to the mathematical treatment of problems in evolution. Evolution by natural selection, stated in its simplest terms, means that individuals with certain genotypes, and therefore certain traits, have more offspring surviving in the next generation than do other individuals with contrasting genotypes. They contribute a proportionately greater percentage of genes to the gene pool of the next generation than do organisms with other genes. We now view the process of evolution as a gradual change in the gene frequencies of a population which occurs when the Hardy-Weinberg equilibrium is upset. This upset may result because mutations occur, because reproduction is nonrandom, i.e., because selection occurs, or because the population is small and chance alone may determine the survival or the loss of a specific allele by a process termed *genetic drift*. This process of *differential reproduction* implies that the conditions of the Hardy-Weinberg equilibrium do not apply to that particular population. The individuals that produce more surviving offspring in the next generation are usually, but not necessarily, those that are best adapted to the given environment. Well-adapted individuals may be healthier, better able to obtain food and mates, and better able to care for their offspring, but the primary factor in evolution is how many of their offspring survive to be parents of the next generation.

2–7 METABOLIC PROCESSES ARE MEDIATED BY ENZYMES

One of the characteristics of living things is their ability to metabolize, to carry on a great variety of chemical reactions. Our present-day generalizations about metabolism had their beginnings in 1780, when Lavoisier and La-Place concluded that respiration is a form of combustion. They reached this conclusion from simple experiments comparing the utilization of oxygen and the production of carbon dioxide by animals and by candles kept in bell jars, even though their thinking was hindered by the then-current, erroneous "phlogiston" theory.

The concept that metabolism in all living organisms is mediated by specific organic catalysts, *enzymes*, synthesized by living cells, has gradually been crystallized since 1815, when Kirchhoff prepared an extract of wheat which would convert starch to sugar. A long argument between Liebig and Pasteur as to whether enzymes (or "ferments" as they were called then) themselves were living was resolved in Liebig's favor in 1897 when Eduard Büchner prepared a cell-free extract of yeast which would convert sugar to alcohol. Intensive research in enzymology has resulted in the isolation of many enzymes, the demonstration that they are all large protein molecules, and the generalization that each enzyme, because of its specific configuration, controls a specific kind of chemical reaction. The substance undergoing a chemical reaction (the *substrate*) unites with the enzyme to form a specific enzyme-substrate complex. In this way enzymes control the speed and specificity of essentially all the chemical reactions of living things.

The metabolic reactions of a wide variety of living things—animals, green plants, bacteria and molds—have been found to be remarkably similar in many respects. The continuation of life requires the expenditure of energy, and the ultimate source of the energy used by all living things is sunlight. This energy, captured by green plants in the process of photosynthesis, is made available to the plants in further reactions. Some of the energy may eventually be used by the animals that eat the plants, or by animals that eat the animals that ate the plants.

Metabolic processes are regulated so as to maintain the internal environment of the cell as constant as possible. Changes in the external environment tend to produce comparable changes in the internal environment of the cell. Extreme changes in the internal environment lead to the death of the cell. Living organisms have a host of elaborate, complex devices that tend to resist the effects of these changes and thus keep the internal environment constant. Many of these devices involve the principle of *"feed-back" control* in which the accumulation of the product of a reaction leads to a decrease in its rate of production or a deficiency of the product leads to an increase in its rate of production. This tendency toward constancy is termed *homeostasis*. In the course of evolution, higher organisms have developed a greater degree of homeostatic control than the lower ones had.

2–8 METABOLIC REACTIONS ARE UNDER GENIC CONTROL

One of the major biological generalizations is the "one gene—one enzyme—one reaction" hypothesis stated by George Beadle and Edward Tatum in 1941. According to this widely accepted theory, each biochemical reaction concerned with the development and maintenance of a particular organism is controlled by a particular enzyme, and the synthesis of the enzyme, in turn, is controlled by a single gene. A change *(mutation)* in the gene will result in a change in the properties of the enzyme, or even in its complete absence. This in turn alters the nature or rate of a particular metabolic step and results in a particular change in the development of the organism. This theory provides the basis for understanding the relationship of a given gene to its specific trait.

2–9 DNA IS THE MAJOR REPOSITORY OF GENETIC INFORMATION

By the early 1950's, Alfred Mirsky and Roger Vendrely had shown that all the cells of the various tissues of a given organism contain the same amount of DNA. The only exception to this are the gametes. Eggs and sperm have only one-half as much DNA per cell as do other cells of the same organism. The inference that DNA was the important part of the gene was obvious. Erwin Chargaff had carried out analyses of the relative amounts of purines and pyrimidines in DNAs from a variety of sources. His analyses showed that although DNAs from different sources may have quite different compositions, certain patterns are evident. The amount of adenine equals the amount of thymine and the amount of guanine equals the amount of cytosine. Studies by Maurice Wilkins in London had shown by x-ray crystallography that the DNA molecule was probably a helix, a giant coil. Using these facts, James Watson and Francis Crick in 1953 proposed a model structure for the DNA molecule (Fig. 29–7), which accounted for the known properties of the gene: its ability to replicate itself exactly, its ability to transmit information, and its ability to undergo mutation.

Watson and Crick suggested that the DNA molecule is a huge intertwined *double helix.* The Watson-Crick model of the DNA molecule pictures two polynucleotide chains wrapped helically around each other, with the sugar-phosphate residues forming a chain on the outside and the purines and pyrimidines on the inside of the helix. The two chains are held together by hydrogen bonds between specific pairs of purines and pyrimidines, e.g., between adenine and thymine as one pair and cytosine and guanine as another pair. Thus these two chains are *complementary* to each other; that is, the sequence of nucleotides in one chain dictates the sequence of nucleotides in the other. The two complementary strands have opposite polarity; that is, they extend in opposite directions and have their terminal phosphate groups at opposite ends of the double helix. When Watson and Crick made an exactly scaled molecular model of this double helix, they found that the combinations adenine and thymine on the one hand and guanine and cytosine on the other hand would fit into the space available, whereas other combinations of purines and pyrimidines would not. The Watson-Crick model in addition provides an explanation of how DNA molecules may undergo replication. The two chains separate, each one brings about the formation of a new chain which is complementary to it, and thus two new double chains are established.

2–10 GENETIC CODING AND PROTEIN SYNTHESIS

Although the Watson-Crick model of the DNA molecule implied that genetic information is transmitted somehow by the sequence of its constituent nucleotides, the exact mechanism was unclear. Since there are only four types of nucleotides, A, T, C and G, in the DNA and twenty or more kinds of amino acids in the peptide chain, it was obvious that there could not be a one-to-one correlation between a nucleotide and an amino acid. Neither would two nucleotides to one amino acid provide a means of coding for 20 amino acids. The various combinations of four symbols taken two at a time would provide only 16 different combinations; however, a *triplet code* of three nucleotides for each amino acid would permit 64 different combinations of the four nucleotides taken three at a time. The mathematical and biological arguments for a triplet code were put forward by Francis Crick in 1961. An enormous amount of research since that time has substantiated his thesis that the genetic code is a triplet one with three adjacent nucleotide bases, termed a *codon,* specifying a particular amino acid (Fig. 29–22). Adjacent codons do not overlap. Each single base is part of a single codon.

A further generalization has emerged from these studies: the genetic code appears to be universal, that is, the codons in the DNA and RNA specify the same amino acid in all the organisms that have been studied from viruses to man. Experimental evidence to support Crick's hypothesis of a triplet code was quickly forthcoming from the experiments of Nirenberg and Matthei regarding the incorporation of specific labeled amino acids into proteins by purified enzyme systems under the direction of artificial polynucleotides of known composition. Thus the genetic code is composed of three-letter units or codons, the three nucleotides of which specify a single amino acid. The sequence of codons in a molecule of DNA in turn specifies the sequence of amino acids in the corresponding polypeptide chain.

In each cell generation the gene, the DNA chain, undergoes replication so that when the cell divides, each of the two daughter cells receives an exact copy of the code. Also, in each cell generation one or more transcriptions of the code may be made by which the genetic information is used to regulate the assembly of a specific enzyme or other protein. This is a two-step process. In the first step, the four-letter code of the nucleotides in the DNA of the genes is transcribed into a similar four-letter code composed of the linear sequence of four ribonucleotides, A, U, C and G. This RNA copy, called *messenger RNA*, is carried to the *ribosomes*, the submicroscopic structures in the cell on which amino acids are assembled to form enzymes and other proteins. In separate processes, the amino acids are activated and attached to another kind of RNA termed *transfer RNA*. This contains a sequence of three nucleotides, called an *anticodon* which binds the amino acid-transfer RNA combination to the appropriate codon on the messenger RNA.

The specificity of any protein, its physical and enzymatic properties, depends on the linear sequence of the amino acids that comprise it. There are some 20 different amino acids, and each protein molecule is composed of several hundred or more amino acids. All, or nearly all, of the various kinds of amino acids are present in each protein. At each step in the transfer of information from DNA to messenger RNA and in the alignment of messenger and transfer RNAs, the process depends upon the specific attraction between complementary pairs of purines and pyrimidines mediated by specific but rather weak **hydrogen bonds.** Thus the biosynthesis of any specific protein involves a specific template and the formation of specific hydrogen bonds between complementary purine and pyrimidine nucleotide pairs. We can summarize this as: the DNA gene, with a four-letter code, located in the chromosome in the nucleus of the cell → messenger RNA, with a four-letter code, made in the nucleus by transcribing (copying) the code of the genes → specific protein, an enzyme or other protein, with its specificity residing in the sequence of the amino acids that are present in its peptide chain. (The arrows indicate the flow of biological information from DNA to RNA to protein.)

2–11 CELLULAR DIFFERENTIATION RESULTS FROM THE DIFFERENTIAL ACTIVITY OF THE SAME SET OF GENES IN DIFFERENT CELLS

The mitotic process ensures the exact distribution of genes to each daughter cell in the organism; however, the various tissues of a multicellular organism do have quantitative and even qualitative differences in the pattern of enzymes and other proteins present. Thus the differences in the pattern of proteins found in different cells must arise by differences in the activity of the same set of genes in different cells (Fig. 2–3). The turning on or off of the synthesis of a specific protein could occur by some process regulating the transcription of DNA to form messenger RNA, by some process involving the combination of messenger RNA with the ribosome, by some process involved in protein synthesis on the ribosome, or by some process in the transformation of the ultimate protein product. Each kind of messenger RNA has a half-life ranging from a few minutes in certain microorganisms to 12 or 16 hours in man and other mammals. Each molecule of RNA template probably can serve to direct the synthesis of many molecules of its protein, but eventually the template RNA is degraded and must be replaced. This provides a means by which a cell can alter the kind of protein that it synthesizes as new types of messenger RNA replace the previous ones. Thus the cell can

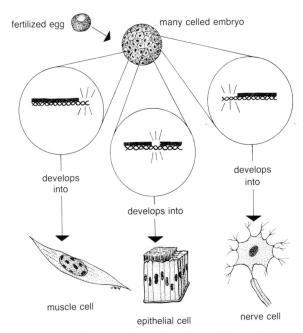

fertilized egg

many celled embryo

develops into

develops into

develops into

muscle cell

epithelial cell

nerve cell

Figure 2–3 The problem of cellular differentiation. The differences between the components of a muscle cell and a nerve cell must arise in some way from differences in the activities of the same set of genes in the two kinds of cells. The diagram postulates that some genes are bound to protein and hence are "silent," whereas others are free and available for transcription to produce the specific proteins characteristic of that cell type.

respond to exogenous stimuli and produce new types of enzymes and other proteins. It has been suggested that the DNA of the genes that are not being transcribed at any given moment is "silent," bound to some kind of protein which makes the DNA unavailable for the transcription system. The process of cellular differentiation is not yet understood and is one of the major unsolved problems in present day biology.

2–12 LIVING CELLS ARE ENERGY TRANSDUCERS

Living organisms and their constituent cells are not heat engines but are *transducers* which convert the chemical energy of foodstuffs—energy ultimately captured by green plants from sunlight—into electrical, mechanical, osmotic or other forms of useful work.

Each living cell is equipped with an efficient and complex series of devices for transforming energy. The radiant energy of sunlight is the major source of energy for all of the life forms on the planet. The first major transformation of energy on the earth is carried out by green plants. These transform the radiant energy of sunlight into chemical energy, which is stored as the energy of the bonds connecting the atoms of molecules such as glucose (Fig. 2–4). This first stage of energy transformation, photosynthesis, is carried out by the pigment chlorophyll. The energy conserved in photosynthesis is used to synthesize carbohydrates and other molecules from carbon dioxide and water.

The second major stage in the flow of energy on this planet occurs in every cell, both plant and animal, in which respiration occurs. In this process, the chemical energy of carbohydrates and other molecules is transformed into a biologically useful kind of energy as these foodstuff molecules undergo oxidation. Cells metabolize foodstuffs such as glucose by a series of enzymic reactions, and the energy present in the chemical bonds of the foodstuffs is transformed and conserved as the energy of *adenosine triphosphate*, ATP.

In the third stage of energy transformation, the chemical energy recovered from foodstuffs in the form of ATP is used by cells to do a variety of kinds of work. The ATP is the source of energy used by cells to transmit nerve impulses, to cause muscle contraction, to carry out the synthesis of complex macromolecules from simpler smaller ones, and to carry on all of the myriad life functions. As these biological functions are performed, the energy eventually flows to the environment and is dissipated as heat. In none of these transformations of energy does the cell act as a heat engine. Neither glucose nor any other molecule is "burned," in the strict sense of the word, within the cell.

The branch of physics that deals with energy and its transformations, *thermodynamics*, consists of a number of relatively simple basic principles which are universally applicable to chemical processes whether they occur in living or nonliving systems. It is well established that the laws of thermodynamics apply to living systems just as they do to nonliving ones.

Under experimentally controlled conditions, the amounts of energy entering and leaving any system may be measured and compared. It is always found that energy is neither created nor destroyed, but only transformed

Photosynthesis:
$H_2O + CO_2 + e \longrightarrow$
glucose $+ O_2$

Oxidative phosphorylation:
Glucose $+ O_2 \longrightarrow$
$CO_2 + H_2O + e$ (ATP)

$ATP \longrightarrow ADP + P + e$
ATP utilized in
muscle contraction

Figure 2–4 The flow of energy (e) from the sun through green plants to animals via photosynthesis, respiration and the utilization of ATP in the performance of work. The radiant energy of sunlight is converted into chemical energy, ATP, and finally is dissipated as heat.

from one form to another. This is an expression of one of the fundamental principles of physics, the *Law of the Conservation of Energy.*

2–13 VITAMINS ARE PRECURSORS OF COENZYMES

The discovery that substances other than salts, proteins, fats and carbohydrates are needed for adequate nutrition—substances called accessory food factors by F. G. Hopkins and *vitamins* by Casimir Funk in 1911—stimulated investigations into the role these substances play in metabolism and why they are needed in the diet of some organisms and not others. The generalization is now amply established that these substances are necessary for the metabolism of *all* organisms—bacteria, green plants, fungi and animals. Many organisms, however, are able to synthesize all of these substances that they require; the ones that cannot synthesize these materials must obtain them in their diet.

The specific roles in metabolism of many of these vitamins are now known. In each instance they are known to become a part of a larger molecule which functions as a *coenzyme*, a partner of the enzyme and substrate which is absolutely necessary for some particular reaction or reactions to occur. The deficiency diseases—pellagra, scurvy, rickets and beri-beri—caused by the lack of the vitamins reflect the impaired metabolism caused by the deficient coenzyme.

2–14 HORMONES REGULATE CELLULAR ACTIVITIES

The term *hormone* was originated by the British physiologist E. H. Starling in 1905 and was defined as "any substance normally produced in the cells of some part of the body and carried by the bloodstream to distant parts, which it affects for the good of the body as a whole." The science of endocrinology can be said to have begun in 1849, when, on the basis of experiments in which testes were transplanted from one bird to another, Berthold postulated that these male sex glands secrete some blood-borne substance essential for the differentiation of the male secondary sex characteristics. This substance, testosterone, was finally isolated and synthesized in 1935.

Our rapidly increasing knowledge of the many different hormones produced by vertebrate and invertebrate animals and by plants and molds has led to the generalization that these are special chemical substances, produced by some restricted region of an organism, which diffuse or are transported to another region where they are effective in very low concentrations in regulating and coordinating the activities of the cells. Hormones thus provide for chemical control and coordination which complement and supplement the coordination resulting from the activities of the nervous system.

2–15 INTERRELATIONS OF ORGANISMS AND ENVIRONMENT

The last major generalization we shall consider, and one of the major unifying concepts of biology today, comes from the field of ecology. From detailed studies of communities of plants and animals in a given area, the generalization has been made that all the living things in a given region are closely related with each other and with the environment. This generalization includes the idea that particular kinds of plants and animals are not found at random over the earth but occur in interdependent *communities* of producer, consumer and decomposer organisms together with certain non-living components. These communities can be recognized and characterized by certain dominant members of the group, usually plants, which provide both food and shelter for many other forms. Why certain plants and animals comprise a given community, how they interact, and how we can control them to our own advantage are major research problems in ecology.

The list of biological principles given here is not intended to be exhaustive but rather to emphasize the fundamental unity of biological science and the many ways in which living things are related and interdependent. These generalizations have been derived from careful observations and experiments made by many biologists over a long period of time. All of them have been tested repeatedly, and many have been revised as new information became available as the result of discoveries made with the aid of new techniques such as the electron microscope, radioactive isotopes for tracer studies, and the many other physical and chemical methods which are being used in biological research. Future studies may result in further revisions of some of these principles.

QUESTIONS

1. If the first interplanetary travelers report that they found living things on Mars, would you expect that the biological generalizations which hold true for living things on this planet would hold equally well for them?
2. If all living things originate from other living things, where do you suppose the very first living things came from?
3. Do you think it possible for the whole to be greater than the sum of its parts, i.e., for a whole organism to have properties which are more than the sum of the properties of its constituent cells?
4. What is meant by "the continuity of the germ plasm"?
5. Contrast the essential features of Darwin's "pangene" hypothesis and the modern gene theory.
6. Why does it follow from Weismann's theory that acquired characters are not inherited?
7. Do you think that modern man evolved by natural selection? What characteristics will make man "fittest to survive" in the future?
8. Define the terms "enzyme" and "hormone." What chemical and biological properties may they have in common?
9. Why may the dietary vitamin requirements of man, mouse and mosquito be different?
10. In what ways do plants and animals interact with their environment?

SUPPLEMENTARY READING

Some of the theories presented here are discussed further in Wightman's *The Growth of Scientific Ideas*, in John Bonner's *The Ideas of Biology* and in Clifford Grobstein's *The Strategy of Life*. Joseph Mazzeo's *The Design of Life* provides an interesting historical analysis of some of the key concepts in biological science. Some of the important ideas in biology, presented by extensive quotations from the original papers, are found in Gabriel and Fogel's *Great Experiments in Biology*. For a physicist's look at some of the basic concepts of biology, read Schroedinger's *What Is Life?*

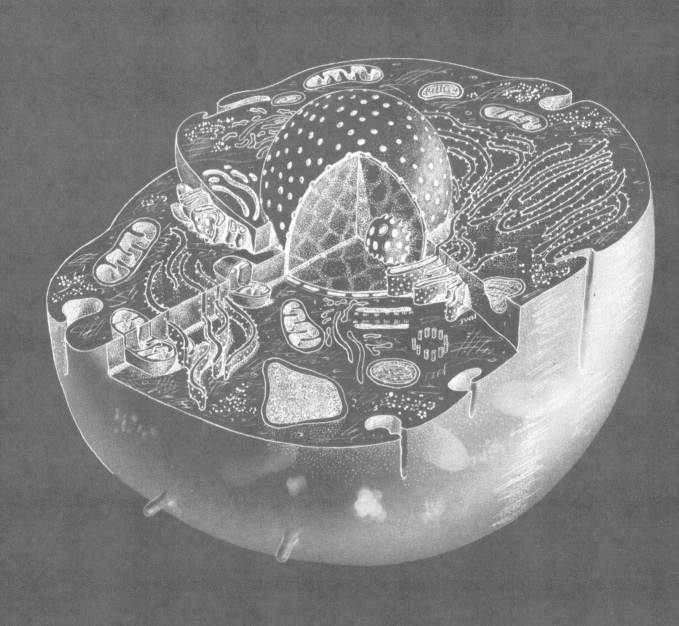

PART 1

CELL STRUCTURES AND FUNCTIONS

INTRODUCTION

The science of biology was originally concerned with the discovery and classification of organisms and with descriptions of their morphology and embryonic development. Gradually biologists have turned to studies of the functions of whole plants and animals and their respective organ systems and subsequently to the structure and function of tissues, cells, subcellular organelles and finally to isolated enzyme systems. This trend toward "molecular biology" has placed increasing emphasis on the physical and chemical aspects of living systems as biologists have explored the energy transformations and enzymatic reactions underlying the manifestations of life. From all of these studies has emerged the realization that the basic chemical organization and the metabolic processes of all living things are remarkably similar despite their morphologic diversity. Further, it is clear that the physical and chemical principles governing living systems are similar to those governing nonliving ones. Thus, an understanding and appreciation of modern biology requires a firm grasp of the chemical and physical principles introduced in Part One.

CHAPTER 3

THE MOLECULAR BASIS OF LIFE

In recent decades biology has grown in all of its ramifications but especially in the areas of biochemistry and biophysics. These branches of biology have scored some notable successes in providing explanations of biological phenomena, and in discussing many areas of biology it will be necessary to refer to physical and chemical principles.

The summary of these principles presented later in this chapter is a necessary introduction to the understanding of basic biological principles. Certain molecules—nucleic acids, proteins, phospholipids, polysaccharides—are uniquely found in living organisms and certain physical phenomena are associated with living organisms.

In differentiating the living from the nonliving, biologists use the word *organism* to refer to any living thing, plant or animal. It is relatively easy to see that a man, an oak, a rosebush and an earthworm are living, whereas rocks and stones are not, but it is more difficult to decide whether such things as viruses are "alive."

To study this fundamental stuff of life we take advantage of the fact that in amoebas and slime molds the living substance is naked and readily visible under the microscope. The living substance of such an organism is colorless, or perhaps faintly yellow, green or pink, and translucent. It has a thick, viscid, syrupy consistency and would feel slimy to the touch. When seen in the light microscope, it may appear to have granules or fibrils of denser material, droplets of fatty substances, or fluid-filled vacuoles, all suspended in the clear, continuous, semifluid "ground substance." The electron microscope has revealed the presence of a remarkable structural complexity in what had appeared in the light microscope to be a more or less homogeneous matrix. The cell membranes and the subcellular constituents have been shown by x-ray diffraction analyses to have an even finer structure, one which can be related to the structure of the large molecules comprising them.

3–1 CHARACTERISTICS OF LIVING THINGS

In discussing the features that separate living from nonliving things, we want to do more than simply list and describe their distinguishing characteristics; we would like ultimately to understand the chemical and physical bases of each. Although the list of properties of living things—organization, metabolism, movement, irritability, growth, reproduction and adaptation — seems specific and definite, the line between the living and nonliving is rather tenuous. Viruses exhibit some but not all of the usual characteristics of living things. When we realize that we cannot answer the question of whether they *are* living or nonliving, but only the question of whether they should be *called* living or nonliving, the problem is put into proper perspective. Even nonliving objects may show one or another of these properties. Crystals in a saturated solution may "grow," a bit of metallic sodium will move rapidly over the surface of water, and a drop of oil floating in glycerol and alcohol may send out pseudopods and move like an amoeba.

Specific Organization. Each kind of living organism is recognized by its characteristic shape and appearance; the adults of each kind of organism typically have a characteristic size. Nonliving things generally have much more variable shapes and sizes. Living things are not homogeneous, but are made of different parts, each with special functions; thus the bodies of living things are characterized by a specific, complex organization.

The structural and functional unit of both plants and animals is the *cell*, the simplest bit of living matter that can exist independently. The processes of the entire organism are the sum of the coordinated functions of its constituent cells. These cellular units may vary considerably in size, shape and function. Some of the smallest animals and plants have bodies made of a single cell; the body of a man or an oak tree, in contrast, is made of countless billions of cells fitted together.

The cell itself has a specific organization and each kind of cell has a characteristic size and shape by means of which it can be recognized. The cell has a *plasma membrane* which separates the living substance from the surroundings, and it contains a *nucleus*, a specialized part of the cell separated from the rest by a nuclear membrane. The nucleus, as we shall learn later, plays a major role in controlling and regulating the activities of the cell.

The bodies of the higher animals and plants are organized in a series of increasingly complex levels: Cells are organized into *tissues*, tissues into *organs*, and organs into *organ systems*.

Metabolism. The sum of all the chemical activities of the cell which provide for its growth, maintenance and repair is called *metabolism.* All cells are constantly taking in new substances, altering them chemically in a variety of ways, building new cellular materials, and transforming the potential energy contained in large molecules of carbohydrates, fats and proteins into kinetic energy and heat as these substances are converted into simpler ones. The never-ending flow of energy within a cell, from one cell to another, and from one organism to another is the essence of life, one of the unique and characteristic attributes of living things. Some kinds of cells — bacteria, for example — have very high metabolic rates. Other kinds, such as seeds and spores, have a barely detectable rate of metabolism. Even within a particular species or person metabolic rates may vary, depending on such factors as age, sex, general health, amount of endocrine secretion, pregnancy and even the time of day. The study of energy transformations in living organisms is termed *bioenergetics.*

It is customary to characterize metabolic processes as anabolic or catabolic. The term *anabolism* refers to those chemical processes in which simpler substances are combined to form more complex substances, resulting in the storage of energy, the production of new cellular materials and growth. *Catabolism* refers to the breaking down of these complex substances resulting in the release of energy and the wearing out and using up of cellular materials. Both types of processes occur continuously; indeed the two are intricately interdependent and difficult to distinguish. Complex compounds may be broken down and their parts recombined in new ways to form different substances. The interconversions of carbohydrates, proteins and fats that occur continuously in human cells are examples of combined catabolic and anabolic processes. Anabolic processes, in general, require energy; therefore some catabolic processes must occur to supply the energy to drive the reactions involved in building up the new molecules.

Both plants and animals have anabolic and catabolic phases of metabolism. Plants, however (with some exceptions), have the ability to manufacture their own organic compounds out of inorganic materials in the soil and air. Animals must depend on plants for their food. Plant cells are simply better chemists than animal cells.

Movement. Living things are characterized further by their ability to move. The movement of most animals is quite obvious—they wiggle, crawl, swim, run or fly. The movements of plants are much slower and less obvious, but occur nonetheless. A few animals—sponges, corals, oysters, certain parasites—do not move from place to place, but most of these have cilia or flagella to move their surroundings past their bodies and thus bring food and other necessities of life to themselves. Movement may result from muscular contraction, from the beating of microscopic hairlike projections of cells called *cilia* or *flagella,* or from the slow oozing of a mass of cell substance *(amoeboid motion)* (Fig. 3–1). The streaming motion of the living material in the cells of the leaves of plants is known as *cyclosis.*

Irritability. Living things are irritable; they respond to stimuli, to physical or chemical changes in their immediate surroundings. Stimuli which are effective in evoking a response in most animals and plants are changes in the color, intensity or direction of light, changes in temperature, pressure or sound, and changes in the chemical composition of the earth, water or air surrounding the organism. In man and other complex animals, certain cells of the body are highly specialized to respond to certain types of stimuli: the rods and cones in the retina of the eye respond to light, certain cells in the nose and in the taste buds of the tongue respond to chemical stimuli, and special cells in the skin respond to changes in temperature or pressure. In lower animals, and in plants, such specialized cells may be absent but the whole organism responds to stimuli. Single-celled organisms, protista, will respond by moving toward or away from heat or cold, certain chemical substances, light, or the touch of a microneedle.

The irritability of plant cells may not be as obvious as that of animal cells, but they are sensitive to changes in their environment. The streaming movements in plant cells may be speeded or stopped by changes in the amount of light. A few plants, such as the Venus flytrap of the Carolina swamps, have a remarkable sensitivity to touch and can catch insects. Their leaves are hinged along the midrib (Fig. 3–2), and the edges of the leaves are covered with hairs. The presence of an insect on the leaf stimulates it to fold, the edges come together and the hairs interlock to prevent the escape of the prey. The leaf then secretes enzymes which kill and digest the insect. The development of fly-trapping is an adaptation which enables these plants to obtain part of the nitrogen they require for growth from the prey they "eat," since the soil in which they grow is deficient in nitrogen.

Growth. Growth, an increase in cellular mass, may be brought about by an increase in the *size* of the individual cells, by an increase

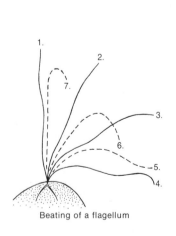

Beating of a flagellum

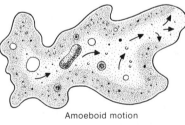

Amoeboid motion

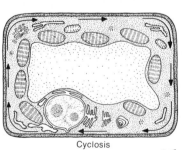

Cyclosis

Figure 3–1 Diagram illustrating several types of cellular movements.

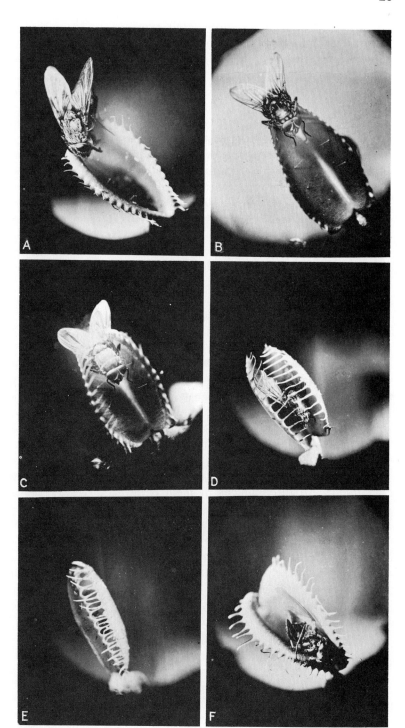

Figure 3–2 A leaf of the Venus flytrap catching and digesting a fly. (Courtesy of General Biological Supply House, Chicago, Ill.)

in the *number* of cells, or both. An increase in cell size may occur simply by the uptake of water, but this swelling is generally not considered to be growth. The term growth is restricted to those processes which increase the amount of living substance of the body, measured by the amount of nitrogen or protein present. Why do you suppose nitrogen or protein is used as the yardstick rather than the amount of carbohydrate, fat, sulfur or sodium? Growth may be uniform in the several parts of an organism, or it may be greater in some parts than in others so that the body proportions change as growth occurs. Some organisms — most trees, for example — will grow indefinitely. Many animals have a definite growth period which terminates in an adult of a characteristic size. One of the remarkable aspects of the growth process is that each organ continues to function while undergoing growth.

Reproduction. If there is any one characteristic that can be said to be the *sine qua non* of life, it is the ability to reproduce. As we shall see (p. 151) the simplest viruses do not metabolize, move or grow, yet because they can reproduce and undergo mutations (p. 706), most biologists regard them as living. Although at one time worms were believed to arise from horse hairs in a water trough, maggots from decaying meat, and frogs from the mud of the Nile, we now know that each can come only from previously existing ones. One of the fundamental tenets of biology is that "all life comes only from living things."

The classic experiment disproving the *spontaneous generation of life* was performed by an Italian, Francesco Redi, about 1680. Redi proved that maggots do not come from decaying meat by this simple experiment. He placed a piece of meat in each of three jars, leaving one uncovered, covering the second with a piece of fine gauze, and covering the third with parchment. All three pieces of meat decayed but maggots appeared on only the meat in the uncovered jar. A few maggots appeared on the gauze of the second jar, but not on the meat, and no maggots were found on the meat covered by parchment. Redi thus demonstrated that the maggots did not come from the decaying meat, but hatched from eggs laid by blowflies attracted by the smell of the decaying

meat. Further observations showed that the maggots develop into flies which in turn lay more eggs. Louis Pasteur, about two hundred years later, showed that bacteria do not arise by spontaneous generation but only from previously existing bacteria. The submicroscopic filtrable viruses do not arise from nonviral material by spontaneous generation; the multiplication of viruses requires the presence of previously existing viruses.

The problem of the original source of life will be discussed later (p. 748), but it is likely that billions of years ago, when chemical and physical conditions on the earth's surface were quite different from those at present, the first living things *did* actually arise from nonliving material.

The process of reproduction may be as simple as the splitting of one individual into two (Fig. 3–3). In most animals and plants, however, it involves the production of specialized eggs and sperm which unite to form the fertilized egg or zygote, from which the new organism develops. Reproduction in certain parasitic worms involves several quite different forms, each of which gives rise to the next in succession until the cycle is completed and the adult reappears.

Adaptation. The ability of a plant or animal to adapt to its environment is the characteristic which enables it to survive the exigencies of a changing world. Each species can become adapted by seeking out an environment to which it is suited or by undergoing modifications to make it better fitted to its present surroundings. Adaptation may involve immediate changes which depend upon the irritability of cells or the responses of enzyme

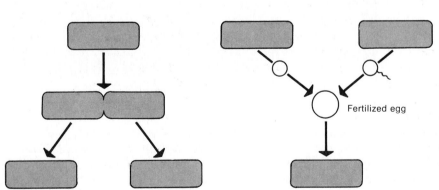

Fertilized egg

Asexual reproduction   Sexual reproduction

Figure 3–3 A comparison of asexual reproduction, in which one individual gives rise to two or more offspring, and sexual reproduction, in which two parents each contribute a gamete and these join to give rise to the offspring.

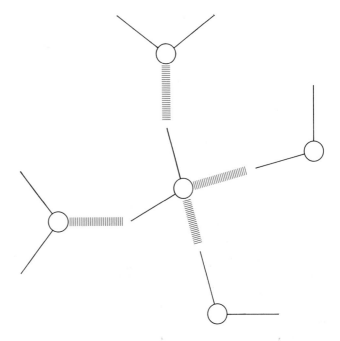

Figure 3–4 Hydrogen bonding of water molecules. Each water molecule tends to form hydrogen bonds with four neighboring water molecules. The hydrogen bonds are indicated by dashed lines. The circles represent the oxygen atoms, and the solid lines the axes of the covalent bonds from the oxygen to the hydrogen atoms.

systems to inducers or repressors (p. 711), or it may be the result of a long-term process of mutation and selection (p. 740). It is obvious that no single kind of plant or animal can adapt to all the conceivable kinds of environment. Hence, there will be certain areas where it cannot survive. The list of factors that may limit the distribution of a species is almost endless: water, light, temperature, food, predators, competitors, parasites, and so on.

3–2 LIFE OCCURS IN AN AQUEOUS PHASE

A large part of each cell is simply water. The percentage of water in human tissues ranges from about 20 per cent in bone to 85 per cent in brain cells. About two thirds of our total body weight is water; as much as 95 per cent of a jellyfish is water. Water serves a number of functions in living systems. Most of the other chemicals present are dissolved in it, and as we shall see, these require a water medium in order to react one with another. Water dissolves the waste products of metabolism and assists in their removal from the cell and the organism. Water has a high heat capacity; that is, it has a great capacity for absorbing heat with only minimal changes in its own temperature. This results from the fact that the neigh-

boring water molecules in ice or in liquid water are held together by hydrogen bonds (Fig. 3–4), and some heat energy is dissipated in breaking these hydrogen bonds. Water thus protects the living material against sudden thermal changes.

Water has the property of absorbing a great deal of heat as it changes from a liquid to a gas, thus enabling the body to dissipate excess heat by the evaporation of water. For example, a football player weighing 100 kg. might lose 2 kg. of water from his body as perspiration in the course of an hour's scrimmage. The heat of vaporization of water is 574 kcal./kg., hence 574 kcal./kg. × 2 kg. = 1148 kcal. If the water had not been vaporized, and if all the heat produced during the scrimmage had remained within his body, his body temperature would have risen 11.5° C, or nearly 20° F! The characteristic high heat conductivity of water makes it possible for heat to be distributed evenly throughout the body tissues. Water serves a further indispensable function as a lubricant, and is present in body fluids whenever one organ rubs against another, and in the joints where one bone moves on another.

Both the liquid within cells and the liquid between cells in man and other multicellular organisms contain a variety of mineral salts, of which sodium, potassium, calcium and magnesium are the chief *cations* (positively charged

Table 3–1 THE IONIC COMPOSITION OF SEA WATER AND THE BODY FLUIDS OF VARIOUS ANIMALS, WITH CONCENTRATIONS EXPRESSED RELATIVE TO SODIUM AS 100

	Na	K	Ca	Mg	Cl	SO$_4$
Sea water (Woods Hole)	100	2.74	2.79	13.94	136.8	7.10
Aurelia, coelenterate	100	2.90	2.15	10.18	113.1	5.15
Strongylocentrotus, echinoderm	100	2.30	2.28	11.21	116.1	5.71
Phascolosoma, sipunculid	100	10.07	2.78	–	114.1	–
Mercenaria, mollusk	100	1.66	2.17	5.70	117.3	5.84
Carcinus, crustacean	100	2.32	2.51	3.70	105.2	3.90
Hydrophilus, insect	100	11.1	0.92	16.8	33.6	0.12
Lophius, fish	100	2.85	1.01	1.61	71.9	–
Frog, amphibian	100	2.40	1.92	1.15	71.4	–
Man, mammal	100	3.99	1.78	0.66	84.0	1.73

ions) and chloride, bicarbonate, phosphate and sulfate are the important **anions** (negatively charged ions). Although the body fluids of terrestrial animals differ considerably from sea water in their total salt content, they resemble it in general in the kinds of salts present and in their relative concentrations (Table 3–1). The total concentration of salts in the body fluids of most marine animals is equal to that in sea water, about 3.4 per cent. Vertebrates, whether terrestrial, fresh water or marine, have less than one per cent of salts in their body fluids. The body fluids of fresh water and terrestrial invertebrates contain 0.3 to 0.7 per cent salts. For life processes to continue, certain salts must be present in relative concentrations that lie within certain limits.

The blood of man and other terrestrial vertebrates is not simply a dilute sea water, but differs in having relatively more potassium and less magnesium and chloride than sea water. Life probably originated in the sea and the cells of those early organisms were adapted to the relative concentration of the salts present (though that early sea probably had a lower total concentration of salts than the present sea). In the course of evolution, animals have evolved with body fluids having generally similar relative concentrations of salts, for any marked difference in the kinds of salts present would inhibit certain enzymes in the cells and place that kind of animal at a marked disadvantage in the competition for survival. Some animals have evolved kidneys and other excretory organs that selectively retain or excrete certain ions, thus leading to body fluids with somewhat different relative concentrations of salts. The concentration of each type of ion is determined by the relative rates of its uptake and excretion by the organism.

The concentration of the various salts is kept extremely constant under normal conditions, and any great deviation from the normal values causes marked effects on cell function, even death. A decreased concentration of calcium ions in the blood of mammals results in convulsions and death. Heart muscle can contract normally only in the presence of the proper balance of sodium, potassium and calcium ions. If a frog heart is removed from the body and placed in a pure sodium chloride solution, it soon stops beating in the relaxed condition. If placed in a solution of potassium chloride or a mixture of sodium and calcium chloride, it will stop in the contracted state. It will continue to beat, however, if placed in a solution containing the proper balance of these three salts. This frog heart method is so sensitive that it can be used to measure the concentration of calcium ions in solutions.

In addition to these specific effects of salts on certain cell functions, mineral salts are important in maintaining the osmotic relationships between the cell and its environment.

3–3 CHEMICAL BONDS

The constituent atoms of a molecule are joined together by forces called chemical bonds. Ionic, covalent and hydrogen bonds are of importance in the molecules present in biologic materials. Ionic and hydrogen bonds are relatively weak and easily broken, but covalent bonds are strong, their formation is *endergonic* (energy must be supplied to form them). Both the formation and the cleavage of covalent bonds are carried out by enzymic reactions.

Ionic bonds are due to the attraction of positively and negatively charged particles. An

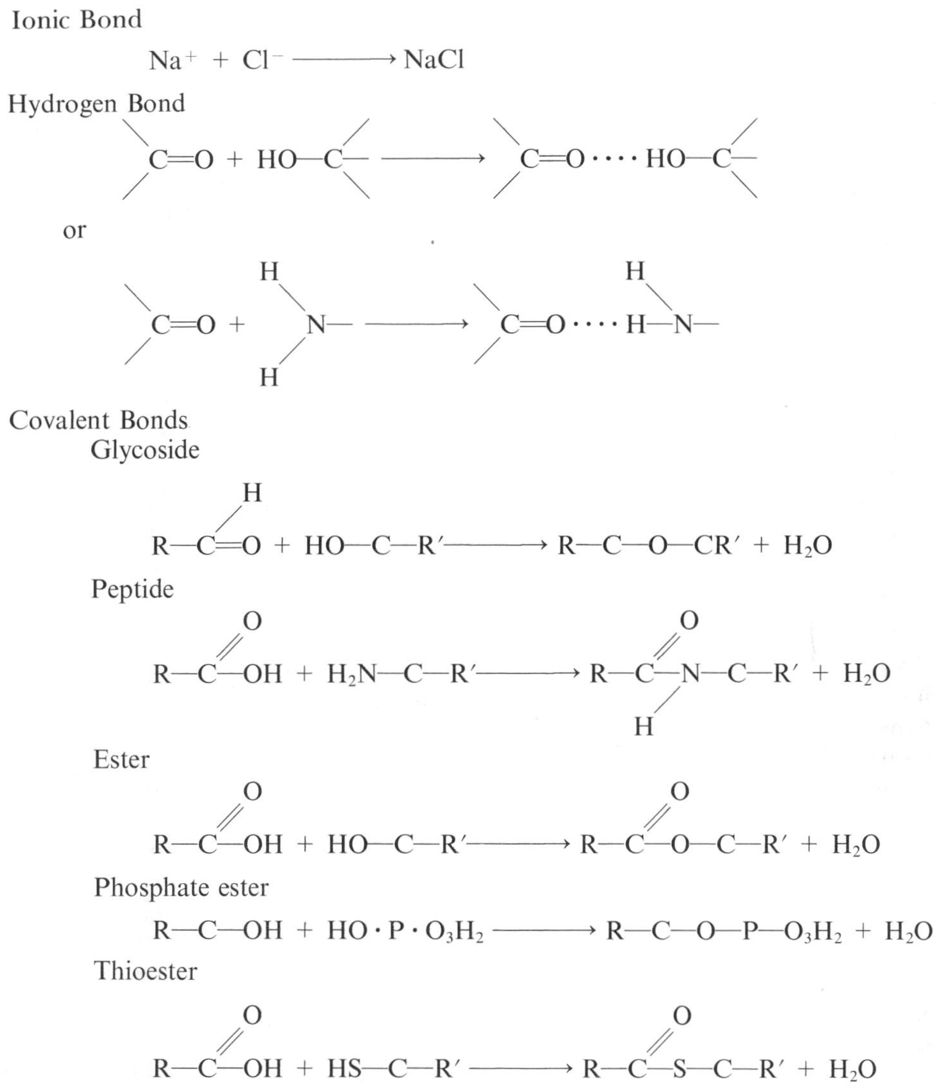

Ionic Bond

$$Na^+ + Cl^- \longrightarrow NaCl$$

Hydrogen Bond

$$\diagdown C{=}O + HO{-}C{\diagdown} \longrightarrow \diagdown C{=}O \cdots HO{-}C{\diagdown}$$

or

$$\diagdown C{=}O + \overset{H}{\underset{H}{\diagup N{-}}} \longrightarrow \diagdown C{=}O \cdots \overset{H}{H{-}N{-}}$$

Covalent Bonds
  Glycoside

$$R{-}\overset{H}{C{=}O} + HO{-}C{-}R' \longrightarrow R{-}C{-}O{-}CR' + H_2O$$

Peptide

$$R{-}\overset{O}{\overset{\|}{C}}{-}OH + H_2N{-}C{-}R' \longrightarrow R{-}\overset{O}{\overset{\|}{C}}{-}\underset{H}{N}{-}C{-}R' + H_2O$$

Ester

$$R{-}\overset{O}{\overset{\|}{C}}{-}OH + HO{-}C{-}R' \longrightarrow R{-}\overset{O}{\overset{\|}{C}}{-}O{-}C{-}R' + H_2O$$

Phosphate ester

$$R{-}C{-}OH + HO \cdot P \cdot O_3H_2 \longrightarrow R{-}C{-}O{-}P{-}O_3H_2 + H_2O$$

Thioester

$$R{-}\overset{O}{\overset{\|}{C}}{-}OH + HS{-}C{-}R' \longrightarrow R{-}\overset{O}{\overset{\|}{C}}{-}S{-}C{-}R' + H_2O$$

Figure 3–5 Types of chemical bonds.

electropositive atom such as a sodium ion, Na^+, unites with an electronegative atom such as a chloride ion, Cl^-, to form sodium chloride. *Hydrogen bonds* are formed when a hydrogen atom is shared between two atoms, one of which is usually oxygen (Fig. 3–5). They tend to form between any hydrogen which is covalently bonded to oxygen or nitrogen and any strongly electronegative atom, usually oxygen or nitrogen in another molecule, or in another part of the same molecule. Hydrogen bonds have a specific length and a specific direction, which is of great importance in their role in determining the structure of macromolecules such as proteins and nucleic acids. Hydrogen bonds are

geometrically quite precise. The water molecules in liquid water are held together largely by hydrogen bonds. This results from the fact that the oxygen atom and two hydrogen atoms in a water molecule form a triangle. The electrons used to bond hydrogen to oxygen are more strongly attracted to the oxygen nucleus than to the hydrogen nuclei, and tend to be located nearer the oxygen atom. Because of this, the two hydrogen atoms have a small local positive charge, and the oxygen atom has a small local negative charge, although the water molecule as a whole is electrically neutral. Molecules which are positive at one end and negative at the other are said to be *polar.* Such molecules are usually soluble in water since the electrostatic attraction of the negative and positive charges tends to align the water molecules around them. When the positively charged hydrogen atom of one water molecule is next to an atom carrying an electronegative charge, such as the oxygen atom in another water molecule, the attraction between them forms a hydrogen bond. **Covalent bonds** are formed when two adjacent atoms *share* a pair of electrons. This is in contrast to the ionic bond in which one atom *transfers* an electron to another atom. Thus sodium transfers an electron to chlorine in the formation of sodium chloride.

Each atom of hydrogen has a nucleus and one electron in its first shell. When two atoms of hydrogen join to form a molecule of hydrogen, each atom shares its electron with the other atom so that, in a sense, each hydrogen has completed its first shell of two electrons. In this instance, the electrons are shared equally by the two atoms and there is no greater probability that the electrons will be nearer one nucleus than the other. Such a bond is said to be a **nonpolar covalent bond.** When two hydrogen atoms are covalently bonded to an oxygen atom to form a molecule of water, the oxygen atom shares electrons with two hydrogen atoms and completes its outer electron circle of eight. At the same time, each hydrogen completes its first shell of two electrons. This covalent bond is a little different from the one between two hydrogen atoms, because when a covalent bond is formed between two different elements, the shared electrons tend to be pulled more strongly to one element than to the other. Such a bond is called a **polar covalent bond.** In a water molecule the electrons tend to lie closer to the oxygen nucleus than to the hydrogen nucleus, giving the oxygen atom a partial negative charge and the hydrogen atoms a partial positive charge. Covalent bonds may have all degrees of polarity, from ones in which the electrons are exactly shared as in the hydrogen molecule, to ones in which the electrons are much closer to one atom than to the other, and the bond is therefore quite polar. In a sense, an ionic bond is simply one extreme of this, in which the electrons are pulled completely from one atom to the other.

Since electrons do not remain in any one position but are constantly moving, the bond may be essentially covalent one instant and ionic another. For this reason a compound that is primarily covalent may be slightly ionized. It is easier to ionize a polar covalent bond than a nonpolar covalent bond. Two other types of weak bonds, van der Waals bonds and hydrophobic bonds, are of particular importance in the structure of protein molecules.

The covalent bonds of greatest importance in joining molecules together are formed by removing an OH group from one molecule and an H group from the other. In biosynthetic reactions, the bond is usually formed not by the actual removal of a molecule of water, but by the substitution of a phosphate group or some other group for an OH group on one molecule and then removing the phosphate and an H from the second molecule to liberate inorganic phosphate and form the bond.

The bonds joining sugar molecules together are **glycoside bonds,** formed by removing an H from an alcohol group on one sugar and an OH from an aldehyde group of the other. The bonds joining the amino acids in proteins are **peptide bonds,** formed by removing an OH from the carboxyl group ($-COOH$) of one amino acid and an H from the amino group ($-NH_2$) of another. The fatty acids and glycerol of neutral fats are joined by **ester bonds,** formed by removing an OH from the carboxyl group of a fatty acid and an H from an alcohol group of glycerol. Other ester bonds of great biologic importance are **phosphate esters,** formed by removing an H from phosphoric acid and an OH group from a sugar, and **thioesters,** which involve the removal of water from an OH of the carboxyl group of an acid and an H from an $-SH$ group rather than an $-OH$ group. Nucleotides have a glycosidic bond between the sugar and the purine or pyrimidine and a phosphate ester bond joining the phosphate to the sugar. Coenzyme A forms thioesters with a variety of substances; acetyl coenzyme A is the thioester of acetic acid and coenzyme A.

3–4 BIOLOGICAL MOLECULES

The major types of organic compounds that cells synthesize and use are carbohydrates, proteins, lipids, nucleic acids and steroids. Some of these are required for the structural integrity of the cell, others to supply energy for its functioning, and still others are of prime importance in regulating metabolism within the cell. Carbohydrates and lipids are the chief sources of chemical energy in almost every form of life; proteins are structural elements but are of even greater importance as catalysts (enzymes) and regulators of cellular processes. Nucleic acids are of prime importance in the storage and transfer of information used in the synthesis of specific proteins and other molecules. The types of substances, and even their relative proportions, are remarkably similar for cells from the various parts of the body and for cells from different animals. A bit of human liver and the cell of an amoeba both contain about 80 per cent water, 12 per cent protein, 2 per cent nucleic acid, 5 per cent fat, 1 per cent carbohydrate and a fraction of 1 per cent of sterols and other substances. Certain specialized cells, of course, have unique patterns of chemical constituents; the mammalian brain, for instance, is rich in certain kinds of fats.

Carbohydrates. Carbohydrates are compounds containing only carbon, hydrogen and oxygen in the ratio of approximately 1C : 2H : 1O. Sugars, starches and cellulose are examples of carbohydrates. Some of the simplest carbohydrates of biological importance are the hexoses, single sugars (monosaccharides) with the formula $C_6H_{12}O_6$. *Glucose* (also called dextrose) and *fructose* are hexoses that differ slightly in the arrangement of their constituent atoms. These different arrangements give them slightly different chemical properties. Compounds having identical molecular formulas but different arrangements of their atoms are termed *isomers.* The arrangements of atoms in molecules are represented by *structural formulas* (Fig. 3–6) in which the atoms are represented by their symbols—C, H, O, etc.—and the chemical bonds, or forces that hold atoms together, are indicated by connecting lines. Hydrogen has one bond to connect to other atoms; oxygen has two, and carbon four. (See Appendix I for a discussion of atomic structure and the nature of the interatomic forces that hold molecules together.)

Carbon atoms can unite with each other as well as with many other kinds of atoms and form an almost infinite variety of compounds. Carbon atoms linked together may form a long chain, as in a fatty acid, or a branched chain, as in certain amino acids. They may form rings, as in purines and pyrimidines, or complex four-ringed systems as in sterol and steroid molecules. Molecules are in fact three-dimensional structures, not simple two-dimensional ones as the formulas on a printed page would suggest. There are more complex ways of representing the third dimension of the molecule. Since the properties of the compound depend in part on its *conformation,* its three-dimensional structure, such three-dimensional formulas are helpful in understanding the intimate relations between molecular structure and biological function. The molecules of glucose and other single sugars in solution are not extended straight chains, as depicted in Figure 3–6 but are present as flattened, boat-shaped or chair-shaped rings formed when a chemical bond connects carbon 1 to the oxygen attached to carbon 5 (or to carbon 4). Glucose in solution typically has a ring of five carbons and one oxygen (Fig. 3–7).

Glucose is the only hexose found in quantity in the body. The other carbohydrates we eat are converted by the liver into glucose. Glucose, an indispensable component of blood, is normally present in the blood and tissues of mammals in a concentration of about 0.1 per cent by weight. A prolonged increase in the concentration of glucose in the blood, as in untreated or poorly controlled diabetes mellitus, can cause metabolic alterations and extensive damage to certain tissues such as the eye and the kidney. A reduced concentration of

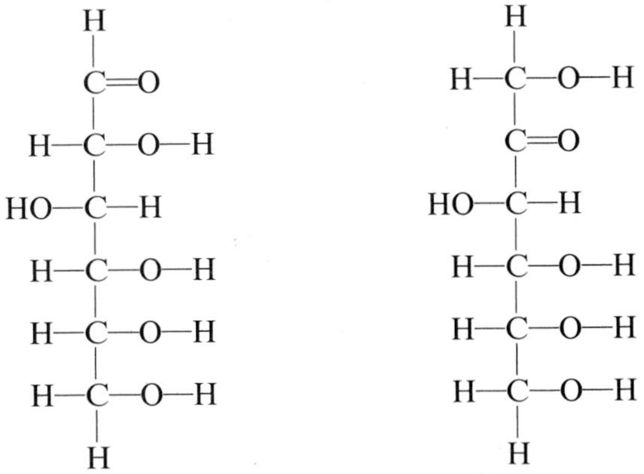

Figure 3–6 Structural formulas of two simple sugars.

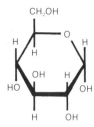

Figure 3–7 Structural formula of glucose depicted as a six-membered ring containing five carbon atoms and one oxygen atom.

Table 3–2 RELATIVE SWEETNESS OF SOME OF THE COMMON SUGARS

Sugar	Relative Sweetness (Sucrose 100)
Lactose	16.0
Galactose	32.1
Maltose	32.5
Glucose	74.3
Sucrose	100.0
Fructose	173.3
Saccharin	55,000

glucose in the blood leads to an increased irritability of certain brain cells, so that they respond to very slight stimuli. As a result of impulses from these cells to the muscles, twitches, convulsions, unconsciousness and death may ensue. Brain cells require glucose for their metabolism and a certain minimal concentration of glucose in the blood is necessary to supply this. A complex physiologic control mechanism, operating like the "feedback" controls of electronic devices, and involving the nervous system, liver, pancreas, pituitary and adrenal glands, maintains the proper concentration of glucose in the blood.

Double sugars (disaccharides), all of which have the formula $C_{12}H_{22}O_{11}$, can be viewed as being made of two single sugars joined together by the removal of a molecule of water. Both cane and beet table sugars are *sucrose,* a combination of one molecule of glucose with one of fructose. Several other double sugars are found in biological systems; all have the formula $C_{12}H_{22}O_{11}$ but differ in the arrangement of their constituent atoms and hence in some of their chemical and physical properties. *Maltose* or malt sugar is composed of two molecules of glucose; *lactose,* or milk sugar, found in the milk of all mammals, is made of one molecule of glucose and one of galactose, a third kind of single sugar. These sugars differ markedly in their sweetness. Fructose is the sweetest of the common sugars. Lactose, the least sweet, is less than one tenth as sweet as fructose. Sucrose is intermediate (Table 3–2). *Saccharin,* a synthetic non-carbohydrate sweetening agent, is much sweeter than any sugar and is used by people who want to sweeten food without using sugar.

The largest carbohydrate molecules are starches and celluloses, composed of a large number of single sugars joined together either in a single long chain (*amylose*) or in a branched chain (*amylopectin*). Since the exact number of sugar molecules joined to make a starch molecule is unknown, and indeed may vary from one molecule to the next, the formula for starch may be written $(C_6H_{10}O_5)_x$, where x stands for the unknown, large number of single sugars which comprise the starch molecule.

Starches vary in the number and kind of sugar molecules present and are common constituents of both plant and animal cells. Animal starch, called *glycogen,* differs from plant starch in being quite highly branched (Fig. 3–8) and more soluble in water. Carbohydrates are stored in plants as starches and in animals as glycogen; glucose could not be stored as such, for its small molecules would leak out of the cells. The larger, less soluble starch and glycogen molecules will not pass through the plasma membrane. In man and other mammals, glycogen is stored especially in the liver and muscles. Liver glycogen is readily converted into glucose by four enzymes working sequentially and the glucose is then carried in the blood to other parts of the body.

Most plants have a strong supporting outer wall of *cellulose,* an insoluble compound sugar (polysaccharide) resembling starch in that it is made of many glucose molecules. However the chemical bonds between the successive glucose molecules of cellulose are β-glycosidic bonds, and are different from the ones in starch or glycogen. They are not split by the enzymes that cleave the bonds in starch.

Carbohydrates serve primarily as a readily available fuel to supply energy for metabolic processes in the cell. Glucose is ultimately metabolized to carbon dioxide and water and energy is released:

$$C_6H_{12}O_6 + 6O_2 \longrightarrow 6H_2O + 6CO_2 + energy$$

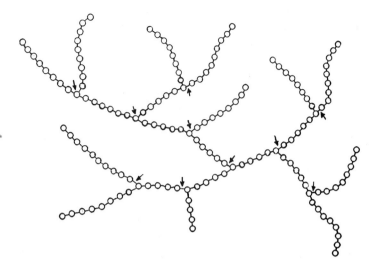

Figure 3-8 A diagrammatic representation of a portion of a glycogen molecule showing its highly branched structure. Each circle represents a glucose molecule bound by a glycoside bond from its carbon 1 to carbon 4 of the adjacent glucose in the straight chain portion of the molecule, or from its carbon 1 to carbon 6 of the adjacent glucose at the branch points indicated by arrows.

Some carbohydrates are combined with proteins (glycoproteins) or lipids (glycolipids) and serve as structural components of cells and cell walls. Glucosamine and galactosamine, nitrogen-containing derivatives of glucose and galactose, are important constituents of cell membranes and of supporting substances such as connective tissue fibers, cartilage and chitin, present in the hard outer shell of insects, spiders and crabs. *Ribose* and *deoxyribose* are five-carbon sugars of great importance biologically as components of ribonucleic acid (RNA) and deoxyribonucleic acid (DNA).

Lipids (Fats). True *fats* are also composed of carbon, hydrogen and oxygen, but have much less oxygen in proportion to the carbon and hydrogen than carbohydrates have. Fats have a greasy or oily consistency; some, such as beef tallow or bacon fat, are solid at ordinary temperatures; others, such as olive oil or cod liver oil, are liquid. Each molecule of fat is composed of one molecule of glycerol and three molecules of fatty acid. All such neutral fats, termed *triacyl glycerols,* contain glycerol but may differ in the kinds of fatty acids present. Fatty acids are long chains of carbon atoms with a carboxyl group (—COOH) at one end. All the fatty acids in nature have an even number of carbon atoms—palmitic has 16 and stearic, 18. Butyric acid, present in rancid butter, has four carbon atoms, and caproic acid, found in goat sweat, has six. Fatty acids with one or more double bonds are called "unsaturated." Oleic acid has 18 carbons and one double bond (and hence has two less hydrogen atoms than stearic). A fat common in beef tallow, tristearin, $C_{57}H_{110}O_6$, has three mole-

cules of stearic acid and one of glycerol (Fig. 3-9). Fats containing unsaturated fatty acids are usually liquid at room temperature, whereas saturated fats, such as tristearin, are solids.

Fats are important as biological fuels and as structural components of cells, especially cell membranes. Glycogen or starch is readily converted to glucose and metabolized to release energy quickly; the carbohydrates serve as short-term sources of energy. Fats yield more than twice as much energy per gram as do carbohydrates and thus are a more economical form for the storage of food reserves. Carbohydrates can be transformed by the body into fats and stored in this form—a restatement of the generally known fact that starches and sugars are "fattening."

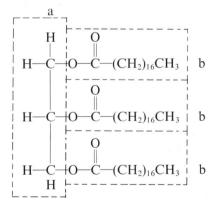

Figure 3-9 Structural formula of tristearin, a fat composed of glycerol (a) and three molecules of stearic acid (b). In the formula, $(CH_2)_{16}$ represents a chain of sixteen carbon atoms joined in a line, —C—C—..., to each of which are attached two hydrogen atoms.

Fats are important structural elements of the body. The plasma membrane around each cell and the nuclear membrane contain fatty substances as important constituents, and the myelin sheath around the nerve fibers (p. 74) has a high lipid content. Animals store fat as globules within the cells of adipose tissue. The layer of adipose tissue just under the skin serves as an insulator against the loss of body heat. Women tend to have a thicker layer of adipose tissue than do men and thus should be more tolerant to cold. Whales, which live in cold water and have no insulating hair, have an especially thick layer of fat (blubber) just under the skin for this purpose. The subcutaneous fat in humans keeps the skin firm in addition to restricting the loss of body heat.

The fat deposits are not simply long-term stores of foodstuff used only in starvation, but are constantly being used and re-formed. Studies with labeled fatty acids showed that mice replace half of their stored fats in seven days.

Besides the true fats, composed of glycerol and fatty acids, lipids include several related substances that contain components such as phosphorus, choline and sugars in addition to fatty acids. The *phospholipids* are important constituents of the membranes of plant and animal cells in general and of nerve cells in particular. The fatty acid portion of the phospholipid molecule is hydrophobic, not soluble in water. The other portion, composed of glycerol, phosphate and a nitrogenous base such as choline, is ionized and readily water-soluble. For this reason, phospholipid molecules in a film tend to be oriented with the polar, water soluble portion pointing one way and the nonpolar, fatty acid portion pointing the other.

Plants contain certain red and yellow pigments called *carotenoids*, which are included with the lipids because they are insoluble in water and have an oily consistency. Carotenoids are found in all plant cells from the lowest to the highest. They play some role in phototropism, the orientation of plants toward light. One of the common carotenoids, *carotene*, is a molecule with a six-carbon ring at each end of a long chain of carbon atoms, linked together with alternating single and double bonds. When the carotene molecule is cut in half it yields a molecule of *vitamin A.* The chemical present in the cells of the retina of the eye that is sensitive to light is termed *retinal.* Retinal is a derivative of vitamin A and undergoes a chemical reaction in the presence of light. It is involved in the actual reception of light stimuli. It is remarkable to realize that photoreceptors or eyes have evolved independently in three different lines of animals—the mollusks, the vertebrates and the insects. These organisms have no common evolutionary ancestor, i.e., the three types of eyes have no common evolutionary source, yet each one of these eyes contains the same chemical compound, retinal, involved in the photoreception process. In contrast to a number of other instances, the fact that retinal is present in each of these types of eyes is not the result of a common evolutionary ancestry, but perhaps is due to some unique fitness of this kind of molecule for the process of light reception.

Steroids. Steroids are complex molecules containing carbon atoms arranged in four interlocking rings, three of which contain six carbon atoms each and the fourth of which contains five (Fig. 3–10). Some steroids of biological importance are vitamin D, the male and female sex hormones, the adrenal cortical hormones, bile salts and cholesterol. Cholesterol is an important structural component of nervous tissue and other tissues, and the steroid hormones are of prime importance in regulating certain phases of metabolism.

Proteins. *Proteins* are compounds containing carbon, hydrogen, oxygen, nitrogen and usually sulfur and phosphorus. All enzymes, certain hormones, and many of the important structural components of the cell are protein.

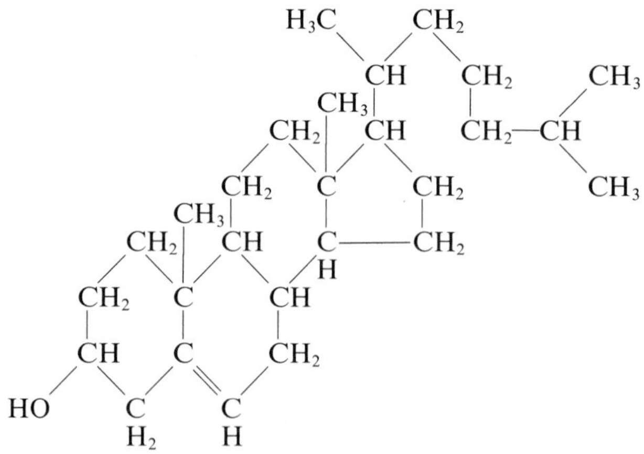

Figure 3–10 Structural formula of cholesterol, a sterol.

Proteins are among the largest molecules present in cells and share with nucleic acids the distinction of having great complexity and variety. A typical protein of great importance to the human body is *hemoglobin,* the red pigment responsible for the color of blood. Some idea of the complexity of the hemoglobin molecule can be gained from its formula: $C_{3032}H_{4816}O_{872}N_{780}S_8Fe_4$. (Fe is the symbol for iron.) Although the hemoglobin molecule is enormous compared with a glucose molecule, it is only a small-to-medium sized protein. A large fraction of the proteins present within cells are *enzymes,* biological catalysts which control the rates at which the many chemical processes of the cell occur.

Protein molecules are made of simpler components, known as *amino acids.* The twenty-odd amino acids commonly found in proteins all contain an amino group ($-NH_2$) and a carboxyl group ($-COOH$) but differ in their side chains. Glycine, the simplest, has an H for its side chain and alanine has a $-CH_3$ group (Fig. 3–11). The amino group enables the amino acid to act as a base and combine with acids; the acid group enables it to combine with bases. Amino acids and proteins in solution serve as *buffers* and resist changes in acidity or alkalinity. Amino acids are linked together to form proteins by *peptide bonds* between the amino group of one and the carboxyl group of another (Fig. 3–11).

Since each protein contains perhaps hundreds of amino acids combined in a certain proportion and in a particular order, an almost infinite variety of protein molecules is possible. Analytic methods have been developed for determining the exact sequence of the amino

acids in a protein molecule. Insulin, the hormone secreted by the pancreas and used in the treatment of diabetes, was the first protein whose structure was elucidated. Ribonuclease, an enzyme secreted by the pancreas, was the first enzyme for which the exact order of its amino acids was known. Not all proteins contain all of the possible amino acids.

It is possible to distinguish several different levels of organization in the protein molecule. The first level is the so-called *primary structure* which depends upon the sequence of amino acids in the polypeptide chain. This sequence, as we shall see, is determined in turn by the sequence of nucleotides in the RNA and DNA of the nucleus of the cell.

A second level of organization of protein molecules involves the coiling of the polypeptide chain into a helix or into some other regular configuration. The polypeptide chains ordinarily do not lie out flat in a protein molecule, but they undergo coiling to yield a three-dimensional structure. One of the common secondary structures in protein molecules is the *α-helix,* which involves a spiral formation of the basic polypeptide chain. The α-helix is a very uniform geometric structure with 3.6 amino acids occupying each turn of the helix. The helical structure is determined and maintained by the formation of hydrogen bonds between amino acid residues in successive turns of the spiral.

A third level of structure of protein molecules involves the folding of the peptide chain upon itself to form globular proteins. Again, weak bonds such as hydrogen, ionic and hydrophobic bonds form between one part of the peptide chain and another part, so that the

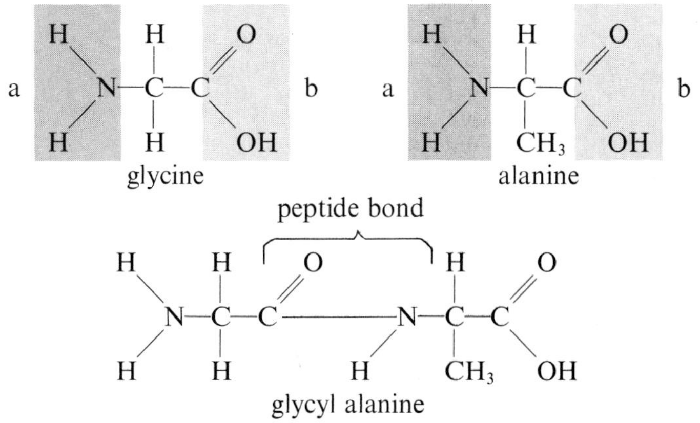

Figure 3–11 The structural formulas of the amino acids glycine and alanine, showing (a) the amino group and (b) the acid (carboxyl) group. These may be joined by a peptide bond to form glycyl alanine by the removal of water.

chain is folded in a specific fashion to give a specific overall structure of the protein molecule. Covalent bonds such as disulfide bonds $(-S-S-)$ are important in the tertiary structure of many proteins. The biological activity of a protein depends in large part on the specific tertiary structure which is held together by these bonds. When a protein is heated or treated with any of a variety of chemicals, the tertiary structure is lost. The coiled peptide chains unfold to give a random configuration accompanied by a loss of the biological activity of the protein. This change is termed "denaturation."

Those proteins composed of two or more subunits have a quaternary structure. This refers to the combination of two or more like or unlike peptide chain subunits, each with its own primary, secondary and tertiary structures, to form the biologically active protein molecule.

Each cell contains hundreds of different proteins and each kind of cell contains some proteins which are unique to it. Every species of plant and animal appears to have certain proteins which differ from those of every other species. The degree of difference in the proteins of two species depends upon the evolutionary relationship of the forms involved. Organisms less closely related by evolution have proteins which differ more markedly than those of closely related forms. Investigations of the similarities and dissimilarities of proteins have been useful in studies of evolution and have added strong confirmatory evidence to ideas of evolutionary relationships derived from other facts. Because of the interactions of unlike proteins, grafts of tissue taken from one species of animal usually will not grow when implanted into a host of different species, but degenerate and are sloughed off by the host. Indeed, even grafts made between members of the same species usually will not grow, but only grafts between genetically identical donors and hosts—identical twins or members of closely inbred strains.

When proteins are eaten, they are hydrolyzed to amino acids before they are absorbed into the bloodstream. The amino acids are carried to all parts of the body to be made into new protein or to be metabolized for the release of energy. When a human eats beef proteins in a steak, they are digested and split to their constituent amino acids. The human tissues then rebuild the amino acids into human proteins.

Pure amino acids have a sweet taste. Monosodium glutamate, or MSG, is the salt of glutamic acid, an amino acid of special importance to metabolism. It is widely used in cooking to add a "meaty" taste.

Proteins are of prime importance as structural components of cells and as the functional constituents of enzymes and certain hormones, but they may also serve as fuel for the liberation of energy. The amino acids first lose their amino group by an enzymatic reaction called *deamination.* The amino group reacts with other substances to form *urea,* and is excreted. The rest of the molecule may be changed, via a series of intermediate steps, into glucose and either used immediately as fuel or stored as glycogen.

Information about the conversion of protein to carbohydrates and to fats has been derived from experiments with substances labeled with isotopes of carbon, hydrogen and nitrogen. In prolonged fasting, after the glycogen and stored fats are exhausted, the cellular proteins may be used as fuel. The evidence available at present indicates that the human body (and animal cells in general) can manufacture some, but not all, of the amino acids if the proper raw materials are present. Those which cannot be made by the animal body must be obtained directly or indirectly from plants as food or perhaps from the bacteria that live in our intestines.

Plants can synthesize all the amino acids from simpler substances. The ones which animals cannot synthesize, but must obtain in their diet, are known as *essential amino acids.* It must be understood that these amino acids are no more essential as components of proteins than are any other ones, they are simply essential *in the diet,* since they cannot be synthesized. Animals differ in their biosynthetic capabilities; what is an essential amino acid for one species may not be essential in the diet of another. Individuals with certain inborn errors of metabolism may require other amino acids in their diet. For example, tyrosine is an essential amino acid for an individual with phenylketonuria (p. 723).

Nucleic Acids. Nucleic acids are complex molecules, larger than most proteins, and contain carbon, oxygen, hydrogen, nitrogen and phosphorus. They were first isolated by Miescher in 1870 from the nuclei of pus cells and gained their name from the fact that they are acidic and were first identified in nuclei. There are two classes of nucleic acid—one containing ribose and called *ribose nucleic acid* or *RNA*, and one containing deoxyribose and called *deoxyribonucleic acid* or *DNA*. There are many different kinds of RNA and of DNA which differ in their structural details and in their metabolic functions. DNA occurs in the chromosomes in the nucleus of the cell and, in much smaller amounts, in mitochondria and chloroplasts. It is the primary repository for biological information. In eukaryotic cells, RNA is present in the nucleus, especially in the nucleolus, in the ribosomes, and in lesser amounts in other parts of the cell.

Nucleic acids are composed of units, called *nucleotides,* each of which contains a nitrogenous base, a five-carbon sugar and phosphoric acid. Two types of nitrogenous bases, purines and pyrimidines, are present in nucleic acids (Fig. 3–12). RNA contains the purines *adenine* and *guanine* and the pyrimidines *cytosine* and *uracil,* together with the pentose, ribose, and phosphoric acid. DNA contains adenine and guanine, cytosine and the pyrimidine *thymine,* together with deoxyribose and phosphoric acid. The molecules of nucleic acids are made of linear chains of nucleotides, each of which is attached to the next by bonds between the sugar part of one and the phosphoric acid of the next. The specificity of the nucleic acid resides in the specific sequence of the four kinds of nucleotides present in the chain. For example, CCGATTA might represent a segment of a DNA molecule, with C = cytosine, G = guanine, A = adenine and T = thymine.

An enormous mass of evidence now indicates that DNA is responsible for the specificity and chemical properties of the *genes,* the units of heredity. There are several kinds of RNA, each of which plays a specific role in the biosynthesis of specific proteins by the cell (p. 702).

Nucleotides and Coenzymes. Related structurally to nucleic acids but with quite different roles in cellular function are several mono- and dinucleotides. Each is composed of phosphoric acid, ribose and a purine or pyrimidine base, like the units comprising the nucleic acids. Each of the bases may form a nucleoside triphosphate, with base, sugar and

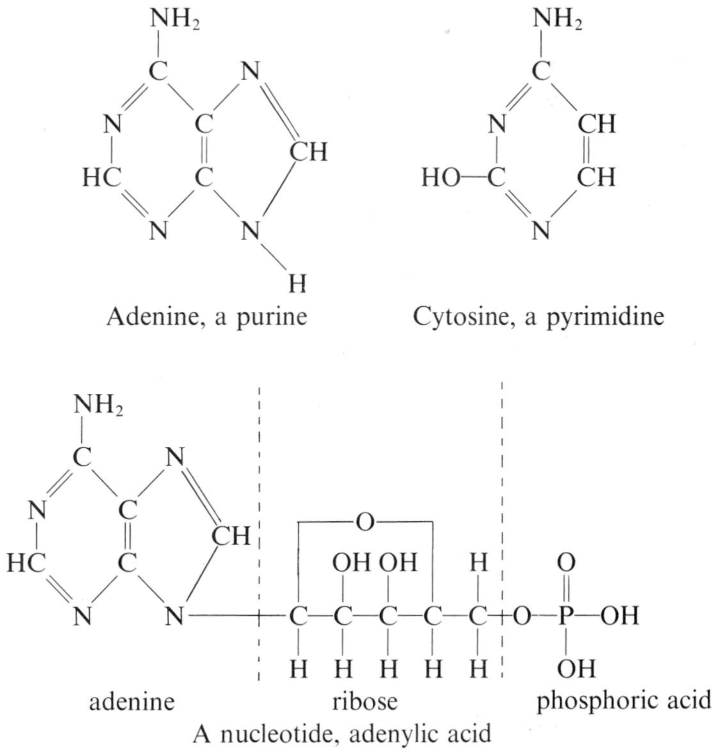

Figure 3–12 Structural formulas of a purine (adenine), a pyrimidine (cytosine), and a nucleotide (adenylic acid).

Adenine, a purine

Cytosine, a pyrimidine

adenine   ribose   phosphoric acid

A nucleotide, adenylic acid

three phosphate groups linked in a row. *Adenosine triphosphate*, abbreviated *ATP*, composed of adenine, ribose and three phosphates, is of major importance as the "energy currency" of all cells. The two terminal phosphate groups are joined to the nucleotide by energy rich bonds, indicated by the ~P symbol. The biologically useful energy of these bonds can be transferred to other molecules; most of the chemical energy of the cell is stored in these ~P bonds of ATP, ready to be released when the phosphate group is transferred to another molecule. *Guanosine triphosphate*, *GTP*, is specifically required in certain steps in the synthesis of proteins; *uridine triphosphate, UTP*, is specifically required in certain steps in carbohydrate metabolism, e.g., glycogen synthesis; and *cytidine triphosphate*, *CTP*, is specifically required for certain steps in the synthesis of fats and phospholipids. All four nucleotide triphosphates are necessary for the synthesis of RNA, and the four deoxyribose nucleoside triphosphates—dATP, dGTP, dCTP and dTTP—are required for the synthesis of DNA.

Completing the list of nucleotides important in metabolic processes are the dinucleotides, NAD, NADP and FAD. *Nicotinamide adenine dinucleotide*, abbreviated *NAD* (and also called diphosphopyridine nucleotide, DPN), consists of nicotinamide, ribose and phosphate attached to an adenine nucleotide, phosphate, ribose and adenine (Fig. 3–13). NAD is of prime importance as a primary electron and hydrogen acceptor in cellular oxidation reactions. Enzymes called *dehydrogenases* remove electrons and hydrogen from molecules such as lactic acid and transfer them to NAD, which in turn passes them on to other electron acceptors. *Nicotinamide adenine dinucleotide phosphate*, abbreviated *NADP* (and also called triphosphopyridine nucleotide,

TPN), serves as electron and hydrogen acceptor for certain other enzymes. NADP is exactly like NAD except it has a third phosphate group attached to the ribose of the adenine nucleotide. *Flavin adenine dinucleotide, FAD*, consists of riboflavin-ribitol-phosphate-phosphate-ribose-adenine and serves as a hydrogen and electron acceptor for certain other dehydrogenases. Notice that these dinucleotides have vitamins—nicotinamide or riboflavin—as component parts. These molecules NAD, NADP and FAD are termed *coenzymes;* they are cofactors required for the functioning of certain enzyme systems but are only loosely bound to the enzyme molecule and are readily removed. When they have accepted electrons and hydrogens, they are changed from their oxidized form, e.g., NAD, to their reduced form, NADH. They are converted back to the oxidized form when they transfer their electrons to the next acceptor in the chain of respiratory enzymes (p. 111).

In summary, the chemical components of the human body include water, inorganic salts and organic substances—nucleic acids, proteins, carbohydrates and fats, as well as many others. Carbohydrates and fats play only a small role in the structure of cells but are important as fuels; carbohydrates are readily available fuel; fats are more permanently stored supplies of energy. Nucleic acids have a primary role in storing and transmitting information. Proteins are structural and functional constituents of cells but may serve as fuel after deamination. The body can convert each of these substances into the others to some extent. The cell's contents comprise a very complex, multicompartmental system. Many of the properties of cells—muscle contraction, amoeboid motion and so on—involve changes in the biophysical state of this complex system.

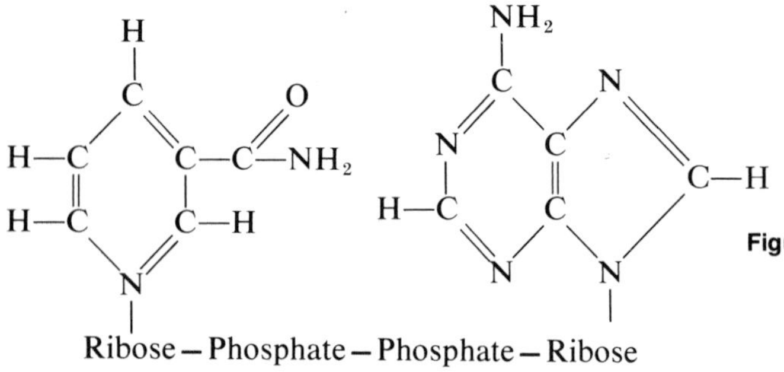

Figure 3–13 Nicotinamide adenine dinucleotide (NAD).

Ribose — Phosphate — Phosphate — Ribose

QUESTIONS

To answer these questions it will be helpful to read the material in Appendix I dealing with fundamental physical and chemical concepts.

1. List and discuss the characteristics of living things. Can you think of any that should be added to or deleted from the list presented in the text?
2. In what ways is the analogy between the cells of the body and the bricks of a house a useful one and in what respects does it fail?
3. Distinguish between an element and a compound.
4. What is the exact meaning of each of the following terms: atom, isotope, ion? Could a single particle of matter be all three simultaneously?
5. What properties of water make it an essential component of living matter?
6. What is the difference between an s orbital and a p orbital?
7. Define a chemical reaction. Does the mixing of alcohol and water constitute a chemical reaction?
8. What is the meaning of the symbol pH? What is the difference in hydrogen ion concentration of two solutions, one of which has a pH of 7.4 and the second a pH of 6.4?
9. Differentiate clearly between acids, bases and salts. What are the functions of salts in living matter?
10. What are the primary functions in the cell of (a) proteins, (b) fats, (c) carbohydrates, (d) nucleic acids and (e) steroids?
11. Distinguish between covalent and ionic bonds.
12. Compare glycosidic, peptide, and ester bonds. What is their importance in biology?

SUPPLEMENTARY READING

Some of the chemical aspects of cellular constituents are discussed in Giese's *Cell Physiology,* Florey's *General and Comparative Animal Physiology* and in *Cell Structure and Function* by A. G. Loewy and P. Siekevitz. *The Living Cell: Readings from the Scientific American,* edited by Donald Kennedy, is a rich collection of articles that have appeared in the Scientific American. The subject of atomic structure is entertainingly explored in George Gamow's *Mr. Thompkins Explores the Atom.* Additional discussions of acids, bases, salts and other chemical compounds can be found in an introductory chemistry text, such as Masterson and Slowinski's *Chemical Principles* or Brescia's *Chemistry: A Modern Introduction.*

In his essay, *Innovation in Biology,* George Wald discusses the concepts from the physical sciences that are needed to explain the unique behavior of living systems. J. A. Ramsay's *The Experimental Basis of Modern Biology,* DuPraw's *Cell and Molecular Biology* and McGilvery's *Biochemical Concepts* provide more detailed and advanced discussions of the complex molecules that make up biological systems.

CHAPTER 4

CELLS AND TISSUES

More than three hundred years ago Robert Hooke, using the newly invented microscope, made the remarkable observation that cork was not a homogeneous material, but consisted of tiny, boxlike cavities which he called **cells.** What he saw, of course, was the cellulose walls of dead cells; the important part of a cell is its contents rather than its cell wall.

The Bohemian physiologist Purkinje introduced, in 1839, the term **protoplasm** for the living contents of the cell. As more has been learned about cell structure and function, it has become clear that the living contents of the cell comprise an incredibly complex system of heterogeneous parts. The term "protoplasm" has no clear meaning in a chemical or physical sense, but it may still be used to refer to all the organized constituents of a cell.

4–1 THE CELL THEORY

Two Germans, Matthias Schleiden, a botanist, and Theodor Schwann, a zoologist, formulated, in 1838, the generalization which has since developed into the *cell theory:* The bodies of all plants and animals are composed of cells. New cells can come into being only by the division of previously existing cells, a generalization first stated by Virchow in 1855. The corollary of this, that all the cells living today can trace their ancestry back to ancient times, was pointed out by August Weismann about 1880. The cell theory includes the concept that the cell is the fundamental unit of both function and structure—the smallest representative bit that shows all the characteristics of living things.

Each cell contains a *nucleus* and is surrounded by a *plasma membrane.* Mammalian red blood cells lose their nucleus in the process of maturation, and skeletal muscles have several nuclei per cell, but these are rare exceptions to the general rule of one nucleus per cell. In the simplest plants and animals, all the living material is found within a single plasma membrane. These organisms may be considered to be unicellular (i.e., single-celled) or acellular (with bodies not divided into cells). However, they may have a high degree of specialization of form and function within this single cell (Fig. 4–1). The single cell may

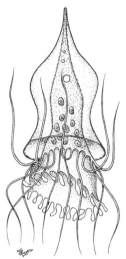

Figure 4–1 *Codonella companella,* a single-celled animal with a high degree of specialization of form and function within the single cell.

be quite large, larger than the whole body of some multicellular organisms. Thus it is wrong to infer that a single-celled animal is necessarily smaller or less complex than a many-celled one.

A single cell, if placed in the proper environment, will grow and eventually divide to form two cells. It is fairly easy to find an environment that will let single-celled plants and animals grow and multiply; for many of these, a drop of pond water will suffice. It is more difficult to prepare a medium that will support the growth and division of cells taken from the body of a man, chick or salamander. This was first accomplished by the American zoologist Ross Harrison, who in 1907 was able to culture salamander cells in an artificial medium outside the body. Since then, many types of plant and animal cells have been cultured *in vitro** and many important discoveries about cell physiology have been made using such systems.

---

*In vitro (Latin, in glass) refers to an experiment performed outside the plant or animal body, typically in a glass vessel of some sort. In contrast to this, *in vivo* refers to an experiment using an intact, living animal or plant body. If we inject some radioactive glucose (i.e., labeled with radioactive carbon) into a rat's vein and then measure the amount of radioactive carbon in the rat's breath and urine, we are performing an *in vivo* experiment. But if we culture some muscle cells in a solution of radioactive glucose in a glass vessel, and do analyses to determine the fate of the radiocarbon, we are performing an *in vitro* experiment.

The cells of different plants and animals and of different organs within a single plant or animal, present a bewildering variety of sizes, shapes, colors and internal structures, but all have certain features in common. Each cell is surrounded by a plasma membrane, and contains a nucleus and several kinds of subcellular organelles—mitochondria, granular endoplasmic reticulum, smooth endoplasmic reticulum, the Golgi complex, lysosomes and centrioles (Fig. 4–2).

4–2 EXCHANGES OF MATERIALS BETWEEN CELL AND ENVIRONMENT

The outer surface of each cell is bounded by a delicate, elastic covering, an integral functional part of the cell, termed the *plasma membrane.* This is of prime importance in regulating the contents of the cell, for all nutrients entering the cell and all waste products or secretions leaving it must pass through this membrane. It hinders the entrance of certain substances and facilitates the entrance of others. Cells are almost invariably surrounded by a watery medium. This might be the fresh or salt water in which a small organism lives, the tissue sap of a higher plant, or the plasma or extracellular fluid of one of the higher animals.

The plasma membrane behaves as though it has ultramicroscopic pores through which certain substances pass. The size of these pores determines the maximal size of the molecule that can pass through the membrane. Factors other than molecular size, such as the electric charge, if any, carried by the particles, the number of water molecules, if any, bound to the surface of the particle, and the solubility of the particle in lipids, may also be important in determining whether or not the substance will pass through the membrane. Infoldings of the plasma membrane may be continuous with channels that extend deep into the interior of the cell, providing paths for the entrance of some materials and for the removal of secretory and excretory products.

All membranes appear to have a basically similar *lipid bilayer* structure. The central region of the membrane consists of two layers of phospholipid, each just one molecule thick, with their hydrophobic (water-repelling) tails

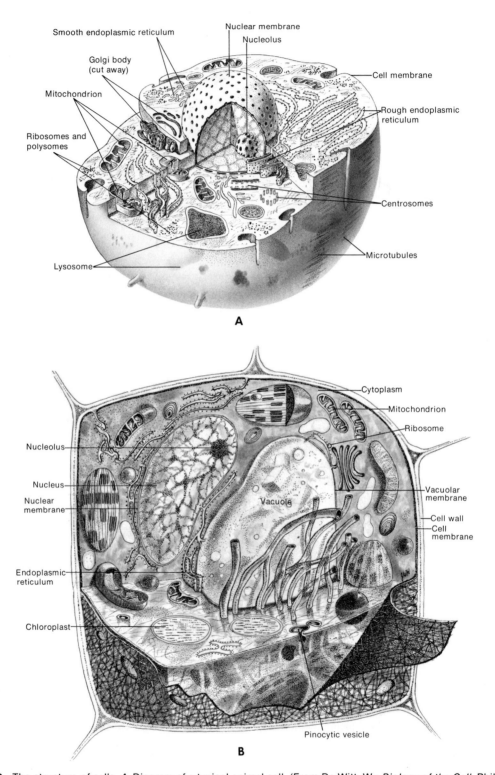

Figure 4–2 The structure of cells. *A*, Diagram of a typical animal cell. (From De Witt, W.: *Biology of the Cell.* Philadelphia, W. B. Saunders Co., 1976.) *B*, Diagram of a typical plant cell.

Illustration continued on opposite page

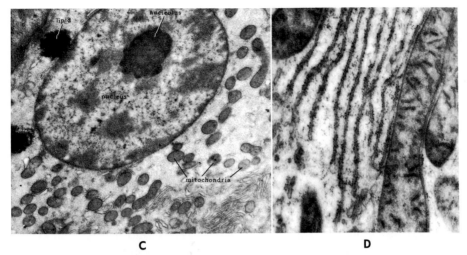

C   D

Figure 4–2 *Continued*   *C,* Electron micrograph of the nucleus and surrounding cytoplasm of a frog liver cell. The spaghettilike strands of endoplasmic reticulum are visible in the lower right corner. Magnification ×16,500. *D,* High-power electron micrograph of a rat liver cell. Granules of ribonucleoprotein are seen on the strands of endoplasmic reticulum, and structures with double membranes are evident within the mitochondria in the upper left corner and on the right. Magnification ×65,000. (*C* and *D* courtesy of Dr. Don W. Fawcett.)

aligned toward each other and with their polar head groups on the outside. The lipid bilayer is a mixture of phospholipid molecules in the fluid state. Special *membrane proteins* are associated with the lipid bilayer, some present only on the outer surface and others only on the inner surface. Some are found only *within* the membrane and still others extend completely through the lipid bilayer. Some of these membrane proteins are enzymes; others are receptors for hormones or other specific compounds. Certain of the proteins can move laterally within the membrane but are unable to rotate (Fig. 4–3).

The plasma membrane is much more than a simple cell envelope that prevents the ready movement of dissolved materials in either direction. It is an active functional structure with enzymatic mechanisms that move specific molecules in or out of the cell against a concentration gradient. Multiple small invaginations of the plasma membrane, termed *microvilli,* are present in certain cells, such as those lining the kidney tubules (p. 452).

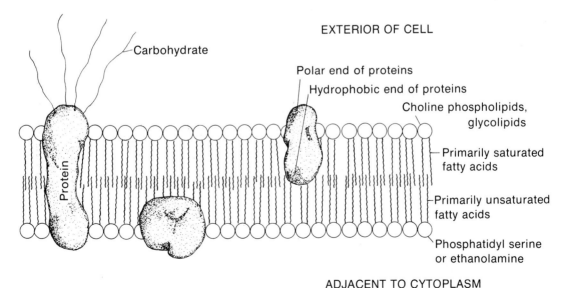

Figure 4–3 Diagram of the molecular architecture of biological membranes such as the plasma membrane or the membrane around the mitochondrion.

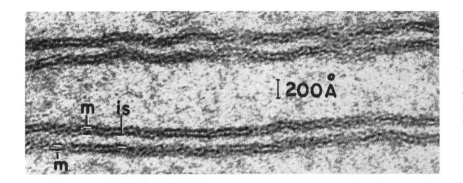

Figure 4-4 Electron micrograph of the cell membranes of intestinal cells showing the three-layered structure. Membrane, *m*; intercellular space, *is*. Magnification ×240,000.

High resolution electron micrographs of the plasma membrane (Fig. 4–4) show a dense-light-dense three-layered structure. The plasma membranes of animal, plant and bacterial cells and the membranes of many subcellular organelles all appear to have a similar three-layered structure.

Nearly all plant cells (but not most animal cells) have a thick *cell wall* made of cellulose, which lies outside the plasma membrane. This cell wall is nonliving, secreted by the cell substance. It is pierced in many places by tiny holes through which the contents of one cell connect with that of the adjacent cells and through which materials can pass from one cell to the next. These tough, firm cell walls provide support for the plant body.

Molecular Motion. What is the nature of the forces driving materials from the cell's environment to its interior and back? All molecules in liquids and gases characteristically tend to move or diffuse in all directions until they are distributed evenly throughout the available space. *Diffusion* may be defined as the movement of molecules from a region of high concentration to one of lower concentration, brought about by the kinetic energy of the molecules (Fig. 4–5). The rate of diffusion is a function of the size of the molecule and the

temperature. The molecules that make up all kinds of substances, even solids, are constantly in motion. The chief difference between the three states of matter—solid, liquid and gas—is simply the freedom of movement of the molecules present. The molecules of a solid are relatively closely packed and the forces of attraction between molecules will allow them to vibrate but not to move around. In the liquid state the molecules are farther apart, the intermolecular forces are weaker, and the molecules move about with considerable freedom. Finally, in the gaseous state, the molecules are so far apart that intermolecular forces are negligible and the movement of the molecules is restricted only by external barriers.

When a drop of water is examined under the microscope, the motion of the water molecules is not evident, but if a drop of India ink (which contains fine particles of carbon) is added, the carbon particles move about continually in aimless and zigzag paths. Each carbon particle is constantly being bumped by water molecules, and the recoil from these bumps gives the carbon particle its motion. This motion of small particles is termed *Brownian movement* after Robert Brown, an English botanist, who first observed it when he looked through the microscope at some tiny pollen

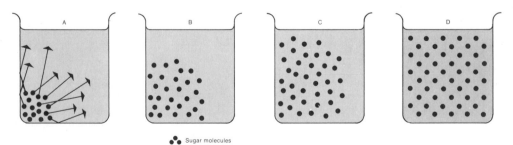

●● Sugar molecules

Figure 4–5 The process of diffusion. When a small lump of sugar is dropped into a beaker of water its molecules dissolve *(left)* and begin to diffuse. The process of diffusion over a long period of time will result in an even distribution of sugar molecules throughout the water in the beaker *(right)*.

grains in a drop of water. Brownian movement provides a model of how diffusing particles move.

The Speed of Diffusion. In the process of diffusion, each individual molecule moves in a straight line until it bumps into something—another molecule or the side of the container—then it rebounds and moves in another direction. The molecules continue to move even when they have become uniformly distributed throughout a given space; however as fast as some molecules move, for example, from left to right, others move from right to left, so that an equilibrium is maintained. Any number of substances will diffuse independently of each other within the same solution. Individual molecules may move at a rate of several hundred meters per second, but each molecule can go only a fraction of a nanometer before bumping into another molecule and rebounding. Thus the progress of any given molecule in a straight line is quite slow. This can be demonstrated by placing a bit of dye at the bottom of a glass cylinder filled with water. As days and weeks go by, the colored substance will gradually move upwards, but it will take months before the dye is uniformly distributed throughout the cylinder. Thus, although diffusion occurs very rapidly over microdistances, it takes a long time for a molecule to travel distances measured in centimeters.

This fact has important biological implications, for it limits the number of molecules of oxygen and nutrients that will reach an organism by diffusion alone. Only a very small organism that needs relatively few molecules per second can survive if it sits in one place and lets molecules come to it by diffusion. A larger organism must either have some means of moving to a new region or some mechanism for stirring up its environment to bring molecules to it. As a third alternative, some organisms live where the environment is constantly moving past it—in a river or in the intertidal zone on the seashore. The larger land plants, shrubs and trees have solved the problem by developing a tremendously branched root system, thus obtaining their raw materials from a large area of the surrounding environment.

Membrane Permeability. Whether or not a membrane will permit the molecules of a certain substance to pass through depends on its structure and on the size of the pores present. A membrane is said to be *permeable* if it will permit any substance to pass through, *impermeable* if it will permit no substance to pass, and *differentially permeable* if it will allow some but not all substances to diffuse through. Permeability, it must be emphasized, is a property of the *membrane*, not of the diffusing substance. All the membranes surrounding cells, nuclei, vacuoles and subcellular structures are differentially permeable.

The diffusion of a dissolved substance through a differentially permeable membrane is termed *dialysis.* To demonstrate the process of dialysis, one can make a pouch of collodion, cellophane or parchment and fill it with a solution of glucose. The pouch is then placed in a beaker of water, and the glucose molecules

Figure 4–6 Diagram illustrating the process of osmosis. *A,* When a 5 per cent sugar solution is placed in a sac made of a differentially permeable membrane such as cellophane and suspended in water, the water molecules diffuse into the sac, causing the column of water in the glass tube to rise. The larger glucose molecules are unable to pass through the pores in the cellophane. *B,* When equilibrium is reached, the pressure of the column of water in the tube just equals, and is a measure of, the osmotic pressure of the sugar solution.

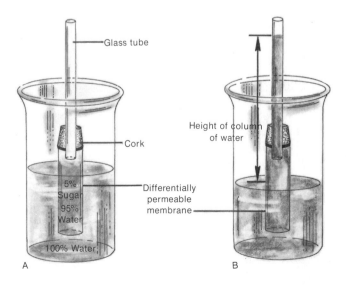

will dialyze through the membrane if the pores are large enough. After an appropriate interval of time, analyses of the glucose content of the liquid inside and outside the pouch will show that they are equal. Molecules will continue to diffuse through the membrane, but there will be no net change in concentration, for the rate of diffusion into the pouch equals the rate of diffusion out of the pouch.

If the pores of the membrane of the pouch are somewhat smaller, so that it is permeable to water molecules but not to glucose molecules, a different phenomenon may be observed. The pouch is fitted with a cork stopper through which passes a glass tube. When this pouch, filled with glucose solution, is placed in a beaker of water, the glucose molecules will be unable to penetrate the membrane and thus remain inside the bag (Fig. 4–6). The water molecules, however, diffuse through the membrane into the pouch. The liquid inside the membrane is 5 per cent glucose, hence it is only 95 per cent water. The liquid outside the membrane is 100 per cent water; therefore, water molecules move from a region of higher concentration (100 per cent, outside the pouch) to a region of lower concentration (95 per cent, inside the pouch). The diffusion of water or solvent molecules through a membrane is termed *osmosis.*

As osmosis continues, water rises in the glass tube. If an amount of water equal to the amount originally inside the pouch were to pass through the membrane, the glucose solution would be diluted to be 2.5 per cent glucose and 97.5 per cent water. The concentration of water on the outside of the pouch would still be higher than that inside, and osmosis will continue. Eventually the water in the glass tube will rise to a height such that the weight of the water in the tube exerts a pressure just equal to that resulting from the tendency of the water molecules to enter the bag. Then there will be no further *net* change in the amount of water within the bag. Osmosis will continue to occur in both directions through the differentially permeable membrane with equal speed.

Osmotic Pressure. The pressure of the column of water is termed the *osmotic pressure* of the sugar solution. The osmotic pressure is brought about by the tendency of the molecules to pass through the membrane and equalize the concentration of water molecules on the two sides. A more concentrated sugar solution would have a greater osmotic pressure, and thus would cause water to rise to a higher level in the tube. A 10 per cent sugar solution would cause water to rise approximately twice as high in the tube as a 5 per cent solution.

Dialysis and osmosis are simply two special forms of diffusion. Diffusion is the general term for the movement of molecules from a region of high concentration to one of lower concentration, resulting from their kinetic energy. Dialysis is the diffusion of dissolved molecules (solutes) through a differentially permeable membrane, and osmosis is the diffusion of solvent molecules through a differentially permeable membrane. In living systems, the solvent molecules are water.

The fluid compartment of every living cell contains dissolved salts, sugars and other substances that give that fluid a certain osmotic pressure. When this cell is placed in a fluid with the same osmotic pressure, there is no net movement of water molecules either into or out of the cell. There is no tendency for the cell either to swell or to shrink. This fluid is said to be *isotonic* or *isosmotic* to the fluid within the cell. Normally the blood plasma and all the body fluids are isotonic. They contain the same concentration of dissolved materials as the cells.

If the concentration of dissolved substances in the surrounding fluid is greater than the concentration within the cell, water tends to pass out of the cell and the cell shrinks. Such a fluid is termed *hypertonic* to the cell. If the surrounding fluid has less dissolved material than the cell, it is said to be *hypotonic.* Water will tend to enter the cell and cause it to swell. A solution of 0.9 per cent sodium chloride, sometimes termed "physiologic saline," is isotonic to the cells of human beings. Red blood cells placed in a solution of 0.6 per cent sodium chloride will swell and burst (Fig. 4–7), and those placed in a 1.3 per cent solution will shrink but those placed in 0.9 per cent sodium chloride will neither swell nor shrink.

Regulation of Intracellular Volume. A cell placed in a solution that is not isotonic to it may adjust to the changed environment by undergoing a change in its water content (by swelling or shrinking), so that it finally reaches the same concentration of solutes as its surrounding fluid. Some cells have the ability to

Figure 4–7 Red cells placed in a hypertonic solution *(left)* will shrink, and red cells placed in a hypotonic solution *(right)* will swell and burst. Red cells placed in an isosmotic solution *(center)* will neither swell nor shrink.

Shrunken, crenated in hypertonic solution

Normal shape in isosmotic solution

Swollen in hypotonic solution

Hemolyzed red cell "ghost"

pump water or certain solutes in or out through the plasma membranes, and in this way can produce an osmotic pressure that differs from that of the surrounding medium. Amoebas, paramecia and other protozoa living in pond water, which is very hypotonic, have evolved *contractile vacuoles,* which collect water from the interior of the cell and pump it to the outside. Plants living in fresh water also have the problem of dealing with the water that enters the cell by osmosis from the surrounding hypotonic environment. Plant cells have no contractile vacuole to pump out the water but their firm cellulose walls prevent undue swelling. As water enters, an internal pressure, called *turgor pressure,* is generated which counterbalances the osmotic pressure and prevents the entrance of additional water molecules. Turgor pressure is characteristic of plant cells in general and is responsible in part for the support of the plant body. A flower wilts when the turgor pressure in its cells has decreased owing to the lack of water.

Many organisms that live in the sea have phenomenal powers to accumulate selectively certain substances from the sea water. Seaweeds can accumulate iodine so that the concentration within the cell is 2,000,000 times that of the sea water. Tunicates, primitive chordates, can accumulate vanadium so that it is also some 2,000,000 times as concentrated within the tunicate cells as it is in the sea water. The pumping of water or solutes in or out of the cell against a concentration gradient is physical work and requires the expenditure of energy. The cell is able to move molecules against the gradient only as long as it is alive and carrying on metabolic activities that yield energy. If the cell is treated with some metabolic poison, such as cyanide, it loses its ability to produce and maintain differences in concentration on the two sides of its plasma membrane.

4–3 THE CELL NUCLEUS

Each cell contains a small, usually spherical or oval, organelle known as the *nucleus.* In some cells the nucleus has a relatively fixed position, usually somewhere near the center. In others it may move around freely and be found almost anywhere in the cell. The nucleus is an important center for the control of cellular processes. It contains the hereditary factors or genes responsible for controlling the traits of the cell and the organism, and it directly or indirectly controls many aspects of cellular activity. The nucleus is separated from the surrounding cytoplasm by a *nuclear membrane,* which regulates the flow of materials into and out of the nucleus. The electron microscope reveals that the nuclear membrane is double-layered and that there are pores in the double membrane (Fig. 4–8) through which the nuclear contents are continuous with the cytoplasm and through which even large molecules of RNA may pass. The outer of the two layers of the nuclear membrane appears to be continuous with the membranes of the endoplasmic reticulum and the Golgi complex.

When a cell is killed by fixation in the proper chemicals and stained with appropriate dyes, several structures within the nucleus become visible. These are difficult to observe in the living cell with an ordinary light microscope, but are evident by the use of a phase microscope. Within the semifluid ground substance, termed the karyoplasm, are suspended a fixed number of extended, linear, thread-like bodies called *chromosomes,* composed of DNA and protein, and containing the hereditary units or genes. In a nondividing cell (Fig. 4–9), the chromosomes usually appear as an irregular network of strands and granules termed *chromatin.* Just prior to nuclear division these strands condense into compact, rod-shaped chromosomes which are subsequently distrib-

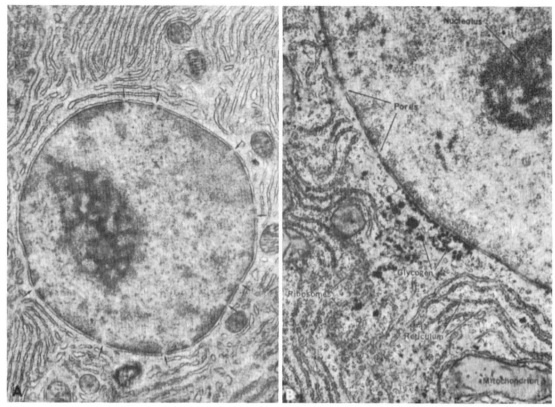

Figure 4–8 Electron micrographs showing pores in the nuclear membrane. The endoplasmic reticulum is evident in both pictures; *B* shows ribosomes on the endoplasmic reticulum. *A*, magnification ×20,000. *B*, magnification ×50,000. (Courtesy of Drs. Don W. Fawcett and Keith R. Porter.)

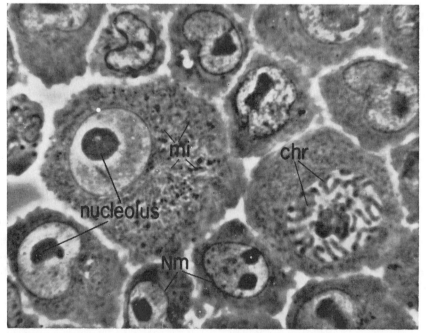

Figure 4–9 A photomicrograph taken by phase contrast microscopy of living cells from an ascites tumor. Chromosomes, *chr;* mitochondria, *mi;* nuclear membrane, *nm.* (Courtesy of N. Takeda.)

uted to the two daughter cells in exactly equal numbers. Each type of organism has a characteristic number of chromosomes present in each of its constituent cells. The fruit fly has eight chromosomes, sorghum has 10, the garden pea, 14, corn, 20, the toad, 22, the tomato, 24, the cherry, 32, the rat, 42, the human, 46, the potato, 48, the goat, 60 and the duck, 80. The somatic cells of higher plants and animals each contain two of each kind of chromosome. The 46 chromosomes in each human cell include two of each of 23 different kinds. They differ in their length, shape and the presence of knobs or constrictions along their length. In most species, including the human, the morphologic features of the different chromosomes are distinctive and permit the cytologist to distinguish the different pairs.

A cell with two complete sets of chromosomes is said to be *diploid.* Sperm and egg cells, which have only one of each kind of chromosome, one full set of chromosomes, are said to be *haploid.* They have just half as many chromosomes as the somatic cells of that same species. When the egg is fertilized by the sperm the two haploid sets of chromosomes are joined and the diploid number is restored.

The *nucleolus,* a spherical body found within the nucleus, is extremely variable in most cells, appearing and disappearing, changing its form and structure. There may be more than one nucleolus in a nucleus, but the cells of any given species of plant or animal usually have a fixed number of nucleoli. The nucleoli disappear when a cell is about to divide, and reappear afterwards. They appear to play a role in the synthesis of the ribonucleic acid constituents of ribosomes. If the nucleolus is destroyed by carefully localized ultraviolet or x-irradiation, cell division is inhibited. This does not occur in control experiments in which regions of the nucleus other than the nucleolus are irradiated.

Experimental Analyses of Nuclear Function. The role of the nucleus can be studied by removing it and observing the consequences. When we remove the nucleus of the single-celled amoeba with a microneedle, the cell will continue to live and move, but it cannot grow and will die after a few days. The nucleus, we conclude, is necessary for the metabolic processes, primarily the synthesis of

nucleic acids and proteins, that provide for growth and cell reproduction.

But, you may object, what if the operation itself, not the loss of the nucleus, caused the ensuing death? We can decide this by a controlled experiment, in which we subject two groups of amoebas to the same operative trauma, but have them differ in the presence or absence of a nucleus. For example, we can stick a microneedle into some of the amoebas, perhaps push the needle around inside to simulate the operation of removing a nucleus, but then withdraw the needle, leaving the nucleus inside. An amoeba treated to such a "sham operation" will recover and will subsequently grow and divide, demonstrating that it was the removal of the nucleus, not the operation, that brought about the death of the first group of amoebas.

A classic series of experiments demonstrating the importance of the nucleus in controlling the growth of the cell was performed by Hämmerling, using the single-celled *Acetabularia mediterranea.* This marine alga, which may be as much as five centimeters long, superficially resembles a mushroom. It has roots and a stalk surmounted by a large, disc-shaped umbrella. The entire organism is a single cell and has but one nucleus located near the base of the stalk.

Hämmerling severed the stalk and found that the lower part could live and would regenerate an umbrella, recovering completely from the operation (Fig. 4–10). The upper umbrella portion, which lacked a nucleus, may live for a considerable time but eventually would die without being able to regenerate a lower part. In *Acetabularia,* as in the amoeba, the nucleus appears to be necessary for those metabolic processes underlying growth. Regeneration is, of course, a form of growth. In further experiments, Hämmerling severed the stalk above the nucleus and made a second cut just below the umbrella. The isolated section of stalk, when replaced in sea water, was able to regenerate a partial or complete umbrella. This might, at first, seem to show that a nucleus is not necessary for regeneration; however, when Hämmerling removed the second umbrella, the stalk was unable to form a third one. From such experiments, Hämmerling concluded that the nucleus produces some substance required for

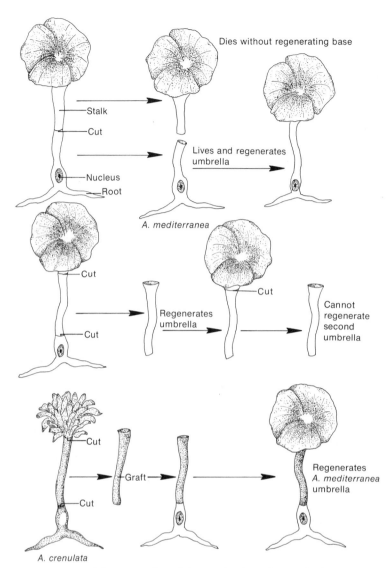

Figure 4–10 Hämmerling's experimental demonstration of the production of an umbrella-regenerating substance by the nucleus of the alga *Acetabularia. Lower line,* When a stalk from *A. crenulata* is grafted onto the base *(white)* of an *A. mediterranea* plant, the stalk regenerates an umbrella whose shape is that characteristic of *A. mediterranea* plants.

the formation of umbrellas which diffuses up the stalk and instigates the growth of an umbrella. In the experiments just described, enough of this umbrella material remained in the stalk after the initial cut to produce one new umbrella. After that had been exhausted in the formation of the first umbrella, no second regeneration was possible in the absence of a nucleus.

A second species of *Acetabularia (A. crenulata)* has a branched instead of a disc-shaped umbrella. Hämmerling grafted a piece of *A. crenulata* stalk that lacked a nucleus onto the

base of an *A. mediterranea* plant, which contained an *A. mediterranea* nucleus. A new umbrella developed at the top of the stalk. However, the shape of the umbrella was dictated not by the species that supplied the stalk but by the species that supplied the base and the nucleus. The nucleus, through the activity of its genes, provided the specific information that controlled the type of umbrella that was regenerated. The nucleus can override the tendency of the stalk to form an umbrella characteristic of its own species. The nucleus controls the activities of the other portions of the cell

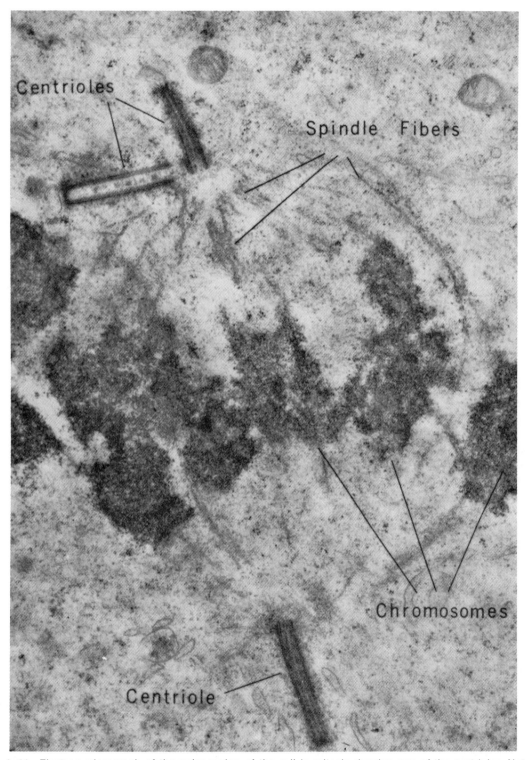

Figure 4–11 Electron micrograph of the polar region of the cell in mitosis showing one of the centrioles. Notice the tubular aspect of the spindle fibers which converge at the centriole. Two chromosomes are evident in the central portion of the figure. Magnification ×80,000.

because it has coded within its chromosomes the information needed for the synthesis of proteins and other materials on which cellular structure and function depend. Each time a cell divides, the entire set of instructions must be replicated, and a duplicate copy must be passed to each daughter cell.

4–4 CENTRIOLES AND SPINDLES

The cells of animals and certain lower plants contain, adjacent to the nucleus, two small, dark-staining cylindrical bodies, the *centrioles.* These play a prominent role at the time of cell division, separating, migrating to opposite poles of the cell, and organizing the spindle between them. With the electron microscope, each centriole appears as a hollow cylinder with a wall in which are embedded nine parallel, longitudinally oriented groups of tubules, with three tubules in each group (Fig. 4–11). The cylinders of the two centrioles are typically oriented with their long axes perpendicular to each other.

When cell division begins, the centrioles move to opposite sides of the cell. From each centriole there extends a cluster of raylike filaments called an *aster,* and between the separating centrioles a *spindle* forms, composed of protein threads with properties similar to those of the contractile proteins in muscle, *actin* and *myosin.* The protein threads of this spindle are arranged like two cones placed together, base to base, narrow at the ends or poles near the centrioles and broad at the equator of the cell. The spindle fibers stretch from equator to pole and comprise a definite structure. One can introduce a fine needle into a cell and push the spindle around. Spindle fibers may be isolated by special techniques (Fig. 4–12) and have been found to contain protein, largely a single kind of protein, together with a small amount of RNA. Some of the spindle fibers are attached to the centromeres of the chromosomes and appear to push or pull the chromosomes to the poles during mitosis. When the spindle fibers are viewed under high magnification in the electron microscope they appear to be fine, straight hollow tubules. During cell division they first lengthen and then shorten, but they do not appear to get thicker or thinner. This suggests

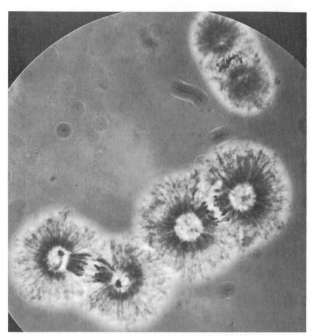

Figure 4–12 Photomicrograph of isolated spindle fibers of dividing cells from a sea urchin embryo. A metaphase figure appears in the upper right and two anaphase figures are seen below. (Courtesy of Dr. Daniel Mazia.)

that rather than stretching or contracting, new material is added to the fiber or removed from it as the spindle changes its size. When a moving spindle fiber was marked by burning a portion with ultraviolet light, the marked spot could be seen to move from a point near the equator to the pole and finally to disappear from the end of the fiber. This suggests that protein material is added at the equator, moves to the pole, and then is removed.

Cells bearing cilia on their exposed surfaces have a structure, termed a *basal body,* at the base of each cilium. The basal body strongly resembles the centriole in containing nine parallel tubules. Each cilium contains nine peripherally located, longitudinal filaments plus two centrally located ones. Like centrioles, basal bodies can duplicate themselves.

4–5 THE CELL CYCLE

The doubling of all the constituents of the cell followed by its division into two daughter cells is generally termed the *cell cycle.* A cell is "born" when its parent cell divides; it undergoes a cycle of growth and division and gives rise to two daughter cells. The doubling and halving processes in a population of cells are approximate rather than exact. For example, the size of connective tissue cells that are about to undergo division varies by about 12 per cent around the mean size. The size of the cell is limited in some way and a nondividing cell stops growing. It had been believed that growth was limited by the relative masses of nucleus and cytoplasm. This theory, proposed by Von Hertwig in 1908, suggested that when the ratio of the masses of nucleus and cytoplasm reaches a certain value the cell becomes unstable and cell division is triggered. It does seem to be true that the size of the cell is limited by the capacity of the single cell nucleus to support growth. If a cell is set up artificially containing two nuclei or two sets of chromosomes in a single nucleus, the cell can grow to twice the normal adult size. Cells exposed to an appropriate dose of x-rays can go through the cell cycle, but are unable to undergo division. The amount of genetic material in the nucleus increases and so does the size of the entire cell.

The cell cycle in a typical plant or animal cell requires about 20 hours for completion. Only an hour or so is devoted to mitosis; the rest of the time is required for interphase growth. Under optimal conditions of nutrition and temperature, the length of the cell cycle for any given kind of cell is constant. Under less favorable conditions it may be slowed, but it has not been possible to speed up the cell cycle and make cells grow faster. From this we infer that the duration of the cell cycle is the time required for carrying out some precise program that has been built into each cell. This program appears to include two parts: one having to do with the replication of the genetic material in the chromosomes and the other involving the doubling of all of the other constituents of the cell involved in growth.

The fertilized eggs of animals are unusual in that they rapidly undergo repeated divisions that may be completed in an hour or less. An egg cell is generously endowed with all of the kinds of molecules produced by growing cells and it divides repeatedly, yielding smaller and smaller cells. Its cell cycle does not involve growth, but simply preparation for division and the division process itself.

It was realized about 1950 that chromosomes undergo replication during the interphase and simply separate during mitotic division. Experiments using tritiated thymidine showed that DNA is synthesized during the interphase. The period of DNA replication during the interphase is termed the *S phase*. The time between mitotic division and the beginning of DNA replication is termed the *G_1 phase*, or "gap 1" phase. Following the completion of the S phase, the cell is usually not ready to divide immediately, but undergoes a second gap phase, the *G_2 phase*. The completion of G_2 is marked by the beginning of mitotic division, the *M phase*. In a typical cell cycle, the G_1 phase occupies about eight hours, the S phase six hours, the G_2 phase four and one-half hours, and the M phase one hour. Although the duration of each phase may vary somewhat, the greatest variation is found in the G_1 phase. Cells with very long cell cycles show most of the prolongation in the G_1 phase. Cells with very short cell cycles, such as eggs, have no measurable G_1 phase. In essentially all mammalian cells the S phase lasts six to eight hours and the G_2 phase three to five hours. The G_1 phase may be as short as a few hours or minutes or as long as several days or weeks. During the G_1 phase certain key processes occur that make it possible for the cell to enter the S phase and become committed to a future cell division (Fig. 4–13). In the S phase there is not simply a doubling of the amount of DNA in the nucleus, but an exact replication of each chromosome. The human cell contains 46 threads of DNA with a total length of 2 meters or more, all of which are stuffed into a nucleus about 5 micrometers in diameter. In the complex process of replication an exact copy is made of

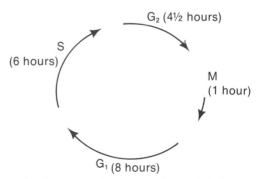

Figure 4–13 Diagram of the cell cycle showing durations of the four phases in a typical cell.

each of these 46 threads. Replication does not simply begin at one end of each thread and travel along to the other end; rather, each thread undergoes replication in many segments according to a definite program. The segments do not replicate in tandem in any one chromosome, nor does any one chromosome complete its replication before the next chromosome begins. When the last segments have been replicated, DNA synthesis shuts down and does not resume until the next cycle. The G_2 phase includes the final steps in the cell's preparation for division, during which there is increased protein synthesis.

Some of the enzymes of the cell are syn-

thesized at a constant rate throughout the cell cycle. A second group, the "step" enzymes, undergo abrupt increases at certain stages in the cell cycle, with different enzymes increasing at different stages. A third group, "peak" enzymes, increase at certain points in the cycle and thereafter decrease as though they were meant to carry out a particular job at a particular time and thereafter disappear. The enzymes involved in DNA synthesis are peak enzymes that appear only during the S phase. These observations imply that within the cell there are different programs for producing enzymes at certain times in the cell cycle. One of the basic dogmas of molecular genetics is that the pro-

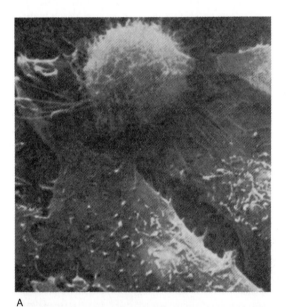

A

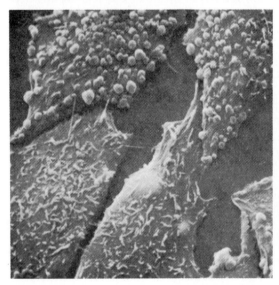

B

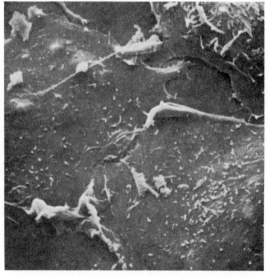

C

Figure 4–14　Hamster ovary cells in mitosis. *A,* The shape of the cell is changing from flat to round. A number of long, thin projections covering the cell surface help to secure the cell to the substrate. *B,* Flattened cells in late G_1 phase; the surfaces are covered with microvilli and blisterlike spheres (blebs). Several cells are establishing contact with each other. *C,* Increased flattening of cells having entered the S phase; no blebs now remain and microvilli are less abundant. "Ruffles" (light areas) appear on the edges of the cells. (Scanning electron micrographs by Keith R. Porter, David M. Prescott and Jearl F. Frye. From Mazia, D.: The cell cycle. Scientific American *230:*54, 1974. © 1974 by Scientific American, Inc. All rights reserved.)

duction of an enzyme is the end result of a read-out of the gene for that enzyme. That there is a program for the production of enzymes during the cell cycle suggests that there is a program for the read-out of the genes for those enzymes. The reproducibility of the time table of the phases of the cell cycle suggests that certain genes determine the progression of the cell cycle from one phase to the next.

Scanning electron microscopy reveals that cells grown in culture have quite different external appearances during the different phases (Fig. 4–14). During the M phase the cells are spherical and not strongly attached to the substrate. As they enter the G_1 phase they begin to flatten, the cell surface shows bubble-like blisters and microvilli appear. Later in the G_1 phase the margins of the cells become thin and active and appear ruffled. During the S phase the cells flatten further and their surface becomes smooth. As the cell enters the G_2 phase it once again thickens, the surface shows ruffles and microvilli, but not as many surface blisters as during the G_1 phase. Two important transition points occur in the cell cycle: one between the G_1 phase and the S phase when the replication of the chromosomes starts, and the other between the G_2 phase and the M phase when the chromosomes condense and mitosis begins. Certain animal viruses, the *Sendai viruses*, produce a change in the cell membrane allowing two or more adjacent cells to fuse into a single cell. The new cell contains within a single membrane the cytoplasm and the nuclei of all the cells undergoing fusion. Quite different kinds of cells—even cells from different species of animals—can be fused by the action of the Sendai viruses, forming hybrids that will not only carry out normal processes but will reproduce as hybrids.

By producing hybrids of cells in different phases of the cell cycle, we can ask whether there are certain signals that move a cell from one phase of the cycle to the next. When cells in the G_1 phase are fused with cells in the S phase, nuclei of the G_1 cells begin to make DNA long before they normally would. Thus a cell in the S phase contains something that triggers DNA synthesis, something that is absent from the cell that has not completed the G_1 phase.

Further insight has come from experiments with hybrids of different cells having cell cycles of different lengths. Hamster cells normally have a shorter G_1 phase than mouse cells, but in hybrids of mouse and hamster cells the mouse nuclei have a short G_1 phase like the hamster nucleus, and both nuclei enter the S phase simultaneously. This permits the inference that whatever turns on DNA synthesis in the hamster nucleus also turns it on in the mouse nucleus. Once the nuclei enter the S phase the nucleus of each species follows its own timetable and its own replication program. Those parts of the chromosomes that replicate first in the parent cell also replicate first in the hybrids, and those that normally replicate last in the parent are last in the hybrid. This suggests that the signal for starting the S phase does not exert any control over what happens once the phase has started. Rather, the replication program characteristic of each species must be contained in the nucleus of that species' cells. Thus there is something that pervades the cell and is responsible for initiating the replication of chromosomes. Moreover, it can induce any nucleus to enter the S phase whether or not it is "ready." This signal for beginning chromosome replication might be some specific molecule or it might be some specific alteration in the internal environment of the cell.

Entrance into the M phase is characterized by the condensation of the chromosomes; they become packed into threads that are visible under the microscope. When cells in the M phase are fused with cells in the G_2 phase, the chromosomes in the G_2 nucleus undergo condensation (Fig. 4–15). The chromosomes are double structures; they had undergone replication in the preceding S phase. However, when an M phase cell is fused with a cell in the G_1 phase, thereby forcing the chromosomes in the G_1 nucleus to condense before replication, the chromosomes are seen to be single rather than double structures. When an M phase cell is fused with a cell in the S phase the chromosomes in the S phase nucleus condense, but they condense in small fragments. Thus something in the M phase cell forces chromosomes to condense even when they are not ready to do so. Whatever it is, it appears to work in any kind of cell, even when fused with the cell of an animal only remotely related. If one cell contains condensed chromosomes in the M phase it can force premature chromosome condensation in the other cell.

An organism such as a higher animal,

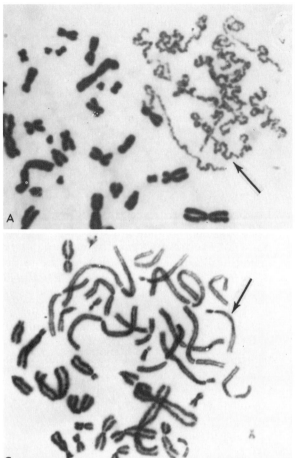

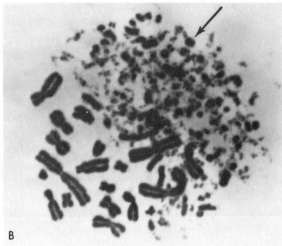

Figure 4–15 Prematurely condensed chromosomes (PCC) of Chinese hamster ovary cells. Cells synchronized in mitosis were fused with interphase cells with the help of U.V. inactivated Sendai virus. Metaphase chromosomes are darkly stained and well condensed; PCC are lightly stained with variable morphology. *A,* G_1 cell was fused with a mitotic cell. G_1-PCC *(arrow)* have a single chromatid and are greatly spiralized. *B,* Heterophasic binucleate cell formed by fusion of a cell in S phase with one in mitosis. S-PCC appear to be fragmented because of uneven condensation of S phase chromatin, dependent upon the state of DNA replication. Some chromosome regions have replicated; others have not yet started replication or are in the active process of replication. Replicated elements consist of double chromatids *(arrow)*; unreplicated regions are seen as single chromatids. Actual site of replication (replicon) in the chromatin does not undergo condensation; hence it appears as a gap between the condensed regions. *C,* Product of fusion between G_2 and mitotic cells. G_2-PCC *(arrow)* are long and slender, with the two chromatids still attached to each other. (Courtesy of Dr. Potu N. Rao, The University of Texas System Cancer Center, M. D. Anderson Hospital and Tumor Institute, Houston, Tex.)

which can be viewed as a society of cells, includes some cells that will reproduce and others that will not. The cells of tissues that perform certain special services—those of the nervous system and the muscular system—do not in general reproduce at all. In tissues such as the skin, the blood-forming system and epithelial linings, new cells are produced at a rate which just compensates for the continued loss of old cells. The cells involved in immune reactions and the healing of wounds multiply at a rapid rate as a bodily response to some external provocation. Finally, in malignant cancer the cells no longer abide by the rules of governance of the organism and go through their cycles in an anarchic fashion. There is nothing special about the cell cycle in cancer cells compared with the cell cycle in normal cells. The tempo, the cycle and its phases are the same, but the difference is that cancer cells repeat the cycle without restraint.

A very simple rule emerges from these studies. Cells that do not divide never enter the S phase, and conversely, cells that enter the S phase almost always complete that phase and go on to divide. Thus replication of the chromosomes is a strong commitment ultimately to undergo division, and the major control of cell division lies in whether or not a cell enters into replication. However, cells that are not going to divide are not cells that have become stuck somehow in the transition between the G_1 and the S phase. Nondividing cells are probably best considered as ones that have not entered the cycle at all. The many kinds of specialized cells in the body are noncycling cells, making no progress at all toward cell division. Many of the cells that have left the cell cycle can be made to reenter it by placing them under specific culture conditions in the laboratory. Presumably agents that cause cancer must in some way cause noncycling cells to enter the cycle. The conversion of lymphocytes from noncyling to cycling cells is of importance in the immune response. The cells begin to grow and divide, producing cells that contribute to the formation of antibodies. Noncycling lymphocytes can be transformed into cycling ones in the laboratory by exposing them to *lectins*, specific kinds of plant proteins. When noncycling lymphocytes are artificially stimulated by exposure to lectins, nearly 24 hours pass before they enter the S phase and begin to replicate the chromosomes. The delay apparently is due to the need of these stimulated lymphocytes to produce the enzymes required for the replication process as a preliminary to entering the S phase. Thus they are not stuck in the G_1 phase, ready to go into the S phase when they are signaled. When stimulated, a noncycling cell must first do the things

that a cycling cell does in the G_1 phase before it can enter the S phase and replicate its chromosomes.

4–6 MITOSIS

The regularity of the process of cell division ensures that each daughter cell will receive exactly the same number and kind of chromosomes that the parent cell had. If a cell receives more or less than the proper number of chromosomes by some malfunctioning of the process of cell division, the resulting cell may show marked abnormalities and may be unable to survive. Although the chromosome appears to split longitudinally into two halves, in fact,

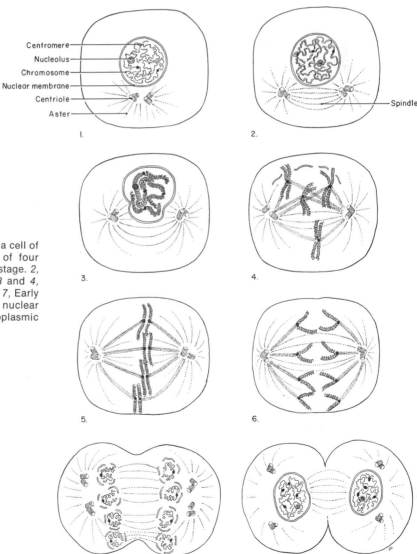

Figure 4–16 Diagram of mitosis in a cell of an animal with a diploid number of four (haploid number = two). *1,* Resting stage. *2,* Early prophase: centriole divided. *3* and *4,* Later prophase. *5,* Metaphase. *6* and *7,* Early and late anaphase. *8* Telophase: nuclear membrane has reappeared and cytoplasmic division has begun.

each original chromosome has brought about the synthesis of an exact replica of itself immediately adjacent to it during the preceding S phase. The old and new chromosomes are identical in structure and function and at first lie so close to one another that they appear to be one. As mitosis proceeds and the chromosomes contract, the line of cleavage between them becomes visible. In each human cell at the time of mitosis each of the 46 chromosomes has produced an exact replica of itself so that for a time there are 92 chromosomes in the cell. Then as cell division is completed, 46 go to one and 46 go to the other daughter cell. A rather complicated mechanism ensures an exactly equal division of the chromosomes between the two daughter cells.

The term mitosis in a strict sense refers to the division of the nucleus into two daughter nuclei and the term *cytokinesis* is applied to the division of the cytoplasm to form two daughter cells, each containing a daughter nucleus. Nuclear division and cytoplasmic division, although almost invariably well synchronized and coordinated, are separate and distinct processes. Each mitotic division is a continuous process, with each stage merging imperceptibly into the next one. For descriptive purposes mitosis may be divided into four stages: *prophase, metaphase, anaphase* and *telophase* (Fig. 4–16). Between mitotic divisions a nucleus is said to be in the *interphase* or resting stage. The nucleus is "resting" only with respect to division, however, for during this time it may be very active metabolically. It is difficult to realize from a description, a diagram or even from a prepared slide of cells undergoing mitosis (Fig. 4–17) just how active this process is. Time lapse movies of the process reveal graphically what an extremely active process mitosis is.

Prophase. *Prophase* begins when the chromatin threads begin to condense and the chromosomes appear as a tangled mass of threads within the nucleus. Initially the chromosomes are stretched maximally and the individual chromomeres are clearly visible. In certain favorable instances the chromomeres differ enough in size and shape so that individual ones can be recognized. When the chromosomes subsequently contract, the chromomeres come to lie close together and individual ones can no longer be distinguished. Each part of the

doubled chromosome is called a *chromatid;* the two chromatids are held together at the centromere, which remains single until the metaphase.

At the beginning of prophase the centriole divides and the two daughter centrioles migrate to opposite sides of the cell. From each centriole there extends a cluster of raylike filaments called an *aster.* Between the separating centrioles a *spindle* forms. While the centrioles have been separating and the spindle has been forming, the chromosomes in the nucleus have been contracting, getting shorter and thicker. Their double nature, which might not have been visible before, can now be clearly seen.

Metaphase. When the chromosomes have contracted fully and are short, dark-staining, rodlike bodies, the nuclear membrane disappears and the chromosomes line up across the equatorial plane of the spindle which has been forming around them (Fig. 4–16). Prophase is complete, and the short period during which the chromosomes are in the equatorial plane constitutes the *metaphase.* At this time the centromere divides and the two chromatids become completely separate *daughter chromosomes.* The division of the centromeres occurs simultaneously in all of the chromosomes, under the control of some as yet unknown mechanism. The daughter centromeres begin to move apart, marking the beginning of anaphase. In the division of human cells, prophase lasts from 30 to 60 minutes and metaphase from two to six minutes. The times vary considerably for different tissues and for different species.

Anaphase. The chromosomes separate, one daughter chromosome going to each pole. The events from the time when the chromosomes first begin to move apart until they reach the poles constitute the *anaphase,* a period which lasts 3 to 15 minutes. The mechanism which moves the chromosomes to the poles is unknown. One theory suggests that the spindle fibers contract in the presence of ATP and pull the chromosomes to the poles. Spindles isolated from cells about to divide can be induced to contract when ATP is added. With the aid of the spindle fibers all of one set of daughter chromosomes are gathered at one pole and all of the other set at the other pole. The chromosomes moving toward the poles usually assume a **V** shape with the centromere at the apex

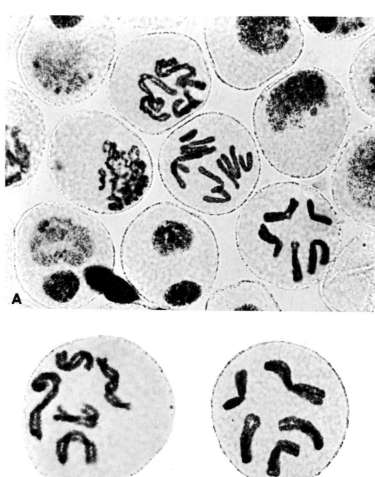

Figure 4–17 Photographs of mitosis in cells of the plant *Trillium erectum. A,* Field of normal microspore cells ×1400. *B,* Late prophase ×2300. C, Normal metaphase cell ×2300. D, Normal anaphase cell ×2300. *E,* Binucleate microspore (early) ×2000. (Courtesy of A. H. Sparrow and R. F. Smith.)

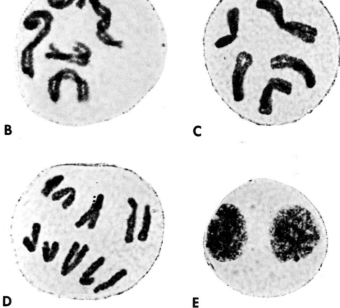

pointing toward the pole. It appears that whatever force moves the chromosome to the pole is applied at the centromere. Evidence from electron microscopy indicates that a spindle fiber attaches at the centromere. Chromosomes that lack a centromere, perhaps as a result of exposure to x-radiation, do not move at all in mitosis.

Telophase. When the chromosomes reach the poles the *telophase* begins. This period is roughly equal in duration to the prophase and lasts 30 to 60 minutes. The chromosomes elongate and return to the resting condition in which only chromatin threads or granules are visible. A nuclear membrane forms around each daughter nucleus. This completes nuclear division *(karyokinesis),* and the division of the cell body *(cytokinesis)* follows. The division of animal cells is accomplished by a furrow which encircles the surface of the cell in the plane of the equator. The furrow gradually deepens and separates the cytoplasm into two daughter cells, each of which has a nucleus. In plants, division occurs by the formation of a *cell plate,* a partition which forms in the equatorial region of the spindle and grows laterally to the cell wall. The cell plate is secreted by the endoplasmic reticulum. Each daughter cell then forms a cell membrane on its side of the cell plate and the cellulose cell walls are finally formed on either side of the cell plate.

Control of Mitosis. The frequency of mitosis varies widely in various tissues and in different species. In human red bone marrow, for example, where 10 million red blood cells are produced per second, 10 million mitoses must occur per second. In other tissues, such as those of the nervous system, mitoses occur very rarely. During the early development of an organism cell divisions take place extremely rapidly, every 30 minutes or so. The factors controlling the stage in development at which the mitotic rate slows and ceases are unknown. Cell divisions in the central nervous system largely cease in the first few months of life, whereas cell divisions in the red bone marrow, the lining of the digestive tract and the lining of the kidney tubules continue until the end of life.

The factors that initiate mitosis are not known exactly, but the ratio between the volume of the nucleus and the volume of the cytoplasm appears to play a role. The increase in the size of a cell involves the synthesis of proteins, nucleic acid, lipids and other cellular components. This requires the transport of substances back and forth through both nuclear and cell membranes. Although the volume of a sphere increases as the *cube* of its radius, the surface of a sphere increases only as the *square* of the radius. A cell is usually not a simple sphere, but comparable relationships hold and the volume of a cell increases more rapidly than the surface of the cell or nuclear membranes. At a critical point the surface of the nucleus becomes inadequate for the exchange of materials between nucleus and cytoplasm necessary to provide for further growth. The division of the cell greatly increases the surface of both nuclear and cell membranes without increasing the volume, and it is believed that this limiting factor in the *nucleoplasmic ratio* somehow initiates mitosis.

This nucleus-cytoplasmic mass theory, proposed by von Hertwig in 1908, requires some changes, for factors other than cell mass are involved in controlling mitosis. Cells can be made to divide before they have doubled in size and the eggs of many organisms grow to very large sizes before dividing. There is some evidence that the two cyclic nucleotides, cyclic AMP and cyclic GMP, play roles in regulating many cellular functions, including cell division. The two have opposite, antagonistic effects on the growth and division of certain cells grown in culture, cGMP stimulating growth and cAMP inhibiting it.

The Biological Significance of Mitosis. The process of mitosis ensures the precise and equal distribution of chromosomes to each of two daughter nuclei so that each cell in a multicellular organism has exactly the same number and kind of chromosomes as every other cell. The chromosomes contain genetic information coded in DNA and the regular and orderly mitotic process ensures that this genetic information is precisely distributed to each daughter nucleus; each cell has all the genetic information for every characteristic of the organism. Thus it is understandable why a single cell from a fully differentiated adult plant can, under suitable conditions in cell culture, develop into an entire plant.

4–7 CYTOPLASMIC ORGANELLES: MITOCHONDRIA

The material within the plasma membrane but outside the nuclear membrane is termed *cytoplasm.* Under the light microscope this appears to be composed of a semifluid ground substance in which are suspended a variety of droplets, vacuoles, granules and rodlike or threadlike structures. The electron microscope, however, reveals that the "cytoplasm" is an incredibly complex maze of membranes and spaces enclosed by membranes (Fig. 4–18). When a thin section of a cell is examined under the electron microscope, these membranes have the appearance of a profusion of spaghetti-like tubular strands termed *endoplasmic reticulum.* When viewed in three dimensions, these strands are sheetlike membranes that fill most of the space in the cytoplasm. The

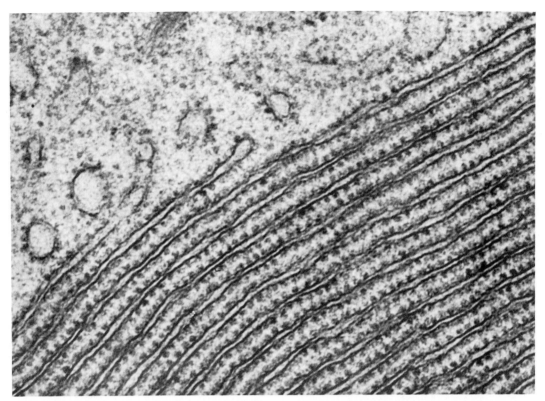

Figure 4–18 Electron micrograph of the granular endoplasmic reticulum of a pancreatic acinar cell. This form of the reticulum consists of parallel arrays of broad flat sacs or cisternae. The outer surface of their limiting membranes is studded with particles (15 nm.) of ribonucleoprotein (ribosomes). Osmium tetroxide fixation, ×85,000.

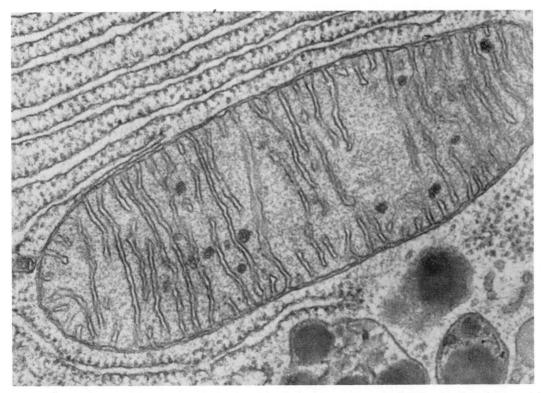

Figure 4–19 Electron micrograph of a typical mitochondrion from the pancreas of a bat showing the cristae, matrix and matrix granules. Endoplasmic reticulum is seen at the upper left and some lysosomes at the lower right. ×79,000. (Courtesy of Keith R. Porter.)

remainder of the space is occupied by other specialized structures with specific functions—mitochondria, the Golgi apparatus, centrioles and plastids.

Both plant and animal cells contain *mitochondria,* organelles ranging in size from 0.2 to 5 microns and in shape from spheres to rods and threads. The number of mitochondria per cell may range from just a few to more than a thousand. When living cells are examined, their mitochondria can be seen to move, to change size and shape, to fuse with other mitochondria to form bigger structures, or to cleave to form shorter ones. They are usually concentrated in the region of the cell with the highest rate of metabolism.

Some of the larger mitochondria are visible in the light microscope, but the details of their internal structure are revealed only in the electron microscope (Fig. 4–19). Each mitochondrion is bounded by a double membrane, the outer layer of which forms a smooth outer boundary and the inner layer of which is folded repeatedly into parallel plates that extend into the center of the mitochondrial cavity (Fig. 4–20). These plates may meet and fuse with folds coming in from the opposite side. Each of the outer and inner membranes is a unit membrane consisting of a central double layer of phospholipid molecules and a layer of protein molecules on each side. The shelf-like inner folds, called *cristae,* contain the enzymes of the electron transmitter system, which are of prime importance in converting the potential energy of foodstuffs into biologically useful energy for cellular activities. The semifluid material within the inner compartment, the matrix, contains certain enzymes of the Krebs citric acid cycle (p. 104). The mitochondria, whose prime function is the release of biologi-

cally useful energy, have been aptly termed the "powerhouses" of the cell.

Biochemists are able to homogenize cells and separate mitochondria from other subcellular organelles by differential high speed centrifugation. These purified mitochondria, when incubated *in vitro,* metabolize carbohydrates and fatty acids to carbon dioxide and water, utilizing oxygen and releasing energy-rich phosphate compounds in the process. Mitochondria swell and contract as they carry out these metabolic functions.

Mitochondria (and chloroplasts) are unusual in that they are under the control of two distinct genetic systems. DNA present in the mitochondria provides the genetic code for about 10 per cent of the mitochondrial protein, especially the hydrophobic polypeptides of the inner mitochondrial membrane. Nuclear DNA provides the genetic code for the remainder of the mitochondrial proteins, which are synthesized, like other proteins, on the ribosomes outside of the mitochondria and subsequently taken up and incorporated into the mitochondrial structure.

Biologists have speculated about the evolutionary origin of mitochondria. The cells of bacteria do not contain mitochondria, but they do contain membranes in which are embedded the enzymes of the electron transmitter system. In some bacteria these membranes are just inside the plasma membrane. Other bacteria, such as certain marine forms, have a complex system of parallel membranous sheets that stretch across the central part of the cell (Fig. 4–21). The enzymes of the electron transmitter system are embedded in these membranes. One can speculate that as cells grew larger and more complex, these membranes underwent folding and finally pinched off to form discrete

Inner membrane
particles

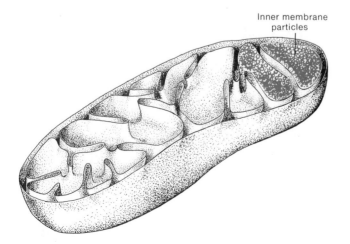

Figure 4–20 Diagram of a mitochondrion illustrating its internal structure and the arrangement of its inner and outer membranes. The inner surface of the inner membrane is covered with regularly spaced polygonal structures connected to the membrane by a narrow stalk. These structures, termed elementary particles, are believed to contain the enzymes involved in oxidative phosphorylation.

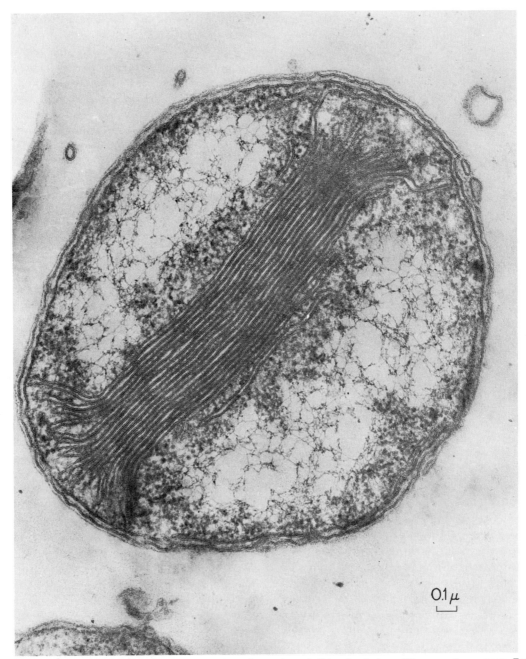

Figure 4–21   An electron micrograph of a thin section of a marine nitrifying bacterium, *Nitrosocystis oceanus*. Extending across the central portion of the cell are the parallel lamellae of the membranous organelle, which contain the enzymes of the electron transmitter system. On either side are light areas containing strands of DNA and near the periphery of the cell are ribosomes, which appear as dark dots. Outside the plasma membrane is a cell wall composed of four dense layers. (Courtesy of Dr. S. W. Watson.)

organelles, which were the precursors of the present-day mitochondria. Other investigators have speculated that entire bacterial cells, complete with their membranes of electron transmitter enzymes, invaded larger cells and took up a symbiotic existence as the mitochondria of the larger cells.

4–8 CHLOROPLASTS

The cells of most plants contain *plastids*, small bodies involved in the synthesis or storage of foodstuffs. The most important plastids, *chloroplasts*, contain the green pigment, chlorophyll, which imparts the green color to

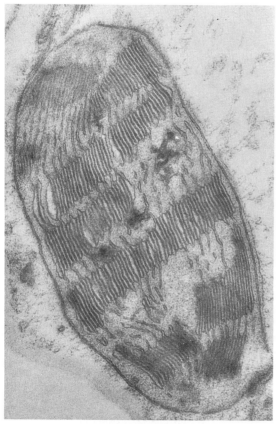

Figure 4–22 An electron micrograph of a chloroplast from a leaf of the tobacco plant, *Nicotiana rustica,* showing the fine structure of the grana. ×30,000. Note the alternate layers of protein and lipid in the grana within the chloroplast, and the membrane separating the chloroplast from the surrounding cytoplasm. (Courtesy of Dr. E. T. Weier.)

plants and is of paramount importance in photosynthesis in trapping the energy of sunlight. The chloroplasts of higher plants are typically disc-shaped structures, some 5 microns in diameter and 1 micron thick. The electron microscope reveals a complex internal structure consisting of a lamellar arrangement of tightly stacked membranes (Fig. 4–22). Each cell has some 20 to 100 chloroplasts which can grow and divide to form daughter chloroplasts. Within each chloroplast are many smaller bodies called *grana,* which contain the chlorophyll (Fig. 4–23).

The chloroplast is not a simple bag of chlorophyll. Indeed, the capacity of chlorophyll to capture light energy depends upon its distribution within the lamellae of the grana.

A layer of chlorophyll molecules and a layer of phospholipid molecules are sandwiched between layers of protein. This arrangement spreads the chlorophyll molecules

over a wide area and the layered structure may be important in facilitating the transfer of energy from one molecule to an adjacent one during photosynthesis. The material surrounding each granum is termed the stroma. The several grana within each chloroplast are connected one with another by sheets of membranes which pass through the stroma.

Other colorless plastids, *leukoplasts,* serve as centers for the storage of starch and other materials. *Chromoplasts,* a third type of plastid, contain pigments and are responsible for the colors of flowers and fruits.

4–9 RIBOSOMES AND THE SYNTHESIS OF PROTEINS

Cells, such as those of the pancreas, that are especially active in protein synthesis, are crowded with the membranous labyrinth of the endoplasmic reticulum (Fig. 4–18). Other cells may have only a scanty supply of such membranes. Two types are found, *granular* or *"rough" endoplasmic reticulum* to which are bound many *ribosomes,* and *agranular* or *smooth endoplasmic reticulum,* consisting of membranes alone. Bound to the granular reticulum or free in the cell matrix are many *ribosomes,* ribonucleoprotein particles on which proteins are synthesized. Both smooth and rough endoplasmic reticulum may be found in the same cell. The agranular endoplasmic reticulum may play some role in the process of cellular secretion. The tightly packed sheets of endoplasmic reticulum may form tubules some 50 to 100 nanometers in diameter. In other regions of the cell, the cavities of the endoplasmic reticulum may be expanded, forming flattened sacs called *cisternae.* The membranes of the endoplasmic reticulum divide the cytoplasm into a multitude of compartments in which different groups of enzymatic reactions may occur. The endoplasmic reticulum serves a further function as a system for the transport of substrates and products through the cytoplasm to the exterior of the cell and to the nucleus.

After mitochondria have been sedimented from homogenized cells by centrifugation a heterogeneous group of smaller particles, termed *microsomes,* can be sedimented by centrifuging at about 100,000 times the force of

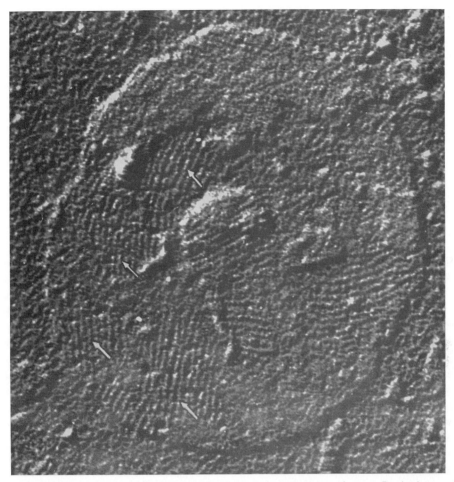

Figure 4–23 Electron micrograph showing the inner surface of one compartment of grana. By shadow-casting, a uniform linear arrangement of spheroids of 10 × 20 nm. called quantosomes, can be demonstrated (*arrows*). (Courtesy of M. Calvin.)

gravity. Ribosomes can be separated from the rest of the microsomal fraction by treatment with appropriate detergents. Isolated ribosomes can then synthesize proteins *in vitro* if supplied with the appropriate instructions in the form of messenger RNA and with an assortment of amino acids, an energy source and the other enzymes and transfer RNAs required (see Chapter 29). Ribosomes are ubiquitous, occurring in all kinds of cells from bacteria to higher plants and animals. Ribosomes contain three kinds of RNA and 55 proteins and are composed of two subunits, shaped like overstuffed chairs, which are combined to form the active protein synthesizing unit (Fig. 4–24). The smaller subunit (molecular weight about 600,000) contains one kind of RNA and 21 kinds of protein; the larger subunit (molecular weight about 1,300,000) contains two kinds of RNA and 34 kinds of protein. The specific role in protein synthesis of certain of these ribosomal proteins has now been discovered.

Ribosomes are synthesized in the nucleus and pass to the cytoplasm where they are active in synthesizing proteins. Ribosomes may be bound to the membranes of the endoplasmic reticulum, or they may be free in the matrix of the cytoplasm. In many cells, clusters of five or six ribosomes, termed *polysomes* (Fig. 29–21), appear to be the functional unit that is effective in protein synthesis. It is estimated that a bacterial cell, such as *Escherichia coli*, contains some 6000 ribosomes and that the rabbit reticulocyte (precursor of the red blood cell) contains some 100,000 ribosomes. Ribosomes are remarkably uniform in their size, structure and composition, whether isolated from bacterial or mammalian sources.

The microsomal fraction contains, in addition to the ribosomes involved in the synthesis of polypeptide chains, several less well characterized particles that contain enzymes involved in the metabolism of other types of compounds.

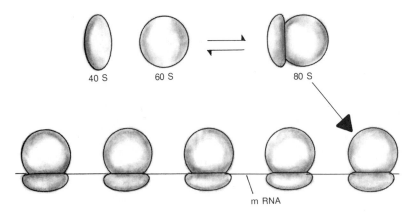

40 S   60 S   80 S

m RNA

Polysome

Figure 4–24 Diagram illustrating the assembly of the two subunits 40 S and 60 S to form a ribosome, 80 S, the subcellular organelle on which proteins are synthesized. The 80 S ribosomes attach to a messenger RNA strand to form a polysome.

4–10 OTHER INTRACELLULAR ORGANELLES

The *Golgi complex* is another cytoplasmic component present in almost all cells except mature sperm and red blood cells. Under the electron microscope, the Golgi complex is seen to consist of parallel arrays of membranes without granules, which may be distended in certain regions to form small vesicles or vacuoles filled with cell products (Fig. 4–25). The complex is usually located near the nucleus surrounding the centrioles and is believed to serve as a temporary storage place for proteins and other compounds synthesized in the endoplasmic reticulum. There is evidence to suggest that proteins made in the cisternae of the endoplasmic reticulum are sealed off in little packets of endoplasmic reticulum and travel to the Golgi complex. There they are repackaged in larger sacks made up of membranes from the Golgi complex. In these packets they travel to the plasma membrane, which fuses with the membrane of the vesicle, opening the vesicle and releasing its contents to the exterior of the cell.

The Golgi apparatus of plant cells may be involved in the secretion of the cellulose of the cell walls (Fig. 4–26). It usually appears as discrete bodies dispersed throughout the cell, each one of which consists of a stack of flattened vesicles that are slightly dilated at the edges.

Many, if not most, cells have hollow cylindrical cytoplasmic subunits termed *microtubules* (Fig. 4–11), which appear to be important in maintaining or controlling the shape of the cell. Microtubules play a role in such cellular movements as the movement of chromosomes by the mitotic spindle (p. 58) and serve as channels for the oriented flow of cytoplasmic constituents within the cell. Microtubules are also the major structural components of cilia and flagella.

Microtubules are composed of several types of proteins, one of which, termed *tubulin*, is present in largest amount. Tubulin is made up of two different subunits, termed α and β tubulin, which differ in their composition and molecular weights. Microtubules are long, hollow cylinders with wall thicknesses of 4.5 to 7 nm and diameters ranging from 20 to 30 nm. The thickness of the walls of microtubules corresponds approximately to the dimensions of the subunits of tubulin; this suggests that the walls are one molecule thick. The microtubules of nerve axons play a role in the rapid transport of proteins and other molecules, such as the hormones of the posterior pituitary and the hypothalamic releasing factors, down the axon to the tip (p. 558).

Solid cytoplasmic *filaments* are present in some cells in addition to hollow microtubules. These protein filaments play additional roles in cell structure and motion. The cytoplasm of skeletal muscle fibers contains many long *myofibrils*, protein filaments that participate in muscle contraction (p. 477).

Another group of intracellular organelles found in animal cells are *lysosomes*. About the size of mitochondria but somewhat less dense, lysosomes consist of a membrane-bounded structure containing a variety of enzymes that can hydrolyze the macromolecular constituents of the cell, proteins, polysaccharides and nu-

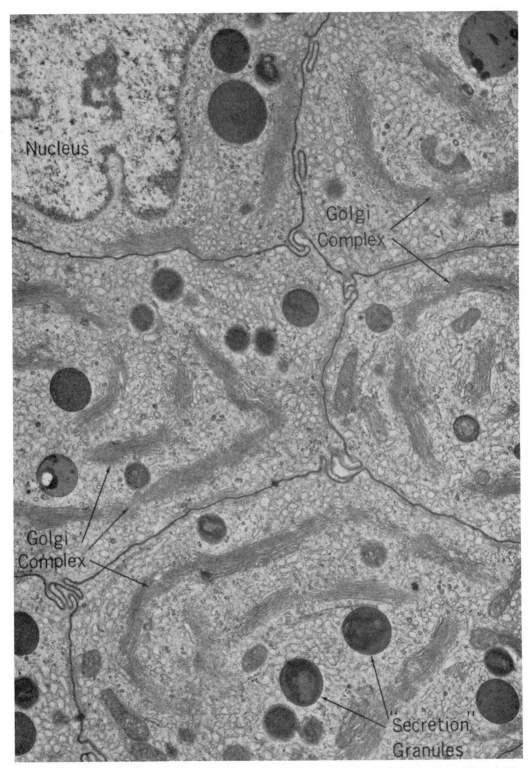

Figure 4–25 Electron micrograph of rabbit epididymis showing the extensive Golgi complex, evident in the parallel arrays of the membranes. Magnification ×9500. (From Fawcett, D. W.: *The Cell*. Philadelphia, W. B. Saunders Co., 1966.)

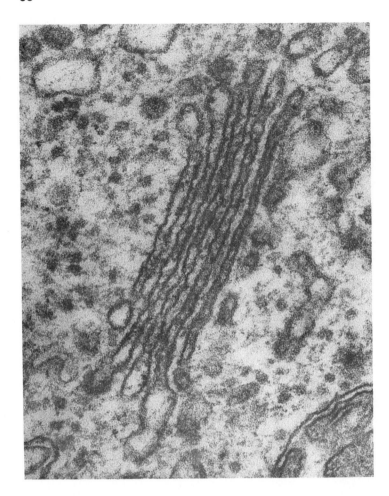

Figure 4–26 The Golgi body of a root cell of garlic, ×270,000. (From Norstog, K., and Long, R. W.: *Plant Biology*. Philadelphia, W. B. Saunders Co., 1976. Photo courtesy of Dr. Laszlo Hanzely.)

cleic acids. These packets serve to segregate these enzymes in the intact cell and presumably prevent their digesting the contents of the cell. Rupture of the lysosomal membrane releases these enzymes and accounts, at least in part, for the lysis of dead or dying cells and the resorption of cells such as those in the tail of a tadpole during metamorphosis. Since the lysosomes contain enzymes that can hydrolyze the major cellular constituents when the lysosomes rupture and release them, they have been termed "suicide bags" by the Belgian biochemist, Christian De Duve.

In addition to these living elements, the cytoplasm may contain *vacuoles*, bubble-like cavities filled with watery fluid and bordered by a vacuolar membrane similar in structure to the plasma membrane. Vacuoles are fairly common in the cells of plants and of lower animals but are rare in those of higher animals. Most protozoa have *food vacuoles*, containing food undergoing digestion, and *contractile vacuoles*, which serve to remove excess water

from the cell. Cytoplasm, in addition, may contain granules of stored starch or protein, or droplets of oil.

4–11 PLANT AND ANIMAL CELLS

Animal and plant cells differ in three major respects. Animal cells, but not the cells of higher plants, have a centriole; plant cells, but not animal cells, have plastids; and plant cells have a stiff cell wall of cellulose which prevents their changing position or shape, whereas animal cells usually have only a thin plasma membrane and thus are able to move about and alter their shape.

Most cells, both plant and animal, are too small to be seen with the naked eye. Their diameters range from 1 to 100 micrometers, and a speck 100 micrometers in diameter is near the lower limit of visibility. A few species of amoebas are a millimeter or two in diameter; some single-celled plants such as *Acetabularia* may be more than a centimeter long. Some of the largest single cells are the egg cells of fishes and birds. The egg cell of a large bird may be several centimeters in diameter. Only the yolk of the egg is the true egg cell; the white of the egg is a noncellular material secreted by the hen's oviduct.

4-12 METHODS OF STUDYING CELLS

Many of the basic principles of biology have been discovered by observations and experiments with single cells. Living cells can be examined in a drop of fluid, using a microscope to reveal the movement of amoebas or the beating of the hairlike projections of cilia that cover the bodies of paramecia.

The discovery of methods of culturing cells removed from the body of a higher animal or plant has enabled investigators to bring to light many new facts about cell function and structure. In culturing animal cells, a special complex nutritive medium, made of blood plasma, an extract of embryonic tissues and added vitamins and other chemicals, is prepared and sterilized.

Animal cells may be grown in "tissue culture" as a sheet of cells on a solid substrate bathed in nutritive medium or in "cell culture" with individual cells suspended in a liquid nutritive medium. Human connective tissue cells (fibroblasts) will survive in culture and undergo a relatively fixed number of mitoses (cell generations) — about 50 — after which they can no longer divide. In contrast, cancer cells can survive in culture and undergo divisions indefinitely. Muscle cells in culture divide a few times and then fuse to form a muscle fiber, becoming contractile.

Details of cell morphology may be studied by using a bit of tissue that has been killed quickly by a chemical "fixative" that does not destroy cell structure and then sliced with a *microtome* and stained with special dyes. The stained slices, mounted on a glass slide and covered with a glass cover slip, are then ready to be observed under the microscope.

The nucleus, mitochondria and other specialized parts of the cell differ chemically, and therefore they combine with different dyes and are stained characteristically. Tissues are prepared for electron microscopy by fixation with osmic acid, mounted in acrylic plastic to be cut into extremely thin sections with a glass or diamond knife and placed on a fine grid to be inserted in the path of the beam of electrons.

4-13 TISSUES

One of the major trends in the evolution of both plants and animals has been that leading to specialization and division of labor of the constituent cells. The cells that make up the body of a tree or a man are not all alike; each is specialized to carry out certain functions. This specialization allows the cells to function more efficiently but also makes the parts of the body more interdependent: the injury or destruction of one part of the body may result in death of the whole. The advantages of specialization, however, outweigh the disadvantages.

A *tissue* may be defined as a group or layer of similarly specialized cells which together perform certain special functions. The study of the structure and arrangement of tissues is known as *histology.* Each kind of tissue is composed of cells which have a characteristic size, shape and arrangement. Tissues may consist of noncellular materials in addition to living cells; blood and connective tissues, for example, contain some nonliving material between the cells.

Biologists differ somewhat in their ideas of how the various types of tissue should be classified and, consequently, of how many types of tissue there are. We shall classify animal tissues in six groups — epithelial, connective, muscular, blood, nervous and reproductive — and the tissues of higher plants in four categories — meristematic, protective, fundamental and conductive.

Epithelial Tissues. *Epithelial tissues* are composed of cells which form a continuous layer or sheet covering the body surface or lining cavities within the body. They may have one or more of the following functions: protection, absorption, secretion and sensation. The epithelia of the body protect the underlying cells from mechanical injury, from harmful chemicals and bacteria, and from drying. The epithelia lining the digestive tract absorb food and water into the body. Other epithelia secrete a wide variety of substances as waste products or for use elsewhere in the body. Finally, since the body is entirely covered by epithelium, it is obvious that all sensory stimuli must penetrate an epithelium to be received. Examples of epithelial tissues are the outer layer of the skin, the lining of the digestive tract, the lining of the windpipe and lungs and the lining of the kidney tubules. Epithelial tissues are divided into six subclasses according to their shape and function.

Squamous epithelium, composed of flattened cells shaped like pancakes or flagstones (Fig. 4–27 A), is found on the surface of the skin and the lining of the mouth, esophagus and vagina. In humans and higher animals, there are usually several layers of these flat cells piled one on top of another, a condition called *stratified squamous epithelium.*

The tissue lining the kidney tubules, composed of cells that are cube-shaped, resembling dice, is known as *cuboidal epithelium* (Fig. 4–27 B).

The cells of *columnar epithelium* are elongated, resembling pillars or columns; the nucleus is usually located near the base of the cell (Fig. 4–27 C). The stomach and intestines are lined with columnar epithelium.

Columnar cells may have *cilia* on their free surface. These small cytoplasmic projections (Fig. 4–27 D) beat rhythmically and move materials in one direction. Most of the respiratory system is lined with *ciliated epithelium.* The cilia assist in removing particles of dust and other foreign material.

Sensory epithelium is composed of cells specialized to receive stimuli (Fig. 4–27 E). The cells lining the nose—the olfactory epithelium, responsible for the sense of smell—are an example.

The cells of *glandular epithelium* (Fig. 4–27 F) are specialized to secrete substances such as milk, wax or perspiration. They are either columnar or cuboidal in shape.

Connective Tissue. The connective tissues, bone, cartilage, tendons, ligaments and fibrous connective tissue, support and hold together the other cells of the body. The cells of these tissues characteristically secrete a large amount of nonliving material, called *matrix,* and the nature and function of each kind of connective tissue is determined largely by the nature of this intercellular matrix. The cells thus perform their functions indirectly by secreting a matrix which does the actual connecting and supporting.

In *fibrous connective tissue* the matrix is a thick, interlacing, matted network of microscopic fibers secreted by and surrounding the connective tissue cells (Fig. 4–28 A). Such tissue occurs throughout the body and holds skin to the muscle, keeps glands in position and binds together many other structures. Tendons and ligaments are specialized types of fibrous connective tissue. *Tendons* are not elastic but are flexible, cable-like cords that connect muscles to each other or to bone. *Ligaments* are somewhat elastic and connect one bone to another. There is an especially thick mat of connective tissue fibers just below the skin of

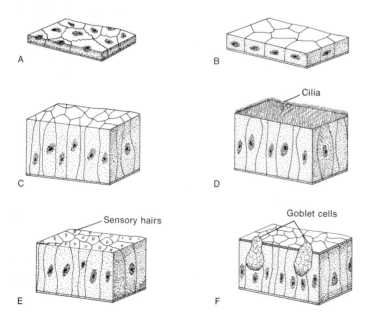

Figure 4–27 Types of epithelial tissue. *A,* Squamous epithelium; *B,* cuboidal epithelium; *C,* columnar epithelium; *D,* ciliated columnar epithelium; *E,* sensory epithelium (cells from the lining of the nose); *F,* glandular epithelium: two single-celled glands (goblet cells) in the lining of the intestine.

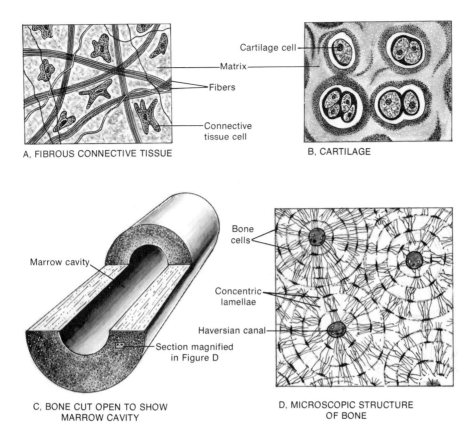

Cartilage cell

Matrix

Fibers

Connective
tissue cell

A, FIBROUS CONNECTIVE TISSUE

B, CARTILAGE

Marrow cavity

Bone
cells

Concentric
lamellae

Haversian canal

Section magnified
in Figure D

C, BONE CUT OPEN TO SHOW
MARROW CAVITY

D, MICROSCOPIC STRUCTURE
OF BONE

Figure 4–28 Types of connective tissues.

most vertebrates (Fig. 4–29). When this is treated chemically (tanned) it becomes leather.

Connective tissue fibers contain a unique protein, *collagen,* rich in the amino acids glycine, proline and hydroxyproline. When the fibers are treated with hot water, collagen is converted into the soluble protein *gelatin.* Because there is so much connective tissue present, about one-third of all the protein in the human body is collagen. The collagen units comprising the fibers consist of a helix made of three peptide chains wound around each other in a supercable and joined by hydrogen bonds.

The supporting skeleton of vertebrates is composed of *cartilage* or *bone.* Cartilage is the supporting skeleton in the embryonic stages of all vertebrates, but is largely replaced in the adult by bone in all but the sharks and rays. In the human body, cartilage may be felt in the supporting structure of the ear flap (pinna) and the tip of the nose. It is firm yet elastic. Cartilage cells secrete this hard, rubbery matrix around themselves and come to lie singly or in groups of two or four in small cavities in the homogeneous, continuous matrix (Fig. 4–28 *B*). The

cartilage cells in the matrix remain alive; some of them secrete fibers which become embedded in the matrix and strengthen it.

Bone cells remain alive and secrete a bony matrix throughout a person's life. The matrix contains calcium salts (the mineral *hydroxyapatite*) and proteins, principally collagen. The calcium salts in the bony matrix make it very hard, and the collagen prevents the bone from being overly brittle. The dense bony matrix enables the skeleton to support the body against the pull of gravity. Contrary to appearance, bone is not a solid structure. Most bones have a large *marrow cavity* in the center (Fig. 4–28 *C*), which may contain yellow marrow, mostly fat, or red marrow, the tissue in which red and certain white blood cells are made.

Extending through the matrix of the bone are *haversian canals,* microscopic channels through which blood vessels and nerves supply and control the bone cells. The bone matrix is secreted in concentric rings (lamellae) around the canals, and the cells become embedded in cavities in these rings (Fig. 4–28 *D*). Bone cells are connected to each other and to

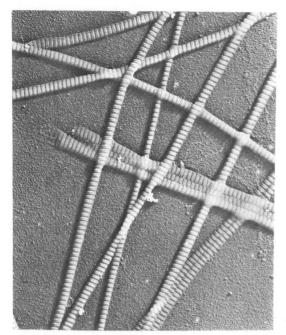

Figure 4–29 Electron micrograph of collagen fibrils teased from a preparation of calf skin. Note the regular periodic striations of the fibrils, which indicate its repeating structural unit. ×33,000. (Courtesy of Dr. Jerome Gross.)

the haversian canals by cellular extensions which lie in minute canals (canaliculi) in the matrix. The bone cells obtain oxygen and raw materials and eliminate wastes by way of these minute canals. Bones contain other cells which can dissolve and remove the bony substance; thus the shape of the bone can gradually change in response to continued stresses and strains.

Muscular Tissue. The movements of most animals result from the contraction of elongate, cylindrical or spindle-shaped cells, each of which contains many small, longitudinal, parallel, contractile fibers called *myofibrils,* composed of the proteins myosin and actin. Muscle cells can perform mechanical work only by contracting, by getting shorter and thicker; they cannot push. Three types of muscle can be distinguished (Fig. 4–30). *Cardiac muscle* is found in the walls of the heart and *smooth muscle* in the walls of the digestive tract and certain other internal organs. *Skeletal muscle* makes up the large muscle masses attached to the bones of the body. Skeletal and cardiac fibers are exceptions to the rule that cells have only one nucleus; each fiber has many nuclei. The nuclei of the skeletal fibers are also unusual in their position: they lie pe-

ripherally, just under the cell membrane; presumably this is an adaptation to increase the efficiency of contraction. Skeletal muscle cells are extremely long, some 2 or 3 centimeters in length. Indeed, some investigators believe that muscle cells extend the entire length of the muscle.

Skeletal and cardiac fibers have alternate light and dark microscopic transverse stripes or *striations.* These stripes appear to be involved in contraction, for they change their relative sizes during contraction, the dark stripes remaining essentially constant and the light stripes decreasing in width. Striated muscles can contract very rapidly but cannot remain contracted; a striated muscle must relax and rest momentarily before it can contract again. Skeletal muscle is sometimes called voluntary muscle, because it is under the control of the will. Cardiac and smooth muscles are called involuntary, because they cannot be regulated by the will. Table 4–1 summarizes the features that distinguish the three types of muscle tissue.

Blood. Blood includes the red and white blood cells and the liquid, noncellular part of the blood, the *plasma* (p. 347). Many biologists classify blood with the connective tissues because they originate from similar cells.

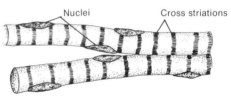

A, SKELETAL MUSCLE FIBERS

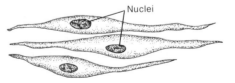

B, SMOOTH MUSCLE FIBERS

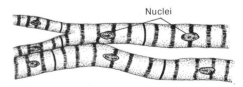

C, CARDIAC MUSCLE FIBERS

Figure 4–30 Types of muscle tissues.

Table 4–1 COMPARISON OF THE TYPES OF MUSCLE TISSUE

	Skeletal	Smooth	Cardiac
Location	Attached to skeleton	Walls of viscera, stomach, intestines, etc.	Wall of heart
Shape of fiber	Elongate, cylindrical, blunt ends	Elongate, spindle-shaped, pointed ends	Elongate, cylindrical, fibers branch and fuse
Number of nuclei per fiber	Many	One	Many
Position of nuclei	Peripheral	Central	Central
Cross striations	Present	Absent	Present
Speed of contraction	Most rapid	Slowest	Intermediate
Ability to remain contracted	Least	Greatest	Intermediate
Type of control	Voluntary	Involuntary	Involuntary

The red cells (*erythrocytes*) of vertebrates contain the pigment hemoglobin, which can combine easily and reversibly with oxygen. Oxygen combined as oxyhemoglobin is transported to the cells of the body in the red cells. Mammalian red cells are flattened biconcave discs without a nucleus (Fig. 4–31 *D*); those of other vertebrates are more typical cells with an oval shape and a nucleus.

Five different kinds of white blood cells—lymphocytes, monocytes, neutrophils, eosinophils and basophils (Fig. 4–31)—are present in human blood. White cells have no hemoglobin but can move around and even slip through the walls of blood vessels and enter the tissues of the body to engulf bacteria.

The fluid part of the blood, plasma, transports many types of substances from one part of the body to the other. Some of the substances transported are in solution, others are bound to one or another of the plasma proteins. In certain invertebrate animals, the oxygen-carrying pigment is not localized in cells but is dissolved in the plasma and colors it red or blue. Platelets are small fragments broken off from large cells in the bone marrow. They play a role in the clotting of blood.

Nervous Tissues. Nervous tissue is made of *neurons*, cells specialized for conducting electrochemical nerve impulses. Each neuron has an enlarged *cell body*, which contains the nucleus, and two or more thin, hairlike nerve fibers extending from the cell body (Fig. 4–32). The nerve fibers, made of cytoplasm and covered by a plasma membrane, vary in width from a few micrometers to 30 or 40 micrometers and in length from a millimeter or two to more than a meter. Those extending from the spinal cord down the arm or leg in the human may be a meter or more in length. The neurons are connected together in chains to pass impulses for long distances through the body.

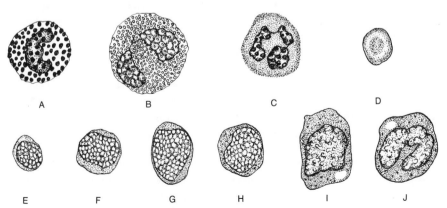

Figure 4–31 Types of white blood cells. *A*, Basophil; *B*, eosinophil; *C*, neutrophil; *D*, a red blood cell drawn to the same scale; *E–H*, a variety of lymphocytes; *I* and *J*, monocytes.

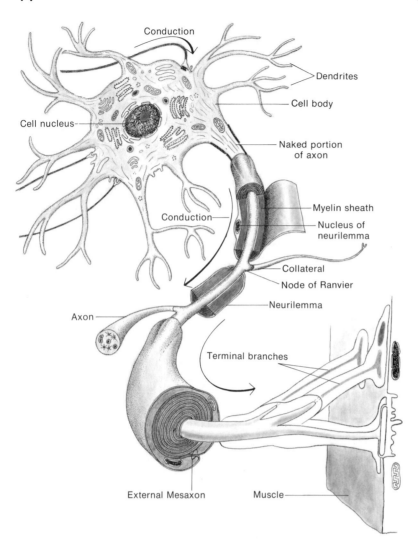

Figure 4–32 Diagram of a single neuron and its parts. The arrows indicate the direction of the normal nerve impulse. (After Warwick, R., and Williams, P. L.: *Gray's Anatomy,* 35th British Ed. London, W. B. Saunders Co., 1973.)

The nerve fibers of the peripheral nervous system are enveloped by a sheath of cells, the *neurilemma.* On some nerve fibers these cells secrete a spiral wrapping of insulating fatty material, *myelin.* There are gaps, the *nodes of Ranvier,* between the neurilemma cells where the fiber is free of myelin.

Two types of nerve fibers, *axons* and *dendrites,* are differentiated on the basis of the direction in which they normally conduct a nerve impulse: axons conduct nerve impulses *away* from the cell body; dendrites conduct them *toward* the cell body. The junction between the axon of one neuron and the dendrite of the next is called a *synapse.* The axon and dendrite do not actually touch at the synapse; there is a small gap between the two. An impulse can travel across the synapse only from an axon to a dendrite; the synapse serves as a valve to prevent the backflow of impulses.

Neurons occur in many sizes and shapes but all have the same basic plan.

Reproductive Tissue. Reproductive tissue is composed of cells modified to produce offspring—egg cells in females and sperm cells in males (Fig. 4–33). Egg cells are usually spherical or oval and are nonmotile. The eggs of most animals, but not of the higher mammals, contain a large amount of *yolk* which serves as food for the developing organism from the time of fertilization until it is able to obtain food in some other way. Sperm cells are much smaller than eggs; they have lost most of their cytoplasm and developed a tail by which they propel themselves. A typical sperm consists of a *head,* which contains the nucleus, a *middle piece* and a *tail.* The shape of the sperm varies in different species of animals (Fig. 28–12). Because eggs and sperm develop from epithelial-like tissue in the ovaries and

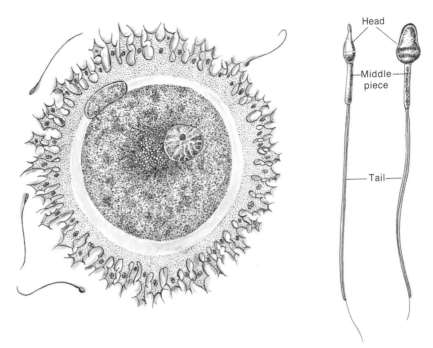

Figure 4–33 *Left,* Human egg and sperm magnified 400 times. Note the large nucleus and nucleolus in the egg. A polar body is visible as a flattened oval body between the egg and the surrounding corona radiata cells. Compare with Figure 27–2. *Right,* Side and top views of a sperm, magnified about 2000 times.

testes, some biologists classify them with those tissues.

Meristematic Tissue. The *meristematic tissues* of plants are composed of small, thin-walled cells with large nuclei and few or no vacuoles (Fig. 4–34). Their chief function is to grow, divide and differentiate into all the other types of tissue. An embryonic plant begins development composed entirely of meristem. As it develops, most of the meristem becomes differentiated into other tissues, but even in an adult tree there are regions of meristem which provide for continued growth. Meristematic tissues are found in the rapidly growing parts of the plant—the tips of the roots and stems, and in the cambium. The meristem in the tips of roots and stems, called *apical meristem,* is responsible for the increase in length of roots and stems, and the meristem in the *cambium,* called lateral meristem, makes possible the increase in diameter of growing stems and roots.

Protective Tissue. *Protective tissues* consist of cells with thick walls that serve to protect the underlying thin-walled cells from drying out and from mechanical abrasions. The epidermis of leaves and the cork layers of stems and roots are examples of protective tissues. The epidermis of leaves secretes a waxy, waterproof material known as *cutin,* which decreases the loss of water from the leaf surface.

On the surface of leaves are specialized epidermal cells, termed *guard cells,* that occur in pairs around each tiny opening, termed a *stoma,* into the interior of the leaf. The turgor pressure (p. 234) in the guard cell regulates the size of the stoma and the rate at which oxygen, carbon dioxide and water vapor pass into or out of the leaf.

Some of the epidermal cells of the roots have outgrowths called *root hairs* that increase the absorptive surface for the intake of water and dissolved minerals from the soil. Stems and roots are covered by layers of cork cells, produced by a separate cork cambium, another lateral meristem. Cork cells are closely packed and their cell walls contain another waterproof material, *suberin.* Since the suberin prevents the entrance of water into the cork cells themselves, they are short-lived and all mature cork cells are dead.

Fundamental Tissue. The *fundamental tissues* make up the great mass of the plant body, including the soft parts of the leaf, the pith and cortex of stems and roots, and the soft parts of flowers and fruits. Their chief functions are the production and storage of food. The simplest of the fundamental tissues, *parenchyma,* consists of cells with a thin wall and a thin layer of cytoplasm surrounding a central vacuole (Fig. 4–35 A). *Chlorenchyma* is a modified parenchyma containing *chloroplasts* in which photosynthesis occurs. The chloren-

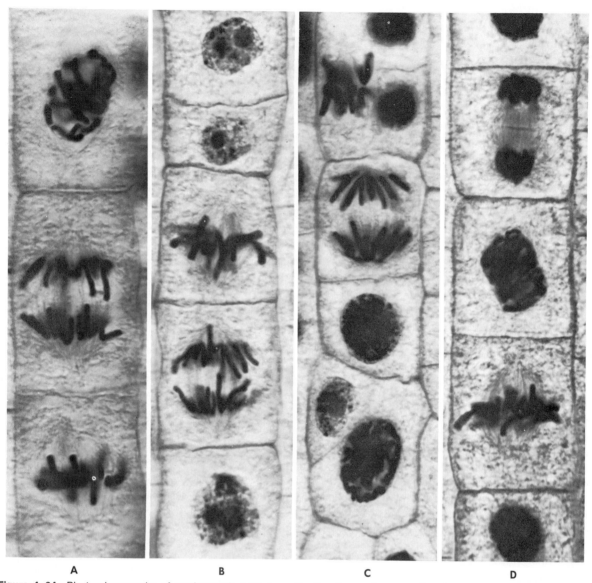

A   B   C   D

Figure 4–34 Photomicrographs of meristematic cells from the tip of an onion root showing stages in mitotic division. Note the characteristic thin walls, large nuclei and absence of vacuoles. *A*, Prophase and anaphase. *B*, Metaphase and anaphase. *C*, Late anaphase. *D*, Telophase and anaphase. (Courtesy of the Carolina Biological Supply Company.)

chyma cells are loosely packed and make up most of the interior of leaves and some stems. They are characterized by thin cell walls, large vacuoles, and the presence of chloroplasts.

In some fundamental tissues the corners of the cells are thickened to provide the plant with support. Such tissue, called *collenchyma* (Fig. 4–35 *B*), occurs just beneath the epidermis of stems and leaf stalks. In *sclerenchyma* (Fig. 4–35 *C*), the entire cell wall becomes greatly thickened. These cells, which provide support and mechanical strength, are found in many stems and roots. They sometimes take the form of long thin fibers. Spindle-

shaped sclerenchyma cells called **bast fibers** are found in the phloem of many stems. Rounded sclerenchyma cells called *stone cells* are found in the hard shells of nuts.

Conductive Tissue. There are two types of conductive tissue in plants: *xylem*, which conducts water and dissolved salts, and *phloem*, which conducts dissolved nutrients such as glucose. In all higher plants, the first xylem cells to develop are long *tracheids*, with pointed ends and thickenings on the walls in a circular, spiral or pitted pattern (Fig. 4–35 *E*). Later, other cells join end to end to form *xylem vessels.* As the vessels develop, the

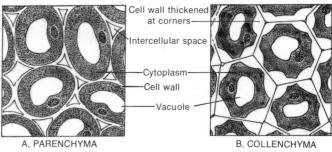

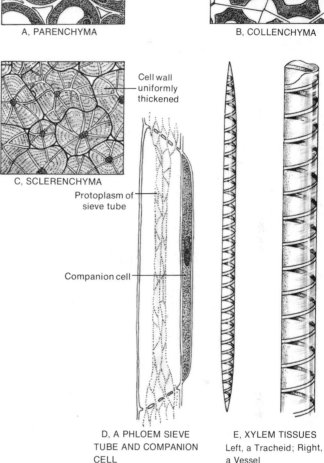

A, PARENCHYMA   B, COLLENCHYMA

Cell wall thickened
at corners

Intercellular space

Cytoplasm

Cell wall

Vacuole

C, SCLERENCHYMA

Cell wall
uniformly
thickened

Protoplasm of
sieve tube

Companion cell

Figure 4–35 Some types of plant tissues. *A,* Parenchyma; *B,* collenchyma; *C,* sclerenchyma; *D,* sieve tube; and *E,* tracheid and vessel.

D, A PHLOEM SIEVE
TUBE AND COMPANION
CELL

E, XYLEM TISSUES
Left, a Tracheid; Right,
a Vessel

end walls dissolve and the side walls thicken, leaving a cellulose tube as much as 3 meters long for the conduction of water. In both tracheids and vessels the cytoplasm eventually dies, leaving the tubes, which continue to function. The thickening, which involves the deposition of *lignin,* the substance responsible for the hard, woody nature of plant stems and roots, enables xylem to act as a supportive as well as a conductive tissue.

A similar end-to-end fusion of cells gives rise to the *sieve tubes* of the phloem (Fig. 4–35 *D*). The ends of the cells do not disappear, but remain as a perforated plate, the *sieve plate.* Unlike the tracheids and vessels of the xylem, mature sieve tubes remain alive and have an abundance of cytoplasm but lose their nucleus. Adjacent to the sieve tubes are nucleated "companion cells" (Fig. 4–35 *D*) which may serve to regulate the functions of the sieve tubes. The streaming movement or cyclosis of the cytoplasm of the sieve tubes is important in speeding up the transport of dissolved foods (translocation, cf. p. 236) by the sieve tubes. Sieve tubes are found in woody stems, in the soft bark just outside the cambium layer.

4–14 ORGAN SYSTEMS

The bodies of single-celled animals and plants are not, of course, organized into tissues and organs; all the life functions are carried on by the one cell. In more complex organisms a division of labor has occurred and special systems have evolved to perform each of the principal life functions. In the human, for example, the circulatory system is made of organs— heart, arteries and veins; the heart is made of several types of tissue—cardiac muscle, fibrous connective tissue, nerves, etc.; and each type of tissue is composed of millions of individual cells.

In the human and other vertebrates eleven organ systems can be distinguished:

The *circulatory system* transports materials around the body.

The *respiratory system* provides a means for the exchange of oxygen and carbon dioxide between the blood stream and the external environment.

The *digestive system* takes in food, secretes enzymes which split the larger molecules of nutrients into smaller molecules, and absorbs them into the blood.

The *excretory system* eliminates the waste products of metabolism.

The *integumentary system* covers and protects the entire body.

The *skeletal system* supports the body and provides for movement and locomotion.

The *muscular system* functions with the skeletal system in movement and locomotion.

The *nervous system* conducts impulses around the body and integrates the activities of the other systems.

The *sense organs* receive stimuli from the outer world and from various regions of the body.

The *endocrine system* serves as an additional coordinator of the body functions.

The *reproductive system* provides for the continuation of the species.

4-15 BODY PLAN AND SYMMETRY

In referring to parts of the body, biologists use the terms *anterior*, referring to the head end of the body; *posterior*, to the tail end of the body; *dorsal*, to the back side; *ventral*, to the belly side; *medial*, to the midline of the body; and *lateral*, to the side. These terms may also be used to indicate relative position. For example, the neck is anterior to the chest; the

ribs are posterior to the collar bone; the spinal cord is dorsal to the body cavity.

Animal and plant bodies may be organized according to one of three types of symmetry. A structure is said to be symmetrical if it can be cut into two equivalent halves. If a body is spherical, and completely homogeneous, like a rubber ball, so that it is possible to cut it in any plane through the center and get two equal halves, it is said to have *spherical symmetry.* Only a few of the lowest plants and animals have this type of organization. In *radial symmetry* two sides are distinguishable, a top and a bottom, as in a starfish or a mushroom. A mushroom can be cut into two equal halves by any plane which includes the line or axis running through the center from top to bottom. Human beings have *bilateral symmetry;* only one special cut will divide the body into two equivalent halves. In bilaterally symmetrical animals, anterior, posterior, dorsal and ventral sides can be distinguished. In man, for example, only a plane from head to foot exactly in the center will divide the body into equivalent halves, right and left. And, as we shall see when we study some of the details of human structure, even the right and left halves of the human body are not exactly equivalent.

In a bilaterally symmetrical animal, three planes or sections can be distinguished:

Transverse planes, which include a dorsoventral axis and a left-right axis, but are at right angles to the anterior-posterior axis.

A *sagittal* plane, which includes the dorsoventral axis and the anterior-posterior axis, but is at right angles to the left-right axis.

A *frontal* plane, which includes the anterior-posterior axis and the left-right axis, but is at right angles to the dorso-ventral axis.

In learning to differentiate between these body planes, it is helpful to make a rough model of some bilaterally symmetrical animal such as a fish, using modeling clay or some similar material. Practice making the various types of sections until you are completely familiar with these terms.

QUESTIONS

1. How would you define a cell? What are the chief parts of a cell?
2. What are the chief differences between plant and animal cells?

3. What advantages may accrue to an organism if its body is organized in cells?

4. Contrast the meanings of the term "cell" in the time of Robert Hooke, in the time of Schleiden and Schwann, and at present.

5. What are the functions of the nucleus? What is the evidence that indicates the role of the nucleus in cell metabolism?

6. What are the functions in the cell of (a) mitochondria, (b) ribosomes?

7. What is meant by *in vitro* and *in vivo* experiments? Give an example of each type.

8. What factors may limit the size of a cell?

9. What are the features of the S phase of the cell cycle? of the M phase?

10. Distinguish between "step" enzymes and "peak" enzymes.

11. Outline briefly the events occurring in each stage of mitosis. Illustrate your discussion with a series of diagrams of mitosis in an organism with a haploid number (n) of four.

12. Differentiate clearly among diffusion, dialysis and osmosis.

13. In what ways do gases, liquids and solids differ?

14. In what ways is the phenomenon of diffusion important to living things?

15. What is the structure of bone? Of tendon? How are the types of muscle tissues differentiated?

16. Which tissues of plants and animals are comparable? Which are peculiar to one or the other?

17. Compare the structures and functions of the components of xylem and phloem.

18. What kinds of tissues make up the following organs: the eyeball, the lung, the intestines, the sweat glands and the liver?

19. How would you define the terms anterior, ventral and medial?

20. What kind of symmetry is exhibited by each of the following organisms: earthworm, snail, pine tree, volvox, clam?

21. How could you describe the position of the hump of a camel? Of a cat's eye in relation to its nose?

22. What difficulties arise in applying the terms "anterior" and "posterior" to an animal, such as the human, whose body is upright? Are these difficulties eliminated by using the terms "superior" and "inferior"?

SUPPLEMENTARY READING

The development of the cell theory is presented in Hall's *A Source Book in Animal Biology* by means of long quotations from some of the original scientific papers. The properties of cells, together with a number of electron micrographs of tissues, are presented in *The Cell* by Fawcett, *Cell Biology* by De Robertis, Nowinski and Saez, *Anatomy of the Cell* by Bjorn Afzelius, *Fine Structure of Cells and Tissues* by K. R. Porter and M. A. Bonneville, and in Bloom and Fawcett's *Textbook of Histology*. The latter is a detailed and technical discussion of the tissues of the human body and includes many fine illustrations of each type of tissue. A briefer survey of the subject is given in *The Cell* by C. P. Swanson. *The Living Cell: Readings from the Scientific American*, edited by Donald Kennedy, contains a readable collection of articles from *Scientific American* on subjects relating to cell structure and function. The nature of biological membranes and their role in cell function are discussed in L. I. Rothfield's *Structure and Function of Biological Membranes*.

CHAPTER 5

CELLULAR ENERGETICS

Energy, defined as the capacity to do work, to produce a change in matter, may take the form of heat, light, electricity, motion, or chemical energy. We can distinguish **potential energy** as the capacity to do work owing to the position or state of a particle, and **kinetic energy** as the energy of a particle in motion. A boulder at the top of a hill has potential energy because of its position. As it rolls down the hill the potential energy is converted to kinetic energy.

One of the major attributes of living cells is the presence within them of complex and efficient systems, such as chloroplasts and mitochondria, for transforming one type of energy into another. The chemical reactions of cells, which provide for their growth, irritability, movement, maintenance, repair and reproduction, are collectively called **metabolism.** The metabolic activities of animal, plant and bacterial cells are all remarkably similar despite the differences in the appearance of these organisms. In all cells simple sugars, such as glucose, are converted by way of a series of intermediate compounds to carbon dioxide and water. During these conversions a part of the energy of the glucose molecule is conserved and made available to the cells to drive other processes.

5–1 BIOENERGETICS: ENERGY TRANSFORMATIONS

The never-ending flow of energy within a cell, from one cell to another, and from one organism to another organism, is the essence of life itself. The study of energy transformations in living organisms is termed *bioenergetics.* Three major types of energy transformations can be distinguished in the biological world (Fig. 5–1). In the first, the radiant energy of sunlight is captured by the green pigment chlorophyll, present in green plants, and is transformed by the process of photosynthesis into chemical energy. This is used to synthesize carbohydrates and other complex molecules from carbon dioxide and water. The radiant energy of sunlight, a form of kinetic energy, is transformed into a type of potential energy. The chemical energy is stored in the molecules

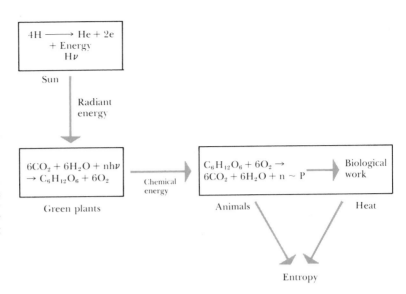

Figure 5–1 Energy transformations in the biological world. (1) The radiant energy of sunlight is transformed in photosynthesis into chemical energy in the bonds of organic compounds. (2) The chemical energy of organic compounds is transformed during cellular respiration into biologically useful energy, the energy-rich phosphate bonds of ATP and other compounds. (3) The chemical energy of energy-rich phosphate bonds is utilized in cells to do mechanical, electrical, osmotic or chemical work. Finally energy flows to the environment as heat in the "entropy sink."

of carbohydrates and other foodstuffs as the energy of the bonds which connect their constituent atoms.

In a second type of energy transformation, the chemical energy of carbohydrates and other molecules is transformed by the process termed *cellular respiration* into the biologically useful energy of energy-rich phosphate bonds. This kind of energy transformation occurs in the mitochondrion. A third type of energy transformation occurs when the chemical energy of the energy-rich phosphate bonds is utilized by the cells to do work—the mechanical work of muscular contraction, the electrical work of conducting a nerve impulse, the osmotic work of moving molecules against a gradient, or the chemical work of synthesizing molecules for growth. As these transformations occur, energy finally flows to the environment and is dissipated as heat. Plants and animals have evolved some remarkably effective *energy transducers,* such as chloroplasts and mitochondria, to carry out these processes, together with efficient control mechanisms to regulate the transducers and enable the cells to adjust to variations in environmental conditions (Table 5–1).

The branch of physics that deals with energy and its transformations, *thermodynamics,* consists of certain relatively simple basic principles which are universally applicable to chemical processes whether these occur in living or nonliving systems.

Under experimentally controlled conditions, the amount of energy entering and leaving any system may be measured and compared. It is always found that energy is neither created nor destroyed but is only transformed from one form to another. This is an expression of the first law of thermodynamics, sometimes called the *Law of the Conservation of Energy:* the total energy of any object and its surroundings (i.e., a system) remains constant. As any given object undergoes a change from its initial state to its final state it may absorb energy from the surroundings or deliver energy to the surroundings. The difference in the energy content of the object in its initial and final state must be just equalled by a corresponding change in the energy content of the surroundings. Heat is a convenient form in which energy may be measured, which is why the study of energy has been called thermodynamics, i.e., heat dynamics.

Table 5–1 ENERGY TRANSFORMATIONS IN CELLS

Transformation	Type of Cell
Chemical energy to electrical energy	Nerve, brain
Sound to electrical energy	Inner ear
Light to chemical energy	Chloroplast
Light to electrical energy	Retina of eye
Chemical energy to osmotic energy	Kidney
Chemical energy to mechanical energy	Muscle cell, ciliated epithelium
Chemical energy to radiant energy	Luminescent organ of firefly
Chemical energy to electrical energy	Sense organs of taste and smell

The unit of energy most widely used in biological systems is the *calorie,* which is the amount of heat required to raise one kilogram of water one degree centigrade (strictly speaking, from 14.5° to 15.5°C). This unit, more correctly called a kilogram-calorie or kilocalorie, is the unit commonly used in calculating the energy content of foods. Other forms of energy—radiant, chemical, electrical, the energy of motion or position—can be converted to heat and measured by their effect in raising the temperature of water.

Nearly every physical or chemical event is accompanied by the delivery of heat to the surroundings or the absorption of heat from the surroundings. When a process occurs with delivery of heat to the surroundings it is said to be *exothermic.* When it occurs with the absorption of heat from the surroundings it is termed *endothermic.* Although heat is a simple and familiar means by which energy is transferred in many of the machines made by man it is not a useful way of transferring energy in biological systems for the simple reason that living organisms are basically *isothermal* (equal temperature). There is no significant temperature difference between the different parts of the cell or between the different cells in a tissue. Stated in another way, cells do not act as heat engines. They have no means of allowing heat to flow from a warmer to a cooler body.

The **Second Law of Thermodynamics** may be stated briefly as "the entropy of the universe increases." *Entropy* may be defined as a randomized state of energy that is unavailable to do work. The second law may be phrased that "physical and chemical processes proceed in such a way that the entropy of the system becomes maximal." Entropy, then, is a measure of randomness or disorder. In almost all energy transformations there is a loss of some heat to the surroundings, and since heat involves the random motion of molecules, such heat losses increase the entropy of the surroundings. Living organisms and their component cells are highly organized and thus have little entropy. They preserve this low entropy state by increasing the entropy of their surroundings. You increase the entropy of your surroundings when you eat a candy bar and convert its glucose to carbon dioxide and water and return them to the surroundings.

The force that drives all processes is the tendency of the system to reach the condition of maximum entropy. Heat is either given up or absorbed by the object to allow the system to reach the state of maximum entropy. The changes of heat and entropy are related by a third dimension of energy termed *free energy.* Free energy may be visualized as that component of the total energy of a system which is available to do work under isothermal conditions; it is thus the thermodynamic parameter of greatest interest in biology. Entropy (S) and free energy (G) are related inversely; as entropy increases during an irreversible process, the amount of free energy decreases. The two are related by the equation $\Delta G = \Delta H - T\Delta S$, in which H is the enthalpy, or total heat content of the system, and T is the absolute temperature. All physical and chemical processes proceed with a decline in free energy until they reach an equilibrium in which the free energy of the system is at a minimum and the entropy is at a maximum. Free energy is useful energy; entropy is degraded, useless energy.

5-2 CHEMICAL REACTIONS

A chemical reaction is a change involving the molecular structure of one or more substances; matter is changed from one substance, with its characteristic properties, to another with new properties, and energy is released or absorbed. Hydrochloric acid reacts with the base, sodium hydroxide, to yield water and the salt, sodium chloride. In the process energy is released as heat. The chemical properties of HCl and NaOH are very different from those of H_2O and NaCl. In chemical shorthand a plus sign connects the symbols of the reacting substances, HCl and NaOH, and the products of the reaction, H_2O and NaCl. An arrow indicates the direction of the reaction:

$$HCl + NaOH \rightarrow NaCl + H_2O + energy\ (heat)$$

Note that the number of atoms of a given element in the products is just equal to the number of atoms of that element in the reactants. Atoms are neither destroyed nor created in a chemical reaction, but simply change partners. This is an expression of one of the basic tenets of physics, the Law of the Conservation of Matter.

Most chemical reactions are reversible; this is indicated by a double arrow, ⇌. The energy relations of the several chemicals involved, their relative concentrations, and their solubility are some of the factors that determine whether or not a reaction will occur, and whether it will go from right to left or left to right.

For each reaction a constant, the thermodynamic equilibrium constant (K), expresses the chemical equilibrium reached by the system. Thus for the reaction A + B ⇌ C + D,

$$K = \frac{[C] \times [D]}{[A] \times [B]}$$

The equilibrium constant for any given reaction is determined by the tendency of the reaction components to reach maximum entropy, or minimum free energy, for the system. The equilibrium constant, K, is related mathematically to the change in free energy of the components of the reaction: $\Delta G = -RT \ln K$. R is the gas constant, T the absolute temperature, and ln K is the natural logarithm of the equilibrium constant. The symbol ΔG represents the standard free energy change in Kcal./mole (i.e., the gain or loss of free energy in calories as one mole* of reactant is converted to one mole of product).

Examination of this equation reveals that when the equilibrium constant, K, is high, the standard free energy change, ΔG, is negative. Such a reaction will proceed with a decrease in free energy. However, when the equilibrium constant is less than 1 (e.g., 0.3) the reaction does not go far in the direction of completion, and the free energy change is positive. Indeed, such reactions are said to "favor the starting material." It is necessary to put energy into the system to transform one mole of reactant into one mole of product. When the equilibrium constant is 1, then the change in free energy is zero and the reaction is freely reversible.

How does one go about measuring the equilibrium constant and free energy change of a biochemical reaction? Glucose-1-phosphate and glucose-6-phosphate are interconverted in the cell by a reaction catalyzed by the enzyme *phosphoglucomutase*. We set up a system with a carefully measured amount of glucose-1-phosphate, add an adequate amount of the enzyme, and measure the amounts of the reaction mixture at succeeding times until no further change occurs, i.e., until equilibrium is reached. At equilibrium there will be 19 times as much glucose-6-phosphate as glucose-1-phosphate, and therefore the equilibrium constant, K, is 19. This number is substituted in the equation $\Delta G = -RT \ln K$, and we calculate that $\Delta G = -1745$ calories per mole. There is a decline in free energy of 1745 calories when one mole of glucose-1-phosphate is converted to one mole of glucose-6-phosphate at 25°C.

The rate at which a chemical reaction occurs is determined by a number of factors, one of which is temperature. Thus we must always state the temperature at which the reaction was carried out (25°C in this case). Each increase of 10°C approximately doubles the rate of most reactions. This is true of biological processes as well as reactions in a chemist's test tube, which indicates again that the chemical reactions of living things are fundamentally similar to those of nonliving ones. Whether or not an enzyme or any other catalyst is present in the system has no effect on either the equilibrium constant, K, or the free energy change, ΔG, undergone. Enyzmes and other catalysts simply speed up the rate at which the system approaches equilibrium but do not change the equilibrium point itself.

Reactions which have a high equilibrium constant, K, and a negative standard free energy change, ΔG, are said to be *exergonic*. A reaction with a very low equilibrium constant and therefore a positive standard free energy change will not go to completion under standard conditions. Such processes are called *endergonic* processes. In biological systems, the endergonic processes must be coupled with some exergonic process so that the exergonic process delivers the required energy to drive the endergonic process. In such coupled systems, the endergonic process can occur only if the decline in free energy of the exergonic process to which it is coupled is larger than the gain in free energy of the endergonic process.

---

*A mole or gram molecular weight of a compound is defined as the amount of that substance equal to its molecular weight in grams. One mole of glucose is 180 grams of glucose.

5–3 CATALYSIS

Many of the substances that are rapidly metabolized by living cells are remarkably inert outside the body. A glucose solution will keep indefinitely in a bottle if it is kept free of bacteria and molds; it must be subjected to high temperature, strong acids or bases before it will break down. Living cells cannot use these extreme conditions to cleave the glucose molecule, for the cell itself would be destroyed long before the glucose. The reactions are brought about by agents called *enzymes,* which belong to the class of substances known as *catalysts.*

A catalyst is a substance which affects the velocity of a chemical reaction without affecting its final equilibrium point and without being used up as a result of the reaction. Almost any substance may serve as a catalyst for some reaction. Water is an excellent one. For example, pure, dry hydrogen gas and chlorine gas may be mixed without result, but if a slight amount of water is present, the hydrogen and chlorine react with explosive violence to form hydrogen chloride.

Finely divided metals—iron, nickel, platinum, palladium, and so on—are catalysts widely used in industrial processes such as the hydrogenation of cottonseed oil to make margarine or the "cracking" of petroleum to make gasoline. A catalyst is not used up in a reaction, but can be used over and over again; a very small amount of catalyst will speed up the reaction of vast quantities of reactants.

There is an energy barrier to almost every chemical reaction, even an exergonic one with a strongly negative ΔG, which prevents the reaction from beginning. This energy barrier is termed the *activation energy.* In a population of molecules of any given kind, some have a relatively high energy content, others have a lower energy content, and the energy content of the entire population of molecules conforms to a bell-shaped curve of normal distribution, like the distribution of the heights of a population of adult men (Fig. 28–11). Only those molecules with a relatively high energy content are likely to react to form the product. To make the reaction go faster we must raise the energy content of more of the population of molecules so that the activation energy barrier is overcome. This can be done by heating the mixture, for the heat absorbed by the molecules increases their internal energy and increases the likelihood that they will collide and react. Alternatively, the activation energy barrier can be overcome by adding a catalyst. This lowers the activation energy of the reaction and allows a larger fraction of the population of molecules to react at any one time (Fig. 5–2). The catalyst does this by forming an unstable intermediate complex with the substrate, which then decomposes to the product and frees the catalyst to react with a second molecule of reactant.

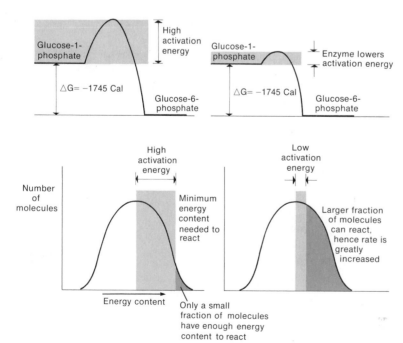

Figure 5–2 The role of a catalyst such as an enzyme in lowering the activation energy of a reaction and in increasing the fraction of the population of molecules with an energy content great enough to react and form the product. The catalyst does this by forming an unstable intermediate complex with the substrate, glucose-1-phosphate, which then decomposes to the product, glucose-6-phosphate, and frees the catalyst to react with a second molecule of substrate.

5–4 ENZYMES AND THEIR PROPERTIES

Enzymes are protein catalysts produced by living cells. They regulate the speed and specificity of the thousands of chemical reactions that occur within cells. Although enzymes are synthesized within cells, they need not be inside a cell to be effective catalysts. Many enzymes have been extracted from cells with their activity unimpaired. They can then be purified and crystallized and their catalytic abilities can be studied. Enzyme-controlled reactions are basic to all the phenomena of life: respiration, growth, muscle contraction, nerve conduction, photosynthesis, nitrogen fixation, deamination, digestion and so on. Enzymes are usually named by adding the suffix *-ase* to the name of the substance acted upon, e.g., sucrose is split by the enzyme *sucrase* to give glucose and fructose. There are group names for enzymes that catalyze similar reactions: *lipases* cleave triacylglycerols, proteinases cleave the peptide bonds in proteins, and dehydrogenases transfer hydrogens from one compound to another. Enzymes are usually colorless, but they may be yellow, green, blue, brown or red. Most enzymes are soluble in water or dilute salt solution, but some, for example the enzymes present in the mitochondria, are bound together by lipoprotein (a phospholipid-protein complex) and are insoluble in water.

The catalytic ability of some enzymes is truly phenomenal. For example, one molecule of the iron-containing enzyme *catalase*, extracted from beef liver, will bring about the decomposition of 5,000,000 molecules of hydrogen peroxide (H_2O_2) per minute at 0°C. The substance acted upon by an enzyme is known as its *substrate;* thus, hydrogen peroxide is the substrate of the enzyme catalase.

The number of molecules of substrate acted upon by a molecule of enzyme per minute is called the *turnover number* of the enzyme. The turnover number of catalase is thus 5,000,000. Most enzymes have high turnover numbers and are very effective catalysts even though present in the cell in relatively minute amounts. Hydrogen peroxide is a poisonous substance produced as a by-product in a number of enzyme reactions. Catalase protects the cell by destroying the peroxide.

Hydrogen peroxide can be split by iron atoms alone, but only at a very slow rate. It would take 300 *years* for an iron atom to split the same number of molecules of H_2O_2 that a molecule of catalase (containing one iron atom) splits in one *second*. This is an example of the evolution of a catalyst and emphasizes one of the major characteristics of enzymes—they are very efficient catalysts.

Enzymes differ in their specificity, in the number of different substrates they will attack. A few enzymes are absolutely specific; *urease*, which decomposes urea to ammonia and carbon dioxide, will attack no other substance. Specific enzymes split each of the three common double sugars, sucrose, maltose and lactose. Other enzymes are relatively specific and will work upon only a few, closely related substances. *Peroxidase* will decompose several different peroxides including hydrogen peroxide. Peroxidase, found in a wide variety of plant and animal tissues, can be demonstrated by mincing some raw potato and adding some hydrogen peroxide. A vigorous bubbling will ensue as peroxidase converts the peroxide to water and oxygen.

Finally, a few enzymes are specific only in requiring that the substrate have a certain kind of chemical bond. The *lipase* secreted by the pancreas will split the ester bonds connecting the glycerol and fatty acids of a wide variety of different fats.

Theoretically, enzyme-controlled reactions are reversible and the enzyme does not determine which way the reaction will go; it simply accelerates the rate at which the reaction reaches equilibrium. One of the classic examples of this is the action of lipase in splitting the ester bonds between glycerol and fatty acids in a triacylglycerol, and its action in joining glycerol and fatty acids to form triacylglycerols. Whether the reaction begins with pure triacylglycerol or with pure glycerol and fatty acids, the same equilibrium mixture of fat, fatty acids, and glycerol is achieved. The equilibrium point depends upon complex thermodynamic principles, a discussion of which is beyond the scope of this book. Since many reactions give off energy when going in one direction, it is obvious that the proper amount of energy in a usable form must be supplied to drive the reaction in the reverse direction.

To drive an energy-requiring reaction, some energy-yielding reaction must occur at about the same time. In most biologic systems, energy-yielding reactions result in the synthesis of energy-rich phosphate bonds, ~ P, such as the terminal bonds of *adenosine triphosphate* (abbreviated as ATP). The energy of these phosphate bonds can then be used by a cell to conduct a nerve impulse, contract muscle, synthesize proteins and so on. The term "coupled reactions" is applied to two reactions which must take place together so that one can furnish energy, or one of the reactants, needed by the other.

Enzymes usually work in teams, with the product of one enzyme-controlled reaction serving as the substrate for the next. We can picture the inside of a cell as a factory with many different assembly lines (and disassembly lines) operating simultaneously. Each of the assembly lines is composed of a number of enzymes, each of which carries out one step such as changing molecule A into molecule B and then passes it along to the next enzyme, which converts molecule B into molecule C, and so on. From germinating barley seeds one can extract two enzymes that will convert starch to glucose. The first, *amylase,* hydrolyzes starch to maltose and the second, *maltase,* splits maltose to glucose. Eleven different enzymes, working consecutively, are required to convert glucose to lactic acid. The same series of eleven enzymes is found in human cells, in green leaves and in bacteria.

Some enzymes, such as pepsin, consist solely of protein. Others consist of two parts, one of which is protein (and called an *apoenzyme)* and the second (called a *coenzyme)* a smaller organic molecule, usually containing phosphate. Coenzymes can be separated from their enzymes and, when analyzed, have proved to contain some vitamin as part of the molecule—thiamine, niacin, riboflavin, pyridoxine, and so on. This finding has led to the generalization that *all vitamins function as parts of coenzymes in cells.*

Neither the apoenzyme nor the coenzyme alone has any catalytic activity; only when the two are combined is activity present. Other enzymes require for activity, in addition to a coenzyme, the presence of some ion. Several of the enzymes involved in the breakdown of glucose require magnesium (Mg^{++}). The *amylase* secreted by the salivary glands requires chloride ion (Cl^-) for activity. Most, if

not all, of the elements needed by plants and animals in very small amounts—the so-called *trace elements,* manganese, copper, cobalt, zinc, iron, etc.—function as such enzyme activators, usually as an integral part of the enzyme molecule.

In summary, enzymes, as catalysts, influence the rate but not the equilibrium of chemical reactions; they are very efficient catalysts; they have a high degree of specificity with respect to their substrates; they are subject to specific activators and inhibitors; and they direct the pathways of chemical reactions. Each enzyme is under the control of a specific gene.

5–5 LOCATION OF ENZYMES IN THE CELL

Many enzymes are simply dissolved in the cytoplasm of the cell. It is possible to make a water extract of ground liver that contains all the enzymes necessary to convert glucose to lactic acid. Other enzymes are tightly bound to certain subcellular organelles. The respiratory enzymes that catalyze the metabolism of lactic acid (and also substances derived from amino acids and fatty acids) to carbon dioxide and water are bound to the mitochondria; in fact, the mitochondrial membranes are in large part made of these enzymes. The enzymes which carry out the synthesis of proteins are integral parts of smaller cytoplasmic particles, the *ribosomes.*

The location and functioning of enzymes within the cell can be studied histochemically. The tissue is fixed and sliced by methods which do not destroy enzyme activity. Then the proper chemical substrate for the enzyme is provided and, after a specified period of incubation, some substance is added which will form a colored compound with one of the products of the reaction mediated by the enzyme. The regions of the cell which have the greatest enzyme activity will have the largest amount of the colored substance (Fig. 5–3).

5–6 MODE OF ACTION OF ENZYMES

An enzyme can speed up only those reactions that would occur to some extent, however

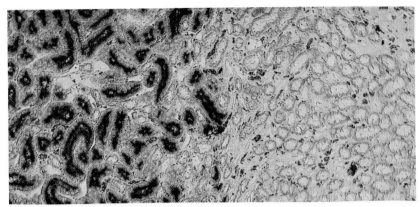

Figure 5-3 Histochemical demonstration of the location of alkaline phosphate within the cells of the rat's kidney. The tissue section is incubated at the proper pH with a naphthyl phosphate. The naphthyl phosphate is hydrolyzed where the phosphatase is located. The naphthol released by the action of the enzyme couples with a diazonium salt to form an intensely blue, insoluble azo dye which remains at the site of the enzymatic activity. The brush borders of the cells of the proximal convoluted tubules in the kidney cortex (left) show thick deposits of azo dye, i.e., intense phosphatase activity. The tubules in the kidney medulla (the loop of Henle) have no activity (right). (Courtesy of Dr. R. J. Barrnett.)

slight, in the absence of the enzyme. Many years ago the German chemist Emil Fischer suggested that the specific relationship of an enzyme to its substrate indicated that the two must fit together like a lock and key.

The idea that the enzyme combines with its substrate to form an intermediate *enzyme-substrate complex,* which subsequently decomposes to release the enzyme and the reaction products, was formulated mathematically by Leonor Michaelis over fifty years ago. By brilliant inductive reasoning, he assumed that such a complex does form, and then calculated how the speed of the reaction should be affected by varying the concentrations of enzyme and substrate. Exactly these relationships are observed experimentally, which is strong evidence that Michaelis' assumption, that an enzyme-substrate complex forms as an intermediate, is correct.

Direct evidence of the existence of an enzyme-substrate complex was obtained by David Keilin of Cambridge University and Britton Chance of the University of Pennsylvania. Chance isolated a brown-colored peroxidase from horseradishes. When he mixed this with hydrogen peroxide, a green-colored enzyme-substrate complex was formed. This was then changed to a second, pale red enzyme-substrate complex which finally split to give the original brown enzyme plus the breakdown products. By following the changes of color, Chance was able to calculate the rate of formation of the enzyme-substrate complex and its rate of breakdown.

Although it is clear that the substrate is much more reactive when part of an enzyme-substrate complex than when it is free, it is not clear *why* this should be true. One current theory postulates that the enzyme unites with the substrate at two or more points, and the substrate is held in a position which strains its molecular bonds and makes them more likely to break.

One approach to the study of enzyme action is to investigate the structure of the enzyme molecule itself. By certain analytic methods it is possible to determine the kinds of amino acids present in a given protein and their relative numbers. By more refined methods it is possible to determine the sequence of the amino acids in the peptide chain or chains comprising the protein. *Ribonuclease,* which consists of 124 amino acids in a single chain that is looped on itself like a pretzel, was the first enzyme for which it was possible to give the exact sequence of amino acids in the entire molecule (Fig. 5-4). The surface of the enzyme may gather together in favorable proximity the several substrates and cofactors involved in the reaction, thereby bringing about a marked increase in their local concentrations. This, of course, would increase the rate of the reaction. The enzyme may also ensure the appropriate orientation in space of the reacting groups so that a specific product results.

For most enzymes it is probable that only a relatively small part of the molecule combines with the substrate. This part is termed the *ac-*

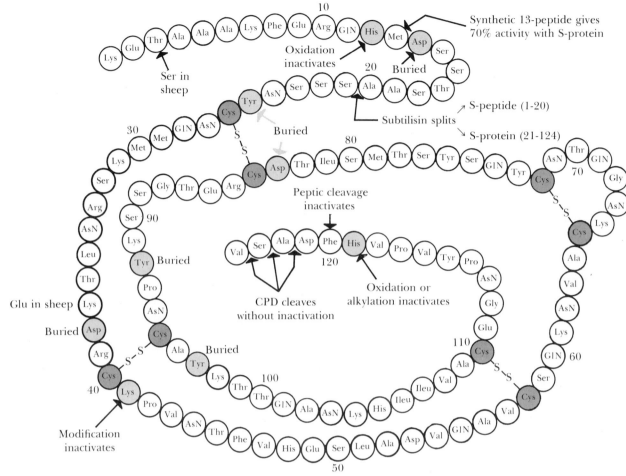

Figure 5–4 The primary structure of ribonuclease. This enzyme, which hydrolyzes ribonucleic acid (RNA), is composed of 124 amino acids, numbered in order beginning with the N-terminal amino acid (lysine) and ending with the carboxyl terminal amino acid (valine). The amino acids are abbreviated in the conventional fashion: *ala,* alanine; *arg,* arginine; *asp,* aspartic acid; *asp:NH₂,* asparagine; *cys,* cysteine; *glu,* glutamic acid; *glu:NH₂,* glutamine; *gly,* glycine; *his,* histidine; *ileu,* isoleucine; *leu,* leucine; *lys,* lysine; *met,* methionine; *phe,* phenylalanine; *pro,* proline; *ser,* serine; *thr,* threonine; *tyr,* tyrosine; *val,* valine.

tive site, and another approach to the study of enzyme action is to determine the location of the active site, and the nature and sequence of the amino acids present in it. This information may throw light on the nature of the interaction between the active site and the substrate and provide an explanation of the mechanism by which the enzyme functions. Studies of the constituents and topography of certain active sites have revealed the presence of certain amino acids or metal ions which actually participate in the enzyme-catalyzed reaction as proton donors or acceptors or as nucleophilic (electron-donating) or electrophilic (electron-attracting) agents. A metal ion present at the active site may bind with certain groups on the substrate molecule and, by forming a chelated intermediate, participate in the strain-producing process.

5–7 FACTORS AFFECTING ENZYME ACTIVITY

Temperature. Enzymes are inactivated by heat—temperatures of 50 to 60°C rapidly inactivate most enzymes. The inactivation is irreversible, for activity is not regained upon cooling. This explains why most organisms are killed by a short exposure to high temperature: their enzymes are inactivated and they are unable to continue metabolism.

A few remarkable exceptions to this rule exist: there are some species of primitive blue-green algae that can survive in hot springs, such as the ones in Yellowstone National Park, where the temperature is almost 100°C. These algae are responsible for the brilliant colors in the terraces of the hot springs. Below the temperature at which enzymes are inactivated (about 40°C) the rates of most enzyme-controlled reactions, like other chemical reactions, are about doubled by each increase of 10°C.

Enzymes are not usually inactivated by freezing; their reactions go on very slowly or not at all at low temperatures but their catalytic activity reappears when the temperature is raised to normal.

Acidity. Enzymes are sensitive to changes in pH, changes in acidity and alkalinity of the reaction medium. *Pepsin*, the protein-digesting enzyme secreted by the stomach lining, is remarkable in that it will work only in a very acid medium, and works optimally at pH 2. *Trypsin*, a protein-splitting enzyme secreted by the pancreas, is an example of an enzyme that works optimally in an alkaline medium, at about pH 8.5. The majority of intracellular enzymes have pH optima near neutrality and will not work well in an acid or alkaline medium; stronger acids or bases will irreversibly inactivate them.

The marked influence of pH on enzymic activity is predictable from the fact that enzymes are proteins. The number of positive and negative charges associated with the protein molecule and the topography of the molecular surface are determined in part by the pH. Probably only one particular state of the enzyme molecule, with a specific number of positive and negative charges, is active as a catalyst.

Concentration of Enzyme, Substrate and Cofactors. If the pH and temperature of an enzyme system are kept constant, and if an excess of substrate is present, the rate of the reaction is directly proportional to the amount of *enzyme* present. Knowing this, one can measure the amount of some particular enzyme present in a tissue extract. If the pH, temperature and enzyme concentration of a system are kept constant, the initial rate of reaction is proportional to the amount of *substrate* present, up to a limiting value. If the enzyme system requires a coenzyme or specific activator ion, the concentration of this substance

may, under certain circumstances, determine the overall rate of the reaction.

Enzyme Poisons. Certain enzymes are particularly susceptible to poisons such as cyanide, iodoacetic acid, fluoride, lewisite, etc.—even a very low concentration of poison inactivates the enzymes. Cytochrome oxidase, one of the enzymes of the electron transmitter system, is especially sensitive to cyanide, and a person dies of cyanide poisoning because his cytochrome enzymes have been inhibited. One of the enzymatic steps in the breakdown of glucose is inhibited by fluoride and another is inhibited by iodoacetic acid; biochemists have used these inhibitors as tools to investigate the properties and sequences of many different enzyme systems.

Enzymes themselves can act as poisons if they get into the wrong place. For example, as little as 1 milligram of crystalline trypsin will kill a rat if it is injected intravenously. Several types of snake, bee and scorpion venoms are harmful because they contain enzymes that destroy blood cells or other tissues.

5-8 ENERGY FLOW IN LIVING SYSTEMS

Human beings, like other animals, derive their energy from the foodstuffs they eat. Our fruits and vegetables are derived directly from plants and our meats, fish and shellfish are products of animals; but these animals in turn derive their energy supply from the plants that they eat. Ultimately all of the food and energy of the animal world comes from the plant world. For growth, plants require water, carbon dioxide, nutrient salts and nitrogen, but, most important, they require an abundant supply of the radiant energy of sunlight. Sunlight is thus the ultimate source of all the biological energy on this planet. The radiant energy of sunlight arises from nuclear energy when, at the very high temperatures that occur in the interior of the sun, hydrogen atoms undergo transformation to helium atoms with the release of energy initially as gamma rays. The reaction is $4H \longrightarrow He^4 + 2e^- + h\nu$; h is Planck's constant and ν is the wavelength of the gamma radiation. The gamma radiation reacts with electrons and the energy is ultimately emitted again as photons of light energy which pass out of the sun.

Only a small fraction of the light energy reaching the earth from the sun is trapped. Large areas of the earth have no plants, and plants can utilize in photosynthesis only some 3 per cent of the incident energy. The radiant energy is converted into the potential energy of the chemical bonds of the organic substances made by the plant. When an animal eats a plant, or when bacteria decompose it and these organic substances are oxidized, the energy liberated is just equal to the amount of energy used in synthesizing the substances (First Law of Thermodynamics) but some of the energy is converted to heat and not useful energy (Second Law of Thermodynamics). When the animal in turn is eaten by another animal, a further decrease in useful energy occurs as the second animal oxidizes the organic substances in the first to liberate energy to synthesize its own cellular constituents. Eventually all the radiant energy originally trapped by plants in photosynthesis is converted to heat and dissipated to outer space.

We can roughly estimate that the total amount of carbon fixed by all the plants on the earth and in the waters of the globe is some 200 billion tons each year. Land plants synthesize about one tenth of this total and the marine plants, mostly microscopic algae, synthesize the remainder. The formation of each mole of glucose—180 grams—requires the input of 686 kilocalories of radiant energy. If we make the simple assumption that the carbon is fixed in the form of glucose, then we can see that it would require biological energy amounting to 10^{16} kilocalories per year. If we make appropriate corrections for friction losses, our estimate for the total biological energy flux rises another hundredfold, or perhaps more, to 10^{18} kilocalories of solar energy captured per year. This, in turn, is only about one thousandth of the total solar energy, about 10^{21} kilocalories per year, that is estimated to fall on the earth in the course of the year. It has been estimated that the activity of green plants leads to a renewal of all the carbon dioxide in the atmosphere and dissolved in the waters of the world every 300 years, and a renewal of all the oxygen in the atmosphere in about 2000 years.

All of the phenomena of life which we discussed previously—growth, movement, irritability, reproduction and others—require the expenditure of energy by the cell. Living cells are not heat engines; they cannot use heat energy to drive these reactions but must use chemical energy, chiefly in the form of energy-rich phosphate bonds, abbreviated ~ P. These bonds have a relatively high *free energy of hydrolysis*, ΔG (i.e., the difference in the energy content of the reactants and the products after the bond is split is relatively high). The free energy of hydrolysis is *not* localized in the covalent bond joining the phosphorus atom to the oxygen or nitrogen atom. Thus the term "energy-rich phosphate bond" is actually a misnomer, but it is so deeply ingrained by long usage that it is not likely to be changed.

5–9 LIGHT AND PHOTOCHEMICAL REACTIONS

To discuss biological processes involving light—photosynthesis, vision, phototropism and bioluminescence—we must first consider the properties of light itself. Light acts as though made of small packets of energy, called *photons* or light *quanta*, that have no electric charge and very little mass. Photons are a class of ultimate physical particles, like protons and electrons. Light is also characterized by its wave motion, and the different colors of light (different regions of the light spectrum) are identified by their wavelength or frequency.

Each photon has an energy content, E, equal to hc/λ, where c is the velocity of light (3×10^{10} cm. per sec.), λ is the wavelength of light and h is Planck's constant that relates frequency to energy. The energy content of a photon, the work it can do, is *inversely* proportional to wavelength. Light of shorter wavelength (violet) has a higher energy content than light of longer wavelength (red). The intensity of light equals the rate at which photons are being delivered. In photochemical reactions, each molecule, by absorbing one quantum of light, is excited to enter into a chemical reaction. One mole of the substance (6×10^{23} molecules) will be excited by 6×10^{23} quanta. The energy content of "one mole" of quanta, 6×10^{23} quanta, is defined as one *einstein*.

By multiplying the appropriate constants, we can find that the energy content of one einstein equals 2.854×10^4 kilocalories divided by the wavelength in nanometers. The energy

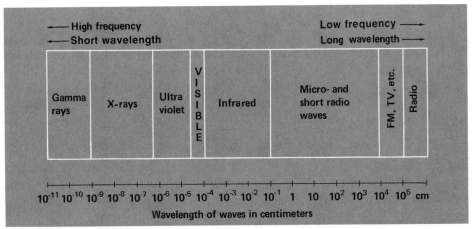

Figure 5–5 The spectrum of electromagnetic waves. (From Jones, M. M., Netterville, J. T., Johnston, D. O., and Wood, J. L.: *Chemistry, Man and Society,* 2nd Ed. Philadelphia, W. B. Saunders Co., 1976.)

content of one einstein of blue light (wavelength, 450 nm.) is 64 kilocalories, that of red light (660 nm.) is 43 kilocalories. The different colors of light represent quanta of different energy, those of blue light having greater energy than those of red light. All the radiations of the electromagnetic spectrum, from very short x-rays, through ultraviolet, visible and infrared light, to very long radio waves, represent a single phenomenon, differing only in wavelength and in the energy of their photons (Fig. 5–5).

When light quanta strike some types of metal plates, electrons are ejected, the number of electrons ejected being proportional to the number of photons striking the plate (this is the principle of an exposure meter for photog-

raphy). When a quantum strikes a molecule, such as a molecule of chlorophyll, an electron may be ejected or moved to an orbital farther from the nucleus (with a higher energy content). Energy must be put into the system to move an electron from an inner orbital to one farther out, because a negatively charged particle is being moved away from the positively charged nucleus. There can be no more than two electrons in any orbital (see Appendix I) and the electrons must be paired—must spin in opposite directions. If one of these electrons absorbs a light quantum and moves to another orbital, the molecule is in an "excited" state (Fig. 5–6). To promote a reaction, light must be absorbed; there is a quantitative relationship between the number of quanta absorbed and the number of

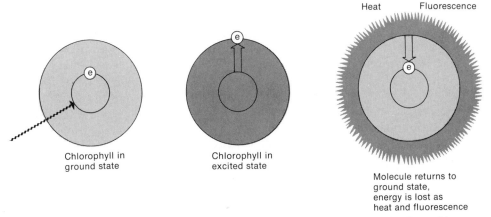

Figure 5–6 Model of a photochemical reaction. A quantum of light energy strikes a molecule, and its energy is used to move an electron to an orbital farther from the nucleus. When the electron returns to the inner orbital, the energy may be released as light energy of a different wavelength, a process termed fluorescence.

molecules activated. *Photochemical reactions* are characterized by the presence of intermediates which are electronically excited atoms or molecules—ones in which an electron has been moved to an outer orbital of higher energy.

The transition of an electron to a new orbital leads to a redistribution of the electronic charge over the entire molecule. The chemical properties of a molecule depend on the energy and on charge density at different sites; thus the excitation of an electron results in the formation of a new molecule, usually one which is chemically more reactive than a molecule in the ground state.

The electrons of greatest significance in photochemical reactions are the *"lone-pair" electrons,* found in molecules with nitrogen and oxygen atoms and not involved in bonding. These two, occupying the same orbital, must be paired, i.e., have opposite spins. When one is activated by absorbing a photon and moves to an outer orbital of higher energy, the spins may be opposite (the *singlet state*) or alike (the *triplet state*).

The probability of a direct transition from the ground state to the triplet state is very small. Usually the triplet state is achieved by internal conversion from the singlet state of the same electronic configuration. The decay of an excited singlet state to the ground state is known as *fluorescence;* it is extremely rapid (10^{-8} second) and is independent of temperature. If the excited singlet state undergoes a transition to the related triplet state, and the latter then decays to the ground state, a different kind of radiation, *phosphorescence*, is emitted. This is much slower, lasting from 10^{-4} to 1 second.

The photochemical reactions most important biologically, however, do not involve the emission of light but are *radiationless transitions* from the triplet to the ground state. Instead of being emitted as light, the energy is transferred to chemicals reacting in another system. The triplet state, with its relatively long half-life, has a much greater probability of reacting chemically than the singlet state. Photosynthesis involves such a radiationless transition from the triplet to the ground state of the unique molecule *chlorophyll.*

The energy required to move an electron from one orbital to another depends on the difference in energy between the two. The energy in a photon must be used completely (one cannot use part of a quantum); hence only photons with the proper energy to move the electron to an allowed orbital will be absorbed. By shining lights of specific wavelengths through a solution of a substance, and measuring the fraction absorbed, one obtains the *absorption spectrum* of that substance. If the absorption peak is at 660 nm. (in the red region), then about 43 kilocalories per mole are required to move the electron from a filled inner orbital to an unfilled outer orbital. Each type of molecule has a characteristic absorption spectrum, and measuring the absorption spectrum can be useful in identifying some unknown substance isolated from a plant or animal cell.

5–10 PHOTOSYNTHESIS

The essence of the photosynthetic process now appears to be the conversion of the radiant energy of sunlight into chemical energy in the form of ATP and reduced nicotinamide adenine dinucleotide phosphate (NADPH). Our theories of photosynthesis have undergone many changes as new evidence has come to light.

The first studies date back to 1630, when van Helmont, a Flemish botanist, showed that plants make their own organic materials and do not simply absorb them from the soil. He weighed a pot of soil and the willow tree planted in it and showed that the tree gained 164 pounds in five years but the soil weighed only two ounces less. Van Helmont concluded that the rest of the substance came from the water he had added; we now know that carbon dioxide removed from the air by the plant contributed some 70 per cent of the mass of the plant material synthesized.

Joseph Priestley showed in 1772 that a sprig of mint would "restore" air that had been "injured" by the burning of a candle. Seven years later Jan Ingenhousz showed that vegetation could restore "bad air" only if the sun was shining, and that the ability of the plant to restore air was proportional to the clearness of the day and to the exposure of the plant to the sun. In the dark, plants gave off air "hurtful to animals."

The next major step in understanding photosynthesis came in 1804, when de Saussure weighed both air and plant before and after photosynthesis and showed that the increase in the dry weight of the plant was greater than the weight of carbon dioxide removed from the air. He concluded that the other substance contributing to the gain in weight was water. Thus, 170 years ago, the broad outlines of the photosynthetic process appeared to be: carbon dioxide plus water plus light energy yields oxygen and organic material.

Ingenhousz suggested that light functions in photosynthesis to split carbon dioxide to liberate oxygen and yield "carbon," which is used in forming plant substance. On this basis, living organisms were divided into green plants, which could use the radiant energy for the "assimilation" of carbon dioxide, and other organisms, without chlorophyll, which could not use radiant energy and could not assimilate CO_2.

This logical division of the living world was upset when Winogradsky discovered (1887) chemosynthetic bacteria, organisms without chlorophyll that could assimilate carbon dioxide—convert it to organic substances—in the dark. It was upset further by Engelmann's discovery (1883) of purple bacteria that carried out a kind of photosynthesis in which no oxygen was liberated. Although not fully appreciated at the time, the discovery of chemosynthetic bacteria that can carry on carbon dioxide assimilation in the dark revealed that this process is not peculiar to photosynthesis. Since 1940, by the use of labeled carbon as a tracer, experiments have shown that all cells, plant, bacterial and animal, can assimilate carbon dioxide—can incorporate it into organic molecules—but differ in the source of the energy that is required for this process.

Another major advance in our understanding of the photosynthetic process was made in 1905 when the British plant physiologist Blackman demonstrated that it includes two successive series of reactions, a rapid *light reaction* and a slower series of steps not affected by light, which he termed the *dark reaction.* Using light of high intensity, he found that photosynthesis proceeded as rapidly when the light was flashed alternately on and off for periods of a fraction of a second as when the light shone continuously, even though the photosynthetic system received less than half as much energy. Only when the length of the dark period was increased considerably was the rate of photosynthesis decreased. In further experiments, he showed that the rate of the dark reaction was markedly increased by increasing the temperature.

The next hypothesis regarding the central chemical process in photosynthesis was supplied by the experiments of C. B. van Niel, who showed in 1931 that bacterial photosynthesis could proceed anaerobically without the evolution of oxygen. He suggested that there is a basic similarity in bacterial and green plant photosynthesis; in the latter, light energy is used to carry out the photolysis of water, H_2O, to yield a reductant (H) which reacts in some way in the assimilation of carbon dioxide, and an oxidant (OH) which was postulated to be the precursor of molecular oxygen. In bacterial photosynthesis, the process is basically similar but a different hydrogen donor, H_2S or molecular hydrogen, is used and there is no evolution of oxygen.

It is now clear that all the reactions for the incorporation of CO_2 into organic materials can occur in the dark (the "dark reactions") and thus perhaps are not in the strict sense part of the process of photosynthesis. They fundamentally constitute a reversal of the reactions by which carbohydrates are broken down. The reactions dependent on light (the "light reactions") are those in which radiant energy is converted into chemical energy, and the first stable, chemically defined products of these reactions are ATP and NADPH.

Before proceeding with the discussion of the chemical reactions by which CO_2 assimilation occurs (which are now fairly well understood) and the reactions by which ATP and NADPH are formed (which are still not completely clear), let us consider the structure of the cell organelles and the properties of the pigments involved in this process.

Chloroplasts. When a bit of leaf is examined under the microscope it can be seen that the green pigment is not uniformly distributed in the cell but is confined to small bodies called *chloroplasts.* Each cell has some 20 to 100 chloroplasts, which can grow and divide to form daughter chloroplasts. The electron microscope reveals that the chloroplast, like the mitochondrion, has a double-layered outer membrane within which are many smaller bodies, called *grana*, which contain the chlorophyll.

Each granum is composed of layers of molecules arranged like a stack of coins. The stacks of flat disks which compose each granum are made of layers of protein molecules alternating with layers containing chlorophyll, carotenes and other pigments and special types of lipids (containing galactose or sulfur but only one fatty acid). These surface-active lipids are believed to be adsorbed between the layers and serve in stabilizing the lamellae composed of alternate layers of protein and pigments. As we shall see, this characteristic layered structure of the grana may be important in permitting the transfer of energy from one molecule to the adjacent one during the photosynthetic process.

Electron microscopy has revealed in the lamellae repeating unit structures termed *quantasomes*, made up of some 230 molecules of chlorophyll. The material within the chloroplast and lying between the grana is called the *stroma;* it contains the enzymes which carry out the "dark" reactions. Isolated chloroplasts, when carefully prepared from leaves, are able to carry out *in vitro* the entire sequence of photosynthetic reactions.

Land plants absorb the water required for the photosynthetic process through their roots: aquatic plants receive it by diffusion from the surrounding medium. The carbon dioxide required diffuses into the plant by way of small holes, *stomata* (p. 75), in the surface of the leaves. Carbon dioxide is used up as a result of the photosynthetic process and its concentration in the cell is always slightly lower than that in the atmosphere. Oxygen is liberated in the process and diffuses out of the cell and finally out of the plant through the stomata. The sugars formed also tend to diffuse away from the site of formation to regions of lower concentration.

Plants need vast quantities of air to carry on photosynthesis, for air contains only 0.03 per cent carbon dioxide — 10,000 liters of air are needed to supply 3 liters of carbon dioxide, enough to make about 4 grams of glucose. Plants generally grow better in air with a higher carbon dioxide content and some greenhouses are maintained with an atmosphere containing 1 to 5 per cent CO_2.

Chlorophyll and Other Photosensitive Pigments. The chlorophyll molecule, made of atoms of carbon and nitrogen joined in a com-

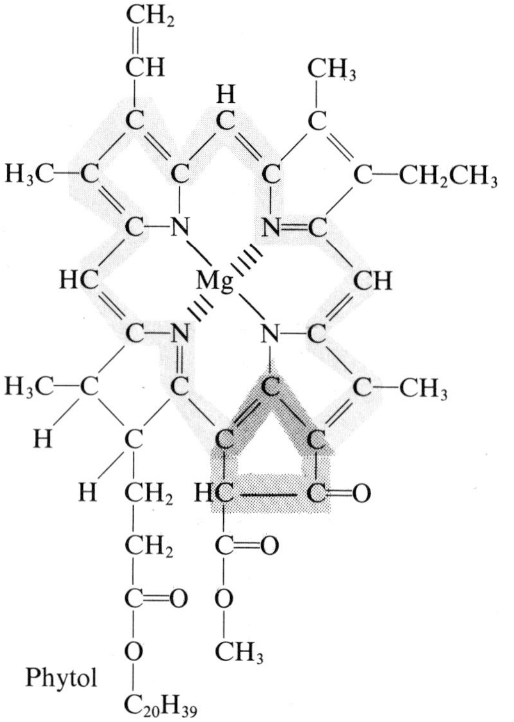

Figure 5–7   Structural formula of chlorophyll A. The light gray tint outlines the conjugated system of single and double bonds present in the molecule. The dark gray tint indicates the 5-carbon ring that may function in the transfer of hydrogen atoms.

plex ring (Fig. 5–7), is strikingly similar to the heme portion of the red pigment hemoglobin present in red blood cells, but contains an atom of magnesium instead of an atom of iron in the center of the ring, bound to two of the four nitrogen atoms. The chlorophyll molecule has a long tail composed of *phytol,* an alcohol containing a chain of 20 carbon atoms.

An examination of the chlorophyll molecule reveals that it is a *conjugated system* of double and single bonds (indicated in light shading) alternating around the ring. Such a conjugated system provides for many rearrangements, many different patterns of single and double bonds in the ring structure. It is a *resonating system,* which includes many different ways of arranging the external electrons without moving any of the constituent atoms. The possibilities of resonance in the ring give the chlorophyll molecule considerable stability. Such systems of conjugated single and double bonds have mobile electrons, called *pi electrons,* which are associated not with a single atom or bond but with the conjugated system as a whole. Only a small amount of energy is needed to raise such pi electrons to an orbital of higher energy.

There are several types of chlorophyll, of which the two most important are called a and b. Chlorophyll b has one oxygen more and two hydrogens less than chlorophyll a. All green plants have chlorophyll a, but many algae and certain other plants lack chlorophyll b. Plants contain, in addition to chlorophyll, many pigments which give them their great variety of colors. Some of them play a role in absorbing light energy and transferring it to chlorophyll to be used in photosynthesis. Most plants have a deep orange pigment, *carotene,* which can be converted in the animal body to vitamin A, and a yellow pigment, *xanthophyll.* The red and blue-green algae contain other pigments, *phycocyanin* and *phycoerythrin,* which are more effective in photosynthesis in these red algae than are the chlorophylls present.

The sunlight reaching the surface of the earth has its maximal intensity in the blue-green and green portions of the spectrum, from 450 to 550 nm., yet it is in this region that the chlorophyll molecule absorbs the smallest fraction of the incident light energy. Chlorophyll a and b both have absorption peaks in the violet region at about 440 nm.; chlorophyll b strongly absorbs light quanta in the red region at about 650 nm. and chlorophyll a at about 675 nm. in the far red region.

Despite its relative inability to absorb the wavelengths of sunlight with the greatest energy content, chlorophyll has other qualities that may account for its being selected in the course of evolution. In addition other pigments, such as the carotenoids, which absorb more strongly in the blue-green or green region, 400 to 550 nm., can transfer the energy absorbed to chlorophyll where it can be used in photosynthesis.

The term *photosynthetic unit* is applied to that group of pigment molecules within which the energy of light, at any wavelength in the visible spectrum from 400 to 700 nm., absorbed anywhere in the unit, can be transferred to the specific chlorophyll molecule that participates in the conversion of light to chemical energy. The excitation energy of the photon migrates within 10^{-9} second to that pigment molecule in the unit with an absorption band at the lowest energy state. If that molecule can readily undergo a singlet-triplet conversion, or can form a charge transfer complex with an adjacent molecule, it can trap excitation energy received anywhere in the unit. Energy can be transferred over distances as great as 10 nm. in 10^{-12} to 10^{-11} second. Each photosynthetic unit serves as a light-receiving apparatus which can collect light energy and deliver it to the chlorophyll molecule that will participate in the photochemical reaction. The absorption of a photon increases the energy of an electron, moving it from an inner orbital to one farther out from the nucleus. The amount of energy absorbed must just equal the difference in energy content of the old and new orbitals. An electron can occupy only a limited number of orbitals, and, unless the photon can raise the energy of the electron by just the right amount for a defined orbital level, the photon will not be absorbed. The unique capacity of chlorophyll to function in photosynthesis depends on its remarkable ability to absorb the energy of visible light with a high degree of efficiency and to transfer the energy to other molecules. Chlorophyll has a chemically reactive site (the five-carbon ring indicated in dark shading in Figure 5–7) which may function in transferring hydrogen atoms.

Carbon Dioxide Assimilation (the "Dark Reactions"). The reactions by which carbon dioxide is incorporated into organic substances have now been worked out in detail by investigators who incubated suspensions of algae in the presence of carbon dioxide labeled with carbon-14. By incubating for varying lengths of time and isolating and identifying the products, they were able to determine that the first compound to become labeled was 3-phosphoglyceric acid; the carbon-14 was present predominantly in the carboxyl carbon. Further experiments revealed that the "dark" reactions of carbohydrate synthesis occur in a cyclic sequence of (1) carboxylative, (2) reductive and (3) regenerative phases (Fig. 5–8).

The five-carbon sugar with a phosphate group attached to carbon number 5, *ribulose-5-phosphate,* is phosphorylated by ATP to yield *ribulose diphosphate.* This is carboxylated by the addition of CO_2, presumably to an intermediate six-carbon substance, but this immediately splits by the addition of a molecule of water to give two molecules of *phosphoglyceric acid.* The phosphoglyceric acid is then reduced by an enzymic reaction

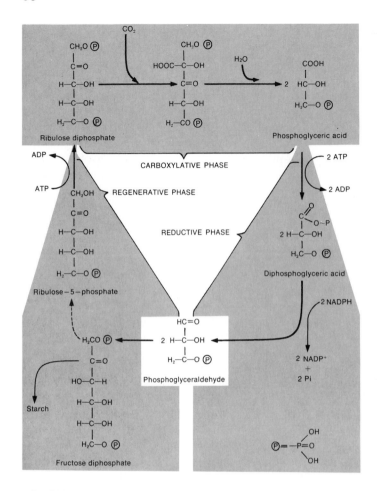

Figure 5–8 Diagram of the "dark" reactions of carbohydrate synthesis by which carbon dioxide is incorporated into sugars. The dashed line from fructose diphosphate to ribulose-5-phosphate indicates several reactions of the pentose phosphate pathway (cf. Figure 5–9).

which requires ATP and reduced nicotinamide adenine dinucleotide phosphate (NADPH) to give *phosphoglyceraldehyde* (a three-carbon sugar, triose, containing a phosphate group). Two of these trioses can then condense to form one hexose molecule, which can be added onto a starch molecule and thus stored.

To enable the cycle to continue, ribulose diphosphate must be regenerated from some of the fructose phosphates produced; this is achieved by certain reactions of the pentose phosphate pathway (Fig. 5–9). The overall reaction for the *net* production of one mole of fructose phosphate is: 6 ribulose-1,5-diphosphate + 6 CO_2 + 18 ATP + 12 NADPH + $12H^+ \rightarrow$ 6 ribulose-1,5-diphosphate + 1 fructose-6-phosphate + 17 P_i + 18 ADP + 12 $NADP^+$. This cycle is termed the Calvin cycle, or "C3 pathway," for the first compound into which carbon dioxide is assimilated is the three-carbon compound, phosphoglyceric acid.

Several observations indicate that there are other pathways for the synthesis of carbohydrates. The carboxylation of acetyl CoA (p. 103) to form pyruvate and the carboxylation

of succinyl CoA (p. 105) to form α-ketoglutarate occur in certain photosynthetic bacteria.

In certain tropical grasses such as sugar cane, the first compounds to be labeled after incubation with $^{14}CO_2$ are the *four*-carbon dicarboxylic acids, oxaloacetate, malate, and aspartate. This "C4 pathway" is present in addition to the C3 pathway in a number of plant orders and apparently has evolved independently several times. The C4 plants evolved in areas of the world with high temperatures, high light intensities and limited amounts of water, and they have a higher temperature optimum, a higher light optimum, high rates of photosynthesis and growth, and less loss of water by transpiration (p. 269) than plants having only the C3 pathways.

The C4 plants have a discrete layer of *bundle sheath cells* surrounding the xylem and phloem of the vascular bundle and a population of *mesophyll cells* surrounding the bundle sheath cells. When these cells were separated and tested, the C3 pathway was found largely or entirely in the bundle sheath cells, whereas the mesophyll cells contained the enzyme phos-

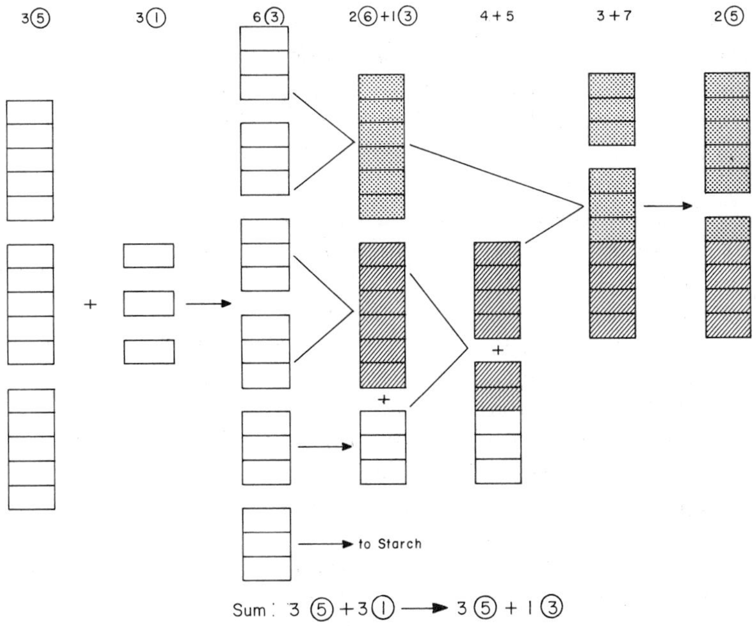

Figure 5-9 Diagrammatic representation of the reactions by which three pentose molecules accept three molecules of carbon dioxide to yield six molecules of triose. One of these trioses can be used in the synthesis of hexose and starch. The other five, by a further series of reactions in which the upper two or three carbons of the molecule are transferred as a unit, eventually regenerate three molecules of pentose so that the cycle can be repeated.

phoenolpyruvate carboxykinase, which catalyzes the reaction CO_2 + phosphoenolpyruvate (three carbons) $\longrightarrow$ oxaloacetate (four carbons). The phosphoenolpyruvate is generated in another reaction catalyzed by enzymes in the mesophyll layer: pyruvate + ATP + P_i $\longrightarrow$ phosphoenolpyruvate + AMP + PP_i. The ATP is generated in the sequence of reactions:

$$AMP + ATP \longrightarrow 2\ ADP$$
$$PP_i \longrightarrow 2\ P_i$$
$$2\ ADP + 2\ P_i \longrightarrow 2\ ATP\ (by$$
$$photophosphorylation$$
Sum: $ATP + AMP + PP_i \longrightarrow 2\ ATP$

The oxaloacetate may undergo transamination to form aspartate or it may be reduced to malate: Oxaloacetate + NADPH $\rightarrow$ malate + $NADP^+$. Within the chloroplasts of the bundle sheath cells is a different enzyme which catalyzes the decarboxylation of malate: Malate + $NADP^+$ $\rightarrow$ pyruvate + CO_2 + NADPH. The role of the C4 cycle is to increase the concentration of CO_2 within the bundle sheath cells so as to drive the C3 (Calvin) cycle there. The operation of the C4 cycle in the mesophyll cells and the transfer of the malate to the bundle sheath cells where it is decarboxylated serves to increase the local intracellular concentration of CO_2 in the bundle sheath cells from tenfold to sixtyfold over that in the cells of plants having only the C3 pathway.

None of these reactions, carboxylative, reductive or regenerative, is unique to photosynthetic cells. The only difference found so far is that the reductive reaction by which phosphoglyceric acid is converted to phosphoglyceraldehyde requires NADPH rather than the usual NADH.

The Light Reactions. The ATP and NADPH needed to drive these dark reactions are derived from the light reactions. It was thought at one time that the light reaction in the chloroplasts makes NADPH and that this might then be used by the mitochondria in the plant cell to synthesize ATP. However, the leaves that are most specialized for photosynthesis have cells containing very few mitochondria, and Arnon and his colleagues have demonstrated that isolated chloroplasts, free of mitochondria, can carry out photosynthesis and thus must be able to make ATP as well as NADPH.

Working with isolated chloroplasts, Arnon demonstrated the complete separability of the light and dark reactions. He first illuminated chloroplasts in the absence of CO_2 but in the presence of ADP and NADP. Under these conditions the chloroplasts accumulated ATP and NADPH and evolved molecular oxygen. He then extracted the enzymes needed for the dark reactions of carbon dioxide assimilation, discarded the green part of the chloroplasts and, using the ATP and NADPH previously made in the light reaction, carried out CO_2 assimilation in the dark; i.e., the system made carbohydrates which could be isolated and identified. In other experiments these enzymes were supplied with NADPH and ATP made by animal cells and they were again able to carry out CO_2 assimilation in the dark.

Photosynthetic Phosphorylation. When Arnon illuminated isolated chloroplasts in the absence of CO_2 and pyridine nucleotide, but in the presence of large amounts of ADP and inorganic phosphate, the chloroplasts could utilize the light energy to synthesize ATP from the ADP and inorganic phosphate. This process has been termed *photosynthetic phosphorylation* to distinguish it from the oxidative phosphorylation that occurs in mitochondria (p. 111).

The chloroplast contains an electron transport system that includes a flavoprotein, two or more cytochromes, an iron-containing protein called *ferredoxin,* which can undergo cyclic oxidation and reduction, and a substance related to ubiquinone called *plastoquinone.* The photosynthetic phosphorylation of ADP to produce ATP can proceed in the absence of oxygen. It has been demonstrated in chloroplasts from a variety of green plants and in the chlorophyll-containing chromatophores of several species of photosynthetic bacteria.

In oxidative phosphorylation, as we shall see (p. 111), the energy to add the inorganic phosphate (P_i) to ADP is obtained when electrons are transferred from an electron donor at one energy level to an electron acceptor at another. This occurs in three steps, in each of which an energy-rich phosphate group ($\sim P$) is made and added to ADP to form ATP. The flow of electrons (in some of the steps the electrons are attached to protons to form hydrogen atoms) from one energy level to another releases energy. Each of these steps has been likened to a water wheel, turned by the "falling" electron, and driving the energy-requiring process of attaching an inorganic phosphate group to ADP. The process is fundamentally one of energy transfer—as the electrons move from one energy level to another, the energy is normally conserved in the form of an energy-rich phosphate group. It is possible, however, to "uncouple" the phosphorylation process from the flow of electrons (p. 112). In oxidative phosphorylation the ultimate electron acceptor is oxygen and the electron donor is sugar or some other organic substance.

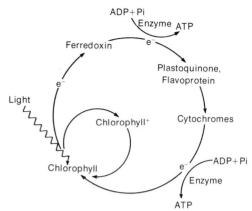

Figure 5-10 Diagrammatic representation of the reactions which compose cyclic photophosphorylation.

Since photophosphorylation uses neither oxygen nor any substrate molecules, only light, the electron donors and acceptors must be *within* the chloroplast itself. Arnon proposed that the chlorophyll molecule can serve in both capacities. Light striking a chlorophyll molecule excites one of the electrons to an energy level high enough to eject it from the molecule (Fig. 5-10). The chlorophyll molecule, having lost an electron, is now ready to serve as an electron acceptor (*chlorophyll*$^+$). If it simply took the same electron back directly it would reemit the light energy as heat and fluorescence (pure chlorophyll does fluoresce in this way). However, if the electron injected is taken up by ferredoxin, it can then return to the chlorophyll by a series of graded steps, somewhat like those in oxidative phosphorylation. The electrons pass from ferredoxin to the flavoprotein, to cytochrome and then to chlorophyll, yielding two $\sim P$ in the process as the energy of the electron is transferred via a phosphorylating enzyme system to $\sim P$. The purpose of this flow of electrons out of the chlorophyll molecule through ferredoxin, flavoprotein and cytochrome back to chlorophyll is to conserve in a chemical form some of the energy taken in as radiant energy rather than have it lost as fluorescence and heat. Only the light-induced production of high energy electrons and the nature of the ultimate electron acceptor, chlorophyll, are peculiar to this photosynthetic process. This process of photosynthetic phosphorylation appears to be a primary reaction of photosynthesis, and it accounts for the production of some, at least, of the ATP needed in the dark reactions of CO_2 assimilation, but it does not account for the NADPH which is also needed nor for the splitting of water into hydrogen and oxygen.

Noncyclic Photophosphorylation. In the evolution of algae and higher plants, the photosynthetic apparatus acquired the ability to use H_2O to reduce pyridine nucleotides, a reaction that requires about 55 kilocal./mole of $NADP^+$. Two einsteins of light at 675 nm. would provide about 84 kilocal. and could drive the process if the system were highly efficient. The mechanism that did evolve is somewhat less efficient; it uses four einsteins of light to drive *two* electron transport systems that operate in series. By using four quanta of light, about 160 kilocal. become available to drive a reaction that requires some 55 kilocal./mole.

Although light of wave lengths longer than 700 nm. is ineffective in driving photosynthesis, a mixture of light at 700 and 680 nm., applied simultaneously or alternately, is considerably more effective than the sum of the two applied alone. This observation suggests that there are two electron transport systems. The photosynthetic unit of system I contains a molecule of *chlorophyll* P_{700} (which absorbs light at 700 nm. and ejects an electron) plus other molecules that serve as electron donors and acceptors to the photoactivated chlorophyll

P_{700}. The photochemical center of system II, chlorophyll P_{680}, has not yet been defined precisely; neither the identity of the chlorophyll at the center nor the electron donors and acceptors in its immediate neighborhood are known. The general pathway of electron flow from H_2O to NADPH by these two systems, and their associated oxidants and reductants, is shown in Figure 5–11.

Examination of Figure 5–12 permits analysis of the electron transport systems of photosynthesis into four components, those leading to and from each of the two photochemical systems. The chain leading to photochemical system I, which must accomplish the four-electron oxidation of 2 OH^- to O_2, is completely unknown; no intermediates in the process have been detected. The specific acceptor of electrons from that system is also unknown. A component designated C_{550}, because it is affected by light at 550 nm., may have this property and transfer the electrons to *plastoquinones*. These serve as connections between systems I and II and transfer electrons to cytochromes b and c, to a copper protein called *plastocyanin*, and then to chlorophyll P_{700}. The passage of electrons along this chain generates a molecule of ATP.

The electron acceptor from chlorophyll

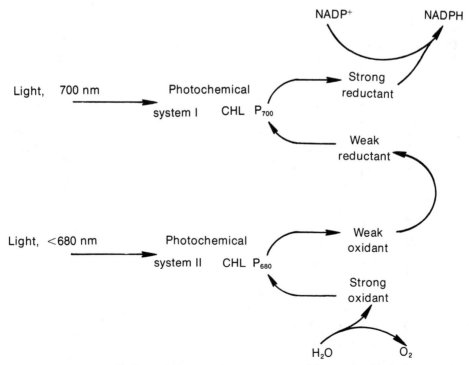

Figure 5–11 Diagram illustrating the flow of electrons from water to pyridine nucleotide in the two electron transport systems of photosynthesis in the chloroplasts of plants.

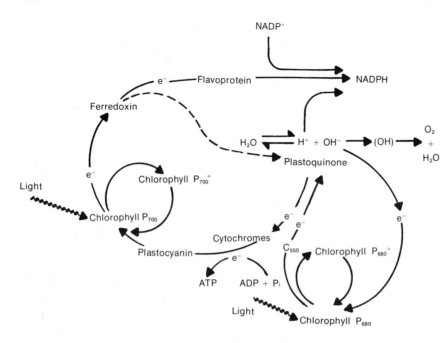

Figure 5–12 Diagram of the reactions that make up non-cyclic photophosphorylation.

P_{700} is first a bound ferredoxin ("ferredoxin reducing substance") and then ferredoxin. The flavoprotein, *ferredoxin-NADP reductase*, then transfers electrons from ferredoxin to $NADP^+$, reducing it to NADPH. System I, with chlorophyll P_{700}, can also carry out cyclic photophosphorylation by transferring electrons from ferredoxin to plastoquinone (the dashed line in Figure 5–12) via a substance termed phosphodoxin, which has not yet been characterized.

Considerable effort has been expended over the years in attempting to define the "quantum yield" of photosynthesis, the number of photons required for the evolution of one molecule of oxygen. The system outlined here would require at least four quanta at each of the two photochemical systems, a total of eight quanta, to achieve the evolution of one O_2 and the formation of 2 moles of NADPH. If the mean energy of the absorbed photons is 40 kilocal./einstein, a total of 8 × 40, or 320 kilocal., would be used per O_2 formed. At a minimum, 12 NADPH and 18 ATP are required for the synthesis of one mole of hexose. The absorption of 6 × 320 kilocal., or 1920 kilocal., would form 12 NADPH and 12 ATP. To obtain the additional 6 ATP required for the formation of one mole of hexose, 12 more quanta, another 480 kilocal., must be absorbed and used in cyclic phosphorylation for a grand total of 2400 kilocal. A mole of glucose yields 686 kilocal. when combusted to CO_2 and H_2O, and an equivalent amount of energy is the minimum

required for the reverse process, the formation of glucose from CO_2 and H_2O. Thus, the efficiency of the overall photosynthetic process can be estimated at 686/2400, or approximately 28 per cent.

Although the general concepts of the photosynthetic process are now reasonably well established, the description of the two photochemical centers system has been developed only in the past few years, and many details remain to be discovered.

The chlorophyll of certain photosynthetic bacteria, such as the red sulfur bacterium, *Chromatium*, can be activated by light to raise electrons to a high energy level. These electrons can then either reduce pyridine nucleotides (as in green plants) or reduce nitrogen gas to ammonia or reduce protons to hydrogen gas.

It seems possible that cyclic photophosphorylation is the more primitive process and perhaps was the first to appear, when the earth's atmosphere contained little or no oxygen gas. It simply provided these primitive organisms with a way of synthesizing ATP in an anaerobic environment other than by the process of fermentation.

Perhaps the next evolutionary step was a noncyclic photophosphorylation, such as the one found in certain bacteria today, in which pyridine nucleotides are reduced (and thus made available for biosynthetic reactions) but the electron donors are molecules such as thiosulfate or succinate. The third step, achieved by the green plants, was the ability to use water molecules rather than thiosulfate or succinate as the electron donor in photophosphorylation. By evolving a system which could obtain electrons for reductive processes from water molecules, green plants were enabled to live nearly everywhere and were not restricted to places where special electron donors such as thiosulfate might be found.

Then, as plants spread and multiplied, they released into the atmosphere oxygen molecules that had previously been locked up in water; this made possible the further biochemical evolution of animals and other organisms that require molecular oxygen.

The overall process of photosynthesis by which organic molecules are built up, represented by

$$6\ CO_2 + 6\ H_2O \longrightarrow C_6H_{12}O_6 + 6\ O_2$$

is a very complex one, as complex as the reactions of cellular respiration by which they are subsequently broken down. In the light reactions, the remarkable properties of chlorophyll enable it to capture the radiant energy of light with a high degree of efficiency and, with the aid of ferredoxin, cytochromes and other compounds in the membranes of the grana, to produce NADPH, ATP and molecular oxygen. Experiments with water and CO_2 labeled with ^{18}O show that all of the oxygen gas evolved in photosynthesis comes from the O in H_2O, while none comes from CO_2.

In the subsequent dark reactions, enzymes in the stroma of the chloroplasts condense carbon dioxide with ribulose diphosphate to yield two molecules of phosphoglyceric acid, which are then converted to phosphoglyceraldehyde in a reaction driven by ATP and NADPH. The two phosphoglyceraldehydes condense to fructose diphosphate, and a complex sequence of enzymatic transfers of two- and three-carbon units from one carbohydrate to another eventually regenerates ribulose phosphate and leads to the net synthesis of glucose.

5–11 BIOLOGICAL OXIDATION AND REDUCTION

The enzymatic processes within each cell by which carbohydrates, fatty acids and amino acids are metabolized ultimately to carbon dioxide and water with the conservation of bio-

logically useful energy are termed cellular respiration. Many of the enzymes catalyzing these reactions are located in the cristae and walls of the mitochondria (Fig. 4–19).

All living cells obtain biologically useful energy by enzymic reactions in which electrons flow from one energy level to another. For most organisms, oxygen is the ultimate electron acceptor; oxygen reacts with the electrons and with hydrogen ions to form a molecule of water. Electrons are transferred to oxygen by a system of enzymes, localized within the mitochondria, called the *electron transmitter system.*

Electrons are removed from a molecule of some foodstuff and transferred (by the action of a specific enzyme) to some *primary acceptor.* Other enzymes transfer the electrons from the primary acceptor through the various components of the electron transmitter system and eventually combine them with oxygen (Fig. 5–13).

The chief source of energy-rich phosphate bonds, $\sim P$, in the cell is from the flow of electrons through the acceptors and the electron transmitter system. This flow of electrons has been termed the "electron cascade," and we might picture a series of waterfalls over which electrons flow, each fall driving a waterwheel, an enzymatic reaction by which the energy of the electron is captured in a biologically useful form—that of energy-rich compounds such as adenosine triphosphate, ATP.

ATP is the "energy currency" of the cell; all of the energy-requiring reactions of cellular metabolism utilize ATP to drive the reaction. Energy-rich molecules do not pass freely from cell to cell, but are made at the site in which they are to be utilized. The energy-rich bonds of ATP that will drive the reactions of muscle contraction, for example, are produced right in the muscle cells.

Processes in which electrons (e^-) are removed from an atom or molecule are termed *oxidations;* the reverse process, the addition of electrons to an atom or molecule, is termed *reduction.* A simple example of oxidation and reduction is the reversible reaction

$$Fe^{++} \rightleftarrows Fe^{+++} + e^-$$

The reaction towards the right is an *oxidation* (the removal of an electron) and the reaction towards the left is a *reduction* (the addition of an electron). Every oxidation reaction (in

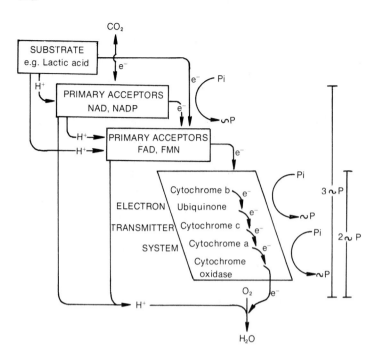

Figure 5–13 Diagram of the reactions of the "electron cascade," the succession of metabolic steps by which electrons are transferred from substrate to oxygen and the energy is trapped in a biologically useful form as energy-rich phosphate bonds, ~P.

which an electron is given off) must be accompanied by a reduction, a reaction in which the electrons are accepted by another molecule; electrons do not exist in the free state.

The passage of electrons in the electron transmitter system is a series of oxidation and reduction reactions termed *biological oxidation.* When the energy of this flow of electrons is captured in the form of ~P, the overall process is called *oxidative phosphorylation.* In most biological systems two electrons and two protons (that is, two hydrogen atoms) are removed together and the process is known as *dehydrogenation.*

The specific compounds of the electron transmitter system that are alternately oxidized and reduced are known as *cytochromes.* There are several of these and each one—a, b and c—is a protein molecule to which is bound a heme group similar to the one present in hemoglobin (p. 349). In the center of the heme group is an atom of iron, Fe, which is alternately oxidized and reduced—converted from Fe^{++} to Fe^{+++} and back—by giving off and taking up an electron:

$$Fe^{++} \rightleftarrows Fe^{+++} + e^-$$

Another component of the electron transmitter system, called *ubiquinone* (it occurs everywhere!), consists of a head, a six-membered carbon ring, which can take up and release electrons, and a very long tail. The tail is composed of ten repeating units, each of which consists of five carbon atoms. The repeating unit, called an isoprenoid group, is the basic unit of molecules of rubber, sterols and steroids.

All of the reactions of biological oxidation are mediated by enzymes and each enzyme is quite specific—it will catalyze the oxidation or reduction of certain compounds but not others. Only certain hydrogen atoms, ones having a certain spatial relationship to the rest of the molecule, can be removed (can undergo dehydrogenation).

5–12 THE OXIDATION OF LACTIC ACID

As an example of biological oxidation, let us consider the oxidation of lactic acid (the acid of sour milk), an important intermediate in metabolism. The reaction catalyzed by the enzyme *lactate dehydrogenase* (Fig. 5–14, reaction 1) is a *dehydrogenation;* the arrangement

$$H—\overset{|}{\underset{|}{C}}—OH$$

is one from which two hydrogens can be removed enzymatically. In this reaction, and in all dehydrogenations, the electrons are transferred to a *primary acceptor.* The primary electron acceptor in this reaction is NAD (p. 38). The product of the dehydrogenation reaction, *pyruvic acid,* does not have a structure suitable for attack by a dehydrogenase and must undergo further enzymatic action to attain a mo-

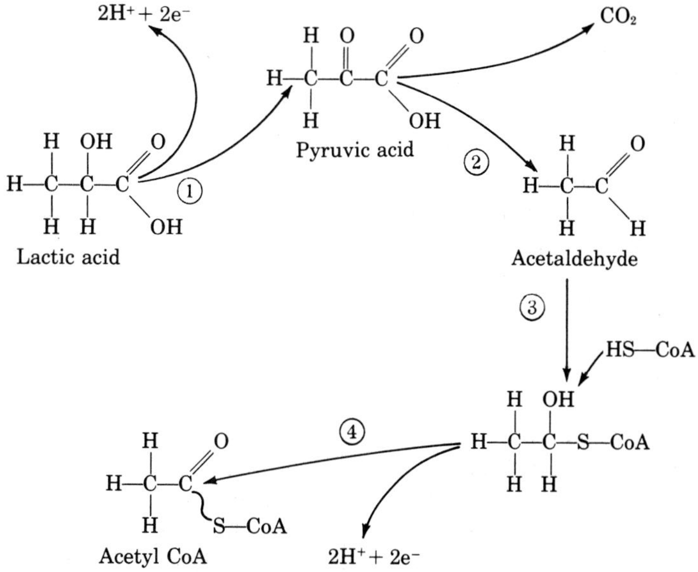

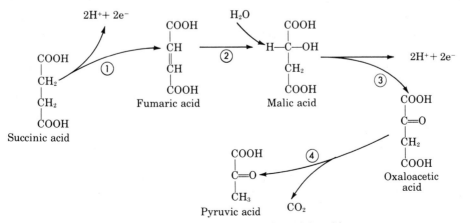

Figure 5–14 The enzymatic conversion of lactic acid to acetyl coenzyme A.

lecular structure which can undergo dehydrogenation. The next reaction of pyruvic acid is a ***decarboxylation,*** the loss of carbon dioxide (Fig. 5–14, reaction 2).

All of the carbon dioxide in the air we breathe out is produced by similar decarboxylation reactions. Carbon dioxide is derived in biological systems only from carboxyl groups (—COOH) by the process of decarboxylation. The product of the decarboxylation reaction undergoes a "make-ready" reaction with a large complex organic molecule, ***coenzyme A*** (Fig. 5–14, reaction 3), which results in an

$$H—\overset{|}{\underset{|}{C}}—OH$$ structure suitable for dehydrogenation. The active end of the coenzyme A molecule is a sulfhydryl group composed of sulfur and hydrogen, abbreviated CoA—SH.

The resulting compound undergoes dehydrogenation (Fig. 5–14, reaction 4) to yield acetyl coenzyme A.

5–13   THE OXIDATION OF SUCCINIC ACID

Another type of dehydrogenation, involving a different molecular structure, is exemplified by the oxidation of succinic acid (Fig. 5–15, reaction 1). The dehydrogenation of molecules having a —CH$_2$—CH$_2$— group involves a ***flavin*** as hydrogen and electron acceptor. The product, fumaric acid, cannot be dehydrogenated directly, but undergoes a make-ready reaction (Fig. 5–15, reaction 2) in which a molecule of water is added. The product of this reaction (malic acid) does have an arrange-

Figure 5–15 The oxidation of succinic acid.

ment, H—C—OH, suitable for dehydrogenation.

Malic acid undergoes dehydrogenation by *malic dehydrogenase*, and the product of the reaction is *oxaloacetic acid* (Fig. 5–15, reaction 3). Oxaloacetic acid may undergo several different reactions, one of which is a decarboxylation to yield pyruvic acid (reaction 4). The further metabolism of pyruvic acid would proceed to acetyl coenzyme A by the reactions just discussed.

In the reactions by which carbohydrates, fats and proteins are oxidized, the cell requires only these three simple types of reactions—dehydrogenation, decarboxylation and makeready. These may occur in different orders in different chains of reactions, as we have seen in the reactions of lactic acid and succinic acid. All of the dehydrogenation reactions are, by definition, oxidations, reactions in which electrons are removed from a molecule. The electrons cannot exist in the free state for any finite period of time but must be taken up immediately by other compounds, *electron acceptors.*

Two of the primary electron acceptors of the cell are pyridine nucleotides, nicotinamide adenine dinucleotide (abbreviated NAD) and nicotinamide adenine dinucleotide phosphate (abbreviated NADP). The functional end of both pyridine nucleotides is the vitamin *nicotinamide.*

The nicotinamide ring accepts one hydrogen ion and two electrons from a molecule undergoing dehydrogenation (e.g., lactic acid) and becomes reduced nicotinamide adenine dinucleotide, NADH, releasing one proton (Fig. 5–16).

NAD and NADP serve as primary electron and hydrogen acceptors in dehydrogenation reactions involving substrates with the

H—C—OH arrangement, such as the dehydrogenation of lactic or malic acid. The two pyridine nucleotides differ from one another in that NAD has two phosphate groups and NADP three phosphate groups in the tail attached to the niacin ring ("R"). Most dehydrogenases specifically require either NAD or NADP as the hydrogen acceptor and will not work with the other; some enzymes are less specific and will work with either one, though they usually work more rapidly with one than the other.

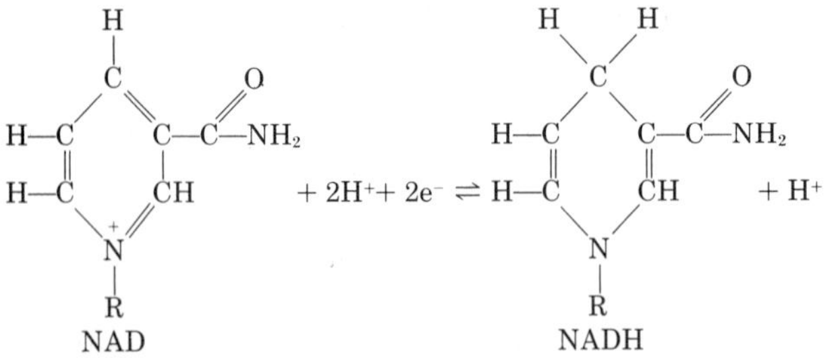

Another primary hydrogen acceptor, *flavin adenine dinucleotide* (abbreviated *FAD*), serves in reactions involving the —CH_2—CH_2— structure as in the dehydrogenation of succinic acid. In some reactions *flavin mononucleotide, FMN,* which consists of part of the FAD molecule, may serve in its place. The FAD of succinic dehydrogenase is bound very tightly to the protein part of the enzyme and cannot be removed easily. Such tightly bound cofactors are termed *prosthetic groups* of the enzyme. The pyridine nucleotide effective in the lactic dehydrogenase system, in contrast, is very loosely bound and is readily removed. Such loosely bound cofactors are termed *coenzymes.*

The reduced pyridine nucleotides, NADH or NADPH, cannot react with oxygen; their electrons must be passed through the intermediate acceptors of the electron transmitter system (the cytochromes) before they can react with oxygen. The flavin primary acceptors usually pass their electrons to the electron transmitter system, but some flavoproteins can react directly with oxygen. When this occurs, hydrogen peroxide, H_2O_2, is produced and no ~P is formed. An enzyme that can mediate the transfer of electrons *directly* to oxygen is termed an *oxidase;* one that mediates the removal of electrons from a substrate to a primary or intermediate acceptor is termed a *dehydrogenase*—e.g., lactic dehydrogenase, malic dehydrogenase or succinic dehydrogenase.

5–14 THE CITRIC ACID CYCLE

Acetyl coenzyme A cannot undergo dehydrogenation directly but undergoes a makeready reaction, combining with oxaloacetic

Figure 5–16 The reduction of nicotinamide-adenine dinucleotide (NAD).

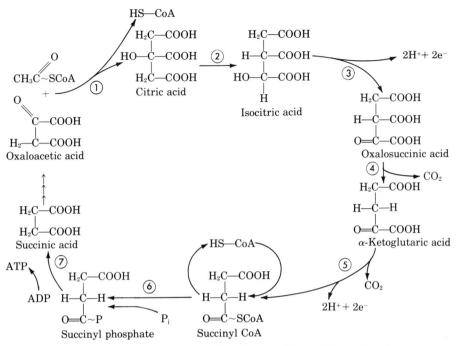

Figure 5–17 The tricarboxylic acid cycle (Krebs citric acid cycle).

acid (which contains four carbons) to yield *citric acid* (six carbons) plus free coenzyme A (Fig. 5–17, reaction 1).

Citric acid has neither an $H—\overset{|}{\underset{|}{C}}—OH$ nor

a —CH₂—CH₂— group and cannot undergo dehydrogenation either. Two additional makeready reactions, in which a molecule of water is removed and then added back (Fig. 5–17, reaction 2), yield *isocitric acid*, which can undergo dehydrogenation at its $H—\overset{|}{\underset{|}{C}}—OH$ group

(Fig. 5–17, reaction 3). The hydrogen acceptor is the pyridine nucleotide, NAD, and the product is *oxalosuccinic acid.* This undergoes decarboxylation to yield *α-ketoglutaric acid.*

Inspection of the formula of α-ketoglutaric acid reveals that the lower part of the molecule is just like pyruvic acid, and α-ketoglutaric acid is metabolized by a set of reactions similar to those of pyruvic acid. A dehydrogenation, a decarboxylation and a make-ready reaction using coenzyme A combine to yield *succinyl coenzyme A* (Fig. 5–17, reaction 5).

Just as pyruvic acid is converted to the coenzyme A derivative of an acid with one less carbon atom, α-ketoglutaric acid (five carbons) is converted to succinyl coenzyme A (four car-

bons). In animal cells the metabolism of pyruvic acid to acetyl coenzyme A and the metabolism of α-ketoglutaric acid to succinyl coenzyme A are complex reactions involving as coenzymes, in addition to NAD and coenzyme A, *thiamine pyrophosphate* (a coenzyme containing another vitamin, *thiamine*) and *lipoic acid.*

The bond joining coenzyme A to succinic acid is an energy-rich one, ~S, like the bond joining coenzyme A to acetic acid in acetyl coenzyme A. The energy of the bond of acetyl coenzyme A was utilized in bringing about the addition of the acetyl group to the oxaloacetic acid. The energy in the ~S bond of succinyl coenzyme A can be converted to an energy-rich phosphate bond, ~P, in ATP. The reaction of succinyl coenzyme A with inorganic phosphate yields succinyl phosphate and free coenzyme A (Fig. 5–17, reaction 6).

The phosphate group is then transferred to ADP (via guanosine diphosphate as an intermediate) to form ATP and free *succinic acid.* This is an example of an energy-rich bond synthesized at the *substrate level* — by reactions not involving the electron transmitter system. Only a small fraction of the energy-rich bonds are synthesized by reactions such as this. Normal cells metabolizing in an atmosphere containing oxygen synthesize most of their ATP by

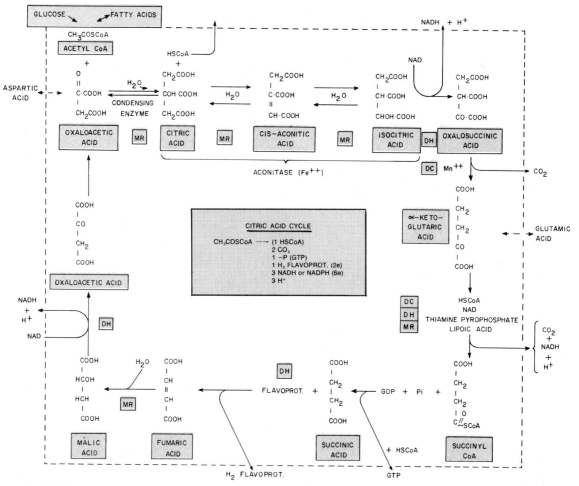

Figure 5–18 Diagram of the cyclic series of reactions, termed the tricarboxylic acid cycle (Krebs citric acid cycle), by which the carbon chains of sugars, fatty acids and amino acids are metabolized to yield carbon dioxide. The reactions are designated as DH, dehydrogenations; DC, decarboxylations; and MR, make-ready. The overall reaction effected by one "turn" of the cycle is summarized in the center.

oxidative phosphorylation in the electron transmitter system.

We have discussed the reactions by which succinic acid is metabolized to fumaric, malic and oxaloacetic acids to complete a cycle of reactions (Fig. 5–18). This cycle of reactions was first described by the English biochemist Sir Hans Krebs and is usually called the *Krebs citric acid cycle* or the *tricarboxylic acid (TCA) cycle.* In this cycle, a *two-carbon unit,* acetyl coenzyme A, combines with a *four-carbon unit,* oxaloacetic acid, to yield a *six-carbon product,* citric acid. By a series of dehydrogenations, decarboxylations and make-ready reactions, citric acid is metabolized back to oxaloacetic acid, which is then ready to combine with another molecule of acetyl coenzyme A. In the course of this cycle, two molecules of CO_2 are released, eight hydrogen atoms are removed,

and one molecule of $\sim P$ is synthesized at the substrate level (Fig. 5–19).

The tricarboxylic acid cycle is the final common pathway by which the carbon chains of carbohydrates, fatty acids and amino acids

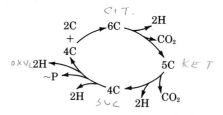

Figure 5–19 The essence of the tricarboxylic acid cycle. A two-carbon unit combines with a four-carbon unit to give a six-carbon compound (citric acid). This undergoes two decarboxylations, four dehydrogenations and several make-ready reactions to regenerate the four-carbon unit (oxaloacetic acid) so that the cycle may continue.

are metabolized. All of these substances are fed into the cycle at one point or another. For example, pyruvic acid undergoes reactions which convert it to acetyl coenzyme A. Acetyl coenzyme A can also be produced from fatty acids by a series of reactions which cleave the long carbon chain into two-carbon units.

5–15 FATTY ACID OXIDATION

Before a fatty acid can be metabolized, it must be "activated" by enzymes which react it first with ATP and then with coenzyme A to yield the fatty acyl coenzyme A. Thus *palmitic acid* (16 carbons) reacts to form palmityl coenzyme A (Fig. 5–20, reaction 1). Palmityl coenzyme A undergoes a dehydrogenation between the second and third carbons of the chain. The group undergoing dehydrogenation is $-CH_2$ $-CH_2-$ and, as you might now expect, the hydrogen acceptor is a *flavin* (Fig. 5–20, reaction 2).

The product of that reaction undergoes a make-ready reaction, the addition of a molecule of water (Fig. 5–20, reaction 3). This molecule contains an $H-\overset{|}{\underset{|}{C}}-OH$ group and can undergo dehydrogenation (Fig. 5–20, reaction 4) with NAD as the hydrogen acceptor.

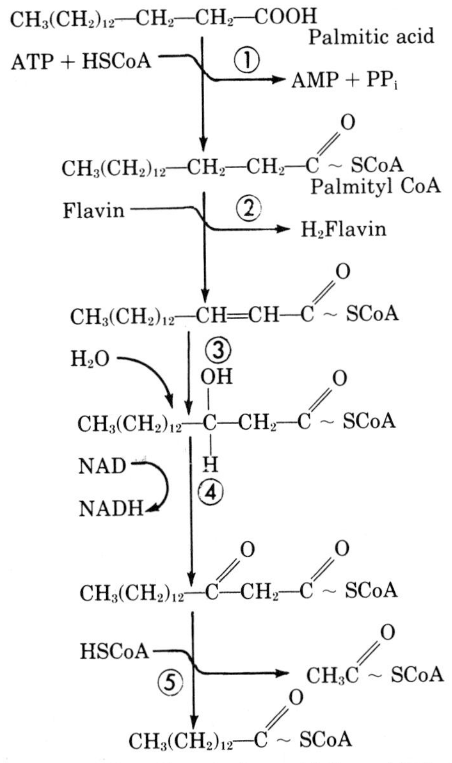

A make-ready reaction with coenzyme A cleaves off a two-carbon unit as *acetyl coenzyme A* and leaves a carbon chain two carbons shorter. This is already activated, it contains a coenzyme A on its carboxyl group, and thus is ready to be dehydrogenated by an enzyme which uses a flavin as hydrogen acceptor. This repeating series of reactions (which includes dehydrogenations and make-ready reactions, but not decarboxylations) cleaves a fatty acid chain two carbons at a time. Seven such series of reactions split palmitic acid to eight molecules of acetyl coenzyme A.

5–16 GLYCOLYSIS

A series of metabolic reactions converts the carbon chain of glucose (and other carbohydrates) to pyruvic acid and then to acetyl coenzyme A (Fig. 5–21). The series of glycolytic (sugar-cleaving) reactions begins (as with fatty acids) with one in which glucose is "activated." The reaction of glucose with ATP to yield glucose-6-phosphate and ADP is catalyzed by the enzyme *hexokinase.* Only the terminal phosphate of ATP is transferred and adenosine diphosphate, ADP, remains. After this first make-ready reaction, additional make-ready reactions finally establish a configuration which can undergo dehydrogenation. A rearrangement yields fructose-6-phosphate, and the transfer of a second phosphate from ATP forms fructose-1,6-diphosphate, with phosphate groups bound at carbons 1 and 6 by ester bonds. Fructose-1,6-diphosphate is split by *aldolase* into two three-carbon sugars *glyceraldehyde-3-phosphate* and *dihydroxyacetone phosphate.* These two are interconverted by the enzyme triose phosphate isomerase.

Glyceraldehyde-3-phosphate reacts with a compound containing an $-SH$ group; not acetyl coenzyme A, as in certain reactions discussed previously, but an $-SH$ group in an amino acid that is part of the glyceraldehyde-3-phosphate dehydrogenase. The combination of the glyceraldehyde-3-phosphate with the $-SH$ group yields an $H-\overset{|}{\underset{|}{C}}-OH$ group that can

Figure 5–20 Reactions in the oxidation of fatty acids.

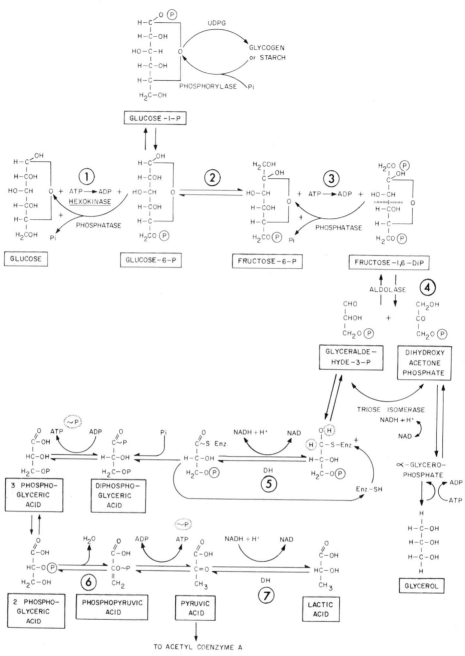

Figure 5–21 Diagram of the series of reactions by which glucose and other sugars are metabolized to pyruvic acid. Note that most of the steps are reversible (indicated by double arrows). Other steps are essentially irreversible: glucose-6-phosphate is converted to glucose by a separate enzyme, a glucose-6-phosphatase, and not by the hexokinase that catalyzes the conversion of glucose to glucose-6-phosphate.

undergo dehydrogenation with NAD as the hydrogen acceptor. The product, *phosphoglyceric acid* bound to the SH group of the enzyme, then reacts with inorganic phosphate to yield *1,3-diphosphoglyceric acid* and free enzyme —SH. The phosphate at carbon 1 is an energy-rich group, which can react with ADP to form ATP. This, like the energy-rich phos-

phate of succinyl phosphate, is one made at the *substrate* level. The resulting 3-phosphoglyceric acid undergoes rearrangement to 2-phosphoglyceric acid and then, in an unusual reaction, an energy-rich phosphate is generated by the removal of water, by a *dehydration*, rather than by the removal of two hydrogens, a *dehydrogenation.*

The product, phosphopyruvic acid, can transfer its phosphate group to ADP to yield ATP and free pyruvic acid. This is the second energy-rich phosphate bond generated at the substrate level in the metabolism of glucose to pyruvic acid. Each glucose molecule yields two molecules of glyceraldehyde-3-phosphate (the second by the conversion of dihydroxyacetone phosphate) and hence a total of four energy-rich bonds are produced as glucose is metabolized to pyruvic acid. However, two energy-rich phosphate bonds are utilized in the process—one to convert glucose to glucose-6-phosphate and the second to convert fructose-6-phosphate to fructose-1,6-diphosphate. The net yield in the process is two $\sim P$ (four $\sim P$ produced minus two $\sim P$ used up in the reactions). Pyruvic acid is then metabolized to acetyl coenzyme A by the reactions described previously.

5–17 ANAEROBIC GLYCOLYSIS

Under anaerobic conditions, without oxygen to serve as the ultimate electron acceptor, the reactions of the electron transmitter system cease when all of the intermediate acceptors have been converted to the reduced condition, when they have taken up all of the electrons possible. Then the metabolism of glucose will lead to the accumulation of pyruvic acid (since it cannot be metabolized to acetyl coenzyme A), and pyruvic acid accepts hydrogens from reduced pyridine nucleotides to yield lactic acid and oxidized pyridine nucleotide (Fig. 5–21, reaction 7), catalyzed by lactic dehydrogenase operating in the opposite direction. The oxidized pyridine nucleotide can then again accept hydrogens from glyceraldehyde-3-phosphate and be reduced to NADH and H$^+$. By the cyclic utilization of NAD in these two reactions glucose is utilized under anaerobic conditions and lactic acid accumulates. See below.

During strenuous exercise our muscles produce pyruvic acid from glucose faster than oxygen can be supplied to metabolize the pyruvic acid to acetyl coenzyme A and to CO_2 and H_2O. Lactic acid accumulates, causing the aches of fatigued muscles and the "oxygen debt" (p. 476) that is eventually repaid by breathing rapidly after the exercise.

The conversion of glucose to lactic acid will yield a net production of two $\sim P$, as we have seen, and cells can obtain in this way a small amount of energy in the absence of oxygen. When glucose is split into two molecules of lactic acid, the free energy change, ΔG, is -52 kilocalories per mole. However, when glucose is enzymatically converted to lactate by anaerobic glycolysis, the decrease in free energy, ΔG, is only -38 kilocalories per mole. The other 14 kilocalories are conserved in the two moles of $\sim P$ formed, some 7 kilocalories per mole.

The reactions by which glucose is metabolized in the absence of oxygen are identical to those by which it is metabolized in the presence of oxygen except for this very last step, at pyruvic acid. In the absence of oxygen, pyruvic acid is converted to lactic acid, which accumulates, and in the presence of oxygen pyruvic acid is metabolized to acetyl coenzyme A and, via the tricarboxylic acid cycle, to carbon dioxide and water. In anaerobic glycolysis the ultimate hydrogen acceptor is some substance other than oxygen—lactic acid in animal cells, ethanol in yeasts, and glycerol or butanol in certain bacteria.

In yeast cells pyruvic acid is converted to acetaldehyde and this can accept hydrogens from reduced NAD to yield oxidized NAD and ethyl alcohol:

$$\begin{matrix} H—C{=}O \\ | \\ CH_3 \end{matrix} + NADH + H^+ \rightarrow \begin{matrix} H_2C—OH \\ | \\ CH_3 \end{matrix} + NAD$$

Acetaldehyde   Ethyl alcohol

If the cells of our bodies had this same enzyme we could, by exercising violently so as to metabolize glucose anaerobically to pyruvate and acetaldehyde, produce ethyl alcohol within our cells and, perhaps, become intoxicated! Mammalian cells, however, metabolize pyruvic acid not to free acetaldehyde but, via a series of intermediates, to acetyl coenzyme A.

5–18 THE PENTOSE PHOSPHATE PATHWAY

The *pentose phosphate pathway* is an alternative route for the oxidation of glucose. A series of enzymes, quite independent of those of the glycolytic sequence, act in concert with several of the glycolytic enzymes and can account for the complete oxidation of glucose to

CO_2. However, the metabolic significance of the pathway is as a source of reducing equivalents for the generation of NADPH and as a means of synthesizing and disposing of pentoses. In contrast to glycolysis, ATP is neither utilized nor generated in the pentose phosphate path, which begins with glucose-6-phosphate.

Glucose-6-phosphate has an $H—\overset{|}{\underset{|}{C}}—OH$ group at carbon 1, which can undergo dehydrogenation (Fig. 5–22). The enzyme catalyzing this reaction, *glucose-6-phosphate dehydrogenase*, specifically requires NADP as its hydrogen acceptor. A make-ready reaction, the addition of water, converts the product to *6-phosphogluconic acid*. This can undergo dehydrogenation (it has an $H—\overset{|}{\underset{|}{C}}—OH$ group) by a second enzyme which also specifically requires NADP as hydrogen acceptor. The product of this reaction undergoes decarboxylation to yield a five-carbon sugar, *ribulose-5-phosphate*, which is of major importance in reactions associated with photosynthesis (p. 95). It is also the source of the five-carbon sugars, ribose and deoxyribose, which are constituents of nucleotides and nucleic acids.

In further reactions, rearrangements occur in which either the upper two or the upper three carbons of the molecule are transferred

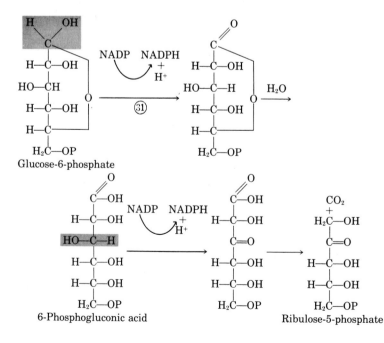

Glucose-6-phosphate

6-Phosphogluconic acid

Ribulose-5-phosphate

Figure 5–22 Glucose-6-phosphate dehydrogenase and 6-phosphogluconate dehydrogenase. The groups undergoing dehydrogenation are shown in tint.

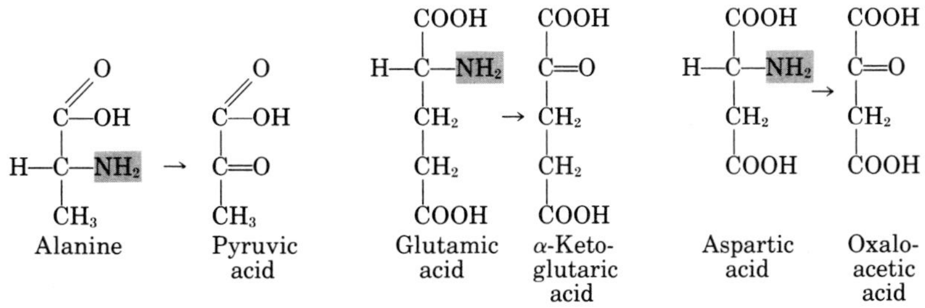

Figure 5–23 The deamination of amino acids to their corresponding α-keto acids. The amino groups are tinted.

Alanine　Pyruvic acid　Glutamic acid　α-Keto-glutaric acid　Aspartic acid　Oxalo-acetic acid

as a unit to another sugar. By such reactions, sugars with three, four, five, six, or even seven carbon atoms are synthesized (Fig. 5–9).

5–19　AMINO ACID OXIDATION

Amino acids are oxidized by reactions in which the amino group is first removed, a process called *deamination,* then the carbon chain is metabolized and eventually enters the tricarboxylic acid cycle. The amino acid *alanine,* for example, yields pyruvic acid when deaminated, *glutamic acid* yields α-ketoglutaric acid, and *aspartic acid* yields oxaloacetic acid (Fig. 5–23).

These three amino acids enter the TCA cycle directly. Other amino acids may require several reactions in addition to deamination to yield a substance which is a member of the TCA cycle, but ultimately the carbon chains of all of the amino acids are metabolized in this way.

5–20　THE ELECTRON TRANSMITTER SYSTEM

Thus far we have considered only the reactions by which electrons are removed from substrate molecules and transferred to the primary acceptors, either a pyridine nucleotide or a flavin. The major reactions by which biologically useful energy is conserved, however, occur when the electrons pass down the "electron cascade" in the electron transmitter system and provide the energy to drive the reactions of oxidative phosphorylation. The electrons entering the electron transmitter system from NADH have a relatively high energy content. As they pass along the chain of enzymes they lose much of their energy, some of which is conserved in the form of ATP.

The enzymes of the electron transmitter system are located within the substance of the mitochondrion in the mitochondrial membranes. They are immediately adjacent to one another, and it seems likely that the electrons in fact flow through a solid phase rather than between enzymes that are in solution.

The enzyme *succinic dehydrogenase* is also located in the mitochondrial membranes. The enzymes that convert pyruvic acid to acetyl coenzyme A and α-ketoglutaric acid to succinyl coenzyme A appear to be located in particles within the mitochondria which are evident in the electron microscope.

The components of the electron transmitter system are given in the order of their oxidation reduction potentials, which range from -0.32 volt for the pyridine nucleotides to $+0.81$ volt for oxygen (Fig. 5–13). However, it is not known whether a particular electron, in passing from pyridine nucleotide to oxygen, must go through each and every one of the intermediates or whether it might skip some of the steps. It probably has to pass through at least three different steps to account for the three $\sim$P which are made for each pair of electrons that pass from pyridine nucleotide to oxygen.

Oxidative phosphorylation is measured by the rate at which inorganic phosphate, P_i, is converted to ATP as NADH or some other substance undergoes oxidation. It is possible to extract mitochondria by homogenizing the cells and then separating the subcellular particles by centrifugation. Such mitochondria, when removed carefully from the cell, will still carry out oxidative phosphorylation. Indeed, it is possible to disrupt mitochondria by ultrasonic vibrations and obtain submitochondrial particles which will carry out oxidative phosphorylation.

In these purified systems NADH will be oxidized and oxygen will be utilized *only* if ADP is present to accept the energy-rich phosphates produced by the flow of electrons. The flow of electrons is *tightly coupled* to the phosphorylation process and will not occur unless phosphorylation can occur, too. This, in a sense, prevents waste, for electrons will not flow unless ~P can be formed.

It is known that oxidation is similarly coupled to phosphorylation when the reactions occur within the intact cell. Certain substances, one of which is the hormone **thyroxine,** can "uncouple" phosphorylation from oxidation so that the flow of electrons occurs without their energy being trapped as ~P; the energy is released as heat, instead. The details of the mechanism by which these reactions convert inorganic phosphate to energy-rich phosphate in the electron transmitter system are under intensive investigation in attempts to determine which of the several alternative hypotheses regarding the mechanism(s) provides the best explanation of the phenomenon.

The flow of electrons all the way from pyridine nucleotide to oxygen, a total drop of 1.13 volts (from -0.32 to $+0.81$ volt), would yield 52 kilocalories per pair of electrons if the process were 100 per cent efficient. This may be calculated from the formula $\Delta G' = -nF\ \Delta E$, where $\Delta G'$ is the change in free energy, n is the number of electrons (2), F is the Faraday (23.04 kilocalories), and ΔE is the difference in the oxidation-reduction potentials of the reactants (1.13 volts). Under experimental conditions, most cells will produce at most three ~P per pair of electrons as the electrons pass from pyridine nucleotide to oxygen (Fig. 5–13). Each ~P is equivalent to about 10 kilocalories, hence the three ~P amount to about 30 kilocalories. The efficiency of the electron transmitter system may thus be calculated at 30/52 or about 57 per cent.

The amount of ATP present in any cell is usually rather small. In muscle cells, in which large amounts of energy may be expended in a short time during the contraction process, an additional substance, **creatine phosphate,** serves as a reservoir of ~P. The terminal phosphate of ATP is transferred by an enzyme, creatine kinase, to creatine to yield creatine phosphate and ADP. The phosphate bond of creatine phosphate is also an energy-rich bond. The ~P of creatine phosphate must be transferred back to ADP, converting it to ATP, to be utilized in some reaction requiring energy, such as muscle contraction.

$$\begin{array}{c} \text{utilization} \\ \uparrow \\ \text{Creatine} + \text{ATP} \rightleftharpoons \text{Creatine} \sim\text{P} + \text{ADP} \\ \uparrow \\ \text{synthesis} \qquad \text{storage} \end{array}$$

The fact that phosphorylation is tightly coupled to oxidation (electron flow) in the electron transmitter system provides the basis for a system of control which can regulate the rate of energy production and adjust it to the rate of energy utilization. In a resting muscle cell oxidative phosphorylation will occur until all of the ADP has been converted to ATP. Then, since there are no more acceptors of ~P, phosphorylation stops. Since oxidation is tightly coupled to phosphorylation, oxidation (i.e., the flow of electrons and utilization of oxygen) will cease.

When the muscle contracts, the energy required is obtained by the splitting of the energy-rich terminal phosphate of ATP:

$$\text{ATP} \rightarrow \text{ADP} + \text{P}_i + \text{energy}$$

The ADP formed can then serve as an acceptor of ~P, phosphorylation begins, and the flow of electrons to oxygen occurs. Oxidative phosphorylation continues until all of the ADP has been converted to ATP. Electric generating systems have an analogous control device which adjusts the rate of production of electricity to the rate of utilization of electricity.

5–21 ENERGY TRANSFORMATIONS IN THE HUMAN BODY

Some interesting calculations of the overall energy changes involved in metabolism in the human body have been made by E. G. Ball of Harvard University. The conversion of oxygen to water requires the participation of hydrogen ions and electrons; thus the total flow of electrons in the human body can be calculated and expressed in terms of amperes. From the average oxygen consumption of an adult 70 kg. man at rest, 264 ml. per minute, and the fact that each oxygen atom requires two hydrogen atoms and two electrons to form a molecule of water, Dr. Ball calculated that 2.86×10^{22} electrons are flowing from foodstuff via dehydrogenases and the cytochromes to oxygen each minute in all the body cells. Since an ampere equals 3.76×10^{20} electrons per minute, this current amounts to 76 amperes. This is quite a bit of current, for an ordinary 100 watt light bulb uses just a little less than 1 ampere.

The flow of electrons from substrate to oxygen involves a potential difference of 1.13 volts (from -0.32 v to $+0.81$ v). In electrical units, volts times amperes equals watts, and $1.13 \times 76 = 85.9$ watts.

The total expenditure of energy can also be calculated from the number of kilocalories used per minute (about 1.27 at rest). Using the appropriate conversion factors this can be shown to be equivalent to about 88 watts, which agrees satisfactorily with the value calculated previously.

The human body, then, utilizes energy at about the same rate as a 100 watt light bulb, but differs from it in having a much larger flow of electrons passing through a much smaller voltage change.

The conversion of glucose to carbon dioxide and water in a calorimeter yields about 4 kcal./gm. In describing the successive steps in the metabolism of glucose by cells, we noted that energy is released in the form of energy-rich phosphate bonds, a form in which it can be used to perform a variety of kinds of work. Let us now dissect the overall equation

$$C_6H_{12}O_6 + 6O_2 \rightarrow 6CO_2 + 6H_2O + energy$$

and review where useful energy is released (Table 5–2).

In glycolysis (reaction 1, Table 5–2), glucose is activated by the addition of 2 ~P and converted to 2 pyruvate + 2 NADH + 4 ~P. Then the two pyruvates are metabolized (reaction 2), to 2 CO_2 + 2 acetyl CoA + 2 NADH. Finally, in the citric acid cycle (reaction 3), the 2 acetyl CoA are metabolized to 4 CO_2 + 6 NADH + 2 H_2FP + 2 ~P.

These reactions can be added together (reaction 4) by eliminating items that are present on both sides of the arrows. Then, since the oxidation of NADH in the electron transmitter system yields 3 ~P per mole, the 10 NADH = 30 ~P. The oxidation of H_2FP yields 2 ~P per mole and the 2 H_2FP = 4 ~P. Summing these we see that the complete aerobic metabolism of one mole (180 gm.) of glucose yields 38 ~P (reaction 5). Each ~P is equivalent to about 7 kilocalories and the 38 ~ P = 266 kilocalories.

When a mole of glucose is burned in a calorimeter some 690 kilocalories are released as heat. The metabolism of glucose in cells releases 266/690, or about 40 per cent of the total energy as biologically useful energy, ~P. The remainder of the energy is dissipated as heat.

5–22 THE MOLECULAR ORGANIZATION OF MITOCHONDRIA

Mitochondrial shapes range from spherical to elongated, sausage-like structures, but an average mitochondrion is an ellipsoid about 3 μm. long and a little less than 1 μm. in diameter. Some very generously endowed large cells, such as the giant amoeba, *Chaos chaos*, have several hundred thousand mitochondria, and an average mammalian liver cell has about one thousand. The mitochondrial protein accounts for about 20 per cent of the total protein of the liver cell.

Each mitochondrion has two membranes, an outer smooth one and an inner one that is folded repeatedly to form the *cristae* or shelf-like projections within the cavity of the mitochondrion (Fig. 5–24). (Some mitochondria have wall-to-wall cristae.) Both the inner and outer mitochondrial membranes are lipid bilayers containing protein molecules. The fluid material within the inner membrane is called the matrix.

Table 5–2 THE ENERGY YIELD FROM THE COMPLETE OXIDATION OF ONE MOLE OF GLUCOSE

(1)	$C_6H_{12}O_6$ + 2 ~P	$\longrightarrow$ 2 Pyruvate + 2 NADH + 4 ~P
(2)	2 Pyruvate	$\longrightarrow$ 2 CO_2 + 2 Acetyl CoA + 2 NADH
(3)	2 Acetyl CoA	$\longrightarrow$ 4 CO_2 + 6 NADH + 2 H_2FP + 2 ~P
(4) Sum	$C_6H_{12}O_6$	$\longrightarrow$ 6 CO_2 + 10 NADH + 2 H_2FP + 4 ~P
(5)	$C_6H_{12}O_6$ + 6 O_2	$\longrightarrow$ 6 CO_2 + 6 H_2O + 30 ~P + 4 ~P + 4 ~P

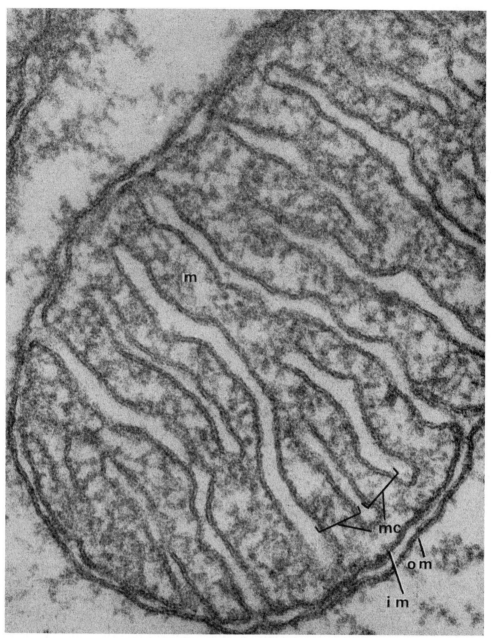

Figure 5–24   Electron micrograph of a mitochondrion from a pancreatic acinar cell. The double-layered unit membrane is evident in the smooth outer membrane (*om*) and in the inner membrane (*im*), which folds to form the cristae (*mc*), between which lies the matrix (*m*). Magnification ×207,000. (From De Robertis, E. D., Nowinski, W. W., and Saez, F. A.: *Cell Biology.* Philadelphia, W. B. Saunders Co., 1970. Courtesy of G. E. Palade.)

The enzymes of the TCA cycle have been found in the soluble matrix within the mitochondrion. The enzymes of the electron transmitter system are tightly bound to the inner membrane. Each group of these electron transmitter enzymes, termed a *respiratory assembly,* is one of the fundamental units of cellular activity. It has been estimated that the mitochondrion of the liver cell contains some 15,000 res-

piratory assemblies and that they make up about one quarter of the mass of the mitochondrial membranes. Thus the mitochondrial membrane is not just a protective skin but an important functional part of the mitochondrion.

High resolution electron micrographs of mitochondria, made by Humberto Fernandez-Moran at the University of Chicago, have demonstrated the presence of small particles on the inner surface of the inner membrane (Fig. 5–25). The particles on the inner membrane typically have a spherical knob-like head, a cylindrical stalk and a base plate and may be sites of some of the enzyme reactions carried out by mitochondria. These "inner membrane" spheres are not the respiratory assemblies, as was suggested when they were first found, but they do contain one or more coupling factors necessary for oxidative phosphorylation.

It seems clear that much of the cell's biologically useful energy, ATP, is generated by enzyme systems located in the *inner* membrane of the mitochondrion, yet most of the energy utilized by the cell is required for processes that take place *outside* of the mitochondrion. ATP is used in the synthesis of proteins, fats, carbohydrates, nucleic acids and other complex molecules, in the transport of substances across the plasma membrane, in the conduction of nerve impulses, and in the contraction of muscle fibers, all of which are reactions occurring largely or completely outside of the mitochondrion in other parts of the cell. We do not yet know how the ~P generated within the mitochondrion becomes available outside of the mitochondrion, for membranes such as those comprising the walls of the mitochondria are largely impermeable to large, charged molecules such as ATP. A carrier pro-

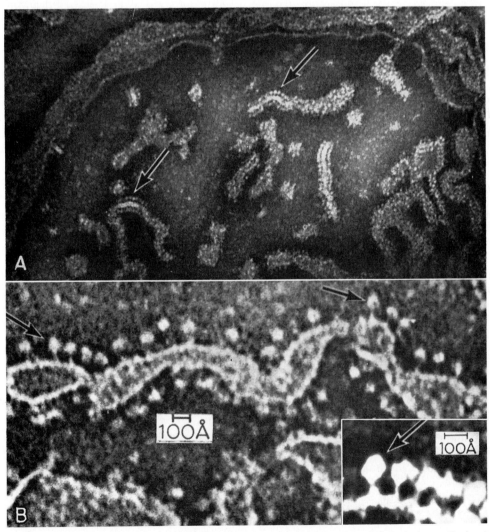

Figure 5–25 Electron micrograph of a mitochondrion swollen in a hypotonic solution and negatively stained with phosphotungstate. *A,* Isolated cristae, ×85,000. *B,* At higher magnification, ×500,000, the elementary particles attached by a stalk to the surface of the cristae are evident. *Insert:* Magnified 650,000×, the polygonal shape of the elementary particle and its slender stalk are clearly visible. (From De Robertis, E. D., Nowinski, W. W., and Saez, F. A.: *Cell Biology.* Philadelphia, W. B. Saunders Co., 1970. Courtesy of H. Fernandez-Moran.)

tein that seems to transfer ADP and ATP across the inner mitochondrial membrane has been identified. It permits an ADP to enter the mitochondrion only if an ATP comes out (a veritable Maxwell's demon!). It thus permits a specific exchange diffusion to occur. The carrier protein has high affinity and specificity for ATP and ADP and exhibits the phenomenon of saturation with its two substrates. It thus has many of the characteristics of an enzyme, catalyzing a two-way exchange across the inner membrane. The outer mitochondrial membrane is much more freely permeable to most low molecular weight solutes, whereas the inner membrane is permeable to water, glycerol, urea and short-chain fatty acids.

A single cell may have a thousand or more mitochondria and the function of each one must be controlled appropriately to generate the amount of energy required by the cell at any given moment. The rate at which ~P is utilized by a cell may vary over a remarkably wide range as the cell becomes active or quiescent. This has been measured in the muscle cells of a frog. One hundred grams of frog muscle utilizes 1.6 μ moles of ~P per minute when quiescent and 3300 μ moles of ~P per minute in a state of tetanus (continuous contraction). The rate of production of ATP in the cell is controlled in large part by the rate of utilization of ATP by the cell. The flow of electrons in the electron transmitter system is tightly coupled to phosphorylation, and oxidative phosphorylation can occur only when there is ADP to be converted to ATP. These facts provide the basis for the system by which a cell's utilization of ATP, which produces ADP, is used to regulate the rate at which ATP is produced. In addition, the structure and biological activity of some of the enzymes involved in glucose oxidation are affected by the concentration of ADP present. In this way an increased concentration of ADP can lead to increased activity of these enzymes and an increased production of ATP. The problem of the nature and interrelations of these various biological control systems is very much in the forefront in biology today.

5–23 THE DYNAMIC STATE OF CELLULAR CONSTITUENTS

The body of a plant or animal appears to be unchanging as days and weeks go by. It was inferred from this that the cells of the body, and the component molecules of the cells, are equally unchanging. Until about 1937 it was believed that the molecules of nutrients not used to increase the total cellular mass were rapidly used to provide energy. It followed from this that two kinds of molecules could be distinguished: relatively static ones which constituted the cellular "machinery," and ones which were rapidly metabolized and thus corresponded to cellular "fuel."

However, experiments in which amino acids, fats, carbohydrates and water, each suitably labeled with some radioactive or heavy isotope, were fed to rats or other animals, have shown that the cellular constituents are in a constant state of flux. Labeled amino acids are rapidly incorporated into body proteins, and labeled fatty acids are rapidly incorporated into fat deposits, even though there is no increase in the total amount of protein or fat. The proteins and fats of the body—even the substance of the bones—are constantly and rapidly being synthesized and broken down. In the adult the rates of synthesis and of degradation are essentially equal so that there is little or no change in the total mass of the body. Thus the distinction between "machinery" molecules and "fuel" molecules becomes much less sharp. The machinery is constantly being overhauled and some of the machinery molecules are broken down and used as fuel.

The one exception to this general rule of molecular flux is provided by the molecules of DNA that constitute the units of heredity, the genes, in the nucleus of each cell. Experiments with labeled atoms have shown that DNA molecules are remarkably stable and are broken down and resynthesized only very slowly if at all. The amount of DNA in the nucleus of each cell of an organism is constant. New DNA molecules must be synthesized before a cell can divide. The stability of the DNA molecules is of importance in ensuring that hereditary characters are transmitted to succeeding generations with as few chemical errors as possible. In contrast, molecules of RNA are constantly undergoing synthesis and degradation; indeed, the amount of RNA per cell may vary within wide limits.

From the rate at which the labeled atoms are incorporated into macromolecules it has been calculated that one half of all the tissue proteins of an adult human are broken down and rebuilt every eighty days. This is an

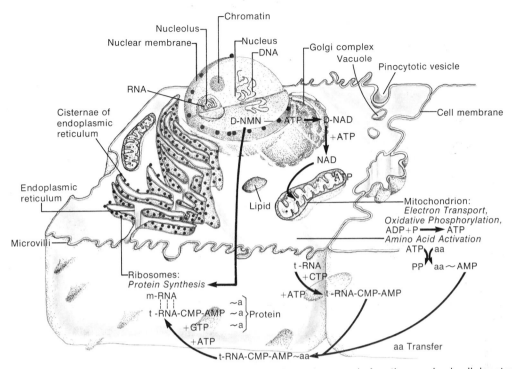

Figure 5–26   Diagram of a cell illustrating the relationship of certain enzymic functions and subcellular structure.

average figure; some proteins are replaced much more rapidly, others more slowly. The proteins of the liver and of blood serum are replaced very rapidly, one half of them every ten days. Some specific enzymes in the liver have half-lives as short as two to four hours. Muscle proteins are replaced more slowly, one half of them every 180 days. You are not the same person, chemically speaking, that you were yesterday!

Some aspects of our present concept of the location within the cell of specific enzyme systems are shown in Figure 5–26. The synthesis of DNA and of RNA occurs in the nucleus. The processes of electron transport and oxidative phosphorylation occur in the mitochondria. The synthesis of proteins occurs on the ribosomes which are either free or located on the membranes of the endoplasmic reticulum. The activation of amino acids for protein synthesis, the process of glycolysis, and many other reactions occur in the soluble cell sap; apparently their enzymes are not attached to any of the subcellular particles.

5–24 BIOSYNTHETIC PROCESSES

Our discussion thus far has dealt with processes that break down molecules of food-

stuffs and conserve their energy in the biologically useful form of ~P. Both plant and animal cells possess a remarkable array of enzymes which catalyze a variety of biosynthetic processes utilizing the energy of ATP and, as raw materials, some of the five-, four-, three-, two- and one-carbon compounds that are intermediates in the metabolism of glucose, fatty acids, amino acids and other compounds.

This whole subject of intermediary metabolism, by which an enormous variety of compounds is synthesized, is much too complex to be discussed in detail here. The enzyme controlling each reaction is a protein molecule whose structure is determined genetically by processes that will be discussed in Chapter 29. The overall maze of enzyme reactions by which any compound is synthesized includes a variety of self-adjusting control mechanisms to regulate and integrate the reactions. Several basic principles of cellular biosynthesis can be distinguished:

1. Each cell, in general, synthesizes its own proteins, nucleic acids, lipids, polysaccharides and other complex molecules and does not receive them preformed from other cells. Muscle glycogen, for example, is synthesized within the muscle cell and is not derived from liver glycogen.

2. Each step in the biosynthetic process is catalyzed by a separate enzyme.

3. Although certain steps in a biosynthetic sequence will proceed without the use of energy-rich phosphate, the overall synthesis of these complex molecules requires chemical energy at various points along the way. Why should this be true? Can you relate this to your understanding of the concept of entropy?

4. The synthetic processes utilize as raw materials relatively few substances, among which are acetyl coenzyme A, glycine, succinyl coenzyme A, ribose, pyruvate and glycerol.

5. These synthetic processes are in general not simply the reverse of the processes by which the molecule is degraded but include one or more separate steps which differ from any step in the degradative process. These steps are controlled by different enzymes and this permits separate control mechanisms to govern the synthesis and the degradation of the complex molecule.

6. The biosynthetic process includes not only the formation of the macromolecular components from simple precursors but their assembly into the several kinds of membranes that comprise the outer boundary of the cell and the intracellular organelles. Each cell's constituent molecules are in a dynamic state—they are constantly being degraded and synthesized. Thus, even a cell that is not growing, not increasing in mass, uses a considerable portion of its total energy for the chemical work of biosynthesis. A cell that is growing rapidly must allocate a correspondingly larger fraction of its total energy output to biosynthetic processes, especially the biosynthesis of protein. A rapidly growing bacterial cell may use as much as 90 per cent of its total biosynthetic energy for the synthesis of proteins.

Many of the steps in biosynthetic processes involve the formation of peptide bonds, glycosidic bonds, and ester bonds, but these bonds are *not* formed by reactions in which water is removed. The biosynthesis of sucrose in the cane sugar plant, for example, which involves the formation of a glycosidic bond, does not proceed via

$$\text{Glucose} + \text{fructose} \rightleftharpoons \text{sucrose} + H_2O$$

This reaction would require energy, some 5.5 kilocalories per mole, to go to the right if all reactants were present in the concentration of one mole per liter. However, the concentration of glucose and fructose in the plant cell is probably less than 0.01 mole per liter, whereas the concentration of water is very high, about 55 moles per liter. Thus, the equilibrium point of the reaction under these conditions would be very far to the left.

Instead, one or more of the reactants is activated by a reaction with ATP. The terminal phosphate is enzymatically transferred to glucose with the conservation of some of the energy of the terminal phosphate bond. The glucose phosphate, with a higher energy content than free glucose, can react with fructose via another enzyme-catalyzed reaction to yield sucrose and inorganic phosphate.

$$\text{ATP} + \text{glucose} \rightarrow$$
$$\text{ADP} + \text{glucose-1-phosphate}$$
$$\text{Glucose-1-phosphate} + \text{fructose} \rightarrow$$
$$\text{sucrose} + \text{phosphate}$$
$$\text{Sum: ATP} + \text{glucose} + \text{fructose} \rightarrow$$
$$\text{sucrose} + \text{ADP} + P_i$$

This reaction proceeds to the right because there is a net decrease in free energy. The 7 kilocalories of the $\sim P$ bond are used to supply the 5.5 kilocalories needed to assemble the glucose and fructose into sucrose; the overall decrease in free energy is 1.5 kilocalories per mole. Since water is not a product of this reaction, the high concentration of water in the cell does not inhibit it.

ATP has two energy-rich bonds and can react in several different ways to transfer energy to another molecule and provide that molecule with the energy needed to carry out a further reaction. Depending on which of the energy-rich bonds of ATP is split, the result can be (a) the transfer of the terminal phosphate group and the release of ADP; (b) the transfer of the final two phosphate groups (called a pyrophosphate group) and the release of AMP, adenosine monophosphate; (c) the transfer of the adenosine monophosphate group with the release of the pyrophosphate; or (d) the transfer of the entire adenosine group and the release of both a pyrophosphate from the terminal two phosphate groups, and an inorganic phosphate from the third phosphate group of ATP. These four types of reactions are shown in Figure 5–27.

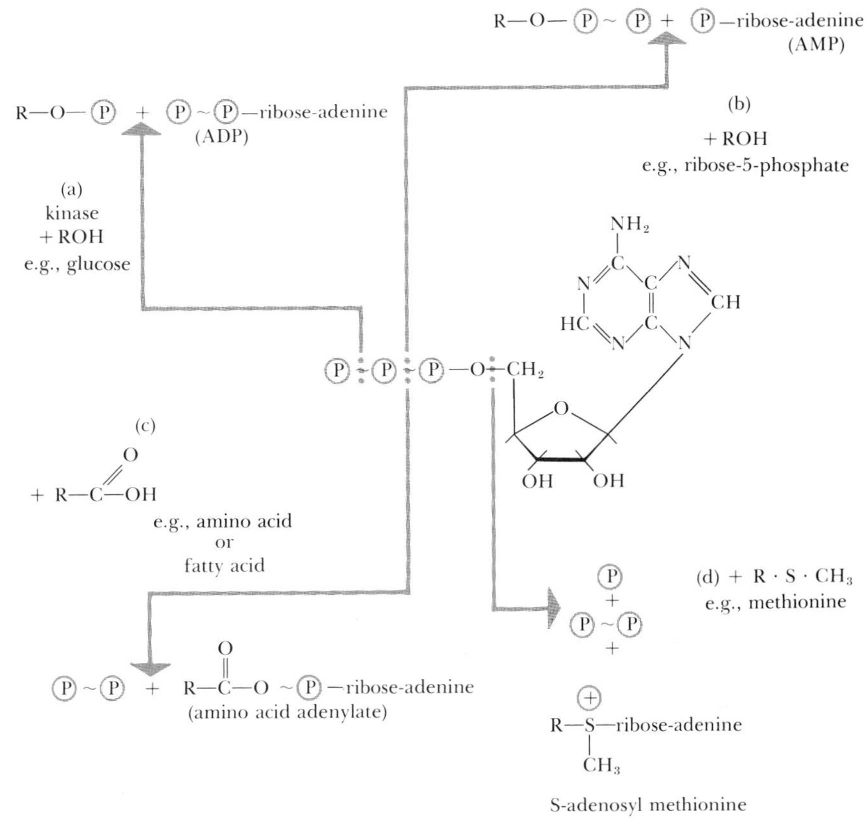

$$R—O—\textcircled{P} \sim \textcircled{P} + \textcircled{P}—ribose\text{-}adenine$$
$$(AMP)$$

$$R—O—\textcircled{P} + \textcircled{P} \sim \textcircled{P}—ribose\text{-}adenine$$
$$(ADP)$$

(b)

+ ROH
e.g., ribose-5-phosphate

(a)
kinase
+ ROH
e.g., glucose

Figure 5–27 Four major types of reactions of adenosine triphosphate: the transfer of phosphate *(a)*, the transfer of pyrophosphate *(b)*, the transfer of adenylic acid *(c)* and the transfer of an adenosyl group *(d)*.

$$\textcircled{P} \vdots \textcircled{P} \vdots \textcircled{P}—O \vdots CH_2$$

(c)

$$O$$
$$\parallel$$
$$+ R—C—OH$$
e.g., amino acid
or
fatty acid

$$\textcircled{P}$$
$$+$$
$$\textcircled{P} \sim \textcircled{P}$$
$$+$$

(d) + R · S · CH$_3$
e.g., methionine

$$\textcircled{P} \sim \textcircled{P} + \begin{matrix} O \\ \parallel \\ R—C—O \end{matrix} \sim \textcircled{P}—ribose\text{-}adenine$$
(amino acid adenylate)

$$R—S—ribose\text{-}adenine$$
$$|$$
$$CH_3$$

S-adenosyl methionine

The most common reaction is type (a). If the terminal phosphate group is transferred to water, the reaction is a hydrolysis of this terminal phosphate group, and the free energy change of this reaction is negative with a ΔG of 7 kilocalories per mole. The fact that the transfer of phosphate onto water is exergonic (it releases energy) explains the transfer potential of the terminal phosphate group of ATP. This terminal phosphate can be transferred from ATP onto hydroxyl groups

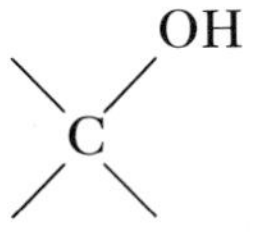

onto carboxyl groups

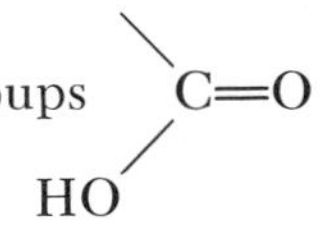

or onto amino groups

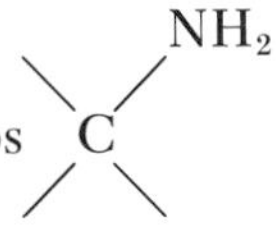

All of these transfer reactions are catalyzed by enzymes called **kinases.** You are already familiar with hexokinase, which transfers a phosphate from ATP onto the hydroxyl group (OH) at carbon 6 of glucose. The phosphate donor for most kinases is ATP; however, in a few re-

actions, uridine triphosphate (UTP), cytidine triphosphate (CTP) or guanosine triphosphate (GTP) may act in the same way.

The second type of reaction, in which a **pyrophosphate** group is transferred, is much rarer than the first type. The reaction of ribose-5-phosphate with ATP provides an example of such a transfer. The two terminal phosphates of ATP are transferred to the ribose-5-phosphate to yield 5-phosphoribosyl-1-pyrophosphate (called PRPP) and free AMP. The PRPP molecule is an important reactant in the synthesis of both purine and pyrimidine nucleotides, the components of both DNA and RNA.

The third type of reaction, the transfer of AMP to another molecule with the release of pyrophosphate, is fairly common. The result is an "activated" compound (R~AMP) with potential for group transfer. This type of reaction occurs in the activation of amino acids to prepare them for protein synthesis, and in the activation of fatty acids to prepare them for metabolism. In the latter, the resulting fatty acid ~AMP compound undergoes a second reaction with coenzyme A to yield free AMP and fatty acid ~coenzyme A.

A curious variation of this type of transfer

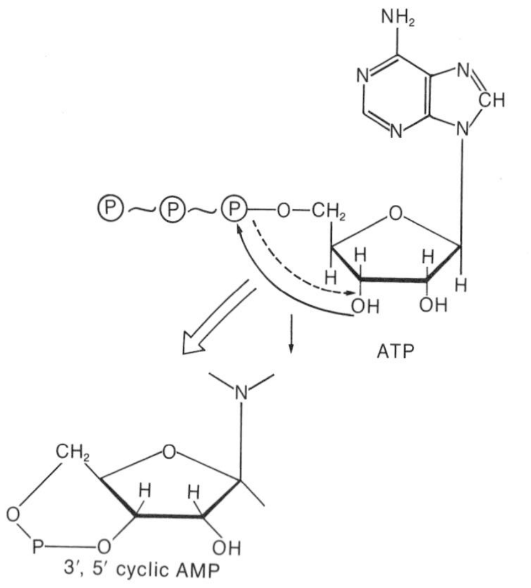

ATP

3′, 5′ cyclic AMP

Figure 5–28 The formation of 3′,5′-cyclic adenylic acid from adenosine triphosphate in the reaction catalyzed by adenyl cyclase.

is seen in the synthesis of *cyclic adenylic acid* (cyclic AMP), adenosine 3,′5′-monophosphate, Figure 5–28. Cyclic AMP is in the center of interest at the present time because of its role in the mechanism of action of many types of hormones. In its synthesis, ATP carries out an intramolecular transfer of AMP onto the 3′-hydroxyl group of the ribose with the elimination of pyrophosphate.

The fourth type of reaction, the transfer of an adenosyl group, with the release of both orthophosphate (P_i) and pyrophosphate (PP_i), occurs in the reaction of ATP with the amino acid *methionine.* The products are inorganic phosphate, pyrophosphate and an adenosine group attached to the sulfur of methionine, a

compound called *S-adenosyl methionine* (SAM). This reaction results in an activation of the methyl group of methionine so that it can be transferred from S-adenosyl methionine to certain acceptor compounds.

Unique Biosynthetic Processes. In addition to the general metabolic activities described above, certain animals and plants have special metabolic activities: green plants can photosynthesize; certain bacteria, molds and animals can produce light enzymically; a few fish such as the electric eel can produce shocking amounts of electricity; certain plants produce a wide variety of substances—flower pigments, perfumes, many types of drugs—and bacteria and molds, perhaps the best chemists of them all, can make everything from deadly poisons to antibiotics.

5–25 BIOLUMINESCENCE

Although the firefly and glowworm are the most conspicuous light-emitting organisms, a number of other animals and some bacteria and fungi, but no green plants, also have this ability (Fig. 5–29). Luminescent animals are found among the sponges, cnidaria, ctenophores, nemerteans, annelids, crustaceans, centipedes, millipedes, beetles, echinoderms, mollusks, hemichordates, tunicates and fishes.

There appears to be no single evolutionary line of luminescent forms; the ability to emit light has appeared independently a number of times in the course of evolution. It is sometimes difficult to establish the fact that an orga-

Figure 5–29 A school of luminescent squid, *Watasenia scintillans.* (Drawn by Miss E. Grace White. From Dahlgren, U.: Phosphorescent animals and plants. Nat. Hist. N.Y. *22*:23, 1922.)

Figure 5–30 Two species of luminescent fish from the waters of the Malay Archipelago: *A, Anomalops katoptron* and, *B, Photoblepharon palpebratus.* The half-moon-shaped luminescent organs just ventral to the eyes are equipped with reflectors. (From Steche, O.: Über die Leuchtorgane von *Anomalops katoptron* und *Photoblepharon palpebratus*, zwei Oberflächenfischen ans dem malayischen Archipel. Ein Beitrag zur Morph. u. Physiol. der Leuchtorgane der Fische. Z. Wiss. Zool. *93*:349, 1909.)

nism is itself luminescent; in a number of instances, the light has been found to be emitted not by the organism but by bacteria. Several unusual East Indian fish have light organs under their eyes in which live luminous bacteria (Fig. 5–30). The light organ contains special long cylindrical cells, well equipped with blood vessels to supply the bacteria with adequate amounts of oxygen. The bacteria emit light continuously and the fish have a black membrane, like an eyelid, that can be drawn up over the light organ to turn off the light. An unsolved mystery is how the bacteria come to collect in the fish's light organ, as they must in each newly hatched fish.

Some species, such as the shrimp, have accessory lenses, reflectors and color filters with the light-emitting organ, so that the whole assembly is like a lantern.

The amount of light produced by some luminescent animals is amazing. Fireflies may produce as much light, in terms of lumens per square centimeter, as fluorescent lamps. Different animals emit lights of different colors—red, green, yellow or blue. The "railroad worm" of Uruguay, the larva of a beetle, is remarkable in being able to produce two different colors: it has a row of green lights along each side of the body, and a pair of red lights at the head end. The light produced by luminescent organisms is entirely in the visible spectrum; no ultraviolet or infrared light is produced. Bioluminescence is sometimes called "cold light" because very little heat is given off.

The production of light is an enzyme-controlled reaction, the details of which differ in different species. Bacteria and fungi produce light continuously if oxygen is available. Most luminescent animals give out flashes of light only when their luminescent organs are stimulated. The names *luciferin* (the substrate) and *luciferase* (the enzyme) are given to the two major components of the light-emitting system, but the luciferin and luciferase from one species of animal may be different from those in another. The luciferins from the crustacean *Cypridina* and from the firefly *Photinus* have been isolated and crystallized and found to be quite different chemically.

The luciferin-luciferase reaction is a form of biological oxidation and can occur only in the presence of oxygen as an electron acceptor. It is possible to extract luciferin and luciferase from a firefly, mix the two in a test tube with added Mg^{2+} and adenosine triphosphate, and get light. The emission of light is an energy-requiring process, and the biologically useful energy to drive the reaction is supplied by ATP. Under certain conditions, the amount of light emitted is proportional to the amount of ATP present. This system can be used to measure the amount of ATP in a tissue extract.

The sequence of reactions involved in firefly bioluminescence has been shown by W. D. McElroy to begin with the reaction of *luciferase* (E) with the reduced form of *luciferin* (LH$_2$) and ATP to yield an intermediate complex of enzyme-luciferyl-adenosine monophosphate (AMP) and release inorganic pyrophosphate (PP$_i$)

$$E + LH_2 + ATP \rightleftharpoons E—LH_2—AMP + PP_i$$

The adenosine monophosphate is attached to a carboxyl group of the luciferin. Then, in the presence of oxygen, light is emitted when $E—LH_2—AMP$ is oxidized to $E—L—AMP$, the combination of enzyme with oxyluciferin (L) and adenosine monophosphate. (It seems appropriate that L—AMP should be the abbreviation for a light-emitting substance!) Finally the $E—L—AMP$ dissociates to yield free luciferase, luciferin and adenosine monophosphate. In this sequence the chemical energy of ATP is converted to light energy.

Two varieties of the fungus *Panus stipticus* are known, an American one which is luminescent and a European one which is not. When the two varieties are crossed, it is found that the ability to luminesce is inherited by a single dominant gene.

What advantage the ability to emit light may confer on an organism is not clear. For deep sea animals, living in perpetual darkness, light organs would conceivably be useful to enable members of a species to recognize each other or to serve as a lure for prey or a warning to would-be predators. It is known that the light emitted by fireflies serves as a signal to bring the sexes together for mating. The light emitted by bacteria and fungi probably serves no useful purpose to the organism, but is simply a by-product of oxidative metabolism, just as heat is a by-product of metabolism in other organisms.

QUESTIONS

1. Define the term catalyst. Give an example of an inorganic catalyst and of an organic catalyst.
2. List the chief properties of enzymes.
3. Outline a method, giving the precautions to be observed, by which an enzyme could be used to measure the amount of a given substance (its substrate).
4. What are the roles of ATP in the cell? How is ~P produced, how is it stored, and how is it utilized?
5. Trace the sequence of energy transformations from sunlight to the heat released in muscle contraction.
6. Distinguish between an apoenzyme and a coenzyme and between a coenzyme and a prosthetic group.
7. How would you define a photon? How does it differ from an electron? From a proton?
8. How is the energy content of light related to its wavelength? What is an einstein?
9. In what respects are photochemical reactions different from other chemical reactions? Distinguish between fluorescence and phosphorescence.
10. Discuss the structure of the chloroplast and the possible relation of its structure to the process of photosynthesis. What is a quantasome?
11. What is meant by the statement that the atoms of the chlorophyll molecule constitute a "resonating system"? Of what importance is this in the process of photosynthesis?
12. How could you prove that oxygen is given off by green plants during photosynthesis?
13. What pigments may be present in plant cells? What are the functions of these pigments?
14. Describe the light reactions of photosynthesis. What are the products of these reactions?
15. Discuss the sequence of reactions that constitutes the "dark reactions" of carbohydrate synthesis. What are the products of these reactions?
16. Compare photosynthetic phosphorylation and oxidative phosphorylation. What are the roles of ferredoxin and plastoquinone?
17. Compare the process of cyclic photophosphorylation and noncyclic photophosphorylation.
18. Discuss the possible sequence of events in the evolution of photosynthesis.
19. List the enzymes known to be present in the mitochondria of the cell and give their prime functions in intermediary metabolism. What enzymes are present in ribosomes and what are their functions? What are some of the enzymes present in the soluble cell sap and what are their functions?

20. Describe the mode of action of enzymes. What factors affect enzyme activity?
21. Define oxidation, reduction, oxidase and dehydrogenase.
22. Compare the processes of oxidative phosphorylation and "substrate level" phosphorylation.
23. Discuss the biological roles of NAD and NADP.
24. Compare the processes by which glucose and fatty acids are "activated."
25. Outline the reactions by which the carbons of a glucose molecule may be converted to the carbons of a molecule of fat.
26. Discuss the validity of the statement, "pyruvic acid is an important crossroads of metabolism."
27. What factors regulate the rate of chemical reactions in a test tube? in a living cell?
28. Does a yeast cell metabolize more efficiently in the presence or in the absence of oxygen? Explain.
29. Compare the types of reactions in which the energy of ATP is made available for biosynthetic processes.
30. What is the evidence that the chemical compounds of a cell are in a "dynamic state"? How would you set up an experiment to determine whether the proteins and fats of a bean leaf are also in a dynamic state?
31. Suppose you discovered a new species of bioluminescent worm. How could you prove that it was the worm itself and not some contaminating bacterium that was producing the light?
32. Calculate the number of molecules of ATP that would be produced by the complete oxidation of a molecule of palmitic acid.
33. List the coenzymes known to be important in the oxidation of carbohydrates, fats and amino acids and for each give the vitamin from which it is derived.

SUPPLEMENTARY READING

Metabolic Pathways, edited by D. M. Greenberg, is a multivolume, multiauthor work that covers in great detail the subjects discussed in this chapter. Especially pertinent are the chapters by J. M. Loewenstein and by D. E. Green and D. H. MacLennan. A briefer, readable discussion of the cellular basis of metabolic reactions is found in *Cell Structure and Function*, by A. G. Loewy and P. Siekevitz. An advanced discussion of enzymes and their properties is presented in *Enzymes* by Dixon and Webb. The subject of isozymes is discussed in breadth and depth in the four-volume multi-authored *Isozymes* edited by C. L. Markert. Baldwin's *Dynamic Aspects of Biochemistry*, McGilvery's *Biochemical Concepts* and Stryer's *Biochemistry* give technical but extremely interesting discussions of the details of cellular metabolism. *Bioenergetics* by Lehninger and *Mechanisms in Bioenergetics* by Ephraim Racker give excellent discussions of the principles of thermodynamics and their application to biological systems. The *Mitochondrion*, also by Lehninger, is a superb blending of the functional and morphologic aspects of this organelle and its role in the cell's economy. A description of the classic experiments that demonstrated the rapid renewal of the chemical constituents of tissues is found in Rudolph Schoenheimer's *The Dynamic State of the Body Constituents*. For more detailed discussions of the subject, consult *Photosynthesis* by G. E. Fogg. The phenomenon of bioluminescence is discussed in the symposium entitled *Light and Life*, edited by W. D. McElroy and B. Glass. L. J. Henderson in his *Fitness of the Environment* advanced the thesis that the environment had to have certain chemical and physical characteristics for life to develop.

PART 2

THE WORLD OF LIFE: PLANTS

INTRODUCTION

Part Two presents a broad picture of the bacteria, fungi and green plants, together with details of the structures and functions of certain key types. To set the stage for this discussion, Chapter 6 begins with an outline of the system by which all living things are classified, continues with a discussion of the distinctions between plants, animals, fungi, protists and monerans—between eukaryotes and prokaryotes—and ends with a consideration of some of the basic principles of ecology that are needed to understand the roles that particular organisms play in their environment. The following chapter (7) describes the morphologic and physiologic characteristics of bacteria and viruses and examines their effects upon humans. Chapter 8 delineates the various kinds of algae and protozoa assigned to the Kingdom Protista and the several kinds of fungi assigned to the Kingdom Fungi. It describes their life cycles and their effects, positive and negative, on human beings. The invasion of the land by green plants and the evolution of mosses, ferns and related plants, together with their fossil ancestors and their life cycles, are described in Chapter 9. This is followed by a presentation of the evolution and morphologic features of the seed plants and their life cycles in Chapter 10. The subjects of plant hormones, tropisms, photoperiodism and the functions of seed plants are discussed in Chapter 11. The final chapter (12) considers further the adaptations of seed plants for terrestrial life and describes the structures of the roots, stems and leaves and their functions, such as transpiration and the transport and storage of nutrients.

CHAPTER 6

BIOLOGIC INTERRELATIONSHIPS

As we turn from the more cellular and molecular aspects of biology to those concerning whole organisms, we will consider some of the enormous variety of plants and animals that exist and the various ways in which they are related. Although the world of living things appears to be made up of a bewildering variety of plants and animals, all quite different and each going its separate way at its own pace, closer inspection reveals that all organisms have the same basic requirements for survival. Each must obtain food for energy, find space in which to live, and survive to produce the next generation. In addition, each must interact with other species in a variety of ways that enhance or decrease its probability of survival.

In solving these problems, plants and animals have evolved into a tremendous number of different forms, each adapted to live in some particular sort of environment. Each has become adapted not only to the physical environment—has acquired a tolerance to a certain range of moisture, wind, sun, temperature, gravity, and so on—but also to the biotic environment, to all of the plants and animals living in the same general region. The study of the interrelations between living things and their environment, both physical and biotic, is known as **ecology.**

In recent years the general public has begun to become aware of the awesome problems that have arisen from our past ignorance of, and disregard for, the principles of ecology. Our concern over the growing pollution of the air and water is encouraging, and suggests that appropriate steps will be taken eventually to alleviate these problems and prevent their recurrence. However, pollution is only a portion of our present ecologic predicament, which has a much broader basis. Human beings must eventually learn how to keep their own numbers in check and how to live on this planet without altering the physical and biotic environment in such a way as to endanger the continued existence of themselves and the other 1,700,000 kinds of plants and animals.

One organism may provide food or shelter for another or produce some substance harmful to the second, or the two may compete for food or shelter. A detailed study of ecology requires knowledge of the structure and functions of a wide variety of plants and animals. In the latter part of this book, after we have considered some of the details of animal and plant physiology, heredity and evolution, we will return to a discussion of ecology, which serves as one of the major unifying concepts of biology.

6-1 THE CLASSIFICATION OF LIVING THINGS

To deal with the myriad forms of life and describe their characters, biologists first had to name and classify them. Animals, for example, were classified by St. Augustine in the fourth century as useful, harmful or superfluous—to man. The herbalists of the Middle Ages classified plants as to whether they produced fruit, vegetables, fibers or wood. The classification of plants and animals by structural similarities was placed on a firm systematic basis by the Swedish biologist Carl von Linné, or Linnaeus. He catalogued and described plants in *Species Plantarum* (1753) and animals in *Systema Naturae* (1758). With the acceptance of the theory of evolution, taxonomists have tried to set up systems of classification based on natural relationships, putting into a single group those organisms which are closely related in their evolutionary origin. Since many of the structural similarities depend on evolutionary relations, the modern classification of organisms is similar in many respects to the one of Linnaeus based on logical structural similarities.

The unit of classification for both plants and animals is the *species.* It is difficult to give a definition of this term which will apply uniformly throughout the living world, but a species may be defined as a population of similar individuals, alike in their structural and functional characteristics, which in nature breed only with each other, and have a common ancestry.

Closely related species are grouped together in the next higher unit of classification, the *genus* (plural, *genera*). The scientific names of plants and animals consist of two words, the genus and the species, given in Latin. This system of naming organisms, called the *binomial* (two name) *system,* was first used consistently by Linnaeus. In accordance with

it, the scientific name of the domestic cat, *Felis domestica,* applies to all the varieties of tame cats—Persian, Siamese, Manx, Abyssinian and plain tabby. All of them belong to the same species and all are able to interbreed. Related species of the same genus are *Felis leo,* the lion, *Felis tigris,* the tiger, and *Felis pardus,* the leopard. The dog, which belongs to a different genus, is named *Canis familiaris.* Notice that in each of these names, the genus is given first, and is capitalized, the species name is given second and is not capitalized (some species names of plants are capitalized). The use of Latin rather than a modern language in naming species is a carryover from the days when Latin was the international language of science.

Why, you may ask, bother to give Latin names to plants and animals? Why call a sugar maple *Acer* (maple) *saccharum* (sugar)? The primary reason is to be accurate and avoid confusion,* for in some parts of America this same tree is called either hard maple or rock maple. The tree most of us call white pine is *Pinus strobus,* but some people also refer to *Pinus*

---

*The scientific names of organisms are not necessarily fixed forever, for new research sometimes shows that the relationships of certain genera and species differ from our previous ideas about them. This may necessitate changing the names of the organisms concerned, much to the distress of other biologists who are used to calling a particular animal by a particular scientific name. George Wald, in a discussion of biochemical evolution (*Trends in Physiology and Biochemistry,* Academic Press, New York, 1952), describes his difficulty in finding out which animals actually were referred to under the names *Cynocephalus mormon* and *Cynocephalus sphinx* in a paper published in 1904: "I have learned since that one is the mandrill, the other the guinea baboon. Since Nuttall wrote in 1904, these names have undergone the following vagaries. *Cynocephalus mormon* became *Papio mormon,* otherwise *Papio maimon,* which turned to *Papio sphinx.* This might well have been confused with *Cynocephalus,* now become *Papio sphinx,* had not the latter meanwhile been turned into *Papio papio.* This danger averted, *Papio sphinx* now became *Mandrillus sphinx,* while *Papio papio* became *Papio comatus.* All I can say to this is, thank heavens one is called the mandrill, the other the guinea baboon."

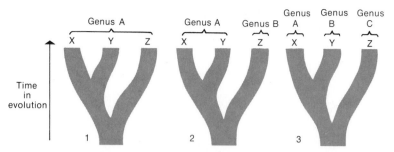

Figure 6–1 A single original stock evolves in the course of time into three different groups of organisms—X, Y and Z. Different biologists looking at these three groups may assign them all to either the same genus (*1*) or to two genera (*2*) or to three different genera (*3*). Some biologists prefer taxonomic systems with a single large classification and many subdivisions; others prefer separate subdivisions for each group that can be recognized. There is no single "correct" way of classifying organisms in any subdivision of the plant or animal kingdom.

flexilis and *Pinus glabra* as white pines and still other people call *Pinus strobus* northern pine, soft pine or Weymouth pine. There are thousands of other instances of confusing common names, but these examples should make it clear that exact scientific names are really necessary and not simply scientific double-talk.

Just as several species may be grouped together to form a genus, a number of related genera constitute a *family*, and families may be grouped into *orders*, orders into *classes*, and classes into *divisions* (plants) or *phyla* (singular, *phylum*) (animals). The latter two are the large, major divisions of the plant and animal kingdoms, as the species are the fundamental small units. A complete classification of a white oak is division Magnoliophyta, class Magnoliopsida, order Fagales, family Fagaceae, genus *Quercus,* species *alba.* Humans are members of the phylum Chordata, subphylum Vertebrata, class Mammalia, subclass Eutheria, order Primates, family Hominidae, genus *Homo,* species *sapiens.*

Many plants and animals fall into easily recognizable, natural groups, and their classification presents no difficulty. Others appear to lie on the borderline between two groups, having some characteristics in common with each, and are difficult to assign to one or the other. The number and inclusiveness of the principal groups vary according to the basis of the classification used and the judgment of the scientist making the classification (Fig. 6–1). Some taxonomists like to group things together in already existing units; others prefer to establish separate categories for forms that do not fall naturally into one of the recognized classifications. Thus, different taxonomists consider that

there are from ten to thirty-three animal phyla and from four to twelve plant divisions.

6–2 TWO KINGDOMS OR FIVE?

Biologists since the time of Aristotle have divided the living world into two kingdoms, plants and animals. The word "plant" suggests trees, shrubs, flowers, grasses and vines—large and familiar objects of our everyday world; and "animal" suggests cats, dogs, lions, tigers, birds, frogs and fish. Further thought brings to mind such forms as ferns, mosses, mushrooms and pond scums, quite different but recognizable as "plants," and insects, lobsters, clams, worms and snails that are definitely animals. But if you have ever had the pleasure of climbing over the rocky shore of the sea coast, looking at the organisms that cling to the rocks or live in a tide pool, you undoubtedly found some things that were difficult to recognize as animals or plants. The one-celled organisms visible under the microscope cannot easily be assigned to the plant or animal kingdom.

The German biologist Ernst Haeckel suggested more than a century ago that a third kingdom, the *Protista,* be established to include all the single-celled organisms that are intermediate in many respects between plants and animals. Some protists are very clearly plantlike and are closely related to other plants. Other protists are animallike; some have characteristics intermediate between animals and plants, and still others have characteristics that are distinctly different from either plants or animals. Even the organisms included by different biologists in the kingdom Protista may differ. Some taxonomists include in the Protista only unicellular forms, whereas others include fungi and multicellular algae as well as bacteria and blue-green algae.

Other biologists have suggested establishing a fourth kingdom, the **Monera,** to include the bacteria and blue-green algae, which have many characteristics in common, such as the absence of a nuclear membrane. The bacteria and blue-green algae are termed *prokaryotes,* to indicate that these cells have no nuclear membrane and only a single "naked" chromosome. Prokaryotes lack membrane-bounded subcellular organelles such as mitochondria or chloroplasts. All the protists, plants and animals are *eukaryotes,* characterized by true nuclei bounded by a nuclear membrane.

A scheme of classification assigning organisms to five kingdoms, suggested by R. H. Whittaker in 1969, has won wide acceptance in biological circles. Whittaker distinguished the fungi as a kingdom separate from the other plantlike forms. The fungi lack photosynthetic pigments but have nuclei and cell walls. Like other proposals for classifying organisms this solves some of the problems regarding the assignment of organisms to the major subdivisions of living things but raises a few others. When we realize that the question is not *what* these organisms are but what we shall *call* them, the argument falls into proper perspective.

There are many fundamental similarities between plants and animals; both are made of cells as structural and functional units and both (cf. Chap. 5) have many metabolic processes in common. But there are some obvious ways and some obscure ways in which they differ.

Plant cells, in general, secrete a hard outer cell wall of *cellulose* which encloses the living cell and supports the plant, while animal cells have no outer wall and hence can change their shape. However, some plants have no cellulose walls, and one group of animals, the primitive chordates known as sea squirts or tunicates, do have cellulose walls around their cells.

Secondly, plant growth generally is indeterminate. Perennial plants keep on growing indefinitely because some of the cells remain in an actively growing state throughout life. Many tropical plants grow throughout the year; those in the temperate regions grow primarily in the spring and summer. In contrast, the ultimate body size of most animals is established

after a definite period of growth. There are, of course, some notable exceptions to this generalization. Alligators, turtles and lobsters are examples of animals which continue to grow for a very long period of time.

Most animals are able to move about, whereas most plants remain fixed in one place, sending roots into the soil to obtain water and salts and getting energy from the sun by exposing broad flat surfaces. A little thought will bring to mind exceptions to both of these distinctions.

The most important difference between the two is their mode of obtaining nourishment. Animals move about and obtain their food from organisms in the environment, but plants are stationary and manufacture their own food. Plants have the green pigment chlorophyll which enables them to carry on photosynthesis, to utilize light energy to split water and reduce carbon dioxide to carbohydrate. There are some species or genetic variants of certain higher plants that have lost the ability to synthesize chlorophyll. These plants, generally white in color, must obtain nutrients from other organisms since they cannot carry out photosynthesis.

Although the reproductive cycles of plants and animals are typically quite different, there are enough exceptions to the generalizations concerning reproductive phenomena that these criteria cannot be used to distinguish plants and animals. Thus, there are no hard and fast rules for distinguishing plants and animals.

Our concepts of the evolutionary relationships between the major phyla of plants and animals are rather vague because the evolutionary events occurred so long ago and the fossil record of these early forms is nearly blank. The evolutionary relationships of viruses and bacteria to other organisms are unknown; there is little evidence regarding the relationships between the major kinds of algae and fungi; and the relationships of the major kinds of protozoa to multicellular animals are unclear.

6-3 MODES OF NUTRITION

Organisms that can synthesize their own food are said to be *autotrophic* (self-nourishing). Autotrophs require only water, carbon dioxide, inorganic salts and a source of energy. Green plants and purple bacteria are *photosynthetic autotrophs*, deriving the energy needed for biosynthetic processes from sunlight. A few bacteria are **chemosynthetic** autotrophs and obtain energy by oxidizing certain inorganic substances such as ammonia or hydrogen sulfide. These bacteria have evolved special enzyme systems that catalyze the oxidation of these substances and couple the oxidation with the generation of energy-rich phosphates. For example, nitrite bacteria (*Nitrosomonas*) oxidize ammonia to nitrites; nitrate bacteria (*Nitrobacter*) oxidize nitrites to nitrates; iron bacteria oxidize ferrous to ferric iron; and still other bacteria oxidize hydrogen sulfide to sulfates. The energy derived from these oxidations is utilized to synthesize all the organic materials necessary to maintain life and grow. The nitrite and nitrate bacteria are also important in the cyclic use of nitrogen, for together they convert ammonia to nitrate, a form readily used by green plants.

Purple bacteria have pigments that can utilize the energy of sunlight to "fix" carbon dioxide as carbohydrate. However, in this reaction oxygen is not produced and the organisms use hydrogen sulfide, hydrogen, or an organic compound such as succinate as the source of hydrogen instead of the water molecules used in photosynthesis by green plants.

In contrast to the autotrophs, heterotrophic organisms are unable to synthesize their own foodstuffs from inorganic materials. Heterotrophs must live at the expense of autotrophs or upon decaying organic matter. All animals, all fungi and most bacteria are heterotrophs.

There are several types of heterotrophic nutrition. When food is obtained as solid particles that must be eaten, digested and absorbed, as in most animals, the process is termed *holozoic* nutrition. Holozoic organisms must constantly find, catch and eat other organisms; to do this animals have evolved a variety of sensory, nervous and muscular structures to find and catch food and several types of digestive systems to convert this food into molecules small enough to be absorbed. Insectivorous plants, such as the Venus flytrap, sundew and pitcher plant, supplement their photosynthetic capabilities by trapping and digesting insects and other small animals (moonlighting in the plant world!). From these the plants obtain amino acids and other nitrogenous compounds for growth.

Herbivorous animals eat plants and obtain their energy-rich compounds from the contents of the plant cells, compounds made by the plant using energy derived from sunlight. Other animals are *carnivores*, eating animals (that ate plants). *Omnivores* are animals that will eat either plant or animal material. All heterotrophic organisms obtain their energy-rich nutrients ultimately from autotrophic organisms that trapped the radiant energy of sunlight to synthesize those compounds.

Yeasts, molds and most bacteria neither make their nutrients by autotrophic processes nor can they ingest solid food. They must absorb their required organic nutrients directly through the cell membrane. This type of heterotrophic nutrition is known as *saprophytic* nutrition. Saprophytes can grow only in places where there are decomposing bodies of animals or plants or masses of plant and animal by-products.

Yeasts are good examples of saprophytic plants. They need only inorganic salts, oxygen and some kind of sugar. From the last they can derive energy and using this energy they can make all the other substances needed for life—proteins, fats, nucleic acids, vitamins, and so on. When plenty of oxygen is available, the yeasts obtain energy by oxidizing the glucose completely to carbon dioxide and water via the tricarboxylic acid cycle. When the supply of oxygen is limited, they ferment the glucose and form alcohol and carbon dioxide. This, as we have seen, yields only about one twentieth as much energy as the complete oxidation of glucose, and therefore yeasts grow very slowly in the absence of oxygen.

Yeasts are used in the manufacture of all alcoholic beverages. Indeed, the only practicable way of obtaining ethyl alcohol is by this action of yeast. Ethanol is used not only in beverages but in many industrial processes as a solvent or as a raw material for the production of plastics and synthetic rubbers. Yeasts have a remarkable resistance to the toxic effects of alcohol and continue to produce it until a concentration of about 12 per cent ethanol is reached in the surrounding environment, at which point the yeast organisms are inhibited. To produce beverages such as brandy or whiskey with a higher alcoholic content, the wine or mash must be distilled. When yeast is mixed with bread dough, it ferments some of the sugar present to form ethanol and carbon dioxide; most of the alcohol is dissipated during the baking process, but the bubbles of carbon dioxide trapped in the dough expand and raise it, making the bread porous.

Another saprophyte, the baker's mold, *Neurospora*, requires the vitamin biotin in addition to salts and sugar; other saprophytes may require many different organic compounds.

A third type of heterotrophic nutrition, found among both plants and animals, is *parasitism.* A parasite lives in or on the living body of a plant or animal (called the *host*) and obtains its nourishment from it. Almost every living organism is the host for one or more parasites. A few plants, such as the mistletoe, are in part parasitic and in part autotrophic, for although they have chlorophyll and make some of their food, their roots grow into the stems of other plants, and they absorb some of their nutrients from their hosts.

Parasites may obtain their nutrients by ingesting and digesting solid particles or by absorbing organic molecules through their cell walls from the body fluids or tissues of the host. Some parasites cause little or no harm to the host. Others produce definite diseases, destroying the host cells or producing toxic substances which interfere with the host's metabolic processes. The *pathogenic* (disease-producing) parasites of man and other animals include viruses, bacteria, fungi, protozoa and an assortment of worms. Most plant diseases are caused by parasitic fungi; a few are due to viruses, worms or insects.

It is curious that parasites are usually restricted to one or a few species of hosts. Most of the organisms that infect man will not infect other animals, or will infect only apes and monkeys which are closely related to man phylogenetically. A few human parasites do have wider host ranges and will infect more distantly related mammals and even birds. Saprophytes such as yeasts or bread molds are easily grown in the laboratory, for they require only inorganic salts, glucose, and perhaps a vitamin or two, and will grow over a considerable range of temperature. But parasitic bacteria usually require a temperature near that of their normal host, and a complex medium containing sugars, amino acids and vitamins. Some, indeed, will grow only if provided with blood, liver or yeast extracts which contain one or more unknown growth factors. Finally, a few parasites, such as rickettsias and viruses, can be grown only in the presence of living cells. Until 1952, the poliomyelitis virus could be grown only by infecting some experimental animal such as a monkey, but in that year John Enders of Harvard discovered methods by which the virus could be grown on tissue cultures of human cells or on the kidney cells of the rhesus monkey. This work, for which Enders and his colleagues received a Nobel Prize in 1954, paved the way for the development of a polio vaccine by Jonas Salk.

6-4 THE CYCLIC USE OF MATTER

The total mass of all the organisms that have lived on the earth in the past two or three billion years is much greater than the mass of carbon and nitrogen atoms present on the planet. The Law of the Conservation of Matter assures us that matter has neither been created nor destroyed over this period. Obviously the carbon and nitrogen atoms must have been used over and over again in the formation of new generations of plants and animals. The earth neither receives any great amount of matter from other parts of the universe nor does it lose significant amounts of matter to outer space. The atoms of each element, carbon, hydrogen, oxygen, nitrogen and the rest, are taken from the environment, made a part of some cellular component and finally, perhaps by a quite circuitous route involving several other organisms, are returned to the environment to be used again. An appreciation of the roles of green plants, animals, fungi and bacteria in this cyclic use of the elements can be gained from considering the details of the more important cycles.

The Carbon Cycle. In the atmosphere over each acre of the earth's surface are about six *tons* of carbon as carbon dioxide. Yet in a single year an acre of luxuriant plant growth, such as sugar cane, can extract as much as 20 tons of carbon from the atmosphere and incorporate it into the plant body. If there were no way to renew the supply, the green plants would eventually, perhaps in a few centuries,

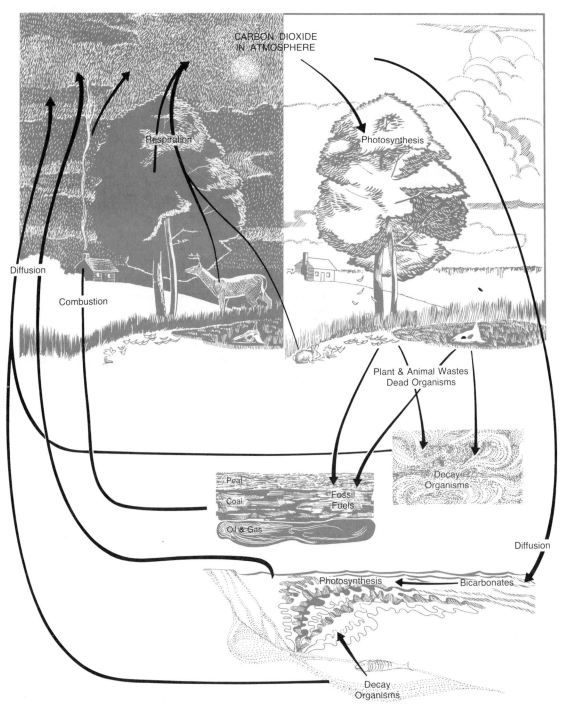

Figure 6–2 The carbon cycle, one of the major cycles of chemical elements in the environment. Through this cycle, carbon dioxide in the air can be utilized by living organisms to build the full range of organic compounds; it is then returned to the atmosphere for use by later generations. Without this cycle, the carbon dioxide content of the air would eventually be depleted. (See page 131 for a more detailed description.)

use up the entire atmospheric supply of carbon. Carbon dioxide fixation by bacteria and animals is another, but quantitatively minor, drain on the supply of carbon dioxide. Carbon dioxide is returned to the air by the decarboxylations that occur in cellular respiration. Plants carry on respiration continuously. Plant tissues are eaten by animals which, by respiration, return more carbon dioxide to the air. But respiration alone would be unable to return enough carbon dioxide to the air to balance that withdrawn by photosynthesis. Vast amounts of carbon would accumulate in the compounds making up the dead bodies of plants and animals. The carbon cycle is balanced by the decay bacteria and fungi which cleave the carbon compounds of dead plants and animals and convert the carbon to carbon dioxide (Fig. 6–2).

When the bodies of plants are compressed under water for long periods of time they are not decayed by bacteria but undergo a series of chemical changes to form *peat*, later brown coal or *lignite*, and finally *coal*. The bodies of certain marine organisms may undergo somewhat similar changes to form *petroleum*. These processes remove some carbon from the cycle but eventually geologic changes or man's mining and drilling bring the coal and oil to the surface to be burned to carbon dioxide and restored to the cycle.

Most of the earth's carbon atoms are present in limestone and marble as carbonates. The rocks are very gradually worn down and the carbonates in time are added to the carbon cycle. But other rocks are forming at the bottom of the sea from the sediments of dead animals and plants, so that the amount of carbon in the carbon cycle remains nearly constant.

The Nitrogen Cycle. The nitrogen for the synthesis of amino acids and proteins is taken up from the soil and water by plants as nitrates (Fig. 6–3). The nitrates are converted to amino groups and used by the plant cells in the synthesis of amino acids and proteins. Animals may eat the plants and utilize the amino acids from the plant proteins in the synthesis of their own proteins and other nitrogenous compounds. When animals and plants die, the decay bacteria convert the nitrogen of their proteins and other compounds into ammonia. Animals excrete several kinds of nitrogen-containing wastes—urea, uric acid, creatinine and

ammonia—and the decay bacteria convert these wastes to ammonia. Most of the ammonia is converted by nitrite bacteria to nitrites and these in turn are converted by nitrate bacteria to nitrates, thus completing the cycle. Denitrifying bacteria convert some of the ammonia to atmospheric nitrogen. Atmospheric nitrogen can be "fixed," converted to organic nitrogen compounds such as amino acids, by some blue-green algae (*Nostoc* and *Anabaena*) and by the soil bacteria *Azotobacter* and *Clostridium*.

Other bacteria of the genus *Rhizobium*, although unable to fix atmospheric nitrogen themselves, can do this when in combination with cells from the roots of legumes such as peas and beans. The bacteria invade the roots and stimulate the formation of root nodules, a sort of harmless tumor (Fig. 6–4). The combination of legume cell and bacteria is able to fix atmospheric nitrogen (something neither one can do alone) and for this reason legumes are often planted to restore soil fertility by increasing the content of fixed nitrogen. Nodule bacteria may fix as much as 50 to 100 kilograms of nitrogen per acre per year, and free soil bacteria as much as 12 kilograms per acre per year.

Atmospheric nitrogen can also be fixed by electrical energy, supplied either by lightning or by the electric company. Although 80 per cent of the atmosphere is nitrogen, no animals and only a few bacteria can utilize it in this form. When the bodies of the nitrogen-fixing bacteria decay, the amino acids are metabolized to ammonia and this is turn is converted by the nitrite and nitrate bacteria to nitrates to complete the cycle.

The Water Cycle. The great reservoir of water is the ocean. The sun's heat vaporizes water and forms clouds. These, moved by winds, may pass over land, where they are cooled enough to precipitate the water as rain or snow. Some of the precipitated water soaks into the ground, some runs off the surface into streams and goes directly back to the sea. The ground water is returned to the surface by springs, by pumps and by the activities of plants (transpiration; see p. 269). Water inevitably ends up back in the sea, but it may become incorporated into the bodies of several different organisms, one after another, en route. The energy to run the cycle—the heat needed to evaporate water—comes from sunlight.

The Phosphorus Cycle. As water runs

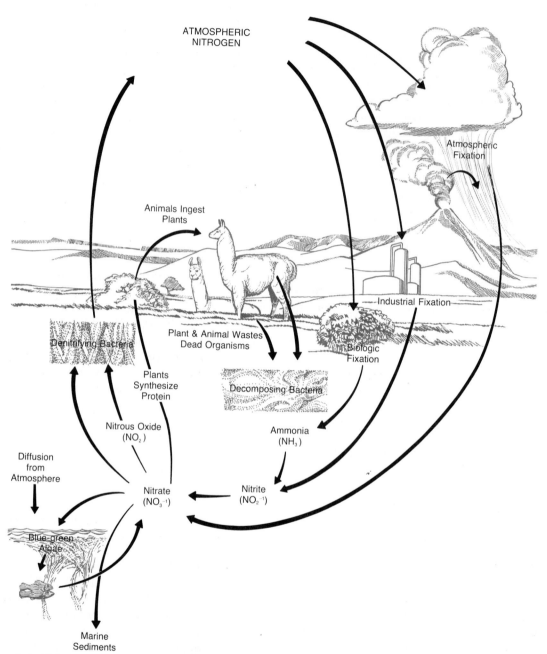

Figure 6–3 The nitrogen cycle, the second major elemental cycle. Certain reactions of the cycle take place in the air, on land and in water. Through this cycle, nitrogen is converted through a variety of biochemical processes into substances that can be used to make components of nucleic acids and other nitrogenous compounds. (See page 133 for a more detailed description.)

Figure 6–4 Roots of a soybean plant with root nodules formed by nitrogen-fixing bacteria. (From Weatherwax, P.: *Botany*, 3rd Ed. Philadelphia, W. B. Saunders Co., 1956.)

over rocks it gradually wears away the surface and carries off a variety of minerals, some in solution and some in suspension. Some of these minerals, such as phosphates, sulfates, calcium, magnesium and others, are necessary for the growth of plants and animals. Phosphorus, an extremely important constituent of all cells, is taken in by plants as inorganic phosphate and converted to a variety of organic phosphates (which are intermediates in the metabolism of carbohydrates, nucleic acids and fats). Animals get their phosphorus as inorganic phosphate in the water they drink or as inorganic plus organic phosphates in the food they eat.

The phosphorus cycle is not completely balanced, for phosphates are being carried into the sediments at the bottom of the sea faster then they are being returned by the actions of marine birds and fish. Sea birds play an important role in returning phosphorus to the cycle by depositing phosphate-rich guano on land. Man and other animals, by catching and eating fish, also recover some phosphorus from the sea. In time, geologic upheavals bring some of the sea bottom back to the surface as new

mountains are raised and in this way minerals are recovered from the sea bottom and made available for use once more.

The Energy Cycle. The cycles of all these types of matter are closed: the atoms are used over and over again. To keep the cycles going does not require new matter but it does require energy, for *the energy cycle is not a closed one.* The Law of the Conservation of Energy, or the First Law of Thermodynamics, states that energy is neither created nor destroyed but only transformed from one kind to another (p. 18). However, the Second Law of Thermodynamics states that whenever energy is transformed from one kind to another, there is an increase in entropy and a decrease in the amount of useful energy. Some energy is degraded into heat and dissipated.

Only a small fraction of the light energy reaching the earth is trapped; considerable areas of the earth have no plants, and plants can utilize in photosynthesis only about 3 per cent of the incident energy. This radiant energy is converted into the potential energy of the chemical bonds of the organic substances made by the plant. When an animal eats a plant (or when bacteria decompose it) and these organic substances are oxidized, the energy liberated is just equal to the amount of energy used in synthesizing the substances (First Law of Thermodynamics), but some of the energy is heat and not useful energy (Second Law of Thermodynamics). If this animal in turn is eaten by another one, a further decrease in useful energy occurs as the second animal oxidizes the organic substances of the first to liberate energy to synthesize its own cellular constituents.

In the successive steps of such a *food chain*—photosynthetic autotroph, herbivorous heterotroph, carnivorous heterotroph, decay bacteria—the number and mass of the organisms in each step is limited by the amount of energy available. Since some energy is lost as heat in each transformation the steps become progressively *smaller* near the top. This relationship is sometimes called a *"food pyramid"* or pyramid of numbers, to emphasize that in each successively higher section of the food chain the number (or more precisely the total mass) of the predators decreases.

Eventually, all the energy originally trapped by plants in photosynthesis is converted to heat and dissipated to outer space and all the carbon of the organic compounds ends up as carbon dioxide. The only important source of energy on earth is sunlight, energy derived from nuclear reactions, largely the conversion of hydrogen to helium, occurring at extremely high temperatures (about 10,000,000° C) in the interior of the sun. When this energy is exhausted and the radiant energy of the sun can no longer support photosynthesis, the carbon cycle will stop, all plants and animals will die and organic carbon will be converted to carbon dioxide. There is no immediate cause for alarm, however; the sun will continue to shine for several billions of years!

6–5 ECOSYSTEMS

In the chapters that follow, we shall discuss the structures and activities of a variety of plants and animals. As we learn more about what each species is and does, it will be apparent that each is not independent of its neigh-

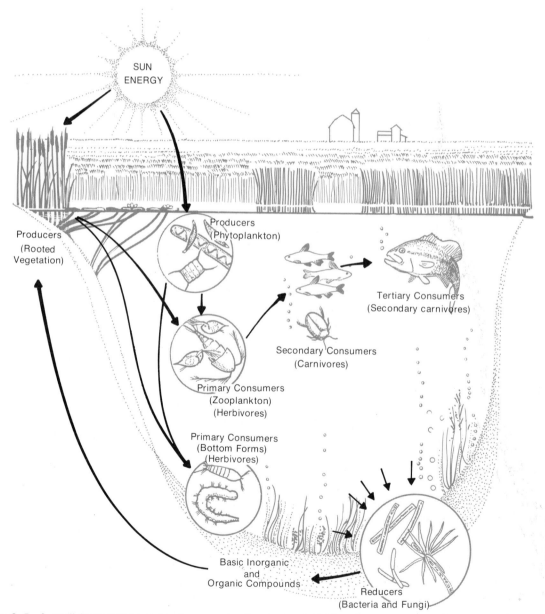

Figure 6–5 A small fresh water pond as an example of an ecosystem. The component parts (producer, consumer, and decomposer or reducer organisms) plus the nonliving parts are indicated.

bors but is part of a system of interdependent and interacting parts which form a larger unit. Ecologists use the term *ecosystem* to indicate a natural unit of living and nonliving parts that interact to produce a stable system in which the exchange of materials between living and nonliving parts follows a circular path. An ecosystem may be as large as the ocean or a forest or one of the cycles of the elements, or it may be as small as an aquarium jar containing tropical fish, green plants and snails. To qualify as an ecosystem, the unit must be a stable system in which the exchange of materials follows a circular path.

A classic example of an ecosystem compact enough to be investigated in quantitative detail is a small lake or pond (Fig. 6–5). The nonliving parts of the lake include the water, dissolved oxygen, carbon dioxide, inorganic salts such as phosphates, nitrates and chlorides of sodium, potassium and calcium, and a multitude of organic compounds. The living organisms may be subdivided into producers, consumers and decomposers according to their specific role in keeping the ecosystem operating as a stable interacting whole.

The *producer* organisms are the green plants that manufacture organic compounds from simple inorganic substances by photosynthesis. In a lake there are two types of producers: the larger plants growing along the shore or floating in shallow water, and the microscopic floating plants, most of which are algae, that are distributed throughout the water, as deep as light will penetrate. These tiny plants, collectively known as *phytoplankton*, are usually not visible unless they are present in great abundance and give the water a greenish tinge. They are usually much more important as food producers for the lake than are the more readily visible plants.

The *consumer* organisms are heterotrophs such as insects and insect larvae, crustacea, frogs, fish, and perhaps some fresh water clams. Primary consumers are the plant eaters and secondary consumers are the carnivores that eat the primary consumers. There might be some tertiary consumers that eat the carnivorous secondary consumers.

The ecosystem is completed by *decomposer* organisms, bacteria and fungi, which break down the organic compounds of cells from dead producer and consumer organisms either into small organic molecules, which they utilize themselves as saprophytes, or into inorganic substances that can be used as raw materials by green plants. Even the largest and most complex ecosystem can be shown to consist of the same components—producer, consumer and decomposer organisms and nonliving components.

6–6 HABITAT AND ECOLOGIC NICHE

In describing the ecologic relations of organisms it is useful to distinguish between *where* an organism lives and what it *does* as part of its ecosystem. The terms habitat and ecologic niche refer to two concepts that are of prime importance in ecology. The *habitat* of an organism is the place where it lives, a physical area, some specific part of the earth's surface, air, soil or water. It may be as large as the ocean or a prairie or as small and restricted as the underside of a rotten log or the intestine of a termite, but it is always a tangible, physically demarcated region. More than one animal or plant may live in a particular habitat.

In contrast, the *ecologic niche* is the status or role of an organism within the community or ecosystem. It depends on the organism's structural adaptations, physiologic responses and behavior. It may be helpful to think of the habitat as an organism's address (where it lives) and of the ecologic niche as its profession (what it does biologically). The ecologic niche is not a physically demarcated space, but an abstraction that includes all the physical, chemical, physiologic and biotic factors that an organism requires to live. To describe an organism's ecologic niche, we must know what it eats and what eats it, its range of movement and its effects on other organisms and on the nonliving parts of the surroundings. One of the important generalizations of ecology is that no two species may occupy the same ecologic niche.

A single species may occupy somewhat different niches in different regions, depending on such things as the available food supply and the number and kinds of competitors. Some organisms, such as animals with distinctly different stages in their life history, occupy different niches in succession. The frog tadpole is a primary consumer, feeding on plants, but an adult frog is a secondary consumer, feeding on insects and other animals. In contrast, young river turtles are secondary consumers, eating snails, worms and insects, whereas the adult turtles are primary consumers and eat green plants such as tape grass.

6–7 INTERSPECIFIC INTERACTIONS

The members of two different species of animals or plants may interact with each other in any of several different ways. If each species is adversely affected by the other in its search for food, space or some other need, the interaction is one of *competition.* Two species may compete for the same space, food or light, or in avoiding predators or disease. They are, in a sense, competing for the same ecologic niche. Competition may result in one species dying off or being forced to change its ecologic niche — to move away or to utilize a different source of food. Careful ecologic studies usually confirm *Gause's Rule:* There is only one species in an ecologic niche. One of the clearest examples of ecologic competition was provided by the classic experiments of Gause with populations of the protozoan *Paramecium.* When either of two closely related species, *Paramecium caudatum* or *Paramecium aurelia,*

was cultured separately on a fixed amount of food (bacteria), it multiplied and finally reached a constant level (Fig. 6–6). But when both species were placed in the same culture with a limited amount of food, only *Paramecium aurelia* was left at the end of sixteen days. The *P. aurelia* had not attacked the other species nor secreted any harmful substance, it simply had a slightly greater growth rate and had thus been more successful in competing for the limited food supply.

Studies in field ecology also generally confirm Gause's Rule. The cormorant and the shag are two fish-eating, cliff-nesting sea birds that seemed at first glance to have survived despite the fact that they occupy the same ecologic niche. However, the cormorant feeds on bottom-dwelling fish and shrimp, whereas the shag hunts fish and eels in the upper levels of the sea. Further study showed that these birds typically choose slightly different nesting sites on the cliffs.

Commensalism, the relationship in which two species habitually live together, one of which (the commensal) derives benefit from the association and the other is unharmed, is especially common in the ocean. Practically every worm burrow and shellfish contain some uninvited guests that take advantage of the shelter and possibly of the abundant food pro-

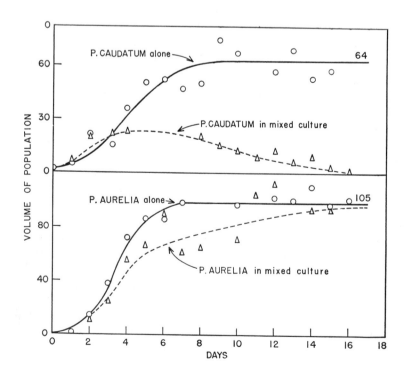

Figure 6–6 An experiment to demonstrate the competition between two closely related species of paramecia that have identical niches. When grown separately in controlled cultures with a fixed supply of food (bacteria), both *Paramecium caudatum* and *Paramecium aurelia* show normal S-shaped growth curves (solid lines). When grown together, *Paramecium caudatum* is eliminated (dashed lines). (From Allee, W. C., et al.: *Principles of Animal Ecology.* Philadelphia, W. B. Saunders Co., 1949. After Gause.)

vided by the host organism but do it neither good nor harm. Some flatworms live attached to the gills of the horseshoe crab and get their food from the scraps of the crab's meals. They receive shelter and transportation from the host but apparently do not injure it. One of the more startling examples of commensalism is that of a small fish that lives in the posterior end of the digestive tract of the sea cucumber (an echinoderm), entering and leaving it at will. These fish are quickly eaten by other fish if removed from their sheltering host.

If both species gain from an association, but are able to survive without it, the association is termed *protocooperation.* Several kinds of crabs put cnidarians of one sort or another on top of their shells, presumably as camouflage. The cnidarians benefit from the association by obtaining particles of food when the crab captures and eats an animal. Neither crab nor cnidarian is absolutely dependent on the other.

When both species gain from an association and are unable to survive separately, the association is called *mutualism.* It seems probable that interspecific associations begin as commensalism and evolve through a stage of protocooperation to one of mutualism. A striking example of a mutualistic association is that of termites and their intestinal flagellates. Termites are famous for their ability to eat wood, yet they have no enzymes to digest it. In their intestines, however, live certain flagellate protozoa that do have the enzymes to digest the cellulose of wood to sugars. Although the flagellates use some of this sugar for their own metabolism, there is enough left over for the termite. Termites cannot survive without their intestinal inhabitants; newly hatched termites instinctively lick the anus of another termite to obtain a supply of flagellates. Since a termite loses all its flagellates along with most of its gut lining at each molt, termites must live in colonies so that a newly-molted individual will be able to get a new supply of flagellates from its neighbor. The flagellates also benefit by this arrangement: they are supplied with plenty of food and the proper environment; in fact, they can survive only in the intestines of termites.

Negative Interactions. In certain types of interspecific associations one of the species is harmed by the other. If one is harmed but the second is unaffected, the relationship is termed *amensalism.* Organisms that produce antibiotics and the species inhibited by the antibiotics are examples of amensalism. The mold *Penicillium* produces *penicillin,* a substance that will inhibit the growth of a variety of bacteria. The mold presumably benefits by having a greater food supply when the competing bacteria have been removed. Human beings, of course, take advantage of this and culture *Penicillium* and other antibiotic-producing molds in huge quantities to obtain bacteria-inhibiting substances to combat bacterial infections. The use of these bacteria-inhibiting agents has had the unexpected effect of increasing the incidence of fungus-induced diseases in humans. These are normally kept in check by the presence of bacteria. When the bacteria are killed off by antibiotics, pathogenic fungi have a golden opportunity to multiply in the host.

We would be quite wrong if we assumed that the host-parasite or predator-prey relationship was invariably harmful to the host or prey *as a species.* This is usually true when such relationships are first set up, but in time, the forces of natural selection tend to decrease the detrimental effects. If the detrimental effects continued, the parasite would eventually kill off all the hosts and, unless it found a new species to parasitize, it would die itself.

Studies of hundreds of different examples of parasite-host and predator-prey interrelations show that in general, where the associations are of long standing, the long-term effect on the host or prey is not very detrimental and may even be beneficial. Conversely, newly acquired predators or parasites are usually quite damaging. The plant parasites and insect pests that are most troublesome to man and his crops are usually those which have recently been carried into some new area and thus have a new group of organisms to attack.

A striking example of the result of upsetting a long-standing predator-prey relationship occurred on the Kaibab plateau, on the north side of the Grand Canyon of the Colorado River. In this area in 1907 there were some 4000 deer and a considerable population of their predators, mountain lions and wolves. When a concerted effort was made to "protect" the deer by killing off the predators, the deer population increased tremendously. By 1925 there were some 100,000 deer on the plateau, far too many for the supply of vegetation. The deer ate everything in reach, grass, tree seedlings and shrubs, and there was marked damage to the vegetation. There was no longer enough vegetation to support the deer population over the winter, and in the next two winters vast numbers of deer starved to death. Finally the deer population fell to about 10,000. The original predator-prey interaction had been maintaining a fairly stable equilibrium, with the number of deer being kept at a level within the available food supply.

The size of the predator population in the wild varies with the size of the population preyed upon. The swings in the size of the predator population lag a bit behind those of the prey.

6–8 INTRASPECIFIC INTERACTIONS

In addition to the associations between the members of two different species just described, aggregations of animals or plants of a single species frequently occur. Some of these aggregations are temporary, for breeding; others are more permanent. Despite the fact that the crowding which accompanies dense aggregations of animals is ecologically undesirable and deleterious, both laboratory experiments and field observations show that such aggregations of individuals may be able to survive when a single individual of the same species placed in the same environment would die. A herd of deer, with many noses and pairs of eyes, is less likely to be surprised by a predator than is a single deer. Wolves hunting in a pack are more likely to make a kill than is a lone wolf. The survival value of intraspecific aggregations is less obvious, but nonetheless real, in some of the lower animals. A group of insects is less likely to dry up and die in a dry environment than is a single insect, and a group of planaria is less likely to be killed by a given dose of ultraviolet light than is a single flatworm. When a dozen goldfish are placed in one bowl and a single one in a second bowl, and the same amount of a toxic agent such as colloidal silver is added to each bowl, the single fish will die, but the group will survive. The explanation for this has proved to be that the slime secreted by the group of fish is enough to precipitate much of the colloidal

silver and render it nontoxic, whereas the amount secreted by a single fish is not.

Such animal aggregations do have survival value for the species. W. C. Allee has called this *"unconscious cooperation."* When genes governing a tendency toward aggregation arise in a species and prove to have survival value, natural selection will tend to preserve this inherited behavior pattern. The occurrence of many fish in schools, of birds in flocks and so on are examples of this "unconscious cooperation," which occurs very widely in the animal kingdom.

From such simple animal aggregations there may evolve complex animal societies, composed of specialized types of individuals, such as the colonies of bees, ants and termites. The human being is another example of a social animal.

In the following survey of plants and animals, you should keep in mind not only how they are classified but also where they live and, even more important, what they do—in what ways they contribute to the ecosystems of which they are members. In this way you can gradually build up a picture of the world of animals and plants, all related closely or distantly by evolutionary descent, and bound together in a variety of interspecific interactions. Without these unifying concepts of evolutionary relationships and ecologic interrelations, the world of living things is simply a bewildering hodgepodge.

QUESTIONS

1. How would you define a species? a class? a phylum? a division?
2. Distinguish between "natural" and "artificial" systems of classifying organisms.
3. What are the major structural and functional differences between plants and animals?
4. What are the advantages of a "five kingdom" system over a "two kingdom" one? List some organisms that are easily assigned to a kingdom. What organisms are especially difficult to assign a place in the taxonomic hierarchy?
5. Discuss the relevant arguments for classifying the blue-green algae as monerans, as protists and as plants.
6. What is meant by the term autotrophic? What types of organisms are autotrophic?
7. Differentiate between the several types of het-

erotrophic nutrition and give an example of each.

8. In what ways do saprophytes differ from parasites? Are parasites necessarily pathogenic?
9. Why are parasites usually restricted to one or a very few host species?
10. In what ways have parasites evolved to become adapted for their parasitic life?
11. Discuss the role of bacteria in (a) the carbon cycle and (b) the nitrogen cycle.
12. In what way does the phosphorus cycle differ from the carbon and nitrogen cycles? In what way does the energy cycle differ from the cycles of matter?
13. Explain why all life on the earth is ultimately dependent on a continued supply of sunlight.
14. What is meant by a "food chain"?
15. Define precisely the terms ecosystem, habitat and ecologic niche.
16. Discuss an aquarium as an example of an ecosystem.
17. List the several types of interactions between species and give an example of each.
18. Differentiate clearly between commensalism, protocooperation and mutualism.

SUPPLEMENTARY READING

The cycles of substances in nature and the kinds of relations between different species of organisms are described in more detail in *Fundamentals of Ecology* by E. P. Odum and in *Energy Exchange in the Biosphere* by D. M. Gates. Much of the September, 1971, issue of *Scientific American* was devoted to considerations of energy relationships in the biosphere and is highly recommended reading. *Life in the Soil* by R. M. Jackson and F. Raw and *Nitrogen Fixation in Plants* by W. D. P. Steward provide extensive discussions of the carbon, nitrogen and other cycles. *Plant Ecology* by Weaver and Clement and *Principles of Animal Ecology* by Allee, Emerson, Park, Park and Schmidt give more detailed discussions of plant and animal ecology respectively.

CHAPTER 7

PROKARYOTES AND VIRUSES

Most laymen associate the words disease and bacteria to such an extent that they cannot hear one without thinking of the other. Yet many diseases are not caused by bacteria and most bacteria do not cause disease. Bacteria were probably first seen by Antonj van Leeuwenhoek (1632–1723), who was a draper in Delft, Holland. With hand lenses he ground himself he examined almost everything at hand—pond water, sea water, vinegar, pepper solutions (he wanted to find out what made them hot), feces, saliva, semen and many other things. He described the objects he saw in letters written to the Royal Society of London. In a letter written in 1683 he described what were unquestionably bacteria—the size, shape and the characteristic motion of the organisms he described leave no doubt that they were bacteria.

The extensive research of Louis Pasteur in the 1870's and 1880's revealed the importance of bacteria as agents of disease and decay. This stimulated Robert Koch, Ferdinand Cohn, Joseph Lister and others, and the science of bacteriology blossomed rapidly in the latter part of the nineteenth century. Pasteur's studies of the "diseases" of souring wine and beer showed that they were caused by microorganisms that entered the wine or beer from the air and brought about undesirable fermentations yielding products other than alcohol. By gently heating (a process now known as **pasteurization**) the grape juice or beer mash to kill the undesirable organisms and then seeding the cooled juice with yeast, he could prevent these diseases.

Another of Pasteur's contributions to bacteriology was his unequivocal demonstration that bacteria cannot arise by spontaneous generation. After his study of the diseases of wine, Pasteur was asked by the French government to investigate a disease of silkworms. When Pasteur found that this, too, was caused by microorganisms he reasoned that many animal and plant diseases might be caused by the invasion of "germs." During his investigations of anthrax, a disease of sheep and cattle, and chicken cholera, he devised a method of treatment, that of **inoculation,** which greatly reduced the death rate from these diseases.

Lord Lister, an English surgeon, was one of the first to understand the significance of Pasteur's discoveries and to apply the germ theory to surgical procedures. He initiated antiseptic techniques by dipping all his operating instruments into carbolic acid and by spraying the scene of the operation with that germicide. In this way he effected a marked decline in the number of fatalities following operations.

In more recent decades research with bacteria has been concerned with their physiological, nutritional, biochemical and genetic properties. Much of our present understanding of the molecular basis of life has been gained from studies of bacteria, especially of *Escherichia coli,* a harmless parasite of the intestines of man and other mammals.

7–1 BACTERIA

There are relatively few places in the world that are devoid of bacteria, for they can be found as much as 5 meters deep in the soil, in fresh and salt water, and even in the ice of glaciers. They are abundant in air, in liquids such as milk, and in and on the bodies of animals and plants, both living and dead. Bacterial cells range in size from less than 1 to 10 micrometers in length and from 0.2 to 1 micrometer in width. There are rod-like bacilli, spherical cocci and spiral forms of bacteria (Fig. 7–1). Most bacterial species exist as single-celled forms, but some are found as filaments of loosely joined cells. Bacteria are clas-

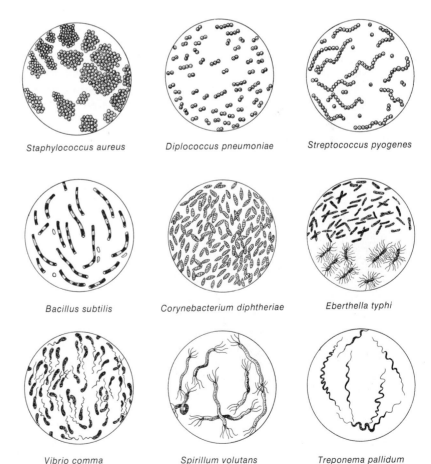

Figure 7–1 Some representative bacteria. *Upper row,* Spherical forms (cocci); *middle row,* rod forms (bacilli); *lower row,* spiral forms.

Staphylococcus aureus   *Diplococcus pneumoniae*   *Streptococcus pyogenes*

Bacillus subtilis   *Corynebacterium diphtheriae*   *Eberthella typhi*

Vibrio comma   *Spirillum volutans*   *Treponema pallidum*

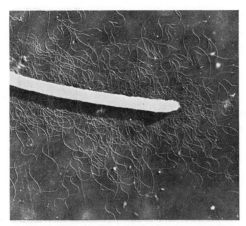

Figure 7–2 An electron micrograph of a bacillus, a rod-shaped bacterium. The bacillus was shadowed with a thin film of gold before being photographed. The thin, whiplike flagella are clearly visible. (Courtesy of Dr. C. F. Robinow and Dr. James Hillier of R.C.A.)

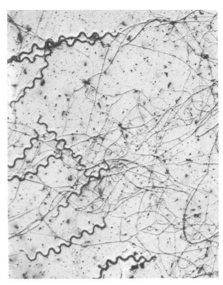

Figure 7–4 Electron micrograph of starting sample of *T. pallidum* extract for zonal centrifuge run. Magnification ×6250. (From Thomas, M., et al.: Applied Microbiology 23:714, 1972.)

sified largely by physiological and biochemical characteristics rather than by morphologic characters, since the various kinds of bacteria have generally similar shapes and cell structures.

Bacilli may occur as single rods (Fig. 7–2) or as long chains of rods joined together (e.g., anthrax bacilli). Diphtheria, typhoid fever, tuberculosis and leprosy are all caused by bacilli. Spherical bacteria occur singly in some species; in groups of two (e.g., gonococcus, the agent causing gonorrhea); in long chains (streptococci); or in irregular clumps like a bunch of grapes (staphylococci), (Fig. 7–3). There are

two types of spiral bacteria: the less coiled comma-shaped spirilla (the organism causing cholera is comma-shaped) and the highly coiled, corkscrew-shaped spirochetes (Fig. 7–4) such as the organism causing syphilis.

The bacterial cell has a cell membrane and is covered by a strong, rigid cell wall which contains *diaminopimelic acid*, an amino acid found only in bacteria and in blue-green algae, and *muramic acid*, a derivative of glucose. With high resolution electron microscopy, the cell walls of bacteria are revealed to be constructed of units 50 to 140 nanometers in diameter, arranged in regular, hexagonal or rectangular patterns (Fig. 7–5). Most bacteria have a slimy *capsule* composed of polysaccharides which lies outside of the cell wall and serves as an additional protective layer. The cytoplasm within the bacterial cell is dense and contains granules of glycogen, protein and fat, but lacks mitochondria and an endoplasmic reticulum (Fig. 7–6). The ribosomes occur free in the cytoplasm and are not bound to an endoplasmic reticulum. The DNA is present in a distinct nuclear region, but no nuclear membrane separates it from the rest of the cytoplasm. Cocci usually have one nuclear region, but rodlike bacilli usually have two or more per cell (Fig. 7–7). Many bacteria have whiplike cellular outgrowths called *flagella*, by means of which they can swim about. Bacterial

Figure 7–3 An electron micrograph of a group of staphylococci, spherical bacteria that occur in bunches like grapes. Magnification ×25,000, reduced ⅕ in printing. (Courtesy of the Department of Physical Chemistry, Lilly Research Laboratories.)

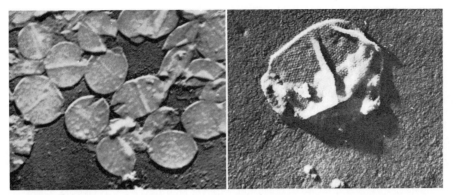

Figure 7–5 Cell walls of bacteria. *Left,* Walls of *Streptococcus fecalis* prepared by grinding the cells; splitting permitted the cell contents to escape. Magnification ×12,000. *Right,* A portion of the cell wall of *Spirillum rubrum* showing the regular pattern of the spherical bodies of which this wall is composed. Magnification ×42,000. (From Salton, M. R. J., and Williams, R. C.: Biochem. Biophys. Acta *14*:455, 1954.)

flagella are curious in that they consist of a single fibril, whereas the flagella of higher organisms are composed of 11 fibrils arranged in a bundle, nine in the periphery and two in the center of the flagellum. Rod- and spiral-shaped bacteria tend to have flagella, whereas spherical ones do not. Some bacteria can travel as much as 2000 times their own length in an hour in this way. (Imagine a man 2 meters tall swimming 4 kilometers per hour simply by swinging a long whip!)

Bacteria generally reproduce asexually by the division of the parental cell into two daughter cells. However, the duplication of the chromosome (the S phase) and the division of the nuclear region (the M phase) can get out of phase with the division of the rest of the cell, so that a given cell may have from one to four

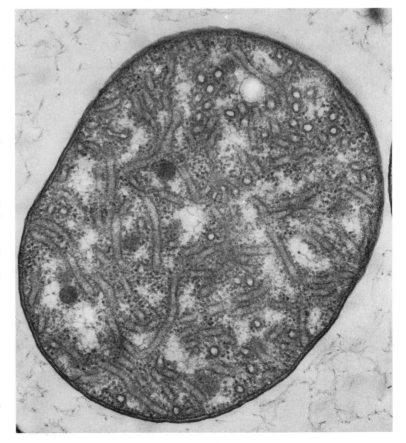

Figure 7–6 An electron micrograph of a thin section of a marine nitrifying bacterium, *Nitrococcus mobilis,* that oxidizes nitrites to nitrates. The micrograph shows both longitudinal and cross sections of the curious tubular membranes present in these cells. These membranes contain the enzymes of the electron transmitter system. The hexagonal bodies scattered through the cell have not been identified but might be viruses. The ribosomes are clearly evident as small electron-dense (black) dots. (Courtesy of Dr. S. W. Watson.)

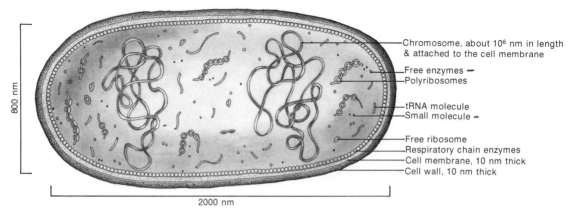

Chromosome, about 10⁶ nm in length
& attached to the cell membrane

Free enzymes
Polyribosomes

tRNA molecule
Small molecule

Free ribosome
Respiratory chain enzymes
Cell membrane, 10 nm thick
Cell wall, 10 nm thick

800 nm

2000 nm

Figure 7–7 Schematic diagram of a bacterial cell of *Escherichia coli,* containing two chromosomes.

or even more nuclear regions (Fig. 7–8). Bacterial cells do not have, nor do they need, a mitotic spindle, for they do not have a set of chromosomes which divide and must be segregated carefully to the two daughter cells. However, the two daughter chromosomes that form when the single chromosome replicates must be regularly segregated to the two daughter cells. This appears to be accomplished by a connection between the chromosome and the plasma membrane, perhaps via a mesosome (a large, irregular, convoluted invagination of the plasma membrane). The mesosome may divide along with the chromosome, and the progeny mesosomes, attached to the plasma membrane,

may migrate in opposite directions, each carrying a daughter chromosome with it, so that the chromosomes are regularly segregated to the daughter cells. Thus the cytoplasmic membrane may serve as a primitive mitotic spindle.

Bacterial cell division can occur with remarkable speed, and some species grown in an appropriately fortified and aerated culture medium can divide every 20 minutes. At this rate, if nothing interfered, one bacterium would give rise to some 250,000 bacteria within six hours. This explains why the entrance of only a few pathogenic bacteria into a human being can result so quickly in the symptoms of disease. Fortunately bacteria cannot reproduce at

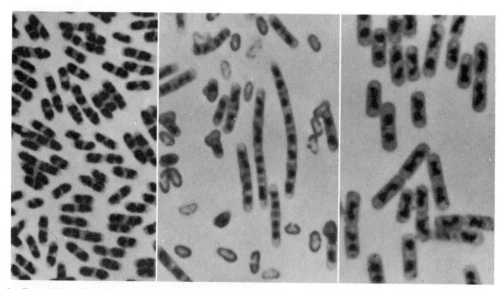

Figure 7–8 Bacterial cells stained to demonstrate nuclear regions. *Left, Escherichia coli,* ×4000; *middle, Bacillus mesentericus,* ×4400 (note spores in various stages of germination); *right, Bacillus cereus,* ×4400. (Photograph of *E. coli* by J. Hillier, S. Mudd and A. G. Smith, S.A.B. LS-239; photographs of *B. mesentericis* and *B. cereus,* S.A.B. LS-266.)

this rate for a very long time, for they are soon checked by lack of food or by the accumulation of waste products.

7-2 BACTERIAL REPRODUCTION

Bacteria have been used extensively in studies of molecular biology and genetics. The phenomenon of bacterial transformation (p. 679) has been of enormous importance in showing not only that genes can be transferred from one bacterium to another, but also that DNA is the chemical basis of heredity. Bacterial genes can be transferred not only by transformation—the uptake of naked DNA—but also by *conjugation*—the mating of two bacterial cells—and by viral infection, called *transduction.*

Both cytologic and genetic evidence indicate that bacterial cells may rarely undergo something similar to sexual reproduction in which two cells come together and genetic material is transferred from one to the other. In the process of conjugation many marker genes are transferred jointly, which suggests that large portions of the genome are transferred rather than small segments, as in the process of transformation. Evidence that conjugation occurs, evidence of genetic recombination, has been found in all of the strains of bacteria that have been carefully studied, and it probably occurs in other species as well. In *Escherichia coli* some cells act as donors and transfer genetic information by direct contact to recipient cells. The ability to transfer genetic material is regulated by the fertility factor, F+ (which can itself be transferred from a donor to a recipient, thereby converting it into a donor). The usual vegetative bacterial cell is haploid, and in conjugation, part or all of the chromosome passes from the donor to the recipient cell, thereby making it partly or completely diploid. Crossing over then occurs between the recipient chromosome and the donor chromosome or fragment, followed by a process of segregation that yields haploid progeny cells.

The process of conjugation requires that the two types of bacterial cells have complementary macromolecules somewhere on their surfaces. The mutual recognition of these leads to pairing. It had been assumed that in conjugation some sort of cytoplasmic bridge was formed between the two cells, a process which would have involved a change in the rigid cell wall. More recent evidence suggests that the F+ gene leads to the formation of hundreds of rodlike *pili* that radiate from the cell surface and that these are involved in the transfer of DNA. The F pili are long and narrow, but have an axial hole with an inner diameter of about 2.5 nanometers, which is similar to the tails of certain bacteriophages through which phage DNA can be transferred. To explain how energy is provided to push the DNA across the conjugation bridge, Jacob and Brenner suggested that the DNA molecule in the donor cell undergoes replication and that one of the new double strands passes across the cytoplasmic bridge while the other one remains inside the cell. Thus the transfer of DNA is believed to require its simultaneous replication.

In the third process of gene transfer, transduction, bacterial genes are transferred within bacteriophage particles. Transduction, like transformation, involves the transfer of only small segments of bacterial DNA from a donor to a recipient. In transformation the DNA transferred is naked, whereas in transduction it is passed to the recipient cell as a packet surrounded by the coat of the bacteriophage. When a bacteriophage particle is formed inside a bacterium, it may incorporate a piece of bacterial DNA into its head in place of some of its own DNA. The bacteriophage particle is released when the parent bacterial cell undergoes lysogeny, and the phage particle may infect a new bacterial cell, carrying into it some bacterial genes from the host. Transduction is a relatively rare event that can be detected only by rather sophisticated genetic techniques. The total length of bacteriophage DNA is a small fraction, about 2 per cent, of the bacterial chromosome. A phage particle can contain only a very small fraction of the total amount of bacterial DNA. The phage particles usually are "defective," with little or no phage DNA in the head, presumably because the bacterial DNA occupies all of the available space.

When the environment of the bacterial cell becomes unfavorable (e.g., when it becomes very dry), many bacteria can become dormant. The cell loses water, shrinks a bit and remains quiescent until water is again available. Other species form *spores* to survive in extremely dry, hot or cold environments (Fig. 7–9). The

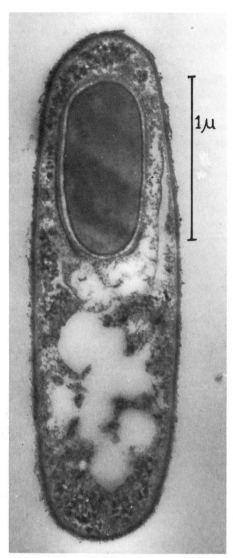

Figure 7-9 Nearly mature spore within a cell of *Bacillus cereus*. Three spore coats are clearly distinguishable. The large, white bodies in the lower end of the cell are vacuole-like inclusions. (Electron micrograph of ultrathin section by G. B. Chapman: J. Bact. *71*:348–355, 1956.)

formation of a spore is not a kind of reproduction, since only one spore is formed per cell. The total number of individuals does not increase as a result of the formation and subsequent hatching of the spore. During the formation of a spore the cell shrinks, rounds up within the former cell membrane, and secretes a new, thicker wall inside the old one. When the environmental conditions are again suitable for growth the spore can absorb water, break out of its inner cell, and become a typical bacterial cell. Spores of anthrax bacilli have been shown to be able to hatch out 30 years after they were formed.

7-3 BACTERIAL METABOLISM

Although a few bacteria are autotrophic and synthesize organic compounds by chemosynthetic or photosynthetic reactions, most bacteria are either saprophytes or parasites. The majority of bacteria, like animals and plants, are aerobic and utilize atmospheric oxygen in cellular respiration. A few bacteria can grow and multiply in the absence of gaseous oxygen. They obtain energy by the anaerobic metabolism of carbohydrates or amino acids, and, in the course of this, accumulate a variety of partially oxidized intermediates, such as ethanol, glycerol and lactic acid. A few bacteria, *obligate anaerobes*, can grow only in the absence of oxygen, and are killed in the presence of molecular oxygen.

The range of organic compounds that can be utilized by one or another kind of bacteria as a source of energy is impressive and includes sugars, amino acids, fats, urea, uric acid and other waste products. One strain of bacteria has become adapted to using penicillin, although many kinds are killed by this antibiotic. The anaerobic metabolism of carbohydrates is termed *fermentation*, and the anaerobic metabolism of proteins and amino acids is termed *putrefaction*. The foul smells associated with the decay of food and plant or animal bodies or wastes are due to nitrogen and sulfur-containing compounds formed in putrefaction. Bacteria play important roles in the carbon, nitrogen, and other cycles of nature, and, in fact, the substances produced by one kind of bacteria may be used as sources of energy by other kinds of bacteria. Without the bacteria and the fungi, all the available carbon and nitrogen atoms would eventually be tied up in the bodies of dead plants and animals, and life would cease because of the lack of raw materials for the synthesis of new cellular components.

A *pure culture* of one species of bacteria can be set up by the method of *serial dilution*. A sample of soil, feces, blood or sputum is mixed with a large volume of nutrient medium and a small portion of this is removed and mixed with another large volume of medium. By repeating this process and making the serial dilutions rapidly enough so that the bacteria cannot reproduce between transfers, you eventually obtain a tube of nutrient medium with but a single bacterial cell, even though the initial sample contained many millions of bacterial cells. When the final tube is incubated for an appropriate period, and the single cell divides repeatedly, the resulting bacterial culture will have arisen by the reproduction of this single cell and will be a pure culture. Another method takes advantage of the fact that bacteria can move only a very short distance on a solid medium. A drop of feces or sputum is mixed with warm liquid *agar* (extracted from a seaweed) that is about to solidify. The mixture is poured in a thin layer on a flat glass dish with a cover. As the medium cools it solidifies and the bacteria are held in

position. The bacteria are spread out on the agar by the mixing process, and when each subsequently multiplies it gives rise to a colony of daughter cells, all of which came from a single initial bacterium and thus is a pure culture (Fig. 7–10). With a sterile wire loop some of the members of this colony are transferred to a fresh dish of agar, mixed and incubated again. In this way, any bacteria contaminating the original colony can be removed.

Bacteria are identified to some extent by their morphologic appearance and the color of the colonies, and by staining with the Gram stain devised by the Danish physician Christian Gram. Bacteria are also identified by their biochemical properties, by testing the kinds of substances they need to grow and the kinds of substances they produce. For example, the

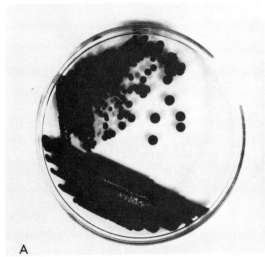

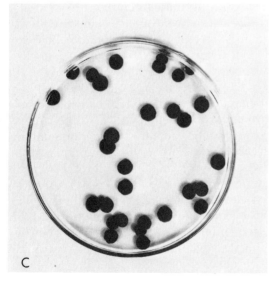

Figure 7–10 *A,* Three-phase discontinuous streak plate; note isolated colonies *(S. marcescens* on nutrient agar). *B,* Turntable for spread plate technique. Bent glass spreading rod is immersed in 95 per cent alcohol and flame-sterilized. Inoculum is spread evenly over the surface of the agar. *C,* Isolated colonies developed from spread plate *(S. marcescens* spread over nutrient agar). (From Frobisher, M., Hinsdill, R. D., Crabtree, K. T., and Goodheart, C. R.: *Fundamentals of Microbiology,* 9th Ed. Philadelphia, W. B. Saunders Co., 1974.)

colon bacillus *Escherichia coli* is a normal inhabitant of the human colon, but typhoid bacilli and *Shigella*, a bacillus that produces food poisoning, are not normal inhabitants. The normal and the pathogenic bacilli cannot be distinguished morphologically, but the colon bacillus can ferment lactose, whereas the pathogenic forms cannot.

Man has used to good advantage the abilities of specific strains of bacteria to produce chemicals such as ethanol, acetic acid, butanol and acetone. Bacteria with specific enzymatic properties are widely used in the synthesis of drugs and other chemicals. Bacterial action is involved in the curing of tobacco, in preparing hides for tanning, and in treating the fibers from flax and hemp to make linen and rope. Bacteria also play a role in the production of butter, cheese, sauerkraut, rubber, cotton, silk, coffee and cocoa. Specific strains of decay bacteria play an important role in the functioning of sewage disposal plants. Sewage is allowed to pass slowly over beds of gravel and sand which contain bacteria. The action of the bacteria converts the raw sewage into solid material that can be dried and used as fertilizer. It removes and kills any disease-producing bacteria that may have been present, thus making the effluent safe as a source of drinking water.

Bacteria not only cause a variety of diseases in man and animals, but also are responsible for certain diseases of plants, such as the *fire blight* of apple and pear trees. The bacteria enter the tree through a wound or through a flower, multiply rapidly and kill the cells so quickly that the tree appears to be burned. The tree typically produces a sticky liquid around the infection which contains bacteria. Other trees become infected when drops of this liquid are transported by insects. Bacteria are responsible for the *soft rot* of a variety of vegetables, the *black rot* of cabbage and the growth of *crown galls* on other plants.

7–4 THE BLUE-GREEN ALGAE

The blue-green algae are similar to the bacteria in many fundamental respects. These algae are probably the most primitive chlorophyll-containing, autotrophic organisms now living. The oldest fossil plants found so far appear to have been blue-green algae. The Latin word *alga* means seaweed, but although most seaweeds are algae, there are many algae that are not seaweeds and some seaweeds that are not algae. Most algae, blue-green as well as the other kinds of algae, live in either fresh or salt water, but a few live on rock surfaces or the bark of trees. The ones living in such relatively dry places usually remain dormant when water is absent. Almost all of the photosynthesis in the oceans and most of that in fresh water is carried on in algae. We do not commonly use algae as food, but a considerable portion of human food is fish, which eat algae or eat other organisms that eat algae.

The blue-greens are quite different from other algae and are structurally similar to bacteria. These similarities are the basis for grouping them with the bacteria in the kingdom **Monera.** Blue-greens, like bacteria, have a diffuse nucleus with no nuclear membrane. The photosynthetic pigments, chlorophyll, xanthophyll, carotene and phycocyanin, are dispersed throughout the cytoplasm in the blue-greens (Fig. 7–11) instead of being concentrated in plastids as they are in other algae. The blue pigment, **phycocyanin,** is found only in the blue-greens; in addition they have a unique cyanophycean starch. Like the bacteria, the blue-greens lack mitochondria, the Golgi complex and other subcellular organelles. The nuclear region contains a single circular chromosome of double-stranded DNA. The blue-green algae secrete a sticky gelatinous outer sheath which covers each cell. Sexual reproduction has never been observed in blue-green algae but it may occur rarely, as in bacteria.

Most of the blue-green algae are found in fresh water pools and ponds, where they may occur in sufficient numbers to color the water and give it an unpleasant taste and smell. A few species are found in hot springs; still others live in the ocean. Some species live on the bark of trees, on dead logs or in other moist environments. Members of the genus *Trichodesmium* contain a red pigment and from time to time occur in such numbers in the Red Sea that they actually color the water. A sudden increase in the numbers of one kind of alga is called a "bloom." As a result of a bloom, the water may become extremely turbid, limiting the penetration of sunlight. The vast numbers of algae may, by respiration at night, remove all of the oxygen from the water and

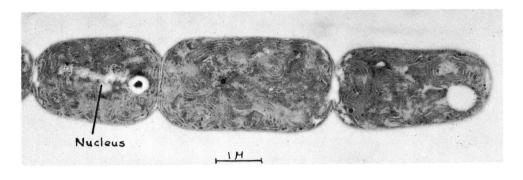

Nucleus

I M

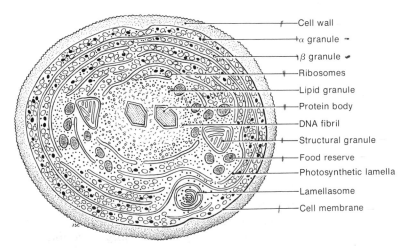

Cell wall
α granule
β granule
Ribosomes
Lipid granule
Protein body
DNA fibril
Structural granule
Food reserve
Photosynthetic lamella
Lamellasome
Cell membrane

Figure 7–11 Blue-green algae. *Upper,* An electron micrograph of *Anabaena.* The low density (pale) material is nuclear substance but no nuclear membrane is present. The membranous structures in the cytoplasm are the functional equivalents of chloroplasts. (From Carpenter, P. L.: *Microbiology,* 3rd Ed. Philadelphia, W. B. Saunders Co., 1972. Courtesy of G. B. Chapman.) *Lower,* Diagram of the structure of blue-green alga. Compare with the diagram in Figure 7–7; note the fundamental similarity of the structure of bacteria and blue-green algae.

cause the death of large numbers of fish. The algal bloom may also result in the production of toxic metabolic products that kill fish or any animals that drink the water.

A few kinds of blue-green algae are unicellular, but most of them occur as globular colonies or as long, multicellular filaments (Fig. 7–12). None of the blue-greens has flagella, but some of the filamentous species are capable of a curious back-and-forth oscillatory movement; others have a slow, gliding motion.

7–5 VIRUSES AND BACTERIOPHAGES

Viruses and bacteriophages are much smaller than bacteria, and indeed are scarcely larger than some very large single molecules of protein or nucleic acid. Too small to be seen with the light microscope, they can be photographed only with an electron microscope.

They resist classification as plants or animals, or even as protists or monerans. In one sense, viruses are not living organisms but rather are large nucleoprotein particles which enter specific kinds of animal, plant or bacterial cells and multiply (or are multiplied) to form new virus particles. When viruses are outside the host cell they are metabolically inert; in fact, some viruses have been crystallized. Each is essentially a bit of genetic material, either DNA or RNA, enclosed within a protective coat of protein which permits it to pass from one cell to the next. The difficulty in deciding whether these forms should be considered living or nonliving simply reflects the difficulty in defining life itself. Viruses and bacteriophages exhibit some, but not all, of the usual characteristics of living things.

Filtrable Viruses. Disease-causing, ultramicroscopic particles small enough to pass through very fine-pored porcelain filters were

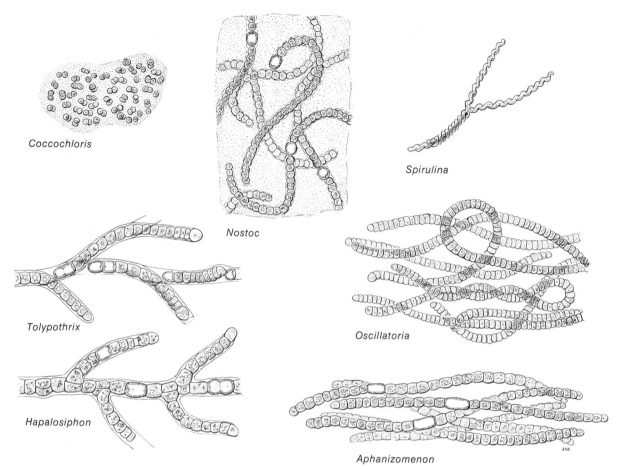

Coccochloris

Spirulina

Nostoc

Tolypothrix

Oscillatoria

Hapalosiphon

Aphanizomenon

Figure 7–12   Some common species of blue-green algae.

discovered by the Russian botanist Iwanowski in 1892. Iwanowski found that a disease of tobacco plants, called mosaic disease because the infected leaves had a spotted appearance, could be transmitted to healthy plants by daubing their leaves with the sap of diseased plants. The sap was effective even after it had been passed through filters fine enough to remove all bacteria.

Viruses are the infective agent in a wide variety of plant and animal diseases and in such human diseases as smallpox, rabies, poliomyelitis, measles, warts, fever blisters and the common cold. There is a strong possibility that viruses may be involved in the etiology of certain types of human cancer. One type of breast cancer in mice has been shown to be caused by

a viruslike agent, but human cancers are not infectious, as one might expect a virus disease to be.

Viruses can undergo duplication only within the complex environment of living cells. Viruses have very few, if any, of the metabolic properties of the cells of higher organisms. The entrance of a virus particle produces profound changes in the metabolic pattern of the host cell, which eventually lead to the production of new virus particles. Thus viruses do not really reproduce themselves, but are reproduced by the enzymatic machinery present in their host cells. This explains why viruses cannot be grown on cell-free culture media. Viruses are commonly cultured for experimental purposes by injecting them into

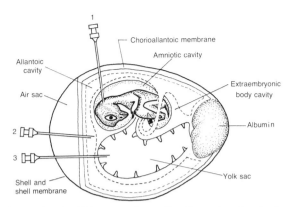

1
Chorioallantoic membrane
Amniotic cavity
Allantoic cavity
Extraembryonic body cavity
Air sac
2
3
Albumin
Shell and shell membrane
Yolk sac

Figure 7–13 Diagrammatic section through a developing chick embryo from 10 to 12 days old indicating how viruses can be inoculated into (1) the head of the embryo, (2) the allantoic cavity or (3) the yolk sac.

fertilized hens' eggs (Fig. 7–13) or by adding them to cells growing in tissue culture. Iwanowski's tobacco mosaic virus was isolated and crystallized by W. M. Stanley in 1935. Since then many other viruses have been obtained as crystals. When these crystals are put back into the appropriate host, they multiply and produce the symptoms of the disease. The tobacco mosaic virus was separated into its component protein and nucleic acid parts by Stanley, who was then able to recombine these parts into an active virus. Fraenkel-Conrat subsequently showed that the nucleic acid portion of the virus is infectious when injected alone, although it is very labile and much less efficient as an infectious agent than the intact virus. Injecting the nucleic acid alone induced the tobacco plant to produce the specific protein of the viral coat, in addition to specific viral nucleic acid, so that the complete virus particle was reconstituted.

Viruses vary widely in size from the psittacosis virus (which is about 275 nanometers in diameter), the cause of a disease transmitted by parrots and other birds, to the virus causing foot-and-mouth disease of cattle, which is only 10 nanometers in diameter. Electron microscopy reveals that some of the viruses are spherical and others are rod-shaped (Fig. 7–14). Although individual virus particles cannot be seen in the light microscope, cells that are infected with viruses frequently contain *inclusion bodies* that are visible by light microscopy. These appear to be huge colonies of virus particles.

Each kind of virus usually attacks some specific part of the host's body. Apparently the virus particles can reproduce only in certain kinds of cells and not in all the cells of the body. The viruses of smallpox, measles and warts attack the skin. Those of poliomyelitis and rabies attack the brain and spinal cord, and those of yellow fever attack the liver. It is fortunate that many of the infections caused by viruses create a lasting immunity against reinfection. Thus inoculations for smallpox, rabies and yellow fever are highly successful.

Interferon. The infection of a cell by one virus interferes with its subsequent infection by a second virus. The infected cell releases a substance which has been named *interferon* and identified as a protein about the size of hemoglobin. Interferon is released when the cell is infected with a live virus or with a heat-killed virus and is produced in amounts which might indeed account for the subsequent resistance of the cell to viral infections. The synthesis of interferon may be stimulated by the presence within the cell of a foreign nucleic acid, either of viral or nonviral origin. Even synthetic polynucleotides such as poly UC are potent stimulators of the production of interferon. Mice infected with influenza virus had the greatest amount of virus in the lungs on the third day, after which the number of viruses decreased. The concentration of interferon in the lungs was greatest on days 3 to 5 and then slowly declined. Antibodies to the virus appeared only on day 7 and increased after that, which suggests that interferon may be more important than antibodies in accounting for the recovery of the animal from the acute infection.

Bacteriophages. Viruses that parasitize bacteria, called *bacteriophages*, were discovered in 1917 by the French scientist d'Herelle, who noticed that some invisible agent was destroying his cultures of dysentery bacilli. Bacteriophages are filtrable and will grow only within bacterial cells which they cause to swell and dissolve. Bacteriophages are found in nature wherever bacteria occur and they are especially abundant in the intestines of man and other animals. There are many varieties of bacteriophages, and usually one kind of phage will attack only one species or one strain of bacteria.

Phages are spherical or comma-shaped or shaped like a Ping-Pong paddle and are some 5

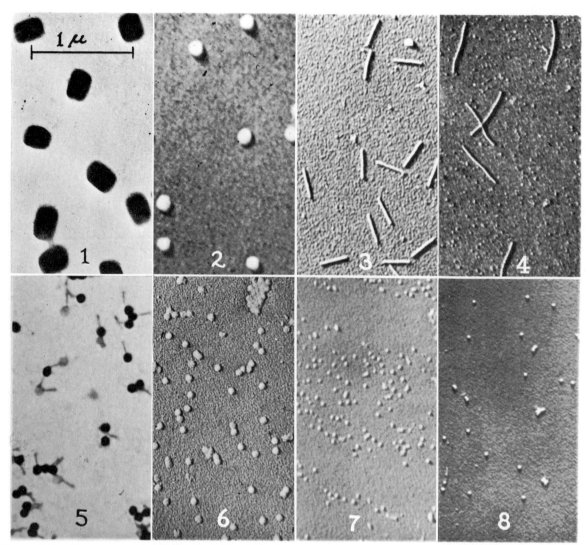

Figure 7-14 Electron micrographs of a variety of viruses. *1*, Vaccinia virus (used in vaccinating for smallpox). *2*, Influenza virus. *3*, Tobacco mosaic virus. *4*, Potato mosaic virus. *5*, Bacteriophages. *6*, Shope papilloma virus. *7*, Southern bean mosaic virus. *8*, Tomato bushy stunt virus. The viruses in *2, 3, 4, 6, 7* and *8* were shadowed with gold before being photographed in the electron microscope. (Courtesy of Dr. C. A. Knight.)

nanometers in diameter (Fig. 7–15). The fact that bacteriophages destroy bacteria, of course, led biologists and medical scientists to attempt to use them to treat patients suffering from bacterial diseases such as dysentery and staphylococcus infections. No preparation of bacteriophages has had any significant effect and this, combined with experimental evidence that bacteriophages are ineffective in the presence of blood, pus or fecal material, has halted attempts to use them for therapeutic purposes.

To obtain a pure strain of bacteriophage, an emulsion of feces, soil or sewage is made and passed through a fine filter. If phages are present, a drop of this filtrate added to a turbid bacterial culture will cause the bacterial cells to swell and dissolve, and the culture becomes clear (Fig. 7–16). When a drop of this clear culture is filtered and added in turn to a second bacterial culture, the latter will also become clear. Serial transfers of the bacteriophage can be made in this way indefinitely.

The virus first attaches to the bacterial cell by means of its protein tail. The tail contains an enzyme which digests part of the bacterial cell wall, and the DNA in the core of the virus is passed into the bacterial cell (Fig. 7–17). By labeling the protein of the virus with radioac-

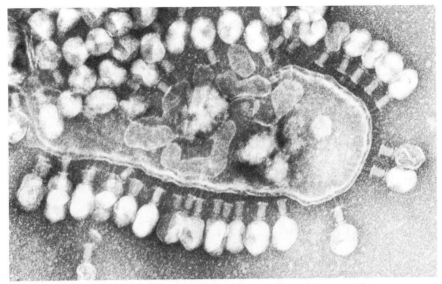

Figure 7–15 Phage infecting an *Escherichia coli* cell. Many phages are attached to the cell wall. The head, tail and base plate of most virus particles are clearly visible. The tail core extending from the base plate to the cell wall is hollow and acts like a hypodermic needle in injecting virus DNA into the cell. The break in the bacterial cell wall is an artifact produced during preparation for viewing under the electron microscope. (From Gerking, S. D.: *Biological Systems,* 2nd Ed. Philadelphia, W. B. Saunders Co., 1974. Courtesy of Dr. Lee D. Simon, Institute for Cancer Research, Philadelphia, Pa.)

tive sulfur (^{35}S) and the nucleic acid with radioactive phosphorus (^{32}P), Hershey and Chase could show that only the DNA of the virus is injected into the bacterial cell (p. 681). The protein coat of the head and tail remains outside. The inference that the DNA contains all the genetic information for the synthesis of the complete virus particle was established. The DNA directs the biosynthetic systems of the host cell to produce both viral DNA and viral protein. For 10 to 15 minutes after the infection of the bacterial cell, no virus particles can be detected within it. Over the next 10 to 15 minutes, increasing numbers of viral particles accumulate, and finally some 30 minutes after the initial infection, the bacterial cell bursts, and several hundred newly formed viral particles are released, ready to attack new bacterial cells.

Bacteria treated with the enzyme *lysozyme* to remove most of the material in the cell wall can then be infected with purified DNA from a bacterial virus, as well as by intact viral particles. When these bacteria are subsequently lysed, *complete* viruses containing both DNA *and protein* are released by a process known as *lysogeny.* This is further evidence that all the information to make a complete virus is present in the DNA and that the protein coat simply plays a role in penetrating the bacterial cell wall.

7–6 RICKETTSIAS

The rickettsias are a group of organisms which are smaller than bacteria but larger than viruses. They are not filtrable and are just barely visible under the light microscope. Some are spherical, and others rod-shaped, and they vary in length from 300 to 2000 nanometers. Their cellular structure is similar to that of bacteria (Fig. 7–18). Their discoverer, Howard Ricketts, died in Mexico in 1910 of typhus fever while studying the organisms that cause it. Rickettsias, like viruses, will multiply only within living cells, and are obligate intracellular parasites. The single exception to this is a nonpathogenic parasite of the sheep tick.

Some 50 different kinds of rickettsias are harmless parasites in the intestinal tracts and salivary glands of insects such as lice, bed bugs and ticks. A few of these rickettsias, when transmitted to man by insect bites, will multiply inside human cells and produce the symptoms of disease. Only six kinds of rickettsias are known to produce human diseases; the

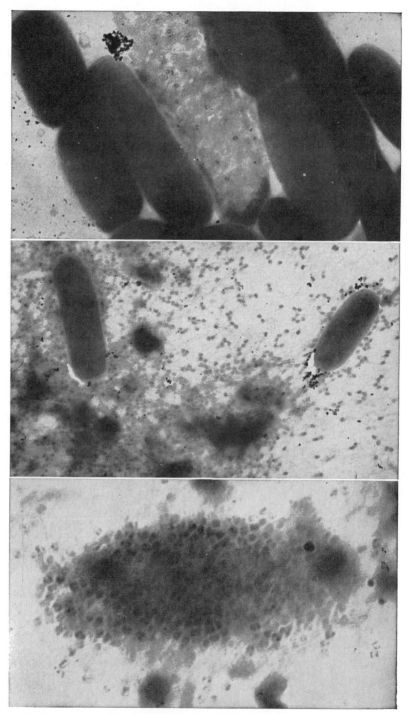

Figure 7–16 Electron micrographs of the destruction of *Escherichia coli* by bacteriophages. *Top,* Dark, sausage-shaped structures are normal bacteria; lighter one in middle has been attacked and destroyed by 'phage. 'Phage particles are evident within the cell. *Center,* Later stage with more 'phage particles visible and more bacteria destroyed. *Bottom,* Dense mass of 'phage particles occupying the space of the bacillus they have destroyed. (From Burrows, W., et al.: *Textbook of Microbiology,* 16th Ed. Philadelphia, W. B. Saunders Co., 1954.)

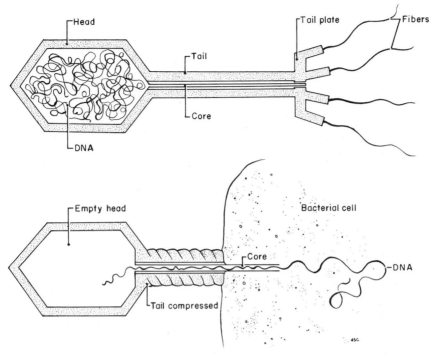

Figure 7–17 *Upper*, Diagram of the structure of a T₂ bacteriophage revealed by electron microscopy. *Lower*, Diagram of the transfer of viral DNA into the bacterial host cell.

principal rickettsial diseases of man are typhus fever and Rocky Mountain spotted fever. Rickettsias can be grown on chick embryos developing inside the egg shell to produce enough organisms to prepare a vaccine.

7–7 EVOLUTIONARY RELATIONSHIPS OF THE BACTERIA

The evolutionary relationships of the bacteria are not at all clear, and because of this their classification has undergone a number of changes. One scheme classifies them as the division Schizomycophyta in the subkingdom Thallophyta, which includes all plants not forming embryos during development, ones commonly called algae and fungi. Other current taxonomic systems place the bacteria in the kingdom Protista along with algae, fungi and protozoa, or in the kingdom Monera together with the blue-green algae. Other biologists simply refer to them as prokaryotes, organisms lacking a nucleus, in contrast to the

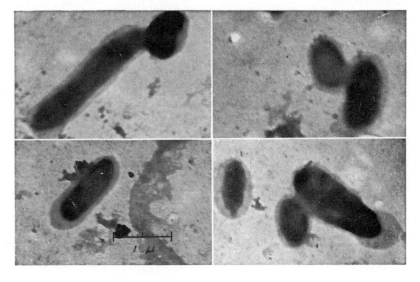

Figure 7–18 Electron micrographs of the rickettsia causing typhus. Note the variations in the size and shape of the particles and the less dense capsule that surrounds each. (Courtesy of Lilly Research Laboratories.)

eukaryotes, organisms with a nucleus, which include the vast majority of all other plants and animals.

One widely held view is that bacteria have evolved from the blue-green algae; after becoming adapted to a saprophytic or parasitic existence, they lost their chlorophyll. This view is based on the general similarity of the cell structure of these two forms. Other investigators believe that the fact that many bacteria have flagella indicates that these organisms descended from simple flagellated forms, perhaps ones that also gave rise to the green algae. Still others believe that the present-day heterotrophic bacteria evolved from autotrophic ones like the present-day iron and sulfur bacteria. Autotrophic bacteria may have appeared before any of the chlorophyll-containing plants. The present-day bacteria could be direct descendants of the original bacterialike heterotrophs that have been suggested as the first types of living cells (p. 750). It is also possible that different groups of bacteria have descended independently from different ancestors and that any two or even all three of these explanations are true.

Our knowledge of the evolutionary relationships of bacteria with higher organisms is just about as unsatisfactory. Bacteria may be a terminal group in evolution that has given rise to no other forms, or, if they did not arise from the blue-green algae, perhaps the blue-green algae have descended from them.

QUESTIONS

1. What contributions to bacteriology have been made by Leeuwenhoek, Pasteur, Koch, Lister, Tatum and Lederberg?
2. What are the structural characteristics of bacteria? In what ways do they resemble animals? In what ways do they resemble green plants? Describe the structure and constituents of the bacterial cell wall.
3. Discuss the roles of bacteria in the carbon and nitrogen cycles.
4. Describe the process of reproduction in bacteria. What is the role of bacterial spores? Compare the processes of transformation, conjugation and transduction.
5. Describe methods for obtaining a pure culture of a given species or strain of bacteria.
6. How are particular species of bacteria identified? What tests are used in the identification of bacteria?
7. In what ways are bacteria of economic importance to man?
8. Distinguish between rickettsias and viruses.
9. Discuss the several theories of the evolutionary origin of bacteria. Which of these seems most reasonable to you?
10. What are bacteriophages? How are they obtained and how are they maintained in culture?
11. What are some of the contributions to our general knowledge of cell metabolism and genetics which have been made using bacteria and bacterial viruses?
12. What is interferon? How was it discovered?

SUPPLEMENTARY READING

The biography of Leeuwenhoek by Clifford Dobell, *Antonj Von Leeuwenhoek and His Little Animals*, includes his description of the discovery of bacteria. William Bullock's *The History of Bacteriology* gives a general picture of the development of theories and knowledge about these interesting organisms. The biographies of Louis Pasteur by Vallery-Radot and by Rene Dubos provide interesting reading. Two fascinating accounts of the great plagues and their influence on historic events are presented by Hans Zinsser in *Rats, Lice and History* and by G. Smith in *Plague on Us*. A popular account of the role of viruses as agents of disease is presented in *The Viruses* by H. Curtis. *General Virology* by Luria and Darnell gives an excellent general account of the viruses. *The Microbial World* by R. Y. Stanier gives a fine overview of the bacteria and related organisms.

CHAPTER 8

THE KINGDOMS PROTISTA AND FUNGI

The more primitive plantlike organisms, the algae and fungi, may collectively be called thallophytes. They neither form embryos during development nor have vascular systems. They are widely distributed in fresh and salt water, on land and as parasites on other plants and animals. The members of this heterogeneous group range in size from microscopic, single-celled organisms to giant seaweeds or kelp that may be as much as one hundred meters long. The body of one of these plants, called a **thallus** (hence the term thallophytes), may show some differentiation of parts but has no true roots, stem or leaves.

Many systems of classifying these primitive, plantlike organisms have been proposed; one currently in vogue is Whittaker's "Five Kingdom" system, which assigns the algae and certain single-celled forms such as protozoa to the kingdom Protista and the slime molds and fungi, which lack chlorophyll and must live as saprobes or parasites, to the kingdom Fungi. The separation of algae and fungi is to some extent an artificial one, for it separates some organisms that are very much alike except for their color, which results from the presence or absence of chlorophyll. Some organisms, such as *Euglena,* that lose chlorophyll and live as saprobes if put in the dark but regain chlorophyll if returned to light strain the classification even further.

Alga is the Latin word for "seaweed," but although most seaweeds are algae, there are many algae that are not seaweeds. Algae are primarily inhabitants of water, fresh or salt, but a few live on rock surfaces or on the bark of trees. The ones living in such comparatively dry places usually remain dormant when water is absent. By virtue of their tremendous numbers, algae are important food producers; almost all the photosynthesis in the seas and most of that in fresh water is carried on by algae. Human beings do not commonly use algae directly as food, but a considerable fraction of human food is fish, which either eat algae or eat other organisms that depend on algae for food.

8–1 LIFE CYCLES

In each species of organism reproducing sexually there is a characteristic sequence of developmental processes by which one generation gives rise to the succeeding one. The events between any given point in one generation and that same point in the next generation constitute a *life cycle.* Sexual reproduction involves two fundamental processes: a special kind of nuclear division, *meiosis,* that reduces the number of chromosomes from 2N to 1N (p. 589), and *fertilization,* the fusion of two 1N gametes (special sex cells) to form a 2N zygote or fertilized egg. The sets of chromosomes are separated in meiosis before the formation of the gametes. Every 2N cell of a plant or animal can be traced to a single cell, the zygote, formed by the union of two gametes, each having the 1N number of chromosomes. The gametes, in turn, were produced by meiosis occurring in a 2N cell or, indirectly, by a series of mitotic divisions of 1N cells which followed meiosis in a 2N cell. In some species the two gametes that fuse are identical in size and structure. Much more commonly, one gamete is larger and nonmotile (the egg or ovum) and the other gamete is smaller and motile (the sperm). When two gametes unite their nuclei fuse to form a single nucleus, but the individual chromosomes within the nuclei remain distinct. The zygote, therefore, has twice as many chromosomes (it is 2N, diploid) as either gamete. Each gamete furnishes one *set* of chromosomes and the zygote has two sets. The number of chromosomes per set varies in different species—from one or two to several hundred—but is constant for any given species.

The union of gametes and the process of meiosis are landmarks dividing the life cycles of plants and animals into two phases. The first, from gamete union to meiosis, is characterized by 2N cells, and the second, from meiosis to gamete union, is characterized by 1N cells (Fig. 8–1). The alternation of 1N and 2N stages in the life cycle, the manner in which these stages are attained, their duration, and their relative size and complexity are of great importance in characterizing the different groups of organisms.

In many kinds of algae and fungi the zygote is the only 2N cell. The zygote divides meiotically and produces vegetative cells with

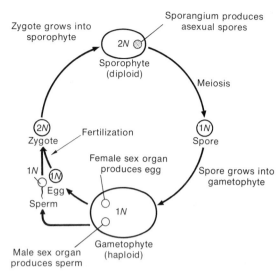

Figure 8–1 Diagram of the alternation of sporophyte (diploid) and gametophyte (haploid) generations in plants.

a 1N chromosome number. These divide repeatedly by mitosis and all the resulting daughter cells maintain the 1N number. Eventually, perhaps as a result of some seasonal change in the environment, some or all of the vegetative cells either become transformed into gametes or produce specialized gametes by some process characteristic of the species. When these gametes combine in fertilization a zygote is again formed and the life cycle has been completed. This type of life cycle has been termed *Life Cycle I* (Fig. 8–2). In a second type, *Life Cycle II,* meiosis occurs as gametes are being produced and the organism spends most of its life cycle in the 2N condition. Life Cycle II is typical of most animals and occurs in a few algae.

In other algae and fungi and in all of the higher plants—the mosses, ferns, conifers and flowering plants—the life cycle is a complex *alternation of generations.* Although the details of this process vary from one group of plants to another, the essentials are similar in all groups. The 2N zygote in these plants divides mitotically and forms a plant with 2N cells. The 2N plant is termed a *sporophyte* and some of the sporophyte cells undergo meiosis to form *meiospores.* These 1N meiospores divide mitotically to form a second plant generation termed a *gametophyte.* The cells of the gametophyte generation are 1N and when a gametophyte matures it produces 1N gametes by mitosis. The union of two 1N gametes in fertilization results

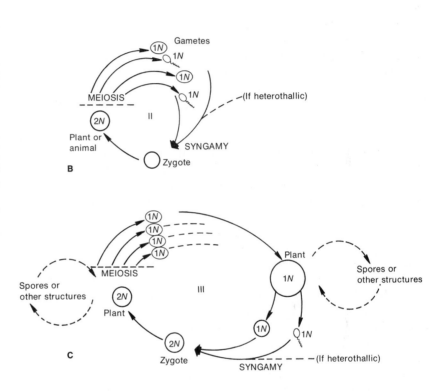

Figure 8–2 Sexual life cycles among algae, fungi and higher plants. *A,* Life cycle without alternation of generations, as seen in many algae and certain fungi, in which meiosis occurs in the zygote (Life Cycle I). *B,* Life cycle without alternation of generations, in which meiosis results directly in the production of gametes; this is seen in some algae and in most animals and is referred to as Life Cycle II. *C,* Life cycle with alternation of generations; this is seen in some algae and fungi and in all higher green plants and is referred to as Life Cycle III. Note that asexual reproduction also may occur in conjunction with various phases of such life cycles. (After Norstog, K., and Long, R. W.: *Plant Biology.* Philadelphia, W. B. Saunders Co., 1976.)

in a 2N zygote, and the cycle is completed. This type of life cycle is termed *Life Cycle III.* In the higher plants the life cycle is characterized by the alternation of a haploid, gamete-producing generation with a diploid, spore-producing generation. In the process of evolution from algae to rosebush the basic pattern of the cycle has remained the same, but there have been tremendous changes in the vegetative and reproductive organs and in the relative sizes and nutritional relationships of the two generations.

8–2 ASEXUAL REPRODUCTION

Asexual reproduction is characterized by the presence of a *single* parent, one that splits, buds, fragments, or produces many spores so as to give rise to two or more offspring. All of the descendants produced asexually from a single parent have the same genetic constitution, the same assortment of genes, as that parent and are termed *clones.* Sexual reproduction, in contrast, involves the cooperation of *two* parents; each supplies one gamete and the two gametes unite to form a zygote. In self-pollinating plants and hermaphroditic animals both "parents" may be located in the same plant or animal body, but reproduction is typically sexual nonetheless.

The advantage of sexual over asexual reproduction is that sexual reproduction makes possible the recombination of the best inherited traits of the two parental types. As a result the offspring may be better suited for survival than either parent. In contrast, offspring produced asexually have exactly the same inherited characteristics as the parent. Evolution by natural selection can proceed much more rapidly and effectively with sexual reproduction than with asexual reproduction.

Bacteria, unicellular algae and fungi reproduce asexually by *fission*—the parental body splits into two more or less equal daughter parts. The cell divisions involved in the splitting process are mitotic and may occur in rapid sequence to yield many new individuals in a short time. Yeasts and certain other protists and plants reproduce asexually by **budding.** A small part of the parent's body separates from the rest and develops into a new individual. The gemmae produced by liverworts (p. 200) are asexual buds; these become separated from the parent plant and may develop into new plants.

One of the major methods of asexual reproduction, important in the life cycles of a great many organisms, is the formation of *spores.* Spores typically have some sort of resistant covering to withstand unfavorable environmental conditions such as heat, cold or desiccation. Each of these special reproductive cells can develop without fertilization into a complete new organism.

Most blue-green algae and bacteria, the prokaryotes, appear to reproduce only by asexual means—by fission. Sexual reproduction has been demonstrated to occur rarely in some bacteria and it may occur in others. Even in the higher plants reproduction may occur asexually by a variety of means, and the farmer and florist take advantage of these in producing food and ornamental plants that will be exactly like the parent plant.

8–3 DIVISION CHLOROPHYTA: THE GREEN ALGAE

About 7000 species of green algae live in a wide variety of habitats ranging from salt to fresh water to damp soil. Most botanists believe that the higher plants probably evolved from algae very similar to the present-day green algae, for the chlorophytes have the greatest number of characteristics in common with the higher plants. The chlorophyte cell

has a distinct nucleus with a nuclear membrane; its pigments, chlorophylls a and b, carotene and xanthophyll, are organized into chloroplasts; food is stored as starch; and a cellulose cell wall is present. The simplest green algae are unicellular and motile; more advanced members of the division have many-celled bodies in the shape of filaments or flat leaflike structures. Even in the more advanced members of the division the cells of the plant body are almost all alike and there is little differentiation of tissues. Many green algae have flagella, but some species are nonmotile. Reproduction may be asexual, by cell division or by the formation of spores, or sexual, by the union of gametes.

A few species of green algae are terrestrial, living on the moist, shady sides of trees, rocks and buildings. One species has become adapted to living on the surface of snow and ice; it has red pigment in addition to chlorophyll and may grow in patches thick enough to give the snow a red tinge. The many-celled green algae living in fresh water include the pond scums; these occasionally grow very thickly in ponds and streams. Among the multicellular marine green algae living near the low tide mark and in the upper 6 meters of water is the sea lettuce, *Ulva*, with a body some 30 centimeters long but only two cells thick. It resembles a crinkled sheet of green cellophane. Some of the tropical marine forms have developed thickened plant bodies the size of a moss or small fern plant, with parts that superficially resemble the roots, stems and leaves of higher plants. The common, nonmotile, single-celled fresh water green algae known as **desmids** have symmetrical, curved, spiny or lacy bodies with a constriction in the middle of the cell. When seen under the microscope they look like snowflakes (Fig. 8–3).

The green algae exhibit an enormous variety of forms and reproductive processes. Asexual reproduction may occur by fragmentation of multicellular colonies. Many species of green algae form asexual spores, termed *mitospores* (because they are produced by mitosis), and zoospores if they are motile. Three variations in sexual reproduction among algae can be discerned. *Isogamy* refers to the union of identical, usually motile, gametes. The union of motile gametes, one of which is much larger than the other, is called *anisogamy,* and

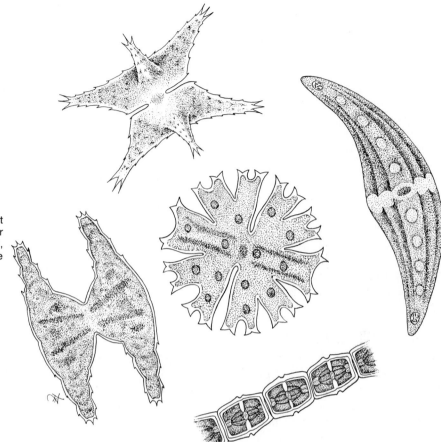

Figure 8–3 Several different species of desmids, unicellular green algae, highly magnified, showing the symmetry of the cells.

the union of a motile male gamete or sperm with a nonmotile egg is known as *oogamy.* Isogamous sexual reproduction usually occurs in the less advanced forms whereas anisogamy and oogamy are characteristic of the more highly evolved plants. In all of these algae, gametes are formed in special cells called *gametangia.* After their release from the gametangia, the gametes, if they are motile, swim about for a short time, then combine to form zygotes. The isogametes of some green algae are similar to the asexual zoospores although generally smaller and physiologically weaker.

Three major groups of green algae may be distinguished on the basis of their cellular organization (Fig. 8–4). One group includes those forms with single motile cells or colonies in which all of the vegetative cells are motile. A second group of algae includes those filamentous species in which the colony is composed of nonmotile cells. A third group contains algae with tubular multinucleate bodies.

Motile Colony Algae. The members of this group are motile and either unicellular like *Chlamydomonas* or colonial like *Volvox* and *Pandorina.* One of the simplest green

algae, and presumably a representative of a primitive type, is the motile, freshwater *Chlamydomonas,* found in pools, lakes and damp soil. The vegetative cell bears two whiplash flagella at its anterior end and is protected by a heavy cellulose wall. Each cell has a single cup-shaped chloroplast containing a pyrenoid involved in the production of starch. Other cellular structures are an eyespot containing a red pigment and two contractile vacuoles near the base of the flagella. When nutrients are abundant and conditions of light and temperature are optimal in quiet standing pools *Chlamydomonas* undergoes rapid asexual reproduction, discoloring the water and resulting in an algal "bloom." The cell divides to form two to eight zoospores within the cellulose wall; these are set free by the rupturing of the parental cell wall, and the zoospores swim away as independent individuals. When environmental conditions worsen, sexual reproduction occurs. The parent cell divides to form 8 to 32 smaller gametes which resemble the zoospores and vegetative cells but are much smaller. Two of these gametes fuse, beginning at the end bearing the flagella, to form a zygote

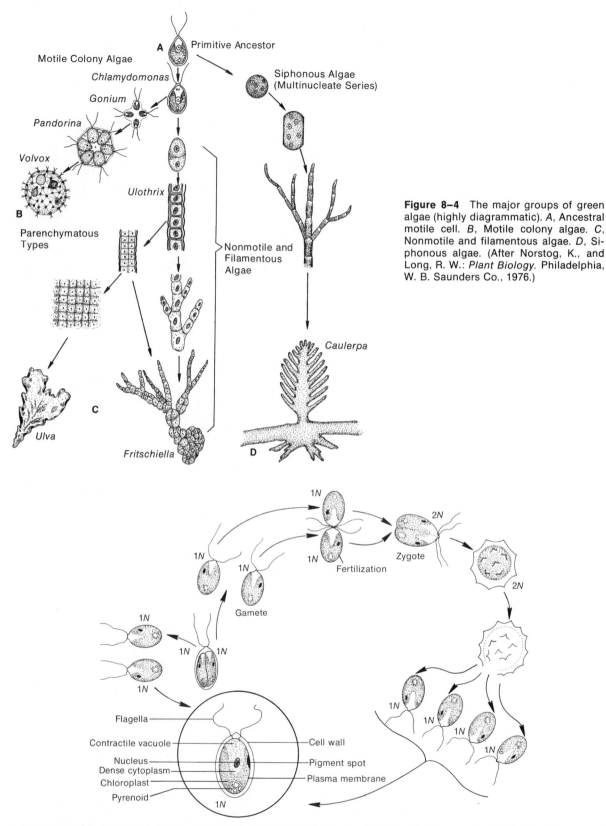

Figure 8–4 The major groups of green algae (highly diagrammatic). *A*, Ancestral motile cell. *B*, Motile colony algae. *C*, Nonmotile and filamentous algae. *D*, Siphonous algae. (After Norstog, K., and Long, R. W.: *Plant Biology*. Philadelphia, W. B. Saunders Co., 1976.)

Figure 8–5 The life cycle of the green alga *Chlamydomonas;* asexual reproduction on the left, stages in sexual reproduction on the right. *Inset:* Enlarged view of a single individual showing body structures. (After Norstog, K., and Long, R. W.: *Plant Biology*. Philadelphia, W. B. Saunders Co., 1976.)

(Fig. 8–5). The fusion of the two isogametes results in a zygote with four flagella, two contributed by each gamete, but in time these are lost. The zygote becomes a dormant *zygospore;* it rounds up and secretes a thick wall which enables it to survive long periods in an unfavorable environment. The isogametes are 1*N* cells and the zygote resulting from their fusion is a 2*N* cell. When favorable environmental conditions return, the zygospore becomes active and the zygospore nucleus divides by meiosis to form four 1*N* zoospores. Zoospores formed directly by meiosis may be termed *meiospores.* The zoospores are motile and swim out of the zygospore when it breaks open, becoming independent vegetative cells reproducing asexually until environmental conditions lead to the generation of a new sexual cycle. In some species of *Chlamydomonas* the isogametes produced by one cell can fuse with another isogamete produced by that same cell. This form of sexual reproduction is called *homothallism.* In other species the isogamete must fuse with a gamete formed by a cell of a different mating type. The two kinds of isogametes are designated by plus and minus signs to indicate that they are of different "mating types." Sexual reproduction of

this type is called *heterothallism.* Both homothallism and heterothallism occur in widely different kinds of algae and fungi.

Pandorina forms a motile colony of 4 to 32 cells arranged as a hollow sphere in a jellylike matrix. Each cell resembles one cell of *Chlamydomonas.* Any of the cells may divide internally and produce from 4 to 32 zoospores. When these are released suddenly they escape as a unit and form a miniature new colony. *Volvox* represents a further evolution of this form and is a colonial alga composed of a hollow ball of cells, each bearing two flagella and connected to its neighbors by fine cytoplasmic strands (Fig. 8–6). Each spherical colony may contain some 500 to 50,000 cells, most of which are alike and function only vegetatively. Small motile sperm cells which bear two flagella are produced only in special sperm-producing structures termed *antheridia.* A large single nonmotile egg is produced within a special egg-producing *oogonium.*

The motile sperm are released and swim to the egg. Their union results in a diploid zygote which forms a thick cell wall and can resist unfavorable conditions. Meiosis occurs during germination of the zygospore and the

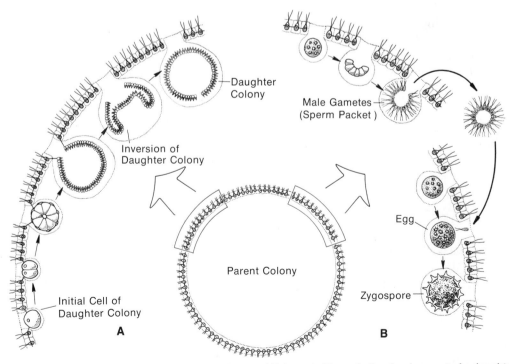

Figure 8–6 Diagrammatic representation of the life history of *Volvox. A,* Stages in the development of a daughter colony. *B,* Sexual reproduction. (From Norstog, K., and Long, R. W.: *Plant Biology.* Philadelphia, W. B. Saunders Co., 1976.)

haploid (1*N*) cells undergo repeated mitotic divisions to give rise to a new colony. In some species of *Volvox* a single colony may have both antheridia and oogonia. In other species a colony has one or the other, but not both, and can be said to be "male" or "female." In these forms sexual reproduction has evolved to a point where there is sex differentiation. *Volvox* may also reproduce asexually; one cell in the colony may enlarge, then divide mitotically to form a mass of cells. These cells form a hollow sphere with their flagella directed inwardly. The sphere, which is complete except for a small opening, then turns itself inside out and floats freely within the parent colony. Several such daughter colonies may occur within a single mature *Volvox* and become free-living when the parental colony breaks apart.

Nonmotile and Filamentous Algae. This group includes some nonmotile unicellular algae and filamentous colonial forms as well as complex, multicellular types that are among the most highly evolved of the green algae. One of the simplest examples of this series is *Chlorococcum*, a nonmotile, unicellular alga that lives in soils and fresh water. It reproduces asexually by the formation of flagellated zoospores which swim for a time but then lose their flagella and become nonmotile. Another nonmotile alga, *Tetraspora*, has nonmotile vegetative cells which adhere together in irregular masses. *Tetraspora* also produces flagellated zoospores and gametes that resemble those of *Chlamydomonas*. From ancestral algae resembling *Tetraspora* may have descended

the filamentous algae and the parenchymatous algae which have flattened, leaflike plant bodies. The fresh water alga, *Ulothrix*, is an example of a filamentous form. In this plant each haploid vegetative cell in the filamentous chain contains a single collar-shaped chloroplast and several pyrenoids. One cell may divide to form 4 to 8 zoospores, each bearing four flagella, which are released and subsequently give rise to a new filament (Fig. 8–7). One of the cells of the filament may undergo several divisions to produce many small gametes resembling zoospores, but with two instead of four flagella. As in *Chlamydomonas*, two of the swimming gametes fuse, forming a zygote which initially has four flagella. Thus in *Ulothrix*, sexual reproduction is isogamous and occurs by the fusion of two identical cells. However, these cells are specialized gametes differing from the usual vegetative cells. The zygote eventually loses its flagella, secretes a thick cell wall, and, like a zygospore, is capable of withstanding cold or drying. It later undergoes meiotic division and gives rise to four cells which are liberated from the zygote wall and develop into new filaments. After swimming, the *Ulothrix* zoospore settles down on a suitable substrate and forms a **holdfast cell** which anchors the plant and initiates growth of the new filament. *Ulothrix* is heterothallic and certain cells produce biflagellate isogametes that are released into the surrounding water where they fuse with an isogamete from another *Ulothrix* filament of a different mating type.

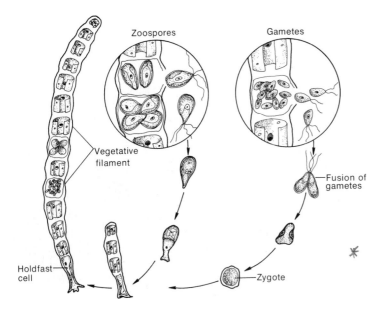

Zoospores

Gametes

Vegetative filament

Fusion of gametes

Holdfast cell

Zygote

Figure 8–7 *Left,* A filament of the green alga *Ulothrix. Center,* An enlarged view of asexual reproduction by zoospores. *Right,* Sexual reproduction by the formation of gametes and the fusion of two gametes to form a zygote.

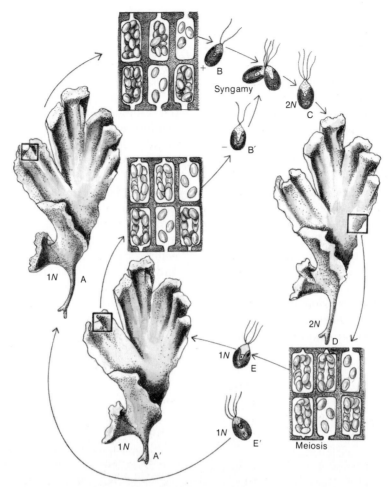

Figure 8–8 Life history of *Ulva* (shown diagrammatically). *A* and *A'*, Plus and minus gametophytes, respectively. *B* and *B'*, Plus and minus gametes. *C*, Zygote. *D*, Sporophyte plant. *E* and *E'*, Meiospores of the two mating types. (After Norstog, K., and Long, R. W.: *Plant Biology*. Philadelphia, W. B. Saunders Co., 1976.)

The parenchymatous green alga, *Ulva*, the sea lettuce, is a flat, membranous plant composed of two layers of cells. *Ulva* produces large numbers of isogametes which, after release, fuse with those from other plants to form zygotes (Fig. 8–8). The zygotes germinate to form 2N *Ulva* plants which are identical in appearance to the 1N *Ulva* plants producing isogametes. When the 2N *Ulva* plant matures it produces 1N zoospores by meiosis. These zoospores are released and settle upon a substrate,

become attached and grow into 1N *Ulva* plants which subsequently produce isogametes. In *Ulva* there is an alternation of generations comparable to that in some of the higher plants, but because it involves plants that look alike it may be called an **isomorphic alternation of generations.** It is categorized as Life Cycle III. One of the types of *Ulva* is a diploid (2N) sporophyte which by meiosis produces 1N zoospores that develop into 1N gametophyte plants. The latter produce 1N gametes that fuse to form a 2N zygote which develops into the 2N sporophyte plant.

The Siphonous Algae. The third group of green algae is characterized by multinucleate cells. Each cell contains several or many nuclei rather than the single nucleus characteristic of other green algae. A multinucleate plant is termed a *coenocyte;* examples of coenocytic organisms are found among both algae and fungi, although none of the higher green plants are coenocytic. Some of the siphonous or multinucleate marine green algae such as *Valonia* are easily seen with the unaided eye even though they are unicellular. *Acetabularia,* the "mermaid's wineglass," (Fig. 8–9) has been used in research on nuclear and cytoplasmic relationships (p. 49). It has but one large nucleus near its base during most of its life cycle but becomes multinucleate later on. Other coenocytic green algae include *Codium,* a marine form which often becomes attached to shellfish and

can be very destructive to oysters, mussels and scallops. *Caulerpa* has leaflike blades, stemlike horizontal parts and rootlike holdfasts (Fig. 8–10). Some of these individual plants may be a meter or more in length but are morphologically a single cell with no cell barriers between the nuclei. These forms are commonly found in shallow marine waters; some secrete particles of calcium carbonate and contribute to the formation of limestone deposits and calcareous sands.

The pond scum, *Spirogyra,* classified in the order Zygnematales, consists of long filaments of haploid nonmotile cells arranged end to end. In the fall when reproduction usually occurs, two filaments come to lie side by side and dome-shaped protuberances appear on the cells lying opposite (Fig. 8–11). These enlarge, fuse and form a *conjugation tube* connecting the two cells. One cell rounds up, oozes through the tube and joins the second. The

Figure 8–9 *Acetabularia crenulata,* the mermaid's wineglass. (From Norstog, K., and Long, R. W.: *Plant Biology.* Philadelphia, W. B. Saunders Co., 1976.)

Figure 8–10 Some siphonous, green, marine algae. *A, Codium. A',* Enlarged drawing of a *Codium* branch. *B, Caulerpa,* with its rhizomatous habit. *B',* Enlarged drawing of a portion of a branch of *Caulerpa.* Both algae have been compared to a single giant cell with many nuclei and continuous cytoplasm. (From Norstog, K., and Long, R. W.: *Plant Biology.* Philadelphia, W. B. Saunders Co., 1976.)

nuclei of the two cells unite and fertilization is complete. The resulting zygote develops a thick wall and is able to survive during the winter. In the spring the zygospore divides meiotically to form four haploid nuclei, three of which degenerate. The fourth remains, and after the thick wall breaks, it divides mitotically to form a new haploid filament. Sexual reproduction in *Spirogyra* involves unspecialized cells: that is, any cell in the filament can fuse with one from the neighboring filament; the two fusing cells are similar (isogamy). *Spirogyra* is characterized by its spiral chloroplasts (Fig. 8–11) and is interesting in that it has dispensed with flagellated cells.

Another filamentous fresh water alga, *Oedogonium,* forms multiflagellated sperm and zoospores; sexual reproduction is oogamous. Any vegetative cell of *Oedogonium* can differentiate into either an egg-forming cell, an

oögonium, or a sperm-forming cell, an *antheridium.* The enlarged, spherical, egg-forming cell shrinks away from the hard cell wall to form a rounded, nonmotile food-laden egg. The sperm-forming cells are produced when a vegetative cell divides several times to produce a series of short disc-shaped cells. Each of these divides to produce two small sperm, each bearing a circle of flagella at the anterior end (Fig. 8–12). Both egg and sperm are $1N$ and their fusion results in a $2N$ zygote. The sperm swims to the egg, apparently attracted by some chemical substance given off by the egg. It enters the egg-containing cell through a crack and fuses with the egg. The zygote secretes a thick cell wall and can survive unfavorable environmental conditions. Eventually the zygote undergoes meiosis to form four haploid cells, each having a circle of flagella at the anterior end and resembling the

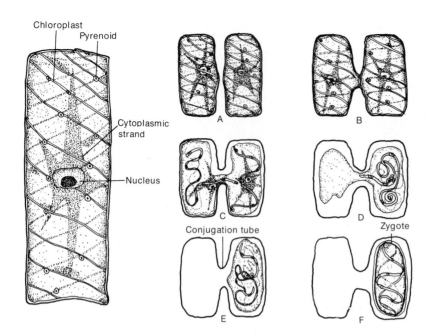

Chloroplast

Pyrenoid

Cytoplasmic strand

Nucleus

Conjugation tube

Zygote

Figure 8–11 *Left,* A single cell of a filament of the green alga *Spirogyra,* showing cell structures. *Right, A-F,* Stages in the sexual reproduction of *Spirogyra;* see text for discussion.

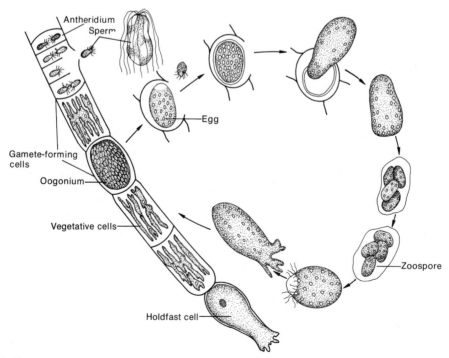

Antheridium
Sperm

Egg

Gamete-forming cells

Oogonium

Vegetative cells

Zoospore

Holdfast cell

Figure 8–12 A filament of the green alga *Oedogonium,* with sexual reproduction involving differentiated eggs and sperm.

asexual reproductive cells, the zoospores. The zoospore, whether formed sexually or asexually, can germinate, settle down on the substrate and differentiate into a holdfast cell. It then divides to form a new *Oedogonium* filament.

The most complicated of the green algae are the **stonewarts** found in fresh water ponds. The fossil record of this group goes all the way back to the Silurian period. One of the living forms, *Chara*, lives submerged at the bottom of clear, fresh water ponds. *Chara* is a comparatively large green alga with whorls of branches, both horizontal and vertical stems, and anchoring structures known as rhizoids. Like the higher plants, *Chara* has an apical growth pattern—that is, an apical cell divides mitotically to form the cells further down the branch that subsequently develop into several kinds of structural and reproductive cells. Most stonewarts have calcified walls and complex gametangia. These multicellular algae resemble miniature trees with structures that superficially look like, and serve the functions of, roots, stems, leaves and seeds, though they are not anatomically like their counterparts in higher plants. In many characteristics they are more advanced than the other green algae, more like higher plants, and some taxonomic systems place the 200 or so species of stoneworts in a separate division.

A number of evolutionary trends can be discerned among the green algae: the change from isogamy to anisogamy to oogamy; and the change from motile unicells to motile colonies or from motile unicells to nonmotile unicells to nonmotile colonies to multicellular filaments. Additional trends include the evolution of parenchymatous forms from multicellular filaments, a change from nonspecialized gametangia to specialized gametangia, and a change from the situation in which the *zygote* undergoes meiosis to one in which the zygote divides to become spores and then the *spores* undergo meiosis to form the gametes.

The green algae have an extremely wide range of ecologic habitats. If light and some moisture are present, green algae will probably be found growing in the vicinity. Species of *Chlamydomonas* can grow on the snow and ice of polar regions and in the alpine regions of high mountains. Many green algae grow in salt lakes as well as in fresh water ponds and the ocean. Species of green algae are also found in some rather unusual places. *Basicladia*, for example, grows on the backs of fresh water turtles while *Desmococcus* is found in the hair of the three-toed sloth, giving the animal a greenish hue. Species of *Chlorella* live within the bodies of the cnidarian *Hydra*, and *Chlorococcum* can be the algal component of lichens (p. 186), which are associations of algae and fungi.

8–4 DIVISION EUGLENOPHYTA

The euglenids are all unicellular flagellated organisms commonly found in fresh water, but also present in soil, brackish water and even salt water. The 800 species that have been described include some of the most primitive organisms known. They have been difficult to classify for they have a mixture of animal and plant characteristics (Fig. 8–13). Botanists call them algae and assign them to a separate division, the Euglenophyta. Zoologists classify them as an order within the class Phytomastigophora in the phylum Protozoa. They are assigned to the kingdom Protista in the Whittaker "Five Kingdom" system of classification. The euglenids are more advanced morphologically than the blue-green algae for they have a definite, easily stained nucleus and the chlorophyll is not scattered in granules but is localized in chloroplasts as in the higher plants. All the euglenids have two flagella (usually one long and one short) by means of which they swim actively. They lack the outer cellulose cell wall present in other algae; instead the protoplast is bounded by a grooved layer called the **pellicle.** The pellicle is quite unlike the plant cell wall, for it incorporates the cell's plasma membrane which is outside a system of spirally arranged rigid pellicle strips made of protein. Euglenids have a gullet near the base of the flagella and an eyespot containing the red pigment **astaxanthin,** found elsewhere only in Crustacea. It is this substance that gives a boiled lobster its red color. The euglenids store their carbohydrates as paramylum, which is chemically distinct from both starch and glycogen.

Reproduction is usually asexual by simple cell division, but sexual reproduction has been observed in at least one genus. Although most euglenids contain chlorophyll they apparently

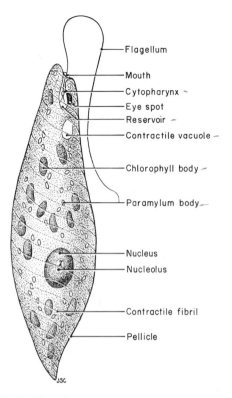

Flagellum
Mouth
Cytopharynx
Eye spot
Reservoir
Contractile vacuole
Chlorophyll body
Paramylum body
Nucleus
Nucleolus
Contractile fibril
Pellicle

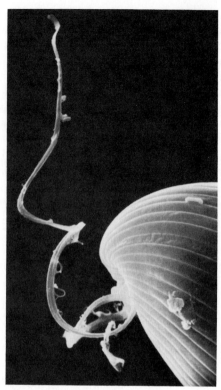

Figure 8–13 Flagellates. *Left,* A diagram of a common simple flagellate, *Euglena. Right,* A scanning electron micrograph of the oral end showing the flagella. (Electron micrograph courtesy of Dr. Eugene Small.)

cannot survive solely by photosynthesis. They will not survive in a medium containing only inorganic salt but will flourish if small amounts of amino acids are added. Some of the 800 species in the division are completely autotrophic, others are saprobic, and some are holozoic, capturing and ingesting other organisms like typical animals. The euglenids with their distinctive mixture of plant and animal characteristics give us an idea of what early living things might have been like when the early autotrophs had evolved from heterotrophs but before plants and animals had evolved separately. Whether they should be called animals, plants or protists is a matter of definition.

8–5 DIVISION CHRYSOPHYTA: GOLDEN-BROWN ALGAE AND DIATOMS

The division Chrysophyta includes a variety of diverse types usually arranged in three classes. The first of these, Bacillariophyceae, contains the **diatoms,** which are microscopic, usually single-celled forms found in fresh and salt water and comprising an important food source for animals. The siliceous cell walls of diatoms are constructed in two overlapping halves and fit together like the two parts of a pillbox. These siliceous walls are ornamented with fine ridges, lines and pores that are characteristic for each species. The markings are either radially symmetrical or bilaterally symmetrical on either side of the long axis of the cell (Fig. 8–14). Many of these markings are at the limit of resolution of the best light microscopes and are used as test objects to determine the quality of the lens. Diatoms are capable of a slow, gliding movement, apparently produced by the streaming of cytoplasm through the grooves on the surface of the cell wall. An alternative hypothesis states that the diatom secretes an adhesive material from the fine groove (raphe) on its lower shell and that this swells and pushes the diatom along. Diatoms store their food as the polysaccharide leucosin and as oil rather than as starch. It is widely believed that petroleum is derived from the oil of diatoms that lived in past geologic ages. Some diatoms are especially sensitive to certain pollutants in water and are widely used as pollution indicators.

The remains of the silica-containing cell walls accumulate as sediments in the oceans. Later geologic uplifts may bring these to the surface and the diatomaceous earth is mined and used in making insulating bricks, as a filtering agent, and as a fine abrasive (several kinds of toothpaste contain diatomaceous earth). Some deposits of diatoms in California are more than 300 meters thick.

Diatoms resemble brown algae in possessing the brown pigment *fucoxanthin.* They are extremely important photosynthesizers: probably three quarters of all the organic material synthesized in the world is produced by diatoms and dinoflagellates.

Diatoms, which are found in truly mind-boggling numbers in certain river waters and marine environments, reproduce sexually or asexually. The presence of the hard silica-containing wall complicates the process of asexual reproduction. The diatom cell divides and forms two cells within the original cell wall, and then two new cell walls form, back to back, between the two cells. Thus each daughter cell ends up with two cell walls, one inherited from the parent and a new one that fits inside the old one. Each successive generation, of course, gets a little smaller because each new cell wall fits inside the old one. Finally, a special cell is formed which discards the old cell walls, en-

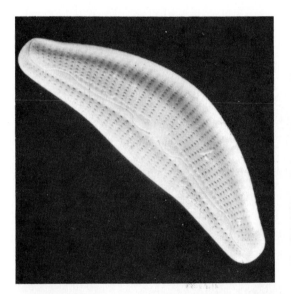

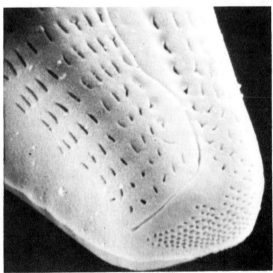

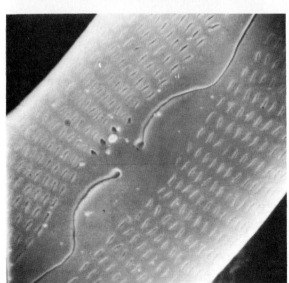

Figure 8–14 Scanning electron micrographs of a pennate diatom. Note the raphe and elaborate series of wall perforations. (From Hufford, T. L., and Collins, G. B.: Some Morphological Variations in the Diatom *Cymbella cistula.* J. Phycol. *8:*192–195, 1972.)

larges and then forms a pair of new, large walls.

The other two classes of Chrysophyta are the yellow-green algae (Xanthophyceae) and the golden-brown algae (Chrysophyceae). Both of these have silica-impregnated, two-shelled cell walls and chloroplasts rich in carotenes and xanthophylls. These pigments give them their characteristic yellow or brown color. Some members of each class have in the course of evolution lost their chlorophylls and other pigments and have become colorless flagellate or ameboid forms that are heterotrophic rather than autotrophic. Some taxonomists consider the yellow-green algae as a separate division, the Xanthophyta, rather than as a class (Xanthophyceae) of the Chrysophyta.

8–6 DIVISION *PYRROPHYTA:* DINOFLAGELLATES AND CRYPTOMONADS

Dinoflagellates are single-celled algae, most of which are surrounded by a shell made of thick, interlocking plates. They are all motile, having two flagella, one projecting from one end and the other running in a transverse groove (Fig. 8–15). Like diatoms, they have fucoxanthin in addition to chlorophyll and food reserves stored as oils as well as polysaccharides. Most dinoflagellates are marine and are important photosynthesizers in the ocean. Occasionally vast numbers of them accumulate in some part of the sea, coloring the water red or brown. Some species of dinoflagellates are

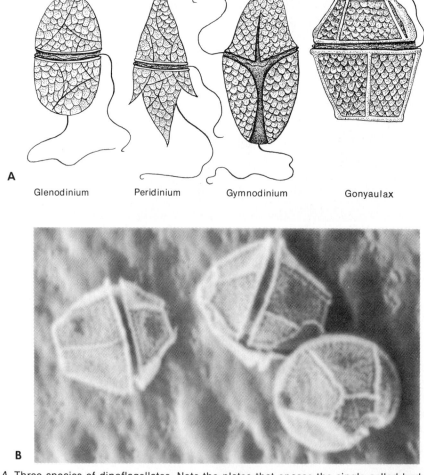

A

Glenodinium   Peridinium   Gymnodinium   Gonyaulax

B

Figure 8–15 *A,* Three species of dinoflagellates. Note the plates that encase the single-celled body and the characteristic two flagella, one of which is located in a transverse groove. *B,* Scanning electron micrograph of *Gonyaulax,* a pyrrophyte. (From Anikouchine, W., and Sternberg, R.: *The World Ocean: An Introduction to Oceanography.* © 1973 by Prentice-Hall, Inc. Published by Prentice-Hall, Inc., Englewood Cliffs, N.J.)

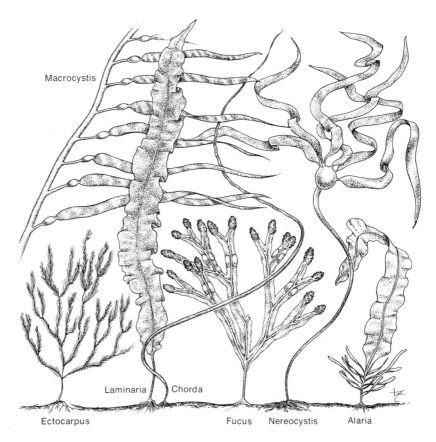

Figure 8–16 Some of the kinds of brown algae or kelps, all of which are multicellular marine plants. The sketches are not drawn to the same scale.

Macrocystis

Laminaria   Chorda

Ectocarpus   Fucus   Nereocystis   Alaria

poisonous to vertebrates and when these accumulate, large numbers of fish in that region are killed. Other dinoflagellates are taken up as food by mussels. The mussels are apparently unharmed by the dinoflagellates, but if humans eat some of these infected mussels they may become seriously ill. Many species of dinoflagellates exhibit bioluminescence, emitting a blue or green light.

Many of the pyrrophytes have nuclei in which the individual chromosomes are visible always rather than just during mitosis or meiosis. The nuclear membrane remains intact during cell division.

The *cryptomonads* (class Cryptophyceae) are a small group including about 100 species of unicellular biflagellate aquatic organisms with bilaterally compressed walls. They have a gullet and yellow-green or brown chloroplasts.

8-7 DIVISION *PHAEOPHYTA:* THE BROWN ALGAE

The brown algae include about 1500 species of multicellular forms ranging in size up to giant kelps, whose bodies may be one hundred meters long. They are the prominent brownish-green seaweeds that usually cover the rocks in the tidal zone and extend out into water 15 or so meters deep. The considerable amount of the golden-brown pigment *fucoxanthin* in these plants tends to mask the chlorophyll present; the color of the plants ranges from light golden to dark brown or black. Some brown algae are large, highly advanced plants with complex body structures and parts that resemble the leaves, stems and roots of higher plants. In the algae, these parts are called *blade, stipe* and *holdfast,* respectively, to indicate that they are not homologous to the corresponding structures of higher plants. Brown algae are found in shallow waters along the coasts of all seas but are larger and more numerous in cool waters. They are both the largest and the most rugged of the algae. They are attached by their holdfasts to the rocks beneath the surface and usually have air bladders to buoy up the free ends.

The plant body, or thallus, of a brown alga may be a simple filament, such as the soft brown tufts of *Ectocarpus,* commonly found on pilings, or tough, ropelike, slimy strands, such as *Chorda,* the "Devil's shoelace," or thick, flattened, branching forms such as *Fucus, Sargassum* or *Nereocystis* (Fig. 8–16). Most phaeophytes have a well-defined alternation of generations. *Ectocarpus,* for example, consists of two kinds of plants that are similar in size and structure but one produces gametes and the other produces spores. The diploid form produces haploid spores (called zoospores) that divide and grow into mature haploid plants. These produce haploid gametes that fuse to produce a diploid zygote. This develops into the diploid plant, completing the life cycle.

Ectocarpus, like the green algae *Ulva,* has a Life Cycle III with an isomorphic alternation of generations. In other brown algae, e.g., *Laminaria,* the larger diploid sporophyte generation is distinguishable from the smaller haploid gametophyte generation. In some, such as *Fucus,* the gametophyte generation is greatly reduced, as in the higher plants.

Brown algae furnish food and hiding places for many marine animals. Some kelps are used as food in oriental countries. Kelps such as *Laminaria* are processed commercially to yield a colloidal carbohydrate known as *algin,* a pectin-like component of the cell wall. This has the property of gelling and thickening mixtures and is widely used in making ice cream, for with it the ice cream manufacturer can use much less real cream and still have a smooth creamy product. Algin is also used in making candy, toothpaste and cream cosmetics.

8–8 DIVISION *RHODOPHYTA:* THE RED ALGAE

The red algae, like the Phaeophyta, are found almost entirely in the oceans. They are usually smaller and have more delicate bodies than the brown algae. A few species are unicellular but most are filamentous or flattened sheets. They are unique among eukaryotes in having the red pigment *phycoerythrin* in addition to chlorophyll (blue-green algae also have phycoerythrin) and are various shades of pink to purple. A second pigment, phycocyanin, provides the bluish component of the purple-colored ones. Red algae can grow at greater depths than other algae and are found as deep as 100 meters beneath the surface. As sunlight penetrates water, first the red, then the orange,

yellow and green rays are filtered out and only the blue and violet rays remain. Phycoerythrin is more effective than chlorophyll is absorbing light energy of these short wavelengths, and hence red algae can live at greater depths than other plants that lack phycoerythrin. Although red algae occur as far up as the low tide line, they reach their greatest development in the deeper tropical waters. Some 4000 species of rhodophytes are known.

Many rhodophytes have lacy, delicately branched bodies that are not as well adapted to survive in the intertidal zone as the tough, leathery brown algae, but they do well in the quieter deep waters (Fig. 8–17). The coralline red algae accumulate calcium from the sea water and deposit it in their bodies as calcium carbonate. *Coralline algae* are abundant in tropical waters and are even more important in the formation of coral atolls than are coral animals. Red algae have complex life cycles corresponding to the Life Cycle III, with a marked alternation of sexual and asexual generations and specialized sex organs.

Several kinds of red algae are used as food. *Porphyra* is considered a great delicacy by the Japanese and is widely cultivated in submarine gardens. Dulse (*Rhodymenia*) is boiled in milk and eaten by the Scottish. *Agar,* used in making culture media for bacteria, is extracted from the red algae *Gelidium* and *Gracilaria.* Agar is extensively used in baking and canning. *Carrageenin,* extracted from Irish moss (*Chondrus*), is used in brewing beer and in the preparation of chocolate milk to keep the chocolate from settling out.

8–9 THE KINGDOM FUNGI

The fungi are an extremely diverse group of eukaryotes that differ widely in their structural characteristics and in their modes of reproduction. They have little in common except the heterotrophic nutrition necessitated by the lack of chlorophyll. Their cells are enclosed in cell walls at least at some stage in the life cycle and they produce some type of spore, usually in large numbers, during the reproductive process. There are two groups of fungi: the *Myxomycota,* or slime molds; and the *Eumycota,* or true fungi. These may be considered as separate divisions in the plant kingdom

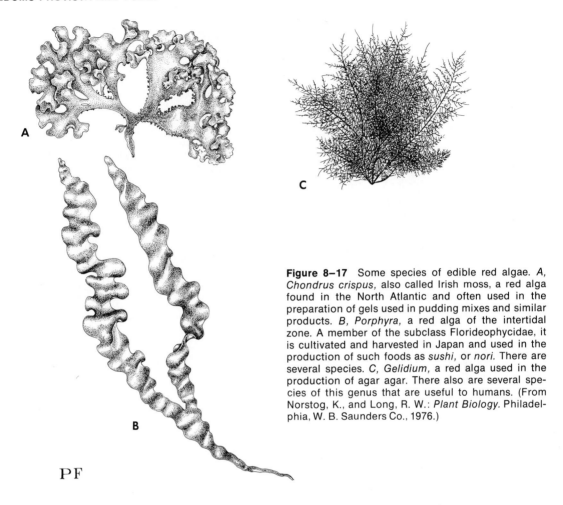

A

B

C

PF

Figure 8–17 Some species of edible red algae. *A, Chondrus crispus,* also called Irish moss, a red alga found in the North Atlantic and often used in the preparation of gels used in pudding mixes and similar products. *B, Porphyra,* a red alga of the intertidal zone. A member of the subclass Florideophycidae, it is cultivated and harvested in Japan and used in the production of such foods as *sushi,* or *nori.* There are several species. *C, Gelidium,* a red alga used in the production of agar agar. There also are several species of this genus that are useful to humans. (From Norstog, K., and Long, R. W.: *Plant Biology.* Philadelphia, W. B. Saunders Co., 1976.)

or the two may be assigned to the separate *kingdom Fungi* as proposed by Whittaker. The Myxomycota include the cellular slime molds and the plasmodial slime molds which differ markedly in their structure and in their means of reproduction. The cellular slime molds (class Acrasiomycetes) are protozoanlike and resemble amoebas through most of their life stages. They have no flagellated cells at any point in their life cycle, and spores are produced not by cytoplasmic divisions as in other fungi but by the formation of walls around individual amoeboid cells.

Ordinarily these cells reproduce asexually by fission, but one of the more curious biological phenomena is the aggregation and fusing of individual, amoebalike slime molds to form a fruiting, spore-producing body (Fig. 8–18). The signal for the aggregation process is the release of cyclic 3′, 5′ adenosine monophosphate (cyclic AMP) which serves as an intermediate in the mechanism of action of several hor-

mones (p. 578) and in the regulation of the cell cycle in higher organisms. If a spore released from the fruiting body falls on some moist surface it will absorb water, split out of its wall and divide to form amoeboid cells. Finally, some of these amoeboid cells serve as gametes, fusing to form a zygote which then divides and grows to become the multinucleate slimy mass, the plasmodium, thereby completing the life cycle.

The plasmodial slime molds (class Myxomycetes) are peculiar organisms found as slimy masses on decaying leaves or lumber; they move by sending out pseudopodia as do amoebas. At successive stages in their life cycle they are single-celled flagellates, single-celled amoebas, multinucleate masses of cytoplasm (the colonial mass is called a *plasmodium*), and finally plantlike bodies with a stalk and fruiting body. The plasmodium has no definite shape, may be brightly colored, and creeps along slowly by the flowing movement

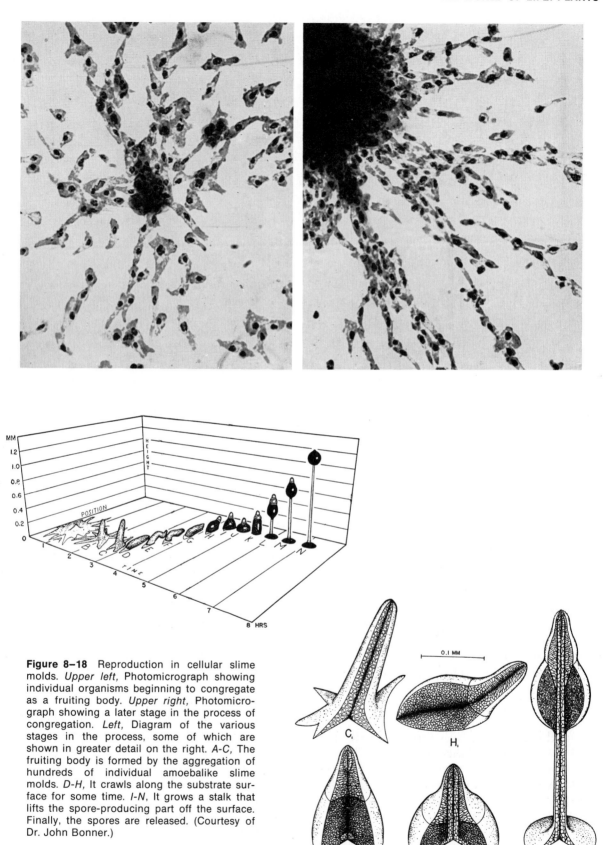

Figure 8–18 Reproduction in cellular slime molds. *Upper left,* Photomicrograph showing individual organisms beginning to congregate as a fruiting body. *Upper right,* Photomicrograph showing a later stage in the process of congregation. *Left,* Diagram of the various stages in the process, some of which are shown in greater detail on the right. *A-C,* The fruiting body is formed by the aggregation of hundreds of individual amoebalike slime molds. *D-H,* It crawls along the substrate surface for some time. *I-N,* It grows a stalk that lifts the spore-producing part off the surface. Finally, the spores are released. (Courtesy of Dr. John Bonner.)

of its cytoplasm, feeding on small organic particles and microorganisms in its path. Plasmodial movements are accompanied by rhythmic and reversible streaming of the cytoplasm. The plasmodium grows by the synthesis of its cytoplasmic components, accompanied by mitotic divisions of the nuclei. Some 450 species of myxomycetes are known and may be distinguished by the size, color and texture of the plasmodium and by the kinds of sporangia and spores that are produced.

In other kinds of slime molds the spores germinate and produce one to four naked motile cells, each bearing two flagella. These swarm cells may divide mitotically or may become amoeboid, but they eventually fuse in pairs to form the zygote. The spores are produced in the fruiting body by a meiotic division; thus the spores and swarm cells are haploid, but the plasmodium has diploid nuclei.

A few slime molds are plant parasites; one causes the disease known as clubroot in cabbages. The slime molds have many characteristics in common with amoebas and flagellates and their evolutionary origin in unclear. They may have evolved from flagellates; however, the formation of a fruiting body is characteristic of other fungi, and for this reason they are usually classified with the fungi.

8–10 DIVISION EUMYCOTA: TRUE FUNGI

The 80,000 or more species of true fungi have a number of characteristics in common with algae and are believed to have arisen from one or more of the algal divisions. The true fungi include the yeasts, molds, mildews, rusts, smuts and mushrooms. A few of the true fungi are unicellular, but most have many-celled bodies made of tubular branching filaments called *hyphae.* The outer wall of the body of the fungus may be composed of cellulose, of chitin, or of a combination of the two. Lignin and the carbohydrate callose may also be present. In some species the hyphae are subdivided by cross walls between successive nuclei and the organism is multicellular; in other species there are no cross walls between adjacent nuclei and the fungus is multinucleate.

One of the distinguishing characteristics of these eumycota is the presence of a *mycelium,* a mass of branching filamentous hyphae. In the common bread mold this mycelium is visible as a cobwebby mass of fibers on the surface of the bread penetrating into its interior. In a fungus such as a mushroom, much of the mycelium is below ground. The mushroom cap that we eat is a fruiting body, a specialized reproductive structure that grows out from the underground, nutritive mycelium.

Fungi grow best in dark, moist habitats; they are either saprobic or parasitic and are found universally wherever organic material is available. Some fungi can grow under apparently very unfavorable conditions. They have a strong resistance to plasmolysis for they can grow in concentrated salt solutions or sugar solutions such as on jelly. As the mycelium branches and comes in contact with organic material it secretes enzymes which hydrolyze proteins, carbohydrates and fats and then absorbs the split products. Fungi are important members of the carbon, nitrogen and other cycles, breaking up the complex organic compounds present in dead leaves and logs and making them once again available for use in the cycle. However, many of these fungi cause serious diseases of man, of his domestic animals, and of his crop plants. They are responsible in large measure for the deterioration of wood, leather, cloth and other material.

Fungal reproduction occurs in a variety of ways—asexually by fission, by budding or by spores; and sexually by means that are characteristic for each of the subgroups. The spores of aquatic fungi typically have flagella whereas the spores of terrestrial fungi are nonmotile cells dispersed by the wind or by animals.

The classes of Eumycota (*Chytridiomycetes, Oomycetes, Zygomycetes, Ascomycetes,* and *Basidiomycetes*) are distinguished on the basis of their means of sexual reproduction. The members of the first three classes have hyphae that are tubular and nonseptate, resulting in a coenocytic condition in which the cytoplasm is continuous within the walls of the hyphae and the organism is multinucleate. The ascomycetes and the basidiomycetes are not coenocytic but have septate hyphae.

Two other classes have been established in the Mycota—the *Deuteromycetes,* or imperfect fungi, and the *Lichens.* These classes are artificial taxonomic groupings of organisms that are incompletely known or are symbiotic associations (lichens) of two different organisms. The Fungi Imperfecti includes a very heterogeneous group of organisms whose status is not completely understood and in which sexual reproduction is unknown. Some fungi may actually have no sexual phase, but others probably have one that has not yet been discovered. When the means of sexual reproduction has been ascertained the fungus can be removed from the Fungi Imperfecti and assigned to the appropriate class. This does not simply represent indecision on the part of the mycologists (specialists in the study of fungi), but is a result of commendable scientific caution in assigning a form to one of the clear-cut classes of fungi only when sufficient evidence is available to justify that classification.

8–11 THE ALGAL FUNGI

The first three classes of Eumycota have in the past been grouped together in a single class, the Phycomycetes, a name that may be translated as the "algal fungi." Many of the 1100 or so species in these three classes resemble some of the simpler filamentous algae in their cytoplasmic organization, in the structure of their sex organs, spores and gametes, and in their aquatic mode of life. These similarities to the algae and to each other may be the result of convergent evolution rather than of close evolutionary relationship.

Chytridiomycetes. Members of the class Chytridiomycetes include protozoanlike forms as well as mycelial types, all of which reproduce by means of uniflagellate zoospores and gametes. The mycelial types have a true alternation of generations and a Life Cycle III. *Allomyces*, a microscopic fungus isolated from soil from various parts of the world, may be grown in culture on boiled hempseed. The fungi develop as a downy growth of hyphal filaments over the hempseeds. The hyphae are coenocytic and branching and develop male and female gametangia. The male gametangia release flagellated male gametes which are attracted by a substance termed *sirenin* released by the larger flagellated female gametes. The two gametes fuse (anisogamy) and the resulting zygotes develop into branching coenocytic organisms resembling those that produced the gametes. These organisms develop mitosporangia and produce motile mitospores (zoo-

spores). These give rise to more sporophyte plants, but as the culture ages, the sporangia begin to develop thick, brownish walls which will resist desiccation. When the resistant sporangia are placed in a moist environment they undergo meiosis and form flagellated meiospores which produce the next gametophyte generation.

Oomycetes. The oomycetes are mycelial forms, many of which are aquatic and algalike. These fungi have oogamous sexual reproduction and flagellated spores. Some of these fungi, the *water molds,* are saprobes and parasites of fish, insect larvae and seeds and attack plant and animal debris. The fish mold, *Saprolegnia,* may parasitize fish and is a common pest in overcrowded aquaria. It appears as an infection in which cottony mycelia form over the fins and eventually over the entire bodies of aquarium fish. Other oomycetes include a group of terrestrial pathogens that cause diseases such as the damping off of seedlings (*Pythium*), the downy mildew of grapes (*Plasmopara*) and the potato blight, caused by the plant pathogen *Phytophthora infestans,* which brought about the tragic potato famine that devastated Ireland in the 1840's. The oomycetes form eggs and reproduce sexually by conjugation between small nonmotile male cells and much larger egg cells. Their spores are biflagellate and motile.

Zygomycetes. The zygomycetes are probably the best known of the algal fungi. One common form is the black bread mold *Rhizopus nigricans.* Bread becomes moldy when a mold spore falls upon it. The spore germinates and grows to form a tangled mass of threads, the mycelium. Some of the hyphae, termed *rhizoids,* penetrate the bread and obtain nutrients; others, termed *stolons,* grow horizontally with amazing speed. Eventually certain hyphae grow upward and develop a sporangium or spore sac at the tip. Clusters of black spherical spores develop within this sac and are released when the delicate spore sac ruptures.

Sexual reproduction occurs when the hyphae of two different plants come to lie side by side. The bread mold is heterothallic and sexual reproduction can occur only between a member of a plus strain and one of a minus strain. This is a sort of physiologic sex differentiation even though there is no morphologic

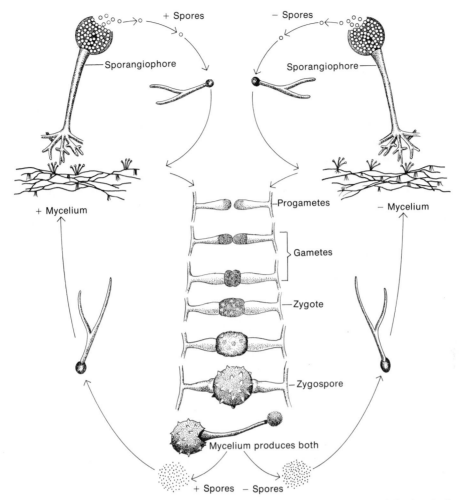

Figure 8-19 The life cycle of the black bread mold *Rhizopus nigricans*. The diagrams of the top indicate the asexual production of mycelia from spores. In the center is a series of stages in sexual reproduction. See text for discussion.

sex differentiation, and we could scarcely call the two strains "male" and "female." The tips of the swellings enlarge and pinch off to form gametes. The two adjacent gametes finally fuse to form a zygote (Fig. 8–19). Only the zygote is diploid; meiosis occurs in the germination of the zygote, and all the hyphae of the next generation are haploid.

8-12 CLASS ASCOMYCETES: THE SAC FUNGI

The members of this largest class of fungi (about 35,000 species) are called sac fungi because their spores are produced in sacs called asci. Each ascus produces two to eight ascospores. The ascomycetes include the yeasts, powdery mildews, the molds appearing on cheese, jelly and fruit, and the edible truffle. The ascomycete molds invading our food may give it an unpleasant taste, but they are not poisonous. Both the unique flavor of cheeses such as Roquefort and Camembert and the antibiotic penicillin are produced by the action of members of the genus *Penicillium*. These appear to be ascomycetes on the basis of their hyphal structure and their conidia but the sexual phase is not yet known and they are assigned to the Fungi Imperfecti.

The bodies of ascomycetes may be unicellular, as in yeasts; many-celled filamentous mycelia, as in the powdery mildews; or thickened and fleshy, as in the truffle. There is much diversity in the class, both in the form of the vegetative cells and in the mode of reproduction. Most ascomycetes are mycelial and have septate hyphae, but the septa characteristically have pores so that there is cytoplasmic continuity from one cell to the next, and nuclei have been observed to pass through the pores. Reproduction is accomplished asexually by budding in yeasts or by spores called *conidia* that develop in sequence at the tips of certain hyphae, and sexually by a fertilization process that occurs in two steps. In the first step, *plasmogamy,* there is a mingling of cytoplasm resulting from the transfer of one or more nuclei and some cytoplasm from a male antheridium into the female gametangium, called an *ascogonium.* The cells of the hyphae that form *before* plasmogamy are uninucleate and those that develop from the ascogonium *after* plasmogamy are binucleate. The second step in fertilization is the fusion of nuclei or *karyogamy,* which takes place in the terminal cells of the binucleate hyphae. The resulting ascus has a 2N nucleus, and meiosis, usually followed by one mitotic division, results in the formation of eight ascospores, each 1N in chromosome number, within the ascus.

Asexual reproduction among ascomycetes occurs by the production of conidia, spores that are budded off in sequence at the tips of certain hyphae rather than formed within sporangia. The conidia, sometimes called "summer spores," are a means of rapidly propagating new mycelia. These occur in various shapes, sizes and colors in different species, and it is the color of the conidia that gives the characteristic black, blue, green, pink or other tint to many of the molds. Although the structures in which asci are produced are often large and fleshy and may superficially resemble true fruits, they have, of course, no relation to them.

The ability of yeasts to produce ethyl alcohol and carbon dioxide from glucose in the absence of oxygen is of great economic importance. The yeasts used in winemaking are usually the wild yeasts normally present on the skins of the grapes. Some of the differences in

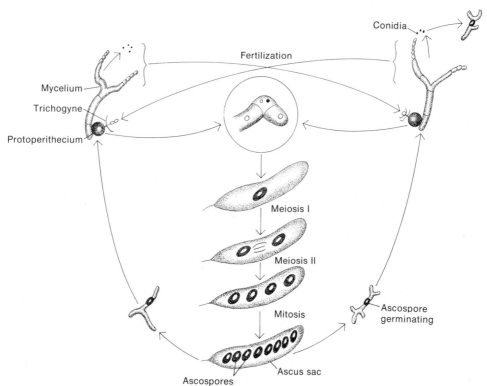

Figure 8-20 Life cycle of the red bread mold *Neurospora crassa.* Sexual reproduction will occur only between hyphae of opposite types. The resulting diploid cell divides by meiosis and then by mitosis to produce eight haploid ascospores. Each ascospore may germinate to form a minute mycelium. Haploid mycelia may also reproduce asexually by producing conidia, which germinate to form additional haploid mycelia.

the flavors of different kinds of wines are due to the kind of yeasts present in the grape-growing region. The yeasts used in baking and in brewing beer are cultivated yeasts carefully kept as pure strains to prevent contamination.

The ascomycete *Neurospora crassa*, a saprobe that occurs on pies and cakes, appearing as a cottony white fluff at first and turning pink as it develops asexual pink spores, has been an important research tool in genetics and biochemistry. As in the black bread mold *Rhizopus* there are two mating types indistinguishable in body form, and sexual reproduction will occur only between hyphae of opposite types (Fig. 8–20). The diploid cell that results from sexual reproduction divides by meiosis and then by mitosis to produce eight 1*N* ascospores within the ascus. Under favorable conditions each ascospore will germinate to produce a new mycelium. It is possible under a microscope to dissect out the individual ascospores and establish pure strains for use in genetic and biochemical research.

A number of ascomycetes are edible and highly prized. These include morels, which are cooked and eaten like mushrooms, and the truffle, an ascomycete that produces underground ascocarps.

Another ascomycete, *Claviceps purpurea*, produces the ergot disease of rye plants and other cereals that results in the poisoning of human beings and of livestock known as ergotism (St. Vitus' dance). Lysergic acid is one of the constituents of ergot and is an intermediate in the synthesis of lysergic acid diethylamide (LSD).

Some characteristics of the ascomycetes resemble those of the oomycetes. However, unlike the oomycetes, none of the ascomycetes produce flagellated cells. In a number of other respects, particularly in their means of reproduction, the ascomycetes resemble the red algae, and some mycologists are inclined to believe that ascomycetes evolved from red algae which became saprobic and lost their photosynthetic pigments.

8–13 CLASS BASIDIOMYCETES: THE CLUB FUNGI

The 25,000 or more species of basidiomycetes include mushrooms, toadstools, puff balls, rusts, smuts and bracket fungi (Fig. 8–21). They derive their name from the fact that they develop a *basidium*, a structure comparable in function to the ascus of ascomycetes. Each basidium is an enlarged, club-shaped, hyphal cell, at the tip of which develop four *basidiospores.* Note that basidiospores develop on the *outside* of the basidium whereas ascospores develop *within* the ascus. The basidiospores are released and develop into new mycelia when they come in contact with the proper environment. The vegetative body of the plant consists of a mycelium made of many-celled hyphae. No motile cells are formed at any stage of the life cycle of basidiomycetes.

The vegetative body of the cultivated mushroom, *Agaricus campestris*, consists of a mass of white, branching, threadlike hyphae that occur mostly below ground. The hyphae are septate but the septa are not perforate as are those of the ascomycetes. After a time, compact masses of hyphae, called "buttons," appear at intervals on the mycelium (Fig. 8–22). The button grows into the structure we ordinarily call a mushroom, consisting of a stalk and an umbrella. On the underside of the umbrella are many thin perpendicular plates called *gills,* extending radially from the stalk to the edge of the cap. The basidia develop on the surface of these gills (Fig. 8–23). Each basidium contains two nuclei which fuse to form a diploid nucleus. This in turn divides by meiosis to form haploid *basidiospores.* Each plant produces millions of basidiospores, each of which can, if it falls in the proper environment, give rise to a new mycelium. When, in the course of its growth, the hypha encounters another hypha of a different mating type they join by plasmogamy and produce binucleate cells, each with one nucleus of each mating type. These hyphae grow extensively and eventually develop basidiocarps, the structure we recognize as a "mushroom." These are composed of intertwined hyphae matted together. In certain regions of the basidiocarp, on the gills of the mushroom, the cells at the tips of the hyphae undergo karyogamy (nuclear fusion) followed by meiotic division to yield haploid basidiospores. Each basidium typically bears four basidiospores but the cultivated mushroom forms only two basidiospores per basidium.

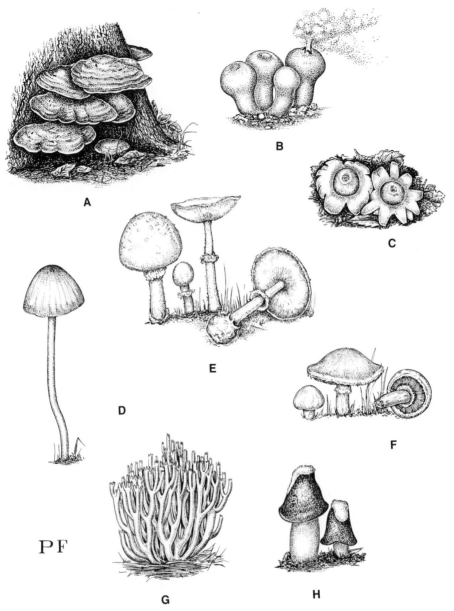

PF

Figure 8–21   Some types of basidiocarps. *A, Fomes,* a bracket fungus. *B, Calvatia,* a puffball. *C, Geaster,* an earth star. *D, Psilocybe,* a sacred mushroom. *E, Amanita verna,* the destroying angel, a poisonous species of gill mushroom. *F, Agaricus campestris,* the edible field mushroom. *G, Clavaria,* a coral mushroom. *H, Phallus,* a stinkhorn. (From Norstog, K., and Long, R. W.: *Plant Biology.* Philadelphia, W. B. Saunders Co., 1976.)

The word "mushroom" does not refer to any particular species of basidiomycetes but simply to the fruiting body of a number of forms. There are some two hundred kinds of edible mushrooms and about twenty-five poisonous ones (poisonous ones are sometimes called "toadstools"). The deadly "destroying angel," *Amanita verna* is exceedingly toxic, and the "sacred mushrooms" of the Aztecs, *Conocybe* and *Psilocybe,* contain dangerous hallucinogens. There is no simple test that distinguishes edible and poisonous mushrooms; they must be identified by an expert.

The evolutionary origin of the basidio-

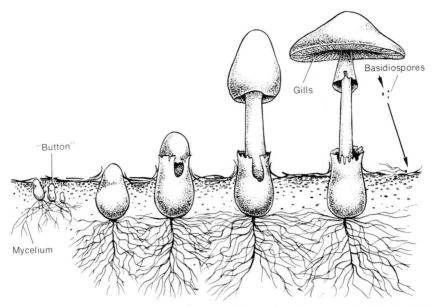

Figure 8–22 Stages in the development of a mushroom from the mycelium, the mass of white, branching threads found underground. A compact "button" appears and grows into the fruiting body, or mushroom. On the undersurface of the fruiting body are "gills," thin perpendicular plates extending radially from the stem (see Figure 8–23). Basidia develop on the surface of these gills and produce basidiospores, which are shed and, if they reach a suitable environment, give rise to new mycelia.

mycetes is shrouded in mystery. They show no relations with any of the algae and it is generally presumed that they are derived from other fungi, possibly from the ascomycetes.

8–14 LICHENS

Although lichens look like individual plants, they are in fact an intimate combination

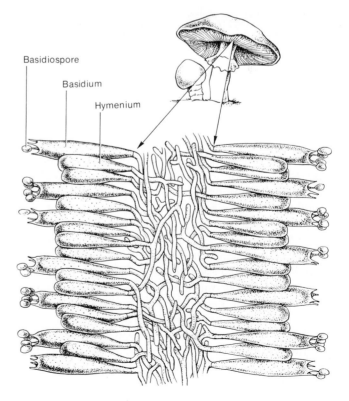

Figure 8–23 Section of a gill from the underside of a mushroom cap, magnified 500 times to show the basidia and their basidiospores.

Figure 8–24 Types of lichens. *Above,* Leafy type growing on the bark of a tree. *Lower left,* The lichen known as "reindeer moss." *Lower right,* An encrusting type growing on the surface of a rock. Flat or cup-shaped fruiting bodies can be seen on some of the plants. (From Weatherwax, P.: *Botany,* 3rd Ed. Philadelphia, W. B. Saunders Co., 1956.)

of an alga and a fungus (Fig. 8–24) and provide a classic example of mutualism (p. 139). The algal component is either a green or blue-green alga, and the fungus is usually an ascomycete; in some lichens from tropical regions the fungus partner is a basidiomycete. The fungal and algal components can be separated and grown separately in appropriate culture media. They can then be reassembled as a lichen but only if they are placed in a culture medium incapable of supporting either the fungus or the alga by itself. If the medium will support the growth of either one independently they will not undergo recombination. The combined fungus and alga can synthesize organic compounds that neither can synthesize alone.

Lichens are resistant to extremes of temperature and moisture and grow everywhere that life can be supported at all. They exist farther north than any other plants of the Arctic

region and are equally at home in the steaming equatorial jungle. The alga, by photosynthesis, produces food for both, while the tough gelatinous mycelium of the fungus protects the alga and provides it with moisture and mineral salts. Lichens play an important role in the formation of soil, for they gradually dissolve and disintegrate the rocks to which they cling. The fungi of some lichens produce colored pigments. One of these, *orchil,* is used to dye woolens, and another, *litmus,* is widely used in chemistry laboratories as an acid-base indicator. The "reindeer moss" of the Arctic is a lichen.

There are some 10,000 species of lichens, and they present a nasty problem in classification—should they be classified according to their algal component, their fungal component, or in some third way? Even though a lichen consists of an alga with one name and a fungus with another name, it is customarily given a third name and placed in a separate class of the division Eumycota.

8–15 ECONOMIC IMPORTANCE OF THE FUNGI

Only a few fungi are used as food by man and only a few are human parasites. The only fungi poisonous to man are the few poisonous mushrooms and the ascomycete *Claviceps*, which causes a disease of rye plants known as *ergot.* If a man eats bread made with flour from diseased plants, he suffers ergot poisoning, characterized by hallucinations, insanity and death. A derivative of ergot, lysergic acid, and the related compound LSD (lysergic acid diethylamide) produce wild hallucinations similar to those of schizophrenia and may

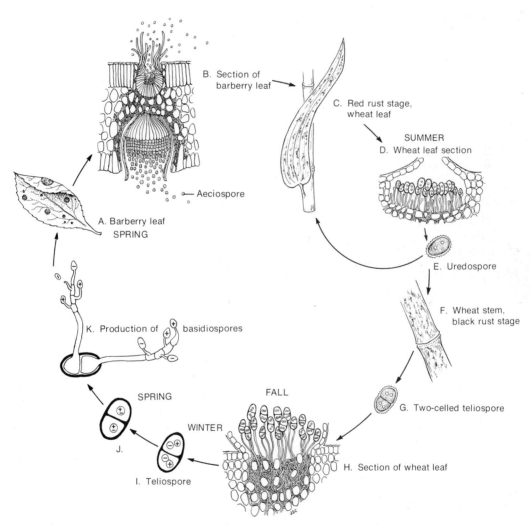

Figure 8–25 Life cycle of the wheat rust, *Puccinia graminis. A,* In the spring basidiospores from infected wheat plants of the previous year infect leaves of the barberry plant, forming pycnia containing clusters of spermagonia on the upper surfaces and cluster cups of aecia on the lower surface of the leaf *(B).* Aeciospores are produced in the aecia; they are binucleate, containing $N+N$ (not a single $2N$ nucleus) chromosomes. In early summer the aeciospores infect the leaves of young wheat plants *(C).* They develop into clusters of red, single-celled uredospores *(D),* producing the "red rust" stage. Uredospores are released *(E)* and infect other wheat plants, producing more uredospores. In late summer uredospores develop into dark-brown, two-celled teliospores *(F)* on the stems and leaf sheaths of wheat plants, forming the "black rust" stage *(G).* A section of the wheat stem *(H)* shows the $N+N$ teliospores, which are thick-walled and remain dormant over the winter *(I).* In the spring $N+N$ nuclei within each cell of the teliospore fuse to form a $2N$ nucleus *(J).* The teliospore, still attached to the wheat plant, germinates and undergoes meiosis, and each teliospore cell produces four basidiospores *(K).* The haploid basidiospores then infect a barberry leaf to complete the cycle.

cause chromosomal damage as well as mental and emotional disturbances.

Fungus-induced plant diseases cause tremendous economic losses. Phycomycetes produce "damping-off disease," which attacks young seedlings of corn, tobacco, peas, beans and even trees. Another phycomycete is the cause of the potato blight. A heavy attack of this in Ireland in 1845 destroyed almost the entire potato crop. The resulting famine led more than a million Irish to migrate to the United States.

Grapes are attacked by a downy mildew introduced into France from the United States. This almost destroyed the French vineyards before an effective fungicide, called Bordeaux mixture, was discovered. Some important plant diseases caused by ascomycetes are chestnut blight, Dutch elm disease, apple scab and brown rot, which attacks cherries, peaches, plums and apricots.

Basidiomycetes include some 700 species of smuts and 6000 species of rusts which attack the various cereals, corn, wheat, oats and so on. In general, each species of smut is restricted to a single host species. Some of these parasites, such as the stem rust of wheat and the white pine blister rust, have complicated life cycles, passed in two or more different plants, and involving the production of several kinds of spores. The white pine blister rust must infect a gooseberry or a red currant plant before it can infect another pine. The wheat rust must infect a barberry plant at one stage in its life cycle (Fig. 8–25).

Since this has been known, the eradication of barberry plants in wheat-growing regions has effectively reduced infection with wheat rust, but the eradication must be complete, for a single barberry bush can support enough wheat rust organisms to infect hundreds of acres of wheat. The spores produced by the wheat rust in the fall, ones with thick walls enabling them to survive a cold winter, will grow only if they land on a barberry. The usual, thin-walled spores produced during most of the summer can infect other wheat plants directly, and infection spreads from plant to plant in a wheat field in this way. If the winter is very mild, some of these thin-walled spores may survive and cause an infection the following year even in the absence of barberry plants. Thus, even the complete eradication of barberry plants does not provide a final solution to the wheat rust problem.

Bracket fungi (Fig. 8–21) cause enormous losses by bringing about the decay of wood, both in living trees and in stored lumber. The amount of timber destroyed each year by these basidiomycetes approaches in value that destroyed by forest fires.

Some of the fungi imperfecti cause important diseases of man. *Candida* causes a throat and mouth disease called "thrush" and also infects the mucous membranes of the lungs and genital organs. The Trichophytoneae infect the skin of man and other animals, causing ringworm, athlete's foot and barber's itch. Other members of the Fungi Imperfecti are parasites of higher plants and cause important diseases of fruit trees and crop plants.

8–16 PROTOZOA

The unicellular protozoa may be classified as a phylum of animals or as a phylum of the Protista. These are functionally complex organisms, even though some may appear to be relatively simple structurally. Typically protozoa have a single nucleus and lead an independent existence, but some have multinucleate cells, and in other species the cells are joined together to form a colony. A colonial protozoan can be distinguished from a multicellular animal, because its cells are quite similar and none is specialized for feeding. Most of the individuals in a population of protozoa are produced by simple cell division of the parent, although sexual reproduction by the mating of two individuals does occur. Protozoa are primarily aquatic and live in fresh or salt water, in small puddles or in the oceans. Some live in damp soils, crawling on the film of water that surrounds each dirt particle. Parasitic protozoa live in the body fluids of animals or in the saps of plants. Some species of protozoa can form inactive spores or cysts that can be dried and distributed with particles of dirt or dust from one habitat to another.

The 25,000 or so species of protozoa are divided into five classes, which differ in their means of locomotion as well as in other respects. The *Flagellata* have one or more long, whiplike flagella; the *Sarcodina* move by forming pseudopods; the *Ciliata* are characterized by the presence of many short hairlike cilia which beat in a coordinated fashion and move the animal along; young *Suctoria* have cilia but the mature animals have tentacles; and the *Sporozoa* are parasites lacking locomotor structures and reproducing by multiple fission.

Metazoan organisms are characterized by some division of labor between cells; certain cells are specialized to carry out nutritive, excretory, locomotor and other functions. In the protozoa these various activities are accomplished by specialized structures, termed *organelles.* Cilia and flagella are examples of locomotor organelles. The cilia beat with an oblique stroke so that the animal revolves as it swims. The coordination of the ciliary beating is good enough so that the animal not only can go forward but can back up and turn around. Coordination is achieved by a system of *neurofibrils* that connect the rows of *basal bodies* at the inner end of each cilium. If the neurofibrils are cut, the beating of the cilia is no longer coordinated. Near the surface of the cells of ciliates are many small *trichocysts*, organelles which can discharge filaments believed to aid the organism in trapping and holding its prey.

Flagellates move rapidly, pulling themselves forward by lashing one or more flagella located at the anterior end. Each *flagellum* is a long, supple filament containing an axial fiber, shown by electron microscopy to be composed of 11 filaments, a sheath of nine surrounding two in the center. The two central filaments give bilateral symmetry to the flagellum and influence its plane of motion. These 11 filaments have a chemical composition similar to that of actomyosin, the contractile protein of muscle (p. 477). Some protozoa can creep along a flat surface with wormlike movements that depend upon a layer of contractile fibers just beneath the surface of the cell which form an organelle comparable to the muscular body wall of a worm.

Although protozoa have no nervous system, they do have conductile organelles such as the basal body and system of neurofibrils just described. The *basal body* stimulates and controls the movement of the flagellum or cilium to which it is attached. The basal bodies may be joined to the centriole by filaments and they appear to be produced by the division and differentiation of centrioles during development. Many flagellates have photosensitive organelles associated with the conductile and locomotor organelles. The photosensitive organelle, or *eye spot*, of *Euglena* consists of a patch of red pigment and a tiny light-sensitive photoreceptor beside the base of the flagellum. The shading of the photoreceptor by the pigment spot enables the organism to determine the direction of the source of light. In other species the photoreceptor may be set in a pigment cup with the opening of the cup directed anteriorly. In a few species the cuticle covering the animal is swollen over the cup to form an optic lens.

A prominent structure in many protozoans is the *contractile vacuole,* a cavity that regularly fills with water from the surrounding cytoplasm and then empties the water into the environment. It is not an excretory organ but a pump to remove the excess water that is constantly entering the cell by osmosis. The cytoplasm of a fresh water protozoan has a higher concentration of dissolved materials—salts, sugars and organic acids—than the surrounding water. Water tends to pass into the cytoplasm by osmosis. Without a pump to remove the excess water, the amoeba or other protozoan would swell and burst, just as human blood cells do when they are placed in distilled water. In contrast, most marine protozoa do not have a contractile vacuole, since the concentration of salts in the sea water is about the same as that in their cytoplasm.

Flagellates. The largest class of protozoa, the flagellates, includes more than half of all living species of protozoa. Flagellates have spherical or elongate bodies, a single central nucleus, and one to many slender whiplike flagella at the anterior end which enable them to move (Fig. 8–26). Some flagellates engulf food by forming pseudopods; others resemble paramecia and have a definite mouth and gullet. Although most flagellates are tiny and difficult to study, a few, such as members of the genus *Euglena*, are large and can be used as representatives of the class. The flagellates with the largest number of flagella and the most specialized bodies are the ones living in the intestines of termites (Fig. 8–27). Some species of flagellates have characteristics in common with other classes of protozoa or with the sponges and suggest some of the intermediate steps by which these forms might have evolved from them.

The body of *Euglena* is covered with a del-

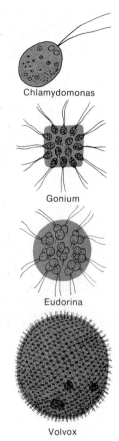

Figure 8–26 Sketches of four representative flagellates.

Chlamydomonas

Gonium

Eudorina

Volvox

viously (p. 165). The cells of the *Volvox* colony are connected to each other by bridges of cytoplasm through which the activities of the different cells can be synchronized.

The more strictly animal-like flagellates are small and rather uncommon. Of special interest, because of their resemblance to the sponges, are the *choanoflagellates.* These sedentary flagellates are attached to the bottom by a stalk and their single flagellum is surrounded by a delicate collar of cytoplasm. There are some parasitic flagellates of medical importance, such as the trypanosomes that produce sleeping sickness and are transmitted from one human to another by the tsetse fly. The flagellates are generally considered to be the basic stock from which evolved not only other kinds of protozoa but also higher plants and animals.

Sarcodinids. The members of the class *Sarcodina*, unlike other protozoa, have no definite body shape (Fig. 8–28). Their single cells are shapeless blobs that change form as they move. The nucleus, contractile vacuole and food vacuoles are shifted about within the cell as the animal moves.

An amoeba moves by pushing out temporary cytoplasmic projections called *pseudopods* (false feet) from the surface of the body (Fig. 8–29). More cytoplasm flows into the pseudopods, enlarging them until all the cytoplasm has entered and the animal as a whole has moved. During amoeboid motion, a stable gel layer at the surface surrounds the central core of liquid, flowing cytoplasm. As a pseudopod forms, the outer layer of cytoplasm becomes a liquid sol momentarily and then returns to the semisolid gel state along the sides of the forming pseudopodial lobe. As the liquid cytoplasm reaches the tip of the pseudopod it is pushed to the side and converted into a gel to form that portion of the wall. At the posterior end of the amoeba, the gel walls are converted to a liquid to be pushed along. The molecular basis of amoeboid motion and the chemical reactions supplying energy for it appear to be fundamentally similar to those of muscular contraction. Human white blood cells also move by amoeboid motion.

The pseudopods are used to capture food, two or more of them moving out to surround and engulf a bit of debris—another protozoan, or even a small metazoan (Fig. 8–30). The food

icate pellicle with spiral thickenings. Beneath the pellicle is a layer of contractile fibrils which permit the organism to change its shape. Scattered inside the cytoplasm are green chloroplasts and transparent, colorless *paramylum bodies* containing stored polysaccharides. Protruding from the gullet is a single flagellum formed by the fusion of two (Fig. 8–13).

A *Euglena* swimming with a single flagellum may be likened to a one-armed man trying to swim. At each stroke the flagellum bends toward the side bearing the pigment spot, not in a simple backward lash but obliquely toward the long axis of the organism. The body not only turns toward one side but rotates a bit. Successive lashes of the flagellum thus move the organism forward in a spiral path with the pigment spot facing the outside of the spiral.

The more plantlike flagellates, dinoflagellates and phytomonads, such as the colonial flagellate *Volvox*, were described pre-

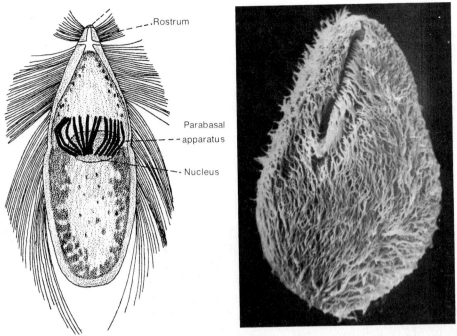

Figure 8–27   *Left,* A diagram of *Trichonympha agilis,* a symbiotic flagellate found in the gut of the termite. *Right,* A scanning electron micrograph of *Nyctotherus ovalis,* a holotrich from the hind gut of the cockroach *Blabarus discoidalis.* (Electron micrograph courtesy of Dr. Eugene Small.)

that has been engulfed is surrounded by a *food vacuole,* and acids and enzymes are secreted from the surrounding cytoplasm into the food vacuole to begin the process of digestion. The digested materials are absorbed from the food vacuole and the latter gradually shrinks as it becomes empty. Any indigestible remnants are expelled from the body and left behind as the amoeba moves along.

The parasitic members of the class *Sarcodina* include the species causing amoebic dysentery in man. Certain free-living species secrete shells around the body or cement sand grains together into a protective layer around their cells. The ocean contains untold trillions

of amoeboid protozoa, the *foraminifera,* which secrete chalky, many-chambered shells with pores through which the animal extends its pseudopods. The dead foraminifera sink to the bottom of the ocean and form a grey mud which is gradually transformed into chalk. Other amoeboid protozoa, the *radiolaria,* secrete elaborate and beautiful skeletons made of silica. These skeletons become mud on the ocean floor and eventually are compressed and converted into siliceous rock.

During the latter part of the Paleozoic era, a group of foraminiferans, the *Fusulinidae,* flourished. Over a relatively brief period (75,000,000 years) many species of fusulinids

Figure 8–28   Some examples of the class Sarcodina. *Difflugia* is a free-living amoeba that builds a protective layer around itself by cementing together grains of sand. *Actinophrys* is a member of the order Heliozoa that lives in fresh water, and *Globigerina* is a marine form, a member of the order Foraminifera.

Difflugia

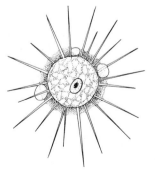

Actinophrys

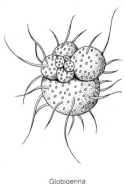

Globigerina

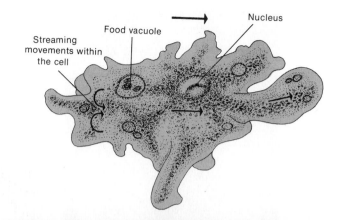

Streaming
movements within
the cell

Food vacuole

Nucleus

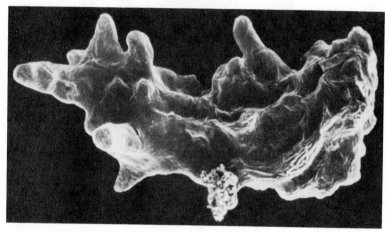

Figure 8–29 *Amoeba proteus. Top,* A diagram illustrating streaming movements. *Bottom,* A scanning electron micrograph of this sarcodinid. The micrograph gives a remarkably clear picture of the surface of the animal and its pseudopodia. (Electron micrograph courtesy of Dr. Eugene Small.)

developed and then became extinct. Some of these were large protozoa, as much as 2 centimeters in diameter, that lay on the bottom of the shallow seas. Their fossils are now found in deposits that have accumulated oil. As an oil well is drilled through the sedimentary rock, the bit passes in rapid succession through these successive species of fusulinids. By analyzing the species present in a specific portion of the core, the driller can estimate how far into the paleozoic deposit he has drilled.

Ciliates. Members of the class Ciliata, typified by *Paramecium,* have a definite, permanent shape due to the presence of a sturdy

flexible outer covering of chitin (Fig. 8–31). The surface of the cell is covered by several thousand fine cytoplasmic hairs, cilia, which extend through pores in the covering and move the animal along. Ciliates and Suctorians differ from other protozoa in having two nuclei per cell, a *micronucleus* which functions in sexual reproduction and a *macronucleus* which controls cell metabolism and growth. Both nuclei divide at each mitosis, but at sexual reproduction the macronucleus disintegrates and the micronucleus gives rise to both nuclei of the offspring. The macronucleus appears to be a compound structure formed by the amalgama-

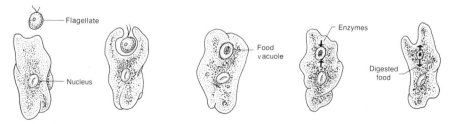

Flagellate

Nucleus

Food vacuole

Enzymes

Digested food

Figure 8–30 Diagram of an amoeba catching a small flagellate and digesting it in a food vacuole.

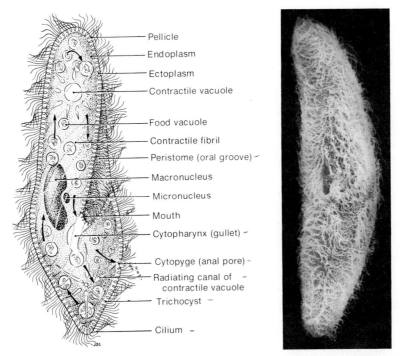

Figure 8–31 Paramecia. *Left,* A diagram of a typical ciliate, *Paramecium caudatum. Right,* A scanning electron micrograph of *Paramecium multimicronucleatum.* (Micrograph courtesy of Dr. Eugene Small.)

Labels (top to bottom): Pellicle — Endoplasm — Ectoplasm — Contractile vacuole — Food vacuole — Contractile fibril — Peristome (oral groove) — Macronucleus — Micronucleus — Mouth — Cytopharynx (gullet) — Cytopyge (anal pore) — Radiating canal of contractile vacuole — Trichocyst — Cilium

tion of many sets of chromosomes. It is simply pulled in half during asexual reproduction, without any mitotic phenomena.

Paramecia have two contractile vacuoles which together can remove a volume of water equal to the total volume of the animal's body within half an hour. In contrast a man excretes an amount of water equal to his body volume in about three weeks. Well-fed paramecia reproduce by division two or three times a day and are ideal subjects for studies of the laws governing population growth. Paramecia and other ciliates may have more than two, and indeed as many as eight, "sexes" or mating types. All the sexes look alike, but an individual of one sex will mate only with an individual of some other sex. The plantlike flagellates, the phytomonads, are haploid organisms, whereas ciliates, like animals, are diploid organisms. During sexual reproduction, two individuals of different sexes conjugate and press together by their oral surfaces. Within each individual the macronucleus disintegrates and the micronucleus undergoes meiosis to form four daughter nuclei. Three of these four degenerate, which is comparable to the degeneration of the polar bodies during oogenesis, leaving one viable haploid nucleus. This di-

vides once mitotically, and one of the two identical haploid nuclei remains within the cell. The other crosses through the oral region into the other individual and fuses with its haploid nucleus. Thus each conjugation yields two fertilizations, and the two new diploid nuclei are identical.

One of the interesting features of certain ciliates is the presence of traits that are transmitted to offspring through the *cytoplasm* rather than through nuclear chromosomes as is usually the case. Some strains of paramecia produce and secrete into the medium **killer particles** which will cause the death of a "sensitive" individual if they come in contact with it. All individuals that are unable to produce killer particles are sensitive, whereas all individuals that do produce these particles are resistant to their effect and do not get killed. These killer particles are manufactured in the cytoplasm from **kappa particles**, granules present in killer animals but not in sensitive ones. A killer paramecium may have in its cytoplasm some 800 kappa particles and will secrete a killer particle every five hours. The kappa particles are nucleoproteins which multiply in the cytoplasm independently of the division of the cell. At each division of the paramecium, the kappa particles are divided randomly between the daughter cells. As long as each daughter cell receives at least one kappa particle it will remain a killer. Under certain conditions the paramecium will divide more rapidly than the kappa particles do, and the number of particles per cell gradually decreases. Ultimately some cells will be formed that lack killer particles and they are then sensitive to them. Sensitive individuals, though unable to produce kappa particles or killer particles, occasionally mate with a killer before encountering a killer particle. During the mating process, kappa particles may be transferred into the sensitive cell. These will subsequently survive, divide and transform the sensitive cell into a killer. The trait will again be transmitted to the offspring as long as the reproduction of the particles keeps pace with the reproduction of the paramecium. Certain strains lack the ability to acquire the trait, because they lack a specific nuclear gene. Hence, even this kind of cytoplasmic inheritance is ultimately determined by nuclear genes.

Another ciliate of general biological interest is the genus *Tetrahymena*, which can be cultured in a chemically defined medium, i.e., one in which the exact amounts and kinds of all chemicals are known. By varying the chemicals present in the medium and by using appropriate radioactive or heavy atoms as labels, biologists have found what materials are needed for growth and maintenance and to what extent they may be converted into other materials, the nature of the enzymes that catalyze these transformations, and how all these processes are under genic control. Since the metabolic pathways in all organisms have many basic similarities, the study of *Tetrahymena* (in which information can be obtained rapidly and easily) is shedding light on similar problems in other organisms, even in the human.

Suctorians. Suctorians, a fourth class of protozoa, are closely related to ciliates and appear to have been derived from them in evolution. The suctorians, like the ciliates, have both a macronucleus and a micronucleus (Fig. 8–32). Young individuals have cilia and swim about, but the adults are sedentary and have stalks by which they are attached to the substrate. The body bears a group of delicate cytoplasmic tentacles, some of which are pointed to pierce

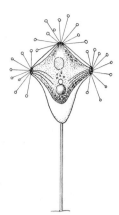

Figure 8–32   Diagram of a suctorian, *Acineta tuburosa*.

their prey, whereas others are tipped with rounded adhesive knobs to catch and hold the prey. The tentacles secrete a toxic material which may paralyze the prey. Although adult suctorians lack cilia they possess basal bodies. During asexual reproduction a bud forms on the suctorian, the basal bodies multiply, become arranged in rows, and develop cilia. After the nucleus has divided, the bud separates from the parent and swims away. When it becomes attached to the substrate, the cilia disappear and tentacles develop.

Sporozoa. The sporozoans comprise a large group of parasitic protozoa, among which are the agents causing serious diseases such as *coccidiosis* in poultry and *malaria* in man. Sporozoa have neither locomotor organelles nor contractile vacuoles. Most sporozoans live as intracellular parasites in the host cells during the growth phase of their life cycle and absorb nutrients through their cell wall. The sporozoan causing malaria, *Plasmodium*, enters the human blood stream when an infected mosquito bites a man (Fig. 8–33). The plasmodia enter the red cells and each divides into 12 to 24 spores, which are released when the red cell bursts. The released spores infect new red cells and the process is repeated. The simultaneous bursting of millions of red cells causes the malarial chill followed by fever as toxic substances are released and penetrate other organs of the body. If a second, uninfected mosquito bites the infected person it will suck up some plasmodian spores along with its drink of blood. A complicated process of sexual reproduction occurs within the mosquito's stomach and new spores are formed, some of which migrate into the mosquito's salivary

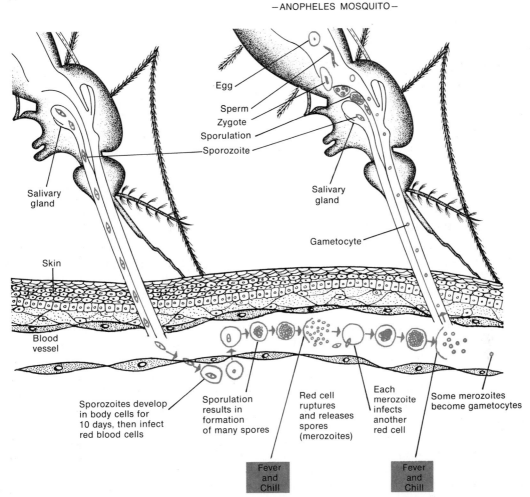

—ANOPHELES MOSQUITO—

Egg
Sperm
Zygote
Sporulation
Sporozoite

Salivary gland

Salivary gland

Gametocyte

Skin

Blood vessel

Sporozoites develop in body cells for 10 days, then infect red blood cells

Sporulation results in formation of many spores

Red cell ruptures and releases spores (merozoites)

Each merozoite infects another red cell

Some merozoites become gametocytes

Fever and Chill

Fever and Chill

Figure 8–33 Diagram of the life cycle of the sporozoan *Plasmodium,* which causes malaria in humans and other mammals. An infected mosquito *(left)* bites a person and injects sporozoites of *Plasmodium* into the bloodstream. These undergo sporulation and reproduce asexually within the red blood cells of the host. Periodically the infected red cells rupture and the new crop of merozoites released then infect other red cells. The bursting of the red cells releases toxic substances that cause the periodic fever and chill. Some merozoites develop into gametocytes, which can infect another mosquito when one bites the infected person. The gametocytes develop into eggs and sperm *(right)* within the mosquito and undergo sexual reproduction. The resulting zygote by sporulation produces a host of sporozoites that migrate to the salivary glands and are ready to be injected when the mosquito bites the next person.

glands to infect the next person bitten. This process of sexual reproduction does not occur within the human.

Relationships among the Protozoa. The flagellates are generally considered to be the basic stock of organisms from which other protozoa and indeed the higher animals and plants arose. The sarcodinids are related to the flagellates through several genera of amoeboid organisms that have flagella and through several forms that resemble typical flagellates when in open water but lose the flagella and creep like amoebas when next to a solid substrate.

The existence of so many intergrades suggests that sarcodinids may have evolved several different times from the flagellates.

The ciliates are a distinct group and probably arose from flagellates only once. Suctorians are clearly related to the ciliates. The evolutionary origin of the macronucleus is unknown. During the conjugation of most ciliates, a bit of cytoplasm is transferred along with the migrating nucleus and its bit of cyoplasm separates in the mouth cavity as a *gamete* with a long tail. The two gametes move past each other to the opposite organism.

The sporozoa may well be a composite group. Some species show affinities with flagellates while others more nearly resemble sarcodinids. The phenomenon of multiple fission may be regarded as an adaptation to parasitism. The number of evolutionary changes needed to develop one class of protozoa from another is not very great, and the possibility that in the course of evolution, the change from one group to another occurred repeatedly cannot be ruled out.

8–17 EVOLUTIONARY RELATIONSHIPS AMONG THE PROTISTS

There is an incredible variety of relatively simple organisms, some autotrophs and some heterotrophs, and neither their evolutionary relationships nor the most appropriate way of classifying them has been resolved to everyone's satisfaction. The several basic similarities in the structure and chemical composition of bacteria and blue-green algae suggest that they lie somewhat more closely together in the evolutionary scheme than do other forms. Some taxonomists classify the blue-greens and the bacteria as members of the same phylum, the Schizophyta, or as a separate kingdom, the Monera. It is widely held that bacteria evolved from the blue-green algae. They became adapted to a saprobic or parasitic existence and subsequently lost chlorophyll. The fact that many bacteria have flagella leads other taxonomists to conclude that these organisms descended from simple flagellated forms, perhaps ones that also gave rise to the green algae. Alternatively the present-day heterotrophic bacteria may have evolved from early autotrophic bacteria similar to the present-day iron and sulfur bacteria. As a further possibility, the present-day bacteria could be direct descendents of the original bacterialike heterotrophs that are generally believed to be the first types of living cells. It is even possible that different groups of bacteria have descended independently from different ancestors and that more than one of these explanations may be true. Perhaps it is more important at this time to realize the extent of this challenging problem in taxonomy rather than attempt to set up any rigid classification for these several disparate groups.

Although it may appear reasonable to put single-celled animal-like organisms and single-celled plantlike organisms together as the Protista, the green, brown and red algae all include some forms that though clearly related to the single-celled forms, are in fact relatively complex multicellular plants. The stoneworts, kelps and coralline algae are even more incongruous as members of the kingdom Protista than as members of the plant kingdom. The term protist may be used to refer in a general way to simple single-celled plants and animals without carrying the implication that it refers to a separate kingdom of organisms distinct from both plants and animals.

The evolutionary relationships of the slime molds and fungi to each other and to other living plants and animals are unclear, although ultimately they may have descended from some simple single-celled flagellated ancestor. The evolutionary relationships of the several classes of fungi have not been established. They may have evolved from one or another of the algae by taking up heterotrophic nutrition and losing chlorophyll. Alternatively they may have descended directly from primitive heterotrophs without ever having passed through an autotrophic stage. The zygomycetes are the most algalike of the fungi and may represent a more primitive descendant of some original algae, possibly the green algae. The characteristics of the ascomycetes are a curious mixture of those of the phycomycetes and the red algae. A number of mycologists believe that the ascomycetes evolved from red algae, which became saprophytic and lost their photosynthetic pigments.

Finally, the evolutionary origin of the basidiomycetes is truly shrouded in mystery, for they show no relations with any of the algae. It is generally presumed that they are derived from some other fungi, perhaps from the ascomycetes.

QUESTIONS

1. Bacteria and blue-green algae are termed prokaryotes and all other organisms, plant and animal, are termed eukaryotes. What do these terms mean?
2. What trends in the evolution of sexual reproduction are evident among the green algae? Have these trends continued in higher plants?

3. It is generally believed that higher plants evolved from the green algae. What are the reasons for this belief?

4. Of what importance to humans are desmids, diatoms, and dinoflagellates?

5. In what ways are red algae adapted to survive in deep water?

6. In what ways are slime molds like true fungi? In what ways do they resemble animals?

7. In what respects does a saprobic organism differ from a parasite?

8. What is a fruiting body? How does it differ from a true fruit?

9. In what ways are fungi economically important to human beings?

10. What measures can you suggest to prevent bread from becoming moldy?

11. How do lichens differ from other plants? How are they classified?

12. What measures should be taken to eliminate white pine blister rust?

13. What are the differences between an ascus and a basidium? Between a hypha and a mycelium?

14. What organisms are responsible for decay?

15. What are the chief characteristics of the protozoa?

16. Compare the methods of movement, nutrition and asexual reproduction in *Euglena, Paramecium* and the amoeba.

17. List the arguments for classifying *Euglena* as an animal, as a plant and as a protist.

SUPPLEMENTARY READING

E. Y. Dawson's *Marine Botany: An Introduction* provides concise coverage of phytoplankton, seaweeds and sea grasses, and marine fungi and bacteria. Norstog and Long's *Plant Biology* gives an excellent discussion of the algae and fungi and their classification. A biochemical approach to the problems of plant parasitism is presented in *Plant Pathology* by G. N. Agrios; the text includes many detailed diagrams of the cycles of disease-causing plants.

CHAPTER 9

THE INVASION OF LAND BY PLANTS

Most biologists believe that land plants and animals evolved from aquatic ancestors. The primitive algae and lower invertebrates living today are aquatic and the assumption that the primitive ancestral forms were also aquatic seems to be well founded. In tracing the evolutionary history of certain plants and animals we may find that, having once become adapted to terrestrial life, they may return to an aquatic habitat—even reemerge later on and once again become terrestrial. But when such evolutionary lines are traced back as far as possible, the primitive ancestral forms are all aquatic, and it is generally believed that life began in a watery environment.

Aquatic plants can survive without many of the specialized structures found in terrestrial plants. The surrounding water keeps them supplied with nutrients, prevents the cells from drying out, buoys up and supports the plant body so that special supporting structures are unnecessary, and serves as a convenient medium for both the meeting of gametes in sexual reproduction and the dispersal of asexual spores. In leaving the friendly water and taking up life on the barren land, the plants had to become adapted by developing new structures to take over the many functions previously served by the surrounding water.

The conquest of the land must have been a long and difficult process, perhaps fraught with many failures. The plants which finally triumphed and became truly terrestrial were able to survive because they evolved the **cuticle,** a waxy outer protective layer protecting the soft, watery tissues underneath; **leaves** extending into the air to absorb light and carry on photosynthesis; **roots** extending into the soil to provide anchorage and absorb water and salts; **stems** supporting the leaves in the sunlight and connecting them with the roots to provide a two-way connection for the transfer of nutrients in the xylem and phloem; and some means of reproduction, such as flowers, pollen and seeds, to permit the union of male and female gametes in the absence of a watery medium and to enable the zygote to begin development protected from desiccation.

9–1 THE FIRST LAND PLANTS

Just as the present-day amphibians of the vertebrate phylum—salamanders, newts and frogs—give us an idea of what the first land vertebrates may have looked like, the *bryophytes*—mosses, liverworts and hornworts—suggest the stages through which aquatic algae may have evolved to become fully terrestrial plants.

The algae have evolved body structures adapted to expose a maximum of surface for absorbing nutrients from the surrounding water. For survival on land, plants need a more compact body to decrease the water lost through the surface. Probably the first land plants lay flat, and exposed only one surface to the air. Plants were successful on land only after they had developed a specialized epidermal tissue, with thickened cell walls impregnated with a waxy, waterproof material. Most land plants, including bryophytes, have an epidermis; it may be slightly thickened and waxy and provided with pores to permit the diffusion of gases.

The bryophytes did not really solve the problem of reproduction in the absence of a watery medium; they evaded the difficulty by evolving reproductive structures that would have a watery medium for the union of gametes (p. 202). All land plants, including the Bryophyta, have evolved a life cycle in which the zygote is retained within the female sex organ. There it obtains food and water from the surrounding parental tissues and is protected from drying while it develops into a multicellular *embryo.* For this reason the bryophytes and the tracheophytes, or higher plants, are classified together in the subkingdom Embryophyta.

9–2 DIVISION BRYOPHYTA

The division Bryophyta is made up of about 25,000 species of *mosses, liverworts* and *hornworts.* The name "moss" is erroneously applied to a number of plants that are not bryophytes. The moss that grows on the bark of a tree may be an alga; *reindeer moss* is a lichen and the *Spanish moss* hanging from trees in the southern states is really a seed plant, a relative of the pineapple. Mosses in general form an in-

significant part of the vegetation even though they are widely distributed. Some of the 15,000 species can live only in damp places; others can survive in a dormant state in dry rocky places where enough moisture for growth is present only during a short part of the year. Liverworts, which are not as well protected against desiccation as mosses, are even more restricted and are found in the deep shade of forests or on the shady side of a cliff; some are true aquatic plants.

Mosses. True mosses (class Bryopsida) are all rather similar in structure, consisting of a filamentous green body, or *protonema,* on or in the soil. From the protonema grows an erect stem to which are attached a spiral whorl of leaves one cell thick. From the base of the stem extend many colorless, hairlike projections called *rhizoids.* Mosses are never more than 15 to 20 centimeters high owing to the inefficiency of the rhizoids as water absorbers and the absence of true vascular and supporting tissues.

The familiar small, green, leafy moss plant is the sexual or *gametophyte* generation. The *sporophyte* or asexual generation develops as a partial parasite on the gametophyte—it can carry on photosynthesis but is dependent on the gametophyte for its supply of water and minerals. Many moss "plants" can be produced asexually by a single protonema.

Mosses, like lichens, can grow on bare places where few other plants could survive. Once mosses have become established and soil begins to accumulate, other plants can follow. An economically important plant is *Sphagnum,* which grows in boggy places. The remains of this plant accumulate under water and form *peat,* used as a fuel in many countries. Dried *Sphagnum* can absorb and retain vast quantities of water and is used as a packing material for live plants.

Liverworts. A second class of bryophytes, Marchantiopsida or the liverworts, is simpler and more primitive than the mosses. Some of the 9000 species of liverworts are simply flat, sometimes branched, ribbonlike structures that lie on the ground, attached to the soil by numerous rhizoids and lacking a stem. Other species tend to grow upright and have a leaflike gametophyte. Still others, the leafy liverworts, have gametophytes that are differentiated into stems, branches and leaves but lack

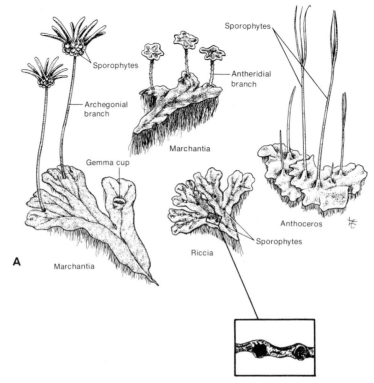

Figure 9–1 *A,* Sketches of some common liverworts, *Marchantia* and *Riccia,* and a hornwort, *Anthoceros.* In each, the flat part of the body is the gametophyte, upon which the sporophyte (*inset*) develops. *B,* Archegonial branch and, *C,* antheridial branch of *Marchantia.* (*B* and *C* courtesy of the Carolina Biological Supply Co.)

vascular tissues. A few liverworts are strictly aquatic. Many of the leafy liverworts are epiphytes and form colonies that grow on the stems, branches and leaves of trees in tropical rain forests. The leafy liverworts are believed to be the most primitive of the bryophytes.

The upper surface of the gametophyte is an epidermis one cell thick, punctured by many pores for the exchange of gases. The lower surface is an epidermis covered with many thin scales and from which grow many long, slender rhizoids. The recognizable plant is the gametophyte generation; as in mosses, the sporophyte grows as a parasite on the gametophyte (Fig. 9–1). The upper surface of some liverwort gametophytes may bear **gemma cups**

(Fig. 9–1). Small, flattened, ovoid gemmae are produced within these cups, separate from the parent plant and grow into new gametophytes, a process of asexual reproduction.

Hornworts. Some 300 species of plants with small, leaflike, irregularly branched gametophytes comprise the class Anthocerotopsida, the hornworts. Hornworts are widely distributed in moist, shaded places in the warmer regions of the world. The genus *Anthoceros* has a sporophyte which grows upward as a slender cylindrical capsule from the foot embedded in the gametophyte (Fig. 9–1).

Mosses and liverworts evolved from algalike ancestors; they have many characteristics in common with the green algae and are generally believed to be derived from them. The protonemata of mosses are remarkably similar to certain filamentous green algae. The pigments present are similar in the two groups: carbohydrates are stored as starch and the hornworts have chloroplasts containing pyrenoids, like the green algae. Most algae, however, have unicellular sex organs whereas bryophytes have multicellular antheridia and archegonia. At one time it was believed that the higher, vascular plants evolved from the bryophytes, from a liverwort or a hornwort. But although there is fossil evidence of true vascular plants in the Silurian period some 360,000,000 years ago, the first evidence of the bryophytes dates from the Pennsylvanian period, which began about 100,000,000 years later. For this and other reasons, botanists are now inclined to believe that vascular plants evolved from the green algae independently of the mosses. The mosses probably represent the end of a separate branch of the evolutionary tree.

9–3 THE LIFE CYCLE OF A MOSS

The trends in the evolution of reproductive processes described in the algae (p. 171) are continued among mosses, ferns and seed plants. These include the evolution of heterogamy, sexual differentiation, and the alter-

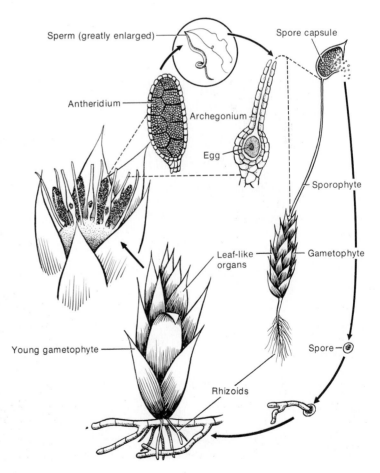

Figure 9–2 The life cycle of a moss plant.

nation of sexual and asexual generations. Each species has a characteristic *life cycle*, the sequence of developmental processes and events occurring between any given point in the life span of one organism and the comparable point in the life span of its offspring. Mosses characteristically have a Life Cycle III (p. 161), with a marked alternation of sporophyte and gametophyte generations.

The familiar, small, green, leafy plants called *mosses* are the haploid gametophyte generation of the plant. The gametophyte consists of a single, central stem, bearing "leaves" arranged spirally, and held in place in the ground by a number of slender rootlets or *rhizoids*, which absorb water and salts from the soil. The leaf cells produce all the other compounds the plant needs for survival, so that each gametophyte is an independent organism. When the gametophyte has attained full growth, sex organs develop at the top of the stem, in the middle of a circle of leaves and sterile hairs called paraphyses (Fig. 9–2).

In some species the sexes are separate; in others, both male and female organs develop on the same plant. The male organs are sausage-shaped structures, *antheridia*. Each antheridium produces a large number of slender, spirally coiled sperm or *antherozoids*, each equipped with two flagella. After a rain or heavy dew the sperm are released and swim through the film of water covering the plant to a neighboring female sex organ, either on the same plant or on another one. The female organ, the *archegonium*, is shaped like a flask and has one large egg in its broad base. This organ releases a chemical substance that attracts sperm, and guided by this, the sperm swim down the neck of the archegonium and into its base, where one sperm fertilizes the egg. The resulting zygote is the beginning of the 2N sporophyte generation.

In contrast to the independent green gametophyte, the sporophyte is a leafless, single, spindle-shaped stalk (*seta*) living as a parasite on the gametophyte and obtaining its nourishment by means of a *foot* which grows down into the gametophyte tissue. At the opposite, upper end of the sporophyte stalk, a sporangium, or *capsule*, forms. The cells comprising the capsule possess chloroplasts and produce some of their food photosynthetically. Within the cylindrical cavity of the capsule each diploid *spore mother cell* undergoes

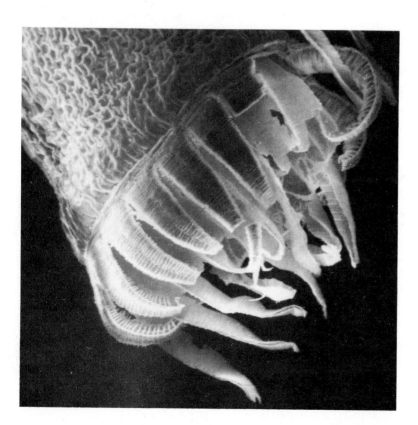

Figure 9–3 Scanning electron micrograph of a moss capsule showing the peristome teeth. (From Norstog, K., and Long, R. W.: *Plant Biology.* Philadelphia, W. B. Saunders Co., 1976. Photo courtesy of Dr. James A. McCleary.)

meiotic divisions to form four haploid spores. These are the beginning of the gametophyte generation.

When the capsule matures, the upper end forms a *lid*, which drops off. In some mosses the opening of the spore capsule is obstructed by one or two rings of wedge-shaped teeth (Fig. 9–3). These bend inward in wet weather and prevent the escape of the spores but bend outward in dry weather, permitting the liberation of spores when they are likely to be dispersed by the wind. When a spore drops in a suitable place, it germinates and develops into a *protonema*, a green, branching filamentous structure. The protonema buds and produces several gametophytes, thereby completing the life cycle. The gametophyte may undergo asexual reproduction by gemmae produced in gemma cups.

When removed and cultured in nutrient media, the lower portion of the capsule of certain species of mosses will grow and develop not into sporophytes (as you might expect, since the cells are $2N$) but into protonemata. These protonemata subsequently develop into mature, functional $2N$ gametophytes. The $2N$ gametophytes may undergo sexual reproduction and produce $4N$ (tetraploid) sporophytes.

9–4 TRACHEOPHYTES: THE VASCULAR PLANTS

The vascular plants compose an ancient and diverse group and include the oldest land plants as well as the dominant land plants of today. All of these have stems with a cortex and stele containing vascular tissues—xylem and phloem—and well-developed meiosporangia. The earliest vascular plants, members of the division *Rhyniophyta*, gave rise to several groups of plants, including the *Equisetophyta* (the scouring rushes) and the *Lycopodiophyta* (the club mosses) which flourished in the Paleozoic era, reached their peak in the Carboniferous period and subsequently declined. There are today no surviving species of Rhyniophyta, some 20 species of Equisetophyta, and about 650 species of club mosses.

The division *Psilotophyta* (the fork ferns) includes two living genera which resemble the ancient rhyniophytes and were classified with them at one time. The true ferns, division

Polypodiophyta, are a large and diverse group with a long evolutionary history; an abundance of fossil ferns has been found in deposits from the Paleozoic era and the ferns are successful modern plants.

The seed plants include the gymnosperms, division *Pinophyta*, and the flowering plants, division *Magnoliophyta*, the dominant land plants of today. The earliest seed plants, the seed ferns (*Lyginopteridopsida*), were gymnosperms that arose some 300,000,000 years ago and are known now only from fossils.

All the tracheophytes, like the mosses, have a life cycle with an alternation of gametophyte and sporophyte plants. However, the sporophyte of the higher plants is a free-living, independent plant and the gametophyte is either a small, independent plant or is contained within the sporophyte.

9–5 DIVISION RHYNIOPHYTA

The most primitive vascular plants known are the Rhyniophyta (Fig. 9–4), which lived in the Devonian period and probably even earlier, in the Silurian. These plants had a creeping horizontal stem or *rhizome* from which grew branching, erect, green stems attaining a height as great as 60 centimeters. The smaller branches were coiled at the tip and probably unrolled as they grew, as present-day ferns do. They had no roots and were either leaflets or had small scalelike leaves. Fossil remains of these plants have been found in Scotland in such a good state of preservation that details of the internal structures were visible. The most ancient vascular plant known is *Cooksonia*, a rhyniophyte, the fossil remains of which have been found in rocks of the Silurian period, which are some 400,000,000 years old.

The center of the stem of the rhyniophytes was a cylinder of vascular tissue composed solely of *xylem*—empty, dead tracheids that serve to conduct water. The other component of the vascular tissues of higher plants, *phloem*, which transports nutrients, cannot be distinguished in the fossil rhyniophytes found so far.

The Rhyniophyta are widely believed to be the ancestors of the other vascular plants. The several fossil species known each suggest the beginning of some sort of specialization found more fully developed in one of the other divisions.

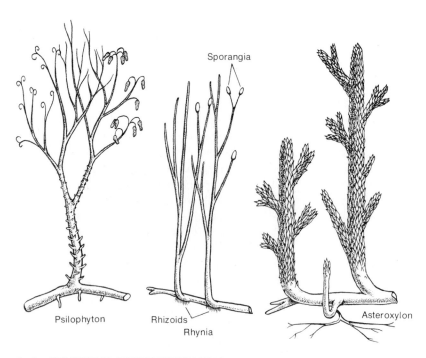

Sporangia

Psilophyton   Rhizoids   Asteroxylon
                   Rhynia

Figure 9–4 Three species of fossil vascular plants that lived during the Devonian period. *Rhynia* and *Psilophyton* are members of the division Rhyniophyta, and *Asteroxylon* is a member of the division Lycopodiophyta.

9–6 DIVISION EQUISETOPHYTA

Like the previous division, the Equisetophyta includes many more fossil plants than living ones. The division originated during Devonian times and developed into a variety of species, some small and some that were gigantic, treelike plants as much as 30 meters tall and 50 centimeters in diameter (Fig. 9–5). The latter flourished especially during the Carboniferous period and their dead bodies, together with those of certain other plants, are the source of our present-day coal deposits (Fig. 9–6).

The present-day equisetophytes, genus *Equisetum*, the "horsetails" or "scouring rushes," are widespread from the tropics to the Arctic on all continents except Australia. These plants, usually less than 40 centimeters tall, are found in both boggy and dry places. The name "horsetail" is appropriate because the multiple-branched, bushy structure of many species resembles a horse's tail. The popular name, "scouring rushes," reflects the fact that deposits of silica in the epidermis give the plants a harsh, abrasive quality. They were used to clean pots and pans for centuries before the invention of steel wool.

The sporophyte of *Equisetum* is made up of a horizontal, branching, underground rhizome from which grow slender, branching roots and jointed aerial stems. The stem contains many vascular bundles arranged in a circle around a hollow center. The stems have conspicuous nodes that divide them into jointed sections; at each node there is a whorl of smaller secondary branches and a whorl of small, scalelike leaves (Fig. 9–7). Some branches develop a conelike structure at their tip which contains numerous structures bearing spore sacs on their inner faces.

The spores released from these structures germinate into green gametophytes with egg- and sperm-producing organs. The zygote formed after fertilization develops into the sporophyte plant; at first this is parasitic on the gametophyte, but it quickly develops its own stem and roots.

9–7 DIVISION LYCOPODIOPHYTA

The club mosses, quillworts and their relatives comprise the division Lycopodiophyta. These plants were widespread in the later Devonian and Carboniferous periods and many of them were tall, treelike forms. Today only five genera remain, all small forms (usually less than 30 centimeters high) that have survived to the present without having undergone any major changes. These inconspicuous plants consist of a creeping stem from which grow true roots and upright stems with thin, flat, spirally arranged true leaves (Fig. 9–8). The stem

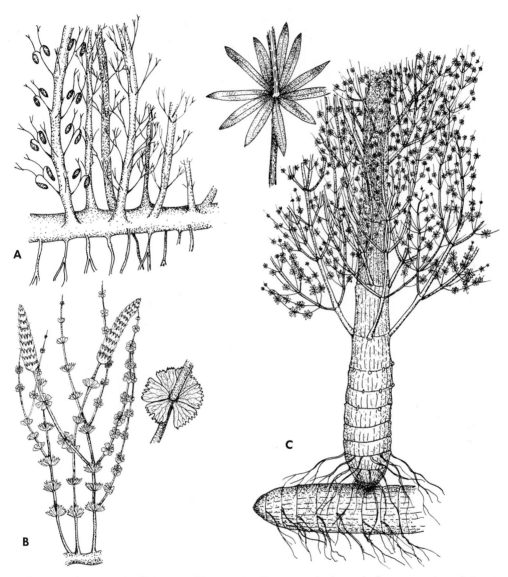

Figure 9–5 Equisetophytes of the Paleozoic Era seen in diagrammatic form. *A,* Reconstruction of *Hyenia,* a Middle Devonian species that had whorls of appendages but lacked cones. *B, Sphenophyllum,* an Upper Devonian and Carboniferous period plant. *C, Calamites,* an Upper Devonian and Carboniferous period plant somewhat resembling the modern-day *Equisetum.* (From Norstog, K., and Long, R. W.: *Plant Biology.* Philadelphia, W. B. Saunders Co., 1976. Redrawn from Smith, G. M.: *Cryptogamic Botany,* Vol. II, 2nd Ed. Copyright © 1955 by McGraw-Hill, Inc. Used by permission of McGraw-Hill Book Company.)

Figure 9–6 Photograph of a calamite, a fossil equisetophyte. The whorls of long, linear leaves are clearly evident. (From Fuller, H. J., and Carothers, A. B.: *The Plant World,* 4th Ed. New York, Holt, Rinehart and Winston, Inc., 1963.)

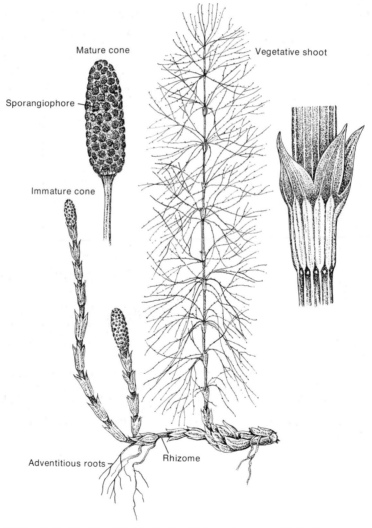

Figure 9–7 Sketch, about half natural size, of a horsetail, *Equisetum.*

Figure 9–8 Sketch of a club moss, *Lycopodium*, about half natural size. This kind of club moss is sometimes used as a Christmas ornament. Note the spirally arranged leaves.

has a core of xylem tissue surrounded by a cylinder of phloem. No cambium is present and hence all of the plant's growth is primary, at the growing tip. However some of the ancient club mosses living in the Carboniferous period did have a cambium and were great treelike plants as much as a meter thick and 20 or more meters tall.

At the tip of the stem are specialized leaves, *sporophylls*, arranged somewhat in the shape of a pine cone. The spore-producing structures, *sporangia*, are born on the sporophylls. One genus of living lycopsids, *Lycopodium*, has a life cycle rather like that of the rhyniophytes. The sporophyte produces spores, all of which are alike and germinate to form bisexual gametophytes. Sex organs develop on the gametophytes and produce eggs and biflagellate sperm. After fertilization, the

developing embryo is, for a time, dependent upon the gametophyte for nutrition. The homosporous lycopods, such as *Lycopodium*, are assigned to the class Lycopodiopsida, and the heterosporous lycopods, *Selaginella* and *Isoetes*, to the class Isoetopsida.

The genus *Selaginella*, commonly called spike mosses or little club mosses, shows an important evolutionary advance in having two types of spores, *megaspores* and *microspores*. The megaspores germinate to produce female megagametophytes, and the *microspores* germinate to produce male microgametophytes (Fig. 9–9). The gametophytes are greatly reduced in size and dependent on the sporophyte for nourishment. When the haploid microspores are released from the sporophyte they may drop near a megaspore. When wet by dew or rain, the microspore wall splits and the

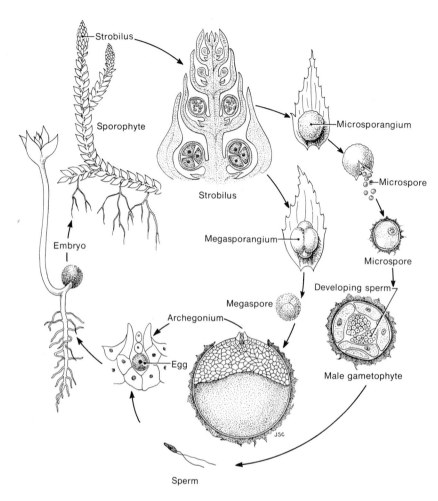

Figure 9–9 Diagram of the life cycle of the "spike moss" *Selaginella*. *Selaginella* produces two kinds of spores in its strobilus: megaspores, produced in megasporangia, which germinate to produce female gametophytes; and microspores, produced in microsporangia, which germinate to produce male gametophytes. Sperm develop within the male gametophyte and, when freed, swim to the archegonium on the female gametophyte and fertilize the egg within. The fertilized egg develops into the new sporophyte.

sperm within are free to swim to the megaspore and fertilize the haploid egg. As in the seed plants, the embryo is produced in the female gametophyte while it is still within the sporophyte. Although the reproductive cycle of these plants foreshadows that of the seed plants, they are not the ancestors of them but are a terminal group. The "resurrection plant" of the Southwest, *Selaginella lepidophylla*, rolls up into a compact ball of apparently dead leaves during the dry season, then unrolls and carries on its normal activities when moisture is present. Investigations of this remarkable phenomenon have shown that the cells may undergo complete dissociation of their cytoplasmic organization, yet within a few hours after a rainfall their chloroplasts are reconstituted and they become photosynthetic.

The **quillworts**, *Isoetes* and *Stylites*, superficially quite different from the club mosses, are deciduous perennial plants living in marshy places with their stems in the ground.

The quillworts are heterosporous and bear microsporangia and megasporangia. Roots project down from the stem and slender, quill-like leaves resembling those of a bunch of garlic project into the air (Fig. 9–10). The leaves are attached by their broad bases to the short stem and have spore-bearing organs at the basal end.

9–8 DIVISION PSILOTOPHYTA

Until recently it was customary to group two living genera of plants, the fork ferns *Psilotum* and *Tmesipteris*, in the same division with the fossil plants, *Rhynia* and *Psilophyton*. There are a number of similarities between the two types, but the sporangia of *Psilotum* and *Tmesipteris* are borne on short lateral stalks rather than terminally as in the Rhyniophytes. The question of whether the upright parts of the fork ferns are to be interpreted as stems or as complex leaves has not yet been resolved. If

and surrounded by an endodermis. The gametophytes are perennial, underground, stemlike plants that undergo branching.

9–9 DIVISION POLYPODIOPHYTA: THE FERNS

The characteristics that distinguish ferns from the other lower, non-seed-bearing, vascular plants include the structure of the leaf, the anatomy of the stem, the location of the sporangia and the pattern of development. Ferns typically have large branching leaves (**fronds**) which develop by uncoiling ("fiddleheads."); their sporangia are borne on the fronds in clusters (**sori**) and their stems have a pith.

Some 9000 species of ferns are widely distributed today in both the tropics and temperate regions. Those in the temperate regions thrive best in cool, damp and shady places. Ferns are abundant in tropical rain forests. Some tropical ferns are tall and superficially resemble palm trees, having an erect, woody, unbranched stem with a cluster of compound leaves or fronds at the top. Some of these *tree ferns* reach a height of 16 meters and have leaves 4 meters long. The common, temperate zone ferns have horizontal rhizomes growing at, or just beneath, the surface of the soil, from which grow hairlike roots. The rhizomes are usually perennial and each year new erect fronds or compound leaves grow from them. The leaves of ferns characteristically are coiled in the bud and unroll and expand to form the mature leaf (Fig. 9–12).

Ferns are remarkably like the seed plants in many respects. The root of a fern has root cap, meristematic elongation and mature zones like the root of a seed plant. Its stem has a protective epidermis, supporting and vascular tissues, and the leaves have veins, chlorenchyma, a protective epidermis and stomata. Ferns differ from the seed plants in that the xylem contains only tracheids—no vessels—and the spores are all alike and are produced in sporangia on the undersurfaces of certain leaves. The sporophyte plant may live for several years and produce a crop of haploid spores each year. These haploid spores are released, carried by the wind, and will germinate into gametophytes on moist soil. Because

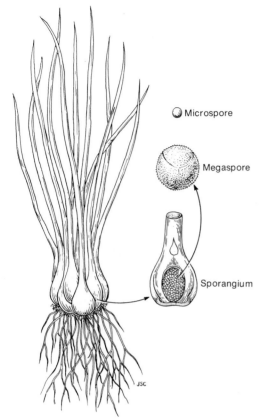

Figure 9–10 Sketch, natural size, of a quillwort, *Isoetes*. *Right,* Sketch of the base of one of the leaves showing the sporangium.

Microspore

Megaspore

Sporangium

they are complex leaves, it would be appropriate to include the fork ferns with the true ferns in the division Polypodiophyta, for ferns characteristically have complex leaves or fronds. Until the question is finally settled to the satisfaction of most botanists it would appear advisable to place these two genera of fork ferns in a separate division, the Psilotophyta. The species of *Psilotum* are small, simple subtropical plants, two species of which grow in the southeastern United States. Some grow as epiphytes among the fibrous roots at the bases of palm trunks (Fig. 9–11). The various species of *Tmesipteris* are small, hanging epiphytes restricted to Australasia and New Zealand.

The fork ferns are homosporous and rootless but have an underground branching stem from which grow unicellular rhizoids and green, photosynthetic upright branches bearing tiny scalelike leaves. *Psilotum* is of great interest to botanists because both its gametophytes and its sporophytes have vascular tissues. The larger and the older gametophytes of *Psilotum* have a stele complete with xylem and phloem

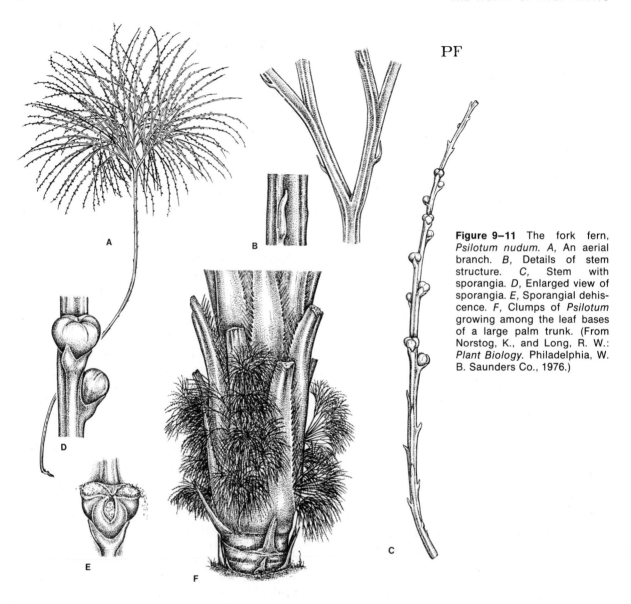

PF

Figure 9–11 The fork fern, *Psilotum nudum. A,* An aerial branch. *B,* Details of stem structure. *C,* Stem with sporangia. *D,* Enlarged view of sporangia. *E,* Sporangial dehiscence. *F,* Clumps of *Psilotum* growing among the leaf bases of a large palm trunk. (From Norstog, K., and Long, R. W.: *Plant Biology.* Philadelphia, W. B. Saunders Co., 1976.)

Figure 9–12 Photograph of young fern plant showing young leaves, which will uncoil as they develop. (From Weatherwax, P.: *Botany,* 3rd Ed. Philadelphia, W. B. Saunders Co., 1956.)

the spores can be carried by the wind for hundreds of miles the gametophyte may develop at a great distance from its parent sporophyte. Gametophytes are better able to withstand the cold than are sporophytes, and the gametophytes of some tropical ferns are found as far north as Ohio, whereas their sporophytes can survive no farther north than Florida.

During the Carboniferous period there were great forests of fern trees. These had tall, slender trunks made of the stem plus an enveloping mass of roots matted together by hairs. The bodies of these fern trees also contributed to our present coal deposits. Another group of fossil plants with fernlike leaves, long considered to be ferns, have been shown to have borne seeds; these fossil *seed ferns* are now classified with the gymnosperms.

9–10 THE LIFE CYCLE OF A FERN

The relatively large, leafy green plant commonly called a fern is the *sporophyte* generation. This consists of a horizontal stem, or *rhizome*, lying just under the surface of the soil and bearing fibrous roots and several leaves or fronds. Each frond is subdivided into a large number of leaflets (Fig. 9–13). On the under surfaces of certain leaflets develop clusters of small, brown spore cases (*sporangia*). Within the sporangia, haploid spores are produced from spore mother cells by meiosis. The sporophyte plant may live several years and produce several yearly crops of spores.

The spores are released at the proper time, fall to the ground, and develop into flat, green, photosynthetic, heart-shaped gametophytes 5 or 6 millimeters in diameter. The gametophyte, called a *prothallus*, grows in moist, shady places, especially on decaying logs and on moist soil and rocks. A number of rhizoids grow from each gametophyte into the soil, anchoring it and absorbing water and salts. The male and female sex organs (*antheridia* and *archegonia*) develop on the under surface of the gametophyte (Fig. 9–13).

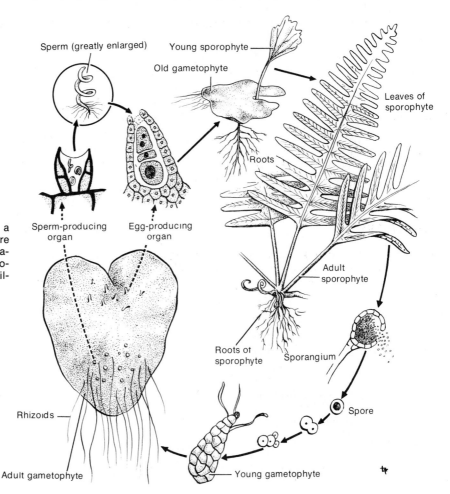

Figure 9–13 The life cycle of a fern plant. As in the mosses, there is an alternation of a sexual (gametophyte) and an asexual (sporophyte) generation; the large, familiar fern plant is the sporophyte.

Sperm (greatly enlarged)

Young sporophyte

Old gametophyte

Leaves of sporophyte

Roots

Sperm-producing organ

Egg-producing organ

Adult sporophyte

Roots of sporophyte

Sporangium

Spore

Rhizoids

Adult gametophyte

Young gametophyte

Each archegonium, usually located near the notch of the heart-shaped structure, contains a single egg. The antheridia, located at the other end of the gametophyte, develop a number of flagellated antherozoids or sperm. These are ovoid in shape and have many flagella on a spiral band at their anterior end. The antherozoids are released after a rain and, attracted by a chemical substance released by the archegonium, swim through the water on the undersurface of the gametophyte to reach the egg. The antheridia usually develop and discharge their antherozoids (in response to a gibberellinlike substance, **antheridogen,** secreted by the oldest gametophyte nearby) before the archegonia of that gametophyte plant have matured. The antherozoid of one plant usually fertilizes the egg of another plant and a new sporophyte develops from the resulting zygote. The fertilized egg begins development within the archegonium into a sporophyte embryo. Initially the sporophyte develops as a parasite on the gametophyte, but it soon develops its own roots, stem and leaves and becomes an independent sporophyte, completing the cycle.

The diploid fern sporophyte is fairly well adapted for terrestrial life: it has conducting and supporting tissues and, in contrast to the mosses, is nutritionally independent of the gametophyte. However, the conquest of land by the ferns has remained incomplete, for the gametophyte generation can survive only where there is plenty of moisture and shade, and the union of eggs and sperm in fertilization requires a watery medium.

QUESTIONS

1. Discuss the adaptations for land life evident in the bryophytes. In what ways do the bryophytes qualify for the title "amphibians of the plant world"?
2. What structural and physiological factors restrict moss plants to a height of about 15 cm.?
3. In what ways do mosses, liverworts and hornworts resemble each other? In what respects are they similar to, and in what respects are they different from, the higher plants?
4. What evidence supports the theory that vascular plants evolved directly from green algae and not from bryophytes?
5. Describe sexual reproduction in a moss. In what ways are bryophytes relatively ill-adapted for terrestrial life?
6. What are spores? How are they produced in a moss? In a fern?
7. Compare the life cycles of *Selaginella* and a fern.
8. What features of the rhyniophytes suggest that they are the ancestors of the other vascular plants?
9. In what ways do ferns resemble seed plants? In what respects do they differ from them?
10. How do the fork ferns differ from true ferns? What unusual feature is present in *Psilotum?*

SUPPLEMENTARY READING

The evolution of plants is discussed in T. Delevoryas's *Plant Diversification,* in A. Cronquist's *Evolution and Classification of Flowering Plants,* and in *The Life of Plants* by E. J. H. Corner, in which a renowned botanist presents his views on the evolution of plants and on the modifications that occurred as plants invaded the land. *The Diversity of Green Plants* by P. R. Bell and C. L. F. Woodstock is a concise, up-to-date survey of the enormous variety of plants. *Plant Anatomy* by Katherine Esau and *Morphology of Plants* by H. C. Bold are beautifully illustrated books providing excellent descriptions of all aspects of plant morphology. Well written treatments of mosses, liverworts and hornworts are to be found in E. V. Watson's *The Structure and Life of Bryophytes* and in W. T. Doyle's *Nonvascular Plants: Form and Function.*

CHAPTER 10

THE SEED PLANTS

The remaining two divisions of the plant kingdom, the gymnosperms, division **Pinophyta,** and the angiosperms, division **Magnoliophyta,** differ from ferns in having no independent gametophyte generation. Their two key characteristics are the formation of **seeds,** structures enclosing the embryo during a resting stage, and the union of male gametes with the egg by **pollination,** by the growth of a pollen tube. Seeds provide for the wide and rapid dissemination of a species and are resistant to desiccation and to high or low temperatures. In the reproductive processes of lower plants the sperm reaches the egg through an external supply of water—the ocean, a pond, a puddle or perhaps simply a film of water. The pollen tube eliminates the need for this and provides a means for the direct union of male and female gametes. These two traits have undoubtedly been responsible in large measure for the success of seed plants as terrestrial organisms.

Seed plants characteristically produce two types of spores, **megaspores** and **microspores.** The megaspores develop into female gametophytes, and the microspores develop into male gametophytes or **pollen.** The female gametophyte is retained within the megaspore and gives rise to a gamete, an egg, that is fertilized there. The resulting zygote develops into an embryo with the rudiments of leaves, stems and roots while still within the seed coat derived from the previous sporophyte. The **seed** thus consists of structures belonging to three distinct generations: the **embryo,** the new sporophyte; the **endosperm,** nutritive tissue derived from the female gametophyte; and the **seed coat,** derived from the old sporophyte.

There are more than 250,000 species of seed plants, adapted to survive in a variety of terrestrial environments and varying in size from the minute duckweed a few millimeters in diameter to giant redwood trees. They also are the plants of greatest value to us as sources of food, shelter, drugs and industrial products. The two general groups of seed plants, the gymnosperms (''naked seeds'') and the angiosperms (''enclosed seeds''), differ in the relationship of the seeds to the structures producing them. The seeds of angiosperms are formed inside a **fruit,** and the seed covering is developed from the wall of the ovule (see p. 224) of the flower. Some angiosperms appear to violate this rule and have exposed seeds—wheat, corn, sunflowers and maples, for example. However, these plants also have enclosed seeds, for the structure commonly called the "seed" actually is a fruit, enclosing the true seed. The seeds of gymnosperms are borne in various ways, usually on **cones,** but they are never really enclosed as are angiosperm seeds. The seed habit has contributed immensely to the success of the seed plants; the stored food nourishes the embryo until it can lead an independent life; the tough outer coat protects the embryo from heat, cold, drying and parasites; and seeds provide a means for the dispersal of the species.

10–1   THE ORIGIN OF THE GYMNOSPERMS

The gymnosperms appear to have evolved from a group of Devonian plants known as *progymnosperms.* These had frondlike branches resembling those of early ferns but in addition had a type of xylem generally associated with gymnosperms—one composed of tracheids with **bordered pits.** A bordered pit is a thin area between adjacent xylem cells, ringed on each side by cell wall material so that it looks superficially like a doughnut. One of the progymnosperms, *Archaeopteris,* was a large tree with a trunk nearly two meters in diameter (Fig. 10–1); its overall appearance may have been similar to that of present-day gymnosperms such as pines and cypresses. In addition to tracheids with bordered pits, *Archaeopteris* had woody stems with extensive amounts of xylem, a characteristic typical of gymnosperms. The nature of the reproductive organs of the progymnosperms is not well understood. Some species had microspores and megaspores; others had spores of only one size and may have been homosporous, but it is also possible that they were pollen-producing plants. Primitive seeds have been found in some deposits associated with fossils of *Archaeopteris* and although the seeds were not attached to the fossils it is possible that some progymnosperms may have borne seeds.

10–2   CLASS LYGINOPTERIDOPSIDA: THE SEED FERNS

The coal beds deposited during the Carboniferous period of the Paleozoic era have

yielded fossil remains of many large, leafy plants with fernlike leaves or fronds (Fig. 10–2). These were initially thought to be ferns, but more recently seeds were found attached to some of these leaves and the plants were named *seed ferns* and classified with the gym-

Figure 10–1   A restoration of *Archaeopteris.* (From Beck, C. B.: Reconstruction of *Archaeopteris,* and further consideration of its phylogenetic position. Amer. J. Bot. *49:*373–382, 1962.)

Figure 10–2 *A,* A reconstruction of a seed fern, *Medullosa noei,* which in life was about 5 meters tall. *B,* A large seed fern seed, attached terminally to a part of a frond. *C,* A microsporangial organ of *Medullosa,* composed of many longitudinally united microsporangia. (From *Morphology and Evolution of Fossil Plants* by Theodore Delavoryas. Copyright © 1962 by Holt, Rinehart and Winston, Inc. Reproduced by permission of Holt, Rinehart and Winston, Inc.)

nosperms. Some of the seed ferns were small trees; other were vinelike plants with large fronds and slender stems. The seeds of the seed ferns, ranging in length from 4 millimeters to 11 centimeters, were attached to the leaves in several ways. In some species the seeds were attached at the tip of the leaf, in others they were attached to the margins of the leaflets, and in others they grew directly from the central extension of the petiole (the ***rachis***) of the frond. Pollen-bearing organs, usually compound structures composed of many

elongate microsporangia, were also attached to the fronds. There had been some speculation about how pollination and fertilization occurred in these fossil plants, but recently fossil pollen tubes in the ovule of a seed fern were described, which indicates that the process resembled that in the group of ancient but still living gymnosperms, the cycads. The seed ferns were especially abundant during the Carboniferous period and made a major contribution to the coal deposits of that period. These ferns have been extinct since the Mesozoic era.

10–3   CLASS CYCADOPSIDA: THE CYCADS

The cycads, found mainly in tropical and semitropical regions, have either short tuberous underground stems or erect cylindrical stems above the ground. With their large compound divided leaves (Fig. 10–3) they resemble ferns or miniature palm trees, for which they are frequently mistaken. The only cycad native to the United States is the sago palm, *Zamia*, found in Florida from Gainesville south. *Zamia* is one of the smallest cycads, having a short fleshy underground stem, a long tap root, and leathery compound leaves, usually less than 1 meter long. In addition to the tap root, cycads have other roots that grow near the surface of the soil, closely branched and containing colonies of the nitrogen-fixing blue-green alga *Nostoc*. It seems likely that the nitrogen requirements of these plants are met at least in part by the blue-green algae. Cycads grow very slowly, but they may live for an exceedingly long time. One very old cycad plant in western Australia is estimated to be about 5000 years old, making this one of the more ancient living organisms in the world.

The pine produces two kinds of cones on the same tree. In contrast, the cycad population consists of two kinds of trees, one producing only cones yielding pollen and the other producing only cones yielding seeds. These are not "male" and "female" trees but are sporophytes that produce male and female gametophytes respectively. The seed cones of cycads are larger than the pollen cones and in some species they may be as much as a meter long and weigh 40 kilograms. Seed cones are usually quite colorful, ranging from yellow to deep red.

The cones of cycads are composed of sporophylls ranged in a helix about a central axis. The microsporophylls have many large microsporangia on their undersurface, and each megasporophyll bears two ovules before fertilization.

The life history of a typical cycad such as *Zamia* (Fig. 10–4) is of interest because of the insight it provides into the methods of reproduction in the earliest seed plants, the seed ferns. The cones begin to appear on the apex of the plant in early summer and microsporangia develop on the lower surfaces of the microsporophyll. These contain microspore mother

Figure 10–3 Photographs of cycads. *Above, Cyas; below left,* a pollen plant of *Zamia;* and *below right,* a mature seed plant of *Zamia.* This plant, found from Florida to South America, is the only cycad native to the United States. (From Weatherwax, P.: *Botany,* 3rd Ed. Philadelphia, W. B. Saunders Co., 1956.)

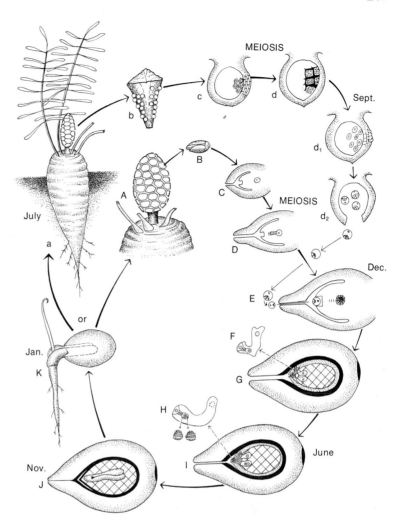

Figure 10–4 Life history of *Zamia. a,* Pollen plant. *b,* Microsporophyll. *c,* Young microsporangium with microspore mother cells. *d,* Microsporangium containing microspore tetrads. *d₁,* Microspores. *d₂,* Three-celled pollen (immature male gametophytes). *A,* Seed cone. *B,* Megasporophyll. *C,* Ovule with megaspore mother cell. *D,* Tetrad of megaspores, one of which is functional. *E,* Pollination, showing pollen in pollen chamber and young megagametophyte. *F,* Developing pollen tube (microgametophyte). *G,* Mature megagametophyte. *H,* Mature microgametophyte. *I,* Fertilization. *J,* Mature seed with embryo. *K,* Germinating seedling. (Redrawn from Norstog, K., and Long, R. W.: *Plant Biology.* Philadelphia, W. B. Saunders Co., 1976.)

cells which produce tetrads of microspores by meiotic divisions. The microspores develop thick walls and divide mitotically to form the three-celled *microgametophytes,* or male gametophytes. The assemblage of microspore wall plus microgametophyte constitutes the pollen grains which are released and dispersed by the wind to the megasporophylls of neighboring seed cones.

During this same time, within each of the two ovules on the megasporophyll a megasporangium produces four megaspores by meiosis. Only one of these survives to form a *megagametophyte,* or female gametophyte. The ovule at the time of pollination consists of an outer layer (the *integument*), the megasporangium, and a developing megagametophyte.

Pollination involves the sticking of the pollen grain to a drop of mucilaginous fluid that fills the micropyle of the megasporangium.

As the sticky fluid dries it shrinks and carries the pollen grains with it, depositing them within a pollen chamber, a region within the megasporangium. The pollen grains germinate and the pollen tubes slowly develop over a period of some five months. The female gametophyte continues to grow and mature at about the same rate as the male gametophyte. At fertilization the female gametophyte occupies most of the interior of the ovule and has a cuplike depression, the fertilization chamber, at its micropylar end. The fertilization chamber contains three to five archegonia which contain large egg cells, nearly 2 millimeters long.

During the growth of the pollen tubes several cell divisions occur which culminate in the formation of two large flagellated sperm cells. Cycads are remarkable among the seed plants in that they have *motile* male gametes within their pollen tubes. Only the cycads and *Ginkgo,* of all the living seed plants, have flagellated sperm. The sperm cells of cycads are enormous, as large as 400 micrometers in one species of *Zamia,* and may be seen with the unaided eye. They are the largest motile male gametes known among higher plants. Not only is the cycad sperm large, but it also has thousands of flagella per cell. Each of these flagella has the typical eukaryote $9 + 2$ pattern of axial filaments. All flagella are attached to a spiral band that encircles the anterior end of the cell. When the sperm become active and move about in the pollen tube, the tubes burst and release them and the sperm swim briefly in the liquid of the fertilization chamber. Each archegonium has four neck cells that project into the fertilization chamber and these open when the sperm are freed from the pollen tubes. Several sperm may enter each archegonium, but only one enters the egg cytoplasm and loses it flagella. The sperm nucleus then combines with the egg nucleus during the first mitosis. Subsequent cell divisions and growth produce an embryo with a long, highly coiled *suspensor* and two seed leaves, or cotyledons. The suspensor apparently plays the role of pushing the embryo deep within the female gametophyte. Nearly a year after pollination has occurred the seed cone breaks apart and the brillant orange-red seeds drop to the ground at the base of the parent plant. The cycads provide a connecting link between the ferns and lycopods (in which the motile male gametes pass through environmental water to the female gamete) and the more advanced seed plants (in which the nonflagellate, nonmotile sperm are transferred simply in the pollen tube).

Cycads are widely used as ornamental plants for landscaping in the tropics. The seeds of *Cycas* are gathered and eaten and a starchy cycad flour is prepared from the fleshy underground stems of *Zamia*. Both the stems and seeds of cycads contain a potent neurotoxin capable of causing paralysis and death. The cycad foliage also contains this, and livestock eating the foliage may be poisoned. A potent carcinogen, *cycasin,* has been isolated from the stems and seeds of cycads. Great care must be taken in preparing the cycad flour to remove these poisons.

An interesting and curious group of extinct Mesozoic plants are the cycadeoids, members of the class Bennettitopsida. These plants were like the cycads in general form, but the cones of at least some of the species produced both pollen and ovules so that they had a general resemblance to a flower. Although some botanists have called these cones "flowers," the cycadeoids were true gymnosperms, for their ovules were exposed and were not enclosed within a fruit. The idea that the flowering plants evolved from the cycadeoids has lost support because of major structural differences between the two groups.

10-4 CLASS PINOPSIDA: THE CONIFERS

The conifers—pine, cedar, spruce, fir and redwood trees—are the most successful, biologically, of the living gymnosperms. Many of the conifers are trees, but some are shrubs; most are evergreen and have needle-shaped leaves. The conifers have a worldwide distribution and are of great importance to us both ecologically and economically. The conifers have no flowers but bear their seeds on the inner sides of scale-like leaves that are usually arranged spirally to form a cone. Most species have both pollen cones and seed cones. Pollen released by the male cone is carried by the wind to the female cone where the eggs are fertilized. In pines, as much as a year may elapse between pollination and fertilization and several more years may elapse between fertilization and the shedding of the seeds.

The conifers are important economically as the source of more than 75 per cent of the wood used in construction and in the manufacture of paper, plastics, rayon, lacquer, photographic film, explosives, and many other materials. Some conifers produce resins used in the production of turpentine, tar and oils. The seeds of a few conifers are used as food, and the berries of junipers produce aromatic oils used for flavoring alcoholic beverages such as gin. The needlelike leaves of the evergreen gymnosperms are well adapted to withstand hot summers and cold winters and the mechanical abrasions of storms. Under the thick, heavily cutinized epidermal layer is a layer of thick-walled sclerenchyma. The stomata are set in deep pits that penetrate the sclerenchyma. Although conifers are found from tropical to subarctic regions the most extensive coniferous forests are found in the Northern Hemisphere in Canada and Siberia where they are the predominant trees of the taiga ecosystem. Some conifers are deciduous, but most are evergreen and retain their leaves for years before shedding them. The conifers are all woody, perennial plants with a characteristic tree form. Their wood is composed of tracheids with bordered pits and their seeds are "naked." It is currently thought that the conifers evolved separately from the seed ferns and the cycads, although all of these probably have a common ancestry among the progymnosperms. Trees that were clearly conifers existed at the same time in the Paleozoic era as the seed ferns. Among them were a number of species of *Cordaites*, all of which are now extinct. The *Cordaites* were large, cone-bearing trees with strap-shaped leaves and stems that were as much as 30 meters tall and a meter in diameter. These formed great forests during the Carboniferous period, and the modern conifers are believed to have descended from *Cordaites* and related forms.

The various species of conifers occur in a variety of habitats ranging from very humid rain forests to dry semideserts. The northern coniferous forest region of the taiga is a relatively dry environment with a low relative humidity during most of the year. Western North America has a most interesting group of conifers including the redwood, *Sequoia;* the California Big Tree, *Sequoiadendron,* the largest living organism; a variety of firs; and many species of pines, spruces and junipers. Fossils of a dawn redwood were found in Japan and in other areas of the Northern Hemisphere and botanists were greatly surprised when living dawn redwoods were found in remote valleys in central China in 1945. Seeds of the dawn redwood were collected and planted and the trees can now be seen at a number of botanical gardens around the world.

10-5 THE LIFE CYCLE OF A GYMNOSPERM

The key characteristics of the life cycles of gymnosperms and angiosperms that have allowed an even better adaptation to terrestrial life than those of mosses and ferns are the production of *seeds* and the formation of a *pollen tube.* The gametophyte has been re-duced to a few cells enclosed within the tissues of the sporophyte and entirely dependent on it for food. The union of sperm and egg is no longer dependent on the presence of a film of water on the gametophyte but is brought about by wind- or insect-dispersed pollen and the growth of the pollen tube. The embryo sporophyte is nourished and protected during early growth not by an independent gametophyte plant, as in mosses and ferns, but by a *seed* with its supply of endosperm and its tough outer coat. The sporophyte of seed plants produces two kinds of spores, larger *megaspores* and smaller *microspores,* and is said to be **heterosporous.** The lycopsid *Selanginella* (p. 208) and a few ferns are also heterosporous.

The reproductive parts of the seed plants were studied and named—stamen, pistil, ovule, and so forth—before the stages in the alternation of sporophyte and gametophyte generations were understood and before the essential parallelism of the life cycles of mosses, ferns and seed plants was recognized.

The life cycles of gymnosperms represent, in several respects, a transition between those of the ferns and those of the angiosperms (Fig. 10–5). A pine tree, a typical gymnosperm, produces two kinds of cones, *staminate,* which are small, less than 3 centimeters long, and *ovulate,* which are the large easily visible cones, as much as 45 centimeters long in some species. The ovulate cone is composed of many scales and on the surface of each scale are two ovules. Within each *ovule* is a diploid *megaspore mother cell;* this divides by meiosis to form four haploid *megaspores.* Only one of these becomes functional and grows into a multicellular *megagametophyte.* On each megagametophyte are two or three female sex organs (archegonia), each containing a single large egg.

On the underside of each scale of the staminate cone are two *microsporangia.* Within each microsporangium are many *microspore mother cells,* each of which divides by meiosis to form four *microspores.* While still within the microsporangium or pollen sac, the microspores divide mitotically to form a four-celled microgametophyte or *pollen grain.* These are released and carried by the wind. When a pollen grain reaches an ovulate cone it enters the ovule through an opening, the micropyle,

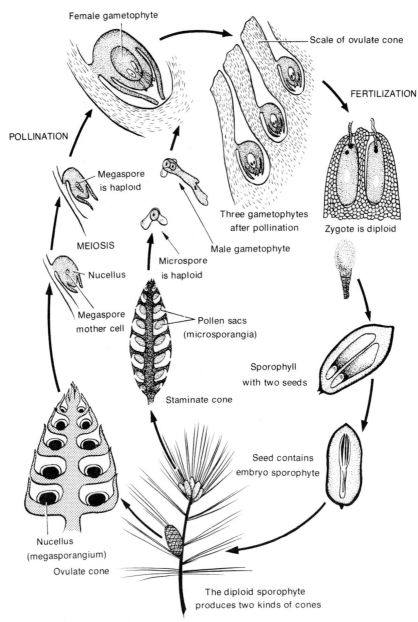

Figure 10–5 The life cycle of a pine tree. See text for discussion.

and comes in contact with the mega-sporangium.

A year or more may elapse before one cell of the pollen grain elongates into a pollen tube and grows through the megasporangium to the megagametophyte. Another cell of the pollen grain divides to form two male *gamete nuclei,* not motile sperm as in lower plants. When the end of the pollen tube reaches the neck of the archegonium and bursts open, the two male nuclei are discharged near the egg. One fuses with the egg nucleus to form the diploid

zygote, and the other disintegrates. After fertilization, the zygote divides and differentiates to produce a sporophyte embryo, surrounded by the tissues of the megagametophyte and those of the parent sporophyte. This entire structure is the *seed.*

The tissues of the megagametophyte provide nutrition for the developing embryo and are called *endosperm.* However, they are haploid cells and are quite different from the triploid (3N) endosperm cells of angiosperms, though both serve the same nutritive function

for their respective embryos. After a short period of growth and differentiation, resulting in the development of several leaflike *cotyledons,* the *epicotyl* and the *hypocotyl,* the embryo remains quiescent until the seed is shed and drops to the ground. When conditions are favorable, it germinates and develops into a mature sporophyte, the pine tree. The epicotyl develops into the stem, and the primary root develops from the tip of the hypocotyl.

10-6 GINKGOS AND GNETOPSIDS

The maidenhair tree, *Ginkgo biloba* (Fig. 10-6), is the only living representative of a once numerous and widespread class with a fossil record extending back some 200 million years to the Triassic period. Ginkgos were cultivated in China and Japan as ornamental trees because of their distinctive fan-shaped leaves which are shed in the fall (Fig. 10-7). These

Figure 10-7 Ginkgo twigs. *Left,* A cluster of mature seeds; *center,* a twig with leaves and young ovules; *right,* a twig with leaves and pollen cones. (Copyright, General Biological Supply House, Chicago.)

"living fossils" are tough and hardy trees under most conditions and studies have shown that they are remarkably resistant to attack by insects and fungi. For this reason the wood, although brittle, has been used to make insect-proof cabinets. Like the cycads there are two kinds of *Ginkgo* trees, one producing only staminate (pollen-producing) cones, and the other only ovulate (egg-producing) cones. *Ginkgo* and the cycads are the only seed plants that produce swimming sperm rather than pollen tube nuclei. When the *Ginkgo* egg is fertilized and matures, the inner seed covering becomes hard while the outer covering becomes soft and pulpy and has a rancid, sour odor. When trees produce many seeds at once the odor can be extremely offensive. For this reason when *Ginkgo* are planted in parks, care must be taken to ensure that all the trees in one region are of the same sex so that no seeds will be produced.

The Gnetopsida. The Gnetopsida is a small class of gymnosperms that includes some very peculiar plants. They are unique among living gymnosperms because they have flower-like compound pollen cones, ovules with two integuments, and vessels in their wood. The latter are tubelike elements in the xylem composed of cylindrical xylem cells arranged end to end with sievelike or open-ended walls. Such vessels are unknown in other gymnosperms but are a characteristic component of the xylem of flowering plants. In other respects the three members of this class, *Gnetum*, *Ephedra* and *Welwitschia*, are as unlike one another as they are unlike other gymnosperms.

Some 40 species of *Gnetum* grow as tropical vines, trees or shrubs in the valley of the Amazon, in West Africa and from India to Malaya. *Gnetum* has a marked resemblance to certain flowering plants: it has opposite leaves that are broad and net-veined and its reproductive organs look somewhat flowerlike, but its ovules are naked as they are in other gymnosperms. Some 35 species of *Ephedra* occur as shrubs in desert regions in many parts of the world. One is found in the southwestern United States and Mexico—a low, much-branched shrub with naked, green, photosynthetic twigs and rudimentary scale leaves (Fig.

Figure 10–8 *Ephedra,* a desert plant. (From Norstog, K., and Long, R. W.: *Plant Biology.* Philadelphia, W. B. Saunders Co., 1976.)

10–8). This *Ephedra* looks superficially like *Equisetum*, one of the horsetails.

There is only one living species of *Welwitschia*, found in the deserts of coastal Southwest Africa. It has an exceedingly deep taproot by means of which it obtains subsurface water. *Welwitschia* consists of a short, broad woody stem that is mostly underground except for a concave, disc-shaped crown that is covered with cork. The mature plant is rather like a large, woody turnip. Some specimens of *Welwitschia* are as much as 2000 years old and have crowns that are over 1 meter in diameter. During its entire life span, each plant produces only a single pair of thick and leathery leaves which grow on the periphery of the crown. These may become quite tattered as they sprawl over the surface of the soil. Although it may not rain in this region of Southwest Africa for periods of four or five years, there are frequent nightly fogs, which apparently supply these plants with much of the water they need. The stomata on the leaves of *Welwitschia* are reported to open at night to take in this water from the fog and then remain closed during the daytime. The resemblances of the Gnetopsida to flowering plants suggested that they may represent an ancestral gymnosperm from which flowering plants evolved. However, most botanists now believe the similarities to flowering plants are coincidental and the result of parallel evolution.

10–7 DIVISION MAGNOLIOPHYTA: THE FLOWERING PLANTS

The true flowering plants of the division *Magnoliophyta* are the largest division in the plant kingdom and include more than a quarter million species of trees, shrubs, vines and herbs adapted to almost every kind of habitat. Some live completely under water, others in extremely arid conditions. The vast majority are autotrophic, but some, such as the Indian pipe and mistletoe, have little or no chlorophyll and are partially or wholly parasitic. A few flowering plants have evolved devices for catching insects and other small animals and hence are holozoic and carnivorous. The flowering plants provide human beings with much of our food, clothing, shelter and drugs and liven the world with the beautiful colors and scents of their flowers.

Flowering plants are assigned to one of two major classes, the **Magnoliopsida**, also called **dicotyledons**, and the **Liliopsida**, or **monocotyledons**. These two classes are distinguished on the basis of their embryo structure, the form of their flowers and the anatomy of their stem and leaf. The cereal plants, wheat, corn and rice, the basic agricultural crops of modern society, are monocotyledons whereas most vegetable and fruit crops are dicotyledons. Other dicotyledons are medicinal plants such as *Digitalis*, which produces a heart stimulant, and *Rauwolfia*, which produces a compound used in the treatment of hypertension. Coffee, tea and cocoa are derived from dicotyledonous plants, *Coffea*, *Thea*, and *Theobroma cacao*, respectively. The Magnoliophyta were called *Angiospermopsida* under an earlier taxonomic system and the term angiosperm has survived as a common name for the group.

Many angiosperms can complete an entire life cycle from the germination of the seed to the production of new seeds within a month, but others require 20 to 30 years to reach sexual maturity. Some live for a single growing season, others live for centuries. The stems, leaves and roots present a bewildering variety of forms, but all angiosperms develop flowers that have a fundamentally similar pattern.

Angiosperms differ from gymnosperms in the abundance and prominence of the xylem vessels (most gymnosperms have only tracheids); in the formation of flowers and fruits; in the presence of sepals, petals or both in addition to the sporophylls; in the formation of a pistil through which the pollen tube grows to reach the ovule and egg (in gymnosperms the pollen lands on the surface of the ovule and the pollen tube grows in directly); and in the further reduction of the gametophyte generation to a few cells completely parasitic on the sporophyte.

Fossil remains of angiosperms have been found in rocks from the Cretaceous period during the Mesozoic era, and although they probably evolved from some primitive gymnosperm, there are no intermediate plants in older rocks to indicate which group of gymnosperms might have been the ancestor of the angiosperms.

The division Magnoliophyta (the angiosperms) includes some 225,000 species of Magnoliopsida (dicotyledons) and 50,000 species of Liliopsida (monocotyledons). The two classes differ in the following respects: (1) The embryo in the monocotyledons has only one seed leaf or cotyledon; the dicot embryo has two. The starch and other foods in the cotyledon of dicots nourish the embryo and seedling until it is capable of making its own food by photosynthesis. The cotyledon of the typical monocot functions as an absorptive organ rather than as a storage organ. (2) The mature seed of the monocots typically has endosperm present whereas endosperm is often absent from the mature seed of the dicots. (3) The leaves of monocots have parallel veins and smooth edges; the veins of dicot leaves branch and rebranch and the edges of the leaves are usually lobed or indented. (4) A cambium is usually absent in the monocot but present in the dicot. Thus most monocots have little or no secondary growth; however, woody monocots such as palms and yuccas have a special thickening meristem. (5) The flower parts of monocots—petals, sepals, stamens and pistils—exist in threes or multiples of three; dicot flower parts usually occur in fours or fives or multiples of these. (6) The monocots are generally herbaceous; a few families have woody plants whereas many families of dicots have genera of woody plants. (7) The vascular bundles of xylem and phloem are scattered throughout the stem of monocots. In dicots the xylem and phloem occur either as a single solid mass extending up the center of the stem or as a ring between the cortex and pith. (8) The roots of monocots are typically fibrous and adventitious whereas the root system of dicots usually consists of one or more primary taproots and secondary roots.

The many different families of monocots and dicots usually take their names from some conspicuous member. Some of the monocot families are the grasses, palms, lilies, orchids and irises. Some important dicot families are the buttercup, mustard, rose, maple, cactus, carnation, primrose, phlox, mint, pea, parsley and aster. The rose family, for example, includes in addition to roses, the apple, pear, plum, cherry, apricot, peach, almond, strawberry, raspberry, hawthorn and other shrubs.

10–8 THE LIFE CYCLE OF AN ANGIOSPERM

Angiosperms, like the lower tracheophytes, exhibit an alternation of gametophyte and sporophyte generations, but the gametophyte is reduced to a few cells lying within the tisssues of the sporophyte *flower.* The sporophyte is the familiar tree, shrub or herb. Not all flowers are easily recognizable as such. The small, inconspicuous green flowers of grasses and certain kinds of trees are quite different from the colorful blossoms we usually think of as flowers.

The Flower. The flower of an angiosperm is a modified stem, but in place of the usual foliage leaves it bears concentric circles of leaves modified for reproduction. A typical flower consists of four concentric rings of parts (Fig. 10–9) attached to the *receptacle,* the expanded end of the flower stem. The outermost parts, usually green and most like ordinary leaves, are called *sepals.* Within the circle of sepals are the *petals,* typically brilliantly colored to attract insects or birds and ensure pollination. Just inside the circle of petals are the *stamens,* the male parts of the flower. Each stamen consists of a slender filament with an *anther* at the tip. The anther is a group of *pollen sacs* (microsporangia). Each pollen sac contains a group of microspore mother cells— *pollen mother cells.* Each diploid pollen mother cell divides meiotically and produces four haploid microspores. A further nuclear division converts each microspore into a microgametophyte or *pollen grain.*

In the very center of the flower is a ring of *pistils* (or a single fused one). Each pistil has a swollen, hollow, basal part, the *ovary,* a long slender portion above this, the *style,* and at the top a flattened part, the *stigma.* In many flowers the stigma secretes a moist, sticky substance to trap and hold the pollen grains that reach it. Different species of angiosperms show great variations in the number, position and shape of the various parts. A flower that has both stamens and pistils is called a *perfect flower;* one that lacks one or the other is called an *imperfect flower.* Flowers with stamens but no pistils are *staminate flowers;* those with pistils but no stamens are *pistillate flowers.* Willows, poplars and date palms are examples of species with two kinds of plants, some bearing only staminate flowers, others bearing only pistillate flowers.

Within the ovary, at the base of the pistils, are one or more *ovules.* An ovule is a megasporangium completely enclosed by one or two integuments. Each ovule typically contains one

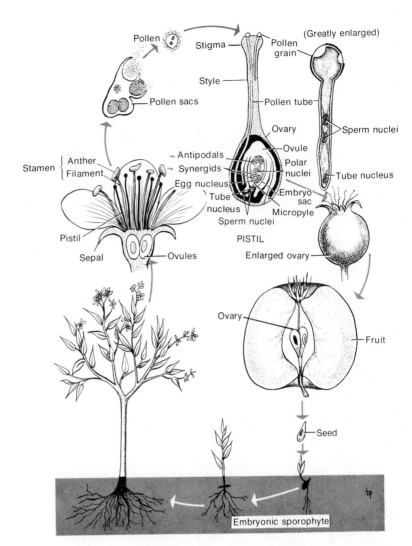

Figure 10–9 The life cycle of an angiosperm. See text for discussion.

megaspore mother cell. Each diploid megaspore mother cell divides by meiosis to form four haploid megaspores. One of the *megaspores* develops into the *megagametophyte;* the other three disintegrate. There is some variation in different species in the details of megagametophyte development, but typically the megaspore enlarges greatly and its nucleus divides. The two daughter nuclei migrate to opposite ends of the cell, then each divides and the daughter nuclei divide again. The resulting *megagametophyte* is a seven-celled structure called an *embryo sac* with three cells at each end and a large binucleate cell in the middle. These two nuclei are called the *polar nuclei* (Fig. 10–10). One of the three cells at one end of the megagametophyte becomes the *egg,* the other two and the three cells at the other end all disintegrate.

The haploid microspore develops into the pollen grain or young microgametophyte while still within the pollen sac. The nucleus of the microspore divides into a larger *tube nucleus* and a smaller *generative nucleus.* Most pollen grains are released while in this state and are carried by the wind, insects or birds to the stigma of the same or a neighboring flower. The pollen grain then germinates; the pollen tube grows out of the pollen grain and down the style to the ovule. The tip of the pollen tube produces enzymes that dissolve the cells of the style, thus making room for the pollen tube to grow. The tube nucleus remains in the tip of the pollen tube as it grows. The generative nucleus migrates into the pollen tube and divides to form two cells, the *sperm.* The mature male gametophyte consists of the pollen grain and tube, the tube nucleus, and the two sperm cells.

When the tip of the pollen tube penetrates

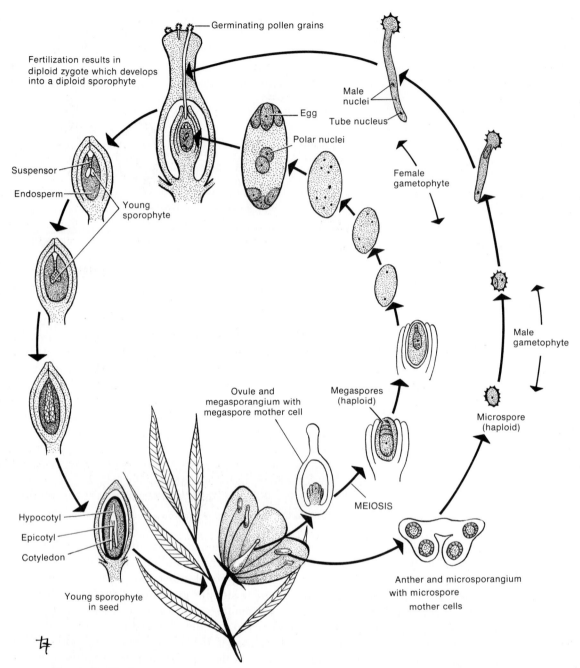

Figure 10–10 Some details of fertilization and seed formation in a dicot. See text for discussion.

the megagametophyte through the micropyle, it bursts and the two sperm cells are discharged into the megagametophyte. One of the sperm cells migrates to the egg nucleus and fuses with it. The resulting diploid cell is the **zygote,** the beginning of the new sporophyte generation. The other sperm nucleus migrates to the two polar nuclei and, in a remarkable and unusual process, all three fuse to form a 3N **endosperm nucleus.** In some species the two

polar nuclei have fused to form a single one before the sperm nucleus arrives. This phenomenon of **double fertilization,** which results in a diploid zygote and a triploid (three sets of chromosomes) endosperm, is peculiar to, and characteristic of, the flowering plants.

After fertilization the zygote undergoes a number of divisions and forms a multicellular embryo. The endosperm nucleus also undergoes a number of divisions and forms a mass of endosperm cells, gorged with nutrients. The endosperm fills the space around the embryo and provides it with nourishment. The sepals, petals, stamens, stigma and style usually wither and fall off after fertilization. The ovule with its contained embryo becomes the seed; its walls become thick and form the tough outer coverings of the seed. The seed consists of the dormant sporophyte embryo plus the stored food of the endosperm, all enclosed in a resistant covering derived from the wall of the ovule. It serves in dispersing the species to new locations and in enabling the species to survive periods of unfavorable environmental conditions (such as winter!) which may kill the mature plants.

10–9 FRUITS

The ovary, the basal part of the pistil containing the ovules, enlarges and forms the *fruit.* The fruit thus contains as many seeds as there were ovules in the ovary. In the strict botanical sense of the word, a fruit is a matured ovary containing seeds—the matured ovules. Although we usually think of only such sweet, pulpy things as grapes, berries, apples, peaches and cherries as fruits, bean and pea pods, corn kernels, tomatoes, cucumbers and watermelons are also fruits, as are nuts, burrs and the winged fruits of maple trees. A *true fruit* is one developed solely from the ovary. If the fruit develops from sepals, petals or receptacle as well as from the ovary it is known as an *accessory fruit.* The apple fruit consists mostly of an enlarged, fleshy receptacle; only the core is derived from the ovary.

The three general types of true and accessory fruits are simple fruits, aggregate fruits and multiple fruits. Simple fruits (e.g., cherries, dates, palms) mature from a flower with a single pistil; aggregate fruits (raspberries and blackberries) mature from a flower with several pistils; and multiple fruits (pineapples) are derived from a cluster of flowers united to form a single fruit. Fruits are also classified as *dry fruits* if the mature fruit is composed of rather hard dry tissues and as *fleshy fruits* if the mature fruit is largely soft and pulpy (Fig. 10–11). Dry fruits represent adaptations for dispersal by the wind or for attachment to animal bodies by hooks. Birds, mammals and other animals eat fleshy fruits and their enclosed seeds. The seeds pass through the animal's digestive tract and are dropped with the feces in a new place. Thus fleshy fruits also represent an adaptation for the dispersal of the species.

In one type of dry fruit, termed a *nut,* the wall of the ovary develops into a hard shell that surrounds the seed. The edible part of a chestnut is the seed within the fruit coat or shell. A

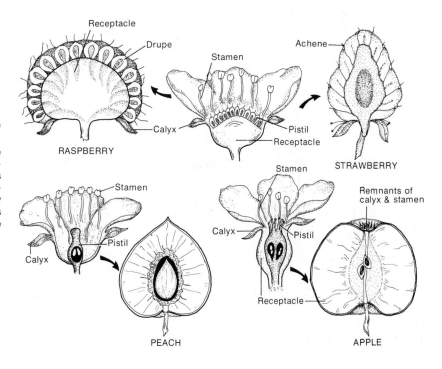

Figure 10–11 The formation of some fleshy fruits from flowers. Raspberries and strawberries are derived from similar types of flowers, but the pistils of raspberries become fleshy drupes and the pistils of strawberries become dry achenes, the yellow seedlike spots scattered over the surface of the fruit.

Brazil nut is really a seed; there are about twenty such seeds borne within a single fruit. An almond is not a "nut" at all, but the seed or "stone" of a fleshy fruit related to the peach.

Grapes, tomatoes, bananas, oranges and watermelons, although superficially quite different, are all examples of fleshy fruits. In these the entire wall of the ovary becomes pulpy; such fruits are technically called **berries.** Peaches, plums, cherries and apricots in contrast are stone fruits or **drupes.** In a drupe the outer part of the ovary wall forms a skin, the middle part becomes fleshy and juicy, and the inner part forms a hard pit or stone around the seed. There are, then, many kinds of fruit, differing in the number of seeds present, in the part of the flower from which they are derived, as well as in color, shape, water and sugar content, and consistency.

Fruits may form, or be induced to form, without the development of seeds. The banana, which has been cultivated for centuries, has only vestigial seeds (the black specks in the fruit) and must be propagated by vegetative means. Plant breeders have been able to develop seedless varieties of grapes, oranges and cucumbers. Many other plants have been induced to form seedless fruits by treatment with plant growth hormones.

10–10 GERMINATION OF THE SEED AND EMBRYONIC DEVELOPMENT

A few seeds germinate shortly after being shed if conditions are suitable, but most seeds remain dormant during the cold or dry season and germinate only with the advent of the next favorable growing season. A prolonged period of dormancy usually occurs only in seeds with thick or waxy seed coats which render them impermeable to water and oxygen. The length of time that a seed will remain viable and capable of germination varies greatly. Willow and poplar seeds must germinate within a few days of being shed or they will not germinate at all. Some seeds remain viable for many years: A long-term experiment in progress at East Lansing, Michigan, showed that seeds of the evening primrose and of yellow dock were able to germinate after 80 years, however at the end of 90 years only those of the evening primrose remained viable. Samples are being tested for germinating ability every ten years; the next ones will be tested in 1979. There are authentic records of lotus seeds preserved in peat germinating more than 200 years after being shed. Grass seeds found in a frozen condition in ancient lemming burrows in the Alaskan permafrost have been dated by radiocarbon measure-

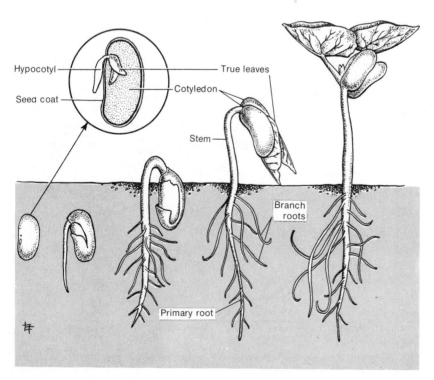

Figure 10–12 Stages in the germination and development of a bean seed. *Inset:* An enlarged view of an opened seed, showing cotyledons; the hypocotyl, from which develops the root; and the epicotyl, from which develop the stem and leaves.

ments as nearly 10,000 years old. When thawed, these have germinated! The ability of a seed to retain its germinating power depends on the thickness of the seed coat, on a low water content, and on the presence of starch rather than fats as the stored food material. Dormant seeds are alive and do metabolize, though at a very low rate.

Germination is initiated by warmth and moisture and requires oxygen. The embryo and endosperm absorb water, swell and rupture the seed coats. This frees the embryo and enables it to resume development. Most seeds do not need soil nutrients to germinate; they will germinate equally well on moist paper.

The cell divisions that the zygote undergoes following fertilization first produce a basal cell and a terminal cell; from the former develops a filament of cells, called the *suspensor*. The terminal cell divides, forming a rounded mass of cells. From this grow (in dicotyledonous plants) two primary leaves or *cotyledons* and a central *axis*. The part of the axis below the point of attachment of the cotyledons is called the *hypocotyl* and the part above it, the *epicotyl* (Fig. 10–12). The embryo is in about this state of development when the seed becomes dormant.

After germination the hypocotyl elongates and emerges from the seed coat. The primitive root or *radicle* grows out of the hypocotyl and, since it is strongly and positively geotropic, it grows directly downward into the soil. The arching of the hypocotyl in a seed such as the bean pulls the cotyledons and epicotyl out of the seed coat, and the epicotyl, responding negatively to the pull of gravity, grows upward. The cotyledons digest, absorb and store food from the endosperm while within the seed. The cotyledons of some plants shrivel and drop off after germination; those of other plants become flat foliage leaves. The cotyledons contain reserves of food that supply the growing seedling until it develops enough chlorophyll to become independent. The stem and leaves develop from the epicotyl.

will develop roots at their tips when placed in moist ground or in water containing a small amount of indoleacetic acid. Willow stems have an almost unbelievable ability to form roots and grow. The amateur gardener who cuts willow poles to support his beans or tomatoes may be dismayed to find that his willows have taken root and are growing better than his beans! A number of commercial plants—bananas, seedless grapes and navel oranges, to mention just a few—have lost the ability to produce functional seeds and must be propagated entirely by asexual means.

Many plants, such as the strawberry, develop long, horizontal stems called *stolons* or *runners*. These grow a meter or more along the surface in a single season and may develop new erect plants at every other node. Other plants spread by means of comparable, but underground, stems called *rhizomes*. Such weeds as witch grass and crab grass are particularly difficult to control because they spread by means of runners or rhizomes. Swollen underground stems or *tubers* also serve as a means of reproduction in plants such as the potato. Some of the cultivated varieties of potato rarely, if ever, produce seeds and must be propagated by planting a piece of tuber containing a bud or "eye."

The stems of raspberries, currants, and wild roses, and the branches of several kinds of trees may droop to the ground. Adventitious roots and a new erect stem may develop at one of the nodes touching the ground. This new plant may lose its connection with the parent plant, a form of asexual reproduction.

Grafting, the uniting of the stem of one plant with the stem or root of another, is not a method of reproduction, since it does not result in an increase in the number of individuals. It is widely used commercially to grow the stem of one variety that produces fine fruit on the root of another variety that has hardy, vigorous roots but produces poor fruit. Most Florida sweet oranges, for example, grow on trees grafted onto the root system of a sour orange variety.

10–11 ASEXUAL REPRODUCTION IN THE SEED PLANTS

Most of the cultivated trees and shrubs are reproduced from the *cuttings* of stems. These

10–12 ECONOMIC IMPORTANCE OF SEEDS

Seeds are used extensively by humans as sources of food, beverages, textiles and oils. Almost all of our carbohydrates are derived from seeds, the major exceptions being potato tubers, sugar cane and sugar beets. Wheat, rye, corn, rice, oats and barley are one-seeded fruits from members of the grass family; beans, peas and peanuts are the seeds of legumes and are rich sources of both proteins and carbohydrates. Coffee and cocoa are beverages derived from seeds, and a variety of spices and seasonings are made from ground seeds. Cotton fibers are produced as epidermal hairs on the seed coats of the cotton plant. Oils derived from seeds may be important industrially, or as foods. Linseed and tung oils are used in the manufacture of paints and varnishes. Oils from peanuts, cotton seeds and soy beans are used to make salad oils and margarine. Cocoanut oil is used in making soaps and shampoos as well as margarine.

10–13 EVOLUTIONARY TRENDS IN THE PLANT KINGDOM

As we glance back over the many types of life cycles that are found from algae to angiosperms, a number of evolutionary trends are evident. One of these is a change from a population that is mostly 1N individuals to one that

is almost entirely 2N (Fig. 10–13). In algae such as *Ulothrix* only one cell in each life cycle, the zygote, is 2N; all the rest are 1N. The haploid phase of the moss is more conspicuous and longer-lived than the diploid phase, but the latter is a complex, multicellular plant. The relative importance of the two phases is reversed in the ferns: The diploid, 2N, phase is the obvious, larger plant, and the haploid gametophyte, though still an independent plant, is small and inconspicuous. The gymnosperms and angiosperms show progressive reductions of the haploid, 1N, phase until, in the angiosperms, the male gametophyte consists of three cells and the female gametophyte of seven. The corollary of this trend toward diploidy is the trend toward the reduction of the gametophyte.

The evolutionary trend toward immobile adult plants, anchored in place by roots, solved several physiological problems, such as the supply of water and salts, but raised the reproductive problem of the means of bringing together the two kinds of gametes. Aquatic plants, like aquatic sessile animals, may have motile sperm that swim to the egg. Such a system persisted in some of the early terrestrial plants but the seed plants evolved pollen and the pollen tube. (The curious persistence of

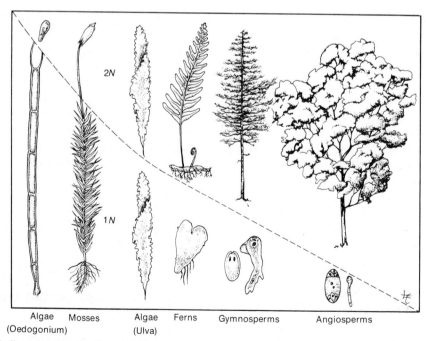

Algae   Mosses   Algae   Ferns   Gymnosperms   Angiosperms
(Oedogonium)   (Ulva)

Figure 10–13 A diagram showing the evolutionary trend toward a greater size and importance of the sporophyte (2N) and a reduction in the size of the gametophyte (1N) generation.

flagellate, motile sperm in the cycads and in *Ginkgo* emphasizes that in the course of evolution there have arisen several solutions to the problem of ensuring the fusion of sperm and egg.) Pollen may be very light and carried by the wind to a height of several thousand meters and for distances of many kilometers. The pollen of other species may be carried by insects and other animals.

Many species of plants are *monoecious*, with both sexes in the same plant, so that self-fertilization is possible. However continued self-fertilization leads to the loss of the advantages of sexual reproduction, the opportunity for genetic recombination. Self-fertilization may be regarded as a sort of safety device to provide for the fertilization of the egg by sperm from the same plant if no other sperm are available.

Immobile plants also have the problem of dispersing the species, spreading out to occupy a larger territory. Algae have evolved motile aquatic spores, fungi and ferns have windborne spores, and seed plants have seeds that may be windborne or carried by "hooking" into the fur of an animal. Some seed plants have attractive fruits that are eaten by certain animals plus seeds that resist digestion; thus the seeds eventually pass through the animal's digestive system, are excreted in the feces, and subsequently germinate.

Several explanations for these evolutionary trends may be advanced. As long as there was an independent gametophyte generation, the transfer of sperm to the egg required a film of water for the sperm to swim in. The evolution of a life cycle in which there is a reduction of the gametophyte to a small group of cells within the sporophyte and in which sperm are transferred to the egg via a pollen tube permitted reproduction to occur in the absence of moisture. The evolutionary advantages of this are obvious. There may be another, less obvious reason for this: a diploid individual can survive despite the presence of deleterious, recessive genes; a haploid individual would be much more susceptible to the effects of such genes. A third explanation is that, since terrestrial life required the development of conducting and supporting tissues, and since these have appeared primarily in sporophyte individuals (the appearance of xylem and phloem in

the gametophyte of the fork fern, *Psilotum,* is a remarkable exception to this rule), evolutionary processes on land favored those plants with longer sporophyte and shorter gametophyte generations.

QUESTIONS

1. What features distinguish asexual and sexual reproduction? What are the evolutionary advantages of the latter?
2. Discuss the various modes of asexual reproduction exhibited by the higher plants.
3. What trends in the evolution of the life cycle are evident from algae to flowering plants?
4. What are the principal characters distinguishing conifers and flowering plants?
5. In what ways do ferns resemble the seed plants? In what ways do they differ from them?
6. What exactly is a seed? What tissues are present and what are their respective functions?
7. Describe the major features of the life cycle of a pine tree. In what ways does the life cycle of an apple tree differ from this?
8. Draw a diagram of a flower and label the parts. What are the functions of each part?
9. Discuss the phenomenon of double fertilization in flowering plants.
10. What is a fruit? Differentiate clearly the several types of fruit.
11. What factors play a role in the germination of a seed?
12. Describe the development of the seed.
13. What are the parts of an embryo of a dicot? What does each become in the seedling?
14. Compare the roles of the cotyledons of dicots and monocots.
15. How are the following plants classified: (a) a ginkgo; (b) a mushroom; (c) a "horsetail"; (d) a pine tree; (e) a club moss; (f) an orchid; (g) a cactus?

SUPPLEMENTARY READING

Further details of the life cycle of plants and of the many different types of fruits and seeds will be found in textbooks of botany such as *Botany* by Wilson and Loomis, *Biology of Plants* by Raven and Curtis, *Plants, An Introduction to Modern Botany* by Greulach and Adams, and *Plant Biology* by Norstog and Long. More specialized books dealing with the morphology and development of the vascular plants include *Evolution and Classification of Flowering Plants* by A. Cronquist, *Comparative Morphology of Vascular Plants* by A. S. Foster and E. M. Gifford, Jr., and *Patterns in Plant Development* by T. A. Steeves and I. M. Sussex.

CHAPTER 11

GENERAL PROPERTIES OF GREEN PLANTS

The unique properties of chlorophyll enable green plants to carry out the process of photosynthesis, discussed earlier (p. 92). In those reactions that are the essence of photosynthesis—the light reactions—the remarkable properties of chlorophyll enable green plant cells to capture the radiant energy of sunlight with a high degree of efficiency and, with the aid of ferredoxin, cytochromes and the other compounds in the membranes of the grana, to produce NADPH, ATP, and molecular oxygen. The process of photosynthesis is, at present, the only significant way in which energy from the sun is made available for life on this planet. A tremendous amount of carbon (estimates place the amount at about 200 billion tons) is converted into organic matter each year by green plants. Land plants synthesize about one tenth of the total, and marine plants, mostly microscopic algae, synthesize the remainder.

In the dark reactions of photosynthesis, enzymes in the stroma of the chloroplasts condense carbon dioxide with ribulose diphosphate to yield two molecules of phosphoglyceric acid, which are converted to phosphoglyceraldehyde in a reaction driven by ATP and NADPH. The two phosphoglyceraldehydes condense to form fructose diphosphate, and a complex sequence of enzymatic transfers of two-carbon and three-carbon units from one carbohydrate to another eventually regenerates ribulose diphosphate and leads to the net synthesis of hexoses.

11–1 THE SYNTHESIS OF ORGANIC COMPOUNDS

The fructose diphosphate synthesized in the dark reactions of photosynthesis is readily converted enzymatically to glucose phosphate and fructose phosphate. Many plants produce large quantities of the disaccharide sucrose (our common table sugar) by combining in an enzyme-catalyzed reaction the phosphorylated forms of glucose and fructose. The reaction does not proceed by the simple removal of a molecule of water from free glucose and fructose. Glucose units can be combined to form macromolecules of starch or other polysaccharides by enzyme systems present in plant cells. These also utilize phosphorylated glucose, not free glucose molecules, as substrate and release inorganic phosphate as the glucose molecules are joined together in a chain. Starch, unlike glucose or sucrose, is relatively insoluble in water and many plants convert hexoses to starch for storage.

The enzyme systems in plant cells that synthesize lipids, proteins, nucleic acids and sterols are basically similar to those in animal cells. Phosphoglyceraldehyde is readily converted to glycerol, and phosphoglyceric acid is metabolized to acetyl coenzyme A (p. 107) from which fatty acids and steroids can be synthesized. Other enzymes synthesize amino acids by adding amino groups to certain of the organic acids that are intermediates in the metabolism of phosphoglyceric acid. The amino acids then react with ATP and transfer RNA and serve as substrates for the synthesis of proteins (p. 702). Still other biosynthetic systems produce vitamins and a host of other chemical substances. Energy, usually in the form of ATP, is required to drive all these biosynthetic reactions; the ATP is produced either by photophosphorylation in the chloroplasts or by oxidative phosphorylation in the mitochondria of the plant cells.

Cells from certain species of plants have enzyme systems that synthesize unique, characteristic substances of great economic importance: rubber, drugs such as quinine and morphine, spices, beverages such as coffee and tea, flavorings and perfumes. Animals, green plants, molds and bacteria have many enzyme systems in common and the general pathways of intermediary metabolism in all of these organisms are remarkably similar. Plant cells, however, have wider biosynthetic capabilities than animal cells, for they can carry on photosynthesis, they can synthesize all of the amino acids

and vitamins they require, and they can synthesize many other substances. Animal cells are unable to synthesize some eight or ten of the twenty-odd amino acids, and must obtain these "essential" amino acids, their vitamins, certain unsaturated fatty acids and an adequate supply of carbohydrates and fats for energy either directly or indirectly from plants.

11–2 CELLULAR RESPIRATION IN PLANTS

The series of enzymic reactions termed *cellular respiration* result in the utilization of oxygen, the release of carbon dioxide, and the transfer of energy from glucose and other substrate molecules to ATP and other forms of biologically useful energy. These occur in green plants as they do in every living cell. The raw materials and products of respiration and photosynthesis are the reverse of each other (Table 11–1).

When a plant is illuminated, the rate of photosynthesis is ten to thirty times greater than the rate of respiration, and the latter process is masked completely. Tracer experiments with ^{18}O show that respiration continues at about the same rate whether or not photosynthesis is occurring. Oxygen is carried to the cells, and carbon dioxide is carried from them by simple diffusion. No special respiratory organs are present. The larger plants have air spaces between the loosely packed cells which facilitate gas diffusion. The roots of plants may be asphyxiated if the surrounding soil is packed too tightly or if it is filled with water as in swampy land.

11–3 THE SKELETAL SYSTEM OF PLANTS

One of the requirements for photosynthesis is sunlight, and plant bodies have evolved

Table 11–1 COMPARISON OF PHOTOSYNTHESIS AND RESPIRATION

	Photosynthesis	Respiration
Occurs in:	Only those cells of green plants and a few bacteria which contain chlorophyll	All living cells
Raw materials:	Water and carbon dioxide	Oxygen and organic substances such as glucose
Time of occurrence:	Only when light shines on the cell	Continuously, night and day
Energy:	Stored by the process	Released by the process
Matter:	Results in an increase in the weight of the plant	Results in a decrease in the weight of the plant or animal
Products:	Oxygen and organic materials	Carbon dioxide and water

in a number of ways to ensure their chlorophyll-containing parts—the leaves—a place in the sun. Plant cells characteristically have a thick cell wall made of cellulose outside of, and in addition to, the plasma membrane. Unlike many animals, plants have no separate skeletal system for support.

Most algae are aquatic and have little need for specialized skeletal structures, for their bodies are generally small and supported by the water. The land plants however do need some structure strong enough to hold their leaves in position to receive sunlight. This has been achieved in two major ways: the cellulose wall can be very thick, as in the woody stems of trees and shrubs, and serve directly for the support of the plant body; or it can be rather thin and provide support indirectly by way of turgor pressure. Trees and shrubs have woody cells, tracheids and vessels, in the xylem for support. These cells secrete a very thick wall of cellulose impregnated with a complex, poorly defined chemical called *lignin.* Phloem, or inner bark, also contains thick fibers to help support the trunk.

11–4 TURGOR PRESSURE

To understand what turgor pressure is, how it provides support for the plant, and why it is decreased when a plant is immersed in salt water requires knowledge of basic plant anatomy and of the physical process of osmosis (p. 46). The plant cell, inside its cellulose wall, has one or more large vacuoles filled with *cell sap.* This sap is an aqueous solution of salts, sugars and other organic molecules. The plasma and vacuolar membranes, which separate the cell sap from the fluid outside the cell, are differentially permeable, and water molecules pass readily through their pores. Inorganic ions and organic molecules pass much less readily across these membranes. When the concentration of solutes in the cell sap is greater than the concentration outside the cell, as it usually is, water tends to enter, moving by diffusion from a region of higher concentration (of water) to a region of lower concentration. This additional water distends the vacuole, pressing the cytoplasm against the cellulose wall (Fig. 11–1). The cellulose wall is slightly

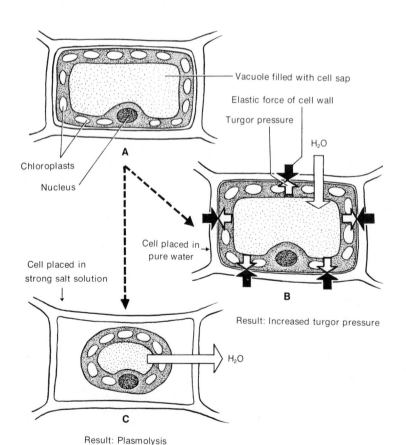

Vacuole filled with cell sap

Elastic force of cell wall

Turgor pressure

H_2O

Chloroplasts

Nucleus

Cell placed in pure water

Cell placed in strong salt solution

A

B

Result: Increased turgor pressure

H_2O

C

Result: Plasmolysis

Figure 11–1 Diagram illustrating the osmotic forces and the movement of water molecules that result in increased turgor pressure *(B)* or in plasmolysis *(C).*

elastic and is stretched by the internal pressure. After a certain amount of water has entered an equilibrium is reached, and the pressure exerted by the stretched cell wall equals the pressure of the cell sap. After this, the number of water molecules entering the vacuole is equaled by the number leaving, and the volume of the cell sap remains constant.

Turgor pressure, the pressure exerted by the contents of the cell against the cell wall, should not be confused with the *osmotic pressure* of the cell sap, the pressure that could be developed if the cell sap were separated from pure water by a membrane completely impermeable to all the solutes in the cell sap. Turgor pressure is less than the osmotic pressure of the cell sap because (1) the fluid outside the cell is usually not pure water but a dilute salt solution; and (2) the cell membranes are permeable to the salts and organic materials of the cell sap. These in time would diffuse through the membrane and reduce turgor pressure if it were not for the active processes of the living cell, which can pump certain substances selectively into the cell and extrude others. In addition, the cell by photosynthesis produces new organic molecules, increasing the concentration of solutes in the cell sap and increasing turgor pressure.

In all nonwoody plants, turgor pressure is important in maintaining the form of the plant body. In young cells, turgor pressure provides the force to stretch cell walls and makes possible cell growth.

11-5   PLASMOLYSIS

If the fluid outside the cell has a higher salt concentration than the cell sap, as when a lettuce leaf is placed in concentrated salt solution, water from the cell sap diffuses out of the cell, passing from a region of higher water concentration to a region of lower water concentration. This decreases the turgor pressure within the cell and the lettuce leaf wilts.

When the volume of the cell sap decreases owing to the loss of water, the cell is no longer pressed against the cellulose cell wall. Instead, it shrinks away from the cell wall, a process called *plasmolysis* (Fig. 11-1). If plant cells are exposed to a hypertonic solution for too long, they die. If exposed for only a short time and

then returned to pure water they can regain their turgidity. To demonstrate this, try putting half a dozen carrots in a beaker of salt solution and removing them one at a time, after varying lengths of time, to pure water.

11-6   PLANT DIGESTION

Plants have no specialized digestive system; their nutrients are either made within the cells or are absorbed through the cell membranes. The nutrients synthesized are either used at once or transported to another part, such as the stem or root, where they are stored for later use. The few insect-eating plants, although without an organized digestive system, do secrete digestive enzymes similar to those secreted by animals.

Plants accumulate reserves of organic materials to be used when photosynthesis is impossible — at night or over the winter. An embryo plant cannot make its own food until the seed has sprouted and the embryo has developed a functional root, leaf and stem system. Most seeds contain rich reserves of carbohydrates and fats; these supply energy for the initial stages of growth.

11-7   PLANT CIRCULATION

The simpler plants, consisting of single cells or small groups of cells, have no circulatory system. Simple diffusion, augmented in certain instances by facilitated diffusion or active transport, suffices to bring in the substances the plant requires: water, carbon dioxide and salts. The circulatory systems of higher plants are simpler than those of higher animals and are constructed on an entirely different plan. Plants have no heart and no blood vessels. Transportation is accomplished by the xylem and phloem systems; a few plants have an additional latex system to assist in circulation. *Latex* is a milky material, rich in food substances (carbohydrates and proteins). The latex of certain plants yields commercially valuable products such as rubber, chicle and opium.

Two kinds of liquid-conducting cells are present in the xylem: the *tracheids* are long, tapered cells with lateral pits; and the *vessel elements* are cylindrical cells, usually having a greater diameter than the tracheids, in which the end walls have disappeared. The vessel elements are arranged end-to-end to form a continuous tube, the *xylem vessel.* The tracheids and the vessel elements die at maturity and function as conductive cells after the cytoplasm has disappeared from within the cell. Their main function is the conduction of water and dissolved minerals.

Four types of cells are present in phloem: the *sieve tube elements, companion cells, phloem fibers* and *parenchyma.* The principal function of the phloem, the transport of nutrients, is carried out by the sieve tube elements. These lack nuclei as mature cells and form continuous columns of cells, the *sieve tubes,* in which the cytoplasm of one cell is continuous with that of the next and extends through perforations in the cell wall called *sieve plate pores.* Adjacent to each sieve tube element is a smaller, nucleated companion cell which is believed to play some role in correlating the conduction of nutrients in the sieve tubes. The phloem of the nonflowering vascular plants has only sieve cells without companion cells. The parenchyma cells of the phloem are associated with the formation of cork in perennial stems and the phloem fibers strengthen the conductive tissue and provide mechanical support. The phloem fibers of plants such as hemp and flax are made into rope and linen, respectively.

Many plant cells are near enough to the surface to obtain their oxygen and carbon dioxide directly from the atmosphere. The rest receive their oxygen from the air spaces or canals that penetrate the deeper parts of the plant; xylem vessels and tracheids and phloem tubes have little or nothing to do with the transport of gases. The xylem vessels and tracheids are chiefly concerned with transporting water and minerals from the roots up the stem to the leaves, while phloem tubes transport food which has been manufactured in the leaves down the stems for storage and use in the stems and roots.

Phloem transports nutrients up the stem as well as down; in the spring, for example, substances pass from their places of storage to the buds to supply energy for growth. The conduction of nutrients in the phloem involves not the flowing of a plant sap, as in the xylem, but rather the circular flowing of the cytoplasm

within each phloem cell. The conduction of water in the xylem and of nutrients in the phloem is termed *translocation.* The two processes have somewhat different molecular bases. Water rises in the vessels and tracheids of the xylem (which are not living cells) by the combined forces of transpiration and root pressure. Nutrients are transported in the living phloem cells with remarkable rapidity and by mechanisms not yet fully understood. Experiments with carbon-14 (^{14}C)–labeled materials have shown that the sugars made photosynthetically are translocated from the leaf to the root at rates as high as several hundred centimeters per hour. Such rates could not be achieved by diffusion alone; translocation involves in addition the circular streaming motion (*cyclosis*) of the cytoplasm in each individual phloem cell. This functions as a sort of "bucket brigade" to pass along the molecules of nutrients from cell to cell.

The connections of the phloem tubes and xylem vessels are fundamentally different from those in arteries and veins. In the latter a large tube branches successively into smaller and smaller tubes. All xylem and phloem tubes are small and occur in bunches called vascular bundles. In the lower part of the stem there are many vessels per bundle; in the upper part there are fewer per bundle, the others having entered the branches of the stem.

11–8 PLANT SAPS

The material in the xylem, phloem and latex tubes of the higher plants, which is collectively called *plant sap,* is somewhat analogous to the blood plasma of humans and higher animals. It is a complex mixture of many substances, organic and inorganic, whose composition varies greatly from one plant to another, from one part of the plant to another, and from season to season. As much as 98 per cent may be water. Other constituents include salts, sugars, amino acids, hormones such as indoleacetic acid, enzymes and other proteins, and organic acids such as citric and malic acids. Citric acid was first isolated from citrus fruits, and malic acid from apples. Plant saps, in contrast to the blood plasma in animals, are usually somewhat acid, with pH's ranging from 7.0 to 4.6.

11–9 PLANT EXCRETION

A striking difference between plants and animals is that plants excrete little or no nitrogenous wastes. Those excreted by animals—urea, uric acid and ammonia—come from the breakdown of proteins, nucleic acids and other substances. Similar nitrogenous compounds are formed by metabolic processes in plants, but instead of being excreted as wastes, they are reutilized in the synthesis of other substances. Plants "recycle" their constituent compounds.

Since plants neither carry on muscular activity nor, with minor exceptions, ingest proteins (the two largest sources of metabolic wastes in animals), the total amount of nitrogenous waste is small and can be eliminated by diffusion as ammonia through the pores of the leaves or by diffusion as nitrogen-containing salts from the roots into the soil. In some plants, a few waste products accumulate and remain as crystals within the cells; for example, spinach leaves contain about 1 per cent oxalic acid. Wastes deposited in the leaves are eliminated when the leaves are shed.

11–10 PLANT COORDINATION

The activities of the various parts of a plant are much more autonomous than are those of the parts of an animal. The coordination between parts that does exist is achieved largely by direct chemical and physical means, since plants have developed no specialized sense organs and no nervous system.

Actively growing plants can respond to a stimulus coming from a given direction by growing more rapidly on one side and hence bending toward or away from the stimulus. Such a *growth* response, called a **tropism,** can occur only in those parts of the plant that are growing and elongating, but the stimulus for it may be received in some distant part of the plant. If an organism is motile, it may respond to a stimulus by moving toward or away from it. An orientation movement in response to a stimulus is called a **taxis.** Taxic responses are limited to cells capable of movement: those of animals and some lower plants, and the male sex cells (antherozoids) of mosses or ferns.

Tropisms and taxes are named for the kind of stimulus eliciting them: **phototropism** (a growth response to light, Fig. 11–2); **geotropism** (a growth response to gravity, Fig. 11–3); **chemotropism** (a growth response to some chemical); and **thigmotropism** (a growth response to contact, Fig. 11–4). Ivy plants are famous for their thigmotropism—they grow so as to maintain contact with a wall, tree or some other supporting object.

Tropisms and taxes may be either positive or negative—the response may be either toward or away from the stimulus. For example, no matter how a seed may be oriented in the ground, the primitive root and shoot of the developing embryo grow so that the root grows downward and the shoot (stem) grows upward (Fig. 11–5). Thus, the root is positively geo-

Figure 11–2 Phototropism in young radish plants. *A,* In the dark or in uniform light, the plants grow straight upward. *B,* When exposed to light coming from a single direction, they quickly bend toward it—that is, they are positively phototropic. The photograph on the right was taken half an hour after the one on the left. (From Weatherwax, P.: *Botany,* 3rd Ed. Philadelphia, W. B. Saunders Co., 1956.)

A   B

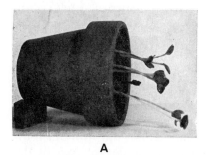

A   B

Figure 11–3 Geotropism in young radish plants. *A,* Pot of straight radish seedlings placed on its side and kept in the dark to eliminate phototropism. *B,* Within 30 minutes the seedlings had bent; since they bend away from the direction of the force of gravity, they are negatively geotropic. (From Weatherwax, P.: *Botany,* 3rd Ed. Philadelphia, W. B. Saunders Co., 1956.)

tropic and the shoot is negatively geotropic. A growing plant can be fooled by fastening it to a disc and spinning the disc, thus imposing a field of centrifugal force on the growing plant. The plant will respond to this force that has replaced gravity and the shoot will grow toward the center of the disc (which corresponds to up) and the roots will grow toward the periphery.

When plants are placed in the dark, they will grow for a while, using the energy stored in the seed, and will form stems and leaves that contain no chlorophyll at all. Such plants, called *etiolated,* will still respond to a beam of light by growing toward it, so we can conclude that the phototropic response is not brought about by chlorophyll. By shining light of different wavelengths on etiolated plants and noting the resulting bending, it was found that blue or violet light was much more effective in eliciting a phototropic response than a green, yellow or red light. (In contrast, red light is most effective in photosynthesis.) This indicated that the pigments responsible for the phototropic response are yellow or orange, perhaps carotenes or xanthophylls.

Figure 11–4 Response of a squash tendril to touch. The straight tendril *(A)* is stroked with a stick *(B)*. Five minutes later *(C)* it is beginning to bend, and within 20 minutes it has completed one coil *(D–F)*. (From Weatherwax, P.: *Botany,* 3rd Ed. Philadelphia, W. B. Saunders Co., 1956.)

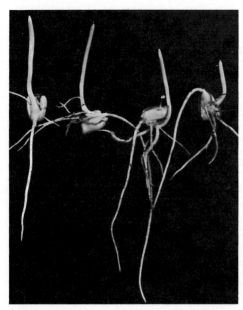

Figure 11–5 Geotropism in the roots and shoots of corn seedlings. The grains were planted in various positions, but the roots tend to grow downward and the shoots upward. (From Weatherwax, P.: *Botany,* 3rd Ed. Philadelphia, W. B. Saunders Co., 1956.)

Since tropistic responses require the actual growth of cells (one side of the plant grows faster than the other) they are necessarily slow and require from an hour to perhaps a week for completion. This differential acceleration of growth, as well as normal uniform growth, depends upon the stimulation of cells by plant growth hormones, or **auxins.** As yet there is no complete explanation as to how light, gravity or any of the other effective stimuli bring about the production of auxins from tryptophan nor how the auxins are distributed differentially to the two sides of the plant. It is known that the movement of auxin from cell to cell is brought about by an oxygen-dependent transport system. The movement is polarized, with the auxin passing from the apex toward the base of the plant. It is obviously of advantage to a plant to grow toward the light and to have its roots grow toward moisture, but no one supposes that the plant *purposefully* grows toward the light because it needs light for photosynthesis.

Light does not react with the auxin itself but reacts with the photoreceptor, bringing about an effect which influences hundreds of auxin molecules, perhaps by an effect on the enzyme system that produces the auxin.

11–11 THE TRANSMISSION OF IMPULSES

All living cells, whether animal or plant, exhibit irritability to some extent and can transmit an excitation, even though only slowly. The excitation produced by the penetration of an egg by a sperm travels across the surface of the egg at a rate of about 1 centimeter per hour. The unspecialized cells of sponges transmit excitations at about 1 centimeter per minute. Plant cells can also transmit excitations, although usually the rate of propagation is so slow that the results are not easily observed. In a few plants, however, responses to stimuli do occur rapidly enough to be readily seen. One of these is the response of the Venus

Figure 11–6 *A,* The "sensitive plant," *Mimosa pudica,* before being disturbed. *B,* The plant five seconds after being touched; note how the leaves have folded and drooped. (Courtesy of General Biological Supply House, Chicago, Ill.)

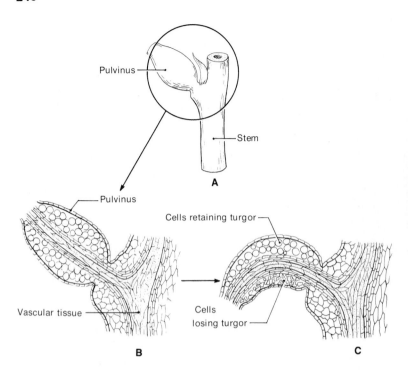

Figure 11–7 The mechanism of response in *Mimosa pudica. A,* The base of the petiole showing the pulvinus. *B,* Section through the pulvinus showing condition of cells when leaf is extended horizontally. *C,* Section through the pulvinus showing cells losing turgor to produce folding of the leaves.

flytrap (Fig. 3–2) to the presence of an insect on the leaf, and another is the response of the "sensitive plant," *Mimosa pudica,* to touch (Fig. 11–6).

Normally the leaves of the latter plant are horizontal, but if one of them is lightly touched, all the leaflets fold within two or three seconds. Touching one leaf sharply causes not only the stimulated leaf but also the neighboring leaves to fold and droop. After a few minutes the leaves return to their original position. The folding of the leaves results from a decrease in the turgor pressure of the cells at the base of the leaf (Fig. 11–7), but the excitation is transmitted along the sieve tubes of the leaves and stems. The rate of transmission in *Mimosa* is about 5 centimeters per second. This is slow when compared to the rate of transmission of a nerve impulse in the higher vertebrates, which is about 120 meters per second, but the nature of the impulse is fundamentally the same. The excitation is accompanied by electrical phenomena, increased permeability of the excited cells, and a temporary change in their metabolism. The response can be altered by various drugs.

11–12 PLANT HORMONES

Plant hormones, like animal hormones, are organic compounds which can produce striking effects on cell metabolism and growth even though present in extremely small amounts. The plant hormones are produced primarily in actively growing tissue, especially by meristem tissue in the growing points at the tip of stems and roots. Like animal hormones, the plant hormones usually exert their effects on parts somewhat removed from the site of production. The plant hormones have many different types of effects on metabolism and cell division: (1) they stimulate the lengthwise growth of individual cells in the growing part of the plant; (2) they initiate the formation of new roots, especially adventitious roots; (3) they initiate the development of flowers and the development of fruit from the flower parts; (4) they stimulate cell division in the cambium; (5) they inhibit the development of lateral buds; and (6) they inhibit the formation of abscission regions (p. 269) and hence prevent the fall of leaves and fruit.

There are several groups of naturally occurring, chemically defined substances that function in the control of growth and development in flowering plants: the *indole auxins, ethylene,* the *cytokinins,* the *gibberellins,* and *abscisic acid.*

Auxins. Some of the first experiments with growth-promoting substances were made by Charles Darwin and his son Francis. It had been known for many years that plants would grow toward the light (were positively photo-

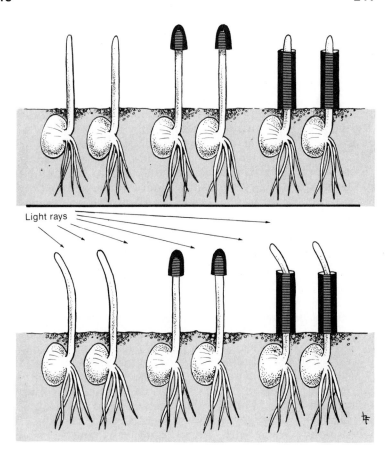

Figure 11-8 Darwin's experiment with canary grass seedlings. *Upper row,* Some plants were uncovered, some were covered only at the tip and others were covered everywhere but at the tip. After exposure to light coming from one direction *(lower row)* the uncovered plants and the plants with uncovered tips bent toward the light; the plants with covered tips *(center)* grew straight up. Darwin's conclusion: The tip of the seedling is sensitive to light and gives off some "influence" that moves down the stem and causes the bending.

Light rays

tropic). To see what part of the plant received the light stimulus, Darwin grew a number of canary grass seedlings, covered the tips of some with black paper caps and covered everything but the tips of others with black paper cylinders (Fig. 11–8). He then put all the seedlings near a window so that they received light from one direction only. By the next day the seedlings with no cover at all and the seedlings with everything but the tip covered were both bent very strongly toward the light, but the capped seedlings had grown straight up. From these experiments Darwin concluded that the light was received by the tip of the plant and that some "influence" moved down the stem from the tip to cause the plant to bend.

The investigations of Boysen-Jensen of Denmark, Frits Went of Holland and others from 1910 until about 1930 clarified the mechanism underlying tropistic responses and showed that this "influence" was a plant growth hormone. These classic experiments were performed using the *coleoptile* of the oat seedling. The coleoptile is a hollow, practically cylindrical organ that envelops the unexpanded leaves like a sheath. After a certain stage of growth, its further increase in length is due almost entirely to cell elongation. In 1910 Boysen-Jensen discovered that if the coleoptile tip is cut off at this stage, the decapitated coleoptile immediately stops elongating. If the tip is replaced, the coleoptile again begins to grow (Fig. 11–9).

Placing a thin sheet of mica between the tip and the rest of the coleoptile also prevented cell elongation; however, a thin sheet of gelatin could be placed between the tip and the rest of the coleoptile without interfering with the growth of the coleoptile. Indeed, the decapitated tip could be placed for a time on a block of gelatin and then be removed, and the bit of gelatin, when placed on the coleoptile, would stimulate it to grow (Fig. 11–9 *D*). These experiments showed that the growth of the coleoptile is controlled by some substance produced in the tip that normally passes downward and stimulates the coleoptile cells to elongate. By leaving a tip on a gelatin block for varying lengths of time it was shown that the amount of growth substance that diffused into the gelatin was proportional to the length of time they were in

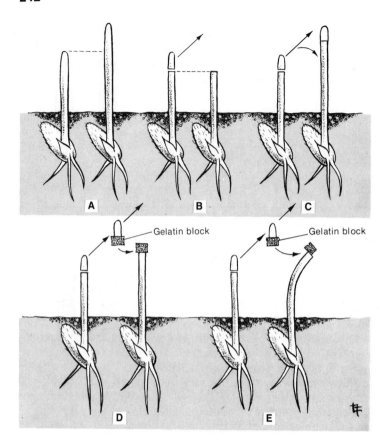

Figure 11–9 A series of experiments that demonstrate the existence and mode of action of plant growth hormones in oat coleoptiles. In each pair of drawings, the figure on the left indicates the experiment performed and the figure on the right the growth after a period of time. *A,* Control: No operation performed, normal growth. *B,* If the tip of the coleoptile is cut off and removed, no growth occurs. *C,* If the tip is cut off and then replaced, normal growth ensues. *D,* If the tip is cut off, placed on a block of gelatin for a time and then the gelatin block, but not the coleoptile tip, is placed on the seedling, growth occurs. *E,* If the tip is placed on a gelatin block for a time and then the gelatin block is placed asymmetrically on the seedling, curved growth results.

contact. The amount of growth substance was measured by its effect on the elongation of the coleoptile.

Further experiments by Frits Went showed that if the gelatin block is placed on one side of the decapitated coleoptile, growth is asymmetrical; the coleoptile bends *away* from the side on which the block was placed; i.e., growth is more rapid in the cells directly under the gelatin block (Fig. 11–9 *E*). This test is extremely sensitive and Went could measure the growth-promoting substance, called **auxin,** in terms of "curvature units"—the number of degrees of bending produced in the coleoptile. This is an example of a **bioassay,** a test in which some chemical substance is measured or assayed in terms of its effect on some biological system.

Auxins are present only in minute quantities even in actively growing tissues. The growing shoot of the pineaple plant contains about six micrograms of indoleacetic acid per kilogram of plant material. As J. P. Nitsch concluded, this is comparable to the weight of one needle in a 22-ton haystack!

Indoleacetic acid, the primary natural auxin, is rapidly metabolized by indoleacetic acid oxidase. This enzyme is inhibited by certain orthodiphenols. The growth-promoting effects of orthodiphenols had been recognized for some time and they were initially believed to be auxins themselves; however, they promote growth by inhibiting indoleacetic acid oxidase and thus raising the effective concentration of endogenous indoleacetic acid.

Indoleacetic acid is synthesized from tryptophan, but how light falling on the growing tip of a plant stimulates the conversion of tryptophan to indoleacetic acid remains unclear. Auxin is transported down the stem at speeds ranging from 0.5 to 1.5 centimeters per hour. The kinetics of auxin transport suggest that this is an active process, driven by metabolic processes; the substance transported must have a certain molecular configuration. The synthetic auxin, **2,4-D** (2,4-dichlorophenoxyacetic acid) is a potent stimulator of plant metabolism but is only poorly transported down the stem and thus does not produce correlated growth of the several parts of the plant as indoleacetic acid does.

The growth response of the coleoptile sheath is a metabolic process, for it will occur only in the presence of oxygen and growth is improved if glucose is supplied as a source of energy. The roots, buds and stems each have growth responses that occur at different concentrations of auxin. Each shows an increasing growth response at low concentrations of auxin and then an *inhibition* of growth at higher concentrations (Fig. 11–10). The auxin concentration producing optimal growth of stems is much higher than the concentration for optimal growth of roots or buds.

The bending of a plant shoot in response to light, or the bending of the root in response to gravity, is due to differential growth resulting from the differential distribution of auxin. Such tropistic responses can be separated into: (1) the perception of the stimulus, light or gravity; (2) the induction of a lateral physiological difference; and (3) a lateral response expressed as differential growth. The perception of light involves the absorption of light energy by pigments such as carotenoids or flavins. Unilateral illumination sets up a transverse gradient of absorbed energy and induces differential physiological activity, the production of auxin. If the tip of a coleoptile is illuminated unilaterally, then cut off and placed on the stump of another coleoptile, the latter will respond with differential growth and will bend toward the source of the light originally reaching the coleoptile tip. Since the side toward the light should absorb more energy, and since the production of auxin is an energy-requiring process, this, at first glance, would seem to account for the bending toward the light. However, a bending toward the light means that the *shaded* side of the plant must grow faster and have more auxin.

Further experiments have shown that the plant redistributes the auxin that is synthesized and transports it laterally to the shaded side. Experiments with ¹⁴C-labeled auxin demon-strated the lateral transport of the auxin following unilateral stimulation by light or gravity. The amount of cellular elongation is proportional to the auxin concentration and hence the growth on the shaded side, with the greater concentration of auxin, is greater than the growth on the lighted side.

The plant's responses to gravity, its **geotropisms,** are also determined by the distribution of auxin. When a plant is placed on its side (Fig. 11–3) auxin accumulates in the cells on the lower side of the stem. This causes greater elongation in those cells, and the stem's growth is curved so that it grows upright. The cells of the root are much more sensitive to auxin than are the cells of the stem (Fig. 11–10) and their cellular elongation is *inhibited* by concentrations of auxin that exceed 10^{-10} M. The increased auxin concentration on the lower side of a horizontal root inhibits elongation whereas the lesser concentration on the upper side stimulates it and the root grows downward.

Auxin may bring about the differentiation of tissues in the parts of the plant through which it is transported or to which it is taken. The differentiation of xylem in a polar fashion results in the production of an integrated system of pipes that pass vertically through the plant from the apical meristem and expanding leaves down through the stem to the root tips. This differentiation is directed in the terminal regions of the stem by the apical meristem. In an interesting series of experiments, Wetmore and Sorokin showed that a bit of lilac callus grown in tissue culture would not form xylem cells unless a piece of meristem was grafted onto it. However, if auxin was applied in a localized position in place of a meristem graft, some xylem cells were formed in the callus.

Figure 11–10 The stimulation of growth in roots, buds and stems produced by varying the concentration of indoleacetic acid in the medium bathing them. Note that each growth is first stimulated, then inhibited, as the concentration of auxin is increased.

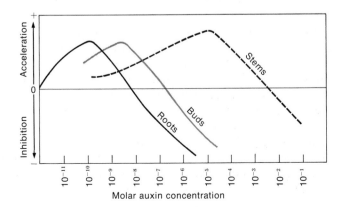

The apical meristem appears to be a source of auxin which passes down into the tissues below and participates in directing them toward differentiation as xylem.

The differentiation of roots is also controlled by auxins. It has long been known that the lower tip of a cut stem placed in water may form roots. Cuttings farther down the stem or from older wood show a lesser ability to form roots; this agrees with the decreasing amount of auxin toward the base of the stem. By placing a cutting in a dilute solution of either natural or synthetic auxin, roots can be readily produced.

Auxins and other plant hormones determine the growth correlations of the several parts of a plant. The terminal bud of the stem normally inhibits the development of lateral buds. If the terminal bud is cut off, the lateral buds, freed of the inhibition induced by the auxin produced in the apical meristem of the terminal bud, may begin to develop. That this inhibition is caused by auxin can be demonstrated by removing the terminal bud and replacing it with an appropriate amount of indoleacetic acid held in place by a suitable material such as fat. The lateral buds of a stem treated in this fashion remain inhibited.

The abscission or shedding of leaves, flowers, fruits and stems from the parent plant is another process controlled by auxin. Auxin produced in leaves passes down the petiole and inhibits the development of the abscission zone. As long as the leaf continues to produce auxin its abscission, or shedding, is inhibited. Abscission is a natural indicator of the decreased auxin formation that normally accompanies aging. Low concentrations of auxin promote abscission and high concentrations inhibit abscission.

Auxins have a variety of practical uses and are of tremendous economic importance in stimulating the growth of roots from cuttings (Fig. 11-11), in producing *parthenocarpic fruits* (ones formed without pollination and hence without seeds), in hastening the ripening of fruit, and in preventing their dropping from the tree before harvest. The synthetic auxin 2,4-D is widely used as a weed killer. Most of the common weeds are dicotyledonous plants and are much more sensitive to stimulation by auxins than are the monocots such as grasses. A lawn sprayed with the proper concentration of 2,4-D, enough to stimulate the weeds but not

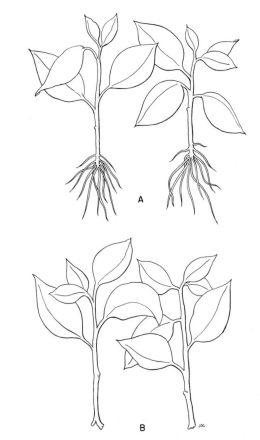

Figure 11-11   The effect of indoleacetic acid in stimulating the formation of roots in cuttings from lemon trees. The cuttings in *A* had been soaked in a dilute solution of indoleacetic acid (500 parts per million) for 8 hours 18 days previously. The control cuttings *(B)* had been soaked in plain water for a similar length of time.

the grass, will be freed of weeds. They take up the 2,4-D and are stimulated to metabolize at a high rate, consuming their cellular constituents and finally dying. Many of the important crop plants—corn, oats, rye, barley and wheat—are monocotyledons and millions of acres of these crop plants are treated each year with 2,4-D to eliminate weeds and to increase the yield of these plants.

The molecular mechanism by which auxins produce their remarkably diverse effects is not yet clear. The elongation of a plant cell in response to auxin requires the uptake of water by the cell and a softening of the cell wall to permit the cell to swell as it takes up water osmotically. Auxin does indeed soften the cell wall and increase its plastic bending in response to a standard force. It might do this by altering the chemical structure of the pectins in the cell wall, and some experimental evidence supports this hypothesis. The finding that actinomycin D, which inhibits the synthesis of RNA, also inhibits the elongation of plant cells induced by auxin indicates that the synthesis of RNA is required for at least some aspect of the cell elongation induced by auxin. An enlarging cell would have more plasma membrane and the RNA synthesis might be involved in producing the structural components of the plasma membrane.

Auxin may be regarded as the most important of the plant hormones, since it has the most marked effects in correlating growth and differentiation so that the normal pattern of development results. Auxin, together with certain other chemical messengers, brings about the differentiation of the dividing plant cells into a true organism instead of into a simple multicellular colony.

Ethylene. It had been known for many years that overripe fruits produce something that hastens ripening in adjacent fruits ("one rotten apple will spoil a barrel"). This fruit-ripening factor has proved to be the simple unsaturated hydrocarbon, *ethylene.* This is a component of manufactured coal gas and it was identified by the hastening of ripening of fruit exposed to accidental leakages of coal gas. Ethylene is a natural plant product which is produced by ripe fruits and acts like a plant hormone. Ripe fruits produce ethylene which stimulates ripening of adjacent fruits (and the production of ethylene by them). Fruits such as bananas that are picked green for transport to market are treated with ethylene so that they will be properly ripe when they reach the market. Premature ripening of fruit in a warehouse can be prevented by ventilation to remove the ethylene and by increasing the content of carbon dioxide in the air, for carbon dioxide counters the effect of ethylene.

Gibberellins. *Gibberellins* increase the length of the stem of some species of plants and increase the size of fruits in others. A unique effect of gibberellins is the stimulation of the germination of seeds. The seeds of wheat, oats, barley and other cereals have two major parts, the *embryo* and the reserve food supply, the *endosperm* (Fig. 11–12). The storage cells of the endosperm appear to be dead but are surrounded by the *aleurone,* a three-layered coat of living cells. During germination the starch in the storage cells is hydrolyzed by *α-amylase,* secreted by the aleurone layer; however, the embryo must be present

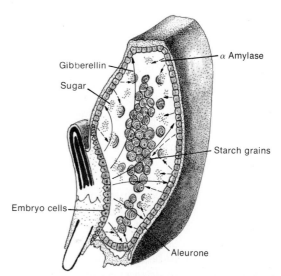

Figure 11–12 The action of gibberellin in the germination of barley grains (highly diagrammatic). Following water inhibition, cells of the embryo produce and secrete gibberellin. The gibberellin in turn activates the synthesis of starch-digesting enzymes (amylases) in the living cells of the aleurone. The amylases are secreted into the starchy endosperm, where the hydrolysis of starch to sugar occurs. (From Norstog, K., and Long, R. W.: *Plant Biology.* Philadelphia, W. B. Saunders Co., 1976.)

before the aleurone layer will secrete α-amylase. If a grain of barley is cut in half and half with the embryo is removed, the starch in the remaining half will not undergo hydrolysis. Gibberellin, a chemical messenger secreted by the embryo, activates the cells in the aleurone layer to produce and secrete α-amylase. Gibberellins activate other enzymes that break down the material in the cells of the seed coats, weakening them so that the growing embryo can burst through. The formation of proteolytic enzymes in the aleurone layer of the endosperm is also stimulated by gibberellin. The resulting hydrolysis of proteins liberates tryptophan, the precursor of indoleacetic acid, in the tip of the coleoptile of the young embryo.

The synthesis of α-amylase by the aleurone layer in response to gibberellin is completely inhibited by actinomycin D. Actinomycin D is a peptide antibiotic which complexes with DNA through the amino group of guanine and displaces DNA-dependent RNA polymerase. Thus the DNA-dependent synthesis of RNA is inhibited. This evidence suggests that gibberellins regulate in some fashion the expression of the genetic information contained in the DNA, perhaps by uncovering a portion of the DNA so that it can be transcribed to produce a specific RNA which in turn results in the formation of the specific enzyme.

More than a dozen different gibberellins have been isolated from plants, each with some slight difference in structure and some difference in biological activity (Fig. 11–13). Some will induce flower formation, others will not. One will produce male sex organs, *antheridia*, on fern gametophytes, but the others will not.

Cytokinins. A third type of plant hormone, the *cytokinins*, stimulate growth of cells in tissue culture or organ culture and have a marked effect in increasing the rate of cell division. *Zeatin*, isolated from young corn seeds, was shown to be a derivative of adenine, 6-(4-hydroxy-3-methyl-*trans*-2-butenyl amino) purine (Fig. 11–14). Cytokinins occur in plants in such small amounts that they were identified only by the technique of mass spectrometry. The side chain attached to the amino group at position 6 of the purine is an isoprenoid and, like the gibberellins, is derived from mevalonic acid. A substance very closely related to zeatin, 6-(γ,γ-dimethylallyl amino) purine was isolated as the ribonucleoside from both serine and tyrosine transfer RNAs from yeast, calf liver, peas and spinach (see p. 700). In both kinds of tRNA this odd base is located just adjacent to the anticodon. Five other transfer RNAs have been tested and shown to lack this curious base.

Cytokinins not only promote cell division but can change the structure of plant cells growing in culture. When the concentration of cytokinin in the culture medium is very low (10^{-9} M or less) only loose, friable tissues appear. At somewhat higher concentrations, 10^{-8} to 10^{-7} M, roots develop on the mass of cells in culture, and finally at somewhat higher concentrations, 3×10^{-6} M, shoots are induced. This experimental demonstration that cyto-

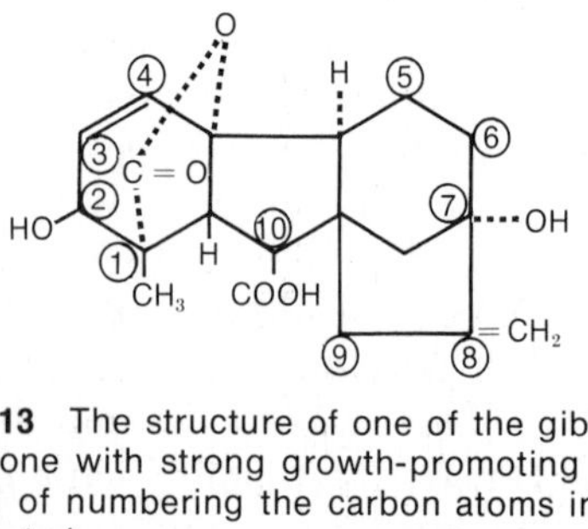

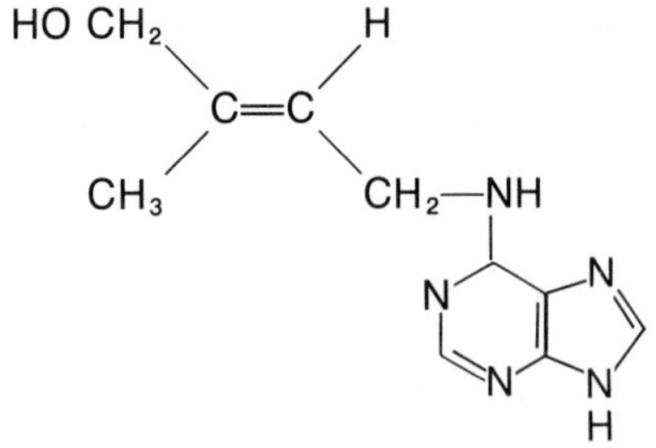

Figure 11–14 The structure of zeatin, a cytokinin.

kinins may control the relative production of shoots and roots in tissue culture preparations suggests that they may perform a similar function in the intact plant.

It is not yet possible to link cytokinins with any specific biochemical reaction and just how they stimulate cell division is unknown. DNA synthesis in dividing tobacco cells is stimulated by the addition of cytokinin.

In the plant, the three major types of plant hormones interact in pairs or all three may interact to regulate specific biological phenomena. The optimal growth of tobacco callus tissue in culture requires specific concentrations of all three factors. Cytokinins and gibberellins have dominant roles in controlling the early phases of growth and development and auxins become dominant later in controlling cell elongation.

Abscisic Acid. Some plant hormones result in an inhibition, rather than a stimulation, of growth and development. One of these, termed *abscisic acid* (Fig. 11–15), causes abscission in cotton and bud dormancy in birch trees. The application of abscisic acid to an actively growing twig of a woody plant results in cessation of elongation of the internodes; some of the leaves develop abscission layers and drop off, young developing leaves form scale leaves instead of foliage leaves, and the terminal bud becomes quiescent. Similar responses in twigs are normally seen at the onset of the winter season and abscisic acid may be described as a "dormancy-inducing hormone." Several synthetic compounds have comparable effects in retarding plant growth without otherwide affecting the health of the plant. One of these, named AMO-1618, suppresses the elongation of stems, either by blocking the action of gibberellins or by inhibiting the movement of auxin in the stem. Such growth retardants are

Figure 11–13 The structure of one of the gibberellins, a plant hormone with strong growth-promoting properties. The system of numbering the carbon atoms in the molecule is indicated.

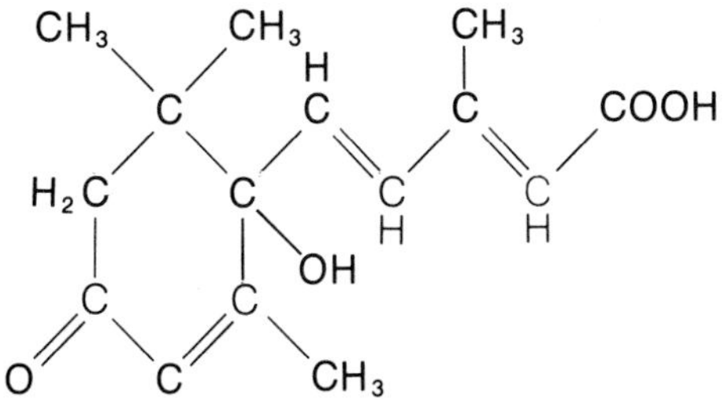

Figure 11–15 Abscisic acid.

used by florists to produce short-stemmed chrysanthemums and other house plants and by farmers to produce bushy rather than tall crop plants with increased production of fruit.

11–13 PHOTOPERIODISM: FLORIGENS AND PHYTOCHROMES

It has, of course, been known for a long time that different kinds of plants flower at different seasons of the year and that the time of flowering can be related to the number of hours of daylight per day, the *photoperiod.* The role of the ratio of daylight and darkness in determining the time of flowering of plants was unknown until 1920, when Garner and Allard, of the United States Department of Agriculture, demonstrated that the time of flowering of tobacco plants could be altered by changing the *photoperiod.*

Some species of plants (e.g., asters, cosmos, chrysanthemums, dahlias, poinsettias and potatoes) are *short-day plants* and will not produce flowers if the photoperiod exceeds a certain critical length, i.e., if there is more than a certain number of hours of daylight per day. Such plants normally flower in the early spring, late summer or fall. Others, termed *long-day plants* (e.g., beets, clover, coreopsis, corn, delphinium and gladiolus), require a photoperiod that *exceeds* a certain critical length for flowering to take place. The length of the critical photoperiod varies from species to species, ranging from 9 to 16 hours, and is not necessarily short in a short-day plant. Indeed, some short-day plants have a longer critical photoperiod than some long-day plants. The important difference is that the photoperiod must be *less* than the critical period in short-day plants to induce flowering but *greater* than the critical period in long-day plants. Short-day plants are actually "long-night" plants, for the controlling factor is the length of the period of uninterrupted darkness. Long-night plants will flower only when exposed to darkness for nine or more hours. They can be made to flower earlier than usual by decreasing their daily exposure to light or by covering them, and they can be kept from flowering by giving them artificial illumination (Fig. 11–16). A long-day (short-night) plant can be prevented from flowering if it is covered for part of each day,

Figure 11–16 Flowering: All petunias received eight hours of daylight each day. Plant on left with small buds got in addition eight hours of fluorescent light, which contains flower-suppressing red light but no infrared. Flowering plant in center was given an extra eight hours of incandescent light, containing both red and flower-stimulating infrared light. (Courtesy of the U.S. Department of Agriculture.)

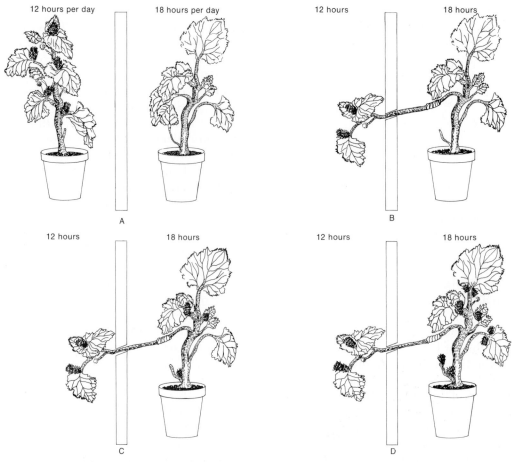

Figure 11–17 An experiment to demonstrate the existence of a flower-inducing hormone. *A,* Two cocklebur plants are grown in pots separated by a light-tight partition, exposed to 12 and 18 hours, respectively, of light per day. The 12-hour plant has flowered; the 18-hour plant has not. *B,* The 12-hour plant is cut off, inserted through a light-tight hole in the partition and grafted to the 18-hour plant. The two parts continue to receive 12 and 18 hours of light respectively. The 18-hour plant gradually develops flowers, first on the twigs nearest the graft *(C)* and eventually on all twigs *(D).* If no graft had been made, the 18-hour plant would not have developed flowers.

thereby reducing its daily exposure to light to less than its critical photoperiod.

Carnations, cotton, dandelions, sunflowers and tomatoes are examples of "*day-neutral*" plants. The flowering of these plants is relatively unaffected by the amount of daylight per day. The time of flowering in short-day, long-day and day-neutral plants is not controlled solely by the photoperiod, for temperature, moisture, soil nutrients and the amount of crowding may also play a role.

How the length of darkness may affect the time of flowering is not known in detail, but the results of some experiments suggest that a flower-producing hormone, named *florigen,* is involved. A typical experiment, using cocklebur, a long-night plant, is as follows (Fig. 11–17): One plant is grown exposed to 12 hours of light per day until it is producing flowers. It is

then grafted to another plant that had been grown exposed to 18 hours of light per day (and thus had been inhibited from producing flowers). The two parts, though grafted, are separated by a light-tight partition and the first part continues to receive 12 hours and the second 18 hours of sunlight. The long-night part of the plant continues to produce flowers, and in time the short-night part of the plant also produces flowers, usually beginning at the point nearest the graft. This is taken as evidence for a diffusible, flower-inducing hormone produced in the leaves and transported in the phloem to the buds. Nothing is known of the chemical composition of this hormone nor of how it might act to induce flowering.

Red light inhibits flowering in short-day plants but induces flowering in long-day plants. Infrared light induces flowering in short-day plants and inhibits flowering in long-day plants. A light-sensitive protein pigment, **phytochrome**, appears to play a fundamental role in this phenomenon of photoperiodism. This protein pigment exists in two forms: one, phytochrome$_{660}$, sensitive to red light (660 nm.); and the other, phytochrome$_{735}$, sensitive to infrared light (735 nm.). P_{660} appears to be the quiescent form in which the plant stores the potentially active compound and P_{735} is the active material. Light converts inactive P_{660} to active P_{735} and infrared light converts P_{735} to P_{660}; these are photochemical reactions that occur as rapidly at 0° C as at 35° C. P_{735} is also converted to P_{660} in the dark by a reaction mediated by enzymes and requiring oxygen. During the day phytochrome exists predominantly in the P_{735} form and during the night it is converted to the P_{660} form enzymatically. This could provide the plant with a means of detecting whether it is light or dark. The rate at which P_{735} is converted to P_{660} provides the plant with a "clock" for measuring the duration of darkness.

A knowledge of the phenomenon of photoperiodism is an intensely practical thing for commercial plant growers, for by altering the amount of light per day in their greenhouses they can speed up or retard the flowering of their plants so that they come into full bloom at just the right time for the Christmas or Easter season.

In general, long-night plants are native to tropical or subtropical regions, where there is no more than 13 or 14 hours of daylight per day. Most short-night plants are native to the higher latitudes. Day-neutral plants may be found almost anywhere.

A number of plant functions other than flowering may also be affected by the daily photoperiod: the formation of tubers by Irish potatoes is accelerated when the daily exposure to light is shortened. Since the growth of the tuber (the part we know as the potato) involves the deposition of starch, the photoperiod must in some way stimulate the transfer of carbohydrates from the leaves to the tuber.

11–14 SLEEP MOVEMENTS

Time-lapse photography shows that plants and their parts are constantly moving. Some of these movements result from differential growth of the several parts of the plant; others are turgor movements which result from changes in the turgor pressure of the cells (p. 234). In addition to the tropistic responses of plants (p. 237)—growth responses to external stimuli that come from some specific direction which determines the direction of the movement—there are **nastic movements** which are *independent* of the direction of the stimulus. Some of the "sleep" movements of plants are nastic movements.

Many plants change the position of their leaves or flower parts in the late afternoon or evening. These parts return to their original position in the morning. The leaves of peas, beans, clover and a number of other plants fold together in darkness and return to an expanded position in daylight. These changes in position have been termed *sleep movements*, although they are in no way related to the sleep of animals. Several kinds of flowers close their petals at night and open them in the morning.

QUESTIONS

1. Compare and contrast the processes of photosynthesis and cellular respiration.
2. What pigments may be present in plant cells? What are the functions of these pigments?
3. Compare the skeletal systems of plants and animals.
4. Describe the physiological mechanism underlying turgor pressure.
5. What are the functions of the xylem, phloem and latex systems?
6. Compare the structure and cellular components of xylem and phloem.
7. What are the major constituents of plant saps?
8. Describe the physiological basis of tropisms. Where in a plant can tropistic responses occur?
9. What is an "etiolated" plant? How have they been useful in studies of phototropism?
10. List the different types of plant hormones and discuss the effects of each. What are some practical applications of plant hormones?
11. Describe an experiment designed to prove that some new synthetic chemical has auxinlike properties.
12. Discuss the interactions, both synergistic and antagonistic, of the several types of plant hormones.
13. What is meant by photoperiodism? How would you determine experimentally whether a particular plant is a long-night, short-night or a day-neutral plant?

14. What are the functions of florigens and phyto-chromes in plant cells?
15. Compare and contrast tropistic responses and nastic movements in plants.

SUPPLEMENTARY READING

Further information about the functions of plant cells will be found in textbooks of general botany such as those by Raven and Curtis, by Greulach, by Ray, by Norstog and Long, and by Galston. More technical discussions of plant physiology will be found in *Introduction to Plant Physiology* by Meyer, Anderson and Böhning, and in Bonner and Galston's *Principles of Plant Physiology. Plants, Life and Man* by Edgar Anderson is concerned with the effects of plants on human existence. Discussions of plant hormones and some of their practical and sci-entific uses are to be found in *Hormonal Regulation in Higher Plants* by A. W. Galston and P. J. Davies, in the *Physiology of Flowering* by W. S. Hillman, and in *Plant Growth and Development* by A. C. Leopold. *Plants and Water* by J. Sutcliffe provides an interesting discussion of the importance of water to plants and its many roles in plant function. *Plants at Work* by F. C. Steward examines the many aspects of plant physiology. *Plant Biochemistry* edited by J. Bonner and J. E. Varner is a collection of articles by leading specialists who discuss structure and functions of plants as well as their chemistry.

The Morphology of Vascular Plants by D. W. Bierhorst is an excellent presentation of the struc-tures and evolutionary relationships of the higher plants. Ledbetter and Porter's *Introduction to the Fine Structure of Plant Cells* contains superior electron micrographs of a variety of plant cells and a discussion of the inferences to be made from these studies.

CHAPTER 12

PROCURING AND DISTRIBUTING
NUTRIENTS IN SEED PLANTS

The functions common to green plant cells described in the previous chapter may all occur within a single cell in the more primitive plants. In the evolution of the more familiar and widespread ferns, conifers and flowering plants, cellular specialization has occurred, and root, stem and leaf (Fig. 12–1) have become differentiated. The evolution of conducting tissues, xylem and phloem, and the specialization of regions of the body have enabled plants to survive on land and to grow to large size. The seed plants have evolved into a great many species which are distributed all over the world, and many of these are of prime importance to man as sources of food, clothing, shelter and other necessities. The adaptations of structure and function evolved by higher plants, described in this chapter, provide an interesting contrast to those evolved by humans and other animals.

12–1 EMBRYOS AND SEEDLINGS

In nearly all multicellular plants, an early step in the development of the mature plant from a single cell—a zygote or a spore—is a cell division that establishes the plant's axis or polarity. In its simplest form the axis determines that the plant has a top and bottom. In the more complex land plants the top and bottom are differentiated into shoot and root. This polarity is established in the early embryo and results in the formation of a short axis, usually termed the **hypocotyl,** having a shoot meristem at one end and a root meristem at the other.

Early in development the shoot meristem becomes clearly different from the root meristem and produces leaves. The first leaves produced are the **cotyledons;** monocots have a single cotyledon and dicots have two cotyledons. The part of the embryo above the cotyledons is the *epicotyl* and that below the cotyledons is the hypocotyl (Fig. 12–2). The meristem at the lower end of the hypocotyl differentiates into the embryonic root, or **radicle.** During germination the hypocotyl may elongate rapidly, lifting the epicotyl and cotyledons above the soil. In other plants the epicotyl elongates and the cotyledons remain buried in the soil. The cotyledons of dicots serve as food storage structures and provide the energy for the germinating plant. In some species the cotyledons carry on photosynthesis. The single cotyledon of monocots serves to absorb food and transfer it to the growing regions. The epicotyl differentiates into the stem of the seedling.

The functional role of the seedling is to re-establish the organization of the plant following sexual reproduction and genetic recombination. The embryo, as part of the seed, serves to disperse the species to new areas. During this dispersal process the embryo remains dormant; it remains inactive for a considerable period of time before it germinates.

The germination of the seed requires moisture, oxygen, light and the proper temperature. The energy needed for the process is

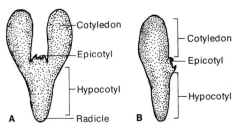

Figure 12–2 Highly diagrammatic view of basic embryo types. *A*, Embryo of a dicotyledon (Magnoliopsida). *B*, Embryo of a monocotyledon (Liliopsida). In the dicot embryo, the cotyledons usually function as food storage organs and sometimes as photosynthetic organs. In the monocot embryo the cotyledon is nonphotosynthetic and absorptive, and the radicle is late in developing. (From Norstog, K., and Long, R. W.: *Plant Biology*. Philadelphia, W. B. Saunders Co., 1976.)

derived from starch and other food stored in the endosperm. The seed absorbs water, becomes hydrated and swells, bursting the seed coat, and the embryo begins to grow into a seedling. The radicle emerges and grows down in the soil and a root system is established. The activity of enzymes in the endosperm is greatly increased; nutrients stored there are digested and transported to the rapidly growing tips of the embryonic axis.

The architecture of the mature plant is determined by the pattern of development of the shoot and root apices formed initially in the embryo. Cells in the apical meristems divide repeatedly, forming new cells which divide for a time, then enlarge and elongate. Behind this elongation region the cells mature and become specialized, giving rise to the various types of tissue present in stems and roots.

In plants, in contrast to animals, all cell division is confined to these meristem regions, at first to the shoot meristem and the root meristem and subsequently to secondary meristems such as the cambium, temporary leaf meristems, and the cork cambium.

The radicle forms the first root, the **primary root** of the young plant. The primary root may continue to grow and form the **taproot** system of the mature plant. In other species, especially in the monocots, the primary root does not persist, and adventitious roots originating from the lower part of the stem form a **fibrous root** system (Fig. 12–3). Roots typically are the principal means by which water and minerals are absorbed from the soil; in addition, they may serve to store reserve food and water and to anchor the plant

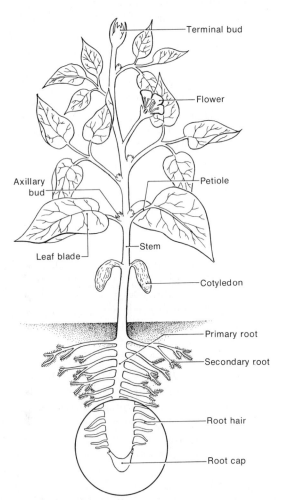

Figure 12–1 A diagram of a young bean plant showing the root, stem and leaf.

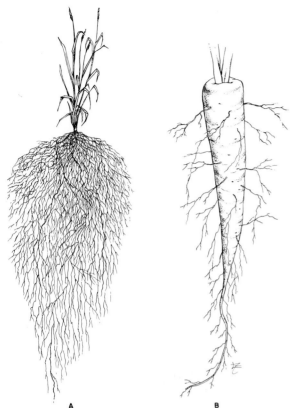

Figure 12–3 Types of root systems in plants. *A*, The diffuse root system of a grass. *B*, The taproot system of a carrot.

in the substrate. They may also provide support for aerial structures in the form of stilt or prop roots.

12–2 THE ROOT AND ITS FUNCTIONS

The most obvious function of the root is to anchor the plant and hold it in an upright position. To do this, roots branch and rebranch extensively through the soil. Although the depth of the root seldom equals the height of the stem, the roots frequently extend farther laterally than the stem's branches and the total surface of the root usually exceeds that of the stem. The second and biologically more important function of the root is to absorb water and minerals from the soil and conduct them to the stem. In some plants—for example, carrots, sweet potatoes, beets—the roots function as storage places for large quantities of food.

The first root formed by a young seedling, the *primary root*, develops in some species into a single large *taproot*, usually growing straight down, from which branch many, much smaller, secondary rootlets. Other species have *fibrous roots*, a system of many slender-branched roots, all about the same size.

Additional roots that grow from the stem or leaf or any structure other than the primary root or one of its branches are termed *adventitious roots*. The prop roots of corn plants and the aerial roots of ivy and other vines attaching the vine to a wall or tree are adventitious roots. Many kinds of flowering plants can be propagated by cutting off bits of their stems and producing adventitious roots at the base of these stem cuttings.

The apical meristem in the tip of each root is covered by a protective *root cap,* a thimble-shaped group of cells which covers the rapidly growing meristematic region (Fig. 12–4). The outer part of the root cap is rough and uneven because its cells are constantly being worn away as the root pushes through the soil. The growing point consists of actively dividing meristematic cells (Figs. 12–5 and 12–6) from which all the other tissues of the root are formed. The growing point also gives rise to new root cap cells to replace the ones worn away. Immediately behind the growing point is the *zone of elongation;* here the cells remain undifferentiated but grow rapidly in length by taking in large amounts of water. The growing point is about 1 millimeter in length and the zone of elongation is 3 to 5 millimeters long; these two regions alone account for the increase in length of the root.

Above the zone of elongation is the *zone of maturation* (Fig. 12–4), characterized externally by a downy covering of whitish *root hairs.* In this zone the cells differentiate into the permanent tissues of the root. Each root hair is a slender, elongated, lateral projection from a single epidermal cell. Any epidermal cell may absorb substances, but the surface exposed by a root hair is much larger and hence most of the water and minerals are absorbed through the root hairs. The delicate, short-lived root hairs are formed just behind the zone of elongation and wither and die as the root elongates (Fig. 12–7). Only a short segment of the root, perhaps 1 to 6 centimeters long, has root hairs.

Some measurements made by Dittmer give an idea of the astonishing amount of root surface available for the absorption of water

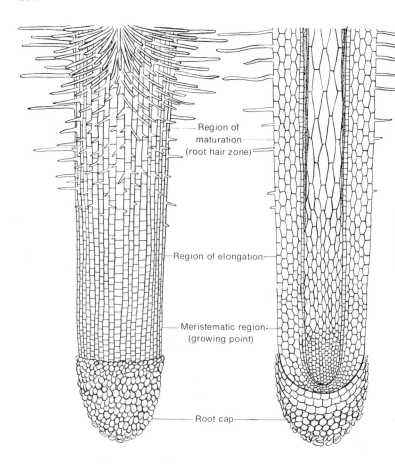

Region of
maturation
(root hair zone)

Region of elongation

Meristematic region:
(growing point)

Root cap

Figure 12–4 An enlarged diagrammatic view of the tip of a young root. *Left,* The surface of the root. *Right,* A longitudinal section of the root showing the internal structure.

and minerals. A single rye plant, grown in a box 30 centimeters square and 55 centimeters deep, reached a height of 50 centimeters in four weeks. Dittmer then carefully washed away the soil and measured the number of roots. This single plant had 625 kilometers of roots with a surface area of 285 square meters. But its root hairs, some 14,500,000,000 of them, were estimated to have a total length of 10,500 kilometers and a surface area of 480 square meters! The total area of roots plus root hairs, through which water could be absorbed, was 765 square meters, or about 6875 square feet (the building lots in some subdivisions have less area than this).

The number and area of the roots on any given plant depend on many factors, such as how close each plant is to its neighbors. Wheat plants grown 3 meters apart each had 70 kilometers of roots, but the same variety grown 15 centimeters apart had less than 1 kilometer of roots. Weed plants in a garden decrease the number and area of the roots on peas, beans and tomatoes and thus decrease productivity.

A cross section of a root in the zone of mat-

uration reveals a complex, highly organized structure (Fig. 12–8). The outer surface, the *epidermis,* is a layer of rectangular, thin-walled cells one cell thick. The epidermis of the root, unlike the epidermis of the stem, usually has no waxy cuticle on its outer surface. A waxy cuticle would undoubtedly interfere with the absorption of water. Each root hair originates as a swelling on an epidermal cell; it increases in size and becomes a hairlike projection as much as 8 millimeters long. There are hundreds of root hairs on each square millimeter of root surface. A branch of the vacuole of the epidermal cell fills most of the volume of the root hair, leaving just a thin film of cytoplasm between the vacuole and the cell wall.

Just inside the epidermis is the *cortex,* a wide area composed of many layers of large, thin-walled, nearly spherical parenchymal cells with many intercellular spaces between them. These cells serve as avenues for the conduction of water and minerals and as storage places for starch and other foods. At the inner edge of the cortex, a single layer of cells, the *endodermis,* separates the loosely packed cor-

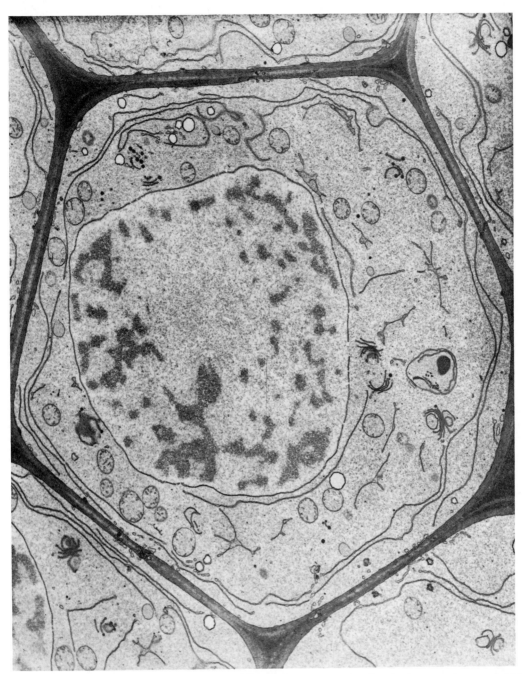

Figure 12–5 Electron micrograph of a cell from the root tip of a corn plant showing the large nucleus, an assortment of mitochondria, endoplasmic reticulum, plastids and Golgi apparatus. The pores in the nuclear membrane are clearly evident. The tissue was fixed with permanganate, a fixative that does not preserve ribosomes or the integrity of the vacuole. (Courtesy of Dr. H. H. Mollenhauer and Dr. W. G. Whaley, Cell Research Institute, University of Texas.)

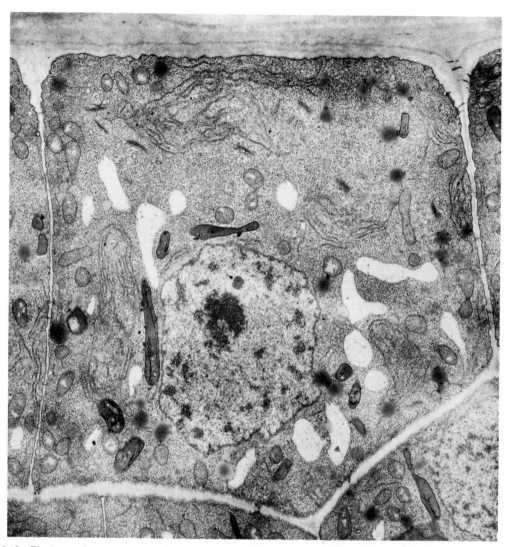

Figure 12–6 Electron micrograph from the root tip of a spinach plant. The tissue was fixed for electron microscopy using glutaraldehyde and osmium, which permits the visualization of the ribosomes, the many small dots seen in the cytoplasm either free or bound to the membranes (rough endoplasmic reticulum), and preserves the vacuoles, the light, empty areas. (Courtesy of Dr. M. Dauwalder and Dr. W. G. Whaley, Cell Research Institute, University of Texas.)

tical parenchyma from the central core of vascular tissue, the *stele.* In younger roots the endodermis is composed of thin-walled cells having a circumferential fatty thickening, called the Casparian strip, in their lateral walls. The Casparian strip, composed of lignin and suberin, seals the otherwise permeable radial walls of the endodermis cells, thus forcing water and solutes to move through the cell's cytoplasm. In the older parts of the root, the endodermis tends to become thick, waxy and lignified and decreases the diffusion of water out of the stele into the cortex. At various points in the endodermis are thin-walled cells having Casparian strips, the *passage cells,*

through which water and minerals pass to the stele. Just inside the endodermis is a single layer of parenchymal cells, the *pericycle.* This can be transformed into meristem to give rise to part of the root's cambium and to lateral roots. That secondary roots originate deep within the primary root ensures a good connection between the vascular tissues of branch and main roots.

The central portion of the stele, surrounded by endodermis and pericycle, is composed of the two vascular tissues, *xylem* and *phloem.* The cells of the xylem, *tracheids* and *xylem vessels,* are usually arranged in the form of a 3- or 4-pointed star or like the spokes of a wheel.

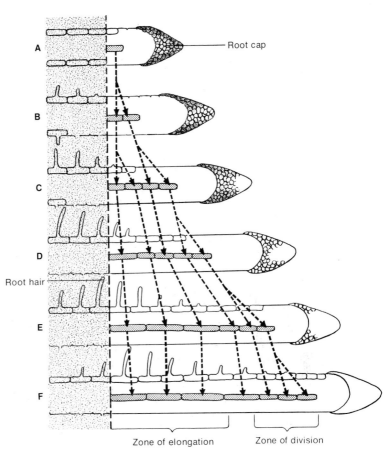

Figure 12–7 Diagram of successive stages in the growth of a root tip. The vertical dashed line is a landmark to indicate the same part of the root as growth occurs. Of the many cells that make up the interior of the root (see Figure 12–4) only one is indicated in A. This divides longitudinally to form two (B) and four (C) cells. The cells nearest the tip continue to divide (D, E, F) and the ones farther back elongate. Root hairs grow out of the epidermal cells in the zone of elongation (B, C, D) and then wither and die as the growing tip moves on (E, F).

Root hair

Zone of elongation   Zone of division

They are thick-walled, cylindrical cells which conduct water. Between adjacent points of the xylem star are bundles of phloem cells, smaller and thinner-walled than the xylem. Some of the phloem cells form specialized *sieve tubes.* The roots of trees, shrubs and other perennial plants typically have a single-celled *cambium* layer between the xylem and

phloem. This divides and gives rise to additional layers of xylem and phloem to provide for increased thickness of the root. A cambium also occurs in woody annuals such as the sunflower.

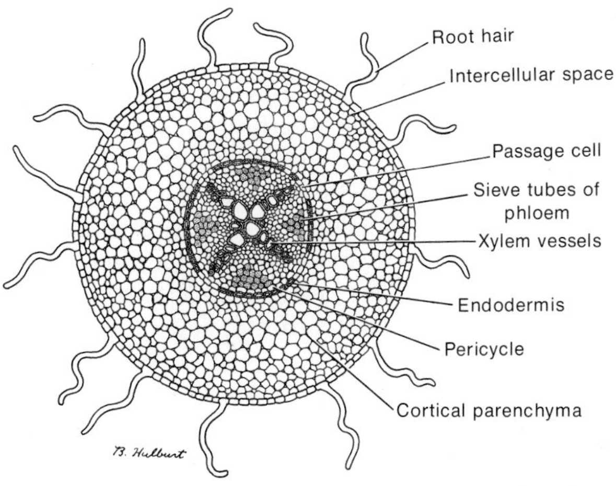

The transfer of water from the soil into the root hairs and across epidermis, cortex and endodermis into the xylem in the stele can be explained on purely physical principles (Fig. 12–9). The water available to plants is present as a thin film loosely held to the soil particles and is called *capillary water.* The roots, especially the root hairs, are in contact with the films of capillary water. The capillary water usually contains some dissolved inorganic salts and perhaps some organic compounds, but the concentration of solutes in capillary water is low and the solution is hypotonic to the fluid within the root hairs. The cell sap in the epidermal cells has a fairly high concentration of glucose and other organic compounds. Water molecules diffuse from a region of higher concentration (the capillary water) to a region of lower concentration inside the root hairs. Water is passing through a differentially per-

Root hair
Intercellular space
Passage cell
Sieve tubes of phloem
Xylem vessels
Endodermis
Pericycle
Cortical parenchyma

B. Hulburt

Figure 12–8 Cross section of a root near its tip where root hairs are present.

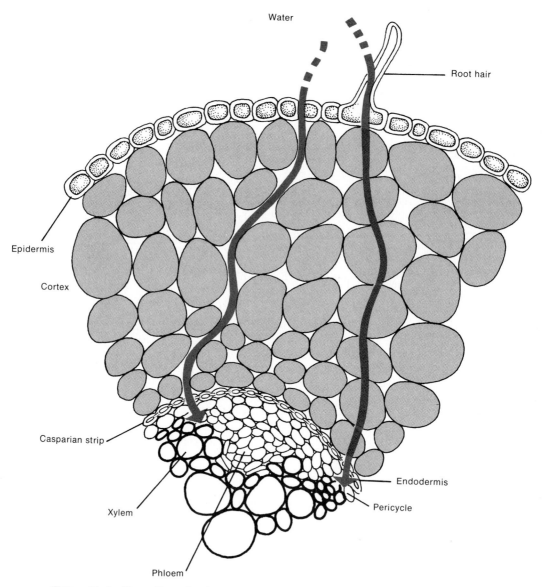

Figure 12–9 The pathway for the transfer of water from the soil to the stele of the root.

meable membrane, the plasma membrane of the root hair cell, by the process of osmosis. The plasma membrane is permeable to water but not to the glucose and other organic compounds in the cell sap.

As the root hair cells take in water, their contents become more dilute and their cell sap becomes hypotonic to that of the adjacent cells in the cortex. Water tends to pass from the epidermal cell to the cortical cell. The outermost cortical cells then have a lower osmotic pressure than the ones farther in, and water tends to diffuse in toward the center. In this way water continues to diffuse inward, finally reaching the ducts of the xylem, and then rises in the xylem to the stem and other parts of the

plant body. The removal of water from the stele via the xylem maintains the concentration gradient of water from epidermis to xylem and permits the continued inflow of water. There is some evidence that the roots of some species are capable of an active, energy-requiring secretion of water inward toward the stele; this may come into play when the concentration gradient between soil and root cells is small or reversed.

Water may pass from epidermis to stele without passing through the intervening cells. The cell walls of all these cells are made of cellulose, which has a strong tendency to imbibe water (the "blotting paper" principle), and water may move inward along these cellulose cell walls to the endodermis. There it enters the stele through the thin-walled passage cells in the endodermis. The cell walls of many plant cells are interrupted by very small holes through which the cytoplasm of one cell may be connected to the cytoplasm of the adjacent one. These connections, called *plasmodesmatata,* may be important as a further avenue for the transport of water, ions, sugars and amino acids from one cell to the next.

The inner layer of the cortex, the endodermis, also functions as an osmotically active membrane system due to the presence of the Casparian strips that tend to direct the flow of water and solutes through the living protoplasts of the endodermal cells. From the endodermis, water moves into the empty, dead xylem vessels and tracheids.

The water entering the xylem vessels increases the pressure within them, just as water entering a semipermeable cellophane membrane containing a glucose solution increases the pressure in the tube to which it is attached (p. 46). This *root pressure* is one of the factors that moves sap upward through the root and stem to the leaves. Water vapor is constantly escaping from the leaves by *transpiration* (p. 269). New molecules of glucose and other organic substances are formed in the leaves by photosynthesis and sent via the phloem to the roots. These two processes maintain the hypertonicity of the sap in the roots, and the absorption of water continues.

The entrance of mineral nutrients in the form of inorganic salts occurs in part by simple diffusion, for any ion present in greater concentration in the capillary water of the soil than in the cell sap of the root hair will tend to diffuse into the root hair. The absorption of inorganic ions is largely independent of the entrance of water, and the absorption of each kind of ion is independent of all others, for each diffuses down its own concentration gradient. The rate of entry of each type of ion is determined by the difference in the concentrations of that ion inside and outside the root and by factors such as the diameter of its hydrated form which determine its rate of penetration of cell membranes.

Some roots, and perhaps all, take in inorganic ions by processes of active transport, requiring the expenditure of energy. An increase in the rate of cellular metabolism can be detected in such roots when they are absorbing inorganic ions against a concentration gradient. The concentration of inorganic ions in plant sap is rather low and large volumes of sap must flow to the stem and leaves to provide the required nutrients. The inorganic ions that enter the root are carried to other parts of the plant and utilized rapidly, thereby decreasing their concentration and maintaining the concentration gradient from soil to root to stem to leaf. The nitrates absorbed from the soil are reduced to amino groups and other nitrogenous compounds. Some of the amino groups are incorporated into amino acids used for protein synthesis and others become constituents of one of a host of nitrogenous organic molecules peculiar to one or another species of plant. These include flavorings, perfumes and drugs such as quinine and morphine.

12–3 THE ENVIRONMENT OF ROOTS: SOIL

The soil provides a solid, yet penetrable foundation for plant growth; it enables plants to become firmly anchored and serves as a reservoir for the water and minerals needed by plants. Plant growth requires (in addition to carbon, hydrogen, oxygen and nitrogen) the minerals calcium, iron, magnesium, potassium, phosphorus and sulfur plus the trace elements boron, copper, cobalt, molybdenum, manganese and zinc. A deficiency of any one of these will limit plant growth even though all the others are present in optimal amounts (Fig. 12–10).

The mineral portion of soils is made up of particles varying in size from large rocks to fine clay, broken off of the rock in the earth's crust by the physical and chemical processes of weathering or by the action of plants and animals. Soils made solely of minerals cannot support plant growth. A productive soil must contain organic material, **humus,** as well. Humus is derived from the decaying remains of plant and animal bodies; it gives soil a brown or black color. Humus, in addition to supplying plant nutrients, increases the porosity of soil so that proper drainage and aeration can occur, and increases the ability of the soil to absorb and hold water.

Soils contain innumerable bacteria and

Figure 12–10 Tobacco plants illustrating the effects of deficiencies of specific elements. The plant in the center (Ck.) received all essential elements; the others were supplied with all essential elements except the one indicated on the label. All plants are the same age and variety. Note the marked growth inhibition resulting from lack of N and K, the green but distorted leaves of the plant lacking Ca, the general chlorosis (except along the veins) of the −Mg plant, and the chlorosis in older leaves only when P was lacking. The other leaves of the −P plant are dark blue-green. The leaves of the −S plant show different degrees of chlorosis, but more commonly the younger leaves are the most chlorotic. Note also the necrosis in the −N, −P and −K plants. (Courtesy of W. R. Robbins, Rutgers University.)

fungi which bring about decay and are of prime importance in keeping the soil in good condition. Soils also contain a wide variety of animals, ranging from microscopic forms to moles, gophers and field mice. The constant burrowing of earthworms turns over the soil and mixes in additional organic material. An acre of soil may contain some 50,000 worms. In the course of a growing season, these ingest some 18 tons of earth, grind it and deposit it on the soil surface.

The soil is an example of a major ecosystem, containing a large number of different kinds of animals, bacteria and plants that compose an interrelated biologic complex. The productivity of a soil depends on such things as its chemical composition and porosity, its content of air and water, and its temperature.

Soils are classified according to the particle size of their mineral constituents, from coarse gravel (over 2 millimeters in diameter) through several classes of gravel, sand and silt to clay (particle size less than 0.002 millimeters in diameter). A soil is called *loam* if it is about one-half sand, one-fourth silt and one-fourth clay. The amount of air and water a soil can contain is determined by its texture. Soil water is largely present as a capillary film on the surface of the mineral particles, hence a clay soil with a large number of small particles will have a high water content. A good porous soil can provide sufficient air for root growth but a tightly packed or wet soil will result in the death of the plants present.

A major conservation goal all over the world is to decrease the amount of valuable topsoil carried away each year by wind and water. Reforestation of mountain slopes, building check dams in gullies to decrease the spread of the run-off water, contour cultivation, terraces and the planting of windbreaks are some of the methods that are currently being used successfully to protect the topsoil against erosion.

About a century ago botanists began experimenting to determine the kinds and amounts of nutrients necessary for plant growth. They attempted to grow plants in water without soil and found that plants could be grown in liquid media if they were supported either by screening or sand. This soilless culture of plants (*hydroponics*) makes possible carefully controlled experiments on the requirements of plants for trace elements, but it is not an economically practicable way to grow plants in quantities. The costs of circulating and aerating the water prohibit any large-scale commercial development of soilless agriculture, but it was used by the Navy in World War II to produce food on certain Pacific islands which could not have supported agriculture otherwise.

12-4 THE STEM AND ITS FUNCTIONS

The stem begins life as an embryonic bud, the *epicotyl,* consisting of an apical dome of cells, the *apical meristem,* surrounded by several immature leaves, the *leaf primordia.* The stem develops from the top down; a cell formed by division in the apical meristem occupies a progressively more basal position as the cells of the apex divide and add more new cells to the top of the shoot. A cell can be followed from its origin in the apical meristem through zones of elongation and maturation to its final position as a fully mature and specialized member of one or another of the plant's tissue systems.

Three fundamental tissues are present in the growing shoot: the *protoderm,* which forms the epidermis; the *ground meristem,* which forms the parenchyma of the pith and cortex; and the *procambium,* which differentiates into the xylem and phloem and, in dicots, into the *cambium* of the vascular bundles. In dicotyledons the vascular bundles are arranged in a circle but in monocotyledons they are rather evenly distributed throughout the stem.

A shoot apex from which the young leaves and bud scales have been removed reveals a domelike mound of unspecialized cells of the apical meristem surrounded by regularly spaced smaller mounds, the leaf primordia. The spacing of the leaf primordia has a characteristic pattern in each species of plant and determines the ultimate arrangement of leaves on the mature stem.

The stems of developing plants consist of a linear series of nodes and internodes. Buds and leaves are attached to the stem at specific points, the *nodes,* and the portion of the stem between adjacent nodes is termed the *internode.* The curvature of the stem and its growth in length are determined by cell elongation and multiplication in the internodes. Both terminal and lateral (axillary) buds are present on young stems. Each bud consists of an apical meristem below which are several nodes, each of which bears a miniature leaf and a partly formed axillary bud. Some buds bear immature flowers as well as leaves; others bear only flowers. Herbaceous plants, which remain soft and green throughout their life, have active buds that continuously grow and produce new leaf, bud and stem growth. Woody plants usually have protected buds that remain dormant during part of the year. Dormant buds often have an outer layer of scale leaves, or *bud scales.* When a bud elongates and forms a leafy stem or branch in the spring, the bud scales are shed, leaving bud scale scars at the base of the newly formed stem.

The shoot that develops from the epicotyl contains only primary tissues at first: the *epidermis, cortex, primary xylem, primary phloem, pith* and *pith rays,* or medullary rays, collectively called the *primary body.* The young stem is usually relatively soft and green and carries on some of the same functions as leaves. Stomata are present on the epidermis and chloroplasts are found in the cells just beneath the epidermis. The epidermis functions to prevent the evaporation of water from the soft tissues within and is generally a single layer of cells covered with a waterproof cuticle composed of a fatty substance, *cutin.* In addition a coating of wax may be secreted in some plants.

The stem, which in a tree includes the trunk, branches and twigs, is the connecting link between the roots, where water and minerals enter the plant, and the leaves, where foodstuffs are synthesized. The vascular tissues of the stem are continuous with those of root and leaf and provide a pathway for the exchange of materials. The stem and its branches support the leaves so that each leaf is exposed to as much sunlight as possible. Stems also support flowers and fruits in the proper position for reproduction to occur. The stem is the source of all the leaves and flowers produced by a plant, for its growing points produce the primordia of leaves and flowers. No seed plant consists solely of roots and leaves, even though all the processes necessary for plant maintenance can occur in these organs. Some stems have cells which contain chlorophyll and carry out photosynthesis; others have cells specialized for the storage of starch and other nutrients.

The beginning student may have difficulty in distinguishing roots from stems, for many kinds of stems grow underground and some roots grow in the air. Ferns and grasses are examples of plants that have underground stems, called *rhizomes.* These grow just beneath the surface of the ground and give rise to aboveground leaves. Thickened underground stems adapted for food storage, called *tubers,* are found in plants such as the potato. An onion bulb is an underground stem surrounded by overlapping, tightly packed scale leaves. Roots and stems are structurally quite different: stems, but not roots, have nodes, points at which leaves are attached. The tip of a root is always covered by a root cap whereas the tip of a stem is naked unless it terminates in a bud. The stem typically contains separate rings of phloem and xylem, with the xylem central to the phloem (Fig. 12–11), whereas in the root the bundles of phloem tubes lie *between* the points of the star-shaped masses of xylem (Fig. 12–8).

Plant stems are either herbaceous or woody. The soft, green, rather thin **herbaceous** stems are typical of plants called **annuals.** Such plants start from seed, develop, flower and produce seeds within a single growing season, dying before the following winter. Other herbaceous plants are biennials, with two-season

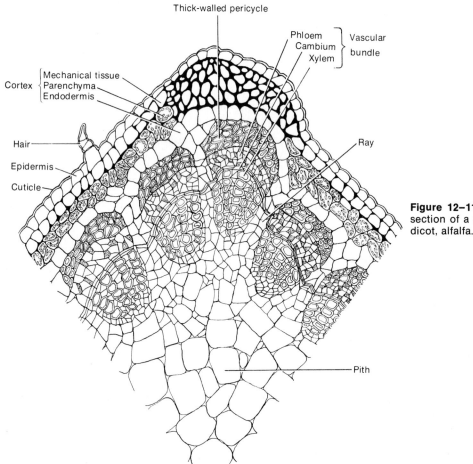

Figure 12–11 A sector of a cross section of a stem from a herbaceous dicot, alfalfa.

Thick-walled pericycle

Phloem
Cambium
Xylem

Vascular bundle

Cortex
Mechanical tissue
Parenchyma
Endodermis

Hair

Epidermis

Cuticle

Ray

Pith

growing cycles. During this first season, while the plant is growing, food is stored in the root. Then the top of the plant dies and is replaced in the second growing season by a second top which produces seeds. Carrots and beets are examples of biennials. Plants whose stems are soft and perishable and supported chiefly by turgor pressure are called *herbs.*

Quite different from the herbaceous annuals and biennials are the *woody perennials,* which live longer than two years and have a thick tough stem, or trunk, covered with a layer of cork. A *tree* is a woody-stemmed perennial that grows some distance above ground before branching and so has a main stem or trunk. *Shrubs,* such as lilacs, oleanders and sagebrush, are woody perennials with several stems of roughly equal size above the ground line.

It is generally believed that woody-stemmed plants are more primitive than the herbaceous ones, for the available evidence indicates that the first true seed plants were woody-stemmed perennials. In past geologic ages such plants grew as far north as Greenland, but with the change in climate toward the end of the Mesozoic era, compounded by the shifting positions of the continents as a result of continental drift, some of these plants were killed by the advancing cold and others were forced to retreat toward the equator. Still other woody plants adapted to the cold by evolving a life cycle in which growth and flowering were completed in the warm summer of a single year and the rigors of the winter were withstood by cold-resistant seeds; i.e., they became herbs.

The tissues of herbaceous stems are arranged around the bundles of xylem and phloem but the details are different in the two main groups of flowering plants, the dicotyledons and the monocotyledons. In the stem of a dicot, such as the sunflower or clover, the circular arrangement of the xylem and phloem bundles subdivides the stem into three concentric regions: the outer *cortex,* the *vascular bundles,* and the central core, or *pith,* composed of colorless parenchyma cells which serve as storage places (Fig. 12–11).

Each vascular bundle has an outer cluster of phloem cells (composed of sieve tube elements, companion cells, phloem fibers and parenchyma) and an inner cluster of xylem

cells (composed of tracheids and vessel elements) separated by a layer of meristematic tissue, the *cambium* (Fig. 12–11). Continued mitotic activity in the cambium produces new phloem cells on its outer margin and new xylem on its inner margin. On the lateral border of the phloem is the *pericycle,* a layer of thick-walled supporting cells. Between the vascular bundles lie groups of cells known as *medullary rays,* which extend radially from the vascular region to both pith and cortex and distribute materials from the xylem and phloem to these inner and outer parts.

The cortex consists of an inner layer, one cell thick, called the *endodermis,* which is immediately adjacent to the pericycle; then a layer of loosely packed, thin-walled parenchyma cells; and finally a layer of thick-walled collenchyma cells, which are supporting tissues. In many stems the endodermis and pericycle are not clearly distinguishable. Immediately surrounding the collenchyma of the cortex is the outermost layer, the *epidermis.* The outer walls of the epidermal cells are thickened and contain cutin.

The stems of woody plants resemble herbaceous ones during their first year of growth but by the end of the first growing season, additional cambium has formed in the medullary rays so that a continuous circle of cambium extends between the vascular bundles as well as through them. In each successive year the cambium forms an additional layer of secondary xylem and secondary phloem. The secondary phloem formed in this way eventually replaces the primary phloem and forms a continuous thin sheath of food-conducting tissue just outside the cambium. The yearly deposits of secondary xylem form the *annual rings* (Fig. 12–12). These can be distinguished because the tracheids and vessels formed in the spring of the year are larger, and hence appear lighter, than those elements of the xylem formed in the summer. Only the youngest, outermost layers of xylem, known as the *sapwood,* carry sap to the leaves; the inner layers of hard nonconducting xylem cells and fibers, known as the *heartwood,* increase the strength of the stem and accommodate the increasing load of foliage as the tree grows.

The width of the annual rings varies according to the climatic conditions prevailing when the ring was formed, so that it is possible

TRANSVERSE SECTION

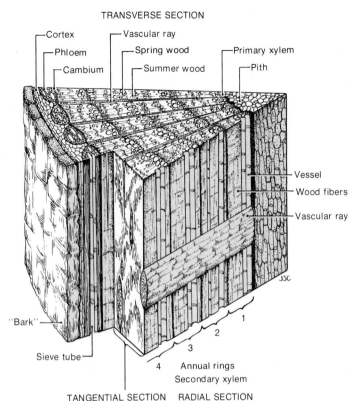

Cortex
Phloem
Cambium
Vascular ray
Spring wood
Summer wood
Primary xylem
Pith

Vessel
Wood fibers
Vascular ray

"Bark"
Sieve tube

1
2
3
4   Annual rings
Secondary xylem

JSC

TANGENTIAL SECTION   RADIAL SECTION

Figure 12–12 Diagram of a four-year-old woody stem, showing transverse, radial and tangential sections and the annual rings of secondary xylem.

to infer what the climate was at a particular time, several hundred or even thousands of years ago, by examining the rings of old trees. An interesting application of this technique is the dating of the time of construction of certain Indian pueblos in the southwestern United States by an analysis of the annual rings in the logs used. By comparing the pattern of thick and thin annual rings in these logs with those of trees whose year of felling is known it has been possible to determine the year of construction of these pueblos and to deduce the weather conditions for the past hundreds of years.

The cambium is important in the healing of wounds. When the stem's outer layer is removed through injury, the cambium grows over the exposed area and differentiates into new xylem, phloem and cambium, each of which is continuous with the same type of tissue in the uninjured part of the plant. Certain cells in the outer cortex of most woody plants become meristematic and form a second, or *cork cambium.* These outer cork cells become impregnated with a waterproof, waxy material and eventually die and fall off, partly under the stress of wind and rain and partly because of

the outward pressure of the growing tissues within. The cork cells and the cork cambium together make up the *periderm.*

The stem of a monocot, such as corn, has an outer epidermis made of thick-walled cells and pierced by openings (stomata) similar to the ones in leaves. The epidermis and the cells of the cortex just beneath the epidermis become thick-walled and lignified and serve as supporting tissues. The vascular bundles are scattered throughout the stem (Fig. 12–13), instead of being arranged in a ring as in the dicots. The bundles are smaller and more numerous in the outer part of the stem. Each bundle contains xylem and phloem but has no cambium; it is usually enclosed in a sheath of supporting sclerenchyma cells. In some monocots, such as wheat and bamboo, the parenchyma cells of the center of the stem disintegrate, leaving a central pith cavity.

The epidermis of a young woody twig has stomata like those of leaves; gases may enter and leave through them. Later, certain cells of the cork cambium divide repeatedly and form masses of cells that rupture the epidermis and form swellings, called *lenticels.* The intercellular spaces in these regions of loosely arranged,

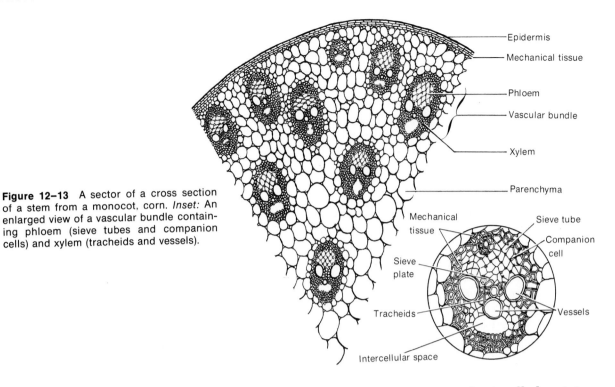

Epidermis

Mechanical tissue

Phloem

Vascular bundle

Xylem

Parenchyma

Mechanical tissue

Sieve tube

Companion cell

Sieve plate

Tracheids

Vessels

Intercellular space

Figure 12–13 A sector of a cross section of a stem from a monocot, corn. *Inset:* An enlarged view of a vascular bundle containing phloem (sieve tubes and companion cells) and xylem (tracheids and vessels).

thin-walled cells are continuous with those of the tissues within and permit the diffusion of gases in and out of the stem. Lenticels are visible as slightly elevated dots or streaks on the bark (Fig. 12–14).

The point on a stem where a leaf or bud develops is called a **node**, and the section of

the stem between two nodes is called an **internode** (Fig. 12–14). Internodes may be quite short or as much as 5 to 10 centimeters in length. In the upper angle of the point of junction of a leaf with the stem, a bud usually appears which is called a lateral or axillary bud to distinguish it from the terminal bud at the tip

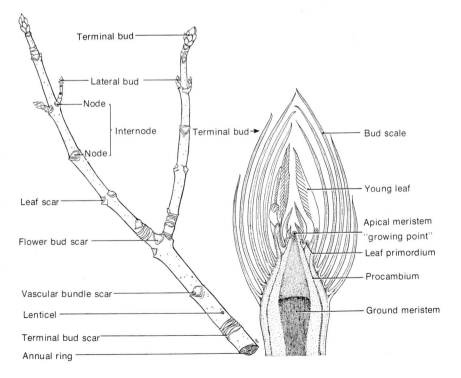

Terminal bud

Lateral bud

Node

Internode

Node

Terminal bud

Bud scale

Young leaf

Apical meristem "growing point"

Leaf primordium

Procambium

Ground meristem

Leaf scar

Flower bud scar

Vascular bundle scar

Lenticel

Terminal bud scar

Annual ring

Figure 12–14 *Left,* A twig from a horse chestnut tree, showing buds and scars. *Right,* An enlarged longitudinal section through a terminal bud from a hickory tree.

of the stem. Terminal buds continue the growth of the main stem; lateral buds give rise to branches. A bud consists of a number of embryonic leaves, a growing point, and (in woody plants) a ring of leaves modified to form outer protective scales (Fig. 12–14). In some species these scales are coated with a waxy secretion or have a dense covering of hairs to increase their protective value. The leaves within the buds may be fairly well developed so that their ultimate shape can be distinguished or they may be shapeless rudiments.

When a terminal bud begins to grow in the spring, its covering scales are forced apart and fall off, leaving a ring of scars (Fig. 12–14). These scars mark the position of the end of the stem at the completion of the growth season. They may remain visible for several years, so that it is possible to determine the age of a twig by counting the number of terminal bud scars.

The overall shape of a tree or shrub is determined by the position, arrangement and relative activity of the terminal and axillary buds. In a tree with a strong terminal bud, such as a pine or poplar, the twig produced by the terminal bud is much more vigorous than those produced by lateral buds, and a single, strong, straight main trunk results. Plants with vigorous lateral buds have strong horizontal branches and a spreading shape. Other factors influencing the shape of a tree are the direction and strength of the prevailing wind, and the presence of other trees nearby.

In addition to terminal and axillary buds, lenticels and bud scars, the surface of a twig may show *leaf scars* (Fig. 12–14), left when the stalk of a leaf breaks away from the twig, and *fruit scars*, produced by the breaking off of fruit.

12–5 THE LEAF AND ITS FUNCTIONS

Each leaf is a specialized nutritive organ whose function is to carry on photosynthesis, a process requiring a continuing supply of water, carbon dioxide and radiant energy. Leaves are generally broad and flat, presenting a maximum of surface to sunlight and having a maximum surface area for the exchange of gases—oxygen, carbon dioxide and water vapor.

The monocots and dicots are distinguished on the basis of leaf structure. The monocots, Liliopsida, typically have simple leaves with parallel venation, and the dicots, Magnoliopsida, typically have branching venation. Leaves range in size from less than 1 centimeter to some 20 meters in some of the large palm trees. The water lily, *Victoria regina*, has enormous floating leaves 2 meters or more in diameter.

Leaves originate as a succession of lateral outgrowths, called *leaf primordia*, from the apical meristem at the tip of the stem. Each outgrowth undergoes cell division, growth and differentiation, and finally a miniature, fully formed leaf is produced within the bud. In the spring, the leaves grow rapidly, forcing apart the bud scales, and largely by the absorption of water, unfold, enlarge, and reach their full size. All leaves are meristematic during their early development. Broad leaves, for example, have lateral meristems that form the blade. Most leaves, however, do not have a long-lived meristem, although the leaves of some ferns do. Most leaves do not live long—a few weeks in some desert plants, a few months for most trees, and up to three or four years for the needle-shaped evergreen leaves.

The leaf of a typical dicot consists of a stalk, the *petiole*, attached to the stem, and a broad *blade*, which may be one simple structure or a compound one, with two or more parts. The petiole may be short and in some species is completely lacking. Like a stem in cross section, it is composed of vascular bundles, attached at one end to those of the stem and at the other end to the midrib of the blade. Within the blade the vascular bundles fork repeatedly and form the *veins*.

A microscopic section through a leaf (Fig. 12–15) shows it to be composed of several types of cells. The outer cells, both top and bottom, make up a colorless, protective *epidermis* which secretes a fatty *cutin*. The epidermal cells—thin, tough, firm-walled and translucent—are well adapted to give protection to the underlying cells and decrease water loss yet admit light. Scattered over the epidermal surface are many small pores, called *stomata*, each surrounded by two *guard cells*. These cells, by changing their shape, can change the size of the aperture and so control the escape of water and the exchange of gases. In contrast to other epidermal cells, guard cells

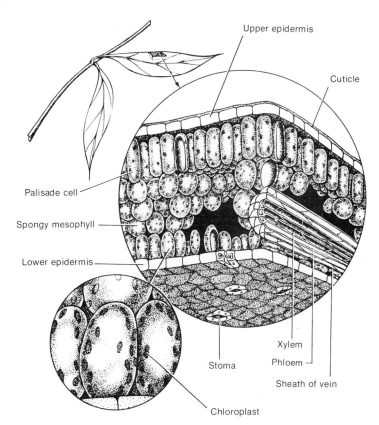

Figure 12–15 A diagram of the microscopic structure of a leaf. Part of a small vein is visible to the right.

contain chloroplasts. There are 50 to 500 stomata per square millimeter of leaf, many more on the lower than on the upper surface in the leaves of most species.

The bean-shaped guard cells have thicker walls on the side toward the stoma than on the other sides. In general, the stomata open in the presence of light and close in the dark; the opening and closing are regulated by changes in the turgor pressure within the guard cells (Fig. 12–16). Increased turgor pressure causes

their outer walls to bulge and the inner walls become curved so that they move apart, creating the stomatal opening between them. When the turgor pressure in the guard cells decreases, the elastic inner walls regain their original shape and the stoma is closed.

The mechanism that increases turgor pressure is complex, involving in part the production of glucose and other osmotically active substances by photosynthesis in the guard cells themselves. Light also initiates a sequence of

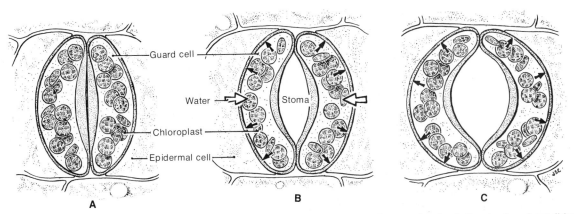

Figure 12–16 Diagrams illustrating the regulation of the size of the stoma by the guard cells. *A*, Nearly closed condition. *B*, When osmotically active substances such as glucose are produced, water enters the guard cells, turgor pressure increases, and the guard cells buckle so as to increase the size of the stoma. *C*, Stoma open.

enzymatic reactions that lead to the conversion of starch stored in the guard cells (starch molecules are large, insoluble and osmotically inactive) to glucose, a small, soluble, osmotically active molecule. Experiments have shown that the turgor pressure in the guard cell can increase threefold or more in the light.

As light increases photosynthesis, carbon dioxide is utilized and the concentration of carbon dioxide in the leaf is decreased. This increases the pH of the system and the enzyme phosphorylase is stimulated by the increased pH to convert starch to glucose-1-phosphate. The glucose-1-phosphate is converted to glucose, increasing the concentration of glucose in the guard cell and increasing the osmosis of water into the cell, thereby raising the turgor pressure in the guard cells and opening the stoma! In the dark the process is reversed.

The open stomata permit the entrance of carbon dioxide and enable the leaf to carry on photosynthesis. In the absence of light, photosynthesis ceases in the guard cells as in all cells, turgor pressure decreases and the stomata close. If, on a hot, dry day, the amount of water supplied by the roots is too small, the guard cells will be unable to maintain the turgid state and hence will close, effectively conserving the decreased supply of water.

Most of the space between the upper and lower layers of the leaf epidermis is filled with thin-walled cells, called *mesophyll cells,* which are full of chloroplasts. The mesophyll layer near the upper epidermis is usually made of cylindrical *palisade cells,* closely packed together and so arranged that their long axes are perpendicular to the epidermal surface. The rest of the mesophyll cells are very loosely packed together, with large air spaces between them. A leaf that is actively photosynthesizing has a high rate of gas exchange through the stomata with the environment. As carbon dioxide is utilized in photosynthesis in a cell its concentration within the cell decreases and carbon dioxide diffuses into the cell from the film of water surrounding it. Other molecules of carbon dioxide pass from the air spaces within the leaf and dissolve in the water film. Still other molecules of carbon dioxide diffuse from the air outside the leaf through the stomata and into the air spaces within the leaf. The molecules of carbon dioxide move from a region of higher concentration to one of lower concentration, driven by the forces of diffusion. Oxygen produced within the cells in the leaf passes from cell to water film to air space and through the stomata to the exterior by diffusion down a chemical gradient (Fig. 12–17).

The veins of a leaf branch and rebranch repeatedly to form an extremely fine network, so that no mesophyll cell is far from a vein. Each vein contains both xylem and phloem tis-

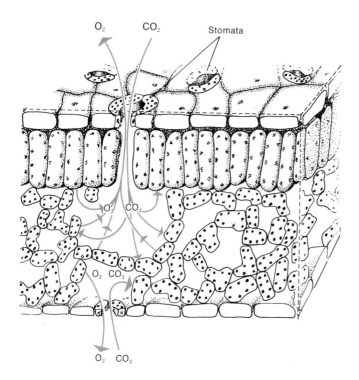

Figure 12–17 An enlarged cross section of a leaf showing its structure. The diffusion of carbon dioxide through the stomata to the interior of the leaf and the diffusion of oxygen from the photosynthetic cells through the stomata to the exterior are indicated by arrows.

sues. The xylem is on the upper side of the vein, the phloem on the lower. In the smallest veins there are only a few xylem tracheids and a few phloem sieve tubes.

The leaves of a number of desert plants are thick and fleshy and serve as storage places for water. The leaves of pond lilies and other aquatic plants have large air spaces to provide buoyancy. A few leaves, such a those of cabbages, store considerable amounts of food. The insect-trapping abilities of the leaves of plants such as the pitcher plant and the Venus flytrap have been discussed.

The fall of leaves in the autumn of the year is brought about by changes at the point where the petiole is attached to the stem. A special *abscission layer* of thin-walled cells, loosely joined together, extends across the base of the petiole, weakening the base of the leaf. The part next to the stem becomes corklike and forms a protective layer which will remain when the leaf falls off. When the abscission layer has formed, the petiole is held on only by the epidermis and the easily broken vascular bundles, so that a high wind will bring about the fall of the leaf. The change in color of the leaves is effected partly by the decomposition of the green chlorophyll, exposing the yellow xanthophyll and orange carotene, previously hidden by the green pigment, and partly by the formation of red and purple pigments—*anthocyanins*—in the cell sap.

12–6 TRANSPIRATION

The leaves of a plant exposed to the air will lose moisture by evaporation unless the air is saturated with water vapor. The sun's heat vaporizes the water from the surfaces of the mesophyll cells, and the resulting water vapor passes through the stomata and escapes. This loss of water, called *transpiration,* may occur in all parts of the plant exposed to the air, but most of it occurs in the leaves. The rate of transpiration is very low during the night when the stomata are usually closed, and the lower temperature decreases the rate of evaporation of water from the mesophyll cells. The stomata also tend to close during the latter part of the afternoon on a hot sunny day. This greatly decreases the rate of transpiration and conserves the plant's supply of water. If the plant

has an adequate supply of water, the stomata remain open and an amazing amount of water is transpired. Only a small fraction—1 or 2 per cent—of all the water absorbed by the roots is used in photosynthesis. All the rest passes through the stomata as water vapor in the process of transpiration. If the plant is not getting sufficient water from its roots, the guard cells around the stomata will become less turgid and the stomata will close, thereby conserving water.

The many small holes of the stomata provide a remarkably effective pathway for the diffusion of water vapor, oxygen and carbon dioxide. Although the total area of the pores is only 1 to 3 per cent of the total area of the leaf's surface, the rate of diffusion through the stomata is from 50 to 75 per cent of the rate through an open surface equal to the area of the leaf. In sunlight an average plant will transpire about 50 milliliters of water per square meter of leaf surface per hour. An average corn plant transpires more than 200 liters of water in the course of a growing season and a medium-sized tree will transpire that much in a single day. The amount transpired varies widely in different plants; for example, it is estimated that an acre of corn will transpire 1,400,000 liters of water in a growing season whereas an acre of cactus in the Arizona desert will transpire no more than 1100 liters in a whole year. The amount of water vaporized from the leaves of trees in a forest is enough to influence significantly the rainfall, humidity and temperature of the region.

Transpiration contributes to the economy of the plant by assisting the upward movement of water through the stem, by concentrating in the leaves the dilute solutions of minerals absorbed by the roots and needed for the synthesis of new cellular constituents, and by cooling the leaves, in a manner analogous to the evaporation of sweat in animals. Although the leaf absorbs some 75 per cent of the sunlight reaching it, only about 3 per cent is utilized in photosynthesis. The rest is transformed into heat and must be removed or it would kill the tissues of the leaf. Some of this heat is removed by the vaporization of water, for 540 kilocalories are required to convert a liter of water to water vapor. The rest of the heat is lost by reradiation and by convection.

When water is lost by evaporation from the

surface of a mesophyll cell, the concentration of solutes in the cell water increases and the cell becomes slightly hypertonic. Water thus tends to pass into it from neighboring cells that contain more water. These cells in turn receive water from the tracheids and vessels of the leaf veins. During transpiration, then, water passes by the purely physical process of osmosis from the xylem vessels of the veins, through the intervening cells, to the mesophyll cells next to the air spaces of the leaf, where it is vaporized. In fact, a continuous stream of water passes from the soil into the vascular system of the roots, up through the stem and petiole to the veins of the leaf blade.

12–7　THE MOVEMENT OF WATER

It was shown experimentally many years ago that water and salts absorbed by the roots move upward in the stem primarily in the tracheids and vessels of the xylem and that sugars and other organic materials are carried primarily in the sieve tubes of the phloem. If a cut is made entirely around a stem, deep enough to penetrate the phloem and cambium but not the xylem, the leaves remain turgid and in good condition for a long time. They must be getting water via the xylem since the phloem has been completely cut. By special techniques it is possible to cut the inner xylem and leave the outer phloem relatively intact. When this is done the leaves wilt and die almost immediately, again showing that water reaches them primarily via the xylem. Although the

route of water transport has been known for some time, the mechanism by which this occurs is still not completely clear. Any acceptable explanation must account for the high rates of water flow (as much as 75 to 100 milliliters per minute) observed in some plants and for the fact that water can be moved to the tops of redwoods and Douglas firs that may be as much as 125 meters high. Although a pressure of about 12 atmospheres would support a column of water 125 meters high, additional pressure would be required to move the water upward against the frictional resistance of the small tubes. Estimates of the pressure required to move water to the top of a tall tree range up to 30 atmospheres. The pressure might be generated at the base of the plant and *push* water upwards; it might be generated at the top of the plant and *pull* water up; or water might be moved by these two forces acting jointly.

Root Pressure. If the stem of a well-watered tomato plant is cut, sap will flow from the stump for some time. If a piece of glass tubing is attached to the stump with a watertight junction, water will rise to a height of 1 meter or more. This is evidence for a positive pressure, termed *root pressure*, at the junction of root and stem, generated by forces operating in the root. A plant growing in well-watered soil under conditions of high humidity such that little water is lost by evaporation from the leaves may force water under pressure out at the ends of the leaf veins, forming droplets along the edges of the leaves (Fig. 12–

Figure 12–18 Guttation. A photograph of a leaf of a strawberry plant showing the drops of water expressed from the veins at the tip of the leaf under pressure, a process termed guttation. (Courtesy of J. Arthur Herrick, Kent State University, Kent, Ohio.)

18). This phenomenon, termed **guttation,** is further evidence that under certain conditions the sap in the xylem may be under pressure generated by the roots.

The sap in the roots is hypertonic to the water in the surrounding soil; this may account at least in part for the generation of root pressure. The movement of water from the soil through the epidermis, cortex, endodermis and pericycle of the root to the xylem, and up the xylem to the stem and leaf, is down a concentration gradient and occurs at least in part by simple diffusion. If roots are killed or deprived of oxygen, root pressure falls to zero, indicating that some active, energy-requiring process is involved. Many plant physiologists have concluded that the cells of the endodermis of the root secrete water inward toward the stele by an active process and that this is responsible for much of the root pressure.

Although early measurements of root pressures indicated that they were small, newer methods of cutting off the stem and fitting a leakproof connection without injuring the stump have demonstrated root pressures of 6 to 10 or more atmospheres even in tomato plants that are raising water less than 1 meter.

In the spring, before leaves have been formed, root pressure is probably the sole force bringing about the rise of sap. Attempts to measure root pressures in conifers have been unsuccessful and perhaps these plants cannot generate this force. If root pressure were the major force causing the rise of sap, you would expect sap to flow out under pressure when a xylem vessel is punctured. Instead you may

hear the hissing sound of air being taken in when the vessel is punctured. Thus root pressure may contribute to the rise of sap in some plants under certain conditions, but it probably is not the principal force causing water to rise in the xylem of most plants most of the time.

Transpiration and the Cohesion Theory. The alternative force that might raise water in a stem is a pull from above rather than a push from below. Water is lost by transpiration or used in photosynthesis in the leaves, and osmotically active substances are produced by photosynthesis; these processes keep the leaf cells hypertonic to the sap in the veins. They constantly draw water from the upper ends of the xylem vessels in the leaves and stem and this tends to lift the column of sap upward in each duct. This pull from above can be demonstrated by attaching a cut branch to a water-filled glass tube by a watertight connection and putting the other end of the tube in a beaker of water (Fig. 12–19). If a small air bubble is introduced into the tube, the rate of movement of the water can be measured by the rate at which the bubble moves.

The water columns in the xylem vessels, being under tension from above, are slightly stretched, but water molecules are joined by hydrogen bonds and have a strong tendency to cling together. The slender column of water in the xylem vessel has a high tensile strength. Transpiration is the major process providing the pull at the top of the column. The tendency of water molecules to stick together transmits this force through the length of the stem and roots and results in the elevation of the whole

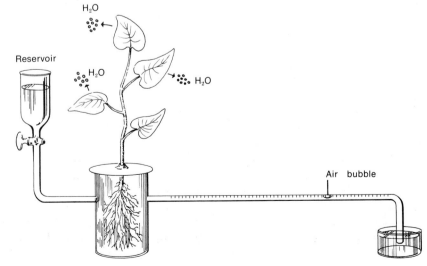

Figure 12–19 A device for measuring the rate of transpiration in a stem. As water is transpired and evaporated from the leaves it is pulled from the reservoir on the right. The rate of transpiration is measured by the movement of the air bubble through the graduated tube. The reservoir on the right can then be refilled from the reservoir on the left and a new measurement begun.

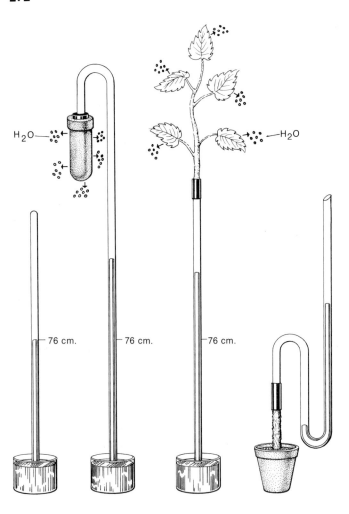

Figure 12-20 A model of the forces operating in transpiration. A porous clay cup *(left)* will draw water up in a glass tube as water is evaporated from its surface. In a similar fashion, water is moved upward in a stem as water molecules evaporate from the leaf *(center).* A device for measuring root pressure is illustrated in the diagram on the right.

column of sap. The idea that water rises in plants owing to a pull at the top resulting from transpiration was first stated by Stephen Hales early in the eighteenth century. The concept that the cohesive properties of water molecules play a role in the ascent of water in the xylem was formulated in 1914 by Dixon and Joly, who predicted that a column of water would have great tensile strength. Some experimental measurements of it have given values considerably greater than the 30 atmospheres necessary to account for the rise of water to the top of a redwood. If water is evaporated from a porous clay tube at the top of a column of water immersed in a beaker of mercury (Fig. 12–20), the mercury can be drawn to a height of 100 centimeters or more, much higher than barometric pressure. For a successful demonstration, the water must be carefully boiled first to remove dissolved gases that might form bubbles when the water column is subjected to tension.

This theory would predict that the diameter of a tree would decrease as transpiration occurs. If the water in xylem ducts is under tension, there should be an inward pull on the wall of each duct. The inward pulls on the walls of all the ducts in a band of sapwood should be great enough to produce a detectable *decrease* in the diameter of a tree during active transpiration. This predicted result was obtained by D. T. MacDougal, who showed that the diameter of the Monterey pine fluctuates daily and is minimal shortly after noon, when transpiration is maximal.

Experiments analyzing the uptake of water by jungle vines also support the Dixon-Joly hypothesis. Some of these vines extend 50 meters or more up into the trees of the jungle. When the base of the vine is severed and placed in a pail of water, water is taken up by the plant at a high rate. If the stem is placed in a sealed container of water it again takes up water rapidly even though it produces a high vacuum in the container. This transpiration-cohesion theory is widely accepted at present and accounts for most of the rise of water by most plants under most conditions. Among the questions yet unanswered is how the column of water becomes established initially.

Plants have become adapted to grow in environments with a wide range of water content, and botanists classify them according to their water preferences as hydrophytes, mesophytes and xerophytes. *Hydrophytes* grow in a very wet environment, either completely aquatic or rooted in water or mud but with stems and leaves above the water. Water lilies, pondweeds and cattails are common hydrophytes. *Mesophytes* are the common land plants that live in a climate with an average amount of moisture—beech, maple, oak, dogwood and birch are typical mesophytic trees. *Xerophytes* are plants such as yuccas and cactuses that are adapted to live where soil water is scarce. By such means as reducing the number of stomata and developing thick stems and leaves to store water and heavily cutinized surfaces to reduce water loss, these plants manage to survive with limited supplies of water. A moment's reflection will reveal that plants living in a saltwater marsh on the seacoast are also xerophytes, for although water is present, it is largely unavailable to the plant. The concentration of salts in the water exceeds that in the plant tissues and the plant is unable to get water by simple diffusion.

12–8 THE TRANSPORT AND STORAGE OF NUTRIENTS

In sunlight a green plant may produce more than twenty times as much food as it is using at the moment. At other times, during the night and over the winter season, it consumes more food than it makes. The accumulated food reserves tide the plant over periods when photosynthesis cannot occur. Food stores may be deposited in leaves, stems or roots. Leaves serve as temporary depots for food but are not suitable for long-term storage for they are too easily and too rapidly lost. The stems of woody perennials serve as storage places for large amounts of food; other plants utilize underground fleshy stems for the purpose. Perhaps the most common storage organs are roots; their underground location protects them from climatic changes and from animals looking for food.

Much of the glucose produced in a day is converted into starch and stored in the leaves.

The starch is subsequently hydrolyzed back to glucose which is translocated in the phloem to the stem and roots. In the root it may pass out of the phloem through the pericycle and endodermis to the cortical parenchyma, where it is reconverted to starch and stored. The flow of liquid in the phloem, though not as large nor as rapid as the flow in xylem, is nonetheless sizable and may attain 1000 centimeters per hour.

The movement of food materials in the sieve tubes of the phloem is dependent on their metabolic activity and is reduced when metabolism is decreased by low temperature, by lack of oxygen or by metabolic poisons. Dissolved nutrients, sugars and amino acids are carried both down and up in the phloem; indeed, two substances may be carried in opposite directions simultaneously. Very little, if any, translocation of solutes occurs in any system other than the phloem. Although the total amount of phloem sieve tubes that are functional in the trunk of a large tree is small, a variety of experimental evidence shows that the phloem is the only system involved in sugar transport and that it does this at a remarkable rate. Not all organic materials are synthesized in the leaves. For example, nitrates taken in through the roots are converted there to amino groups and incorporated into amino acids and other nitrogenous compounds.

Transport in the phloem is through living, active cells, which, however, are highly modified as a result of certain irreversible changes including loss of the nucleus and disintegration of the vacuoles. In contrast, transport in the xylem occurs through hollow tubes, the dead remains of cell walls without cytoplasm. The phloem sieve tube elements contain cytoplasm and are connected end-to-end by cytoplasmic threads that penetrate the small pores (*sieve plates*) in the cell walls.

According to one hypothesis, materials pass into one end of a sieve tube element through the sieve plate and are picked up by the cytoplasm which streams up one side of the tube cell and down the other (Fig. 12–21). At the other end of the sieve tube cell, the material passes across the sieve plate to the next adjacent tube cell by physical diffusion, by facilitated diffusion or by active transport. The *cyclosis,* or cytoplasmic streaming within successive cells, and diffusion or active transport between cells could move sugars and other materials over long distances. The system could account for the simultaneous transport of two substances in opposite directions.

Another widely held theory, termed the "pressure-flow" theory, suggests that water containing the solutes flows through the phloem under pressure along a gradient of turgor or osmotic pressure resulting from a gradient in solute concentrations. Cells in the phloem in the leaf contain a high concentration of sugars and other products of photosynthesis and water thus tends to pass into them from the xylem ducts (Fig. 12-22). This increases the pressure within the phloem cells and tends to push fluid from one cell to the adjacent one down the phloem tube. As the fluid passes down the stem and root, sugars are removed and stored as starch, which is insoluble and exerts no osmotic effect. This withdrawal of solutes lowers the osmotic pressure in the phloem, water passes out of the phloem and the turgor pres-

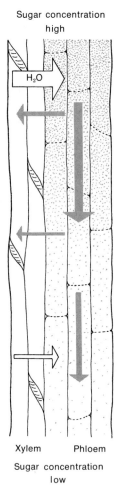

Sugar concentration
high

Xylem   Phloem

Sugar concentration
low

Figure 12-22 Diagram illustrating the pressure-flow theory of fluid transport in plants.

sure falls. The difference in turgor pressure is believed to bring about the mass flow of the contents of the sieve tubes from a region of high turgor pressure — such as the leaves when photosynthesis is occurring — to regions of lower turgor pressure — such as the stem or root where the materials are stored or used. This theory predicts that the contents of the sieve tubes should be under pressure, and this can be shown experimentally to be true. When a sap-sucking insect such as an aphid punctures a phloem tube with its stylet, the sap flows into the insect without assistance. If the body of the insect is removed but the stylet is left in place, the sugar-laden sap will continue to exude through the stylet for days. Although this theory is highly regarded by some botanists it does not readily account for the simultaneous transport of materials in both directions in the phloem. In addition, since it is based on physical phenomena it should be relatively unaffected by changes in the rate of metabolism in

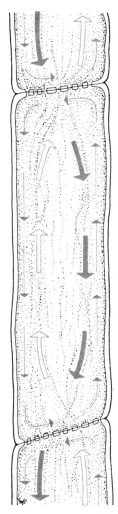

Figure 12-21 Cytoplasmic streaming in a phloem tube. The light and dark arrows illustrate that cytoplasmic streaming could account for the simultaneous transport of two different substances in opposite directions.

the phloem sieve tubes, but translocation is markedly affected by changes in metabolic rate.

The evolution of xylem and phloem systems has enabled tracheophytes to adapt to a land environment and to grow to great heights. A land plant exposing broad, chlorophyll-containing surfaces to the sun is subject to marked loss of water to the dry air. It must procure a supply of water not only as a raw material in photosynthesis but to replace the water lost by transpiration. The structure of the xylem adapts it to carry water absorbed from the soil to the leaf. The roots and stems in turn require a supply of nutrients to remain alive, to keep the plant anchored in the soil and to hold the leaves so they can be exposed to sunlight. The sieve tubes of the phloem transport these nutrients from the leaves to the stem and root and back again as required.

When a plant sheds its leaves each fall, it must put away a reserve of food in the stem or roots to carry it through the winter and to provide energy for the growth of new leaves the following spring. The stored food must be in the form of some insoluble substance to prevent its diffusing away, and the usual form of storage is starch. Plants also deposit rich stores of food in their seeds to provide energy for the development of the embryo until the new plant has developed a functional root, stem and leaf. Such seeds, rich in proteins, fats and starches, are an important source of food for many animals and human beings.

12–9 THE ECONOMIC IMPORTANCE OF PLANTS

We are completely dependent upon plants, both directly, for food, and indirectly, for the food they supply to the animals we use as food. We obtain foods, seasonings, beverages, perfumes and a wide variety of drugs from all parts of plants: roots (radishes, sarsaparilla, sweet potatoes, carrots, tapioca); stems (garlic, sugar cane, white potatoes); leaves (lettuce, spinach, cabbage); flowers (artichokes); seeds (nuts, cocoa, coffee, nutmeg, mustard); and fruits (berries, squash, apples, oranges, eggplant).

The list of nonfood products derived from plants and used in everyday life—including such things as lumber, paint, rubber, soap, cork, cotton and resins derived from a wide va-

riety of plant organs and secretions—is almost endless.

QUESTIONS

1. Draw a diagram of a plant embryo. How is the polarity of the developing embryo established?
2. Draw a three-dimensional diagram of a root. What are the functions of roots? What are the functions of adventitious roots?
3. Describe the processes by which a root absorbs water and salts from the surrounding soil.
4. What are the constituents of a good rich soil? What is the role of each of these constituents in plant nutrition? What measures can be taken to prevent the loss of topsoil?
5. Describe the three-dimensional structure of a woody stem. What are the functions of stems? How are stems and roots differentiated? What structures are uniquely present in roots and in stems?
6. Differentiate between herbaceous and woody plants and between annual and perennial plants. Give an example of each.
7. What life cycle patterns are characteristic of annual, biennial and perennial plants respectively?
8. What are the features that distinguish dicots from monocots?
9. What are the functions of the (a) cambium, (b) stomata, (c) heartwood, (d) lenticels, (e) abscission layer, (f) cutin, (g) meristem, (h) Casparian strip?
10. Describe an experiment by means of which you could show that water is transported both up and down the stem in the xylem.
11. Discuss the mechanism by which the guard cells regulate the size of a stoma.
12. What are the functions of leaves? What is the role of transpiration in the plant? What factors affect the rate of transpiration?
13. In what ways are leaves specially adapted for the manufacture of food substances?
14. What processes in the plant are responsible for the phenomenon of guttation?
15. What experiments could be devised to test the "cohesion theory" of water transport in plants?

SUPPLEMENTARY READING

The structure and function of seed plants is discussed in greater detail in *Biology of Plants* by P. Raven and H. Curtis, in *Growth and Organization in Plants* by F. C. Steward, in A. W. Galston's *The Green Plant* and in *Anatomy of Seed Plants* by K. Esau. Three fine books on plant development are J. G. Torrey's *Development in Flowering Plants*, C. W. Wardlaw's *Morphogenesis in Plants* and T. A. Steeves and I. M. Sussex's *Patterns in Plant Development*.

Plant Physiology by Salisbury and Ross is an up-to-date, general text with an excellent discussion of water movement and translocation. Richardson's *Translocation in Plants* provides an excellent survey of the current theories of the mechanisms involved in the transport of nutrients and water in plants.

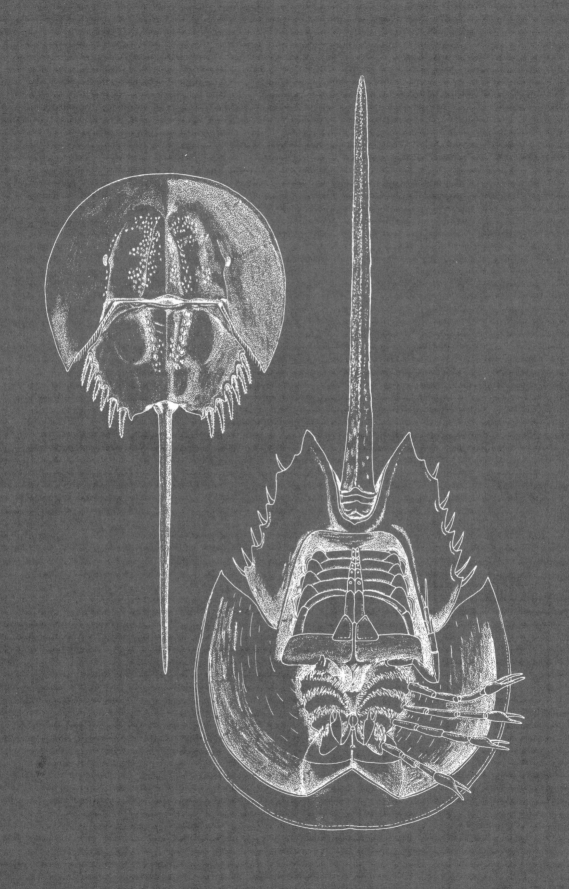

PART 3

THE WORLD OF LIFE: ANIMALS

INTRODUCTION

Somewhat more than one million species of animals (including some 800,000 insects!) have been identified and probably several million more, mostly small ones, remain to be named. In this section only a small fraction of the many different types of animals known can be sampled. The ones described were chosen because of their evolutionary or ecologic significance, because they illustrate some general biological principle, or because of their prevalence. An appreciation of the diversity of animal types, of their special adaptations, their colors and shapes, can be further sharpened by trips to the woods, to the seashore, to aquaria, to zoos and to museums.

A survey of the simpler, generally smaller invertebrate animals—sponges, coelenterates, flatworms and a variety of "worms"—is presented in Chapter 13. The larger, more complex invertebrates—mollusks, annelids, arthropods, and echinoderms—are described in Chapter 14. Chapter 15 summarizes the animals of the phylum Chordata and includes a detailed discussion of the structure of the frog, an animal widely used in laboratory sections of general biology.

CHAPTER 13

THE ANIMAL KINGDOM: LOWER INVERTEBRATES

The traditional division of the animal kingdom into those animals that have a backbone—the vertebrates—and those that lack this structure—the invertebrates—is artificial and difficult to justify. The invertebrates are a diverse, incredibly heterogeneous assemblage of animals, and except for the absence of a backbone, there is no single morphologic or developmental characteristic common to all. In certain respects some invertebrate animals are more closely related to the vertebrates than to the other invertebrate groups.

In discussing and comparing the many different groups of animals, it is convenient to use terms such as "primitive" and "advanced" or "lower" and "higher" or "simple" and "specialized." The terms "advanced," "higher" or "specialized" do not imply that these animals are better or more nearly perfect than others; rather, they are used in a comparative sense to describe the evolutionary relationships of members within a particular group of animals. The terms "higher" and "lower" usually refer to the level at which a particular group has diverged from certain main lines of evolution. It is customary, for example, to refer to sponges and cnidarians as "lower" invertebrates since they are believed to have originated near the base of the phylogenetic tree of the animal kingdom. Neither sponges nor cnidarians are primitive in all morphologic or physiologic characteristics. Each has become specialized in certain respects in the course of evolution. Furthermore, the terms "lower" or "higher" do not necessarily imply that the higher groups have evolved directly from or through the lower groups.

Two phyla, the cnidarians and the ctenophores, are radially symmetrical and are included in the *Radiata.* All of the other animals are bilaterally symmetrical and may collectively be called the *Bilateria.* Some of the higher animals, such as the echinoderms, have secondarily become radially symmetrical as adults, but their embryos are bilaterally symmetrical and are therefore classified as Bilateria.

Comparisons of the patterns of embryonic development provide evidence to support the division of the bilateral animals into two main lines of evolution. The first line, including flatworms, annelids, mollusks and arthropods as well as a number of smaller groups, is termed the **Protostomia.** The second line, the **Deuterostomia,** includes the echinoderms, the chordates and several smaller phyla. The deuterostomes appear to have diverged from the main protostome line at a point in evolution considerably after the flatworms. The protostomes have a basic plan of development that is characteristic and distinct from that of the deuterostomes.

In order to survive, all animals have had to evolve solutions to the same basic biological problems; they must obtain food and oxygen, remove metabolic wastes, maintain water balance, detect and respond appropriately to changes in their environment, and reproduce their kind. It follows that there is a certain basic unity of life and that a discovery about one form may have broad, even universal, application. It seems obvious that rats, rabbits and guinea pigs are similar enough to the human that experiments on their digestive, circulatory and excretory systems contribute to our understanding of the corresponding systems in us. But it is not so obvious that much of our knowledge of nerve action can and does come from experiments on the nerves of the squid and earthworm, that experiments on the heart of the horseshoe crab can give us information about the human heart, and that observations of the toadfish kidney are useful in providing understanding of the human excretory system.

13-1 THE BASIS FOR ANIMAL CLASSIFICATION

In determining relationships, biologists are careful to distinguish between homologous structures and analogous structures. *Homologous structures* are those which develop from similar embryonic rudiments, are similar in basic structural plan and development, and hence reflect a common genetic endowment and evolutionary relationship. In contrast, *analogous structures* are simply superficially similar and serve a similar function, but have quite different basic structures and developmental patterns. The presence of analogous structures does not imply an evolutionary relationship in the animals bearing them. For example, the arm of a man, the wing of a bird, and the pectoral or front fin of a whale are all homologous, with basically similar patterns of bones, muscles, nerves, and blood vessels, and similar embryonic origins, though rather different functions (Fig. 13-1). The wing of a bird and the wing of a butterfly, in contrast, are simply analogous; both enable their possessors to fly, but they have no developmental processes in common. The wings of birds and the wings of bats have a similar structural plan and development and are anatomically homologous; however, they evolved independently as adaptations for flying and are thus analogous in terms of their functions. As more has been learned about the molecular structure of cellular constituents, it has become clear that these terms can be applied at that level. The hemoglobins of different animals, the cytochrome c's present in different vertebrates, or the lactic dehydrogenases present in birds and mammals may be termed homologous proteins. The hemoglobins in different species, for example, have very similar sequences of amino acids; these again reflect a common genetic pattern and evolutionary relationship. In contrast, hemoglobin and hemocyanin may be termed analogous molecules, since they have similar functions (oxygen transport) but quite different molecular structures.

Animals are distinguished by the type of symmetry present, whether their body plan shows spherical, radial or bilateral symmetry.

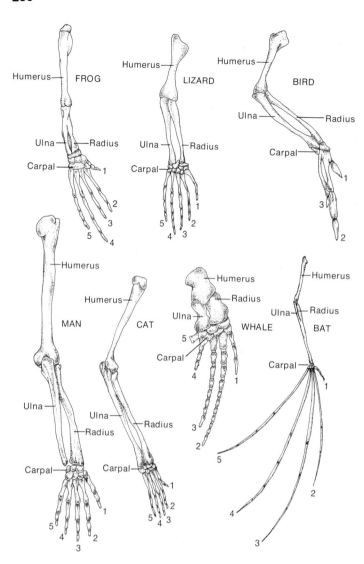

Figure 13–1 The bones of the forelimbs of a frog, lizard, bird, man, cat, whale, and bat, showing the arrangement of the homologous bones in these superficially different structures.

Animal groups are also distinguished by certain features of early development—the patterns of cleavage and gastrulation, the formation and fate of the blastopore, and so on. Some of the metazoa have only two embryonic cell layers or germ layers—an outer *ectoderm* and an inner *endoderm* lining the digestive tract. Others have these two plus a third, the *mesoderm,* extending between the ectoderm and endoderm and making up the rest of the body. In the simpler many-celled animals, the body is essentially a double-walled sac surrounding a single "gastrovascular" cavity with a single opening to the outside—the mouth. The more complex animals have two cavities and bodies constructed on a tube-within-a-tube plan. The inner tube, the digestive tract, is lined with endoderm and is open at both ends—the mouth and the anus. The outer tube or body wall is covered with ectoderm. Between the two tubes is a second cavity, termed a *coelom* if it lies within the mesoderm and is lined by it, or a *pseudocoelom* if it lies between the mesoderm and endoderm. The members of several phyla are characterized by bodies composed of a row of segments, each having the same fundamental plan, with or without some variation. In most of the vertebrates, the segmental character of the body is largely obscured. In humans, the bones of the spinal column—the vertebrae—are among the few parts of the body that are still clearly segmented.

A few structures are found exclusively in members of a single phylum and help distinguish them from all other animals. For example, the cnidarians are unique in having stinging capsules or *nematocysts;* echinoderms have a peculiar *water vascular system* found in no other phylum; and only the chordates have a dorsally located, hollow nerve cord. The appendix lists 21 animal phyla; some systematists recognize more phyla, and some recognize fewer ones. Some 10 of these 21 are major phyla that include most of the animals, both living and extinct.

The protozoa (p. 188) are included in the animal kingdom by some biologists, for they are heterotrophic and most are motile. However, at the same time protozoa are unicellular (or acellular), their evolutionary history has been separate from that of the metazoa, and their complexity has developed in a different fashion.

13–2 SPONGES

The *Porifera,* the phylum of animals commonly called sponges, have porous body walls and internal cavities lined with *choanocytes* (Fig. 13–2). The cells are strikingly similar to the choanoflagellates. Much of the body is composed of a jellylike matrix, containing a skeleton made of protein, calcium carbonate or silica. There seems to be no nervous system present. Sponges are organized on a cellular level. Instead of a single cell carrying on all the life activities as in protozoa, there is a division of labor, with certain cells specialized to perform particular functions such as nutrition, support or reproduction. The sponge shows cellular differentiation but little or no coordination of cells to form tissues. The cells are very loosely organized and cell relations can be disrupted by passing the sponge through silk bolting cloth without damaging cellular integrity. The cells reaggregate and form a structure similar to the initial one. Because of their many distinctive morphologic features, the sponges are usually considered to be a side branch in the evolution of the me-

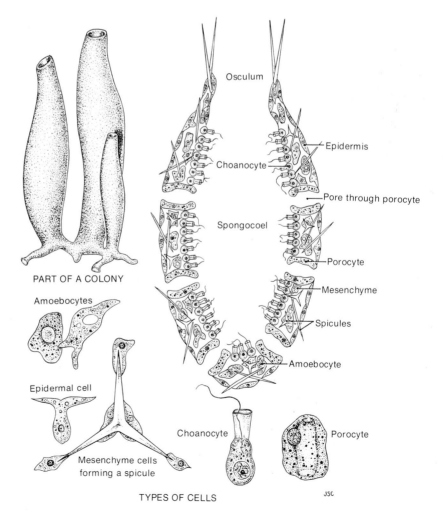

Figure 13–2 Sponges. *Upper left,* Sketch of part of a colony of sponges. *Upper right,* Diagram of a section through a simple sponge showing its cellular organization. *Lower,* Sketches of the types of cells found in a sponge.

PART OF A COLONY

Osculum

Epidermis

Choanocyte

Pore through porocyte

Spongocoel

Porocyte

Mesenchyme

Spicules

Amoebocyte

Amoebocytes

Epidermal cell

Mesenchyme cells forming a spicule

Choanocyte

Porocyte

TYPES OF CELLS

JSC

tazoa. In fact they were classified as plants until 1765! They evolved from flagellates independently of the other metazoa and have not given rise to any other phylum.

Living sponges may be drab colored or bright green, orange, red or purple. They are usually slimy to the touch and have an unpleasant odor. Sponges are sedentary organisms ranging in size from 1 to 200 centimeters in height and varying in shape from flat, encrusting growths to balls, cups, fans and vases. Most sponges are marine; only one family occurs in fresh water.

Sponges make their living by filtering water, straining out microscopic organisms which they use for food. Characteristic of sponges is the presence of choanocytes, or collar cells, with flagella that beat, creating the currents of water to bring in food and oxygen and to carry away carbon dioxide and wastes. The choanocytes of some complex sponges can pump a volume of water equal to the volume of the sponge each minute!

The collar of the choanocyte is composed of fine parallel fibers between which water passes. The food particles are trapped and pass down to the base of the collar where they are taken in by phagocytosis and either digested within the choanocyte or transferred to an amoebocyte for digestion. Wandering through the gelatinous matrix of the sponge body are numerous amoebocytes, which collect food from the cells lining the pores, secrete the matrix and the protein, calcium carbonate or silica of the skeleton, collect wastes and become converted to epidermal cells as needed. Each cell of the body is irritable and can react to stimuli, but there are no sense cells or nerve cells which would enable the animal to react as a whole.

The three classes of sponges, made up of about 10,000 species, are distinguished on the basis of the kind of skeleton present: *Calcispongiae* have a skeleton of calcium carbonate spicules, *Hexactinellida* have a skeleton of six-rayed siliceous spicules, and the *Demospongia* have a skeleton of spongin fibers or of siliceous spicules that do not have six rays (Fig. 13–3).

The bath sponges are found in warm shallow waters on rocky bottoms. Sponge fishermen hook them from the ocean bottom by poles with a pronged fork on the end. They are kept out of water until they die, then are left

Figure 13–3 Photograph of the skeleton of the glass sponge, *Euplectella*. The hexagons are fused to form intersecting girders. (Courtesy of the American Museum of Natural History.)

lying in shallow water until the flesh is decayed. After being beaten, washed and bleached in the sun they are ready for the market. All that remains is the spongin network whose many interstices enable it to soak up a large amount of water.

All sponges appear to be diploid and have the usual metazoan processes of oögenesis and spermatogenesis. The eggs are retained just beneath the choanocytes where they are fertilized by sperm from another sponge brought in with the current.

The pattern of development in the sponge is quite different from that in any other metazoan. The fertilized egg cleaves to form a blastulalike structure that is inside out compared to the blastulas of other animals. The nuclei lie toward the inner ends of the cells, and the flagella that appear on the cells project inward instead of outward. The embryo is also peculiar in having a mouth at the vegetal pole through which food is taken from the parent. When fully developed the embryo turns inside out through its mouth, penetrates the maternal choanocyte layer and escapes into the channels of the parent sponge and finally to the open ocean as a free-swimming larva. The eversion process brings the flagella to the outside of the larva and by means of these the larva swims. It finally becomes attached to the bottom by its anterior end and the flagellated half invaginates into the posterior half to form a two-layered structure. The flagellated cells become the inner choanocytes, whereas the outer layer forms the remainder of the sponge.

The remarkable ability of the sponges to reorganize was demonstrated by H. V. Wilson in 1907. He squeezed sponges through fine silk cloth into a dish, thus disaggregating the sponge cells into minute clumps. The choanocytes liberated in this fashion swim about on the bottom and the amoebocytes crawl. When the cells come in contact they remain together. The bottom of the dish becomes covered with balls of cells, each of which develops into a tiny sponge if it includes both choanocytes and amoebocytes. The cells are species specific and if two kinds of sponges, one yellow and one orange, are disaggregated in the same dish, the clumps of cells that aggregate will be either all yellow or all orange, but not a mixture of the two.

Sponges are believed to be an evolutionary side line, they may have evolved from a different group of flagellates from those which are thought to be ancestral to the rest of the many-celled animals. According to one theory, sponges evolved from an early choanoflagellate; another theory maintains that they are derived from a hollow, free-swimming colonial flagellate. Sponge larvae resemble such flagellate colonies. There is no evidence to suggest that any of the higher animals evolved from the sponges.

13-3 JELLYFISH AND COMB JELLIES

Cnidaria

In addition to fish, whales and porpoises, which swim actively, open ocean waters contain many organisms (termed *plankton*) floating passively with the water current. They may swim but they cannot swim strongly enough to travel in a horizontal direction or to remain in

one place against the current. Radiolaria, foraminifera and a variety of algae are planktonic protists.

The largest of the plankton are jellyfish, often seen from shipboard as vast swarms in the upper meter or so of water. The common name *jellyfish* is applied to a heterogeneous group of organisms with soft bodies of a jellylike consistency, members of the phylum *Cnidaria.* These organisms swim weakly by muscular contractions of their umbrella-shaped bodies. They characteristically have a radially symmetrical body plan organized as a hollow sac. The interior is a digestive cavity opening to the outside by a mouth; hence the former name of the phylum, Coelenterata (*coel* = hollow; *enteron* = gut). The mouth is surrounded by a circle of tentacles bearing *cnidoblasts,* stinging cells, containing nematocysts, or stinging capsules. The present name for the phylum, Cnidaria, derives from the characteristic presence of these cnidoblasts in this group.

The Cnidaria are believed to have evolved from the same stock as all the higher animals, because, like the latter, they have this central digestive cavity connected to the outside by a mouth (the sponges, in contrast, lack it). The tissues of cnidarians fall into the same general categories as those of higher animals: epithelial, connective, muscular, nervous and reproductive. The cells lining the digestive cavity are the *gastrodermis* (endoderm); those covering the outside of the body are the *epidermis* (ectoderm). In contrast to the higher animals, the cnidarians have no mesoderm cells between these two; the space is filled with *mesoglea,* a gelatinous matrix containing a very few scattered cells. The cnidarian, or coelenterate, phylum also includes a number of sessile organisms attached to the pond or ocean bottom—hydras, sea anemones and corals—and floating colonies such as the Portuguese man-of-war. The Cnidaria are grouped in three classes, the *Hydrozoa,* the hydras, the hydroids such as *Obelia,* and the Portuguese man-of-war; the *Scyphozoa,* the jellyfish; and the *Anthozoa,* the sea anemones, true corals and alcyonarians (sea fans, sea whips and precious corals).

Hydrozoa. The body plan of the cnidarians is typified by a tiny animal, *Hydra,* found in ponds and appearing to the naked eye

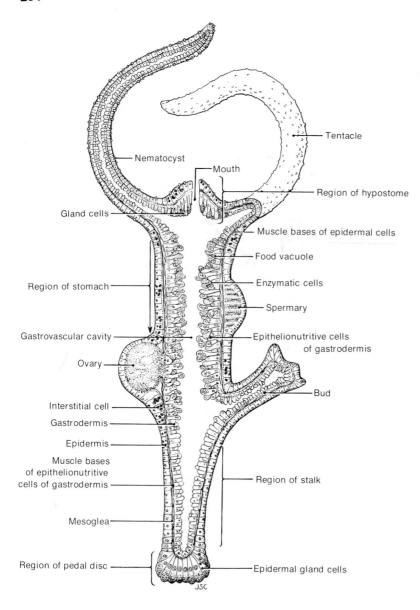

Figure 13–4 Diagram of a *Hydra* cut longitudinally to reveal its internal structure. Both a spermary and an ovary are illustrated, but they occur on separate individuals; no animal has both an ovary and a spermary. Asexual reproduction by budding is represented on the right.

like a bit of frayed string (Fig. 13–4). This animal takes its name from the multiheaded monster of Greek mythology with the remarkable ability to grow two new heads for each head cut off. The cnidarian hydra has a comparable ability to regenerate, and when it is cut into several pieces, each one may grow all the missing parts and become a whole animal. The hydra's body, seldom more than 1 centimeter long, consists of two layers of cells enclosing a central gastrovascular cavity (Fig. 19–11). The outer epidermis serves as a protective layer; the inner gastrodermis is primarily a digestive epithelium. The bases of the cells of both layers are elongated into contractile muscle fibers; those of the epidermis run lengthwise,

and those in the gastrodermis run circularly. By the contraction of one or the other, the hydra can shorten, lengthen or bend its body. Throughout its life the animal lives attached to a rock, twig or leaf by a disc of cells at its base. At the other end is the mouth, connecting the gastrovascular cavity with the outside and surrounded by a circlet of tentacles. Each tentacle may be as much as one and a half times as long as the body itself. The tentacles, composed of an outer epidermis and an inner gastrodermis, may be hollow or solid.

The Cnidaria are unique in producing "thread capsules," or *nematocysts* (Fig. 13–5) within stinging cells (cnidoblasts) in the epidermis. The nematocysts, when appropriately

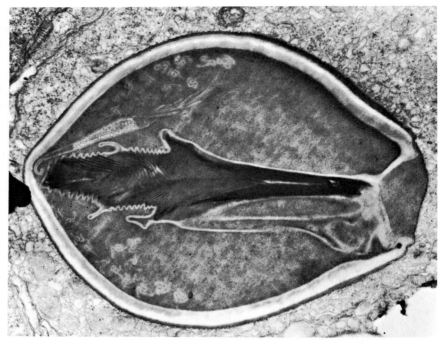

Figure 13–5 Electron micrograph of an undischarged nematocyst of hydra (sagittal section). (Courtesy of G. B. Chapman, Cornell University Medical College. From Lenhoff, H. M., and Loomis, W. F. (eds.): The Biology of Hydra. Coral Gables, University of Miami Press, 1961.)

stimulated, can release a coiled, hollow thread. Some types of nematocyst threads are sticky; others are long and coil around the prey; a third type is tipped with a barb or spine and can inject a protein toxin that paralyzes the prey. The nematocyst is shaped like a balloon, with a long tubular neck that develops tightly coiled and inside out within the cavity of the balloon. Each stinging cell has a small projecting trigger on its outer surface that responds to touch or to chemicals dissolved in the water ("taste") and causes the nematocyst to fire its thread. A nematocyst can be used only once; when it has been discharged, it is discarded and replaced by a new one, produced by the cnidoblasts. The tentacles encircle the prey and stuff it through the mouth into the gastrovascular cavity, where digestion begins. The partially digested fragments are taken up by pseudopods of the gastrodermis cells, and digestion is completed within food vacuoles in those cells.

Respiration and excretion occur by diffusion, for the body of a hydra is small enough that no cell is far from the surface. The motion of the body as it stretches and shortens circulates the contents of the gastrovascular cavity, and some of the gastrodermis cells have flagella whose beating aids in circulation. The hydra has no other circulatory device.

The first true nerve cells in the animal kingdom are found in the Cnidaria. These animals have many nerve cells in the form of an irregular network connecting the sensory cells in the body wall with muscle and gland cells. The coordination achieved thereby is of the simplest sort; there is no aggregation of nerve cells to form a "brain" or spinal cord, and an impulse set up in one part of the body passes in all directions more or less equally.

Hydra reproduce asexually by budding during periods when environmental conditions are optimal, but are stimulated to form sexual forms, males and females, when the pond water becomes stagnant. W. L. Loomis showed that the stimulus for the formation of sexual forms is an increased carbon dioxide tension in the water. Loomis showed further that males always bud to form other males and females give rise by budding only to other females.

The Portuguese man-of-war (*Physalia*) (Fig. 13–6) superficially resembles a jellyfish but is actually a colony of hydroids and medusae. The long tentacles of this form are equipped with stinging capsules that can paralyze a large fish and wound a human severely. The colony is supported by a gas-filled float of vivid, iridescent purplish green.

Scyphozoa. Besides the hydra and hydralike organisms, the phylum Cnidaria includes such superficially different forms as jellyfish (Fig. 13–6), corals and sea anemones among its

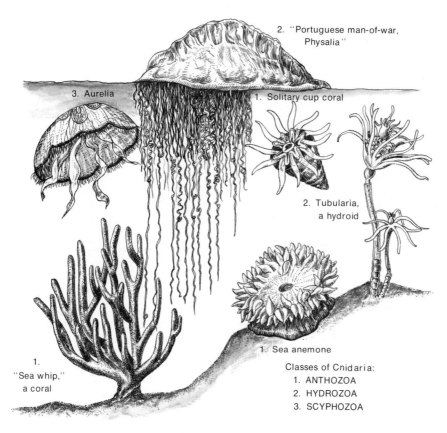

Figure 13-6 Some common representatives of the three classes of the phylum Cnidaria (Coelenterata).

Classes of Cnidaria:
1. ANTHOZOA
2. HYDROZOA
3. SCYPHOZOA

10,000 different species. Both jellyfish and hydroids have bodies composed of an outer epidermis and inner gastrodermis, with a nonliving jelly (mesoglea) layer between them. In the hydroids, the mesoglea layer is thin, whereas in the jellyfish it is thick and viscous, giving firmness to the body. The fundamental similarity between the two is illustrated in Figure 13-7. A jellyfish is like a hydra turned upside down, with a greatly increased mesoglea layer. The hydroids and jellyfish are, then, two ramifications of the same fundamental plan, one adapted for an attached life, the other for a free-swimming life.

Some of the marine cnidarians are remarkable for an alternation of sexual and asexual generations analogous to that in plants. Many species of jellyfish reproduce sexually to give

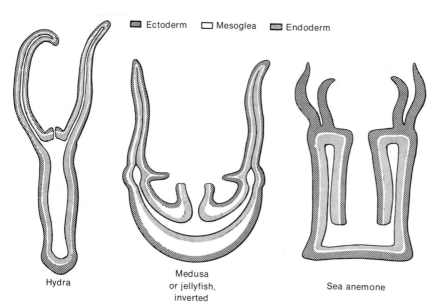

Ectoderm   Mesoglea   Endoderm

Figure 13-7 Comparison of a hydra, an inverted jellyfish and a sea anemone to show their fundamental similarity of structure.

Hydra

Medusa or jellyfish, inverted

Sea anemone

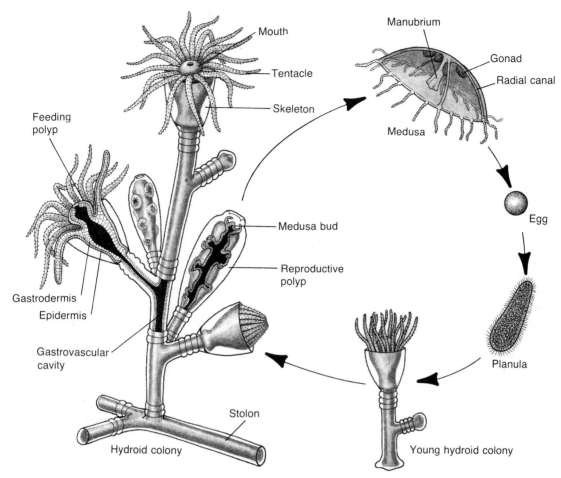

Figure 13–8   Life cycle of *Obelia,* showing structure of hydroid colony. (From Barnes, R. D., *Invertebrate Zoology,* 3rd Ed. Philadelphia, W. B. Saunders Co., 1974.)

rise to *planula* larvae which develop into sessile, sac-shaped, hydralike animals (*polyps*). These, in turn, reproduce asexually to form new free-swimming jellyfishes, *medusae,* shaped like inverted bowls (Fig. 13–8). This alternation of generations differs from that of plants in that both sexual and asexual forms are diploid; only sperm and eggs are haploid. Many of the marine cnidarians form colonial organizations of hundreds or thousands of individuals. A colony begins with a single individual that reproduces by budding, but instead of separating from the parent, the buds remain attached and continue to bud themselves. Several types of individuals may arise in the same colony, some specialized for feeding, others for reproduction.

The largest jellyfish, *Cyanea,* may be 4 meters in diameter and have tentacles 30 meters long. These orange and blue monsters, among the largest of the invertebrate animals,

are a real danger to swimmers in the North Atlantic Ocean.

Anthozoa.   The sea anemones and corals have no free-swimming jellyfish stage, and the polyps may be either individual or colonial forms. They differ from hydras in that the gastrovascular cavity is divided by a series of vertical partitions into a number of chambers, and the surface ectoderm is turned in at the mouth to line a gullet (Fig. 13–7). The partitions in the gastrovascular cavity increase the digestive surface, so that an anemone can digest an animal as large as a crab or fish.

In warm shallow seas almost every square meter of the bottom is covered with coral or anemones, most of them brightly colored. The extravagant reefs and atolls of the South Pacific are the remains of billions of microscopic, cup-shaped calcareous structures, secreted during past ages by coral colonies and by coralline plants. Living colonies occur only in the uppermost part of such reefs, adding their own secretions to the mass.

Ctenophora

The **Ctenophores,** or comb jellies, are similar in many ways to cnidarians although the 100 or so species of the group are usually placed in a separate phylum. They may be as small as a pea or as large as a tomato and they consist of two layers of cells enclosing a mass of jelly. The outer surface is covered with eight rows of cilia, resembling combs. The coordinated beating of the cilia in these combs moves the animal through the water (Fig. 13–9). At the upper pole of the body is a sense organ containing a mass of limestone particles balanced on four tufts of cilia connected to sense cells. When the body turns, these particles bear more heavily on the lower cilia, stimulating the sense cells. This causes the cilia in certain of the combs to beat faster and bring the body back to its normal position. Nerve fibers extending from the sense organ to the cilia control the beating. If they are cut, the beating of the cilia below the incision is disorganized. Ctenophores differ from cnidarians in several respects: they are biradially symmetrical, they have a markedly different type of larval development, they lack stinging capsules and they have only two tentacles instead of many. Many ctenophores are luminescent.

Both cnidarians and ctenophores have remarkable powers of regeneration; a half, quarter or even smaller piece is able to grow into a whole animal. These animals also have a marked ability to return disarranged structures to their normal relationships. It is possible to turn a hydra inside out by pulling the base out through the mouth. The hydra, though unable to turn itself inside in, does restore the normal relations of epidermis and gastrodermis by the migration of the individual cells to their proper position. A fascinating question is, how do the individual cells know where to go?

13–4 FLATWORMS

Flatworms, members of phylum Platyhelminthes, live in either fresh or salt water, or they may be terrestrial, creeping over rocks, debris and leaves. Like the hydra, flatworms have a single gastrovascular cavity (Fig. 19–11)—sometimes extensively branched—connected to the outside by a single opening, the mouth, located on the middle of the ventral surface. In addition to an outer ectoderm and an inner endoderm, the flatworm has a third, middle layer, the mesoderm, which comprises most of the body and forms many of the organs. The flatworms are the simplest animals that have well developed **organs,** functional units made of two or more kinds of tissue. They have several simple organs—a muscular pharynx for taking in food, eyespots and other sense organs on the head, a brain ganglion and a pair of interconnected ventral nerve cords and complex reproductive organs. Like the higher animals, and in contrast to the cnidarians, the flatworms are bilaterally symmetrical and have a definite anterior and posterior end. The development of a distinct anterior end with a concentration of nervous tissue, termed **cephalization,** is typical of free-living, motile, bilaterally symmetrical animals. Flatworms always keep the ventral surface down toward the substratum as they crawl along. Locomotion is achieved partly by means of cilia on their ventral surface, and partly by undulatory muscular contractions, similar to those of the earthworm. Movement is

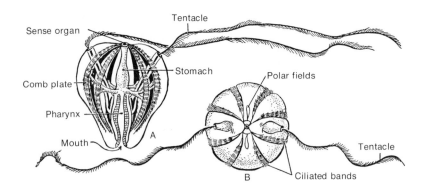

Figure 13–9 A ctenophore, *Pleurobrachia. A,* Side view; *B,* top view. (From Hunter, G. W., and Hunter, F. R.: *College Zoology.* Philadelphia, W. B. Saunders Co., 1949.)

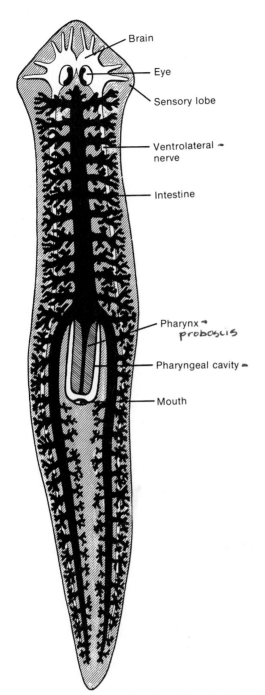

Figure 13–10 labels:
- Brain
- Eye
- Sensory lobe
- Ventrolateral nerve
- Intestine
- Pharynx — *proboscis*
- Pharyngeal cavity
- Mouth

Figure 13–10 The common American planarian *Dugesia*. (From Villee, C. A., Walker, W. F., Jr., and Barnes, R. D.: *General Zoology*, 4th Ed. Philadelphia, W. B. Saunders Co., 1973.)

facilitated by a slime secreted by gland cells in the ventral epidermis.

The commonest free-living flatworms are the planarians, found in ponds and quiet streams all over the world. The common Amer-ican planarian is *Dugesia*, about 15 millimeters long, with what appear to be crossed eyes and flapping ears (Fig. 13–10). Planarians are carnivorous and feed on living and dead small animals (Fig. 13–11). Flatworms can survive without food for months, gradually digesting their own tissues and growing smaller as time passes. As in the cnidarians, respiration takes place by diffusion. To secrete waste products, the flatworm has a branching network of fine tubes opening to the surface by pores and ending in branches known as *flame cells* or *protonephridia* (Fig. 21–9). Each flame cell is a single, hollow cell containing a tuft of cilia. The beating of the cilia resembles a flame. The motion of the cilia drives the excreted fluid along the tubes and out the pores. Planarians living in fresh water have the same problem of getting rid of excess water that the fresh water protozoa face. The flame cells, like the contractile vacuole, solve it. Some flatworms confiscate intact nematocysts from the hydras they eat, incorporate them in their own epidermis and use them for defense, discharging them when appropriately stimulated.

Tapeworms and Flukes. Besides the free-living flatworms such as *Dugesia*, which compose the class **Turbellaria**, there are two classes of parasitic platyhelminths, the **Trematoda**, or flukes, and the **Cestoda**, or tapeworms, both of which lack a ciliated epidermis.

The flukes are structurally like the free-living flatworms, but differ in having one or more suckers with which to cling to the host and a thick outer layer, the cuticle, in place of cilia. The organs of digestion, excretion and coordination are like those of the other flatworms, but the mouth is anterior rather than ventral. The reproductive organs are extremely complex. The flukes parasitic in human beings are the blood flukes, widespread in China, Japan and Egypt, and the liver flukes, common in China, Japan and Korea. Both of these parasites go through complicated life cycles, involving a number of different forms, alternation of sexual and asexual generations, and parasitism on one or more intermediate hosts such as snails and fishes (Fig. 13–12).

Tapeworms are long, flat, ribbonlike animals, some species of which live as adults in the intestines of probably every kind of vertebrate, including humans. The anterior end of a tapeworm (Fig. 13–13) lacks eyes but is

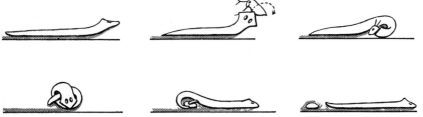

Figure 13-11 Hunting and feeding in *Dugesia*. A small crustacean *(Daphnia)* is captured and eaten, its tough exoskeleton remaining as an empty shell. (From Villee, C. A., Walker, W. F., Jr., and Barnes, R. D.: *General Zoology*, 4th Ed. Philadelphia, W. B. Saunders Co., 1973.)

equipped with suckers, and some species attach themselves to the lining of the host's intestine by a circle of hooks. Behind the head a growing region constantly gives rise by budding to new body sections, called *proglottids.* The rest of the body consists of a series of these sections, each containing little more than a complete set of reproductive organs. Each proglottid mates with itself or with an adjacent proglottid, becomes filled with fertilized eggs (each in its own capsule), breaks off and passes out of the host's body. The fertilized egg is eaten by another host, the larva hatches out of its capsule and continues development. Most cestodes must invade two or three host species in succession to complete their life cycles.

Tapeworms have no mouth and no trace of a digestive system; they live by absorbing the digested materials present in the intestine of their host.

13–5 THE ORGAN SYSTEM LEVEL OF ORGANIZATION

The Nemertea or Proboscis Worms. This relatively small group of animals (550 species) is important to us only as an evolutionary landmark, for the proboscis worms are the simplest living animals illustrating the organ system level of organization (Fig. 13-14). None of them is parasitic and none is of economic importance. Almost all of them are marine, although a few inhabit fresh water or damp soil. They have long narrow bodies, either cylindrical or flattened, varying in length from 5 centimeters to 20 meters. Some of them are a vivid orange, red or green with black or colored stripes. Their most remarkable organ—the *proboscis,* from which they take their name—is a long, hollow, muscular tube which they evert from the anterior end of the body and use for

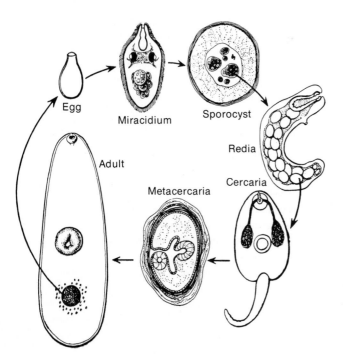

Figure 13-12 Life cycle of a trematode. The miracidium burrows into a snail, rounds up and becomes a sporocyst. Embryos within the sporocyst develop into another form, called redia, which escape from the sporocyst and feed on the tissues of the host snail. Reproductive tissue within the redia develops into embryos, which grow into redia or into cercariae, which escape from the redia and are miniature flukes, complete with a tail. The cercaria leaves the snail, swims to the next host (crayfish, clam, fish), bores into the new host, becomes surrounded by a cyst and develops into a metacercaria. When this is eaten by the final host it develops into the adult fluke and migrates to the liver or lungs of the final host. There it lays eggs which are fertilized and develop into the ciliated larval form, the miracidium, to complete the cycle. (From Villee, C. A., Walker, W. F., Jr., and Barnes, R. D.: *General Zoology,* 4th Ed. Philadelphia, W. B. Saunders Co., 1973.)

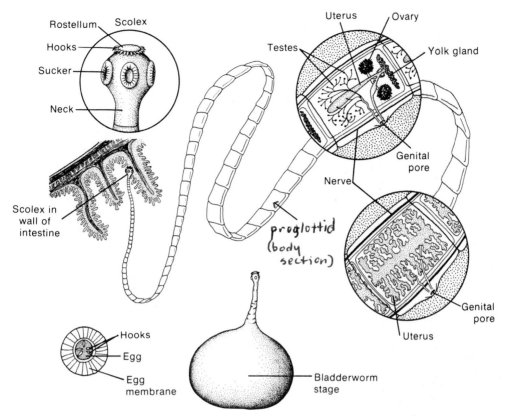

Figure 13–13 The pork tapeworm, *Taenia solium.* Insets show the head, an immature and a mature section of the body. (From Villee, C. A., Walker, W. F., Jr., and Barnes, R. D.: *General Zoology,* 4th Ed. Philadelphia, W. B. Saunders Co., 1973.)

seizing food. The proboscis secretes mucus, helpful in catching and retaining the prey. The proboscis of certain species is equipped with a

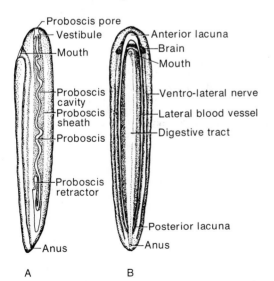

Figure 13–14 Diagrams of the structure of a typical proboscis worm or nemertean, *A,* Lateral view of the digestive tract and proboscis. *B,* Dorsal view of the digestive, circulatory and nervous system. (Redrawn from Villee, C. A., Walker, W. F., Jr., and Smith, F. E.: *General Zoology,* 3rd Ed. Philadelphia, W. B. Saunders Co., 1968.)

hard point at the tip and poison-secreting glands at the base of this point. This proboscis is thrust outward by the pressure of the surrounding muscular walls on the contained fluid; a separate muscle inside the proboscis retracts it.

The important advances displayed by the nemerteans are, first, a complete digestive tract, with a mouth at one end for taking in food, an anus at the other for eliminating feces, and an esophagus and intestine in between. This is in contrast to the cnidarians and planarians, whose food enters and wastes leave by the same opening. In the proboscis worm, water and metabolic wastes are eliminated from the body by flame cells (protonephridia), as they are in the flatworms.

A second advance exhibited by the nemerteans is the separation of digestive and circulatory functions. These animals are the most primitive organisms to have a separate circulatory system. It is, of course, rudimentary, consisting simply of three muscular tubes—the blood vessels—extending the length of the body and connected by transverse vessels.

These primitive forms have red blood cells filled with **hemoglobin,** the same red pigment that transports oxygen in human blood. Nemerteans have no heart, and the blood is circulated through the vessels by the movements of the body and the contractions of the muscular blood vessels. There are no capillaries. The nervous system is more highly developed than it is in the flatworm; there is a "brain"* at the anterior end of the body, consisting of two groups of nerve cells (ganglia) connected by a ring of nerves extending around the sheath of the proboscis; two nerve cords extend posteriorly from the brain.

The Nematoda. The phylum Nematoda, the roundworms, has a great many members (about 10,000 species), all of them remarkably similar in general body pattern. Some live in the sea, others in fresh water, in the soil, or in other plants or animals as parasites. There are some 50 species of nematodes which are human parasites, including the hookworm, trichina worm (Fig. 13–15), ascaris worm (Fig. 13–16), filaria worm and guinea worm. A mi-

---

*The word "brain" is loosely applied to the aggregation of nerve cells at the anterior end of the nerve cord, which acts as a reflex center. It should not be inferred that anything like thought processes occur in any of the lower animals.

croscopic examination of a shovelful of earth from almost anywhere in the world will reveal a number of tiny white worms which thrash around, coiling and uncoiling. Their elongate, cylindrical, threadlike bodies, pointed at both ends, are covered with a tough cuticle. A feature of nematode anatomy is the presence of a primitive body cavity, the **pseudocoelom,** between the body wall and gut wall. It is a derivative of the embryonic blastocoele and lies between the endoderm and mesoderm. A true coelom is a body cavity lying within tissues of mesodermal origin and lined with a simple epithelium of mesodermal origin. In contrast to the nemerteans, which have cilia all over the epithelium and the lining of the digestive tract, none of the nematodes has any cilia at all. Nematodes have no circularly arranged muscle fibers, only longitudinal ones. They can only bend, and they swim poorly despite vigorous thrashing movements.

With the evolution of a complete digestive system, a separate circulatory system, and a nervous system composed of a "brain" and nerve cords, as illustrated in the proboscis worm, the essential structures of higher animals were established. The proboscis worm is believed to resemble animals that might be the ancestors of both the higher animals and itself. Evolution beyond this point branched out in a

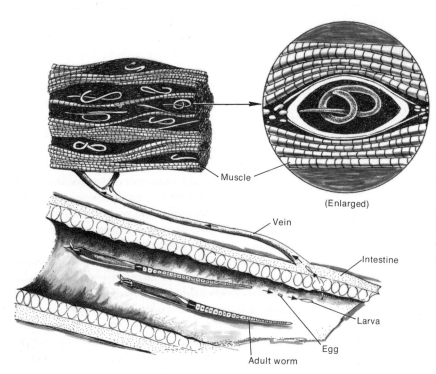

Muscle

(Enlarged)

Vein

Intestine

Larva

Egg

Adult worm

Figure 13–15 The roundworm *Trichinella spiralis,* which causes trichinosis. When a piece of pork infected with *Trichinella* is eaten, the larvae are released and grow rapidly to maturity in the intestine. After fertilization, the females produce tiny larvae which burrow into the blood vessels and are carried to the muscles, where they encyst *(inset, upper right).*

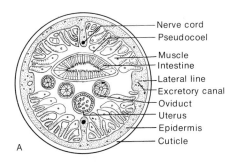

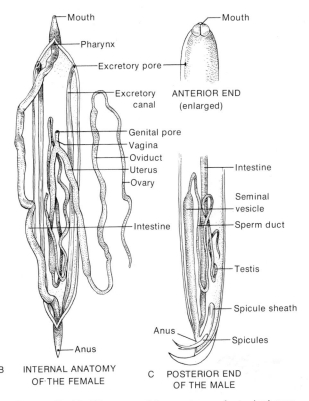

Figure 13–16 Diagrams of the anatomy of a typical parasitic nematode, *Ascaris lumbricoides*. A, Transverse section showing arrangement of internal organs. B, Female worm dissected open to show reproductive tract. C, Posterior end of male worm dissected open to show reproductive organs.

great many different directions, and the more advanced animals cannot be arranged in a single series of progressively higher and more complex forms. One main branch of evolution led to the vertebrates, another to the arthropods, and another to the clam, squid and other mollusks.

— **The Rotifera.** Among the more obscure invertebrates are the "wheel animals" of the phylum Rotifera. These aquatic, microscopic worms, although no larger than many of the protozoa, have many-celled bodies with a complete digestive tract including a *mastax*, a mus-

cular organ for grinding food; a pseudocoelom; an excretory system made up of flame cells and a bladder; a nervous system with a "brain" and sense organs; and a characteristic crown of cilia on the anterior end which gives the appearance of a spinning wheel (Fig. 13–17).

Rotifers and gastrotrichs are "cell constant" animals: Each member of a given species is composed of exactly the same number of cells; indeed, each part of the body is made of a precisely fixed number of cells arranged in a characteristic pattern. Cell division ceases with embryonic development and mitosis cannot subsequently be induced; growth and repair are impossible. One of the challenging problems of biology is the nature of the difference between such nondividing cells and the dividing cells of other animals. Do rotifers never develop cancer?

— **The Gastrotricha.** Another phylum of microscopic, aquatic animals, resembling rotifers in many respects but lacking the crown of cilia, is the gastrotrichs. Some of the fresh water gastrotrichs are peculiar in that they apparently consist entirely of females which reproduce by parthenogenesis; no males of this group have ever been found.

— **The Bryozoa.** The bryozoa, or moss animals, comprise two phyla of colonial animals, the *Entoprocta* and the *Ectoprocta*, that superficially resemble the colonial hydroids (Fig. 13–18). The colonies of some species, delicately branched and beautiful, are sometimes mistaken for seaweed; the colonies of other species appear as thin, lacy encrustations on rocks. Each animal secretes about itself a protective case made of calcium carbonate or of a horny protein into which it can withdraw when danger threatens. This case may be shaped like a vase, a box or a tube. Around the animal's

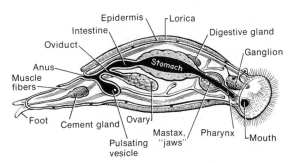

Figure 13–17 Diagram of the anatomical structures of a rotifer, seen in lateral view.

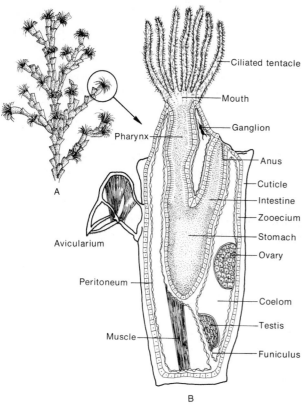

A

Ciliated tentacle

Mouth

Ganglion

Pharynx

Anus

Cuticle

Intestine

Zooecium

Avicularium

Stomach

Ovary

Peritoneum

Coelom

Testis

Muscle

Funiculus

B

Figure 13–18 *A,* A colonial ectoproct bryozoan, *Bugula.* *B,* Diagram of one animal cut open longitudinally to reveal its internal structure.

mouth is a circular or horseshoe-shaped ridge bearing a set of ciliated tentacles. An adaptation to living in a "vase" is the **U**-shaped digestive system. In the Entoprocta, the anus lies within the ring of tentacles, called a tentacular crown, and the body cavity is a pseudocoelom. In the other phylum of bryozoa, the Ectoprocta, the anus lies outside the ring of tentacles, called a lophophore, and the body cavity is a true coelom. In both phyla, new members of the colony arise by budding from the other ones; new colonies may also arise as the result of sexual reproduction during certain seasons.

The ectoproct colony has specialized members, *avicularia,* resembling the head of a bird. These organisms are in constant motion from side to side, and, as they move, a peculiar organ, shaped like a bird's lower beak and operated by muscles, frequently snaps open and shut. The avicularia function not to catch food—other members do that—but to keep small animals from settling on the colony. The evolutionary relationships of both ectoproct and endoproct bryozoa are uncertain.

The Brachiopoda. Members of another phylum characterized by a lophophore, and commonly known as lampshells, superficially resemble clams. Brachiopods have two clam-like shells, usually calcareous, that can be opened and closed by muscles. Unlike the clam, whose two shells are on the right and left sides of the body, the brachiopod shells lie above and below the animal, the bottom shell being attached to a rock or other object by a sturdy, muscular stalk. All brachiopods live in the sea. Although only about 200 species of this extremely ancient phylum exist today, there once were more than 3000 species. Because of their great age and well-preserved hard shells, the fossil brachiopods are useful to geologists in determining the age of rocks. Fossils obtained from rocks more than 500,000,000 years old are almost exactly like the brachiopods living today. The genus *Lingula,* represented by both fossil and living forms, is the oldest known genus that has living members.

Some of the less familiar invertebrates—members of the phyla Ectoprocta, Brachiopoda and Phoronida—as well as others to be discussed presently, are characterized by a body cavity, or *coelom,* lying between the body wall and the wall of the digestive tract. This cavity arises during embryonic development from a split in the mesoderm layer, and hence is lined with mesoderm. The development of the coelom freed the digestive tract from the body wall, and permitted the two sets of muscles to contract independently; this was an important step toward the development of the higher animals.

QUESTIONS

1. Distinguish between homologous and analogous organs. Why can the former but not the latter be used as criteria in classification?
2. How do sponges obtain food and water?
3. Compare the body plans of a jellyfish and a hydra. List the adaptations each shows to its mode of life.
4. Compare the alternation of generations exhibited by *Obelia* and by a moss plant.
5. How would you distinguish a multicellular organism from a colonial organism? Do you think the sections of a tapeworm constitute a single individual or a colony of individuals comparable to a cnidarian colony?
6. What are flame cells? What is their role and how do they function?

7. In what ways are planarians and hydras similar? How do they differ?
8. An animal is multicellular, has no digestive system, and its body is perforated with pores. What is it?
9. Compare mesoglea and mesoderm. In what types of organisms is each found?
10. Discuss the ways in which physical diffusion is important to the survival of cnidarians.
11. What are the distinguishing features of each of the classes of the phylum Platyhelminthes?
12. Define the terms coelom, pseudocoelom, cnidoblast, choanocyte.

SUPPLEMENTARY READING

Prime source books for the details of the structure of invertebrate animals are *The Invertebrates,* in six volumes, by Libbie Hyman and *Invertebrate Zoology,* in three volumes, by A. Kaestner. The standard college text, and an excellent general reference, is Robert Barnes' *Invertebrate Zoology.* Another recommended text is *Invertebrate Zoology* by P. A. Meglitsch. *Animals Without Backbones,* by Ralph Buchsbaum, is a well-written book with outstanding illustrations. A popular book on animals is R. W. Hegner's *Parade of the Animal Kingdom.* More details about flatworms, tapeworms and cestodes can be found in Noble and Noble's *Parasitology.* Many of the invertebrate animals are marine and descriptions of them and their modes of life will be found in E. F. Ricketts and J. Calvin's *Between Pacific Tides,* Douglas P. Wilson's *They Live in the Sea,* and G. E. MacGinitie's *Natural History of Marine Animals.* More popular accounts of marine life are found in Rachel Carson's *The Sea Around Us* and in C. M. Yonge's *A Year on the Great Barrier Reef. The Sex Gas of Hydra* by W. F. Loomis describes the discovery of the chemical mechanism which leads to the production of eggs and sperm in this lower invertebrate.

CHAPTER 14

THE HIGHER INVERTEBRATES

The "higher" invertebrates—the annelids, arthropods, mollusks and echinoderms—all have a separate mouth and anus, a muscular gut, a well developed circulatory system and a true *coelom,* a cavity within the mesoderm lined by peritoneum. The coelom is formed during development either by a splitting within originally solid masses of mesoderm (a *schizocoelom,* typically found in mollusks, annelids and arthropods) or by pouches that bud off from the original gut cavity (an *enterocoelom,* typically found in echinoderms and chordates), but there are notable exceptions to both of these generalizations.

Of these four phyla, only the arthropods are very successful terrestrial animals. It is true that the earthworm is a terrestrial animal, but most annelids are marine; there are a few land snails, but most mollusks live in the sea; all the echinoderms are marine. Of the five classes of arthropods, one, the Crustacea (crabs, lobsters, etc.), includes largely marine forms, but the other four, Insecta (insects), Arachnida (spiders, mites, etc.), Chilopoda (centipedes) and Diplopoda (millipedes), are mostly terrestrial. From the fossil record we know that the first air-breathing land animals were scorpionlike arachnids that came ashore in the Silurian, some 410,000,000 years ago. The first land vertebrates, the amphibians, did not appear until the latter part of the Devonian, some 60,000,000 years later.

14–1 ADAPTING TO TERRESTRIAL LIFE

In evolving to become adapted to terrestrial life, animals, like plants (Chapter 9), had certain problems to solve to allow survival in the absence of a surrounding watery medium. The chief problem of all land organisms is that of preventing desiccation. Reproduction poses a second problem: Aquatic forms can shed their gametes into the water and fertilization will occur there; the delicate embryos that result are protected by the surrounding water as they begin development. Land plants by pollination transfer sperm nuclei to the egg in the absence of a watery medium, and the developing embryo is protected by the tissues of the parent gametophyte or by seed coverings. Some land animals, including most amphibians, return to the water for reproduction, and the young forms—larvae or tadpoles—develop in the water. Earthworms, insects, snails, reptiles, birds and mammals transfer sperm from the body of the male directly to the body of the female by *copulation;* the sperm are surrounded by a watery medium or semen. The fertilized egg either is covered by some sort of tough, protective shell secreted around it by the female or it develops within the body of the mother.

The problem of supporting a body against the pull of gravity in the absence of the buoyant effect of water is not too acute for small animals that burrow in the ground. Larger burrowing animals and those living on the surface of the earth generally need some sort of skeleton. The arthropods and mollusks evolved one on the outside of the body (called an *exoskeleton*) and the vertebrates have one within the body (an *endoskeleton*). Land forms are subject to much wider variations in temperature than marine organisms and had to evolve suitable adaptations to survive. The ocean acts as a great, constant temperature bath. The deep water varies only a few degrees from summer to winter and even a small lake has less drastic temperature fluctuations than air does.

With all these disadvantages it might seem incredible that any land forms *did* evolve. However, one of the major tendencies in evolution is for organisms to become diversified and to spread into new types of environment. Wherever the environment can support life at all, some form of life, suitably adapted for survival there, will eventually evolve. A land environment is not without some advantages, however. Once land plants evolved, the land offered the first animals an environment with a plentiful supply of food, no predators and few competitors.

The tough exoskeleton that evolved in the arthropods and mollusks serves several functions—it provides stiffening, enabling the body to stand against the pull of gravity; it serves as a point of attachment for muscles; it provides protection against desiccation; and it serves as a coat of armor to protect the animal against predators. Thus the evolution of an exoskeleton solved many of the problems of survival on land.

14-2 THE MOLLUSCA

This phylum, with its 128,000 living species and 35,000 fossil species, is the second largest of all the animal phyla. It includes the oysters, clams, octopuses, snails, slugs and the largest of all the invertebrates—the giant squid, which achieves a length of 16 meters, a circumference of 6 meters and a weight of several tons.

Molluscan Body Plan. The adult body plan of these animals is remarkably different from that of any other group of invertebrates, but the more primitive mollusks have a characteristic larval form, called a *trochophore,* similar to that of certain marine annelids. This suggests that mollusks and annelid worms arose from a common ancestral type; the worms, however, evolved a segmented body plan, while the mollusks evolved a unique body plan without segmentation. The sluggish marine animals, the *chitons* (Fig. 14-1), members of the class *Amphineura,* live by scraping algae off the rocks of the seashore. Their relatively simple structural characteristics provide clear illustrations of the basic molluscan traits: a broad, flat muscular *foot* for creeping across rocks; a *visceral mass* above the foot, containing most of the organs of the body; a *mantle,* or fold of tissue which covers the visceral mass and projects laterally over the edges of the foot; and a hard, calcareous *shell,* secreted by the upper surface of the mantle as eight separate plates. Like the outer covering of the arthropods, this shell gives protection but has the disadvantage of making locomotion difficult.

The molluscan digestive system is a single tube, sometimes coiled, consisting of a mouth, pharynx, esophagus, stomach, intestine and anus. The pharynx characteristically contains a rasplike structure, the *radula,* operated by a set of muscles. This can drill a hole in another animal's shell or break off pieces of a plant. The bivalves are the only mollusks that lack a radula; they are filter feeders, obtaining their food by straining sea water. The circulatory system is well developed and consists of a pumping organ which sends blood through a system of branched vessels and open spaces containing the body organs. Two "kidneys" (metanephridia), lying just below the heart, extract metabolic wastes from the blood and dis-

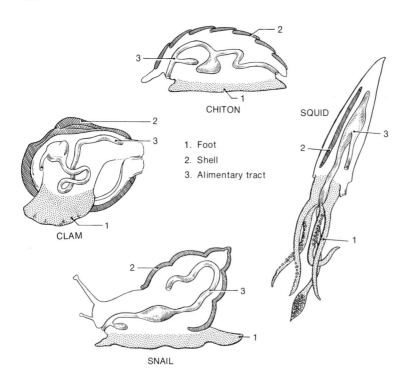

CHITON

SQUID

CLAM

1. Foot
2. Shell
3. Alimentary tract

Figure 14–1 Variations in the basic molluscan body plan in chitons, snails, clams and squid. Note how the foot (1), shell (2) and alimentary tract (3) have changed their positions in the evolution of the several classes.

SNAIL

charge them through pores located near the anus. The nervous system consists of two pairs of nerve cords, one going to the foot, another to the mantle. The ganglia of these are connected around the esophagus, at the anterior end of the body, by a ring of nervous tissue, thus forming the "brain." Many mollusks have no well developed sense organs, but snails have a pair of simple eyes, usually located on stalks extending from the head, and squids and octopuses have well developed, image-forming eyes.

One usually thinks of snails as having a spirally coiled shell, and many of them do, yet many members of the same class (*Gastropoda*), such as the limpets and abalones, have shells like flattened dunce caps, and others, such as the garden slugs and some marine snails, the *nudibranchs*, have no shell at all. At a particular stage in the development of each gastropod there occurs a unique, sudden, permanent twisting of the body so that the anus is brought around and comes to lie above the head (Fig. 14–2). Subsequent growth is dorsal and usually in a spiral coil. The twist limits space in the body, and typically the gill, heart, kidney and gonad on one side are absent. The viscera of the shell-less slugs and nudibranchs undergo a similar twisting during development.

The members of the class *Pelecypoda* (meaning hatchet-foot), commonly called *bivalves*, developed two shells, hinged on the dorsal side and opening ventrally. This arrangement allows the hatchet-shaped foot to protrude for locomotion and the long, muscular siphon, containing two tubes for the intake and output of water, to be extended. Some bivalves, such as oysters, are permanently attached to the substratum; others, such as clams, burrow slowly through sand or mud by means of the foot. A third type burrows through rock or wood, seeking protected dwellings. (The

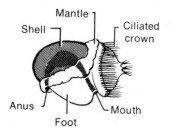

Shell

Mantle

Ciliated crown

Anus

Foot

Mouth

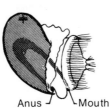

Anus

Mouth

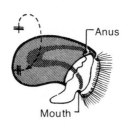

Anus

Mouth

Figure 14–2 Embryonic torsion in the gastropod *Acmaea* (a limpet). (After Boutan, 1899).

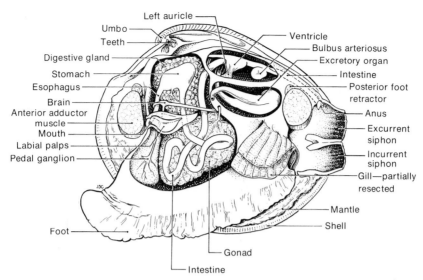

Figure 14-3 Longitudinal section of the marine clam *Mercenaria mercenaria* showing major organ systems.

shipworm, *Teredo*, which damages dock pilings and other marine installations, is just looking for a home.) Finally, some bivalves, such as scallops, swim with amazing speed by clapping their two shells together by the contraction of a large adductor muscle (the only part of the scallop that is eaten).

Clams (Fig. 14-3) and oysters obtain food by straining the sea water brought in over their gills by the siphon. The water is kept in motion by the beating of cilia on the surface of the gills, and food particles trapped in the mucus secreted by the gills are carried to the mouth. An average oyster filters about 3 liters of sea water per hour.

The innermost, pearly layer of the bivalve shell, made of calcium carbonate (mother-of-pearl), is secreted in thin sheets by the epithelial cells of the mantle. If a bit of foreign matter gets between the shell and the epithelium, the epithelial cells are stimulated to secrete concentric layers of calcium carbonate around the intruding particle; in this way, a pearl is formed.

In contrast to other mollusks, squids (Fig. 14-1), nautiluses and octopuses, composing the class *Cephalopoda*, are active, predatory animals. They have evolved a specialized, complex head-foot with a large, well developed "brain" and two big eyes. These are strikingly like vertebrate eyes in structure, but develop as a folding of the skin, rather than as an outgrowth of the brain. The independent evolution of similar structures with similar func-

tions in two different, unrelated animals is known as **convergent evolution.** The foot of the squid and octopus is divided into ten and eight long tentacles, respectively, covered with suckers for seizing and holding the prey. Besides having a radula, the animals have (in the mouth) two strong, horny beaks used to kill the prey and tear it to bits. The mantle is thick, muscular and fitted with a funnel. By filling the mantle cavity with water and ejecting it through this funnel, the animals attain rapid jet propulsion in the opposite direction.

Cephalopods are equipped with an **ink sac** which produces a thick black liquid. This is released when the squid or octopus is alarmed. The ink distracts the pursuer (perhaps a moray eel); MacGinitie has shown that the octopus ink paralyzes the chemoreceptors of the animals pursuing the octopus.

The shell of a nautilus is a flat, coiled structure, consisting of many chambers built up year by year; each year the animal lives in the latest and largest chamber of the series. By secreting a gas resembling air into the other chambers the nautilus is able to float. The shell of the squid is reduced to a small "pen" in the mantle, and the octopus has no shell at all.

Small octopuses survive well in aquaria and have been shown to have a high degree of intelligence. They can make associations among stimuli and generally show an adaptability of behavior that resembles that of the vertebrates more closely than it does the more stereotyped patterns seen in other invertebrates. Octopuses feed on crabs and other arthropods, catching and killing them with a poisonous secretion of their salivary glands. They live among rocks, taking shelter in small caves. Their motion is incredibly fluid and gives little hint of the considerable strength in their eight arms. They usually hide during the day and hunt for food in the evening.

The striking similarities in the development of mollusks and annelids—the process of spiral cleavage and the appearance of a trochophore larva—had suggested that these two phyla were related in evolutionary origin and had a common coelomate ancestor. This view was supported by the discovery, in 1952 and subsequently, of specimens of a primitive mollusk, *Neopilina*, in material dredged from a deep trench in the Pacific off Costa Rica. These animals are about 2.5 centimeters long and have some characteristics in common with gastropods and amphineurans. Their most remarkable feature is the segmental arrangement of certain internal organs—they have five pairs of retractor muscles, six pairs of nephridia and five pairs of gills. This has been interpreted by some zoologists as evidence of the segmental character of their ancestors, evidence that the mollusks, like the annelids, have a basically metameric body plan.

14-3 THE ANNELIDA

Annelid Body Plan. Among the most familiar invertebrate animals are the earthworms, members of the phylum Annelida (Figs. 14–4 and 21–10). This word means "ringed" and refers to the series of rings, or *segments*, that compose the bodies of the

members of this phylum. Both the internal organs and the body wall are segmented. The body is a bilaterally symmetrical tube composed of about 100 more or less similar units. One or a pair of organs of each system may be present in each segment. The segments are separated from each other by transverse, bulkheadlike partitions, the *septa*. The chief evolutionary advance shown by the annelids over the lower forms is this development of segmentation, for each segment constitutes a subunit of the body that may be specialized to carry on a particular function. The dividing of the body into segments is thus similar, on a larger scale, to the original division of the animal body into cells to provide for local specialization. In the earthworm the individual segments are almost all alike, but in many of the segmented animals—the arthropods and chordates—the specialization of different segments reaches a point at which the segmentation of the body plan is obscured.

The earthworm's body is protected from desiccation by a thin, transparent cuticle, secreted by the cells of the epidermis or outer layer of the body wall. Mucus secreted by the glandular cells of the epidermis forms an additional protective layer over the skin. The body wall contains an outer layer of circular muscles and an inner layer of longitudinal muscles. Each segment except the first bears four pairs of bristles, *chaetae*, supplied with small muscles that can move the chaeta in and out and change its angle. The earthworm moves forward by contracting its circular muscles to elongate the body, grasping the ground or walls of the burrow with its chaetae, and then contracting its longitudinal muscles to draw the posterior end forward; locomotion occurs in waves.

The body cavity of the annelids is a large and well developed true *coelom;* the body consists essentially of two tubes, one within the other. The outer tube is the body wall, and the inner tube, the wall of the digestive tract. The coelomic cavity is filled with fluid; this bathes the internal organs and transports gases, nutrients and wastes between the circulatory system and the individual cells of the body.

The digestive system of the earthworm shows several advances over that of the proboscis worm: there is a muscular pharynx for swallowing food, an esophagus, and a stom-

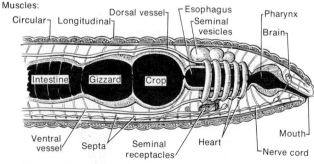

— **Figure 14–4** Diagrammatic longitudinal section of the anterior part of an earthworm showing internal structures.

ach of two parts—a thin-walled *crop*, where food is stored, and behind it, a thick-walled muscular *gizzard*, where food is ground to bits. The rest of the digestive system is a long, straight intestine, where digestion and absorption take place. The posterior end of the intestine, the anus, opens to the exterior.

The circulatory system, more complex and efficient than that of the proboscis worm, consists of two main vessels. One, just dorsal to the digestive tract, collects blood from numerous segmental vessels. It is contractile and pumps blood anteriorly. Blood flows posteriorly in the other, located just below the digestive tract, and is distributed to the various organs. In the region of the esophagus, five pairs of muscular tubes, called "hearts," propel the blood from the dorsal to the ventral vessel. There are other smaller lateral and ventral distributing vessels and tiny capillaries in all the organs as well as in the body wall.

The excretory system is composed of paired organs repeated in almost every segment of the body. Each individual organ, called a *metanephridium*, consists of a ciliated funnel, opening into the next anterior coelomic cavity and connected by a tube to the outside of the body (Fig. 21–10). Wastes are removed from the coelomic cavity partly by the beating of the cilia and partly by currents set up by the contraction of muscles in the body wall. The tube of the excretory organ is surrounded by a capillary network, so that wastes are removed from the blood stream as well as from the coelomic cavity. The metanephridia, open at both ends, are quite different from the protonephridia of the lower invertebrates, which are blind tubules opening only to the exterior. The adults of higher invertebrates typically have metanephridia, but larval forms usually have protonephridia as excretory organs—these typically have a single long flagellum rather than a tuft of cilia. This is consistent with the concept that the higher invertebrates evolved from forms similar to the lower invertebrates.

Reproductive System. Earthworms (and all oligochaete worms) are *hermaphroditic;* each individual contains both male and female reproductive organs. Segments 10 and 11 each contain a pair of *testes* located in isolated coelomic cavities, sperm reservoirs. These have three pairs of prominent lateral pouches, the *seminal vesicles* (Fig. 14–4), that extend into segments 9, 10 and 11. Sperm produced in the testes are stored in the reservoirs and vesicles. Two pairs of *sperm funnels* collect sperm from the reservoirs and pass them through a pair of sperm ducts to the *male pores* on the ventral side of segment 15.

A single pair of tiny *ovaries,* in segment 13, shed their eggs into the coelomic cavity. They are collected by a pair of *egg funnels* into short *oviducts* that open via *female pores* on the ventral side of segment 14. Two pairs of

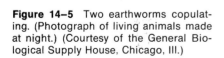

Figure 14–5 Two earthworms copulating. (Photograph of living animals made at night.) (Courtesy of the General Biological Supply House, Chicago, Ill.)

seminal receptacles, in segments 9 and 10, store sperm received during copulation.

During copulation two worms, heading in opposite directions, press their ventral surfaces together (Fig. 14–5), which become glued together by thick mucous secretions of the *clitellum,* a thickened ring of epidermis in segments 32 to 37. Sperm from one worm pass posteriorly to its clitellum and are stored in the seminal receptacles of the second worm. The worms then separate and the clitellum secretes a membranous *cocoon* containing an albuminous fluid. As the cocoon is slipped over the worm's head, eggs are laid into it from the female pores and sperm are added as the cocoon passes the seminal receptacles. When the cocoon is free, its openings constrict so that a spindle-shaped capsule is formed, and the eggs develop into tiny worms within the cocoon. This complex reproductive pattern is an adaptation to terrestrial life.

Nervous System. The nervous system is more advanced than that of the proboscis worm. It consists of a large, two-lobed aggregation of nerve cells (called the *brain*) located just above the pharynx in the third segment, and a *subpharyngeal ganglion* just below the pharynx in the fourth segment. A ring of nerve fibers around the pharynx connects the two ganglia. From the lower ganglion a nerve cord (actually two closely united cords) extends beneath the digestive tract to the posterior end of the body. In each segment there is a swelling of the nerve cord, a *segmental ganglion,* from which nerves extend laterally to the muscles and organs of that segment. The segmental ganglia coordinate the contraction of the muscles of the body wall, so that the worm can creep along. The nerve cord contains a few *giant axons* which transmit nerve impulses more rapidly than ordinary fibers. When danger threatens, these stimulate the muscles to contract and draw the worm back into its burrow. Giant axons from annelids, squids and certain arthropods have been extensively used in studies of the mechanism of nerve conduction (p. 490).

The activities of the earthworm are governed by the brain and subpharyngeal ganglion. Removal of the brain results in *increased* bodily activities and removal of the subpharyngeal ganglion eliminates all spontaneous movements. This is evidence of functional specialization of the nervous system; the brain is in part an *inhibitory center* and the subpharyngeal ganglion is a *stimulatory center.* Living a subterranean life, the earthworm has no well developed sense organs, but some of its sea-dwelling relatives, such as the clamworm, *Nereis,* have two pairs of eyes and organs sensitive to touch and to chemicals in the water.

Annelid Classes. The phylum Annelida contains some 10,000 species divided into four classes: the Polychaeta, Oligochaeta, Archiannelida and Hirudinea. The *Polychaeta* ("many bristles") include marine worms which swim freely in the sea, burrow in the sand and mud near shore, or live in tubes formed by secretions from the body wall. Each segment of their bodies typically has a pair of thickly bristled paddles (called *parapodia*) extending laterally (Fig. 14–6). The anterior end of the body is a well developed "head," or *prostomium,* and may bear eyes, antennae, tentacles, bristles or palps. Most species of poly-

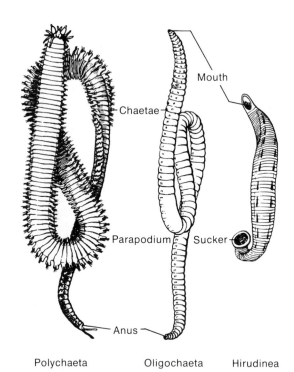

Mouth

Chaetae

Parapodium   Sucker

Anus

Polychaeta   Oligochaeta   Hirudinea

Figure 14–6 Classes of the phylum Annelida. Polychaeta: *Nereis virens,* the clamworm. Oligochaeta: *Lumbricus terrestris,* the earthworm. Hirudinea: *Hirudo medicinalis,* the medicinal leech. (From Villee, C. A., Walker, W. F., Jr., and Smith, F. E.: *General Zoology,* 3rd Ed. Philadelphia, W. B. Saunders Co., 1968.)

chaetes are predators and all have separate sexes.

The eggs and sperm of polychaetes are released into the sea water and fertilization occurs there. Many polychaetes have evolved behavioral patterns that ensure fertilization. By responding to certain rhythmic variations, or cycles, in the environment nearly all of the males and females of a given species release their gametes at the same time. The *seasonal cycles* produce variations in temperature, length of day and amount of food; *lunar cycles* produce variations in the height of tides, strength of currents, the relation between time of tide and hour of the day, and the amount of light at night; and *diurnal cycles* produce great variations in light from day to night. Coordinated by the combined effects of these cyclic events in the environment, over 90 per cent of the *Palolo worms,* a species of polychaete living on coral reefs in the South Pacific, shed their eggs and sperm within a single two-hour period on one night of the year. The seasonal rhythm limits the reproductive period to November; the lunar rhythm, to a day during the last quarter of the moon when the tide is unusually low; and the diurnal rhythm, to a few hours just after complete darkness. The posterior half of the Palolo worm, loaded with gametes, actually breaks off from the rest, swims backward to the surface and eventually bursts, releasing the eggs or sperm so that fertilization may occur.

The 2000 species of the class *Oligochaeta* (to which the earthworm belongs) have few bristles per segment and are found almost exclusively in fresh water and in moist terrestrial habitats.

The *Archiannelida,* a small group of simple marine worms, are not segmented externally and do not have bristles.

The leeches, class *Hirudinea,* (Fig. 14–6), are provided with stout muscular suckers at the anterior and posterior ends for clinging to their prey. They differ from other annelids in having neither chaetae nor appendages. Most leeches feed by sucking the blood of vertebrates; they attach themselves by their suckers, bite through the skin of the host, and suck out a quantity of blood, which is stored in pouches in the digestive tract. An anticoagulant (*hirudin*), secreted by glands in its crop, ensures the leech a full meal of blood. Their meals may be

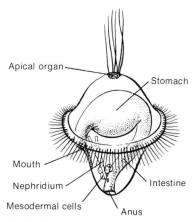

Figure 14–7 The trochophore larva of a polychaete.

infrequent, but they can store enough food from one meal to last a long time. The so-called "medicinal leech," a fresh water worm about 10 centimeters long, was used by physicians for bloodletting in the seventeenth and eighteenth centuries, when the humoral theory of disease was in vogue.

The archiannelids and polychaetes appear to represent one branch of annelid evolution and the oligochaetes and leeches another. The development of polychaetes and archiannelids is characterized by a *trochophore* larva (Fig. 14–7). The remarkable resemblance of the trochophore and the larva of mollusks is one of the bases for the theory that the annelids and mollusks arose from a common ancestor.

14–4 THE ONYCHOPHORA

The theory that the annelids and arthropods developed from a common, segmented ancestor is substantiated by the existence of a curious animal called *peripatus* (Fig. 14–8), found in the moist, tropical forests of Africa, Australia, Asia and South America. This caterpillarlike creature, 5 to 8 centimeters in length, appears to be a connecting link between the two phyla. It is not believed that peripatus *is* the ancestor of the present-day arthropods, but perhaps it is a descendent which has not changed much from the original ancestor of both annelids and arthropods. The anatomical characters of peripatus are a mixture of those of annelids and arthropods. It has many pairs of legs, each with a pair of claws at

Figure 14–8 Peripatus, a member of the phylum Onychophora, with structural features intermediate between those of the Annelida and the Arthropoda. (Courtesty of Ward's Natural Science Establishment.)

the tip. Its excretory, reproductive and nervous systems are similar to those of annelids, but its circulatory system and respiratory system, which consists of air tubes (tracheal tubes), are arthropodlike. Most zoologists classify peripatus and related species as a separate phylum, **Onychophora;** others classify them with the annelids or with the arthropods.

Peripatus provides a possible link between annelids and the terrestrial arthropods, but it is quite unlike the trilobites (Fig. 14–9). In 1930, 11 well preserved fossils from Cambrian deposits were found; these marine animals, named *Aysheaia,* resembled peripatus in many respects. Perhaps peripatus and *Aysheaia* are two representatives of a varied and widespread ancient group that had developed many arthropodlike characteristics before the arthropods evolved.

14–5 THE ARTHROPODA

The animals that make up this phylum are, without doubt, the most successful, biologically, of all animals. There are more of them—about 1,000,000 species, of which some 800,000 are insects. They live in a greater variety of habitats and can eat a greater variety of food than the members of any other phylum.

The term "arthropod" refers to the paired, jointed appendages characteristic of these animals. These function as swimming paddles, walking legs, mouth parts or accessory repro-

ductive organs for transferring sperm. An important factor in the evolutionary success of the arthropods is the hard, chitinous, armorlike exoskeleton, or cuticle, that covers the entire segmented body and appendages. The cuticle has an outer, waterproof, waxy layer, the epicuticle, composed of proteins and lipids; a rigid middle exocuticle; and a flexible, inner endocuticle. The principal constituent of the two inner layers is **chitin,** a polysaccharide composed of units of acetyl glucosamine. The rigid layer is thin in certain regions, such as the joints of the legs and between the body segments, allowing the cuticle to be bent. The exoskeleton provides protection against excessive loss of moisture and predators and gives support to the underlying soft tissues. But it has disadvantages, too: body movement is somewhat restricted, and in order to grow, the arthropod must shed the outer shell periodically and grow another larger one, a process which leaves it temporarily vulnerable. Arthropods have distinct muscle bundles attached to the inner surface of the exoskeleton. These act upon a system of levers that permit the extension and flexion of parts at the joints.

Arthropod Body Plan. The bodies of most arthropods are divided into three regions: the **head,** always composed of exactly six segments; the **thorax;** and the **abdomen,** both of which are composed of a variable number of segments. In contrast to most annelids, each arthropod has a fixed number of segments which remains the same throughout life. The incredible range of variations in body plan and in the shape of the jointed appendages in the numerous species begs description.

The nervous system of the more primitive arthropods, like that of the annelids, consists of a ventral nerve cord connecting segmental ganglia, but in the more complex arthropods the successive ganglia usually fuse together. Arthropods have a variety of well developed sense organs: complicated eyes, such as the compound eyes of insects; organs in the antennae sensitive to touch and chemicals; organs of hearing; and touch cells on the surface of the body.

The true coelom is small and made up chiefly of the cavities of the reproductive system. The large body cavity is not a coelom but a *hemocoel*, a blood cavity, which is part of the circulatory system. The latter includes, in addition to the enclosed vessels, open spaces throughout the body in which the organs are bathed; a pumping organ, or "heart," in the dorsal part of the body stirs the blood around in these spaces. Most of the aquatic arthropods have a system of gills for external respiration. The land forms, in contrast, usually have a system of fine, branching air tubes, or *tracheae*, which conduct air to the internal organs. The digestive system typically is a simple tube like that of the earthworm, lined in part with a cuticle similar to the outer covering of the body. In insects and some other forms the excretory system consists of tubules emptying into the digestive tube. These metabolic wastes then pass out of the body with the feces, through the anus.

Classes of Arthropods

The Trilobitomorpha. The most primitive arthropods, members of the subphylum *Trilobitomorpha*, are extinct marine arthropods that were once abundant and widely distributed in the Paleozoic seas. From fossil remains, some 3900 species of trilobites have been described; most of these lived on the sea bottom and walked on or dug into the sand and mud. They ranged in length from a millimeter to nearly a meter, but most were between 3 and 10 centimeters long. Their body was a flattened oval divided into three parts (Fig. 14–9)—an anterior *cephalon* of four fused segments bearing a pair of antennae and a pair of compound eyes; a *thorax* consisting of a varying number of segments; and a posterior *pygidium* composed of several fused segments. The body was divided further into a median lobe and two lateral lobes by two furrows extending from the anterior to the posterior end of the animal body. Each segment of the body had a pair of segmented biramous (two-branched) appendages. Each appendage consisted of an inner walking leg (telopodite) and an outer branch (pre-epipodite) bearing gills.

It is remarkable that fossil evidence has

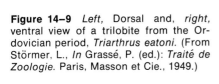

Figure 14–9 *Left*, Dorsal and, *right*, ventral view of a trilobite from the Ordovician period, *Triarthrus eatoni*. (From Störmer, L., *In* Grassé, P. (ed.): *Traité de Zoologie*. Paris, Masson et Cie., 1949.)

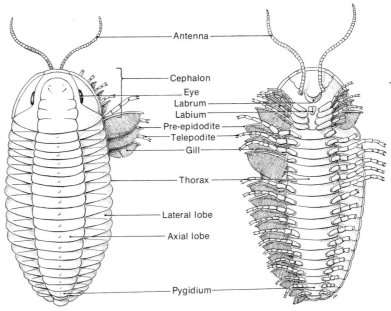

yielded information not only about the structure of the adult but also about the developmental stages of the trilobites. The trilobites passed through three larval periods; during each one the larvae underwent several molts. As the successive molts occurred additional segments were added to the body and the body structure became more complex. The trilobites have a number of characteristics in common with the crustacea and others in common with the arachnids and horseshoe crabs. They may have been ancestors of both of these groups, but the exact evolutionary relationship of the three is not clear.

The Mandibulata. The crustaceans, insects, millipedes and centipedes all have a pair of jawlike *mandibles* as the first post-oral appendages and are classified in the subphylum *Mandibulata.* The centipedes, *Chilopoda,* and the millipedes, *Diplopoda,* are similar in having a head and an elongated trunk with many segments, each bearing legs (Fig. 14–10). All of them are terrestrial; they are typically found beneath stones or wood or in the soil in both temperate and tropical regions. The centipedes have one pair of legs on each segment behind the head. Most centipedes have many fewer than a hundred legs—the common numbers being in the thirties—although there are a few species with enough legs to merit the term "centipede." The legs of centipedes are long, enabling them to run rapidly. Centipedes are

carnivorous and feed upon other animals, mostly insects, but the larger centipedes have been known to eat snakes, mice and frogs. The prey is captured and killed with poison claws located just behind the head on the first trunk segment. A pair of poison glands at the base of the claws empty into ducts that open at the tip of the pointed, fanglike claw. Centipedes breathe by means of a series of air tubes, or *tracheae,* that open to the exterior through openings called *spiracles.*

The millipedes or "thousand-leggers" are also secretive animals that live beneath leaves, stones and logs. The distinguishing feature of the class is the presence of doubled trunk segments resulting from the fusion of two original somites. Each double segment has two pairs of legs and two pairs of ganglia. The most anterior three or four segments have only a single pair of legs. The body of the millipede tends to be cylindrical, whereas the body of the centipede tends to be flattened. Diplopods are not as agile as chilopods, and most species can crawl only slowly over the ground. Millipedes are generally herbivorous and feed on both living and decomposing vegetation. Millipedes also breathe by tracheae that open through spiracles. In both chilopods and diplopods eyes may be completely lacking or the animals may have simple eyes (ocelli). A few species of centipedes have eyes that are similar to the compound eyes of insects, being composed of a group of as many as one hundred optical units on each side of the head.

THE CRUSTACEA. Members of the class *Crustacea* differ from other arthropods in having two pairs of *antennae,* or sensory feelers, a pair of *mandibles,* and two pairs of *maxillae* on their heads. They usually have compound eyes. The 26,000 species of crustacea include some familiar animals such as crabs, shrimp, lobsters and crayfish (Fig. 14–11), plus thousands of less familiar species. There are myriads of tiny crustaceans living in seas, lakes and ponds that play an important role in the aquatic food chains. The principal food of some of the largest whales is "krill," marine crustacea less than 25 millimeters long. The crustacea are the only class of arthropods that are primarily aquatic. Most of the crustaceans are marine, but some live in fresh water. There are a few species—the hermit crab of the Caribbean islands, for example—that survive for

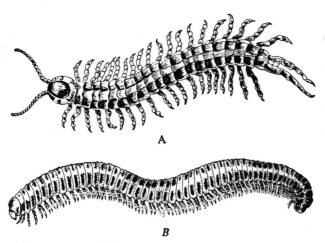

A

B

Figure 14–10 *A,* Centipede, a member of the class Chilopoda. Centipedes have one pair of appendages per segment. *B,* Millipede, a member of the class Diplopoda. Millipedes have two pairs of appendages per segment. (From Hunter, G. W., and Hunter, F. R.: *College Zoology.* Philadelphia, W. B. Saunders Co., 1949.)

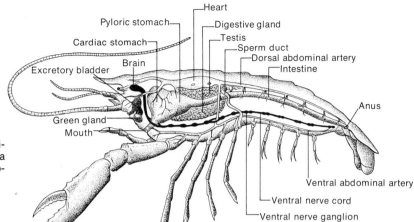

Pyloric stomach — Heart
Cardiac stomach — Digestive gland
Testis
Sperm duct
Dorsal abdominal artery
Excretory bladder — Brain — Intestine
Anus
Green gland
Mouth
Ventral abdominal artery
Ventral nerve cord
Ventral nerve ganglion

Figure 14–11 Diagrammatic longitudinal section of a male crayfish, a fresh water animal similar to a lobster.

extended periods of time on land in a moist environment. The crustaceans are either carnivores, scavengers or filter feeders. Certain appendages in the latter have fine hairs (setae) that function as a filter to collect small particles of food, which are then removed from the setae by other hairs and transferred to the mouth parts.

One of the more familiar crustaceans is the lobster, a decapod or "ten-legged" form. The six segments of the lobster's head and the eight segments of the thorax are fused into a *cephalothorax*, covered on its top and sides by a shield, the *carapace*, composed of chitin impregnated with calcium salts. The two pairs of antennae are the sites of chemoreceptors and tactile sense organs; the second pair of antennae are especially long. The mandibles are short and heavy with opposing surfaces used in grinding and biting food. Behind the mandibles are two pairs of accessory feeding appendages, the first and second maxillae. The appendages of the first three segments of the thorax, the *maxillipeds*, aid in chopping up food and passing it to the mouth. The fourth segment of the thorax has a pair of large *chelipeds*, or pinching claws, and segments 5 through 8 have pairs of *walking legs*. The appendages of the first abdominal segment are part of the reproductive system and function in the male as sperm transferring structures. On the following four segments of the abdomen are paired *swimmerets*, small paddlelike structures used by some decapods for swimming and by the females of all species for holding eggs. The uropods and the telson on the last two segments make up a fan-shaped tail used for swimming backwards.

Respiration in the crustacea is carried out generally by gills, which are usually attached to the proximal segment of certain appendages. Crustaceans have an open circulatory system with a contractile heart and arteries that end in the hemocoel, a large blood-filled space that ramifies through most parts of the body. The blood of the lobster contains a bluish pigment, *hemocyanin*, for the transport of oxygen.

The order *Decapoda*, the largest order of crustaceans, contains some 8500 species of lobsters, crabs, crayfish and shrimp. Most decapods are marine but a few, such as the crayfish, certain shrimp and a few crabs, live in fresh water. The crustaceans in general and the decapods in particular show in a striking way the specialization and differentiation of parts in the various regions of the animal. The segments of the trilobites and perhaps of the earliest crustaceans bore appendages that were very similar. In the lobster, no two of the nineteen pairs of appendages are identical, and the appendages in the different parts of the body differ markedly in form and function.

Although lobsters, crabs and shrimps are the most familiar of the crustaceans, they are not the most important in the over-all economy of nature. There are countless billions of microscopic crustaceans that swarm in the ocean and form the food of many fish and other marine forms such as the whales. The subclass *Branchiopoda* includes a variety of small shrimplike forms—fairy shrimps, tadpole shrimps, water fleas—found mostly in fresh water (Fig. 14–12). The subclass *Ostracoda* (mussel shrimps) includes other minute crustaceans found in the sea and in fresh water. The ostracods are characterized by the pres-

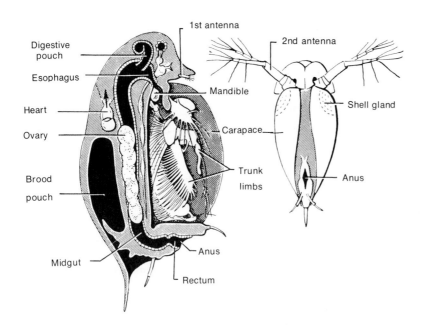

Figure 14–12 *Daphnia*, the water flea. Side view *(left)* with one side of the carapace removed to show enclosed body organs. Ventral view *(right)* with trunk appendages omitted. (From Villee, C. A., Walker, W. F., Jr., and Smith, F. E.: *General Zoology*, 3rd Ed. Philadelphia, W. B. Saunders Co., 1968.)

ence of two round or elliptical protective shells, in addition to the usual cuticle, which look like miniature clam shells and are impregnated with calcium carbonate.

Another group of very small crustaceans, the subclass *Copepoda*, are marine or fresh water dwellers (Fig. 14–13). In addition the group contains many species that are parasitic on other marine or fresh water animals. Copepods are also important in the diet of whales and fishes. The free-living copepods have bodies that are typically short and cylindrical.

The barnacles (subclass *Cirripedia*) are the only sessile crustaceans. They differ markedly in their external anatomy from other crustacea, and it was only in 1830, when the larval stages

were investigated, that the relationship between the barnacles and other crustaceans was recognized. The barnacles are exclusively marine and secrete complex calcareous cups within which the animal lives. The larvae of barnacles are free-swimming forms that go through several molts and eventually become sessile and develop into the adult form. Barnacles were described many years ago by Louis Agassiz as "nothing more than a little shrimplike animal standing on its head in a limestone house and kicking food into its mouth."

Two numerous groups of crustaceans are small animals that look rather like bugs and live either in the ocean, in fresh water or in

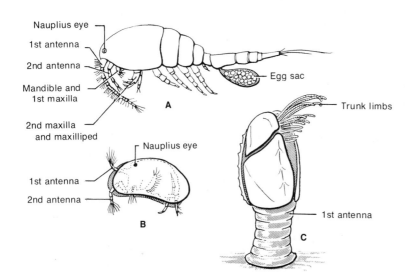

Figure 14–13 Additional orders of small Crustacea. *A,* Order Copepoda. *B,* Order Ostracoda, with a hinged carapace enclosing head and body. *C,* Order Cirripedia, the barnacles, attached by an enormous first antenna, with the body enclosed in calcareous plates. (From Villee, C. A., Walker, W. F., Jr., and Smith, F. E.: *General Zoology*, 3rd Ed. Philadelphia, W. B. Saunders Co., 1968.)

damp places on land. Members of the orders *Isopoda* and *Amphipoda* are commonly called "pill bugs," "wood lice," "beach fleas," "sow bugs" and so on. Although they superficially look like insects, they are crustaceans. The isopods have bodies that are typically flattened from top to bottom and the amphipods have bodies that are typically compressed from side to side.

Molting. Crabs, lobsters, crayfish and the other crustacea molt many times during development. The newly hatched animals pass by successive molts through a series of larval stages and finally reach the body form characteristic of the adult. The lobster, for example, molts seven times during the first summer; at each molt it gets larger and resembles the adult more. After it reaches the stage of a small adult, additional molts provide for growth. Just before molting the glands in the epidermis se-

crete a *molting fluid* which contains enzymes to digest the chitin and proteins of the inner layers of the cuticle. A soft, flexible, new cuticle, folded to allow for growth, is formed under the old one. The digested remains of the old cuticle are absorbed by the body; some substances (e.g., the calcium salts) are stored for reuse. The animal may swallow air or water to aid in swelling up and bursting the old cuticle. It extricates itself from the old cuticle, swells to stretch the folded new cuticle to its full size, and then the epidermis secretes enzymes that harden the cuticle by oxidizing some of the compounds and by adding calcium carbonate to the chitin. Additional layers of cuticle are secreted subsequently. The endocrine regulation of molting and metamorphosis is discussed later (p. 554).

THE INSECTA. The class *Insecta* is the largest, most successful and most diverse of all

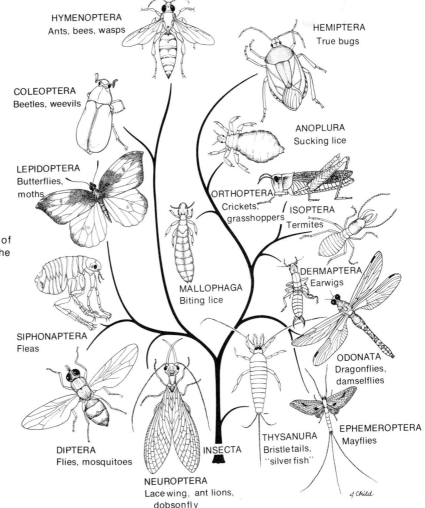

Figure 14–14 Representatives of some of the important orders of the class Insecta.

HYMENOPTERA
Ants, bees, wasps

HEMIPTERA
True bugs

COLEOPTERA
Beetles, weevils

ANOPLURA
Sucking lice

LEPIDOPTERA
Butterflies, moths

ORTHOPTERA
Crickets, grasshoppers

ISOPTERA
Termites

MALLOPHAGA
Biting lice

DERMAPTERA
Earwigs

SIPHONAPTERA
Fleas

ODONATA
Dragonflies, damselflies

DIPTERA
Flies, mosquitoes

INSECTA

THYSANURA
Bristletails, "silverfish"

EPHEMEROPTERA
Mayflies

NEUROPTERA
Lacewing, ant lions, dobsonfly

J. Child

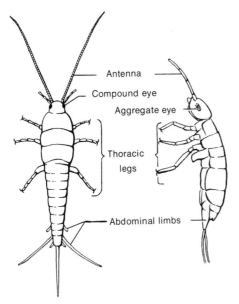

Figure 14–15 Primitive wingless insects (Apterygota): a silverfish *(left)* and a springtail *(right)*. (From Villee, C. A., Walker, W. F., Jr., and Barnes, R. D.: *General Zoology,* 4th Ed. Philadelphia, W. B. Saunders Co., 1973.)

thorax is separated from the abdomen. The appendages of one of the head segments are the sensory antennae, and the other appendages are the complex mouthparts. In various species they are adapted for biting, for sucking or for piercing. The thorax consists of three segments fused together. Each segment has a pair of legs (hence the total of six legs characteristic of the insects). Insects typically have two pairs of wings on the last two thoracic segments. The abdomen has up to eleven segments, all usually lacking appendages. Respiration is carried out by tracheae opening to the outside by spiracles; the circulatory system is open, without capillaries or veins. Insects have a variety of sense organs, including simple and compound eyes, chemoreceptors and receptors for sound waves. Insects are classified into 20 to 25 orders, each representing adaptations to a wide range of habitats and environments (Fig. 14–14). There are four orders of primitive, wingless insects placed in a separate category of the class Insecta, or in a separate class by some taxonomists. These *Apterygota* (Fig. 14–15), such as silverfish and springtails, have small appendages on their abdominal segments but have compound eyes similar to those of the winged insects.

The larger group of winged insects, the *Pterygota*, is divided into Paleoptera and Neoptera. In the Paleoptera the wings are held permanently at right angles to the body. The

animals. Insects are primarily terrestrial animals; some species live in fresh water, a few have become adapted to living along the shore between the tides and a few are truly marine. In contrast to the crustaceans, the insect head, composed of six completely fused segments, is clearly separated from the thorax, and the

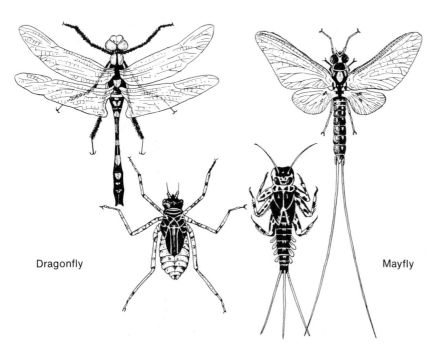

Dragonfly

Mayfly

Figure 14–16 Living Paleoptera. Orders Odonata *(left)* and Ephemeroptera *(right)*. Adults above, and nymphs below. (From Villee, C. A., Walker, W. F., Jr., and Smith, F. E.: *General Zoology,* 3rd Ed. Philadelphia, W. B. Saunders Co., 1968.)

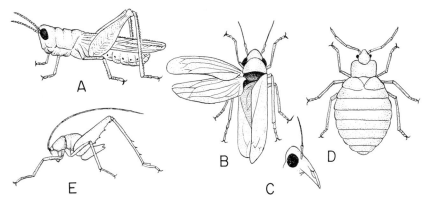

Figure 14–17 Representative orders of the Exopterygota. *A,* Orthoptera (grasshopper). *B,* Hemiptera (leafhopper). Hemipterans have sucking mouthparts *(C).* Wingless forms in each order include the bedbug *(D)* and the camel cricket *(E).* Other orders include the Blattaria (cockroach) and the Isoptera (termite). (From Villee, C. A., Walker, W. F., Jr., and Smith, F. E.: *General Zoology,* 3rd Ed. Philadelphia, W. B. Saunders Co., 1968.)

group includes several orders of extinct insects plus dragonflies, damselflies and mayflies (Fig. 14–16). The Neoptera include insects with wings that can be folded back over the body when not in use. These in turn are subdivided into Exopterygota (Fig. 14–17), which have external wing buds and incomplete metamorphosis, and the Endopterygota (Fig. 14–18), which have internal wing buds and complete metamorphosis.

Insect Flight. Insects are the only invertebrates to have developed wings (though not all species have done so). These structures are simply analogous, not homologous, to the wings of the vertebrates. Insects usually have two pairs of wings; flies and other Diptera have one pair of wings plus a pair of balancers (*hal-*

teres) which evolved from the second pair of wings. The halteres beat up and down rapidly during flight and apparently serve as gyroscopes. In most insects, both sets of wings are functional as flying organs, but in grasshoppers and beetles the anterior pair are simply stiffened protective devices for the functional posterior pair.

Unlike birds, most insects do not have flight muscles attached to the wings. Instead, the wings are attached to the body wall over a fulcrum in such a way that slight changes in the shape of the thorax cause the wings to beat up and down (Fig. 14–19). The contraction of the vertical muscles pulls down the *tergum,* a plate on the upper surface of the thorax, but raises the wings, on the opposite side of the

Figure 14–18 The major orders of the Endopterygota. The Coleoptera (beetles) have thick, rigid forewings. The Lepidoptera (butterflies and moths) have scales on the wings and sucking mouthparts. The Hymenoptera (bees, ants, etc.) have membranous wings with few veins. The Diptera (flies) have two wings, the hindwings being reduced to balancing organs. (From Villee, C. A., Walker, W. F., Jr., and Smith, F. E.: *General Zoology,* 3rd Ed. Philadelphia, W. B. Saunders Co., 1968.)

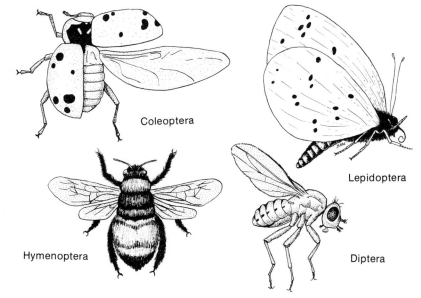

Coleoptera

Lepidoptera

Hymenoptera

Diptera

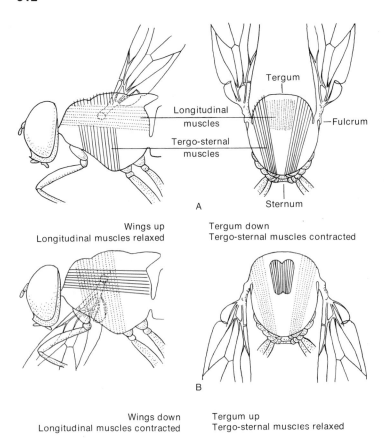

Wings up
Longitudinal muscles relaxed

Tergum down
Tergo-sternal muscles contracted

Wings down
Longitudinal muscles contracted

Tergum up
Tergo-sternal muscles relaxed

Figure 14–19 Diagram of the arrangement of the flight muscles of an insect. The contraction of the longitudinal muscles forces the tergum up and the wings down. Contraction of the tergosternal muscles forces the tergum down and the wings up.

fulcrum. The contraction of the longitudinal muscles causes the tergum to bulge upward and the wing is pulled downward. The movements of the body wall are barely perceptible, but the length of the lever on the two sides of the fulcrum is very different and the distance moved by the tips of the wings is several hundred times as great. For flight the wings must move forward and backward as well as up and down. Other muscles produce the back and forth motion and change the angle of the wings to provide both lift and forward thrust.

In many insects—among them butterflies and moths—the frequency of the wing beat is correlated with the frequency of the nerve impulses to the flight muscles. The impulses to the two sets of muscles are staggered so that rhythmic up and down movements of the wings occur. The rate of these movements ranges from about 8 beats per second in large moths to about 75 beats per second in some smaller insects. In other insects—flies and bees, for instance—wing beats are not correlated with the number of nerve impulses. When nerve impulses reach the muscles with a

frequency greater than a certain minimal threshold, the wings beat but at a higher frequency. The frequency is set by the muscles themselves and is a function of the tension in the two sets of opposing muscles. It may reach several hundred beats per second.

Social Insects. Certain species of bees, ants and termites exist not as single individuals but as colonies or societies made up of several different types of individuals, each adapted for some particular function. In this they resemble a cnidarian or bryozoan colony, but they differ in that they are not joined together anatomically as are these lower forms; instead they constitute a *social colony.* A termite colony (Fig. 14–20), for example, contains reproductives (the queen and king), soldiers and workers. The king and queen give rise to all other members of the colony. The soldiers, strong-jawed, heavily armored termites, protect the colony from enemies. The workers gather food, build the nest and care for the young. Both soldiers and workers are sterile, and neither reproductives nor soldiers can feed themselves. Thus the members of the colony are

Figure 14–20   Model of a royal cell of the termite *Constrictotermes cavifrons,* from Guyana. The queen, with an enlarged abdomen, occupies the center of the chamber with her head toward the right. The king is at the lower left. Most of the individuals are workers. A few soldiers with "squirt gun" heads and reduced mandibles are at the left. (Courtesy of the Buffalo Society of Natural Sciences.)

completely dependent on each other. Each year new reproductives develop in a colony as winged forms that leave the group, mate, and form a new colony. A queen termite may lay as many as 6000 eggs per day every day in the year for years. She is simply a specialized egg-laying machine and must be fed and cared for by the workers.

A honey bee colony consists of a single queen, a few hundred drones or males, and thousands of workers, sterile females. The queen bee mates during one or more "nuptial flights," and stores the sperm in a sperm sac in her body. During copulation the male's reproductive organs are "exploded" into the female and he dies. The queen returns to the hive and thereafter can lay either fertilized or unfertilized eggs. The latter develop parthenogenetically into haploid male drones, and the fertilized eggs develop into diploid females. If female larvae are fed "royal jelly" for about six days they develop into queens; if fed this for two or three days and then a mixture of nectar and pollen for three days, they develop into workers. Young adult workers serve as "nurse bees" to feed the larvae and prepare brood cells. Other adults are "house bees" that stand guard at the entrance of the hive, receive and store nectar and pollen, secrete wax for new

cells and keep the hive clean. The oldest adults are "field bees" that fly from the hive and forage for water, pollen and nectar.

Insect Communication.   The members of the colony communicate with each other by "dances" and by pheromones (p. 580). When a worker bee has found a source of food, it collects a sample and flies back to the hive. It transmits information to other members of the colony by the kind of "dance" it does on a vertical surface of the hive. If the food is near the hive, the honey bee circles first in one direction and then in the other in a "round dance" (Figure 14–21). The other worker bees then fly out and search in all directions near the hive. If the food is located at a greater distance from the hive, the bee goes through a half circle, then moves in a straight line, waggling its abdomen from side to side, and finally moves in a half circle in the other direction. During the straight portion of the dance, the bee produces a series of sounds. The angle that the straight, waggling portion of the dance makes with the vertical is the same as the angle of flight to the food relative to the sun (Fig. 14–22). Information on the distance to be traveled in that direction is also transmitted by the pattern of the "waggle dance." There is a correlation between the number of waggle dances executed in a given time and the distance of the food from the hive. An even closer correlation, however, exists between distance and the sound pulses the dancing bee produces. Bees can sense the vibrations of the substratum, and sound would seem to be a good way to communicate in the dim interior of the hive.

The distance to be traveled is "calculated" by the bee on the basis of the flight *to* the food source, not the return trip. Moreover, the instructions apparently include corrections for wind and large obstacles. The dancing bee also indicates the richness of the food source by how long and vigorously it repeats the dance. Other bees gain further information about the food source by the smell of the flowers the bee has recently visited. The more primitive social bees, the stingless bees, lay down an odor trail, a series of chemical marks, to the food source; in addition, the collector bee guides other individuals along the trail.

Fire ants communicate information about food sources by a pheromone produced in a special gland, Dufor's gland, located in the abdomen of the worker ant. When one ant has located a source of food, it gathers what it can and starts back toward the nest. As it moves along the ground, it periodically extrudes its

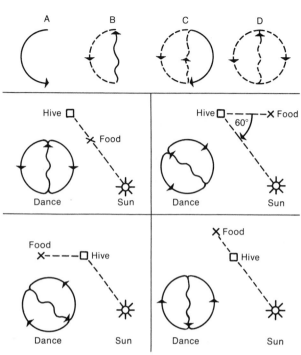

Figure 14–22 Indication of direction by the waggle dance. (From Villee, C. A., Walker, W. F., Jr., and Barnes, R. D.: *General Zoology*, 4th Ed. Philadelphia, W. B. Saunders Company, 1973.)

stinger and places a small amount of pheromone on the substratum. When another fire ant comes across this trail of chemical dashes it becomes very active, follows the trail and is led to the source of the food.

Insects can affect human life in many ways, either positively by pollinating our domesticated plants or negatively by destroying plants and stored crops. The bites of mosquitoes, fleas, bedbugs and flies can contribute directly to human misery. Some insects serve as vectors of human diseases or of diseases of our domesticated animals. Mosquitoes can transmit malaria, elephantiasis and yellow fever. The tsetse fly transmits sleeping sickness, lice transmit typhus, fleas transmit bubonic plague and the house fly can transmit typhoid fever and dysentery. A great deal of money and effort is expended each year to control insect pests which greatly reduce the high agricultural yields necessary to support the burgeoning human population of the world. However, overzealous use of pesticides can raise other environmental problems.

The Chelicerata. Another large and important group of arthropods is the chelicerates

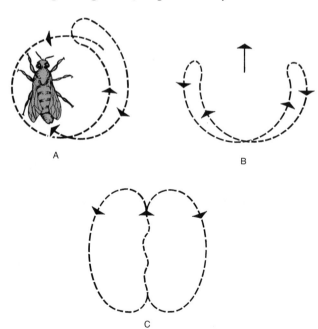

Figure 14–21 Three kinds of communication dances performed by honey bees. *A*, The round dance; *B*, the sickle dance; *C*, the waggle dance. (From von Frisch, K., and Lindauer, M.: *In* McGill, T. (ed.): *Readings on Animal Behavior.* New York, Holt, Rinehart and Winston, Inc., 1965.)

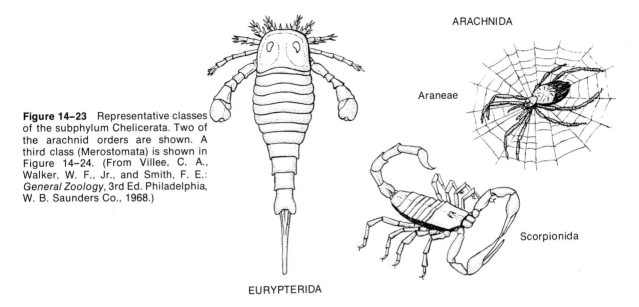

ARACHNIDA

Araneae

Scorpionida

Figure 14–23 Representative classes of the subphylum Chelicerata. Two of the arachnid orders are shown. A third class (Merostomata) is shown in Figure 14–24. (From Villee, C. A., Walker, W. F., Jr., and Smith, F. E.: *General Zoology,* 3rd Ed. Philadelphia, W. B. Saunders Co., 1968.)

EURYPTERIDA

(subphylum Chelicerata), members of the class *Arachnida* and the class *Merostomata.* The chelicerate body consists of a fused cephalothorax and an abdomen. The first pair of appendages on the head are the *chelicerae,* the pinching claws of the Merostomata and typically the poison-injecting claws in spiders. The Arachnida have four pairs of walking legs, and the Merostomata five.

The class Arachnida includes the spiders, harvestmen, scorpions, mites and ticks (Fig. 14–23). Except for a few groups that are secondarily aquatic, arachnids are terrestrial chelicerates. They are believed to have evolved from water scorpions, the second, now completely extinct, group of Merostomata. Arachnida have, in addition to the chelicerae, a pair of *pedipalps* and four pairs of legs, but no antennae. Most arachnids are carnivorous and prey upon insects and other small arthropods. Arachnids respire either by book lungs, by tracheae, or by both. *Book lungs* are internal and occur in pairs on the ventral side of the abdomen. An arachnid may have as many as four pairs. The eyes are simple rather than compound. In addition arachnids have tactile hairs and slit sense organs that may serve as chemoreceptors.

Horseshoe crabs, or king crabs (Fig. 14–24), relics of a formerly numerous class, Merostomata, are living fossils clearly related to the arachnids although having a hard calcareous exoskeleton like the crabs. The present species of horseshoe crabs have survived essentially

unchanged for 350 million years or more. Other fossil chelicerates, the eurypterids, were abundant in the Paleozoic era and included some species that were as much as 3 meters long. These giant water scorpions had compound eyes, a pair of chelicerae, four pairs of walking legs and a pair of large paddles for swimming. They were carnivorous and preyed on the earliest vertebrates, the ostracoderms (p. 325).

There are three other relatively minor classes of arthropods: the *Pycnogonida,* or sea spiders, the *Symphyla* and the *Pauropoda* (Fig. 14–25). Pauropods are very small (0.5 to 2.0 mm. long), soft-bodied animals frequently abundant in forest litter. Symphylans are small (usually less than 10 mm. long), centipedelike, active animals living in soil and leaf mold. Their mouth parts and six-jointed legs are similar to those of insects. The symphylans have been proposed as possible ancestors of the insects. Pycnogonids are also usually small (1 to 10 mm. in length), although a few deepwater forms may attain much greater proportions. This class is entirely marine.

14–6 THE ECHINODERMATA

The phylum Echinodermata ("spiny-skinned") includes the classes Asteroidea (sea stars), Echinoidea (sea urchins), Holothuroidea (sea cucumbers), Ophiuroidea (serpent stars) and Crinoidea (sea lilies) (Fig. 14–26). These

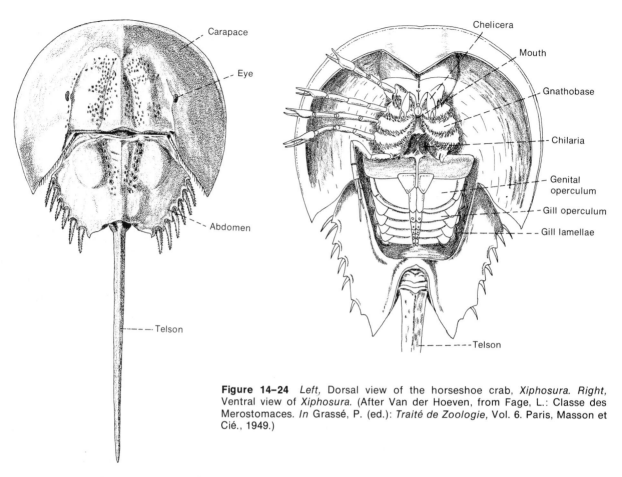

Figure 14–24 *Left,* Dorsal view of the horseshoe crab, *Xiphosura. Right,* Ventral view of *Xiphosura.* (After Van der Hoeven, from Fage, L.: Classe des Merostomaces. *In* Grassé, P. (ed.): *Traité de Zoologie,* Vol. 6. Paris, Masson et Cié., 1949.)

animals are radically different from all other invertebrates and appear to be more closely related to the chordates. The larvae of the echinoderms and of the hemichordates have many features in common (Fig. 14–30). The larvae are bilaterally symmetrical, but the adults secondarily achieve radial symmetry. All of the 6000 or so species of this phylum are marine.

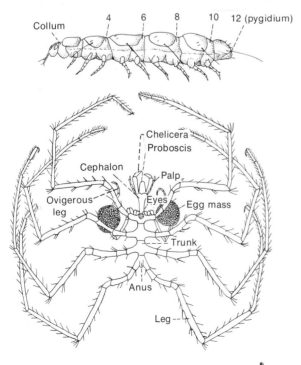

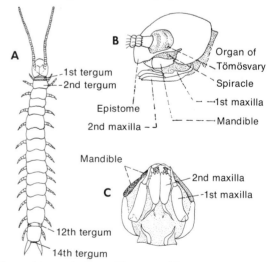

Figure 14–25 *Left top,* The sea spider, *Nymphon rubrum.* (After Fage, L.: Classe des Pycnogonides. *In* Grassé, P. (ed.): *Traité de Zoologie,* Vol. 6. Paris, Masson et Cié., 1949.) *Left bottom,* Lateral view of the pauropod, *Pauropus silvaticus. Right,* Symphylans. *A,* Dorsal view of *Scutigerella immaculata; B,* lateral view of the head of *Hanseniella; C,* ventral view of the head of *Scutigerella immaculata.* (*Left bottom* and *right* after Snodgrass, R. E.: *A Textbook of Arthropod Anatomy.* Ithaca, N.Y., Cornell University Press, 1952.)

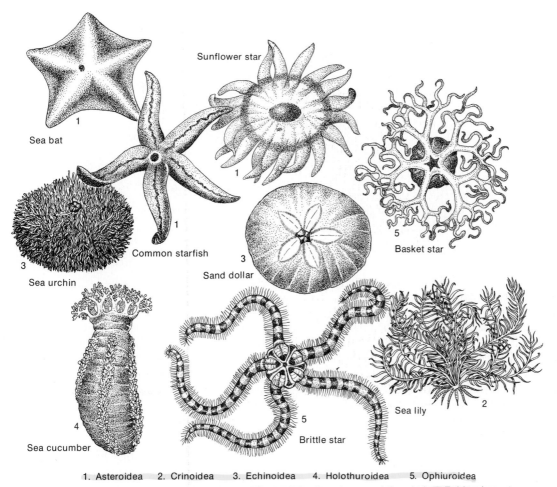

Sunflower star

Sea bat

1

Common starfish

1

Basket star

5

Sea urchin

3

Sand dollar

3

Sea cucumber

4

Brittle star

5

Sea lily

2

1. Asteroidea   2. Crinoidea   3. Echinoidea   4. Holothuroidea   5. Ophiuroidea

Figure 14–26 Sketches of some representatives of the five classes of the phylum Echinodermata.

A starfish, or sea star (Fig. 14–27), consists of a central *disc* from which radiate five to twenty or more *arms.* On the underside of the disc in the center is the mouth. The skin of the entire animal is embedded with tiny, flat bits of calcium carbonate, some of which give rise to spines. A number of these spines are movable. Around the base of some, especially near the delicate skin gills used in respiration, are still tinier, specialized spines, in the form of pincers. Operated by muscles, these keep the surface of the animal free of debris. Echinoderms are equipped with a unique hydraulic arrangement called a *water vascular system,* which provides for locomotion. The undersurface of each arm is equipped with hundreds of pairs of tube feet—hollow, thin-walled, sucker-tipped muscular cylinders—at the base of which are round muscular sacs, called *ampullas.* To extend the feet, these sacs contract, forcing water into the feet. The feet are with-

drawn by the contraction of muscles in their walls which forces the water back into the ampullas. The cavities of the tube feet are all connected by radial canals in the arms, and these, in turn, are connected by a circular canal in the central disc. The circular canal is connected by the stone canal to a button-shaped *madreporite* on the upper (aboral) surface of the central disc. The madreporite has many pores, as many as 250, by which the cavity of the stone canal opens to the exterior.

To attack a clam or oyster, the sea star mounts it, assuming a humped position as it straddles the edge opposite the hinge. Then, with its tube feet attached to the two shells, it begins to pull. The clam, of course, reacts by closing its shell tightly. The starfish pulls hard enough to produce a small gape between the two shells of the clam, then passes its everted cardiac stomach through the gape. The cardiac stomach of some species can pass through a

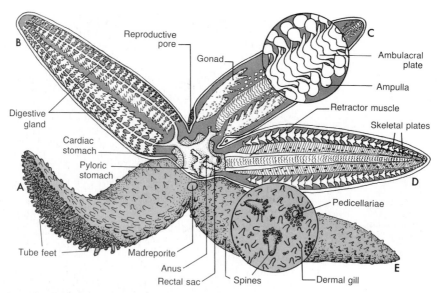

Figure 14–27 The starfish *Asterias* viewed from above with the arms in various stages of dissection. *A,* Arm turned to show the lower side. *B,* The upper body wall removed. *C,* The upper body wall and the digestive glands removed and a magnified detail of the ampullas and ambulacral plates. *D,* All the internal organs removed except the retractor muscles, showing the inner surface of the lower body wall. *E,* The upper surface, with a magnified detail showing the features of the surface.

gape as small as 0.1 millimeters and apparently is not damaged if the clam's valves press together. Enzymes secreted by the stomach gradually digest the clam's adductor muscles and the gape increases. The soft parts of the clam are digested to the consistency of a thick soup and passed into the body for further digestion in glands located in each arm. After the clam has been digested, the starfish retracts its stomach into the interior of the disc. The water vascular system does not enable the starfish to move rapidly, but since it usually preys upon slow-moving or stationary clams and oysters, speed of attack is not necessary as it is for most predators. Starfish occasionally catch and eat small fish (Fig. 14–28).

There are no special respiratory or circulatory systems in these animals; both functions are accomplished by the fluid which fills the large coelomic cavity and bathes the internal organs. Nor is there any special excretory system — metabolic wastes pass to the outside by diffusion. The nervous system consists of a ring of nervous tissue encircling the mouth and a nerve cord extending from this into each arm. There is no aggregation of nerve cells which could be called a brain.

The brittle stars, or serpent stars, members of the class Ophiuroidea, also have a central disc but their arms are long and slender, en-

abling them to move rapidly. The arms are discarded and replaced when injured.

Sea urchins, members of the class Echinoidea, look like animated pincushions, bearing on their spherical bodies long, movable spines between which the tube feet protrude. In these creatures the calcareous plates have fused, forming a spherical shell, and in the center of the undersurface of this is the mouth. The tube feet, arranged in five bands on the surface of the shell, are longer and more slender than those of sea stars, but the water vascular system is otherwise similar.

Sea cucumbers, members of the class Holothuroidea, are appropriately named, for many of them are green and about the size and shape of a gherkin. Like the members of several other phyla, these animals have a circle of tentacles around the mouth, and, in common with the starfish, they have a water vascular system; some species have external tube feet. The bodies of sea cucumbers are flexible, hollow, muscular sacs. Whenever environmental conditions are unfavorable, because of high temperatures, lack of oxygen or excessive irritation, the sea cucumber contracts violently and ejects part or all of its digestive tract. When conditions are again favorable the cucumber grows a new digestive tract.

The crinoids, or sea lilies, are sessile crea-

Figure 14–28 The common starfish, *Asterias,* eating a fish. Transparent lobes of the cardiac stomach can be seen surrounding the body of the fish. A number of tube feet are being used to hold the starfish to the side of the aquarium. (Courtesy of Robert S. Bailey.)

tures rather like a starfish turned mouth side up, with a number of arms extending upward. A stalk attaches the animal to the sea bottom. There are many more fossil than living species of crinoids.

14–7 THE HEMICHORDATA

The hemichordates, or acorn worms, are a small group of wormlike marine animals that burrow in sand or mud. The anterior section is a

Figure 14–29 The acorn worm *Saccoglossus,* a member of the phylum Hemichordata. *Left,* External view showing external features. *Right,* A diagrammatic section through the anterior part of the body showing some of the internal organs. A lateral fold subdivides the pharynx into a ventral channel along which the sand passes and a dorsal channel containing the gill slits.

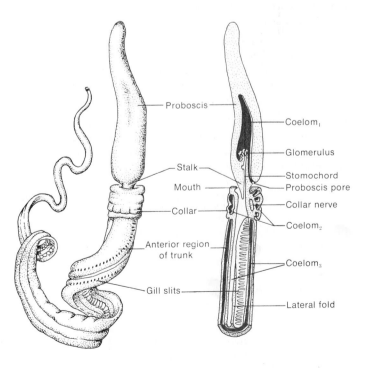

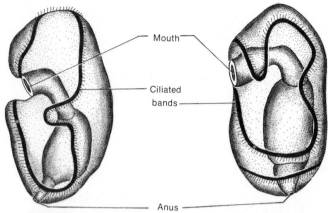

Figure 14–30 *Left,* The bipinnaria larva of a starfish and, *right,* the tornaria larva of an acorn worm. Note the striking similarities between the two.

short, conical, muscular *proboscis* (this apparently resembled an acorn in the eyes of early biologists) connected by a narrow stalk to the *collar,* behind which extends a long, cylindrical, rather flaccid, wormlike body (Fig. 14–29). The mouth is on the lower side of the body at the base of the collar. Just behind the collar and perforated by many gill slits, is the pharynx, through which water passes. As it burrows along, the animal feeds on organic matter in the sand. The nervous system is a diffuse network over most of the body, but concentrated into dorsal and ventral nerve cords in the anterior region. Only the dorsal nerve cord extends into the collar, where it becomes thick and hollow. There is a short, rodlike outgrowth from the anterior end of the digestive tract called a "stomochord," extending into the cavity of the proboscis. The larva of some acorn worms is much like that of some echinoderms (Fig. 14–30) and is often mistaken for it, but of course the later development of the two forms is quite different. This is taken as evidence of the evolutionary relationship of the two phyla. The hemichordates may have some evolutionary affinity with the chordates (only these two phyla have pharyngeal clefts).

The foregoing description does not exhaust the great variety of animals. In addition to these phyla, there are other groups of invertebrates sometimes put in phyla of their own, sometimes classified under other phyla, which are not important enough to be discussed here.

QUESTIONS

1. What are the fundamental differences between a tapeworm, an earthworm and a roundworm such as *Ascaris?*
2. Why do you suppose echinoderms are found only in the sea?
3. Discuss the various means utilized by land plants and land animals to bring about the union of an egg and sperm in the absence of a watery medium.
4. What are the advantages and disadvantages of an exoskeleton? How have some of the disadvantages of the exoskeleton been overcome by the arthropods? By the mollusks?
5. What are the advantages of a segmented body plan?
6. In what ways are the bodies of earthworms and *Nereis* adapted to their habitats?
7. What are the basic features of the molluscan body plan? How do they differ in the several classes of mollusks?
8. Describe the peripatus. What is its evolutionary significance? How would you classify it—with the annelids, with the arthropods, or by itself?
9. What characteristics of arthropods have been of primary importance in their evolutionary success?
10. What is the importance of *Aysheaia?* of *Neopilina?*
11. Discuss the similarities and differences in the body plans of a tick and a lobster.
12. What are the distinguishing features of the several classes of arthropods?
13. What are the distinguishing features of the subclasses of insects?
14. Compare the factors important to the survival of a social colony such as those of bees or termites with those important to the survival of a colonial organism such as *Obelia.*
15. Why do you suppose the horseshoe crab managed to survive whereas the trilobites became extinct?

SUPPLEMENTARY READING

The Invertebrates, a multivolume work by Libbie Hyman, is an up-to-date, definitive treatise on the invertebrates and is an excellent primary reference book. An excellent one-volume textbook is Robert Barnes's *Invertebrate Zoology.* A useful paperback field guide to the common invertebrates of the seashore is *Seashores* by H. S. Zim and L. Ingle. Many interesting aspects of invertebrate physiology are presented in E. Florey's *General and Comparative Animal Physiology.* Crustaceans by W. L. Schmitt is an interesting general discussion of this important group of invertebrates, and *The Insects: Structure and Function* by R. F. Chapman gives a detailed account of insect anatomy and physiology. Excellent reviews of the behavior, physiology and ecology of echinoderms are to be found in *Physiology of Echinodermata,* edited by R. A. Boolootian, and in *Echinoderms* by D. Nichols. *Spiders, Scorpions, Centipedes and Mites* by J. L. Cloudsley-Thompson provides a broad discussion of the structure and ecology of these groups of arthropods. *Molluscs* by J. E. Morton gives a useful general account of this important and diverse group of invertebrates, and *Annelids* by R. P. Dales is a brief, clearly written treatment of the polychaetes and other annelids. A superb general biology of the social insects is *The Insect Societies* by E. O. Wilson.

CHAPTER 15

THE PHYLUM CHORDATA

The last great phylum of animals, that to which we belong, is the phylum Chordata, whose members are distinctive in having a *notochord,* a dorsal, hollow *nerve cord* and paired, *pharyngeal pouches* and *gill slits.* The latter are present in all chordate embryos but are not evident in adult higher vertebrates. In addition to the fishes, amphibia, reptiles, birds and mammals, which make up the classes of the subphylum *Vertebrata* — characterized by a cartilaginous or bony vertebral column — the phylum Chordata includes two subphyla of curious soft-bodied marine animals which show the chordate characteristics to some extent, and are of interest as possible connecting links between vertebrates and invertebrates.

The notochord is a dorsal longitudinal rod composed of a fibrous sheath enclosing vacuolated cells. The turgidity of these cells makes the notochord firm yet flexible. The notochord, by preventing the body from shortening when the longitudinal muscles in the body wall contract, facilitates the lateral undulatory movements involved in the swimming motions of fishes. The nerve cord dorsal to the notochord differs from that of invertebrates not only in its position but in its structure, for it is single rather than double and a hollow tube rather than a solid cord. Chordates also have pharyngeal pouches extending laterally from the anterior part of the digestive tract toward the body wall, perhaps breaking through as gill slits. The earliest chordates apparently were filter-feeders and this arrangement of pouches and slits permitted water to escape from the digestive system and concentrated the small food particles in the gut. Chordates share other characteristics with certain invertebrates. They are bilaterally symmetrical, they have three germ layers, and they have a tube-within-a-tube body plan, with a true coelom separating the gut from the body wall.

The vertebrates are less diverse and much less numerous and abundant than the insects but rival them in their adaptation to a variety of modes of existence and excel them in the ability to receive stimuli and react to them. Vertebrates are generally active animals and show a high degree of *cephalization*, the accumulation of nerves and sense organs in the head.

15–1 SEA SQUIRTS OR TUNICATES: UROCHORDATA

The adult *sea squirts* or *tunicates* (Fig. 15–1) are barrel-shaped, sessile, marine animals unlike the other chordates; indeed, the primitive members are often mistaken for sponges or cnidarians. The larval tunicates, however, are typically chordate, superficially resembling a tadpole. Their expanded bodies have a pharynx with gill slits, and the long muscular tails contain a notochord and dorsal nerve cord. The larva eventually becomes attached to the sea bottom and loses its tail, notochord and most of its nervous system. In the adult, only the gill slits suggest that it is a chordate.

The adult develops a *tunic,* quite thick in most species and covering the entire animal, which is composed (curiously enough) principally of a kind of cellulose. The tunic has two openings, the *incurrent siphon,* through which water and food enter, and the *excurrent siphon,* through which water, waste products and gametes pass to the outside.

Tunicates are filter-feeders, removing plankton from the stream of water passing through the pharynx. The food particles are trapped in mucus secreted by the *endostyle* and carried by the beating of the cilia down into the esophagus and digestive gland.

15–2 CEPHALOCHORDATES

In the *Cephalochordata,* the second chordate subphylum, all three chordate characteristics are highly developed. The *notochord* extends from the tip of the head to the tip of the tail, a large pharyngeal region contains many pairs of *gill slits,* and a hollow, dorsal *nerve cord* stretches the entire length of the body (Fig. 15–2). The cephalochordates are small, translucent, fishlike, segmented animals, 5 to 10 centimeters long and pointed at both ends. They are widely distributed in shallow seas, either swimming freely or burrowing in the sand near the low tide line. Like the tunicates, they feed by drawing a current of water into the mouth (by the beating of cilia) and straining out the microscopic plants and animals. Food particles are trapped in the pharynx by mucus secreted by the endostyle and are then carried

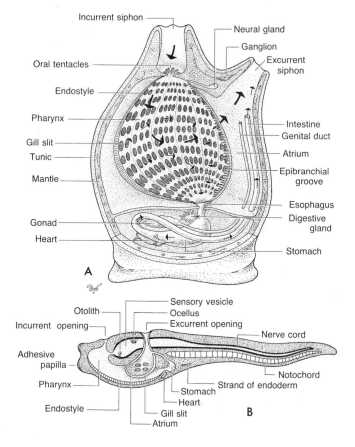

Figure 15–1 Diagrammatic lateral views of an adult *(A)* and a larval *(B)* tunicate showing major internal organs. The large arrows in *A* represent the course of the current of water, and the small arrows represent the path of the food. The stomach, intestine and other visceral organs are embedded in the mantle.

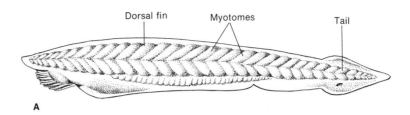

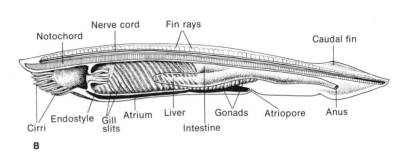

Figure 15–2 External view and a longitudinal section of Amphioxus, a member of the subphylum Cephalochordata.

back to the intestine. The water passes through the gill slits into the *atrium,* a chamber lined with ectoderm. This has a ventral opening, the atriopore, just anterior to the anus. Metabolic wastes are excreted by segmentally arranged, ciliated *protonephridia* that open into the atrium. Although superficially similar to fishes, they are much more primitive, for they lack paired fins, jaws, sense organs and a brain. It is generally believed that the cephalochordate *Amphioxus* is rather similar to the primitive ancestor from which the vertebrates evolved. In contrast to the invertebrates, the blood of this animal flows anteriorly in the ventral vessel and posteriorly in the dorsal vessel.

15–3 THE VERTEBRATES

The vertebrates are distinguished from these lower chordates by the presence of an internal skeleton of cartilage or bone that reinforces or replaces the notochord. The notochord is the only skeletal structure present in the lower chordates, but in the vertebrates segmental bony or cartilaginous *vertebrae* surround the notochord. In the higher vertebrates the notochord is visible only early in development; later the vertebrae replace it completely. Vertebrates have a bony or cartilaginous brain case, the *cranium,* which encloses and protects the brain, the enlarged anterior end of the dorsal, hollow nerve cord.

Vertebrates have a pair of eyes that de-

velop as lateral outgrowths of the brain. Invertebrate eyes, such as those of insects and cephalopods, may be highly developed and quite efficient, but they develop from a folding of the skin. Another vertebrate characteristic is a pair of ears, which are primarily organs of equilibrium in the lowest vertebrates. The *cochlea,* containing the cells sensitive to sound vibrations, is a later evolutionary development.

The circulatory system of vertebrates is distinctive in that the blood is confined to blood vessels and is pumped by a ventral, muscular heart. The higher invertebrates such as arthropods and mollusks typically have hearts but they are located on the dorsal side of the body and pump blood into open spaces in the body, called *hemocoels.* Vertebrates are said to have a *closed circulatory system;* arthropods and mollusks have an open circulatory system, for the blood is not confined solely to tubular blood vessels.

The classes of the subphylum Vertebrata are as follows: the *Agnatha,* the jawless fishes such as the lamprey eels; the *Placodermi,* earliest of the jawed fishes, known only from fossils; the *Chondrichthyes,* the sharks and rays with cartilaginous skeletons; the *Osteichthyes,* the bony fishes; the *Amphibia,* frogs and salamanders; the *Reptilia,* lizards, snakes, turtles and alligators plus a host of fossil forms like the dinosaurs; the *Aves,* birds; and the *Mammalia,* the warm-blooded, fur-bearing animals that suckle their young. The Agnatha, Placodermi, Chondrichthyes and Osteichthyes compose the superclass *Pisces,* and the Amphibia, Rep-

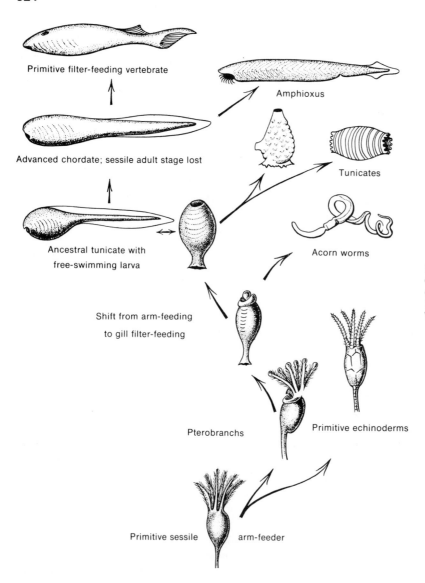

Primitive filter-feeding vertebrate

Amphioxus

Advanced chordate; sessile adult stage lost

Tunicates

Ancestral tunicate with
free-swimming larva

Acorn worms

Figure 15–3 A diagram illustrating
one hypothesis of the evolution of
chordates. (After Romer, A. S.: *The
Vertebrate Body*, 4th Ed. Philadelphia,
W. B. Saunders Co., 1970.)

Shift from arm-feeding

to gill filter-feeding

Pterobranchs

Primitive echinoderms

Primitive sessile   arm-feeder

tilia, Aves and Mammalia comprise the superclass **Tetrapoda.**

There is no clear fossil record of the ancestors of the chordates, but whatever they were,
they were undoubtedly small and soft-bodied.
An impression of an Amphioxus-like animal,
Jamoytius, has been found in rocks of the
Silurian period in Scotland. Some paleontologists have interpreted it as a primitive vertebrate.

Theories of the origin of the chordates
must depend on other types of evidence. The
most widely held theory at present is that the
echinoderms, hemichordates and chordates
have a common evolutionary origin (Fig. 15–3).
This is based on the striking similarity of the
tornaria larva of the hemichordate and the
bipinnaria larva of the starfish (Fig. 14–29)

plus the generally similar modes of formation
of the mesoderm and coelom in the three
phyla.

15–4 JAWLESS FISHES

The Agnatha, or jawless fishes, include the
ostracoderms, the earliest known fossil chordates (Fig. 32–3), and the living lamprey eels
and hagfishes (Fig. 15–4). These have cylindrical bodies up to a meter long, with smooth
scaleless skin and no jaws or paired fins. Lampreys and hagfishes have a circular sucking
disc around the mouth, which is located on the
ventral side of the anterior end. They attach
themselves by this disc to other fish, and using
the horny teeth on the disc and tongue, bore

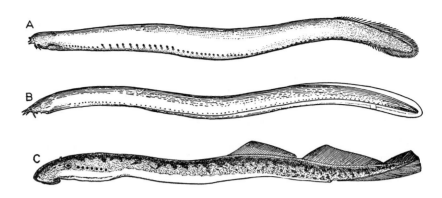

Figure 15–4 The three main types of cyclostomes. *A,* A slime hag; *B,* a hagfish; *C,* a lamprey. Note the absence of jaws and paired fins. (From Romer, A. S.: *The Vertebrate Body,* 4th Ed. Philadelphia, W. B. Saunders Co., 1970.)

through the skin and get blood and soft tissues to eat. They are the only parasitic vertebrates; hagfishes may bore their way completely through the skin and come to lie within the body of the host. Both are of great economic importance because of their destruction of food fish such as cod, flounder, lake trout and whitefish. The trout of the Great Lakes have been killed off in great numbers by sea lampreys that apparently came up from the St. Lawrence via the Welland Canal. Lampreys leave the ocean or lakes and swim upstream to spawn. They build a nest, a shallow depression in the gravelly bed of the stream, into which eggs and sperm are shed. The fertilized eggs develop into **ammocoetes** larvae in about three weeks. These larvae probably resemble the ancestral primitive vertebrate more closely than any living adult vertebrate does. They drift downstream to a pool and live as filter-feeders in burrows in the muddy bottom for several years. They then undergo a metamorphosis, become adult lampreys, and migrate back to the ocean or lake.

The earliest vertebrates were the ostracoderms, fossils found in rocks of the Ordovician period. These were small, armored, jawless, bottom-dwelling, filter-feeding, fresh water fishes. The head was covered with thick bony plates and the trunk and tail were covered with thick scales. Ostracoderms had median fins and some species had paired pectoral fins.

15–5 THE EARLIEST JAWED FISHES

During the Silurian and Devonian periods, some descendants of the ostracoderms evolved jaws and paired appendages and changed from filter-feeding bottom-dwellers to active predators. The earliest jawed fishes, the spiny-skinned sharks, are placed in the subclass Acanthodii. Another group of primitive jawed fishes living in the Paleozoic era were the placoderms, small, armored, fresh water fish with a variable number (as many as seven) of paired fins (Fig. 15–5). The placoderms, all of which are now extinct, were probably the ancestors of both cartilaginous and bony fishes. One of the best known placoderms is *Dunkleosteus*, a monster that attained a length of 3 meters. The head and anterior part of the trunk had a bony armor, but the remainder of the body was naked. The evolution of jaws from a portion of the gill arch skeleton enabled the placoderms and their descendants to become adapted to new modes of life. The success of the jawed vertebrates undoubtedly contributed to the extinction of the ostracoderms.

15–6 CARTILAGINOUS FISHES

The ostracoderms and placoderms were primarily fresh water fishes; only a few ventured into the oceans. The cartilaginous fishes evolved as successful marine forms in the Devonian and most have remained as ocean-dwellers; only a few have secondarily returned to a fresh water habitat.

The Chondrichthyes—sharks, rays and skates—have a skelton of cartilage which may or may not be calcified. The cartilaginous skeleton of these fishes represents the retention of an embryonic condition, not a primitive one, for the adult ancestors had bony skeletons. The dogfish is commonly used in biology classes because it demonstrates the basic vertebrate characteristics in a simple, uncomplicated form. All the Chondrichthyes have paired jaws and two pairs of fins. The skin contains scales composed of an outer enamel and an inner dentine layer. The lining of the mouth contains larger but essentially similar scales which

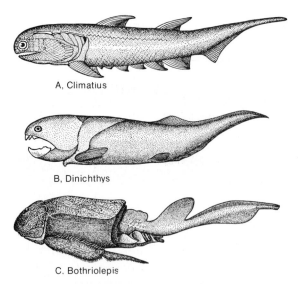

A, Climatius

B, Dinichthys

C. Bothriolepis

Figure 15–5 Acanthodians and placoderms from the Devonian period. *A, Climatius,* a "spiny-skinned shark," with large fin spines and five pairs of accessory fins between the pectoral and pelvic pairs. *B, Dinichthys,* a giant arthrodire that grew to a length of 10 meters. Its head and thorax were covered by bony armor, but the rest of the body and tail were naked. *C, Bothriolepis,* a placoderm with a single pair of jointed flippers projecting from the body. (From Romer, A. S.: *The Vertebrate Body,* 4th Ed. Philadelphia, W. B. Saunders Co., 1970.)

serve as teeth. The teeth of the higher vertebrates are homologous with these shark scales.

All fishes, from lampreys to the highest bony fishes, have highly vascular *gills* with a large surface for the transfer of oxygen and carbon dioxide. Cartilaginous fishes have five to seven pairs of gills. The gills of some fishes also secrete salts to maintain osmotic equilibrium between the blood and the surrounding water. A current of water enters the mouth, passes over the gills and out the gill slits, constantly providing the fish with a fresh supply of dissolved oxygen.

Unlike the bony fishes, cartilaginous fishes have no swim bladders and their bodies are denser than water; hence they tend to sink unless they are actively swimming. The large pectoral fins give a lift component to their forward motion and the sculling action of the tail provides additional lift.

The whale shark, which reaches a length of 16 meters, is the largest fish known, but it feeds on microscopic crustacea and other plankton. Sharks are elongate, streamlined predators, that swim actively and catch other fish.

Rays and skates are sluggish, flattened creatures, living partly buried in the sand and feeding on mussels and clams. The undulations of its enormous pectoral fins propel the ray or skate along the bottom. The stingray has a whiplike tail with a barbed spine at the tip, which can inflict a painful wound (Fig. 15–6). The electric ray has electric organs on either side of the head; these modified muscles can discharge enough electricity to stun fairly large fishes. Shark skin is tanned and used in making shoes and handbags, and shark liver oil is an important source of vitamin A. Some sharks and rays are used for food. The swordfish and certain other fishes can take up mercury from the surrounding sea water and concentrate it in their tissues.

15–7 BONY FISHES

The Osteichthyes include some 20,000 species of fresh and salt water fishes, ranging in size from guppies to sturgeons weighing over a ton. The fossil evidence now available indicates that bony fishes evolved from placoderms independently of, and at about the same time as, the cartilaginous fishes. They did not evolve from cartilaginous fishes.

The bony fishes originated in fresh water but subsequently entered the oceans and became dominant there, too. Lungs evolved in the primitive fresh water bony fishes, presumably as an adaptation to environmental conditions. There were frequent seasonal droughts in the Devonian period during which ponds became stagnant or dried up completely. Fishes with some adaptation for breathing air, such as lungs, had a tremendous advantage for survival under those conditions. By the middle of the Devonian, the bony fishes had evolved into three major groups—lungfishes, lobe-finned fishes and ray-finned fishes. All three groups possessed lungs and an armor of bony scales. A few lungfishes have survived to the present in the rivers of tropical Africa, Australia and South America. The ray-finned fishes evolved slowly during the late Paleozic era, then ramified greatly in the Mesozoic and gave rise to the modern bony fishes, the teleosts. The lobe-finned fishes, generally credited as being the ancestors of the land vertebrates, were almost extinct by the end of the Paleozoic.

Figure 15-6 *Upper,* the sawfish *Pristis.* *Lower,* the stingray *Dasyatis.* (Courtesy of Marine Studios.)

A few specimens of marine lobe-finned fishes, *coelacanths,* nearly 2 meters longs (Fig. 15–7), have been caught in the deep waters off the east coast of Africa near the Comoro Islands.

Most of the bony fishes have beautifully streamlined bodies and swim by contracting the body and tail muscles which move the tail back and forth laterally in a sculling motion. The fins are used chiefly for steering. Bony fishes typically have a *swim bladder,* a gas-filled sac located in the dorsal part of the body cavity (Figs. 15–8 and 18–8). By secreting gases into the bladder or by absorbing them from it, the fish can change the density of its body and so hover at a given depth of water. The gills of bony fishes are covered by a hard, bony protective flap, the *operculum.* The skeleton is composed of bone rather than cartilage, and the head is encased in many bony plates which form a *skull.* Bony fishes have protective, overlapping bony scales in the skin which differ from those found in sharks.

Many of the bony fishes, particularly those in tropical waters, are brightly and beautifully colored—red, orange, yellow, green, blue and black. Some fishes, such as flounders, can change color and pattern to conform to the color and pattern of the background, and thus render themselves inconspicuous to predators. The teleosts have undergone a remarkable adaptive radiation (p. 743), evolving into a great variety of types adapted to many different ecologic niches. Fish evolution has led to a tremendous variety of sizes, shapes and colors, and to a number of curious adaptations—the sea horse male, which has a brood pouch in which eggs are carried until they hatch; the deep sea forms that have evolved luminescent

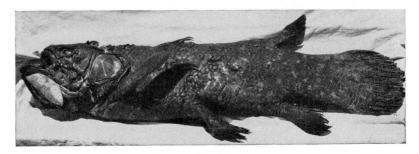

Figure 15-7 A photograph of a coelacanth 1.5 meters in length caught in the Indian Ocean off South Africa in 1952. Note the thick, lobe-shaped fins. (Courtesy of LIFE, © TIME, Inc.)

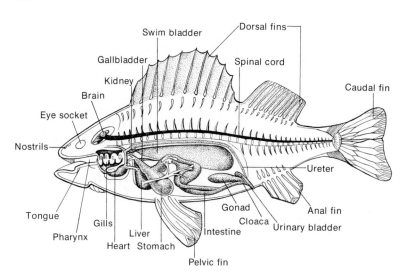

Figure 15–8　A diagram of the structure of a perch, a bony fish.

structures as lures for their prey; the male stickleback, which builds a nest of sticks held together by threads which he secretes and then guards the eggs in the nest; the true eels, which live as adults in streams in North America or Europe but which migrate to the Atlantic near Bermuda to spawn, and so on. Certain muscles of several fish, such as the

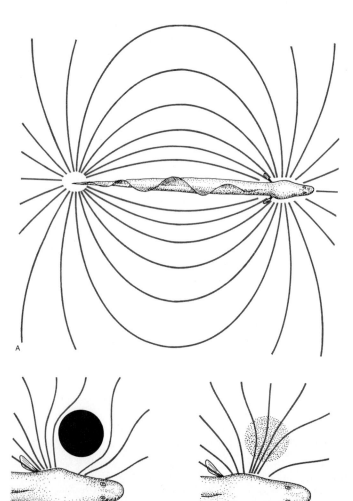

Figure 15–9　Electrolocation in the fish *Gymnarchus.* The fish generates an electric field, its tail negative to the head *(A).* Objects that conduct less *(B)* or more *(C)* electricity than the surrounding water distort the electric field, and this is detected by special sense organs along the sides of the fish. (Courtesy of H. W. Lissmann.)

electric eel, have evolved into electric organs capable of emitting pulses strong enough to stun or kill their prey. In other fishes weak pulses are emitted continuously and used to generate an electric field in the surrounding water. This serves as a guidance system, a biological "radar" for the fish to detect the disturbances in the electric field produced by rocks and other objects in the water (Fig. 15–9).

The lungfishes, order **Dipnoi**, were once thought to be the ancestors of land vertebrates, but in the arrangement of the bones of the skull, the type of teeth, the pattern of fin bones and types of vertebrae, the lobe-finned fishes resemble the primitive amphibians closely and the lungfishes do not.

15–8 THE AMPHIBIA

The four-legged land vertebrates, the amphibia, reptiles, birds and mammals, are placed together in the superclass **Tetrapoda.** Not all the tetrapods have four legs (e.g., the snakes) but they evolved from four-legged ancestors. Not all the tetrapods now live on land (e.g., whales, seals), but they evolved from terrestrial ancestors.

The first successful land vertebrates were ancient amphibians, the *labyrinthodonts* (Fig. 15–10 A), clumsy, salamanderlike, ancient amphibians with short necks and heavy muscular tails. They closely resembled their ancestral lobe-finned fishes but had evolved limbs strong enough to support the weight of the body on land. These earliest arms and legs were five-fingered, a pattern that has generally been kept by the higher vertebrates. There were many different kinds of ancient amphibia, all of which became extinct in the first part of the Mesozoic. The labyrinthodonts ranged in size from small, salamander-sized animals up to ones as large as crocodiles. They gave rise to other primitive amphibians, to modern frogs and salamanders and to the earliest reptiles, the *cotylosaurs,* or stem reptiles (Fig. 15–10 B). The modern amphibia, the frogs and salamanders, appeared in the latter part of the Mesozoic. The salamanders and water dogs more closely resemble the ancient amphibia; frogs and toads are highly specialized for hopping.

Although some adult amphibia are quite successful as land animals and can live in comparatively dry places, they must return to water

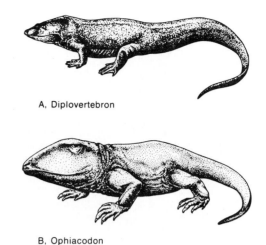

A, Diplovertebron

B, Ophiacodon

Figure 15–10 *A, Diplovertebron,* a primitive Paleozoic amphibian (labyrinthodont). *B, Ophiacodon,* an early Permian pelycosaur. Although the pelycosaurs were primitive reptiles, they had characteristics indicating that they represented a first stage in the evolution of the mammals. (From Romer, A. S.: *The Vertebrate Body,* 4th Ed. Philadelphia, W. B. Saunders Co., 1970.)

to reproduce. Eggs and sperm are generally laid in water and the fertilized eggs, nourished at first by the yolk, develop into larvae, or *tadpoles.* These breathe by means of gills and feed on aquatic plants. After a time the larva undergoes *metamorphosis* and becomes a young adult frog or salamander, with lungs and legs. Like the metamorphosis of insects and crustacea, that of amphibia is under hormonal control. Amphibia undergo a single change from larva to adult in contrast to the four or more molts involved in the development of arthropods to the adult form. Amphibian metamorphosis is regulated by thyroxine, the hormone secreted by the thyroid gland, and can be prevented by removing the thyroid or the pituitary, which secretes a thyroid-stimulating hormone.

During metamorphosis the forelegs grow out of the fold of skin that had enveloped them, the gills and gill slits are lost, the tail is resorbed, the digestive tract shortens, the mouth widens, a tongue develops, the tympanic membrane and eyelids appear, and the shape of the lens changes. In addition, a host of biochemical changes occur to provide for the change from a completely aquatic life to an amphibious one. The kind of light-sensitive pigment in the retina of the eye changes at the time of metamorphosis under the control of thyroxine.

Several species of salamanders, such as the mud puppy, *Necturus,* do not undergo meta-

Figure 15–11 *A,* Leopard frog adapted to a light background, and *B,* one adapted to a dark background. (From Turner, C. D.: *General Endocrinology,* 4th Ed. Philadelphia, W. B. Saunders Co., 1966.)

morphosis but grow to be very large "larvae" and reproduce in the larval state. The thyroid glands of these animals produce thyroxine, but their tissues fail to respond to the hormone, perhaps because they have lost the ability to synthesize the receptor for thyroxine.

Adult amphibia do not depend solely on their primitive lungs for the exchange of respiratory gases; their moist skin, plentifully supplied with blood vessels, also serves as a respiratory surface. The skin of salamanders and frogs has no scales and may be brightly colored. Frogs especially have the ability to change color, from light to dark, by increasing or decreasing the size of the melanocytes, the pigment-containing cells of the skin (Fig. 15–11). The change in color is controlled by the melanocyte-stimulating hormone, MSH, secreted by the intermediate lobe of the pituitary (p. 551). Salamanders, but not frogs, have a marked ability to regenerate lost legs and tails. In some species of frogs and toads the fertilized eggs do not develop in the water but are kept on the back of the female, in the mouth of the male, or in a string wrapped around the male's hind legs.

A number of frogs, toads and salamanders have skin glands that secrete poisonous substances. These may serve as a means of protection for the species, by discouraging would-be predators.

A small group of tropical, legless, worm-like amphibia (the Apoda) burrow in moist earth.

15–9 THE FROG

The frog is neither the simplest nor the most representative vertebrate, but it is readily available, easily studied, and is frequently chosen as an example of vertebrate structure and function in biology courses. Frogs are highly specialized and have short trunks, no tail, and powerful, enlarged hind legs with webbed feet, features that adapt them for their mode of life. On the frog's head are a large mouth, a pair of external nostrils or *nares,* a pair of eyes and a pair of round *tympanic membranes,* or eardrums.

The forelegs (*pectoral appendages*) are much shorter than the hind legs (*pelvic appendages*) and have four instead of five digits on the hand—the thumb is missing. The most medial digit (which corresponds to our index finger) is thicker in males than in females, especially during the breeding season, and helps the male grasp the female during amplexus. The skin is soft, smooth and moist, and serves for gas exchange as well as for protection and the reception of sensory stimuli. It is greenish in color, but contains no green pigment. The green color is due to the reflection of yellow and blue light by certain pigment and refractive granules lying just below the epidermis.

The skeleton of the frog (Fig. 15–12) serves as the supporting framework of the body, provides a point of attachment for most of the muscles, and encases and protects much of the nervous sytem. The *axial skeleton* includes the vertebral column, sternum and skull, and the *appendicular skeleton* includes the girdles and limbs. The *urostyle* represents several caudal vertebrae fused together and specialized for the attachment of the powerful pelvic muscles. Some details of the ventral aspect of the skull and of the *hyoid apparatus,* which supports the floor of the mouth, are

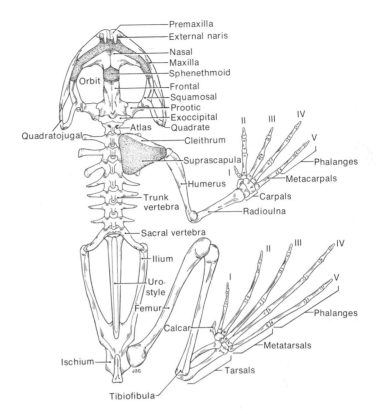

Figure 15-12 Dorsal view of the skeleton of a frog. The roman numerals refer to the digit numbers.

shown in Figure 15-13. The **sternum,** part of the axial skeleton, together with the pectoral girdle, forms an arch of bone and cartilage that nearly encircles the front of the trunk. The girdle consists of an anterior **clavicle** and posterior **coracoid** extending laterally from the

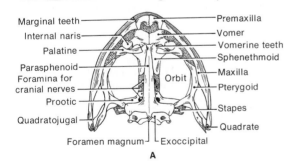

A

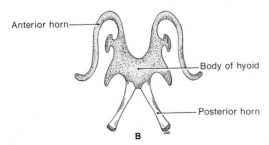

B

Figure 15-13 *A,* Ventral view of the skull of a frog. *B,* Ventral view of the hyoid apparatus.

sternum and a **scapula** that extends dorsally from their lateral ends (Fig. 15-14). The **glenoid fossa,** the concavity where these three bones meet, articulates with the humerus, the bone of the upper arm. The pectoral girdle is bound to the trunk only by muscles; there is no bony connection between the girdle and the vertebral column. In contrast, the pelvic girdle, composed of **ilium, ischium** and **pubis,** is attached to the sacral vertebra. The **femur,** the bone of the upper part of the hind leg, articulates with a cavity, the **acetabulum,** located where pubis, ischium and ilium join.

The superficial skeletal muscles, shown in Figure 15-15, illustrate the general pattern of vertebrate musculature. Muscles can produce movement of parts of the body only by contracting or shortening, hence they are arranged in **antagonistic sets;** one moves a part in one direction and the antagonist moves it in the other. Thus there are **flexors** and **extensors, adductors** and **abductors,** and **protractors** and **retractors.** Most muscles are attached to bones by **tendons.** The **origin** of the muscle is the relatively more fixed end of the muscle, the end that moves less when the muscle contracts, and the **insertion** is the end that moves.

The internal organs protrude into the

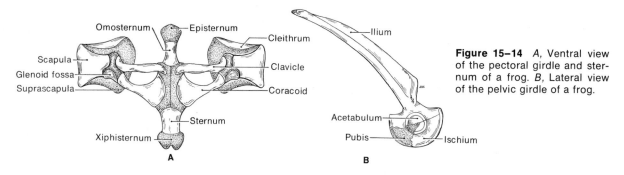

Figure 15–14 *A,* Ventral view of the pectoral girdle and sternum of a frog. *B,* Lateral view of the pelvic girdle of a frog.

coelomic cavity, which contains a small amount of coelomic fluid and is lined with a thin epithelial layer, the *peritoneum.* Most of the coelomic cavity is occupied by the organs of the *digestive system* (Fig. 15–16), which are generally similar to the human organs (Fig. 19–2). The tongue of the frog is attached anteriorly and can be protruded from the mouth to catch insects and other prey. The *respiratory system* of the frog includes not only the lungs but also the skin and the mucous membranes lining the mouth and pharynx. All of these are moist, vascular, differentially permeable membranes through which gases can diffuse. Air is moved into the lungs by the pumping action of muscles in the floor of the mouth. During inspi-

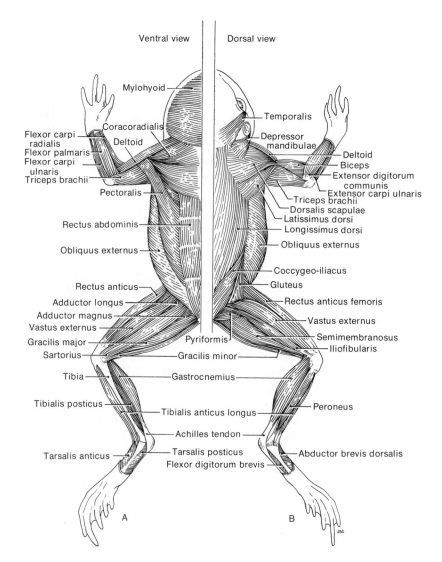

Figure 15–15 Superficial skeletal muscles of the frog seen in ventral view *(left)* and dorsal view *(right).*

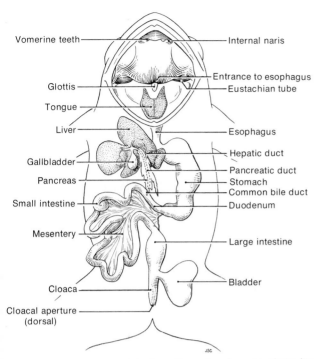

Vomerine teeth

Glottis

Tongue

Liver

Gallbladder

Pancreas

Small intestine

Mesentery

Cloaca

Cloacal aperture
(dorsal)

Internal naris

Entrance to esophagus

Eustachian tube

Esophagus

Hepatic duct

Pancreatic duct

Stomach

Common bile duct

Duodenum

Large intestine

Bladder

Figure 15–16 Ventral dissection of a frog to show its digestive system. The lobes of the liver have been pulled forward to reveal the gallbladder.

ration the floor of the mouth is lowered and air is drawn into the mouth through the external nares. These are then closed by pressure of the lower jaw on the movable *premaxillary bones,* the floor of the mouth is raised and air is forced through the *glottis* into the lungs.

The pattern of the major arteries and veins, shown diagrammatically in Figure 15–17, differs in several respects from the human pattern (Figure 17–9). Much of the blood from the hind legs and back enters a pair of *renal portal veins* which lead to capillaries within the kidneys. These are drained by *renal veins* which empty into the posterior vena cava. The frog's heart (Fig. 15–18) consists of a thin-walled *sinus venosus* which receives blood from the venae cavae and passes it on to the right *atrium.* The left atrium receives blood from the lungs via the pulmonary veins. Both empty into a single *ventricle,* from which blood is forced by muscular contraction of the ventricular wall out through the *conus arteriosus* (which has a peculiar spiral valve in its interior) into the *truncus arteriosus.* The oxygenated blood from the left atrium is mixed to a considerable extent in the ventricle with the blood from the right atrium, which is only partially oxygenated.

Some waste products of metabolism are removed through the skin and lungs but most are excreted by the *kidneys.* These elongate organs lie just dorsal to the coelomic cavity (Figs. 15–19 and 15–20). The *adrenal glands* are evident as irregular, yellowish bands on the ventral surface of each kidney. The *wolffian ducts,* a pair of which drain the kidneys, are functionally, but not embryologically, com-

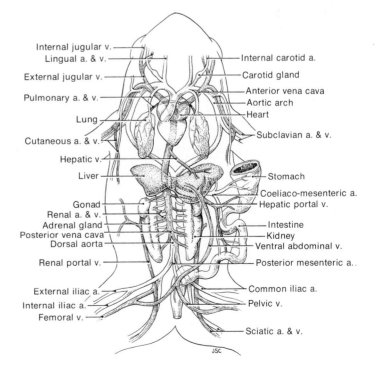

Figure 15–17 Ventral view of the major arteries and veins of the frog. Veins are shaded, arteries are white.

Internal jugular v.

Lingual a. & v.

External jugular v.

Pulmonary a. & v.

Lung

Cutaneous a. & v.

Hepatic v.

Liver

Gonad

Renal a. & v.

Adrenal gland

Posterior vena cava

Dorsal aorta

Renal portal v.

External iliac a.

Internal iliac a.

Femoral v.

Internal carotid a.

Carotid gland

Anterior vena cava

Aortic arch

Heart

Subclavian a. & v.

Stomach

Coeliaco-mesenteric a.

Hepatic portal v.

Intestine

Kidney

Ventral abdominal v.

Posterior mesenteric a.

Common iliac a.

Pelvic v.

Sciatic a. & v.

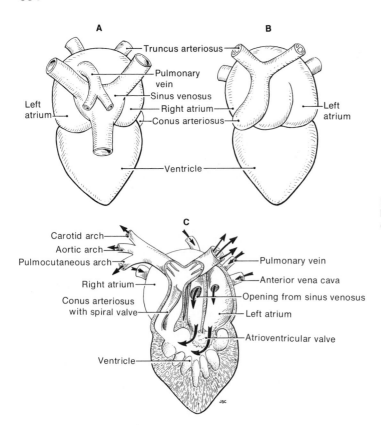

Figure 15–18 Views of the frog's heart. *A,* Dorsal aspect; *B,* ventral aspect; *C,* ventral view of a dissection of the heart.

parable to the ureters of higher vertebrates. These ducts empty into the *cloaca,* and urine may be discharged directly from the cloaca or may be stored for a time in the *urinary bladder* attached to the ventral surface of the cloaca.

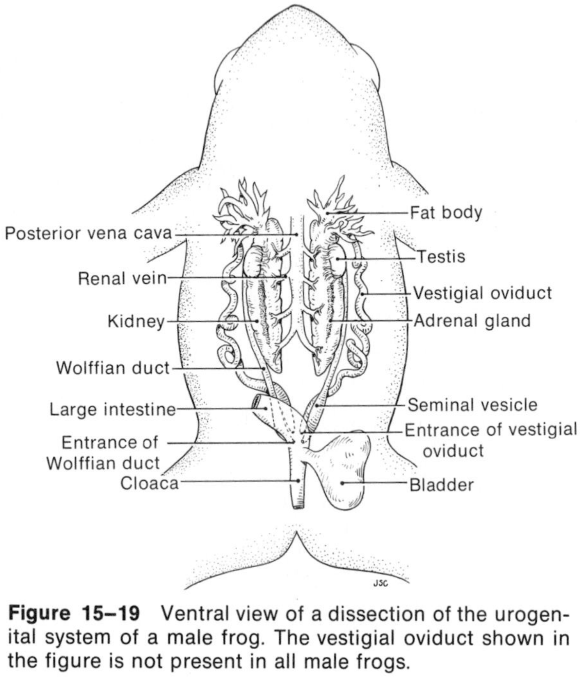

Figure 15–19 Ventral view of a dissection of the urogenital system of a male frog. The vestigial oviduct shown in the figure is not present in all male frogs.

Some water may be reabsorbed from the urine in the urinary bladder.

The *reproductive system* includes the *gonads* and the ducts that convey the gametes produced in the gonads to the exterior. The small, yellowish, bean-shaped testes (Fig. 15–19), suspended by mesenteries from the kidneys, produce the sperm. A fingerlike *fat body,* large in the fall and small in the spring, is attached to the anterior end of the testes and serves as a reserve supply of food over the winter. Sperm pass via microscopic ducts, the *vasa efferentia,* lying in the mesentery connecting the testis to the kidney, to certain kidney tubules whence they pass, via the wolffian duct, to the cloaca and the exterior.

The *ovary* in the spring is filled with hundreds of large, ripe eggs and occupies most of the coelomic cavity. The eggs are forced out of the ovary by the contraction of smooth muscles in the wall of the sac (*follicle*) surrounding each ovary. This is stimulated by a *gonadotropic hormone* secreted by the pituitary gland (p. 556). The eggs pass into the coelomic cavity and are carried to its anterior end by the beating of cilia. There they enter the funnel-shaped openings (*ostia*) of the paired *oviducts* (Fig. 15–20). As the eggs are carried down the ovi-

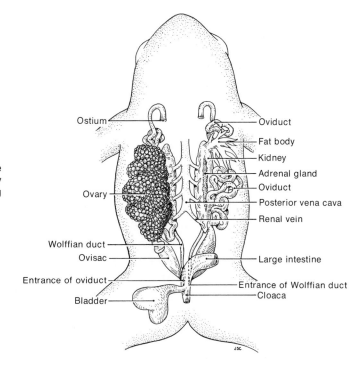

Figure 15–20 Ventral view of a dissection of the urogenital system of a female frog. The left ovary has been removed to reveal the structures lying dorsal to it.

Labels for Figure 15–20: Ostium, Ovary, Wolffian duct, Ovisac, Entrance of oviduct, Bladder, Oviduct, Fat body, Kidney, Adrenal gland, Oviduct, Posterior vena cava, Renal vein, Large intestine, Entrance of Wolffian duct, Cloaca

ducts by the beating of cilia, they are covered with several layers of a jellylike substance. Near its connection with the cloaca, each oviduct expands into a thin-walled *ovisac*, where the eggs may be stored for short periods until mating occurs.

During mating, the male grasps the female with his forelimbs, an embrace known as *amplexus,* and as she discharges her eggs into the water, he sheds his sperm. Fertilization occurs externally and the eggs develop to the larval stage within the jelly coat, then emerge as free-swimming larvae, or *tadpoles.*

The eyes of the frog (Fig. 15–21 *A*) are

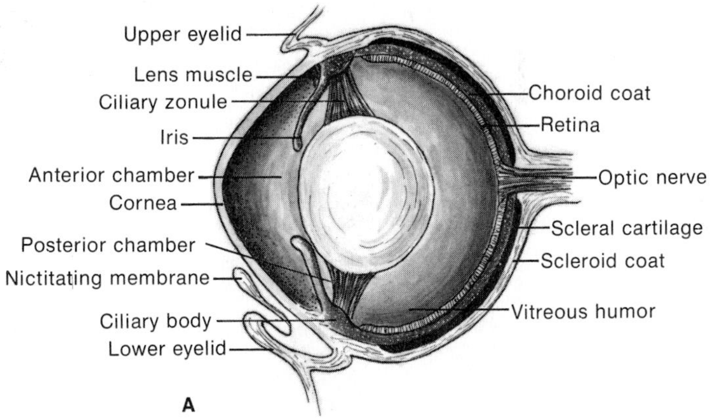

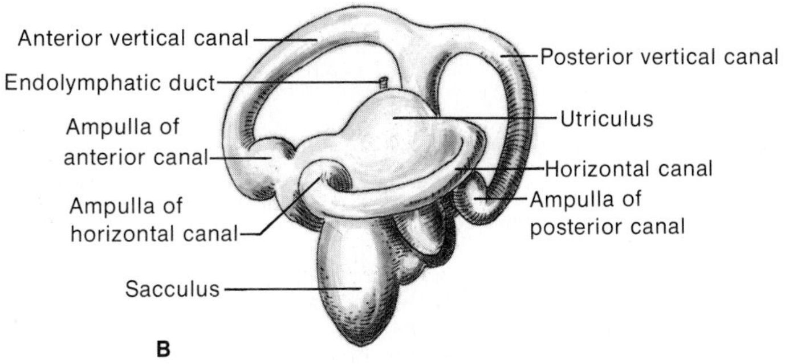

Figure 15–21 *A*, Sagittal section through the eye of a frog. *B*, Dissection of the canals and ducts of the inner ear of a frog.

Labels for A: Upper eyelid, Lens muscle, Ciliary zonule, Iris, Anterior chamber, Cornea, Posterior chamber, Nictitating membrane, Ciliary body, Lower eyelid, Choroid coat, Retina, Optic nerve, Scleral cartilage, Scleroid coat, Vitreous humor

A

Labels for B: Anterior vertical canal, Endolymphatic duct, Ampulla of anterior canal, Ampulla of horizontal canal, Sacculus, Posterior vertical canal, Utriculus, Horizontal canal, Ampulla of posterior canal

B

basically similar to human eyes (Fig. 24–12), but change focus for near and far vision by moving the lens forward and back (as we do in focusing a camera) instead of by changing the *shape* of the lens as in the mammalian eye. Light rays are focused by the lens onto the light sensitive cells in the *retina* at the back of the eye.

The ears are structures concerned with the sense of equilibrium as well as with sound detection. The inner ear (Fig. 15–21 *B*) consists of three fluid-filled canals located at right angles to each other whereby motion in any plane can be perceived by the motion of the liquid in the canals and sacs. Sound vibrations cause the tympanic membrane in the skin just behind the eye to vibrate. The vibrations are transmitted across the *middle ear cavity,* just under the tympanic membrane, by a rod-shaped bone, the *stapes.* The middle ear cavity is homologous to a gill pouch in a fish and is connected to the pharynx by a duct, the *eustachian tube.* The inner end of the stapes fits into an *oval window* in the inner ear and vibrations are thus transmitted to the liquid-filled canals and

sacs of the inner ear (Fig. 24–19). The vibrations stimulate specific groups of cells in these canals to initiate impulses in the acoustic nerve which pass to the brain.

The brain of the frog (Fig. 15–22) differs from the human brain (Figs. 23–15 and 23–16) not only in its absolute size, but in the relative sizes of the various regions. The *cerebral hemispheres* and *cerebellum* are small and the *medulla* is relatively large. The frog has only ten pairs of *cranial nerves* (not twelve as in higher vertebrates). The spinal cord, like the entire trunk region of the frog, is short and there are only ten pairs of *spinal nerves* (Fig. 15–23). Each spinal nerve is attached to the cord by a dorsal and ventral root. The ventral root transmits efferent motor impulses, and the dorsal root afferent sensory impulses. Each spinal nerve has a *ramus communicans* extending to the *sympathetic trunk,* which lies parallel to the spinal cord. The two are connected at an enlargement, the *sympathetic ganglion.* The two sympathetic trunks lie on either side of the dorsal aorta. The motor fibers in the sympathetic trunks and nerves, together with the

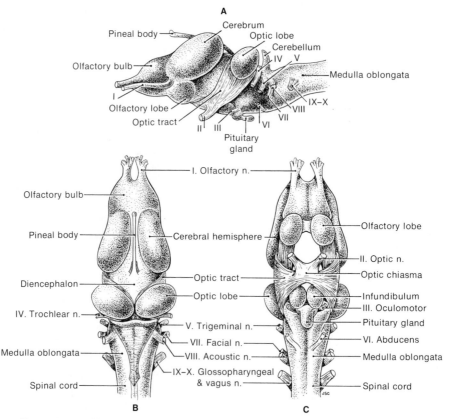

Figure 15–22   Views of the brain of the frog. *A*, Lateral; *B*, dorsal; and *C*, ventral views.

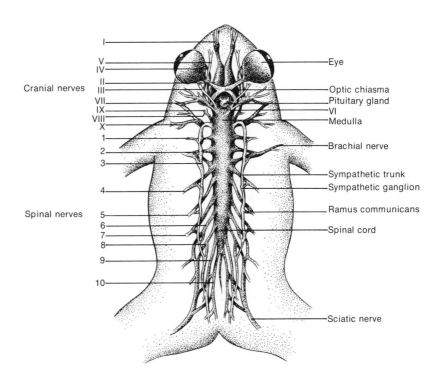

Figure 15–23 Ventral view of a dissection of the spinal cord, spinal nerves and sympathetic trunks of the frog.

Cranial nerves

I
V
IV
II
III
VII
IX
VIII
X

Spinal nerves

1
2
3
4
5
6
7
8
9
10

Eye
Optic chiasma
Pituitary gland
VI
Medulla
Brachial nerve
Sympathetic trunk
Sympathetic ganglion
Ramus communicans
Spinal cord
Sciatic nerve

motor fibers in the vagus and certain other cranial nerves, constitute the *autonomic nervous system* (Fig. 23–25), which innervates visceral organs, blood vessels and glands.

In addition to the adrenal glands and gonads noted previously, the endocrine system of the frog includes the *pituitary, thyroid* and *pancreas.* The pituitary secretes MSH, which is involved in the control of skin color, *growth hormone* and *gonadotropins. Thyroxine,* secreted by the thyroid, is necessary for metamorphosis from the larval to the adult form and for the maintenance of the proper level of metabolism in the adult.

15–10 REPTILES

The class Reptilia has many more extinct than living species. Reptiles are true land forms and need not return to water to reproduce, as amphibians must. The embryo develops in a watery medium within the protective leathery egg shell secreted by the female. Since a sperm cannot penetrate this shell, fertilization must occur *within* the body of the female *before* the shell is added. This necessitated the evolution of some means of transferring sperm from the body of the male to that of the female. Reptiles were the first to evolve a

male copulatory organ, the *penis,* for this purpose.

The bodies of reptiles are covered with hard, dry, *horny scales* which protect the animal from desiccation and from predators. They breathe by means of lungs, for the dry scaly skin cannot serve as an organ of respiration. Reptiles are able to survive in dry terrestrial regions such as deserts because, in evolving, the kidney tubules have been modified so that less water is removed from the blood and more of the water in the glomerular filtrate (p. 451) in reabsorbed in the distal tubule and urinary bladder. Reptiles excrete nitrogenous wastes as uric acid; amphibia excrete them as urea. Uric acid is less toxic and less soluble than urea and is excreted as a watery paste.

Like fish and amphibia, reptiles do not have a mechanism for regulating body temperature and therefore have the same temperature as their surroundings. In hot weather their body temperature is high, metabolism occurs rapidly, and they can be quite active. In cold weather their body temperature is low, their metabolic rates are low and they are very sluggish. Because of this they are much more successful in warm than in cold climates. The reptiles living today are turtles (order Chelonia), alligators (order Crocodilia), snakes and lizards (order Squamata), and the tuatara of

Figure 15–24   Adaptive radiation among lizards. *A*, The Old World chameleon has grasping feet and a prehensile tail with which to climb about the trees. *B*, The gecko climbs by means of digital pads. *C*, The horned toad, *Phrynosoma*, is a ground-dwelling species that often burrows. *D*, The glass snake, *Ophisaurus*, also burrows. *E*, The Gila monster, *Heloderma*, and a related Mexican species are the only poisonous lizards in the world. *F*, The tuatara, *Sphenodon*, is one of the most primitive of living reptiles. (Courtesy of the New York Zoological Society.)

New Zealand (Fig. 15–24). The many types of reptiles that flourished in the Mesozoic are discussed in Chapter 32.

15–11   BIRDS

The members of the class Aves are characterized by the presence of feathers, which are modified reptilian scales; these decrease the loss of water through the body surface, decrease the loss of body heat, and aid in flying by presenting a plane surface to the air. Birds and mammals are the only animals with a *constant* body temperature. They are sometimes called *warm-blooded*, and other animals cold-blooded, but this is inaccurate, for a frog or snake may have a higher body temperature on

Figure 15–25   Some neognathous birds. *A,* Courtship in the Adelie penguin. *B,* An American robin. *C,* A kestrel, a predator of small rodents. *D,* A mockingbird attacking a great horned owl. (From Welty, J. C.: *The Life of Birds,* 2nd Ed. Philadelphia, W. B. Saunders Co., 1975. *A* by Expeditions Polaires Francaises; *B* by G. R. Austing; *C* by E. Hosking; *D* by G. R. Austing.)

a hot day than a bird or mammal. Birds and mammals independently evolved mechanisms to keep body temperature constant despite wide fluctuations in the environmental temperature. The constant body temperature permits metabolic processes to proceed at constant rates and enables these animals to remain active in cold climates.

The earliest known species of bird, *Archaeopteryx,* was about the size of a crow, had rather feeble wings, jawbones armed with reptilianlike teeth (so this early bird could get its worm?) and a long reptilian tail covered with feathers (Fig. 33–14). Birds did not evolve from the flying reptiles, the pterosaurs, but from a group of primitive reptiles called **thecodonts.** Cretaceous rocks have yielded fossils of two other early birds: *Hesperornis,* an aquatic diving bird with powerful hind legs and vestigial wings, and *Ichthyornis,* a flying bird about the size of a tern.

Like reptiles, birds lay eggs and have internal fertilization. Birds have reptilian scales on their legs, and, as mentioned, the earliest birds, known only from fossils, had teeth. Adaptation to flight has involved the evolution of hollow bones and the presence of *air sacs* — extensions of the lungs that occupy the spaces between the internal organs. Not all birds fly; some, such as penguins, have small, flipperlike wings used in swimming (Fig. 15–25). Others, such as the ostrich and cassowary, have vestigial wings but well developed legs. Birds have become adapted to a variety of environments, and different species have very different types of beaks, feet, wings and tails.

People have long been fascinated by the colors, songs and behavior of birds and these have been studied extensively. One of the most fascinating aspects of bird behavior is the annual migration that many birds make. Some birds, such as the golden plover and arctic tern, fly from Alaska to Patagonia and back each year, flying perhaps 25,000 miles en route. Others migrate only a few hundred miles south each winter and some, such as the bobwhite and great horned owl, do not migrate at all. The stimulus for the northward spring migration of certain birds that winter in California has been shown to be the increasing amount of daylight per day. This in some way stimulates the secretion of hypothalamic "releasing factors" which stimulate the pituitary gland to secrete gonadotropic hormones. These in turn initiate the growth of the ovary or testes and increase its production of sex hormones. The increased amount of sex hormones circulating in the blood, testosterone in the male and estradiol in the female, initiates the behavior patterns of migration, mating and nesting.

How birds navigate, especially on long over-water flights, remains an intriguing but incompletely solved mystery. Theories have been advanced suggesting the earth's magnetic fields, the sun and stars, visual landmarks and other features of the environment as cues used in navigation by birds, but no single theory explains all the facts.

The services rendered to humans by birds include the destruction of harmful rodents, insects and weed seeds and the dispersal of the pollen and seeds of many plants. The guano (excrement) deposited by sea birds in certain regions is valuable fertilizer. Birds also provide us with food and with feathers that serve a variety of purposes.

15–12 MAMMALS

The distinguishing features of mammals are the presence of hair, **mammary glands** and sweat glands, and the differentiation of the teeth into incisors, canines and molars. Mammals have a constant body temperature, and the covering of hair serves as insulation to aid in thermoregulation. Mammals evolved from a group of reptiles called **therapsids** (Fig. 32–9), probably during the Triassic period. At an early time in the evolution of mammals one line of descent branched off from the major line and led to the egg-laying animals called **monotremes.** Only two monotremes have survived to the present — the Australian duck-billed platypus and the spiny anteater (Fig. 32–15). The young, after hatching from the egg, are nourished by milk secreted by mammary glands.

The second subclass of mammals, the **marsupials** or pouched mammals, are also found largely in Australia — kangaroos, koalas and wombats. The opossum more closely resembles the primitive ancestral marsupials than the Australian marsupials do and is one of the few found outside of Australia. Marsupials do not lay eggs; the young are born alive in a very immature state and are transferred to a pouch on the mother's abdomen where they feed on milk secreted by the mammary glands and complete their development.

The subclass Eutheria, or placental mammals, includes all the other mammals, all characterized by the formation of a *placenta* for the nourishment of the developing embryo while within the uterus (womb) of the mother. The placenta is formed in part from tissues derived from the embryo and in part from maternal tissues. The embryo receives nutrients and oxygen and eliminates wastes via the placenta. The young are born alive, in a more advanced state of development than the newborn marsupials. Some of the principal orders of placental mammals are the following:

(1) *Insectivora*—moles, hedgehogs and shrews. These are insect-eating animals, considered to be the most primitive placental

Figure 15–26 High-speed stroboscopic photograph of a little brown bat. *Myotis lucifugus,* about to catch a falling meal worm. (Courtesy of Mr. Frederic Webster.)

mammals and the ones closest to the ancestors of all the placentals. The shrew is the smallest living mammal; some weigh less than 5 grams.

(2) *Chiroptera*—bats. These mammals are adapted for flying; a fold of skin extends from the elongated fingers to the body and legs, forming a wing. They eat insects and fruit or suck the blood of other animals. Bats are guided in flight by a sort of biologic sonar (Fig. 15–26): they emit high frequency squeaks and are guided by the echoes from obstructions. Blood-sucking bats may transmit diseases such as yellow fever and paralytic rabies.

(3) *Carnivora*—cats, dogs, wolves, foxes, bears, otters, mink, weasels, skunks, seals, walruses and sea lions. These are all flesh eaters, with sharp, pointed canine teeth and shearing molars.

(4) *Rodentia*—squirrels, beavers, rats, mice, porcupines, hamsters, chinchillas and guinea pigs. These numerous mammals have sharp, chisel-like incisor teeth.

(5) *Edentata*—sloths, anteaters and armadillos. Mammals with few or no teeth.

(6) *Primates*—lemurs, monkeys, apes and humans. These mammals have highly developed brains and eyes, nails instead of claws, opposable great toes or thumbs, and eyes directed forward.

(7) *Artiodactyla*—cattle, sheep, pigs, giraffes and deer. Herbivorous hooved animals with an even number of digits per foot.

(8) *Perissodactyla*—horses, zebras, tapirs and rhinoceroses. Herbivorous hooved animals with an odd number of digits per foot.

(9) *Proboscidea*—elephants, mastodons and woolly mammoths. Animals with a long muscular proboscis or trunk, thick, loose skin, and incisors elongated as tusks. These are the largest land animals, weighing as much as 7 tons.

(10) *Sirenia*—sea cows, dugongs and manatees. These are herbivorous aquatic animals with finlike forelimbs and no hind limbs. They are probably the basis for most tales about mermaids.

(11) *Cetacea*—whales, dolphins and porpoises. These are marine mammals with fish-shaped bodies, finlike forelimbs, no hind limbs, and a thick layer of fat, called blubber, covering the body. The blue whale, the largest animal that has ever existed, attains a length of 30 meters and a weight of 135 metric tons.

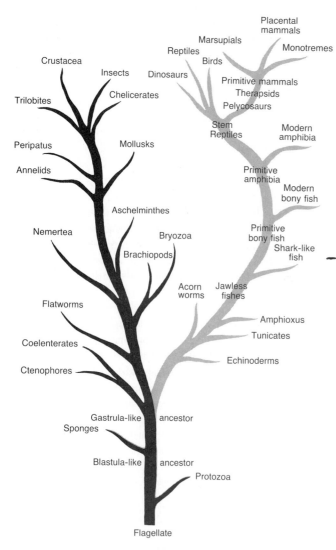

Figure 15–27 A diagram illustrating current theories of evolutionary relationships in the animal kingdom. The branch including the Protostomia is indicated in black and the branch including the Deuterostomia is indicated in gray.

The various members of the animal kingdom cannot be placed on a single scale ranging from "lowest" to "highest," for evolution has occurred in the manner of a branching tree, rather than in a single continuous series (Fig. 15–27). We cannot say, for example, that the starfish is "higher" or "lower" than the oyster; the two forms are simply representatives of the two main trunks of the evolutionary tree, the deuterostomes and the protostomes. Between the two groups are deep-lying differences of structure and development which will be discussed further in Chapter 32.

QUESTIONS

1. What is the evidence for the belief that vertebrates are more closely related to echinoderms than to any other invertebrate phylum?

2. What are the three chief characteristics of phylum Chordata? How are these evident in a tunicate larva? In an adult tunicate? In humans?

3. What characteristics distinguish the vertebrates from the rest of the chordates?

4. Two quite different structures are both termed "atrium." Identify these structures. Look up the derivation of the word and discuss why it is appropriate to call each of these structures an "atrium."

5. How do lampreys and hagfishes differ from other fishes? Of what economic importance are they?

6. What are the functions of the gills of bony fishes?

7. Compare the skins of sharks, frogs, turtles and mammals.

8. Where are the following found and what are their functions: air sac, swim bladder, placenta?

9. Give the phylum, subphylum and class to which humans belong and name three other animals which are included in each division.

10. An animal is multicellular, has a dorsal, hollow

nerve cord, many pairs of gills, a notochord longer than the nerve cord, and no vertebrae. What is it?

11. Why are monotremes considered to be more primitive than marsupials? One group of paleontologists considers them to be therapsid reptiles rather than mammals; give the arguments for and against their theory.

12. Which are more specialized, birds or mammals?

13. Make a diagram of a generalized vertebrate showing the arrangements of the major organs.

14. Make a diagram of a cross section of a frog showing the relationship of the internal organs to the coelom, peritoneum and mesenteries.

15. List the major regions of the frog's brain.

SUPPLEMENTARY READING

For further information on the structure and function of the vertebrates, consult *The Vertebrate Body* by A. S. Romer or *Vertebrate Biology* by R. T. Orr. The former is the standard text of comparative anatomy of the vertebrates, and the latter is a valuable reference book covering many aspects of vertebrate life such as territoriality, dormancy and population dynamics. Two excellent general books on birds are J. C. Welty's *The Life of Birds* and A. A. Allen's *The Book of Bird Life; Bird Navigation* by G. N. Matthews deals in depth with this fascinating phenomenon. There are many books dealing with the natural history of particular chordate groups such as *The Biology of the Hemichordates and Protochordata* by E. J. W. Barrington, *The Life of Fishes* by N. B. Marshall, *The Life of Reptiles* by A. Bellairs, *Herpetology* by K. R. Porter, D. M. Cochrane's *Living Amphibians of the World* and C. G. Hartman's *Possums*. J. Z. Young's two books, *The Life of Vertebrates* and *The Life of Mammals*, are fascinating accounts of the evolution and adaptation of these animals. T. A. Vaughan's *Mammology* is an excellent exposition of the origins, ecology, zoogeography, behavior and physiology of mammals. H. T. Andersen's *The Biology of Marine Mammals* has fascinating accounts of diving, echolocation and other aspects of the life of cetaceans and other marine mammals. Many facts about the various methods used by animals, vertebrate and invertebrate, to carry on their basic life functions are presented in Prosser and Brown's *Comparative Animal Physiology* and in E. Florey's *General and Comparative Animal Physiology* and in *How Animals Work* by K. Schmidt-Nielsen.

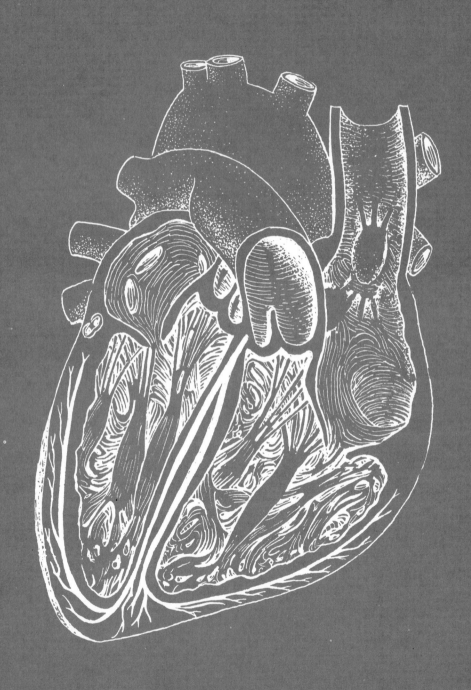

PART 4

THE ORGANIZATION OF THE BODY

INTRODUCTION

In the preceding parts, we have surveyed some of the many different kinds of plants and animals, both simple and complex, that now inhabit the earth. We will next examine in some detail how the body of a single animal — the human — is put together, how it functions, and the nature of the complex control mechanisms that regulate and integrate the activities of its several parts. The section begins with the consideration of the transport system of the body, the blood and blood vessels that supply all cells with nutrients and remove the waste products of metabolism. Every large and active animal requires an efficient transport system to supply each of its cells with all of the materials it needs to function. The respiratory system together with the circulatory system supplies oxygen to each cell of the body and removes an important waste product of metabolism, carbon dioxide. The section continues with an exploration of the role of the digestive system in supplying the other cells of the body with their nutrients and the role of the liver in intermediary metabolism. The kidney and other organs of the excretory system not only remove wastes but regulate the amounts of the various constituents of tissue fluids and intracellular fluids. The skeleton and integument protect and support the body, and the muscles move the various parts of the body one on another.

CHAPTER 16

BLOOD CELLS, PLASMA AND THE IMMUNE RESPONSE

The metabolic processes of all cells require a constant supply of nutrients and oxygen and a constant removal of waste products. This is accomplished simply by diffusion in small plants and animals living in a watery environment, but all the larger animals, including the human have developed some system of internal transport, a circulatory system. The human circulatory system includes the heart and blood vessels, the lymph vessels, and the blood and lymph. Blood fits our definition of a tissue—it is a group of similar cells specialized to perform certain functions. Many of the studies on the relations of cells to their immediate environment have utilized blood cells, for they can be readily obtained with a hypodermic needle and syringe. The blood cells are not injured when carefully collected and can be studied in their normal state.

The volume of blood depends upon body weight; a person weighing 70 kilograms has about 5 liters of blood (just the quantity of oil in the crankcase of most cars). Blood transports nutrients and oxygen to cells and removes wastes from them; it transports hormones, the secretions of the endocrine glands; it has a role in regulating the amount of acids, bases, salts and water in cells; it is important in regulating body temperature, cooling organs such as the liver and muscles where an excess of heat is produced and heating the skin where heat loss is greatest; its white cells are a major defense against bacteria and other disease organisms; and its clotting mechanism helps prevent the loss of this valuable fluid.

Although blood appears to be a homogeneous crimson fluid as it pours from a wound, it is composed of a yellowish liquid, called *plasma,* in which float the *red blood cells, white blood cells* and *blood platelets.* The latter are small cell fragments, important in initiating the clotting process, that are derived from large cells in the bone marrow. The formed elements make up about 45 per cent of whole blood; the remaining 55 per cent is plasma. The loss of water in profuse sweating may reduce the plasma volume to 50 per cent of the blood, and drinking a lot of water or beer may increase it to 60 per cent. The formed elements, with a specific gravity of 1.09, are heavier than plasma, which has a specific gravity of 1.03, and the two may be separated by centrifuging. Blood is constantly mixed as it circulates in the blood vessels so that the plasma and blood cells do not separate.

16–1 PLASMA

Plasma is a complex mixture of proteins, amino acids, carbohydrates, lipids, salts, hormones, enzymes, antibodies and dissolved gases. It is very slightly alkaline, with a pH of 7.4. The two chief constituents are water (90 to 92 per cent) and proteins (7 to 8 per cent). The plasma is in dynamic equilibrium with the *interstitial fluid* bathing the cells and the *intracellular fluid* within the cells. Plasma constantly takes up and gives off substances as it passes through the capillaries, yet its composition remains remarkably constant. Any change in its composition initiates responses on the part of one or more organs of the body to restore the normal equilibrium.

Plasma contains several kinds of proteins, each with specific properties and functions—fibrinogen; alpha, beta and gamma globulins; albumin; and lipoproteins. *Fibrinogen* is one of the proteins involved in the clotting process; *albumin* and *globulins* regulate the water content of cells and body fluids (p. 389). The *gamma globulin* fraction, rich in antibodies, provides immunity to certain infectious diseases such as measles and infectious hepatitis. Purified human gamma globulin is used in the treatment of these diseases. The presence of these proteins makes blood about six times as viscous as water. The plasma protein molecules, too large to pass readily through the walls of the blood vessels, exert an osmotic pressure and play an important role in regulating the distribution of water between the plasma and tissue fluids. The plasma proteins and the hemoglobin in the red cells are important acid-base buffers that keep the pH of the blood and body cells within a narrow range.

In many acute infections the synthesis of fibrinogen and of globulin is increased, and the increased amount of these two in the plasma speeds up the rate of sedimentation of red cells. Measuring the sedimentation rate of red cells is a commonly used test to follow the course of a disease and the patient's recovery from it.

The normal functioning of nerves, muscles and other tissues requires the proper balance in the concentrations of sodium, potassium, magnesium and calcium ions. Plasma contains these, together with chloride, bicarbonate and phosphate ions, in a total concentration of about 0.9 per cent in mammals and slightly less in lower vertebrates. The transport of these ions and the regulation of the amount of each present in the tissues is an important function of the blood.

The concentration of glucose in plasma ranges from 0.08 to 0.14 per cent, and averages about 0.1 per cent. It is carried in the blood from the intestines, where it is absorbed into the body, to the liver, where it is stored as glycogen, and eventually to all the cells of the body, where it is metabolized to release energy. Brain cells are especially dependent upon a constant supply of glucose for fuel. If the concentration in the blood falls below 0.04 per cent, the irritability of certain brain cells is greatly increased and muscular twitching and convulsions occur. If the blood glucose concentration remains low, the brain cells become unable to function and coma and death ensue. This may occur if a diabetic patient takes an excessive dose of insulin, which reduces the concentration of glucose in the blood below the critical value. Diabetics are taught to recognize the signs and symptoms of too little glucose in the blood (hypoglycemia) and to take some candy or sugar to prevent disaster.

16–2 ERYTHROCYTES

Floating in the plasma are red blood cells, *erythrocytes,* which are biconcave discs 7 to 8

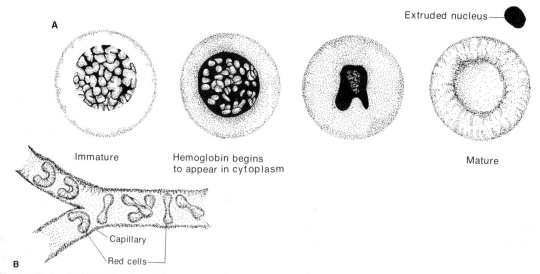

Extruded nucleus

Immature

Hemoglobin begins
to appear in cytoplasm

Mature

Capillary

Red cells

Figure 16-1   Red blood cells. *A,* Stages in the formation of human red cells. *B,* Bending and twisting of red cells as they pass through capillaries. Their supporting elastic framework enables them to resume their normal disc shape after tension is released.

micrometers in diameter and 1 to 2 micrometers thick (Fig. 16-1). The mature mammalian erythrocyte has no nucleus; it is lost in the course of development from the precursor cell, the erythroblast. An internal elastic framework maintains the disc shape and permits the cell to bend and twist as it passes through blood vessels smaller than its diameter. Erythrocytes are moved about by the pumping action of the heart. There are, on the average, about 5,400,000 red cells per cubic millimeter of blood in the adult male, and about 5,000,000 per cubic millimeter in the adult female. Newborn infants have 6 to 7 million per cubic millimeter; this number decreases after birth and the adult number is reached at about three months. Each red cell contains some 265,000,000 molecules of hemoglobin, the red pigment that gives the red cell its color and is responsible for the transport of oxygen.

16-3   TRANSPORT IN THE BLOOD: HEMOGLOBIN

Materials are transported in the blood in one of three states—in simple solution in the plasma, bound to one of the plasma proteins, or carried in the red cell in combination with hemoglobin. A small amount of the oxygen and carbon dioxide is simply dissolved in the plasma water, but most of the oxygen and much of the carbon dioxide are bound to hemoglobin in the red cell. The anions chloride, bicarbonate and phosphate and the cations sodium, potassium, magnesium and about half of the calcium are carried in plasma in solution. The remaining half of the calcium is bound to a plasma protein. Glucose, amino acids and organic acids such as lactic and citric are present in solution in the plasma, as are a variety of metabolic waste products such as urea, uric acid, ammonia and creatinine and very small amounts of certain vitamins.

Many other compounds in the plasma are bound to proteins. Neither free fatty acids nor triacylglycerols are soluble enough in water to dissolve in the plasma. Free fatty acids are carried in the blood largely in combination with serum albumin, and triacylglycerols are carried from the intestine to the liver and adipose tissue as *chylomicrons,* droplets of fat that are stabilized by a thin coat of proteins, phospholipids and cholesterol esters. The liver synthesizes and secretes several kinds of *lipoproteins,* which are about half peptide chain and half lipids—phospholipids, cholesterol and its esters, and triacylglycerols. One type of lipoprotein is especially rich in triacylglycerols and is the form in which these fats are transported from the liver to skeletal muscle, adipose tissue and other tissues.

Some of the hormones, those of the anterior and posterior lobes of the pituitary, for example, are carried in the blood dissolved in the plasma. Others are bound to plasma proteins, either to albumin or to some unique protein that specifically binds that hormone. For instance, thyroxine is bound either to a specific *thyroxine-binding globulin* or to albumin, from which it is slowly released over a period of days. This helps ensure a steady, slow supply of thyroxine to the tissues. Most of the insulin present in plasma is bound to globulins. Some of the steroid hormones, such as cortisol, are carried in plasma bound to specific proteins, e.g., *transcortin.*

The transport of oxygen and carbon dioxide in the blood depends largely on the hemoglobin present in the red cell. Human blood contains about 15 grams of hemoglobin per 100 milliliters. If blood were simply water, it could carry only about 0.2 milliliters of oxygen and 0.3 milliliters of carbon dioxide in each 100 milliliters. The properties of hemoglobin enable whole blood to carry some 20 milliliters of oxygen and 30 to 60 milliliters of carbon dioxide per 100 milliliters and serve simultaneously as an acid-base buffer to minimize changes in the *p*H of the blood. The protein portion of hemoglobin is composed of four peptide chains, typically two α chains and two β chains, to which are attached four heme (porphyrin) rings. An iron atom is bound in the center of each heme ring (Fig. 16–2).

Hemoglobin has the remarkable property of forming a loose chemical union with oxygen; the oxygen atoms are attached to the iron atoms in the hemoglobin molecule. In the respiratory organ, the lung or gill, oxygen diffuses into the red cells from the plasma and combines with hemoglobin (Hb) to form *oxyhemoglobin* (HbO$_2$): $Hb + O_2 \rightleftarrows HbO_2$.

The reaction is reversible and hemoglobin releases the oxygen when it reaches a region where the oxygen tension is low, in the capillaries of the tissues. The combination of oxygen with hemoglobin and its release from oxyhemoglobin are controlled by the concentration of oxygen and, to a lesser extent, by the concentration of carbon dioxide. Carbon dioxide reacts with water to form carbonic acid, H_2CO_3; hence an increase in the concentration of carbon dioxide results in an increased acidity of the blood. The oxygen-carrying capacity of hemoglobin decreases as blood becomes more acid. Thus the combination of hemoglobin with oxygen is controlled indirectly by the amount of carbon dioxide present. This results in an extremely efficient system: In the capil-

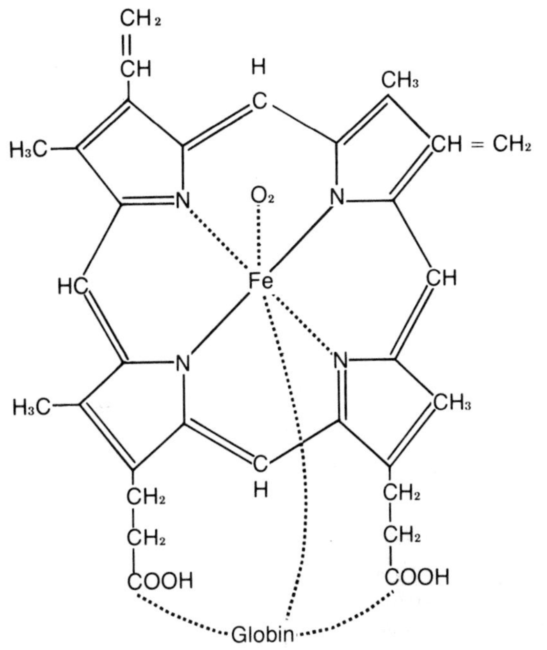

Figure 16–2 One of the subunits of the hemoglobin molecule, a combination of a porphyrin ring, ferrous iron (Fe^{++}), and a peptide chain, globin. Four subunits, two α chains and two β chains, compose the hemoglobin molecule.

laries of the tissues, the concentration of carbon dioxide is high and a large amount of oxygen is released from hemoglobin by the combined action of the low oxygen tension and the high carbon dioxide tension. In the capillaries of the lung or gill, carbon dioxide tension is lower and a large amount of oxygen is taken up by hemoglobin owing to the combined effect of the high oxygen tension and low carbon dioxide tension.

At the oxygen tension of arterial blood, 100 millimeters Hg, each 100 milliliters of blood contains about 19 milliliters of oxygen. At the oxygen tension of venous blood, 40 millimeters Hg, each 100 milliliters contains 12 milliliters of oxygen. The difference, 7 milliliters, represents the amount of oxygen delivered to the tissues by each 100 milliliters of blood. Since some 5 liters of blood are delivered to the tissues each minute, this means that 350 milliliters of oxygen can be supplied per minute. At rest, the cells of the body need about 250 milliliters of oxygen per minute and with exercise this requirement may increase 10- or 15-fold. Hence, during exercise, muscles must obtain a major fraction of their total energy requirement from anaerobic metabolism.

The cells of the body produce, at rest,

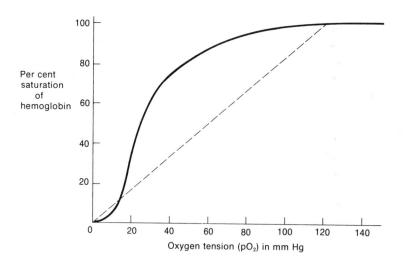

Figure 16–3 The combination of oxygen with hemoglobin as a function of oxygen tension. The dashed line shows the relation that would exist if hemoglobin bound oxygen as a linear function of oxygen tension. Because of the interactions between the four heme groups in a single hemoglobin molecule, the binding of oxygen to hemoglobin follows the **S**-shaped curve shown in black.

about 200 milliliters of carbon dioxide per minute. If this were simply dissolved in plasma (which can transport in solution only 4.3 milliliters of carbon dioxide per liter), blood would have to circulate at a rate of 47 liters per minute instead of 4 or 5. This amount of carbon dioxide dissolved in water would yield a pH of 4.5. The unique properties of hemoglobin enable each liter of blood to transport some 50 milliliters of carbon dioxide from tissue to lung with only a few hundredths of a unit difference in the pH of arterial and venous blood. Some carbon dioxide is carried in a loose chemical union with hemoglobin as ***carbaminohemoglobin***, and a small amount is present as carbonic acid, but most of it is transported as bicarbonate ion, HCO_3^-. The CO_2 produced by cells dissolves in the tissue fluid to form H_2CO_3, a reaction catalyzed by ***carbonic anhydrase***, and the carbonic acid is neutralized to bicarbonate by the sodium and potassium ions released when oxyhemoglobin is converted to hemoglobin. Oxyhemoglobin is a stronger acid than reduced hemoglobin, hence some cations are released when HbO_2 is converted to Hb. In the course of evolution this one molecule has been endowed with all the properties needed for the transport of large amounts of oxygen and carbon dioxide, with a minimal change in the pH of the blood while these gases are being transported.

The properties of the heme pigments are such that the amount of oxygen taken up by the pigment is not directly proportional to the oxygen tension. A graph of the relationship gives an **S**-shaped curve (Fig. 16–3). Blood is a more effective transporter of oxygen than it would be if the oxygen content were a simple linear function of oxygen tension. The effect of carbon dioxide (really the change in pH

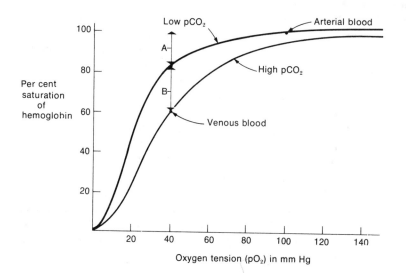

Figure 16–4 The effect of carbon dioxide tension ((pCO_2) on the delivery of oxygen to the tissues. The arrow *A* indicates the amount of oxygen released from hemoglobin as the pO_2 falls from that of arterial blood (100 mm. Hg) to that of venous blood (40 mm. Hg). The arrow *B* indicates the additional amount of oxygen released because of the greater pCO_2 in venous blood.

brought about by the change in the carbon dioxide content) on the combination of oxygen with hemoglobin is shown in Figure 16–4. The oxygen dissociation curves for arterial blood, with low carbon dioxide tension, and for venous blood, with high carbon dioxide tension, illustrate how much more oxygen is delivered to the tissue by a given amount of blood as carbon dioxide is taken up in the tissue capillaries. The properties of the heme proteins in different species of animals are quite different and are generally adapted to the amount of carbon dioxide present. This is generally low in water-breathing animals and high in air-breathing animals. This points up the generalization that the evolution of air-breathing animals from water-breathing ones involved marked changes not only in the structure of the respiratory organs but also in the chemical properties of the heme proteins serving as blood pigments.

The red cells of the fetus contain a slightly different kind of hemoglobin, called *fetal hemoglobin.* This gradually disappears after birth and by twenty weeks has been replaced by the adult type. Cells containing fetal hemoglobin can take up and give off oxygen at lower oxygen tensions than adult cells. This may be important to the fetus, for as it develops in the uterus, it has less oxygen available than the adult has.

Carbon monoxide, present in commercial illuminating gas and in the exhaust gas of automobiles, forms a compound with hemoglobin; in fact, it has a much *greater* affinity for hemoglobin than oxygen does. Carbon monoxide and oxygen both are attached to hemoglobin by the iron atoms; thus, when hemoglobin has combined with carbon monoxide it is unable to combine with oxygen. When the air breathed in contains only 0.5 per cent carbon monoxide, more than half of the hemoglobin in the blood has combined with this gas and only half is left to transport oxygen. This has the same effect as the sudden loss of half of one's red cells. The union of hemoglobin with carbon monoxide, like its union with oxygen, is reversible, but the decomposition of carbon monoxide hemoglobin is a slow process, and several hours in carbon monoxide-free air are required to free the blood of it.

In certain diseases, such as blackwater fever, a complication of malaria, red blood cells are destroyed and the hemoglobin is released into the plasma. It is then excreted in the urine, giving it a dark color. Other diseases decrease the amount of hemoglobin in the blood, a condition known as anemia.

All mammals have red cells similar to ours—non-nucleated, biconcave discs containing hemoglobin. Birds, reptiles, amphibians and fishes have nucleated oval red cells containing hemoglobin. The red cells of the lower vertebrates are typically larger than those of mammals; frog's erythrocytes are approximately 35 micrometers on the long axis.

Hemoglobin is found in all the major groups of animals above the flatworms. In some, the hemoglobin is simply dissolved in the plasma, while in others it is present in blood cells. Mollusks, crustacea and certain other animals have a blue-green blood pigment, *hemocyanin,* which contains copper instead of iron.

The respiratory enzymes of all cells, both plant and animal, the *cytochromes,* which catalyze the transfer of electrons from substrate to oxygen and the concomitant transfer of energy to ATP (Fig. 5–13), are heme proteins closely related to hemoglobin in their molecular structure.

16–4 REGULATION OF RED CELL NUMBER

Red cells are constantly being destroyed and new ones made, yet the total number remains remarkably constant. Red cells originate in the *red bone marrow,* located in the central hollow spaces of certain bones. The red marrow consists of a network of connective tissue cells and of thousands of small blood vessels, from the lining of which develop the red cells. The unspecialized precursor cells, hemocytoblasts, have a nucleus and no hemoglobin (Fig. 16–1), and divide to form more precursor cells. After its last division each precursor cell is gradually transformed into a mature red cell by the loss of the nucleus, the manufacture of hemoglobin and the assumption of the biconcave disc shape.

The average life span of the human red cell is about 127 days, demonstrated by labeling the cells with radioactive iron or some other isotope. The *spleen,* an oval organ about

12 centimeters long, lying to the left of the stomach, is connected only to the blood stream and serves as a reservoir for red cells. In the walls of the blood vessels in the spleen and liver are cells with the ability to engulf, or *phagocytize,* red cells and thus destroy them. The hemoglobin molecules recovered from old red cells are dismantled in the spleen and liver. The iron atoms are recovered and returned to the red bone marrow to be used in the synthesis of new molecules. The heme portion of the molecule undergoes chemical degradation and is excreted by the liver in the bile as *bile pigments.* These substances undergo further reactions by the bacteria in the intestine and pass from the body in the feces. The bile pigments are primarily responsible for the color of the feces; if the bile duct is blocked, by a gallstone, for example, the bile pigments cannot pass into the intestine and the feces are a grayish clay color.

From the total number of red cells in the body and their average life span, it is simple to calculate that about 2.5 million red cells are made, and an equal number destroyed, each second throughout the day and night every day. Since there are 265×10^6 molecules of hemoglobin per cell, some 650×10^{12} molecules of hemoglobin are made each second. Experiments by Dr. Howard Dintzis indicate that about 90 seconds are required for the assembly of the 574 amino acids that compose each hemoglobin molecule. Thus at any given moment $90 \times 650 \times 10^{12}$ or 6×10^{16} molecules of hemoglobin are in the process of being assembled!

The constancy of the number of red cells provides us with an excellent example of a *dynamic equilibrium.* Under normal circumstances the rate of formation of new cells just equals the rate of destruction of old ones and the total number of circulating red cells remains constant. The rate of production of red cells is increased by any factor which decreases the amount of oxygen delivered to the tissues. The loss of red cells by hemorrhage decreases the capacity of the blood to transport oxygen and leads to increased red cell production (Fig. 16–5). The stimulus is not simply the decreased number of red cells, for if a person with a normal number of red cells moves to a very high altitude for a few weeks, the number of red cells will increase to 6 or 7 million per

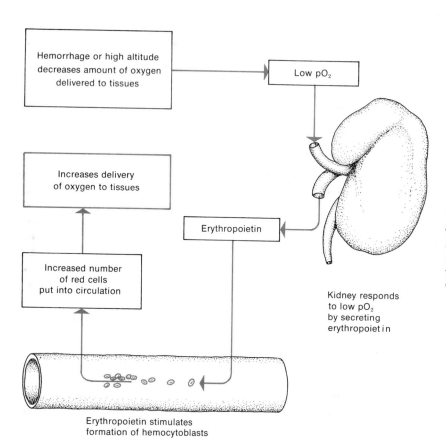

Hemorrhage or high altitude decreases amount of oxygen delivered to tissues

Low pO₂

Increases delivery of oxygen to tissues

Erythropoietin

Increased number of red cells put into circulation

Kidney responds to low pO₂ by secreting erythropoietin

Erythropoietin stimulates formation of hemocytoblasts

Figure 16–5 The mechanism by which erythropoietin is released from the kidney in response to low *p*O₂ and passes to the bone marrow, where it stimulates the formation of hemocytoblasts, thereby increasing the number of red cells in the circulation and the delivery of oxygen to the tissues.

cubic millimeter. At high altitudes there is less oxygen in the air and consequently less oxygen is delivered to the tissues. The number of red cells remains elevated as long as the individual is exposed to air with a low oxygen content. The same increase can be produced experimentally at sea level by keeping animals in a chamber the atmosphere of which has a low oxygen content but a total pressure equal to that of air at sea level.

Lack of oxygen in the red bone marrow itself does not stimulate red cell production. Instead, in response to lowered oxygen tension, the kidneys and perhaps the liver and other tissues secrete *erythropoietin*, a glycoprotein with a molecular weight of 60,000. This passes in the blood to the bone marrow and stimulates the initial stage in red cell production, the formation of hemocytoblasts from primitive stem cells. When the number of red cells has returned to normal, the oxygen tension in the tissues is returned to normal and the stimulus for the production of erythropoietin is removed. The lower amount of erythropoietin results in a decreased rate of production of red cells; this prevents an excess of red cells. In sum, the number of circulating red cells is regulated automatically by the production of erythropoietin, and this in turn is regulated by the requirements of the tissues for oxygen.

The synthesis of hemoglobin and the production of red cells are not necessarily correlated. A deficiency of iron, for example, decreases hemoglobin synthesis, but red cells are produced at the normal rate or even at an elevated rate in response to the stimulus of decreased delivery of oxygen to the tissues. The cells produced have less hemoglobin than normal (they are said to be hypochromic) and, of course, are less effective than normal red cells in transporting oxygen.

16–5 LEUKOCYTES

Human blood contains five kinds of white cells, *leukocytes*, all having nuclei but no hemoglobin. All can move actively by amoeboid movement. Leukocytes can move against the current of the blood stream and even slip through the walls of blood vessels and enter the tissues.

White cells are much less numerous than red ones; there are about 7000 white cells per cubic millimeter on the average. The number fluctuates from 500 to 9000 or 10,000 in different persons and even in the same person at different times of day. The white count is lowest early in the morning and highest in the afternoon. Poorly nourished individuals have

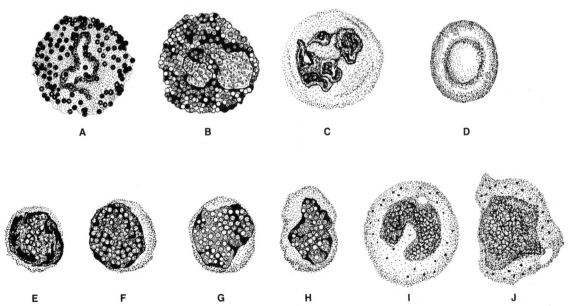

Figure 16–6 Types of white blood cells. *A*, Basophil; *B*, eosinophil; *C*, neutrophil; *E–H*, a variety of lymphocytes; *I* and *J*, monocytes. *D* is a red blood cell drawn to the same scale. (After Bloom, W., and Fawcett, D. W.: *A Textbook of Histology*, 10th Ed. Philadelphia, W. B. Saunders Co., 1975.)

fewer white cells than normal and lower resistance to infection and disease; a drop to 500 or fewer white cells per cubic millimeter is fatal.

Two of the types of white cells, *lymphocytes* and *monocytes* (Fig. 16–6), are produced in lymphoid tissue such as the spleen, thymus and lymph nodes. The other three, *neutrophils, eosinophils* and *basophils*, are produced in the bone marrow along with the red cells. All three contain cytoplasmic granules differing in size and staining properties. Neutrophils normally make up 60 to 70 per cent of the population of circulating white cells; they can move actively by amoeboid motion and slip through the wall of the capillary between adjacent cells, and they are important in taking up bacteria and other infective agents by phagocytosis. The neutrophils also phagocytize the remains of dead tissue cells. Neutrophils and other white cells are guided to points of infection by chemicals released by the inflamed and infected tissues.

Monocytes are the largest white cells, reaching 20 micrometers in diameter. Like the smaller lymphocytes, they have no cytoplasmic granules. Monocytes also move actively by amoeboid motion and phagocytize bacteria. After several hours in the tissue space, monocytes tend to enlarge and become *macrophages*, which can move quite rapidly and engulf 100 or more bacteria. Neutrophils are of prime importance in resisting acute bacterial infections; monocytes become of greater importance in countering long-term infections. Macrophages can ingest large particles of cellular debris and are important in cleaning up an infected area after the bacteria have been eliminated (Fig. 16–7). Eosinophils, which have large granules that stain with eosin and other acidic dyes, are amoeboid and phagocytic. They increase greatly in number during allergic reactions and during infections with parasites such as trichina worms.

Lymphocytes are remarkably interesting cells which have the potential of being converted into many other types of cells in the body. The lymphocyte can swell and become a monocyte, then enter connective tissues and other spaces and become a macrophage. Lymphocytes can also enter the bone marrow and develop into either the precursors of red cells or the precursors of the granulocytic white cells such as neutrophils. Lymphocytes in tissue can develop into *fibroblasts* and secrete collagen fibers, elastic fibers and other elements of connective tissue. Stem lymphocytes from the bone marrow may also develop into T, or thymus, lymphocytes and B, or bursal, lymphocytes, which are involved in the development of cellular immunity and humoral immunity respectively (p. 363). T lymphocytes develop into *lymphoblasts,* and B lymphocytes into *plasma cells,* both of which produce and secrete antibodies, of prime importance in the immune process.

The proportion of white cells of each type is determined by a *differential white cell count.* A drop of blood is smeared thinly and evenly on a glass slide, stained with Wright's stain, and examined under the microscope. Several hundred white cells are counted and classified. The average values for a normal person are: 60 to 70 per cent neutrophils, 25 to 30 per cent lymphocytes, 5 to 10 per cent monocytes, 1 to 4 per cent eosinophils and 0.5 per cent basophils.

One final group of formed elements circulating in the blood are blood platelets, or *thrombocytes,* important in initiating the clotting of the blood. In most vertebrates other than mammals, the blood contains thrombocytes, small, oval, pointed cells with a nucleus. Mammalian blood contains tiny, spherical or disc-shaped *blood platelets,* lacking a nucleus. These are formed by the pinching off of protrusions from the surface of giant cells, *megakaryocytes,* in the red bone marrow. Some platelets are formed from phagocytic cells in the lungs.

16–6 PROTECTIVE FUNCTIONS OF WHITE CELLS

The white cells protect the body against disease organisms. Neutrophils and monocytes destroy invading bacteria by ingesting them (much as an amoeba ingests a particle of food). The phagocytized bacteria are digested by enzymes secreted by the white cell. White cells continue to ingest particles until they are killed by the accumulated breakdown products. Neutrophils have been observed to phagocytize 5 to 25 bacteria, and monocytes, as many as 100 bacteria before dying.

When bacteria enter the tissues of the

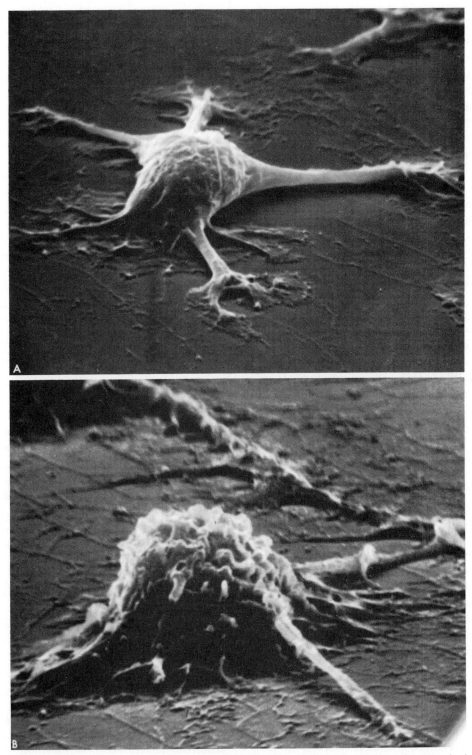

Figure 16–7 Scanning electron micrographs of macrophages recovered from the peritoneal cavity of mice. The upper photograph *(A)* is of an unstimulated macrophage. The macrophage in the lower picture *(B)* was taken from a mouse that had received an injection of mineral oil in the peritoneal cavity a few days earlier. The mineral oil acts as an irritant, producing "angry" macrophages with greatly increased metabolic rates, folding of the plasma membrane and ability to ingest bacteria. The irregular surface and many lateral microprojections of the "angry" macrophage are clearly evident. The macrophages were grown on coverslips in tissue culture medium for a brief period before photography. Magnifications: *A,* ×14,400: *B,* ×16,200. (From Albrecht et al.: Exp. Cell Res., *70:*230–232, 1971.)

body they either attack cells directly or produce toxins. The blood vessels of the affected region dilate and bring in an increased supply of blood, causing the characteristic reddening and increased temperature known as *inflammation.* The walls of the blood vessels become more permeable, fluid enters the tissue from the blood stream, and swelling results. The white cells, and the neutrophils in particular, migrate through the walls of the blood vessels and phagocytize the invaders and the remains of any destroyed tissues. The aggregation of dead tissue cells, bacteria, and living and dead white cells forms a thick yellowish fluid called *pus.* The white cells are guided to points of infection by chemicals released by the invading organisms and by the inflamed tissues.

After the bacteria have been destroyed, the lost tissue is replaced. Some tissues have the ability to regenerate by the multiplication of neighboring cells; others have a very limited ability to regenerate and are replaced by connective tissue cells which secrete fibers to form *scar tissue.* The lymphocytes are believed to be active in this process, for they tend to accumulate in areas where healing is occurring. Lymphocytes grown outside the body in sterile media can become connective tissue cells; presumably this same process can occur within the body to facilitate the processes of repair.

The number of circulating white cells is increased in most infections; the white count may rise to 20,000 or more per cubic millimeter in appendicitis or pneumonia. Inflamed tissues are believed to liberate a substance (leukocyte promoting factor) which passes via the blood to the bone marrow, where it stimulates the production and release of white cells, especially neutrophils. The number of white cells in the blood is a reflection of the severity of the infection, and successive white cell counts are useful in gauging a patient's recovery. Certain diseases are characterized by an increase in a particular kind of white cell, and a differential white cell count is helpful in diagnosis. The number of lymphocytes is increased in whooping cough and pernicious anemia, by living at high altitudes or in the tropics, by sun tan and by chronic diseases such as tuberculosis. Typhoid fever and malaria usually effect an increase in the number of monocytes, and pneumonia, appendicitis and other acute bacterial infections typically increase the number of neutrophils. A marked increase in eosinophils occurs with infections of trichina worms, tapeworm, hookworm and other animal parasites, and in scarlet fever, asthma, allergic conditions and some skin diseases.

16–7 THE CLOTTING OF BLOOD

Animals have evolved elaborate mechanisms to prevent the accidental loss of blood. In some animals blood loss is prevented by the powerful contraction of muscles in the wall of the blood vessel when the vessel is severed. In man and other vertebrates, and in many invertebrates as well, blood loss is prevented by a series of chemical reactions in which a solid *clot* is formed to plug the broken vessel. Clotting is essentially a function of the plasma, not of the blood cells, and involves the enzymatic conversion of *fibrinogen,* a soluble plasma protein, to *fibrin,* an insoluble protein.

Blood drawn from a blood vessel into a test tube changes from a liquid to a semisolid gel in about six minutes (the clotting time ranges from four to ten minutes). Red cells and white cells trapped by the fibrin fibers contribute to the solidity of the clot but are not essential for the clotting process. The clot later shrinks and squeezes out a straw-colored fluid, *serum,* which is similar to plasma in most respects but is unable to clot because it lacks fibrinogen.

Blood does not clot, as many people think, because it is exposed to air or because it stops flowing; if it is carefully drawn into a paraffin-lined vessel it will not clot even though it is stationary and exposed to air. The clotting mechanism is actually quite complex, involving many different substances in the plasma which interact in three sequential reactions. Each of the first two reactions produces an enzyme required for the succeeding reaction (Fig. 16–8).

The first step, the production of *thromboplastin,* is initiated when a blood vessel is cut. Traumatized tissues release a lipoprotein called thromboplastin which interacts with calcium ions and several protein factors in the blood plasma (proaccelerin, proconvertin) to produce *prothrombinase,* the enzyme that catalyzes the second step. Prothrombinase can also be synthesized by the interaction of factors

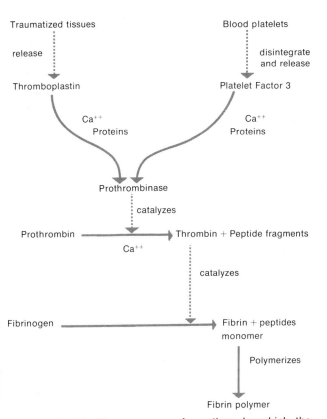

Figure 16–8 The sequence of reactions by which the clotting of blood occurs. Each reaction is catalyzed by an enzyme that is produced by the previous reaction. The process is initiated by the release of thromboplastin from damaged tissue or by the release of a specific factor when platelets disintegrate.

released from platelets, calcium ions and other plasma globulins. One of these, termed *antihemophilic factor,* is present in normal plasma but is absent from the plasma of individuals with *hemophilia,* or "bleeder's disease." The prothrombinase (made either by the system involving thromboplastin released from traumatized tissue or by a comparable factor released by platelets) catalyzes a reaction in which *prothrombin,* a plasma globulin made in the liver, is split into several fragments, one of which is *thrombin.* This reaction also requires calcium ions. Finally, the thrombin acts as a proteolytic enzyme to cleave two peptides from fibrinogen and form an active fibrin monomer, which polymerizes to form long threads of insoluble fibrin. The network of fibrin threads traps red cells, white cells and platelets to form a clot. The conversion of inactive precursors to enzymatically active clotting agents is similar to the activation of digestive enzymes by the removal of peptides (p. 425). Indeed, the amino acid sequence of

thrombin is homologous with sequences of the pancreatic proteinases; perhaps prothrombin originated by duplication of the genes that regulate the synthesis of the zymogens.

This mechanism, involving a cascading sequence of enzymatic reactions, is admirably adapted to provide for rapid clotting when a blood vessel is injured and yet prevent clotting in an intact blood vessel. Although normal blood may contain a small amount of thromboplastin, it also contains a strong anticoagulant, *heparin,* produced in mast cells in the lungs and liver. Heparin inhibits the conversion of prothrombin to thrombin. The synthesis of prothrombin in the liver requires an adequate supply of vitamin K and anything that interferes with the dietary supply of vitamin K or its absorption from the intestine may lead to deficient clotting.

The blood that flows during menstruation does not coagulate, either because its fibrinogen has been removed in the uterus or because it has already clotted and the fibrin has subsequently been destroyed by proteolytic enzymes.

Some bacteria attack the walls of blood vessels, and the clotting chemicals aid in the repair of such weak spots. The destruction of the cells and the accumulation and disintegration of the platelets on the roughened wall liberate thromboplastin and cause the formation of a clot across the weakened area. The clot alone is not very effective but connective tissue cells from the arterial wall migrate in and secrete a tough, fibrous connective tissue to form a scar and reinforce the clot. This is not without danger, for the clot may completely occlude the blood vessel and prevent the passage of blood.

Whether caused by bacterial infection or other factors, the pathological clotting of blood within blood vessels, especially in the arteries supplying the heart, brain and lungs, is one of the major causes of illness and death. Many organs are supplied by several blood vessels and an intravascular clot, called a *thrombus,* would probably cause little harm to these organs. However, other organs that are served by a single artery would have no blood supply at all if that artery were occluded by a thrombus. A thrombus formed in one vessel may break loose and be carried by the blood to some other vessel, obstructing it. A thrombus or any other particle carried by the blood stream which blocks a blood vessel is called an *embolus.*

16–8 DISEASES OF THE BLOOD

Anemia. Anemia is not a single, specific disease but rather a condition which may have many different causes. Anemia is characterized by a decreased number of red cells in the blood, by a decreased amount of hemoglobin per red cell, or both. The number of red cells may drop to 4, 3, or even 1 million per cubic millimeter of blood. Anemia may result from severe loss of blood by hemorrhage or from the destruction of red cells. In several inherited conditions (e.g., sickle cell anemia, p. 743) the red cells are unusually fragile and are rapidly destroyed. Red cells are destroyed by rattlesnake venom, malaria, burns and certain chemicals. The bone marrow is stimulated to increase red cell production and releases immature, nucleated red cells into the bloodstream. Anemias harm the patient by decreasing the transport of oxygen, thereby impairing the metabolism of all the tissues. In addition, the decreased number of red cells leads to a decreased viscosity of the blood, and this, indirectly, causes the heart to beat faster. The increased work load on the heart is one of the major ill effects of anemia.

Anemia may also result from injury to the bone marow, liver or spleen. The marrow tissue may be destroyed by tumors or by substances such as benzol or lead; industrial workers continually exposed to these substances may develop anemia. A fourth cause of anemia is a deficiency of some substance essential to the manufacture of red cells—iron, *cobalamin* (vitamin B_{12}) or folic acid. Pregnant women frequently become anemic because of the demands of the fetus upon the mother's supply of iron and vitamins for red cell manufacture. The fetus must store up enough iron to carry it through the first year or so of life. Milk is quite deficient in iron, and a prolonged diet of milk will produce anemia. Most of the iron of hemoglobin is salvaged and reused; the daily requirement for iron, about 0.01 gram, is small and easily supplied by the average diet.

A severe and formerly fatal disease is *pernicious anemia,* characterized by red cells that are immature, very fragile and decreased in number. The bone marrow of patients with this disorder does not receive enough cobalamin, vitamin B_{12}, to make normal red cells. Cobalamin is present in the diet in adequate amounts but the lining of the stomach of these patients does not secrete the substance called intrinsic factor necessary for the absorption of cobalamin. When dogs were made anemic by repeated bleeding, and then fed various diets, Dr. Whipple of the University of Rochester noticed that liver was most effective in promoting recovery. Physicians at Harvard then tried feeding liver to patients with pernicious anemia and found that the red cell count began to rise after a few days and reached the normal level in a few weeks. It remained normal as long as the patients ate large amounts of liver. Potent liver extracts were subsequently prepared, and in 1948 the active substance, cobalamin, an organic compound containing the element cobalt, was isolated and identified.

Polycythemia. An increase in the number of circulating red cells—the number may reach 11 to 15 million per cubic millimeter—is called *polycythemia.* Diarrhea, by decreasing the fluids of the body and hence the total blood volume, leads to a temporary increase in the number of red cells *per cubic millimeter,* for the total number of red cells remains the same. True polycythemia results from an overproduction of red cells; the blood becomes very viscous and tends to plug the blood vessels.

Leukemia. The leukemias are diseases of the cells which produce the white cells. They multiply too rapidly, and release large numbers of white cells, many of which are immature, into the blood stream. There are several types of leukemia, each characterized by an increase in one particular kind of white cell. Leukemia is a type of cancer, characterized by the abnormally rapid growth of one kind of cell. It is treated, as other cancers are, by x-rays or by the radiation from a radioactive element such as radiophosphorus, or by antivitamins such as *aminopterin,* which is an antagonist of folic acid. The leukemic cells, by filling up the bone marrow and displacing the normal cells which produce red cells, frequently cause anemia.

16–9 BLOOD TYPES AND TRANSFUSIONS

Attempts to replace human blood lost by hemorrhage date back to 1667, when the

Addition of Blood Type: Type A Serum   Type B Serum

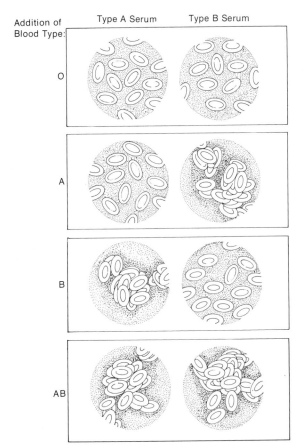

Figure 16–9 Agglutination tests for the several types of human blood groups. On a glass slide is placed a drop of type A serum (containing b agglutinin) and a separate drop of type B serum (containing a agglutinin). When a drop of type O blood is added to each drop (O blood has neither A nor B agglutinogen) there is no agglutination (clumping) of the red cells in either drop. If a drop of type B blood is added, agglutination occurs with type A serum; type A red cells are agglutinated by type B serum, and type AB red cells are agglutinated by both sera.

transfer of animal blood into human veins was tried. Such transfusions were uniformly unsuccessful and evoked severe reactions and often death. The transfusion of blood from one person to another was sometimes successful but occasionally led to the *agglutination* (clumping) of the patient's red cells. Agglutination, caused by an antigen-antibody reaction, must

not be confused with coagulation, the clotting of blood caused by the reaction of thrombin and fibrinogen.

The mystery of why transfusions were sometimes unsuccessful was solved in 1900 by Landsteiner, who found that the blood of different persons may differ chemically and that agglutination occurs when the bloods of donor and recipient are incompatible (Fig. 16–9). The four chief blood groups, designated *O, A, B* and *AB,* are distinguished by the presence of *agglutinogens* (a type of antigen) A or B in the red cells and by *agglutinins* (a type of immunoglobulin, p. 363) a or b in the plasma. The inheritance of the blood groups and the chemical nature of the agglutinogens is discussed later (p. 669). The characteristics of the blood groups and the types of transfusions possible are summarized in Table 16–1. Normally, of course, no blood agglutinates itself, because the corresponding agglutinogen and agglutinin (A and a, for example) are not present together.

In making a transfusion the physician tries to find a donor of the same group as the patient and tests the two bloods by mixing some serum from the patient with some red cells from the prospective donor to make sure that they are compatible. In addition to the four main groups there are two subgroups to be considered. In an emergency, if no donor of the patient's type is available, blood of another type may be used, provided that the plasma of the recipient does not agglutinate the red cells of the donor. For example, a type B individual may receive blood from an O donor, because the O red corpuscles have no agglutinogens and will not be clumped by any kind of plasma. The O plasma contains both a and b agglutinins, but in the transfusion they are gradually diluted by the patient's plasma and do not agglutinate the patient's red cells. Type O individuals are called universal donors because their blood can be given to any person, and type AB people are known as universal recipients, because they can receive blood of any type. Blood groups

Table 16–1 THE HUMAN BLOOD GROUPS

Blood Group	Agglutinogen in Red Cells	Agglutinin in Plasma	Can Give Blood to Groups	Can Receive Blood from Groups
O	None	a and b	O, A, B, AB	O
A	A	b	A, AB	O, A
B	B	a	B, AB	O, B
AB	A and B	None	AB	O, A, B, AB

are inherited (see p. 669) and remain constant throughout life.

Some notable differences have been found in the relative *frequency* of the different types in different races. The proportion of blood types in a population remains constant from one generation to the next as long as there is no intermarriage with other groups. Gypsies, who originally came from India, have lived in Hungary for several hundred years, but because there is little mating between them and the native Hungarians, the proportion of blood types among the Gypsies is similar to that of the Hindus and quite different from that of the

Hungarians. Descendants of the Germans who migrated to Hungary about 1700 still have, in addition to their German language and customs, a blood type frequency characteristic of the Germans in Germany. With the development of techniques for determining the blood types of mummies, and even of skeletons, the use of blood tests in anthropology has been considerably broadened. Candela tested the blood types of thirty prehistoric Aleutian mummies and found eight of them to be B or AB. Previously the Aleuts had been thought to have been derived from an Eskimo-Indian cross, but both Eskimos and Indians have rela-

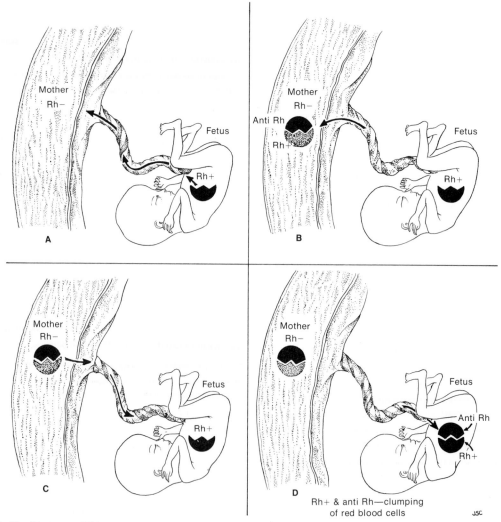

Figure 16–10 Diagram of the sequence of events leading to erythroblastosis fetalis, a condition in which the red cells of the fetus clump within the uterus.

A, Red cells pass from an Rh⁺ fetus to its Rh⁻ mother through some defect in the placenta.

B, Maternal white cells produce anti-Rh⁺ antibodies.

C, During a subsequent pregnancy anti-Rh⁺ antibodies pass from the maternal blood to the fetal blood stream.

D, The reaction of Rh⁺ cells with anti-Rh⁺ antibodies causes clumping of the fetal red cells.

tively low frequencies of B and AB groups, whereas B is common in Asia. These findings, indicating that the Aleuts were of Asiatic origin, have been substantiated by other evidence.

Red cells contain, in addition to the A and B agglutinogens, a second pair known as **M** and **N**, inherited independently of the A and B pair. These, together with other less important blood types, provide additional means of identifying blood.

A third set of hereditary factors causes the presence or absence of another agglutinogen, the **Rh factor**, so called because it was first found in the blood of rhesus monkeys. About 85 per cent of white people are Rh positive, i.e., have the Rh antigen in their red cells, and 15 per cent are Rh negative, with no Rh antigen. If a woman is Rh negative and her husband Rh positive, the fetus may be Rh positive, having inherited the factor from its father. Blood from the fetus may pass through some defect in the placenta into the maternal blood stream and stimulate the formation of antibodies to the Rh factor by the white cells of the mother (Fig. 16–10). Then, when this woman becomes pregnant a second time, some of these antibodies may pass through the placenta into the child's blood stream and cause the clumping of its red cells, a condition known as erythroblastosis fetalis. In extreme cases, so many red cells are destroyed that the fetus dies before birth; more frequently it is born alive but dies after birth. Newborn children with erythroblastosis fetalis are now given massive blood transfusions to replace essentially all of their red cells. Studies revealed that the major transfer of fetal red cells to the mother occurred during the process of childbirth. Because of this, Rh negative women are treated just after childbirth (or abortion) with an anti-Rh gamma globulin preparation to block the formation of maternal antibodies.

The O, A, B and AB groups have been demonstrated in chimpanzees, orang-utans and gorillas; thus, these blood group substances arose before the primates had finally evolved into different types. Substances similar to, but not identical with, the human A and B substances have been found in the bloods of many mammals and birds. M and N agglutinogens, similar to those of humans, have been found in chimpanzees, orang-utans and some monkeys, but in no other animals. The M and N substances of the chimpanzee are most like those of humans, confirming the generally accepted idea that chimpanzees are the most humanlike of the arthropoid apes.

The transfusion of whole blood, plasma or plasma fractions is extremely important in saving lives. Blood drawn from a suitable donor is treated, usually with sodium citrate (which binds the calcium ions) to prevent coagulation. Care must be taken to prevent infection, and the blood must be introduced into the recipient's veins at the proper temperature and rate. The heart will be overloaded if blood is transfused too rapidly. Blood can be stored for weeks in a "blood bank" by adding citric acid and glucose and keeping the blood at a temperature of 4 to 6° C. By separating the red cells from the plasma, and then suspending the cells in purified albumin, cells can be kept for as long as three months.

The search for substitutes for whole blood to be used in transfusions dates back to 1878, when cow's milk was tried. Plasma and certain plasma fractions, which can be stored much longer than the whole blood, are effective substitutes for whole blood in many clinical conditions such as shock. Dried plasma or plasma fractions, prepared by freezing and drying and placed in a sealed sterile container, can be kept even without refrigeration for a long time. To be used, the plasma is mixed with the proper quantity of sterile distilled water and injected. In preparing plasma, bloods from 16 different people of assorted blood types are pooled so that the different agglutinins are diluted below their effective concentration and will not agglutinate the red cells of the recipient.

The polysaccharide **dextran**, a glucose polymer made by bacteria, can also be used as a blood substitute. It can be prepared inexpensively in large amounts, does not cause agglutination of red cells, gives fewer toxic reactions than any other substitute tried, and eliminates the possibility always present with the transfusion of blood or plasma of transmitting the virus of serum hepatitis. Gelatin, pectin, gums and albumin from cow's blood have all been tried as substitutes but none has been satisfactory.

16-10 IMMUNITY AND IMMUNOGLOBULINS

In addition to distributing nutrients and hormones, the blood and lymph play important roles in the body's defense against pathogenic bacteria and viruses. The study of the defenses of the host against the continuous assaults of the myriad pathogenic microorganisms in our environment is part of the science of immunology. The phenomenon of *immunity* was first recognized centuries ago when it became apparent that a person who had recovered from one of the communicable diseases that were the scourges of the time was unlikely to have it again—he was "immune" to it. In the late eighteenth century Edward Jenner noticed that dairy workers who handled cows with *cowpox* never contracted the much more serious human disease *smallpox.* When he scratched some serum from the pustules on a cow's udder into human skin, the individual had a mild disease with a single, localized pox at the site of injection. Anyone so vaccinated never acquired smallpox. We know that cowpox and smallpox are caused by two distinct but closely related viruses. Inoculation with cowpox virus (*vaccinia*) stimulates the production of antibodies that will also react with the smallpox virus (*variola*). Since then, vaccination procedures have been developed for many other diseases; these have been remarkably effective in improving the general health of the world's population. Untold millions of people have been protected against debilitating and fatal diseases such as poliomyelitis, measles and whooping cough.

More recently, it has been realized that the mechanisms of immunity to disease are simply one aspect of a very broad and general biologic phenomenon. Certain vertebrate cells not only can react to microbial proteins and polysaccharides, but can recognize essentially all proteins foreign to them and respond with reactions that will remove the foreign proteins or neutralize their effects. Many of these defense mechanisms involve either the white blood cells and their derivatives, or antibodies that are present in the blood plasma.

Humans and other vertebrates have several lines of defense against invading pathogenic microorganisms. The tough protective layer of skin, with its content of fatty acids and its low pH, is an effective barrier against bacteria as long as it is intact. The mucous membranes are constantly flushed and produce antimicrobial substances which assist in preventing the entrance of microorganisms. If microorganisms manage to penetrate the skin or mucous membranes, a second line of defense, consisting of the several types of phagocytic cells in the body which engulf and destroy pathogens, comes into play. Finally, the specific immunoglobulins or antibodies produced by certain cells of the body provide a third line of defense against the pathogens that pierce the other two.

Neutrophils and monocytes defend the body against invading microorganisms primarily by phagocytizing them. The phagocytic cells in the blood are backed up by relatively large phagocytic *macrophages* found principally in reticular connective tissue. Some of the macrophages wander around in the tissue spaces; others are fixed in one place. Together the two types constitute the *reticuloendothelial system* of phagocytes, an important defense mechanism of the body. Lymphocytes have relatively little phagocytic activity, but by producing antibodies they provide another important defense mechanism.

Actively acquired immunity, such as that produced in response to vaccination with cowpox, depends upon the production of specific protein *antibodies* in response to the presence of some foreign substance (an *antigen*) and their subsequent release into the blood and tissue fluids. An antigen can be defined as any substance capable of provoking an immune response, such as the production of antibodies. A typical antigen is a protein or polysaccharide or a complex containing both substances. A bacterial cell contains many different proteins and polysaccharides, each of which could act as an antigen to produce a specific antibody. In general, it is not the entire protein molecule which acts as an antigen, but rather some segment of the antigen's peptide chain with some unusual surface configuration, which acts as an *immunologically specific determinant group.* A single protein may bear several of these determinant groups and evoke the production of several antibodies differing in their specificities for the several determinant groups.

Antigens are able to combine with the antibodies that they elicit; the antigen-antibody combination can be demonstrated by precipitin tests. The response of the antibody-forming cell to an antigen is a reflection of the ability of the cell (*self*) to recognize and react against anything that is *nonself.* Anything not derived from that cell or from a genetically identical cell is nonself and hence a foreign substance. If a human is injected with some rabbit serum the human cells recognize the rabbit proteins as nonself and respond by producing antibodies to reject them. In a similar fashion, if a piece of an organ from some animal were transplanted to a human being, it would be quickly rejected by the body cells of the recipient because the transplant is nonself; even if the tissue came from another human being it would be rejected as nonself. The rejection of tissue from another member of the same species is called the *homograft reaction.* Usually skin grafts are successful only if the patient's own skin (or skin from his identical twin) is used. Even certain tissues of an individual's own body (e.g., certain nervous and connective tissues and the protein of the eye lens) may be rejected as nonself. Substances that exhibit antigenic activity in the same body are termed *autoantigens.* Certain types of arthritis, multiple sclerosis and other diseases, which have such reactions as their basic cause, are termed *autoimmune diseases.*

During fetal and neonatal life the lymphocytes that will subsequently react to antigens have not yet begun to function. If an antigenic substance is injected during this time the antigen is not recognized as nonself and no antibodies are made. The remarkable thing is that when this individual becomes mature he will still be "tolerant" to these substances and will recognize them as self. This adaptation during early life to an antigenic substance is called *acquired tolerance* or *induced tolerance.*

The Immune Response. The evolution of the ability to produce antibodies was a giant biological step forward in improving the defense mechanisms of the host. The ability to produce antibodies appeared relatively late in evolution and is limited to the vertebrates and certain echinoderms. The primitive cyclostomes are immunologically competent, but their immune responses are not as highly developed as those of other vertebrates.

The key cell in the recognition of an antigen and the subsequent immune response is the lymphocyte. Two populations of lymphocytes can be distinguished, both derived from bone marrow. The *T,* or *thymus, lymphocytes* are primarily responsible for cellular immunity, and the *B,* or *bursal, lymphocytes* are primarily responsible for the production of circulating specific antibodies, i.e., for humoral immunity. B cells have a greater density than T lymphocytes and the two types can be separated on density gradients.

Stem cells from the bone marrow find their way in the blood stream either to the thymus or to those bursal lymphoid systems associated with the digestive system—the tonsils, the appendix and certain groups of lymphoid cells in the wall of the intestine (Fig. 16–11). The stem cells which mature in the thymus and become T lymphocytes eventually leave the thymus and enter the circulation again, finally colonizing other lymphoid tissues such as the spleen and lymph nodes (Fig. 16–12). The T lymphocytes are long-lived cells which, in the presence of antigens, become transformed into *lymphoblasts,* cells which do not secrete antibodies but are responsible for cellular immunity. The stem cells which mature in the bursal tissues emerge as B lymphocytes and colonize other portions of the spleen and lymph nodes. The B lymphocytes, upon contact with an antigen, become transformed into *plasma cells* which produce immunoglobulins; that is, they are the source of humoral immunity. Both populations of lymphocytes have immunological memory; they retain the ability to recognize foreign substances as nonself and to take some action to rid the body of those substances. Depending upon the nature of the antigen, the length of exposure and the dosage, either B or T lymphocytes or both may respond to the presence of the antigen. Usually both B and T cells are involved in the immunity that an individual develops to an infectious microorganism.

The Nature of Antibodies. Antibodies are immunoglobulins, proteins present in blood, in lymph and in certain secretions of the body such as mucus. The Nobel Prize in medicine was awarded to Rodney Porter and Gerald Edelman in 1972 for their studies revealing the structure of immunoglobulins. Five classes of immunoglobulins can be distinguished on the basis of their size and chemical composition, their antigenic specificity and the type of im-

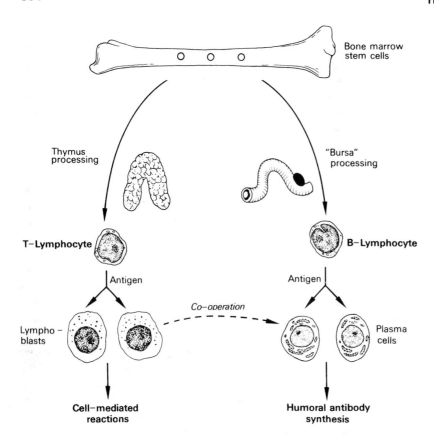

Figure 16–11 Bone marrow stem cells undergoing maturation in the thymus or the bursal tissue to become T- or B-lymphocytes, respectively. (From Frobisher, M., Hinsdill, R., Crabtree, K., and Goodheart, C.: *Fundamentals of Microbiology,* 9th Ed. Philadelphia, W. B. Saunders Co., 1974.)

munologic phenomena in which they participate. Some 80 to 85 per cent of the total circulating antibodies are *IgG immunoglobulins* with molecular weights of 150,000. IgG is composed of four subunits, two *heavy chains* with molecular weights of 53,000 and two *light chains* with molecular weights of 22,000 (Fig. 16–13). In humans the IgG (but not the other immunoglobulins) pass across the placenta to the fetus and give passive protection to the newborn against many infectious agents for the first six months or so of life. The immunological specificity of the IgG immunoglobulins lies in the unique spatial configuration at the tips of the Y-shaped molecule where the heavy and light chains come together. The Y-shaped molecule has two branches and is "divalent," that is, it contains two combining sites per molecule. This allows the antibody to bind two antigenic molecules, thus forming a complex.

All the other immunoglobulins are also composed of light and heavy chains. Each molecule has two identical light chains containing some 220 amino acids with a molecular weight of 22,000. Each molecule also has two identical heavy chains with molecular weights ranging from 53,000 (IgG) to 75,000 (IgE). The poly-peptide chains are held together and their configurations are stabilized by disulfide (−S−S−) linkages and by noncovalent bonds. Comparisons of IgG molecules from different individuals reveal that they share some determinant groups but have other unique ones. The other immunoglobulins, *IgM, IgA, IgD* and *IgE,* reveal similar variations in determinant groups. This led to the generalization that the immunoglobulins are quite heterogeneous with respect to their determinant groups.

Both light and heavy chains have "constant" and "variable" regions. The constant regions, which have identical sequences of amino acids in all the immunoglobulins, may determine the antigenic specificity of the immunoglobulin and may be responsible for some of the biological properties which determine whether it can be secreted, whether it will bind to macrophages and so on. The variable portions of the four polypeptide chains in each antibody molecule are believed to be responsible for its antibody specificity. The sequence of amino acids in these portions determines the spatial configuration of the tip, which will just fit some particular antigenic determinant group.

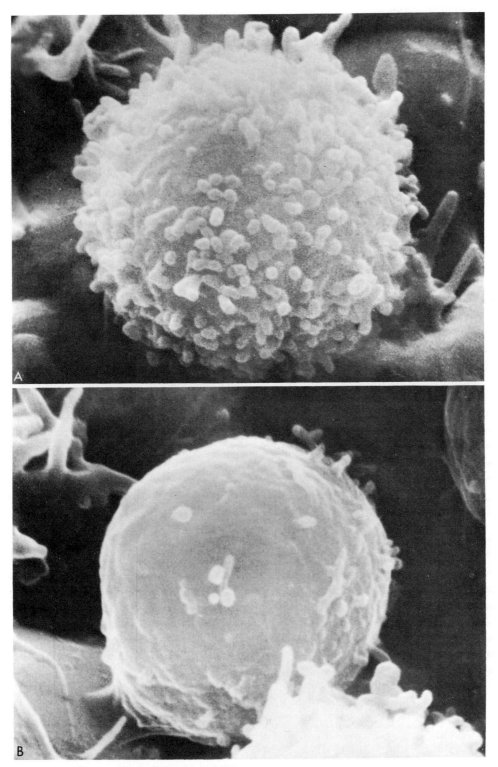

Figure 16–12 *A,* Scanning electron micrograph of a typical circulating lymphocyte with its fimbriated surface. This surface is frequently associated with B-derived lymphocytes. *B,* Another cell with a relatively smooth surface with few microvilli; this surface is frequently found in thymic lymphocytes. (From Polliack, A., Lampen, N., Clarkson, B. D., DeHarven, E., Bentwich, Z., Siegal, F. P., and Kunkel, H. G.: Identification of human B and T lymphocytes by scanning electron microscopy. J. Exp. Med. *138*:607, 1973.)

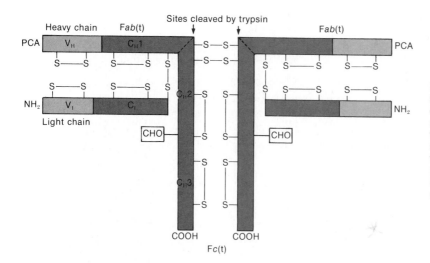

Figure 16–13 A diagram of the structure of the immunoglobulin molecule IgG, composed of two light chains and two heavy chains joined together by disulfide bonds. The constant (C) and variable (V) regions of the chains are indicated. The carbohydrate chains are attached to the heavy chains at the points indicated—CHO.

The IgA immunoglobulins make up about 10 per cent of the immunoglobulins in serum, but they are the principal ones found in secretions such as saliva and tears. In secretions IgA is generally a dimer, composed of two units, and has an attached *transport piece*, a β-globulin with a molecular weight of some 50,000. This transport piece becomes attached to the IgA as the immunoglobulin passes through the glandular cell producing the secretion. It is believed that IgA is of special importance in defending the membranes of the mouth, vagina and other body cavities against infections.

IgM immunoglobulin has a molecular weight of about one million and is a pentamer, composed of five subunits. The IgM immunoglobulins have a different content of carbohydrate (12 per cent versus 3 per cent in IgG) and heavy chains with a molecular weight of about 70,000. All types of immunoglobulins have antigenically identical light chains but antigenically distinct heavy chains. The IgM immunoglobulins are among the earliest antibodies to appear in response to an antigen; somewhat later in the immune response they are replaced by IgG antibodies.

It is remarkable that the antibodies produced in response to a given antigen are not necessarily homogeneous; in fact they may differ not only in their specificity and in their avidity for the antigen, but also in such physical-chemical properties as size, shape, net charge and amino acid sequence. This makes the problem of relating their structure to their specificity as antibodies even more difficult to solve. The differences between different anti-bodies must be quite subtle—probably even more subtle than the differences between enzymes.

IgD immunoglobulins were first discovered in patients suffering from multiple myeloma, a disease of the lymphoid tissues. Their biological function is not known, but they are present in certain intestinal fluids of normal individuals. The IgE immunoglobulins are present only in very low concentrations in the serum and their possible function in the serum is not clear. The IgE immunoglobulins may be more harmful than beneficial for those individuals who are allergic or hypersensitive to a substance. The IgE immunoglobulins become bound to cells, sensitizing them to the allergen and playing a role in the allergic response. When transferred from a sensitized individual to a normal one, the IgE sensitizes the latter to the allergen.

Antibodies are produced in plasma cells derived from B lymphocytes and in certain large lymphocytes located in lymph nodes and other lymphoid tissue. Plasma cells have an extensively developed endoplasmic reticulum (Fig. 16–14) for the synthesis of proteins. Apparently only a small portion of the large immunoglobulin molecule—perhaps some 15 to 30 amino acid residues—is involved in the immunologically active site. The differences between antibodies appear to reside in small variations in the shape of the protein that are caused by differences in the sequence of amino acids in the variable regions. The antigen and antibody are believed to have some sort of complementary geometrical configuration so that they fit together like a lock and key.

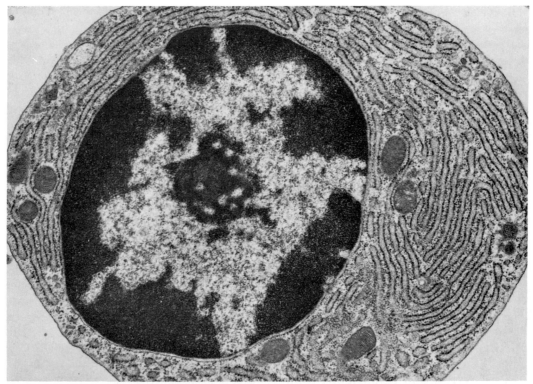

Figure 16-14 An electron micrograph of a guinea pig plasma cell showing the greatly hypertrophied endoplasmic reticulum. (From Bloom, W., and Fawcett, D.: *Textbook of Histology*, 10th Ed. Philadelphia, W. B. Saunders Co., 1975.)

Primary and Secondary Responses. Following the injection of an antigen there is a latent period of nearly a week before any antibody appears in the blood. The titer of antibodies rises slowly to a low peak, called the *primary response*, and then decreases (Fig. 16-15). A second injection of antigen—several days, weeks or even months later—will induce a rapid production of antibodies after a shorter

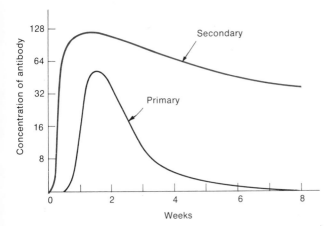

Figure 16-15 Primary and secondary responses of antibody formation to successive doses of antigens.

latent period. This is termed the *secondary response*. The titer of the antibody in the secondary response reaches a higher level and decreases more slowly. Further injections will induce additional secondary responses until a maximal titer is achieved. This usually declines with time and periodic reimmunizations ("booster shots") help maintain a satisfactory level of immunity. A secondary response may also be induced when a previously immunized individual is exposed to the natural infectious agent. An antibody is usually produced rapidly enough to prevent the appearance of the disease symptoms.

Although any foreign protein, and indeed many smaller molecules combined to a protein, may act as an antigen and activate the plasma cells to produce a specific antibody, the more common antigens are the proteins present in certain bacteria and viruses that produce infectious diseases. Antibodies can subsequently combine with the bacteria or viruses, neutralizing them and causing them to clump or agglutinate, thereby effectively preventing further penetration of the host. Antibodies may also cause the microorganisms to undergo *lysis*, to break up and dissolve, or they may make the invaders more susceptible to phagocytosis.

All of us have a **natural immunity** to certain infectious diseases that affect other organisms but not humans. The virus for canine distemper, which kills about half the dogs it infects, does not infect human beings. The virus **herpes simplex** is lethal to a rabbit but usually causes only a "fever blister" in humans. Natural immunity is apparently due to the presence in the plasma of antibodies that can combine with the effective agent. Vertebrates in general can form antibodies to a variety of infectious agents and develop specific immunity to them.

A person who has contracted a disease and needs antibodies immediately to combat its antigens may be given a **serum,** a solution of antibodies produced by an animal that had been infected with the disease and recovered from it. The antibodies in the serum can protect the patient until his own plasma cells can develop enough antibodies to the infective agent. The method of passively acquiring immunity is immediately effective, but the immunity disappears in a few weeks. Since it involves giving the patient a serum from another animal it may raise an immune response in the patient to the plasma proteins of the donor animal.

To prepare a serum for a given disease, a pure strain of the infectious agent is grown and injected in increasing doses into an animal, such as a goat or a horse, which responds by synthesizing large amounts of the specific antibody and secreting them into its blood. Blood is drawn from the animal at intervals and processed to remove blood cells and the proteins other than the desired antibody.

Theories of Antibody Formation. Among the questions facing immunologists at present are three: (1) How can plasma cells make so many different kinds of antibodies and so many proteins with different amino acid sequences? (2) How can they make a specific antibody on demand when a specific antigen is present? (3) Why do an animal's plasma cells not make antibodies to the animal's own native proteins?

Among the many theories of antibody formation that have been proposed, the one favored about 1940 stated that the antigen supplies the information necessary for the production of its specific antibody. The **instructive** theory held that the antigen acts as a template, and as the globulin molecule is formed it is molded around the antigen and thus develops a complementary conformation at its active sites. This theory assumed that the specificity of the antibody was due not to its specific amino acid sequence but to its conformation; it was ruled out by the discovery that different antibodies do have different amino acid sequences and that the specificity of the antibody is a function of its amino acid sequence.

The most widely accepted theory at present is that of **clonal selection**, advanced by Burnet in the late 1950's. According to this theory, the lymphoid cells in the primary lymphoid tissues undergo differentiation during fetal development. Many different kinds of cells result, each with one or more recognition sites on its surface. After a short time thousands of cells, each with one or more different recognition sites, are present in the cell population. As the cells migrate through the developing fetus some of the cells encounter fetal tissue or soluble antigens which are complementary to their own recognition sites. The two combine and the lymphoid cell dies. In this way any lymphoid cell carrying recognition sites for "self" proteins will be destroyed during fetal life.

The cells that survive this procedure can recognize and react only to nonself proteins; all the cells that respond to self proteins have been eliminated. The surviving cells are then free to become located in the lymphoid tissue and each can divide to produce generations of daughter cells exactly like itself. The daughter cells derived from a single initial cell compose a **clone.** It has been calculated that by the time an individual becomes immunologically competent, a large number of clones of lymphoid cells, with 10,000 to 100,000 different kinds of recognition sites, will be present in his tissues. These, it is believed, will enable the host to recognize almost any nonself antigen to which it may be exposed. Then, when an antigen enters the body and makes contact with a B lymphocyte or a T lymphocyte bearing the proper recognition site, the lymphocyte is stimulated to reproduce rapidly and the size of that clone increases. Depending on which cell line they represent, its progeny become either **plasma cells** (and produce humoral antibodies) or **lymphoblasts** (and activate cellular immunity). The greater the antigenic stimulation, the greater will be the proliferation of the clones of cells possessing corresponding recognition sites and the greater the production of antibodies or phagocytic cells. How would you explain the secondary response to antigens on the basis of this clonal selection theory?

It should be emphasized that according to the clonal selection theory the antigen does not provide information to the antibody-forming cells but simply selects those cells that are making antibodies that fit it and stimulates them to proliferate and make large quantities of antibodies. The antibodies are not tailor-made to fit the antigen (instructive theory) but are simply selected from among the ready-mades on the rack if they fit (clonal theory).

If the antigen molecule is small and contains only one combining site—i.e., if it is univalent—only one antibody can combine with each antigen molecule. However, two antigen molecules may combine with each antibody since antibody molecules are typically divalent. If the antigen is a complex molecule with several or many determinant groups, each of these will evoke an antibody molecule specific for that site. It follows that although such a multivalent antigenic protein is a pure substance, the antibodies produced in response to it will be a mixture of antibody molecules, each of which is specific for a different determinant group on the antigen molecule. When all of the determinant group sites are saturated with the specific antibodies, a large, stable, insoluble colloidal complex forms and precipitates.

To summarize, the cells that synthesize any given kind of antibody are believed to represent the descendants of a single progenitor cell (a clone) that had acquired, by some random process, the genetic capacity to synthesize an immunoglobulin with a specific amino acid sequence and consequently the ability to react with a specific antigen. In the absence of the specific antigen the clone consists of relatively few cells. The presence of the antigen stimulates the proliferation of the members of the clone that can make the appropriate antibody and hence stimulates the production of the specific immunoglobulin. This theory is satisfying in that it supposes that the synthesis of immunoglobulins involves the production of peptides with specific amino acid sequences by a ribo-

somal mechanism identical to that for the synthesis of other proteins. However, it is not yet clear how a specific antigen and the cells of the appropriate clone of plasma cells recognize each other, nor how this recognition leads to the proliferation of the cells of that clone.

Immunogenetics. The clonal selection theory would require only a few structural genes for the light and heavy chains of the immunoglobulin to be transmitted by the germ cell. Subsequent somatic mutations in different clones would provide the genetic diversity needed to code for all the different antibodies a given individual can make. This theory does not readily account for the fact that the ability to synthesize antibodies of certain specificities appears during development in a fixed time sequence and not at random. The clonal selection hypothesis implies that any single plasma cell would make only one kind of antibody, but several experiments have shown that a single plasma cell may make more than one kind of antibody. Thus the concept that there are different clones of cells and that each of these responds to only one or a few antigens has not yet been established. A related hypothesis suggests that there is one C gene that regulates the synthesis of the constant region of the molecule that is common to all immunoglobulin chains and that one or more V genes code for the variable region of the chain that differs in different antibodies. The recombination of a specific V gene with the C gene yields a VC gene that is read out and forms the specific immunoglobulin.

The antibodies discussed so far are all humoral antibodies produced by plasma cells and present in the plasma and secretions. In addition, some antibodies are synthesized and remain within cells, providing the basis for the cellular immunity involved in the phenomena of hypersensitivity and immunologic tolerance.

Allergy and Hypersensitivity. Although the immune system in general has a protective role and is beneficial to the host organism, some immunologic reactions result in one or more pathologic effects. The adverse responses, such as allergy and hypersensitivity, are of two major types—those that are mediated by humoral antibodies and those that are mediated by cellular responses.

The several types of hypersensitivity reactions vary considerably as to the mechanisms involved and the symptoms that result, but they have two features in common. First, the individual must be sensitized by being exposed to the allergen (the sensitizing dose), and some time must elapse during which the immune response develops. Secondly, the individual must be exposed to an eliciting dose of the same allergen to produce the hypersensitive reaction. The most severe type of allergic reaction, *anaphylaxis,* is systemic and involves the whole organism. Its clinical manifestations are respiratory distress, obstruction of the bronchial tubes and the larynx, vascular collapse, itching, the formation of blisters (urticaria) and gastrointestinal upsets, such as abdominal pain, diarrhea and vomiting. The anaphylactic reaction is caused by the release from the mast cell of histamine, *SRS-A* (slow reacting substance of anaphylaxis) and *ECF-A* (eosinophil chemotactic factor of anaphylaxis), which cause the anaphylactic reactions.

The mast cell has receptors which concentrate IgE antibodies. The IgE are trace antibodies, not normally present in the fluid phase of the blood and interstitial fluid. Two IgE antibodies bound side by side to the mast cell form a recognition site to which the allergen can bind and activate the mast cell to secrete the anaphylactic agents. Some of these are stored in the mast cell and some are synthesized there when the cell is activated. Histamine causes changes in vascular permeability and constricts the proximal airways; SRS-A constricts the distal airways. SRS-A belongs to a new class of compounds; it has a molecular weight of 400 and contains an acidic sulfate ester. The ECF-A is an acidic peptide with a molecular weight of 500.

When an individual has been sensitized to a substance, such as penicillin or the material in a bee sting, the anaphylactic reaction can appear and death can ensue very rapidly following the eliciting dose of the allergen. Symptoms may appear within 30 seconds and death may follow within 16 to 120 minutes after the administration of penicillin or following a bee sting. The administration of antihistamines can counteract the effect of histamine, but the SRS-A and ECF-A continue to operate. These responses are not affected by antihistamines. Some of the symptoms of anaphylaxis result from the contraction of smooth muscle fibers induced by the presence of histamine. During anaphylaxis the blood vessels are damaged, resulting in dilation and the escape of fluid into the tissues, causing swelling (edema) and shock.

Some 10 per cent of the human population suffer from allergies, which we might consider as localized anaphylactic reactions. Allergens such as pollens, animal danders, house dust or mold spores contact the cell-bound IgE immunoglobulin present in the nasal mucosa, the bronchial tissues or the conjunctiva. As a result, histamine and other mediators are released and produce the symptoms of hay fever (tears, redness of the eyes, nasal secretion) or of asthma (constriction of the bronchial tubes).

Some individuals show hypersensitivity to certain foods. Shortly after the allergenic food has been ingested the individual breaks out in hives (generalized urticaria). Apparently food allergens are absorbed intact from the gut and make their way to the skin by way of the blood stream. Desensitization of the hypersensitive individual by the repeated injection of small amounts of the allergen is often attempted but has variable results. When desensitization does work it is not clearly understood *how* it works. One theory suggests that injections of the antigen may elicit the formation of IgG, and circulating IgG reacts with the allergen before the latter can react with the cell-bound IgE. It should be understood that only the *recognition* of the allergen by the IgE bound to the mast cell is an immunologic phenomenon. The subsequent production and secretion of histamine, SRS-A and ECF-A by the stimulated mast cells are biochemical and pharmacologic phenomena. Cyclic AMP provides an important control of the secretion of these mediators; increasing the cyclic AMP content of the cell causes a decrease in the release of histamine and decreases the synthesis and release of SRS-A. In contrast, cyclic GMP increases the synthesis and release of SRS-A.

Immunologic Tolerance and Transplantation Immunity. When cells from one adult animal, the donor, are injected into or grafted onto a second adult, the host, the transplanted cells may grow for a short time but will be rejected and slough off in a few weeks. If a second graft is transplanted from the same donor to the same host it will be rejected in a much shorter time. This accelerated rejection is termed the *second set reaction,* and it is specific for the donor. If a skin graft is made from a different, genetically unrelated individual into this host, it will undergo a *first set reaction* and be rejected only after some weeks. The immunologic basis of the second set reaction, the capacity to reject the transplant from another member of the same species, is acquired when the host tissue is exposed to the donor's tissue. It can be acquired not only by a previous skin graft but also by the injection of a suspension of spleen cells or other cells from the donor. Only if transplants are made between identical twins or between members of a highly inbred, genetically, homogeneous strain will the graft be successful and survive in the host body.

If embryonic cells from a mouse of strain A are infused into a newborn mouse of strain B, the latter will develop normally and, when adult, will be able to accept transplants of skin or other tissues from mice of strain A. It will have developed *immunologic tolerance* by being exposed to foreign tissues before it developed immunologic competence. If an animal is exposed to an antigen before it has developed the capacity to react to it, that is, to produce the specific antibodies to it, then the development of that capacity is delayed. In the continued presence of antigen the development of the capacity to synthesize antibodies can be postponed indefinitely. However, this same animal will respond quite normally to the presence of other antigens and will form antibodies to them. A newborn rabbit injected with bovine serum albumin will not develop antibodies to it. Several months later it can be injected with more bovine serum albumin, but it still will not form antibodies to it. Immunologic tolerance can be induced only in fetal or neonatal animals. If the *thymus gland* (a large gland in the neck of the fetus or newborn) is removed, the animal's lymph nodes remain small and the animal is deficient in cellular immunity; it will accept tissue grafts from other animals that differ from it genetically.

The genes that determine the antigens responsible for the response of the host to transplanted tissues are termed *histocompatibility genes.* Genetic experiments with mice have shown that there are at least 14 independently segregating loci for transplantation antigens, and the number in humans is undoubtedly at least as large.

There is a rapidly growing interest in the possibility of using healthy tissues and organs, such as hearts and kidneys, to replace diseased ones. Four major types of transplantations can be recognized. In the first, tissue is simply removed from one location and grafted onto another in the same individual. Such *autografts* are usually successful. When the donor and recipient of the graft are identical twins or are members of the same purebred strain of mice, for example, such *isografts* are usually successful. When donor and recipient are of the same species, but of different genetic constitution, these *homografts* are usually rejected. Finally, when donor and recipient are members of different species, such *heterografts* are unsuccessful. Because the donor and recipient are less closely related and therefore have a greater number of different histocompatibility genes, the probability for success of the transplant greatly decreases. The present limited success of homograft transplants depends on how well the tissues of donor and recipient can be matched, just as individuals are matched for blood transfusions. The lower the number of different histocompatibility genes, the greater the possibility of successful transplants. Efforts are made to minimize the immune response of the recipient toward those foreign antigenic groups that cannot be avoided, but measures taken to paralyze the immune response of the recipient are not without considerable danger.

QUESTIONS

1. List seven functions of the blood in man. How are these functions carried out in animals that do not have a circulatory system?
2. Distinguish between plasma and serum and between lymph and tissue fluid.
3. How would you measure the volume of blood in a man's body without draining it out?
4. What is oxyhemoglobin? What is its significance in the body?
5. Why is carbon monoxide poisonous?
6. Where and how are red cells manufactured? What happens to old red cells?
7. Why does the number of red cells in the human body increase at high altitudes? (Note that this "why" refers to the physiologic mechanism involved and not to a teleologic explanation of any possible advantage to the man of having more red cells at higher altitudes.)
8. List the main structural and functional differences between white blood cells and red blood cells.

9. What are antibodies? Where are they produced?
10. Distinguish between the primary response and the secondary response to an antigen.
11. What is the physiologic basis of allergies? What measures can be taken to decrease the effect of allergies?
12. Distinguish between humoral immunity and cellular immunity.
13. Compare the "instructive" and the "clonal selection" theories of antibody specificity.
14. Describe briefly the series of reactions involved in the clotting mechanism. What prevents clotting from occurring within the blood vessels?
15. Differentiate between clotting and agglutination.
16. What substances have been found useful as substitutes for whole blood in transfusions?
17. Citric acid forms a tight union with calcium ions so that the calcium is unable to react with other substances. In view of this, why is sodium citrate added to whole blood immediately after it is drawn from a blood donor?
18. List the major components of blood plasma and give their respective functions.
19. Describe two different ways that leukocytes act to protect the body from microorganisms.
20. What factors would have to be considered when giving a blood transfusion to an Rh negative woman who has had several Rh positive children?

SUPPLEMENTARY READING

Further discussions on the subjects of red cells, white cells, blood clotting and immunology will be found in A. C. Guyton's *Function of the Human Body* and in *Human Physiology* by A. J. Vander, J. H. Sherman and D. S. Luciano. Further details of these subjects will be found in J. W. Harris's *The Red Cell*, L. O. Jacobson and M. Doyle's *Erythropoiesis* and in J. W. Rebuck's *The Lymphocyte and Lymphocytic Tissue*. Further details of the process of blood coagulation are described in *Human Blood Coagulation and Its Disorders* by E. R. Biggs and R. G. MacFarlane. Excellent source books in immunology and related subjects are J. F. Ackroyd's *Immunological Methods*, W. C. Boyd's *Introduction to Immunochemical Specificity* and P. L. Carpenter's *Immunology and Serology*.

CHAPTER 17

INTERNAL TRANSPORT AND CIRCULATION

The circulatory system, frequently termed the system of internal transport, carries nutrients and oxygen to all the tissues of the body, removes the waste products of metabolism, carries hormones from endocrine glands to their target organs and equalizes body temperatures. It includes the heart, the blood vessels and the lymph vessels, in addition to the blood, lymph, cerebrospinal fluid and tissue fluid.

The cells of the human body are bathed with tissue fluid and the concentration of substances within each of our cells is regulated in part by the concentration of that substance in the tissue fluid. The concentration in the tissue fluid is in turn regulated by the concentration in the blood and this is regulated by the kidneys, lungs, liver and intestines.

17–1 BLOOD VESSELS

The circulatory systems of humans and other vertebrates include three types of blood vessels: *arteries, veins* and *capillaries.* Arteries and veins are distinguished from each other by the direction of the flow of blood in them, not by the kind of blood (aerated or nonaerated) that they contain. Arteries carry blood from the heart *to* the tissues of the body; veins return blood *from* the tissues to the heart. Capillaries, which connect the arteries and veins, are microscopic, thin-walled vessels located in the tissues. Only the walls of the capillaries are thin enough to permit the exchange of nutrients, gases and wastes between blood and tissues. The walls consist of a single layer of cells, the *endothelium,* continuous with the endothelial lining of the artery and vein on either side (Fig. 17–1 *C*). Some capillaries are so small that the red cells are bent in passing through them.

Arterial walls have an outer coat of connective tissue, a middle coat of smooth muscle cells and an inner coat of endothelium and connective tissue (Fig. 17–1 *A*). The tough fibrous tissue in the outer coat makes the artery resistant to internal pressures while permitting it to expand and contract with each heart beat. The arterial walls are supplied with two sets of nerves; impulses carried by one set cause the smooth muscles to contract, those carried by the other set effect a relaxation. By contracting and relaxing, and thus regulating the size of

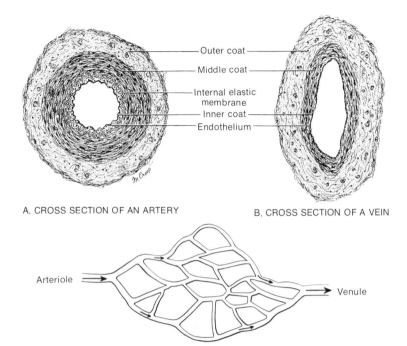

A, CROSS SECTION OF AN ARTERY

B, CROSS SECTION OF A VEIN

Figure 17–1 Arteries and veins compared. The diagram of the capillary network, *C,* is drawn at a much greater magnification.

Arteriole

Venule

C. DIAGRAM OF A CAPILLARY NETWORK

the arterial lumen, the smooth muscle in the middle layer regulates the amount of blood passing to a particular organ. In addition to the endothelial lining, the inner coat of most arteries contains a strong *internal elastic membrane* to give additional strength to the walls. The largest artery, the *aorta,* is about 25 millimeters in diameter near the heart, and has a wall about 3 millimeters thick.

The walls of veins are thinner than those of arteries, but the same three coats are present (Fig. 17–1 *B*). The outer coat has fewer elastic fibers, the middle muscular layer is thinner than the corresponding arterial layer, and most veins have no internal elastic membrane. Veins, but not arteries, are supplied with *valves* along their length to prevent the back-flow of blood.

Blood does not come in direct contact with the cells of the body; instead the cells are surrounded by and bathed with *interstitial fluid,* part of the extracellular fluid of the body. Substances must diffuse from the blood through the wall of a capillary and across the space filled with interstitial fluid to get to the cells (Fig. 17–2). Professor A. Baird Hastings characterized interstitial fluid, lymph and blood plasma as "the sea within us." An adult 70 kilogram man has about 10^{15} cells, bathed by only 14 liters of extracellular fluid. An equivalent

number of protozoan cells living in the sea would require 10,000,000 liters of sea water to provide them with the gases and nutrients they need. The efficient lungs, liver, intestines and kidneys continually replenish the oxygen and foodstuffs of these fluids and remove their wastes, enabling the cells of the body to survive even though they have relatively little extracellular fluid.

Capillaries. Each drop of blood passing

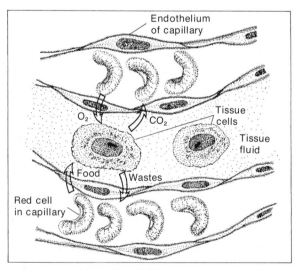

Figure 17–2 Diagram of the diffusion of materials between capillaries and the cells of the body, by way of the tissue fluid which bathes the cells.

through a capillary network is exposed to a large surface area through which substances may diffuse or be transported to the interstitial fluid bathing the cells. It has been estimated that each cubic centimeter of blood is exposed to 7000 square centimeters of capillary surface. The number of capillaries in the human body is almost beyond calculation. In tissues with a high metabolic rate, such as skeletal muscle, they are close together, the distance between adjacent ones being about twice the diameter of the capillary (Fig. 17–2). One investigator places the number of capillaries in muscle tissue at about 240,000 per square centimeter. Less active tissues are not so well supplied—fatty tissue has few capillaries, and the lens of the eye has none at all. Ordinarily, only a fraction of the capillaries of any organ are filled with blood and functioning (Fig. 17–3), but during periods of intense activity, all or nearly all the capillary networks are full.

Each capillary bed has certain "thoroughfare" channels through which some blood flows continually from *arterioles* (the smallest arteries) to the *venules* (smallest veins). Small muscular *precapillary sphincters,* located at the arteriolar end of capillaries branching off the thoroughfare channels, open or close other parts of the capillary bed to meet the varying metabolic requirements of the tissue. These sphincters, plus the smooth muscles in the walls of the arteries and arterioles, regulate the supply of blood to each organ and its subdivisions. Blood flows in the capillaries in intermittent spurts, not at a continuous rate, owing to the contraction and relaxation of the muscles in the arterioles and the precapillary sphincters.

The contraction and relaxation of the muscles in the wall of the arteriole are primarily under nervous control, responding to stimuli reaching them through the sympathetic nervous system. In contrast, the muscles of the precapillary sphincters are controlled by local conditions in the tissues. For example, decreased oxygen tension in the tissue causes these muscles to relax, which increases the flow of blood through the capillary bed and increases the oxygen supply to the tissue.

17–2 THE HEART AS A PUMP

The heart is a powerful muscular organ located in the chest directly under the breast bone. Its walls are composed of cardiac muscle cells held together by strands of connective tissue. Enclosing it is a tough, connective tissue sac, the *pericardium.* The inner surface of this sac and the outer surface of the heart are covered by a smooth layer of epithelium-like cells, and the cavity contains a fluid which reduces friction to a minimum as the heart beats.

The muscle fibers branch and fuse to form a complex network throughout the heart wall, across which nerve impulses can be transmitted. The contraction of the heart follows the "all-or-none" law (see p. 473): if a nerve impulse is sufficiently strong to make the heart beat at all, it responds with a maximal contraction. The heart and all the blood vessels are lined with a layer of smooth, thin, flattened cells, the *endothelium,* which prevents blood from clotting within the circulatory system. Any disease or injury that roughens the endothelium may lead to the formation of a thrombus within the vessel.

The human heart and those of other mammals and birds are divided into four chambers (Fig. 17–4): the upper right and left *atria,** and

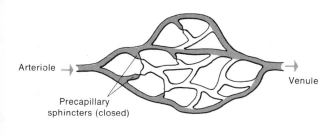

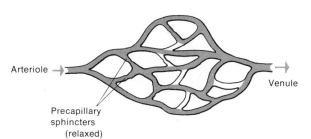

Figure 17–3 Changes in blood flow through a capillary bed as the tissue becomes active. *A,* The tissue at rest; only the thoroughfare channels are open in the capillary bed. *B,* In an active tissue the decreased oxygen tension in the tissue brings about a relaxation of the precapillary sphincters, and more capillaries become open. This increases the blood supply and the delivery of oxygen to the active tissue.

---

*The term auricle is sometimes used as a synonym of atrium, but strictly speaking the auricle is only one part of the atrium, the lateral pouchlike appendage.

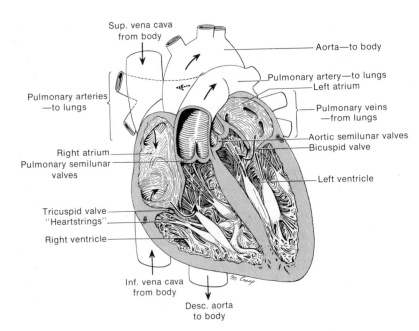

Figure 17–4 Diagram of the human heart showing chambers, valves and connecting vessels.

the lower right and left *ventricles.* The atria have relatively thin walls, receive blood from the veins and push it into the ventricles. The latter, with much thicker walls, pump the blood out of the heart and around the body. There is an opening for blood to pass from the right atrium to the right ventricle, and a second opening from the left atrium to the left ventricle, but no connection between the right and left atria or right and left ventricles. The heart is really two separate pumps, sometimes called the right heart and the left heart.

For its function as a pump, the heart is supplied with valves that close automatically and prevent blood from flowing in the wrong direction. The valve between the right atrium and ventricle, with three flaps or cusps, is called the *tricuspid* valve; that between the left atrium and ventricle, with two cusps, is called the *bicuspid* valve. These two valves are held in place and prevented from being pushed back into the atria when the ventricles contract by stout cords, or "heart-strings," attached to the valves and the walls of the ventricles (Fig. 17–4). At the bases of the pulmonary artery and aorta, which leave the right and left ventricles, respectively, are the two half-moon–shaped *semilunar* valves, pouches opening away from the heart. When blood passes out of the ventricle, the pouches are pushed aside and offer no resistance. But when the ventricles are relaxing and filling with blood from the atria, the blood pressure in the arte-

ries is higher than that in the ventricles. Then blood fills the pouches, stretching them across the cavity of the pulmonary artery or aorta and thus preventing a flow of blood back into the ventricle (Fig. 17–5). There is no valve at the openings of the large veins into the right atrium or at the openings of the pulmonary veins into the left atrium, and some blood is forced back into the veins when the atria contract.

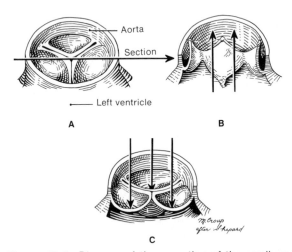

Figure 17–5 Diagram of the operation of the semilunar valves. *A,* Arrangement of the three pouches of the semilunar valves. The aorta has been cut across just above its point of attachment to the ventricle to expose the valves. *B,* When the ventricle contracts, the expelled blood (*arrows*) pushes the pouches aside and passes into the aorta. *C,* When the ventricle relaxes, blood from the aorta fills the pouches (*arrows*), causing them to extend across the cavity and prevent the leakage of blood back into the heart.

The right atrium receives blood from all parts of the body (except the lungs) by way of two large veins. The *superior vena cava* drains the head, arms and upper part of the body, and the *inferior vena cava* drains the legs and the lower part of the body. The contraction of the right atrium pushes open the flaps of the tricuspid valve, pumping blood into the right ventricle. The contraction of the right ventricle then closes the tricuspid valve, opens the semilunar valve, and pushes the blood out via the pulmonary artery to the lungs. Blood returning from the lungs in the pulmonary veins enters the left atrium, and is pumped by its contraction through the bicuspid valve into the left ventricle. The contraction of the left ventricle closes the bicuspid valve, opens the semilunar valve, and sends blood spurting out the aorta to all parts of the body. Every drop of blood entering the right atrium must go through the lungs before it can get into the left ventricle and be pumped around the body. The muscular walls of the left ventricle are thicker than those of the right ventricle. The muscle fibers in the ventricle are spirally arranged and their contraction "wrings" the blood out of the ventricular cavity. The elastic recoil of ventricular relaxation reduces the pressure within the chamber and blood enters from the atria.

In certain diseases the valves are attacked and their ability to function is greatly reduced. In rheumatic fever, which results from an immune reaction to the toxin of streptococci, antibodies formed against the toxin attack the valves, causing small, cauliflowerlike growths on their edges which erode the valves and cause the ingrowth of fibrous tissue. The valve may become so constricted and hardened that it cannot close properly, or the valve opening may be so greatly narrowed by scar tissue (valvular stenosis) that the rate of blood flow through the hole is much reduced. If the valve is eroded so that the flaps cannot close tightly, blood leaks back during diastole (valvular regurgitation), reducing the efficiency of the heart beat. A valve may be both leaky and narrowed, reducing cardiac function in two ways.

The heart muscle is not nourished by the blood within its chambers; its walls are too thick for nutrients and oxygen to diffuse through. Instead, the cardiac muscle is supplied by *coronary arteries* branching from the aorta at the point where that vessel leaves the heart. These cardiac arteries ramify through the heart muscle. Veins from the heart tissue drain into the right atrium, except for a few small veins that drain from the heart wall directly into the ventricle.

When one of the coronary arteries is blocked, the area of heart muscle served by that artery is deprived of oxygen and nutrients. It stops contracting, becomes flaccid, and dilates outward when the rest of the heart is contracting. When the blood supply to a sizeable fraction of the heart is affected, the heart can no longer function. Coronary occlusion is the cause of about one third of all deaths. It usually is the result of *atherosclerosis,* the deposition of fatty material (including cholesterol, phospholipids and triacylglycerols) in the wall of the arteries. Fibrous tissue grows in and around the deposit, and calcium combines with the fat to form solid, hard, platelike deposits ("hardening of the arteries"). In addition to this gradual process of occlusion, which occurs in all of us as we grow older, arteries may be suddenly occluded by a blood clot or by a bit of fatty deposit which breaks off from one location and is carried in the blood to another vessel, which it plugs. This occlusion results in a sudden decrease in cardiac function called a *"heart attack."* A severe occlusion causes immediate death; a smaller occlusion may cause a temporary weakening of the heart that can be repaired in part when new blood vessels grow into the area formerly served by the occluded artery. Diets rich in fats and smoking increase the probability of heart attack.

17–3 THE HEART BEAT

Beating is an inherent capacity of the heart, exhibited early in embryonic development and continuing without pause throughout life. All tissues need the constant supply of oxygen provided by the circulating blood. Unconsciousness results if the heart stops beating for a few seconds, and death ensues if it stops for a few minutes. The heart of a resting human pumps about 5 liters of blood per minute, or about 75 milliliters per beat. This means that a quantity of blood equal to the total amount contained in the body passes through the heart each minute. Not all the blood actually goes through the heart once a minute: some of the blood on the shorter circuits will return to the heart in less than a minute, whereas that on the longer circuits will take more than a minute to complete the trip. In man's three score and ten years his heart beats about 2,600,000,000 times, and pumps at least 155,000,000 liters (about 150,000 tons) of blood. It has been estimated that the work done by the heart is enough to raise a weight of 10 tons to a height of 10 miles—a truly remarkable performance for an organ that weighs approximately 300 grams! During exercise both the number of beats per minute and the amount of blood pumped per beat are greatly increased. Physical training enables the heart to increase its volume per beat, making it possible for an athlete to increase the total amount of blood pumped without as great an increase in the heart rate as would be necessary in an untrained person.

The heart continues to beat normally after its nerves have been severed; it is not dependent upon stimuli from the brain. If kept in the proper liquid medium, it will continue to beat when entirely removed from the body. Even a few muscle fibers dissected from the heart retain this ability. The rate of the fundamental, innate contraction is regulated by the *nodal tissue* within the heart and by two sets of nerves from the brain.

17–4 NODAL TISSUE

The heart beat is initiated and regulated by *nodal tissue,* made up of specialized muscle

fibers called *Purkinje tissue* (Fig. 17–6). In the lower vertebrates, such as fish and frogs, a separate chamber of the heart, the **sinus venosus,** into which the veins empty, passes blood to the right atrium. In higher forms this has disappeared except for a vestigial mass of nodal tissue called the **sinoatrial node** located at the point where the superior vena cava empties into the right atrium (Fig. 17–6). A second node, lying between the atria just above the ventricles, is known as the **atrioventricular node.** From the latter a bundle of branching Purkinje fibers ramifies to all parts of the ventricles. The sinoatrial node initiates the heart beat and regulates its rate of contraction. For this reason it is called the *pacemaker.* At regular intervals an action potential is initiated in the sinoatrial node and passes through the atrial muscle. When it arrives at the atrioventricular node, the impulse is transmitted to the ventricles and throughout the ventricular tissue by bundles of Purkinje fibers. There is no muscular connection between the atria and ventricles; their beating is correlated solely by the specialized nodal tissue, which conducts impulses about ten times as fast as ordinary muscle does (the rates are approximately 5 meters and 0.5 meter per second, respectively). Conduction by the nodal tissue ensures that all parts of the ventricle contract almost simultaneously. If conduction in the ventricles were by way of ordinary muscle tissue, the muscles near the base of the ventricles would contract first, causing the uncontracted apex to bulge and possibly injuring it.

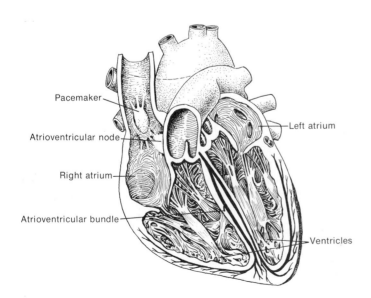

Pacemaker

Atrioventricular node

Right atrium

Atrioventricular bundle

Left atrium

Ventricles

Figure 17–6 Diagram of the heart showing the location of the pacemaker (the sinoatrial node), the atrioventricular node and the atrioventricular bundle, which regulate and coordinate the beating of the parts of the heart.

That the sinoatrial node regulates the rate of the heart beat can be demonstrated by warming the node, which results in an increased rate, whereas cooling it causes a decreased rate. Heat and cold usually have similar effects on other physiologic reactions, but heating or cooling other parts of the heart does not affect the rate of its beat. The acceleration of the beat accompanying fever is caused by the stimulation of the sinoatrial node by the warmer blood. When the sinoatrial node is destroyed by injury or disease, the atrioventricular node takes over its function as pacemaker.

Each heart beat consists of a contraction, or *systole*, of the heart muscle, followed by its relaxation, or *diastole*. At the normal rate of seventy beats per minute, each complete beat occupies about 0.85 second. The atria and ventricles do not contract simultaneously: atrial systole occurs first, occupying about 0.15 second, followed by ventricular systole, which takes approximately 0.30 second. During the other 0.40 second all chambers rest in the relaxed state.

17–5 THE HEART CYCLE

The action of the heart in pumping blood follows a cyclic pattern. The successive stages of the cycle, beginning with atrial systole, are as follows:

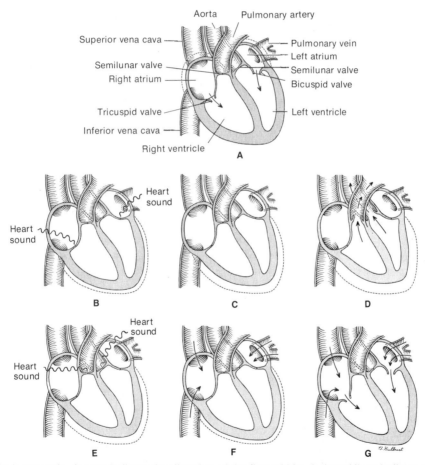

Figure 17–7 The heart cycle. Arrows indicate the direction of the flow of blood; dotted lines indicate the change in size as contraction occurs. *A,* Atrial systole; atria contract, blood is pushed through the open tricuspid and bicuspid valves into the ventricles. The semilunar valves are closed. *B,* Beginning of ventricular systole; ventricles begin to contract, pressure within ventricles increases and closes the tricuspid and bicuspid valves, causing the first heart sound. *C,* Period of rising pressure. *D,* The semilunar valves open when the pressure within the ventricles exceeds that in the arteries, and blood spurts into the aorta and pulmonary artery. *E,* Beginning of ventricular diastole. When the pressure in the relaxing ventricles drops below that in the arteries, the semilunar valves snap shut, causing the second heart sound. *F,* Period of falling pressure. Blood flows from the veins into the relaxed atria. *G,* The tricuspid and bicuspid valves open when the pressure in the ventricles falls below that in the atria, and blood flows into the ventricles.

(a) Atrial systole (Fig. 17–7 A). A wave of contraction, stimulated by the sinoatrial node, spreads over the atria, forcing blood into the ventricles. The ventricles are already partly filled with blood, owing to the fact that the pressure is lower there than in the atria, and the tricuspid and bicuspid valves are open. The conduction of the impulse through the atrioventricular node is slower than in other parts of the nodal tissue. This accounts for the brief pause after atrial systole before ventricular systole begins.

(b) Beginning of ventricular systole (Fig. 17–7 B). The muscles of the ventricular wall, stimulated by the impulse carried by the bundle of nodal tissue from the atrioventricular node, begin to contract, causing a rapid increase of pressure in the ventricles. The bicuspid and tricuspid valves close immediately, producing part of the first heart sound.

(c) Period of rising pressure (Fig. 17–7 C). The pressure in the ventricles mounts rapidly, but until it equals the pressure within the arteries, the semilunar valves remain closed and no blood flows into or out of the ventricles.

(d) When the intraventricular pressure exceeds that in the arteries, the semilunar valves open (Fig. 17–7 D) and blood spurts into the pulmonary artery and aorta. As the ventricles complete their contraction, blood is ejected more slowly until it finally stops. Ventricular diastole then follows.

(e) Beginning of ventricular diastole (Fig. 17–7 E). As the ventricles relax, the pressure within them decreases until it is less than the pressure within the arteries, and the semilunar valves snap shut. This causes the second heart sound.

(f) Period of falling pressure (Fig. 17–7 F). After the semilunar valves close, the ventricular walls continue to relax, and the pressure within the ventricles continues to decrease. The tricuspid and bicuspid valves, which closed during the previous ventricular systole, remain closed because the pressure within the ventricles still exceeds that within the atria. No blood is flowing into or out of the ventricles at this time, although some blood is flowing from the veins into the relaxed atria.

(g) The continued relaxation of the ventricular walls finally lowers the intraventricular pressure below the pressure within the atria. At that moment the tricuspid and bicuspid valves open (Fig. 17–7 G) and blood flows rapidly from the atria into the ventricles. No contraction of the heart causes this; it is due simply to the fact that the pressure within the relaxed ventricle is less than that in the atria and veins. The ventricles may be half filled before the atria undergo systole.

17–6 THE HEART SOUNDS

The beating heart produces characteristic sounds which can be heard by placing the ear against the chest or by using a stethoscope, an instrument which magnifies sounds and conducts them to the ear. In most normal persons, two sounds are produced per heart beat, one of which is low-pitched, not very loud, and of long duration. This is caused partly by the closure of the tri- and bicuspid valves and partly by the contraction of the muscle in the ventricle (all muscles make a noise when they contract). The first sound, which marks the beginning of ventricular systole, is followed quickly by a second which is higher-pitched, louder, sharper and shorter in duration. It is the result of the closure of the semilunar valves and marks the end of ventricular systole. The two sounds have been described by the syllables "lubb-dup," and their quality indicates to the physician the state of the valves. When the semilunar valves are injured, a soft hissing noise ("lubb-shhh") is heard in place of the second sound. This is known as a heart murmur and may be caused by syphilis, rheumatic fever or any other disease which injures the valves and prevents their closing tightly, so that blood can leak back from the arteries into the ventricles during diastole. Damage to the bicuspid or tricuspid valve affects the quality of the first heart sound.

17–7 ELECTROCARDIOGRAMS; ELECTRICAL CHANGES ACCOMPANYING THE HEART BEAT

When any tissue becomes active—when a muscle contracts, a gland secretes or a nerve conducts an impulse—it becomes electrically negative with respect to the surrounding tissues. The slight *action currents* accompanying each beat of the heart are detectable even at

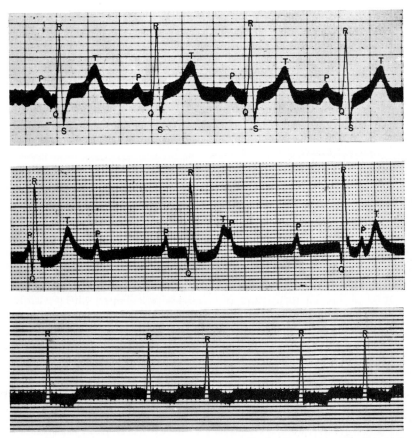

Figure 17–8 Electrocardiograms. *Above,* Tracing from a normal heart. The P wave corresponds to the contraction of the atria, the QRS complex to the contraction of the ventricle, and the T wave to the relaxation of the ventricle. *Middle,* Tracing from a man with a complete block of the atrioventricular node, so that the atrium and ventricle beat independently, each at its own rate. Note that P waves appear at regular intervals and QRS and T waves appear at regular but longer intervals, but that there is no relation between the P and QRS waves. *Below,* Tracing from a man with atrial fibrillation. The individual muscle fibers of the atrium twitch rapidly and independently. There is no regular atrial contraction and no P wave. The ventricle beats independently and irregularly, causing the QRS wave to appear at irregular intervals. (Courtesy of Dr. Lewis Dexter and the Peter Bent Brigham Hospital, Boston, Mass.)

the surface of the body by means of an ***electrocardiograph,*** which records them as a complex, curved line (Fig. 17–8). Malfunctioning of the heart causes aberrant action currents, and the recording from a heart with a pathologic condition will be distinctly different from normal.

Normally, when one impulse has passed along the membranes of the fibers of heart muscle, a second impulse cannot spread along the same membranes until some 0.3 second later, after the ***refractory period*** (p. 474) is over. When an impulse enters from the atrium it spreads to the end of the ventricle in about 0.06 second. The entire ventricle is then in the refractory state and the impulse stops. Under abnormal conditions the impulse may continue on around and around the heart for long periods of time in a "circus movement."

This eliminates the pumping action of the heart, for pumping requires that the muscle relax as well as contract. During a circus movement the muscles of the entire heart neither relax nor contract simultaneously and hence the alternate filling and squeezing action of the heart cannot occur. Circus movements around the atria cause atrial flutter, with the atria fluttering rapidly but unable to pump blood. The circus movements may pass in oddly shaped patterns around the atrium, causing ***atrial fibrillation,*** minute fibrillatory movements of the muscle. Circus movements developing in the ventricle result in ***ventricular fibrillation,*** in which the heart muscle contracts continuously in fine, rippling fibrillatory movements. Such ventricles are incapable of pumping blood and the person dies very rapidly. Ventricular fibrillation may be initiated by a shock with 60 cycle alternating current, which causes impulses to go in many directions at once in the heart and establishes odd-shaped patterns of impulse transmission.

17–8 ADAPTATION OF THE HEART BEAT TO BODY ACTIVITY

Active tissue requires several times as much oxygen and nutrients as the same tissue at rest. Both heart and blood vessels participate in the adjustments necessary to provide the required amounts. During periods of intense exercise the heart can pump seven or eight times as much blood as normal by increasing the number of beats per minute, or by increasing the blood volume per beat, or both. The heart normally pumps about 75 milliliters of blood per beat, but it can pump as much as 200 milliliters. The following stimuli, singly or together, can effect this increase:

(a) *A rise in the carbon dioxide content of the blood.* During exercise, the rate of production of biologically useful energy (~P) is increased, more carbon dioxide is produced in tissues and diffuses into the bloodstream, and this stimulates the heart to increase the volume per beat.

(b) *Stretching of the heart muscle.* During exercise the pressure in the veins is increased and more blood flows into the heart chambers before they contract, stretching the muscular walls. The contractile power of a muscle, within limits, is increased by the tension it is under when it begins to contract; hence the greater the volume of blood in the heart at the beginning of systole, the greater will be the quantity of blood pumped per heart beat.

An increase in the rate of the heart beat from the normal 70 to as much as 200 beats per minute is possible during exercise. Again, several factors may be involved:

(a) *Increased temperature.* Enough heat is produced by exercising to raise the body temperature a few degrees. This affects the sinoatrial node (just as fever does), and the heart rate is increased.

(b) *Hormones.* Both epinephrine, produced in increased amounts by the adrenal glands during emergencies, and thyroxine, produced by the thyroid gland, accelerate the heart beat. When thyroxine is injected into an experimental animal, however, acceleration of the heart does not occur until three or four hours have elapsed—a reaction too slow to be effective in bringing about the rapid adjustment that the heart must make continuously, although it can affect the long-term responses of the heart to the general condition of the body.

(c) *Nerves.* Nervous control of the heart rate is localized in the *vasomotor center* in the medulla of the brain. Two sets of motor nerves pass from this center to the heart; one, via the sympathetic nerve trunk, accelerates heart rate, and the second, via the vagus nerve, decreases the heart rate. Both sets of nerves end in the sinoatrial node and increase or decrease the frequency with which impulses are initiated. Sensory receptors in the walls of the vena cava and right atrium are stimulated when the vessels are distended with blood and the wall is stretched. These initiate impulses that pass to the vasomotor center and result in a faster heart beat. Sensory nerves in the walls of the aorta and carotid arteries, stimulated by their distention, conduct impulses to the center that lead to a slower heart rate.

The action of these "feedback" controls increases the heart rate during exercise. The contraction of muscles during exercise increases venous return and distends the vena cava, stimulating the stretch receptors there. Impulses from these stimulate the vasomotor center of the brain. As the heart rate increases, the amount of blood in the aorta and carotid arteries increases, and the distention of their walls stimulates the stretch receptors there. Impulses from these increase the impulses from the vasomotor center via the vagus nerve and result in a slowing of the heart rate. This complex control system adjusts the heart rate quickly to the metabolic demands of the body yet prevents any serious over-response, for an increased heart rate leads to a stimulation of the aortic stretch receptors and these, in turn, lead to a slowing down of the heart rate.

17–9 ROUTES OF THE BLOOD AROUND THE BODY

The head and brain are supplied with blood by the *carotid arteries* and are drained by the *jugular veins* (Fig. 17–9). In addition, the brain is served by a second pair—the *vertebral arteries and veins* (not shown in Fig. 17–9)—lying close to the spinal cord. At the base

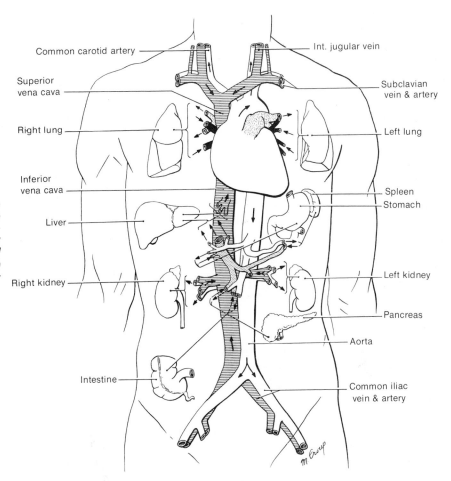

Common carotid artery

Int. jugular vein

Superior vena cava

Subclavian vein & artery

Right lung

Left lung

Inferior vena cava

Spleen

Stomach

Liver

Right kidney

Left kidney

Pancreas

Aorta

Intestine

Common iliac vein & artery

Figure 17–9 Diagram of the major arteries and veins of the human circulatory system; arteries (*white*), pulmonary artery (*stippled*), veins (*horizontal lines*), pulmonary veins (*black*) and hepatic portal system (*dotted*).

of the brain are interconnections between the carotid and vertebral arteries, so that if one of the vessels is cut or occluded, the brain is still adequately supplied with blood.

One exception to the rule that all veins carry blood to the heart is the *hepatic portal system,* which collects blood from the spleen, stomach, pancreas and intestines and conducts it to the liver. There the portal vein breaks up into capillaries, which in turn unite to form the hepatic vein. This vein drains blood from the liver into the inferior vena cava. Because of this arrangement, all blood from the spleen, stomach, intestines and pancreas must pass through the liver before it reaches the heart. Thus, food absorbed in the intestines is carried directly to the liver for storage.

17–10 THE RATE OF FLOW OF BLOOD

The rate of flow through the blood vessels depends primarily on the *pressure,* the force that drives the blood through the vessels, and the *resistance* offered by the smaller vessels to the flow of blood through them. The rate of blood flow to any specific organ at any given moment is remarkably well adjusted to the requirements of that organ for oxygen and nutrients. A set of control mechanisms regulates the flow of blood to each tissue in proportion to its needs from moment to moment.

The amount of blood pumped by the heart, the *cardiac output,* is about 5 liters per minute at rest and as much as 30 liters per minute during vigorous exercise. The amounts flowing to the various parts of the body are summarized in Table 17–1. Under basal conditions the liver, kidneys and brain receive 27, 22 and 14 per cent of the cardiac output, respectively, whereas skeletal muscles get only 15 per cent. During exercise almost all of the increased cardiac ouptut goes to the skeletal muscles, and they receive as much as 70 per cent of the total cardiac output.

In its course through the body, blood does

Table 17–1 Blood Flow to Regions of the Human Body under Basal Conditions and During Strenuous Exercise

	Basal Conditions*		Exercise	
	ML./MIN.	PER CENT OF TOTAL	ML./MIN.	PER CENT OF TOTAL
Brain	700	14	750	4.2
Heart	200	4	750	4.2
Bronchi	100	2	200	1.1
Kidneys	1100	22	600	3.3
Liver	1350	27	600	3.3
Via portal vein	(1050)	(21)		
Via hepatic artery	(300)	(6)		
Skeletal muscles	750	15	12,500	70.3
Bone	250	5	250	1.4
Skin	300	6	1,900	10.7
Thyroid gland	50	1	50	0.3
Adrenal glands	25	0.5	25	0.2
Other tissues	175	3.5	175	1.0
	5000	100.0	17,800	100.0

*Basal conditions data from Guyton, A. C.: *Function of the Human Body,* 4th ed. Philadelphia, W. B. Saunders Co., 1974. Based on data compiled by Dr. L. A. Sapirstein.

not flow at a constant speed. The flow is rapid in the arteries (about 30 centimeters per second in the larger ones), a little less rapid in the veins (about 8 centimeters per second in the larger ones), and slow in the capillaries (less than 1 millimeter per second). The differences in the rate of flow depend upon the total cross-sectional area of the vessels. If a fluid passes from one tube into another of larger size, the rate of flow is less in the larger tube. When the blood flows through a series of tubes of different sizes, connected together end-to-end, its velocity is always inversely proportional to the cross-sectional area of whatever tube it happens to be in.

The circulatory system is constructed in such a way that one large artery (the aorta) branches into many, intermediate-sized arteries. These in turn branch into thousands of small arteries (called *arterioles*), each of which gives rise to many capillaries. Although the individual branches of the aorta are smaller than the vessel itself, there are so many of them that the *total* cross-sectional area is greater, and the rate of flow correspondingly less. It has been estimated that the total cross-sectional area of all the capillaries of the body is about 800 times that of the aorta. Therefore, the rate of flow in the capillaries is about 1/800 as great as in the aorta. At the other end of the capillary network, the capillaries join to form small veins, *venules,* which combine to form increas-

ingly larger veins. As this occurs, the total cross-sectional area decreases and the rate of flow increases.

The length of a capillary, from arteriole to venule, is about 0.5 to 1.0 millimeter. With the blood moving at about 0.4 millimeter per second, the blood remains in the capillary for only one or two seconds. Despite this, the dynamics of exchange between blood and interstitial fluid are such that a large fraction of the substances in the blood can be transferred through the capillary wall.

Since the heart pushes blood into the arteries only during ventricular systole, arterial blood moves spasmodically—rapidly when the ventricles contract, slowly at other times. When the semilunar valves are closed, the blood in that part of the aorta nearest the heart is stationary, but the blood in the arteries farther away from the heart does not stop completely between systoles. In the arterioles the alternation in speeds is less marked; in the capillaries the flow of blood is almost constant, and the transfer of materials can occur continuously. This conversion of the intermittent flow in the arteries to the steady flow in the capillaries is made possible by the elasticity of the arterial wall. The force of the contracting ventricles pushes the blood forward, and distends and elongates the walls of the arteries (Fig. 17–10). During diastole the stretched walls contract and squeeze the blood along. Blood is prevented from flowing backward by the closure of the semilunar valves. The contraction of the arterial wall next to the heart distends the next section of the aorta or pulmonary artery, which in turn contracts and distends the next section, and so on. This alternate stretching and contracting passes along the arterial wall at the rate of 7 to 8 meters (25 feet) per second, and is known as the *pulse.* The blood inside the artery flows at a much slower rate, about 30 centimeters per second.

The heart is assisted in moving blood through the veins by the movements of the skeletal muscles and the motion of the body in breathing. Most of the veins are surrounded by skeletal muscles, which, when they contract, cause the veins to collapse (Fig. 17–11). As the muscles relax, the collapsed section again fills with blood, which must come from the direction of the capillaries. This mechanism by which muscle contractions "milk" blood along

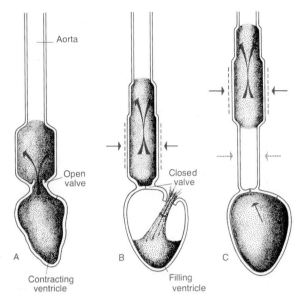

Figure 17–10 The movement of blood from the ventricle through the elastic arteries. For simplicity only one ventricle and artery are shown and the amount of stretching of the arterial wall is exaggerated. *A,* As the ventricle contracts, blood is forced through the semilunar valves, and the adjacent wall of the aorta is stretched. *B,* As the ventricle relaxes and begins to fill for the next stroke, the semilunar valve closes and the expanded part of the aorta contracts, causing the adjacent part of the aorta to expand as it is filled with blood. *C,* The pulse wave of expansion and contraction is transmitted to the next adjoining section of the aorta.

the veins is especially important in returning blood to the heart from the legs against the pull of gravity. If one stands upright quietly for a time, tissue fluid tends to collect in the legs, and swelling (*edema*) results. In walking, the contractions of the leg muscles force the blood along the veins, and the feet and ankles are less likely to swell. In breathing, the chest muscles and diaphragm contract, increasing the space inside the chest and lowering the pressure within the chest cavity below that outside the body, causing air to flow into the lungs. The pressure within the veins in the chest also is lowered as one inhales. Blood moves into the chest veins and atria for the same reason that air moves into the lungs.

These two factors are important in enabling the circulatory system to respond to the increased need for blood during exercise. At that time both the "milking" action of the muscles on the veins and the breathing movements are greatly increased, sending more blood to the atria. You will recall that as the volume of blood entering the heart increases, the heart muscle is stretched more and beats more forcibly, ejecting a greater volume per beat. Because of this, the muscular contractions during excitation, which bring about increased requirements for food and oxygen, are in part instrumental in causing the circulatory system to satisfy the increased requirements.

The rate at which blood is delivered to each part of the body is regulated by smooth muscle fibers present in the walls of arteries and arterioles (Fig. 17–12). By contracting or relaxing, these muscles can change the diameter of the artery over a three- to five-fold range. Since resistance to blood flow is inversely proportional to the *fourth power* of a vessel's diameter, this can be changed 100-fold to 1000-fold by the change in diameter.

The smooth muscles are innervated by two sets of nerves, one of which causes the muscles to contract, decreasing the size of the arterioles and lessening the supply of blood to that organ or part of the body. An increase in the number of nerve impulses in the other set causes the

Figure 17–11 The action of skeletal muscles in moving blood through the veins. *A,* Resting condition. *B,* Muscles contract and bulge, compressing veins and forcing blood toward heart. The lower valve prevents backflow. *C,* Muscles relax, and the vein expands and fills with blood from below; the upper valve prevents backflow.

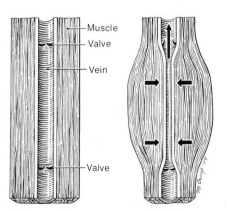

A   B   C

Muscle
Valve
Vein
Valve

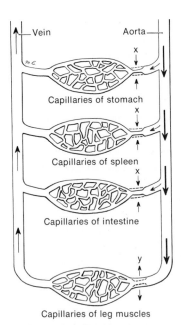

Figure 17–12 Diagram of the reactions of the circulatory system during exercise. Contraction of smooth muscles in the walls of the arterioles of the stomach, spleen and intestine *(X)* decreases the diameter of the arterioles and the flow of blood to those organs. Relaxation of smooth muscles in walls of arterioles of leg and other muscles *(Y)* increases the diameter of these vessels and the flow of blood to the skeletal muscles.

muscles to relax, increasing both the size of the arterioles and the flow of blood to that organ. These muscles usually are partially contracted because of a balance between the two nerve impulses. The arterioles thus act as valves to regulate the amount of blood each organ in the body receives. The smooth muscles of the walls of the arterioles are regulated by the local concentration of carbon dioxide and by the hormone epinephrine. When the brain is metabolizing at a high rate, the increased amount of carbon dioxide acts directly on the smooth muscles, causing them to relax, and thus increases the delivery of blood to the active tissue. In most other active tissues it is the decreased amount of oxygen rather than the increased amount of carbon dioxide that brings about an increased blood flow. Epinephrine causes a relaxation of the walls of the arterioles serving the skeletal muscles, but a contraction of the walls of the arterioles serving the internal organs—the stomach, intestines and liver—which results in a greatly increased flow of blood to the skeletal muscles. These chemical effects are independent of the nerves and

act equally well on normal arterioles and on those with severed nervous connections.

17–11 BLOOD PRESSURE

The force of the heart beat, the volume of blood in the circulatory system and the *peripheral resistance* (the state of constriction or relaxation of the blood vessels) determine the blood pressure. Blood pressure increases with increased force of the heart beat, with increased blood volume and with constriction of the blood vessels. It decreases with their opposites. Blood pressure rises with each contraction and falls with each relaxation of the ventricles. The highest pressure, due to systole of the heart, is called *systolic pressure*; the lowest, due to diastole, is called *diastolic pressure* (Fig. 17–13).

The pressure in an artery can be measured directly by inserting a tube into it and measuring the height to which the column of blood rises. An English clergyman, Stephen Hales, made the first such measurement in 1733. He found the blood pressure in the carotid artery of a mare sufficient to cause a column of blood to rise 9½ feet (285 cm.) in a glass tube.

It is not practicable to puncture an artery in a human being to measure blood pressure. Instead, measurements are made with a *sphygmomanometer,* a cuff tied around the arm and containing a rubber bag, to which is attached a rubber bulb and a device for measuring pressure. The pressure of the blood pushing against the sides of the arteries (Fig. 17–14 A) is measured by determining how much pressure in the rubber bag is required to collapse the artery.

In humans, systolic pressure is about 120 millimeters of mercury (i.e., equal to the pressure of a column of mercury 120 millimeters high), and diastolic pressure about 75 millimeters. The difference between systolic and diastolic pressures—the change in pressure that occurs each time the heart beats—is called the *pulse pressure*. The blood pressure decreases along the circulatory system from the aorta to the veins, being greatest in the aorta (as high as 140) and lowest in the veins near the atria, where it approaches or falls below zero, i.e., atmospheric pressure (Fig. 17–15). The decrease in pressure is caused by the fric-

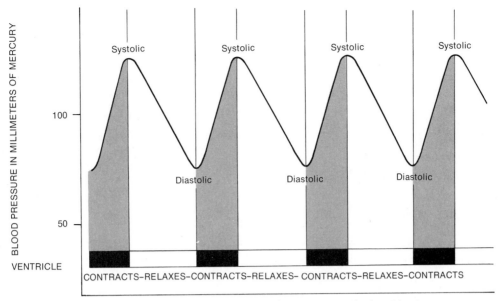

Figure 17–13 The changes in pressure in an artery as the heart beats.

tion of the blood rubbing against the walls of the blood vessels and is especially marked in the arterioles and capillaries, because those vessels are small and friction is greatest. The gradual decrease in pressure is necessary to keep the blood flowing; if the pressure were the same throughout the circulatory system, blood would not move.

17–12 HYPERTENSION

Hypertension, elevated arterial blood pressure, occurs in about one person in eight. It can place excessive strain on the heart, causing it to fail, and can cause blood vessels in the brain, kidneys and other vital organs to rupture. An elevation of blood pressure can be

Figure 17–14 The measurement of blood pressure by the sphygmomanometer. A, Blood pushes against the arterial walls. B, When the pressure in the rubber cuff exceeds 120 mm. of mercury, the artery is collapsed and no blood passes. C, With the pressure in the cuff just below 120, the artery is collapsed except for a small period during systole, when a small amount of blood squirts through, producing a sound audible in the stethoscope. D, With the pressure in the cuff at 95, the artery is open for a longer time during systole, a greater amount of blood passes through, and a louder noise is produced. E, When the pressure in the cuff drops below 75, the artery is open continuously, blood passes through continuously, and no noise is heard.

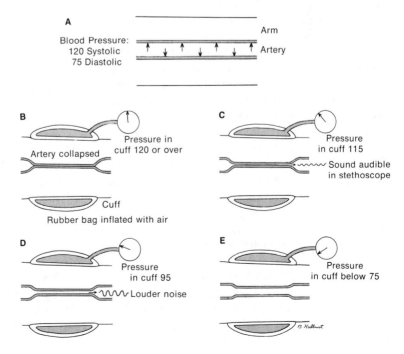

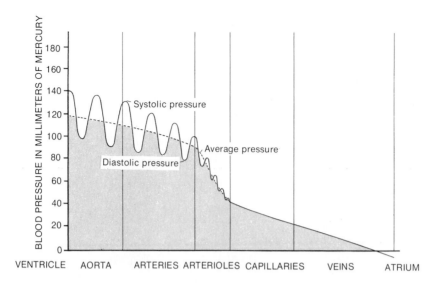

brought about by an increase in the total blood volume, by the constriction of the arteries, particularly the very small arteries, or by both mechanisms. The increased blood volume results from the effect of certain hormones of the adrenal cortex (mineralocorticoids) which act on the tubules of the kidney to increase the reabsorption of sodium. This, in turn, causes the increased retention of water and a resulting increase in blood volume. Even a 5 or 10 per cent increase in blood volume may have marked hypertensive effects. A tumor of the adrenal gland can secrete large amounts of *aldosterone* (the principal mineralocorticoid) and patients with such tumors have hypertension.

Vasoconstriction is brought about by many mechanisms, but one of the most important involves a protein, *angiotensin II.* This protein not only causes a constriction of arteries, but it also stimulates the adrenal gland to make and secrete more aldosterone. Thus angiotensin II promotes both vasoconstriction and increased blood volume, and both of these contribute to the elevation of blood pressure. Angiotensin II is formed enzymatically from an inactive precursor, angiotensin I, in the blood. Angiotensin I is formed from *angiotensinogen* (a glycoprotein made in the liver), and the conversion is catalyzed by a proteolytic enzyme (*renin*) formed in the kidney. Characteristically, damage to a kidney brings about an outpouring of renin which, via angiotensin II, causes an elevation of the blood pressure. Angiotensin II persists in the blood for a minute or two and

then is inactivated by enzymes collectively called angiotensinase.

Hypertension caused by adrenal tumors or obvious kidney damage accounts for only about 10 per cent of all cases of high blood pressure. The other 90 per cent or so involve "*essential hypertension,*" for which the etiology is still unknown. There are clear hereditary aspects to hypertension, for almost all such patients have parents or grandparents who also had essential hypertension. There seems little doubt that dietary factors such as salt intake may also play an important role. The intricate interrelationships of genetic, nutritional and hormonal factors which predispose to hypertension are being studied actively at the present time.

17–13 THE EXCHANGE OF MATERIALS ACROSS THE CAPILLARY WALL

The capillary membrane behaves as though it had much larger pores than those in the plasma membranes of cells. The pores in the capillary wall are about 8 nanometers in diameter (in contrast to the diameter of 0.8 nanometer of the pores in the plasma membrane) and glucose, amino acids and urea, and sodium, chloride and other ions readily pass through the capillary membrane by diffusion.

As blood passes through a capillary, tremendous numbers of water molecules and solutes pass back and forth through the pores in the capillary wall, providing continuous mixing between the interstitial fluid and the plasma. Although the pores make up less than 0.1 per cent of the surface area of the capillary, the rate of thermal motion is so great that the rate at which water molecules diffuse through the capillary membrane is nearly 40 times as great as the rate at which the plasma flows linearly through the capillary. Minute quantities of plasma proteins and other molecules too large to pass through the pores may be transferred across the membrane by pinocytosis in the endothelial cells.

An important aspect of capillary dynamics is the means by which plasma is retained within the capillary and not permitted to pass through the capillary wall into the interstitial fluid. Plasma contains about 7 grams of protein per 100 milliliters, and the interstitial fluid has a much smaller protein content, about 1.5 grams per 100 milliliters. As a result, the blood plasma has a higher osmotic pressure (termed *colloid osmotic pressure*) than the fluid bathing the tissues, and water tends to pass from the interstitial fluid into the capillaries. If two solutions are separated by a semipermeable membrane that is permeable to water but not to the solute molecules, water will tend to pass by osmosis from the solution with the greater concentration of water (and lesser concentration of solute molecules) to the other. The osmotic pressure of the 7 grams per cent protein solution in the blood pasma is about 28 mm. Hg and the osmotic pressure of the 1.5 gram per cent protein solution in the interstitial fluid is about 4 mm. Hg. The difference, 28 − 4 = 24 mm. Hg, is the force tending to move water molecules into the plasma (Fig. 17–16).

This is countered by the **hydrostatic pressure** in the capillary that results from the beating of the heart. This is difficult to measure directly but is estimated to be 18 mm. Hg on the average. Estimates of the pressure in the interstitial fluid (i.e., the pressure in the fluid just outside the capillaries) give values of −6 mm. Hg, just slightly less than atmospheric pressure. The algebraic sum of these two, +18 on the inside of the capillary and −6 on the outside, gives 24 mm. Hg as the total pressure difference between the two sides of the capillary membrane, which tends to push water *out* of the capillaries. This just equals the colloid osmotic pressure tending to move water *into* the capillaries. The balance between these two forces keeps the blood volume remarkably constant despite the higher hydrostatic pressure within the capillaries than in the interstitial fluid.

Although the colloid osmotic pressure is the same, or nearly the same, in both the arterial and venous ends of the capillary, the hydrostatic pressure within the capillaries is higher (about 30 mm. Hg) at the end next to the arteriole and lower (about 10 mm. Hg) at the end adjacent to the venule. These differences in pressure bring about an actual *net flow* of water molecules out of the capillary at the arterial end and back into the capillary at the venous end (Fig. 17–17). The sum of the forces acting at the arterial end gives a net *filtration pressure* of some 12 mm. Hg moving water molecules out of the capillary. At the venous end, the sum of the forces gives a net *absorption pressure* of about 8 mm. Hg moving water molecules back into the capillaries. The two forces are not equal, but part of the difference is counterbalanced by the fact that the venous ends of the capillaries are larger and there is half again as much surface area as at the arterial end.

Some of the fluid that passes out of the arterial end of the capillary, about 10 per cent of the total, is not reabsorbed at the venous end but passes into the lymphatic system and returns to the circulation by that route. Thus under normal conditions the blood volume remains constant as equal quantities of water

Figure 17–16 Diagram of the hydrostatic and osmotic pressures which are responsible for the exchange of materials between capillaries and tissue fluid.

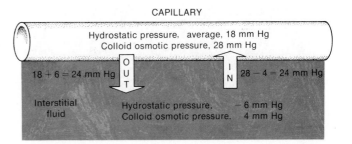

CAPILLARY

Hydrostatic pressure, average, 18 mm Hg
Colloid osmotic pressure, 28 mm Hg

18 + 6 = 24 mm Hg   OUT   IN   28 − 4 = 24 mm Hg

Interstitial fluid

Hydrostatic pressure, − 6 mm Hg
Colloid osmotic pressure, 4 mm Hg

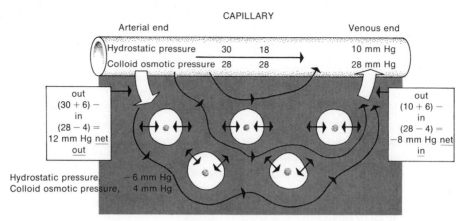

Figure 17–17   The path by which water flows in and out of capillaries to the interstitial fluid.

pass in and out of the capillaries, yet the flow of water in and out is important in supplying nutrients to the tissues.

After the loss of blood in a hemorrhage, the decreased blood volume causes a decrease in blood pressure and hence in the filtration pressure. However, the amount of protein per milliliter of plasma remains the same and the colloid osmotic pressure is not reduced. This increased net pressure returning water to the capillaries rapidly restores blood volume at the expense of the interstitial fluid (and intracellular fluid). In "shock," which may occur after burns or accidents, the permeability of the capillary walls is increased, some of the plasma proteins escape into the interstitial fluid and the colloid osmotic pressure is decreased. Fluids tend to escape from the capillaries into the tissues, causing *edema* (accumulation of fluid in the tissue spaces).

17–14   THE LYMPH VASCULAR SYSTEM

In addition to the blood vascular system, vertebrates have a second, independent group of vessels, the *lymphatic system.* These small, thin-walled vessels originate as minute, dead-end *terminal lymphatic capillaries,* present in nearly all tissue spaces (Fig. 17–18). The cells at the tips of the lymph capillaries are not joined directly but the edge of one overlaps the edge of the adjacent cell to form a flap that serves as a microvalve to prevent the backflow of fluid from the lymph into the interstitial fluid. The lymph capillaries join and form successively larger lymph vessels or veins, which, like the veins of the blood vascular system, have valves to prevent backflow. The final large lymphatic vessels drain into the blood vascular system at the junction of the internal jugular and subclavian (shoulder) veins.

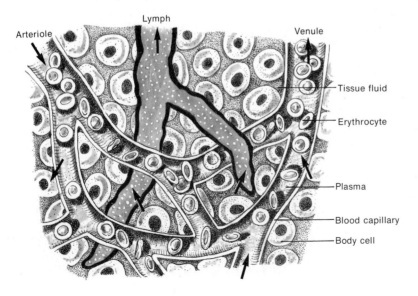

Figure 17–18 Diagram of the relation of blood and lymph capillaries to tissue cells. Note that blood capillaries are connected at both ends whereas lymph capillaries, outlined in black, are "dead-end streets," and contain no erythrocytes. The arrows indicate direction of flow.

The lymphatic vessels are an auxiliary system for the *return* of fluid from the tissue spaces to the circulation; there are no lymphatic arteries. The lymph capillaries, closed at one end, are very permeable and proteins and other large particles readily pass into them along with the tissue fluid. The fluid in the lymph capillaries, called **lymph,** has essentially the same composition as the interstitial fluid and, like it, has much less protein than the blood. Thus, tissue fluid passes into the lymph capillaries, becomes lymph, and then is carried to the junction with the blood vascular system and is mixed with blood.

At the junctions of lymph vessels are aggregations of cells, the **lymph nodes,** which produce one kind of white cell, the **lymphocytes,** and filter out bacteria and other particulate matter so that they do not enter the blood vascular system. Lymph flows very sluggishly through the minute, tortuous channels in the lymph nodes, and invading bacteria are trapped and phagocytized by the cells of the lymph node. Some bacteria may get past the first node and be caught in the second or third. In a massive infection the bacteria may penetrate all the lymph nodes and invade the blood stream. The presence of these phagocytized bacteria causes the lymph nodes to become swollen and tender—the lymph nodes of the neck may become noticeably swollen in individuals with sore throats. The lymph nodes in the lungs of heavy smokers are filled with particles of smoke and become a dark grey or black. These particles may eventually interfere with the functioning of the lymph nodes and reduce resistance to lung infections.

Although some of the lower animals, such as the frog, have four lymph "hearts" which pulsate and squeeze lymph along, the human lymph is moved by contractions of the adjacent skeletal muscles which compress the lymph vessels (the valves prevent backflow) and by the breathing movements of the chest. The lymphatics play an important role in the body's economy by returning fluid, and especially proteins, to the blood vascular system. There is a continual slow loss of protein from the circulation to the interstitial fluid; about 4 per cent of the total protein in the plasma passes out each hour. The lymphatics are the only means by which proteins can be returned to the circulation. Without lymphatics the concentration of protein in the interstitial fluid would soon equal the concentration in the blood capillaries with a consequent fall in colloidal osmotic pressure and decrease in blood volume. In addition, the lymph system produces lymphocytes, filters out bacteria and particulate matter, and plays a role in transferring fats absorbed from the intestinal villi into the blood vascular system (p. 423).

The rate at which lymph flows is slow and variable, but the total lymph flow is about 100 milliliters per *hour*—very much slower than the 5 liters per *minute* in the blood vascular system.

17–15 CIRCULATION IN OTHER ORGANISMS

All organisms have the same problem of transporting substances from one part of the body to another. The human heart, with its remarkable automatic devices for keeping the blood flowing and for adapting to changing conditions, is the result of a long evolutionary process.

The unicellular organisms have no special system for bringing about circulation of substances; foods, wastes and gases simply diffuse through the cytoplasm and eventually reach all parts of the cell. In most protozoa this process is aided by movements of the cytoplasm. As an amoeba moves along, the cytoplasm streams from the rear to the front of the body, distributing substances throughout the cell (Fig. 17–19 A). In other types of protozoa (e.g., *Paramecium*), which have a firm, outer tunic and do not change shape as they move, substances are distributed by a rhythmic, circular movement of the cytoplasm in the direction indicated by the arrows in Figure 17–19 B. Food is taken in through a "mouth" and gullet on one side of the animal. Food vacuoles form at the base of this gullet, break off and move through the cell as they digest and give off food. Gases and waste products of metabolism are moved similarly.

The central cavity of the cnidaria (Fig. 19–11) serves as both a digestive and circulatory organ. When the tentacles have captured their prey they stuff it through the mouth

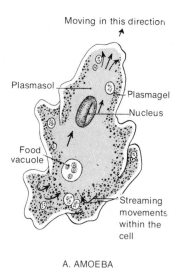

Moving in this direction

Plasmasol

Plasmagel

Nucleus

Food
vacuole

Streaming
movements
within the
cell

A. AMOEBA

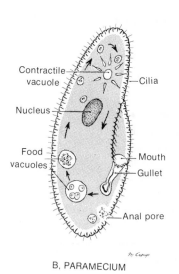

Contractile
vacuole

Cilia

Nucleus

Food
vacuoles

Mouth

Gullet

Anal pore

B. PARAMECIUM

Figure 17–19 Circulation in protozoa by means of streaming movements of the cytoplasm.

into the cavity, where digestion occurs. The digested food substances then pass to the cells lining the cavity and through them, by diffusion, to the outer layer. The movement of the animal's body, as it alternately stretches and contracts, stirs up the contents of the central cavity and aids circulation.

The flatworm planaria (Fig. 19–11) resembles the hydra in having a single, central cavity connected to the outside by a single mouth opening. But besides the inner and outer layers of cells found in hydra, the planaria has a third layer of cells, loosely packed between the other two. The spaces between these cells are filled with a tissue fluid somewhat similar to the tissue fluid of human beings. Food enters through the mouth, is digested in the central cavity, diffuses through the inner layer of cells and passes through the tissue fluid to the other cells. As in the cnidaria circulation is aided by the muscles in the body wall which agitate the fluid in the central cavity and the tissue fluid.

The earthworm and similar forms have a definite closed circulatory system, consisting of plasma, blood cells and blood vessels, although the latter are not specialized as arteries, veins and capillaries. The circulating blood remains within blood vessels. There are two main blood vessels, one on the ventral side, in which blood flows posteriorly (Fig. 17–20), and one on the dorsal side, in which it flows anteriorly. Connecting these in each part of the body are small tubes which serve the intestine, skin and other organs. In the anterior part of the worm

are five pairs of "hearts" or pulsating tubes which conduct blood from the dorsal to the ventral vessel to complete the circuit. The contractions of the muscles of the body wall aid the hearts in circulating the blood.

Mollusks and arthropods, such as clams, squids, crabs and insects, have an "open" circulatory system consisting of a heart, blood vessels, blood spaces, plasma and blood cells. The heart is different from a vertebrate heart, consisting in most forms of a single muscular sac. The blood vessels from the heart open into large spaces, enabling the blood to bathe the body cells. From these spaces, blood is collected by other vessels and returned to the heart. The details are different in various animals, but in all, the circulatory system supplies the cells of the body with oxygen and nutrients, and removes metabolic wastes.

Although both the earthworm and the human have the same red pigment, hemoglobin, for the transport of oxygen, the hemoglobin in human blood, and the blood of vertebrates in general, is present within red blood cells; whereas that in the blood of earthworms and many other invertebrates is present dissolved in the plasma, and any cells in the blood are colorless. In earthworms and other animals in which the oxygen-transport protein is dissolved in the plasma, the protein is of high molecular weight (400,000 to 6,700,000) and composed of many heme-protein subunits (not just four as in the human), each of which binds one O_2 per heme. What adaptive significance do you think this may have had in evolu-

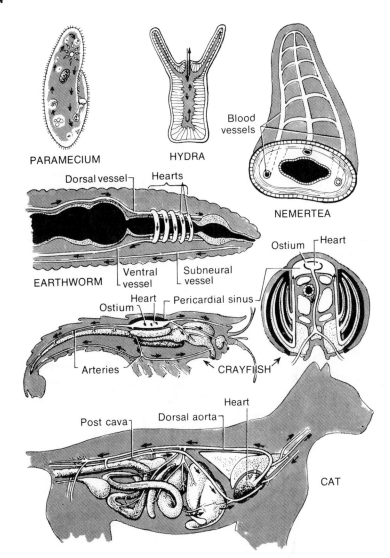

Figure 17–20 The circulatory systems of paramecium, hydra, nemertea, earthworm, crayfish and cat.

tion? What evolutionary advantage accrued to those animals that enclosed their oxygen-carrying proteins within erythrocytes?

The circulatory systems of all vertebrates are fundamentally the same, from fish and frogs through lizards to birds and man. All have a heart and an aorta, as well as arteries, capillaries and veins, organized on a similar basic plan. Because of this similarity, students can learn much about the human circulatory system from the dissection of a dogfish or frog.

In the evolution of the higher vertebrates, from the lower, fishlike forms, the principal changes in the circulatory system occurred in the heart, and are correlated with the change in the respiratory mechanism from gills to lungs. The fish heart consists of four chambers in a row: sinus venosus, atrium, ventricle and conus (Fig. 17–21 *A*). Blood from the veins drains into the sinus venosus, while blood from the

conus is pumped through the ventral aorta to the gills, where it takes up oxygen. It then goes to the dorsal aorta and is distributed to all parts of the body. Blood passes through the fish heart only once each time it makes a circuit through the body.

In the particular group of fish from which the land vertebrates evolved, a number of changes occurred in the heart and blood vessels which are evident in present-day frogs (Figs. 15–18, 17–21 *B*). A partition developed down the middle of the atrium, dividing it into right and left halves. The sinus venosus shifted its connection so that it emptied into the right atrium. Veins from the lungs emptied in the left atrium, while pulmonary arteries to the lungs grew out of the vessels which originally served the most posterior pair of gills. Thus, in the frog, blood should pass from the veins to the sinus venosus, to the right atrium, to the

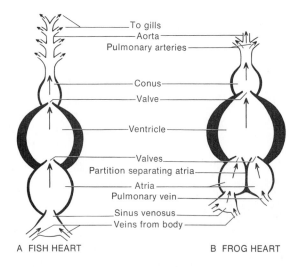

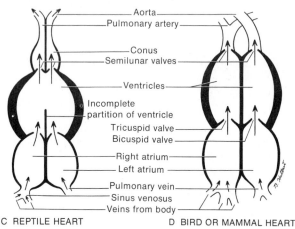

Figure 17–21 Diagram illustrating the evolution of the heart in vertebrates.

cause they are full of other blood, and so passes via the aorta to the body cells. Because aerated and nonaerated blood may mix in the ventricle, blood may pass through the heart once, twice or even more times for each circuit it makes through the body.

In the evolution of reptiles from their ancestral amphibians, one partition developed down the center of the ventricle, and one down the conus (Fig. 17–21 C). The ventricular partition is incomplete in all the reptiles except alligators and crocodiles, and there is still some mixing of aerated and nonaerated blood, although less than in the frog. The sinus venosus is small, foreshadowing its disappearance in the mammalian heart.

The hearts of birds and mammals (Fig. 17–21 D) have completely separate right and left sides. The interventricular partition is complete, precluding any mixture of blood from the right and left hearts. The conus has split and become the base of the aorta and pulmonary artery. The sinus venosus has disappeared as a separate chamber, although a vestige remains as the sinoatrial node. The absolute separation of right and left hearts forces the blood to pass through the heart twice each time it makes a tour of the body. As a result, the blood in the aorta of birds and mammals contains more oxygen than that in the aorta of the lower vertebrates. Hence the tissues of the body receive more oxygen, a higher metabolic rate can be maintained, and the warm-blooded condition is possible. Birds and mammals can maintain a constant, high body temperature in cold surroundings.

ventricle, to the aorta, the pulmonary arteries, the lungs, the pulmonary veins, the left atrium, the ventricle, the aorta, and then to the cells of the body. Of course, there is some mixing of aerated and nonaerated blood in the ventricle, and some blood from the sinus venosus may get into the aorta instead of the pulmonary artery, while some from the left atrium may be pumped into the pulmonary artery. There is less mixing than one might suppose, however. Blood from the right atrium enters the ventricle ahead of that from the left atrium, and so lies nearer the exit. As the ventricle contracts, nonaerated blood from the right atrium leaves first and enters the arteries branching off from the aorta—the pulmonary arteries to the lungs. The aerated blood from the left atrium leaves the ventricle toward the end of the contraction, is unable to enter the pulmonary arteries be-

QUESTIONS

1. Describe in detail the means by which nutrients and oxygen are transported to and enter the cells of the brain.
2. Describe the route of the blood through the heart. What part do the semilunar valves play in the passage of the blood?
3. Discuss the mechanisms that regulate the heart rate in humans and other mammals.
4. Describe the cyclic series of events that occurs when the human heart beats.
5. How does the developing mammalian embryo obtain food and oxygen?
6. In what part of the circulatory system does the blood flow most slowly? What is the physical explanation for this?

7. Calculate the rate of blood flow in the aorta expressed in miles per hour.
8. How can you explain the fact that the pulse beat passes along an artery more rapidly than the blood itself passes through the vessel?
9. What is meant by systolic and diastolic blood pressure?
10. What physiologic changes occur during shock?
11. What is lymph? How does it differ from blood? Where and how is it produced?
12. By what pathways may fluids return to the heart in the human?
13. Compare the manner in which nutrients and oxygen are transported to the body cells in an amoeba, a hydra, a planarian, a crab and a frog.
14. What is the principal difference between the circulatory system of a fish and that of the human?
15. Trace the path of a red cell from a vein in the leg to the kidney in the human.
16. Trace the path of a molecule of sugar from the capillaries of the intestine to a brain cell where it is metabolized and then trace the path of the carbon dioxide formed until it is finally excreted by the lungs.
17. What factors supplement blood pressure in bringing about the return of venous blood?
18. Describe the forces that are involved in the exchange of water and solute molecules between the capillaries and tissue fluid.
19. How does the heart adjust its rate and its output per beat to the increased venous return that occurs during increased bodily activity?
20. Compare the circulation through the heart of a mammalian fetus with that in an adult mammal.

SUPPLEMENTARY READING

John Fulton's *Selected Readings in the History of Physiology* is a collection of 87 passages from as many authors in which the most significant contributions to physiology—the landmarks in the history of physiologic thought—are given. The passages are arranged chronologically by subject so that one can trace the sequence of ideas about the circulation of the blood in Chapter Two and about capillaries in Chapter Three. William Harvey's description of the action of the valves of the veins in preventing the flow of blood away from the heart, one chapter of his *De Motu Cordis,* is presented. The entire text of Harvey's book, translated by Chauncy D. Leake, is also available. Further details of circulation and the many mechanisms that control the rate and volume of blood flow can be found in physiology texts such as those by Davson, Guyton, Ruch and Patton, and Vander, Sherman and Luciano. Carl J. Wiggers' *Scientific American* article, *The Heart,* describes some of the peculiar properties of cardiac muscle and the general functions of the heart. The properties of the microscopic vessels through which blood flows from arteries to veins, the capillaries, are discussed by Benjamin Zweifach in *The Microcirculation of the Blood.* Many aspects of the transport of substances across membranes are discussed in *Membrane Transport* by L. E. Hokin.

CHAPTER 18

RESPIRATION: GAS EXCHANGE

The energy for all the myriad activities of plants and animals is derived from reactions of **biologic oxidations** (Chapter 5). The essential feature of these reactions is the transfer of hydrogen atoms from hydrogen donors to hydrogen acceptors. In most animals and plants there is a series of compounds, each of which accepts hydrogen (or its electron) from the preceding one and donates it to the subsequent one. The ultimate hydrogen acceptor in the metabolism of most plants and animals is oxygen, which is converted to water. Since only small amounts of oxygen can be stored (as oxyhemoglobin in blood or as the comparable oxymyoglobin in muscle), the continuation of metabolism depends upon an uninterrupted supply of oxygen to each cell. Most cells die quickly if deprived of oxygen; mammalian brain cells are especially sensitive and are damaged beyond repair if their supply is cut off for only a few minutes.

Carbon dioxide is removed from certain substrate molecules in other reactions. **Decarboxylation**, the removal of carbon dioxide from a larger molecule, can proceed independently of oxygen utilization. Yeasts, for example, can metabolize sugar to alcohol and carbon dioxide without utilizing oxygen. In most animals and plants the utilization of oxygen and the splitting off of carbon dioxide proceed together. The carbon dioxide must be removed from the cell and body fluids as it is produced, for it reacts with water to form carbonic acid, H_2CO_3.

The term **respiration** is used to refer to those processes by which animal and plant cells utilize oxygen, produce carbon dioxide, and convert energy into biologically useful forms such as ATP. The term respiration has had three different meanings in biology. It originally was synonymous with breathing, and meant inhaling and exhaling; "artificial respiration" refers to this usage of the term. Then, as it became clear that the important process was the exchange of gases between the cell and its environment, the term respiration was applied to this. Finally, as the details of cellular metabolism became known, the term respiration was used to denote those enzymatic reactions of the cell which are responsible for the utilization of oxygen. Thus the cytochromes are called "respiratory enzymes."

18-1 DIRECT AND INDIRECT RESPIRATION

The exchange of gases is a fairly simple process in small, aquatic animals such as paramecia or hydras. Dissolved oxygen from the surrounding pond water diffuses into the cells, carbon dioxide diffuses out, and no special respiratory system is needed. Such gas exchange is called *direct respiration;* the cells of the organism exchange oxygen and carbon dioxide directly with the surrounding environment.

As animals evolved into larger, more complex forms, it became impossible for each cell to exchange gases directly with the external environment. Some form of *indirect respiration* involving a structure of the body specialized for gas exchange with the environment was necessary. This specialized structure had to be thin-walled (the membrane of the wall must be differentially permeable), so that diffusion could occur easily; it had to be kept moist, so that oxygen and carbon dioxide were dissolved in water; and it required a good blood supply. For indirect respiration, fishes, crabs, lobsters and many other animals developed gills; the higher vertebrates, reptiles, birds and mammals, developed lungs; the earthworm uses its moist skin; and insects use tracheal tubes, canals running all through the body, connected by pores with the external environment (Fig. 18–1).

In indirect respiration an external and an internal phase can be distinguished in the exchange of gases between the body cells and the environment. *External respiration* is the exchange of gases by diffusion between the external environment and the bloodstream, by means of the specialized respiratory organ—for example, the lung in mammals. *Internal respiration* is the exchange of gases between the bloodstream and the cells of the body (Fig. 18–2). Between these phases the gases are transported by the circulatory system.

18-2 THE HUMAN RESPIRATORY SYSTEM

The respiratory system in man and other air-breathing vertebrates includes the lungs and the tubes by which air reaches them (Fig. 18–3). Air enters the body through the *external nares,* or nostrils, which open into the *nasal chamber,* a large cavity dorsal to the mouth cavity and ventral to the brain. This cavity contains the sense organs of smell, and is lined with mucus-secreting epithelium. Air is warmed, filtered and moistened as it passes through the nasal chamber. When the capillaries in this chamber dilate excessively and too much mucus is secreted the nose becomes "stopped up" and we have the symptoms of a cold.

Air passes via the *internal nares* to the *pharynx* where the paths of the digestive and respiratory systems cross. Food passes from the

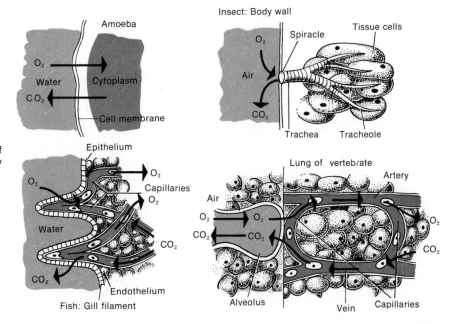

Figure 18–1 A diagram of some of the types of respiratory organs present in animals.

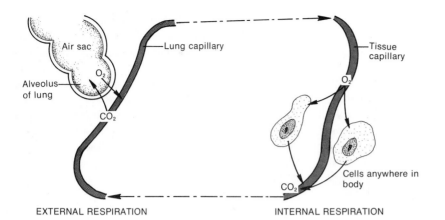

Figure 18–2 Diagram illustrating the exchanges of gases in external and internal respiration.

pharynx to the stomach by way of the esophagus, and air passes to the lungs by way of the larynx and trachea. To prevent food from entering the larynx and trachea and injuring the delicate membranes lining them, a flap of tissue, the *epiglottis*, folds over the opening of the larynx whenever food is swallowed. This is done automatically; we do not have to remember to close the epiglottis each time we swallow. Occasionally the automatic mechanism fails and food goes down the "wrong throat."

The **larynx**, or voice box, evident externally as the Adam's apple, contains the vocal cords, folds of epithelium which vibrate as air passes over them, thereby producing sounds. Muscles adjust the tension of the cords to produce sounds of varying pitch. The **trachea**,

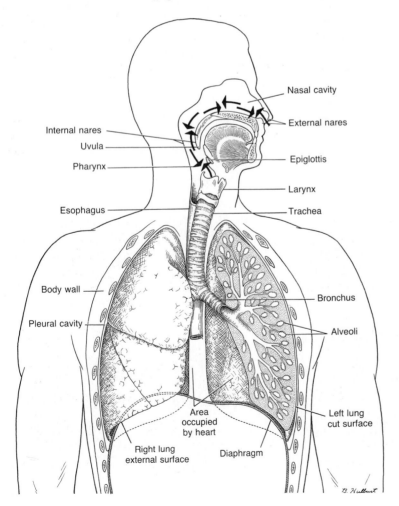

Figure 18–3 Diagram of the human respiratory system.

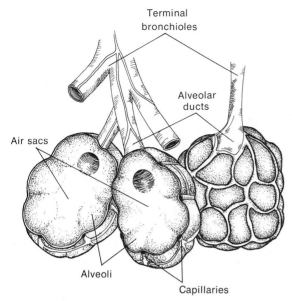

Terminal
bronchioles

Alveolar
ducts

Air sacs

Alveoli

Capillaries

Figure 18–4 Diagram of a small portion of the lung, highly magnified, showing the air sacs at the end of the alveolar ducts, the alveoli in the walls of the air sacs, and the proximity of the alveoli and the pulmonary capillaries containing red blood cells.

or windpipe, may be distinguished from the esophagus by the rings of cartilage embedded in its walls which hold it open. During inspiration the pressure of air in the trachea is less than atmospheric pressure, and without the cartilaginous rings the trachea would collapse.

At the level of the first rib the trachea branches into two cartilaginous *bronchi,* one extending to each lung. Inside the lung, each bronchus branches into bronchioles, which in turn branch repeatedly into smaller and smaller tubes leading to the ultimate cavities, the *air sacs.* In the walls of the smaller vessels and the air sacs are minute, cup-shaped cavities, *alveoli,* kept moist and supplied with a rich network of capillaries (Fig. 18–4). Molecules of oxygen and carbon dioxide diffuse readily through the thin, moist walls of the alveoli. The total alveolar surface across which gases may diffuse has been estimated to be greater than 100 square meters—more than fifty times the area of the skin.

The wall of the trachea or bronchus consists of an inner layer of epithelium, an outer layer of connective tissue, and a middle layer containing the cartilaginous rings and smooth muscle fibers. In a person suffering from asthma these smooth muscle fibers contract abnormally and reduce the size of the passage,

thus making breathing difficult. The lining epithelium secretes mucus and contains ciliated cells. These cilia beat constantly in one direction, so that when bacteria or particles of dust land on the moist surface, they are trapped in the mucus and carried by the beating of the cilia back up to the pharynx.

As the bronchioles and sub-branches of the respiratory passages become smaller, the walls become thinner, the layer of cartilage disappears, and the ciliated cells are replaced by squamous epithelium. The walls of the alveoli are composed of just one layer of flat epithelial cells. Electron micrographs have shown that there are always two membranes, the alveolar epithelium and the endothelium of the capillary, separating pulmonary air from the blood. In between the alveoli and holding them in place are strands of elastic connective tissue. Lungs are so elastic that when they are freshly removed from an animal they may be blown up like a balloon through the trachea. When the pressure is released, the elasticity of the stretched lung deflates it and expels the air.

Each lung, as well as the cavity of the chest in which the lung rests, is covered by a thin sheet of smooth epithelium called the *pleura.* The pleura is kept moist, enabling the lung to move in the chest cavity during breathing without undue friction. The pressure in the pleural cavity (the space between the two sheets of pleura) is generally less than that of the outside atmosphere. The elasticity of the lungs tends to make them pull away from the chest wall, setting up a partial vacuum in the pleural cavity. When these pleural linings become inflamed, they secrete a fluid which accumulates in the cavity between the lung and the chest wall, a condition known as pleurisy.

The chest cavity is closed and has no communication with the outside atmosphere or with any other body cavity. It is bounded on the top and sides by the chest wall, containing the ribs, and on the bottom by a strong, dome-shaped sheet of skeletal muscle, the *diaphragm* (Fig. 18–3).

18–3 THE MECHANICS OF BREATHING

It is necessary to keep clear the distinction between respiration and breathing. Respiration refers to the exchange of gases between a cell and its environment. In man this consists of three phases: external respiration, transportation by the bloodstream and internal respiration. *Breathing* is simply the mechanical process of taking air into the lungs (inspiration) and letting it out again (expiration). The lung capillaries constantly remove oxygen from, and put carbon dioxide into, the air in the alveoli. Thus, the need for replacing the air in the lungs is obvious. In the resting human being the breathing cycle of inspiration and expiration is repeated about fifteen to eighteen times a minute.

In human beings and other mammals the ribs, chest muscles and diaphragm are easily movable, and the volume of the chest cavity can be increased or decreased at will. During inspiration, the rib muscles contract, drawing the front ends of the ribs upward and outward, an action made possible by the hingelike connection of the ribs with the vertebrae (Fig. 18–5 *A*). The floor of the chest cavity, the diaphragm, contracts, decreasing its convexity and consequently enlarging the cavity (Fig. 18–5 *B*). Because the space is closed, this increase in volume results in a lowering of the pressure in the lungs. When it falls below atmospheric pressure, air from the outside rushes in through the trachea and its branches to the air sacs and alveoli.

Air is expelled from the lungs in expiration by the elasticity of the lungs and by the weight of the chest wall. During inspiration the lungs are distended as they become filled with air. When the rib muscles relax, the ribs are permitted to return to their original position, and the simultaneous relaxation of the diaphragm permits the abdominal organs to push it back up to its previous convex shape. These factors decrease the chest volume and allow the distended, elastic lungs to contract and expel the air which had been inhaled.

During muscular exercise this passive expiration of the relaxing rib and diaphragm muscles is not rapid enough to expel the air before the next inspiration must start, and the size of the chest cavity is reduced by muscular contraction. In addition to the muscles which raise the ribs for inspiration, another set with fibers at right angles to the first lowers the front ends of the ribs and thus decreases the thoracic volume. The muscles of the abdominal wall may also contract, squeeze the abdominal organs up against the diaphragm, and further hasten the elastic contraction of the lungs. The chest walls never squeeze the lungs or forcibly expel the air; the decrease in the size of the thoracic cavity simply permits the lungs to contract by means of their own elasticity.

The cough reflex is essential to keep the passageways of the lungs free of foreign material. In response to irritation of the bronchi or trachea by foreign matter the reflex begins with the intake of about 2.5 liters of air. The vocal cords and the epiglottis close tightly to trap the air, and the abdominal muscles and muscles of the chest wall contract, raising the pressure within the lungs to about 100 mm. Hg. Then the vocal cords and epiglottis open suddenly and the air under pressure within the lungs

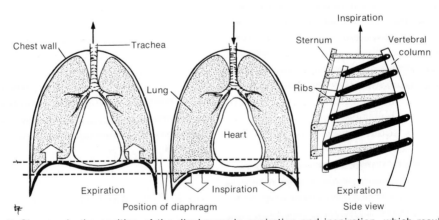

Figure 18–5 *Left,* Changes in the position of the diaphragm in expiration and inspiration, which result in changes in the volume of the chest cavity. *Right,* Changes in the position of the rib cage in expiration and inspiration. The elevation of the front ends of the ribs by the chest muscles causes an increase in the front-to-back dimension of the chest, and a corresponding increase in the volume of the chest cavity. These two factors, by increasing the volume of the chest cavity, result in the intake of a corresponding amount of air.

explodes outward at velocities as great as 120 km. per hour. This rapidly moving air carries with it foreign matter in the trachea and bronchi. The sneeze reflex is similar to the cough reflex but begins in response to irritation in the nasal passages. The intake and expulsion of air are similar to those of the cough reflex, but the uvula is depressed so that a large volume of air passes rapidly through the nose and mouth, helping to clear the nasal passages.

Between breaths the pressure in the lungs is the same as atmospheric pressure. As inspiration starts, there is a slight decrease in the pressure within the lungs to 2 or 3 mm. of mercury below atmospheric. This causes air to enter the lungs. Toward the end of inspiration the newly entered air has equalized the pressure. As expiration begins, the elasticity of the lungs compresses the air within the lung cavities and the pressure rises to 2 or 3 mm. of mercury above atmospheric; consequently, air passes out of the lungs. Of course, by the end of expiration, pressure within the lungs is back to atmospheric pressure.

18–4 THE QUANTITY OF AIR RESPIRED

An adult human at rest breathes in and out about 500 ml. of air with each breath. When these 500 ml. have been expelled, however, another 1.5 liters or so can be expelled by contracting the abdominal muscles. After this there still remains about 1 liter of air that cannot be expelled. During normal breathing, therefore, a reserve of some 2.5 liters of air remains in the lungs, with which the 500 ml. are mixed. After a normal inspiration of 0.5 liter, it is possible, by inspiring deeply, to take in as much as 3 liters more, and during exercise one can increase the amount of air inspired and expired with each breath from 0.5 liter to 5 liters. But even in strenuous exercise the full tenfold increase is seldom used; instead, the rate of breathing is increased. By having an individual breathe in as deeply as possible and then breathe out as completely as possible into some device for measuring air volume, you can measure his *vital capacity.* The average vital capacity of the young adult male is about 4.5 liters and of the young adult female about 3.2 liters. A tall, thin person usually has a higher vital capacity than an obese person, and a well developed athlete may have a vital capacity that is 35 to 55 per cent above normal — that is, 6 to 7 liters. In certain diseases of the heart and lungs the vital capacity may be reduced considerably below normal.

Although about 500 ml. of air are breathed in with each inspiration, only some 350 ml. actually reach the alveoli. The last 150 ml. inhaled remain in the larger air passages, where no exchange of gases between lungs and blood stream can occur. This air is the first to be pushed out with the next expiration. The last 150 ml. expelled from the alveoli with each breath also remains in these tubes, and this air, although laden with carbon dioxide, is the first to be drawn into the alveoli on the next breath. With each breath then, only about 350 ml. of new air reach the lungs to mix with the 2.5 liters already there. The 150 ml. of space in the air passages is known as "dead space." If the dead space is increased (by breathing through a long tube, such as a garden hose) the air reaching the lungs will soon be depleted of oxygen and this can have fatal results.

Because the lungs are not completely emptied and filled with each breath, alveolar air contains less oxygen and more carbon dioxide than atmospheric air (Table 18–1). A sample of essentially alveolar air can be collected at the end of a maximal expiration. Expired air has had less than one quarter of its oxygen removed and can be breathed over again.

The amount of new air moved into the respiratory passages with each breath is called the *tidal volume.* The normal tidal volume of a young adult male is about 500 ml. His normal respiratory rate is about 12 breaths per minute. The product of these two numbers — about 6 liters per minute — is termed the "minute respiratory volume." During strenuous exercise the minute respiratory volume may rise to as much as 100 to 120 liters per minute, and for brief periods (about 15 seconds) a young adult male can force himself to breathe as much as 150 to

Table 18–1 PERCENTAGE COMPOSITION OF ATMOSPHERIC AND ALVEOLAR AIR

	Atmospheric Air	Alveolar Air
Nitrogen	79.0	75.3
Oxygen	20.96	13.2
Carbon dioxide	0.04	5.3
Water vapor	Variable	6.2

170 liters per minute. Thus, the respiratory system has a marked reserve and can increase its minute respiratory volume 20- to 25-fold.

18–5 ARTIFICIAL RESPIRATION

Individuals who have stopped breathing because of drowning, smoke inhalation or electric shock may be sustained by artificial respiration until their own breathing reflexes can be initiated again. A variety of methods of artificial respiration have been practiced over the years, and the manual method presently advocated by the American Red Cross is the Holger-Nielsen method. The individual is placed on his face, and his arms are folded in front of his head. The operator kneels in front of the patient's head and places his hands on the back of the patient over the lower half of the shoulder blades and pushes with about 40 pounds pressure for 3 seconds, causing expiration. He then removes pressure and grasps the subject under his upper arms and lifts, causing inspiration.

Another effective method of artificial respiration is mouth-to-mouth breathing in which the operator takes a deep breath and then breathes rapidly into the mouth of the subject. In the past the method was not highly favored because it was believed that the expired air of the operator would not be beneficial to the subject. However, this is not true, because normal expired air still has an adequate amount of oxygen and the carbon dioxide present in the expired air may be helpful in stimulating the respiratory center of the subject. In any method of artificial respiration it is important to maintain a free airway through the mouth or nose to the trachea. The tongue of a patient has a tendency to fall into the back of the throat and obstruct the airway and it is important to check that the tongue has been pulled forward before beginning artificial respiration.

18–6 EXCHANGE OF GASES IN THE LUNGS

Oxygen passes from the alveoli to the pulmonary capillaries, and carbon dioxide passes in the reverse direction by the physical process of diffusion; each gas passes from a region of higher concentration to one of lower concentration. The extremely thin alveolar epithelium offers little resistance to the passage of the gases, and since there is normally a greater concentration of oxygen in the lung alveoli than in the blood entering the lungs in the pulmonary artery, oxygen diffuses from the alveoli into the capillaries. Similarly, the concentration of carbon dioxide in the blood in the pulmonary artery is normally higher than in the lung alveoli, so that carbon dioxide diffuses from the lung capillaries into the alveoli. In contrast to the cells lining the intestine, which can take a substance from the intestinal cavity and pass it into the blood where the concentration of that substance may be higher, the alveolar epithelium is unable to move either oxygen or carbon dioxide against a diffusion gradient.

Whenever the concentration of oxygen in the alveoli drops below a certain value, the blood passing through the lungs cannot take up enough to meet the body's needs and the symptoms of altitude sickness—nausea, headache and delusions—appear. Altitude sickness begins to occur at about 5000 meters, or even lower in some persons. People can become acclimated to living at high altitudes by increasing the number of red cells in the blood, but no one can live much above 6000 meters without supplementary oxygen. At about 13,000 meters the pressure is so low that even when a man breathes pure oxygen he cannot get enough to supply his body. Airliners are made airtight and provided with pumps to maintain the air pressure in the cabin at about that of sea level, 760 mm. Hg.

In the capillaries of the tissues throughout the body, where internal respiration takes place, oxygen moves by diffusion from the capillaries to the cells, and carbon dioxide from the cells to the capillaries. The continuous metabolism of glucose and other substances in the cells results in the continuous production of carbon dioxide and utilization of oxygen. Consequently, the concentration of oxygen is always lower and the concentration of carbon dioxide is always higher in the cells than in the capillaries.

Throughout the system, from lungs to blood to tissues, oxygen moves from a region of high concentration to one of lower concentration and is finally used in the cells; carbon

OUT OF DATE

dioxide moves from the cells, where it is produced, through the blood to the lungs and out, always toward a region of lower concentration.

18-7 TRANSPORT OF OXYGEN BY THE BLOOD

At rest, the cells of the human body need about 300 liters of oxygen every 24 hours, or 250 ml. per minute. With exercise or work this requirement may increase as much as ten or fifteen times. If oxygen were delivered to the tissues simply dissolved in plasma, blood would have to circulate through the body at a rate of 180 liters per minute to supply enough oxygen to the cells at rest, for oxygen is not very soluble in plasma. Actually, the blood of a man at rest circulates at about 5 liters per minute and supplies all the oxygen his cells need. The reason why the body requires only 5 liters per minute rather than 180 is the action of hemoglobin.

Hemoglobin, the pigment in red blood cells, transports nearly all the oxygen and most of the carbon dioxide. Blood in equilibrium with alveolar air can take up in solution only 0.25 ml. of oxygen and 2.7 ml of carbon dioxide per 100 ml., but by the actions of hemoglobin, 100 ml. of blood can carry about 20 ml. of oxygen and 50 to 60 ml. of carbon dioxide. Human blood contains about 15 grams of hemoglobin per 100 ml.

Approximately 2 per cent of the oxygen in the blood is dissolved in the plasma; the rest is carried in combination with hemoglobin. After oxygen enters the capillaries in the lungs, it diffuses into the red cells from the plasma and unites with hemoglobin (Hb) to form oxyhemoglobin (HbO_2). The oxygen atoms are attached to the iron atoms in the heme rings of hemoglobin:

$$Hb + O_2 \rightleftarrows HbO_2$$

The arrows indicate that the reaction is reversible. Hemoglobin would, of course, be of little value to the body if it could only take up oxygen and not give it off where needed. The reaction goes to the right in the lungs, forming oxyhemoglobin, and to the left in the tissues, releasing oxygen. The difference in color between arterial and venous blood is due to the fact that oxyhemoglobin is a bright scarlet, whereas hemoglobin is purple.

The combination of oxygen with hemoglobin and its release from oxyhemoglobin are controlled by the concentration of oxygen present and, to a lesser extent, by the concentration of carbon dioxide present. In the lungs the concentration of oxygen is relatively high, and oxyhemoglobin is formed. After leaving the lungs the blood passes through the heart and arteries, where there is little change in the oxygen concentration, to the tissues, where the oxyhemoglobin is exposed to an environment with little oxygen. It consequently dissociates, releasing the oxygen to diffuse to the tissue cells.

Carbon dioxide reacts with water to form carbonic acid, H_2CO_3; hence an increase in the concentration of CO_2 increases the acidity of the blood. The oxygen-carrying capacity of hemoglobin decreases as blood becomes more acid; thus the combination of hemoglobin with oxygen is controlled in part by the amount of CO_2 present. This results in an extremely efficient transport system. In the capillaries of the tissues, CO_2 concentration is high and oxygen is released from hemoglobin by the combined effects of low oxygen tension and high CO_2 tension. In the capillaries of the lung (or the gill in fishes), CO_2 tension is lower, and oxygen is taken up by hemoglobin by the combined effects of high oxygen tension and low CO_2 tension. The important thing to realize is that the more carbon dioxide the blood contains, the more acid it is, and that the oxygen-carrying capacity of hemoglobin is less in an acid solution.

The factor that actually determines the direction and rate of diffusion is the pressure, or "tension," of the particular gas. In a mixture of gases, each one exerts, independently of the others, the same pressure it would exert if it were present alone. In air at sea level, where the total pressure is about 760 mm. of mercury, oxygen exerts one fifth of the pressure, i.e., the partial pressure, or tension, of oxygen in the atmosphere is 150 mm. of mercury. The alveolar air contains less oxygen than atmospheric air; the tension of oxygen in the alveoli is about 105 mm. Blood passes through the lung capillaries too rapidly to become completely equilibrated with the alveolar air, so that the oxygen tension in the arterial blood is about 100 mm.

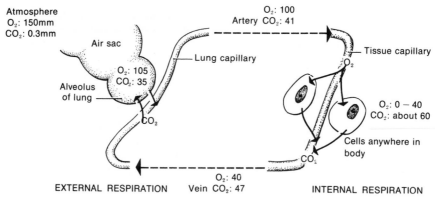

Figure 18–6 Diagram illustrating the diffusion gradients responsible for the transfer of oxygen from lungs to tissues and for the transfer of carbon dioxide from tissues to lungs. The oxygen and carbon dioxide tensions are expressed in millimeters of mercury.

(Fig. 18–6). The oxygen tension in the tissues varies from 0 to 40 mm., with the result that oxygen diffuses out of the capillaries into the tissues. Not all the oxygen leaves the blood, however; blood passes through the capillaries too rapidly for equilibrium to be reached, and the tissues usually have some residual oxygen. The venous blood returning to the lungs has an oxygen tension of about 40 mm. of mercury. At the oxygen tension of arterial blood (100 mm.), each 100 ml. contains about 19 ml. of oxygen. At the oxygen tension of venous blood (40 mm.), each 100 ml. contains 12 ml. of oxygen. The difference of 7 ml. represents the amount of oxygen *delivered to the tissues* by each 100 ml. of blood. Thus the 5 liters of blood in the body can deliver 350 ml. of oxygen on each circuit.

18–8 TRANSPORT OF CARBON DIOXIDE BY THE BLOOD

The transport of carbon dioxide poses a special problem to the body, because when carbon dioxide dissolves, it reacts reversibly with water to form carbonic acid:

$$CO_2 + H_2O \rightleftarrows H_2CO_3$$

The cells of the body produce, at rest, about 200 ml. of carbon dioxide per minute. If this were simply dissolved in plasma (which can transport in solution only 4.3 ml. of carbon dioxide per liter), blood would have to circulate at a rate of 47 liters per minute instead of 4 or 5. This amount of carbon dioxide dissolved

in water would yield a pH of 4.5, and cells are able to survive only within a narrow range on the alkaline side of neutrality (between about pH 7.2 and 7.6). The unique properties of hemoglobin enable each liter of blood to transport about 50 ml. of carbon dioxide from tissues to alveoli, with only a few hundredths of a pH unit difference in the acidity of arterial and venous blood. Some of the carbon dioxide is carried in a loose chemical union with hemoglobin (*carbaminohemoglobin*), and a small amount is present as carbonic acid, but most of the latter is transported as bicarbonate ion, HCO_3^-. The CO_2 produced by cells dissolves in the interstitial fluid to form H_2CO_3, a reaction catalyzed by *carbonic anhydrase*. The H_2CO_3 is neutralized by the sodium or potassium ions released when oxyhemoglobin is changed into hemoglobin. Oxyhemoglobin is a stronger acid than hemoglobin and some cations, sodium and potassium, are released when oxyhemoglobin dissociates into oxygen and hemoglobin. It is amazing that in the course of evolution a single chemical, hemoglobin, has been produced that has all the necessary characteristics to allow the transport of large amounts of oxygen and carbon dioxide with minimal change in the pH of the blood while these gases are being transported.

Carbon dioxide passes from tissues to blood to lungs by diffusing from a region of high tension to one of lower tension. The carbon dioxide tension in the tissues is about 60 mm. of mercury; in the venous blood, about 47 mm.; and in the alveoli, about 35 mm. Arterial blood has a CO_2 tension of about 41 mm. of mercury, so that the blood contains a great deal of carbon dioxide after it has passed through the lungs.

Any condition (such as pneumonia) that interferes with the removal of carbon dioxide by the lungs leads to an increased concentration of carbon dioxide in the blood (really an increased concentration of bicarbonates and carbonic acid). This condition is called *respiratory acidosis*, but the term does not imply that the blood is actually acid (it is still on the alkaline side of neutrality). There is a decrease in the alkaline reserves of the blood (chiefly sodium). Acidosis also occurs in diabetes. Here, however, the trouble is not a respiratory acidosis due to a failure to remove the carbon dioxide in the lungs, but a metabolic acidosis due to an overproduction of acids by the tissues as a result of impaired carbohydrate metabolism.

18-9 ASPHYXIA

Asphyxia results whenever there is an interruption in the delivery of oxygen to the tissues or a failure in the utilization of oxygen by the tissues. Therefore, the cause of asphyxia may lie in the lungs, blood or tissues. In drowning, the lung alveoli become filled with water, and in pneumonia they become filled with tissue fluid, bringing about asphyxia from lack of oxygen. In carbon monoxide poisoning, asphyxia results because the hemoglobin of the blood unites with carbon monoxide instead of oxygen and so is unavailable for transporting oxygen to the cells. In cyanide poisoning, asphyxia is caused by the inactivation of cytochrome oxidase, an important link in the chain of enzymes responsible for the utilization of oxygen by the tissues.

18-10 DISEASES OF THE RESPIRATORY SYSTEM

The thin, delicate membranes of the lungs, exposed as they are to the air, are subject to invasion by disease organisms and to the deleterious effects of many environmental influences. Tuberculosis, caused by tubercle bacilli, results in the walling off of the infected areas of the lung by the growth of fibrous tissue as a protective response, but the walled off region reduces the area of the lung available for gas exchange and reduces vital capacity. Pneumonia, the filling of the lung alveoli with fluid, may result from infection by pneumococci or by viruses or as a result of severe pulmonary edema. The surface available for gas exchange is reduced as the infection spreads from one lobe of the lung to the next. Red and white blood cells, as well as fluid, may pass into the alveoli, and a large area of the lung may become filled with fluid and cellular debris.

The prime deleterious environmental influence attacking the lungs is, of course, tobacco smoke, which can cause chronic emphysema as well as lung cancer. Emphysema, difficulty in breathing, results from the obstruction of the air flow through the terminal bronchioles and from destruction of the lung tissue itself, both of which greatly reduce the effective area of respiratory membranes. The obstruction of the bronchioles results in increased airway resistance and increased work of breathing, especially on expiration. The loss of lung tissue decreases the number of lung capillaries through which blood can pass, increases pulmonary vascular resistance, and results in pulmonary hypertension and failure of the right side of the heart. Prolonged exposure of the lung cells to the carcinogens in tobacco smoke can result in lung cancer, in which the normal cells change into abnormal cancerous ones that grow and may break off to be carried in the lymph system to other parts of the body where they again grow. Such secondary growths are called metastases.

The alveolar cells secrete a material called *surfactant,* a complex of protein and phospholipid, which lowers the surface tension within the alveoli some five to ten fold and prevents collapse of the lungs. In a fraction of newborn infants, especially those born prematurely, the lungs do not secrete an adequate amount of surfactant and it may be impossible for their respiratory forces to overcome the surface tension in their unexpanded alveoli and thus expand the lungs. The infants die of suffocation in what is called *respiratory distress syndrome.* The adult lungs may lose the ability to secrete surfactant after they have been perfused during open heart surgery and this may result in collapse of the alveoli in parts of the lung.

18-11   THE CONTROL OF BREATHING

Since the needs of the body for oxygen are different at rest and at work, the rate and depth of breathing must change automatically to meet varying conditions. During exercise, oxygen consumption by muscles and other tissues may increase four or five times.

Breathing requires the coordinated contraction of a great many separate muscles, achieved by the *respiratory center*, special groups of cells in the medulla and pons of the brain. From this center, volleys of nervous impulses pass out rhythmically to the diaphragm and rib muscles, resulting in their regular and coordinated contraction every four or five seconds. Under ordinary conditions the breathing movements are automatic and occur without our voluntary control. When the nerves to the diaphragm (the phrenic nerves) and to the rib muscles are cut or destroyed, breathing movements stop at once. Of course, we can voluntarily change the rate and depth of breathing. We can even hold our breath for a while, but we cannot hold it very long—the automatic mechanism takes over and brings about an inspiration.

The question naturally occurs: Why does the respiratory center give off this volley of impulses periodically? Through a series of experiments it has been determined that if the connections of the respiratory center with all other parts of the brain are cut—that is, if the sensory nerves and those from the higher brain centers are severed—the center sends out a constant stream of impulses, and the breathing muscles contract and remain contracted. The respiratory center, then, if left to its own devices, causes a complete contraction of the breathing muscles. If either the sensory nerves or the nerves from the higher centers of the brain are left intact, however, the breathing movements continue in normal fashion. This means that for normal breathing to occur, the respiratory center must be *inhibited* periodically so that it stops sending out impulses which cause contraction of the muscles. Further experiments have revealed that the *pneumotaxic center* in the anterior part of the pons, together with the medullary respiratory center, forms a "reverberating circuit" which provides for the basic control of the respiratory rate (Fig. 18–7).

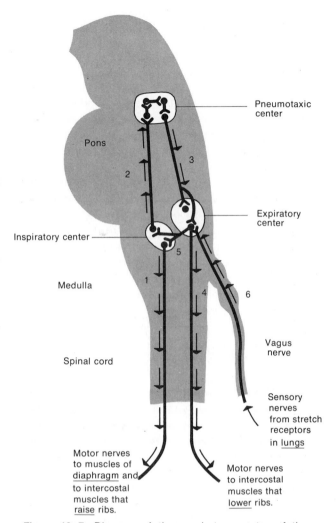

Figure 18–7 Diagram of the respiratory center of the brain. *(1)* Nerve impulses from inspiratory center in medulla stimulate muscles of diaphragm to contract and intercostal muscles to raise ribs. *(2)* Other impulses pass to pneumotaxic center in pons, around its neuronal circuits and eventually *(3)* to expiratory center in medulla. Expiratory center is excited and sends impulses *(4)* to intercostal muscles that lower ribs. Other impulses *(5)* pass to inspiratory center to inhibit it momentarily. When impulses from pneumotaxic center die out inspiration begins again and cycle is repeated. In addition, sensory nerve endings in lungs stimulated by stretching during inspiration send impulses via vagus nerves *(6)* that stimulate the expiratory center and inhibit the inspiratory center. This reflex from the stretch receptors in the lungs provides for a second feedback mechanism to regulate the respiratory cycle.

The stretching of the walls of the alveoli during inspiration stimulates pressure-sensitive nerve cells in their walls which send impulses to the brain to inhibit the respiratory center and bring about the following expiration.

Many other nervous pathways connected with the respiratory center carry either stimulating or inhibiting impulses. Severe pain in any part of the body causes a reflex acceleration of breathing. Also, both the larynx and the pharynx have receptors in their linings which, when stimulated, send impulses to the respiratory center to inhibit breathing. These are valuable protective devices. When an irritating gas, such as ammonia or acid fumes, passes down the respiratory tract, it stimulates the receptors in the larynx, which, by sending impulses to the respiratory center to inhibit breathing, bring about an involuntary "catching of the breath." This prevents the harmful substance from entering the lungs. Similarly, when food accidentally passes into the larynx, it stimulates receptors in the lining of that organ to send inhibitory impulses to the respiratory center. This momentarily stops breathing so that the food particles do not enter the lungs and injure the delicate lining epithelium.

During exercise both the rate and depth of breathing increase to meet the increased needs of the body for oxygen and to prevent the accumulation of carbon dioxide. The concentration of carbon dioxide in the blood is the prime factor controlling respiration. An increased concentration of carbon dioxide in the blood flowing to the brain increases the excitability of both the respiratory center and the pneumotaxic center. Increased activity of the former increases the strength of contraction of the muscles of respiration and increased activity of the latter increases the rate of breathing. When the concentration of carbon dioxide returns to normal, these centers are no longer stimulated and the rate and depth of breathing return to normal.

This mechanism also works in reverse. If a person voluntarily takes a series of deep inhalations and exhalations, he reduces the carbon dioxide content of his alveolar air and blood to such a degree that when he stops breathing deeply, all breathing movements cease until the carbon dioxide in the blood again builds up to a normal level. The first breath of a newborn child is initiated largely by this mechanism. Immediately after a baby is born and separated from the placenta, the carbon dioxide content of its blood increases, stimulating the respiratory center to send nerve impulses to the diaphragm and rib muscles to contract in the first breath. When a newborn infant has difficulty in taking its first breath, air containing 10 per cent

carbon dioxide may be blown into its lungs to set off this mechanism.

Experiments have shown that an increase in the carbon dioxide content of the blood, rather than a decrease in the oxygen content, is primarily effective in stimulating the respiratory center. If a person is placed in a small, airtight chamber so that he breathes and rebreathes the same air, the oxygen in the air gradually decreases. If a chemical is placed in the chamber to absorb the carbon dioxide as fast as it is given off so that its concentration in the lungs and blood does not increase, the man's breathing accelerates only slightly, even if the experiment is continued until the oxygen content is greatly reduced. If, however, the carbon dioxide is not absorbed but is allowed to accumulate, breathing will be greatly accelerated, causing discomfort and a choking sensation in the subject. When he is supplied with air that contains the normal amount of oxygen but an increased amount of carbon dioxide, there is again an acceleration of breathing. Obviously, it is primarily the accumulation of carbon dioxide, not a deficiency of oxygen, that stimulates the respiratory center.

As additional protection against the failure of the body to respond properly to changes in the carbon dioxide and oxygen content of the blood, still another control has evolved. At the base of each internal carotid artery is a small swelling, the *carotid sinus*, containing receptors sensitive to changes in the composition of the blood. If the carbon dioxide content increases, or the oxygen decreases, these receptors are stimulated to send nerve impulses to the respiratory center in the medulla, increasing its activity.

18–12 HIGH FLYING AND LOW DIVING

The barometric pressure decreases as one goes to progressively higher altitudes. The fraction of air that is oxygen, about 21 per cent, remains the same; hence the partial pressure of oxygen decreases along with the barometric pressure. At an altitude about 3000 meters the barometric pressure is about 525 mm. Hg and the partial pressure of oxygen is 110 mm. Hg. Under these conditions the hemoglobin in our arterial blood is only 90 per cent saturated with oxygen. At 6000 meters the barometric pressure is about 350 mm. of mercury, the partial pressure of oxygen is 75 mm. Hg., and arterial oxygen saturation is 70 per cent. At 10,000 meters the barometric pressure is about 225 mm. Hg, the partial pressure of oxygen is 50 mm. Hg, and arterial oxygen saturation is only 20 per cent. Thus hypoxia becomes an ever-increasing problem at higher altitudes. At 10,000 meters a person breathing pure oxygen would be breathing a gas with a partial pressure of oxygen of 225 mm. Hg and would have hemoglobin fully saturated with oxygen. Above 13,000 meters, however, the barometric pressure is so low that even breathing pure oxygen would not enable the individual to keep his arterial hemoglobin fully saturated with oxygen. The arterial saturation becomes 50 per cent (below which a person ordinarily loses consciousness) at an altitude of 7000 meters if you are breathing air and at an altitude of 14,200 meters if you are breathing pure oxygen. All high-flying jets have cabins that are airtight and pressurized to an altitude of about 2000 meters.

Hypoxia, a deficiency of oxygen, results in impaired night vision, drowsiness, decreased mental proficiency, euphoria, nausea, and, ultimately, coma. If a jet was flying at 11,500 meters and underwent sudden decompression, the pilot, even if breathing pure oxygen, would lose consciousness in about 30 seconds and become comatose in about 1 minute.

In addition to the problems of hypoxia, a rapid decrease in barometric pressure can cause bubbles of gas to form in the blood and other body fluids, resulting in *decompression sickness*, with great pain and even death. Whenever the barometric pressure drops below the total pressure of all the gases dissolved in the body fluids, the dissolved gases tend to come out of solution into the gaseous state and form bubbles.

Decompression sickness is even more common in deep sea diving than in high altitude flying. As a person descends below the sea pressure around him increases tremendously—1 atmosphere for each 10 meters. To prevent the collapse of his lungs he must be supplied with air under pressure, thereby exposing his lungs to very high alveolar gas pressures. Workers digging tunnels in caissons often must work in a pressurized area to prevent the tunnel from collapsing. Divers breath-

ing compressed air suffer from nitrogen narcosis when they spend more than an hour at depths below 45 to 50 meters. They first become euphoric as though mildly intoxicated, then drowsy, clumsy, and finally unconscious. The nitrogen apparently dissolves in the lipid membranes of the neurons and reduces their excitability, i.e., their ability to conduct nerve impulses.

Breathing oxygen under high pressure can cause oxygen toxicity resulting in epilepticlike convulsions and coma. The oxygen may be toxic for certain of the oxidative enzymes of the cells, it may decrease blood flow through the brain, or it may cause production of oxidizing free radicals that oxidize certain essential cell components, thereby damaging the cell's metabolic machinery.

A major problem faced by anyone breathing air under high pressures for a long time is the large amount of nitrogen that becomes dissolved in the body fluids. At sea level an adult human has about one liter of nitrogen dissolved in his body, with about half in his fat and half in the body fluids. After a diver's body has been saturated with nitrogen at a depth of 100 meters, the body fluids contain some 10 liters of nitrogen. If the diver suddenly returns to the surface, large quantities of bubbles of nitrogen gas develop in the extracellular fluids and in the fluid phase within the cells, causing pain, dizziness, paralysis, and even unconsciousness. The formation of bubbles within the cells of the central nervous system can disrupt pathways in the spinal cord and brain and result in permanent paralysis. To prevent decompression sickness the diver must be brought to the surface gradually, with stops at certain levels on the way up, so that nitrogen can be expelled normally through the lungs. Alternatively, the diver can be brought to the surface rapidly and then placed in an airtight tank where he is subjected to air under pressure. The pressure in the tank is gradually reduced to sea level and the diver may then emerge. The total decompression time required depends upon the depth the diver reached and the time the diver remained at that depth.

18–13 EVOLUTION OF THE LUNGS

The earliest suggestion of a lung is found in certain types of fish. Some ancestral fish developed an outgrowth of the anterior end of the digestive tract, and in the line of fishes that eventually gave rise to the land vertebrates, this outgrowth became a lung. In other fishes it became a *swim bladder,* an organ that facilitates floating, though it may function as a respiratory organ (Fig. 18–8). The swim bladder is usually single, although it may be paired and may exhibit great variations in size and shape in different species. In some fish it has lost its connection with the digestive tract. The cells at the anterior end of the swim bladder have the ability, found nowhere else in the animal kingdom, to secrete oxygen from the blood into the cavity of the bladder. Other cells at the posterior end remove oxygen from the cavity and return it to the blood stream. By thus "pumping" oxygen in or out of these swim bladders, the fish can maintain a certain depth in the water without muscular effort. Certain

species are even equipped with a series of bones that connect this organ with the internal ear and presumably act as a depth gauge.

The swim bladder may also serve as an organ for the production of sounds. A few fish, such as the toadfish and sea robin, are able to make a noise by contracting the muscles attached to the swim bladder, causing it to vibrate.

Close relatives of the fish that gave rise to the land vertebrates are the *lung fish,* a few of which have survived to the present time in the headwaters of the Nile and the Amazon, and in certain Australian rivers. These animals live in streams that dry up periodically, and during the dry seasons they remain in the mud of the stream bed, breathing by means of their swim bladders. They are also equipped with gills, used in external respiration when swimming. The swim bladders of these fish are simple sacs, single in some forms, paired in others. In

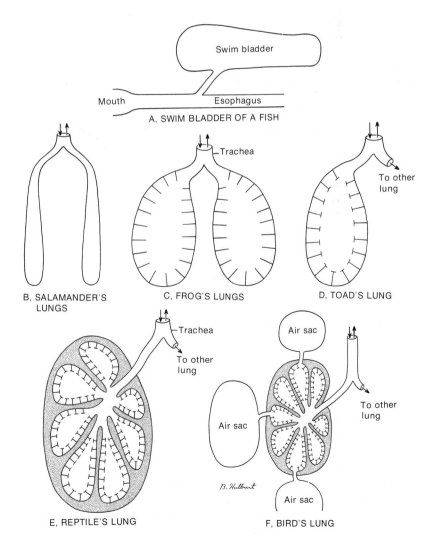

Figure 18–8 Some stages in the evolution of the lungs.

contrast to those of other fish, they are equipped with a pulmonary artery. The lungs of the most primitive amphibians, the mud puppies, are two long simple sacs, covered on the outside by capillaries. Frogs and toads have folds on the inside of the lung sac which increase the respiratory surface (Fig. 18–8). Since frogs have no diaphragm or rib muscles, their method of breathing is quite different from that of humans and depends on the action of valves in the nostrils and muscles of the throat. With the nose valves open, the floor of the mouth cavity is drawn down and air is taken into the mouth. Then the nose valves close, and the throat muscles contract, decreasing the size of the mouth cavity and forcing the air back into the lungs. A frog cannot breathe with his mouth open!

The trend in subsequent evolution was toward a greater subdivision of the lung into smaller and smaller sacs, so that the lung structure became increasingly complex in reptiles, birds and mammals. The lungs of some lizards—for example, the chameleon—have supplementary air sacs that can be inflated, enabling the animal to swell up, which perhaps is a protective device to frighten would-be predators. Birds have rather similar sacs at several points throughout the body (Fig. 18–8) by which air can be drawn all through the lungs and completely renewed on each inspiration. In addition, when the bird is flying and the chest wall must be held rigid to form an anchor for the flight muscles, the air sacs act as a bellows to move air in and out of the lungs. The air sacs, lying between certain flight muscles, are squeezed and relaxed on each stroke of the wing; thus, the faster the bird flies, the

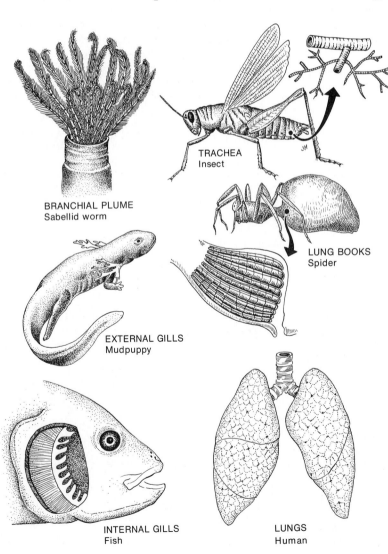

BRANCHIAL PLUME
Sabellid worm

TRACHEA
Insect

LUNG BOOKS
Spider

EXTERNAL GILLS
Mudpuppy

INTERNAL GILLS
Fish

LUNGS
Human

Figure 18–9 Diagrams of the types of respiratory organs found in animals.

more rapid is the circulation of air through the lungs.

18–14 GAS EXCHANGE IN OTHER ANIMALS

A primitive form of external respiration not involving any specialized respiratory organ is found in both invertebrates (e.g., worms) and vertebrates (e.g., amphibia). In these the moist skin serves as a respiratory organ. The membranes lining the mouth cavity and pharynx may also serve as a respiratory organ.

External respiration in most aquatic animals is carried on by specialized structures called *gills*. Fish, mollusks (oysters, squids) and many arthropods (shrimps, crabs, spiders, and so forth, but not insects) have these organs (Fig. 18–9). Every animal with gills has some arrangement for keeping a current of water flowing over them. A fish opens its mouth, takes a gulp of water, then closes its mouth and forces the water out past its gills by contracting its mouth cavity. Gills, like lungs, have thin walls and are moist and well supplied with blood capillaries. Oxygen dissolved in the water diffuses through the gill epithelium into the capillaries, and carbon dioxide diffuses in the reverse direction. The amount of oxygen dissolved in sea water is relatively constant, but the amount in fresh water ponds and rivers may fluctuate widely. Fish suffocate in water lacking sufficient dissolved oxygen, such as that in stagnant ponds.

Insects have quite a different system for getting oxygen to the cells. Each segment of the body has a pair of holes, called *spiracles*, through which air passes by a system of branched air ducts, *tracheal tubes*, extending to all of the internal organs (Fig. 18–9). The ducts terminate in microscopic fluid-filled *tracheoles*; oxygen and carbon dioxide diffuse through these into the adjacent cells. The body walls of insects pulsate, drawing air into the trachea when the body expands, and forcing air out when the body contracts. Grasshoppers, for example, draw air into the body through the first four pairs of spiracles when the abdomen expands and expel it through the last six pairs of spiracles when the abdomen contracts. Thus, in contrast to a fish or crab, in which blood is brought to the surface of the body to be aerated in a gill, the tracheal system conducts air deep within the insect body, near enough to each cell so that it can diffuse in through the wall of the tracheal tube. The tracheal system is efficient; oxygen reaches the cells and carbon dioxide is removed by diffusion. The insect need not maintain a rapid flow of blood as vertebrates must to supply their cells with oxygen.

Animals differ enormously in their rates of utilization of oxygen. A resting mouse uses some 2500 mm.³ of oxygen per hour for each gram of body weight and as much as 20,000 mm.³ per hour per gram when active. An earthworm uses perhaps 60 mm.³ per hour per gram, and a sea anemone uses only 13 mm.³ per hour per gram.

QUESTIONS

1. List the parts of the human respiratory system. How is each adapted for its particular functions?
2. Differentiate clearly between "breathing" and "respiration."
3. What is meant by the "vital capacity" of a person? In what conditions is it increased or decreased?
4. What function does oxygen serve in the body?
5. Discuss the role of hemoglobin in the transport of oxygen. In the transport of carbon dioxide.
6. What is meant by "oxygen tension"?
7. What is acidosis? What factors may cause acidosis?
8. Where is the respiratory center and what are its functions? Describe briefly the neural control of breathing.
9. Describe the sequence of events that occurs when there is an accumulation of carbon dioxide in the cells.
10. Describe the major trends in the evolution of the lungs.
11. How do gills operate?
12. How is respiration carried on in insects?
13. Trace the path of an anesthetic, such as ether, from the ether cone over the nose to the cells in the brain. List each structure passed and the processes involved in its passage.
14. Differentiate between direct and indirect respiration and between external and internal respiration.
15. Why does one experience difficulty in breathing at high altitudes?
16. Why is it that alveolar air differs in its composition from atmospheric air? Of what significance is this fact?
17. What physiologic mechanisms bring about an increase in the rate and depth of breathing during exercise? Why is such an increase necessary?

18. In what group of vertebrates and under what environmental conditions did lungs first arise in evolution?

SUPPLEMENTARY READING

Some of the classic experiments on respiration by Hooke, Mayow, Priestley, Lavoisier and Barcroft are quoted in J. F. Fulton, *Selected Readings in the History of Physiology*, Chapter Four. Chapter Ten of W. B. Cannon's *The Wisdom of the Body* includes discussions of some of the quantitative aspects of human breathing. The human respiratory system is described in greater detail in the texts on physiology by Davson, Guyton, Ruch and Patton, and Vander, Sherman and Luciano. Recommended for further reading on the subject of the comparative aspects of respiration are chapters in Florey's *General and Comparative Animal Physiology*, Prosser and Brown's *Comparative Animal Physiology* and Hoar's *General and Comparative Physiology*.

CHAPTER 19

DIGESTION

All animals are heterotrophic and require carbohydrates, fats, proteins, vitamins, water and minerals for the synthesis and maintenance of the many compounds present in their constituent cells. The food eaten by animals is composed of proteins, fats, polysaccharides, polynucleotides and other complex molecules which must be digested, or hydrolyzed to simpler subunits, to be taken up by the cells.

Each molecule of protein, fat or carbohydrate is held together largely by bonds formed by the removal of a molecule of water, or by an equivalent reaction. The splitting of the peptide bonds of proteins, the glycosidic bonds of carbohydrates and the ester bonds of fats involves **hydrolytic cleavage,** the addition of a molecule of water across the bond to split it (Fig. 3–5). Enzymes are highly specific and a different enzyme is required for the hydrolytic cleavage of each type of bond. The process of **digestion,** from amoeba to man, involves the same or a very similar series of enzymes, but differs in where the enzymes are produced, where they act and how the process is controlled. Digestion may be **intracellular**—food particles are taken into the cell by phagocytosis and digestive enzymes act within the cell; or **extracellular**—the enzymes are secreted by the cells that produce them into some cavity, typically that of the gut, where hydrolytic cleavage takes place.

Protozoa, sponges and hydras take food into **food vacuoles** within the cells, and digestion occurs there. In the course of evolution, the more complex animals have developed special organs for obtaining and digesting food. The digestive tract of man is essentially a long tube, composed of several organs and adapted for ingestion, digestion and absorption. The term **ingestion** refers to the mechanical taking in of food, chewing and swallowing. The passage of substances through the wall of the digestive tract, called **absorption,** can occur only after the food molecules have undergone digestion. The wall of the digestive tract is a differentially permeable membrane which will permit only relatively small molecules to pass.

19–1 THE MOUTH CAVITY

The mouth cavity is supported by jaws and is bounded on the sides by the teeth, gums and cheeks, on the bottom by the tongue, and on the top by the *palate.* The last separates the mouth cavity from the nasal cavity and includes an anterior bony part, the *hard palate,* and a posterior fleshy part, the *soft palate.* The latter plays a role during swallowing in preventing food from passing up into the nasal cavity. The tongue, the teeth, and the salivary glands, which have ducts emptying into the mouth cavity, play roles in ingestion or digestion. The tongue and teeth in the human have assumed an additional function, that of *speech.*

The tongue consists of several sets of striated muscles oriented in different planes. The contractions of these muscles move the tongue in or out, up or down, or from side to side. To stick the tongue out, the muscles running up and down and from side to side contract, while the muscles extending from front to back relax. Food is pushed by the tongue between the teeth, to be chewed and then shaped into a spherical mass, called a *bolus,* to be swallowed. Swallowing is initiated when the tongue pushes a bolus into the pharynx. The epithelium covering the tongue contains groups of sensory cells, *taste buds,* stimulated by substances in solution.

The Teeth. The teeth of vertebrates break food up into smaller particles; they vary in size and shape according to the diet of the particular species. The teeth of different mammals, although superficially different, have a common pattern (Fig. 19–1). The part of the tooth projecting above the gum is called the *crown,* that surrounded by the gum is called the *neck,* and the part below the neck is the *root,* embedded in a socket in the jawbone. Each tooth is composed of several layers, a hard outer *enamel,* an inner layer of *dentin* that is not quite so hard, and an innermost pulp cavity, filled with blood vessels, nerves and soft tissue, the *pulp.* The enamel covers only the crown and upper part of the neck of the tooth. The tooth is fastened to the jawbone by a substance called *cement.* Dentin resembles bone in its composition and hardness. It consists of about 72 per cent inorganic matter (primarily calcium phosphate) and 28 per cent organic matter. Enamel, the hardest substance

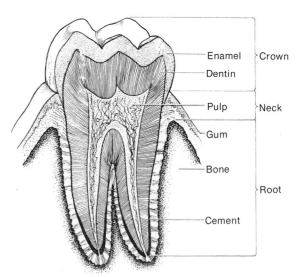

Figure 19–1 Diagram of a section through a human molar tooth.

in the body, consists of almost 97 per cent inorganic matter.

Unlike the simple, pointed, conical teeth of fish, amphibians and reptiles, the teeth of mammals are specialized to perform particular functions. The eight front, chisel-shaped *incisors* are used for biting and are especially large in gnawing animals such as rats, squirrels and beavers. The four cone-shaped *canine* teeth, one in each front corner of the mouth, are used for tearing food. Flesh-eating animals such as wolves and lions have large canine teeth or fangs (they are called canines because they are so large in dogs). In man, each jaw on each side has, behind the canines, two *premolars* and three *molars,* with flattened surfaces adapted for crushing and grinding food. The molars of flesh-eating animals are knife-shaped and are used in shearing flesh. Herbivores, such as horses and cows, have large, flat molars for grinding food and well developed incisors for cutting off grass. Upper incisors never develop in ruminants; these animals crop grass by pulling it with their tongue and upper lip across the cutting edge of the lower incisors.

Our ancestors were omnivorous for millions of years, and human teeth are relatively unspecialized. The last molar, or wisdom tooth, frequently fails to erupt, or if it comes through the gum, it is often crooked and useless. This reflects a trend in the evolution of modern man toward a shortening of the jaws, with a resulting crowding of the teeth which leaves inadequate space for the last molar. In primitive

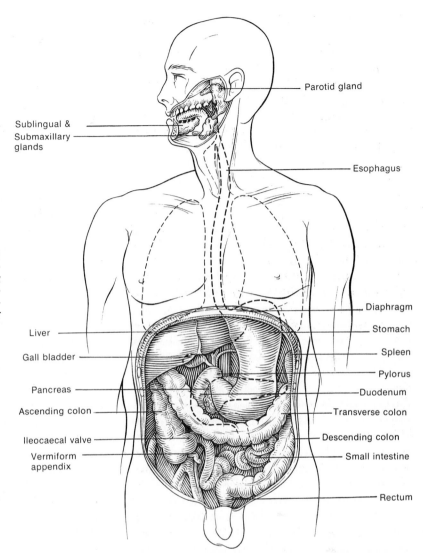

Parotid gland

Sublingual & Submaxillary glands

Esophagus

Figure 19–2 Diagram of the human body showing the parts of the digestive system. The liver, which normally covers part of the stomach and duodenum, has been folded back to reveal these organs and the gallbladder on its undersurface.

Diaphragm

Liver

Stomach

Gall bladder

Spleen

Pylorus

Pancreas

Duodenum

Ascending colon

Transverse colon

Ileocaecal valve

Descending colon

Vermiform appendix

Small intestine

Rectum

races, such as the Australian aborigines, this has not occurred. It is quite possible that in another hundred thousand years, humans may not have wisdom teeth.

The Salivary Glands. To assist in moving food down the throat, and to begin its chemical breakdown, two kinds of saliva are secreted by three pairs of salivary glands. One type of saliva is watery, to dissolve dry foods, and the other contains mucus, a slimy mucoprotein that enables particles of food to stick together in a bolus for swallowing and that lubricates the passage of the bolus down the esophagus. Saliva also protects the lining of the mouth against drying out, cleans it, and facilitates speech by moistening the tongue so that it does not stick to the roof of the mouth. The three pairs of glands produce about 1.5 liters of

saliva each day. The ***parotid glands*** (Fig. 19–2), located in the cheek just in front of the ear, produce only a watery saliva; the ***submaxillary glands,*** in front of the angle of the jaw, produce a mixture of watery and mucous saliva, as do the ***sublingual glands,*** located on the floor of the mouth under the tongue. Each glandular mass weighs about 30 grams and is connected to the mouth cavity by a duct.

Salivary amylase hydrolyzes starch to maltose, and ***salivary maltase*** splits maltose to glucose. Saliva is normally slightly acid, with a pH of 6.5 to 6.8, and amylase works best in this range. In the very acid condition of the stomach, amylase is denatured and inactivated, but because food is swallowed in masses, amylase retains its activity until the acid penetrates the mass of food.

19–2 THE PHARYNX

Food passes from the mouth cavity into the **pharynx**, the cavity behind the soft palate where the digestive and respiratory passages cross. It has no less than seven tubes connecting with it: the two internal nares from the nasal cavity, the connection from the mouth, the glottis opening into the trachea, the esophagus connecting to the stomach, and two eustachian tubes to the middle ear cavity that equalize pressure on the two sides of the ear drum (see p. 540).

Swallowing. The movement of food from the mouth to the stomach is aided by a series of reflexes. The first part of the swallowing act is under voluntary control: The tongue is raised against the roof of the mouth, and the bolus of food between the tongue and palate is pushed into the pharynx by a wavelike movement of the tongue. When swallowing begins, breathing is stopped momentarily by a reflex mechanism to prevent the passage of food into the larynx or trachea. Once the food is in the pharynx, there are four possible exits for it, only one of which is desirable. Normally, the reflex closing of the other three forces the food down the esophagus when the pharynx contracts. The opening to the nasal cavity is closed by the reflex elevation of the soft palate (Fig. 19–3), while the tongue is held against the roof of the mouth, preventing the food from returning. The opening into the larynx is closed by the contraction of muscles which raise the entire larynx, bringing the opening, the **glottis**, under the fold of tissue called the **epiglottis**. This action completely closes the glottis and prevents food from going down the trachea. At the same time it enlarges the opening of the esophagus to facilitate the passage of the bolus. The raising of the larynx can be observed in the bobbing up of the Adam's apple (the projection of the larynx) each time you swallow.

19–3 THE WALLS OF THE DIGESTIVE TRACT

The walls of all parts of the digestive system, from the esophagus to the rectum, have a similar structure and are composed of the same three layers: an inner mucous membrane or **mucosa**, a muscular middle layer, and an outer layer of connective tissue (Fig. 19–4). The inner lining of the mucosa, next to the cavity of the tract, is composed of epithelial cells, usually columnar, some of which secrete the viscous lubricating mucus. The mucosa in the stomach and intestines is greatly folded to increase the secreting and absorbing surface of the tube. The glands of the digestive tract are formed as outpocketings of the mucosal lining.

The muscular layer is composed of smooth muscle except in the upper third of the esophagus, where striated muscle is present. Most of the digestive tract has two distinct layers of muscle: an inner one with the fibers arranged circularly, and an outer one with the fibers arranged longitudinally. By contracting these layers alternately or in unison, the digestive organs can perform a variety of peristaltic movements to churn the food and move it along.

The outermost layer of the digestive tract

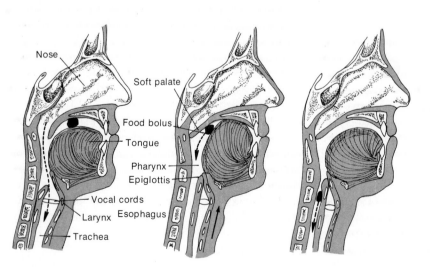

Nose
Soft palate
Food bolus
Tongue
Pharynx
Epiglottis
Vocal cords
Larynx
Esophagus
Trachea

Figure 19–3 Diagram of the position of the tongue and epiglottis during breathing *(left)* and swallowing *(center* and *right).* Note how a food bolus is pushed from the mouth into the pharynx by the tongue to initiate swallowing *(center).*

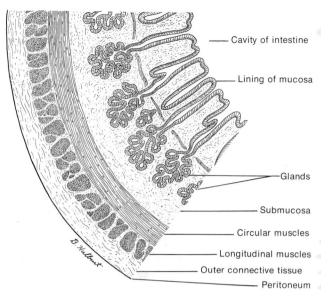

Cavity of intestine

Lining of mucosa

Glands

Submucosa

Circular muscles

Longitudinal muscles

Outer connective tissue

Peritoneum

Figure 19-4 Cross section of the human intestine.

is composed of strong, flexible connective tissue fibers covered by a smooth sheet of *peritoneum.* The fluid secreted by the peritoneum lubricates the surface of the stomach and intestines and reduces friction as the parts of the digestive tract move and rub against each other and the abdominal wall. The esophagus, buried in the muscles of the neck and chest, has no peritoneal covering.

The walls of the digestive tract are richly supplied with nerves that coordinate the actions of the various parts, and with blood and lymph vessels to supply the cells with nutrients and oxygen, to drain away wastes and to carry the absorbed food to a place of storage.

19-4 THE ESOPHAGUS

The esophagus is a muscular tube leading directly downward from the pharynx, passing between the lungs and behind the heart, and penetrating the diaphragm to reach the stomach. The contraction of the muscles in the wall of the pharynx and the presence of the bolus in the upper part of the esophagus cause a single, powerful, rhythmic wave of muscular contraction in the wall of the esophagus, called *peristalsis.* This pushes the bolus down to the stomach. This wave is preceded by one of relaxation, which dilates the tube to make room for the food. Similar peristaltic waves

move the contents through all the organs of the digestive tube. The wave of contraction is rapid in the esophagus, and it takes only about six seconds for solid food to reach the stomach from the mouth. If some of the food escapes the first wave of contraction and remains in the esophagus, it stimulates another muscular contraction to move it to the stomach. An emotional upset, excessive smoking or food swallowed too hastily may cause the muscles of the esophagus to contract in a spasm when no food is present, resulting in the sensation of a "lump in the throat."

The opening from the esophagus to the stomach is controlled by a ring of smooth muscle, called a *sphincter.* Normally, this is closed; it opens reflexly only when a wave of contraction in the esophagus has pushed a bolus of food against it. As liquids are swallowed, they may pass through the esophagus to the sphincter faster than the accompanying wave of contraction of the esophagus muscles, but the ring of muscle does not open until the peristaltic wave reaches it. Similar sphincters control the movement of food at three other places in the digestive tract: the opening of the stomach into the small intestine, the opening of the small intestine into the large intestine, and the opening at the anus where the digestive tract ends.

19-5 THE STOMACH

The stomach, a thick-walled, muscular sac on the left side of the body just beneath the lower ribs, is divided into an upper *cardiac region,* nearest the heart; a deep part below this, called the *fundus;* and the section extending to the opening of the small intestine, called the *pyloric region* (Fig. 19-5). The muscular layers of the stomach wall are exceptionally thick and include a diagonal layer of fibers in addition to the circular and longitudinal fibers found elsewhere in the digestive tract. Millions of microscopic *gastric glands* in the mucosal layer secrete mucus and gastric juice containing enzymes and hydrochloric acid. Pure *gastric juice* is extraordinarily acid, having a pH of about 1, but the stomach contents, in which the gastric juice is mixed with food, are less acid, with a pH of about 3.

The size of the stomach varies, of course,

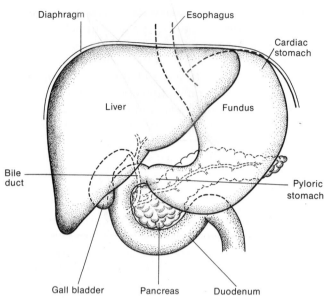

Figure 19–5 Diagram of the relations of the stomach, liver, pancreas and duodenum.

as a meal is eaten and digested. The capacity of the average person's stomach is about 2.5 liters. As swallowing occurs, the stomach relaxes reflexly to make room for the food. Soon after food reaches the stomach, peristaltic waves begin in the pyloric region, passing from left to right, toward the opening into the intestine. The rest of the stomach, containing most of the food, remains quiescent at this stage. As digestion proceeds, the waves originate farther and farther to the left, and finally the entire stomach wall, from cardiac to pyloric end, is swept by deep, powerful peristaltic waves. These mix the contents and mechanically break the larger bits of food into smaller ones until the contents achieve the consistency of cream soup. At intervals, the pyloric sphincter relaxes, and a small amount of *chyme* (as the contents of the stomach and small intestine are called) is pushed into the small intestine by the contraction of the stomach. The opening of the pyloric sphincter at the proper time is regulated by an enterogastric reflex from the duodenum to the stomach and by a hormone. In response to a surfeit of chyme, especially one containing fatty acids, the duodenum releases a hormone, *enterogastrone,* which passes via the blood stream to the stomach, where it inhibits peristaltic activity and slows down the rate of emptying of the stomach. In one to four hours, depending upon the amount

and kind of food eaten, the stomach is emptied. Foods rich in carbohydrates leave the stomach more rapidly than proteins, and proteins more rapidly than fats. When the stomach is empty it may continue to contract. The squeezing of the empty stomach stimulates nerves in the wall, and may cause hunger "pangs."

Vomiting. Occasionally something may be taken into the stomach which should be ejected. To make this possible, most mammals—with the exception of rabbits and rodents—have a *vomiting reflex.* Vomiting may be initiated by mechanical irritation of the pharynx (sticking a finger in the throat is used to induce it when some poisonous substance has been swallowed) or by disturbances in the semicircular canals of the ears, as in motion sickness. It is controlled by the vomiting center in the midbrain, which coordinates the contraction of the stomach and of the muscles of the abdominal wall, the closing of the pyloric sphincter, the opening of the cardiac sphincter and the closing of the glottis.

Ulcers. A peptic ulcer occurs when the gastric mucosa is damaged by the proteolytic enzymes and hydrochloric acid secreted in the gastric juice. The ulcer results when there is an excessive secretion of gastric juice or too little secretion of the protective mucus. Ulcers may be caused by prolonged psychic tension and anxiety, perhaps via stimulation of the vagus nerves. Ulcer patients have a high rate of gastric secretion even between meals when the stomach is empty. A patient with an ulcer is treated by feeding him a bland diet in six or more small meals per day to keep food in the stomach most of the time. This tends to neutralize the acid and dilute the gastric juice. Ulcers that do not respond to dietary measures may be treated surgically. The ulcerous area together with a large portion of the stomach may be removed and the remaining stump joined to the jejunum. Alternatively, the vagus nerve to the stomach may be severed to block secretion by the stomach. Vagotomy often results in almost complete loss of gastric motility and great difficulty in emptying the stomach after a meal unless the opening from the pylorus to the small intestine is enlarged. An untreated ulcer may continue to eat away at the wall of the stomach and ultimately perforate into the abdominal cavity, causing massive bleeding and sometimes death.

DIGESTION

INTESTINAL TRACK LENGTH / DIGESTION - 19

DIET   LENGTH
(1) PLANTS   LONG
(2) MEAT   SHORT
(3) BOTH   INTERMEDIATE

419

19-6 THE SMALL INTESTINE

The small intestine, into which the chyme passes by the force of the peristaltic waves in the stomach, is a coiled tube, about 7 meters long and approximately 2.5 centimeters in diameter. The greater part of enzymatic digestion and almost all absorption occur here. Only alcohol and a few poisons are absorbed through the stomach wall.

The length of the intestine is correlated with the type of diet: Plant-eating animals have a long small intestine; meat-eating animals have a short one; and omnivorous ones, like man, have one of intermediate length. The frog larva or tadpole is herbivorous and has a long small intestine, but the adult frog is carnivorous and has a much shorter one.

The first segment of the intestine, about 25 centimeters long, is called the **duodenum.** It occupies a fixed position in the abdominal cavity and is held in place by ligaments connecting it to the liver and stomach, as well as to the dorsal body wall. The rest of the small intestine (and most of the large intestine) is attached only to the dorsal body wall by a thin, transparent membrane called the **mesentery,** and it is able to move about with considerable freedom. Nerves and blood vessels pass from the body wall to the intestine in the mesentery. Two important juices, bile from the liver and pancreatic juice from the pancreas, are added to the contents of the duodenum. Millions of tiny **intestinal glands** in the intestinal mucosa secrete the intestinal juice containing a number of enzymes. These three juices are mixed in the small intestine and complete the digestive process begun in the mouth and stomach.

Intestinal Movements. After a meal, peristaltic activity in the small intestine is greatly increased by the **gastroenteric reflex.** The distention of the stomach by food stimulates nerve impulses that pass to the walls of the small intestine, increasing the excitability of the intestine and resulting in increased motility and secretion of intestinal juices.

There are two types of intestinal movements: **peristaltic contractions** move the chyme along, and **churning movements** simply mix the intestinal contents (Fig. 19–6). A single peristaltic wave does not move far in the intestine; usually after 10 centimeters or so it is dis-

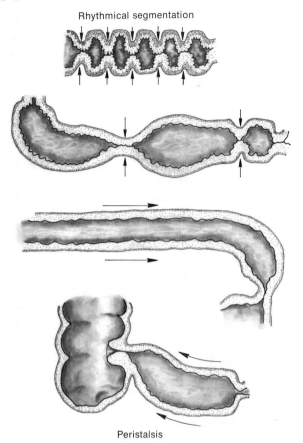

Rhythmical segmentation

Peristalsis

Figure 19–6 Diagrams to illustrate the churning action of rhythmical segmentation in the intestine and the movement of food through the digestive tract by peristalsis.

sipated, though occasionally rapid movements, called **peristaltic rushes,** sweep along for considerable distances. The churning movements, caused by alternate contractions and relaxations of successive segments of the intestine, are repeated about ten times a minute. These movements complete the mechanical breaking up of the intestinal contents, mix them with the various digestive juices, and ensure that all parts of the intestinal contents will be brought in contact with the intestinal wall so that the digested foods may be absorbed into the blood stream. In each part of the intestine these churning movements continue for a time; then a peristaltic wave carries the contents to the next section, and the churning movements begin again. The contents are carried through the small intestine to the large intestine in about eight hours. By the time the remains of the food pass from the small intestine, digestion has been completed and the molecules of

DUODENUM → CHOLECYSTOKININ → RELAXES SPHINCTER
(HORMONE)
RELEASES BILE IN THE PRESENCE
OF FATS

nutrients have been absorbed. The materials passing into the large intestine consist of indigestible matter and large quantities of water derived from the food taken in or from the digestive juices.

19–7　THE LIVER AND PANCREAS

The liver and pancreas are large glandular outgrowths from the small intestine. The liver, one of the largest organs of the human body, continually secretes *bile* (some 600 to 800 ml. per day) which passes by a system of ducts to the *gallbladder.* Bile does not enter the intestine immediately, for a sphincter at the intestinal end of the bile duct remains closed until food enters the intestine. Stimulated by cholecystokinin (a hormone secreted by the duodenum in response to the presence of fats in the duodenal contents), the sphincter relaxes, the wall of the gallbladder contracts and bile is forced out into the intestine. In the gallbladder, bile is concentrated by the removal of water and salts.

Bile contains no digestive enzymes but is alkaline and aids in digestion by neutralizing the acid chyme from the stomach. The enzymes secreted by the pancreas and intestinal glands have *p*H optima in the neutral or slightly alkaline range. The *bile salts* in bile act as detergents, emulsifying the fats in the intestine, increasing the surface area of the fat droplets, and promoting the action of *lipase.* In addition, bile salts combine with lipids and promote their absorption through the intestinal mucosa. When the bile duct is obstructed and bile salts are absent from the intestine, both the digestion and absorption of fats are impaired and much of the fat eaten is excreted in the feces. The bile salts are carefully conserved by the body; they are reabsorbed in the lower part of the intestine and transported back to the liver via the blood stream to be secreted again. Another constituent of the bile, *cholesterol,* is very sparingly soluble in water. The removal of water from the bile in the gallbladder may concentrate the cholesterol to the point where it precipitates, producing hard little pellets called *gallstones.* These may obstruct the bile duct and stop the flow of bile.

The color of bile results from the presence of *bile pigments* (green, yellow, orange or red

OPTIMUM
?H
neutral-
alkaline

in different species of animals), which are excretory products derived from the degradation of hemoglobin in the liver. The bile pigments undergo further chemical reactions by the intestinal bacteria and are converted to the brown pigments responsible for the color of feces. If their excretion is prevented by a gallstone or some other obstruction of the bile duct, the bile pigments are reabsorbed by the liver and gallbladder and accumulate in the blood and tissues, giving a yellowish tinge to the skin, a condition called *jaundice.* The absence of the pigments from the intestinal contents gives the feces a clay-colored appearance.

The *pancreas* is an irregular, diffuse mass of tissues lying between the stomach and the duodenum (Fig. 19–5). Its secretion, the pancreatic juice, contains enzymes that hydrolyze proteins, fats, nucleic acids and carbohydrates. It passes into the duodenum by way of the pancreatic duct. In addition, certain cells of the pancreas, called the *islets of Langerhans,* secrete the hormones *insulin* and *glucagon* (p. 552) into the blood stream. These two secretions are entirely separate and unrelated. It just happens that in man, and most vertebrates, the two types of cells occur together in the same gland; in certain types of fish the two types are spatially separated into two different glands.

The pancreatic juice is a clear, watery alkaline fluid with a *p*H of about 8.5. It is an important factor in neutralizing the acid chyme. The enzymes secreted by the pancreas and the intestinal wall will not work in an acid medium; hence the need for a neutralizing agent after food has been received from the stomach. The average human secretes about 1 or 1.5 liters of pancreatic juice each day. If the pancreatic duct becomes blocked so that the pancreatic enzymes are unable to reach the intestine and act on the food, the person develops a tremendous appetite and eats a great deal. In spite of this he loses weight, a striking demonstration of the importance of the enzymes of the pancreatic juice in normal digestion.

19–8　THE ABSORPTION OF NUTRIENTS

After the digestive enzymes have cleaved the large molecules of proteins, polysaccharides, nucleic acids and lipids into their

FOLDS
VILLI
MICROVILLI

constituent subunits, the products are absorbed through the wall of the intestine, especially the small intestine. The intestines of man and other vertebrates are greatly folded to increase the surface area through which absorption may occur. In addition, countless small, fingerlike projections, called *villi,* cover the entire surface of the intestinal mucosa (Fig. 19–7). Each villus contains a network of blood capillaries and a lymph capillary in its center, into which the nutrients are transferred. A third adaptation for increasing the surface area is the presence of countless, closely packed, cylindrical processes called *microvilli* on the surface of each epithelial cell of the intestine (Fig. 19–8). The folds, villi and microvilli together provide an enormous area through which absorption may occur. Some vertebrates have one or more blind pouches, *caeca,* which are attached to the intestine and increase the area available for absorption. The area of the mucosal surface of the intestine of sharks is increased by an epithelial

sharks

fold (the spiral valve) attached to the side in a spiral fashion and shaped like a corkscrew. Food passing through the intestine must follow the spiral fold and thus is exposed to a greater area.

Absorption is a complex process, occurring in part by simple physical diffusion, in part by facilitated diffusion and in part by active transport. The various hexoses are absorbed by active transport, by a process which requires the expenditure of energy to move the molecules against a chemical gradient. The several hexoses—glucose, fructose and galactose—are absorbed at quite different rates. Galactose is absorbed more rapidly than glucose, and glucose more rapidly than fructose. The hexoses pass through the intestinal wall by an energy-requiring process which is essentially unidirectional and which occurs without phosphorylation or any other chemical change of the hexose molecule.

Amino acids are also absorbed by a process

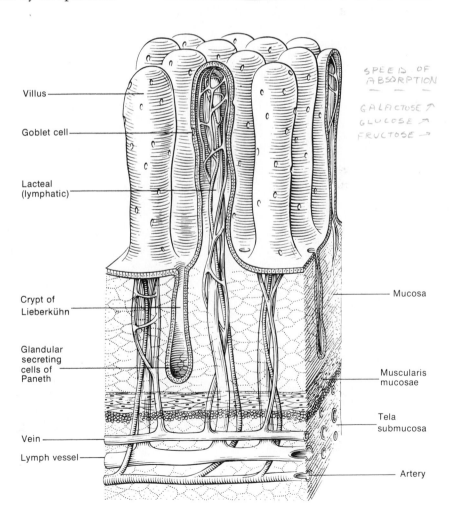

Villus

Goblet cell

Lacteal
(lymphatic)

Crypt of
Lieberkühn

Glandular
secreting
cells of
Paneth

Vein

Lymph vessel

Mucosa

Muscularis
mucosae

Tela
submucosa

Artery

SPEED OF
ABSORPTION

GALACTOSE ↗
GLUCOSE ↗
FRUCTOSE →

Figure 19–7 Details of the structure of some intestinal villi showing their blood and lymph supply. Semischematic representation of jejunal intestinal villi, ×35 (approx.).

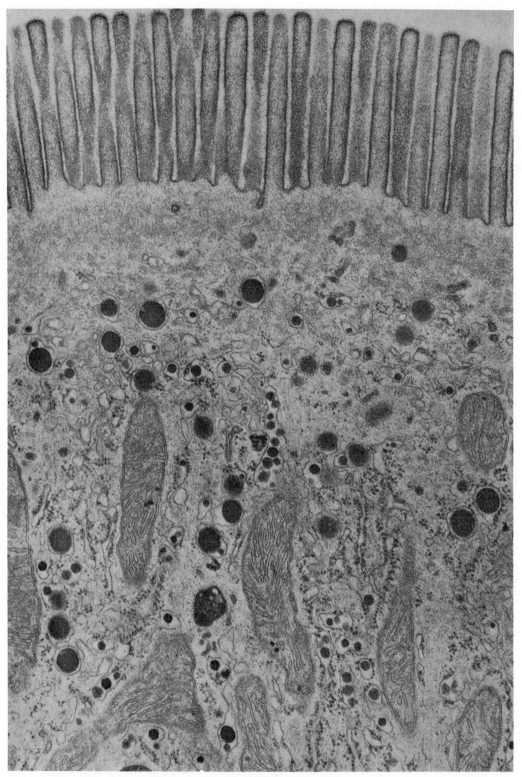

Figure 19–8 Electron micrograph illustrating the microvilli on the free surface *(top)* of the epithelial cells of the intestine.

of active transport into the blood capillaries and transported to the liver for short-term storage and subsequent distribution to the rest of the body.

The products of the complete or partial hydrolysis of lipids enter the body by a different process and a different route. Bile salts play an important role in enhancing the absorption of fatty acids, monoacylglycerols and diacylglycerols, and of other lipid-soluble substances, such as the fat-soluble vitamins. The lipids need not be completely hydrolyzed to glycerol and fatty acids to be absorbed. The short-chain fatty acids are absorbed into the blood capillaries, whereas the longer chain fatty acids are taken up into the lymph capillaries. As the products of lipid hydrolysis pass through the epithelial cells of the mucosa, they are resynthesized into molecules of fat. These subsequently aggregate into *chylomicrons,* fine globules of fat with a thin coat of protein, and enter the lymph. During the absorption of a meal rich in fat, the lymphatic vessels of the intestine have a milky color owing to this fat emulsion. The resynthesis of the fat occurs in the intestinal mucosal cells and requires the conversion of the free fatty acids to fatty acyl-coenzyme A compounds, which can then react with monoacylglycerols and diacylglycerols to form triacylglycerols. The intestinal lymph vessels empty into the great thoracic duct and then into the subclavian vein. The chylomicrons appear in the blood stream and may give the blood itself a turbid, "milky" appearance after a meal rich in fat. Cholesterol is absorbed through the wall of the intestine in the free state, then is esterified in the mucosal cells and passes to the lymph capillaries.

A portion of the water in the intestinal contents is absorbed in the small intestine, but most of it is absorbed in the large intestine or *colon.* The colon absorbs water and converts the wastes to a semisolid state for defecation. The feces contain an enormous number of bacteria; half of the mass of feces may be bacteria. The intestinal bacteria synthesize a variety of vitamins and other nutrients that are absorbed and used by their vertebrate host. Bacteria have enzymes for digesting cellulose walls of plant cells and play an important role in the digestive processes of herbivores such as cattle and rabbits.

19–9 THE LARGE INTESTINE AND RECTUM

The material remaining after the nutrients have been absorbed passes from the small intestine into the U-shaped large intestine, or *colon,* which is larger in diameter and has thicker walls than the small intestine. The small intestine empties into the side of the colon a short distance from its end, leaving a blind sac, the *cecum,* at the tip of which is a small projection about the size of the little finger, the *appendix.* The cecum and appendix were probably larger in our remote ancestors and functioned in the digestion of vegetable materials. Herbivores such as rabbits and guinea pigs have a large, functional cecum. From the junction of the small and large intestines, the ascending colon extends up the right side of the body to the level of the liver, makes a right-angle turn, and, as the transverse colon, extends across the abdominal cavity just below the liver and stomach (Fig. 19–2). At the left side of the body it makes another right turn, becomes the descending colon, and passes down the left side of the body to the rectum.

Most of the nutrients have been removed from the material reaching the colon, but it is still liquid. Some water is absorbed in the small intestine, but almost as much is added by the bile and pancreatic juice. The colon absorbs water and reduces the wastes to a semisolid state. Both churning and peristaltic movements occur in the colon, although both are ordinarily slower and more sluggish than those in the small intestine. Periodically, more vigorous peristaltic movements force the contents along until they finally reach the rectum. These occur especially after eating, because of a reflex mechanism whereby the filling of the stomach stimulates an emptying of the colon. This *gastrocolic reflex* is responsible for defecation usually occurring after a meal.

Defecation is partly voluntary, depending upon the contraction of the abdominal wall muscles and the diaphragm and the relaxation of the outer ring of muscle (sphincter) of the anus, and partly involuntary, depending on the relaxation of the inner anal sphincter and the contraction of the muscles of the large intestine and rectum to force the feces out through the anus. It is the distention of the rectum and the consequent stimulation of nerves in its walls that bring about the desire to defecate. If this signal is ignored, the rectum tends to adapt to the new size and the stimulus diminishes and finally disappears.

Twelve to twenty-four hours are required for the waste products of digestion to pass through the colon and rectum. The end product, the *feces,* contains indigestible remnants of the food, certain substances secreted by the body, such as bile pigments and heavy metals, and large quantities of bacteria.

If the lining of the colon is irritated, as in an infection such as dysentery, peristalsis is increased and the intestinal contents pass through rapidly, with only a small amount of the water removed from them. This condition, known as *diarrhea,* results in frequent defecation and watery feces. The opposite condition, *constipation,* results when the contents pass through too slowly, so that an abnormally large amount of water is removed, and the feces become excessively hard and dry. Because of their irritating effect on the colon, the repeated use of cathartics leads to a condition known as cathartic constipation. The muscles of the colon become incapable of their normal churning and peristaltic movements and remain contracted. Constipation can be avoided by eating foods that contain sufficient indigestible cellulose fibers ("roughage") to give bulk to the intestinal contents.

The headaches and other symptoms that usually accompany constipation are not caused by absorption of "toxic substances" from the feces, but are due to the distention of the rectum. If the rectum is packed with some inert substance such as cotton, the same symptoms appear.

19–10 CHEMICAL ASPECTS OF DIGESTION

Polysaccharides such as starch and glycogen form an important part of the food ingested by man and most animals. The glucose units of these large molecules are joined by *glycosidic bonds* linking carbon 4 (or carbon 6) of one glucose molecule with carbon 1 of the adjacent molecule. These bonds are hydrolyzed by *amylases.* These enzymes will digest polysaccharides to the disac-

charide, *maltose,* but will not split the bond between the two glucose units of maltose. The amylases will split the α-glycosidic bonds present in starch and glycogen but not the β-glycosidic bonds present in cellulose. The garden snail's digestive juice contains β-*glycosidases* that can hydrolyze cellulose. In most vertebrates, amylase is secreted only by the pancreas; in man and certain other mammals, amylase is secreted by the salivary glands as well.

Different individuals have salivas containing different amounts of amylase; thus the time required for the salivary digestion of a given amount of starch will vary. You can test the activity of your own saliva by this simple experiment. An estimate of the amount of starch in solution can be obtained by adding a standard amount of iodine solution, which yields a blue-colored complex with starch but not with sugar. Place 10 ml. of a dilute boiled starch solution in a test tube and add 1 ml. of saliva. Mix thoroughly and at the end of each minute after the addition of the saliva, remove a drop and add it to a drop of iodine solution. At first the resulting solution will be blue; subsequently, samples will turn violet or red and eventually remain yellow, indicating that all the starch has been digested. The red color results from the reaction of iodine with certain intermediate substances formed in the breakdown of starch to sugar. All enzymes are proteins and are denatured, i.e., rendered inactive, by heating. You can prove that the digestion of starch by saliva is mediated by enzymes by repeating the test using saliva that has been boiled.

Disaccharides are cleaved to monosaccharides by enzymes that are specific for the particular disaccharide (Table 19–1). Maltose is split by *maltases* present in saliva and in the intestinal juice secreted by intestinal glands. The intestinal juice also contains *sucrase,* which splits the disaccharide sucrose to its constituents, glucose and fructose, and *lactase,* which splits lactose (milk sugar) to glucose and galactose. The ultimate products of the digestion of carbohydrates—the hexoses, glucose, fructose and galactose—are absorbed into the blood stream through the intestinal wall.

There are several kinds of hydrolases that attack the peptide bonds of proteins; each is specific for peptide bonds in a specific location

Table 19-1 ENZYMES IMPORTANT IN DIGESTION

Enzyme	Source	Optimum pH	Type of Bond Split	Product
Salivary amylase	Saliva	Neutral	α-glycoside	Maltose
Maltase	Saliva	Neutral	α-glycoside	Glucose
Pepsin	Stomach	Acid	Peptide bonds within chain and adjacent to tyrosine or phenylalanine	Peptides
Rennin	Stomach	Acid	Peptide bonds in casein	Coagulated casein
Trypsin	Pancreas	Alkaline	Peptide bonds within chain adjacent to lysine or arginine	Peptides
Chymotrypsin	Pancreas	Alkaline	Peptide bonds within chain adjacent to tyrosine or phenylalanine	Peptides
Lipase	Pancreas	Alkaline	Ester bonds of fats	Glycerol, fatty acids, mono- and diacylglycerols
Amylase	Pancreas	Alkaline	α-glycoside	Maltose
Ribonuclease	Pancreas	Alkaline	Phosphate esters of RNA	Nucleotides
Deoxyribonuclease	Pancreas	Alkaline	Phosphate esters of DNA	Nucleotides
Carboxypeptidase	Intestinal glands	Alkaline	Peptide bond adjacent to free carboxyl end	Free amino acids
Aminopeptidase	Intestinal glands	Alkaline	Peptide bond adjacent to free amino end	Free amino acids
Enterokinase	Intestinal glands	Alkaline	Peptide bonds of trypsinogen	Trypsin
Maltase	Intestinal glands	Alkaline	α-glucoside of maltose	Glucose
Sucrase	Intestinal glands	Alkaline	α-glucoside of sucrose	Glucose and fructose
Lactase	Intestinal glands	Alkaline	β-galactoside of lactose	Glucose and galactose

in a polypeptide chain (Fig. 19-9). *Exopeptidases* cleave the peptide bond joining the terminal amino acids to the peptide chain. *Carboxypeptidase* splits the peptide bond joining the amino acid with the free terminal carboxyl group to the chain, and *aminopeptidase* removes the amino acid with a free terminal α-amino group. Other hydrolases, the *endopeptidases,* will cleave only peptide bonds within a peptide chain. *Pepsin,* secreted by the chief cells in the gastric mucosa, and *trypsin* and *chymotrypsin,* secreted by the pancreas, are endopeptidases but differ in their requirements for specific amino acids adjacent to the peptide bond to be cleaved. Pepsin requires tyrosine or phenylalanine adjacent to the bond to be split; trypsin requires lysine or arginine; and chymotrypsin requires tyrosine, phenylalanine, tryptophan, methionine or leucine at the site of cleavage. These endopeptidases split peptide chains into smaller fragments which are then cleaved further by exopeptidases. The combined action of the endopeptidases and exo-

peptidases results in splitting the protein molecules to free amino acids, which are then absorbed through the intestinal wall into the blood stream by active transport. Some dipeptides and tripeptides are also absorbed and subsequently cleaved within the cells lining the small intestine.

These powerful proteolytic enzymes would constitute a serious threat to the tissues secreting them. However, pepsin, trypsin and chymotrypsin are not secreted as such, but are secreted in the form of inactive precursors—pepsinogen, trypsinogen and chymotrypsinogen. This prevents their digesting the proteins of the cells that produce them. In the gut each is activated by the removal of part of the precursor molecule to yield the active enzyme and an inactive fragment. *Pepsinogen,* with a molecular weight of 42,500, is converted to pepsin, with a molecular weight of 34,500, by the high concentration of H⁺ in the gastric juice and by pepsin itself. *Trypsinogen* is converted to trypsin by *enterokinase,* an enzyme

Figure 19-9 Formula of a peptide, indicating points of attack of pepsin (P), trypsin (T), chymotrypsin (C), aminopeptidase (AP) and carboxypeptidase (CP).

$$H_2N—gly—ala—leu—tyr—ala—asp—lys—val—glu—gly—COOH$$

AP   C C or P   T   CP

secreted by glands in the wall of the intestine, or by trypsin itself. The conversion of *chymo-trypsinogen* to chymotrypsin is mediated by trypsin but not by chymotrypsin. As a further protection to the pancreas, which secretes trypsinogen and chymotrypsinogen, it also secretes a small protein called "trypsin inhibitor," which will combine with and inactivate any molecules of free trypsin that may be formed accidentally in the pancreas. If the pancreas is damaged or if the duct is blocked, large amounts of pancreatic enzymes accumulate, the trypsin inhibitor is overwhelmed and the enzymes may digest the pancreas, resulting in acute pancreatitis, which is frequently fatal.

The amount of protein in the various digestive juices is not inconsequential but composes a significant fraction of the total protein hydrolyzed in the gut and is the source of a significant fraction of the amino acids absorbed through the intestine. The pancreas alone secretes about one quarter of its total protein content each day. It is estimated that about 100 grams of protein is secreted each day in the digestive juices of humans. These proteins are subsequently hydrolyzed and their amino acids are reabsorbed. This amount is comparable to the total dietary intake of proteins by many people!

The digestion of fats is catalyzed by *esterases* that split the ester bond between glycerol and fatty acid. The principal mammalian esterase is *lipase,* secreted by the pancreas. Like other proteins, lipase is water-soluble, but its substrates are not. Thus the enzyme can attack only those molecules of fat at the *surface* of a fat droplet. The *bile salts* are detergents that reduce the surface tension of fats, breaking the large droplets of fat into very fine ones; this greatly increases the surface area of fat exposed to the action of lipase and increases the rate of digestion of lipids. Conditions in the intestine are usually not optimal for the complete hydrolysis of lipids to glycerol and fatty acids. The products of digestion include glycerol and free fatty acids plus monoacylglycerols, diacylglycerols and some triacylglycerols, undigested fats.

The pancreas also secretes *ribonuclease,* an esterase which splits the phosphate ester bonds linking adjacent nucleotides in ribonucleic acids, and *deoxyribonuclease,* which splits the phosphate ester bonds linking adjacent nucleotides in deoxyribonucleic acids.

Enzymes that complete the cleavage of nucleic acids are secreted by the intestinal mucosa. *Phosphodiesterase* removes nucleotides one at a time from the end of a polynucleotide chain. The nucleotides in turn are attacked by *phosphatases* which remove the phosphate group and leave the nucleosides, which are absorbed.

It should be noted that these hydrolytic enzymes do *not* usually serve for the formation of the bonds they normally split. Starches are not synthesized by amylases, proteins are not synthesized by pepsin or trypsin, and fats are not usually synthesized by lipase. The synthesis of a polysaccharide chain does not involve the removal of water from molecules of glucose but rather the removal of phosphate from molecules of glucose-phosphate. As a general rule organic compounds are synthesized in the cell by enzyme systems that differ, at least in certain key enzymes, from those that take the compound apart.

19–11   CONTROLLING THE SECRETION OF DIGESTIVE ENZYMES

The salivary glands are controlled entirely by the nervous system. Either smelling or tasting food stimulates nerve cells in the nose or mouth to send impulses to the *salivation center* in the medulla of the brain; these are relayed to the salivary glands, causing them to secrete saliva. The mere presence in the mouth of tasteless, odorless objects, such as pebbles, stimulates other cells in the lining of the mouth which act similarly to cause salivation. Impulses may come from the higher centers of the brain; simply seeing or thinking of food can bring about salivation. Salivary glands respond to chemical, mechanical or psychic stimuli.

The classic experiments on the functions of the human stomach were carried out by the Army surgeon William Beaumont. In 1822 a trapper, Alexis St. Martin, was shot in the stomach and Beaumont treated the wound. St. Martin recovered nicely but the wound healed in such a way that there was an opening from the abdominal wall to the lumen of the stomach. Beaumont in succeeding years made a series of observations of the effects of dietary and emotional factors on the secretion of gastric juice and on the other activities of the stomach lining.

Much of our knowledge of the mechanisms controlling the secretion of gastric juice we owe to the Russian physiologist Pavlov, who devised many experimental techniques (such as the Pavlov pouch) and performed many critical experiments. One of these was to sever the esophagus of a dog and bring the two cut ends to the surface of the neck so that when the dog was fed, the food, instead of going to the stomach, went out through the hole in the neck. Although the food never reached the stomach, such a "sham feeding" caused a flow of gastric juice, about one quarter of the normal amount (the average human secretes between 400 and 800 ml. of gastric juice per meal). This quarter of the normal flow is stimulated by nerve impulses originating in the taste buds, or in the eye, and passing to the brain, whence they go to the stomach. The flow is completely abolished when the nerves to the stomach are cut. By putting food into the cut end of the esophagus leading to the stomach, and preventing the dog from seeing, smelling or tasting it, about half the normal flow was stimulated when the food reached the stomach. This flow occurs, although in reduced amount, even when the nerves to the stomach are cut; it depends in part on nervous stimulation of the gastric glands by impulses from cells in the stomach lining, and in part on a hormone called *gastrin*. This peptide hormone was isolated in pure form by Gregory at the University of Liverpool in 1966 and the sequence of its fifteen constituent amino acids was determined.

The cells of the mucosal layer of the pyloric part of the stomach secrete gastrin into the blood stream whenever partly digested food comes in contact with them. If extracts of the cells are injected into an animal, the gastric glands begin to secrete within a short time. The final proof of the existence and action of this hormone was given by cross circulation experiments, in which the blood system of one dog was connected by tubes with that of another. When food was placed in the pyloric region of one dog, the gastric glands of the other began to secrete. Since there were no nerve connections between the two dogs, the secretion of gastric juice by the second dog must have been caused by a substance carried by the blood, the hormone gastrin.

Some secretion of gastric juice is caused by the presence of food in the intestine. Perhaps amino acids, absorbed into the blood from the small intestine, are responsible for this, or it may be caused by some as yet unrecognized reflex or hormone. The operation of so many different mechanisms enables the stomach to provide the proper amount of gastric juice for the amount and type of food eaten. A meal rich in protein causes the flow of a copious amount of gastric juice; a meal with little protein and much carbohydrate causes a moderate flow; while one with a high fat content causes a small flow.

The presence of food in the small intestine stimulates the cells of the duodenal mucosa to secrete two hormones that regulate pancreatic function. *Secretin,* released by the presence of acid food, stimulates the secretion by the pancreas of large amounts of fluid with a high concentration of bicarbonate ion, HCO_3^-, that neutralizes the acidity. The second hormone, *pancreozymin,* is released by the presence of polypeptides in the chyme. This hormone is also carried by the blood stream to the pancreas and stimulates the release of digestive enzymes such as trypsin and chymotrypsin (Fig. 19–10). Secretin also causes an increased production of bile by the liver. Pancreozymin, which has also been called cholecystokinin, stimulates the gallbladder to contract and release bile into the bile duct.

When the nerves to the pancreas are stimulated, little secretion results; and when they are cut, there is little or no suppression of the flow, indicating that its secretions are almost entirely dependent on secretin and pancreozymin.

19–12 DIGESTIVE SYSTEMS OF OTHER ANIMALS

Some animals (sponges, clams, tunicates and even some whales) are *filter feeders,* making a living by filtering out of sea water or pond water the microscopic plants and animals that are present. The choanocytes of sponges take up these small particles by phagocytosis and digest them intracellularly. Certain parasites, such as tapeworms, live in the gut cavity of man and other vertebrates and absorb "predigested" nutrients from the surroundings. Protozoa take food, either living microscopic plants or animals or dead bits of organic matter,

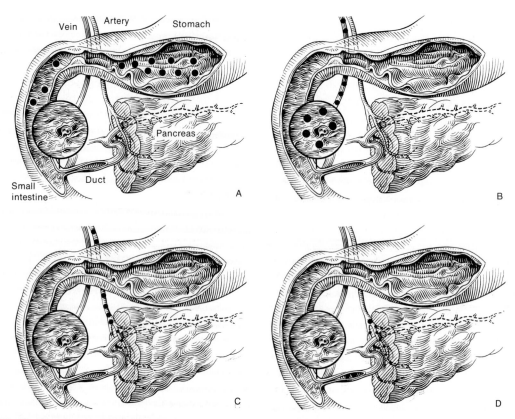

Figure 19-10 The control of the release of pancreatic juice by the hormone secretin. *A,* Hydrochloric acid *(black dots)* is secreted by the glands in the wall of the stomach and passes through the pylorus to the duodenum. *B,* Some of the hydrochloric acid diffuses into the wall of the duodenum and causes cells there to secrete the hormone secretin *(black dots),* which passes into the adjacent capillaries. *C,* Secretin is distributed by the blood vessels to all parts of the body and some of it is carried via the pancreatic artery to the pancreas. *D,* The secretin stimulates the pancreas to secrete pancreatic juice, which is visible in the pancreatic duct. The duct carries the pancreatic juice to the small intestine where its enzymes are important in digestion.

into their cells by *phagocytosis.* An amoeba engulfs a ciliate or flagellate and forms a food vacuole around it (Fig. 19–11 *A*). Hydrolytic enzymes synthesized in the cytoplasm are secreted through the vacuolar membrane into the cavity, and digestion occurs within the vacuole as it circulates in the cell. The products of digestion are absorbed through the vacuolar membrane and utilized for the production of biologically useful energy or as substrates for the synthesis of macromolecules. Any undigested remnants are expelled and left behind as the animal moves on.

In cnidarians and platyhelminthes there has evolved a *gastrovascular cavity,* lined with a *gastrodermis* (Fig. 19–11 *B* and *C*). Small animals are taken in through the mouth, and digestion begins within the gastrovascular cavity. The gastrodermis secretes digestive enzymes into the cavity and the prey is partially

digested there. The bits are absorbed into the gastrodermis cells where digestion is completed within food vacuoles. Digestion is partly extracellular, occurring in the gastrovascular cavity, and partly intracellular, within food vacuoles in the gastrodermis. The gastrovascular cavity of the flatworms may be greatly branched, ramifying through most of the body and facilitating the distribution of digested food. Neither cnidarians nor flatworms have an anal aperture; undigested wastes are excreted through the mouth.

In most of the rest of the invertebrates, and in all the vertebrates, the digestive tract is a tube with two apertures; food enters by the mouth and undigested residues leave by the *anus* (Fig. 19–11 *D*). The digestive tract may be short or long, straight or coiled, and may be subdivided into specialized organs. Even though they may have similar names in dif-

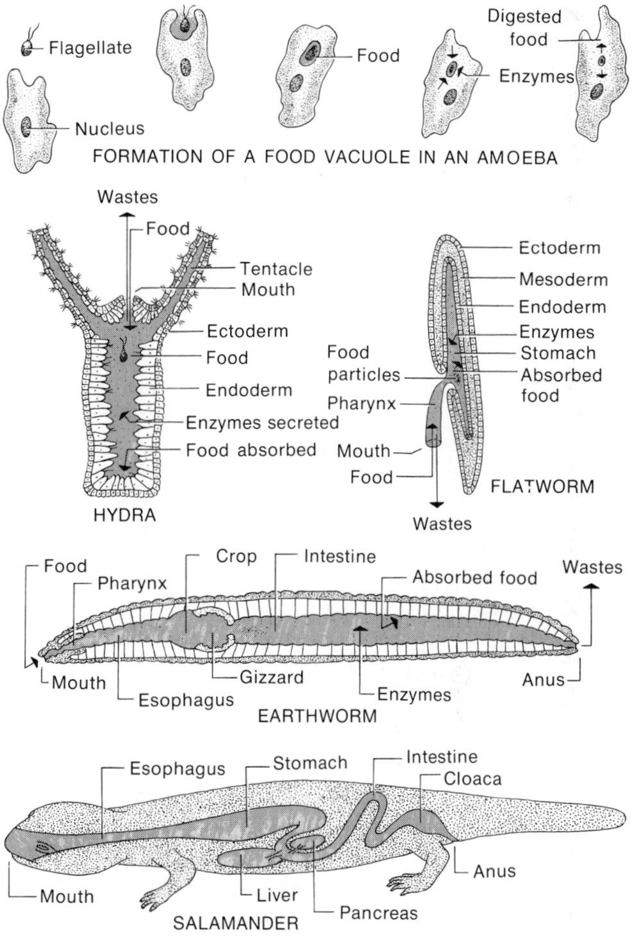

FORMATION OF A FOOD VACUOLE IN AN AMOEBA

Figure 19–11 The structural basis of the process of digestion in amoeba, hydra, flatworm, earthworm and salamander, illustrating the similarities and differences in the digestive systems of these widely different animal forms.

ferent kinds of animals, these organs may be quite different and may have different functions. The digestive system of the earthworm, for example, includes a *mouth,* a muscular *pharynx* which secretes a mucous material to lubricate food particles, an *esophagus,* a thin-walled *crop* where food is stored, a thick, muscular *gizzard* where food is ground against small stones, and a long, straight *intestine* in which extracellular digestion occurs. The products of digestion are absorbed through the intestinal wall by diffusion, by facilitated diffusion or by active transport, and the undigested residues pass out through the anus. Some invertebrates—worms, squid, crustacea and sea urchins—have hard, toothed mouthparts which can tear off and chew bits of food.

As the vertebrates evolved, the digestive system was gradually elaborated and organs added, resulting in the complex human mechanism we have just discussed. The digestive systems of the vertebrates from fish to humans are similar, and in all animals, from lowest to most complex, the chemistry of digestion and the enzymes involved are much alike.

QUESTIONS

1. List the organs of the human digestive system in order and give the functions performed by each.
2. What keeps the food moving through the digestive tract?
3. What is the role of bile in digestion? Where is it manufactured and how does it reach the food undergoing digestion? What controls the secretion of bile?
4. What are gallstones and how are they formed?
5. What are the islets of Langerhans? What is their function?
6. Discuss the role of enzymes in digestion in humans.
7. Discuss the absorption of glucose and amino acids. How does this differ from the mechanism of absorption of glycerol and fatty acids?
8. In certain abnormal conditions the stomach does not secrete hydrochloric acid. What effect might this have on the digestive process?

9. What prevents the stomach from being digested by its own secretions?
10. What controls the secretion of the digestive enzymes?
11. Compare the process of digestion in paramecium, hydra, flatworm and earthworm.
12. How does digestion in plants differ from digestion in animals?
13. Describe the path followed by a molecule of sugar from the time it enters the mouth as part of a molecule of starch until it reaches the liver.
14. Compare the teeth of mammals with those of lower vertebrates.
15. Describe the mechanism that normally prevents food from going down "the wrong throat" when we swallow.
16. Suppose one were to eat a ham sandwich. In what part of the digestive system and by what means would the several components of the sandwich be digested? How would the products of digestion be absorbed?

SUPPLEMENTARY READING

William Beaumont was a military physician stationed in upper Michigan. On June 6, 1822, a trapper, Alexis St. Martin, received a gunshot wound that opened a hole from his stomach to the outside. Beaumont treated St. Martin and made unique observations on the movement of the stomach during digestion, the stimuli effective in evoking gastric secretion, the normal appearance of the gastric mucosa, and so on. His book, *Experiments and Observations on the Gastric Juice and the Physiology of Digestion,* describing these observations (reprinted in 1929) is a beautiful example of the contributions to basic science that can be made by careful clinical observation. Part of this book, plus experiments by Reaumur, Spallanzani, Prout, Pavlov and Bayliss, and Starling's discovery of secretin are given in John F. Fulton's *Selected Readings in the History of Physiology,* Chapter Five. Further details regarding digestive enzymes and the mechanisms controlling the secretion of digestive juices are given in Guyton, *Basic Human Physiology;* Davson, *Textbook of General Physiology;* and *Animal Function* by M. S. Gordon. *Protein-Digesting Enzymes* by Hans Neurath relates enzyme structure and function and discusses the basis of enzymic specificity. The comparative physiology texts by Florey, by Prosser and Brown and by Hoar provide further discussions of digestion in a variety of animals.

CHAPTER 20

METABOLISM AND NUTRITION

Food may be defined as any substance taken into the body that can be utilized for the release of energy, for the building and repair of tissues, or for the regulation of body processes. This broad classification includes carbohydrates, fats, proteins, water, mineral salts and vitamins. The first three are sources of energy; the latter three, though not energy producers, are equally essential to life.

After being taken into the body, the molecules of food participate in a variety of enzymatic reactions, and these, plus all the other chemical activities of the organism, are termed **metabolism.** The presence of metabolic processes is one of the outstanding characteristics of living things. The nutrients may serve as raw materials for the synthesis of new macromolecules or they may be oxidized to provide energy. Some of this energy is required for the continual synthesis of new cellular components, some for the functioning of the cells (transmission of nerve impulses, contraction of muscles, secretion of enzymes, and so forth), and some is released as heat.

There are many ways of subdividing the general field of metabolism. We may study the metabolism of a single tissue, e.g., liver metabolism, or we may study the chemical reactions undergone by a particular kind of molecule or ion. Carbohydrate metabolism, for example, includes all the chemical reactions that starches and sugars undergo from the time they are taken in, until, after digestion and absorption, they are stored, converted into other cellular constituents, or oxidized for energy and leave the body as carbon dioxide and water.

Cells can obtain biologically useful energy from any of the three types of fuel. The complete oxidation of a gram of carbohydrate or protein yields about 4 kilocals., and the oxidation of a gram of fat yields about 9 kilocals. Foods such as whipped cream, mayonnaise and butter have more calories per unit weight, and hence are more "fattening," than fruits, meat or bread because of their high content of fat. In the average American diet about one half of the energy used daily comes from carbohydrates, one third from fats and one sixth from proteins.

20–1 THE BASAL METABOLIC RATE

The daily expenditure of energy varies widely from person to person, depending on activity, age, sex, weight, body proportions, and hormonal state. Metabolic rates are measured under standardized conditions; the subject is tested after he has had a night of restful sleep, at least 12 hours after his last meal, and after he has been reclining at complete rest for at least 30 minutes at an ambient temperature between 18 and 26° C. Under these conditions he uses energy to keep his heart beating, to breathe, to conduct a vast number of nerve impulses, and to maintain the constancy of his body fluids and his body temperature. The amount of energy expended by the body just to keep alive, when no food is being digested and no muscular work is being done, is called the *basal metabolic rate.* The basal metabolic rate for young adult men is about 1600 kilocals. per day and for women it is about 5 per cent less; in other words, if a young adult remained in bed for 24 hours without eating or moving he would expend about 1600 kilocals. in keeping alive. From thousands of determinations of basal metabolic rates in different people, tables have been established giving the normal basal metabolic rate for a given age, sex and total body area. The rate is proportional to the surface area of the body, which can be calculated from the height and weight. (Body surface area = weight$^{0.425}$ × height$^{0.725}$ × 0.007184.) A normal young adult uses 40 kilocals. per square meter of body surface per hour. If a young adult male has a basal metabolic rate of 45 kilocals. per square meter of body surface per hour, his rate is $\frac{45-40}{40}$ × 100, or 12.5 per cent above normal, expressed as BMR = +12.5.

Since chemical reactions occur more rapidly at higher temperature, the basal metabolic rate increases about 5 per cent for each degree of rise in body temperature. This is the primary reason why weight is lost during feverish illnesses (another is that we tend to eat less when we don't feel well).

An individual's basal metabolic rate can be determined directly by measuring the heat given off. The subject is placed in an insulated chamber surrounded by water, and the increase in the temperature of the air in the chamber and the surrounding water is mea-

Table 20–1 ENERGY EXPENDITURE PER HOUR OF A 70 KILOGRAM MAN DURING DIFFERENT TYPES OF ACTIVITY

Form of Activity	Kilocalories per Hour
Sleeping	65
Awake, lying still	77
Sitting at rest	100
Standing relaxed	105
Dressing and undressing	118
Tailoring	135
Typewriting rapidly	140
"Light" exercise	170
Walking slowly (4.2 km. per hour)	200
Carpentry, metal working, industrial painting	240
"Active" exercise	290
"Severe" exercise	450
Sawing wood	480
Swimming	500
Running (8.5 km. per hour)	570
"Very severe" exercise	600
Walking very fast (8.5 km. per hour)	650
Walking up stairs	1100

From Guyton, A. C.: *Textbook of Medical Physiology,* 5th Ed. Philadelphia, W. B. Saunders Co., 1976. Data compiled by Professor M. S. Rose.

sured. A simpler method measures the person's oxygen consumption over a short period of time. Since the release of energy and the production of heat depend on the oxidation of glucose and other foods, the amount of heat produced can be calculated from the amount of oxygen consumed.

Energy Requirements. If a person remains in bed for 24 hours and receives nourishment, he will expend about 1800 kilocals. The additional 200 kilocals. are required for the movements of the muscles of the digestive tract, the synthesis and secretion of the digestive juices, and the active uptake of the products of digestion. A person living a sedentary life uses about 2500 kilocals. per day, and one doing heavy physical work may expend 6000 or more kilocals. per day (Table 20–1). Most adults achieve a balance between the intake and utilization of calories, and their weight remains remarkably constant for years. There is a tendency for middle-aged people to gain weight because physical activity, but not appetite, decreases with age. An excess of only 10 kilocals. per day over the energy requirement would lead to an increase in weight of about one kilogram in the course of a year.

When the caloric intake is less than the daily energy requirement, the body must draw on its stored materials. The first to be used are the carbohydrates, stored as glycogen in the liver and muscles. Next, fat is withdrawn from storage in the fat deposits of the body and metabolized to supply energy. The average adult male has about 9 kilograms and the average adult female about 11 kilograms of stored fat (about 15 per cent and 21 per cent, respectively, of the total weight). The calories from the stored fat will supply energy for five to seven weeks of life. During prolonged starvation, the cells metabolize their own enzymes and structural proteins, first from the skeletal muscles and then from the heart, internal organs and brain, until death occurs.

20–2 CELLULAR FUELS

Carbohydrates. Sugars and starches are the principal sources of energy in the ordinary human diet, but they are not essential to the body. We could obtain energy as well from a mixture of proteins and fats. Foods rich in carbohydrates are usually the cheapest ones available, and the economic factor plays a role in determining the percentage of carbohydrates in an individual's diet. The citric acid of citrus fruits and the malic acid of apples and tomatoes may serve as sources of energy.

Fats. Fats and oils are the most concentrated foods, since they not only supply more than twice as many calories per gram as carbohydrates and proteins but also contain less water than these substances. They are digested and absorbed more slowly than other foods, so that one does not become hungry after a meal rich in fats as soon as after one of proteins and carbohydrates.

Fats are hydrolyzed to yield glycerol and fatty acids. The human body is able to synthesize most fatty acids, but not the "polyunsaturated" ones containing two or more double bonds. These, termed "essential" fatty acids, must be present in the diet. The amount of essential fatty acids required is small and is provided by almost any diet. The fact that they are essential was discovered only when animals were raised on highly purified diets from which fat had been removed chemically. Fats and oils are also important as sources of fat-soluble vitamins.

Proteins. Foods rich in proteins are usually the most expensive, and the amount of protein in the diet may be determined in part by the person's income. Since all the protein constituents of the body are constantly undergoing degradation and replacement, there is a continued requirement for a certain minimum of protein in the diet, even for adults whose growth has ceased. Growing children, pregnant women and people recovering from wasting diseases (all persons whose cells are carrying on a net synthesis of proteins) have increased requirements for proteins in the diet. It is difficult to say just how much protein is necessary in the diet to maintain health, since that depends on the kind eaten and the amount of other substances in the diet.

Proteins differ widely in the number and kind of amino acids they contain. When the body cells are synthesizing a particular type of protein, all the specific amino acids that comprise it must be available. If even a single amino acid is absent, the protein cannot be made. Animal cells can manufacture certain amino acids, but not all of them, and the ones that cannot be synthesized, called "essential" amino acids, must be supplied in the diet. "Essential" amino acids are no more essential for protein synthesis than other amino acids, but since they cannot be synthesized they are essential *in the diet.* There are ten amino acids required by man, and proteins that contain all of them in adequate amounts are called "adequate proteins." Milk, meat and eggs contain biologically adequate proteins, but the major protein of corn kernels (zein) has only small amounts of two of the essential amino acids. An experimental animal raised on a diet in which corn is the sole source of protein would lose weight and eventually die if the diet was not supplemented with tryptophan and lysine. Several recently developed strains of corn, with the genes *opaque-2* or *floury-2*, have kernels with twice as much lysine as ordinary corn and hence are a more valuable food source. Most people eat a diet containing many different proteins and are in no danger of suffering from a deficiency of any of the essential amino acids.

20–3 THE METABOLISM OF CARBOHYDRATES, FATS AND PROTEINS

In the previous chapter we traced the passage of foods from their entrance into the mouth to their absorption through the wall of the small intestine—proteins and carbohydrates being absorbed into the capillaries of the villi, and fats by way of the lymph vessels of the villi. After absorption the amino acids and simple sugars are carried to the liver by the hepatic portal vein. Perhaps originally the liver was important in digestion only, but during the evolutionary process it assumed many other functions and is now a chemical jack-of-all-trades. It protects the other cells of the body by detoxifying certain harmful substances; it is active in the storage and interconversion of carbohydrates, fats and proteins; it is important in the metabolism of hemoglobin; it stores certain vitamins; it manufactures substances necessary for the coagulation of blood; and it converts some of the harmful waste products produced by the metabolism of other body cells to less harmful, more soluble ones which can be excreted by the kidneys.

Carbohydrate Metabolism. Three kinds of simple sugars—glucose, fructose, and galactose—are derived from the hydrolytic cleavage of double sugars and are absorbed from the digestive tract. They pass to the liver, which converts the other simple sugars to glucose and stores them all as *glycogen.* The metabolic pathways of these interconversions are outlined in Figure 20–1. Glycogen is a highly branched

polysaccharide of high molecular weight, composed of glucose units linked by α-glycosidic bonds.

The role of the liver in storing carbohydrates was discovered by the French physiologist Claude Bernard. He analyzed the glucose content of blood entering and leaving the liver just after a meal and found a much higher concentration of sugar in the blood entering the liver than in that leaving it. Analysis of the liver showed that new glycogen appeared simultaneously. Between meals the liver glycogen is reconverted to glucose and the concentration of glucose in the blood leaving the liver is greater than in that entering it. In this way Bernard discovered that the liver maintains the glucose concentration of the blood at a more or less constant level throughout the day.

The liver can store enough glycogen to supply glucose for about 12 to 24 hours; after that the normal concentration of glucose in the blood is maintained by the conversion of other substances, principally amino acids, into glucose.

Glucose is the primary source of energy for all cells and its concentration in the blood must be maintained above a certain minimal level, about 60 mg. per 100 ml. of blood. The brain is

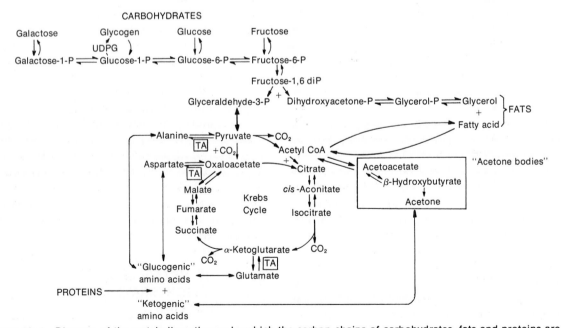

Figure 20–1 Diagram of the metabolic pathways by which the carbon chains of carbohydrates, fats and proteins are interconvertible. Transamination reactions, by which amino acids are converted to keto acids, are indicated by TA. Reactions requiring separate enzyme systems for the reverse direction are indicated by ⇄.

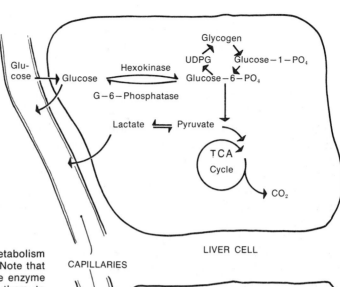

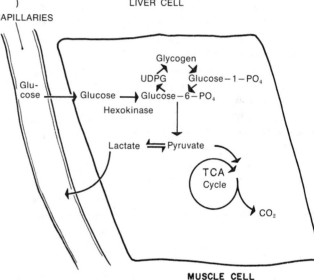

Figure 20–2 Diagram comparing the metabolism of glucose by liver and skeletal muscle. Note that liver cells, but not muscle cells, have the enzyme glucose-6-phosphatase, which enables them to secrete glucose into the blood stream.

the first organ to suffer when the concentration falls below this. In contrast to most other cells of the body, brain cells are unable to store any appreciable quantity of glucose as glycogen and they have only a limited ability to use fatty acids or amino acids as sources of energy. When the glucose level is low, and the brain is not adequately supplied with fuel, symptoms resembling those accompanying a lack of oxygen appear—mental confusion, convulsions, unconsciousness and death. Whenever the brain cells are deprived of either glucose or oxygen, they cannot carry on the metabolic processes that yield energy for their normal functioning.

Muscle cells also can change glucose into glycogen for storage, but muscle glycogen serves only as a local fuel deposit, available for muscular work, and it is not available for regulating the concentration of glucose in the blood.

Liver cells, but not muscle cells, contain the enzyme glucose-6-phosphatase, which converts glucose-6-phosphate to free glucose to be secreted into the blood stream (Fig. 20–2).

In addition to being stored as glycogen or oxidized for energy, glucose may be transformed into fat for storage. Whenever the supply of glucose exceeds the immediate needs, it is converted into fat in the liver and adipose tissue and is used for energy at some later time. It has been known for many years that eating large amounts of starches or sugars is fattening; the starch of corn or wheat eaten by cattle and pigs is converted into the fat of butter and bacon. With the use of radioactive or stable isotopes, it is possible to demonstrate that a particular carbon or hydrogen atom that enters the body as carbohydrate can be recovered as fat in adipose tissue or liver. The metabolic pathways by which carbohydrates are converted to fats are outlined in Figure 20–1. Both the glycerol and the fatty acids of the lipid molecule can be synthesized from the carbon chain of glucose.

The functioning of the liver in carbohydrate metabolism is regulated by a complex interaction of four hormones: insulin from the pancreas, epinephrine from the adrenal medulla, cortisol from the adrenal cortex and growth hormone from the pituitary.

Lipid Metabolism. Each species of animal or plant deposits fat containing a certain proportion of the different kinds of fatty acids. When we eat beef fat or olive oil, it must be changed, largely by the liver, into the type of fat characteristic of human beings. The fat in adipose tissue, besides being available as a source of energy when needed, serves as a supporting cushion for certain internal organs and as an insulating layer under the skin, preventing too rapid heat loss. The role of adipose tissue in thermal insulation is especially evident in aquatic mammals such as whales, which have a thick layer of fat-laden cells, *blubber*, just under the skin.

The oxidation of fatty acids does not proceed properly unless oxaloacetic acid, derived primarily from carbohydrate metabolism, is available to condense with the acetyl coenzyme A formed from the fatty acids (Figs. 5–18 and 20–1). Diabetics, whose carbohydrate metabolism is interfered with, have abnormal lipid metabolism also, and certain intermediate products (called acetone bodies; Fig. 20–1) tend to accumulate in their blood and be excreted in the urine. In addition, large amounts of lipid collect in the liver; fatty liver is a symptom occurring in certain abnormalities of liver function.

Lipids, as well as proteins, are important structural components of the nuclear, mitochondrial and plasma membranes.

The metabolism of fats is controlled by hormones from the pancreas, pituitary and adrenals, and to some extent by sex hormones. Any severe disturbance of liver function results in the almost complete absence of fat from the usual adipose tissues, indicating that fat must be acted upon in some way by the liver before it can be stored or metabolized.

Protein Metabolism. Most of the amino acids entering the liver from the hepatic portal vein are removed from the blood and stored temporarily. Later, some of them are returned to the blood and carried to other cells to be incorporated into new proteins. Experiments using amino acids labeled with ^{15}N, or "heavy" nitrogen, have shown that the body proteins are constantly and rapidly being torn down and rebuilt.

If there are more amino acids in the diet than are necessary for the synthesis of cell proteins, enzymes in the liver remove the amino group from amino acids, a process called *deamination.* Other enzymes combine the split-off amino groups with carbon dioxide to form a waste product, *urea* (Fig. 20–3), carried by the bloodstream to the kidneys and eliminated in the urine.

The parts of the amino acids remaining after deamination are simple organic acids. The carbon skeleton of certain amino acids, called "glucogenic" amino acids, can be converted (Fig. 20–1) into glucose or glycogen. The carbon skeletons of other amino acids yield acetone bodies and these are termed "ketogenic" amino acids. There is little or no storage of proteins as such in the body; the proteins the body draws upon when carbohydrates and fats are exhausted are not stored proteins, but are the actual enzymes and structural proteins of the cells.

The hormonal control of the metabolism of proteins and amino acids is even more complex than that of lipids. Since growth is essentially the deposition of new protein, the growth hormone of the pituitary is an important regulator of protein metabolism. Insulin, the sex hormones and cortisol from the adrenal cortex are also involved in the control of protein metabolism.

20–4 OTHER COMPONENTS OF THE DIET

Minerals. Some fifteen elements are known to be essential as mineral salts in the diet. A few of these are needed only in trace amounts. The daily requirements of some of these are as follows: sodium chloride, 2 to 10 grams; potassium, 1 to 2 grams; magnesium, 0.3 grams; phosphorus, 1.5 grams; calcium, 0.8 grams (more during growth, pregnancy or lactation); iron, 0.012 grams; copper, 0.001 grams; manganese, 0.0003 grams; iodine, 0.00003 gram. The constant loss of mineral salts from

Figure 20–3 The urea cycle is a sequence of enzymic reactions by which the urea molecule is assembled from carbon dioxide and ammonia. The amino acids ornithine, citrulline and arginine are utilized in the cycle, but neither ammonia nor carbon dioxide reacts as such with them. First, carbamyl phosphate is synthesized from ammonia and carbon dioxide by a complex reaction which requires 2 ATPs. The carbamyl phosphate condenses with the terminal amino group of ornithine to form citrulline. This in turn reacts with aspartic acid, another amino acid, to form the intermediate argininosuccinic acid, a reaction that requires another ATP to drive it. The argininosuccinic acid is cleaved to yield arginine and fumaric acid; thus the amino group of aspartic acid is transferred to arginine. The arginine is hydrolyzed by arginase to yield urea and ornithine, which can be used in the next cycle. The energy to drive the cycle and synthesize urea is provided by the two ~P used in the synthesis of carbamyl phosphate and the ATP converted to AMP (adenosine monophosphate) and PP_i (inorganic pyrophosphate) in the synthesis of argininosuccinic acid.

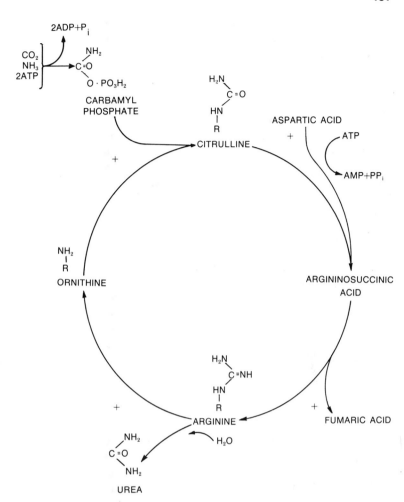

the body (about 30 grams per day) in the urine, sweat and feces must be balanced by the intake of equivalent amounts in the food. A diet that contains no minerals is more rapidly fatal than no food at all, because the excretion of wastes from the metabolism of carbohydrates, fats and proteins requires the simultaneous excretion of a certain amount of salt (to keep the pH of the blood constant). Thus, a salt-free diet actually exhausts the body reserve of salts. Deficiencies of minerals are rather rare, since meat, cheese, eggs, milk and vegetables are rich sources. Human diseases resulting from deficiencies of iron, calcium, copper, phosphate and iodine do occur.

Blood and other body fluids are about 0.9 per cent salt, and most of this is sodium chloride. The sodium and chloride ions play an important role in maintaining osmotic balance and acid-base balance in the body fluids. They are major components of the secretions of the digestive tract—the hydrochloric acid of the stomach and the pancreatic and intestinal juices. The salts in these secretions are reabsorbed and used over again, so that the loss of salts via the digestive tract is negligible. The daily requirement for sodium chloride varies widely and depends on the amount lost in perspiration. Men doing heavy work in hot places—for example, "sand hogs" digging tunnels—may drink salt water instead of plain water to prevent a decrease in the amount of salt in the blood, which would result in muscular cramps and heat exhaustion.

Potassium and magnesium are required for muscle contraction and for the functioning of many enzymes.

Calcium and phosphorus are the chief constituents of bones and teeth, and a deficiency in childhood of either one (or of vitamin D, required for their absorption and metabolism) produces rickets. Phosphorus has an important role in metabolism; DNA, RNA, and the nucleotides so important in intermediary metabolism—NAD, NADP, ATP, etc.—all contain phosphorus. In addition, sugars and fatty acids must be phosphorylated before they can be metabolized to provide biologically useful energy.

Trace Elements. A number of elements are required only in trace amounts. These serve, in general, as metal components of specific enzyme systems.

Iodine is a constituent of the hormone of the thyroid gland, and if the diet is deficient in this substance, the gland is unable to make thyroxine and enlarges to form a *goiter* (p. 561). Iodine is abundant in sea water and sea foods but is rare elsewhere, and formerly it was common for people living in inland regions to suffer from goiter. Now, most table salt is fortified with small amounts of potassium iodide to prevent this.

Iron is a constituent of hemoglobin and of the cytochromes. This iron is used over and over, and as long as there is no loss of blood the amount of iron needed daily in the diet is negligible. Because women lose a considerable amount of blood each month by menstruation, their iron reserves are typically very small and they are more likely than men to become anemic owing to an iron deficiency.

Small amounts of copper are necessary in the diet to bring about the proper utilization of iron for normal growth and as a component of certain enzymes. Traces of manganese, molybdenum, zinc and cobalt are required for normal growth and as activators of certain enzymes. Zinc is a constituent of carbonic anhydrase, alcohol dehydrogenase, and many other enzymes. Traces of fluorine in the drinking water are remarkably effective in preventing dental decay.

Water. Water makes up about two thirds of the human body and is an essential component of every cell. It is the fluid part of blood and lymph and is the medium in which the other chemicals are dissolved and in which all chemical reactions occur. It is indispensable for digestion, since the splitting of carbohydrates, proteins and fats requires a molecule of water for each pair of sugar molecules or amino acid molecules separated. Water dissolves metabolic wastes, distributes and regulates the body heat, and, as perspiration, cools the body surface. The amount of water lost daily averages about 2 liters, although it varies with individual activities and the climate. The loss must be replaced promptly; men can live weeks without food but only a few days without water. All foods contain some water, and some, such as green vegetables and fruits, may contain as much as 95 per cent water. Aquatic animals have no problem in obtaining water; indeed, their problem is to prevent the osmotic inflow of water and the consequent bursting of their cells. Certain desert animals live indefinitely without drinking water, obtaining it from the foods they eat and from the oxidation of their food.

Condiments and Roughage. Pepper and other spices, collectively known as condiments, have little or no food value themselves but are important for making foods more palatable. By stimulating the appetite they help ensure that a sufficient quantity of food is eaten.

We have already discussed the importance of "bulk," or undigested matter, in stimulating the movements of the intestines and preventing constipation. The diet should contain some indigestible matter for this purpose, such as the cellulose from vegetables and fruits.

20–5 VITAMINS

One of the most notable biochemical achievements since the turn of the century has been the discovery of vitamins and the analysis of their properties and functions in metabolism. Vitamins are relatively simple organic compounds, which, though present in such scanty amounts that they cannot be used as sources of energy, are absolutely essential to life. The various vitamins are quite different chemically, but are similar in that they cannot be synthesized in adequate amounts by the animal body and therefore must be present in the diet. There are two major groups of vitamins: those that are readily soluble in fats or lipid solvents, the *fat-soluble vitamins* (A, D, E and K) and those that are readily soluble in water, the *water-soluble vitamins* (C and the B complex). When the amount of any one in the diet is inadequate, a specific pathologic condition or deficiency disease occurs, curable only by administration of the specific vitamin; for example, only vitamin C is effective in curing scurvy.

In 1912, investigators found that animals could not survive on a diet of purified carbohydrates, proteins and fats, but that accessory growth factors, or vitamins, were necessary. At first the chemical structure of these substances was unknown, and they were referred to as vitamins A, B and C, which prevented night blindness and rickets, beriberi, and scurvy, respectively. Today the chemical structures of nearly all of them are known, and most of them have been made synthetically. The original vitamins A and B have been shown to be complexes of several vitamins, A being subdivided into A, D and E, while vitamin B consists of almost a dozen different ones. The vitamins of known chemical structure are usually referred to by their chemical names, e.g., thiamine rather than vitamin B.

The distinction between vitamins and such things as essential amino or fatty acids is not clear cut; the latter are also simple organic substances, essential for life, which cannot be made by the animal body and must be taken in with the food. They are needed in much larger amounts, however; the term vitamin is reserved for the substances required by the body in very small quantities.

The average adult eating a normal varied diet has no need to take vitamin pills; he will obtain the necessary kinds and amounts of vitamins from his food. Infants and younger children, whose diets are more restricted, may need supplementary amounts of certain vitamins, especially A and D. The vitamin requirements of different animals are not the same; most animals do not require vitamin C in the diet but synthesize it from glucose. Only the human, monkey and guinea pig need it in the diet. Insects require only cholesterol and the B complex vitamins in their diet. Thus, what is a vitamin for one animal is not a vitamin for another, although it is probable that all animals and plants require all or nearly all the known vitamins.

The function of almost all of the vitamins has been discovered; each one has been found to serve as an integral part of a coenzyme for one or more of the fundamental enzyme reactions common to all living things. Clear evidence that plants require the same vitamins as animals, but that they normally synthesize all the vitamins they require, has been provided by the experiments with the mold *Neurospora* done by Beadle and Tatum and their collaborators. The ordinary "wild" strain of this mold requires in its culture medium only a single vitamin, biotin, in addition to salts and a sugar of some sort. By exposing mold organisms to x-rays or ultraviolet light, thus causing a gene mutation which interfered with some step in the synthesis of a vitamin, these investigators produced mutant strains which would grow only when one additional vitamin was added to the culture medium. Now there are strains that require each of the vitamins necessary for growth in animals. This is evidence that the mold cells (and probably other plant cells) need these substances for growth just as much as animal cells do, but that ordinarily they synthesize all the vitamins they need except biotin.

20–6 FAT-SOLUBLE VITAMINS

Vitamin A. Vitamin A, or *retinol*, occurs in animal products such as butter, eggs and fish liver oils. Plants contain a yellowish substance, *carotene*, which can be split to yield two molecules of vitamin A in animal cells. Vitamin A itself is fat-soluble and can be stored in the body, especially in the liver. The daily requirement for adults is about 1.5 mg (5000 International Units), for a child under three about 0.6 mg., and for older children an intermediate amount.

This vitamin, necessary for the growth and maintenance of the epithelial cells of the skin, eye, and digestive and respiratory tracts, is stored in the liver. In vitamin A deficiency these cells become flat, brittle and less resistant to infection than normal (vitamin A is sometimes called the "anti-infection vitamin"). A deficiency of vitamin A and consequent disturbance of the epithelial cells can cause the ducts of all types of glands to become blocked, leading to atrophy of the glands. Atrophy of the germinal epithelium of the testis causes sterility in the male. Skeletal growth ceases in individuals with vitamin A deficiency, apparently because of defective synthesis of chondroitin sulfate.

In advanced cases of vitamin A deficiency the eye epithelium forms a dry and horny film over the cornea, resulting in a characteristic type of blindness called *xerophthalmia* (Fig. 20–4). Vitamin A is necessary also for the maintenance of normal nerve tissue and for the growth of bone and the enamel of teeth. It is

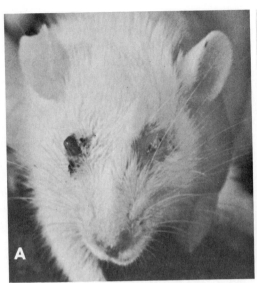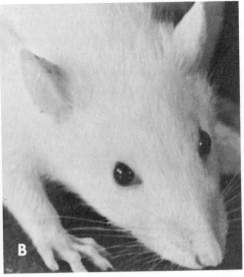

Figure 20–4 *A,* Typical eye condition produced by lack of vitamin A. *B,* The eyes restored to normal by the feeding of 3 units (about 0.001 mg.) of vitamin A daily. (Courtesy of E. R. Squibb and Sons.)

involved in the chemistry of vision, so that *night blindness,* inability to see in a dim light, may result from vitamin A deficiency. The rods in the retina of the eye contain a substance called *rhodopsin* (visual purple), made up of retinal, a derivative of vitamin A, and a protein, opsin. Rhodopsin is split into retinal and opsin by a chemical reaction triggered by light, which stimulates the receptor cell to send an impulse to the brain, resulting in the sensation of sight. Ordinarily, rhodopsin is quickly re-synthesized in an energy-requiring reaction, but in vitamin A deficiency the resynthesis of rhodopsin is retarded and night blindness re-sults. Deficiencies severe enough to produce xerophthalmia are rare in the United States, but night blindness is more common. During World War II, the pilots of night fighter planes were given diets particularly rich in vitamin A to prevent this. Toxic symptoms due to an overdose of vitamin A have been observed in human beings; some of the first cases occurred in people who had eaten polar bear liver, which is very rich in vitamin A!

The chemical form in which vitamin A exerts its effects on growth is not known. *Re-tinoic acid* can partially replace retinol in the rat diet and promote the growth of bone and soft tissues and sperm production, but it cannot be used in the visual process. Retinoic acid is converted in the body to some unknown form that is several times more active than the parent compound.

Vitamin D. Vitamin D, *cholecalciferol,* is antirachitic. It leads to the mobilization of cal-cium and phosphate from bone and stimulates the transport of calcium across the intestinal mucosa. Cholecalciferol can be formed in the skin from 7-dehydrocholesterol by the action of ultraviolet light, which cleaves the B ring of the precursor molecule (Fig. 20–5). Thus cho-lecalciferol is a "vitamin" only if a person is not exposed to an adequate amount of sunlight. The disease *rickets* (Fig. 20–6), in which bones do not form properly, probably first appeared when our ancestors started wearing clothes and living in houses. By greatly reducing the amount of ultraviolet irradiation of the skin, the conversion of 7-dehydrocholesterol to cholecal-ciferol was reduced and, as a result, the latter compound became a vitamin, something that was required in the diet. At one time rickets was widely prevalent in northern Europe and North America, but the discovery of vitamin D in 1922 and the further discovery that sterols could be converted to vitamin D by ultraviolet irradiation provided a ready source of the ac-tive material. Rickets has essentially been com-pletely eliminated in the western world by the addition of vitamin D to milk and other foods.

One of the first clues that suggested that a metabolite of vitamin D, rather than the vita-min itself, might be the biologically active form, came from the observation that there is a lag period of 10 hours or more between the ad-ministration of vitamin D and its effect on cal-

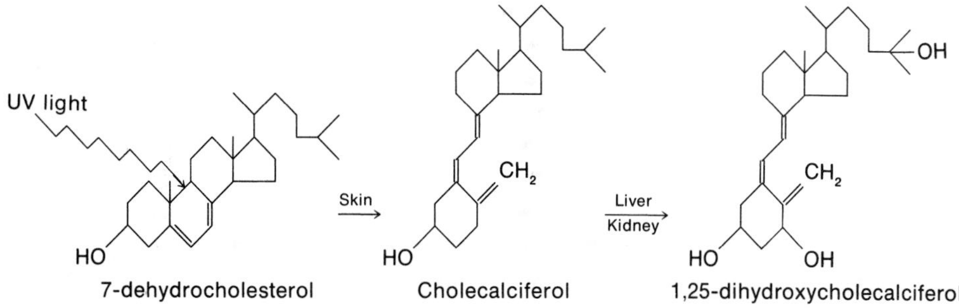

UV light

Skin →

Liver
Kidney →

HO

7-dehydrocholesterol

HO   CH₂

Cholecalciferol

HO   OH   CH₂

OH

1,25-dihydroxycholecalciferol

Figure 20–5 The synthesis of cholecalciferol from its precursor, 7-dehydrocholesterol, and its conversion to 1,25-dihydroxycholecalciferol, the hormonelike, biologically active compound that stimulates the uptake of calcium from the lumen of the intestine.

cium transport. Two hypotheses were advanced to account for this; the first suggested that the lag period might represent the time required for the synthesis of a carrier protein for calcium transport, and the second suggested that the time was required for the transformation of the vitamin into an active molecule. Both of these working hypotheses have now been found to be correct. Vitamin D is converted to 25-hydroxycholecalciferol by an enzyme in the liver and then to 1,25-dihydroxycholecalciferol by an enzyme in the kidney. The active, hormonelike molecule, 1,25-dihydroxycholecalciferol, stimulates the synthesis of a calcium transport protein in the intestinal mucosa and this substance is responsible for the increased uptake of calcium from the intestinal contents. Excessive doses of vitamin D are toxic, causing hypercalcemia and the deposition of calcium in soft tissues.

Vitamin E (Alpha-tocopherol). Experimental studies on rats, chicks and ducks have shown that alpha-tocopherol, or vitamin E, is necessary to prevent sterility. If it is absent from the diet, male animals undergo degenerative changes in the testes and become sterile. Vitamin E-deficient females are unable to complete pregnancy successfully, since the embryos die and are resorbed. Eggs from vitamin E-deficient hens fail to hatch. It has not been shown that a deficiency of vitamin E is responsible for human sterility, but this is possible. No figure can be given for the daily human requirement of the substance, but it is so widespread in both vegetable and animal oils that a deficiency in any normal diet is almost impossible.

Vitamin E serves as an antioxidant and protects certain labile cellular components from being oxidized. It also plays some role as a constituent of the electron transport system, but the exact nature of this role is unclear. Some investigators believe that the effects of tocopherol deficiency can be traced to the accumulation of fatty acid peroxides, which react with and destroy other cellular components. A deficiency of vitamin E produces progressive deterioration and paralysis of the muscles, presumably by degeneration of the nerves (just as the destruction of nerves in infantile paralysis leads to the wasting of muscles and paralysis). Certain types of paralysis in human beings have been treated with vitamin E preparations, with beneficial results.

Vitamin K. A number of similar substances, referred to as vitamin K, play a role in the normal coagulation of blood by promoting the synthesis in the liver of prothrombin and proconvertin, two components of the blood-

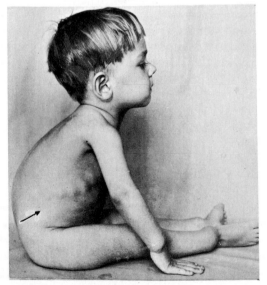

Figure 20–6 A child with rickets. A deficiency of vitamin D decreases the body's ability to absorb and use calcium and phosphorus and produces soft, malformed bones. These are most clearly evident in the ribs *(arrow)* and in the wrists and ankles. (Courtesy of Dr. Niilo Hallman.)

clotting system. These chemicals occur in a variety of foods and are manufactured by the bacteria in the human intestine, so that vitamin K deficiency in man usually results from some abnormality in its absorption rather than from its lack in the diet. It can be absorbed only in the presence of bile salts (vitamins A, D and E also require bile salts to be absorbed), and hence an obstruction of the bile duct results in vitamin K deficiency, no matter how much is present in the diet or made by the intestinal bacteria. Patients with vitamin K deficiencies are poor surgical risks because of the likelihood of hemorrhages after the operation. The administration of vitamin K (and bile salts, if necessary) before operation removes this danger and has saved many lives. Before they acquire their quota of intestinal bacteria, newborn infants are likely to be deficient in vitamin K, and administration of this substance to the mother in the last few days of pregnancy helps prevent the hemorrhages that often occur in infants after delivery. No estimate can be given of the daily requirement of this vitamin, but in vitamin K deficiencies, 1 to 5 mg., administered daily, brings the clotting time of the blood back to normal.

20–7 WATER-SOLUBLE VITAMINS

Vitamin C. The deficiency disease *scurvy*, resulting from a lack of vitamin C, is one of the principal noninfectious plagues of history. It is characterized by bleeding gums, bruised skin, painful swollen joints and general weakness. Scurvy occurs whenever people are deprived of fresh fruits, vegetables and meat for long periods of time, as they were on extended sailing voyages and during the long northern winters.

The earliest report of a cure for scurvy is found in the records of Jacques Cartier's expedition to Canada in 1536. His crew suffered severely from scurvy and were cured by an extract of fir needles prescribed by the Indians. The scurvy-preventing vitamin was isolated in 1933 and proved to be ascorbic acid, a substance that had been known for many years but whose antiscorbutic properties had not been suspected. Ascorbic acid is rather unstable and is readily destroyed by cooking. The best sources of it are fresh fruits and juices, although modern freezing and canning processes preserve most of the ascorbic acid content of other foods. Ascorbic acid plays some part in cellular oxidations, particularly in the oxidation of tyrosine. It also plays a role in the hydroxylation of the amino acids proline and lysine to form hydroxyproline and hydroxylysine, two of the constituents of collagen. In vitamin C deficiencies the capillaries become exceedingly fragile and easily ruptured, resulting in hemorrhages under the skin and in the joints. The development of bones and teeth is also abnormal. Normal human adults require between 75 and 100 mg. of ascorbic acid daily, an amount supplied by an 8-ounce glass of orange juice.

The Vitamin B Complex. The original vitamin B was characterized as the anti-beriberi factor, but from the same extracts of liver, yeast or rice hulls that yield an anti-beriberi substance, nine other materials with specific biologic effects have been isolated. Some of these substances were once given separate alphabetical designations: riboflavin was called vitamin G, and biotin was called vitamin H, but they are now all grouped together as members of the B complex, not because they are similar chemically or in their effects, but because they tend to occur together. Individual members of the B complex are referred to by their chemical names.

Thiamine (Vitamin B_1). This substance, the first to be separated from the rest of the complex, prevents beriberi. It is a white, crystalline material with an odor like that of yeast, found in small quantities in a wide variety of foods. Yeast, liver, nuts, pork and whole grain cereals are the best sources of all the vitamin B complex. Since the average American diet is somewhat deficient in thiamine, flour, bread and breakfast cereals are now enriched with it. The daily requirement varies with the body weight, the number of calories eaten and the proportion of carbohydrate in the diet (the more carbohydrate, the more thiamine needed), but the amount needed daily by the average person is from 2 to 3 mg. Thiamine and the other B complex vitamins are not stored in the body to any great extent, and evidence of a deficiency appears within a few weeks. Most diets contain enough thiamine to prevent the appearance of beriberi but may not contain enough for maximum health.

Thiamine pyrophosphate is the coenzyme

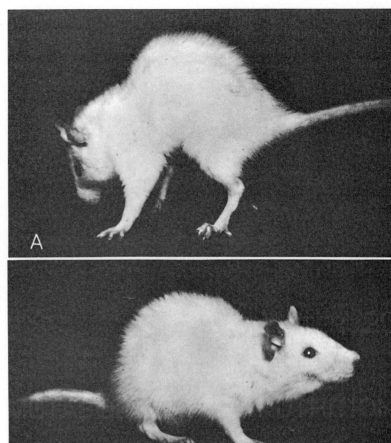

Figure 20–7 *A,* Polyneuritis (beriberi) in a rat raised on a diet deficient in thiamine. Note that the back is arched and the hind legs are stretched and far apart. Such animals have a peculiar halting gait and are particularly awkward in turning, readily losing their balance. When rotated, they have great difficulty in regaining equilibrium, probably because of degeneration of the nerves to the semicircular canals. *B,* The same rat eight hours after receiving an adequate dose of thiamine; the back and hind legs are normal, and the animal readily regains equilibrium when spun. (Courtesy of the Upjohn Company.)

for the oxidative decarboxylation of pyruvic and α-ketoglutaric acids. It is also the coenzyme for transketolase, one of the enzymes of the pentose phosphate pathway. When thiamine deficiency interferes with carbohydrate metabolism, a number of characteristic symptoms appear: in mild deficiencies there is fatigue, loss of appetite, weakness and muscular cramps; in more marked deficiencies these symptoms are accentuated, and there is also a painful degeneration of the nerves and a secondary wasting of the muscles resulting in paralysis. This condition, known as *beriberi* (Fig. 20–7), disappears rapidly when thiamine is given. Any diet deficient in thiamine is likely to be deficient in the other B complex vitamins as well, so that cases of thiamine deficiency alone are rare.

Riboflavin (Vitamin B$_2$ or G). Riboflavin is a yellow pigment found in both plant and animal tissues; it occurs most abundantly in foods rich in thiamine: yeast, liver, wheat germ, meat, eggs and cheese. Riboflavin forms part of flavin mononucleotide (FMN) and *flavin adenine dinucleotide* (FAD), coenzymes in certain cellular oxidative processes in the metabolism of glucose and amino acids. One to two milligrams of riboflavin per day is required to maintain health in man. A deficiency of riboflavin is marked by the appearance of cracks in the corners of the mouth, a characteristic purplish red color of the tongue and stunted growth. In experimental riboflavin deficiencies in rats there is a failure of growth, loss of hair, cataract, inflammation of the eyes and, eventually, death (Fig. 20–8).

Niacin or Nicotinic Acid. Niacin is a component of two coenzymes, nicotinamide adenine dinucleotide (NAD) and nicotinamide adenine dinucleotide phosphate (NADP), important as coenzymes for many dehydrogenases. They serve as hydrogen acceptors and donors in many reactions. Niacin was known as an organic compound for over 50 years before its function as a vitamin was recognized. It is found in yeast, fresh vegetables, meat and

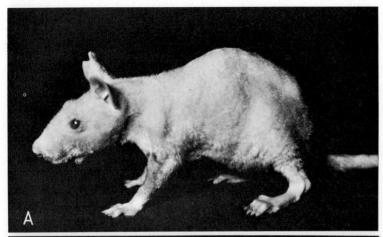

Figure 20–8 *A,* Riboflavin-deficient rat, with stunted growth, general inflammation of the skin (note the open sore on the left front leg), scanty hair, and inflammation of the eyes. *B,* The same rat after two months of treatment with riboflavin: no signs of the deficiency are visible; growth has been resumed, and the lesions of the skin and eyes are cured. (Courtesy of the Upjohn Company.)

beer. Corn meal has an unusually low niacin content, and wherever this food forms a large part of the diet, the deficiency disease *pellagra* is fairly prevalent. Pellagra is characterized by dermatitis (reddened inflammation of the skin, especially in those parts of the body exposed to light), diarrhea and dementia. Niacin plays a role in the maintenance of the epithelia of the skin and digestive tract and normal nerve functioning—processes that depend upon its action as the coenzyme of one or another enzyme. The recommended daily allowance of niacin is about 20 to 25 mg., but a considerable part of the human requirement is synthesized by the intestinal bacteria. When a person is treated with sulfa drugs for some infection, the intestinal bacteria are killed, and deficiencies of a number of vitamins, including niacin, may occur.

The amino acid tryptophan can be metabolized by human tissues to yield niacin (nicotinic acid), hence the dietary requirement of niacin depends on the amount of tryptophan in the diet.

Pyridoxine (Vitamin B₆). This vitamin occurs in a wide variety of foods—meat, eggs, nuts, whole grain cereals and beans—so that a clear-cut deficiency of pyridoxine in man has not been found. *Pyridoxal phosphate* is the coenzyme for glycogen synthetase and for many different enzymatic reactions involving amino acids, such as transamination and decarboxylation to amines. Experimental animals fed a diet deficient in pyridoxine fail to grow, become anemic and have atrophied lymph tissue, resulting in a lack of white blood cells and antibodies, with a consequent lowered resistance to infection. The daily requirement is about 1 to 2 mg., but it varies with the amount of protein in the diet.

Pantothenic Acid. This vitamin is necessary for the maintenance of normal nerves and skin, and experimental deficiencies of it cause failure of growth, dermatitis, gray hair and

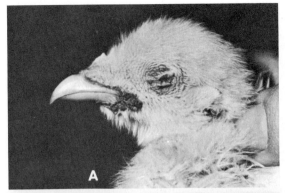

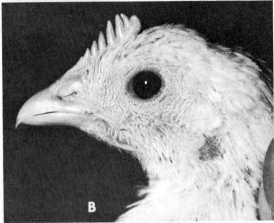

Figure 20–9 *A,* Chick after being fed a diet deficient in pantothenic acid. The eyelids, corners of the mouth, and adjacent skin are inflamed. The growth of feathers is retarded, and the feathers are rough. *B,* The same chick after three weeks on a diet with pantothenic acid: the lesions are completely cured. (Courtesy of the Upjohn Company.)

damage to the adrenal gland (Fig. 20–9). The "burning foot" syndrome suffered by some prisoners in prison camps during World War II responded to treatment with pantothenic acid. Almost any normal diet will provide the 20 mg. required daily by human beings. Especially rich sources of it are eggs, meat, sweet potatoes and peanuts. It forms part of *coenzyme A,* important in a number of steps in the metabolism of carbohydrates, fats and proteins, and in the transfer of energy.

Biotin. This was first discovered as a factor indispensable for the growth of yeast. It has since been shown to be necessary in the diet of mammals, though only in extremely small amounts. Some rich sources of it are molasses, egg yolk and liver. Egg white contains a protein called *avidin,* which combines with biotin in the intestine and prevents its absorption. Avidin is destroyed by heat, however, so that

cooked egg white does not interfere with the absorption of biotin, and much more than the amount of raw egg white in an eggnog or two is needed to cause biotin deficiency in experimental animals (Fig. 20–10). One of the few cases of human biotin deficiency occurred in a man who lived almost entirely on raw eggs and wine; he suffered from an inflammation of the skin, which cleared up when biotin was administered.

Biotin is a coenzyme for reactions in which carbon dioxide is added to an organic molecule (carbon dioxide fixation) and for the carboxylation of acetyl coenzyme A to form malonyl coenzyme A, the first step in the biosynthesis of fatty acids.

Folic Acid, Vitamin B$_{12}$, Choline, Inositol and Para-aminobenzoic Acid. Folic acid and vitamin B$_{12}$ (*cobalamin*) are necessary to prevent anemia and are used in conjunction with liver extract in treating pernicious anemia. Folic acid appears in the coenzyme tetrahydrofolic acid, required in the reactions by which one-carbon compounds are transferred from one molecule to another, and in biopterin, the coenzyme for the conversion of phenylalanine to tyrosine. Cobalamin is a complex molecule composed of a porphyrin ring containing cobalt ion, cyanide, the sugar ribose, and other constituents. The role of folic acid and cobalamin in preventing anemia is apparently that of facilitating the synthesis of nucleic acids involved in the production of red cells. Cobalamin is also known to serve as coenzyme in the interconversion of certain organic acids such as succinate and methyl malonate. Cobalamin is synthesized by bacteria but not by higher plants or animals. Thus this is a "vitamin" for green plants as well as for animals.

Choline is a growth factor, the absence of which causes hemorrhages in the kidneys and a bone deformity called perosis in chicks. It is important in the metabolism of fats and proteins, not as a coenzyme as many other B vitamins are but as a source of methyl groups to be used in the synthesis of certain essential substances. An adult requires about 2000 mg. of choline daily.

Lipoic acid is an eight-carbon fatty acid, containing two sulfur atoms, that functions as a cofactor in the oxidative decarboxylation of pyruvic and α-ketoglutaric acids, along with thiamine pyrophosphate. It has not been shown to be needed in the diets of humans and

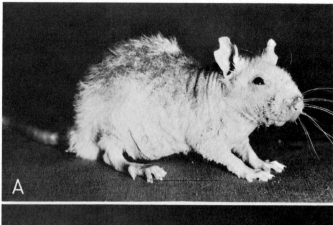

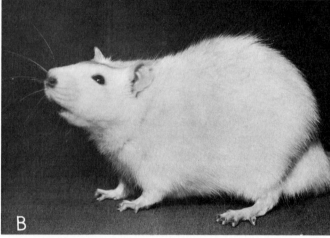

Figure 20–10 *A,* Rat after being fed a diet deficient in biotin, to which raw egg white was added. Growth has been retarded, and there is generalized inflammation of the skin. *B,* The same rat after three months on a diet containing adequate amounts of biotin: growth is normal and the skin lesions are completely healed. (Courtesy of the Upjohn Company.)

other animals, but it is a growth factor for certain microorganisms. Carnitine, needed as a growth factor by certain insects, is found in mammalian cells and plays a role in the transport of fatty acids across the mitochondrial membrane.

Inositol and para-aminobenzoic acid have been reported as important in preventing the loss of hair and the graying of hair, respectively. Both are necessary for normal growth of rats and presumably of other animals, including humans. Para-aminobenzoic acid forms part of folic acid. These B vitamins are also synthesized by the intestinal bacteria.

20–8 ANTIMETABOLITES

D. D. Woods found in 1940 that para-aminobenzoic acid reversed the action of the "sulfa drug" sulfanilamide on bacteria. Sulfanilamide is bacteriostatic—it prevents the multiplication of bacteria and thus aids the body defenses in dealing with invading bacteria.

This observation suggested the theory that sulfanilamide interferes with bacterial growth by acting as a competitive inhibitor of some bacterial enzyme for which para-aminobenzoic acid forms an integral part of the coenzyme. Sulfanilamide is quite similar in chemical structure to para-aminobenzoic acid, similar enough to fool the enzyme and be taken into the enzymatic mechanism but different enough so that the enzymatic mechanism becomes jammed. This theory set off a search for other substances (called *antimetabolites*) that are like, but slightly different from, ordinary vitamins, to serve as inhibitors of bacteria or of the growth of cancer cells. Aminopterin, an antimetabolite of folic acid, has been successful in alleviating certain kinds of leukemia.

20–9 DIET

Human beings, unlike many other animals, can adapt to a variety of diets. We are able to operate quite well on one made up chiefly of protein with only small amounts of fats and carbohydrates, on one composed mainly of carbohydrate with small amounts of protein and fats, or even on one that is primarily fat with small amounts of the others. The diet of Eskimos is an example of the latter.

The most important nutritional problem in the United States at present is *obesity.* Surveys show that some 25 per cent of the population are overweight. Obesity predisposes to a number of diseases, such as diabetes, and materially decreases life expectancy.

Dr. Clive McCay of Cornell University has shown that a low calorie diet, particularly during the early part of life, will double the life span of rats and dogs. The animals fed diets restricted in calories were healthier, spryer and more fertile than the control animals fed *ad libitum.* The statistics produced by these experiments prove what animal breeders and trainers have long known—an animal looks and behaves better when it is slightly underfed. There is every indication that what is true for rats and dogs is true for human beings. Many physiologists, performing experiments on themselves, have found that a restricted diet produced beneficial psychological effects: they felt more alert, happier and better able to stick to tedious, problem-solving tasks.

An adequate diet must supply water, salts and vitamins, sufficient calories to balance the daily expenditure of energy (unless one wishes to lose weight), and enough fats and proteins for tissue repair.

QUESTIONS

1. List the major types of foodstuffs. Which of these yield biologically useful energy?
2. Define a calorie. Compare the caloric value of a kilogram of carbohydrate, a kilogram of fat and a kilogram of protein.
3. What is meant by the term basal metabolic rate? What are two conditions in which this is increased?
4. What is meant by a "biologically adequate" protein?
5. What is meant by the "specific dynamic action" of proteins? What is the physiologic explanation of this phenomenon?
6. What role does glycogen play in cell metabolism?
7. Compare the pattern of carbohydrate metabolism in liver and skeletal muscle.
8. What are the functions of the liver in nutrition? In metabolism?
9. List the minerals most essential to the body and give the function of each in cell metabolism.
10. Define the term vitamin. Give the diseases that may result from vitamin deficiencies and list the vitamin which may cure each of these conditions.
11. What vitamins are present in milk? Eggs? Green vegetables? Pork chops?
12. What should be the caloric content and the specific constituents of a diet that would be adequate for an active young man. How should the diets of a pregnant woman, a 10 year old boy and a 65 year old man differ from this?
13. How can carbohydrates be converted into fats in the body? How can proteins be converted into carbohydrates in the body?
14. Compare a person's requirement for (a) vitamins, (b) essential amino acids and (c) essential fatty acids.

SUPPLEMENTARY READING

Quotations from papers by F. G. Hopkins and by Casimir Funk on the discovery of vitamins are found, together with a description by James Cook of the measures taken to prevent scurvy in his crew, in Chapter Eight of J. F. Fulton's *Selected Readings in the History of Physiology.* Hopkins' discovery of the accessory food factors is described in J. Needham and E. B. Baldwin's *Hopkins and Biochemistry.* The discovery of biotin and of some of its roles in metabolism are discussed by John D. Woodward in *Biotin.* More extensive discussions of intermediary metabolism and the roles of vitamins as enzymes are presented in textbooks of biochemistry such as those by Lehninger, by McGilvery and by Stryer.

CHAPTER 21

HOMEOSTASIS AND EXCRETION

Although the removal of wastes from body fluids is an important function of the kidney, its major role is regulating the volume, pH and composition of the blood and body fluids. Cells can survive and function only within a limited range of conditions. By excreting certain substances and conserving others, the kidneys maintain the constant environment in the blood and body fluids required by cells for their continued normal functioning. The term **homeostasis** was introduced by Walter Cannon to refer to the tendency of organisms to maintain constant the conditions in their internal environment. The term was originally applied to the capacity of the body to regulate the volumes of blood and extracellular fluids and their concentrations of solutes. However, it has gradually been broadened to include the many regulatory processes that maintain constant, or minimize fluctuations in, essentially all of the physiologic functions of the body.

The terms **defecation, excretion** and **secretion** are sometimes confused. Defecation refers to the elimination of wastes and undigested food, collectively called feces, from the anus. Undigested food has never entered any of the body cells and has not taken part in cellular metabolism; hence these are not metabolic wastes. Excretion refers to the removal from the cells and blood stream of substances which are of no further use in the body. The excretion of wastes by the kidneys involves an expenditure of energy by the cells, but the act of defecation requires no such effort of the cells lining the intestine. Secretion is the releasing from a cell of some substance that is utilized elsewhere in some body process—for example, the salivary glands secrete saliva, used in the mouth and stomach for digestion. Secretion also involves cellular activity and requires the expenditure of energy by the secreting cell.

The human excretory system includes more than the kidneys and their ducts; the skin, lungs and digestive tract have excretory and regulatory functions too. Water and carbon dioxide, important metabolic wastes, are excreted by the lungs; bile pigments, the breakdown products of hemoglobin, are excreted by the liver; and certain metals, such as iron and calcium, are excreted by the colon. The sweat glands of the skin are primarily concerned with the regulation of body temperature, but they also serve in the excretion of 5 to 10 per cent of all metabolic wastes. Sweat contains the same substances (salts, urea and other organic compounds) as urine but is much more dilute, having only about one eighth as much solid matter. The volume of perspiration varies from about 500 ml. on a cool day to as much as 2 or 3 liters on a hot one. While doing hard work at high temperatures, a man may excrete from 3 to 4 liters of sweat in an hour!

21-1 THE KIDNEY AND ITS DUCTS

The human kidneys are paired, bean-shaped structures about 10 cm. long, one of which is located on each side of the mid-dorsal line of the abdominal cavity, just below the level of the stomach (Fig. 21-1). On the medial, concave side of each kidney is a funnel-shaped chamber called the *pelvis.* The urine, excreted by the kidney in a continuous trickle, collects in the pelvis and passes down the *ureters* by peristaltic waves of contraction of the ureteral walls to the *urinary bladder.* This hollow, muscular organ located in the lower, ventral part of the abdominal cavity (Fig. 21-1), distends as its muscular walls relax to make room for the urine as it accumulates. Valvular flaps of tissue at the openings of the ureters into the urinary bladder prevent the back flow of urine and keep any bacteria that may be in the blad-

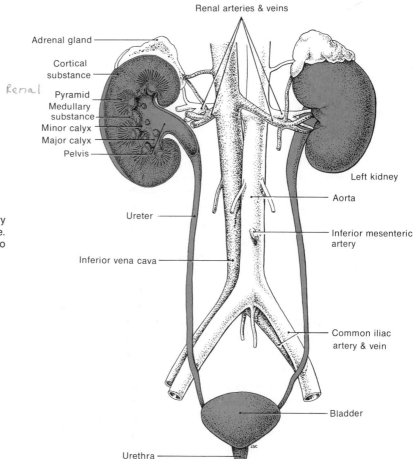

Renal arteries & veins

Adrenal gland

Cortical substance

Renal

Pyramid
Medullary substance
Minor calyx
Major calyx
Pelvis

Left kidney

Aorta

Ureter

Inferior mesenteric artery

Inferior vena cava

Common iliac artery & vein

Bladder

Urethra

Figure 21-1 The human urinary system, seen from the ventral side. The right kidney is shown cut open to reveal the internal structures.

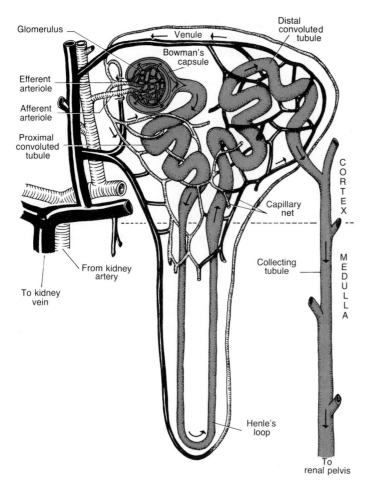

Figure 21–2 Diagram of a single kidney tubule and its blood vessels.

der from ascending to the kidney. As the volume of urine in the bladder increases, the distention of the muscular walls stimulates nerve endings in the bladder walls to send impulses to the brain, producing the sensation of fullness. *Micturition,* the expulsion of urine from the bladder, requires that nerve impulses from the brain cause a contraction of the muscles in the wall of the bladder and a relaxation of the sphincter guarding the opening from the bladder to the urethra.

The kidney consists of masses of microscopic *tubules* arranged in an inner core, the *medulla,* and an outer layer, the *cortex.* The functional unit of the mammalian kidney, the kidney tubule, or *nephron,* consists of a double-walled hollow sac of cells, *Bowman's capsule,* which surrounds a spherical tuft of capillaries, a *glomerulus* (Fig. 21–2), and the coiled, looped *tubules* which reabsorb some substances into the blood but not others. Branches of the renal artery ramify to all parts of the kidney; each ultimate arteriole passes to

the end of one kidney tubule and supplies its glomerulus. The inner wall of Bowman's capsule consists of flat epithelial cells which adhere closely to the capillaries of the glomerulus, permitting ready diffusion of substances from the capillaries into the cavity of Bowman's capsule. Each kidney contains about 10^6 nephrons, each a separate, independent unit for excreting wastes and regulating the composition of the blood. Each filters the blood and then reabsorbs certain substances but not others as the filtrate passes through the tubule.

21–2 THE FORMATION OF URINE

The three processes of *filtration, reabsorption* and *tubular secretion* enable the kidney to remove wastes while conserving the useful components of the blood. Filtration occurs at the junction between the glomerular capillaries and the wall of Bowman's capsule. The blood is "filtered" as it passes through the capillaries so that water, salts, glucose, urea and other small molecules pass from the blood into the cavity of Bowman's capsule to become the *glomerular filtrate* (Fig. 21–3). The blood cells and macromolecules, such as the plasma proteins, are retained in the blood. The total volume of blood passing through the kidneys is about 1200 ml. per minute, or about one quarter of the entire cardiac output. The plasma passing through the glomerulus loses about 20 per cent of its volume as the glomerular filtrate; the rest leaves the glomerulus in the efferent arteriole. The mechanism underlying this process is the purely physical one of *pressure filtration,* and it results from the fact that the arteriole entering the glomerulus, the *afferent arteriole,* is larger than the vessel leaving it, the *efferent arteriole.*

The pressure driving fluid out of the glomerulus and into Bowman's capsule is the pressure of the blood in the glomerular capillaries, 70 mm. Hg. The pressure tending to move fluid in the reverse direction is the sum of the hydrostatic pressure in Bowman's capsule, about 14 mm. Hg, and the colloid osmotic pressure of the plasma in the glomerular capillaries, about 32 mm. Hg. (Why is the colloid osmotic pressure greater here than in the capillaries elsewhere in the body?) Thus the net force moving fluid out of the glomerulus, the *filtration pressure,* is 70 − (32 + 14), or 24 mm. Hg. Most of the fluid filtered across the glomerular membrane is subsequently reabsorbed

from the tubules into the capillaries surrounding them.

By introducing a fine glass syringe into the Bowman's capsule of a frog's kidney and collecting and analyzing some of the glomerular filtrate, A. N. Richards of the University of Pennsylvania showed that it has the same concentration of urea, salts, glucose and so forth as the plasma but lacks its proteins. The cells of Bowman's capsule are thin and unable to move materials from the capillaries; the work of pushing the filtrate from the plasma into the capsule is done by the heart. It can be shown experimentally that the rate at which fluid passes from the glomerulus into Bowman's capsule, the *glomerular filtration rate,* rises and falls with the blood pressure and consequently the filtration pressure. The normal glomerular filtration rate is about 125 ml. per minute, which amounts to 180 liters per day. This is four and one half times the amount of fluid in the entire body!

The amount filtered is also regulated by the constriction or dilation of the arterioles leading to and from the glomerulus. The amount filtered is increased by the constriction of the efferent arterioles and dilation of the afferent arterioles. A rise in arterial blood pressure increases, in turn, glomerular pressure, glomerular filtration rate and the total amount of urine excreted. The increased loss of fluid from the blood decreases the blood volume and consequently the blood pressure. A fall in arterial blood pressure leads, by a comparable

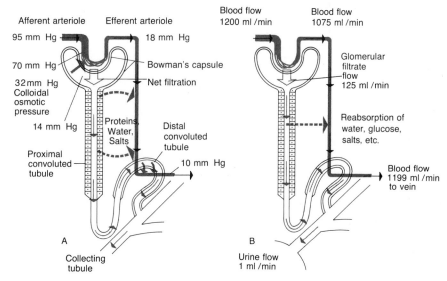

Figure 21–3 *A,* Diagram of a kidney tubule showing the pressure gradients that move fluids from blood to glomerular filtrate (filtration pressure). Substances are reabsorbed, largely in the proximal convoluted tubules, by processes of active transport. *B,* Diagram illustrating the total fluid movement in all the tubules of the kidneys.

sequence of events, to a decrease in the amount of urine excreted. The decreased loss of fluid from the blood increases the blood volume and the blood pressure. The kidney in this way provides a mechanism by which the blood pressure is automatically regulated.

Reabsorption. If the composition of the urine ultimately excreted were like that of the glomerular filtrate, excretion would be a wasteful process, and a great deal of water, glucose, amino acids and other useful substances would be lost; however, the concentrations of substances in the urine are quite different from those in the plasma and glomerular filtrate. From each Bowman's capsule, located in the cortex, the filtrate passes first through a *proximal convoluted tubule* (also in the cortex), then through a long *loop of Henle* passing deep into the medulla and back into the cortex, then through a *distal convoluted tubule,* and empties at last into a *collecting tubule,* through which it passes again through the medulla into the pelvis (Fig. 21–2). There is no further change in the composition of the urine as it passes from the pelvis of the kidney through

the ureters, bladder and urethra to be voided; the changes in composition occur when the filtrate passes from the Bowman's capsule through the long, coiled tubules and collecting tubule to the pelvis.

The walls of the kidney tubules are made of a single layer of cuboidal or flat epithelial cells. The cells making up the walls of the proximal convoluted tubules are richly endowed with mitochondria, and their inner border is a *brush border* (Fig. 21–4) composed of many fine, hairlike processes extending from the cells into the lumen of the tubule. As the filtrate passes through, these reabsorb much of the water and virtually all the glucose, amino acids and other substances needed by the body and secrete them back into the blood stream.

The efferent arteriole does not pass directly to a vein but connects with a second network of capillaries around the proximal and distal convoluted tubules (Fig. 21–2). Thus the route of blood in the kidney is unique—it passes through *two* sets of capillaries in sequence in passing from the renal artery to the renal vein. The ability of the kidney to reg-

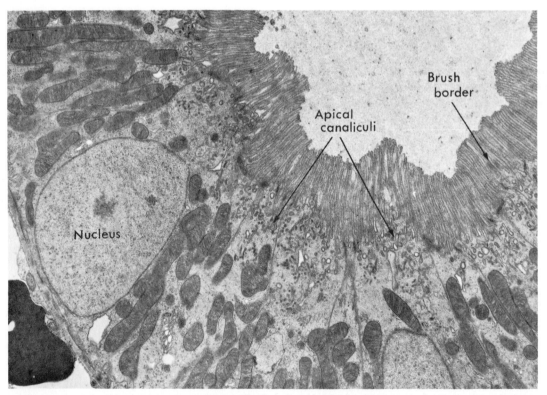

Figure 21–4 An electron micrograph of the cells of the proximal tubule of the kidney showing the brush border of the lumen, ×6000. (From Bloom, W., and Fawcett, D. W.: *Histology,* 10th Ed. Philadelphia, W. B. Saunders Co., 1975.)

ulate the composition of the blood depends upon this structural feature.

Substances are reabsorbed into the blood stream selectively and the rate is regulated in part by the momentary requirements of the body. The cells lining the tubules must expend energy, utilize ATP and do work to secrete these substances by the process of *active transport* back into the blood, usually against a diffusion gradient. A given amount of kidney tissue consumes more oxygen per hour than an equivalent weight of heart muscle, indicating that the kidneys work harder than the heart. The energy for this work is derived from biological oxidations within the mitochondria in their cells; when the kidney is deprived of oxygen, reabsorption, but not filtration, ceases. This reemphasizes the fact that filtration is a *passive* process driven by the blood pressure and not by the metabolic activities of the glomerular cells. The substance reabsorbed in greatest amount is sodium chloride. Each day, our kidney tubules reabsorb about 1200 grams of sodium chloride—a bit more than 2.5 pounds! Sodium ions are actively reabsorbed by a *sodium pump,* and glucose and amino acids are reabsorbed by selective active-transport mechanisms. These result in a decreased concentration of solutes in the tubular fluid and an increased concentration of solutes in the interstitial fluid surrounding the tubule. Water is reabsorbed osmotically, moved by the concentration gradient of water.

The human kidney produces about 125 liters of filtrate for every liter of urine formed; the other 124 liters of water are reabsorbed. In this way, the waste products, urea, uric acid and creatinine, are greatly concentrated as the filtrate passes down the tubules. The concentration of urea in the urine is about 65 times that in the glomerular filtrate and would be even higher but for the fact that some urea is reabsorbed in the tubules. Urea, uric acid and creatinine are not actively reabsorbed by the tubules, but small amounts of them pass by diffusion from the lumen of the tubule back into the capillaries surrounding the tubules. The quantity of water reabsorbed depends on the body's current need for it and is regulated by the *antidiuretic hormone,* ADH, secreted by the posterior lobe of the pituitary (p. 551). If a large quantity of water or beer is drunk, less water is reabsorbed and a copious, dilute urine

is excreted. If water intake is restricted, a maximum amount of water is reabsorbed by the cells of the tubules, conserving water, and a scanty, concentrated urine is excreted.

Tubular Secretion. The cells of the kidney tubules not only remove substances from the filtrate and secrete them into the capillaries but also excrete additional wastes from the blood stream into the filtrate by active transport mechanisms. This process of *tubular secretion* probably plays only a minor role in the function of human kidneys, but in animals like the toadfish, whose kidneys lack glomeruli and Bowman's capsules, secretion by the tubules is the only method available. When the blood pressure, and consequently the filtration pressure, drop below a certain level, filtration ceases, although urine is still formed by tubular secretion. Dyes injected into experimental animals can be seen to pass from the blood stream into the urine through the cells of the tubules. Drugs such as penicillin and atabrine are removed from the blood and excreted by the process of tubular secretion. There is no doubt that tubular secretion can occur in human kidneys and in those of other animals, but how large a role it normally plays in the process of excretion is unclear.

When the fluid reaches the end of the collecting tubule, and some substances have been reabsorbed and others added, the glomerular filtrate becomes *urine.*

21–3 THE REGULATION OF THE GLOMERULAR FILTRATION RATE

The amount of water, salts and sugars reabsorbed from the tubules depends to a large extent on the rate at which the glomerular filtrate passes through the tubules. If the rate is too rapid, important quantities of essential materials are lost in the urine because the fluid passes through the tubules before the reabsorption process is completed. If the rate is too slow, nearly everything, including urea and other waste products, is reabsorbed. Thus there is an optimal glomerular filtration rate which will ensure that water and salts are reabsorbed but not urea and other wastes.

The glomerular filtration rate in each nephron is regulated automatically by the concentration of certain substances in the distal

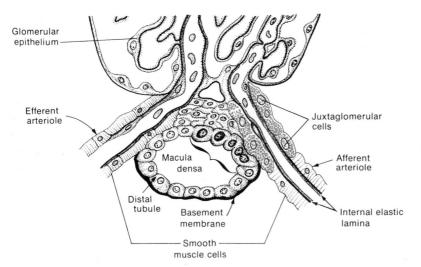

Glomerular epithelium

Efferent arteriole

Macula densa

Juxtaglomerular cells

Afferent arteriole

Distal tubule

Basement membrane

Internal elastic lamina

Smooth muscle cells

Figure 21–5 Diagram illustrating the relationship of the macula densa and the juxtaglomerular cells, muscle cells in the wall of the afferent arterioles of the kidney. The cells of the macula densa monitor the concentration of substances in the glomerular filtrate and stimulate the juxtaglomerular cells to contract or relax, thereby controlling the glomerular filtration rate.

convoluted tubule. The distal convoluted tubule comes to lie very close to the afferent arteriole serving the glomerulus of that tubule. At the point of contact the cells of the distal tubule become dense and increased in number, forming a structure called the *macula densa* (Fig. 21–5). The smooth muscle cells in the wall of the arteriole adjacent to the macula densa, called the *juxtaglomerular cells,* are swollen and filled with granules. By microinjection under the microscope it is possible to inject sodium chloride solutions directly into the distal convoluted tubule and show that the afferent arteriole immediately becomes constricted. Thus the composition of the fluid in the distal tubule is monitored, presumably by the cells in the macula densa, and this regulates the degree of constriction of the afferent arteriole, presumably via the juxtaglomerular cells. The glomerular filtration rate is of course controlled by the filtration pressure, which depends on the degree of constriction of the walls of the afferent arteriole. In this way, each nephron, by continuously assaying the concentration of salts in the tubular fluid entering the distal tubule from the loop of Henle, regulates its glomerular filtration rate to the optimal value.

21–4 RENAL THRESHOLDS AND RENAL CLEARANCES

Although glucose is present in the glomerular filtrate, there normally is little or none in the urine because it has been reabsorbed by the cells of the tubules. But if the concentration of glucose in the blood, and consequently in the glomerular filtrate, is very high, not all of it can be reabsorbed as the filtrate passes through the tubules, and some glucose will appear in the urine. The concentration *in the blood* of a substance such as glucose at the point where it just begins to appear in the urine is termed its *"renal threshold."* The threshold for glucose is about 150 mg. glucose per 100 ml. blood. When this value is exceeded, glucose begins to "spill" into the urine. There are comparable renal thresholds for many other substances; the concentration at which the substance begins to appear in the urine is different for each.

Since a major function of the kidney is to "clear" the extracellular fluids of the body, renal physiologists have adopted the concept of *renal clearance* to express quantitatively the kidney's ability to eliminate any given substance from the blood. As a given volume of plasma passes through the glomerulus, and glomerular filtrate is formed and passes through the tubules, a certain amount of the substance appears in the filtrate, and some of this may be reabsorbed into the cells of the tubules. The relationship between the amount of fluid reabsorbed and the amount of the substance reabsorbed is expressed as the number of milliliters of plasma that is "cleared," i.e., completely freed of that substance, per minute. By taking simultaneous samples of blood and urine and measuring the concentration in each of the substances in question, you can calculate the quantity of that substance in each milliliter of blood and the quantity of that substance appearing in the urine each minute. Plasma clearance is defined as

$$\frac{\text{amount secreted in urine per minute}}{\text{amount in each milliliter of plasma}}$$

If the concentration of urea proved to be 0.2 mg. per ml. and the amount appearing in the urine is 12 mg. per minute, the clearance is 12 mg. per minute/0.2 mg. per ml. = 60 ml. per minute. Thus, 60 ml. of plasma are said to be cleared of urea per minute. Measures of renal clearance are commonly made to assess general kidney function.

21–5 HOW THE KIDNEY EXCRETES A CONCENTRATED URINE: THE COUNTER-CURRENT MODEL

To become successful on land, the higher vertebrates had to evolve a mechanism to excrete a concentrated urine and thus conserve body water. This ability appears to depend on certain properties of the *loop of Henle.* The glomeruli and proximal and distal tubules are located in the outer part of the kidney, the *cortex,* whereas the loop of Henle extends deeply into the central *medulla* (Fig. 21–2). The peritubular capillaries also form long loops, *vasa recta,* that extend down into the medulla. Blood passes down into the medulla and then back up to the cortex in these vasa recta before emptying into the renal veins. The

collecting tubules, into which the distal tubules empty, pass through the medulla and empty into the pelvis of the kidney.

This anatomic arrangement permits the kidney to excrete a urine that is *hypertonic* to the blood. As tubular fluid passes through the ascending loop of Henle, sodium is actively pumped into the interstitial fluid, chloride ions go along passively, and the concentration in the interstitial fluid increases. Some of the sodium and chloride ions diffuse passively back into the descending loop, and the cycling of sodium from ascending limb to interstitial fluid to descending limb results in the establishing of a concentration gradient of sodium and chloride in the tissue fluid surrounding the loop, with the lowest concentration near the cortex and the highest concentration deep in the medulla (Fig. 21–6).

As blood flows into the medulla in the vasa recta, sodium and chloride diffuse into it, but as the blood flows back up and out of the medulla, sodium and chloride diffuse out of the blood into the interstitial fluid. This "countercurrent" flow of blood prevents the loss of sodium and chloride from the medulla and permits the concentration gradient in the interstitial fluid to be maintained. The active transport of sodium out of the tubular fluid in the ascending loop of Henle is so powerful that the fluid reaching the distal convoluted tubule actually has a lower concentration of sodium than the fluid in the glomerular filtrate. The walls of the ascending limb of the loop of Henle appear to be impermeable to water, for water does not

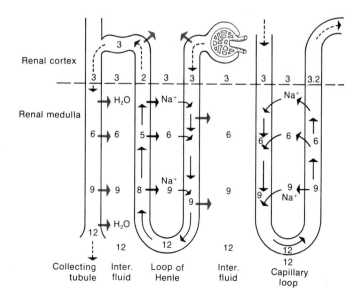

Figure 21–6 Diagram of countercurrent flow mechanism of the mammalian kidney. The general direction of fluid movement is shown by dashed arrows; active sodium transport by thick arrows; passive sodium transport by thin arrows and the movement of water by gray arrows. The numerals refer to the relative concentrations of osmotically active solutes. When two zeros are added, they refer to the concentration of solutes in milliosmoles per liter.

diffuse out as the sodium is pumped out. As the filtrate passes through the loop of Henle it loses a lot of sodium but very little water. Then the urine enters the collecting tubules and flows down through the medulla, through an ever-increasing concentration of sodium and chloride in the interstitial fluid. The walls of the collecting tubule are permeable to water and water moves by osmosis from the dilute urine in the collecting tubule to the interstitial fluid with a high concentration of solutes. The urine finally passing into the pelvis is nearly as concentrated as the interstitial fluid deep in the medulla and is quite a bit more concentrated than the initial glomerular filtrate.

The steroid hormone *aldosterone,* secreted by the adrenal cortex, acts on the cells in the ascending loop of Henle and increases the active reabsorption of sodium ions. As you might guess, the rate at which the adrenal cortex secretes aldosterone is in turn regulated by the concentration of sodium ions in the blood.

The permeability of the walls of the collecting tubules to water can be varied and is controlled by a hormone from the posterior lobe of the pituitary, *antidiuretic hormone (ADH).* This, by a mechanism mediated by *cyclic AMP* (p. 120), greatly increases the permeability of the collecting tubules to water so that water is removed from the urine and a concentrated urine is excreted. In the absence of antidiuretic hormone, the permeability of the collecting tubules to water is very low, and the dilute fluid entering the collecting tubule from

the loop of Henle passes through the collecting tubule nearly unchanged and is excreted as a very dilute urine. In *diabetes insipidus,* in which there is a deficiency of ADH, the output of urine may reach 30 to 40 liters per day instead of the normal 1.2 to 1.5 liters.

21–6 OSMORECEPTORS

The complex counter-current flow mechanism in the loop of Henle, by which the kidney can vary the amount of water reabsorbed from the tubular filtrate, is controlled by the amount of antidiuretic hormone secreted by the posterior lobe of the pituitary. This in turn is under the control of *osmoreceptors* (Fig. 21–7) in the supraoptic nuclei of the hypothalamus (p. 501). These monitor the concentration of solutes in the blood and increase or decrease the secretion of ADH to correct any change in the *osmolarity,* the total concentration of solutes, in the blood.

Osmoreceptors are specialized neurons that are believed to contain small fluid chambers. These swell when the concentration of solutes in the blood falls and decrease in size when the concentration of solutes in the blood rises. When contracted by the increased osmolarity of the blood they initiate impulses which pass through nerve fibers to the posterior lobe of the pituitary and cause the release of ADH (Fig. 21–7). This passes in the blood to the kidneys where it increases the permeabil-

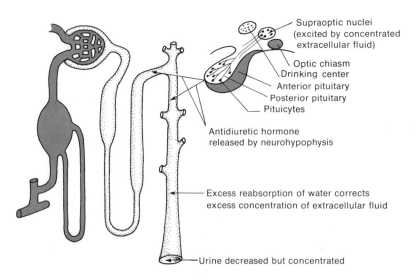

HYPOTHALAMUS

Supraoptic nuclei (excited by concentrated extracellular fluid)
Optic chiasm
Drinking center
Anterior pituitary
Posterior pituitary
Pituicytes
Antidiuretic hormone released by neurohypophysis

Excess reabsorption of water corrects excess concentration of extracellular fluid

Urine decreased but concentrated

Figure 21–7 The release of antidiuretic hormone in response to stimulation of osmoreceptors in the supraoptic nuclei by increased concentration of solutes in the extracellular fluid. The antidiuretic hormone passes to the collecting tubules of the kidney and causes increased reabsorption of water, which reduces the concentration of solutes in extracellular fluid.

osmolarity - total concentration of solutes
osmoreceptors

ity of the collecting tubules to water, and a greater fraction of the water passing through the collecting tubules is reabsorbed. This leads to the retention of water, while solutes continue to pass into the urine and the osmolarity of the body fluids decreases. If the osmolarity of the body fluids decreases, the osmoreceptors emit fewer or no impulses, the secretion of ADH is decreased or stopped, and the kidneys excrete a dilute urine because less water is reabsorbed in the collecting tubules.

When you drink several glasses of water or beer it is absorbed from the gut in 15 or 20 minutes and dilutes the blood. The osmoreceptors detect the decreased osmolarity of the blood, decrease the secretion of ADH, and the collecting tubules of the kidney become less permeable to water, less water is reabsorbed from the urine, and a large quantity of dilute urine is passed from the kidneys. The rate of urine flow will rise from a normal value of about 2 ml. per minute to 7 or 8 ml. per minute until the osmolarity of the extracellular fluids is returned to normal.

21-7 KEEPING THE pH CONSTANT

The cells of humans and most animals have become adapted to surviving only within a relatively small range of hydrogen ion concentrations, and several mechanisms have evolved to maintain the pH within these limits. The role of hemoglobin in transporting carbon dioxide and its role with the plasma proteins in serving as acid-base buffers was described previously (p. 404). The lungs, by removing carbon dioxide from the blood as fast as it is formed, play a role in maintaining the pH constant, for CO_2 combines with water to form carbonic acid.

The kidneys regulate the hydrogen ion concentration of the intra- and extracellular fluids by excreting acidic or basic constituents when these deviate from normal, thus restoring the normal balance of the two. They do this by exchanging hydrogen ions, derived from carbonic acid within the tubular cells, for sodium ions in the tubular fluid. The hydrogen ions pass into the urine and combine with the bicarbonate there to form carbonic acid; this dissociates to form CO_2 and water. The CO_2 is reabsorbed into the blood from the tubules and is

eliminated through the lungs. The decrease in the concentration of bicarbonate ions in the tubular fluid represents a net excretion of hydrogen ions. Ordinarily the amount of hydrogen ions exchanged for sodium in the distal tubules is just equivalent to the amount of bicarbonate ions in the tubular fluid. However, if the extracellular fluids become very acidic, the quantity of bicarbonate ions in the filtrate decreases and hydrogen ions combine with other buffers in the tubular fluid such as phosphate and are excreted.

If the extracellular fluids become too alkaline, the amount of bicarbonate in the glomerular filtrate exceeds the amount of hydrogen ions secreted by the tubules. The bicarbonate that has not reacted with hydrogen ions to form carbonic acid is simply excreted as bicarbonate along with the excess sodium ions, and the loss of the sodium bicarbonate makes the body fluids more acidic, returning the acid-base balance to normal.

The kidney has an additional mechanism for coping with an excess of acids: it can secrete ammonium ions, NH_4^+, derived from the hydrolysis of glutamine in the kidney by the enzyme *glutaminase.* The NH_4^+ is exchanged for Na^+ ions in the tubular filtrate, conserving sodium ions and excreting an acid anion, such as Cl^-, in combination with the NH_4^+ ions.

An individual who is fasting, eating a high fat diet, or suffering from diabetes mellitus has an increased content of acid *ketone bodies* in his blood and extracellular fluids. The kidney compensates for this by increasing the exchange of hydrogen ions for sodium ions in the distal convoluted tubules, thereby conserving sodium, and by increasing the output of ammonium ions to be exchanged for additional sodium ions.

21-8 REGULATING THE VOLUMES OF BODY FLUIDS

The volumes of the blood, extracellular fluid and intracellular fluid are all regulated by automatic feedback controls. The volume of blood in an adult man is kept within a few hundred milliliters of the normal total of 5 liters by two mechanisms, both of which depend on the fact that an increase in blood volume increases the blood pressure. One operates in the capillaries of all tissues and the other in the glomeruli of the kidneys.

An increase in blood volume raises the blood pressure and will increase the pressure in the capillaries above the normal value of about 18 mm. Hg. Thus, the pressure that tends to move fluid out of the capillaries into the interstitial fluid becomes greater than the colloid osmotic pressure, the force that tends to move fluid into the capillaries from the interstitial fluid. The result is a net movement of fluid from the capillaries into tissue spaces until the blood pressure and blood volume are returned to normal. A decrease in blood volume and hence of blood pressure has the opposite effect, and fluid tends to move from the tissue spaces into the capillaries to restore normal pressures and volumes.

An increased blood volume increases the pressure within the glomeruli of the kidney, filtration pressure is increased and the volume of urine formed increases. This is reinforced by another mechanism that is based on the presence of *baroreceptors* in the walls of the arteries in the chest and neck (Fig. 21–8). These are sense organs that detect the degree of stretching of the walls of the arteries. An increase in the blood pressure stretches the walls of the arteries and stimulates the baroreceptors to send impulses to the vasomotor center in the medulla of the brain. The *vasomotor center* initiates impulses to the smooth muscles in the walls of the renal afferent arterioles, causing them to relax, which allows the arterioles to dilate and increase the flow of blood into the glomeruli.

An increase or decrease in the volume of interstitial fluid will lead to a corresponding change in its pressure and will alter the pressure relations between capillaries and interstitial fluid. The interstitial fluid normally has a slight negative pressure (p. 389), which is one of the forces moving fluid out of the capillaries. An increase in volume will change this pressure toward zero, or make it positive, moving fluid into the blood. The increased blood volume in turn would increase the output of urine, and the excess interstitial fluid would be removed from the body via the blood and then the urine.

Each cell maintains its own fluid volume, largely by carrying in certain electrolytes and excluding others; it maintains an osmotic equi-

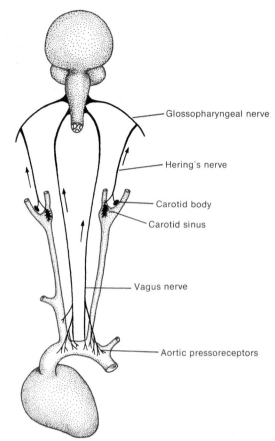

Figure 21–8 Diagram of the baroreceptors involved in regulating pressures within the circulatory system.

librium with the extracellular fluid. The kidneys do not *directly* regulate intracellular fluid volume but do this *indirectly,* by maintaining constant the volume and constituents of the extracellular fluids.

One further center that plays a role in regulating fluid volumes is the *drinking center,* located in the hypothalamus, immediately posterior to the osmoreceptors in the supraoptic nuclei (Fig. 21–7). Stimulation of this center, when the body fluids are too concentrated, leads to the sensation of thirst, and the individual is stimulated to find and drink water, which will dilute the body fluids and turn off the stimulation of the drinking center.

21–9 KIDNEY DISEASE AND THE ARTIFICIAL KIDNEY

The kidney is an extremely complex organ, and malfunction, failure or obstruction of any

baroreceptors
vaso motor center
pyelonophrites

of its parts can cause serious and sometimes fatal disease. *Acute glomerular nephritis* results from an antigen-antibody reaction occurring within the glomeruli. Large numbers of white cells accumulate in the glomeruli, and both the cells of Bowman's capsule and the endothelium of the capillaries become inflamed and swollen, thereby partially or completely blocking the glomeruli and capsule. This inflammation occurs one to three weeks after a streptococcal infection somewhere else in the body, for example, scarlet fever or a streptococcal sore throat. The damage to the kidney results not from the bacterial toxin itself but from the antigen-antibody reaction that occurs after the body has produced antibodies to the streptococci. The acute inflammation usually subsides in about two weeks; the nephrons may return to normal function or they may be permanently damaged or destroyed.

In tubular necrosis, the epithelial cells of the kidney tubules are destroyed either by poisons such as mercuric ion, by the shutting off of the blood supply to the kidney or to a portion of it, or by a "transfusion reaction" in which red cells burst and release their hemoglobin when improperly matched blood is transfused from a donor into a patient. The hemoglobin itself may block the glomeruli or it may have associated with it some substance that causes constriction of the blood vessels in the glomeruli.

Renal insufficiency may result either from arteriosclerosis, in which constriction of the arteries greatly reduces the flow of blood to the kidneys, or from infection and inflammation of the kidney tubules *(pyelonephritis)* by any of many different types of bacteria. The bacterial infection results in progressive destruction of renal tubules and glomeruli and loss of functional kidney tissue. Loss of kidney function results in *uremia* (accumulation of urea in the blood), *edema* due to the retention of water, *acidosis,* and increased potassium concentration in the blood. After a week or so of renal shutdown the patient becomes disoriented and ultimately comatose because of acidosis and the accumulation of urea in the blood.

Patients with severe renal insufficiency were first treated by dialysis in an *artificial kidney* some 30 years ago and the treatment has become more effective as the design of the artificial kidney has been improved. It is now possible to keep people alive almost indefi-nitely by repeated dialysis even though they have no functioning kidney at all. In an artificial kidney, the patient's blood is passed through a system of very fine tubes bounded by thin membranes. The other side of the membrane is bathed by a dialysis fluid into which waste products pass from the blood by diffusion. Blood passes from one of the patient's arteries, is pumped through the artificial kidney, and then returned into one of the patient's veins. Great care must of course be taken to maintain the sterility of the artificial kidney. The total amount of blood in the artificial kidney at any moment is 400 to 500 ml. and the rate of flow is several hundred milliliters per minute over a diffusion surface of 10,000 to 20,000 cm.^{2} Usually heparin is added to the blood as it enters the artificial kidney to prevent coagulation, and an antiheparin compound is added to the blood as it is returned to the patient's vein to prevent internal bleeding. The composition of the dialysis fluid is carefully calculated so that normal electrolytes are not lost but the unwanted blood constituents are removed as rapidly as possible. An artificial kidney can clear urea from the blood at a rate of about 200 ml. of plasma per minute; the urea clearance of the two normal kidneys is only about 70 ml. per minute. Thus the artificial kidney can be very effective, but it can be used for only about 12 hours every 3 or 4 days because of the danger of internal bleeding from the heparin that must be added to the blood during the dialysis procedure.

A *diuretic* is a compound that will increase the rate of urine output when it is administered to a patient. Diuretics are usually given to reduce the total amount of fluid in the body of a patient suffering from edema, but diuretics are usually of little value in treating the edema caused by kidney disease because the abnormal kidneys cannot respond to the diuretic. A diuretic may function by increasing the glomerular filtration rate, by increasing the osmotic load in the kidney tubules, by diminishing the active reabsorption of substances in the tubular epithelial cells, or by inhibiting the secretion of antidiuretic hormone.

21–10 EXCRETORY DEVICES IN OTHER ANIMALS

Every organism has had to solve the problem of getting rid of metabolic wastes. In amoebae and paramecia, the wastes simply diffuse through the cell wall into the outside environment where the concentration is lower. A major nitrogenous waste is ammonia, which is very toxic and inhibits many enzymatic processes but readily diffuses out of the cell before reaching a dangerous concentration. Protozoa living in fresh water have a special problem of getting rid of water, because their cellular contents are hypertonic to pond water and tend to absorb it continuously. To control this situation, which would otherwise result in the swelling and bursting of the cell, they have a *contractile vacuole,* a small vesicle which fills with fluid from the surrounding cytoplasm and then ejects the fluid to the exterior. A protozoan in fresh water is like a leaky boat that must be bailed constantly to stay afloat. The contractile vacuole probably plays no significant role in ridding the cell of nitrogenous wastes, for these pass out readily by diffusion. Marine protozoa that live in an environment that is isotonic to their cell contents usually have no contractile vacuole.

Sponges and cnidarians have no specialized excretory organs and their wastes pass by diffusion from the intracellular fluid to the external environment. The vast majority of both sponges and cnidarians are marine organisms living in an isotonic environment and have no

special problems of an excess of water intake. There are a very few fresh water sponges and a few fresh water cnidarians such as *Hydra* that live in a medium that is very hypotonic to their intracellular fluid. No contractile vacuoles have been observed in these animals and the means by which they prevent the inflow of water or pump it back out again remain a mystery.

The simplest animals with specialized excretory organs are the flatworms and nemerteans, which have *flame cells* (protonephridia) (Fig. 21–9) equipped with flagella. A branching system of excretory ducts connects the protonephridia with the outside. The flame cells lie in the fluid which bathes the cells of the body, and wastes diffuse into the flame cells and from there into the excretory ducts. The beating of the flagella, which suggests a flickering flame when seen under the microscope, presumably moves fluid in the ducts out through the excretory pores. Like the contractile vacuoles of the protozoa, the chief role of the flame cell is to regulate the water content of the animal. Metabolic wastes pass by diffusion through the skin or through the lining of the gastrovascular cavity. The number of protonephridia in a planarian is adjusted to the salinity of the environment. Planaria grown in slightly salty water develop few flame cells, but the number quickly increases if the con-

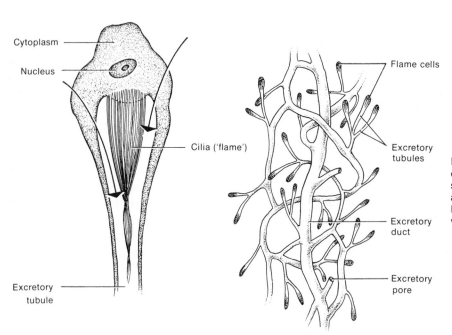

Cytoplasm

Nucleus

Cilia ('flame')

Excretory tubule

Flame cells

Excretory tubules

Excretory duct

Excretory pore

A   B

Figure 21–9 The excretory organs of the flatworm. *A,* A single flame cell. *B,* Flame cells are connected by excretory tubules and ducts to the pore, which opens to the outside.

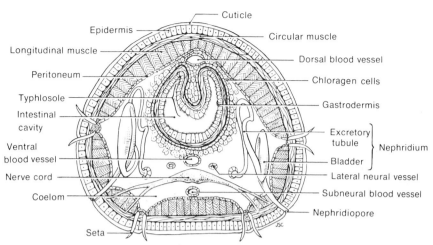

Figure 21–10 Diagrammatic cross section of the common earthworm. *Lumbricus terrestris,* showing the excretory organs, the paired metanephridia. Each consists of a ciliated funnel opening into the coelomic cavity, a coiled tubule, and a pore opening to the outside.

centration of salt in the environment is reduced.

Earthworms have in each segment of their bodies a pair of specialized organs, called *metanephridia,* which function in excretion. The metanephridium, in contrast to the flame cells of flatworms, is a tubule open at both ends, the inner end connecting to the coelom by a ciliated funnel (Fig. 21–10). Around each tubule is a coil of capillaries, which permits the removal of wastes from the blood stream. As the body fluid, moved by the beating of the cilia in the funnel, passes through the metanephridium, water and substances such as glucose are reabsorbed, while the wastes are concentrated and passed out of the body. The earthworm excretes a very dilute, copious urine at a rate of about 60 per cent of its total body weight each day.

The excretory organs of crustacea are the *green glands,* a pair of large structures located at the base of the antennae and supplied with blood vessels. Each gland consists of a coelomic sac, a greenish glandular chamber with folded walls, and a canal which leads to a muscular bladder. Wastes from the blood pass to the coelomic sac and glandular chamber; the fluid in them is isotonic with the blood. Urine collects in the bladder and then is voided to the outside through a pore at the base of the second antenna.

The excretory system of insects consists of organs called *malpighian tubules,* which lie within the hemocoel and empty into the diges-

tive tract. Each tubule has a muscular wall, and its slow writhing movements assist the passage of the wastes down its lumen to the gut cavity. The tubules are bathed in blood in the hemocoel, and their cells transfer the wastes by diffusion or by active transport from the blood to the cavity of the tubule. Water is reabsorbed into the hemocoel both from the tubule and from the digestive tract. The major waste product, *uric acid,* is very sparingly soluble in water, and as the water is reabsorbed the uric acid precipitates and is excreted as a dry paste. This adaptation helps conserve the insect's body water.

The urinary systems of all the vertebrates are essentially similar. Each is composed of units called *kidney tubules,* or *nephrons,* which remove wastes from the blood, but the number and arrangement of the kidney tubules differ in different vertebrates. In the lower vertebrates the kidney tubules open into the body cavity instead of into the hollow ball of cells called *Bowman's capsule,* the arrangement found in the higher vertebrates. These *pronephric tubules* represent a type of excretory organ intermediate between the metanephridia found in annelids and the mesonephric and metanephric tubules of higher vertebrates.

Vertebrates living in or on the sea have evolved special means for coping with salt. The marine bony fishes have blood and body fluids that are hypotonic to the sea water. They tend to lose water osmotically and, like Coleridge's "Ancient Mariner," are in danger of

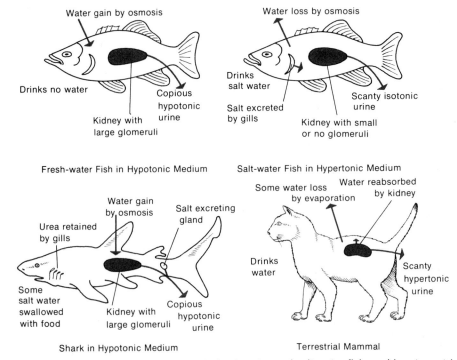

Figure 21-11 Water and salt balance in sharks, in fresh water and salt water fish, and in a terrestrial mammal.

"drying up" even though surrounded by water. Some bony fish have, in the course of evolution, lost their kidney glomeruli, and the resulting *aglomerular kidney* filters very little water from the blood. They compensate further by drinking sea water constantly, retaining the water and secreting the salts by specialized salt secreting glands in their gills (Fig. 21-11).

The elasmobranchs have evolved a different solution to this problem. These animals convert the ammonia from their nitrogenous wastes into *urea* and retain the urea in their blood and tissues in a concentration high enough to render these fluids slightly hypertonic to sea water. Their body fluids may take up some water osmotically and they are able to excrete a hypotonic urine.

Vertebrates evolved originally in fresh water and the ancestral pronephric kidney was adapted for filtration so that it could eliminate the water that kept pouring in from the hypotonic environment. The fresh water fish seldom drink; they absorb salts by active transport across the gills and excrete a copious, dilute urine to eliminate the water taken in osmotically through the gills and lining of the mouth.

Marine turtles and sea gulls have specialized salt-secreting glands in the head which

can excrete the salts from the sea water they drink. The ducts from these salt glands empty either into the nasal cavity or onto the surface of the head. Marine mammals apparently eliminate their excess salt through their kidneys.

Amphibians have kept a primitive kind of kidney tubule with large renal corpuscles that produce a copious, dilute urine. A frog can lose through its skin and urine an amount of water equivalent to one third of its body weight in a day. Reptiles conserve water by having a dry horny skin and kidneys with small glomeruli. Less water is removed from the blood by this type of glomerulus than by the larger glomeruli of fresh water fish and amphibia. Birds and mammals have moderate-sized glomeruli and have evolved *loops of Henle* in which water is reabsorbed. This makes possible the excretion of a hypertonic urine. Mammals that live in the desert and must operate on very limited water supplies have evolved exceptionally long loops of Henle and can remove a greater fraction of water from the urine than other animals can. Toads and some reptiles can reabsorb water from the urinary bladder, but in most animals urine is not changed after it leaves the kidney.

The land vertebrates, reptiles, birds and mammals developed a third kind of kidney, a *metanephros,* with tubules that have two highly coiled regions and a long loop of Henle that extends deep into the medulla of the kidney. These long portions of the tubule function in the reabsorption of water, and their ability to produce a concentrated hypertonic urine was a major factor in enabling their possessors to become efficient land animals.

The evolution of the urinary system has been complicated by the fact that in many animals the reproductive system has come to share some of its structures, and several organs play a dual role. This relationship is so close that the two systems are frequently considered together as the *urogenital system.*

QUESTIONS

1. What is the difference between excretion, secretion and defecation?
2. List the organs that perform excretory functions in the human body.
3. Discuss in detail how metabolic wastes are eliminated from the body by the kidneys.
4. Trace the path of a molecule of urea from its formation in the liver to its excretion in the urine.
5. Describe the role of the kidneys in regulating the osmotic pressure of the body fluids.
6. What is meant by the terms kidney threshold and renal clearance?
7. What substances are present in normal urine? What substances may be present in greater or lesser amounts than normal in the urine of ill people?
8. What is edema? Why may edema result when the kidneys are malfunctioning?
9. What is the cause of acute glomerular nephritis and what effect does this condition have on the patient?
10. By what methods do single-celled animals excrete metabolic wastes?
11. Describe the functioning in excretion of the metanephridia of earthworms.
12. Artificial kidneys have been devised for patients critically ill with kidney disease. What problems had to be solved in the design and use of artificial kidneys?
13. What chemical and physical processes may be involved in reabsorption and tubular secretion in the cells of the kidney tubules?

SUPPLEMENTARY READING

More detailed discussions of the mechanisms of excretion and the regulation of the internal environment in man and mammals are found in the textbooks of physiology by Guyton, by Davson, and by Vander, Sherman and Luciano. Homeostatic mechanisms in other animals are discussed in Prosser and Brown's *Comparative Animal Physiology,* in Florey's *General and Comparative Animal Physiology* and in Hoar's *Comparative Physiology.* The evolutionary significance of the kinds of nitrogenous waste materials excreted by members of the different classes of vertebrates is discussed in Ernest Baldwin's *An Introduction to Comparative Biochemistry.* A helpful discussion of the biochemical mechanisms involved in maintaining constant composition of body fluids is to be found in *Body Fluids and Acid Base Balance* by H. N. Christensen. *The Physiology of the Kidney and Body Fluids* by R. S. Pitts is a useful reference book for students interested in kidney function. H. W. Smith's definitive monograph of his life-long study of kidney function is his *Principles of Renal Physiology.* An *Introduction to Body Fluid Metabolism* by A. V. Wolfe and M. A. Crowder is an advanced account of kidney function and the role of other organs in the regulation of body fluids. J. P. Merrill's article, *The Artificial Kidney,* describes how these devices work in supplementing the functioning of kidneys in patients with kidney disease.

CHAPTER 22

SKIN, BONES AND MUSCLES: PROTECTION AND LOCOMOTION

The skin covers the body, the bony framework supports it, and the muscles move the bony framework. Each of these is an organ system, a group of organs that act together to perform one of the primary life functions. The integumentary, skeletal and muscular systems each function independently of the others, but since they serve as protective devices for the body, and because they determine the shape and symmetry of the body, we shall consider them in the same chapter.

In human beings, as in most animals, the ability to move depends upon a group of specialized contractile cells, the muscle fibers. Humans and most vertebrates are quite muscular animals; almost half of the mass of the human body consists of muscle tissue. In the vertebrates three types of muscle fibers have evolved to perform various kinds of movements: *skeletal muscle* is attached to and moves the bones of the skeleton; *cardiac muscle,* present in the walls of the heart, beats to move blood through the circulatory system; and *smooth muscle* makes up the walls of the digestive tract and certain other internal organs and moves material through the internal hollow organs (Fig. 4–30). All three types of muscle have the ability to shorten when stimulated, and ordinarily this stimulation reaches the muscle fibers by a nerve. Both cardiac and smooth muscle can contract in the absence of nervous stimulation, and both the heart and the digestive tract function almost normally even when all the nerves leading to them have been cut. In contrast, when the nerves to skeletal muscle are severed or blocked, the muscle is completely paralyzed; for a few weeks it will respond to artificial stimulation, such as an electric shock applied to the overlying skin, but even this ability is gradually lost.

The drug *curare,* the chief ingredient of the arrow poison of certain South American Indians, blocks the junction between nerve and muscle so that impulses can no longer pass. This produces the same effect as the cutting of the nerves to all the muscles of the body. The muscles of a curarized animal will still respond to direct electric stimulation, however, demonstrating that muscle is "independently irritable" and need not receive its stimulation through a nerve.

22–1 THE SKIN

All multicellular animals are covered externally by a skin or *integument,* consisting of one or many layers of cells. The skin is much more than merely an outer wrapping for the animal; it is an important organ system and performs many diverse functions. Perhaps the most obvious and vital of these is to protect the body against a variety of external agents and to help maintain a constant internal environment. It shields the underlying cells from mechanical injuries caused by pressure, friction or blows. The skin is practically germ-proof as long as it is unbroken, and it protects the body against disease-producing organisms. Its waterproof quality protects the body from excessive loss of moisture, or, in aquatic animals, from the excessive intake of water. Important, too, is the protection it affords the underlying cells from the harmful ultraviolet rays of the sun, by virtue of the pigment it can produce (melanin, the substance that makes your skin look tan).

The skin also functions as a thermostatically controlled radiator, regulating the elimination of heat from the body. Heat is constantly being produced by the metabolic processes of the cells and distributed by the blood stream. A certain amount of heat must be lost all the time to maintain a constant temperature within the body. Some heat leaves the body in the expired breath and some in the feces and urine, but approximately 90 per cent of the total is lost through the skin. When the external temperature is low, temperature-sensitive nerve endings in the skin are stimulated and the arterioles in the skin are reflexly contracted, thereby decreasing the flow of blood through the skin and decreasing the rate of heat loss. In a warm environment the reverse occurs: the arterioles expand, and the increased flow of blood causes the skin to appear flushed, resulting in increased heat loss. In very warm environments this mechanism cannot eliminate enough heat from the body, and the sweat glands of the skin are stimulated to increase the amount of perspiration secreted. The evaporation of sweat from the surface of the skin lowers the body temperature by removing from the body the heat necessary to convert the liquid sweat into water vapor—540 kilocalories are required to convert a liter of water to water vapor.

The skin contains several types of sense receptors responsible for our ability to feel pressure, temperature and pain (Fig. 24–5). Human skin and the skin of other mammals contains sweat, oil and mammary glands. Some 2½ million sweat glands occur all over the body, but are most numerous on the palms of the hands, the soles of the feet, in the arm pits and on the forehead. Oil glands are especially numerous on the face and scalp. The oil secreted keeps the hair moist and pliable and

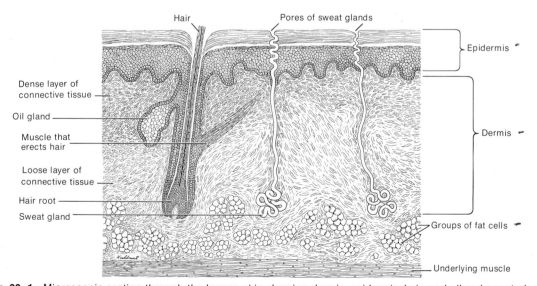

Figure 22–1 Microscopic section through the human skin showing dermis, epidermis, hair, and oil and sweat glands.

Hair   Pores of sweat glands

Epidermis

Dense layer of connective tissue

Oil gland

Muscle that erects hair

Loose layer of connective tissue

Hair root

Sweat gland

Dermis

Groups of fat cells

Underlying muscle

prevents the skin from drying and cracking. The mammary glands are also derivatives of the skin, specialized for the secretion of milk.

Parts of the Skin. Human skin is composed of a comparatively thin, outer layer, the *epidermis,* which is free of blood vessels, and an inner, thicker layer, the *dermis,* which is packed with blood vessels and nerve endings (Fig. 22–1). The several layers of cells in the epidermis vary in number in different parts of the body. The thickness of the skin varies considerably from one part of the body to another. It is thickest on the soles of the feet and the palms of the hands, where the epidermal surface is thrown into the countless tiny ridges forming the fingerprint patterns. These patterns are unique in each person and remain constant throughout life. The columnar cells in the layer of epidermis next to the dermis undergo frequent cell division to give rise to the layers above. The outer layers of the skin are constantly sloughing off and being replaced by cells from beneath. As each cell is pushed outward from the bottom layer, it is compressed into a flat, lifeless, scalelike epithelial cell. Dandruff consists of the flaky particles of dead, outer epidermal cells of the scalp.

The dermis is much thicker than the epidermis and is composed largely of connective tissue fibers and cells. The outer part, made of thickly matted connective tissue fibers, is tanned to make leather. Below this, and connected with the underlying muscles, is a layer composed of many fat cells and a more loosely woven network of fibers. This part of the dermis is one of the principal depots of body fat. The fat helps prevent excessive loss of heat and acts as a cushion against mechanical injury. The dermis is richly supplied with blood and lymph vessels, nerves, sense organs, sweat glands, oil glands and hair follicles.

The color of the skin depends on three factors: the yellowish tinge of the epidermal cells, their translucent quality which allows the pink of the underlying blood vessels to show through, and the kind and amount of pigment—red, yellow or brown—present in the inner layer of epidermal cells.

Outgrowths of the Skin. Human hair and nails and the feathers, scales, claws, hoofs and horns of other vertebrates are derivatives of the skin. With the exception of the palms of the hands and the soles of the feet, the entire skin is equipped with countless hair follicles—cellular inpocketings from the inner layer of the epidermis. These cells undergo division and give rise to the hair cells just as the inner layer of the epidermis gives rise to the outer layers. The hair cells die while still in the follicle, and the hair visible above the surface of the skin consists of tightly packed masses of their remains. Hair grows from the bottom of the follicle, not from the tip. Its color (and that of feathers and fur) depends on the amount and kind of pigment present, on the number of air bubbles, and on the nature of the surface of the hair, which may be smooth or rough.

Fingernails and toenails also develop from inpocketings of cells from the inner layer of the epidermis, and the growth of nails is similar to that of the hair. The translucent, densely packed dead cells of the nails allow the underlying capillaries to show through and give the nails their normal pink color.

Oil and sweat glands are derived from the inner layer of the epidermis by inpocketings which go deep into the dermis. Each hair follicle is associated with an oil gland (Fig. 22–1).

22–2 THE SKELETON

The most obvious function of the skeleton is to give support and "shape" to the body. For an animal to rise off the ground and move around, some hard, durable substance is needed to maintain the soft tissues against the pull of gravity and act as a firm base for the attachment of muscles. The skeleton protects the underlying organs, such as the brain and lungs, from injury. The marrow tissue, found within the cavity of the bones (Fig. 4–28), performs the special task of manufacturing all red cells and some kinds of white ones.

The skeletal system includes, in addition to bone, a variety of connective tissue fibers important in holding cells, tissues and organs together and in place. Two specialized kinds of connective tissue fibers, *ligaments* and *tendons,* attach bones to bones and muscles to bones, respectively, thereby playing an indispensable role in locomotion.

The bones of a dead animal are much more likely to be preserved than its soft parts. Our knowledge of the kinds of animals that lived in the past and their evolutionary relationships has come, in large part, from intensive study of their fossil bones. Much can be learned about an animal from studying the size and shape of the bones, the points at which muscles were attached, and so on.

Types of Skeletons. The skeleton of an animal may be located on the outside of the body (an *exoskeleton*) or inside the body (an *endoskeleton*). The hard shells of lobsters, crabs, oysters, clams and snails, are examples of exoskeletons. The advantage of an exoskeleton as a protective device is obvious, but a serious disadvantage is the attendant difficulty of growth. Snails and clams meet the difficulty by secreting additions to their shells as they grow. Lobsters and crabs have evolved a complicated solution to this problem whereby the outer shell is first softened by the removal of some of

its salts, so that it can be split down the back. The animal then crawls out of the old shell, grows rapidly for a short time, and then produces a new, larger shell which hardens by the redeposition of mineral salts. During this molting process the animal is weak and barely able to move and thus easily falls prey to its enemies.

Human beings and the other vertebrates characteristically have an endoskeleton. The skeleton of sharks and rays is made of cartilage, but in the bony fishes and other vertebrates most of the cartilage has been transformed to bone. The human skeleton consists of approximately 200 bones; the exact number varies at different periods of life as some of the bones, at first distinct, gradually become fused. Most of the bones are hollow and contain bone marrow cells. Bony scales and plates are present in the

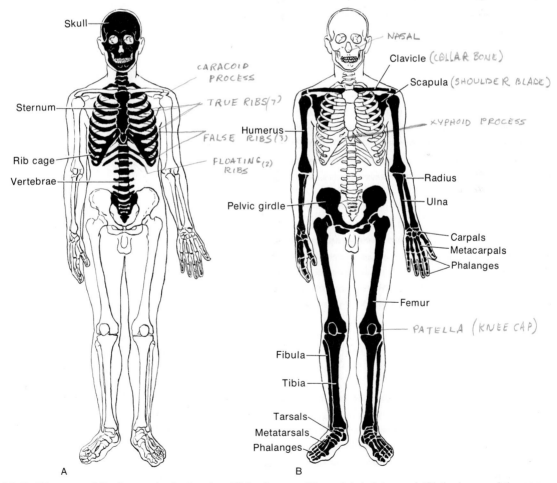

Figure 22–2 Diagrams of the human body showing *(A)* the bones of the axial skeleton and *(B)* the bones of the appendicular skeleton.

dermis of many vertebrates; some of these become associated with, and an integral part of, the skull and pectoral girdle.

Parts of the Skeleton. The vertebrate skeleton may be divided into the *axial skeleton* (the bones and cartilages in the middle or axis of the body) and the *appendicular skeleton* (the bones and cartilages of the fins or limbs) (Fig. 22–2). The axial skeleton includes the *skull*, *backbone* (vertebrae), *ribs*, and *breast bone* (sternum). Compare the features of the human skeleton with those of the frog's skeleton (Figs. 15–12, 15–13 and 15–14).

A number of bones fuse together to make up the skull: the *cranium*, or bony case immediately around the brain, and the bones of the face. The lower vertebrates have gill arches, cartilages or bones supporting the gill pouches. In the higher vertebrates these have disappeared or have been converted into other structures, such as the small bones in the middle ear—the *hammer*, *anvil* and *stirrup*. These transmit sound waves from the ear drum to the inner ear. In the course of evolution, certain gill arches of fishes have been transformed into parts of the epiglottis and larynx of higher vertebrates.

The human backbone is made of 33 separate *vertebrae* which differ in size and shape in different regions of the spine. A typical vertebra (Fig. 22–3) consists of a basal portion, the *centrum*, and a dorsal ring of bone, the *neural arch*, which surrounds and protects the delicate spinal cord. Different vertebrae have different projections for the attachment of ribs and muscles and for articulating (joining) with neighboring vertebrae. The first vertebra, the *atlas* (named for the mythical Greek who held the world on his shoulders), has rounded depressions on its upper surface into which fit two projections from the base of the skull.

The rib basket, composed of a series of flat bones, supports the chest wall and keeps it from collapsing as the diaphragm contracts. The ribs are attached dorsally to the vertebrae, each pair of ribs being attached to a separate vertebra. Of the twelve pairs of ribs in the human, the first seven are attached ventrally to the breastbone, the next three are attached indirectly by cartilages, and the last two, called "floating ribs," have no attachments to the breastbone.

The bones of the appendages (the arms

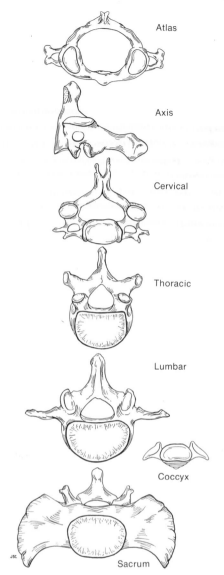

Figure 22–3 Types of human vertebrae. The axis is shown in side view; the others are seen from above.

and legs) and the *girdles* which attach them to the rest of the body make up the appendicular skeleton. The *pelvic girdle* consists of three fused hipbones, and the *pectoral girdle* consists of the two collarbones, or *clavicles*, and the two shoulder blades, or *scapulas*. The pelvic girdle is securely fused to the vertebral column, whereas the pectoral girdle is loosely and flexibly attached to it by means of muscles. Both pectoral and pelvic fins of fishes are simple structures, adapted for paddling. Paleontologic evidence shows that fins evolved into limbs adapted for moving on land, and that these in turn evolved into wings, hoofs, and the flippers of whales.

The appendages of human beings are comparatively primitive, terminating in five digits—the fingers and toes; whereas the more specialized appendages of other animals may be characterized by four digits (as in the pig), three (as in the rhinoceros), two (as in the camel) or one (as in the horse). The diagrams in Figure 22–4 illustrate the arrangement of the bones in the appendages. The *patella* or kneecap is a separate bone of the leg; it has no counterpart in the arm.

The Joints. The point of junction between two bones is called a *joint.* Some joints, such as those between the bones of the skull, are immovable and extremely strong, owing to an intricate dovetailing of the edges of the bones. Other joints, such as the articulation of the humerus to the scapula or of the femur to the hipbone, are in the form of a ball and socket, permitting free movement in several directions. Both the scapula and the hipbone contain rounded, concave depressions to accommodate the rounded, convex heads of the humerus and femur, respectively. Between these two extremes are joints with moderate freedom of motion, such as the hinge joint at the knee in which movement is restricted to a single plane; or the pivot joints at the wrists and ankles, with freedom of movement intermediate between that of the hinge and the ball and socket types (Fig. 22–5).

Wherever two bones move on one another, the ends of the bones are covered with smooth, slippery *cartilage* which reduces friction. The bearing surfaces are completely enclosed in a liquid-tight capsule made of ligaments, and the joint cavity is filled with a liquid lubricant secreted by the membrane lining the cavity. This liquid is similar to lymph or tissue fluid but contains a small amount of mucoproteins. During youth and early maturity the lubricant is replaced as needed, but in middle and old age the supply is often decreased, with resulting joint stiffness and difficulty of movement.

22–3 TYPES OF LOCOMOTION

Animals differ as to the part of the foot they put on the ground in walking and running. Men and bears walk flat on the sole of the foot. This type of locomotion, adapted for a comparatively slow gait, is called *plantigrade.* To increase the effective limb length, and thus the running speed, animals such as dogs and cats have become adapted to running on their digits, a type of locomotion called *digitigrade.* Speed is increased still further in the hoofed animals, horses and deer, by the lengthening of the limb bones and the raising of the wrist and ankle still farther from the ground. The animal runs upon the tips of one or two digits of each limb. This is known as *unguligrade* locomotion, and the animals are known as *ungulates.*

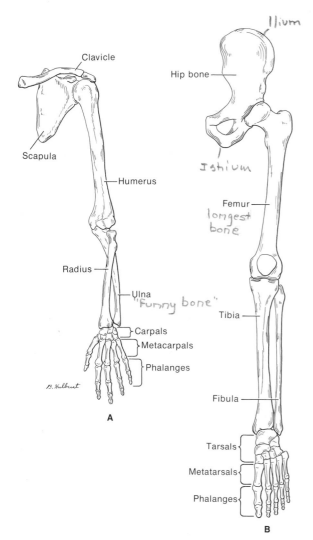

Figure 22–4 The bones of the left arm *(A)* and left leg *(B)*, as seen from the front. Note the fundamental similarity of pattern in the arrangement of the bones in the two limbs. Human limbs have the primitive pentadactyl (five-fingered) arrangement.

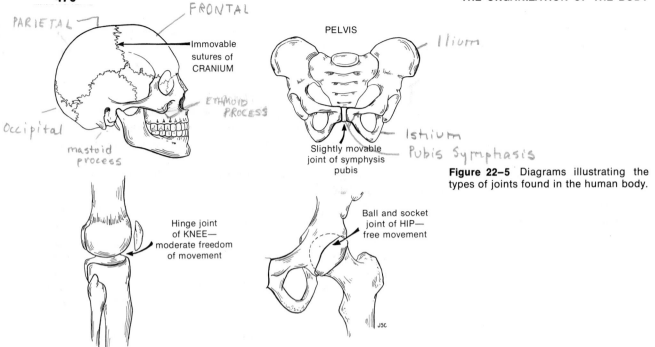

PARIETAL

FRONTAL

Immovable sutures of CRANIUM

ETHMOID PROCESS

Occipital

mastoid process

PELVIS

Ilium

Ishium

Pubis Symphasis

Slightly movable joint of symphysis pubis

Figure 22–5 Diagrams illustrating the types of joints found in the human body.

Hinge joint of KNEE— moderate freedom of movement

Ball and socket joint of HIP— free movement

22–4 SKELETAL MUSCLES

The principal effectors of all the multicellular animals, which provide for movements in response to stimuli, are muscles composed of specialized contractile cells. A typical skeletal muscle of a vertebrate is an elongated mass of tissue composed of millions of individual *muscle fibers* bound together by connective tissue fibers. The entire structure is surrounded by a tough, smooth sheet of connective tissue so that it can move freely over adjacent muscles and other structures with a minimum of friction. The two ends of the muscle in vertebrates are typically attached to two different bones, and the contraction of the muscle draws one bone toward the other, with the joint between the two acting as the fulcrum of the lever system. A few muscles pass from a bone to the skin or, as in the muscles of facial expression, from one part of the skin to another. The end of the muscle that remains relatively fixed when the muscle contracts is called the *origin;* the end that moves is called the *insertion;* and the thick part between the two is called the *belly* (Fig. 22–6). The origin of the *biceps* is on the shoulder and its insertion is on the bone in the forearm called the radius; when the biceps contracts the shoulder remains fixed and the elbow is bent.

Muscles typically contract in groups rather than singly. You cannot, for example, contract the biceps alone; you can only bend the elbow, which involves the contraction of a number of other muscles in addition to the biceps. Mus-

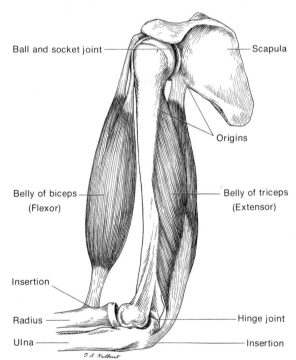

Ball and socket joint

Scapula

Origins

Belly of biceps (Flexor)

Belly of triceps (Extensor)

Insertion

Radius

Hinge joint

Ulna

Insertion

Figure 22–6 Muscles and bones of the upper arm, showing origin, insertion and belly of a muscle and the antagonistic arrangement of the biceps and triceps.

cles can exert a pull but not a push, and hence muscles are typically arranged in *antagonistic pairs:* one pulls a bone in one direction and the other pulls it in the reverse. The biceps, for example, bends or flexes the arm and is termed a *flexor.* Its antagonist, the *triceps,* straightens or extends the arm and is termed an *extensor.* Similar pairs of opposing flexors and extensors are found at the wrist, knee, ankle and other joints. When a flexor contracts, the opposing

extensor must relax to permit the bone to move; this requires proper coordination of the nerve impulses going to the two sets of muscles. Other antagonistic pairs of muscles are *adductors* and *abductors,* which move parts of the body toward or away from the central axis of the body; *levators* and *depressors,* which raise and lower parts of the body; *pronators,* which rotate body parts downward and backward, and *supinators,* which rotate them up-

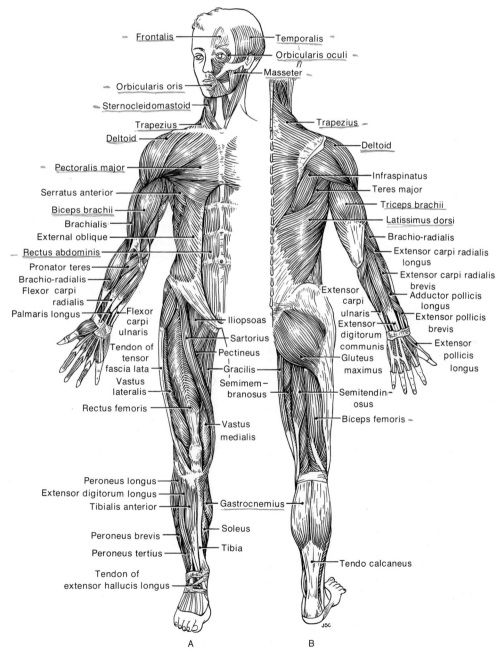

Figure 22–7   *A,* A ventral and, *B,* a dorsal view of the superficial muscles of the human body.

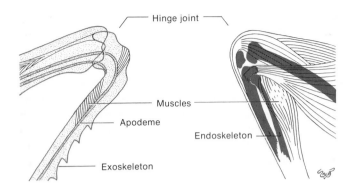

Figure 22–8 A comparison of the vertebrate endoskeleton and joint with the arthropod exoskeleton and joint. Note the difference in the arrangement of the muscles and skeletal elements at the two types of joints.

ward and forward; and *sphincters* and *dilators*, which decrease and enlarge the size of an opening. Some of the superficial muscles of the human body are shown in Figure 22–7.

Musculo-skeletal Relationships. The vertebrate *endoskeleton* is a bony or cartilaginous framework lying within the body and surrounded by muscles. The arthropod *exoskeleton* is a chitinous framework on the outside of the body surrounding the muscles. The basic differences in the mechanical arrangement of vertebrate and arthropod joints are illustrated in Figure 22–8. The muscles of the vertebrate surround the bones; one end of each is attached to one bone and the other end to the second bone. The contraction of the muscle moves one bone with respect to the other. The muscles of the arthropod lie *within* the skeleton and are attached to its inner surface. The arthropod exoskeleton has *joints*, regions in which the exoskeleton is thin and flexible so that it can bend. A muscle may stretch across the joint so that its contraction will move one part on the next; or a muscle may be located entirely within one section of the body or appendage and be attached at one end to a tough *apodeme*—a long, thin, firm part of the exoskeleton extending into that section from the adjoining one.

The hydra and other cnidarians, the planarian and other flatworms, and the earthworm and other annelids move by the same basic principle of antagonistic muscles even though they have no hard exo- or endoskeleton to anchor the ends of the muscles. Instead, the noncompressible fluid contents of the body cavity serve as a *hydrostatic skeleton.* Such animals typically have a set of circular muscles, which decrease the diameter and increase the length of the animal, and a set of antagonistic longitudinal muscles, which decrease the

length and increase the diameter of the animal (Fig. 22–9). Some marine worms have additional diagonally arranged muscles that permit more complex movements of the body and of the paddlelike *parapodia* that extend laterally from the body wall. Many marine worms live in tubes and the movements of the parapodia are important not only in locomotion but in moving currents of water laden with oxygen and nutrients through these tubes.

Posture and Movement. When a muscle is not contracting to effect movement, it is not completely relaxed. As long as an individual is conscious, all his muscles are contracted slightly, a phenomenon termed **tonus.** Posture is maintained by the partial contraction of the muscles of the back and neck and of the flexors and extensors of the legs. When a person stands, both the flexors and extensors of the thigh must contract simultaneously so that the body sways neither forward nor backward on the legs. The simultaneous contraction of the flexors and extensors of the shank locks the knee in place and holds the leg rigid to support

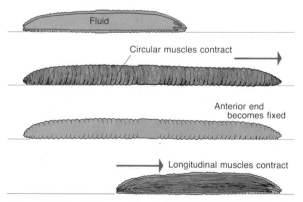

Figure 22–9 A diagram illustrating the hydrostatic skeleton of an earthworm. The fluid within the animal's body enables the alternate contraction of the circular and longitudinal muscles to move the animal along.

the body. When movement is added to posture, as in walking, a complex coordination of the contraction and relaxation of the leg muscles is required. It is not surprising that learning to walk is such a long and tedious process.

Some of the larger muscles of the human body are remarkably strong. The **gastrocnemius** muscle in the calf of the leg has its origin at the knee and its insertion, by the tendon of Achilles, on the heel bone. Because the distance from the toes to the ankle joint is at least six times that from the ankle joint to the heel, the gastrocnemius is working against an adverse lever ratio of 6:1. Thus when a man weighing 70 kg. stands on one leg and rises on his toes, the one gastrocnemius muscle must exert a force of 420 kg. When one ballet dancer holds another in his arms and rises on the toes of one leg, his gastrocnemius is exerting a force of nearly 1000 kg.

Normal nervous coordination prevents muscles from contracting maximally, but in certain diseases in which nervous control is impaired, a muscle may contract forcefully enough to rip tendons and break bones.

22–5 THE PHYSIOLOGY OF MUSCLE ACTIVITY

The functional unit of vertebrate muscles, the **motor unit**, consists of a single motor neuron and the group of muscle cells innervated by its axon, all of which will contract when an impulse travels down the motor neuron to the **motor end plate.** The human body, it is estimated, has some 250,000,000 muscle cells but only some 420,000 motor neurons in spinal nerves. Obviously, some motor neurons must innervate more than one muscle fiber. The degree of fine control of a muscle is inversely proportional to the number of muscle fibers in the motor unit. The muscles of the eyeball, for example, have as few as three to six fibers per motor unit, whereas the leg muscles have as many as 650 fibers per unit.

If a single motor unit is isolated and stimulated with brief electric shocks of increasing intensity, beginning with stimuli too weak to cause contraction, there will be no response until a certain intensity is reached at which point the response is maximal. This phenomenon is termed the *"all-or-none" effect.* In contrast, an entire muscle, composed of many individual motor units, can respond in a graded fashion depending on the number of motor units contracting at any given time. Although an entire muscle cannot contract maximally, a single motor unit can contract *only* maximally or not at all. The strength of contraction of an entire muscle composed of thousands of motor units depends upon the number of its constituent motor units that are contracting and upon whether the motor units are contracting simultaneously or alternately.

Muscles retain their ability to contract for some time after they have been removed from the body. The muscle usually used for experimental purposes is the gastrocnemius muscle of the frog, and if care is taken to keep it moist it will contract for hours. To make a record of these contractions, the muscle is mounted with its origin attached to a fixed hook and its insertion connected by means of another hook to a lever with a pointed stylus at its tip (Fig. 22–10). This stylus is in contact with a cylinder covered with recording paper and revolved by a motor. Each contraction of the muscle raises the stylus and its vigor and duration are recorded. Additional styluses can be used to record an appropriate time scale and to mark the point when the muscle is stimulated.

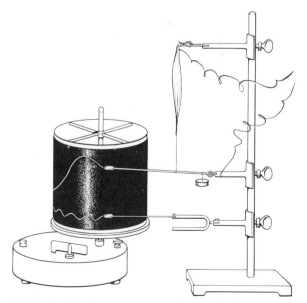

Figure 22–10 The apparatus used to study the contraction of an isolated muscle. The stylus attached to the insertion of the muscle writes a record of the contraction on the rotating cylinder, the kymograph. The contraction is timed by the vibrating tuning fork, whose stylus draws a wavy line on the kymograph record.

The Single Twitch. When a muscle is given a single stimulus, a single electric shock, it responds with a single, quick twitch lasting about 0.1 second in a frog's muscle and about 0.05 second in a human muscle. A record of a single twitch (Fig. 22–11) reveals that it consists of three separate phases: (1) the *latent period,* lasting less than 0.005 second, an interval between the application of the stimulus and the beginning of the visible shortening of the muscle; (2) the *contraction period,* about 0.04 second in duration, during which the muscle shortens and does work; and (3) the *relaxation period,* the longest of the three, lasting 0.05 second, during which the muscle returns to its original length.

Muscle fibers, like nerve fibers, have a *refractory period,* a very short period of time immediately after one stimulus during which they will not respond to a second stimulus. The refractory period in skeletal muscle is so short (about 0.002 second) that muscle can respond to a second stimulus while still contracting in response to the first. The super-

position of the second contraction on the first results in a greater than normal shortening of the muscle fiber, an effect known as *summation.*

The first event after the stimulation of a muscle is the initiation and propagation of an *action potential* (Fig. 22–11), followed by changes in the structure of the contractile proteins *actin* and *myosin,* evident as a change in the *birefringence* of the muscle. After a twitch, the muscle consumes oxygen and gives off carbon dioxide and heat at a rate greater than during rest, marking a *recovery period* in which the muscle is restored to its original state. This recovery period lasts for several seconds, and if a muscle is stimulated repeatedly so that successive contractions occur before the muscle has recovered from the previous ones, the muscle becomes fatigued and the twitches grow feebler and finally stop. If the fatigued muscle is allowed to rest for a time, it regains its ability to contract.

Tetanus. Muscles do not normally contract in single twitches but rather in sustained contractions evoked by volleys of nerve impulses reaching them in rapid succession. Such a sustained contraction is called *tetanus,*° and during a tetanic contraction the stimuli occur so rapidly (several hundred per second) that relaxation cannot occur between the contractions of successive twitches. In most tetanic contractions the individual muscle fibers are stimulated in rotation rather than simultaneously, so that although individual muscle fibers contract and relax, the muscle as a whole remains partially contracted. From personal experience you know that any muscle of your body can contract to different degrees. This gradation of contraction is controlled through the nervous system: in a weak contraction only a small percentage of the muscle fibers is stimulated at one time; for a stronger contraction a larger percentage of muscle fibers contracts simultaneously.

Tonus. The term tonus, or "tone," refers to the state of sustained partial contraction present in all normal skeletal muscles as long

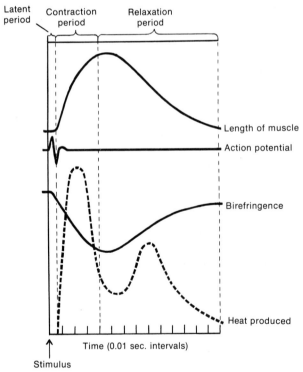

Latent period

Contraction period

Relaxation period

Length of muscle

Action potential

Birefringence

Heat produced

Time (0.01 sec. intervals)

↑
Stimulus

Figure 22–11 The changes that occur in a muscle during a single twitch. Note the temporal relationship between the action potential, the shortening of the muscle and the production of heat.

°This term should not be confused with tetany, the muscular spasms occurring in deficiencies of the parathyroid hormone, or with the disease tetanus ("lockjaw"), characterized by abnormal muscular contractions and caused by the tetanus bacillus.

as the nerves to the muscle are intact. Although cardiac and smooth muscles exhibit tonus even after their nerves are cut, severing the nerve to a skeletal muscle eliminates tonus immediately. Each muscle is normally stimulated by a continuous series of nerve impulses, causing a constant, slight contraction or tonus. Tonus is a mild state of tetanus, present at all times and involving only a small fraction of the fibers of a muscle at any moment. It is believed that the individual fibers contract in turn, working in relays, so that each fiber has an opportunity to recover completely while others fibers are contracting before it is called upon to contract again. A muscle under slight tension can react more rapidly and contract more strongly than one that is completely relaxed.

22-6   THE BIOCHEMISTRY OF MUSCULAR CONTRACTION

A steam engine can convert only about 10 per cent of the heat energy of its fuel into useful work (the rest is wasted as heat); but muscles are able to use between 20 and 40 per cent of the chemical energy of glucose in the mechanical work of contraction. The remainder is converted into heat but is not wholly wasted since it is used to maintain the body temperature. If one refrains from contracting the muscles, the heat produced elsewhere in the body is insufficient to keep it warm in a cold place. In these circumstances the muscles contract involuntarily (one "shivers"), and heat is thereby produced to restore and maintain normal body temperature.

The problem of how a muscle can exert a pull has been attacked enthusiastically by physiologists and biochemists for many years, and great gains have been made in our understanding of the process. However, some of the chemical and physical events involved in muscle contraction are still a matter of conjecture.

About 80 per cent of muscle is water; the remainder is mostly protein. Small amounts of fat and glycogen, and two phosphorus-containing substances, *phosphocreatine* and *adenosine triphosphate,* are also present. The actual contractile part of a muscle fiber is a protein chain which apparently shortens by the sliding together of its parts. Two proteins, *myosin* and *actin,* have been extracted from muscle, nei-

ther of which is capable of contracting alone. When they are combined to form a thread of actomyosin, and potassium, calcium and adenosine triphosphate are added, the thread undergoes contraction. This demonstration of contraction in a test tube, made by Albert Szent-Györgyi, is one of the most exciting discoveries yet made in biochemistry.

As a first step in analyzing the biochemical mechanism and the sequence of events involved in muscle contraction, we can measure what substances are used up in the process. Glycogen, oxygen, phosphocreatine and adenosine triphosphate decrease in amount during contraction, and carbon dioxide, lactic acid, creatine, adenosine diphosphate and inorganic phosphate increase. The fact that oxygen is used up and carbon dioxide is formed suggests that muscular contraction is an oxidative process. But this oxidation is not essential, for a muscle can twitch a good many times even when completely deprived of oxygen—for example, when it is removed from the body and placed in an atmosphere of nitrogen. Such a muscle becomes fatigued, however, much more rapidly than one contracting in an atmosphere of oxygen. Furthermore, although we breathe faster during muscular exertion, the accelerated breathing continues for some time after the physical work has ceased. This suggests that oxidation is involved not in muscular contraction but in the recovery from contraction.

The disappearance of glycogen and the formation of lactic acid are related, for in the absence of oxygen, the amount of lactic acid formed is just equivalent to the glycogen that disappears. Since the breakdown of glycogen to lactic acid requires no oxygen, and since it liberates energy rapidly, it was once thought that this reaction was directly responsible for muscle contraction. When oxygen is present, the muscle oxidizes about one fifth of the lactic acid to carbon dioxide and water, and the energy released by this oxidation is used to reconvert the other four fifths of the lactic acid to glycogen. This explains why lactic acid does not accumulate as long as muscle has sufficient oxygen, and why a muscle becomes fatigued more rapidly (uses up its glycogen and accumulates lactic acid) when it contracts in the absence of oxygen.

About 1930 it was found that a muscle

poisoned with iodoacetate, which inhibits the chemical reactions by which glycogen breaks down to lactic acid, can still contract, although it is capable of twitching only 60 to 70 times, instead of the 200 or more times achieved by a muscle deprived of oxygen. But the fact that it can twitch at all when the breakdown of glycogen is prevented shows that this is not the primary source of energy for contraction.

The other change that can be detected chemically during contraction is a splitting off of inorganic phosphate from phosphocreatine and adenosine triphosphate, accompanied by the release of energy. It is now believed that this is the immediate source of energy for contraction. The reactions whereby glucose and other substances are metabolized to yield energy-rich phosphate compounds such as adenosine triphosphate were described in Chapter 5. Phosphocreatine serves as a reservoir of $\sim P$ in the muscle; its energy-rich phosphate must be transferred to ADP to form ATP before it can be utilized in contraction. After a muscle has contracted the breakdown of glycogen to lactic acid and the oxidation of lactic acid in the tricarboxylic acid cycle provide energy for the resynthesis of adenosine triphosphate and phosphocreatine. In summary, muscle contraction involves the following chemical reactions:

(1) Adenosine triphosphate → inorganic phosphate + adenosine diphosphate + energy (used in actual contraction)

(2) Phosphocreatine + ADP ⇌ creatine + ATP

(3) Glycogen ⇌ intermediates ⇌ lactic acid + energy ($\sim P$, used in resynthesis of organic phosphates)

(4) Part of lactic acid + O_2 → CO_2 + H_2O + energy ($\sim P$, used in resynthesis of rest of lactic acid to glycogen and in resynthesis of ATP and phosphocreatine)

Myosin is an enzyme as well as a contractile protein and can catalyze the cleavage of ATP to ADP and inorganic phosphate. The enzyme creatine kinase catalyzes the transfer of $\sim P$ from ATP to creatine (reaction 2).

It is estimated that the energy from organic phosphates alone could sustain maximal muscular contraction for only a few seconds; a runner might complete a 50 meter dash with this. By calling upon all the sources of energy available in the absence of oxygen, a man might be able to continue maximal muscle contractions for 30 to 60 seconds.

The Oxygen Debt. The fact that the actual contraction, and part of the recovery from contraction, occur without oxygen is extremely important. Our muscles are often called upon to do great spurts of work, and although both the rate of breathing and the heart rate increase during exertion, oxygen cannot be supplied in sufficient quantities to permit these exertions. During violent exercise, such as running the 100 meter dash, glycogen breaks down to lactic acid faster than the lactic acid can be oxidized, so that the latter accumulates. In such circumstances the muscle is said to have incurred an *oxygen debt*, which is afterward repaid by our rapidly breathing enough extra oxygen to oxidize part of the lactic acid, which furnishes energy for resynthesizing the rest to glycogen. In other words, during short spurts of extreme muscular activity, muscles use energy from sources that do not require utilization of oxygen. After the activity has ceased, the muscles and other tissues pay off the oxygen debt by utilizing an extra amount of oxygen to restore the energy-rich phosphate compounds and glycogen to their normal condition. During a long race a runner may reach an equilibrium in which he gets a "second wind," and, because of the increase in breathing and heart rate, takes in enough oxygen to oxidize the lactic acid formed at that moment, so that the oxygen debt is not increased.

Fatigue. A muscle that has contracted many times, exhausted its stores of organic phosphates and glycogen and accumulated lactic acid is unable to contract any more and is said to be *fatigued*. Fatigue is primarily induced by this accumulation of lactic acid, although animals feel fatigue before the muscle reaches the exhausted condition.

The exact spot most susceptible to fatigue can be demonstrated experimentally if a muscle and its attached nerve are dissected out and the nerve stimulated repeatedly by electric shocks until the muscle no longer contracts. If the muscle is then stimulated directly, by placing the electrodes on the muscle tissue, it will respond vigorously. With the proper apparatus for detecting the passage of nerve impulses, it can be shown that the nerve leading to the muscle is not fatigued; it is still capable of conduction. The point of fatigue, then, is the *junction* between the nerve and the muscle, where nerve impulses instigate muscle contraction.

22-7 THE BIOPHYSICS OF MUSCLE CONTRACTION

Electron micrographs show that muscle myofibrils are made of longitudinal filaments, the *myofilaments*. These are of two kinds: thick *primary filaments* (10 nm. thick and 1.5 μm. long) and thin *secondary filaments* (5 nm. thick and 2 μm. long) (Fig. 22–12). By selectively extracting the proteins and by histochemical and immunochemical staining it has been possible to show that the primary filaments consist of myosin and the secondary filaments of actin. The primary and secondary filaments are arranged in such a fashion that, seen in cross section, each primary filament is surrounded by six secondary filaments, which in turn are shared with six surrounding primary filaments.

The alternating light and dark bands seen in the light microscope (Fig. 4–30) consist of dense A bands and light I bands (Fig. 22–13). Each unit consists of an A band bounded on each side by an I band and separated from the adjacent unit by a Z line, a thin, dense line through the center of the I band. The central portion of the A band is somewhat less dense and is called the H zone. Electron micrographs reveal that the thick primary myosin filaments are found only in the A band and that the I band contains only thin secondary actin filaments. The actin filaments, however, are not limited to the I band but extend for some distance into the A band, interdigitating with the myosin filaments. Thus, at either end of the A band are both myosin and actin filaments interdigitating, but the central part of the A band (the H zone) contains only myosin filaments. The actin filaments appear to be smooth but the myosin filaments appear to have minute spines every 6 to 7 nm. along their length which project toward the adjacent actin filament. These spines look like bridges connecting the two sets of filaments.

Since there are six actin filaments arranged hexagonally around the myosin filament (Fig. 22–12), the cross-bridges are repeated six times

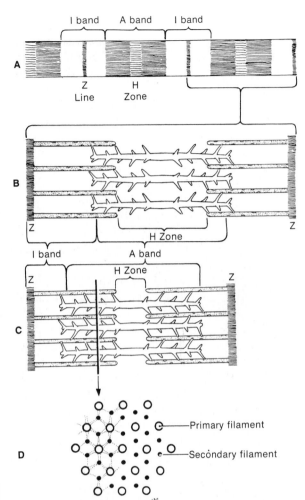

Figure 22–12 Diagrams illustrating the sliding filament hypothesis of the mechanism of muscle contraction. *A*, Diagram of part of a single myofibril showing the pattern of light (I) and dark (A) bands. *B*, Longitudinal view of the arrangement of thick and thin filaments within a myofibril in the relaxed state. *C*, Longitudinal view of the arrangement of thick and thin filaments in a contracted myofibril, showing that the I band decreases in thickness. Note that the two types of filaments appear to slide past one another during contraction. *D*, Transverse view through *C* at arrow, showing each thick primary filament surrounded by six thinner secondary filaments. (After Huxley, H. E.: The mechanism of muscular control. Science *164*:1357, 1969.)

around the circumference of the myosin filament. Actin filaments do not have projections; each filament is composed of two long actin molecules wrapped helically about each other, with a complete turn every 70 nm. Every 40 nm. along the actin filament there are reactive sites that interact in some way with the ends of the myosin cross-bridges to provide the force to draw the actin filaments along the myosin filaments.

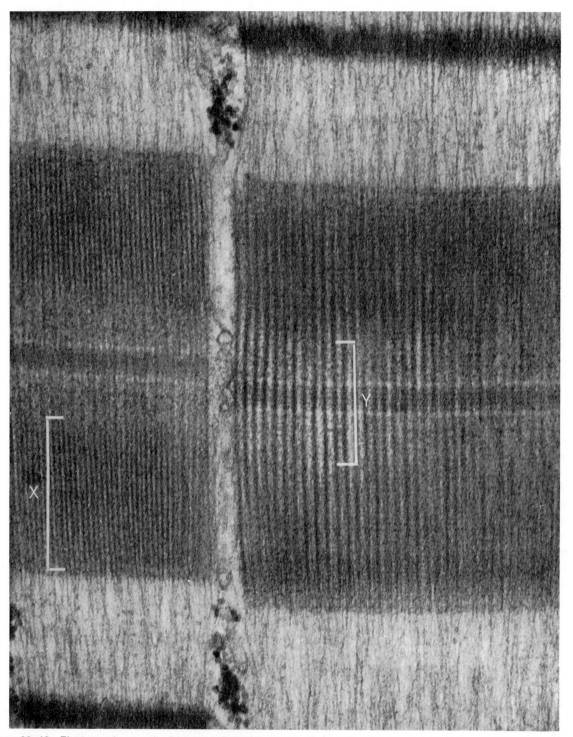

Figure 22–13 Electron micrograph of one muscle unit (sarcomere) extending from one dense Z line above to the next Z line below. The central A band, of medium density, is composed of myosin filaments 10 nm. thick. The less dense I bands, above and below the A band, are made of thinner (6 nm.) actin filaments extending from the Z band. The actin filaments extend into the A band and occupy the spaces between the myosin filaments. Actin and myosin filaments interdigitate in the region indicated by X. The region marked Y, the central portion of the A band, contains only myosin filaments. Papillary muscle of the cat heart, ×78,000. (From Fawcett, D. W.: *The Cell.* Philadelphia, W. B. Saunders Co., 1966.)

During contraction the length of the A band remains constant but the I band shortens and the length of the H zone within the A band decreases. Huxley and others have proposed that during contraction the filaments maintain their initial length but the primary and secondary filaments *slide* past each other. During contraction the thin secondary filaments of actin extend farther into the A band, decreasing the central H zone and narrowing the I band as the ends of the myosin filaments approach the Z band.

The physicochemical mechanism by which this sliding may occur is not yet clear; perhaps the bridges are broken and then re-formed further along. The energy of the ~P may be utilized in forming new cross-bridges between the myosin and actin filaments. If we assume, as Huxley did, that the formation of one new cross-bridge requires the energy of one ~P band, we can calculate from the rate of contraction and the total number of cross-bridges per fiber how much ATP would be utilized per second by the fiber. The resulting figure is of the same order of magnitude as the rate of ATP utilization determined experimentally.

The contraction of a muscle is initiated by nervous stimuli that arrive at the specialized ends of motor nerves, the *motor end plates*, and cause the release of acetylcholine. The acetylcholine presumably depolarizes the membrane of the muscle fiber, triggering an action potential which changes ion concentrations (increases calcium ion concentration) and causes myosin to hydrolyze ATP.

A recently discovered component of the actin filament is *troponin*, a small globular protein, molecular weight 50,000, which normally binds to and inhibits the actomyosin ATPase. Troponin molecules are distributed along the actin fiber at intervals of about 40 nm. and occupy the "reactive sites" described previously. Troponin has a very strong affinity for calcium ions, and when the concentration of calcium within the muscle fiber increases following an action potential, the calcium is tightly bound by troponin. The binding of troponin to calcium releases its binding to actomyosin and its inhibition of actomyosin so that contraction can occur, i.e., so that the sliding filaments can move back and forth. Normally the intracellular concentration of calcium is very low and

calcium is sequestered within the sarcoplasmic reticulum. The injection of calcium ions into a muscle fiber leads to its contraction. The stimulation of the muscle by an action potential causes the release of calcium from the sarcoplasmic reticulum and this binds troponin, releasing actomyosin from its previous inhibition, permitting contraction to occur.

Muscle relaxation requires that the intracellular calcium ion concentration be lowered and this is accomplished by sequestering the calcium back into the sarcoplasmic reticulum. This demands energy, and ATP is used to *pump* the calcium back into the sarcoplasmic reticulum. Thus it is now clear that metabolic energy is required for *relaxation* of the muscle rather than for its contraction. If anything interferes with energy metabolism in the muscle, calcium ions will leak out of the sarcoplasmic reticulum and will not be pumped back in, and this will lead to muscle contraction. The contraction of all the muscles of the body that occurs shortly after death in the human and other animals is termed *rigor mortis.*

The relationship between electrical stimulation, the release of calcium ions, and muscle contraction were demonstrated in experiments carried out by Ashley and Hoyle, which provide a beautiful example of the advantages of comparative biology. For these studies Hoyle chose the muscles of the giant barnacle, *Balanus nubilus,* composed of muscle fibers as much as 2 mm. thick and 4 cm. long. To measure the concentration of calcium within the fiber, these investigators used a remarkable protein, *aequorin,* produced by the jellyfish *Aequorea.* Aequorin has the property of emitting a bluish light when it reacts with calcium ion; it is involved in the natural phosphorescence of the jellyfish. The reaction is specific for calcium and very sensitive. Light is emitted at calcium concentrations as low as 10^{-8} mole per kilogram. A giant muscle fiber from the barnacle was suitably mounted in a holder so that the dim blue light emitted could be measured in a spectrophotometer, aequorin was injected into the fiber, and then an action potential was applied. By timing the reaction with an oscilloscope (Fig. 22–14), Ashley and Hoyle showed that first there was a change in membrane potential, then calcium ion was freed within the muscle fiber, as shown by the emission of the blue light, and lastly, the mus-

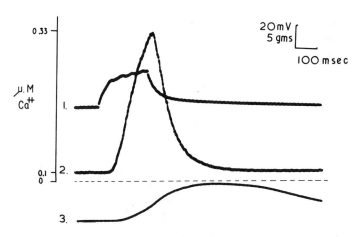

Figure 22-14 Sequence of events in E-C coupling in a single barnacle muscle fiber. Trace 1, the passive electrical response on the outer membrane; trace 2, the change in calcium inside the fiber *(calcium transient)* as monitored by aequorin; and trace 3, the tension response. (From Ashley, C. C.: Single twitch in a single barnacle muscle fiber. Endeavour *30*:18, 1971.)

cle underwent contraction. The emission of light rose to a peak intensity and then fell almost immediately after the electrical stimulus was removed. This showed that calcium did not remain free within the fiber, for light would have been emitted continually if this was the case, but was again sequestered after contraction had occurred. Other experiments showed that the calcium was taken up in the sarcoplasmic reticulum.

When a muscle contracts it becomes shorter and fatter but there is no change in its total volume. This has been shown experimentally by dissecting out a muscle, placing it in a glass vessel with a narrow neck, and filling the vessel with water. The muscle is then stimulated electrically, and as it contracts and relaxes there is no change in the water level in the neck of the vessel.

22-8 CARDIAC AND SMOOTH MUSCLE

The muscles of the heart and internal organs, though resembling skeletal muscle in a general way, have certain distinctive characteristics. They both contract much less rapidly then skeletal muscle: skeletal muscle fibers contract and relax in 0.1 second, but cardiac muscle requires from 1 to 5 seconds, and smooth muscle, from 3 to 180 seconds. All the phases of contraction are prolonged.

Smooth muscle exhibits wide variations in tonus; it may remain almost relaxed or tightly contracted. It also, apparently, can maintain the shortened condition of tonus without the expenditure of energy, perhaps owing to a reorganization of the protein chains making up the fibers.

Each beat of the heart represents a single twitch. Cardiac muscle has a long **refractory period,** the period following one stimulus when it is unable to respond to any other. Consequently, it is unable to contract tetanically, since one twitch cannot follow another quickly enough to maintain a contracted state.

The basic mechanism of the contraction of cardiac and smooth muscles is probably very similar to the sliding filament mechanism that operates in skeletal muscle. Both cardiac and smooth muscles contain actin and myosin and the contraction process involves the hydrolysis of ATP and an interaction between actin and myosin initiated by calcium ions.

A unique feature of cardiac muscle is its inherent rhythmicity; it contracts at a rate of about 72 times per minute even when denervated and removed from the body. Cardiac muscle discharges its membrane potential each time it has built up to a certain level. After each impulse has passed, the membrane becomes repolarized but then suddenly becomes permeable again, initiating the transmission of the next action potential.

22-9 THE MUSCLES OF LOWER ANIMALS

From the flatworm to the human, all animals have muscles that are similar in their elongate, cylindrical or spindle shape and in their content of contractile protein filaments. Even the cnidarians, which lack true muscle fibers, have cells that can contract. Most of the invertebrates have only smooth muscle, whereas arthropods have only striated muscle.

In contrast to the pattern in the vertebrates of a single neuron innervating relatively few muscle fibers to form a motor unit, a single axon in an arthropod may not only innervate *all* the fibers of one muscle but may innervate those of another muscle as well. Furthermore, most muscles receive just a few axons, perhaps only two, but each axon has a different effect upon muscular contraction. In a three-axon system, one axon produces a strong brief contraction, another a weak sustained contraction and the third inhibits the action of the other two. By varying the frequency of stimulation among the axons, the strength and duration of muscular contraction can be varied considerably. This relatively simple pattern of connections between nerves and muscles permits remarkably fine control of activity.

Electrical phenomena, action potentials, are associated with all types of muscle contraction (Fig. 22–11). In general, muscles are arranged with their fibers in parallel so that the voltage difference in a large muscle is no greater than that of a single fiber. In the electric organ of the electric eel, however, the electric plates are modified muscle cells, motor end plates, arranged in series. Although each plate has a potential difference of about 0.1 volt, the discharge of the entire organ, made of several thousand plates, may amount to several hundred volts. Records of the electric eel show that it can produce a potential of 400 volts or more, enough to stun or kill the fish on which it preys and give quite a jolt to a human.

QUESTIONS

1. What is the primary function of the skin of human beings? Name four other functions it performs. Draw a diagram of a cross section of the skin, labeling all parts.
2. How are hairs formed? What is their microscopic structure? What structures are associated with each hair follicle?
3. What kinds of tissue comprise the skeletal system? How is each type of tissue adapted for its role in supporting the body?
4. Differentiate between the axial and appendicular skeletons. What is the basic difference between an exoskeleton and an endoskeleton? In what vertebrates are exoskeletons present?
5. How do bones grow?
6. What characteristics would enable an expert to determine whether a skeleton was that of a man or a woman? What characteristics of the skeleton would enable him to determine the approximate age of the person at death?
7. What is a joint? Why may they become stiffened in old age?
8. How is the skeleton of a horse adapted for running? Compare the method of walking of a bear, a cat and a deer.
9. Contrast the skeletons of a human and a frog.
10. What are the distinguishing characteristics of the three types of muscle fibers?
11. Make a diagram of a typical skeletal muscle and label its parts.
12. Differentiate between tonus and tetanus.
13. Discuss the "all-or-none" law. Does the muscle as a whole operate according to this principle? If not, what does control the degree of contraction of a muscle?
14. What produces shivering? Of what use to the body is shivering?
15. What is meant by the term oxygen debt? Describe the chemical mechanisms involved in muscle contraction and in muscle relaxation. What physical or chemical changes are associated with fatigue?
16. What substances are responsible for the contractile quality of a muscle fiber? How are they organized structurally within the muscle fiber? Discuss the Huxley theory of muscle contraction.
17. Describe the muscular and neural mechanisms involved in the maintenance of an erect posture.
18. Which muscles contract and which relax when you hold your right arm out at the side? When you throw a ball?
19. Compare the muscle fibers of a mammal with the contractile fibers of a cnidarian and with a cilium of a paramecium.
20. List the structural and functional differences between a muscle and an electric organ.

SUPPLEMENTARY READING

The anatomy of the vertebrate skeleton and its evolution is discussed in detail in *The Vertebrate Body* by A. S. Romer. D'Arcy Thompson's interesting book, *On Growth and Form*, discusses the skeleton as an example of the application of engineering principles.

Two classic books on muscular exercise and the effects of exercise and training are A. V. Hill's *Muscular Movement in Man* and F. A. Bainbridge, A. V. Bock and D. B. Dill's *The Physiology of Muscular Exercise*. More recent developments in this field are summarized in *Textbook of Work Physiology* by P. Astrand and K. Rodahl. Quotations from some of the classic papers on muscle physiology are found in Chapter Six of J. F. Fulton's *Selected Readings in the History of Physiology*. A discussion of how the muscle may convert chemical energy to mechanical work is presented in *The Mechanism of Muscular Contraction* by H. E. Huxley.

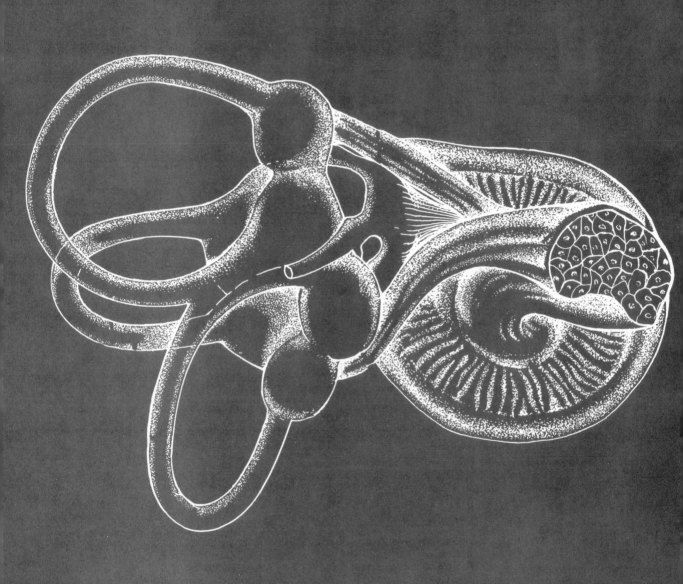

PART 5

THE BIOLOGICAL
BASIS OF BEHAVIOR

INTRODUCTION

Behavior is what an organism does and how it does it. Behavior depends on information received from the environment through the organism's receptors, on the integration of this information in its nervous system and on the appropriate responses by its effectors. Behavior patterns may be quite rigid and undeviating; or they may be flexible, modulated by hormonal factors, by conditioning, by the behavior of other organisms or by learning. The study of behavior may be approached from either the physiologic or the behavioral viewpoint. Physiologists have a primary interest in the neural mechanisms that underlie behavior and they attempt to provide an explanation of behavior in terms of the functions of the receptors, integrative systems and effectors involved. The psychologic school of behaviorists have been attempting to establish general principles of behavior, and especially of learning, by studying animal behavior in a controlled laboratory setting. B. F. Skinner's studies on training rats and pigeons to do fairly complex tasks, which employed immediate reward or reinforcement when the task was accomplished, have provided the experimental basis for programmed learning. In contrast, the ethologic school of behaviorists, led by Lorenz and Tinbergen, carefully watch animals carry out their normal behavior in the natural environment and attempt to identify and evaluate the many factors controlling behavior. The two approaches are obviously complementary and both are needed for a full understanding of the behavior of an animal in a given set of circumstances.

The discussions of neural integration, sensory receptors and hormonal integration in Part 5 provide the physiologic basis for an understanding of animal behavior. Reflex and operant conditioning, learning, imprinting and other aspects of behavior are discussed in Chapter 23. The behavior of an animal is one of its means of adapting to changes in the environment, and the animal's success and survival depend in part on the quality of its behavioral mechanisms.

CHAPTER 23

CONTROL SYSTEMS: NEURAL INTEGRATION

The orderly and efficient functioning of a complex multicellular organism like the human body requires that the various parts act in concert. To this end there must be means for monitoring the activities of the different parts and for providing a flow of information among parts or between parts and centers that receive, collate, integrate, store and retrieve information and subsequently issue appropriate commands. Two systems working in intimate association provide these functions: the **endocrine system** provides relatively slow and long-lasting controls of events, whereas the **nervous system** provides very rapid controls. The human nervous system, composed of brain, spinal cord and nerve trunks, connects the eyes, ears and other sense organs (the **receptors**) with the muscles and glands (the **effectors**). The connectors function in such a way that when a given receptor is stimulated the proper effector responds appropriately.

The chief functions of the nervous system are the **conduction** of impulses and the **integration** of the activities of the parts of the body. Integration means a putting together of generally dissimilar things to achieve unity from diversity. The coordinating activities of the nervous system, together with those of the endocrine system (Chapter 25), and the controls intrinsic in the enzyme systems within each cell (inhibition and stimulation of enzymatic activity, induction and repression of enzymes) are all factors tending toward homeostasis, the maintenance of a constant internal environment.

23–1 NEURONS

The structural and functional unit of the nervous system of all multicellular animals is the **neuron** (Fig. 23–1). The average neuron is slightly less than 0.1 mm. in diameter, but it may be several meters long. Traditionally the neuron is pictured as having three parts: axon, cell body and dendrite. The long **axon** emerges from one end of the **cell body** and bushy **dendrites** emerge from the other. However, there are so many exceptions that the usefulness of

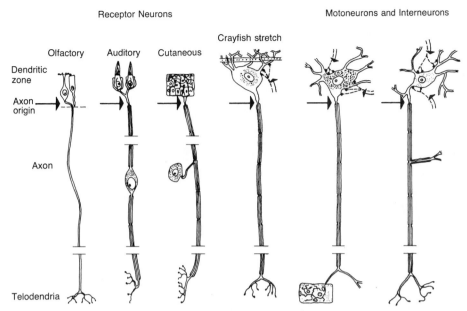

Figure 23–1 Diagram of a variety of receptor and effector neurons, arranged to illustrate the idea that impulse origin, rather than cell-body position, is the most reasonable focal point for the analysis of neuron structure in functional terms. Thus, the impulse conductor or axon may arise from any response generator structure, whether transducing receptor terminals or synapse-bearing surfaces (dendrites, cell body surface, or axon hillock). The interior of the cell body is conceived of as related primarily to the outgrowth of axon and dendrites and to metabolic functions other than membrane activity. Thus, the position of the cell body in the neuron is not critical with respect to the "neural aspects" of neuron function — namely, response generation, conduction and synaptic transmission. Except for the stretch neuron of the crayfish, the neurons shown are those of vertebrates. (Redrawn from Bodian, D.: The generalized vertebrate neuron. Science *137*:323–326, 1962.)

this anatomical subdivision is limited. Functional distinctions are more accurate; thus there are three functional parts. The dendrites constitute that part of the neuron specialized for receiving excitation, whether from environmental stimuli or from another cell. The axon is that part specialized to distribute or conduct excitation away from the dendritic zone. It is generally long and smooth but may give off an occasional collateral. Inside the central nervous system it is surrounded by non-nervous cells called **neuroglia;** outside the central nervous system it is wrapped in **Schwann cells.** It ends in a distribution or emissive apparatus, the **telodendria.** The cell body is concerned with metabolic maintenance and growth and may be situated anywhere with respect to other parts. In the sensory nerves from the skin, for example, it is situated on an offshoot of the axon. The variety in structure of neurons is enormous. Different functional and anatomical types of neurons characterize different parts of the nervous system.

It is the axon that is responsible for the tremendous length of some neurons. The axon

of a **sensory** cell barely 0.1 mm. in diameter located in the toe of a giraffe traverses a distance of several meters before ending in the spinal cord. The bundling together of many axons makes up the nerves and nerve trunks observed in gross anatomical dissection. A common connective tissue sheath surrounds the nerves.

Neurons are classified as **sensory, motor** or **interneurons** on the basis of their functions. Sensory neurons, or **afferent neurons,** are either receptors (olfactory receptors) or connectors of receptors (taste receptors) that conduct information to the central nervous system. Motor neurons, or **efferent neurons,** conduct information away from the central nervous system to the **effectors** (muscles, glands, electric organs, light organs). **Interneurons,** which connect two or more neurons, usually lie wholly within the central nervous system. In contrast the sensory and motor neurons have one of their endings in the central nervous system and the other close to the external environment or to the animal's internal environment.

The cell bodies of neurons are usually

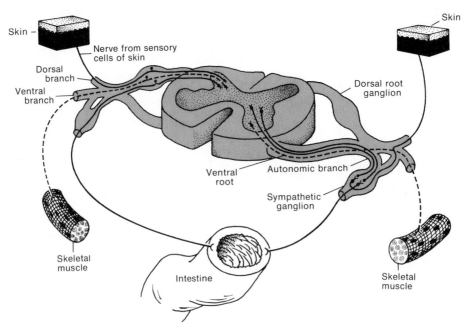

Figure 23–2 Diagram of the primary types of sensory and motor neurons of the spinal nerves and their connections with the spinal cord. For convenience, the sensory neurons are shown on the left and the motor neurons on the right, though both kinds are found on each side of the body.

grouped together in masses called *ganglia.* In the simplest sense a ganglion is any aggregation of neural cell bodies. Examples of ganglia are the *dorsal root ganglia* of vertebrates (Fig. 23–2), which are merely a collection of cell bodies of sensory neurons, and the *autonomic ganglia* of vertebrates, which are groups of cell

bodies of motor neurons. More commonly a ganglion is an aggregation of neural cell bodies plus interneurons. It is a place where different neurons connect with one another and where much integration may occur. The human brain and the brains of other animals are fusions of many ganglia.

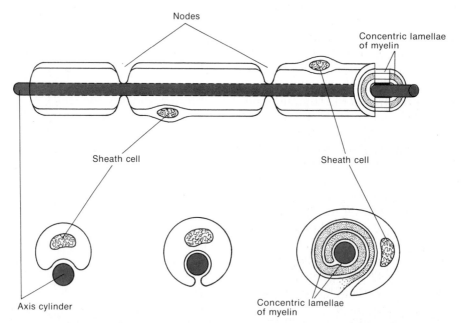

Figure 23–3 Sheath cells on neuron. *Upper,* Dissection of a myelinated nerve fiber. *Lower,* Envelopment of axis cylinder by a sheath cell. (After B. B. Geren, 1954; from Ballard, W. W.: *Comparative Anatomy and Embryology.* New York, The Ronald Press Co., 1967.)

In the central nervous systems of all animals the cellular part of the neurons and the fibrous part are separated into two zones. In vertebrates the **gray matter** (usually on the inside, but also on the outside in higher brain centers) contains cell bodies plus axons and dendrites. The **white matter** consists exclusively of axons plus their myelin sheaths. In invertebrate nerve cords the outside, or rind, consists solely of cell bodies, while the inside (core) consists of fibers.

A neuron consists of the usual cellular components: a nucleus, cytoplasm that extends to the outermost branches, and a cell membrane enclosing all. Enveloping the axon that is outside the central nervous system is a cellular sheath, the **neurilemma,** composed of the Schwann cells. These cells, migrating in from the mesenchyme, line up along axons and wrap around them. On some axons the Schwann cell lays down within its folds a spiral wrapping of insulating fatty material called **myelin** (Figs. 23–3, 23–4). Between adjacent cells there are

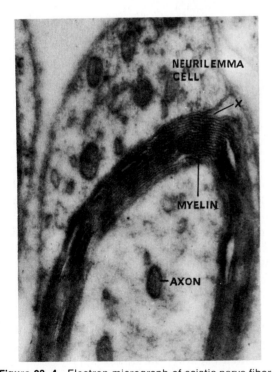

Figure 23–4 Electron micrograph of sciatic nerve fiber of a seven-day-old mouse showing the development of the myelin sheath by the folding of the cell walls of the neurilemma. The spiral infolding of the cell membrane of the neurilemma cell is visible as it is forming the thick, compact, many-layered myelin sheath. Note that at X the layers of myelin are continuous with the cell membrane of the neurilemma. Magnification ×83,000. (Courtesy of Dr. Betty G. Uzman.)

gaps. At these gaps, or **nodes,** the axon is free of myelin. Axons lying within the brain and spinal cord have no neurilemma sheath, and their myelin is provided by satellite cells (**oligodendrocytes**) rather than by Schwann cells. Nerves consisting of heavily myelinated fibers (e.g., those in the brain and spinal cord; those to skin and skeletal muscle) are white in appearance; those with little or no myelin are gray.

The roles of the neurilemma and myelin are not completely understood; however, there is clearly an interdependence between the neuron and its sheath cells. When an axon is separated from its cell body by a cut, it soon degenerates. A hollow tube of Schwann cells remains, but myelin eventually disappears. As long as the cell body of the neuron has not been injured it is capable of regenerating a new axon. Sprouting begins within a few days following cutting (Fig. 23–5). The growing axon enters the old sheath tube and proceeds along it to its final destination in the central nervous system or periphery. Axons can grow in the absence of sheaths if some conduit is provided for them. They can, for example, be made to grow within sections of blood vessels or extremely fine plastic tubes. The length of time required for regeneration depends on how far the nerve has to grow and may require as much as two years. When cuts occur within the spinal cord or brain, regeneration is very feeble and usually totally absent.

It is a remarkable fact that each regenerating axon of a cut nerve finds its way back to its former point of termination whether this be a specific connection in the central nervous system or a specific muscle or sense organ in the periphery. If, during the early stages of development of an amphibian, one transplants an extra limb bud next to the normally developing limb, both will grow to maturity. The extra limb then moves synchronously with the normal one. Anatomical examination reveals that the nerve that innervates the normal limb sends out branches to the extra one. Clearly the extra limb exerts some stimulating influence on the growing nerves to produce more branches, and some directive influence as well.

Similarly, in the optic system, after the optic nerve is cut the axons of retinal cells regenerate and make proper connections within the central nervous system. Behavioral observations confirm this: When a frog with a regen-

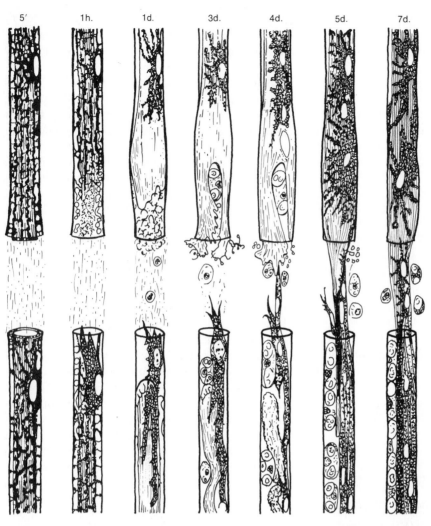

Figure 23–5 Distal part of nerve fiber five minutes after cutting shown at the left. Retraction of the central end of the cut axon one hour later. Budding at third and fourth day. Slender axon grows into the distal degenerated stumps at days 5 and 7. (Courtesy of J. Z. Young. From Young, J. F.: Factors influencing the regeneration of nerves. Adv. Surg. *1*:165, 1949.)

erated optic nerve sees a fly, it strikes at it accurately. This visual-motor performance indicates that there is topological correspondence between points of the retinal area and points in the visual area of the central nervous system. When an excised eye is rotated 180 degrees before reimplantation, the frog strikes 180 degrees off target. This aberration is explained by the fact that the retinal cells regenerate to their proper destinations in the central nervous system but the retina itself is upside down with respect to the central nervous system.

The principal known function of the myelin sheath is to provide for a special kind of nerve conduction (p. 494), but it may have other functions as well. The neuron clearly has an effect on it. When the distant nerve cell body is cut off from its axon, the myelin in the axon begins degenerating within a few minutes. The neuron obviously has some trophic function that is necessary for the well-being of the Schwann cell. As a matter of fact, it is now realized that the nervous system has, in addition to its role of transmitting impulses, important trophic relations with all the organs it innervates. The presence of nerves is essential to the regeneration of amputated amphibian limbs, the normal maintenance of taste buds and the continued functional integrity of muscles. Something other than impulses is obviously transported along the long cellular extensions of the neurons, and it has been demonstrated that there is active flow of cytoplasm away from the cell body.

23–2 SYNAPSES

Since the nervous system is composed of discontinuous units, the neurons, but behaves like a continuous transmission system, there

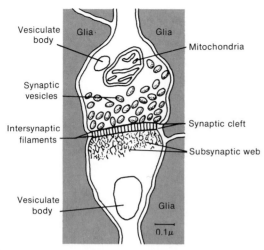

Figure 23–6 A common form of synapse in the mammalian brain. The axonal (presynaptic) side above; the dendritic (postsynaptic) side below. (From DeRobertis, E. D. P.: Ultrastructure and chemical organization of synapses in the central nervous system. *In* Brazier, M. A. B. (ed.): *Brain and Behavior.* Amer. Ass. Adv. Sci., 1962.)

are obviously functional connections between neurons. These functional junctions were termed *synapses* by Sherrington. A synapse is a region where one cell (the *presynaptic*) comes into contact or near contact with, and influences, another cell (the *postsynaptic*) (Fig. 23–6). When there is a space between the two cells it seldom exceeds 50 nm. in width. The synaptic connection involves only limited areas of the neurons involved. In the vertebrate nervous system many synapses are between the telodendria of axons and the cell body of the postsynaptic neuron. In invertebrates the majority of synapses are between axon telodendria and dendrite arborizations. Most synapses transmit excitation in one direction only, i.e., from the pre- to the postsynaptic cell. Characteristically the synaptic region of the presynaptic axon is packed with small rounded bodies (*synaptic vesicles*) 10 to 20 nm. in diameter in vertebrates (Fig. 23–13). The vesicles contain specific chemicals which are released when the axon is excited and transmit the excitation to the postsynaptic cell.

Transmission across the synapse is considerably slower than transmission along the nerve. Impulses normally pass in one direction only: Those in sensory neurons pass from the sense organs to the spinal cord and brain; those in motor neurons, from the brain and spinal cord to the muscles and glands. The synapse controls this, because only the tip of the axon is capable of secreting the chemical which stimulates the next neuron. Any individual nerve fiber can conduct an impulse in either direction; if it is stimulated electrically in the middle, two impulses will be set up, one going in one direction and one in the other (these can be detected by appropriate electrical devices), but only the one going *toward* the *axon* tip can stimulate the next neuron in line. The one going toward the dendrite will stop when it arrives at the tip.

23–3 ACTION POTENTIALS

The study of the nature of the nerve impulse has been fraught with special difficulties because nothing visible occurs when an impulse passes along a nerve. Only with the development of microchemical techniques was it possible to show that the nerve fiber expends more energy, consumes more oxygen and gives off more carbon dioxide and heat when an impulse is transmitted than it does in the resting state. This indicates that metabolic, energy-requiring processes are involved in the conduction of an impulse, or in the recovery of the nerve after conduction, or both.

There is a difference in electrical potential between the inside and the outside of all cells. This can be measured by placing one electrode, insulated except at the tip, inside the cell and a second electrode on the outside surface and connecting the two with a suitable recording instrument such as a galvanometer. The potential difference across the plasma membrane of the neuron is about 60 millivolts, with the inside being negative with respect to the outside. This potential difference is called the *resting potential.* If both electrodes are placed on the outside surface of the neuron, no potential difference between them is registered; all points on the outside are at equal potential.

If the neuron is stimulated electrically, chemically or mechanically, the resting potential changes. When the neuron is unstimulated it is said to be *polarized;* when stimulated it becomes *depolarized.* When the neuron is fully depolarized, the galvanometer would detect no difference in electric potential between the inside and the outside. If the two electrodes are placed at different points on the surface of a

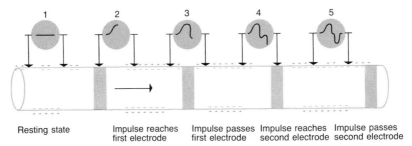

Resting state　　Impulse reaches　Impulse passes　Impulse reaches　Impulse passes
　　　　　　　　first electrode　first electrode　second electrode　second electrode

Figure 23-7 The electrical changes in a nerve fiber as an action potential passes toward the right. Five oscilloscope faces are drawn to illustrate the successive changes when the impulse passes two externally placed electrodes. *1,* Resting state; both electrodes recording same potential, no deflection of line. *2,* Impulse reaches first electrode; depolarization at this point, this electrode negative with respect to other. *3,* Impulse passes, area under first electrode repolarized, second electrode not yet depolarized, line on oscilloscope returns to zero since both electrodes are at equal potential. *4,* Impulse reaches second electrode; this electrode now negative with respect to first since current is flowing in opposite direction. *5,* Areas under both electrodes now repolarized, so line returns to zero. This is a double, or diphasic, action potential.

neuron (or a muscle fiber, or a giant algal cell) and the cell is in the resting state with its membrane polarized, the potential on the outside of the cell is the same all over the surface; no voltage difference is evident between the two electrodes. When the cell is stimulated, the resting potential at the site of stimulation decreases. A voltage difference appears and the electrode near the stimulus becomes negative relative to the one farther away (Fig. 23-7). This is so because the second electrode is on a portion of the neuron that is still polarized (its outside is positive with respect to the inside), while the first electrode is on a spot that has become depolarized; the inside and outside of the neuron are more nearly equal electrically and thus negative with respect to the unstimulated area. This local state of depolarization, the *generator potential,* is the starting point of the *nerve impulse.* If the generator potential is not large enough, enzymatic processes in the neuron repolarize the membrane; the potential differences slowly disappear and nothing is transmitted along the nerve.

However, if the depolarization reaches a critical magnitude within a short enough time, the axon on either side of the area of depolarization becomes stimulated. As a result of this local stimulation, current flows from the two inactive areas to the point of depolarization. The two newly depolarized areas in turn stimulate the inactive areas adjacent to them, and the process passes along the neuron in a chain reaction. This self-generating depolarization reaction passes along the neuron; the propagated wave of depolarization is the *action*

potential. The action potential is fast, lasting about two milliseconds, and is all-or-none; that is, it either fires at maximum voltage or not at all. After the change in potential has occurred, a finite period of recovery, the *absolute refractory period,* must elapse before a second action potential can be generated. The absolute refractory period lasts from 0.5 to 2 milliseconds. Within a neuron an action potential can travel in either direction from its point of origin, but the normal direction is from the dendrite toward the cell body, and away from the cell body in the axon.

Although action potentials have been studied most intensively in neurons and muscle fibers, they have been detected in plant cells, fertilized eggs and a variety of other cells. The height of the *action potential,* its duration and its rate of propagation may vary quite a bit in different kinds of cells. In the giant axon of the squid the action potential is about 90 millivolts, its duration is about 1 millisecond, and it progresses at about 1 meter per second (Fig. 23-8). The rate of propagation of the impulse is 1 cm. or less per second in plant cells, and as much as 120 meters per second in vertebrate neurons. Even the latter is slow, however, compared with the velocity of an electric current in a solution of electrolytes. The velocity of propagation of the action potential is proportional to the distance across which one local response, one change in membrane permeability, can trigger the next adjacent one.

The electrical and chemical processes involved in the transmission of a nerve impulse are similar in many ways to those involved in

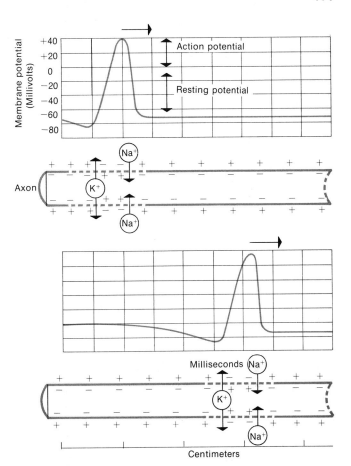

Figure 23–8 The movement of an action potential along a nerve membrane. The wave of depolarization in the membrane is the initial event as the nerve impulse is transmitted along the neuronal membrane.

muscle contraction. Compared with a contracting muscle, however, a transmitting nerve expends little energy; the heat produced by 1 gram of nerve stimulated for 1 minute is equivalent to the energy liberated by the oxidation of 10^{-6} grams of glycogen. This means that if a nerve contained only 1 per cent glycogen to serve as fuel, it could be stimulated continuously for a week or more without exhausting the supply. Nerve fibers are practically incapable of being fatigued as long as an adequate supply of oxygen is available. Whatever "mental fatigue" may be, it is not due to the exhaustion of the energy supply of the nerve fibers.

Changing the concentration of sodium ions on the outside of the membrane leads to changes in the action potential. For example, a decrease in the concentration of sodium ions on the outside of the membrane is accompanied by a decrease in the height of the action potential. To explain these observations, Hodgkin suggested that the change in membrane potential is due to the sudden penetration of sodium ions into the interior of the cell. The

action potential is related to some physical change in the plasma membrane that results in an increased permeability to sodium ions. Direct measurements of Na^+ and K^+, using isotopically labeled ions, showed that the *sodium flux* inward increases some twentyfold during stimulation and the *potassium flux* outward increases about ninefold. Thus the nerve impulse is accompanied by a sudden influx of sodium down an electrochemical gradient, followed by an efflux of potassium down an electrochemical gradient, which abolishes or reverses the resting potential. The actual number of ions that move through the membrane during a 1-millisecond pulse is very small and does not change significantly the concentration of Na^+ or K^+ ions on the two sides of the membrane. The change is detected primarily by its effect on the membrane potential.

By removing the entire cytoplasmic content of a squid axon and filling the axon with solutions of known ionic composition, Baker showed that these axons would respond to stimuli with action potentials of 90 to 130 millivolts for as long as five hours, during which they transmitted several hundred thousand impulses. The axon's only requirement was for potassium salts in the internal medium and sodium salts in the external medium. Further studies showed that as a nerve impulse passes a given point there is first an inward flow of current due to the inward penetration of sodium ions. This is followed by a second, more gradual flow of current outward through the membrane due to the outward diffusion of potassium ions.

To summarize, the spike of an action potential results from the following sequence of events: (1) a sudden increase in the permeability of the membrane to sodium; (2) an inflow of sodium ions driven by the electrochemical gradient (the concentration of sodium outside a neuron is about 137 milliequivalents and the concentration of sodium inside is 10 milliequivalents); (3) this flow causes the inside of the membrane to become positively charged relative to the outside and appears as the ascending portion of the spike; (4) the membrane becomes less permeable to sodium and more permeable to potassium; (5) an outflow of potassium ions along an electrochemical gradient follows (the concentration of potassium inside the human neuron is 141 milliequivalents and its concentration outside is 4 milliequivalents); (6) the outflow of potassium appears as the descending curve of the spike potential.

23–4 THE MEMBRANE THEORY OF NERVE CONDUCTION

Our present knowledge of the nature of the nerve impulse has been derived in large part from experiments using the large axons found in squid, crayfish and certain worms. The giant axon of the squid is large enough, nearly 1 mm.' in diameter, so that investigators can introduce microelectrodes and micropipettes into the substance of the nerve fiber and measure the electrical potential across the nerve membrane.

The *membrane theory* of nerve conduction states that the electrical events in the nerve fiber are governed by the *differential permeability* of the neuronal membrane to sodium and potassium ions, and that these permeabil-

ities are regulated by the *electric field* across the membrane. The interaction of these two factors, differential permeability and electric field, leads to the requirement for a critical threshold of change for excitation to occur. Excitation is a regenerative release of electrical energy from the nerve membrane, and the propagation of this change along the fiber is the *action potential,* the brief, all-or-none electrochemical depolarization of the membrane.

The resting neuron is a long cylindrical tube whose plasma membrane separates two solutions of different chemical composition, though they have the same total number of ions. In the external medium, sodium and chloride ions predominate; whereas within the cell, potassium and various organic ions predominate. In the giant axon of the squid, on which most experiments have been done and which is the prototype for all axons, sodium is some 10 times more concentrated outside the membrane than inside, chloride is 5 to 10 times more concentrated outside than inside, and potassium is 30 to 40 times more concentrated inside than out. The membrane, which is about 5 nm. thick, has high electrical resistance, low selective ionic permeability, and high electrical capacity. Potassium and chloride ions diffuse relatively freely across this membrane, but the permeability to sodium ions is low. Potassium tends to leak out of the neuron and sodium tends to leak in, but because of the selective permeability of the membrane, potassium tends to leak out faster than sodium leaks in. This, plus the fact that the negatively charged organic ions within the cell cannot get out, causes an increasing negative charge on the inside of the membrane. As the inside becomes more negative, it impedes the exit of potassium ions. Ionic conditions would eventually change and come to a new equilibrium if something was not done to counteract this leakage of ions. The steady state is maintained by the *sodium pump,* which actively transports sodium ions from the inside to the outside against a concentration and electrochemical gradient. The sodium pump requires energy, utilizing ATP derived from metabolic processes within the nerve cell. The pump can be turned off experimentally by poisoning the nerve cell with a metabolic inhibitor such as cyanide. The differential distribution of ions on the two sides of the membrane results in a potential difference of 60 to 90 millivolts across the membrane, the *resting membrane potential.*

The extrusion of sodium ions is accompanied by the entrance of potassium ions, and although the details of the process are not yet clear, there appears to be an exchange of cations at the cell surface, with a potassium ion entering for each sodium ion extruded. The relatively low ionic permeability of the membrane is such that even when the pump is poisoned with cyanide, many hours elapse before the concentration gradients of sodium and potassium ions across the membrane disappear.

Electrical studies of the *cable properties* of the nerve fiber show that the axon could hardly serve as a passive transmission line (like a copper wire), because its cable losses are enormous. When a weak signal is applied to the fiber, one too small to excite its usual relay mechanism, the signal fades out within a few millimeters of its origin. The nerve impulse could not be propagated over the long distances in the nerve unless there were some process to boost the signal. The excitatory process regenerates and reamplifies the signal at each point along the nerve fiber. The cable properties of the nerve allow a change in electrical potential to spread along the nerve fiber for a short distance and stimulate the excitatory process in the adjacent portion of the nerve.

The electrochemical potential energy of the resting potential forms the basis for the generation and transmission of the action potential (Fig. 23–9). The permeability of the membrane depends upon the magnitude of the transmembrane electrical potential. Although the permeability of the membrane to sodium is very low at the usual resting membrane potential, the permeability increases as the membrane potential decreases (Fig. 23–9). This permits the leakage of sodium ions down an electrochemical gradient into the interior of the nerve. This further decreases the mem-

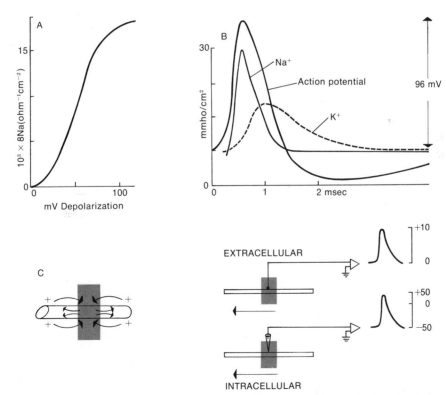

Figure 23–9   Generation of the nerve impulse. *A,* Increase in Na$^+$ permeability as membrane depolarizes. *B,* Na$^+$ and K$^+$ movement across membrane during action potential. *C,* Diagram illustrating flow of electrotonic current in vicinity of action potential (*gray areas*), and examples of extracellular and intracellular recording of action potentials. With a penetrating microelectrode the entire action potential is recorded; whereas with an extracellular electrode only the positive overshoot is detected. (*A* and *B* adapted from Katz, B.: The Croonian Lecture: The transmission of impulses from nerve to muscle, and the subcellular unit of synaptic action. Proc. Roy. Soc. Biol. *155:*455, 1962.)

brane potential and further increases the permeability to sodium. The process thus is self-reinforcing and progressive, resulting in the upward deflection of the action potential. The entering sodium ions drop the transmembrane potential to zero and beyond, to about -40 or -50 millivolts. After one or two milliseconds, the permeability of sodium decreases and potassium begins to move out. This movement leads to the restoration of the resting potential, that is, to repolarization in the membrane. When the membrane is completely repolarized, the permeability to potassium becomes normal and the excess sodium that entered during the process is now slowly removed by the sodium pump. The actual quantities of ions that move in and out during the passage of an action potential are so small that there is no detectable change in the concentration of either ion in the fiber during an impulse.

Because of the changes in permeability that accompany the depolarization of the nerve membrane, the fiber cannot immediately transmit a second impulse. This period of inexcitability, the **absolute refractory period,** is brief and lasts until normal permeability relations have been restored. The nerve impulse, therefore, is a wave of depolarization that passes along the nerve fiber. The change in membrane potential in one region renders the adjacent region more permeable, and the wave of depolarization is transmitted along the fiber. The entire cycle of depolarization and repolarization requires only a few milliseconds.

Propagation of the action potential depends upon electrotonic currents that flow ahead of the nerve impulse. The process has some similarities to the charging and discharging of a condenser. At any point where an action potential has been generated, current flows ahead within the neuron, then through the membrane, and returns on the outer surface (Fig. 23–9C). The circuit is completed by current flowing in the solution that bathes the nerve. If the nerve is immersed in mineral oil, so that only a thin film of conductive fluid bathes the nerve, the external resistance is greatly increased and the rate of conduction of impulses is greatly decreased. The effectiveness of the electrotonic currents in propagating the impulse depends upon the magnitude of the current and the resistances of the neuronal membrane, the cytoplasm and the

surrounding medium. These factors determine at what distance from the active site the membrane permeability will be sufficiently increased to start the regenerative sodium entry. The concept that the action potential is propagated by electrotonic currents moving ahead of it is termed the **local circuit theory of propagation.**

The rate of conduction of impulses increases as the diameter of the axon increases, because the internal resistance (R) decreases. Because of this, large nerve fibers conduct impulses more rapidly than small ones do. Giant nerve fibers have evolved in various members of the animal kingdom, although not in the higher vertebrates. The giant axons of the squid and earthworm conduct impulses very rapidly and have been studied intensively. They generally serve the purpose of conducting danger signals, and for this, speed rather than detailed information is critical.

In vertebrates a high rate of conduction is achieved by a different evolutionary development. A **myelin sheath** surrounds the neuron, and this highly insulating covering prevents the flow of current between the fluid external to the sheath and the fluid within the axon. At the successive **nodes of Ranvier** there are gaps in the sheath where there is no insulation. At these points, free ionic communication between the inside and outside of the membrane is possible. Impulses are generated only at the nodes, and nerve impulses leap from one node to the next (Fig. 23–10). This type of transmission is known as **saltatory conduction.** A myelinated nerve that is only a few micrometers in diameter can conduct impulses at velocities of up to 100 meters per second, compared with velocities of 20 to 50 meters per second in the largest unmyelinated axons, which are 1 mm. in diameter.

Neuronal Functions. The idea that a nerve fiber is simply some sort of cytoplasmic telephone wire has rapidly changed as we have come to appreciate that neurons are very dynamic and metabolically active cells. These cells, many of which are very long, carry out the intracellular transport of materials over long distances by active cytoplasmic streaming and undulations of the axons. Neurons may, in addition, be true secretory cells, secreting hormonal **neurohumors,** such as **vasopressin, oxytocin** and the hypothalamic **releasing factors.**

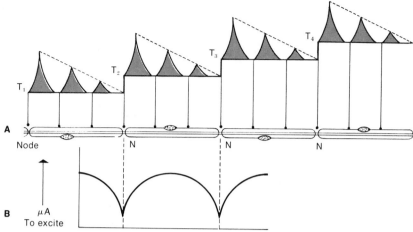

Figure 23–10 *Saltatory conduction in a myelinated axon. A*, Recording at and between successive nodes of Ranvier demonstrates electrotonic transmission between nodes, as impulse appears instantaneously at all points within a node while diminishing in magnitude. At the node a time delay occurs and the impulse regains initial magnitude, showing the node to be the site of the active, impulse-generating process. *B*, Support for the idea that electrotonic current flowing from a previous active site exits and excites at the next node comes from a demonstration that current (μA = microamperes) required to excite the axon is least at the nodes. (From Case, J.: *Sensory Mechanisms.* New York, The Macmillan Company, 1966.)

The regeneration of amputated limbs in salamanders requires specific neuronal influences and will not occur if the nerves are removed. In humans and other mammals, when an axon is severed a new cytoplasmic process grows out of its stump and advances by an amoeboid tip at the rate of 3 or 4 mm. per day. This tip carries on pinocytosis and shows a remarkable ability to find and recognize its proper connection. After this tip has made contact with its proper distal portion, the mass of the axon is brought back to normal by a complex synthetic process which is accompanied by striking morphological changes in the cell body of the neuron. The *cell body* contains an abundant accumulation of endoplasmic reticulum and ribosomes, detected by light microscopy in the nineteenth century and termed *Nissl substance.* The Golgi apparatus of neurons is also abundant and well developed. The cytoplasm of the axon contains longitudinally oriented *neurofibrils* and *neurotubules,* which are associated with the intracellular transport of molecules.

23–5 SYNAPTIC TRANSMISSION

The nervous systems of invertebrate and vertebrate animals are composed of individual discontinuous neurons. This requires some mechanism for the transfer of the neural message from the axon of one neuron to the dendrite of the next neuron or, at the neuromuscular junction, to the muscle. The junction between the axon of one neuron and the dendrite of the next is termed a *synapse.* At some specialized synapses transmission is accomplished electrically. The arrival of the action potential at the end of the axon of the presynaptic neuron sets up electric currents in the external fluid of the synaptic gap, and these currents, in turn, stimulate the dendrite of the postsynaptic cell to generate an action potential. Thus the transmission and the generation of the action potential take place in basically the same way in the postsynaptic cell as in a single nerve fiber. Electrical synapses of this sort are found in parts of the nervous systems of crayfish and fish. Transmission across the giant synapse in an abdominal ganglion of the cord of the crayfish is accomplished by electrical means. The membrane contact of this special synapse is able to act as a rectifier and allows current to pass easily in one direction from the axon of a connector neuron to the dendrite of a motor neuron, but not in the reverse direction.

At most synapses, however, a gap of some 20 nm. separates the two plasma membranes, and the impulse is transmitted across this gap by special chemical transmitters. The classic

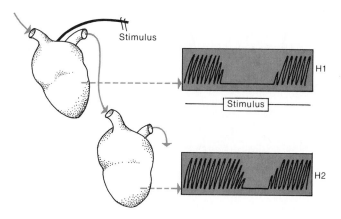

Figure 23–11   A demonstration of junctional chemical transmission. Paired heart experiment showing stimulation of vagus nerve of one heart arrests its beat, as shown in the kymograph recording (*H1*). Action of the vagus nerve must have involved a diffusible substance (subsequently determined to be acetylcholine), because the beat of a second heart (*H2*) is arrested even though its only connection is by way of the perfusion fluid flowing.

paired heart experiment of Loewi provided a striking demonstration of the production of a diffusible substance by a stimulated nerve, which, in turn, can affect muscle. When two isolated hearts were arranged in such a manner that the blood leaving one heart entered the other (Fig. 23–11), stimulation of the vagus nerve of one heart arrested the beating of both hearts. The diffusible substance was subsequently identified as *acetylcholine.* Acetyl-

choline is a potent stimulant causing a local depolarization of the membrane of the muscle cell, which sets up propagated impulses in the membrane and causes the contraction of the muscle fiber (Fig. 23–12). *Curare* prevents the transmission of impulses from nerve to muscle, specifically at this type of synapse, by combining with the receptors for acetylcholine and preventing their normal reaction with it.

Experiments similar to those of Loewi

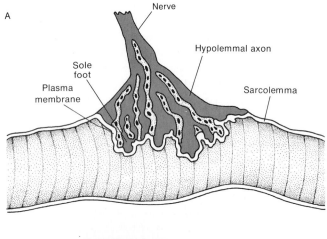

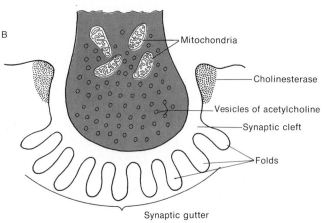

Figure 23–12   *A*, Diagram of a neuromuscular junction. *B*, Enlarged view showing the vesicles of acetylcholine in the sole foot of the nerve which penetrates into the membrane of the muscle fiber. This membrane becomes extensively folded to form the "synaptic gutter" in the plasma membrane of the muscle cell. The transmission of a nerve impulse from one neuron to a second neuron across a synapse is believed to involve a similar structural and functional junction.

have shown that the sympathetic postganglionic fibers accelerate the heart rate by releasing norepinephrine. Such fibers are termed *adrenergic,* whereas those that secrete acetylcholine are termed *cholinergic.* In the synaptic area are potent enzymes: acetylcholinesterase, which specifically hydrolyzes and inactivates acetylcholine, and monoamine oxidase, which oxidizes and inactivates norepinephrine. These enzymes prevent the continuous stimulation of the dendrite or muscle by the neurotransmitter material.

Chemical transmission at the synapse involves the two processes of *neurosecretion* and *chemoreception.* The arrival of a nerve impulse at the tip of the axon stimulates the release of the specific neurotransmitter from its storage place in the tip of the axon into the narrow synaptic space between the adjacent neurons. The specific transmitter substance is then attached to specific molecular sites in the dendrite, producing a change in the properties of the cell membrane so that a new nerve impulse is established. The chemical transmitter passes from axon to dendrite by simple diffusion. Over the short distance involved, diffusion is rapid enough to account for the speed of transmission observed at the synapse.

It has been shown that acetylcholine is released by motor nerves in discrete, tiny packets, each of which contains about 1000 molecules. The mechanism that releases acetylcholine requires calcium ions but is inhibited by magnesium ions. The transmitter substance is stored within the nerve endings in small intracellular structures which discharge their entire contents to the surface. Electron micrographs (Fig. 23-13) of the tips of the neuron at the synapse reveal masses of *synaptic vesicles,* which appear to be the sites of storage of the neurotransmitters. Thus the arrival of the nerve impulse leads to the liberation of the contents of one or more of these vesicles into the synaptic space. How an action potential induces the release of a packet of acetylcholine molecules from its vesicle is still unknown, but the existence of these vesicles provides a satisfactory explanation for the polarity of the synapse (i.e., for the fact that a nerve impulse will travel in one direction but not in the reverse between axon and dendrite). Norepinephrine is also concentrated in synaptic vesicles of adrenergic fibers and is released by the arrival of an action potential. Other neurons may have other neurochemical transmitters, such as serotonin or dopamine.

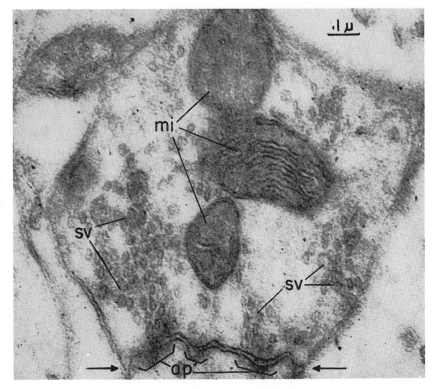

Figure 23-13 Electron micrograph of a synaptic ending in the stimulated olfactory bulb of the rat. The synaptic ending contains three mitochondria (*mi*) and several synaptic vesicles (*sv*). The zone of contact between the two neurons is indicated by the two arrows. The nerve membranes appear to be thickened at "active points" (*ap*) in this zone of contact. (From De Robertis, E., and Pellegrino De Iraldi, A.: Anat. Rec. *139*:299, 1961.)

23–6 SYNAPTIC RESISTANCE

The synaptic junction is a point of resistance to the flow of impulses in the nervous system, and not every impulse reaching the synapse is transmitted to the next neuron. The resistance varies in different synapses, so that they are important in determining the route of impulses through the nervous system and the response of the organism to a specific stimulus.

The entire nervous system is a functional unit, and an impulse arising in any receptor can be transmitted to every effector in the body. Consider, for example, the effects of burning a finger: The muscles of the arm contract to pull the finger away from the heat, a sensation of pain is produced in the brain, a cry may be emitted, the heart beat, digestion and breathing may be altered—in fact, it is conceivable that every muscle and gland in the body may be affected temporarily. Our sense organs receive a constant stream of stimuli, but selective resistance at the synaptic junction prevents the uncontrolled, continuous contraction of muscles and secretion by glands. The drug *strychnine* decreases synaptic resistance; in a person suffering from strychnine poisoning, the slightest stimulus sets off the secretion of all glands and the convulsive contraction of all muscles of the body.

The amount of synaptic resistance can be modified by nerve impulses. One impulse might cancel out the effect of another, a process known as *inhibition.* The opposite condition, whereby one impulse strengthens another, is called *facilitation.* These two processes are of prime importance in effecting integration of body activities. We have seen that all the muscles of the body are in a state of constant, slight contraction, known as tonus, owing to the constant volley of nerve impulses reaching them. But for one muscle, such as the triceps, to contract, its antagonist, the biceps, must relax. This is achieved by impulses that inhibit the volley of impulses to the biceps and by impulses that reinforce the volley to the triceps. Inhibition and facilitation can occur only at the synapse, since once an impulse starts along a neuron it can be neither stopped nor accelerated. Information transmitted along the axon as a spike potential is coded and decoded at the synapse by processes involving neurosecretion and chemoreception, graded

thresholds, and the facilitation and summation of inhibitory and excitatory impulses. These are the substrata of animal behavior and the basis of the complex human processes of learning, memory and intelligence.

23–7 THE CENTRAL NERVOUS SYSTEM: SPINAL CORD

The ten billion or so neurons that make up the nervous system of man are divided into two main parts: those belonging to the **central nervous system,** which make up the brain and spinal cord, and those belonging to the **peripheral nervous system,** which make up the cranial and spinal nerves.

The tubular **spinal cord,** surrounded and protected by the neural arches of the vertebrae, has two important functions: to transmit impulses to and from the brain and to act as a reflex center. In cross section, two regions are evident: an inner, butterfly-shaped mass of **gray matter,** made up of nerve cell bodies; and an outer mass of **white matter,** made up of bundles of axons and dendrites (Fig. 23–14). The whiteness of these bundles is due to the myelin sheaths of the axons and dendrites; the ends of the axons and dendrites, present in the central gray matter, have no myelin sheath. The "wings" of the gray matter are divided into two dorsal horns and two ventral horns. The latter contain the cell bodies of motor neurons, whose axons pass out through the spinal nerves to the muscles; all the other neurons in the spinal cord are interneurons.

The axons and dendrites of the white matter are segregated into bundles with similar functions: the **ascending tracts,** transmitting impulses to the brain; and the **descending tracts,** carrying impulses from the brain to the effectors. Neurologists have carefully noted the symptoms of patients with injured spinal cords and later correlated these with the particular tracts found to be destroyed when the patient's nervous system was examined after death. From these observations they have been able to map out the location and functions of the various tracts (Fig. 23–14). For example, the dorsal columns of the white matter transmit impulses originating in the sense organs of muscles, tendons and joints, by means of which we

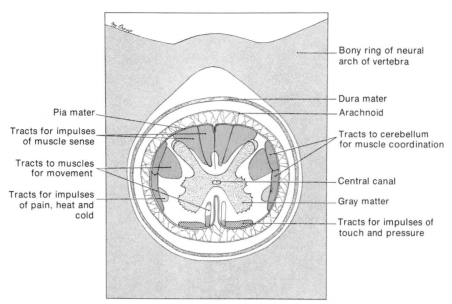

Figure 23-14 Cross section of the spinal cord surrounded by the bony vertebra, showing the meninges (dura mater, pia mater and arachnoid), the gray matter, and some of the important nerve tracts of the white matter.

are aware of the position of the parts of the body. In advanced syphilis these columns may be destroyed, so that the patient cannot tell where his arms and legs are unless he looks at them, and he must watch his feet in order to walk.

One curious fact, still not satisfactorily explained, emerged from these studies of the location and function of the fiber tracts. All the fibers in the spinal cord *cross over* from one side of the body to the other somewhere along their path from sense organ to brain, or from brain to muscle. Thus the right side of the brain controls the left side of the body and receives impressions from the sense organs of the left side. Some fibers cross in the spinal cord itself; others cross in the brain.

In the center of the gray matter is a small canal that extends the entire length of the neural tube and is filled with *cerebrospinal fluid.* This fluid is similar to plasma but contains much less protein. The spinal cord and brain are wrapped in three sheets of connective tissue, known as *meninges.* Meningitis is a disease in which these wrappings become infected and inflamed. One of these sheets (*dura mater*) is fastened against the bony neural arches of the vertebrae; another (*pia mater*) is located on the surface of the spinal cord, and the third (*arachnoid*) lies between. The spaces between the meninges are filled with more cerebrospinal fluid, so that the spinal cord (and

the brain) floats in this liquid and is protected from bouncing against the bone of the vertebrae or the skull with every movement.

23–8 THE CENTRAL NERVOUS SYSTEM: THE BRAIN

The *brain* is the enlarged anterior end of the spinal cord. In the human the enlargement is so great that the resemblance to the spinal cord is obscured, but in the lower animals the relationship of brain to spinal cord is clear. Embryologically the brain develops from three primary swellings in the anterior end of the neural tube. These give rise to the forebrain, midbrain and hindbrain. The fore- and hindbrains further subdivide, so that the adult brain has six major regions: the medulla, pons and cerebellum in the hindbrain; the midbrain; and the thalamus and cerebrum in the forebrain (Fig. 23–15).

The most posterior part of the brain, lying next to the spinal cord, is the *medulla.* Here the central canal of the spinal cord enlarges to form a cavity called the *fourth ventricle* (three other ventricles, or cavities, lie farther up in the brain). The roof of the fourth ventricle is thin and contains a cluster of blood vessels which secrete part of the cerebrospinal fluid; the rest of this fluid is secreted by similar clusters of blood vessels in the other ventricles. In the roof of

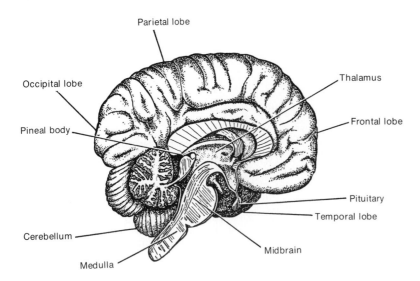

Parietal lobe

Occipital lobe

Pineal body

Thalamus

Frontal lobe

Cerebellum

Medulla

Pituitary

Temporal lobe

Midbrain

Figure 23–15 The parts of the human brain seen in a median sagittal section.

the fourth ventricle are three tiny pores through which cerebrospinal fluid escapes into the meningeal spaces. The walls of the medulla are thick and made up largely of nerve tracts connecting with the higher parts of the brain. The medulla also contains a number of clusters of nerve cell bodies, the *nerve centers.* These reflex centers control respiration, heart rate, the dilatation and constriction of blood vessels, swallowing and vomiting.

Above the medulla is the *cerebellum,* consisting of a central part and two hemispheres extending laterally, which resemble pine cones in shape. Its gray surface is made up of the cell bodies of neurons, and beneath is a mass of white tissue composed of fiber tracts connecting with the medulla and with the higher parts of the brain. The size of the cerebellum in different animals is roughly correlated with the amount of their muscular activity. It regulates and coordinates muscle contraction and is proportionately large in extremely active animals, such as birds. Removal or injury of the cerebellum is accompanied not by paralysis but by impairment of muscle coordination. A bird without a cerebellum is unable to fly and its wings thrash about jerkily. When the human cerebellum is injured by a blow or disease, all muscular movements are uncoordinated. Any activity requiring delicate coordination, such as threading a needle, is impossible.

Running crosswise on the ventral side of the brain just below the cerebellum is a thick bundle of fibers known as the *pons,* or bridge, which carries impulses from one hemisphere of the cerebellum to the other, thus coordinat-

ing muscle movements on the two sides of the body.

In front of the cerebellum and pons lies the thick-walled *midbrain.* It contains a small central canal connecting the fourth ventricle of the medulla with the third ventricle in the thalamus. In the thick walls of the midbrain are certain reflex centers and the main fiber tracts leading to the thalamus and cerebrum. On the upper side of the midbrain are four low, rounded protuberances (*corpora quadrigemina*) containing centers for certain visual and auditory reflexes. The reflex constriction of the pupil when light shines on the eye and the pricking up of a dog's ears in response to a sound are controlled by reflex centers there. The midbrain also contains a cluster of nerve cells regulating muscle tonus and posture.

In front of the midbrain the central canal again widens and becomes the *third ventricle,* the roof of which contains another cluster of blood vessels secreting cerebrospinal fluid. The thick walls of the third ventricle are called the *thalamus.* This is a relay center for sensory impulses; fibers from the spinal cord and lower parts of the brain synapse here with other neurons going to the various sensory areas of the cerebrum. The thalamus appears to regulate and coordinate the external manifestations of emotions; thus, by stimulating the thalamus, a sham rage can be produced in a cat—the hair stands on end, the claws protrude, the back becomes humped, and other signs of anger are evinced. However, as soon as the stimulation stops, the appearance of rage ceases.

In the floor of the third ventricle (the

hypothalamus) are centers regulating body temperature, appetite, water balance, carbohydrate and fat metabolism, blood pressure and sleep. Curiously, the front part of the hypothalamus prevents a rise in body temperature and the rear part prevents a fall. The hypothalamus controls certain functions of the anterior lobe of the pituitary, such as the secretion of gonadotropins, by producing "releasing factors" (p. 567); it also produces the hormones oxytocin and vasopressin, released in the posterior lobe of the pituitary.

The parts of the brain considered so far have to do with unlearned, automatic behavior determined by the fundamental structure of these parts, a structure which is essentially the same from fish to human. The *cerebral hemispheres*, the largest and most anterior part of the human brain, have a basically different function, that of controlling learned behavior. The complex psychologic phenomena of consciousness, intelligence, memory, insight and the interpretation of sensations have their physiologic basis in the activities of the neurons of the cerebral hemispheres. The importance of the cerebrum to different animals can be investigated by removing it surgically. A cerebrumless frog behaves almost exactly like a normal one, and a pigeon whose cerebral cortex has been removed can fly and balance on a perch, but tends to remain quiet for hours. When stimulated, it moves about, though in a random, purposeless way, and it fails to eat when given food. A dog whose cerebral cortex has been removed can walk and will swallow food placed in its mouth, but shows no signs of fear or excitement. Human infants occasionally are born whose cerebral cortex fails to develop, and although they can carry out the vegetative functions of breathing and swallowing, they are incapable of learning and make no voluntary movements. They usually die soon after birth.

The cerebrum contains slightly more than half of the ten billion neurons of the human nervous system. The cerebral hemispheres develop as outgrowths of the anterior end of the brain. In human beings and other mammals they grow back over the rest of the brain and hide it from view. Each hemisphere contains a cavity, the first and the second ventricle, and each of these is connected to the third ventricle of the thalamus by a canal. These ventricles, like the others, contain clusters of blood vessels which secrete cerebrospinal fluid. The cerebrum is made of both gray and white matter; the latter, composed of tracts of nerve fibers, is on the inside of the cerebrum, while the gray matter, made of nerve cell bodies, lies on the surface, or *cortex*, of the cerebrum. Deep in the substance of the cerebral hemispheres lie other masses of gray matter, nerve centers which act as relay stations to and from the cortex. The lower vertebrates, with little gray matter, have smooth cerebral cortices, but in man and other mammals the surface of the

Figure 23–16 The right cerebral hemisphere of the human brain, seen from the side. The stippled areas are regions of special function; the light areas are "association areas." *Inset:* Enlarged view of the sensory and motor areas adjacent to the fissure of Rolando, showing the location of the nerve cells supplying the various parts of the body.

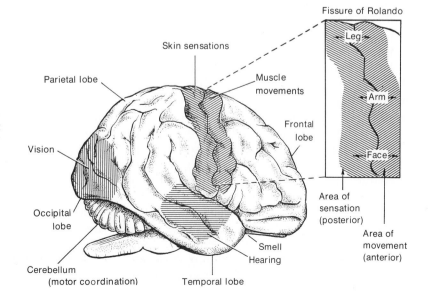

cerebral hemispheres is convoluted. Ridges separated by furrows increase the amount of space available for the cortical gray matter. The pattern of these convolutions is quite constant even in humans of widely different degrees of intelligence, and the geography of the cerebral cortex has been carefully studied (Fig. 23–16). The idea that certain parts of the brain have special functions is an old one, the "science" of phrenology having been based on the premise that functions were localized in the brain. A specially gifted person was believed to have an enlargement in a particular area of the brain and a corresponding bump on the skull. It was believed that an analysis of such bumps would indicate what an individual was best fitted for.

Experimental evidence has established that there is a considerable amount of localization of function in the cortex. By surgically removing particular regions of the cortex from experimental animals, it has been possible to localize many functions exactly; and by observing the paralysis or loss of sensation in a patient with a brain injury or tumor and then examining the brain after death to see where the injury was located, it has been possible to map the human brain. During operations on the brain, surgeons have electrically stimulated small regions and observed which muscles contracted, and since brain surgery can be carried on under local anesthesia, the patient could be asked what sensations were felt when a particular region was stimulated. Curiously, the brain itself has no nerve endings for pain, so that stimulation of the cortex is not painful. Brain activity may be studied by measuring and recording the electrical potentials or "brain waves" given off by various parts of the brain when active.

By combining the data obtained in several ways, investigators have been able to locate many functions of the brain (Fig. 23–16). The posterior part contains the visual center; its removal causes blindness, and stimulation of it, even by a blow on the back of the head, causes the sensation of light. Removal of the region from one side of the brain causes blindness in half of each eye, for the nerves from each eye split, and half go to each side of the brain. The center for hearing is located on the side of the brain, above the ear. Its stimulation by a blow causes a sensation of noise. Although removal of both auditory areas causes deafness, removal

of one does not cause deafness in one ear but a decrease in the auditory acuity of both ears.

Running down the side of the cerebral cortex is an easily recognizable, deep furrow, called the *fissure of Rolando.* This separates the motor area, controlling the skeletal muscles, from the area just behind the furrow, which is responsible for the sensations of heat, cold, touch and pressure from stimulation of sense organs in the skin. In both areas there is a further specialization along the furrow from the top of the brain to the side—neurons at the top of the cortex control the muscles of the feet; the neurons next in line control those of the shank, thigh, abdomen, and so on; and the neurons farthest around to the side control the muscles of the face. The size of the motor area in the brain for any given part of the body is proportional not to the amount of muscle but to the elaborateness and intricacies of movement; thus there are large areas for the control of the hand and face. There is a similar relationship between the parts of the sensory area and the region of the skin from which it receives impulses. Thus, in the connections between the body and the brain, there is not only a twisting of the fibers so that one side of the brain controls the opposite side of the body, but a further "reversal" makes the uppermost part of the cortex control the lower extremities of the body.

When all the areas of known function are plotted, they cover almost all of the rat's cortex, a large part of the dog's, a moderate amount of the monkey's, but only a small part of the total surface of the human cortex. The rest, known as *association areas,* is made up of neurons that are not directly connected to sense organs or muscles but supply interconnections between the other areas. These regions are responsible for the higher intellectual faculties of memory, reasoning, learning and imagination and for personality. In some way, the association regions integrate into a meaningful unit all the diverse impulses constantly reaching the brain, so that the proper response is made. They interpret and manipulate the symbols and words by means of which our thought processes are carried on. When disease or accident destroys the functioning of one or more association areas, the condition known as *aphasia* may result in which the ability to recognize certain kinds of symbols is lost. For example, the names of objects may be forgotten, although their functions are remembered and understood.

23–9 BRAIN WAVES

Metabolism is invariably accompanied by electrical changes, and the electrical activity of the brain can be recorded by a device known as the *electroencephalograph.* To obtain recordings, electrodes are taped to different parts of the scalp and the activity of the underlying parts of the cortex is measured. The electroencephalograph shows that the brain is continuously active. The most regular manifestations of activity, called *alpha waves,* come from the visual areas at the back of the brain when the subject is resting quietly with his eyes shut. These waves occur rhythmically at the rate of 9 or 10 per second and have a potential of about 45 microvolts (Fig. 23–17). When the eyes are opened, the waves disappear and are replaced by more rapid, irregular waves. That the latter are produced by objects seen can be demonstrated by presenting the eyes with some regular stimulus, such as a light blinking at regular intervals, and observing that brain waves with a similar rhythm appear. Sleep is the only normal condition in which the brain waves are drastically altered. During sleep the waves become slower and larger (have a greater potential) as the subject falls into deeper and deeper unconsciousness. The dreams of a sleeping subject are mirrored in flurries of irregular waves.

Certain brain diseases alter the character of the waves; epileptics, for example, exhibit a distinctive, readily recognizable wave pattern, and even people who have never had an epileptic attack, but might under certain conditions, show similar abnormalities. The location of brain tumors can be detected by noting the part of the brain showing abnormal waves.

23–10 SLEEP

The neural mechanisms involved in sleep are unknown and investigators are still trying to discover why sleep is necessary. Sleep is characterized by decreased electrical activity of the cerebral cortex (Fig. 23–17) and this might be correlated with its recuperative effect on the nervous system. Only the higher vertebrates with fairly well developed cerebral cortices sleep, and those with larger hemispheres seem to require more sleep than others. Fatigue is popularly considered the cause of sleep, but there is no experimental evidence to verify this belief. An important sleep-inducing factor is the absence of stimuli; it is

Figure 23–17 Electroencephalograms made while the subject was excited, relaxed, and in various stages of sleep. Recordings made during excitement show brain waves that are rapid and of small amplitude, whereas in sleep the waves are much slower and of greater amplitude. The regular waves characteristic of the relaxed state are called alpha waves. (From Jasper. *In* Penfield and Erickson: *Epilepsy and Cerebral Localization.*)

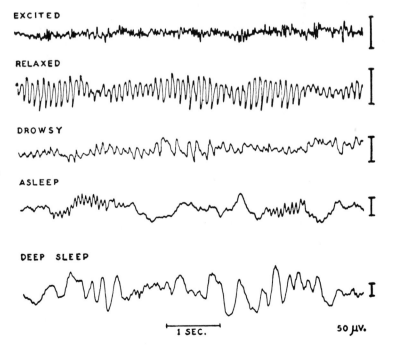

EXCITED

RELAXED

DROWSY

ASLEEP

DEEP SLEEP

1 SEC.

50 μV.

easy to go to sleep, even when one is not particularly tired, if there is nothing interesting to occupy the mind. But although we tend to be wakeful in the presence of attention-holding stimuli, there is a limit beyond which sleep is inevitable. For all the higher animals, life is characterized by a basic rhythm of sleep alternating with wakefulness, a pattern regulated by the hypothalamus. There is a sleep center in the anterior part and a wakefulness center in the posterior part of the hypothalamus. It is believed that the change from wakefulness to sleep and back is controlled by "feedback" circuits involving these two centers.

A sleep-inducing substance is synthesized by the brain, apparently by the thalamus, when an animal is deprived of sleep. It has been extracted both from cerebrospinal fluid and from the brains of sleep-deprived animals. The sleep-inducing substance appears to be a short peptide, with a molecular weight of about 500.

23–11 NEUROSES AND PSYCHOSES

Certain types of brain derangements have an easily understood basis in damage to the brain tissue produced by disease or a wound. If the pores in the roof of the fourth ventricle become clogged, trapping the cerebrospinal fluid within the ventricles, the pressure of this fluid inside the brain will gradually destroy the tissue there. Or a blood vessel in the meninges covering the brain may rupture and the pressure of the accumulated blood will destroy parts of the brain. Tumors and infectious diseases such as syphilis can damage tissue; the actual symptoms—paralysis, loss of sensation or of other functions—depend on which part of the brain is affected.

The causes of other types of disorders, the so-called functional disorders—psychoneuroses and psychoses—are more baffling, for they occur without any structural or chemical change in the brain that pathologists have so far been able to detect. Typically these involve emotional disturbances rather than changes in intelligence.

Psychoneuroses are comparatively common, mild disorders with a great variety of symptoms: anxiety, fear, shyness and hypersensitivity. The emotional upsets may actually produce organic disorders, such as irregular heart beat or digestive disturbances. The cause of this type of mental disorder is not positively known, and there is evidence for believing that the cause differs from person to person and is complex in every instance. The most widely accepted theory at present states that psychoneuroses are due to deep-seated emotional conflicts. Heredity, present environment, past experience and general health may play a part in causing the condition. The patient is usually completely unaware of the cause or causes of his unhappiness. There is no single cure for the various psychoneuroses; many of them respond to psychiatric treatment, by means of which the reason for the sense of anxiety, guilt, conflict or fear is brought to the patient's attention. Other psychoneuroses disappear gradually for no apparent reason, others become increasingly worse, and a few develop into the more serious psychoses.

Psychoses are the severe mental diseases, which usually require hospitalization because the patient may be a danger to himself or to others. There are three main types of psychoses, each of which represents an exaggeration of normal tendencies. *Manic-depressive psychoses* are characterized by an alternation of excessive elation and depression, sometimes accompanied by delusions and hallucinations. Most manic-depressives are normal for most of their lives but suffer recurrent episodes of insanity. *Paranoia* is a psychosis characterized by delusions typically of grandeur or persecution. *Dementia praecox*, or *schizophrenia*, is marked by a withdrawal from the everyday world into a world of daydreams which to the patient become reality. Most psychotic patients have symptoms which represent combinations of these types.

Psychoses are more difficult to cure than psychoneuroses, but encouraging results have been obtained in recent years with psychotherapy. One of the most drastic therapeutic methods is shock treatment, based on the theory that psychotics actually can be jolted back into sanity. Violent fits or comas are produced by the injection of insulin or by the application of electric currents. The drawbacks to such treatments are many, and the neural mechanisms underlying the results are not clearly understood, but a number of patients have been cured by some variation of shock treatment. Treatment with tranquilizing drugs such as chlorpromazine has been successful in many cases and is gradually supplanting shock treatment.

23–12 THE PERIPHERAL NERVOUS SYSTEM

Emerging from the brain and spinal cord and connecting them with every receptor and effector in the body are the paired cranial and spinal nerves; these make up the peripheral nervous system. Cranial and spinal nerves are made of bundles of nerve fibers—axons and dendrites. The only nerve cell bodies present in the peripheral nervous system are those of the sensory neurons, aggregated into clusters known as *ganglia,* near the brain or spinal cord, and those of certain motor neurons of the autonomic system discussed further on.

Cranial Nerves. Twelve pairs of nerves originate in different parts of the human brain and innervate primarily the sense organs, muscles and glands of the head. The same twelve pairs, innervating similar structures, are found in all the higher vertebrates—reptiles, birds and mammals; fish and amphibia have only the first ten pairs. Like all nerves, these are com-

posed of neurons; some have only sensory neurons (nerves I, II, and VIII), some are composed almost completely of motor neurons (III, IV, VI, XI and XII), and the others are made up of both sensory and motor neurons. The names and structures innervated by the cranial nerves are given in Table 23–1. One of the most important cranial nerves, the *vagus,* forms part of the autonomic system and innervates the internal organs of the chest and the upper abdomen.

Spinal Nerves. All the spinal nerves are mixed nerves, having motor and sensory components in roughly equal amounts. In human beings they originate from the spinal cord in 31 symmetrical pairs, each of which innervates the receptors and effectors of one region of the body. Each nerve emerges from the spinal cord as two strands or roots which unite shortly to form the spinal nerves. All the sensory neurons *enter* the cord through the *dorsal root* and all motor fibers *leave* the cord through the *ventral root* (Fig. 23–2). If the dorsal root is severed, the part of the body innervated by that nerve suffers complete loss of sensation without any paralysis of the muscles. If the ventral root is cut, there is complete paralysis of the muscles innervated by that nerve, but the senses of touch, pressure, temperature, kinesthesis and

Table 23–1 THE HUMAN CRANIAL NERVES

Number	Name	Origin of Sensory Fibers	Effector Innervated by Motor Fibers
I	Olfactory	Olfactory mucosa of nose (smell)	None
II	Optic	Retina of eye (vision)	None
III	Oculomotor	Proprioceptors of eyeball muscles (muscle sense)	Muscles that move eyeball (with IV and VI); muscles that change shape of lens; muscles that constrict pupil
IV	Trochlear	Proprioceptors of eyeball muscles (muscle sense)	Other muscles that move eyeball
V	Trigeminal	Teeth and skin of face	Some of muscles used in chewing
VI	Abducens	Proprioceptors of eyeball muscles (muscle sense)	Other muscles that move eyeball
VII	Facial	Taste buds of anterior part of tongue	Muscles of the face; submaxillary and sublingual glands
VIII	Auditory	Cochlea (hearing) and semicircular canals (senses of movement, balance and rotation)	None
IX	Glossopharyngeal	Taste buds of posterior third of tongue, lining of pharynx	Parotid gland; muscles of pharynx used in swallowing
X	Vagus	Nerve endings in many of the internal organs—lungs, stomach, aorta, larynx	Parasympathetic fibers to heart, stomach, small intestine, larynx, esophagus
XI	Spinal accessory	Muscles of shoulder (muscle sense)	Muscles of shoulder
XII	Hypoglossal	Muscles of tongue (muscle sense)	Muscles of tongue

pain are unimpaired. The size of each spinal nerve is related to the size of the body area it innervates; the largest in man is one of the pairs supplying the legs. Shortly beyond the junction of the dorsal and ventral root, each spinal nerve divides into three branches: the *dorsal branch,* serving the skin and muscles of the back; the *ventral branch,* serving the skin and muscles of the sides and belly; and the *autonomic branch,* serving the viscera (Fig. 23–2).

23–13 REFLEXES AND REFLEX ARCS

A reflex is an innate, stereotyped, automatic response to a given stimulus that depends only on the anatomic relationships of the neurons involved. A reflex typically involves part of the body rather than the whole. Flexion of the leg in response to a painful stimulus and constriction of the pupil in bright light are typical reflexes. Reflexes are the functional units of the nervous system, and most of our activities are the result of them. We have already seen how important reflexes are in controlling heart rate, blood pressure, breathing, salivation, movements of the digestive tract and so forth. When we step on something sharp or come in contact with something hot, we do not wait until the pain is experienced by the brain and then after deliberation decide what to do; our responses are immediate and automatic. The foot or hand is being withdrawn by reflex action *before* pain is experienced. Many of the more complicated activities of our daily lives, such as walking, are regulated to a large extent by reflexes. Those present at birth, and common to all human beings, are called *inherited reflexes;* others, acquired later as the result of experience, are called *conditioned reflexes.*

The minimum anatomical requirements for reflex behavior are a sensory neuron with a receptor to detect the stimulus, connected by a synapse to a motor neuron, which is attached to a muscle or some other effector. This, the simplest type of *reflex arc,* is termed *monosynaptic* because there is only one synapse between the sensory and motor neurons. Most reflex arcs include one or more interneurons between the sensory and motor neurons (Fig. 23–18). A *simple reflex,* in which stimulation of a receptor produces contraction of a single muscle, is typified by the "knee jerk." When the tendon of the knee cap is tapped, and thereby stretched, receptors in the tendon are stimulated, an impulse travels over the reflex arc up to the spinal cord and down again, and the muscle attached to the tendon contracts, resulting in a sudden straightening of the leg.

Many reflexes involve the connections of many interneurons in the spinal cord. Not only does the spinal cord function in the conduction of impulses to and from the brain; it also plays an important role in the integration of reflex behavior. Its importance in this respect is readily demonstrated in a "spinal" animal, an animal whose brain has been destroyed or removed. An experiment demonstrating this consists of removing the brain of a frog while leaving the spinal cord intact, and then applying a piece of acid-soaked paper to the animal's back. No matter how many times the piece of paper is placed on the skin, one leg will invariably come up and flick it away. This response, involving many muscles working in a coordinated fashion, is purely reflex and clearly demonstrates one of the chief characteristics of a reflex: fidelity of repetition. A frog with a brain might make the response two or

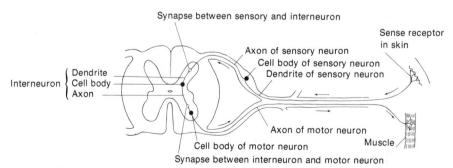

Figure 23–18 Diagram of a reflex arc showing the pathway of an impulse, indicated by arrows.

three times, but eventually it would do something else — perhaps hop away. Most reflexes have some survival value to the animal; the anatomic configuration responsible for the reflex was selected in evolution because of this survival value.

The degree of complexity of a reflex varies with the number of spinal segments involved. The entire reflex arc may be contained in a single spinal segment. Many reflexes, however, require interneuronal connections between two or more segments of the spinal cord. The scratch reflex, the righting reflex and walking reflexes involve the input of sensory information from many segments of the cord. Complex reflexes may even involve certain parts of the brain, namely the midbrain and medulla.

23–14 REFLEXES AND BEHAVIOR

The behavior of a newborn child is determined largely by his innate reflexes, but as he grows older, the relationship between stimulus and response may be altered, and a new stimulus substituted for the old one in eliciting a response. The pattern of behavior is determined by the capabilities of the organism's receptor, effector and nervous systems, which have in turn been determined by evolution. What happens in response to a change in the environmental variable, the stimulus, depends upon how the receptor, effector and nervous systems are interconnected and integrated. Their organization may be relatively direct so that the stimulus triggers a simple set of muscle actions that proceed automatically once

started. As muscular action proceeds, it may be constantly monitored by other stimuli that it has generated in the body, which therefore steer and guide it. Alternatively, the stimulus may trigger a complex response that is completely programmed genetically in the central nervous system. The response mechanism in other behavioral systems may involve both learned and genetically programmed components.

In the simplest sort of response, the stimulus triggers a muscle or a set of muscles, which respond in a predictable, unvarying fashion. This behavior is determined only by the presence of the appropriate sense organs connected with a particular set of muscles. In the knee jerk or *stretch reflex,* a tap on the tendon below the kneecap stretches the attached muscle. The stretching stimulates the spindle organ (Fig. 23–19). Impulses from this receptor neuron are conducted along its axon to the spinal cord, where they pass by a synapse to a motor neuron that stimulates the muscle to contract, producing the knee jerk. This, an example of a *monosynaptic reflex arc,* is behavior dependent simply upon the presence of a stretch receptor connected with a specific muscle.

The nature of the reflex response may be altered by the characteristics of the synapses connecting the incoming and outgoing segments of the circuit. There is delay at the synapse, and the stronger the stimulus, the shorter is the delay. This relates the onset of the behavior to the strength of the stimulus. In addition, the phenomenon of *after-discharge* may appear: the synapse may continue to be

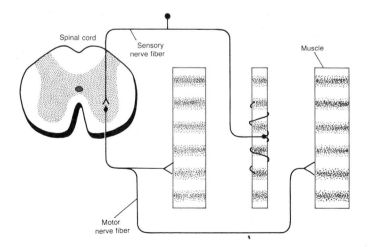

Figure 23–19 A diagram of the anatomy of the stretch reflex, a monosynaptic reflex arc. (After Van der Kloot, W. G.: *Behavior.* New York, Holt, Rinehart and Winston, Inc., 1968.)

active after the stimulus has ceased. A cockroach stimulated to run in response to a touch on its anal cercus may continue to run for some seconds after the stimulus has ceased. The continued running is due in part to after-discharge at a synapse in the ventral nerve cord.

Synapses also exhibit the phenomenon of *temporal summation.* If an investigator applies to a fly's mouthparts a sugar solution that is concentrated enough to stimulate taste hairs but too dilute to evoke a behavioral response, the proboscis is not extended. However, if this stimulation is quickly followed by another weak stimulation (which would not itself evoke a response), proboscis extension does occur.

When two or more neurons synapse with the same postsynaptic neuron, there may be *spatial summation.* Two incoming impulses, one in each presynaptic neuron and each too weak to trigger the discharge of the motor neuron, may combine and be strong enough to be effective.

Both temporal and spatial summation may be operative in the scratch reflex of a dog. When any part of a saddle-shaped area of the skin on a dog's body is rubbed, shocked or mechanically stimulated, the dog will raise the hind leg on the appropriate side and scratch its

back. The leg moves back and forth rhythmically four times per second regardless of the frequency of the stimulus. If a weak electric shock is repeated 18 times per second, there is no response. However, after 44 stimuli have been received, a response occurs—an example of temporal summation. If a shock too weak to elicit a response (a subthreshold stimulus) is given, and several other weak shocks are applied simultaneously to other parts of the skin, a response will occur. This is an example of spatial summation.

The reflex arc underlying the flexion reflex involves two synapses—one between the sensory neuron and interneuron and one between interneuron and motor neuron (Fig. 23–20). The interneuron is inhibitory, and when stimulated by sensory input from one muscle, it inhibits activity in the antagonistic muscle. Considerable complexity of behavior can be developed when the number of synapses in the reflex arc is increased. For example, in the flexion reflex, different responses can be elicited as the strength of the stimulus is increased. In experiments with dogs in which the connection between the spinal cord and brain has been cut, a weak noxious stimulus applied to the foot pad results in withdrawal of the foot.

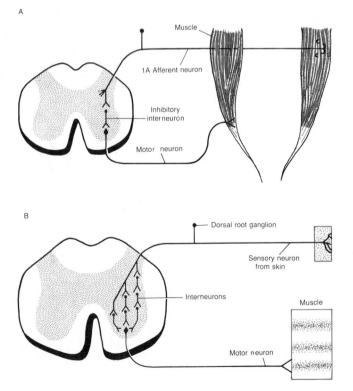

Figure 23–20 *A,* The most direct nerve pathway between a stretch receptor and a motor neuron innervating the antagonist muscle. *B,* A polysynaptic reflex arc between sensory neurons in the skin and motor neurons. (After Van der Kloot, W. G.: *Behavior.* New York, Holt, Rinehart and Winston, Inc., 1968.)

An increase in stimulus strength causes a strong flexion of the lower leg. A stronger stimulus elicits flexion of the upper leg as well. This spreading of response with increase in stimulus strength is termed *irradiation.*

Although one might, in theory, construct a responsive animal simply out of reflexes, reflexes in the higher animals account for only a small portion of total behavior. In some of the simpler animals, however, reflexes play a very important role, and consequently the behavior of these animals is relatively simple, rigid and stereotyped. Much of the behavior of starfish can be explained in terms of reflex arcs. They can extend and retract their tube feet and make postural movements associated with ambulation. The extension and retraction of the tube feet are unoriented reflex responses. The apparent coordination of all the tube feet retracting in unison when a wave washes over the animal is simply the sum of individual responses to a common stimulus. Numerous aspects of crustacean behavior are primarily reflex responses — the withdrawal of the eye stalk, the opening and closing of the claws, and the movements associated with escape, defense, feeding and copulation.

When reflex arcs are connected together, a capacity for the constant monitoring of behavioral responses is introduced. In a complex system the component units of a series of reflex actions may depend on a succession of different stimuli, as illustrated by the catching of prey by the praying mantis. When a fly appears in the visual field of a mantis in ambush, the mantis turns its head to face the fly, for its eyes are immovable in its head. The rest of the body is then usually brought into line with the head, although this alignment is not necessary for a successful strike. When a fly is close enough, either through its own action or through careful stalking by the mantis, the mantis makes a lightning strike with its front legs. The strike is completed in 10 to 30 milliseconds; hence, all the information necessary to direct a strike must be acquired before the action takes place. The mantis must have information about the position of the fly relative to its head and the position of its head relative to its prothorax. The first is provided by the eyes and the second by proprioceptors in the neck. A normal mantis actually hits about 85 per cent of the flies at which it strikes. If information from the proprioceptors is eliminated by cutting the sensory nerve, hitting accuracy drops to about 25 per cent. If the head is fastened to the thorax in the median position, performance remains normal; but if the head is turned to one side and fixed, the prey is missed toward the other side. If the proprioceptors are eliminated on one side and the head is fixed in a turned position, the hitting errors are compounded. The alignment of the head preceding the strike is steered by the difference between the optic center message, a function of the angle between the prey and fixation line, and the proprioceptive center message, a function of the angle between the head and the body axis (Fig. 23–21). When fixation movements cease, the direction of the strike is set principally by the optic center messages, and to some extent by the proprioceptive center messages. The inborn genetic basis of these actions is demonstrated by the fact that mantids that have been hand-fed from the time they hatched and have never had to catch a fly are able to perform accurately at their first opportunity.

23–15 LEARNING

Learning is a relatively long-lasting adaptive change in behavior resulting from experience. It is usually defined by exclusion; that is, it is a modification of behavior that cannot be accounted for by sensory adaptation, central excitatory states, endogenous rhythms, motivational states, or maturation. Learning exhibits many forms and is not a unitary phenomenon. Laboratory experiments have shown that the members of almost every animal phylum can undergo learning phenomena, and field observations have shown that learning is important in a variety of natural situations with a wide variety of animals. Learning in different animal species has different characteristics and may involve different mechanisms. It is convenient to divide learning into several categories: habituation; classic reflex conditioning; operant conditioning, including trial and error learning; insight learning; and imprinting.

Habituation. This is perhaps the simplest form of learning, for it does not involve the acquisition of new responses but rather the loss of old ones. An animal gradually stops

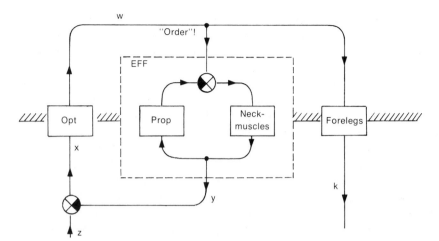

Figure 23-21 The reflex localization of prey in mantids. The control pattern of subsystem EFF is shown to be a loop, which turns the head into a position determined by the order (*w*). (From Mittelstaedt, H.: Ann. Rev. Entom. 7:183, 1962.)

responding to stimuli that have no relevance to its life, ones that are neither rewarding nor punishing. Birds soon ignore the scarecrow which put them to flight when it was first placed in a field. The taming of animals represents a common form of habituation. Habituation is distinguished from fatigue and sensory adaptation by its relatively long persistence. Tube-dwelling marine polychaete worms rapidly pull back into their tubes when a shadow passes over them. When the stimulus is repeated many times, the response becomes irregular, and finally the worm no longer pulls back into its tube in response to the shadow. The curves of habituation to visual and tactile stimuli are different, emphasizing that the decrease in response is not due to muscle fatigue. The nature of the processes underlying habituation is obscure, but it must be a property of the central nervous system and not of the sense organ.

Habituation plays an important part in the development of the behavior of young animals. Threatened by many kinds of predators, young animals initially show escape responses to any large moving object. They quickly become habituated to neutral stimuli such as leaves blowing in the wind.

Classic Reflex Conditioning. All other types of learning consist of strengthening responses that are significant to the animal, and the simplest of these is *classic reflex conditioning,* or Pavlovian conditioning. Pavlov's classic experiments with conditioning in dogs frequently dealt with the salivation reflex. If food is placed in a dog's mouth, the dog salivates. By placing a tube in the salivary duct, Pavlov

could collect and measure all the saliva produced in response to a given stimulus. He then gave the dog standard stimuli by puffing known amounts of meat powder into its mouth through a tube. A standard amount of meat powder caused the secretion of a certain amount of saliva. Pavlov then associated the unconditioned stimulus, the meat powder, with another stimulus, the ringing of a bell or the ticking of a metronome. Initially the second stimulus caused no response, but after a number of pairings of the bell and meat powder stimuli, the saliva began to drip from the tube when the bell was rung and before the meat powder was administered. Pavlov called the second stimulus the conditioned stimulus and the response to it, in this case salivation in response to a bell or metronome, the *conditioned reflex.* Pavlov's experiments showed that almost any stimulus could act as a conditioned stimulus provided it did not produce too marked a response itself.

Conditioned reflexes of this type have been observed in many kinds of animals, from earthworms to chimpanzees. Birds become conditioned to avoid the black and orange caterpillars of the cinnibar moth, which have a very bad taste. The birds associate the bad taste with the orange and black color pattern and avoid not only the caterpillars but certain wasps and other black and orange colored insects as well.

Operant Conditioning. In *operant conditioning* a particular act is rewarded (or punished) when it occurs, thus increasing (or decreasing) the probability that the act will be repeated. A central feature of operant conditioning is the concept that the animal must play a role in bringing about the response which is rewarded or punished. In the simplest type, a rat presses a bar in his cage and is rewarded with food. Initially the pressing of the bar occurs by chance, but on successive trials the rat learns that pressing the bar provides him with food. The rat might at first press the bar with its nose or foot, but the experimenter can determine which of these will be rewarded. He can arrange matters so that the rat will receive the reward of food only when he presses the bar with his left front foot. The experimenter can indeed set up his experimental cage in such a way that the rat is punished, perhaps with an electric shock, if he presses the bar with anything other than his left front foot. Thus operant training may be either reward training (obtaining food) or escape or avoidance training (avoiding the electric shock). The *trial and error learning* that occurs in nature is probably more like operant conditioning than classic conditioning. In both types of training only the *probability* of the occurrence of a certain act under a given set of conditions is changed. Young chicks, for example, initially peck at almost any small object, but they gradually learn not to peck at inedible things.

Maze running is a more complicated form of conditioning because the animal is required to make successive discriminations at the numerous choice points. Correct responses are rewarded and incorrect responses are punished. In other types of trial and error experiments animals are placed in boxes where they have to manipulate a variety of bars or locks to escape. Trial and error is probably one of the most common forms of learning experienced by animals in nature.

Some generalizations can be made about associative learning, as exemplified by these conditioning experiments. First, the time relations are critical. If the unconditioned stimulus precedes the conditioned stimulus, there is little or no conditioning. If the conditioned stimulus precedes the unconditioned stimulus by more than a second, little or no conditioning results. Thus, *contiguity* of conditioned and unconditioned stimuli is required for associative learning. Another relevant feature is *repetition*. The more often the conditioned stimulus and the unconditioned stimulus are paired, the stronger is the acquired conditioned response. The amount of saliva produced by Pavlov's dogs in response to the bell or metronome increased with each trial. A rat learning to run a maze makes fewer and fewer mistakes with each repetition. The rate at which an animal learns is described by a *learning curve*, generated by plotting against the number of trials the time taken to complete the task at each trial or the number of errors committed at each trial. Repetition finally produces a degree of learning beyond which there is no improvement; however, training beyond this point does make the response more resistant to extinction. Extinction is the decay of learning in the absence of reinforcement. Without reinforcement the conditioned response may disappear completely.

Two other concepts associated with conditioning are generalization and discrimination. *Generalization* refers to the common observation that an animal conditioned to one stimulus will also be conditioned to closely related stimuli. The closer the two resemble each other, the better the response will be to the new stimulus. If a dog is conditioned to respond in some way to a 1000 cps tone, it will respond somewhat to a 500 cps tone or to a 1500 cps tone; its response to a 100 cps or a 2000 cps tone would be poorer. The opposite process is termed *discrimination*. Dogs naturally discriminate between different sounds or they would salivate equally in response to all tones. This discrimination can be sharpened if only one tone is followed by a reward and other tones are followed by a punishment. This conditioned discrimination method has been used to test the sensory capacities of animals. The investigator trains the animal to one particular stimulus—color, shape, texture, sound, smell, weight and so on—and then tests to see how well the animal can discriminate this stimulus from others. The stimuli are presented, but only the responses to the "correct stimulus" are rewarded, and the others are given some slight punishment. The two stimuli are made increasingly similar until a point is reached beyond which the animal is unable to discriminate the correct and incorrect stimuli. This technique has been used to measure the honey bee's ability to discriminate different colors, the touch sensitivity of the octopus and the chemical senses of fish.

A type of learning in which an animal learns without reinforcement and later puts the information to good use has been termed *latent learning.* An example of this is exploration. Most animals spend a lot of time exploring new environments, familiarizing themselves with their surroundings. Ants, bees and wasps make *orientation flights* around a nest they have built to learn its position. Some wasps have been shown to be able to learn the essential landmarks around their nest in an orientation flight lasting only nine seconds.

Insight Learning. This type, sometimes called insight reasoning, is considered to be the highest form of learning. It is defined as the ability to combine two or more isolated experiences to form a new experience tailored to a desired goal. If an animal is blocked from obtaining food he can see and selects an appropriate detour to it on his first try, he has probably used insight to solve the problem. Only monkeys and chimpanzees seem to be able to succeed on their first attempt in situations such as this; even dogs and raccoons usually fail on their first attempt. One of the classic examples of insight learning is that of the chimpanzee who piled up boxes or fitted two bamboo poles together to get some bananas hung out of reach. The chimpanzee figured out the solution without being taught.

There are several other types of problem-solving tests used in measuring an animal's ability to learn. These are called *learning set tests.* In one, *oddity principle learning,* the animal must pick out the odd member of a set of three no matter what the objects. In another, the *delayed reaction set,* the animal is allowed to see which of two cups has food hidden under it. Later, after a delay, it has to lift the correct cup. In a third type, the *learning set test,* an animal is presented with two dissimilar objects, one of which always has food under it. Repeated trials are necessary before the animal immediately goes to the object concealing the reward. This discrimination problem is repeated with many different pairs of objects. Although each problem is just as difficult as the first, the animal usually performs with increasing accuracy on repeated trials. Finally, when presented with two objects the animal lifts one. If the food is there, it chooses that object on each subsequent presentation. If the food is not there, it lifts the other object and then always chooses that one first on subsequent presentations. The animal finally gets the idea, or forms a learning set, that the food is always under a particular object. If it picks the object correctly the first time, there is no need to look further. The rates at which various mammals form discrimination learning sets vary tremendously (Fig. 23–22).

Imprinting. This form of learning was

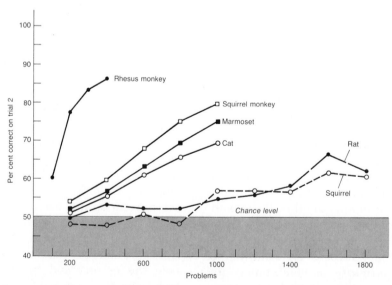

Figure 23–22 The rate at which various mammals can form discrimination learning sets. With each new problem the animal's choice on the first trial has to be random, but if it has learned the principle behind the problems, trial 2 should be correct. Note how long it takes before the scores of rats or squirrels on trial 2 become better than chance, or 50 per cent. Many monkeys reach almost 100 per cent within 400 problems. (From Warren: *The Behavior of Nonhuman Primates,* vol. 1. New York, Academic Press, 1965.)

first described in birds but is now known to occur in sheep, goats, deer, buffalo and other animals whose young are able to walk around at birth. It is a phenomenon whereby a young animal becomes "attached" to the first moving object it sees (or hears, or smells) and reacts to it as it would toward its mother. Konrad Lorenz, who investigated this phenomenon extensively in the 1930's, described imprinting as a unique learning process whereby the young of precocial birds (those able to walk just after hatching) form an attachment with a mother figure. Normally this is their actual mother, but they will become attached to almost any moving object they see during a brief critical period shortly after hatching. This "following" response is of considerable adaptive value in keeping the young bird close behind its parent and well within the parent's protective range. Lorenz found that the young bird's choice of a mother figure during this imprinting period also affected its choice of a sexual partner when it matured. Young birds reared by a foster mother of another species cease to follow her when they develop and become independent. However, when they became sexually mature, they court and attempt to mate with birds of the foster mother's species.

Imprinting occurs in mammals as well as in birds. Orphan lambs reared by humans follow them about and show little interest in other sheep. Harlow's studies with rhesus monkeys show very clearly that the development of normal sexual and social responses depends on their being reared with their mother or with siblings during infancy. There is little doubt that something similar to imprinting occurs in humans; human infants are extremely sensitive to maternal deprivation from about 18 months to 3 years of age.

The Physical Basis of Learning and Memory: The Engram. Learning in some way involves the storage of information in the nervous system and its retrieval on demand. Many of the recent studies of learning have dealt with attempts to discover the physical basis of memory and learning. Somewhere in the nervous system there must be stored a more or less permanent record of what has been learned that can be recalled on future occasions. This record has been termed the memory trace, or *engram*.

Electrical stimulation of the cerebral cortex of a patient undergoing brain surgery can cause vivid recollection of long-forgotten events. From this it was inferred that the items of memory, the engrams, were filed away somewhere in specific places in the brain. Some years ago, Karl Lashley investigated the retention of maze learning in rats by removing portions of the cortex after the rats had learned to solve various problems. The essence of Lashley's results was that the extent of the memory removed by the operation was a function of how much of the cortex was removed and not of what specific part of the cortex was removed. Lashley concluded that the cortex was equipotential and that engrams were not sorted out at specific cortical sites but were in some way present throughout its substance. He speculated that memory might be some system of impulses in reverberating circuits.

Other investigations of the role of the cerebral hemispheres in memory have used the "split-brain" technique of R. W. Sperry. The cerebral hemispheres are separated by a cut down the midline which severs the *corpus callosum*, a large band of transverse fibers that passes ventrally and links the two hemispheres (Fig. 23–23). If a subject learns something using one eye (the other eye is covered), there is no problem when the eye coverings are reversed—the subject can make discriminations with its "untrained" eye. Fibers from the left and right halves of the retina cross over in the optic chiasma, and the visual cortex in

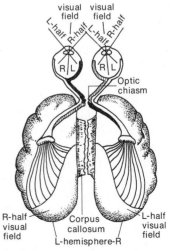

Figure 23–23 Split brain technique. The right and left halves of the brain have been completely separated by cutting the corpus callosum and the optic chiasm. (Redrawn from Sperry, R. W.: Science *133*:1749–1757, 1961.)

each cerebral cortex receives information from both eyes. If the optic chiasma is cut, something learned by the right eye is still recognized by the left eye as long as the corpus callosum is intact. However if the corpus callosum is cut also, half of the brain no longer knows anything learned visually by the other half of the brain following the operation. These split-brain animals are of great interest to psychologists for they are in effect animals with two separate brains. It is possible to train the two halves of the brain to make diametrically opposite discriminations. The left half may learn that the red circle conceals the food, whereas the right half learns that the green circle conceals the food. These experiments show that one function of the corpus callosum is to transfer engrams (or information used in forming engrams) from one hemisphere to the other and that storage is eventually bilateral. The two cerebral cortices may be linked by fiber tracts lower in the brain, but they apparently are not effective in transferring engrams.

Investigations of the memory system of mammals have shown that memories of recent events, *"short-term" memory,* have properties that differ somewhat from memories of events further in the past, *"long-term" memory.* People suffering from concussion are often unable to recall the events that immediately preceded their accident but have normal recall of events in the past. When their memory returns, events are remembered roughly in the order of their occurrence. This phenomenon has been termed *retrograde amnesia.* It appears that each learning trial sets up a process in the brain which relies on the continuous activity of certain neurons. This *consolidation process* results in the formation of an engram and its storage somewhere in the cortex. Concussion, electroshock and similar treatments can interfere with the consolidation process that forms the engrams, but they have no effect on fully formed engrams. The *hippocampus,* on the lower inner margin of each cerebral hemisphere, is connected to the hypothalamus and to the cortex via the reticular formation of the forebrain. If both hippocampi are damaged, human beings have severe defects in their short-term memory system.

Studies of memory in the octopus by J. Z. Young and B. B. Boycott suggest that it, too,

has different mechanisms for short- and long-term memory and that the vertical lobe of the octopus brain is concerned only with long-term memory. The engrams of visual learning become located in the visual lobes themselves and a short-term memory is established there after each presentation of stimulus, punishment or reward. This short-term memory quickly fades unless the vertical lobe is intact. The vertical lobe, however, appears not to be the actual site of long-term storage of engrams but rather part of the mechanism by which short-term memories are converted to long-term memories and then are stored elsewhere.

Investigations of the nature of the engram have been pursued actively for many years, and although they have provided a fair amount of evidence about what the engram is *not,* they have given relatively little positive information. The areas of brains known to be the sites of engram storage—the mammalian cerebral cortex and the optic lobes of the octopus—typically have many self-exciting or reverberating circuits. In these, neurons are arranged in a loop so that impulses conducted along the axon eventually reach the dendrite or cell body. One earlier theory suggested that engrams were represented by continuous activity within such reverberating circuits. These loops may well be involved in the consolidation process, but it is unlikely that they could be the basis of long-term storage. Nerve cells are efficient transmitters over brief periods, but the accurate repetition of a very specific pattern of impulses could hardly continue over many minutes, much less days or years. Clear evidence to the contrary was provided by experiments in which rats were cooled down to 0°C for periods of one hour. At this temperature all electrical activity in the brain ceased, yet when the animals were rewarmed they retained their memory of events prior to the cooling as well as normal, noncooled rats did.

Another theory suggests that memory is not the continuous activity of the special circuit, but rather that the electrical activity associated with a learning trial "wears a path" which will facilitate transmission along that specific path the next time the events occur.

Several features of the engram suggest that it is some sort of permanent growth process, which in turn suggests that protein synthesis may be involved in some way in the conversion of short-term memory into the long-term engram.

Experiments with both goldfish and mice show that puromycin, which inhibits the synthesis of proteins, interferes with the consolidation of memory but not with short-term learning. Established long-term memories can be normal in puromycin-treated animals; hence, it is unlikely that the drug affects the mechanism by which memories are recalled. These experiences suggest that short-term memory, memory immediately following training, is stored by neural activity, but this is shortly followed by a storage phase involving protein synthesis. This storage phase, which may last for several days, converts the short-term memory into the stable, long-term memory. This conclusion can be only tentative, for puromycin is not specific and may affect processes that are necessary for the formation of memories but only indirectly connected with it.

The deciphering of the genetic code led some investigators to the theory that memory might have a basis in the synthesis of RNA—that engrams might consist of specific sequences of nucleotides in RNA molecules. There have been reports of the transfer of specific conditioned reflexes in planaria by the administration of RNA extracts. Studies by Hyden and others suggest that there are changes in the structure of RNA in specific neurons that are involved in specific kinds of learning—for example, in the neurons of the vestibular nuclei of rats that have undergone training in a task involving balance.

Figure 23-24 Photograph (A) and diagram (B) of a salmon fixed in place with an electrode in its olfactory lobe so that a recording of its action potentials can be made as water from different rivers is placed in its nostril. (Photograph and diagram courtesy of Dr. Kiyoshi Oshima, Kyoto University.)

The homing of salmon to the specific tributary in which they were born depends primarily on olfactory cues; the salmon smells some distinctive aroma of his home river, remembers it, and subsequently homes in on it. Salmon have been prepared in the laboratory with electrodes in their olfactory lobes to record neuronal activity and with a catheter in the nasal orifice (Fig. 23–24). When water from the salmon's home river is perfused through the fish's nose, a clear burst of nerve impulses is detected in the olfactory lobe. When water from other rivers is perfused through the nose, little or no electrical activity is detected in the electrodes buried in the olfactory lobe. The experimenters prepared an RNA extract of the olfactory lobe of a salmon that homed in on river A and injected that into the olfactory lobe of a salmon that homed in on river B. Within 48 hours the olfactory lobe of this salmon showed clear bursts of electrical activity, recorded by the indwelling electrode, when water from river A was perfused into the salmon's nostril! This suggests that some sort of memory was transferred from one salmon to another via the RNA extract. There have been similar reports of the transfer of training from one animal to another by the injection of RNA extracts, but negative reports in other instances. There still is no conceptual framework as to how memory might be coded in RNA, in protein, or in any other kind of molecule, but the subject is being attacked experimentally with considerable vigor.

23–16 THE AUTONOMIC NERVOUS SYSTEM

The heart, lungs, digestive tract and other internal organs are innervated by a special set of peripheral nerves, collectively called the *autonomic nervous system.* This system in turn is composed of two parts: the *sympathetic* and the *parasympathetic* nerves.

The autonomic system contains only motor nerves and is distinguished from the rest of the nervous system by several features. There is no willful control by the cerebrum over these nerves; we cannot voluntarily speed up or slow down the heart beat or the action of the muscles of the stomach or intestines. Another important characteristic of the autonomic system is that each internal organ receives a *double* set of fibers, one set coming via the sympathetic nerves and one set via the parasympathetic nerves. Impulses from the sympathetic and parasympathetic nerves always have antagonistic effects on the organ innervated. Thus, if one speeds up an activity, the other decreases it. These effects are summarized in Table 23–2.

Still another peculiarity of the autonomic system is that the motor impulses reach the effector organ from the brain or spinal cord not by a single neuron, as do those to all other parts of the body, but by a *relay* of two or more neurons. The cell body of the first neuron in the chain, termed the *preganglionic neuron,* is located in the brain or spinal cord; that of the second neuron (the *postganglionic neuron*) is located in a ganglion somewhere outside the central nervous system (Fig. 23–2). The cell bodies of the postganglionic neurons of sympathetic nerves are close to the spinal cord; those of the parasympathetic nerves are close to, or actually within, the walls of the organs they innervate. Afferent fibers from the internal organs enter the central nervous system along with the somatic nerve fibers.

The Sympathetic System. The sympathetic system consists of nerve fibers whose preganglionic cell bodies are located in the

Table 23–2 ACTIONS OF THE AUTONOMIC SYSTEM

Organ Innervated	Action of Sympathetic System	Action of Parasympathetic System
Heart	Strengthens and accelerates heart beat	Weakens and slows heart beat
Arteries	Constricts arteries and raises blood pressure	Dilates arteries and lowers blood pressure
Digestive tract	Slows peristalsis, decreases activity	Speeds peristalsis, increases activity
Urinary bladder	Relaxes bladder	Constricts bladder
Muscles in bronchi	Dilates passages, making breathing easier	Constricts passages
Muscles of iris	Dilates pupil	Constricts pupil
Muscles attached to hair	Causes erection of hair	Causes hair to lie flat
Sweat glands	Increases secretion	Decreases secretion

lateral portions of the gray matter of the spinal cord. Their axons pass out through the ventral roots of the spinal nerves, in company with the motor neurons to the skeletal muscles, and then separate from these and become the autonomic branch of the spinal nerve going to the sympathetic ganglion. These ganglia are paired, and there is a chain of 18 of them on each side of the spinal cord from the neck to the abdomen (Fig. 23–25). In each ganglion the axon of the preganglionic neuron synapses with the dendrite of the postganglionic neuron. The cell body of this neuron is located within the ganglion, and its axon passes to the organ innervated.

In addition to the fibers going from each spinal nerve to each ganglion, there are fibers passing from one ganglion to the next. The axons of some of the postganglionic neurons pass from the sympathetic ganglion back to the spinal nerve and through it to innervate sweat glands, the muscles that make the hair stand erect, and the muscles in the walls of blood vessels. The axons of other postganglionic neurons pass from the sympathetic ganglia of the neck up to the salivary glands, the iris of the eye, and the pineal gland. The sensory neurons (fibers) innervating the organs served by the autonomic system are located within the same nerve trunks as the motor neurons but enter the spinal cord by way of the dorsal root, together with other sensory nerves of the nonautonomic system.

The Parasympathetic System. The parasympathetic system consists of fibers originating in the brain and emerging via the third, seventh, ninth and especially the tenth, or vagus, nerve, and of fibers originating in the pelvic region of the spinal cord and emerging by way of the spinal nerves in that region (Fig. 23–25). The vagus nerve arises from the medulla and passes down the neck to the chest

Figure 23–25 Diagram of the autonomic nervous system. The parasympathetic system is shown on the right, the sympathetic system on the left. Roman numerals refer to the numbers of the cranial nerves.

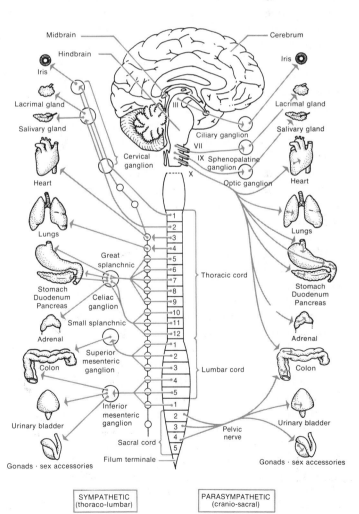

and abdomen, innervating the heart, respiratory system and digestive tract as far as the small intestine. The large intestine and the urinary and reproductive systems are innervated by parasympathetic nerves from the pelvic spinal nerves. The iris of the eye, the sublingual and submaxillary glands, and the parotid gland are innervated by the third, seventh and ninth cranial nerves, respectively. These nerves all contain the axons of preganglionic neurons in the chain; the ganglia of the parasympathetic system are located in or near the organs innervated, so that the axons of the postganglionic neurons are all relatively short.

23–17 THE NERVOUS SYSTEMS OF LOWER ANIMALS

The amoeba manages to exhibit some simple responses, such as movement toward food or away from a needle point, without any specialized structure for integration. The paramecium coordinates the beating of its cilia so that it can move by a system of tiny *neuromotor fibers,* which stretch from the anterior end of the animal to all the cilia.

The sponges have no nervous system. Neurons, synapses and neuromuscular junctions are first found in cnidarians, and have changed relatively little in the course of evolution. The neurons and synapses present in invertebrates are essentially identical with those of human beings. Evolutionary development of the nervous system has included an enormous *multiplication* of units, *diversification* of form and *specialization* of function; an increasing *complexity* of interneuronal connections; a gathering together of neurons with common or associated functions into groups; and a *centralization* and cephalization of neural tissue.

In the cnidarians there is no central nervous system. Neurons tend to be evenly distributed throughout the body, usually in the form of a net; however, there is some tendency for neurons to be gathered into cords in the jellyfish and sea anemones. The cnidarian nerve net is composed of many cells whose branched processes connect, so that an impulse starting in one part of the body can spread in all directions to every part. All the neurons are usually similar, but in the more complex cni-

darians there is some differentiation into sensory, ganglion and motor nerves.

Most of the invertebrates have a highly centralized nervous system. This is especially true of the arthropods (the insects, spiders, crabs and lobsters), the mollusks (squids and snails), and the annelids, or segmented worms. These animals typically have a brain and a longitudinal nerve cord. The earthworm has a true nervous system with axons and dendrites arranged in definite nerve cords and fibers. There is a differentiation into central and peripheral systems with separate sensory neurons, interneurons and motor neurons joined by synapses so that nerve impulses can travel in one direction only. This permits the central nervous system to act as an integrator, selecting certain incoming sensory impulses and passing them on to effectors, while inhibiting or suppressing others. The earthworm has a central nerve cord extending the entire length of the body which enables its separate segments to move in a coordinated fashion. Each segment has a pair of ganglia, a collection of nerve cell bodies. The most anterior part of the cord is an enlarged ganglion sometimes referred to as the "brain," which sends impulses down the cord to coordinate movements. After removal of this "brain," the animal can move almost as well as before, but it persists in futile efforts to go ahead, instead of turning aside, when it

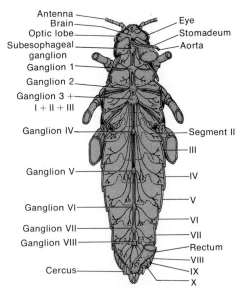

Figure 23–26 Ventral nervous system and brain of a grasshopper, *Dissosteira carolina.*

comes to some obstacle. The "brain," therefore, is necessary for adaptive movements. The nervous systems of the other higher invertebrates are rather similar to that of the earthworm. In all, the nerve cord is ventral to the digestive system and is solid. Although the annelids and more primitive arthropods have a pair of ganglia in each segment of the body, the ganglia of the higher arthropods tend to be fused (Fig. 23–26).

Despite its small size, the invertebrate nervous system is enormously complex. The crayfish system, for example, contains 97,722 neurons. Its sense organs—the compound eye, taste organs, and the mechanical sense organs concerned with postural relationships and tactile discrimination—are highly developed. Its synaptic connections are incredibly complex. The brain and ganglia are jammed with integrating interneurons.

In the vertebrates the nerve cord is dorsal to the digestive system and is hollow, with a central cavity. The nervous systems of all the vertebrates are fundamentally alike; the differences are primarily in the development of the various regions of the brain and in the size of the brain relative to the spinal cord.

QUESTIONS

1. List the primary functions of the nervous system. What other systems serve similar functions?
2. Distinguish between a neuron and a nerve.
3. How many kinds of neurons are there? What functions are performed by each type?
4. Discuss the chemical and physical events associated with the transmission of the nerve impulse.
5. What is the evidence for the theory that the neurilemma sheath plays a role in the regeneration of severed nerves?
6. Compare the nature of the nerve impulse in the optic nerve stimulated by a beam of light with that in a sensory nerve from the skin stimulated by a hot poker.
7. Compare the chemical and physical processes involved in transmission at the synapse with those involved in transmission along a neuron.
8. What are nerve centers, where are they located, and what are their functions?
9. Make a diagram of the brain, labeling the principal parts, and list the functions carried on by each.
10. Describe and give an example of a reflex arc.
11. What is meant by the autonomic nervous system? What are its subdivisions and to which of the larger divisions of the entire nervous system does it belong?
12. Pilocarpine is a drug that stimulates the nerve endings of parasympathetic nerves. What effect would you expect this drug to have on (a) the digestive tract, (b) the iris of the eye, (c) the heart rate?
13. Atropine blocks the action of the parasympathetic system and thus produces the equivalent of a stimulation of the sympathetic system. What effects would you expect atropine to have on (a) the digestive tract, (b) the iris of the eye, and (c) the heart rate?
14. List the cranial nerves of humans. Give the types of neurons found in each and the organs innervated. Which of these cranial nerves are not present in the frog?
15. What metabolic processes are responsible for the brain waves evident in the electroencephalograph? What abnormal conditions are known to affect the character of the brain waves?
16. What is meant by the "facilitation" of nerve impulses?
17. Compare classic reflex (Pavlovian) conditioning with operant conditioning. Which of these more closely resembles the trial and error learning that occurs in nature? What roles do generalization and discrimination play in the conditioning process?
18. What is meant by imprinting? In what kinds of animals does this occur?
19. Design an experiment to test memory using the "split brain" technique.
20. Compare and contrast short-term memory and long-term memory.

SUPPLEMENTARY READING

The role of the nervous system in maintaining body conditions at a constant level is discussed in Chapters 15 and 16 of the *Wisdom of the Body* by W. B. Cannon and in more detail in *The Integrative Action of the Nervous System* by C. S. Sherrington. A comprehensive account of his early experiments in conditioning is given in Pavlov's book *Conditioned Reflexes*. The development of psychology and psychiatry is described in *A History of Medical Psychology* by G. Zilboorg and some of the fundamental investigations in the field are described in *Great Experiments in Psychology* by H. G. Garrett.

A detailed account of the structure of the brain and nerves is presented in *The Anatomy of the Nervous System* by F. W. Ranson and F. L. Clark. *The Waking Brain* by H. W. Magoun, *The Brains of Animals and Man* by R. Freedman and J. E. Morris and *Brain Mechanisms and Mind* bu Keith Oatley give fascinating accounts of the function of this remarkably complex organ. Some classic experiments in neurophysiology are presented in Chapters 6 and 7 of J. F. Fulton's *Selected Readings*.

Discussions of the nature of the nerve impulse

and some of the experiments that have led to our present theories regarding the nerve impulse are presented in B. Katz, *Nerve, Muscle and Synapse,* in A. L. Hodgkin, *The Conduction of the Nerve Impulse* and in D. J. Aidley, *The Physiology of Excitable Cells.* An excellent reference text at an advanced level is *Neurophysiology* by G. C. Ruch, H. D. Patton, W. Woodbury and A. L. Towe.

There is an exceptionally rich literature in the field of animal behavior, and the titles selected here are only a very few of the many excellent sources available. *Animal Behavior* by R. A. Hinde is a good general text combining ethologic and psychologic viewpoints. *Mechanisms of Animal Behavior* by P. Marler and W. J. Hamilton is an excellent general text, giving detailed accounts of animal behavior. Two shorter introductions to the subject, written from different points of view, are *An Introduction to Animal Behavior* by A. Manning and *Animal Behavior* by V. G. Dethier and E. Stellar. Two multi-author books that provide samples of current thought and research on animal behavior are *Readings in Animal Behavior,* edited by T. E. McGill, and *Psychobiology,* edited by J. L. McGaugh, N. M. Wein-

berger and R. E. Whalen. One of the classic books in ethology is N. Tinbergen's *Social Behavior of Animals. Social Behavior and Organization Among Vertebrates,* edited by William Etkin provides both a description of social behavior from the ecological and evolutionary point of view and some experimental analyses of animal behavior. *The Behavior of Arthropods* by J. D. Carthy is a fascinating summary of studies of the mechanisms controlling the responses of insects and other arthropods to stimuli. Two interesting books of reproductive behavior are the classic *Hormones and Behavior* by Frank Beach and *Courtship: A Zoological Study* by M. Bastock. Other books recommended for their coverage of specific aspects of behavior are J. A. Deutsch, *The Structural Basis of Behavior;* J. Gaito, *Macromolecules and Behavior; The Behavior of Nonhuman Primates,* edited by A. M. Schrier, H. F. Harlow and F. Stollnitz; *Behavior-Genetic Analysis,* edited by J. Hirsch; K. Lorenz, *On Aggression;* C. T. Morgan, *Physiological Psychology;* W. Slucken, *Imprinting and Early Learning;* W. H. Thorpe, *Learning and Instinct in Animals;* and M. J. Wells, *Brain and Behavior in Cephalopods.*

CHAPTER 24

RECEPTORS AND EFFECTORS

In order to survive, each organism has evolved means whereby it can make appropriate, meaningful, adaptive responses to specific changes in the environment. This requires that it have **receptors** (sense organs) to detect environmental changes, systems of nerves and endocrine organs to integrate and coordinate the information received and to trigger the response, and **effectors** to carry out the responses. The major effectors in the human body are the skeletal, smooth and cardiac muscles, which are discussed in Chapter 22. Other effectors include the digestive glands, sweat glands, accessory sex glands, mammary glands and, in certain animals, melanocytes, nematocysts, electric organs and luminescent structures.

Single-celled organisms are sensitive to many different kinds of stimuli, as evidenced by their negative responses to bright lights, certain chemicals, electric currents, and so on. For survival on a higher, more complex level of existence, the metazoa have evolved a variety of specialized receptor cells, each sensitive to one type of stimulus. These sense organs enable their possessors to search for food, find and attract a mate, escape enemies, and so on, and are of great importance in the survival of the individual and of the species. The receptors in these sense organs are remarkably sensitive to the appropriate stimuli: the eye is stimulated by an extremely faint beam of light, whereas only a very strong light can stimulate the optic nerve directly. The negligible amount of vinegar that can be tasted or the amount of vanilla that can be smelled would have no effect if applied directly to a nerve fiber. The sense organs of some animals are sensitive to stimuli that are completely ineffective in humans: dogs and cats can hear high-pitched whistles that are inaudible to us. Bats emit and hear very high-pitched noises and are guided in flight by the echoes that rebound from objects in their path. They can even catch insects, guided by the echoes from their small prey.

24–1 THE FUNCTIONS OF SENSE ORGANS

Traditionally, humans are said to have five senses: touch, smell, taste, sight and hearing. Some of these, however, can be divided into several distinct senses; for example, touch, pain, pressure, cold and heat are all included in the general sense of touch. The ear contains an organ for sensing equilibrium and rotation as well as the organ for sensing sound waves. There are other senses—perhaps more vague and more generalized but nonetheless important—for determining the internal state of the body. We can sense the tension of muscles, the movements of joints, and internal conditions such as thirst, hunger, nausea, pain and orgasm. The receptors for such senses are located in the muscles, joints, viscera, throat and other places.

Sense organs have the dual function of detecting change and then transmitting information concerning the nature of the change to the central nervous system. A sense organ would make very little useful information available to an organism if it responded indiscriminately to all kinds of environmental changes. In the course of evolution, sense organs have become highly specific for one kind of environmental stimulus: one organ detects light, another detects mechanical pressure, a third detects certain kinds of chemicals, and so on. No sense organ would be very useful if it responded only to gross changes in the environment. However, if it were so sensitive that it responded to every moving molecule or electron, it would transmit only "noise." Each sense organ has thus developed specificity and *optimal* (not maximal) sensitivity so that it maintains an optimal ratio of signal to noise. Sense organs must, in addition, have the capacity for discrimination and for recording not only "on" and "off" but also rate, magnitude and direction of change.

Each *sense organ* is a specialized structure consisting of one or more *receptor cells* and *accessory tissues.* For instance, the receptor cells of the human eye are the *rods* and *cones;* the accessory structures are the *cornea, lens, iris* and *ciliary muscles* (Fig. 24–12). While some of the capacities enumerated earlier are intrinsic to the receptor cell itself, others are conferred upon the sense organ by the accessory structures. Without a lens, an eye would be incapable of detecting many events. In this case the lens enhances the versatility of the organ. Accessory structures may also act as filters that limit the performance of receptors. For example, ultraviolet light is not seen by the human eye because it is filtered out before reaching the retina; the retina itself is not insensitive to light of this wavelength. In contrast the eyes of insects have no filters for ultraviolet light and

the detection of ultraviolet is important in the lives of these organisms.

The importance of accessory structures is especially evident in the mechanical senses, in which they modify sensitivity and determine the nature of the effective stimulus. Among invertebrates, for example, one mechanoreceptor may be sensitive only to gross touch because it is associated with a stout rigid spine, while another may actually be an auditory receptor because it is connected to a long filamentous hair that is moved by sound waves. The hairs on the cerci (paired posterior appendages) of crickets are examples of this type of auditory organ.

Receptors are generally nerve cells, the axons of which extend directly into the central nervous system or connect synaptically with one or more interneurons connected in turn with the central nervous system. Most receptors are of this type. Some receptors (e.g., the receptors in the human taste buds) are modified epithelial cells connected to one or more nerve cells (Fig. 24–8).

24–2 THE STIMULUS-RECEIVING PROCESS

The sense organ performs two functions: it *detects* and it *transmits* information to the central nervous system. In those sense organs in which the receptor is a *primary neuron,* this neuron both detects and transmits. In organs in which the receptor is an *epithelial cell* it detects, but the information is transmitted by the associated neuron.

In its capacity as detector or sensor, a receptor receives a small amount of energy from the environment. Each kind of receptor is specialized to receive one particular form of energy more efficiently than another. Rods and cones absorb the energy of photons of certain specific energies. Temperature receptors respond to radiant energy transferred by radiation, conduction or convection. Electricity is detected by the energy of electrons. Tastes and smells are detected by the potential energy in the mutual attraction and repulsion of molecules. Each receptor clearly possesses some degree of intrinsic specificity.

The various kinds of environmental energy act as triggers, causing the receptors to perform

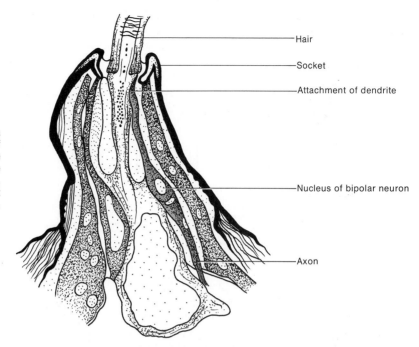

Hair

Socket

Attachment of dendrite

Nucleus of bipolar neuron

Axon

Figure 24–1 A tactile hair from a caterpillar, showing the attachment of the dendrite of the bipolar neuron (the mechanoreceptor) at the point where the shaft of the hair enters the socket. (From Hsu, F.: Étude cytologique et comparée sur les sensilla des insectes. La Cellule *47*:1–60, 1938.)

biological work. This work transforms metabolic energy into electric energy. These relationships are best exemplified by a very simple sense organ, the tactile hair of an insect. This hair plus its associated cells is a complete sense organ (Fig. 24–1). The **bipolar neuron** at its base is the receptor. Its dendrite is attached to the base of the hair near the socket; its axon passes directly to the central nervous system without synapsing. In its unstimulated state this neuron maintains a steady resting potential, i.e., there is a potential difference between the inside and outside of the neuron. This potential difference exists because the ionic compositions of the fluids on each side of the semipermeable cell membrane are different. The difference is maintained by the sodium pump and by metabolic work performed by the cell. When the hair is touched, its shaft moves in the socket and mechanically deforms the dendrite. Deformation (the stimulus) increases the permeability of the neuronal membrane to ions with the result that the potential difference between the two sides of the membrane decreases, disappears or increases. If it decreases or disappears, the cell is said to be **depolarized;** if it increases, the cell is said to be **hyperpolarized.** The state of depolarization caused by the stimulus is called the **receptor potential.** It spreads relatively slowly down the dendrite, decaying exponentially as it goes. When a special area of the cell near the axon

(the **axon hillock**) becomes depolarized, **action potentials** are generated. The action potentials then travel along the axon to the central nervous system. The primary receptor thus performs all the essential functions of a sense organ: it detects an event in the environment (a force acting on the hair); it generates electrical energy at the expense of its metabolic energy (the receptor potential); and it transmits information (action potentials) to the central nervous system. With minor variations this is how all receptors operate.

The relations between the stimulus, the receptor potential and the action potentials are summarized in Figure 24–2. The amplitude and duration of the receptor potential are related to the **strength** and **duration** of the stimulus. Thus a strong stimulus causes a greater depolarization of the receptor membrane than does a weak one. The action potentials are **repetitive** and the frequency at which they are generated is related to the magnitude of the receptor potential. The strength of a stimulus is reflected in the *frequency* of the action potentials. The amplitude of each action potential bears no relation to the stimulus; it is characteristic of the particular neuron under the usual recording conditions.

It must be remembered that the action potential is an all-or-none phenomenon. The receptor potential, in contrast, is a **graded response.** Once a stimulus has triggered a

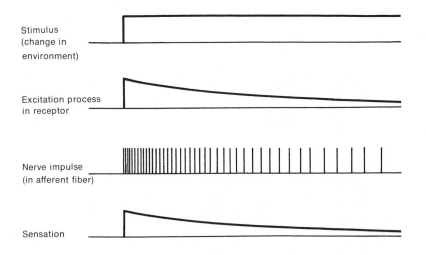

Stimulus (change in environment)

Excitation process in receptor

Nerve impulse (in afferent fiber)

Sensation

Figure 24–2 A diagram showing the relations among the stimulus, the receptor potential, the action potential and sensation. (From Adrian, E. D.: *The Basis of Sensation.* London, Chatto & Windus Ltd., 1949.)

receptor to generate action potentials, the stimulus has no further control over them. The situation is analogous to lighting a fuse. The heat of the match is the stimulus. When it raises the end of the fuse to the combustion point, the fuse begins to burn, and utilizing its own energy, it ignites adjacent parts of itself. In this way the "message" travels the length of the fuse independently of the temperature of the match flame.

Even though the stimulus may continue unabated, neither the receptor potential nor the action potentials continue unchanged. The receptor potential gradually falls, and the frequency of the action potentials decreases (Figs. 24–2 and 24–3). This is the phenomenon of *adaptation.* Some receptors adapt very rapidly and completely, but others do so more slowly (Fig. 24–3).

24–3 SENSORY CODING AND SENSATION

The stimulation of any sense organ initiates what might be considered a coded message, composed of action potentials transmitted by the nerve fibers and decoded in the brain. Impulses from the sense organ may differ in: (1) the total number of fibers transmitting, (2) the specific fibers carrying action potentials, (3) the total number of action potentials passing over a given fiber, (4) the frequency of the action potentials passing over a given fiber, or (5) the time relations between action potentials in specific fibers. These are the possibilities in the "code" sent along the nerve fiber; how the sense organ initiates different codes and how

the brain analyzes and interprets them to produce various sensations are not yet understood. It is important to remember that *all action potentials are qualitatively the same.* Light of the wavelength 400 nanometers (blue), sugar molecules (sweet), and sound waves of 440 hertz (A above middle C) all cause action potentials to be sent to the brain via the appropriate nerves; these action potentials are identical. How can the organism assess its environment accurately? The qualitative differentiation of stimuli must depend either upon the sense organ itself, upon the brain, or upon both. In fact, it depends upon both. Primarily, our ability to

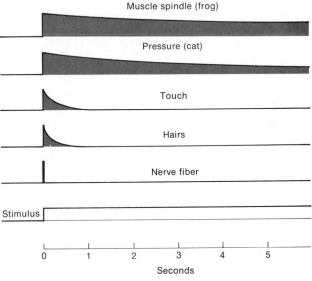

Muscle spindle (frog)

Pressure (cat)

Touch

Hairs

Nerve fiber

Stimulus

0 1 2 3 4 5

Seconds

Figure 24–3 A diagram showing the relation between the stimulus and the different rates of adaptation for different receptors and a nerve fiber. The heights of the curves indicate the rates of discharge of action potentials. (From Adrian, E. D.: *The Basis of Sensation.* London, Chatto & Windus Ltd., 1949.)

discriminate red from green, hot from cold or red from cold is due to the fact that particular sense organs and their individual sensitive cells are connected to specific cells in particular parts of the brain.

The frequency of the repetitive action potential codes the intensity of the stimulus. Since each receptor *normally* responds to but one category of stimuli (i.e., light, sound, taste and so forth), a message arriving in the central nervous system along this nerve is interpreted as meaning that a particular stimulus occurred. Interpretation of the message and, in the case of human beings, of the quality of sensation depend upon which central interneurons receive the message. Sensation, when it occurs, occurs in the brain. Rods and cones do not see; only the combination of rods, cones and centers in the brain see. Furthermore, many sensory messages never give rise to sensations. For example, chemoreceptors in the carotid sinus and the hypothalamus sense internal changes in the body but never stir our consciousness.

Since only those nerve impulses that reach the brain can result in sensations, any blocking of the impulse along the nerve fibers by an anesthetic has the same effect as removing the original stimulus entirely. The sense organs, of course, will continue to initiate impulses, which can be detected by the proper electrical apparatus, but the anesthetic prevents them from reaching their destination.

Spatial localization of stimuli impinging on the body, especially mechanical and pain stimuli, also depends upon the destination of specific nerves in the brain. The importance of the brain in localization and in making sensa-

tions possible is emphasized by the phenomenon of "misreference," which occurs occasionally in connection with pain. A well known example of this is the experience of people suffering from heart pains who complain of pain in the shoulder, upper chest or medial side of the left arm. Actually the stimuli originate in the heart, but the nerve impulses terminate in the same part of the brain as impulses genuinely originating in the shoulder, chest or arm.

Cross-fiber patterning, another method of coding information, is probably the one used in olfactory organs. It is unlikely that the olfactory organ contains a specific receptor for each of the thousands of individual odors that can be recognized. There is, in fact, evidence that there are a limited number of categories of receptors, each of which responds to a spectrum of odors. There is no rigid specificity because the spectra overlap. Perception of different characteristic odors probably depends, therefore, on the pattern of response of all fibers responding together.

The *temporal pattern* of action potentials generated in a single neuron may serve as a code for different stimuli. Single taste receptors of flies, for example, generate action potentials at an even, regular frequency when the stimulus is salt, but generate irregular frequencies when the stimulus is acid.

In invertebrates it is usual for the axon of a sensory neuron to extend all the way to the central nervous system without synapsing. In these circumstances the message generated at the periphery arrives unaltered. However, in the compound eye, as in most vertebrate sense organs, many interneurons are interposed between the receptor and the central nervous sys-

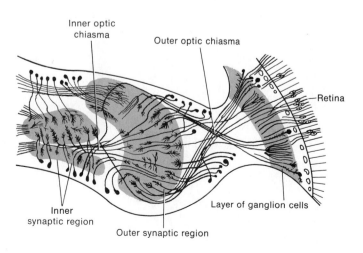

Figure 24–4 An example of neural circuitry in the optic lobes of insects. This is a diagrammatic representation of the optic lobe of the blowfly *Calliphora vomitoria.* (From Cajal, S. R., and Sanchez, D.: Contribucion al conocimiento de los centros nerviosos de los insectos. Trabajas del Lat. de Investig. Biologicas [University of Madrid] *13*:1–167, 1915.)

Inner optic chiasma

Outer optic chiasma

Retina

Inner synaptic region

Outer synaptic region

Layer of ganglion cells

tem (Fig. 24–4). The vertebrate retina or olfactory bulb has an exceptionally complicated neural circuitry. As a consequence of all these synaptic connections, the original message is altered and may lose or gain some of its information. The message that finally arrives at the brain has been well censored by interneurons and bears even less resemblance to the original stimulus than did the action potentials from the receptor.

24–4 MECHANORECEPTORS

Receptors are customarily classified according to the nature of their effective stimuli. The human being, like most animals, is equipped with mechanoreceptors, chemoreceptors, photoreceptors and temperature receptors. Some fish are equipped with electroreceptors.

Mechanoreceptors are sensitive to stretch, compression or torque imparted to tissues by the weight of the body, the relative movement of parts, the gyroscopic effects of moving parts and the impact of the substratum or the surrounding medium (air or water). Mechanoreceptors are concerned with enabling an organism to maintain its primary body attitude with respect to gravity (for us, anterior end up and posterior end down; for a dog, dorsal side up and ventral side down; for a tree sloth, ventral side up and dorsal side down). They are also concerned with maintaining postural relations (i.e., the position of one part of the body with respect to another), information that is essential for all forms of locomotion and for all coordinated and skilled movements from spinning a cocoon to completing a reverse one and a half dive with twist. Mechanoreceptors provide, in addition, information about the shape, texture, weight and topographical relations of objects in the external environment. Finally, mechanoreception is necessary for operation of some of the internal organs. They supply, for example, information about the presence of food in the stomach, feces in the rectum, urine in the bladder, or a fetus in the uterus. Mechanoreceptors may be divided into the following categories: tactile, proprioceptive and auditory.

The Tactile Sense. Among the simplest tactile receptors are the *tactile hairs* of invertebrates. The tactile hair of an insect is a *phasic* receptor, i.e., it responds only when the hair is moving. When the hair is displaced, a receptor potential develops and a few action potentials are generated, but all activity ceases when motion ceases, even though the hair is maintained in the displaced position.

The remarkable tactile sensitivity of the human, especially in the fingertips and lips, is due to a large and diverse number of sense organs in the skin (Fig. 24–5). By making a careful point by point survey of a small area of skin, using a stiff bristle to test for touch, a hot or cold metal stylus to test for temperature and

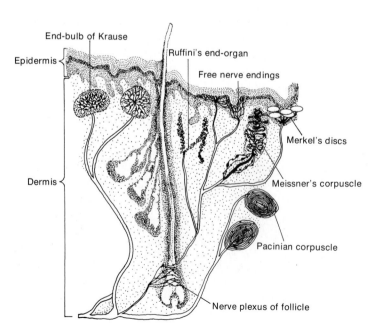

Figure 24–5 Diagrammatic section through the skin showing the types of sense organs present. The sense organs respond to the following stimuli: cold—end-bulbs of Krause; warmth—Ruffini's end-organs; touch—Meissner's corpuscles and Merkel's discs; deep pressure—pacinian corpuscles; and pain—free nerve endings.

a needle to test for pain, it has been found that receptors for each of these sensations are located at different spots. By comparing the distribution of the different types of sense organs and the types of sensations produced, it has been found that the free nerve endings are responsible for pain perception, that basket nerve endings around hair bulbs (Meissner's corpuscles and Merkel's discs) are responsible for touch, that the end-bulbs of Krause and Ruffini's endings are responsible for sensations of cold and warmth and that pacinian corpuscles mediate the sensation of deep pressure.

The pacinian corpuscle has been particularly well studied. The bare axon is surrounded by lamellae interspersed with fluid. Compression causes displacement of the lamellae which provides the deformation stimulating the axon. Even though the displacement is maintained under steady compression, the receptor potential rapidly falls to zero and action potentials cease. This is a phasic receptor responding to velocity.

Proprioception (Kinesthesis). Among invertebrates the sense organs most commonly concerned with relaying postural information are hairs, plates (campaniform organs) and other modified cuticular structures. These are *tonic* (static) sense organs. Unlike phasic receptors, the receptor potential is maintained (though not at constant magnitude) as long as the stimulus is present, and action potentials continue to be generated. Thus there is continued information about the position of the organ concerned.

Each human muscle, tendon and joint is equipped with *proprioceptors* sensitive to muscle tension and stretch. By means of these sense organs we can, even with our eyes closed, perform manual acts such as dressing or tying knots. Impulses from the proprioceptors are also extremely important in ensuring the harmonious contraction of different muscles involved in a single movement; without them, complicated skillful acts would be impossible. Impulses from these organs are also important in the maintenance of balance. Proprioceptors are probably more numerous and more continuously active than any of the other sense organs, although we are less aware of them than of any of the others. The existence of this sense was discovered only a little more than one hundred years ago. One obtains some idea

of what life without proprioceptors would be like when a leg or arm "goes to sleep"—a feeling of numbness results from the lack of proprioceptors.

The Mammalian Muscle Spindle. The mammalian muscle spindle is certainly one of the more versatile stretch receptors and illustrates beautifully how sensory performance may be modified by the consequences of its own action. It is an example of a *feedback* mechanism. In the muscles of higher vertebrates there are in addition to the regular striated muscle fibers (*extrafusal* fibers) special *intrafusal* fibers associated with sensory nerve endings. A bundle of intrafusal fibers together with their sensory endings is called a *muscle spindle* (Fig. 24–6). The intrafusal fiber is striated except in the region of the nucleus. Here there are two kinds of nerve endings: flower-spray endings belonging to a thin sensory (afferent) nerve, and annulospiral endings belonging to a thick sensory (afferent) nerve. In the region of attachment of the extrafusal muscle fibers to the tendon there is another sense organ, the **Golgi** or *tendon organ.* There are two sets of motor neurons to the muscle: the *alpha efferents* innervate the ordinary (extrafusal) muscle fibers; the *gamma efferents* innervate the intrafusal fibers.

For violent muscular contraction, commands from the central nervous system come mostly via the alpha efferents (the "emergency" pathway). If, as a consequence, a muscle is stretched excessively, the Golgi organ and the muscle spindle are stimulated. Messages from the Golgi organ pass up the sensory nerve to a point where they synapse with the alpha efferent. They inhibit the alpha efferent and the muscle stops contracting. Thus tension is kept within bearable limits and the muscle is kept at constant length under a specific load.

Under "ordinary" circumstances (i.e., in the production of slow voluntary movements), commands from the central nervous system descend the gamma efferents to the intrafusal fibers, which begin a slow, graded contraction. As a consequence the muscle spindle is stimulated and sends impulses to the synapse with the alpha efferent, which is excited to cause the extrafusal fibers to contract. Since the intrafusal fibers are connected in parallel with the other fibers rather than in series, as are the

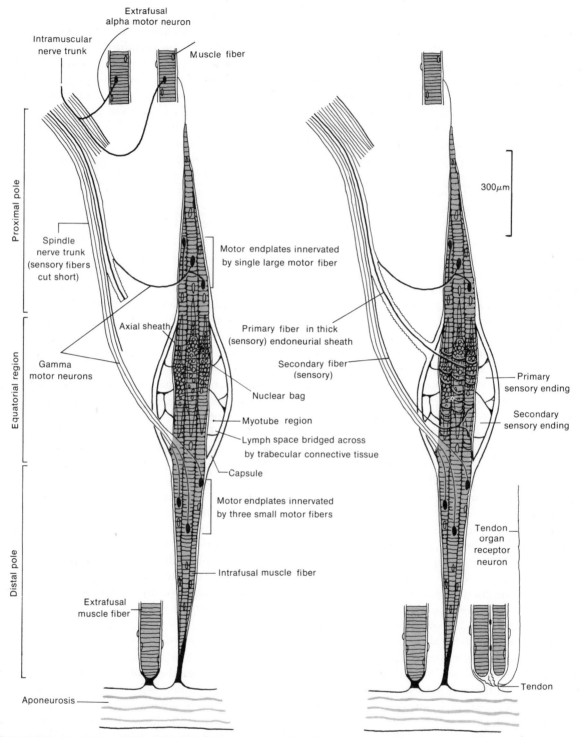

Figure 24-6   A diagram of the muscle receptor system showing the relations between the intrafusal and extrafusal fibers, the tendon organ, the spindle receptors and the motor innervation.

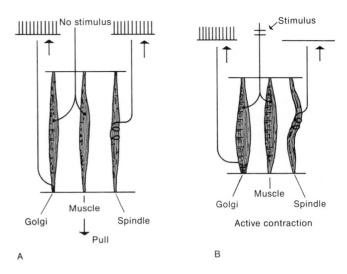

Figure 24–7 Golgi spindle receptors have a series relationship to the muscle fibers. Spindles are in parallel. Pull on the muscle increases the rate of firing of both receptors. *A*, Active contraction of the muscles (*B*) will cause an increase in discharge of the Golgi tendon organ and a decrease in rate of discharge from the spindle. (From Ochs, S.: *Elements of Neurophysiology.* New York, John Wiley & Sons, Inc., 1965.)

Golgi organs, they become slack (Fig. 24–7). The spindle then stops exciting the alpha efferent, and the muscle ceases contracting. The net result is to cause the muscle to come to a new state of tension. Thus muscle tone is maintained and precise voluntary movements are made.

While the muscle has been contracting it has been stretching its antagonist. Naturally the spindle of the antagonist excites it to contract. If it continues to do so in the face of the pull being exerted upon it, the excessive strain stimulates its Golgi organ. This inhibits its contraction. Thus are antagonistic muscles prevented from "fighting" each other.

Visceral Sensitivity. The sensations associated with the receptors located in the internal organs, which are extremely important in regulating the activities of the viscera, seldom reach the level of consciousness. They bring about reflex control of the functioning of the internal organs by way of reflex centers in the medulla, midbrain or thalamus. A few of the impulses from these receptors do get to the cerebrum, however, and give rise to sensations such as thirst, hunger and nausea. The sensation of thirst originates in receptors in the lining of the throat; when this lining becomes dry, the receptors send impulses to the brain which we interpret as a feeling described as "being thirsty."

The wall of the stomach also contains receptors. When the stomach is empty, a series of strong, slow, muscular contractions sweeps over the walls, stimulating these receptors and resulting in the feeling of hunger. By having a subject swallow a rubber balloon connected by a tube to a recording device, it was found that hunger pangs were closely correlated with these characteristic contractions. However, since patients who have had the entire stomach removed surgically still feel hunger, other stimuli, such as the decreased concentration of glucose in the blood, are involved. The feeling of nausea may originate from receptors in the stomach, but the contractions responsible for it move up, instead of down, the tract as in normal peristalsis.

Still other receptors are located in the mesenteries holding the internal organs in place. When these mesenteries become inflamed or stretched by unusual movements of the organs to which they are attached, sensations of pain result. There are other nerve endings for pain in the linings of the organs themselves.

The sense of fullness and the urge to defecate and urinate depend upon receptors in the walls of the rectum and urinary bladder, respectively, which are stimulated by the tension resulting when these hollow organs are distended by their contents.

Many other, less well defined visceral sensations are felt during sexual activity, illness or emotional crises.

24–5 CHEMORECEPTION: TASTE AND SMELL

Throughout the animal kingdom many sexual, reproductive, social and feeding activities are initiated, regulated or influenced in some way by specific chemical aspects of the environment. Insects, for example, use a great number of chemicals in communication, for defense from predators and for the recognition of specific foodstuffs. Many vertebrates employ chemical secretions to mark territory, to attract their sexual partners, or to defend themselves. Chemoreception is also involved in the tracking of prey in carnivores and in the detection of carnivores by intended prey.

Sensitivity to chemicals may be very specific in that only certain compounds act as stimuli and then in low concentrations. The sensitivity of other receptors, such as those in the skin of the frog, may be gross and nonspecific. As most beginning physiology students know, a frog will scratch its back when dilute acid or concentrated solutions of inorganic salts are applied to the skin. Free nerve endings are the chemoreceptors involved. This *common chemical sense* is widely distributed among aquatic animals. Among mammals it is restricted to moist areas of the body. Recall how your eye smarts and waters in the presence of ammonia fumes or a peeled onion and how a broken blister stings if touched by a nonphysiologic solution.

The specific and highly sensitive chemoreceptive systems comprise the senses of *taste* and *smell* (olfaction). These are easily distinguishable in ourselves and other terrestrial organisms. As one examines aquatic organisms, and especially those lower on the phylogenetic scale, it becomes increasingly difficult to decide what is taste and what is olfaction.

Human beings depend primarily on visual or auditory cues; we use our chemical senses much less than other mammals do, and we tend to minimize the importance of the chemical senses. Perhaps for this reason, we still know relatively little about their mode of action.

24–6 THE HUMAN SENSE OF TASTE

The organs of taste are budlike structures located predominantly on the tongue and soft palate. They are situated in papillae, of which there are four kinds: *circumvallate, foliate, fungiform* and *filiform*. The cells of the *taste buds* were originally classified on the basis of

size and histology as gustatory cells and supporting cells. Recent studies with the electron microscope reveal that this concept is too simple, that there are all gradations between the two. There is also a very rapid substitution of cells: every 10 to 30 hours the cells are completely replaced. Each cell has at its free surface a border of microvilli, many of which project into a tiny pore connecting with the fluids bathing the surface of the tongue. There are no taste hairs as once believed.

Each taste cell is an epithelial cell and is the receptor. The connections with the nerve cells are complicated. Each taste cell is innervated by more than one neuron. Furthermore, some neurons may connect with one taste cell and others with many. This complexity of connections renders interpretation of taste-sensory physiology difficult.

Traditionally there are four basic tastes: sweet, salt, sour and bitter. To this must now be added water. While it is true that the greatest sensitivity to each of these tastes is restricted to a given area of the human tongue (Fig. 24–8), not all papillae are restricted in their sensitivity to a single taste modality; some indeed are specific to salt, acid or sugar, but the majority respond to two or more categories of taste solutions. Nor is a single taste bud restricted in its sensitivity to a single type of chemical; a single taste cell may respond to more than one category of taste. Thus the de-

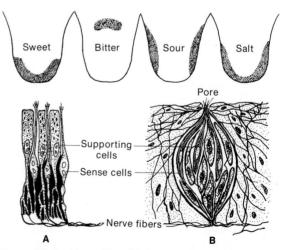

Figure 24–8 *Above,* The distribution on the surface of the tongue of taste buds sensitive to sweet, bitter, sour and salt. *Below: A,* Cells of the olfactory epithelium of the human nose. *B,* Cells of a taste bud in the epithelium of the tongue.

tection and processing of information in the taste organs of the tongue are very complex. Taste discrimination probably depends on a code that consists of cross-fiber patterning; i.e., each receptor responds to more than one kind of chemical, but no two respond exactly alike, so that the total pattern of messages going to the brain is different for different solutions.

Flavor does not depend on the perception of taste alone. It is compounded of taste, smell, texture and temperature. Smell affects flavor because odors pass from the mouth to the nasal chamber via the internal nares.

24–7 THE SENSE OF TASTE IN INSECTS

One of the most thoroughly studied organs of taste is the taste hair of the fly (Fig. 24–9). The terminal segments of the legs and the mouthparts of flies, moths, butterflies and a number of other insects are equipped with very sensitive hairs and pegs. In the fly, each one of these contains four taste receptors and a tactile receptor. All are primary neurons. One taste receptor is more or less specific to sugars, one to water and two to salts. If water is placed on one hair of a thirsty fly, action potentials generated by the water cell pass directly to the central nervous system and cause the fly to respond by extending its retractible proboscis and drinking. Similarly, sugar on one hair stimulates the sugar receptor and causes feeding. Salt causes the fly to reject the solution.

24–8 THE SENSE OF SMELL (OLFACTION)

The sense of smell of terrestrial vertebrates is served by primary neurons located in the *nasal epithelium* in the upper part of the nasal cavity (Fig. 24–10). Each of these neurons has a short axon that passes through the *cribriform plate* and immediately synapses with other neurons. The complexity of this

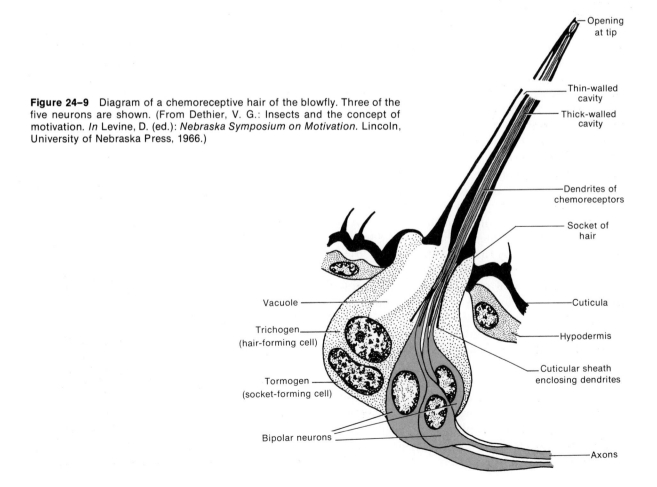

Figure 24–9 Diagram of a chemoreceptive hair of the blowfly. Three of the five neurons are shown. (From Dethier, V. G.: Insects and the concept of motivation. *In* Levine, D. (ed.): *Nebraska Symposium on Motivation.* Lincoln, University of Nebraska Press, 1966.)

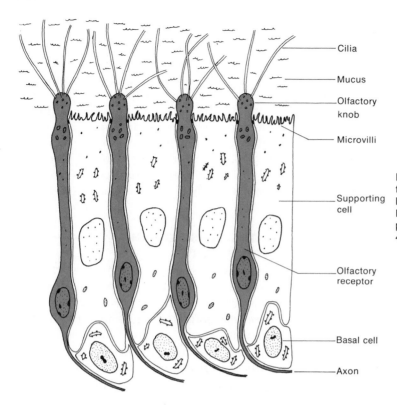

- Cilia
- Mucus
- Olfactory knob
- Microvilli
- Supporting cell
- Olfactory receptor
- Basal cell
- Axon

Figure 24–10 Simplified diagram of olfactory epithelium indicating the various cellular components. (From Moulton, D. G., and Beidler, L.: Structure and function in the peripheral olfactory system. Physiol. Rev. 47:4, 1967.)

neural circuitry is illustrated in Figure 24–11. In the rabbit, for example, there are 10^8 receptors. Twenty-six thousand of these synapse with each *glomerulus,* and each glomerulus connects with 24 *mitral cells* and 68 *tufted cells.* The possibilities for processing olfactory data generated by the receptors before they even reach the brain are enormous.

In contrast to the sensations of taste, the various odors cannot be assigned to specific classes; each substance has its own distinctive smell. The olfactory organs respond to remarkably small amounts of a substance. The synthetic substitute for the odor of violets, ionone, can be detected by most people when it is present to the extent of one part in more than thirty billion parts of air. The sense of smell is rapidly fatigued, and air originally having a powerful stimulus may seem odorless after a few minutes. This fatigue is specific for the particular substance producing it; thus receptors that have become insensitive to one substance will react to another quite normally. This suggests that there are many different kinds of sense cells, each specific for a particular chemical. Some people either completely lack a sense of smell, or are able to smell some substances but not others.

24–9 VISION

Light-sensitive cells exist in almost all organisms. Even protozoa respond to changes in light intensity, usually moving away from the source of light. Most plants orient their leaves and flowers toward the sun, although they have no special light-sensitive structures. In most of the higher animals this light-sensitivity is localized in certain cells and is highly developed. The human eye is an excellent example of an extremely sensitive, specialized organ for perceiving light. When fully dark-adapted, it can detect as little as 6 to 10 quanta of light. Just as matter consists of atoms, light consists of units called *photons,* and, by definition, the energy of 1 photon is 1 *quantum.* The light reaching the eye from a candle 14 miles away is just at the limit of visibility of a normal dark-adapted eye and is about 6 or 7 quanta of light.

Some protozoa have "eye spots" which are more sensitive to light than the rest of the cell, but the most primitive light-sensitive organs in the evolutionary scale are those of flatworms. These organs are *photoreceptors* rather than eyes—they cannot form images. They are bowl-shaped structures containing black pigment, at the bottom of which are clusters of

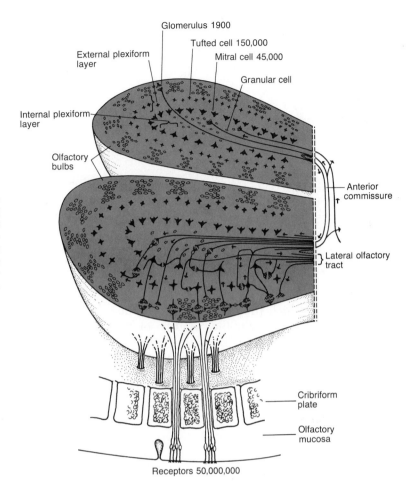

Glomerulus 1900
External plexiform layer
Tufted cell 150,000
Mitral cell 45,000
Granular cell
Internal plexiform layer
Olfactory bulbs
Anterior commissure
Lateral olfactory tract
Cribriform plate
Olfactory mucosa
Receptors 50,000,000

Figure 24–11 Structure of the olfactory bulbs and their relations to the nerves and mucosa. The figures are estimates of the numbers of each type of cell in an olfactory bulb and in the olfactory mucosa lining one nasal cavity of the rabbit. (From Moulton, D. G., and Tucker, D.: Electrophysiology of the olfactory system. Ann. N.Y. Acad. Sci. *116*:380–428, 1964.)

light-sensitive cells. These are shaded by the pigment from light coming from all directions except above and slightly to the front. This arrangement enables the planarian to detect the direction of the source of light. Planaria have other light-sensitive cells over the entire body surface and will continue to react to light after their "eyes" (the photoreceptors) have been removed, though more slowly and less accurately.

A necessary first step in the evolution from photoreceptor to true eye was the development of a *lens* to concentrate light on a group of photoreceptors. As better lens systems evolved, the photoreceptor became able to form images, and an eye in the strict sense of the word evolved. The most highly developed eyes are found in arthropods (insects, crabs, lobsters, and so on), in cephalopods (squids and octopuses) and in vertebrates. Two fundamentally different types of eyes have evolved: the *camera eye* of the vertebrates and cephalopods and the *compound* or *mosaic eye* of the arthropods.

24–10 THE HUMAN EYE

The squid or octopus eye is rather like a simple box camera equipped with slow, black and white film, whereas the human eye is like a de luxe Leica loaded with extremely sensitive color film.

The analogy between the human eye and a camera is complete: the eye (Fig. 24–12) has a *lens* which can be focused for different distances; a diaphragm (the *iris*) which regulates the size of the light opening (the *pupil*); and a light-sensitive *retina* located at the rear of the eye, corresponding to the film of the camera. Next to the retina is a sheet of cells filled with black pigment which absorbs extra light and prevents internally reflected light from blurring the image (cameras are also painted black on the inside). This sheet, called the *choroid coat,* also contains the blood vessels which nourish the retina.

The outer coat of the eyeball, called the *sclera,* is a tough, opaque, curved sheet of con-

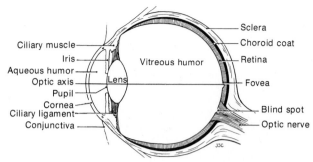

Figure 24–12 Diagrammatic section of the human eye. The retina contains the light-sensitive rods and cones; the lens and cornea focus light rays on the retina; and the iris regulates the amount of light entering the eye by changing the diameter of the pupil. The human eye changes focus for near and far vision by changing the shape of the lens; certain birds change the curvature of the cornea; fish change the position of the lens in the eye; and mollusks shorten the eye, bringing the retina nearer the lens for distant vision.

nective tissue which protects the inner structures and helps maintain the rigidity of the eyeball. On the front surface of the eye this sheet becomes the thinner, transparent *cornea,* through which light enters.

A transparent, elastic ball, the *lens,* located just behind the iris, bends the light rays coming in, bringing them to a focus on the retina. The lens is aided by the curved surface of the cornea and by the refractive properties of the liquids inside the eyeball. The cavity between the cornea and the lens is filled with a watery substance, the *aqueous humor;* the larger chamber between the lens and the retina is filled with a more viscous fluid, the *vitreous humor;* both fluids are important in maintaining the shape of the eyeball. They are secreted by the *ciliary body,* a doughnut-shaped structure which attaches the ligament holding the lens to the eyeball.

The eye accommodates, or changes focus for near or far vision, by changing the curvature of the lens. This is made possible by the stretching and relaxing of the lens by the *ciliary ligament,* which attaches the lens to the ciliary body. Because of the pressure of the fluids within, the eyeball is under tension transmitted by the ciliary ligament to the lens. Tension on the ligament flattens the lens and focuses the eye for far vision, the condition of the eye at rest. Just in front of the ciliary body, and attached to the ciliary ligament, are ciliary muscles which, when contracted, take up the strain on the ligament and lens, leaving the latter free to assume the more spherical shape for near vision.

As people grow older, the lens becomes less elastic and thereby less able to accommodate for near vision. When this occurs, spectacles with one portion ground for distant vision and one portion ground for near vision (bifocals) are worn to accomplish what the eye can no longer do.

The amount of light entering the eye is regulated by the *iris,* a ring of muscle which appears as blue, green or brown, depending on the amount and nature of pigment present. The structure is composed of two sets of muscle fibers, one arranged circularly, which contracts to decrease the size of the pupil, and one arranged radially, which contracts to increase the size of the pupil. The response of these muscles to changes in light intensity is not instantaneous, but requires 10 to 30 seconds; thus when one steps from a light to a dark area, some time is needed for the eyes to adapt to the dark, and when one steps from a dark room to a brightly lighted street, the eyes are dazzled until the size of the pupil is decreased.

Each eye has six muscles stretching from the surface of the eyeball to various points in the bony socket. These enable the eye as a whole to move and be oriented in a given direction. These muscles are innervated in such a way that the eyes normally move together and focus on the same area.

The only light-sensitive part of the human eye is the *retina,* a hemisphere made up of an abundance of receptor cells, called, according to their shape, *rods* and *cones.* There are about 125,000,000 rods and 6,500,000 cones. In addition, the retina contains many sensory and connector neurons and their axons. Curiously enough, the sensitive cells are at the *back* of the retina; to reach them, light must pass through several layers of neurons. The eye develops as an outgrowth of the brain, and folds in such a way that the sensitive cells eventually lie on the farthermost side of the retina (Fig. 24–13). At a point in the back of the eye, the individual axons of the sensory neurons unite to form the optic nerve and pass out of the eyeball. Here there are no rods and cones. This area is called the "blind spot" since images falling on it cannot be perceived. Its existence can be demonstrated by closing the left eye and focusing the right one on the + in Figure 24–14. Starting with the page about 13 cm. from the eye, move it away until the circle disappears. At that position the image of the

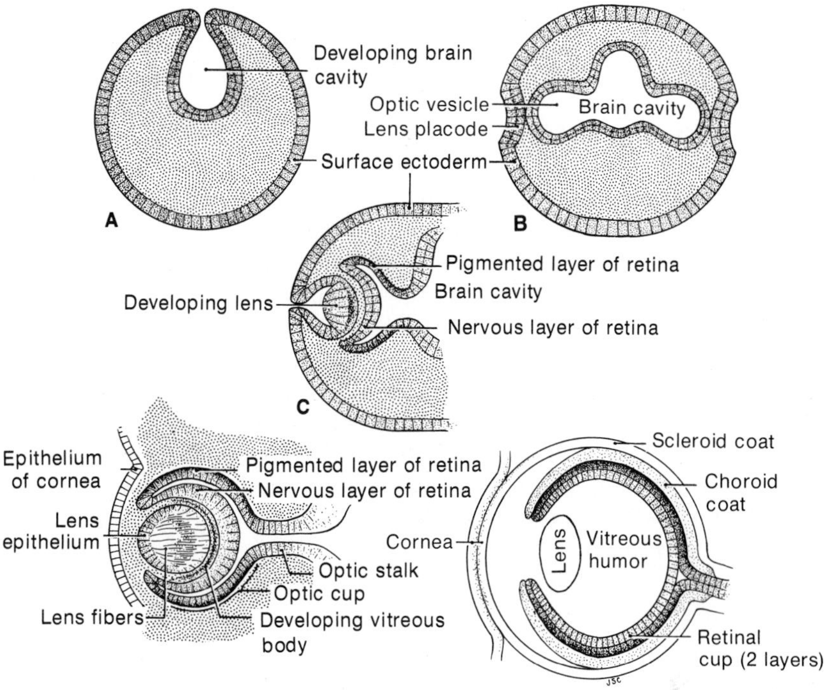

Figure 24–13 Diagrams illustrating some of the steps in the development of the eye in humans. From the developing brain cavity (A), the optic vesicles grow laterally (B) to meet the lens placode. This folds in (C), pushing part of the optic vesicle in with it to form the optic cup (lower left) and the final structure of the eye (lower right).

Developing brain cavity

Optic vesicle
Lens placode
Surface ectoderm

A

B

Brain cavity

Pigmented layer of retina
Brain cavity

Developing lens

Nervous layer of retina

C

Epithelium of cornea

Lens epithelium

Pigmented layer of retina
Nervous layer of retina

Cornea

Optic stalk
Optic cup

Lens fibers

Developing vitreous body

Scleroid coat

Choroid coat

Lens

Vitreous humor

Retinal cup (2 layers)

circle is falling on the blind spot and so is not perceived.

In the center of the retina, directly in line with the center of the cornea and lens, is the region of keenest vision, a small depressed area called the *fovea.* Here are concentrated the light-sensitive cones, responsible for bright light vision, for the perception of detail and for color vision.

In normal vision the eyes are constantly in motion; there are small involuntary movements even when the eye is fixed on a stationary object. Thus the image on the retina is in constant motion, drifting away from the center of the fovea and flicking back to it. Superimposed on these movements is a high speed tremor of the eye. When, by some suitable device, the image is fixed immovably on the retina it fades and disappears, later reappearing completely or in part. Such experiments by Donald Hebb and his colleagues have shown that what a subject "sees" is determined at least in part by whether or not it makes sense to him.

The other light-sensitive cells, the rods, are more numerous in the periphery of the retina, away from the fovea. These function in twilight or dim light and are insensitive to colors. One is not ordinarily aware that only those objects more or less directly in front of the eyes can be perceived in color, but this can be demonstrated by a simple experiment. Close one eye and focus the other on some point straight ahead. As a colored object is gradually brought into view from the side, you will be aware of its presence and of its size and shape before you are aware of its color. Only when the object is brought closer to the direct line of vision, so that its image falls on a part of the retina containing cones, can its color be determined. The rods are actually more sensitive in dim light than are the cones. Since the rods are located not in the center but in the periphery of the retina, it is a curious fact that you can see an object better in dim light if you do not look at it directly (for then its image will fall on the cones in the center of the retina),

Figure 24–14 Demonstration of the blind spot on the retina. See text for details.

but slightly to one side of it, so that its image falls on the rods in the periphery of the retina.

24–11 THE CHEMISTRY OF VISION

All *visual pigments* have a common plan—a *chromophore (retinal$_1$)* bound to a protein (*opsin*). The combination is known as *rhodopsin* or *visual purple.* It is the visual pigment in the rods of most vertebrates and in the retinal cells of insects. The cones contain *iodopsin,* a visual pigment consisting of the same chromophore (retinal$_1$) but a different protein. It is found in humans, chickens, cats, snakes, frogs and crayfish. Fresh water fish have different visual pigments. Their rod pigment (*porphyropsin*) consists of *retinal$_2$* plus opsin. The cone pigment cyanopsin consists of retinal$_2$ plus a different protein. Cyanopsin is found in the eyes of fresh water fish, tortoises and tadpoles. It is interesting that euryhaline eels, salmon and trout have both rhodopsin and porphyropsin, but the one (porphyropsin) commonly associated with spawning (in fresh water) predominates. The sea lamprey has

mostly rhodopsin on its downstream migration to the sea and mostly porphyropsin when going upstream as a sexually mature adult. Amphibians change from porphyropsin to rhodopsin when they change from tadpole to adult.

Retinal is the aldehyde of Vitamin A (*retinol*). It is formed from retinol by an oxidation catalyzed by *alcohol dehydrogenase.* Retinal combines with the different opsins, as just mentioned, to form the various kinds of visual pigments.

Quanta of light striking the rods or cones trigger the emission of a nerve impulse by the receptor cell. The nerve structures are ready to discharge, having been charged with the requisite energy by internal chemical reactions. Thus the phenomenon of vision is basically different from the phenomenon of photosynthesis (p. 92), in which the light supplies the energy to drive the chemical reactions.

The eye is called upon to respond to an enormous range of light intensities, and the rhodopsin system is peculiarly adapted to provide for this wide range of responses. The eyes of mollusks, arthropods and vertebrates, which arose quite independently in the course of

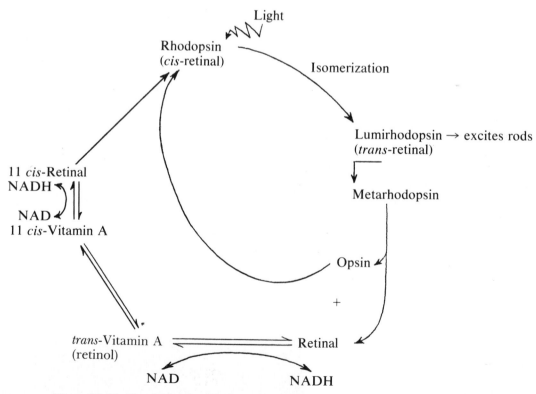

Figure 24–15   The rhodopsin-retinal system underlying the sensitivity of rods to light.

evolution, all utilize the same basic chemical reaction, the conversion by light of the *cis* form of retinal to the *trans* form. This involves a simple rearrangement of the molecular structure, which can occur very rapidly. Light energy converts rhodopsin (with the *cis* form of retinal) into **lumirhodopsin,** an unstable compound containing the *trans* form of retinal. This decays first into metarhodopsin and then into free retinal and opsin (Fig. 24–15). The sensation of vision, the stimulation of nerve impulses by the rods, occurs when light strikes rhodopsin and triggers the isomerization of *cis*-retinal into the all-*trans* form. The all-*trans* form of retinal does not fit onto the opsin molecule as neatly as does the bent *cis* form and hence the lumi- and metarhodopsin are more easily hydrolyzed. The cleavage of retinal from opsin is a much slower reaction than the isomerization, and is too slow to account for the high speed of the visual process.

The resynthesis of rhodopsin from retinal and opsin requires that the retinal first be reisomerized to the *cis* form. It can then be recombined with opsin to form rhodopsin and provide for further excitation of the rods. Rhodopsin is involved in a cyclic process: it is continually synthesized, and after the light-induced isomerization of *cis*-retinal to the *trans* form of retinal in lumi- and metarhodopsin, it is hydrolyzed to yield free retinal and opsin. The free all-*trans*-retinal is then converted to retinol (vitamin A), reisomerized to the *cis* form, and reoxidized to *cis*-retinal before it can be combined with opsin to yield new rhodopsin.

It has been shown that a single quantum of light can be absorbed by a single molecule of rhodopsin and lead to the excitation of a single rod. When the eye is exposed to a flash of light lasting only one millionth of a second, the eye sees an image of light that persists for nearly one tenth of a second. This is the length of time that the retina remains stimulated following a flash, and this presumably reflects the length of time that lumirhodopsin persists in the rods. This persistence of images in the retina enables your eye to fuse the successive flickering images on a motion picture or television screen and you have the impression of seeing a continuous picture.

The ability to see an exceedingly faint light depends on the amount of rhodopsin present in the retinal rods, and this is turn depends on the relative rates of synthesis and breakdown of rhodopsin. In bright light, much of the rhodopsin is broken down to free retinal and opsin. The synthesis of rhodopsin is a relatively slow process and the concentration of rhodopsin in the retina is never very great as long as the eye is exposed to bright light. When the eye is suitably shielded from light, the breakdown of rhodopsin is prevented and its concentration gradually builds up until essentially all of the opsin has been converted to rhodopsin. The sensitivity of the eye to light, a function of the amount of rhodopsin present, can increase 1000-fold if the eye is dark-adapted for a few minutes and can increase about 100,000-fold if the eye is dark-adapted for as much as an hour.

24–12 COLOR VISION

The chemistry of the cones and of color vision is less well understood, but the cones contain an analogous light-sensitive pigment, **iodopsin,** composed of retinal and a different opsin. The cones are considerably less sensitive to light than are rods and cannot provide vision in dim light. The prime function of the cones is to perceive colors. The evidence from certain psychological tests is consistent with the hypothesis that there are three different types of cones, which respond respectively to blue, green and red light. This has been substantiated recently by the demonstration that from human and monkey retinas can be extracted three kinds of color receptors—red, green and blue. Each type can respond to light with a considerable range of wavelengths; the green cones, for example, can respond to light of any wavelength from 450 to 675 nanometers (i.e., blue, green, yellow, orange and red light), but they respond to green light more strongly than to any of the others. Intermediate colors, other than blue, green and red, are perceived by the simultaneous stimulation of two or more types of cones. According to this theory, yellow light (i.e., light with a wavelength of 550 nanometers) stimulates green and red cones to an approximately equal extent, and this is interpreted by the brain as "yellow color." Color blindness (p. 674) results when one or more of the three types of cones is absent because of the absence of the gene that is necessary for the formation of that type of cone.

24–13 DEFECTS IN VISION

The most common defects of the human eye are nearsightedness (*myopia*), farsightedness (*hypermetropia*) and *astigmatism.* In the normal eye (Fig. 24–16 A), the shape of the eyeball is such that the retina is the proper distance behind the lens for the light rays to con-

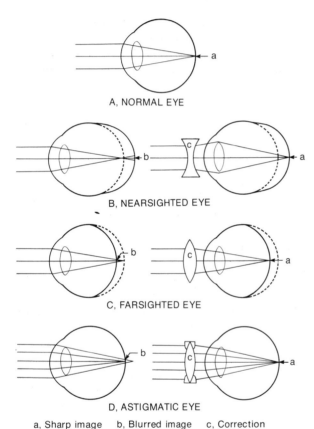

A, NORMAL EYE

B, NEARSIGHTED EYE

C, FARSIGHTED EYE

D, ASTIGMATIC EYE

a, Sharp image   b, Blurred image   c, Correction

Figure 24–16 Diagram illustrating common abnormalities of the eye. *A,* Normal eye, in which parallel rays coming from a point in space are focused as a point on the retina. *B,* Nearsighted eye, in which the eyeball is elongated so that parallel light rays are brought to a focus in front of the retina (on dotted line, which represents the position of the retina in the normal eye) and so form a blurred image on the retina. This situation is corrected by placing a concave lens in front of the eye, which diverges the light rays, making it possible for the eye to focus these rays on the retina. *C,* Farsighted eye, in which the eyeball is shortened and light rays are focused behind the retina. A convex lens converges the light rays so that the eye focuses them on the retina. *D,* Astigmatic eye, in which light rays passing through one part of the eye are focused on the retina, while light rays passing through another area of the lens are not focused on the retina, owing to unequal curvature of the lens or cornea. A cylindrical lens will correct this by bending light rays going through only certain parts of the eye.

verge in the fovea. In a near-sighted eye (Fig. 24–16 B), the eyeball is too long, and the retina is too far from the lens. The light rays converge at a point in front of the retina, and are again diverging when they reach it, resulting in a blurred image. In a far-sighted eye (Fig. 24–16 C), the eyeball is too short and the retina too close to the lens. Light rays strike the retina before they have converged, again resulting in a blurred image. Concave lenses correct for the near-sighted condition by bringing the light rays to a focus at a point farther back, and convex lenses correct for the far-sighted condition by causing the light rays to converge farther forward.

In astigmatism the cornea is curved unequally in different planes, so that the light rays in one plane are focused at a different point from those in another plane (Fig. 24–16 D). To correct for astigmatism, lenses must be ground unequally to compensate for the unequal curvature of the cornea.

In old age the lens may lose its transparency, become opaque, and interfere with the transmission of light to the retina, causing blindness. The only cure for this is surgical removal of the lens. This restores sight, but removes the ability to focus, so that a lensless person must wear special spectacles as a substitute for the lens.

24–14 BINOCULAR VISION AND DEPTH PERCEPTION

The position of the eyes in the head of humans and certain other higher vertebrates permits both of them to be focused on the same object. This **binocular vision** is an important factor in judging distance and depth. To focus on a near object, the eyes must converge (become slightly cross-eyed). The proprioceptors in the eye muscles causing this convergence are stimulated by this contraction to send impulses to the brain; hence part of our judgment of distance and depth depends upon impulses originating when the sensory fibers in those muscles are stimulated. In addition, the eyes, being a little over 5 cm. apart, see things from slightly different angles and thus get slightly different views of a close object. Depth perception is also made possible by the differential size of near and far objects on the retina, by perspective, by overlap and shadow, by distance over the horizon, and by the increasing dimness of distant objects.

24–15 THE COMPOUND EYE

The compound eye of arthropods is made up of hundreds to thousands of *monopolar neurons* that are gathered together in groups of seven or eight, each group being provided with its own cornea and lens (Fig. 24–17). The *cornea* is a transparent thickened area of the body cuticle, and the *lens* is formed by special epithelial cells. Each unit consisting of cornea, lens and neurons is called a *facet,* or *ommatidium.* It has a sheath of pigmented, enveloping cells. In nocturnal and crepuscular insects and many crustacea, this pigment is capable of migrating proximally and distally (Fig. 24–17). When the pigment is in the proximal position, each ommatidium is shielded from its neighbor and only light entering directly along its axis can stimulate the receptors. When the pigment is in the distal position, light striking at an angle may pass through several ommatidia and stimulate many retinal units. Thus in dim light, sensitivity of the eye is increased, and in bright light, the eye is protected from excessive stimulation. Pigment migration is under neural control in insects and under hormonal control in crustacea. In some species it follows a circadian rhythm.

The perception of form by compound eyes is poor. Although the lens system of each ommatidium is adequate to focus a small inverted image on the retinal cells, these images are physiologically unimportant. Each ommatidium in gathering light from a narrow sector of the visual field is in fact sampling a mean intensity from that sector and projecting it as a point of light on the retinal field. All of these points of light taken together form a *mosaic* picture. To appreciate the nature of this mosaic picture one need only look at a newspaper photograph through a magnifying glass. It is a mosaic of many dots of different intensities. The clearness and definition of the picture will depend upon how many dots there are per unit area—the more dots the better the picture. So it is with the compound eye.

Although the compound eye forms only coarse images, it compensates for this by being able to follow *flicker* to high frequencies. Flies are able to detect flickers up to about 265 per second; in contrast, the human eye can detect flickers of 45 to 53 per second. Because flickering lights fuse above these values, we see motion pictures as smooth movement and the ordinary 60-cycle light in the room as steady. To an insect, both motion picture and light must flicker horribly. Because the insect has such a high critical flicker fusion rate, any movement of prey or enemy is immediately detected by one of the eye units. Hence the compound eye is peculiarly well suited to the arthropod's way of life.

Compound eyes are superior to our eyes in two other respects. They are sensitive to different wavelengths of light from the red into the ultraviolet, and they are able to analyze the *plane of polarization* of light. Accordingly an insect can see well in the ultraviolet, and its world of color is much different from ours. Since different flowers reflect ultraviolet to different degrees, two flowers that appear iden-

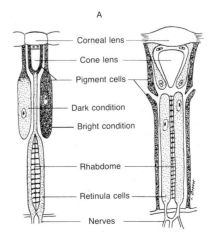 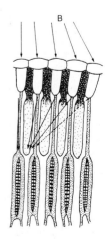

Figure 24–17 *A,* Insect ommatidia, showing a diurnal type (*left*) and a nocturnal type (*right*). In the diurnal type, the pigment is shown in two positions, adapted for very dark conditions on the left side, and for relatively bright conditions on the right. *B,* Nocturnal type of eye adapted for dark conditions, showing how light can be concentrated upon one rhabdome from several lenses. If the pigment moved downward, light from peripheral lenses would be screened out.

Corneal lens
Cone lens
Pigment cells
Dark condition
Bright condition
Rhabdome
Retinula cells
Nerves

Figure 24–18 *A,* Ultraviolet video-viewing of marsh marigolds, *Caltha palustris,* in the field. *B,* The marsh marigold appears to be uniformly yellow to the human eye. *C,* When viewed with an ultraviolet camera the flowers have dark, light-absorbing centers. (From Eisner, T.: Science 28:1172, 1969.)

tically white to us may appear strikingly different to insects. How the world appears to an insect with ultraviolet vision can be appreciated by viewing the landscape through a television camera with an ultraviolet-transmitting lens (Fig. 24–18). A sky that appears equally blue to us in all quadrants reveals quite different patterns to an insect, because the plane of polarization of the light is not the same in all parts of the sky, and the insect's eye can detect the difference. Honey bees and some other arthropods employ this ability as a navigational aid.

24–16 THE EAR

The organs of two different senses, hearing and equilibrium, are located in the ear. These mechanoreceptor organs are buried deep in the bone of the skull, and a number of accessory structures are needed to transmit sound waves from the outside to the deep-lying sensory cells.

The outer ear consists of two parts, the skin-covered cartilaginous flap, or *pinna,* and the *auditory canal* leading from it to the middle ear.

The pinnas, or visible ears, are of some slight use in man for directing sound waves into the canal, but in other animals such as the cat, the larger, movable pinnas are very important. At the junction of the auditory canal and the middle ear is stretched a thin membrane of connective tissue, the *eardrum* (Fig. 24–19), set vibrating by the sound waves.

The middle ear is a small chamber containing three tiny bones, the *hammer, anvil* and *stirrup* (so called because of their shapes). These are connected in series and transmit sound waves across the middle ear cavity. The hammer is in contact with the eardrum, and the stirrup is in contact with the membrane of the opening into the inner ear, called the *oval window.* The narrow *eustachian tube* connects the middle ear to the pharynx and serves to equalize the pressure on the two sides of the eardrum. If the middle ear were completely closed, any variation in atmospheric pressure would cause a pronounced and painful bulging or caving in of the eardrum. A valve at the pharyngeal end of the eustachian tube is normally closed to prevent one from becoming unpleasantly aware of his own voice. This valve is opened during yawning or swallowing, and during an abrupt ascent or descent in an elevator or airplane such acts help prevent the cracking sensation of the eardrums produced by the changes in atmospheric pressure accompanying changes in altitude. Unfortunately, the eustachian tube also provides a path for organisms to invade the middle ear cavity and cause infections, which can result in the fusing of the middle ear bones and loss of hearing.

The inner ear consists of a complicated group of interconnected canals and sacs, often

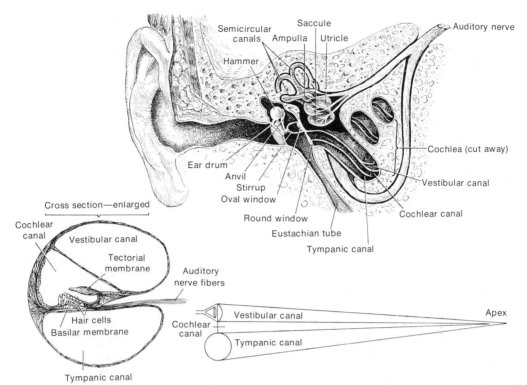

Figure 24–19 *Upper right,* The coiled cochlea, shown dissected out of the skull and cut open to reveal the vestibular and tympanic canals. *Lower right,* A diagram of the cochlea as though it were uncoiled and drawn out in a straight line. *Lower left,* A cross section through the cochlea to show the organ of Corti resting on the basilar membrane and covered by the tectorial membrane. Vibrations transmitted by the hammer, anvil and stirrup set the fluid in the vestibular canal in motion; these vibrations are transmitted to the basilar membrane and the organ of Corti. The hair cells of the organ of Corti are the receptor cells for hearing and are innervated by branches of the auditory nerve.

referred to, most appropriately, as the *labyrinth.* The part of the labyrinth concerned with hearing is a spirally coiled tube of two and a half turns, resembling a snail's shell, called the *cochlea.* If the cochlea were uncoiled, as in Figure 24–19, it could be seen to consist of three canals separated from each other by thin membranes and coming almost to a point at the apex. The oval window is connected to the base of one of these tubes, the *vestibular canal.* At the base of the *tympanic canal* is another opening covered by a membrane, the *round window,* also leading to the middle ear. These two canals are connected with each other at the apex of the cochlea and are filled with a fluid known as the *perilymph.* Between the two lies a third, the *cochlear canal,* filled with a fluid called *endolymph* and containing the actual organ of hearing, the *organ of Corti.* This structure consists of five rows of cells extending the entire length of the coiled cochlea; each cell is equipped with hairlike projections

extending into the cochlear canal. Each organ of Corti contains about 24,000 of the hair cells. These cells rest upon the *basilar membrane* separating the cochlea from the tympanic canal. Overhanging the hair cells is another membrane, the *roof* (or tectorial) *membrane,* attached along one edge to the membrane on which the hair cells rest, and with its other edge free. The hair cells initiate impulses in the fibers of the auditory nerve.

For a sound to be heard, sound waves must first pass down the auditory canal and set the eardrum vibrating. These vibrations are transmitted across the middle ear by the hammer, anvil and stirrup, which are so arranged that they decrease the amplitude but increase the force of the vibrations. The stirrup transmits the vibrations, via the oval window, to the fluid in the vestibular canal. Since fluids are incompressible, the oval window could not cause a movement of the fluid in the vestibular canal unless there were an escape valve for the pressure. This is provided by the round window at the end of the tympanic canal. The pressure wave presses upon the membranes separating the three canals, is transmitted to the tympanic canal, and causes a bulging of the round window. The movements of the basilar membrane produced by these pulsations are believed to rub the hair cells of the organ of Corti against the overlying roof membrane, thus stimulating them and initiating nerve impulses in the dendrites of the auditory nerve, lying at the base of each hair cell.

Since sounds differ in pitch, intensity and quality, any theory of hearing must account for the ability to discriminate such differences. Microscopic examination of the organ of Corti reveals that the fibers of the basilar membrane are of different lengths along the coiled cochlea, being longer at the apex and shorter at the base of the coil, thus resembling the strings of a harp or piano. Sounds of a given frequency (and pitch) set up resonance waves in the fluid in the cochlea that cause a particular section of the basilar membrane to vibrate. The vibration stimulates the particular group of hair cells in that section. Thus the pitch of a sound is sensed by the particular hair cells stimulated. Loud sounds cause resonance waves of greater amplitude and lead to a more intense stimulation of the hair cells and to the initiation of a greater number of impulses per second passing over the auditory nerve to the brain.

When the ear is subjected to intense, continuous sound, the organ of Corti is injured. This was demonstrated by an experiment in which guinea pigs were exposed to continuous pure tones for a period of several weeks. When their cochleas were examined microscopically after death, it was found that the guinea pigs subjected to high-pitched tones suffered injury only in the lower part of the cochlea, while those subjected to low-pitched tones suffered injury only in the upper part of the cochlea. Boiler-makers and other workers subjected to loud, high-pitched noises over a period of years frequently become deaf to high tones because of injury of the cells toward the base of the organ of Corti. The nerve impulses produced by particular sounds have the same frequency as those sounds; thus the brain may recognize particular pitches by the frequency of the nerve impulses reaching it, as well as by the identity of the nerve fibers conducting the impulses.

The auditory nerves transmit two kinds of nerve impulses: ordinary nerve impulses like those of any other nerve, and a different type called *microphonic.* The energy for the latter is not derived from the metabolism of the nerve fiber, as is the energy for the former; instead, the cochlea acts as a microphone to convert the mechanical energy of the sound vibrations into electrical energy. For this reason the wave form of the electrical potential from the cochlea closely resembles that of the stimulating sound wave. In fact, Wever and Bray placed electrodes on the auditory nerve of a decerebrated cat, and then, listening with a telephone receiver to the amplified signals of the nerve, were able to hear not only musical tones, but actual words spoken to the cat. The hair cells of the organ of Corti are believed to be responsible for this conversion of mechanical to electrical energy, the upper and lower ends of the cochlea responding to low and high tones, respectively. It is still a disputed question, however, whether these microphonics have anything to do with the actual sensation of hearing in the normal animal.

Variations in the quality of sound, such as are produced when an oboe, a cornet and a violin play the same note, depend upon the number and kinds of *overtones,* or *harmonics,* present. These stimulate different hair cells in addition to the main stimulation common to all three; thus, differences in quality are recognized by the *pattern* of the hair cells stimulated. Careful histologic work has shown that the nerve fibers from each particular part of the cochlea are connected to particular parts of the auditory area of the brain, so that certain brain cells are responsible for the perception of sensations of high tones, others for low tones.

The human ear is equipped to register sounds of frequencies between about 20 and 20,000 cycles per second, although there are great individual differences. Some animals—dogs, for example—can hear sounds of much higher frequencies. The human ear is more sensitive to sounds between 1000 and 2000 cycles per second than to higher or lower ones. Within this range the ear is extremely sensitive; in fact, when compared with the energy of light waves necessary to produce a sensation, the ear is ten times more sensitive than the eye.

The normal human ear is just about as efficient a hearing device as anything could possibly be, for, like the eye, it has evolved to the point where any further increase in sensitivity would be useless. If it were more sensitive, it would pick up the random movement of the air molecules, which would result in a constant hiss or buzzing. If the eye were more sensitive, a steady light would appear to flicker because the eye would be sensitive to the individual photons (light particles) impinging on it.

There is little fatigue connected with hearing. Even though it is constantly assailed by noises, the ear retains its acuity, and fatigue disappears after a few minutes. When one ear is stimulated for some time by a loud noise, the other ear also shows fatigue, i.e., loses acuity, indicating, not unexpectedly, that some of the fatigue is in the brain rather than in the ear itself.

Deafness may be caused by injuries or malformations of either the sound-transmitting mechanisms of the outer, middle or inner ears, or of the sound-perceiving mechanism of the latter. The external ear may become obstructed by wax secreted by the glands in its wall; the middle ear bones may become fused after an infection; or, more rarely, the inner ear or auditory nerve may be injured by a local inflammation or the fever accompanying some disease.

Relatively few animals have a sense of hearing. The vertebrate ear began as an organ of equilibrium, the cochlea being a later evolutionary outgrowth of the saccule that reached full development only in mammals. The human ear is indeed a curious evolutionary hodgepodge: the cells sensitive to sound are apparently adaptations of cells sensitive to the motion of liquids; the middle ear and eustachian tube were originally part of the respiratory apparatus of fish; the stirrup was originally a structure which attached the jaws of primitive fishes to the cranium, and the hammer and anvil are the remnants of the lower and upper jaws, respectively, of our ancestral fish. In the jawless fish ancestral to these, the structures were part of the support for the gills. Thus, respiratory organs became, first, eating organs, and then organs for hearing. This is an example of one of the fundamental patterns of evolution—the reshaping of old organs to perform new functions, rather than the creation of completely new structures.

24–17 EQUILIBRIUM

Besides the cochlea, the labyrinth of the inner ear consists of two small sacs—the *saccule* and *utricle*—and three *semicircular canals* (Fig. 24–20). These structures are filled with endolymph and float in a pool of perilymph. Destruction of these structures leads to a considerable loss of the sense of equilibrium. A pigeon in which these organs have been destroyed is unable to fly, but in time it can relearn to maintain equilibrium using visual stimuli.

Equilibrium in the human depends upon the sense of vision, stimuli from the proprioceptors, and stimuli from cells sensitive to pressure in the soles of the feet, as well as upon stimuli from the organs in the inner ear. In certain types of deafness the equilibrium organs of the inner ear, as well as the cochlea, are inoperative, yet the sense of equilibrium remains unimpaired.

The utricle and saccule are small, hollow sacs lined with sensitive hair cells and containing small ear stones, or *otoliths,* made of calcium carbonate. Normally, the pull of gravity causes the otoliths to press against particular hair cells, stimulating them to initiate impulses to the brain via sensory nerve fibers at their bases. When the head is tipped, the otoliths

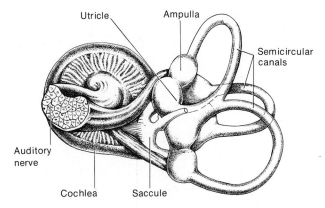

Figure 24–20 The right semicircular canals and cochlea of an adult human, shown dissected free of surrounding bone and enlarged about five times, seen from the inner and posterior side. Note that the plane of each semicircular canal is perpendicular to those of the other two.

press upon the hairs of other cells and stimulate them.

Many invertebrates, such as the crayfish and lobster, have similar organs. An ingenious experiment was performed to demonstrate the action of these organs in the crayfish; it depended on the fact that as the crayfish molts—sheds its skin and grows another, larger one—it also develops new organs of equilibrium and supplies them with grains of sand picked up from the environment. By supplying the molting crayfish with particles of iron, the experimenters could subsequently cause the animals to respond to a magnet. When the magnet was placed directly over the animal, pulling the iron filing against the hair cells on the top of its equilibrium organ, the crayfish thought that "up" was "down" and responded by turning over and swimming on its back.

The labyrinth of each ear has three semicircular canals, each consisting of a semicircular tube connected at both ends to the utricle. The canals are so arranged that each is at right angles to the other two. At one of the openings of each canal into the utricle is a small, bulblike enlargement (the *ampulla*) containing a

clump of hair cells similar to those in the utricle and saccule, but lacking otoliths. These cells are stimulated by movements of the fluid (endolymph) in the canals. When the head is turned, there is a lag in the movement of the fluid within the canals, so that the hair cells in effect move in relation to the fluid and are stimulated by its flow. This stimulation produces not only the consciousness of rotation but also certain reflex movements in response to it, movements of the eyes and head in a direction opposite to the original rotation. Since the three canals are located in three different planes, a movement of the head in any direction will stimulate the movement of the fluid in at least one of the canals.

By irrigating the canal of the outer ear with warm or cold water, convection currents can be established, causing movements in the fluid of the canals without movement of the head. Sensations of rotation and dizziness result. Human beings are used to movements in the horizontal plane, which stimulate certain semicircular canals, but we are unused to vertical movements parallel to the long axis of the body. Such movements, the motion of an eleva-

A

B

Figure 24–21   *A,* Ventral view and *B,* dorsal view of a model of the left haltere of the blowfly *Lucilia sericata.* Rows of sense organs are visible at the base. Magnification ×130. (Courtesy of J. W. S. Pringle.)

tor or of a ship pitching in a rough sea, stimulate the semicircular canals in an unusual way and may produce the sensation of nausea and the vomiting of sea or motion sickness. When one lies down, the movement stimulates the semicircular canals in a different way, and nausea is less likely to occur.

24–18 EQUILIBRIUM IN FLIES: HALTERES

Millions of years before man invented the gyroscope, flies had evolved a balancing organ to stabilize flight. Any flying machine must maintain stability if it is to be controllable in the air. Flies must be able to control lift and to stabilize in all three planes of rotation; i.e., they must correct for pitch, roll, and yaw. They accomplish this with information derived from the *halteres,* a pair of marvelously modified hind wings. Each is a heavy mass of tissue on a thin stalk (Fig. 24–21) and resembles an Indian club. The base is folded and articulated in a complicated fashion and is equipped with about 418 mechanoreceptors. These respond to strains produced in the cuticle by *gyroscopic torque* generated by the beating of the halteres. These oscillating masses generate forces at the base of the stalk as the whole fly rotates. They probably do not act as stabilizing gyroscopes of the sort placed in ships to offset their movement. Their action is indirect in that their mechanoreceptors signal the central nervous system to make the necessary corrections in flight.

QUESTIONS

1. What is meant by a proprioceptor? What is its function in the mammalian body?
2. How do anesthetics act to reduce or eliminate the sensation of pain?
3. Draw a diagram of the human eye, labeling all parts. How are rods and cones distributed in the retina?
4. Discuss the mechanism by which the photoreceptors are stimulated by light. What is the function of lumirhodopsin? How is it regenerated?
5. What produces (a) myopia, (b) hypermetropia and (c) astigmatism?
6. How is the human eye regulated for far and near vision and for seeing in bright and dim lights? In what respects does the frog eye differ in these regulatory mechanisms?
7. Draw a diagram of the ear, labeling all parts.
8. Discuss the mechanism by which the sensory cells of the ear are stimulated by sound waves.
9. What are otoliths and what is their role in maintaining equilibrium?
10. The philosopher Berkeley, along with many others, believed that there is no material world independent of our sense perceptions, and that what we call perceptions of the external world are really ideas transferred from God's mind to our own. Do you think there are any facts known today which absolutely disprove this theory?
11. Do you think there is any justification for the often repeated statement that one person's taste is as good as another's? What do you think is, or should be, the basis for esthetic judgment? Could there be esthetic standards completely independent of the human race?

SUPPLEMENTARY READING

A number of the original papers describing the perception of sensory stimuli are found in *Readings in the History of Psychology* by W. Dennis. The role of the brain in the perception of sensory stimuli is discussed in Edgar Adrian's *The Physical Background of Perception* and in R. L. Gregory's *Eye and Brain: The Psychology of Vision.* Interesting accounts of the functioning of the eye are found in the three Nobel laureate lectures by R. Granit, H. K. Hartline and George Wald, published in *Science,* Vol. 168, 1968. A general discussion of the perception of sensory stimuli is presented by M. Alpern, M. Lawrence and B. Wolsk in *Sensory Processes.* An interesting theory of the chemical basis of the perception of odors is presented in *The Stereochemical Theory of Odor* by J. E. Amoore, J. W. Johnston, Jr. and M. Rubin. A well illustrated account of the sense of hearing is given in *Sound and Hearing* by S. Stevens and F. Warshossky. Vincent Dethier's classic studies of the sensory systems of insects are wittily summarized in *The Hungry Fly.*

CHAPTER 25

CONTROL SYSTEMS: HORMONAL INTEGRATION

The activities of the various parts of the bodies of higher animals are integrated by the nervous system and by the hormones of the endocrine system. The swift responses of muscles and glands, measured in milliseconds, are typically under nervous control. The hormones secreted by the endocrine glands diffuse or are transported by the bloodstream to other cells of the body and regulate their activities. The responses controlled by hormones are in general somewhat slower—measured in minutes, hours or even weeks—but longer lasting than those under nervous control. The long-term adjustments of metabolism, growth and reproduction are typically under endocrine control. Endocrines play a major role in maintaining constant the concentrations of glucose, sodium, potassium, calcium, phosphate and water in the blood and extracellular fluids.

The activities of the several parts of the higher plants are also coordinated and integrated by hormones, although plants do not have distinct, well defined endocrine glands. Plant hormones are mostly concerned with the regulation of growth and differentiation and are secreted by actively growing meristematic tissues in the growing points at the tips of stems and roots. They move through the vascular tissues of the plants, primarily in the phloem, to other regions where they regulate certain cellular activities.

25–1 ENDOCRINE GLANDS

Endocrine glands secrete their products into the blood stream, rather than into a duct leading to the outside of the body or to one of the internal organs. For this reason they are sometimes referred to as ductless glands or glands of internal secretion. The thyroid, parathyroids, pituitary and adrenals function only in the secretion of hormones and are truly ductless glands. The pancreas, ovaries and testes have both external secretions, via ducts, and internal secretions, carried by the blood stream. The pancreas is really two functionally separate organs combined in the same structure; it produces digestive enzymes as well as hormones. In some of the lower vertebrates the two parts of the pancreas are anatomically separate.

Practical knowledge of endocrinology—exemplified by the castration of both men and animals—has existed for several thousand years, but modern endocrinology usually is said to date from 1849. From the results of experiments in which he transplanted testes from one bird to another, Berthold postulated that these male sex glands secrete into the blood stream some substance essential for the differentiation of the male secondary sex characteristics. The first attempt at endocrine therapy was made in 1889, when the French physiologist Brown-Séquard injected himself with testicular extracts and claimed that the injections were beneficial and rejuvenating. It is uncertain whether the extracts he used had any effect other than a psychologic one, but his claims stimulated a great deal of research.

To determine whether a gland suspected of producing a hormone is in fact an endocrine gland, an investigator usually begins by removing the gland surgically and observing the effect on the animal. Next he replaces the gland with one transplanted from another animal and determines whether the changes induced by removing the gland can be reversed by replacing it. In replacing the gland he is careful to ensure that the new gland becomes connected only with the vascular system of the recipient and not with a duct of some sort. He might next try feeding dried glands to an animal from which the gland has been removed to determine whether the active substance, the hormone, can be replaced in this way. Finally he extracts the gland with a variety of solvents to determine the solubility characteristics of the active material and gain a clue as to its chemical nature. By making an extract of the gland, or perhaps by making an extract of blood or urine, and purifying the extract by a sequence of appropriate chemical and physical methods, he finally obtains the pure compound and can determine its exact chemical structure. At each step of the extraction and purification procedure, he tests his material by injecting it into a test animal from which the gland has been removed. In this way he finally determines what specific chemical in the original extract will

reverse the changes caused by removing the gland.

Methods for investigating plant hormones are complicated by the fact that plant hormones are not secreted by well defined glands but by all of the growing tips. This has challenged investigators to devise means of removing the growing tips, extracting them to obtain the pure hormones, and finding an appropriate test tissue to measure hormonal activity.

Hormones are remarkably effective substances and a very small quantity produces a marked effect on the structure and function of one or another part of the body. Only small amounts of a hormone are secreted at any one time by the endocrine gland and the concentration in the circulating blood is very small. Even the amount excreted each day in the urine is not very large. Because of this the isolation of a pure hormone can be a difficult job indeed. To obtain a few milligrams of pure estradiol, one of the female sex hormones, more than 2 tons of pig ovaries were extracted! When the hormone secreted by a given gland has been extracted and its chemical structure has been determined, it can then be prepared synthetically to be used in treating diseases resulting from a deficiency of that hormone. Synthetic thyroxine, epinephrine, oxytocin and ACTH are widely used, and it is possible to synthesize insulin and other peptide hormones, although this has not yet been carried out on a commercial scale. Synthetic glucocorticoids are used in the treatment of Addison's disease and as antiinflammatory agents. Synthetic estrogens and progestins are widely used as constituents of oral contraceptive pills.

25–2 WHAT IS A HORMONE?

The word hormone does not describe some particular class of chemical compound, as do the words protein, lipid and carbohydrate. It is an operational term defined by Bayliss as "a substance secreted by cells in one part of the body that passes to another part where it is effective in very small concentrations in regulating the growth or activity of other cells." Hormones are typically carried in the blood from the site of production to the site of action, but neurohormones may pass down an axon, and prostaglandins may be transferred in the se-

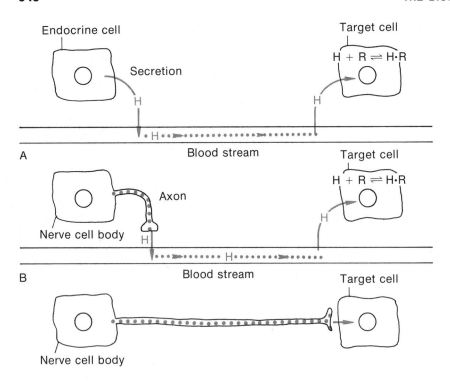

Figure 25–1 Diagram illustrating the differences between endocrine secretion (*A*), neurosecretion (*B*) and neurotransmission (*C*). Hormone (*H*); receptor (*R*).

minal fluid. This definition of hormones includes some remarkably diverse chemical compounds—amino acids and amines, peptides, proteins, fatty acids, purines, steroids and gibberellins. It does not seem likely that all of these diverse compounds would affect cell function by the same or similar mechanisms. Indeed, it is becoming apparent that some, perhaps many, hormones have several independent mechanisms of action by which they regulate cellular activities.

The impulses traveling in sensory neurons, motor neurons and interneurons are all similar simple action potential spikes. The type of information transmitted depends on where the impulse originates and where it ends. The kind of perception generated in the brain depends on where the sensory stimulus originated—in the eye, the ear, the nose, the skin or in the internal organs. In the endocrine system the information transferred depends not only on where the information originates and where it ends but also on the kind of material being transported. In any endocrine system we can distinguish three parts—the secreting cell, the transport mechanism and the target cell—each characterized by a greater or lesser degree of specificity (Fig. 25–1). In general, each type of hormone is synthesized and secreted by a specific kind of cell. Some hor-

mones are transported in the blood stream in solution, but most are bound to some protein component of the serum. Some are bound nonspecifically to albumin; others are selectively bound to specific high affinity proteins.

Most hormones characteristically elicit a detectable response only in certain cells of the body. With some hormones only a few types of cells respond; with others a broader spectrum of response can be observed. An engrossing question at the present is "How does the target cell recognize its appropriate hormone?" For many hormones a specific protein, a *receptor,* has been found in the target cell and the receptors have been at least partially purified. Apparently the first interaction of the hormone with any component of the target cell is with the receptor protein, and it is the specificity of the latter that determines which hormone is effective in that cell. The blood bathing each cell in the body contains the entire spectrum of hormones, but the specific receptor enables the cell to pick out one specific hormone and ignore all the others.

Amino Acid Derivatives. *Epinephrine* and *norepinephrine* are amines derived from the amino acid tyrosine by hydroxylation and decarboxylation (Fig. 25–2). They are synthesized and stored in the chromaffin cells of the adrenal medulla and released when the cells

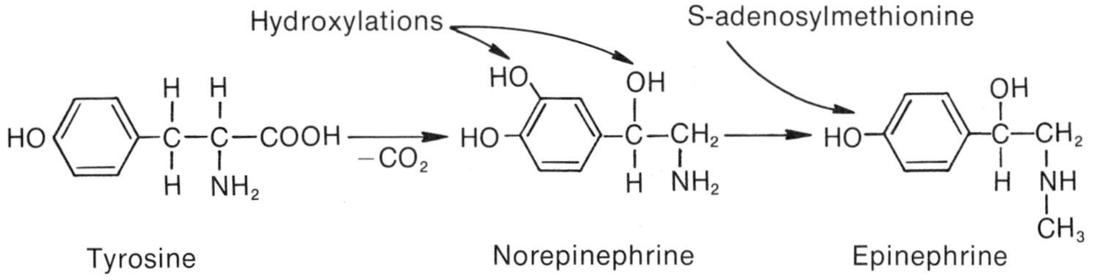

Figure 25-2 The synthesis of epinephrine and norepinephrine from tyrosine.

are stimulated by impulses transmitted by the sympathetic nervous system. Norepinephrine is also a neurotransmitter produced at the tips of the axons of adrenergic nerves and involved in the transmission of the nerve impulse across the synapse to the adjacent neuron.

Thyroxine, synthesized in the thyroid gland, is another metabolic derivative of tyrosine. The thyroid has a remarkably effective iodide pump that can accumulate iodide from the blood stream and concentrate it many fold. Tyrosine residues are iodinated to form diiodotyrosine, and two of these are coupled to form thyroxine (Fig. 25-3). The iodination of the tyrosine and the joining of the iodinated tyrosines occur while the amino acids are part of a large protein molecule, *thyroglobulin.* Thyroxine contains four atoms of iodine, and *triiodothyronine,* which is even more hormonally active than thyroxine, has three atoms of iodine. The anterior pituitary secretes thyroid stimulating hormone (TSH) which stimulates the several processes in the thyroid gland involved in the production and secretion of thyroxine—the uptake of iodide, the addition of iodine to tyrosine, and the release of thyroxine and triiodothyronine from thyroglobulin.

Indoleacetic acid, the primary growth hormone or *auxin* of plants, is synthesized from the amino acid tryptophan by transamination and decarboxylation (Fig. 25-4) in the growing points at the tips of stems and roots. Only small amounts of auxin are present in a plant at any given moment, for it is converted by enzymatic reactions to the inactive compound indoleformaldehyde.

Purine Derivatives. A second plant growth hormone, *cytokinin,* is a purine, a derivative of adenine. It has a five-carbon isoprenoid chain attached to the amino group at carbon 6 in the purine ring (Fig. 25-5). *Zeatin,* the cytokinin found in young corn seeds, is 6-(4-hydroxy-3-methyl-*trans*-2-butenylamino) purine. A purine very similar to this was found to be a constituent of both serine transfer RNA and tyrosine transfer RNA (p. 700). This is 6-(γ, γ-dimethylallyl amino) purine, which differs from zeatin only in its lack of a hydroxyl group on one of the methyl groups in the side chain.

Fatty Acid Derivatives. There is great interest at present in the *prostaglandins,* derivatives of 20-carbon polyunsaturated fatty acids. Prostaglandins have a characteristic five-carbon ring in the middle of the chain and hydroxyl or ketone groups on certain carbons (Fig. 25-6). Many different prostaglandins have

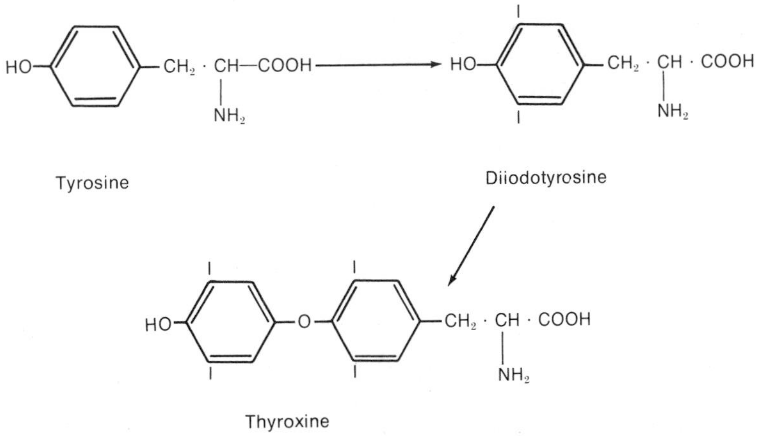

Figure 25-3 Diagram of the reactions by which thyroxine is synthesized from tyrosine.

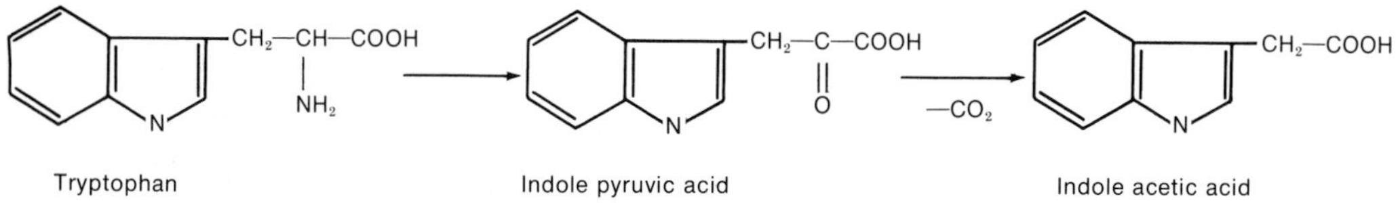

Figure 25-4 Diagram of the reactions by which indoleacetic acid (auxin) is synthesized from tryptophan.

been discovered or synthesized and their chemistry is quite complex. Although prostaglandins were first found in semen and are produced by the seminal vesicle, it is now clear that they are produced by a great many types of tissues. They are synthesized from polyunsaturated fatty acids such as *arachidonic,* a 20-carbon fatty acid chain with four double bonds. The production rate in the human, the amount synthesized per day, was measured by Samuelson in Sweden. In four normal adult males the production rate ranged from 109 to 226 μg. per 24 hours, and in two normal adult females it ranged from 23 to 48 μg. per 24 hours. These experiments indicate that there is a sex difference in the rate of production of prostaglandins and that only a small fraction of the polyunsaturated fatty acids consumed in the diet (about 10 grams per day) is converted to prostaglandins. The high concentration of prostaglandin in human semen appears to be essential for normal fertility.

Prostaglandins have a variety of effects on smooth muscles, the nervous system and blood pressure, and have been implicated in the control of many different types of biological events—perhaps by regulating the production of cyclic AMP by adenyl cyclase. Prostaglandins decrease arterial blood pressure, but they increase the motility of uterine muscles and are used in increasing uterine contractions at the time of childbirth. They are also used to induce abortions earlier in pregnancy. Prostaglandins also increase the motility of intestinal muscles and may cause severe cramps, nausea, vomiting and diarrhea. In a number of species

certain prostaglandins have a sedative tranquilizing effect on the animal.

A second hormone derived from fatty acids is *juvenile hormone,* involved in the control of insect metamorphosis. The juvenile hormone isolated from cecropia moths is methyl 10-epoxy-7-ethyl-3,11-dimethyltrideca-2,6-dienoate (Fig. 25–7); this methyl ester of an unsaturated branched-chain fatty acid resembles an intermediate in the synthesis of cholesterol. The hormone ecdysone induces molting accompanied by metamorphosis. Juvenile hormone secreted by the corpora allata permits molting but inhibits metamorphosis (p. 309), ensuring that each larva will molt several times and reach a large size before pupating. It is not produced during the last larval stage; hence, pupation can occur at the next molt. It has been found that the leaves of yew trees and certain other plants produce substances with ecdysone-like activity. The trees may have evolved this as a protection against moths, for a larva eating the leaves and thereby ingesting these substances will undergo premature molting and die. Other plants and trees secrete substances with juvenile hormone-like activities

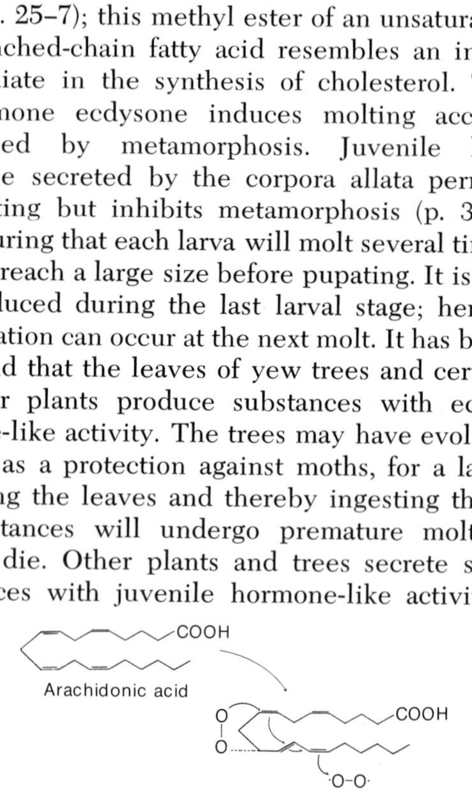

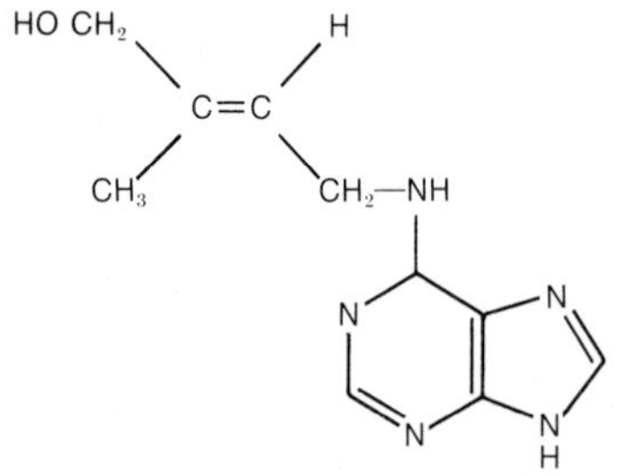

Figure 25-5 The structure of zeatin, a cytokinin.

Figure 25-6 The biosynthesis of prostaglandins from arachidonic acid, a polyunsaturated fatty acid with a 20-carbon chain. (From Page, E. W., Villee, C. A., and Villee, D. B.: *Human Reproduction,* 2nd Ed. Philadelphia, W. B. Saunders Co., 1976.)

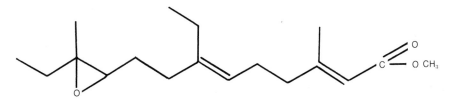

Figure 25–7 Juvenile hormone (methyl-10-epoxy-7-ethyl-3, 11-dimethyl-2, 6-tridecadienoate).

and these, by preventing a larva from becoming adult and reproducing, would protect the trees against insects. In a few field tests, foliage has been sprayed with synthetic juvenile hormones, and these have shown promise as effective and relatively specific insecticides.

Short Peptides. *Oxytocin* and vasopressin, or *antidiuretic hormone* (Fig. 25–8), are both short peptides composed of nine amino acids, seven of which are identical in the two hormones. The two substances have quite different physiologic properties even though they differ in only two of the nine amino acids. They are synthesized in neurosecretory cells in the supraoptic and paraventricular nuclei of the hypothalamus and then pass down the axons of those cells to the posterior lobe of the pituitary where they are stored and subsequently released. In a number of animals, gene mutations have occurred which led to the substitution of one or two amino acids in the chain, yielding a compound with somewhat different biological properties. A lysine vasopressin is found in the pig and the hippopotamus; isotocin is found in certain fishes; and arginine vasotocin is the hormone regulating water balance in amphibians.

Melanocyte stimulating hormone (MSH), a peptide containing 13 amino acids, is secreted by the anterior lobe of the pituitary. In frogs and other amphibians, MSH has a physiologic role in bringing about the darkening of the skin by causing the melanocytes in the skin to expand. The human pituitary produces MSH and human skin does respond somewhat to the hormone, but the physiological role of MSH in humans and other mammals is far from clear.

Long Peptides. Insulin, glucagon, adrenocorticotropic hormone (ACTH) and calcitonin are somewhat longer peptides, with about 30 amino acids in the chain. ACTH, which is secreted by the anterior lobe of the pituitary, is a single peptide chain of 39 amino acids, the first 13 of which have a sequence identical to the peptide chain in MSH. ACTH is synthesized rapidly, little is stored in the pituitary and it is rapidly removed from the plasma. Its biological half-life, the time required for half of a given amount produced to be removed from the plasma, is 20 minutes or less. It stimulates the growth of the adrenal cortex and the production of adrenocortical steroids by that gland. The ACTHs from various species have similar, but not absolutely identical, peptide chains. The substitutions of amino acids in the ACTHs of different species are restricted to the region from the twenty-fifth to the thirty-second residue. A synthetically produced polypeptide chain consisting of the first 23 amino acids is fully active in stimulating the secretion of adrenal corticoids, but it does not react with antibodies developed in the rabbit against the ACTH of the pig. Thus it appears that the usual immunologic determinants (p. 362) of ACTH are located in the carboxyl tail of the peptide.

Insulin, secreted by the β cells of the pancreas, consists of two peptide chains joined by disulfide bonds. The pioneering work of Fred Sanger and his colleagues at the University of Cambridge determined the sequence of the 21 amino acids in one chain and the 30 amino acids in the other. Insulin was the first protein whose amino acid sequence was determined, and these studies earned Sanger a Nobel Prize,

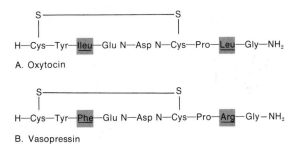

A. Oxytocin

B. Vasopressin

Figure 25–8 The structure of oxytocin and vasopressin (antidiuretic hormone). Note that the two hormones have amino acid sequences that differ only in two amino acids.

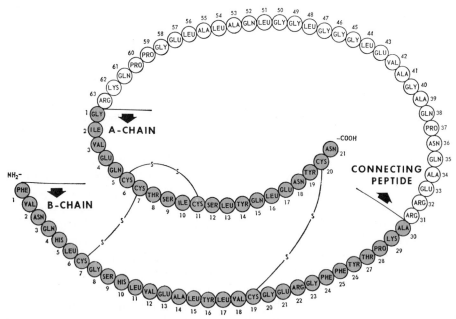

Figure 25–9 The molecular structure of proinsulin, a single polypeptide chain, and its conversion to insulin, composed of two peptide chains. This is achieved by the formation of three disulfide bonds and the removal of a section of the peptide chain between peptide A and peptide B. (From McGilvery, R. W., *Biochemistry*. Philadelphia, W. B. Saunders Co., 1970.)

for they not only clarified the structure of insulin but also established methods by which the amino acid sequence of other proteins could be determined. Insulin is synthesized in the β cells of the pancreas as a *single peptide* composed of 84 amino acids. This compound, *proinsulin,* undergoes folding, three disulfide bonds are formed, and a C peptide, containing 33 amino acids, is removed from the center of the original chain by enzymatic cleavage. This leaves two peptide chains joined by the disulfide bridges (Fig. 25–9).

Proinsulin is an example of a *prohormone.* It has been discovered that several other peptide hormones are synthesized within cells as prohormones which undergo partial cleavage to yield smaller peptides with full hormonal activity. Proinsulin does have some insulinlike activity when tested.

Glucagon, secreted by the α cells of the pancreatic islets, is a short peptide containing 29 amino acids. *Secretin,* a hormone produced by duodenal mucosa, is a peptide containing 27 amino acids, 16 of which are identical in sequence with those in glucagon.

An even longer peptide, *parathyroid hormone* (PTH), is secreted by the parathyroid glands and plays a role in the regulation of calcium metabolism. This hormone is a single peptide chain containing 84 amino acids. *Calcitonin,* secreted by the parafollicular cells embedded in the thyroid, has an action antagonistic to that of parathyroid hormone. Calcitonin is a peptide chain containing 32 amino acids with a seven-membered ring formed by an intrachain disulfide bridge between two cysteines.

Proteins. The gonadotropins thyrotropin and growth hormone, secreted by the anterior pituitary, are proteins with molecular weights of 25,000 or more. The growth hormone of primates has a molecular weight of about 25,000, whereas the growth hormones of the sheep and ox have molecular weights of about 45,000. Growth hormone from the pituitary of the ox or sheep will not stimulate growth in humans, but both human and beef growth hormone will stimulate growth in the rat. Human growth hormone consists of 191 amino acids in a single peptide chain. In contrast to ACTH, growth hormone does accumulate in the pituitary and as much as 10 per cent of the dry weight of the pituitary may be growth hormone. It is curious that a man in his eighties has about as much growth hormone in his pituitary as does a rapidly growing child! Growth hormone is rapidly degraded after it has been secreted into the blood; its biological half-life is about 25 minutes.

Thyrotropin (TSH) is a basic glycoprotein with a molecular weight of about 25,000. It contains several kinds of carbohydrates: N-acetyl-galactosamine, mannose and fucose. Follicle-stimulating hormone (FSH) has a molecular weight of 32,000 and contains about 8 per cent carbohydrate, including some sialic acid. This apparently is required for biological activity, for FSH is no longer hormonally effective after the sialic acid has been removed by treating the hormone with the enzyme neuraminidase. Luteinizing hormone (LH) is a slightly smaller glycoprotein with a molecular weight of about 26,000. The LHs from different species have carbohydrate contents ranging from 4.5 per cent in human LH to 11 per cent in sheep LH. The third gonadotropin, called prolactin or luteotropic hormone, is a protein with a single peptide chain of 198 amino acids but no carbohydrate components.

The three glycoprotein hormones of the pituitary, TSH, FSH and LH, are each composed of two subunits termed α and β. The α subunits of all three are very similar peptides with 96 amino acids in the chain, whereas the β subunits are distinctive and appear to be responsible for the biological specificity of the hormone. The hormones can be separated into their subunits, which have little or no biological activity, and then recombined to give full activity. One can combine a TSH α subunit with an FSH β subunit and obtain a protein with FSH activity.

Steroids. The adrenal cortex, testis, ovary and placenta secrete steroids synthesized from cholesterol and composed of 21, 19 or 18 car-

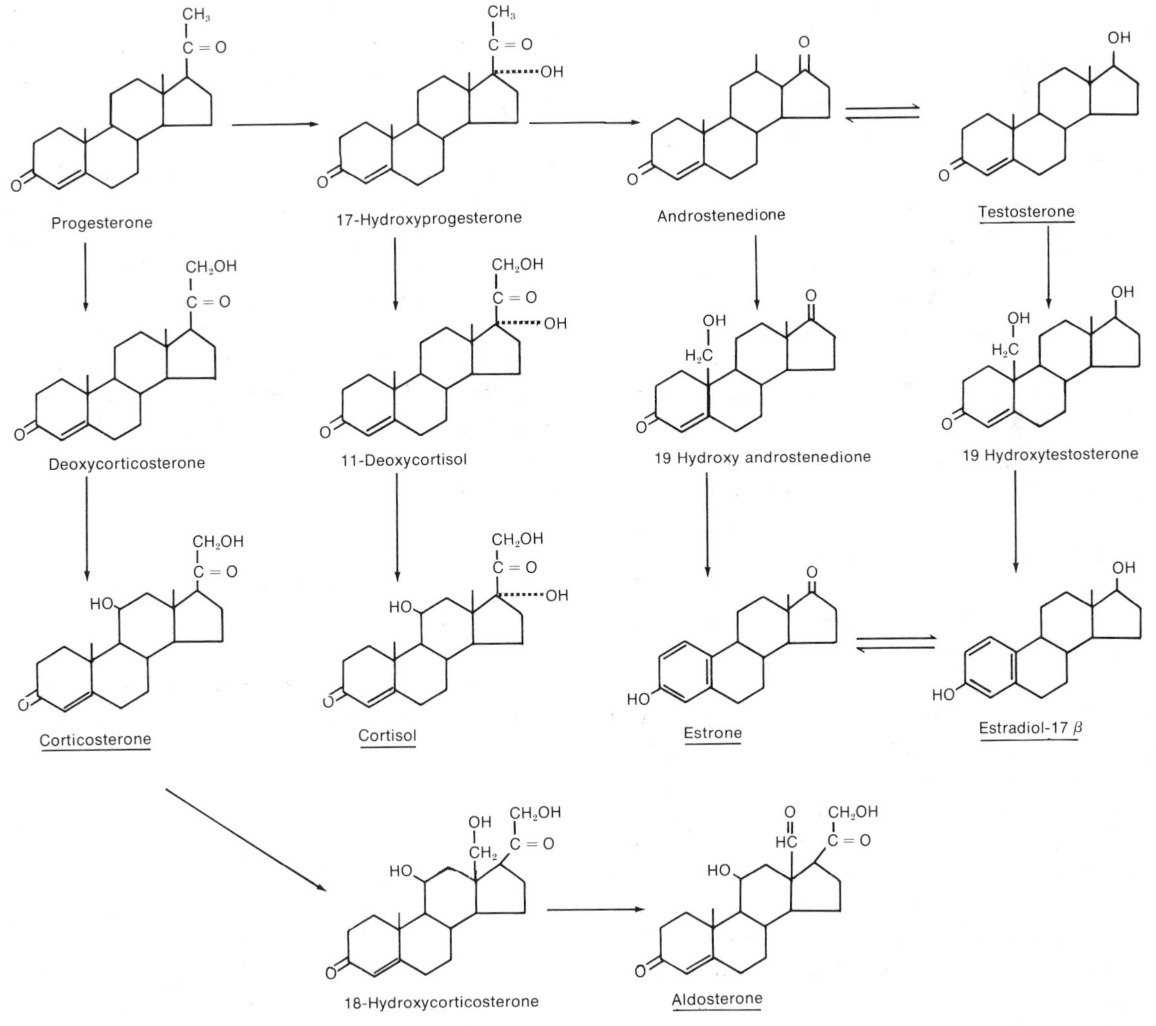

Figure 25-10 The sequence of reactions by which androgens, estrogens and corticoids may be synthesized from progesterone.

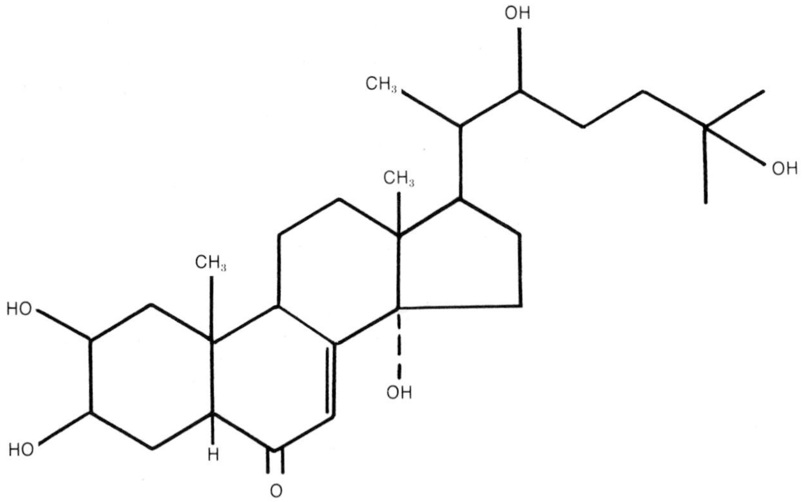

Figure 25-11 The molting hormone of insects, α-ecdysone.

bons arranged in four connected rings; three of the rings have six carbons and the fourth has five. The primary female sex hormone, *estradiol*, with 18 carbons, is synthesized from the male sex hormone *testosterone*, which has 19 carbons (Fig. 25-10). This in turn is synthesized from the second type of female sex hormone, *progesterone*, which has 21 carbons. Progesterone is a precursor in the adrenal cortex of both glucocorticoids and mineralocorticoids. *Glucocorticoids*, such as cortisol, stimulate the conversion of proteins to carbohydrates, whereas *mineralocorticoids*, such as aldosterone, regulate the metabolism of sodium and potassium. The adrenal cortex of both men and women produces dehydroepiandrosterone and adrenosterone, 19 carbon steroids with slight male sex hormone activity, together with a small amount of the potent male sex hormone testosterone.

The hormone stimulating molting in insects, *ecdysone*, is a steroid with 27 carbons arranged in four rings and a tail like the cholesterol molecule (Fig. 25-11). The parent cholesterol molecule is assembled enzymatically by putting together five-carbon "isoprenoid units." Six of these are joined to yield lanosterol with 30 carbons, then three methyl groups are removed enzymatically to yield cholesterol with 27 carbons. Insects synthesize ecdysone from cholesterol, but they cannot synthesize cholesterol; it must be a constituent of their diet. In a sense, cholesterol is a vitamin for insects.

Gibberellins. The final group of compounds with hormonal activity are *gibberellins*, which have specific growth-promoting effects

in plants. In investigating a disease of rice plants that causes excessive elongation of the shoot, Kurosawa found that the active substance was secreted by a fungus that infected the plant. This fungus was *Gibberella fujikuroi*, and so Kurosawa named the substance gibberellin. Some nine or more kinds of gibberellins have been identified and found to be normal constituents of green plants (Fig. 25-12). Like the steroids, the gibberellins are derived from isoprenoid units, but by way of a 20-carbon (diterpene) intermediate instead of a 30-carbon (triterpene) intermediate. The growth-promoting effect of these compounds was strikingly shown by Radley, who found he could make dwarf pea plants grow tall by providing them with a gibberellin extracted from normal tall pea plants.

25-3 THE ENDOCRINE GLANDS OF HUMANS

The glands of the human body known to secrete hormones are shown in Figure 25-13.

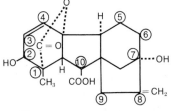

Figure 25-12 The structure of one of the gibberellins, a plant hormone with strong growth-promoting properties. The system of numbering the carbon atoms in the molecule is indicated.

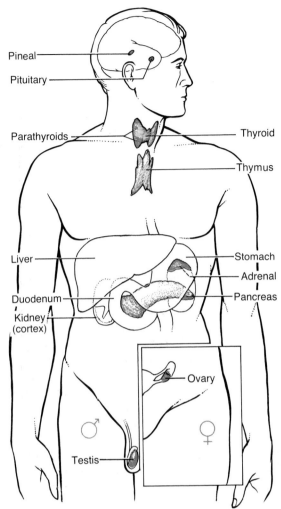

Figure 25–13 The approximate locations of the endocrine glands in men and women.

energy-releasing processes in most of the body tissues. When an extra amount is given, the body uses more oxygen, produces more metabolic wastes, and gives off more heat than it normally does. Thyroxine increases the activity of a variety of enzymes involved in carbohydrate metabolism and in oxidative phosphorylation. There is a latent period of 24 or more hours following the injection of thyroxine into a human being before any effect on metabolic rate can be detected. The maximal effect is achieved only some 12 days after the single dose.

When the supply of thyroxine is inadequate, the basal metabolic rate falls to as little as 600 to 900 kilocalories per day, or between 30 and 50 per cent of the normal amount. Individual tissue slices from an animal with thyroid deficiency also show a metabolic rate lower than normal when incubated *in vitro*. By its effect on metabolism, thyroxine has a marked effect on growth and differentiation. Removal of the thyroids of a young animal produces decreased body growth, retarded mental development and delayed or decreased growth of the genitalia.

In addition to the follicular cells that secrete thyroxine, the thyroids contain parafollicular cells that secrete *calcitonin.* This hormone acts with parathyroid hormone to regulate the concentration of calcium in the blood. Its effects oppose those of parathyroid hormone; it inhibits bone resorption and leads to a decrease in the concentration of calcium in the blood and body fluids.

The *parathyroids* are four masses of tissue, each about the size of a small pea, attached to or embedded in the substance of the thyroid gland. The cells of the parathyroids are arranged in a compact mass and not in follicles like those of the thyroid. Like the thyroids, the parathyroids originate from outgrowths of the pharynx and are evolutionary remnants of the third and fourth gill pouches.

Scattered among the acinar cells of the pancreas that secrete the digestive enzymes are a million or more islands of endocrine tissue called the *islets of Langerhans.* These contain two types of cells which can be readily distinguished in histological sections. The β cells secrete insulin and the α cells secrete glucagon (Fig. 25–14).

The paired *adrenal glands* situated at the

Table 25–1 summarizes the source and major physiological effects of the hormones produced by these glands.

The bilobed *thyroid gland,* located in the neck on either side of the trachea just below the larynx, has an exceptionally rich blood supply. The two lobes are joined by a narrow isthmus of tissue which passes in front of the trachea. The thyroid develops as a central outgrowth of the floor of the pharynx; however, its connection with the pharynx is lost early in development. The gland is composed of cuboidal epithelial cells arranged in hollow spheres one cell thick. The cavity of each follicle is filled with a gelatinous colloid secreted by the epithelial cells lining it (Fig. 25–14).

Thyroxine and triiodothyronine (another product of the thyroid) speed up the oxidative,

Table 25–1 VERTEBRATE HORMONES AND THEIR PHYSIOLOGIC EFFECTS

Hormone	Source	Physiologic Effect
Thyroxine	Thyroid gland	Increases basal metabolic rate
Parathyroid hormone (PTH)	Parathyroid glands	Regulates calcium and phosphorus metabolism
Calcitonin	Parafollicular cells of thyroid	Antagonist of PTH
Insulin	Beta cells of islets in pancreas	Increases glucose utilization by muscle and other cells, decreases blood sugar concentration, increases glycogen storage and metabolism of glucose
Glucagon	Alpha cells of islets in pancreas	Stimulates conversion of liver glycogen to blood glucose
Secretin	Duodenal mucosa	Stimulates secretion of pancreatic juice
Pancreozymin	Duodenal mucosa	Stimulates release of bile by gallbladder and release of enzymes by pancreas
Epinephrine	Adrenal medulla	Reinforces action of sympathetic nerves; stimulates breakdown of liver and muscle glycogen
Norepinephrine	Adrenal medulla	Constricts blood vessels
Cortisol	Adrenal cortex	Stimulates conversion of proteins to carbohydrates
Aldosterone	Adrenal cortex	Regulates metabolism of sodium and potassium
Dehydroepiandrosterone	Adrenal cortex	Androgen; stimulates development of male sex characters
Growth hormone	Anterior pituitary	Controls bone growth and general body growth; affects protein, fat and carbohydrate metabolism
Thyrotropin (TSH)	Anterior pituitary	Stimulates growth of thyroid and production of thyroxine
Adrenocorticotropin (ACTH)	Anterior pituitary	Stimulates adrenal cortex to grow and produce cortical hormones
Follicle-stimulating hormone (FSH)	Anterior pituitary	Stimulates growth of graafian follicles in ovary and of seminiferous tubules in testis
Luteinizing hormone (LH)	Anterior pituitary	Controls production and release of estrogens and progesterone by ovary and of testosterone by testis
Prolactin (LTH)	Anterior pituitary	Maintains secretion of estrogens and progesterone by ovary; stimulates milk production by breast; controls "maternal instinct"
Oxytocin	Hypothalamus, via posterior pituitary	Stimulates contraction of uterine muscles and secretion of milk
Vasopressin	Hypothalamus, via posterior pituitary	Stimulates contraction of smooth muscles; antidiuretic action on kidney tubules
Melanocyte-stimulating hormone (MSH)	Anterior lobe of pituitary	Stimulates dispersal of pigment in chromatophores
Testosterone	Interstitial cells of testis	Androgen; stimulates development and maintenance of male sex characters
Estradiol	Cells lining follicle of ovary	Estrogen; stimulates development and maintenance of female sex characters
Progesterone	Corpus luteum of ovary	Acts with estradiol to regulate estrous and menstrual cycles
Prostaglandins	Seminal vesicle and other tissues	Stimulate uterine contractions
Chorionic gonadotropin	Placenta	Acts together with other hormones to maintain pregnancy
Placental lactogen	Placenta	Has effects like prolactin and growth hormone
Relaxin	Ovary and placenta	Relaxes pelvic ligaments
Melatonin	Pineal gland	Inhibits ovarian function

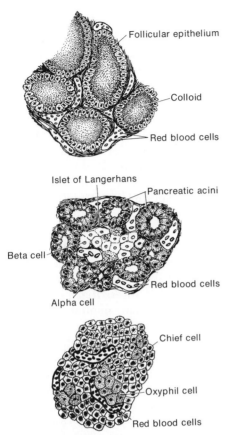

Figure 25–14 Diagrams of the microscopic structure of the thyroid (*above*), the islets of Langerhans in the pancreas (*center*) and the parathyroid (*below*). (Redrawn from Guyton, A. C.: *Function of the Human Body,* 4th Ed. Philadelphia, W. B. Saunders Co., 1974.)

upper end of each kidney are combinations of two entirely independent glands; the adrenal medulla secretes epinephrine and norepinephrine, and the adrenal cortex secretes the adrenal cortical steroids. The two parts have different embryonic origins and different cellular structures. In some of the lower vertebrates the cells corresponding to the medulla and cortex are spatially separated.

The dark reddish-brown adrenal medulla is derived embryologically from the same source as the nervous system and its cells look like modified nerve cells. The medulla secretes large quantities of epinephrine when the human or animal is frightened or angered. Epinephrine promotes several responses, all of which are helpful in coping with emergencies: the blood pressure rises, the heart rate increases, the glucose content of the blood rises, the spleen contracts and squeezes out a reserve store of blood, the clotting time of blood is decreased, the pupils dilate and the muscles which erect the hairs contract, providing a thicker protective mat in those mammals with fur and gooseflesh in the human.

The pale yellowish-pink outer adrenal cortex, composed of three layers of cells, secretes glucocorticoids such as *cortisol,* which stimulate the conversion of amino acids to glucose; mineralocorticoids such as *aldosterone,* which regulate the content of sodium and potassium in extracellular fluids by promoting the reabsorption of sodium by the kidney tubules; and androgens such as *dehydroepiandrosterone, androsterone* and *androstenedione.*

The *pituitary,* also called the hypophysis, lies in a small depression on the floor of the skull just below the hypothalamus. The human pituitary is a double gland about the size of a pea. Its anterior lobe forms in the embryo as an outgrowth of the roof of the mouth. The posterior lobe grows down from the floor of the brain. The two parts meet and the anterior lobe grows partly around the posterior one. The anterior lobe loses its connection with the mouth

Figure 25–15 Section of a human testis showing parts of several seminiferous tubules containing cells in various stages of spermatogenesis. Between the seminiferous tubules are evident the interstitial cells that synthesize and secrete testosterone.

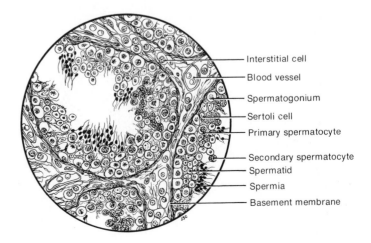

but the posterior lobe retains its connection with the hypothalamus. Certain cells in the hypothalamus secrete *releasing hormones,* which bring about the secretion of specific pituitary hormones. The first of these to be chemically characterized was *thyrotropin-releasing hormone* (TRH), shown to be a tri-peptide, pyroglutamyl-histidyl-proline amide.

The anterior lobe contains at least five different types of cells that differ in their shape, size, staining properties and the kind of granules present in the cytoplasm. It seems likely that each type produces and secretes a different kind of hormone. The cell secreting growth hormone has been identified as a rounded cell whose cytoplasm is packed with dense, round, acidophilic granules. These have been isolated and shown to have a high content of growth hormone. The cells secreting prolactin stain deeply with carmine stain and contain granules that are larger and more ovoid than the granules in the growth-hormone-secreting cells.

The posterior lobe of the pituitary is composed of nonmyelinated nerve fibers and branching cells which contain brownish cytoplasmic granules. Its two hormones, oxytocin and vasopressin, are not produced in the posterior lobe but are secreted by neurosecretory cells in the supraoptic and paraventricular nuclei of the brain, pass down the axons of the

hypothalamic-hypophyseal tract and are stored and released by the posterior lobe. Investigators have found that there are two types of neurons in these nuclei: one produces only oxytocin and the other only vasopressin. This is the basis of the "one neuron–one hormone" hypothesis: a neuron produces oxytocin or vasopressin but not both. Both the supraoptic and paraventricular nuclei have about equal numbers of the two kinds of neurons.

In between the seminiferous tubules that produce the sperm are *interstitial cells* (Fig. 25–15), which produce and secrete the male sex hormones (androgens) such as *testosterone.* If the testes remain in the abdominal cavity instead of descending into the scrotal sac, the seminiferous tubules degenerate and the man is sterile, but his interstitial cells are normal and secrete a normal amount of testosterone. The sperm-forming cells are particularly susceptible to heat, and the scrotal sac, which is about three degrees cooler than the abdominal cavity, provides an environment in which the spermatocytes can develop. Androgens stimulate the development of the male secondary sex characters: the beard, the growth and distribution of hair on the body, the deepened voice, the enlarged and stronger skeletal muscles, and the development of the accessory sex glands, the prostate and seminal vesicles. Testosterone plays a role in determining male sex-

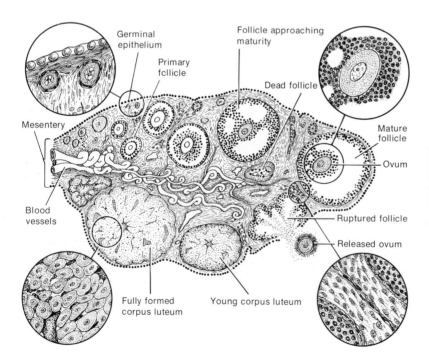

Figure 25–16 Diagram of the stages in the development of an egg, follicle and corpus luteum in a mammalian ovary. Successive stages are depicted clockwise, beginning at the mesentery. The insets show the cellular structure of the successive stages.

tual behavior and the sex urge. The Sertoli cells of the seminiferous tubules can convert testosterone to estradiol. Although the function of estradiol in the male is not clear, it is well established that males produce female sex hormones and that females produce androgens. One of the richest sources of female sex hormones is the urine of stallions.

In the human, the major sources of the female sex hormones are the cells lining the ovarian follicles and the cells of the corpus luteum formed from these cells after ovulation has occurred (Fig. 25–16). The follicular cells of the ovaries secrete *estradiol* and the luteal cells secrete *progesterone.* Estradiol regulates the body changes that occur in the female at the time of puberty or sexual maturity: the broadening of the pelvis, development of the breasts, growth of the uterus, vagina and external genitalia, growth of pubic and axillary hair, change in voice quality and the onset of the menstrual cycle. Progesterone is required for the completion of each menstrual cycle, for the implantation of the fertilized egg in the uterus and for development of the breasts during pregnancy.

The *pineal gland,* a small round structure on the upper surface of the thalamus, lying between the cerebral hemispheres, is derived embryologically as an outgrowth of the brain. It secretes *melatonin,* a methoxy indole synthesized from tryptophan, which is converted to 5-hydroxy tryptamine (serotonin). Serotonin is acetylated at the amine nitrogen and a methyl group is added to the 5-hydroxy group by *hydroxy indole-O-methyl transferase,* an enzyme found only in the pineal. The enzyme is stimulated by norepinephrine, released by the tips of the sympathetic nerves that extend to the pineal from a cervical sympathetic ganglion. Light falling on the retina of the eye increases the synthesis of melatonin by the pineal. A small nerve, the *inferior accessory optic tract,* passes from the optic nerve through the medial forebrain and connects with the sympathetic nervous system. The pineal secretion inhibits ovarian functions either directly or via an effect on the pituitary. Girls blind from birth undergo puberty earlier than normal, apparently because they lack the inhibitory effect of melatonin on ovarian function.

Although the *placenta* is primarily an organ for the support and nourishment of the developing fetus, it is also an endocrine organ and produces estradiol, progesterone and at least three protein hormones. Enzymes present in the placenta can synthesize progesterone from cholesterol and cholesterol from two-carbon precursors and can convert testosterone to estradiol. However, the placenta lacks the enzymes to convert progesterone to testosterone; these are present in the fetal adrenal.

Chorionic gonadotropin (hCG) is similar in many respects to the luteinizing hormone secreted by the pituitary—it is a glycoprotein with α and β subunits. *Human placental lactogen* is a polypeptide of 191 amino acids without a carbohydrate moiety. Its amino acid composition and structure are similar to those of growth hormone and prolactin, and the hormone has both growth-promoting and lactogenic effects. The placenta also synthesizes a protein hormone with thyrotropin-like activities, but its physiologic importance to mother or fetus is not yet clear.

During pregnancy large amounts of chorionic gonadotropins are produced and passed into the urine of the mother. This is the basis for one of the classic pregnancy tests. Formerly an extract of urine was tested for its effect on the ovaries of a rabbit or a rat or on sperm production in a frog or a toad. Now much more sensitive *radioimmunoassay tests* make possible the detection of hCG and the diagnosis of pregnancy as soon as implantation has occurred.

The placenta produces large amounts of steroid hormones, especially in the last month or two of gestation—as much as 250 mg. of progesterone and 30 mg. of estrogen per day. The placentas of the rabbit and some other animals, but not the human placenta, produce *relaxin,* a protein hormone which relaxes the ligaments of the pelvis just before birth to facilitate the passage of the young through the birth canal. Relaxin is effective only after the connective tissue of the pubic symphysis has been sensitized by prior treatment with estradiol.

The lining of the digestive tract produces hormones that stimulate or inhibit the secretion of digestive juices—*gastrin,* secreted by the mucosal cells in the pyloric region of the stomach, and *secretin, pancreozymin* and *enterogastrone,* secreted by the mucosal cells of the duodenum.

25-4 TARGET CELLS

All the hormones produced by the endocrine glands in humans and other vertebrates are secreted into the blood stream and carried by the blood to all parts of the body. A few hormones, such as thyroxine and growth hormone, affect metabolism in every cell in the body; every cell responds to the presence of these hormones and will show an altered metabolic state when deprived of them. Most hormones, however, affect only certain cells in the body, despite the fact that the blood stream carries them to all parts of the organism. For instance, only the pancreas responds to the secretin circulating in the blood. The cells that respond to a given hormone are called the *target cells* of that hormone. The follicle cells of the thyroid are the target cells of thyrotropin (TSH) secreted by the pituitary, and the ovary and testis contain the target cells of follicle-stimulating hormone (FSH) and luteinizing hormone (LH), pituitary gonadotropins. Some hormones, such as estradiol, have marked effects on their primary target cells—the uterus, vagina and mammary glands—but lesser effects on other characteristics—the voice, the distribution of hair on the body, bone growth and so forth—and even smaller effects on certain other tissues.

The ability of a tissue to respond to estradiol is correlated with the presence in its cells of a *receptor* protein that specifically takes up and binds estradiol. Certain cells in the uterus, vagina, pituitary and hypothalamus contain an estrogen receptor and can accumulate estradiol many-fold from the blood. This protein plays a role in transporting estradiol from the cytoplasm into the nucleus, where it produces its effect. There are comparable protein receptors in the oviduct and uterus which specifically take up and bind progesterone, and receptors in the prostate specific for the androgen dihydrotestosterone. There are similar receptors for the peptide and protein hormones, but these are present primarily on the plasma membrane of the target cell rather than in the cytoplasm. Instead of conducting the hormone to the nucleus, the binding of the hormone to the receptor occurs on the cell membrane and results in the stimulation of adenyl cyclase on the inner side of the plasma membrane.

As a result of the process of cellular differentiation (about which we know very little), certain cells become competent to respond to a certain hormone, whereas others do not. Certain cells develop the ability to secrete the hormone; others do not. Thus both *secreting* cells and *responding* cells must develop their special potentialities in order to have an effective system of chemical coordination. A striking example of this is seen in the inherited human condition called **testicular feminization.** Individuals with this condition are genetic males, XY, and secrete normal male levels of testosterone, yet they are phenotypically female. At puberty female breast development occurs, but pubic and axillary hair is scanty. External genitalia are female and a small vagina is present, but there is no uterus, fallopian tubes or ovaries. Testes are present either within the abdomen or in the inguinal canal. After puberty these individuals develop typically feminine proportions, voice and habitus, and are usually quite attractive. Some could readily qualify for a *Playboy* centerfold. They are completely feminine psychologically and have a strong maternal instinct. The gene that regulates the production of the androgen receptor is located on the X chromosome. Individuals with testicular feminization have a mutant gene which does not permit the synthesis of the receptor. Patients with this syndrome produce no androgen receptors and fail to respond to androgens at any time in life. They develop as females under the stimulus of the estrogens secreted by the testis and adrenal.

25-5 ENDOCRINE ABNORMALITIES

The functions of the endocrine glands are of more than theoretical interest to humans, for the underactivity or overactivity of these glands may produce marked effects on the body, conditions sometimes called functional diseases to distinguish them from infectious diseases, caused by bacteria and other agents, and from deficiency diseases, caused by the lack of a vitamin or similar substance. Indeed, the existence of certain endocrine glands was first discovered when the causes of myxedema, diabetes and tetany were investigated.

Myxedema. A deficiency in the amount of thyroxine secreted by the thyroids in an adult results in *myxedema*,* characterized by a low metabolic rate and decreased heat production. The body temperature may drop to several degrees below normal, so that the patient constantly feels cold. His pulse is slow and he is physically and mentally lethargic. His appetite usually remains normal, however, and since the food consumed is not used up at the normal rate, there is a tendency toward obesity. The skin becomes waxy and puffy, owing to the deposition of mucous fluid in the subcutaneous tissues, and the hair usually falls out. Myxedema responds well to the administration of thyroxine or dried thyroid gland. Since thyroxine is not digested appreciably by the digestive juices, it can be given by mouth.

Myxedema is caused by underactivity or degeneration of the thyroid gland itself. Another type of hypothyroidism results when the diet does not contain enough iodine for the synthesis of thyroxine. The gland tends to compensate for the insufficiency by increasing in size. The resulting enlargement, known as a simple *goiter,* may be a small swelling, barely detectable by touching the neck, or a large, disfiguring mass, weighing more than a kilogram. The symptoms accompanying the goiter resemble those of myxedema, but are milder. This type of goiter occurs in areas where the soil lacks iodine, or in regions remote from the sea where seafood (rich in iodine) is unobtainable. The incidence of simple goiter has been greatly reduced by modern packing and shipping methods which permit the transport of seafood to all parts of the country, and by the addition of iodine (as potassium iodide) to table salt. The efficacy of this preventive measure was demonstrated in Detroit, where in seven years the incidence of goiter in school children was reduced from 36 to 2 per cent.

Hypothyroidism present from birth is known as *cretinism.* Children suffering from the disease are dwarfs of low intelligence who never mature sexually. If treatment with thyroxine is begun early, normal growth and mental development can be effected.

Exophthalmic Goiter. Hyperthyroidism results either from the overactivity of a normal-sized gland or from an increase in the size of the gland itself. In both cases the basal metabolic rate increases to as much as twice the normal amount. The excessively rapid heat production causes the hyperthyroid person to feel uncomfortably warm and to perspire profusely. Because the food he eats is used up quickly, he tends to lose weight even on a high caloric diet. High blood pressure, nervous tension and irritability, muscular weakness and tremors are symptomatic of the condition, but probably the most characteristic symptom is the protrusion of the eyeballs, called *exophthalmos,* which gives the patient a wild, staring expression. The swelling of the gland as the result of hyperthyroidism is known as *exophthalmic goiter,* to distinguish it from simple goiter caused by insufficient iodine. Identical symptoms can be caused by feeding thyroid substance or thyroxine to normal people.

Hyperthyroidism can be treated by surgically removing some of the thyroid gland or by killing the cells with x-rays. Hyperthyroidism may also be treated successfully by administering the drug thiouracil, which inhibits the synthesis of thyroxine, or by injecting radioactive iodine, ^{131}I. The thyroid accumulates the radioactive iodine, and the radiation given off by it is concentrated in the gland and destroys the cells.

Tetany. The primary function of parathyroid hormone (PTH) is to regulate the calcium and phosphate content of the blood and tissue fluids. It promotes the absorption of calcium from the lumen of the intestine, the release of calcium from the bones and the reabsorption of calcium from the glomerular filtrate in the kidney tubules. It also inhibits the reabsorption of phosphate in the kidney tubules and thus promotes the excretion of phosphate in the urine. When the parathyroids are removed, the animal suffers muscular tremors, cramps and convulsions in response to stimuli which in a normal animal produce no response or only a slight twitch. This condition, called *tetany,* is due to an increased irritability of muscles and nerves caused by a decrease in the calcium content of the blood and tissue fluids. The calcium content of the blood of a parathyroidectomized animal falls to about half the normal amount. If a solution of a calcium salt is injected into a vein of an animal in tetanic convulsions, the convulsions cease within

---

*Any deficiency of a gland secretion is indicated by the prefix "hypo," while an oversecretion is designated by the prefix "hyper." Hence myxedema is a type of hypothyroidism.

a minute and further convulsions can be prevented by the repeated injection or feeding of calcium. The amount of phosphorus in the blood increases as the calcium decreases and is decreased by the injection of parathyroid hormone.

Hyperfunction of the parathyroids is brought about by tumors or enlargements of the glands and is characterized by a high blood calcium level. Since the calcium comes at least partly from the bones, hyperparathyroidism is characterized by soft bones which are easily bent and fractured. The muscles are less irritable than they are normally, and they may become atrophied and painful. As the level of calcium in the blood increases, the mineral is deposited in abnormal places, such as the kidney. The disease can be treated by removing the excess parathyroid tissue surgically, or by destroying it with x-rays.

Diabetes Mellitus. In 1886 two German investigators, Minkowski and von Mehring, were studying the role of the pancreas in digestion by removing the gland surgically from dogs and noting the subsequent digestive disturbances. The caretaker in charge of the animals noticed that their urine attracted swarms of flies to the cages. Upon analysis, large amounts of sugar were found in the urine, and the resemblance to diabetes was recognized. Diabetes had been known since the first century A.D., but its cause was unknown, and no treatment was effective. A short time before Minkowski and von Mehring performed their experiments, the cure for myxedema—the feeding of thyroid gland—had been discovered, and it was hoped that a similar feeding of pancreas or the injection of pancreatic extracts would cure diabetes. After 1892, when the discoveries were published, many scientists tried to prepare effective extracts of pancreas, but none of the preparations was very good and many were toxic. The digestive enzymes of the pancreas destroyed the hormone before it could be extracted and purified. Finally, in 1922, two Canadians, Banting and Best, obtained an active substance by making extracts from pancreases in which the ducts had been tied several weeks previously so that the enzyme-producing cells had degenerated. Since the islet cells develop before the enzyme-producing cells in embryonic animals, they were also able to obtain active extracts from fetal pancreas.

Insulin increases the rate at which glucose is taken from the blood into certain cells, especially skeletal muscle cells, and converted to glucose-6-phosphate. This decreases the concentration of glucose in the blood, increases the storage of glycogen in muscles, and increases the metabolism of glucose to carbon dioxide and water. A deficiency of insulin decreases the utilization of glucose, and the alterations in carbohydrate metabolism which result secondarily produce changes in the metabolism of proteins, fats and other substances.

The hypofunction of the pancreas in diabetes causes impaired glucose utilization, high concentration of glucose in the blood and the excretion of large amounts of glucose in the urine because the concentration in the blood exceeds the renal threshold (p. 454). Extra water is required to excrete this sugar, urine volume increases, and the patient becomes dehydrated and thirsty. The tissues, unable to get enough glucose from the blood, convert protein into carbohydrate. Much of this is also excreted and there is a progressive loss of weight. Fat deposits are mobilized and the lipids are metabolized. The increased oxidation of fats leads to an accumulation of incompletely oxidized fatty acids. These **ketone bodies** are volatile and have a sweetish smell, which gives to the breath of diabetics its peculiar and characteristic odor. Ketone bodies are acidic and must be excreted in the urine, causing an acidosis (decrease in the alkaline reserve of the body fluids).

Untreated diabetes is ultimately fatal because of the acidosis, the toxicity of the accumulated ketone bodies, and the continuous loss of weight. The injection of insulin alleviates all the diabetic symptoms: the patient is enabled to utilize carbohydrates normally, and the other symptoms disappear. However, since the action of insulin persists for a short time only, repeated injections are necessary. Insulin does not "cure" diabetes, since the pancreas does not begin secreting its hormone again, but continued injections of it prevent the appearance of the symptoms and enable the diabetic to lead a normal life. Why the pancreas stops secreting adequate amounts of insulin is not known, but there appears to be a hereditary basis for diabetes.

If in the treatment of diabetes too much insulin is injected, the blood sugar level falls drastically and shock results. The nerve cells of the brain require a certain level of glucose for their normal functioning; if this level is not maintained, they become overirritable, and convulsions, unconsciousness and death may follow. There are rare cases of enlarged pancreases due to tumors which produce so much insulin that the patients suffer from recurring attacks of convulsions and unconsciousness. These symptoms can be relieved by eating candy, but the condition is cured only by the surgical removal of part of the pancreas.

Addison's Disease. The human disease resulting from decreased secretion of adrenocortical hormones was first described in 1855 by the English physician Thomas Addison. *Addison's disease* is usually caused by a tubercular or syphilitic infection of the cortex which destroys its cells. It is characterized by low blood pressure, muscular weakness, digestive upsets and increased excretion of sodium and chloride in the urine; an increased concentration of potassium in body fluids and cells; and a peculiar bronzing of the skin caused by the deposition of the pigment *melanin.* Addison's disease is treated by the regular oral or intravenous administration of a natural adrenal steroid, such as cortisol, or of a synthetic steroid, such as dexamethasone or 2α-methyl-9α-fluorocortisol acetate. The latter is extremely potent in both mineralocorticoid and glucocorticoid activities.

Diabetes Insipidus. Injury of the posterior lobe of the pituitary or of its nerve tracts may result in a hormonal deficiency causing *diabetes insipidus.* This disease, characterized by the failure of the kidney to concentrate urine, results in the patient's excreting as much as 30 or 40 liters of urine daily, and hence suffering from excessive thirst. Injection of vasopressin does not cure the disease, just as the injection of insulin does not cure diabetes mellitus, but it does relieve all the symptoms, and by repeated injections of vasopressin the patient can live a normal life.

Hypo- and Hyperpituitarism. The importance of the pituitary in the body's economy is demonstrated by the symptoms that result when the gland is removed experimentally (Fig. 25–17). Young animals whose pituitary is removed stop growing immediately and never reach sexual maturity. If adults are operated upon, both males and females show a regression of the reproductive organs and an accompanying atrophy of both thyroid glands and the adrenal cortex. When pituitary extracts are injected into normal young animals, growth is stimulated and they reach sexual maturity at an early age; in addition, the adrenal cortex, the thyroid and the gonads respond by growing abnormally large and oversecreting.

Giants have been known since the beginning of history, but it was not until 1860 that excessive growth was correlated with an en-

Figure 25–17 Hypophysectomy of the prepuberal rat. The two animals are littermate brothers. The smaller one (experimental) was hypophysectomized at 28 days of age, and each weighed 72 grams at that time. At the time of this photograph, the two animals were 10 months of age; the hypophysectomized animal weighed 81 grams and the control 465 grams. The hypophysectomized animal never developed adult pelage, and testicular descent failed to occur. The juvenile hair was lost rapidly, and bald areas frequently appeared on the back, particularly near the base of the tail. When rats are hypophysectomized as adults, the coarse adult hair is gradually replaced by the soft, fluffy hair characteristic of juveniles. (From Turner, C. D., and Bagnara, J. T.: *General Endocrinology.,* 6th Ed. Philadelphia, W. B. Saunders Co., 1976.)

largement of the pituitary. The first hormone from this gland to be discovered was the *growth-stimulating hormone,* finally isolated as a pure protein from extracts of beef pituitary in 1944. This hormone controls general body growth, and especially the growth of the long bones. Consequently, when the pituitary is overactive during the growth period, there is a general acceleration of the process, resulting in a very tall though fairly well proportioned person (Fig. 25–18). Most circus giants are of this type. If the pituitary is underactive during the growth period, a small, normally proportioned person, known as a *midget,* is the result. Oversecretion of growth hormone after normal growth has been completed, however, produces a condition known as *acromegaly.* Since by this time most of the parts of the body have lost their capacity for growth, only the hands, feet and bones of the face develop. The hands and feet become grossly enlarged, the jaws grow abnormally long and broad, and the bony

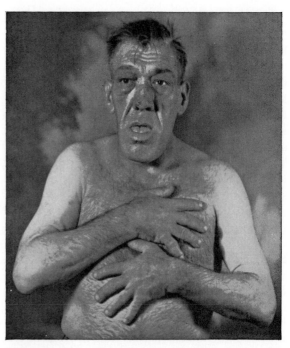

Figure 25–19 A case of acromegaly. The oversecretion of the growth hormone of the pituitary in the adult results in an overgrowth of those parts of the skeleton that are still able to respond. Note the enlargement of the lower jaw and hands and the thickening of the nose and ridges above the eyes. (From Turner, C. D.: *General Endocrinology.* Philadelphia, W. B. Saunders Co., 1948.)

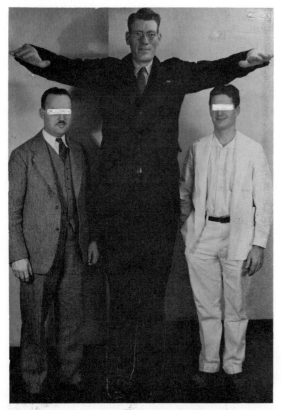

Figure 25–18 A hyperpituitary giant. The giant, approximately 7 feet, 3 inches tall, had a large pituitary tumor and exhibited some of the characteristics of acromegaly—note the enlarged lower jaw and hands. (Courtesy of Dr. E. Perry McCullagh.)

ridges over the eyes and the cheekbones enlarge (Fig. 25–19).

Growth hormone has a host of effects on the tissues of the body, promoting the transfer of amino acids into cells, and increasing protein synthesis. It increases the synthesis of DNA and RNA in the liver and muscles, and the synthesis of collagen and mucopolysaccharides in connective tissue and skin. A deficiency of growth hormone increases the sensitivity of the individual to insulin, and a given dose of insulin produces a greater than normal decrease in the concentration of glucose in the blood. The concentrations of both urea and amino acids in the blood are decreased by growth hormone, reflecting the greater uptake of amino acids and their utilization in protein synthesis. Growth hormone decreases the rate of conversion of amino acid nitrogen to urea. It stimulates the mobilization of fat from adipose tissue and increases the concentration of fatty acids in the plasma. The effects of growth hormone often counter those of insulin.

Many effects of growth hormone are mediated by *somatomedin,* a stable, neutral peptide synthesized in the liver. Somatomedin, with a molecular weight of about 8000, increases the incorporation of thymidine into DNA and the incorporation of sulfate into chondromucoprotein. Purified somatomedin also has some insulinlike effects: it increases the incorporation of amino acids into proteins; it increases glucose utilization; and, like insulin, it inhibits adenyl cyclase activity in adipose tissue and liver cells.

25–6 ENDOCRINES AS SIMPLE TRANSMITTERS OF INFORMATION

Perhaps the simplest model of hormone action is the kind of "information transmitter" exemplified by secretin (Fig. 19–10). The presence of acidic food in the duodenum stimulates the release of *secretin* from the cells of the duodenal mucosa, and the secretin passes via the blood to all cells in the body. The specific arrangement of amino acids in the 27-amino acid peptide means little or nothing to any cell except the acinar cells of the pancreas, and to them it means "secrete pancreatic juice." The duodenal mucosal cells continue to secrete secretin as long as acid food is present. They also secrete *pancreozymin,* another transmitter of information, which stimulates the gallbladder to contract and release bile and stimulates the pancreas to secrete a pancreatic juice rich in enzymes. The bile released from the gallbladder passes down the bile duct to the duodenum and plays a role in neutralizing its contents. It thus turns off the stimulus for the secretion of both hormones.

25–7 ENDOCRINES AS LIMIT CONTROLS

A somewhat more complex model of endocrine function is one in which two hormones act together to set upper and lower limits for some physiological function. The amount of glucose in the blood, for example, is regulated by the hormones *insulin* and *glucagon* (Fig. 25–20). After a meal rich in carbohydrates or after an injection of glucose, the concentration of glucose in the blood rises (1). Some glucose is taken up by the liver and stored as glycogen, but the rise in the concentration of glucose is the signal for the secretion of insulin (2). An increase of only a few milligrams per 100 ml. in the concentration of glucose in the blood will trigger insulin release within 60 seconds. Insulin release can also be triggered by glucagon or by certain amino acids, such as leucine and arginine. Hormones secreted by the digestive tract—secretin and pancreozymin—may stimulate the release of insulin as well as the release of pancreatic enzymes. When the β cell is stimulated by glucose, insulin is released in a burst, then the rate declines and secondarily rises again. From this observation it has been inferred that the insulin is present in the pancreas in two pools, one readily available and one held in reserve. The major effect of insulin is its dramatic increase in the rate of transport of glucose into skeletal muscle and adipose tissue. This leads to a decrease in the concentration of glucose in the blood (3).

The secretion of glucagon by the α cells of the pancreas is also controlled by the concentration of glucose in the blood; it is inhibited by high concentrations of glucose and stimu-

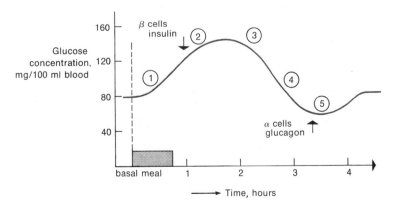

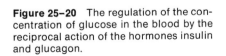

Figure 25-20 The regulation of the concentration of glucose in the blood by the reciprocal action of the hormones insulin and glucagon.

lated by low concentrations. Thus if glucose concentrations fall below the optimal level (4), the release of glucagon is stimulated (5). Glucagon, by activating the glycogen phosphorylase system in the liver, stimulates the conversion of glycogen to glucose-1-phosphate. This in turn is converted to free glucose by way of glucose-6-phosphate and secreted. The glucose secreted by the liver returns the concentration of glucose in the blood to normal and serves to define the lower limit.

The pancreas continually monitors the concentration of glucose in the blood passing through it and secretes either insulin or glucagon as required. An increased concentration of glucose in the blood is the stimulus for a system—the secretion of insulin by the β cells of the islets in the pancreas—which will return it to normal. A decreased concentration of glucose in the blood serves as a stimulus to the opposing system—the secretion of glucagon by the α cells of the pancreatic islets—which will return it to normal.

Another example of a system of limit controls is the regulation of the concentration of calcium in the blood by *parathyroid hormone,* which increases it, and *calcitonin,* which decreases it. Parathyroid hormone, secreted by the parathyroid gland, causes the release of calcium from the bones and teeth and an increased concentration of calcium in the blood. The parathyroid gland in some way senses the concentration of calcium in the blood bathing it and responds to a decreased calcium concentration by releasing parathyroid hormone. By causing a dissolution of bone mineral, parathyroid hormone brings about a release of calcium and phosphate and an increased concentration of calcium in the blood. Calcitonin is secreted by the parafollicular cells in the thyroid in response to an increase in calcium concentration in the blood. It stimulates the deposition of calcium phosphate in the bones. Thus calcitonin regulates the upper limit of calcium concentration and parathyroid hormone regulates its lower limit.

25–8 ENDOCRINES AS RECIPROCAL OR NEGATIVE FEEDBACK CONTROLS

The major glucocorticoid produced by the human adrenal cortex is *cortisol.* This hormone promotes the mobilization of amino acids from skeletal muscle and other peripheral tissues and increases the conversion, in the liver, of the carbon chains of these amino acids into glucose and glycogen, a process termed *gluconeogenesis* (the new formation of glucose). It decreases the utilization of glucose in peripheral tissues and favors the mobilization of lipids and the formation of ketone bodies. Its effects in general are opposite to those of insulin.

The synthesis of cortisol in the adrenal cortex is stimulated by ACTH. This pituitary hormone, which causes growth of the adrenal cortex by stimulating the synthesis of RNA and protein, also specifically stimulates the production of cortisol from cholesterol by increasing the activity of one or more of the enzymes concerned in this conversion. Stimulation of the adrenal cortex by ACTH leads to an increased production of cortisol and an increased concentration of cortisol in the blood (Fig. 25–21). This increased concentration of cortisol inhibits the secretion of ACTH by the pituitary. Cortisol may do this directly by inhibiting the synthesis of ACTH in the pituitary or indirectly by decreasing the production of corticotropin-releasing hormone, CRH, by the hypothalamus.

The primary stimulus that elicits the secretion of cortisol is some sort of physical stress —an injury, a burn, a painful disease, exposure to heat or cold—that sends impulses to the brain that are forwarded to the hypothalamus. The hypothalamus secretes the corticotropin-releasing hormone, which passes along a special portal system of blood vessels directly to the anterior lobe of the pituitary. The CRH stimulates the appropriate cells in the pituitary to secrete ACTH, and this is carried in the blood to the adrenal cortex where it stimulates the production and release of cortisol. The mobilization of amino acids and lipids from peripheral tissues and the process of gluconeogenesis in the liver provide substrates for the repair of the damage and decrease the stimulus which led to the production of the releasing factor. The increased concentration of cortisol in the blood acts in the hypothalamus or pituitary or both to decrease the production and release of ACTH.

An inherited defect of any one of the several enzymes involved in the synthesis of cortisol from pregnenolone and cholesterol may lead to enlargement of the adrenal cortex. The

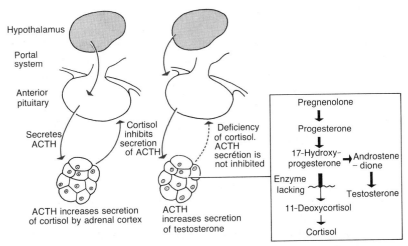

Figure 25–21 *Left,* The normal control of adrenal function by ACTH. A releasing factor from the hypothalamus stimulates the pituitary to secrete ACTH, which increases the synthesis of cortisol in the adrenal cortex. The cortisol then inhibits the secretion of ACTH by the pituitary. *Right,* In one type of adrenal cortical virilism, the adrenals lack an enzyme and the conversion of progesterone to cortisol is greatly reduced. The production of cortisol cannot be increased by ACTH, and ACTH secretion is not inhibited. The secretion of ACTH remains high; this results in increased adrenal size and the secretion of androgens, such as testosterone, instead of cortisol.

commonest defect is a deficiency of the enzyme catalyzing the insertion of a hydroxyl group at carbon 21 of the steroid. Such a defect leads to an accumulation of intermediates in the biosynthetic pathway. Since they cannot be converted to cortisol, some of these are converted to androgens such as **androstenedione** (Fig. 25–21). This may be converted either in the adrenal or elsewhere in the body into testosterone, the most potent androgen. The failure of the adrenal cortex to secrete cortisol results in the oversecretion of ACTH, since there is nothing to shut it off. The negative feedback control cannot operate, the adrenal grows larger and secretes even more androgens, and the individual becomes virilized. A female fetus lacking the enzyme has external genitalia that are masculinized to varying degrees, but a male fetus with the enzymatic defect may show no abnormality at birth. After birth, virilization progresses in both males and females, manifested by enlargement of the phallus, early development of pubic and axillary hair, lowering of the voice and other effects of androgens. Patients with this condition, **adrenal cortical hyperplasia,** can be treated by injecting cortisol to turn off the production of ACTH by the pituitary.

A similar feedback system regulates the synthesis of TSH (thyrotropin) by the pituitary and of thyroxine by the thyroid gland. The an-

terior pituitary secretes TSH, which stimulates several processes in the thyroid gland involved in the production and secretion of thyroxine. An increased concentration of thyroxine in the blood leads to a decreased secretion of TSH, whereas a decreased concentration of thyroxine in the blood leads to an increased secretion of TSH. This reciprocal feedback relationship maintains fairly constant the concentrations of both hormones in the blood.

The release of TSH from the pituitary can be increased by a TSH-releasing hormone secreted by the hypothalamus in response to neural stimulation triggered by lowered body temperature. The increased releasing hormone raises the output of TSH; this in turn increases the secretion of thyroxine, and thyroxine increases the metabolic rate in all body cells and raises body temperature.

The existence of these feedback controls raises certain problems in the treatment of a patient with an endocrine abnormality. Administering thyroxine to a patient depresses the output of TSH just as endogenous thyroxine does, and this will decrease the output of thyroxine by the patient's own thyroid gland. Similarly, giving cortisol to a patient decreases the secretion of ACTH by the pituitary, which decreases the size of the adrenal cortex and its secretion of cortisol. When a patient has been treated for a long time with thyroxine and then the treatment is stopped abruptly, the sudden decrease in the concentration of the thyroxine in the blood may cause a greater than normal secretion of TSH by the pituitary and result in undue stimulation of the thyroid.

25–9 COMPLEX CONTROL SYSTEMS: REGULATION OF THE ESTROUS AND MENSTRUAL CYCLES

For a mammalian egg to be fertilized and develop, it must be released from the ovary when sperm are likely to be present in the oviduct and when the lining of the uterus, the *endometrium*, is in the proper condition to permit implantation of the fertilized egg. The coordination of these events involves some half-dozen hormones in addition to neural mechanisms.

A key event in the estrous cycles of lower mammals and the menstrual cycles of primates is *ovulation*, the release of a mature egg from a follicle in the ovary. Ovulation is triggered by a surge of luteinizing hormone (LH) secreted by the pituitary in response to gonadotropin-releasing hormone (GnRH) secreted by the hypothalamus. This in turn is triggered by a surge of estradiol which operates by *positive* feedback control to induce the release of GnRH and LH. Prolonged treatment of the female with a constant amount of either estradiol or progesterone can block ovulation; it thus appears that a sharp rise in estradiol must occur to affect the positive feedback center in

the hypothalamus so that the pituitary is induced to secrete LH. This theory is supported by experiments in which a single dose of estrogen induced ovulation in animals whose ovaries had been primed with suitable doses of FSH (follicle-stimulating hormone) and LH.

The pattern of gonadotropin release is cyclic in the female but noncyclic in the male. In the rat the male-female patterns of gonadotropin release are determined in early postnatal life. The hypothalami of both genetic male (XY) and female (XX) rats will develop the female cyclic pattern of release unless exposed to testosterone early in postnatal life. However, human females with congenital adrenal cortical hyperplasia, who have been exposed to large amounts of androgen during fetal life, subsequently develop normal ovulatory cycles if treated with cortisol. Evidence suggests that both men and women have a "tonic" center in the arcuate nucleus, the cells of which produce gonadotropin-releasing hormone at a relatively constant rate. Other neurons in the suprachiasmatic nucleus of women constitute the "cyclic" center; this responds to the estrogen surge and secretes the gonadotropin-releasing hormone, which regulates the LH surge that causes ovulation.

GnRH is released at the tips of the axons located in the median eminence. From there it passes into capillaries of the hypophyseal portal vessels (Fig. 25–22), which pass down the pituitary stalk to the anterior pituitary. The vessels break up into a second network of sinusoids and capillaries and the releasing hor-

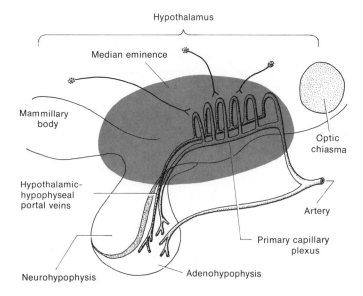

Figure 25–22 Diagram of the relations of the blood vessels supplying the pituitary and hypothalamus, indicating the hypothalamic-hypophyseal portal veins by which hormones are transferred from the hypothalamus to the pituitary.

mone passes to the cells of the pituitary. GnRH has been isolated and identified as a decapeptide with the sequence pyro-glu-his-trp-ser-tyr-gly-leu-arg-pro-gly NH_2. A synthetic decapeptide with this sequence has properties identical to the natural hormone in regulating the release and synthesis of both LH and FSH by pituitary tissue. Some investigators believe that there are separate FSH-releasing and LH-releasing hormones, but others conclude that a single GnRH regulates the release of both FSH and LH.

Sexual Maturation. An interesting question is how this complex system initially becomes established during adolescence. Experiments indicate that the sensitivity of the hypothalamus to sex hormones changes with age. Before puberty the hypothalamus is very sensitive to estradiol and the normally very low concentration of circulating estradiol inhibits gonadotropin release by negative feedback control. A decrease in the sensitivity of the hypothalamus to negative feedback control by estrogen triggers the increased release of gonadotropin and the onset of puberty. During childhood the pituitary secretes a small amount of gonadotropin, mostly FSH, and the ovary or testis responds by releasing small amounts of estrogen or androgen. The hypothalamic set point is low and the low concentration of sex hormones is sufficient to maintain the secretion of gonadotropin below the level required to stimulate the maturation of the gonads. As puberty approaches, the hypothalamus becomes less sensitive to negative feedback control and produces more of the gonadotropins, which cause the growth of the ovaries and testes. The concentration of gonadotropins in the blood increases in girls at about age 10 and in boys at about age 12.

Gonadotropins. Follicle-stimulating hormone (FSH) from the pituitary causes the ovary to grow and stimulates a few of its primary follicles to begin development (Fig. 25–23). Cells surrounding the ovum proliferate rapidly and begin to secrete estradiol. Luteinizing hormone (LH) causes further growth of the follicle, and a surge of LH, perhaps coupled with a surge of FSH, causes the follicle to rupture and release the egg (ovulation).

The remaining follicular cells, under the stimulation of LH and prolactin, increase in size and develop a yellow, fatty appearance, forming a *corpus luteum* in the space previously occupied by the follicle. The cells of the corpus luteum secrete progesterone and some estrogens. The corpus luteum in the human persists for about two weeks if fertilization does not occur, at which point it then stops secreting steroids and degenerates.

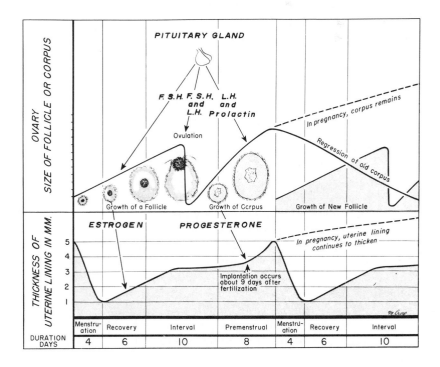

Figure 25–23 The menstrual cycle in the human female. The solid lines indicate the course of events if the egg is not fertilized; the dashed lines indicate the course of events when pregnancy occurs. The actions of the hormones of the pituitary and ovary in regulating the cycle are indicated by arrows.

Steroid Sex Hormones. The primary estrogenic hormone secreted by the ovary, *estradiol,* stimulates the development of the characteristic structural features of the female and plays a role in regulating the cyclic changes of the menstrual cycle. Estradiol stimulates the growth of the uterus at puberty by causing the muscle cells to increase in number and size. The growth of the vagina, the development of the labia, clitoris and other external genitalia, the growth of pubic hair, the broadening of the hips and changes in the structure of the pelvic bones, the growth of the breasts, the proliferation of glandular cells in the breast, and the deposition of fat in hips and thighs characteristic of adult females are all brought about by estradiol. The growth of the lining of the uterus, the *endometrium,* during the first part of the menstrual cycle (the proliferative phase) is also controlled by estradiol (Fig. 25–23).

Progesterone has little effect on the development of the female sex characteristics, but it plays a major role in stimulating the development of the endometrium during the second, or secretory, phase of the menstrual cycle to a condition suitable for the implantation of a fertilized egg. Glycogen and fats accumulate within the endometrial cells, and glands develop which secrete nutrient fluid. The blood vessels in the endometrium grow, becoming long and coiled.

Estrous and Menstrual Cycles. The females of most species of mammals show cyclic periods of the sex urge and will permit copulation only during periods of *estrus,* or "heat," when conditions are optimal for the fertilization of the egg. Most wild animals have one estrous period per year, the dog and cat have two or three, and rats and mice experience estrus every five days. Estrus is characterized by heightened sex urge, ovulation and changes in the lining of the uterus and vagina. Following estrus the endometrium thickens and its glands and blood vessels develop to provide an optimal environment for the implanting embryos. In contrast, the menstrual cycle of the primates is characterized by periods of vaginal bleeding, called *menstruation* (Latin *menstrualis*: monthly), resulting from the degeneration and sloughing of the endometrial lining of the uterus. Primates, unlike other mammals, show little or no cyclic change in the sex urge and permit copulation at any time in the menstrual cycle.

Reflex and Spontaneous Ovulators. A key point in both estrous and menstrual cycles is ovulation, the release of an ovum from the ovary. The rabbit, cat, ferret, mink and certain other mammals are *reflex ovulators;* i.e., the nervous stimulation of mating acts reflexly to bring about ovulation. Nerve impulses from the lining of the vagina pass up the spinal cord to the brain and stimulate the hypothalamus to secrete gonadotropin-releasing hormone (GnRH). The stimulus for ovulation may be a single copulation, as in a rabbit, or a minimum of 19 copulations per day, as required by the female short-tailed shrew. Direct electrical stimulation of the appropriate regions of the hypothalamus, the *tuber cinereum* and *preop-*

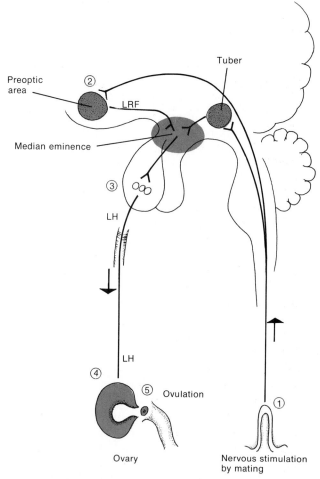

Figure 25–24 Diagram of the nervous pathways involved in the reflex ovulation in an animal such as the rabbit. The stimulation of receptors in the vagina by the mating act generates impulses that pass to the hypothalamus and bring about the secretion of releasing factors. These pass to the pituitary and cause the release of luteinizing hormone, which goes via the blood stream to the ovary and initiates ovulation, the release of the egg.

tic regions, can cause ovulation in the rabbit, cat or monkey. In the rabbit the neural stimulation of coitus causes the hypothalamus to produce GnRH, which passes via the portal system to the anterior pituitary and causes a surge in release of LH from the pituitary. This passes to the ovary and causes the follicle to rupture, thereby releasing the ovum (Fig. 25–24).

Spontaneous ovulators compose a second and larger group of mammals. Ovulation is not stimulated by coitus, but the timing and frequency of ovulation may be influenced by environmental factors. A laboratory rat kept under normal conditions of day and night lighting ovulates early in the morning between 1:00 and 2:30 a.m. If rats are kept for two or more weeks under artificial conditions in which the periods of light and dark are reversed, the time of ovulation is shifted 12 hours. Rats exposed to continuous light 24 hours a day eventually stop ovulating and go into constant estrus, with persistent vaginal cornification. Primates undergo spontaneous ovulation. There is ample evidence that the rhythm of the human menstrual cycle and ovulation is influenced by environmental factors. Nurses on night duty and airline stewardesses who travel long distances east or west to different time zones frequently report changes in their menstrual cycles. Spontaneous ovulators appear to have some sort of light-dependent hypothalamic "clock," which provides the neural stimulation for the release of the hypothalamic-releasing hormones. The hypothalamic neurons end in the median eminence, where they secrete the releasing hormone, which passes by way of the portal vessels to the pituitary and stimulates the release of a surge of LH, initiating ovulation (Fig. 25–25).

The pituitaries of males as well as of females produce and secrete both FSH and LH, which control the development and function of the testis. The pituitary of a male transplanted into a hypophysectomized female will support a normal estrous or menstrual cycle. An ovary transplanted into a castrated male will develop ripe follicles but will not undergo ovulation, for there is no cyclic release of an LH surge from the male pituitary. Thus, although the pituitaries of the male and female are equivalent, the hypothalami are distinctive. The difference between the two sexes—the development of a hypothalamic clock—appears at a critical

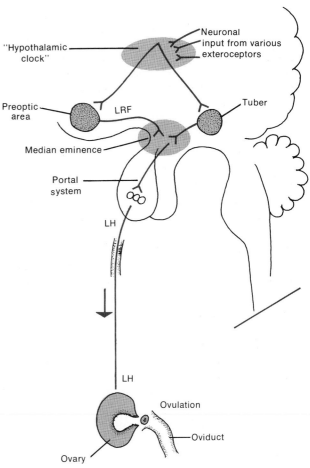

Figure 25–25 Spontaneous ovulation in humans and other primates. The secretion of releasing factors from the hypothalamus is stimulated not by nerve impulses coming from the vagina but by some sort of endogenous biological clock in the hypothalamus.

stage early in the development of the central nervous system. In the developing male, testosterone inhibits the development of the cyclic center. Young females lack testosterone and develop a hypothalamic clock, which regulates the rhythmic release of gonadotropins, controlling the sexual cycles. If testosterone is injected into a female rat anytime between the second and fifth day after birth, the activity of the cyclic center in the hypothalamus is permanently abolished. She never ovulates but remains in a state of constant estrus, like a rat kept under constant illumination.

Feedback Controls. Two types of feedback mechanisms are involved in regulating the estrous and menstrual cycles. Just as thyroxine and cortisol decrease the pituitary's secretion of TSH and ACTH, respectively, so the concentrations of estradiol and proges-

terone in the blood affect the output of gonadotropins by the pituitary, either directly or by an
effect on the hypothalamus. In addition, there
is some evidence that pituitary gonadotropins
have a negative feedback effect on the hypothalamus and decrease the output of releasing
factors.

The Human Menstrual Cycle. The development of sensitive, precise methods for measuring FSH and LH by radioimmunoassays
and for measuring estradiol and progesterone
by specific protein binding methods has made
it possible to measure each of these in a small
sample of blood taken daily during a menstrual
cycle (Fig. 25–26). Such assays show that the
concentrations of LH and FSH rise abruptly
and then fall over a period of two to three days
in the middle of the menstrual cycle. The concentration of FSH in the blood is also elevated
during the first week or so of the proliferative
phase, beginning during the previous menstruation and continuing after the menstrual flow
stops. The concentration of estradiol in the
blood is low during the first ten days or so of
the proliferative phase, but it rises sharply and
reaches a peak, the *estrogen surge,* at about the
time when the concentration of LH begins to

rise. The concentration of estradiol then falls
and has nearly reached the low basal level at
the time the LH concentration reaches its
peak. A second peak of estradiol concentration,
generally broader and lower than the first peak,
is typically seen during the secretory phase of
the cycle. The concentration of progesterone in
the blood begins to rise at about the time of the
LH peak, reaching *its* peak about six days
later; it remains elevated until the end of the
cycle, then falls. The withdrawal of progesterone precipitates a decreased blood flow to
the endometrium, resulting in the death and
sloughing of the cells of the endometrial lining. The menstrual flow, made up of dead endometrial tissue, blood that oozes from the ruptured ends of the endometrial blood vessels,
and tissue fluid from the uterine surface, is
gradually expelled from the uterus over a three
to five day period. The basal portion of the
endometrium remains intact during menstruation and is the source of the new epithelium
and glands that develop subsequently under
the influence of estradiol.

The regulation of the menstrual cycle appears to involve the following sequence of
events (Fig. 25–27).

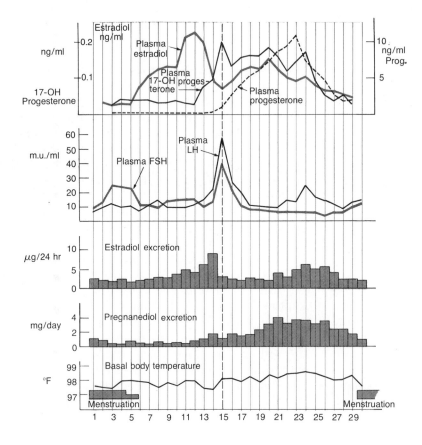

Figure 25–26 Diagram illustrating the concentrations of gonadotropins, estrogens and progestins in the plasma during a single human menstrual cycle. The urinary excretion of estradiol and pregnanediol and the changes in basal body temperature are also shown.

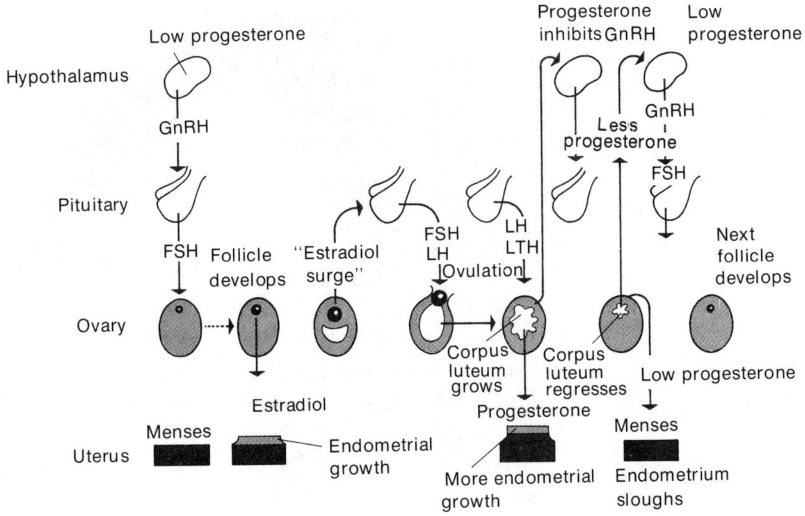

Figure 25–27 Interrelations of hypo-thalamic, pituitary and ovarian hor-mones in regulating the events of the menstrual cycle in women.

1. Following the previous menstruation, the withdrawal of progesterone removes its inhibitory influence, and gonadotropin-releasing hormone is released from the hypothalamus, stimulating the release of FSH from the pituitary. This is reflected in the increased level of FSH circulating in the blood early in the cycle.

2. The FSH causes one or more of the follicles in the ovary to enlarge rapidly and begin to secrete estradiol. The estradiol secreted by the follicular cells causes the proliferation of the endometrium. When the follicle reaches a certain size it secretes a surge of estradiol, which rises and falls and triggers the release of gonadotropin-releasing hormone, which in turn triggers a surge of LH, causing ovulation.

3. The peak of LH, together with prolactin secreted by the pituitary, causes the follicular cells to undergo luteinization, form a corpus luteum and secrete progesterone.

4. The concentration of progesterone in the blood rises and remains high during most of the secretory phase of the cycle. Progesterone causes the continued growth of the endometrial lining and stimulates the endometrial glands to secrete a nutrient fluid.

5. Progesterone has an important function in inhibiting the release of FSH and preventing the development of any additional follicles and eggs. Eventually the corpus luteum begins to regress, the concentration of progesterone in the blood decreases, and the hypothalamus,

freed of its inhibitory effect, releases gonadotropin-releasing hormone and a new cycle begins.

If the egg has been fertilized and implants in the endometrium, the trophoblast cells in the developing placenta secrete *chorionic gonadotropin*. This has luteinizing and luteotropic properties which maintain the corpus luteum and stimulate the continued secretion of progesterone. By the sixteenth week of pregnancy in the human, the placenta produces enough progesterone so that the corpus luteum is no longer necessary and it undergoes involution.

The nature of the reproductive process determines to a considerable extent the organization of animal societies, their migrations and behavior. It also plays a role in the organization of human society. It is interesting to contemplate what kind of human society might have evolved if women were sexually receptive only in April and October, or how human behavior and family units might have been altered if women emitted some visual or olfactory stimulus at the fertile period or (heaven forbid!) if they ovulated reflexly after each coitus.

Oral Contraceptives. Estrogen and progesterone block ovulation not by a direct effect on the ovary but by preventing the secretion of gonadotropin-releasing hormone by the hypothalamus. If a woman were to ovulate after she is pregnant and if this second egg were also fertilized, she would then have two fetuses in her uterus—one several months younger than the other. At parturition both would be ejected and the younger fetus would be at a severe disadvantage in surviving. This has been prevented by the evolution of a hormonal system which prevents ovulation once conception has occurred. The high levels of progesterone and estrogen that are maintained through pregnancy prevent the release of an LH surge.

The oral contraceptives contain synthetic estrogens and progestins which prevent ovulation in a similar manner. Natural estradiol and progesterone are rapidly metabolized in the body, but synthetic hormones, with slight changes in their molecular structure that decrease the rate at which they are destroyed, persist for a longer time. The oral contraceptives, like the natural hormones, inhibit the secretion of releasing hormone and LH and thus prevent ovulation. A woman taking "the pill" has no midcycle surge of LH and FSH and does not ovulate.

The oral contraceptives in general use are "combination pills," containing both an estrogen and a progestin. These are taken daily for a period of 21 days followed by a break of 7 days during which uterine bleeding (menstruation) occurs. Gonadotropin secretion is suppressed, preventing ovulation; in addition, the oral contraceptive prevents the normal maturation of the endometrium and produces an altered cervical mucus which is hostile to penetration by sperm. Another type of pill, containing only progestin, is taken every day throughout the cycle and, although ovulation may occur, fertilization is prevented by the changed nature of the cervical mucus.

After an egg has been released from the ovary and is passing down the oviduct, it retains the ability to be fertilized for only a relatively short time—about 24 hours. Sperm deposited in the female reproductive tract during intercourse retain their ability to fertilize an egg for up to 48 hours. The period of maximal fertility in human beings thus narrows down to a few days midway between two successive menstrual periods. It is difficult to determine precisely when this period of maximal fertility occurs in an individual woman because of the variability of the time of ovulation in the menstrual cycle.

25–10 COMPLEX CONTROL SYSTEMS: REGULATING THE DEVELOPMENT AND FUNCTION OF THE MAMMARY GLAND

The growth and development of the mammary glands after puberty and the production and secretion of milk after parturition are controlled by a complex sequence of hormonal events. The initial development of the breast and the proliferation of glandular elements is stimulated by estradiol and requires the presence of insulin and growth hormone. Progesterone stimulates the further development of the glands at and after puberty. During pregnancy the additional development of the mammary glands is stimulated by estradiol and by

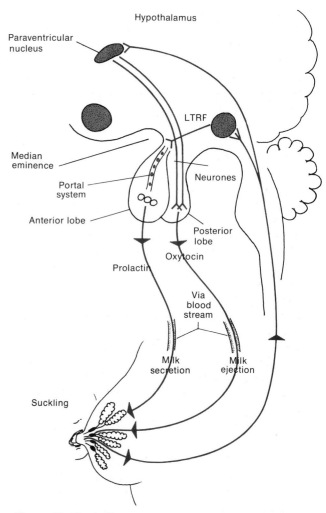

Figure 25–28 A diagram of the hormonal controls that stimulate the production and secretion of milk in the mammary gland.

progesterone produced primarily in the placenta. The alveoli and ducts of the mammary glands continue to develop and become secretory; however, progesterone *inhibits* the production of milk, and it is only after parturition and the sudden decrease in progesterone in the blood that occurs with the loss of the placenta that milk production can begin. Throughout pregnancy the production of pituitary gonadotropins is inhibited by the placental gonadotropins, but with the expulsion of the placenta at birth, the pituitary begins to secrete large quantities of prolactin, which stimulate the breast to produce milk (Fig. 25–28). If the breast is regularly emptied of milk, the pituitary continues to secrete prolactin; however, if milk is not removed, the pituitary stops secreting prolactin. Nerve impulses from the nipple to the hypothalamus stimulate the secretion of prolactin-releasing hormone, which passes to the pituitary and increases the production of prolactin. Other hormones secreted by the anterior pituitary—ACTH, TSH and growth hormone—may also play a role in controlling both the growth of the mammary glands and their secretion of milk.

The *secretion* of milk from the alveolar glands into the milk ducts is under the control of prolactin. The *transport* of milk from the alveolus to the nipple, where it can be removed by the suckling infant, is triggered by the milk ejection reflex, a neurohormonal reflex. Shortly after the infant is put to breast the mammary gland suddenly seems to fill with milk, which comes under pressure and may spurt from the nipple. Milk flow can occur in anticipation of the suckling reflex, and in contrast, stress or discomfort can inhibit this flow so that much less milk is obtained by the suckling child. The rapid increase in pressure within the mammary gland is due to the sudden expulsion from the alveoli of milk that has been synthesized previously. The movement of milk results from the contraction of myoepithelial cells that squeeze the alveoli and expel their contents. The act of suckling (or the milking of a cow) triggers nerve impulses from receptors in the nipple, which pass up the spinal cord to the hypothalamus where they cause the release of oxytocin from the posterior lobe of the pituitary (Fig. 25–28). In the mammary gland, oxytocin causes the myoepithelial cells to contract, increasing intramammary pressure and bringing about the ejection of milk. Thus the milk ejec-

tion response is a reflex; unlike ordinary reflexes, however, the afferent arc is nervous, but the efferent arc is hormonal. Like neural reflexes the response can be conditioned, which explains the occurrence of milk ejection before milking in response to stimuli associated with nursing or milking. Insulin and cortisol play minor roles not yet clearly defined in regulating the development of the breasts and the secretion of milk.

During lactation the release of gonadotropins is suppressed so that ovulation occurs only occasionally and menstrual periods are usually irregular.

25–11 HORMONAL REGULATION OF METABOLIC RATES

The survival of a complex organism such as the human requires, among other things, that order be maintained among the multitude of possible metabolic reactions. Each of these reactions must increase or decrease in an appropriate fashion in response to specific changes in the environment. In part, these changes result from the kinetic properties of the enzymes; the rates increase and decrease as the concentrations of substrates and cofactors rise and fall. For example, glucose-6-phosphate can be converted in the liver to glucose-1-phosphate, to fructose-6-phosphate, to 6-phosphogluconate, or to free glucose (Fig. 25–29). Each reaction is catalyzed by a specific enzyme and the rate of each is controlled by the **affinity** of the enzyme for glucose-6-phosphate and by the **maximum rate** of the enzyme when it is fully saturated with substrate. Merely changing the intracellular concentration of glucose-6-phosphate will change the *direction* in which metabolism flows and not simply the *rate* at which metabolism occurs. At low concentrations of glucose-6-phosphate, **glucose-6-phosphate dehydrogenase,** which has the highest affinity for the substrate, is the only enzyme saturated with substrate and most of the small amount of glucose-6-phosphate present will be converted to 6-phosphogluconic acid. At this low concentration of substrate, **glucose-6-phosphatase,** which has a very low affinity for substrate, will hardly operate at all and little or none of the glucose-6-phosphate will be converted to free glucose. At high concentrations of glucose-6-

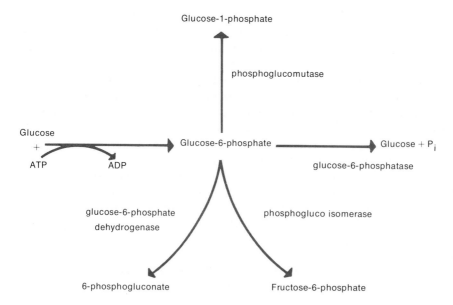

Figure 25–29 Diagram illustrating the possible metabolic fates of glucose-6-phosphate in a mammalian liver cell.

phosphate, when all the enzymes are saturated with substrate, the factors determining the direction of metabolic flow are the relative maximum rates of the four enzymes. Glucose-6-phosphatase has a high maximum rate, and at high concentrations of glucose-6-phosphate most of the substrate is converted to glucose.

In addition to the effects of the concentrations of substrate and cofactors, the activity of many enzymes is *modulated* by the presence of other molecules. The activity of *phosphofructokinase,* for example, which converts fructose-6-phosphate to fructose-1, 6-diphosphate, is increased by adenylic acid, AMP, but is inhibited by ATP (Fig. 25–30).

The enzyme which catalyzes the conversion of fructose-1, 6-diphosphate back to fructose-6-phosphate, *fructose diphosphatase,* is inhibited by AMP. The opposite effects of AMP on the two enzymes prevent the wasteful cyclic hydrolysis of ATP which would occur if both enzymes were active at the same time. The subject of enzyme modulation and the details of these changes in enzyme activity are quite complex and beyond the scope of this book, but they have been introduced to help the student appreciate the many kinds of metabolic controls that may operate within cells.

Superimposed on the controls that depend on the innate properties of the enzymes and

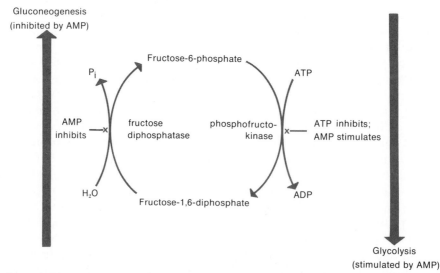

Figure 25–30 Diagram illustrating the modulation of the enzymes fructose-1,6-diphosphatase and phosphofructokinase by adenylic acid (AMP) and adenosine triphosphate (ATP).

their responses to metabolites that modulate their activity are other changes induced by specific hormones. Several hormones—thyroxine, insulin, growth hormone and cortisol—have broad effects on the metabolism of a wide variety of tissues. These controls also help maintain order among the many possible metabolic reactions in those tissues.

Most vertebrates, for example, will not attain their normal adult form and dimensions in the absence of thyroxine. Thyroxine and growth hormone appear to act synergistically in promoting the normal growth of the skeleton. The effect of thyroxine in controlling amphibian metamorphosis is well known, and thyroxine affects the growth and differentiation of other vertebrates, though not in such a marked fashion. Thyroxine and triiodothyronine accelerate the general metabolic rate and oxygen consumption of nearly every organ and tissue in the body. The activities of more than 100 enzymes have been reported to be increased following the administration of thyroxine. Since thyroxine increases oxygen consumption and oxygen is utilized primarily in the mitochondria, it was natural to look for a primary effect of thyroxine on mitochondrial structure and function. Isolated mitochondria do indeed undergo swelling when exposed to thyroxine. Although very high concentrations of thyroxine can "uncouple" oxidative phosphorylation so that energy is released as heat rather than being retained in energy-rich phosphate bonds ($\sim$P), this phenomenon probably bears no relation to the physiological effects of thyroxine. The increased basal metabolic rate of a hyperthyroid individual cannot be explained in terms of an effect on any single metabolic process. In addition to its effects on mitochondrial metabolism, thyroxine increases the synthesis of RNA in the nucleus and the synthesis of proteins on the ribosomes. Thyroid hormones play an especially important role in the maturation of brain and bone, and a deficiency of thyroxine during development results in markedly retarded mental development and bone growth.

Insulin regulates the metabolism of proteins and lipids as well as carbohydrates. Insulin facilitates the entry of glucose into skeletal muscle and adipose tissue; it also increases the activity of a number of enzymes such as glucokinase and glycogen synthetase. Insulin is firmly bound to a specific receptor protein in the plasma membranes of muscle and adipose tissue cells. The uptake of amino acids into cells and the incorporation of amino acids into proteins are both increased by insulin. Insulin increases the conversion of glucose to fatty acids and inhibits the hydrolysis of triacylglycerols in adipose tissue.

Cortisol has metabolic effects that oppose those of insulin. It brings about a mobilization of amino acids from peripheral tissues and accelerates *gluconeogenesis,* the conversion of the carbon chains of the amino acids to glucose and glycogen. Cortisol inhibits the utilization of glucose by skeletal muscle and other peripheral tissues and accelerates the mobilization of fatty acids, increasing the rate of production of ketone bodies.

25–12 MOLECULAR MECHANISMS OF HORMONE ACTION

Any theory of the molecular mechanism by which a given hormone produces its effects in specific tissues must account for the high degree of specificity of many hormones and for the remarkable degree of biological amplification inherent in hormonal processes. Hormones circulate in the blood stream in very low concentrations—steroid hormones at concentrations of about 10^{-9}M and peptide and protein hormones at concentrations of about 10^{-12}M. The several current theories regarding the mechanism of hormone action are alike in suggesting that the hormone first combines with a specific *receptor protein.* The receptors for steroid hormones are soluble proteins, located in the cytoplasm, which combine with the steroid and transport it into the nucleus (Fig. 25–31). In contrast, the receptors for peptide and protein hormones are located on the plasma membranes of the target cell. The peptide hormone is bound to the receptor on the surface of the cell and it is believed that this combination of hormone with receptor stimulates in some fashion the adenyl cyclase on the inner side of the plasma membrane, resulting in an increased production of cyclic AMP. An important intracellular function of cyclic AMP is to activate one or more of the protein kinases within the cell. The protein kinases in turn phosphorylate other proteins which convert an

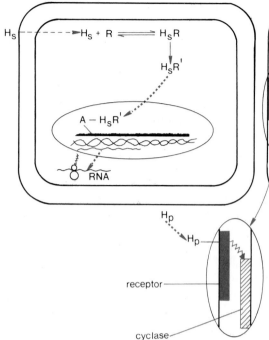

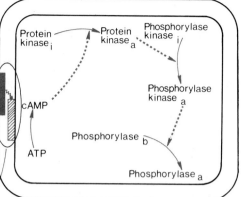

Figure 25–31 A comparison of current concepts of the receptors for steroid hormones (H_s) and peptide hormones (H_p). Note that the steroid hormone enters the cell, binds with a soluble receptor, passes into the nucleus and is attached to the chromatin, stimulating RNA synthesis. In contrast, the peptide hormone remains outside of the cell, binding to a receptor on the plasma membrane. This activates a membrane-bound adenyl cyclase, producing cyclic AMP. The latter stimulates a cascade of protein kinases, which phosphorylate an enzyme, thereby activating (or inactivating) it.

inactive protein to an active enzyme or perhaps an active enzyme to an inactive phosphorylated form. For example, phosphorylase kinase is *activated* when phosphorylated, whereas glycogen synthetase becomes *inactivated* when phosphorylated.

The effects of hormones in facilitating the entrance of certain substrates into the cell, as for example in the uptake of glucose by muscle cells stimulated by insulin, have suggested that

the hormone combines with some protein or other component of the cell membrane. This leads to a change in the molecular architecture of the membrane and hence in its permeability to specific substrates.

Another hypothesis states that the hormone increases the enzymatic activity of a protein. The stimulation of **adenyl cyclase** and the increased production of **cyclic 3′,5′ adenylic acid** from ATP induced by several protein hor-

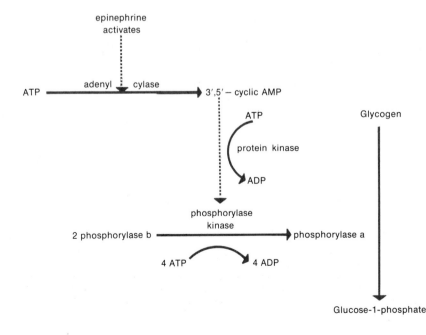

Figure 25–32 The sequence of enzymatic events by which epinephrine or glucagon stimulates adenyl cyclase and brings about the synthesis of 3′,5′-adenosine monophosphate (cyclic AMP). This in turn activates a protein kinase that phosphorylates phosphorylase kinase, which in turn phosphorylates and activates phosphorylase. This finally brings about the cleavage of glycogen and the secretion of glucose.

mones are examples of this. The cyclic AMP is regarded as an intracellular "second messenger" that mediates the effect of the hormone. Epinephrine, for example, stimulates the adenyl cyclase of liver cells (Fig. 25–32). The resulting cyclic AMP activates a protein kinase that transfers a phosphate group from ATP to a third enzyme, phosphorylase kinase, and activates it so that it in turn can convert an inactive fourth enzyme, phosphorylase b, to active phosphorylase a. The latter then catalyzes the production of glucose-1-phosphate from glycogen. At each of these successive steps in the enzymatic cascade there is an amplification of 10- to 100-fold, just as in the sequence of enzymes involved in blood clotting (p. 356). The cascade effect permits a very small amount of epinephrine to lead to the production of a very large amount of glucose-1-phosphate.

A third general hypothesis suggests that the hormone enters the nucleus and activates certain genes that were previously repressed (Fig. 25–33). This leads to the production of additional messenger RNA or new kinds of messenger RNA which code for the synthesis of specific new proteins. The messenger RNA undergoes processing in the nucleus and a long tail of polyadenylic acid is added. The polyadenylic acid messenger RNA passes out of the nucleus to the ribosomes and there serves as a template for the synthesis of proteins. This theory also accounts for the marked amplification of hormonal effects, because a small amount of hormone, by turning on the transcription of a specific gene, could result in the production of many molecules of messenger RNA and many, many more molecules of protein. The generalization emerging from the many studies of hormonal action is that protein and peptide hormones bind to receptors on the

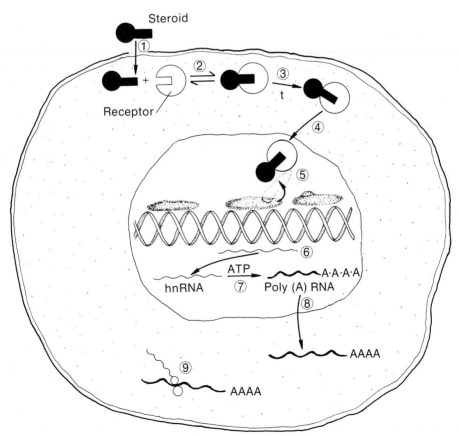

Figure 25–33 Successive steps in the postulated mechanism by which a steroid hormone stimulates the synthesis of specific proteins in its target cells. The steroid enters the cell (1) and binds to a specific receptor protein (2). The steroid-receptor complex undergoes a temperature-dependent transformation process (3), enters the nucleus (4) and binds to the chromatin (5). This in some way facilitates the transcription of the DNA in that section of the chromatin (6). The resulting "heterogeneous nuclear RNA" undergoes processing with ATP (7), and a long polyadenylic acid tail is added. The poly (A) rich mRNA leaves the nucleus (8) and serves as the template for the synthesis on the ribosome (9) of the specific protein.

cell membrane and stimulate adenyl cyclase. The product of adenyl cyclase (cyclic AMP) stimulates protein kinases which activate enzymes by phosphorylating them. All of these phenomena occur in the cytoplasm. In contrast, steroid hormones are bound to cytoplasmic receptors and transferred as a receptor-steroid complex into the nucleus, where they activate the transcription of certain genes. The resulting polyadenylate messenger RNA codes for the synthesis of specific proteins, specific enzymes which are responsible for the physiologic changes resulting from hormonal action. Evidence supporting the latter hypothesis is derived from studies of the effects of estradiol on the uterus, of estradiol and progesterone on the oviduct, and of dihydrotestosterone on the seminal vesicle and prostate.

It is not clear whether hormones are typically used up as they carry out their regulatory action in the target tissue. Estradiol, however, is not used up or changed chemically as it stimulates the growth of the uterus. Hormones bound to their receptors appear to be relatively stable, but hormones circulating in the blood have relatively short biological half-lives. They are inactivated and eliminated from the body and must be replaced by new hormone molecules synthesized in the endocrine glands.

It seems unlikely that all hormones have a common molecular mechanism by which their effects are produced. Indeed, there is evidence that certain hormones produce their effects not by a single mechanism but by several different mechanisms acting in parallel within a single cell. The theories current at any given moment tend to reflect our general knowledge of cellular and molecular biology. Theories implicating effects of hormones on cell membranes have given way to theories implicating an effect on altering the activity of an enzyme, and these have been replaced in turn by theories involving effects of the hormone on the genetic mechanism which result in the synthesis of specific kinds of RNA and proteins.

25–13 PHEROMONES

In recent years it has been appreciated that the behavior of animals may be influenced

not only by hormones—chemicals that are released into the internal environment by endocrine glands and that regulate and coordinate the activities of other tissues—but also by *pheromones*—substances that are secreted by *exocrine* glands and released into the *external* environment and that influence the behavior of other members of the same species. We are used to thinking that information can be transferred from one animal to another by sight or sound; pheromones represent a means of communication, of transferring information, by smell or taste. Pheromones evoke specific behavioral, developmental, or reproductive responses in the recipient; these responses may be of great significance for the survival of the species.

Some pheromones act in some way on the recipient's central nervous system and produce an immediate effect on its behavior. Among these releaser pheromones are the *sex attractants* of moths and the *trail pheromones* and *alarm substances* secreted by ants. Other pheromones, termed primer pheromones, act more slowly and trigger a chain of physiological events in the recipient which affect its growth and differentiation. These include the regulation of the growth of locusts and control of the numbers of reproductives and soldiers in termite colonies.

The sex attractants of moths provide some of the more spectacular examples of pheromones. Among the ones that have been isolated and identified are *bombykol,* a 16-carbon alcohol with two double bonds, secreted by female silkworms, and *gyplure,* 10-acetoxy-Δ^7-hexadecanol, secreted by female gypsy moths. The male silk moth has an extremely sensitive device in his antennae for sensing the attractant. When an investigator records the nerve impulses coming from the antennae, he finds that these electro-antennagrams show specific responses to bombykol and not to other substances. The male silk moth cannot determine the direction of the source by flying up a concentration gradient because the molecules are nearly uniformly dispersed except within a few meters of the source. Instead he responds to the stimulus by flying *upwind* to the source. With a gentle wind, the bombykol given off by a single female moth covers an area several thousand meters long and as much as 200 meters wide. An average silkworm contains some 0.01 mg. of bombykol. It can be shown experimentally that when as little as 10,000 molecules of attractant are allowed to diffuse from a source 1 cm. from a male he responds appropriately. He can have received only a few hundred of these molecules, perhaps less. Thus the amount of attractant in one female could stimulate more than one billion males! The attractants, generally hydrocarbons, contain 10 to 17 carbons in the chain, which provides for the specificity of the several kinds of attractants.

The sex attractant of the American cockroach is not a long chain alcohol like bombykol or gyplure but has a central three-carbon ring to which methyl groups and a propanoxy group are attached. Sex attractants have been tested as possible specific insecticides. By putting sex attractant on stakes placed every 10 meters in a large field, investigators could blanket the air with sex attractant, thus confusing the males and greatly decreasing the probability of their finding females and mating with them.

When returning to the nest after finding food, the fire ants secrete a "trail pheromone" which marks the trail so that other ants can find their way to the food. The trail pheromone is volatile and evaporates within two minutes, so that there is little danger of ants being misled by old trails. Ants also release alarm substances when disturbed and this (rather like ringing the bell in a firehouse) in turn transmits the alarm to ants in the vicinity. These alarm substances have a lower molecular weight than the sex attractants and are less specific, so that members of several different species respond to the same alarm substance.

Worker bees, on finding food, secrete *geraniol,* a 10-carbon, branched chain alcohol, to attract other worker bees to the food. This supplements the information conveyed by their wagging dance (p. 314). Queen bees secrete 9-ketodecanoic acid which, when ingested by worker bees, inhibits the development of their ovaries and their ability to make royal cells in which new queens might be reared. This substance also serves as a sex attractant to male bees during the queen's nuptial flight.

In colonial insects, such as ants, bees and termites, pheromones play an important role in regulating and coordinating the composition and activities of the population. A termite colony includes morphologically distinct queen, king, soldiers, and nymphs or workers. All develop from fertilized eggs; however, queens, kings and soldiers each secrete inhibitory substances, pheromones, that act on the corpus allatum of the nymphs and prevent their developing into the more specialized types. If the queen dies there is no longer any "antiqueen" pheromone released and one or more of the nymphs develop into queens. The members of each colony will permit only one queen to survive, devouring any excess ones. Similarly the loss of the king termite or a reduction in the number of soldiers permits other nymphs to develop into the specialized castes to replace them. Males of migratory locusts secrete a substance from the surfaces of their skin which accelerates the growth of young locusts.

There are examples of primer pheromones in mammals as well as in insects. When female mice are placed four or more per cage there is a greatly increased frequency of pseudopregnancy. If their olfactory bulbs are removed this effect disappears. When more females are placed together in a cage their estrous cycles become very erratic. However if one male mouse is placed in the cage his odor can initiate and synchronize the estrous cycles of all the females (the "Whitten effect") and reduce the frequency of reproductive abnormalities. Even more curious is the finding (the "Bruce effect") that the odor of a strange male will block pregnancy in a newly impregnated female mouse. The nerve impulses from the nose pass to the hypothalamus and block the output of prolactin-releasing hormone. The subsequent lack of prolactin leads to regression of the corpora lutea and the failure of the fertilized ova to implant.

The question of whether there are human pheromones remains unanswered, but of interest in this respect is the observation of the French biologist J. LeMagnen that the odor of 15-hydroxypentadecanoic acid is perceived clearly only by sexually mature females and that it is perceived most sharply at about the time of ovulation! Males and immature females are relatively insensitive to this substance but male subjects become more sensitive to it after an injection of estrogen.

Analyses of the menstrual cycles of the students at an American women's college showed a statistically significant tendency for the increasing synchronization of menstrual cycles among roommates and close friends. The study ruled out several possible explanations for the phenomenon but suggested some pheromonal effect between girls who were together much of the time. The study also demonstrated a significant shortening of the menstrual cycles of women who were with male companions three or more times per week; those who dated twice a week or less had longer and more irregular cycles. This appears to be analogous to the Whitten effect, but whether it involves a pheromone is unknown.

QUESTIONS

1. Compare the integrative actions of the nervous system and the endocrine system.
2. Define a hormone. How would you go about proving that a particular gland secretes a particular hormone?
3. Define a pheromone. Distinguish between hormones and pheromones as to (a) chemical nature, (b) site of production and (c) mode of action.
4. Distinguish between a hormone and a vitamin; a hormone and an enzyme.
5. What kinds of experiments might be used to determine whether a newly discovered gland in a vertebrate secretes a hormone?
6. Where is the thyroid gland located? What does it secrete?
7. Differentiate between myxedema, simple goiter and cretinism. What radioactive substance is especially useful in studying the functioning of the thyroid? Why?
8. What is the function of the parathyroid glands? What role does parathyroid hormone play in regulating metabolism?
9. What is the importance of insulin in the body? Explain why the feeding of thyroid gland may cure myxedema and why the feeding of pancreas does not cure diabetes.
10. List the hormones secreted by each part of the adrenal gland. What is the function of each? What is Addison's disease? What reasons can you advance to explain why it is that Addison's disease is not cured by the feeding of adrenal glands?
11. What are oxytocin and vasopressin? What effects do they have on the mammalian body?
12. Why is the pituitary sometimes called "the master gland"? Do you think this term is justified?
13. In view of the newer evidence of the source of the hormones of the posterior lobe of the pituitary, what function might be ascribed to the pineal body? Outline a series of experiments that might provide evidence for your hypothesis.
14. Describe the feedback mechanism that regulates the production of thyroxine and thyrotropin.
15. What is ovulation? When does it occur in the human? What hormones are involved in controlling ovulation? How do the oral contraceptives act to prevent ovulation?
16. What are the functions of progesterone? Of luteinizing hormone?
17. What tissues secrete prostaglandins? What tissues are affected by them?
18. How can you account for the fact that one of the best sources of estrogens is the urine of the stallion?
19. Describe the feedback mechanism that regulates the events of the menstrual cycle.
20. Make a list of all the structures in the body known to be endocrine glands and list the hormones secreted by each.
21. List and describe the effects of all the hormones that are required for the normal completion of pregnancy.
22. Describe the hormonal interrelations that control the development and functioning of the breasts.
23. What hormones stimulate adenyl cyclase? What is the role of cyclic AMP in the functioning of the endocrine system?
24. To what extent do you think an individual's personality is affected by his endocrine glands? Discuss the ways in which knowledge of physiology and psychology can help in establishing fair laws regarding behavior.

SUPPLEMENTARY READING

General Endocrinology by C. D. Turner and J. T. Bagnara is an excellent introductory text covering the basic biological aspects of endocrinology. The more clinical aspects of the subject are discussed in R. H. Williams' *Textbook of Endocrinology* and in Dorothy B. Villee's *Human Endocrinology*. Two classic treatments of the comparative and evolutionary aspects of endocrine systems in vertebrate and invertebrate animals are *Textbook of Comparative Endocrinology* by E. J. W. Barrington and *A Textbook of Comparative Endocrinology* by A. Gorbman and H. A. Bern. The more recent discoveries in comparative endocrinology are described in *Invertebrate Endocrinology and Hormonal Heterophylly* edited by Walter J. Burnette. *Recent Progress in Hormone Research* is a series of books published annually; each volume contains the papers presented at the annual Laurentian Hormone Conference and a transcript of the discussion following each presentation. The series is a mine of information on the latest discoveries in endocrinology. *Recent Advances in Enzyme Regulation*, edited by George Weber and published annually, is another excellent summary of the latest research in endocrine control of enzymes. *Neuroendocrinology* by E. Scharrer and B. Scharrer is an excellent account of the process of neurosecretion with examples drawn from both insect and vertebrate species. The two-volume, multi-author book *Sex and Internal Secretions*, edited by W. C. Young, includes chapters by experts in the many fields relating to the hormonal control of cycles and behavior. The metabolic effects of the mammalian hormones are presented clearly in J. Tepperman's *Metabolic and Endocrine Physiology*. Clear discussions of how the oral contraceptives work and what their major side effects may be are presented in V. A. Drill's *The Oral Contraceptives* and Robert Kistner's *The Pill*.

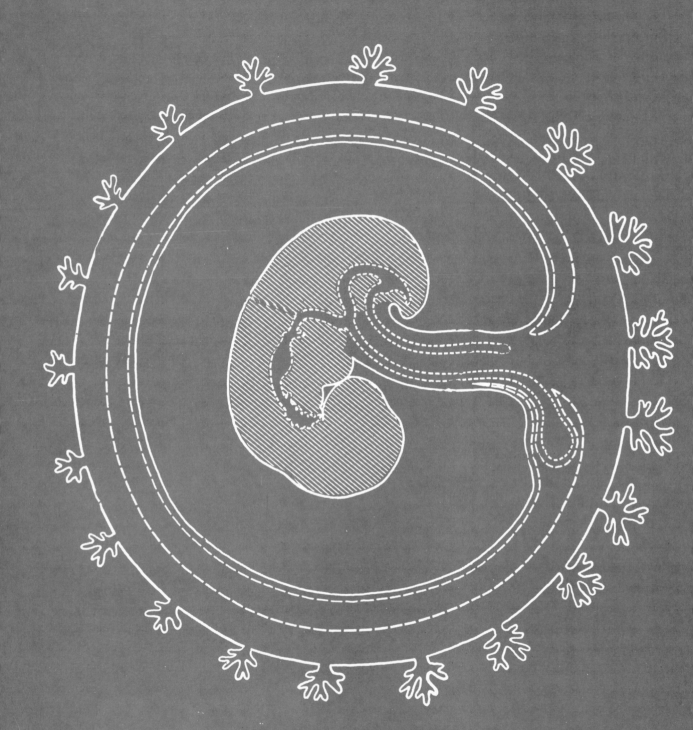

PART 6

REPRODUCTIVE PROCESSES

INTRODUCTION

If there is any one feature of a living system that qualifies as the "essence of life" it is the ability to reproduce and perpetuate the species. The survival of each species of plant or animal requires that its individual members multiply, that they produce new individuals to replace the ones killed by predators, parasites or old age. Reproduction at the molecular level is a function of the unique capacity of the nucleic acids for self-replication, which depends on the specificity of the relatively weak hydrogen bonds between pairs of nucleotides. At the level of the whole organism, reproduction ranges from the simple fission of bacteria and other unicellular organisms (a process that does not involve sex at all) to the incredibly complex structural, functional and behavioral processes of reproduction in the higher animals. Human reproduction involves not only the genetic transfer of biological information from one generation to the next but also the endocrine regulation of the development of the genital tracts, oögenesis, ovulation and spermatogenesis, as well as the intricate behavior patterns that ensure that eggs and sperm are released at the same time and the same place so that they can meet to form a fertilized egg or zygote. This is followed by the complex processes of development and differentiation by which a zygote becomes an adult organism. The reproductive process is a cyclic one, and a study of reproduction could begin at any point in that cycle.

CHAPTER 26

REPRODUCTION

The details of the reproductive process vary tremendously from one kind of organism to another, but we can distinguish two basically different types—asexual and sexual. In **asexual reproduction** a single parent splits, buds or fragments to give rise to two or more offspring that have hereditary traits identical with those of the parent. Even higher animals may reproduce asexually; indeed, the production of identical twins in humans by the splitting of a single fertilized egg is a kind of asexual reproduction.

In contrast, **sexual reproduction** involves two parents: each contributes a specialized gamete, an egg or sperm, and these fuse to form the fertilized egg. The egg is typically large and nonmotile, with a store of nutrients to support the development of the embryo which results when the egg is fertilized. The sperm is usually small and motile, adapted to swim actively to the egg by beating its long, whiplike tail. Sexual reproduction has the biological advantage of making possible the recombination of the inherited traits of the two parents; thus the offspring may be better able to survive than either parent. Evolution can proceed much more rapidly and effectively with sexual reproduction than it can with asexual reproduction.

26–1 SEXUAL AND ASEXUAL REPRODUCTION

Perhaps the simplest form of asexual reproduction is the splitting of the body of the parent into two more or less equal parts, each of which becomes a new independent whole organism (Figs. 26–1 and 26–2). This form of reproduction, termed *fission,* occurs chiefly among the protists, the single-celled plants and animals. The cell division involved is mitotic.

Hydras and yeasts reproduce by **budding,** in which a small part of the parent's body separates from the rest and develops into a new individual. It may split away from the parent and take up an independent existence or it may remain attached and become a more or less independent member of the colony.

Salamanders, lizards, starfish and crabs can grow a new tail, leg, arm or certain other organs if the original one is lost. When this ability to regenerate a part is carried to an ex-

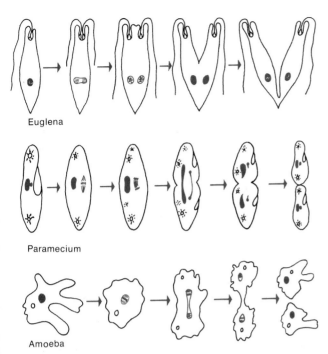

Euglena

Paramecium

Amoeba

Figure 26–1 Asexual reproduction in *Euglena, Paramecium* and *Amoeba.* In each, one cell divides mitotically to give rise to two cells.

blood cells. Inside the red cell each *Plasmodium* grows and then divides by multiple fission into from 12 to 24 spores, each of which enters the blood stream when the red cell, after swelling, disintegrates. The spores released infect new red cells and the process is repeated. The simultaneous breakdown of billions of red cells causes the malarial chill, which is followed by fever as the toxic substances released penetrate to other organs of the body. If a second, uninfected mosquito bites the infected person, it becomes infected by sucking up some of the plasmodia along with the blood. Inside the mosquito's stomach a complicated process of sexual reproduction occurs, involving the formation of eggs and sperm, fertilization, and multiple fission of the resulting zygote to yield new infective spores. Some of these then migrate into the mosquito's salivary glands, where they remain ready to infect the next person bitten.

treme, it becomes a method of reproduction. The body of the parent may break into several pieces and each piece then regenerates the missing parts and develops into a whole animal. Such reproduction by *fragmentation* is common among the flatworms. Starfish have the ability to regenerate an entire new starfish from a single arm. If you try to kill starfish by chopping them in half and throwing the pieces back into the sea you will simply double the number of starfish preying on your oyster beds!

Many plants and some animals reproduce asexually by *spores*, which are special cells with resistant coverings adapted to withstand unfavorable environmental conditions such as excessive heat, cold or desiccation. During the growth phase of their life cycle the parasitic protozoa, the Sporozoa, typically live as parasites within the cells of the host and reproduce by spores. Sporozoa cause some serious diseases such as malaria in humans and coccidiosis in poultry. The malaria organism, *Plasmodium,* has a complex life cycle involving both humans and the *Anopheles* mosquito (Fig. 26–3). Through the bite of an infected mosquito the malaria organisms are introduced into a person's blood stream and enter the red

Figure 26–2 An electron scanning micrograph of an early stage of binary fission in the protozoan *Didinium nasutum.* This type of asexual reproduction is common among protozoa. (Courtesy of Dr. Eugene B. Small.)

—ANOPHELES MOSQUITO—

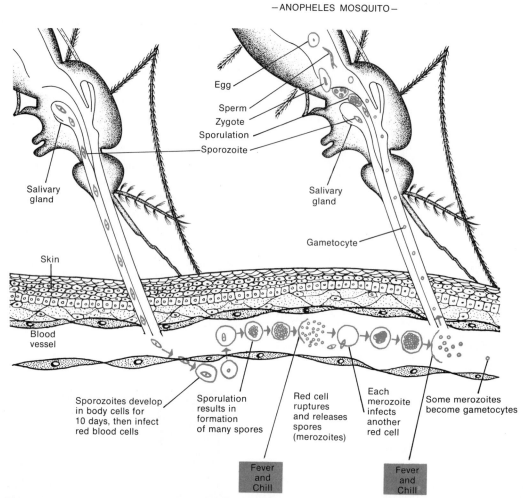

Egg
Sperm
Zygote
Sporulation
Sporozoite

Salivary
gland

Salivary
gland

Gametocyte

Skin

Blood
vessel

Sporozoites develop
in body cells for
10 days, then infect
red blood cells

Sporulation
results in
formation
of many spores

Red cell
ruptures
and releases
spores
(merozoites)

Each
merozoite
infects
another
red cell

Some merozoites
become gametocytes

Fever
and
Chill

Fever
and
Chill

Figure 26–3 A diagram of the life cycle of the malaria parasite, *Plasmodium*. An infected mosquito (*left*) bites a person and injects some *Plasmodium* sporozoites into his blood stream. These reproduce asexually by sporulation within the red blood cells of the host. The infected red cells rupture and the new crop of merozoites released then infects other red cells. The bursting of the red cells releases toxic substances which cause the periodic fever and chill. In time some merozoites become gametocytes which can infect a mosquito if one bites a person with malaria. The gametocytes develop into eggs and sperm (*right*) and undergo sexual reproduction in the mosquito, and the zygote, by sporulation, produces sporozoites which migrate to the salivary glands.

26–2 REPRODUCTIVE SYSTEMS

Although the details of the reproductive process vary tremendously from one organism to another, it is possible to make some generalizations about animal reproductive systems that may be helpful in understanding the variations. The basic components of the male reproductive system are the male gonad, or *testis*, in which sperm are produced, and the *sperm duct* for the transport of sperm to the exterior of the body. Parts of the sperm duct, or of areas adjacent to it, may be modified for specific functions. A part of the duct may be given over for sperm storage; such a part is often called a *seminal vesicle*, but the human organ termed the seminal vesicle does *not* store sperm. There may be glandular areas for the production of seminal fluid, which serves as a vehicle for the sperm and may also activate, nourish and protect them. The terminal part of the sperm duct may open onto or into a copulatory organ, a *penis,* which provides for the transfer of sperm to the female.

The basic parts of the female system are the female gonad, or *ovary,* and the *oviduct,* a tube for the transport of eggs to the exterior. Like the sperm duct, the oviduct may be modified in any of several ways in different species. It may have a glandular portion for the secretion of an egg shell, a case or a cocoon. A section of the oviduct, a *seminal receptacle,* may be modified for the storage of sperm following their transfer from the male. Another portion of the oviduct may be modified as a *uterus* for egg storage or for the development of the fertilized egg within the body of the female. The terminal portion of the oviduct may be adapted as a *vagina* for receiving the male copulatory organ.

The male and female systems of most animals do not possess all these modifications and they cannot be considered as part of any kind of progressive sequence of evolutionary changes. Rather, the presence or absence of some specific adaptation of the reproductive system is correlated with the circumstances of reproduction—whether the animal lives in the sea, in fresh water or on land; whether fertilization is external or internal; whether the eggs are liberated singly into the water to develop or are deposited onto the bottom within envelopes; and whether or not the developing individual passes through a larval stage.

Most species of fishes utilize a rather primitive form of reproduction in which the female sheds into the water enormous numbers of eggs, which may or may not be fertilized by sperm from some passing male. Thereafter, both male and female fishes ignore the eggs. In such fishes, the uterus is missing and the male does not have a copulatory organ. In other fishes, however, internal fertilization takes place before egg-laying. In the sea catfish, for example, the male's anal fin is used as an intromittent organ. The fertilized eggs are deposited in a clump that the male picks up and carries in his mouth during most of their incubation (presumably fasting and yielding not to temptation!).

26-3 MEIOSIS

The process of mitosis, which ensures that each daughter cell will receive exactly the same number and kind of chromosomes that the parent cell had, was discussed earlier (p. 57) together with the features of the cell cycle. The constancy of the chromosome number in the cells of successive generations of organisms is ensured by the process of *meiosis,* which occurs during the formation of eggs or sperm in animals and of spores in plants (p. 160). The process of meiosis involves a pair of cell divisions during which the chromosome number is reduced to one half, so that each gamete receives only half as many chromosomes as are contained in the cells in the body of the parent. When two gametes unite in fertilization, the fusion of their nuclei reconstitutes the $2N$ number of chromosomes. In meiosis the members of each pair of chromosomes separate and pass to different daughter cells. As a result of this, each gamete contains one and only one of each kind of chromosome; in other words, it contains one complete set of chromosomes. This is accomplished by the pairing, or *synapsis,* of the like chromosomes and a separation of the members of the pair, with one going to each pole of the dividing cell. The like chromosomes that pair during meiosis are called *homologous chromosomes.* They are identical in size and shape, have identical chromomeres along their length, and contain similar hereditary factors, or genes. A set of one of each kind of chromosome is called the $1N$, or *haploid number.* A set of two of each kind is called the $2N$, or *diploid number.* The haploid number for the human is 23 and the diploid number is 46. Gametes—eggs and sperm—have the haploid number. Fertilized eggs (zygotes) and all the cells of the body developing from the zygote have the diploid number. A fertilized egg gets exactly half its chromosomes and half its genes from its mother and the other half from its father. Only the last two cell divisions, which result in mature functional eggs or sperm, are meiotic; all other cell divisions are mitotic.

The process of meiosis consists of two cell divisions, the first and second meiotic divisions, which occur in succession (Figs. 26-4 and 26-5). Each of these includes prophase, metaphase, anaphase and telophase stages, but there are important differences between mitosis and meiosis, especially in the prophase of the first meiotic division. In this the chromosomes appear as long thin threads, becoming shorter and thicker as in mitosis. The homologous chromosomes undergo synapsis while

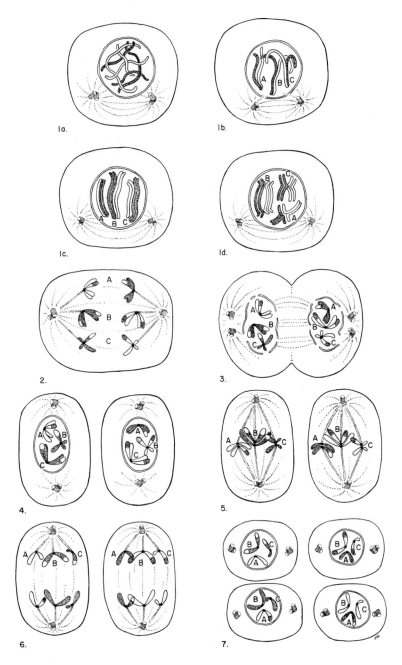

Figure 26–4 Diagrams illustrating the process of meiosis in an animal with a diploid number of six. *1a,* Early prophase; *1b,* later prophase, synapsis beginning; *1c,* apparent doubling of the synapsed chromosomes to form tetrads; *1d,* late prophase of first meiotic division. *2,* Anaphase and, *3,* telophase of first meiotic division. *4,* Prophase of second meiotic division. *5,* Metaphase and, *6,* anaphase of second meiotic division. *7,* Mature gametes, each of which contains the haploid number (three) of chromosomes, one of each kind of chromosome.

they are still elongate and thin. The homologous chromosomes pair longitudinally; they come to lie close together side by side along their entire length and are twisted around each other. After synapsis, or pairing, has occurred, the chromosomes continue to shorten and thicken. Each one becomes visibly double, consisting of two chromatids as in mitosis. This doubling has occurred in the S phase before meiosis begins. At the end of the first meiotic prophase the chromosomes have doubled and undergone synapsis to yield a bundle of four

homologous chromatids called a **tetrad.** Each pair of chromosomes gives rise to a bundle of four, so the number of tetrads equals the haploid number of chromosomes. In human cells there are 23 tetrads (and a total of 92 chromatids) at this stage. The centromeres have not divided and there are only two centromeres for the four chromatids.

While these events are occurring, the two centrioles go to opposite poles, a spindle forms between the centrioles, and the nuclear membrane dissolves. The tetrads line up around the

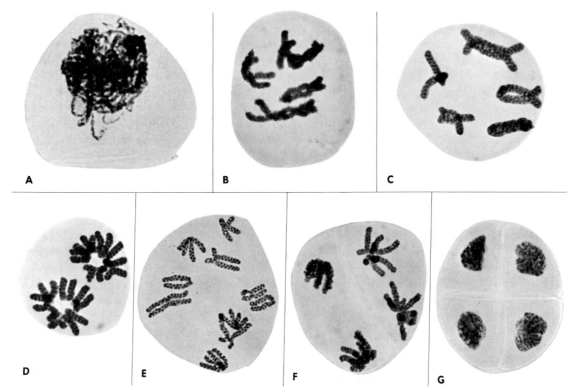

Figure 26–5 Photographs of meiosis in the plant *Trillium erectum*. *A*, Early prophase of the first meiotic division. *B*, Later prophase of the first meiotic division. *C*, Metaphase. *D*, Anaphase of the first meiotic division. *E*, Metaphase of the second meiotic division. *F*, Anaphase of the second meiotic division. *G*, Four daughter cells (quartet). Magnification ×2000. (Courtesy of A. H. Sparrow and R. F. Smith.)

equator of the spindle and the cell is said to be in metaphase. In the anaphase of the first meiotic division, the daughter chromatids formed from each chromosome, still united by their centromere, separate and move toward opposite poles. Thus the homologous chromosomes of each pair, but not the daugher chromatids of each chromosome, are separated in anaphase I. This differs from mitotic anaphase in which the centromeres do divide and the daughter chromatids pass to opposite poles.

In humans there are 23 double chromosomes at each pole in the telophase of the first meiotic division. Cytoplasmic division follows, but as in most animals and plants, there is no clear interphase between the two meiotic divisions. The chromosomes do not divide into daughter chromatids and there is no synthesis of DNA, as there is in the S phase between mitotic divisions. The chromosomes do not form chromatin threads; instead the centriole divides, a new spindle forms in each cell (at right angles to the spindle of the first division), and the haploid number of double chromosomes

lines up on the equator of the spindle. The telophase of the first meiotic division and the prophase of the second meiotic division are usually of rather short duration. The lining up of the double chromosomes on the equator of the spindle constitutes the metaphase of the second meiotic division. The metaphases of the first and second meiotic divisions can be distinguished, because in the first the chromosomes are arranged in bundles of four and in the second the chromosomes are arranged in bundles of two. The centromeres divide and the daughter chromatids, now chromosomes, separate and move to opposite poles. Thus in the telophase of the second meiotic division in humans, 23 chromosomes, one of each kind, arrive at each pole. The cytoplasm then divides, nuclear membranes form, and the chromosomes gradually elongate and become chromatin threads.

The two successive meiotic divisions yield four nuclei, each of which has one and only one of each kind of chromosome, a haploid set. The members of the homologous pairs of chromosomes are segregated into separate daughter cells. The four cells resulting from the two meiotic divisions are now mature gametes and do not undergo any further mitotic or meiotic divisions.

Fundamentally the same process occurs in the meiotic divisions in the testis which result in sperm and in the meiotic divisions in the ovary which result in eggs, but there are some differences in detail.

26–4   SPERMATOGENESIS

The human testis is made up of thousands of cylindrical sperm tubules, in each of which millions of sperm develop. The walls of these tubules are lined with primitive, unspecialized germ cells, *spermatogonia.* Throughout embryonic development and during childhood the spermatogonia divide mitotically, giving rise to additional spermatogonia to provide for the growth of the testis. After sexual maturity, some of the spermatogonia undergo *spermatogenesis,* the formation of mature sperm, while

others continue to divide mitotically and produce more spermatogonia for later spermatogenesis. In most wild animals there is a definite breeding season, either in spring or fall, during which the testis increases in size and spermatogenesis occurs. Between breeding seasons the testis is small and contains only spermatogonia. In man and most domestic animals, spermatogenesis occurs throughout the year once sexual maturity is reached.

Spermatogenesis begins with the growth of the spermatogonia into larger cells known as *primary spermatocytes* (Fig. 26–6). These divide (first meiotic division) into two equal-sized *secondary spermatocytes,* which in turn undergo the second meiotic division to form four equal-sized *spermatids.* The spermatid, a spherical cell with a generous amount of cytoplasm, is a mature gamete with the haploid number of chromosomes. A complicated process of growth and change (though not cell division) converts the spermatid into a functional sperm. The nucleus shrinks in size and becomes the head of the sperm (Fig. 26–7), while the sperm sheds most of its cytoplasm.

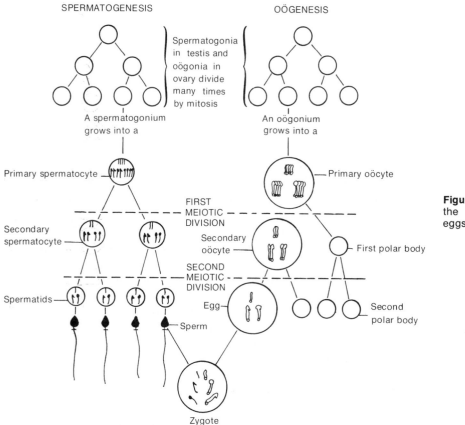

Figure 26–6 Comparison of the formation of sperm and eggs.

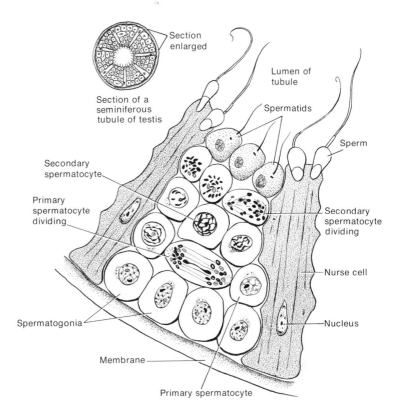

Figure 26–7 Diagram of part of a section of a human seminiferous tubule to show the stages in spermatogenesis and in the transformation of a spermatid into a mature sperm.

Some of the Golgi bodies congregate at the front end of the sperm and form a point (the acrosome) which may aid the sperm in puncturing the egg cell membrane.

The two centrioles of the spermatid move to a position just in back of the nucleus. A small depression appears on the surface of the nucleus, and one of the centrioles, the proximal centriole, takes up a position in the depression at right angles to the axis of the sperm. The second or distal centriole, just behind the proximal centriole, gives rise to the axial filament of the sperm tail (Fig. 26–8). Like the axial filament of flagella, it consists of two longitudinal fibers in the middle and a ring of nine pairs, or doublets, of longitudinal fibers surrounding the two.

The mitochondria move to the point at which head and tail meet, and form a small middle piece which provides energy for the beating of the tail. Most of the cytoplasm of the spermatid is discarded as residual bodies, which are taken up by phagocytosis by the Sertoli cells. The mature sperm retains only a thin sheath surrounding the mitochondria in the middle piece and the axial filament of the tail.

The spermatozoa of various animal species may be quite different. There are great varia-tions in the size and shape of the tail and in the characteristics of the head and middle piece (Fig. 26–9). The sperm of a few animals (e.g., the parasitic roundworm *Ascaris*) have no tail and move instead by amoeboid motion. Crabs and lobsters have a curious tail-less sperm with three pointed projections on the head, which stick to the surface of the egg, holding the sperm securely in place. The middle piece uncoils like a spring, and pushes the nucleus of the sperm into the egg cytoplasm, thus accomplishing fertilization.

26–5 OÖGENESIS

The ova or eggs develop in the ovary from immature sex cells, *oögonia.* Early in development the oögonia undergo many successive mitotic divisions to form additional oögonia, all of which have the diploid number of chromosomes. In many animals, notably the vertebrates, the oögonia and oöcytes are surrounded by a layer of follicle cells. In the human this occurs early in fetal development, and by the third month the oögonia begin to develop into *primary oöcytes* (Fig. 26–10). When a human female is born, her two ovaries contain some

400,000 primary oöcytes, which have attained the prophase of the first meiotic division. These primary oöcytes remain in prophase until the woman reaches sexual maturity; then as each follicle matures, the first meiotic division resumes and is completed at about the time of ovulation (15 to 45 years after meiosis began!).

The events occurring in the nucleus—synapsis, the formation of tetrads, and the separation of the homologous chromosomes—are the same as those occurring in spermatogenesis, but the division of the cytoplasm is unequal, resulting in one large cell, the *secondary oöcyte*, which contains the *yolk* and nearly all the cytoplasm, and one small cell, the first *polar body*, which consists of practically nothing but a nucleus (Fig. 26–6). It was named a polar body before its significance was understood because it appeared as a small speck at the animal pole of the egg.

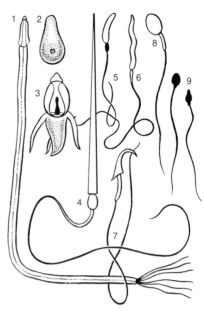

Figure 26–9 Spermatozoa from different species of animals, illustrating the differences in size and shape. *1*, Gastropod. *2*, Ascaris. *3*, Hermit crab. *4*, Salamander. *5*, Frog. *6*, Chicken. *7*, Rat. *8*, Sheep. *9*, Man.

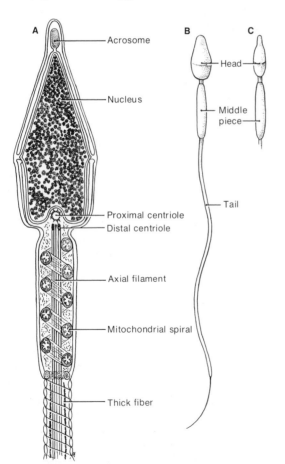

Figure 26–8 *A*, Diagram of the head and middle piece of a mammalian sperm, greatly enlarged, as seen in the electron microscope. *B* and *C*, Top and side views of a sperm as seen by light microscopy.

In the second meiotic division, which proceeds as the ovum enters the fallopian tube, the secondary oöcyte again divides unequally into a large *oötid* and a small second polar body, both of which have the haploid chromosome number. The first polar body may divide into two additional second polar bodies. The oötid then becomes a mature ovum. The three small polar bodies soon disintegrate, so that each primary oöcyte gives rise to just one ovum, in contrast to the four sperm formed from each primary spermatocyte. The unequal cytoplasmic division ensures that the mature egg will have enough cytoplasm and stored yolk to survive, if fertilized. The primary oöcyte in a sense puts all its yolk in one ovum; the egg has neatly solved the problem of reducing its chromosome number without losing the cytoplasm and yolk needed for development after fertilization.

The union of one haploid set of chromosomes from the sperm with another haploid set from the egg, which occurs in *fertilization*, reestablishes the diploid chromosome number. Thus the fertilized egg, or zygote, and all the body cells developing from it by mitosis, have the diploid number of chromosomes. In each individual, exactly half of the chromosomes and half of the genes come from the mother

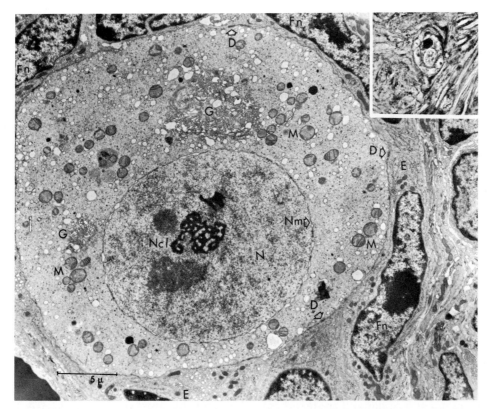

Figure 26–10 Electron micrograph of a young oöcyte of a guinea pig showing Golgi material. *Inset,* Oöcyte in a similar stage, as seen by light microscopy. Nucleus (*N*); nucleolus (*Ncl*); nuclear membrane (*Nm*); Golgi material (*G*); mitochondria (*M*); "desmosomes," denser material connecting plasma membranes of oöcyte and follicle cells (*D*); nuclei of follicle cells (*Fn*); endoplasmic reticulum in follicle cells (*E*). (From Balinsky, B. I.: *An Introduction to Embryology,* 4th Ed. Philadelphia, W. B. Saunders Co., 1975. Courtesy of Professor E. Anderson.)

and the other half from the father. All of the phenomena of Mendelian genetics depend upon these simple facts. Because of the nature of gene interaction, the offspring may resemble one parent more than the other, but the two parents make equal contributions to its inheritance.

26–6 THE HUMAN REPRODUCTIVE SYSTEM: MALE

The paired testes develop within the abdominal cavity but in human beings and some other mammals they descend shortly before or after birth into the *scrotal sac,* an outpocketing of the body wall covered by a loose pouch of skin. The cavity of the scrotal sac is part of the abdominal cavity and is connected to it by the *inguinal canal.* After the testes have descended, this canal is usually closed by the growth of connective tissue. The normal descent of the testes into the scrotal sac is necessary for the production of sperm. If the testes remain in the abdominal cavity the slightly

higher temperature there prevents the formation of sperm.

Each testis consists of about 1000 highly coiled *seminiferous tubules,* with a total length of about 250 meters, in which sperm are produced. Between the tubules lie the *interstitial cells,* which produce male sex hormones (Fig. 25–15). The lining of the seminiferous tubules consists of *spermatogonia,* derived from the primordial sex cells, and *Sertoli cells,* which nourish the sperm as they develop from rounded cells into mature, tailed forms. The formation of sperm proceeds in waves along the tubules. The seminiferous tubules are connected via fine tubes (the *vas efferentia,* derived from the rete testis) to the *epididymis,* a single, complexly coiled tube (as much as 6 meters long in the human) in which sperm are stored. From each epididymis a duct, the *vas deferens,* passes from the scrotum through the inguinal canal into the abdominal cavity and over the urinary bladder to the lower part of the abdominal cavity, where it joins the urethra.

The *urethra* is a tube connecting the urin-

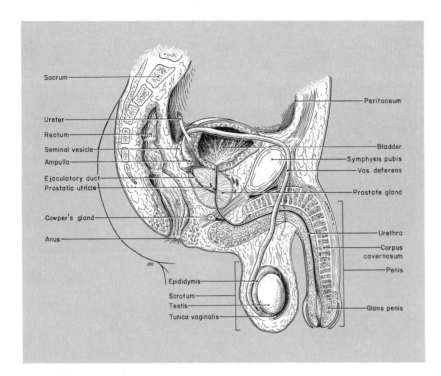

Sacrum

Ureter

Rectum

Seminal vesicle

Ampulla

Ejaculatory duct
Prostatic utricle

Cowper's gland

Anus

Epididymis

Scrotum
Testis
Tunica vaginalis

Peritoneum

Bladder
Symphysis pubis
Vas deferens

Prostate gland

Urethra
Corpus cavernosum
Penis

Glans penis

Figure 26-11 Schematic sagittal section of the pelvic region of the human male showing the organs of the reproductive tract.

ary bladder to the exterior. In the male it passes through the penis and is flanked by three columns of *erectile tissue* (Fig. 26–11) that become engorged with blood during periods of sexual excitement. The engorgement of the erectile tissue and the subsequent erection of the penis are caused not by constriction of the venous outflow but rather by arterial dilatation and increased blood flow at unchanged arterial pressure.

The sperm, suspended in seminal fluid, are transferred to the vagina of the female during copulation. The seminal fluid, amounting to from 2 to 5 ml. per ejaculation, is produced by three different glands. The paired *seminal vesicles* empty into the vasa deferentia just before they join the urethra. Around the urethra, near its source in the urinary bladder, are the paired *prostate glands* (which in the human are fused to form a single prostate). The prostates secrete their contribution to the seminal fluid into the urethra via two sets of short, thin ducts. Farther along the urethra, at the base of the erectile tissue of the penis, lies a third pair of glands, *Cowper's glands,* which contribute the final component of the seminal fluid. Mucous alkaline secretions are provided by the seminal vesicles and Cowper's glands and a thin milky fluid with a characteristic odor is added by the prostate. Seminal fluid may

contain glucose and fructose which are metabolized by the sperm, acid-base buffers and mucous materials that lubricate the passages through which sperm travel.

An operation to tie off and cut the vas deferens, termed a *vasectomy,* is widely used as a contraceptive measure for men who no longer want to be fathers. Cutting the vas does not render a man sterile immediately, for there may be enough sperm remaining in the lower part of the vas for as many as ten ejaculations. Cutting the vas does not stop the production of sperm in the testis; sperm continue to be formed and pass into the epididymis where they die and are resorbed. This may lead to the production of antibodies to the sperm antigens. If the man should subsequently have the vas deferens reconnected he may not regain fertility because of these antibodies to his own sperm.

26-7 THE HUMAN REPRODUCTIVE SYSTEM: FEMALE

The *ovaries,* each about 3 cm. long and the shape of a shelled almond, are held in place by ligaments within the lower part of the abdominal cavity. The egg is released by ovulation from the ovary into the abdominal cavity whence it passes into one of the two *fallopian tubes* through the funnel-shaped ostium at its end (Fig. 26–12). The egg is directed into the ostium by the beating of cilia in the epithelial lining of the fallopian tubes. Very rarely an egg may be fertilized within the abdominal cavity and begin to develop attached to an organ such as the liver or kidney. Usually when this occurs development cannot be completed and the embryo must be removed; there have been instances reported of such embryos completing development and being "born" by surgical removal.

The two fallopian tubes empty into the upper corners of the pear-shaped *uterus*, in which the embryo develops until the time of birth. The uterus is in the central part of the lower abdominal cavity, just behind the urinary bladder. About the size of a clenched fist, it has thick walls of smooth muscles and a mucous lining richly supplied with blood vessels. The uterus terminates in a muscular ring, the *cervix*, which projects a short distance into the vagina. The *vagina*, a single muscular tube, extends from the uterus to the exterior and serves both as a receptacle for sperm during coitus and as the birth canal when the fetus completes development.

The external female sex organs, collectively known as the *vulva*, include the *labia majora*, two folds of fatty tissue covered by skin richly endowed with hair and sebaceous glands, which extend back and down, enclosing the opening of the urethra and vagina and merging behind it. The *labia minora*, thin, pink folds of tissue devoid of hair, lie within the folds of the labia majora and are usually concealed by them. At the junction of these two in front is the *clitoris,* a sensitive erectile organ about the size of a pea. In most women the clitoris, which is homologous to the male penis, is completely covered by the prepuce. Like the penis, the clitoris contains spongy tissue which becomes engorged with blood during sexual excitement; nerve endings in the clitoris and the labia minora respond to erotic stimulation. Behind the clitoris is the opening of the urethra, which in the human female has only a urinary function, and behind the urethra lies the opening of the vagina. The vaginal opening is partly occluded by the *hymen,* a thin membrane composed of elastic and collagenous connective tissue, which is ruptured by the first sexual intercourse. At the junction of the thighs and torso, just above the clitoris, is a small mound of fatty tissue, the *mons veneris,* covered in the adult with pubic hair.

26–8 HUMAN REPRODUCTION: THE SEX ACT

Human reproduction is accomplished sexually by the union of ova and sperm, by the introduction of the erect penis of the male into the vagina of the female. At the height of the male's sexual excitement, seminal fluid is ejected into the upper end of the vagina, around the cervix of the uterus, from which sperm are carried through the cervical canal and uterus to the upper part of the fallopian tubes, where fertilization occurs. The complex structures of the reproductive systems in male and female and the complex physiological, en-

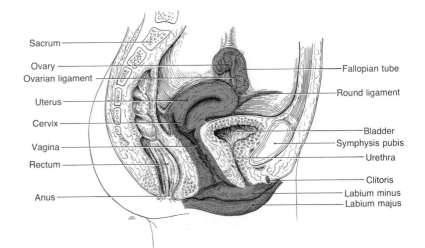

Figure 26–12 Schematic sagittal section of the pelvic region of the human female showing the organs of the reproductive tract.

Sacrum — Ovary — Ovarian ligament — Uterus — Cervix — Vagina — Rectum — Anus — Fallopian tube — Round ligament — Bladder — Symphysis pubis — Urethra — Clitoris — Labium minus — Labium majus

docrine and psychological phenomena associated with sex have just one purpose: to ensure the successful union of the egg and sperm and the subsequent development of the fertilized egg into a new individual.

The first phase of sexual intercourse, termed *excitement,* may last from a few minutes to a few hours. The essential feature of this phase in the male is erection of the penis and psychic tension. Erection of the penis results when the corpora cavernosa become engorged with blood. The excitement phase in the female is characterized by erection of the clitoris and labia minora. The several phases of intercourse merge into each other and the erection of the clitoris and labia may not occur until the second or *plateau* phase. During this second phase the breasts may enlarge as much as 20 per cent in volume and the nipples become erect. The skin from the breast and neck up into the face becomes flushed. During the excitement phase, fluid passes through the walls of the vagina and moistens the vaginal canal and entrance. The tissues surrounding the entrance to the vaginal canal undergo tumescence to form what has been termed the "orgasmic platform" (Fig. 26–13).

The movement of the penis in and out of the vagina, lubricated by mucoid secretions from the male urethra and from the paired Bartholin's glands just inside the vagina, massages the clitoris and labia, causing psychic stimulation of both female and male and leading to the climax or *orgasm.* The male orgasm is marked by ejaculation of seminal fluid, brought about by the contraction of muscles in the walls of the epididymis, vasa deferentia and seminal vesicles and by the contraction of the bulbocavernosus muscle at the proximal end of the urethra. During the female climax the orgasmic platform contracts rhythmically, accompanied by contractions of the uterus and fallopian tubes.

Following orgasm there is a detumescence of penis, labia and clitoris, a decrease in the size of the breasts and a general decrease in muscle tone in both male and female. Not every woman achieves orgasm at every intercourse, but orgasm is not a prerequisite for fertilization. There is intense physical and psychological tension during intercourse, which is relieved by the orgasm. During the excitement and orgasmic phases the respiratory rate may reach 40 per minute, the pulse rate may reach 170 per minute and the blood pressure may increase 30 to 80 mm. Hg. during systole and 20 to 40 mm. Hg. during diastole. Orgasm is followed by a *resolution* phase of muscular relaxation and a feeling of physical and mental lassitude. The male cannot immediately return to another orgasm but requires a latent period of some minutes before beginning again. The female, in contrast, requires no such latent period between climaxes.

Sperm become motile only after they have been in contact with the secretions of the epididymis; sperm removed directly from the testis are nonmotile. Sperm can survive only briefly in the acidic environment of the vagina (pH about 4.5), but the vaginal pH rises to 7.2 within seconds of the deposition of semen because of the buffering capacity of the seminal fluid. Sperm appear in the ovarian end of the oviduct within three minutes after being deposited in the vagina. It has been shown experimentally that even killed sperm are moved rapidly to the upper end of the oviduct. Sperm apparently are transported through the cervix and uterus and up into the fallopian tube by muscular contractions of the uterus and tubes. There is a slight rise in intrauterine pressure at orgasm, followed by a slight negative pressure that might serve to "aspirate" the sperm from the vagina. However, as mentioned before,

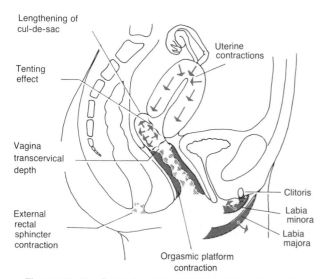

Lengthening of cul-de-sac

Tenting effect

Uterine contractions

Vagina transcervical depth

External rectal sphincter contraction

Orgasmic platform contraction

Clitoris

Labia minora

Labia majora

Figure 26–13 Sagittal section of the pelvic region of the human female showing the structural and functional changes that occur during the orgasmic phase. (After Masters, W. H., and Johnson, V. E.: *Human Sexual Response.* Boston, Little, Brown and Company, 1966.)

pregnancy can occur whether or not the female has experienced an orgasm.

Prostaglandins, secreted by the seminal vesicles into the seminal fluid, increase the activity of uterine muscle and may play a role in furthering the union of egg and sperm.

The inability of a man to have an erection of the penis is termed *impotence*. This is to be contrasted with male *sterility*, which is present in 3 to 4 per cent of men. Some sperm have abnormalities readily seen under the microscope, such as two tails or two heads. Male sterility usually occurs when the total number of sperm in a single ejaculation falls below 150,000,000. A deficiency of prostaglandins has been found to be a cause of male sterility.

Some 14 per cent of women are sterile because their fallopian tubes have been blocked by infection, because the ovary is covered with a thick capsule, or because their pituitary secretes inadequate amounts of gonadotropins. The cervical mucus secreted by some women resists the passage of sperm (sometimes only the sperm of certain men), apparently by some sort of antigen-antibody reaction.

26-9 HUMAN REPRODUCTION: FERTILIZATION

Sperm deposited in the vagina during intercourse travel up the vagina and into the uterus largely by the force of the muscular contractions of the walls of these organs. Most of the sperm become lost on the journey, but a few hundred find their way to the openings of the fallopian tubes and swim up them. Although sperm motility may play little or no role in transporting sperm from the vagina to the oviduct, it probably is important in penetrating the egg. If the egg is fertilized, this usually occurs in the upper third of the fallopian tube. Only one of the hundreds of millions of sperm deposited at each ejaculation fertilizes a single egg (Fig. 26-14).

Each human egg is surrounded by a layer of cells derived from the follicle and termed the *corona radiata*. By the time the ovum has reached the fallopian tube it has completed the first meiotic division and extruded the first polar body. It is probable, though not proven, that hyaluronidase from the seminal plasma and the hydrolytic enzymes of the lysosomes in

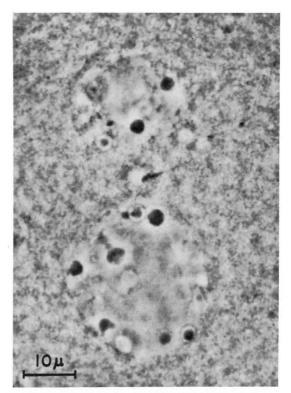

Figure 26-14 Photomicrograph of living human ovum recovered from the oviduct showing the male and female pronuclei before their fusion. The sperm tail is evident in the center of the photomicrograph. This ovum had been fertilized but its pronuclei had not yet fused and the first cleavage division had not begun. It is the earliest stage of development recorded for the human. (Courtesy of Dr. Robert Noyes.)

the sperm head play a role in the penetration of the sperm through the corona radiata and the vitelline membrane. Once one sperm has entered the egg there is some sort of change in the surface layer of the ovum which prevents the entrance of other sperm. The sperm either leaves its tail outside the egg or sheds it shortly after entering the egg's cytoplasm; only the nuclear material of the head and the centriole remain (Fig. 26-15). The ovum completes the second meiotic division and extrudes the second polar body. The head of the sperm swells to form the *male pronucleus* and the nucleus of the ovum becomes the *female pronucleus.* The fusion of the haploid male pronucleus with the haploid female pronucleus forms the nucleus of the fertilized egg or zygote and restores the diploid number of chromosomes.

In view of the many factors working against fertilization it may seem remarkable that it ever does occur, and indeed, humans are relatively infertile animals (yet fertile enough

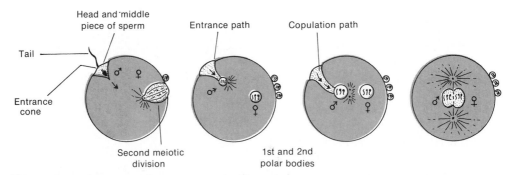

Figure 26-15 Diagram of the successive stages in the union of egg and sperm during fertilization.

so that overpopulation is the major problem facing the world today). Sperm remain alive and retain their ability to fertilize an egg from 24 to at most 48 hours after having been deposited in the female tract, and the egg loses its ability to be fertilized about 24 hours after ovulation. Sperm cells are delicate, their cytoplasm contains meager resources of food, and they are sensitive to heat, slight changes of pH and so on. The leukocytes of the vaginal epithelium which engulf millions of sperm are another hazard. However, the frequency of copulation and the large number of sperm deposited at each ejaculation enable the human race not only to maintain itself but to increase at an alarming rate.

26-10 HUMAN REPRODUCTION: DEVELOPMENT AND IMPLANTATION OF THE BLASTOCYST

The first cleavage of the fertilized egg occurs about 30 hours after insemination and the succeeding mitoses occur every 10 hours or so. By the time the developing ovum reaches the uterus, perhaps 3 to 7 days after fertilization, it is a tight ball of some 32 cells, called a *morula.* If the fertilized egg passes through the fallopian tube too rapidly and reaches the uterus prematurely it cannot be embedded in the wall. One type of contraceptive device, the *intrauterine coil,* may stimulate muscular contractions of the fallopian tube and uterus so that the fertilized ovum reaches the uterine cavity prematurely and dies before it can undergo implantation.

When the developing ovum reaches the uterine cavity it begins to differentiate into a *blastocyst* (Fig. 26-16), composed of an outer

envelope of cells, the *trophoblast,* and an *inner cell mass,* a ball of cells at one pole within the trophoblast which is the precursor of the embryo. This stage implants in the endometrial lining of the uterus by secreting enzymes which erode the cells of the endometrium, permitting the embedded blastocyst to establish close contact with the maternal blood stream (Fig. 26-17). The cells of the trophoblast grow and divide rapidly; they and the adjacent cells of the uterine lining, the *decidua,* form the placenta and fetal membranes. The cells of the endometrium heal over the site of entry of the blastocyst so that it lies wholly within the endometrium and out of the uterine lumen.

The embedding reaction of the endometrium can be elicited by pricking with a glass needle the uterine lining of a female suitably primed with estradiol and progesterone. This

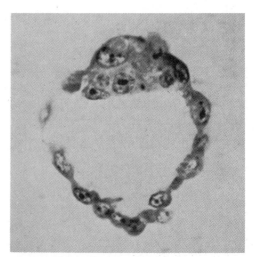

Figure 26-16 Photomicrograph of a human blastocyst; the inner cell mass is visible at the upper pole. (From Reid, D. E.: *Principles of Obstetrics.* Philadelphia, W. B. Saunders Co., 1962.)

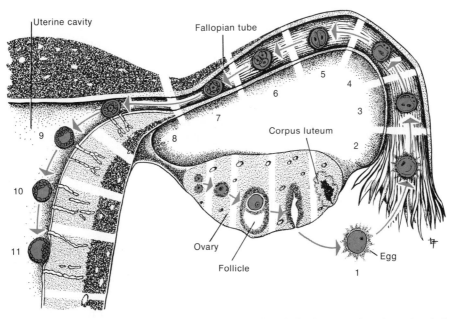

Figure 26-17 Schematic diagram of the maturation of an egg in a follicle in the ovary, its release (ovulation) (*1*), fertilization in the upper part of the oviduct (*2*), cleavage of the egg as it descends the oviduct or fallopian tube (*3–7*), stages in the development of the embryo in the uterus before implantation (*8–10*), and implantation of the embryo in the wall of the uterus (*11*).

stimulus leads to *"pseudopregnancy"* and the uterus develops for a short period just as though an embryo were present.

The trophoblast initially consists of two layers of cells, an inner *cytotrophoblast* composed of individual cells and an outer *syncytiotrophoblast* composed of a multinucleate syncytium. The trophoblastic cells digest and phagocytize materials in the endometrium that were stored prior to implantation. The trophoblast soon is bathed in and nourished by the maternal blood. Normally menstruation occurs about 14 days after ovulation in the nonpregnant woman, and to prevent this the embedded embryo must signal the maternal organism in some way. Because of the time required for traversing the oviduct, and because the fertilized egg remains free in the lumen of the uterus for some days, about 11 days elapse between ovulation and implantation. Thus the embryo has only a very few days to provide the signal which will prevent menstruation. Fairly frequently the signal does not arrive in time and menstruation sweeps out the fertilized egg. The woman has been pregnant in the sense that she had a fertilized egg in her reproductive tract, but was never aware of it and menstruated at her usual time. One of the major contributions of the trophoblast is its secretion of *chorionic gonadotropin,* probably by the cells of the syncytiotrophoblast. Chorionic gonadotropin has properties similar to those of luteinizing hormone and luteotropic hormone of the pituitary; it prevents the corpus luteum from involuting. The secretion of chorionic gonadotropin begins the day the trophoblast is embedded in the endometrial lining.

The process of implantation has implicit in it two questions of general biological interest. Why does the trophoblast generally cease invading the endometrial lining once it has formed a connection with the maternal blood? That is, why doesn't it continue to invade as a group of cancer cells would? And why, since the cells of the trophoblast have the genotype of the developing fetus, a genotype different from that of the mother, do the maternal cells not react as though the trophoblast were a transplant and reject it, as an animal rejects a skin graft from another, genetically different member of the same species?

26-11 NUTRITION OF THE EMBRYO

After implantation in the uterine lining, the embryo continues to develop, obtaining its nourishment at first by enzymatically breaking down the cells of the uterine lining immediately around it, and later by extracting nutrients from the blood stream of the mother via the blood vessels of the placenta.

The new human being develops from only the inner cell mass, the cells lying along one side of the hollow ball originally implanted in the uterus; the other cells form membranes which nourish and protect the developing child. The problem of supplying the embryo with food during development has been solved in different ways by the several groups of vertebrates.

Fishes and amphibia produce relatively large eggs, containing yolk to supply the necessary proteins, fats and carbohydrates. These eggs are laid and develop in water, whence they obtain oxygen, salts and water itself. The embryos of these animals have a pouchlike outgrowth of the digestive tract, the *yolk sac*,

which grows around the yolk, digests it and makes it available to the rest of the organism.

The eggs of reptiles and birds are usually laid on land and are enclosed within a shell that protects them from excessive drying. They have additional membranes for the protection and nutrition of the embryo. The familiar "white" of the hen's egg is an extra store of protein and water to support the embryo until the time of hatching. Both the shell and the white of eggs of reptiles and birds are secreted by glands located in the wall of the oviducts while the egg is passing down these tubes.

26–12 THE EMBRYONIC MEMBRANES

Several embryonic membranes have evolved to enfold, protect, support and nourish the embryos of reptiles, birds and mammals. These membranes, the *amnion*, the *chorion*

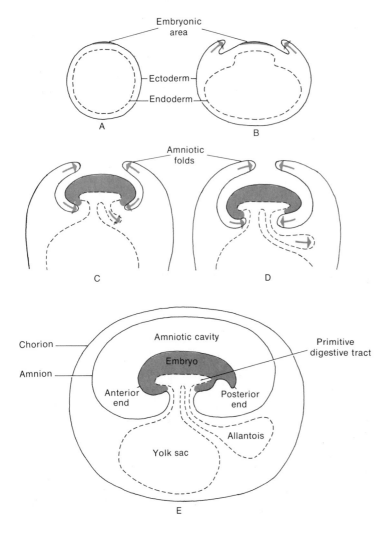

Figure 26–18 Steps in the formation of the embryonic membranes—amnion, chorion, yolk sac and allantois—in a typical mammal, such as a pig. Arrows indicate direction of growth and folding.

and the *allantois,* are sheets of living tissue growing out of the embryo itself—the amnion and chorion out of the body wall to enfold the embryo (Fig. 26–18), and the allantois out of the digestive tract to function in the absorption of food.

The formation of the amnion is a complex process that differs in detail in different species, but it is essentially an outfolding of the body wall of the embryo which grows around the embryo to meet and fuse above it (Fig. 26–18). The space between the embryo and the amnion, known as the *amniotic cavity,* becomes filled with a clear, watery fluid secreted by both the embryo and the membrane. Embryos of the higher vertebrates reach the birth stage enclosed in a small pool within the shell or uterus. The amniotic fluid prevents desiccation of the embryo and acts as a protective cushion that absorbs shocks and prevents the amniotic membrane from sticking to the developing embryo while permitting the organism a certain freedom of motion. During the birth process of humans and other mammals, pressure of the contracting uterus is transmitted via the amniotic fluid and helps to dilate the neck of the uterus; later, shortly before the fetus is born, the amnion normally ruptures, releasing about a liter of amniotic fluid, the so-called waters. Sometimes it fails to burst and the child is born with the amnion still enveloping its head. The amnion is then popularly known as a "caul," and is the source of many odd superstitions.

The amnion develops from the inner part of the original fold from the body wall; the outer part forms a second membrane, the *chorion.* In the eggs of reptiles and birds, this membrane rests in contact with the inner surface of the shell, and in mammals, it is established next to the cells of the uterine wall.

The *allantois,* like the yolk sac, is an outgrowth of the digestive tract. It grows between the amnion and chorion, and in animals like the chick in which it is a large and functional membrane, it fills almost all the space between these two. The allantois of the reptilian and avian egg serves as a depot for nitrogenous wastes. The products of nitrogen metabolism are excreted as uric acid by the kidney of the developing embryo. The poorly soluble uric acid is deposited as crystals in the cavity of the allantois and is discarded along with the allantois when the young hatch out of the egg shells.

The allantois fuses with the chorion to form a compound chorioallantoic membrane, full of blood vessels, by means of which the avian embryo takes in oxygen, gives off carbon dioxide, and excretes waste products. Since the embryo "breathes" through the shell, it will suffocate if the shell is coated with wax. In the human, the allantois is small and nonfunctional except for furnishing blood vessels to the placenta, and the yolk sac is completely nonfunctional. When the chick hatches from the shell, or when the child is born, most of the allantois and all the other membranes are discarded. But the base of the allantois, the part connected to the digestive tract originally, remains within the body and is converted into part of the urinary bladder.

As the human embryo grows, the region on the ventral side from which the folds of the amnion, the yolk sac and the allantois grew becomes relatively smaller, and the edges of the amniotic folds come together to form a tube which encloses the other membranes. This tube, the *umbilical cord,* contains in addition to the yolk sac and allantois the large blood vessels through which the embryo obtains nourishment from the wall of the uterus. The umbilical cord, about 1 cm. in diameter and about 70 cm. long at birth, is composed chiefly of a peculiar jellylike material found nowhere else. It is usually twisted spirally, and in its contortions before birth, the fetus* sometimes passes through a loop of the cord and actually ties a knot in it.

26–13 THE PLACENTA

The outer surface of the chorion in humans and the higher mammals is thin over most of its surface, but at the outer extremity of the umbilical cord it develops a number of fingerlike projections, known as *chorionic villi,* which grow into the tissue of the uterus. These villi, plus the tissues of the uterine wall in which they are embedded, make up the organ known

---

*As soon as the zygote or fertilized egg begins to divide, it is called an embryo. After the embryo has begun to resemble a human being (some two months after fertilization), it is referred to as a fetus until the time of birth.

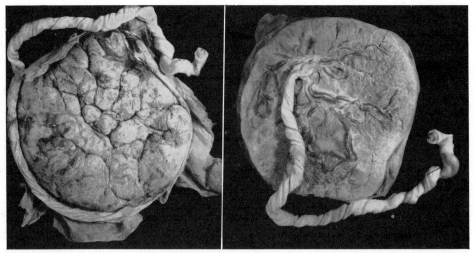

Figure 26–19   Photographs of the maternal (*left*) and fetal (*right*) surfaces of a human placenta at the end of a normal pregnancy. (From Greenhill, J. P., *Obstetrics.* 13th Ed. Philadelphia, W. B. Saunders Co., 1965.)

as the *placenta,* by means of which the developing embryo obtains nutrients and oxygen and gets rid of carbon dioxide and metabolic wastes (Fig. 26–19). There are many capillaries in the villi that receive blood from the embryo by way of one of the two *umbilical arteries* and return it to the embryo by way of the *umbilical vein.* The uterine lining becomes thickened and highly vascularized, forming a mass of spongy tissue filled with maternal blood. *The bloods of the mother and fetus do not mix at all in the placenta or any other place.* The blood of the fetus in the capillaries of the chorionic villi comes in close contact with the mother's blood in the tissues between the villi, but they are always separated by a membrane, through which substances must diffuse or be transported by some active, energy-requiring process. As the embryo develops, the placenta necessarily grows too. At the time of birth, it is a thick, circular disc 15 to 20 cm. in diameter and 2 to 3 cm. thick, weighing about 500 grams. The placenta is a very active tissue with high rates of blood flow and oxygen consumption. At term, about 600 ml. of maternal blood passes each minute through placental spaces that total about 140 ml. and have about 11 square meters of surface area. The fetal blood flow, entering via the two umbilical arteries, is about 300 ml. per minute. The oxygen consumption of the placenta, about 10 μl. per gram of tissue per minute, is twice that of the fetus. Besides serving as the nutritive, respiratory and

excretory organ of the fetus, the placenta is an important endocrine gland (see p. 559).

The uterus increases in size as the fetus grows, and by the end of 9 months its mass is 24 times as great as at the beginning of pregnancy. After 6 months of fetal development the upper end of the uterus is on a level with the navel; by 8 months it is as high as the lower edge of the breastbone. Within the uterus the fetus assumes a characteristic "fetal" position with elbows, hips and knees bent, arms and legs crossed, back curved, and the head bowed and turned to one side. At birth the fetus usually is turned head downward so that its head emerges first, but occasionally the buttocks or feet are presented first, making delivery more difficult.

26–14   THE BIRTH PROCESS

The human *gestation period,* the duration of pregnancy, is normally 280 days, from the time of the last menstrual period to the birth of the baby. Babies born as early as 28 weeks or as late as 45 weeks after the last menstrual period may survive. The factors that actually initiate the process of birth, or *parturition,* after the period of gestation is complete are unknown. Childbirth begins with a long series of involuntary contractions of the uterus, experienced as "labor pains." Labor may be divided into three periods. During the first, which lasts about 12 hours, the contractions of the uterus move the fetus down toward the cervix and cause the latter to dilate, enabling the fetus to pass through. At the end of this period the amnion usually ruptures, releasing the amniotic fluid, which flows out through the vagina. In the second period, which normally lasts between 20 minutes and an hour, the fetus passes through the cervix and vagina and is born, or "delivered" (Fig. 26–20). The fetus is expelled from the uterus by the combined forces of uterine contractions plus the contractions of the muscles of the abdominal wall. With each uterine contraction the woman holds her breath and bears down.

After the baby is born and before the umbilical cord is cut, the contractions of the uterus squeeze much of the fetal blood from the placenta back into the infant. After the pulsations in the umbilical cord cease, the cord is tied and cut, severing the child from the mother. The stump of the cord gradually shrivels until nothing remains but the depressed scar, the *navel.* During the last stage of labor, which lasts 10 or 15 minutes after the birth of the child, the placenta and the fetal membranes are loosened from the lining of the uterus by another series of contractions and expelled. At this stage they are called collectively the afterbirth. In the human and in certain other mammals in which the placenta forms a very tight connection with the uterine lining, the expulsion of the placenta is accompanied by some loss of blood. In other mammals in which the connection between the fetal membranes and uterine wall is not so close, the placenta can pull away from the uterine wall without causing bleeding. After birth the size of the uterus decreases, and its lining is rapidly restored.

The women of developed countries appear to have more difficulty during childbirth than the women of primitive cultures do, and they frequently require help in being delivered. The obstetrician may administer drugs such as oxytocin or prostaglandins to increase the contractions of the uterus or he may have to assist the child into the world by pulling it through the birth canal with special forceps. In some women the aperture between the pelvic bones, through which the vagina passes, is too small to permit the passage of the baby and so the child must be delivered by an operation in which the abdominal wall and uterus are cut open from the front. This operation is now called a cesarean delivery. It seems unlikely that Julius Caesar was born this way (his mother was still alive when he was a grown man). The term is probably derived from the Latin *caedere,* to cut, or from the Roman law, *lex caesarea,* that required the cutting open of a dying woman in an attempt to save her unborn child.

In a little more than 20 per cent of all known pregnancies the infant is born before it is able to carry on an independent existence. When this occurs, the birth is called an *abortion* or *miscarriage.* Such premature births may

Figure 26–20   Models showing the process of birth. (From the Dickinson-Belski series, the Maternity Center Association.) *Left,* Head passing through the dilated cervix of the uterus into the upper part of the vagina. *Right,* Head passing through opening of vagina. (From Patten, B. M.: *Human Embryology,* 2nd Ed. New York, McGraw-Hill Book Co., 1953.)

be caused by improper implantation of the embryo, by faulty functioning of the placenta, by fetal abnormalities or by injury to the mother. The term abortion is applied to the termination of pregnancy before the time when the fetus can survive successfully outside the uterus. Abortions may be either spontaneous or induced by some surgical or medical procedure to terminate an unwanted pregnancy. Since the momentous *Doe* and *Roe* decisions of the United States Supreme Court in January, 1973, which declared restrictive state abortion laws unconstitutional, the number of legal abortions in this country has progressively increased. More than one million induced abortions were performed in this country during 1975, most of them early in pregnancy by uterine aspiration, a quick and relatively safe procedure.

Among mammalian species there are great differences in the condition of the young at birth. The newborn of some species, such as the rat, are blind, hairless and helpless, while others, such as the guinea pig, are well developed and able to walk and eat solid food. There is also great variation in the weight of the newborn in comparison to the mother's weight: a newborn polar bear weighs 0.1 per cent of its mother's weight; a newborn human weighs about 5 per cent as much as its mother; and a newborn bat may weigh as much as 33 per cent of its mother's weight.

26–15 INFANT NUTRITION

During pregnancy both the glands and ducts of the mammary glands grow, stimulated by estradiol and progesterone. The volume of the breast also increases because of the deposition of fat and the engorgement with blood. Each breast increases in volume by about 200 ml. on the average, but the increased volume may be as much as 800 ml. The onset of milk secretion occurs under the stimulus of hormones from the ovary and the anterior and posterior pituitary, but the continued secretion of milk depends upon the presence of a suckling child (Fig. 25–28).

As everyone knows, milk is an excellent food, containing proteins, fats and carbohydrates. However, it is deficient in some things, especially iron and vitamins C and D, so that at the present time an infant's diet is usually sup-

plemented after the first month or so with eggs to supply iron and orange juice and cod liver oil to supply the vitamins. The milks of various species differ in their nutrient content, and to raise a human infant on cow's milk, the latter must be diluted and sugar must be added to approximate the content of human milk.

The widely prevalent assumption that women are infertile after the delivery of a baby is only relatively true. A woman who does not breast feed her child may ovulate within six weeks after delivery. Lactation does not necessarily inhibit either ovulation or menstruation, but the return of ovulation tends to occur later in women who are breast feeding. Even in the latter, ovulation may occur by the twelfth week postpartum.

26–16 CONTROLLING THE SIZE OF THE HUMAN POPULATION

With the rapidly rising world population has come the danger that the human population will outstrip the resources of the world needed to maintain a reasonable standard of life. Many of the other problems facing mankind at present, such as the pollution of the environment, stem directly from the vastly increased number of people on this planet. Even if ways are found to increase the food supply to feed the seven billion people expected on the earth in the year 2000 (compared to the 4.0 billion living now), the number of human beings may well be limited by the supply of air pure enough to breathe, the supply of water pure enough to drink, or the availability of places to put accumulated trash and garbage.

The methods now available for restricting the number of infants born include: (1) suppressing the formation and release of gametes; (2) preventing the union of gametes in fertilization; (3) preventing the implantation of a fertilized egg; and (4) abortion—premature delivery of the implanted embryo. Several types of contraceptive pills are now in use; each typically consists of a small amount of estrogen combined with a larger amount of a progesteronelike compound. The pills are taken daily from the fifth to the twenty-fifth day of the cycle and act on the hypothalamus and pituitary to inhibit the surge of luteinizing hormone required for ovulation. The pills may decrease somewhat the amount of glandular tissue in the endometrium and change the nature of the cervical mucus so that it is more difficult for sperm to enter the uterus. Despite many attempts, no comparably safe and effective method for suppressing spermatogenesis in the male has been found.

The pills most commonly used today are the ones just described, combination pills containing both estrogen and progesterone and taken for 21 days of the cycle. Another type, the sequential pill, contains only estrogen for the first 14 days and estrogen plus progesterone for the remaining seven days. A third type, the so-called minipill, contains only progesterone and is taken daily throughout the year without stopping during menstruation. Progesterone may inhibit the LH surge, thereby preventing ovulation, and suppress the growth of endometrium, thereby preventing implantation, but it mainly alters the state of the cervical mucus, making it more viscous and less readily penetrated by sperm. Experimental field trials carried out with "once-a-month pills" containing a long-lasting estrogen and a progesterone derivative indicate that this is a feasible method of contraception. Other experiments have shown that contraceptive levels of estrogens and progestins can be administered by placing the steroids in silicone capsules that are inserted surgically beneath the skin. The capsules continuously release a small amount of a contraceptive progesterone analogue; they must be replaced with new capsules every four to eight months when the supply of progestin is exhausted. Some family planning clinics have been injecting a long-acting progesterone analogue which remains effective for two to three months before an additional injection is required.

Another method of contraception is to prevent fertilization by interposing a barrier between the gametes. A permanent barrier can be raised by tying or removing a section of the fallopian tubes or of the vasa deferentia. Temporary barriers are provided by a condom on the penis or a diaphragm that stretches across the vault of the vagina and occludes the opening of the cervix. Spermicidal jellies and foams, used alone or in combination with a diaphragm, also reduce the probability of fertilization.

Intrauterine devices placed in the lumen of the uterus, where they remain for months, provide another method of contraception. Several types of IUD's have been devised and, although it was formerly thought that they operated by increasing the motility of the fallopian tubes, the evidence now suggests that they act by producing a local inflammatory response in the lining of the uterus. This produces lysozyme, which destroys the blastocyst. Thus the IUD does not prevent the union of egg and sperm, but prevents the implantation of the fertilized egg. IUD's containing a coil of copper wire apparently have an increased antifertility effect. Another effective IUD is one that releases progesterone in the lumen of the uterus. The IUD has certain drawbacks: it causes severe cramping in some women and so cannot be tolerated, and it may be expelled from the uterus so that the woman is no longer protected against conception.

The fourth method of birth control, abortion, has been widely practiced in all parts of the world. Emptying the contents of the uterus may be brought about in any one of several ways and is a safe procedure when carried out by qualified physicians under sterile conditions in a hospital. It can be a very dangerous procedure under any other conditions. The cervix can be dilated and the lining of the uterus scraped with a special curette, or the implanted embryo can be removed with a special suction device. In another procedure, a needle is inserted through the abdominal wall and the wall of the uterus to the amniotic cavity and some of the amniotic fluid is withdrawn and replaced by hypertonic saline. This kills the fetus and the uterus gradually goes into labor and empties its contents during the next two or three days. Prostaglandins injected intravenously or placed in the vagina or uterus stimulate uterine contraction and cause ejection of the fetus.

The demographic unit for measuring the efficacy of any method of birth control is the number of pregnancies that occur per "100 woman years," i.e., in a group of 100 women using that method of contraception for one year, or for 1200 to 1300 ovulations, assuming that all the women ovulate about once every 28 days. With no attempt to restrict conception, between 50 and 80 pregnancies per 100 woman years will result. Restricting intercourse to the "safe period" results in about 24 pregnancies per 100 woman years. Using the condom or diaphragm results in 14 and 12 pregnancies, respectively, per 100 woman years. With intrauterine devices the number of pregnancies is reduced to about 2 per 100 woman years, and women using the contraceptive pills have between 0 and 1 pregnancies per 100 woman years. The few failures of the last two methods probably result not from any defect inherent in the method but in the way the method is used.

Some ten million or more women in the United States use oral contraceptives and initially there was concern over possible long-term effects of the contraceptive pills. There is a relationship between the amount of estrogen in the pill and the risk of thromboembolism (the formation of clots within the blood stream), but even with the higher levels of estrogen the risk is increased by only about 1 per 100,000. The risk has been minimized by decreasing the estrogen content of the pill and now the most commonly used pills contain only 50 micrograms of estrogen. The same studies in England that showed an increased, although small, risk of thromboembolism showed a considerably greater *decrease* in the risk of breast cancer among pill users as compared with women not taking oral contraceptives. Since this is a much more prevalent disease (about 1 woman in 20 has breast cancer), any decrease in this rate would be greatly desirable. The risk of breast cancer is increased the longer a woman, after puberty, undergoes repeated monthly estrogen surges. The tendency to breast cancer is decreased either by early pregnancy or by the oral contraceptive, both of which eliminate the monthly preovulation estrogen surge.

Demographers dream of a day when it might be feasible to add an antifertility sub-stance to the drinking water, much as we now add fluorides for the prevention of dental caries. When a couple wanted a child, they would then apply to the government to use the appropriate antidote. In light of what we know, this appears likely to remain a dream. It would be totally impractical and unwise to deliver any hormonal agent or any cytotoxic agent in the drinking water or food since the dosage could not be regulated. An amount of estrogen appropriate to inhibit ovulation in an adult woman would cause serious problems in a man or in an immature female or male.

26–17 SEXUAL REPRODUCTION IN LOWER ANIMALS

A few animals, such as the cnidarians, have alternate sexual and asexual generations reminiscent of the life cycles of plants, but both generations of cnidarians are diploid organisms. Most animals reproduce solely by sexual means and have permanent sex organs.

Some of the protozoa, such as paramecia, have a complicated process of sexual reproduction in which two diploid individuals come together, fuse on their oral surfaces, and exchange nuclear material (Fig. 26–21). The original micronucleus of each divides meiotically, and three of the four nuclei degenerate. The remaining haploid nucleus migrates across the cytoplasmic bridge to the other paramecium and fuses with one of its haploid nuclei. Thus, two fertilizations result and the two new diploid nuclei are identical. The paramecia subsequently separate; further nuclear and cytoplasmic divisions result in four individuals, each with a micronucleus and a macronucleus.

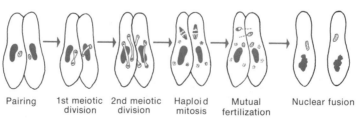

Pairing   1st meiotic division   2nd meiotic division   Haploid mitosis   Mutual fertilization   Nuclear fusion

Figure 26–21 Sexual reproduction in paramecium. Two individuals with diploid micronuclei undergo conjugation (*left*). After meiosis (*second and third figures*) three of the nuclei degenerate and the fourth nucleus divides by mitosis (*fourth figure*). The organisms undergo mutual fertilization; one nucleus passes from each organism to the other (*fifth figure*). This is followed by fusion of the haploid nuclei to form a new diploid nucleus in each of the two organisms (*last figure*). The original macronuclei disappear. The two new diploid nuclei divide several times by mitosis and eventually establish both the new micronuclei and the new macronuclei.

Hermaphroditism. Many of the lower animals are *hermaphroditic;* both ovaries and testes are present in the same individual and it produces both eggs and sperm. Some hermaphroditic animals (the parasitic tapeworms, for example) are capable of self-fertilization. (Does this violate the generalization that sexual reproduction involves two different individuals?) Since a host may be infected with but one parasite, this ability is an important adaptation for the survival of the species. Most hermaphrodites, however, do not reproduce by self-fertilization; rather, two animals copulate and each inseminates the other. This method is used by earthworms. In other species, self-fertilization is prevented by the development of the testes and ovaries at different times. Oysters, although hermaphroditic, cannot undergo self-fertilization, for the ovaries and testes of each individual produce gametes at different times.

Parthenogenesis. A rather rare modification of sexual reproduction, common among honey bees, wasps and certain other arthropods, is *parthenogenesis,* the development of an unfertilized egg into an adult animal. Some species of arthropods consist entirely of females which reproduce in this way. More commonly, parthenogenesis occurs for several generations only, after which males develop, produce sperm, and mate with the females to fertilize the eggs. The queen honey bee is inseminated by a male just once in her entire lifetime, during the "nuptial flight." The sperm she receives are stored in a little pouch connected with her genital tract and closed off by a muscular valve. As the queen lays eggs, she can either open this valve, permitting the sperm to escape and fertilize the eggs, or keep the valve closed, so that the eggs develop without fertilization. Fertilization usually occurs in the fall, and the fertilized eggs are quiescent during the winter. The fertilized eggs become females (queens and workers); the unfertilized eggs become males (drones). Some species of wasps give birth alternately to a parthenogenetic generation and a generation developing from fertilized eggs.

Eggs from species that do not normally undergo parthenogenetic development may be stimulated artificially to develop without fertilization by changing the temperature, pH or salinity of the surrounding water, or by chemical or mechanical stimulation of the egg. Frog eggs can be so stimulated by pricking them with a fine needle; other eggs can be made to divide by shaking them, by adding chemicals to the water in which they are lying, or, in marine eggs, by changing the concentration of salt in the water. It has been possible to stimulate rabbit eggs to begin development parthenogenetically. The cleaving egg that results must be placed in the uterus of a hormonally primed female to complete development.

Fertilization. Most aquatic animals simply liberate their sperm and eggs into the water, and their union occurs by chance. No accessory structures are needed, except the ducts that transport the gametes to the outside of their bodies. Thus, the most rudimentary and uncertain method of uniting the gametes, is called, for obvious reasons, *external fertilization.*

Other animals, especially those living on land, have accessory sex organs for transferring the sperm from the body of the male to that of the female, so that fertilization occurs within the latter. This method, called *internal fertilization,* requires the cooperation of the two sexes, and many species have evolved elaborate behavior patterns to ensure that the two sexes are brought together, mate at the most appropriate time and take care of the resulting offspring. A variety of secondary sex characteristics have evolved and serve as stimuli to attract the opposite sex and elicit a mating response.

An unusual pattern of mating behavior is exhibited by the salamander: the male clasps the female and rubs her nose with his chin. He then dismounts in front of her and deposits a *spermatophore,* a sac containing a large number of sperm. The female picks up the sac and places it in her cloaca. Here the sac breaks, releasing the sperm to fertilize her eggs.

Many breeding habits are dangerous or even fatal to the animals performing them. Salmon swim hundreds of miles upstream to spawn and die. The male spider is frequently eaten by the female after he has performed the necessary act of inseminating her. Yet the survival of the species requires that the individual make such sacrifices in the performance of these acts.

Fertilization involves not only the penetration of the egg by the sperm but also the union of egg and sperm nuclei and the activation of

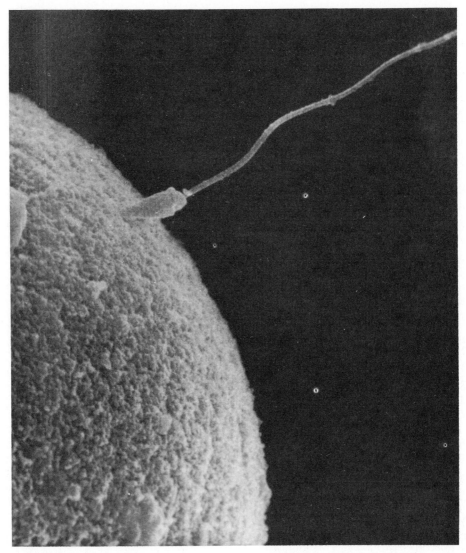

Figure 26–22 Scanning electron micrograph of the sperm of a sea urchin (*Arbacia punctulata*) penetrating the egg. (Courtesy of Dr. Don W. Fawcett and Dr. Everett Anderson.)

the egg to undergo cleavage and development (Fig. 26–22). The egg may be in any stage from primary oöcyte to mature ovum when the sperm enters it, but the fusion of egg and sperm nuclei can occur only after the egg nucleus has completed the two meiotic divisions and has become a mature *ovum.* The eggs of some species secrete *fertilizin,* an important constituent of the jelly coat surrounding the egg. Fertilizin causes sperm to clump together and stick to the surface of the egg. Other substances extracted from the jelly coat (which may be identical with fertilizin) stimulate sperm motility and respiration and prolong sperm viability.

After one sperm enters the egg, a *fertilization membrane* forms around the eggs of some species. This prevents the entrance of other sperm and the possibility that more than one sperm nucleus will fuse with the egg nucleus. The presence of two sperm nuclei may lead to the formation of tripolar spindles and abnormal development.

The females of all birds, amphibians and teleosts, most insects, and many aquatic invertebrates lay eggs from which the young eventually hatch; such animals are said to be *oviparous* (egg-bearing). The small eggs of mammals are kept in the uterus and provided with nutrients from the mother's blood until development has proceeded to the stage when they can live independently; such animals are termed *viviparous* (live-bearing). The females of sharks, lizards and some insects and snakes are *ovoviviparous;* they produce large, yolk-filled eggs which remain in the female reproductive tract for considerable periods after fertilization. The developing embryo usually forms no close connection with the wall of the oviduct or uterus and receives no nourishment from maternal blood.

26-18 MATING BEHAVIOR AND SYNCHRONIZATION OF SEXUAL ACTIVITY

Since most animals breed only during relatively brief seasons of the year, the production and release of eggs and sperm must be synchronized if fertilization is to occur. Typically, the males and females are triggered by some environmental cue such as a change in the photoperiod, or the ambient temperature, seasonal rainfall, or specific relations of tidal and lunar cycles (cf. the Palolo worm, p. 303). In some species the males and females not only must be brought to full sexual activity at the same time but also must be induced to move or migrate to specific mating and breeding grounds. The migration of salmon upstream to breed, the migration of eels to a specific breeding ground in the Central Atlantic, the migration of the gray whale to Baja California and the migration of turtles and birds are examples of such movements. Animals that live a solitary life during most of the year—seals, penguins, certain sea birds—come together for brief periods of mating at specific times and places.

The specific reproductive synchronization of a particular male for a particular female frequently involves some sort of *courtship behavior* (Fig. 26-23). The courtship, usually initiated by the male, may be a very brief ceremony or, in certain species of birds, may last for many days. The courtship behavior serves two additional roles: it tends to decrease aggressive tendencies and it establishes species and sexual identification, i.e., it identifies a member of the same species but of the opposite sex. The members of many species normally fight whenever they meet and

Figure 26-23 Courtship in one species of bower bird. The male builds a bower of sticks (*A*) and decorates it with pebbles and berries (*B*). The female bower bird enters the construction (*C*) and, after the male displays, she sits (*D*) and the birds mate (*E*). The female leaves, builds the nest (*F*) and rears the young birds by herself. (Courtesy of E. T. Gilliard.)

some special cue is needed if two animals are to avoid fighting long enough to mate.

The song patterns of birds and the mating calls of certain fishes, frogs and insects provide effective cues for the discrimination of the species. Female frogs are attracted to calling males of their own species but not to calling males of related species.

26–19 CARE OF THE YOUNG

The evolution of more efficient methods for bringing about fertilization has been accompanied by the evolution of special behavior patterns for the care of the young. In general, the number of eggs produced by a female of a given species and the chance that any particular egg will survive to maturity are inversely related. In the evolution of the vertebrates from fish to mammal, the trend has been toward the production of fewer eggs and the development of better parental care of the young. Fish such as the cod or salmon produce millions of eggs each year but only a small number of these ever become adult fish. At the other end of the spectrum mammals produce relatively few eggs but take good care of their offspring, so that a large fraction of the total attain maturity. Reptiles lay considerable numbers of eggs, usually in the sand or mud, where they develop without parental care, warmed only by the sun. Birds lay relatively few eggs in nests and incubate them by sitting on them. The parents take care of the newly hatched helpless chicks for several weeks until they are abe to survive independently. The mammalian egg develops within the mother's uterus, where it is safe from predators and from most harmful environmental factors. Most mammalian females have a strong "maternal instinct" to take care of the newborn until it can shift for itself. This pattern of behavior is regulated largely by hormones secreted by the pituitary and ovary.

QUESTIONS

1. Define hermaphroditism, parthenogenesis.
2. Compare the devices used by different vertebrates to ensure fertilization.
3. What trends are discernible in the evolution of reproduction in animals?
4. Are the bodies of human beings capable of undergoing regeneration?
5. In what ways are eggs and sperm adapted structurally for their respective functions in reproduction?
6. Distinguish between copulation, insemination and fertilization.
7. How and where is the seminal fluid produced in the human male?
8. Describe the movements of the egg from the time it is released from the ovary until it is implanted in the wall of the uterus.
9. Trace the path of the sperm from the testis to its union with the egg.
10. Describe the various types of nutrition of the human embryo from the time of fertilization until the time of birth.
11. What are the amnion, chorion and allantois? What is the function of each?
12. What is the amniotic cavity?
13. Describe the formation of the umbilical cord in the human.
14. Compare the mode of action, the efficacy and the safety of the several currently employed methods of contraception.
15. What structures are included in the afterbirth?
16. Describe the structure and the functioning of the placenta in the human.

SUPPLEMENTARY READING

An excellent source book on differences in reproductive cycles and processes in mammals, from the aardvark to the zebu, is *Patterns of Mammalian Reproduction* by S. A. Asdell. *Perspectives in Reproduction and Sexual Behavior*, edited by M. Diamond, is a collection of papers analyzing a wide variety of problems relating to the endocrine control of reproduction and sexual behavior. Interesting accounts of mating behavior in a variety of animals are found in *The Mating Instinct* by L. J. Milne and M. J. Milne. The physiologic aspects of human mating are discussed in *The Human Sexual Response* and *Human Sexual Inadequacy*, both by W. H. Masters and V. F. Johnson. Interesting related topics are presented in *The Nature and Evolution of Female Sexuality* by M. J. Sherfing. *The Control of Ovulation*, edited by C. A. Villee, includes a series of papers presented at a symposium that deal with the process of ovulation and its control plus a transcript of the discussion that followed each paper. *Fertilization* by C. R. Austen is a brief text summarizing our knowledge of this field. A series of papers covering a variety of aspects of the fertilization process in invertebrate and vertebrate forms is included in *Fertilization*, edited by C. B. Metz and A. Munroy. A general source book for interested students is *Human Reproduction* by E. W. Page, C. A. Villee and D. B. Villee.

CHAPTER 27

EMBRYONIC DEVELOPMENT

The division, growth and differentiation of a fertilized egg into the remarkably complex and interdependent system of organs which is the adult animal is certainly one of the most fascinating of all biologic phenomena. Not only are the organs complicated, and reproduced in each new individual with extreme fidelity of pattern, but many of the organs begin to function while still developing. The fetal heart, for instance, begins to beat during the fourth week, long before its development is complete.

The pattern of cleavage, blastula formation and gastrulation is seen, with various modifications, in all multicellular animals. The details of later development are quite different in animals of different phyla but are similar in more closely related forms. The main outlines of human development can be discerned by studying the embryos of rats or pigs, or even chicks or frogs. The development of humans and other primates is characterized by the very early development of the embryonic membranes, the amnion, chorion and allantois.

27-1 STAGES IN DEVELOPMENT

The development of a new individual begins with the formation of eggs and sperm in members of the parental generation. During gametogenesis (p. 592), meiosis reduces the number of chromosomes from the diploid to the haploid condition and there is a random selection of the specific genes that will be united in the new individual and guide its development. The egg cell undergoes growth and accumulates in its cytoplasm a variety of substances that will be used in the early stages of development of the fertilized egg. The sperm matures and changes from a spherical cell to one adapted to move actively to the egg and fertilize it.

The fertilization of the egg, the second stage of development, begins with the various processes that bring the parents together and stimulate them to release their gametes simultaneously, and includes the penetration of the egg by the sperm, followed by the activation of the fertilized egg so that it begins to develop. During the activation process the positions of the several kinds of organ-forming material within the egg's cytoplasm may undergo marked changes.

During the third stage, the fertilized egg undergoes a rapid succession of mitotic divisions termed *cleavage;* these result in a compact mass of cells or in a hollow ball of cells, a *blastula,* with a single layer of cells, the *blastoderm,* surrounding a cavity, the *blastocoele.* During the cleavage divisions, the size of the embryo does not increase, and the cleavage cells or *blastomeres* become successively smaller with each division. Since glycogen and other nutrients stored in the cytoplasm of the egg are metabolized to provide energy for the cleavage division, the mass of the blastula (dry weight) may actually be less than the mass of the unfertilized egg.

The fourth phase of development, *gastrulation,* results in the appearance of two or more layers of cells, the *germinal layers,* from the single-layered blastoderm. The bodies of higher animals consist of several layers of tissues and organs, all of which can be traced back to the three germinal layers, *ectoderm, mesoderm* and *endoderm.* The outer ectoderm forms the epidermis of the skin and the nervous system; the innermost layer, the endoderm, forms the lining of the gut and the digestive glands; and the middle layer, the mesoderm, is the source of the muscles, the vascular system, the lining of the coelomic cavity, the reproductive organs, the excretory organs and the bones and cartilages of the internal skeleton (Table 27–1). The cavity of the double-walled gastrula formed by invagination is called the *archenteron,* or primitive gut, and the opening from the archenteron to the exterior is the *blastopore.*

The archenteron, or a portion of it, eventually gives rise to the cavity of the digestive tract. The blastopore develops into the oral opening in cnidarians and becomes the anal opening in echinoderms and chordates. In the mollusks, annelids and arthropods, the blastopore becomes divided into two openings, one of which becomes the mouth and the other the anus. The central portion of the digestive tract is formed from endoderm, but typically the ectoderm invaginates at both oral and anal ends and fuses with the endoderm. Thus the most anterior portion of the digestive tract, the *stomodeum* (mouth cavity), and the most posterior portion of the digestive tract, the *proctodeum* (adjoining the anus), are lined with ectoderm rather than endoderm.

The next stage of development is *organogenesis,* during which the three germ layers split into smaller masses of cells, each destined to form a specific organ or portion of the adult. Each organ is first discernible as a group of cells, a *rudiment,* which becomes segregated from the other cells of the embryo. Organogenesis is followed by growth and differentiation of the cells of the organ rudiments. Eventually the cells acquire the specific structure and the chemical and physiological properties that enable them to perform their functions so that the developing animal can take up an independent existence.

The young emerging from the egg may be a miniature copy of the adult or may lack certain organs and differ to some extent from the adult in form as well as in size. In other animals the form emerging from the egg has special organs, not present in the adult, that are needed for the survival of the young in its particular mode of existence. Such a young animal, termed a *larva,* may lead a mode of life quite different from the adult. Eventually the larva undergoes a relatively rapid and marked change of form, *metamorphosis,* and achieves the adult form. During metamorphosis the spe-

Table 27–1 CONTRIBUTIONS OF THE THREE GERM LAYERS TO THE DEVELOPMENT OF A MAMMAL

Ectoderm	Endoderm	Mesoderm
Epidermis of the skin	Lining of gut	Muscles—smooth, skeletal and cardiac
Hair and nails	Lining of trachea, bronchi and lungs	Dermis of skin
Sweat glands	Liver	Connective tissue, bone and cartilage
Entire nervous system—brain, spinal cord, ganglia, nerves	Pancreas	Dentin of teeth
Receptor cells of sense organs	Lining of gallbladder	Blood and blood vessels
Lens of eye	Thyroid, parathyroid and thymus glands	Mesenteries
Lining of mouth, nostrils and anus	Urinary bladder	Kidneys
Enamel of teeth	Urethra lining	Testes and ovaries

cial larval characteristics are lost and the adult characteristics appear. The young and adult animals continue to undergo changes with time, which we call *aging,* and which must be considered part of the development process.

27–2 TYPES OF EGGS

The amount of yolk, or stored nutrients, present in an egg plays an important role in determining the pattern of development that ensues following fertilization. Yolk granules vary in composition from one species to another but usually contain varying amounts of protein, phospholipids and neutral fats. Most invertebrates and the lower chordates have small eggs with relatively small amounts of yolk. The yolk granules in the egg of the sea urchin, *Arbacia,* for example, make up about 27 per cent of the total volume of the egg. Eggs such as these, with small amounts of yolk present in fine granules distributed fairly uniformly throughout the cytoplasm, are termed *isolecithal* (equal yolk) (Fig. 27–1).

The eggs produced by fishes, amphibia, reptiles and birds have a large amount of yolk,

Figure 27–1 Photographs of eggs of the sea urchin *Arbacia punctulata* subjected to centrifugation to separate the lighter nucleate half from the heavier yolk-laden anucleate half. *Upper left,* Mild gravitational force stratifies constituents of the egg, which from top to bottom are: oil cap (dark), circular nucleus, clear area, mitochondria, dark yolk granules. Further centrifugation first elongates egg (*lower left*), then separates it into nucleate half (*upper right*) and anucleate half (*lower right*). (Courtesy of Grant Patton and Laurens Mets.)

usually present in large granules called *yolk platelets.* The yolk is massed toward the lower or *vegetal pole* of the egg, while the cytoplasm is concentrated toward the upper or *animal pole* of the egg. Such eggs are termed *telolecithal* (end yolk). Two proteins found in the yolks of both amphibian and avian eggs are phosvitin and lipovitellin. *Phosvitin* is a curious protein; about half of its amino acid residues are phosphorylated serine. Lipovitellin is a much larger protein molecule that contains about 17 per cent lipid. In amphibians and birds, phosvitin is synthesized in the maternal liver and transported in the blood to the ovary. Once taken up by the egg it may be phosphorylated further, which makes it less soluble and causes it to precipitate in the oöcyte. A protein kinase has been found in the frog ovary that can catalyze the incorporation of phosphate from ATP into partially phosphorylated phosvitin. Yolk makes up about 45 per cent of the weight of an amphibian egg and as much as 90 per cent or more of the eggs of birds, reptiles and bony fishes. In the latter three, the cytoplasm is restricted to a thin layer on the surface of the egg; a thicker cap of cytoplasm at the animal pole contains the nucleus. The entire center of the egg is filled with yolk platelets.

The eggs of arthropods, especially insects, have a different pattern of yolk distribution and are termed *centrolecithal.* The yolk is concentrated in the center of the egg and the cytoplasm is present as a thin layer on the surface. In addition, there is an island of cytoplasm in the center of the egg that contains the nucleus.

The duck-billed platypus and the echidna have small, yolk-filled eggs, comparable in size to those of a small lizard, which are laid and develop outside the mother's body. The eggs of other mammals are small and relatively free of yolk (Fig. 27–2). They superficially resemble isolecithal eggs but their pattern of development is more nearly like that of telolecithal eggs, probably because mammals evolved from reptiles, which laid the latter kind of egg.

27–3 CLEAVAGE AND GASTRULATION

The entrance of the sperm into the egg initiates a rapid series of changes—the completion of the meiotic divisions, the fusion of male

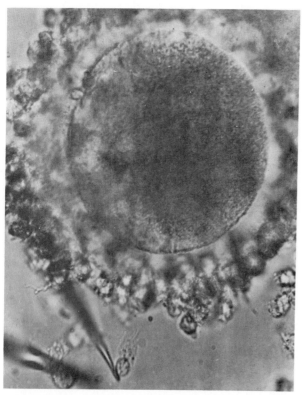

Figure 27–2 A human egg recovered from an ovarian follicle. The large egg cell is surrounded by the many small cells that make up the corona radiata—cells that were originally part of the follicle. The photograph was made in the course of an experiment in which microneedles were used to dissect away the corona radiata cells. Two needles can be seen in the lower corner of the picture dissecting the nucleus out of one of the corona radiata cells. Magnification ×600. (Courtesy of Dr. William R. Duryee.)

lated in protein synthesis. Only later, at about the time of gastrulation, is the genome of the embryo transcribed to form "embryonic" messenger RNA which codes for proteins that play a role in the gastrulation process and in later development. The problem of whether fertilization does lead to the "unmasking" of previously masked, inactive, "maternal" messenger RNA (and if so, how this is achieved) remains unsolved, but the hypothesis has attracted widespread interest and has stimulated many types of experimental approaches.

Cleavage in Isolecithal Eggs. The first cleavage division of an isolecithal egg passes through both animal and vegetal poles and splits the egg into two equal cells (Fig. 27–3). The second cleavage division also passes through both poles of the egg but at right angles to the first, and separates the two cells into four equal cells. The third division is horizontal, at right angles to the other two, and separates the four cells into eight—four above and four below the line of cleavage. Further divisions result in embryos containing 16, 32, 64, 128 cells, and so on until a hollow ball of cells, the *blastula*, is formed. The wall of the blastula is a single layer of cells, the *blastoderm*, surrounding the *blastocoele*, the cavity in the center.

The single-layered blastula is converted into a double-layered sphere, a *gastrula*, by the invagination of a section of one wall of the blastula. This eventually meets the opposite wall and obliterates the original blastocoele. The new cavity of the gastrula is the *archenteron*, and the opening of the archenteron to the exterior is the *blastopore*, which marks the site of the invagination that produced the gastrula. The outer of the two walls of the gastrula is the *ectoderm*, which eventually forms the skin and nervous system. The inner layer of cells, lining the archenteron, is mainly the presumptive endoderm, which will form the digestive tract and its outgrowths such as the liver, pancreas and lungs, plus the presumptive notochord and the presumptive mesoderm, which will form the remaining organs of the body. As the gastrula elongates in the anteroposterior axis the presumptive notochord is stretched into a longitudinal band of cells occupying the mid-dorsal part of the inner layer. The presumptive mesoderm forms two longitudinal bands of cells, one on each side of the presumptive notochord. The remainder of the

and female pronuclei, complex movements of the egg's cytoplasmic constituents, and the beginning of the cleavage divisions. After fertilization, the rates of oxygen consumption and protein synthesis in the eggs of certain species rise sharply. Some evidence supports the hypothesis that the RNA present in the fertilized egg, which directs protein synthesis during the early stages of cleavage, was synthesized in the oöcyte before fertilization and hence is a product of the maternal genotype rather than the genotype of the embryo. The RNA is "masked" in the unfertilized egg, perhaps by combination with a protein, and is unavailable for translation into a peptide chain. The ribosomes of the unfertilized egg are generally single and polyribosomes appear only after fertilization has occurred. One of the effects of fertilization thus appears to be an "unmasking" of the maternal messenger RNA so that it can be trans-

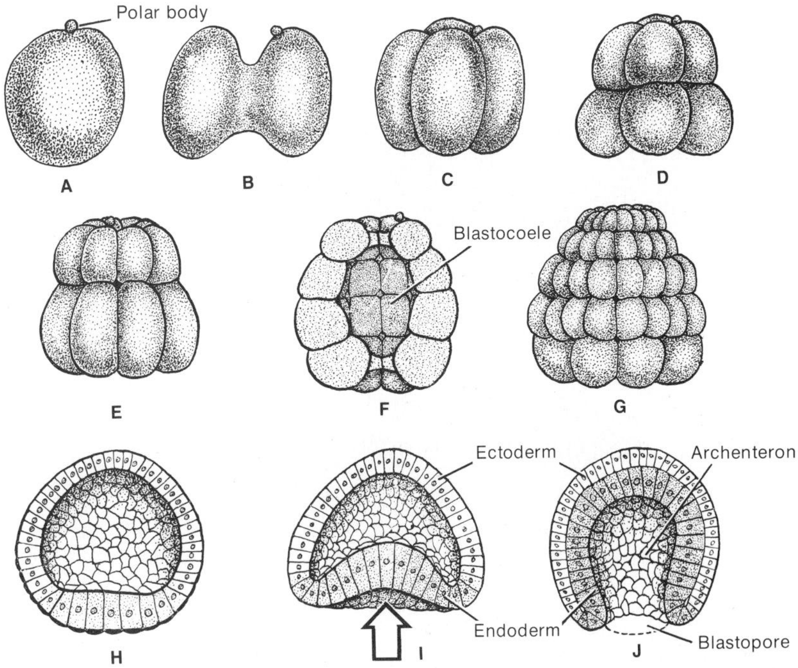

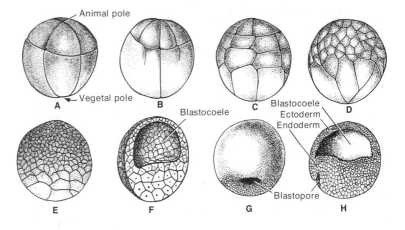

Figure 27-3 Isolecithal cleavage and gastrulation in Amphioxus viewed from the side. *A,* Mature egg with polar body. *B–E,* Two-, four-, eight- and sixteen-cell stages. *F,* Thirty-two-cell stage cut open to show the blastocoele. *G,* Blastula and, *H,* blastula cut open. *I,* Early gastrula showing beginning of invagination at vegetal pole *(arrow). J,* Late gastrula, invagination completed and blastopore formed.

lateral, ventral and anterior parts of the inner layer are presumptive endoderm cells.

Cleavage in Telolecithal Eggs. The processes of cleavage and gastrulation are markedly modified in telolecithal eggs by the large amount of yolk present. The cleavage divisions of the cells originating from the lower part of the frog's egg are slowed by the inert yolk so that the blastula consists of many small cells at the animal pole and a few large cells at the vegetal pole (Fig. 27–4). The lower wall is much thicker than the upper one and the blastocoele is displaced upward.

In eggs with a larger amount of yolk, such as a hen's egg, cleavage occurs only in the small disc of cytoplasm at the animal pole (Fig. 27–5). At first, all the cleavage planes are ver-

tical and all the blastomeres lie in a single plane. The cleavage furrows separate the blastomeres from each other but not from the yolk; the central blastomeres are continuous with the yolk at their lower ends, and the blastomeres at the circumference of the disc are continuous both with the yolk beneath them and with the uncleaved cytoplasm at their outer edge. As cleavage continues more cells become cut off to join the ones in the center, but the new blastomeres are also continuous with the uncleaved underlying yolk. The central blastomeres eventually become separated from the underlying yolk either by cell divisions with horizontal or tangential cleavage planes or by the appearance of slits beneath the nucleated portions of the cells. Horizontal cleav-

Figure 27-4 Successive stages in telolecithal cleavage and gastrulation in the frog, viewed from the side. *A–D,* cleavage; *E,* blastula; *F,* blastula cut open; *G,* early gastrula; *H,* early gastrula cut open.

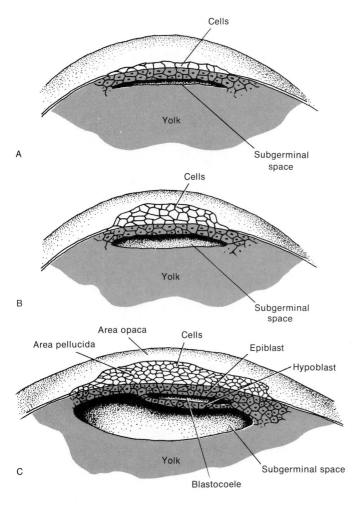

Figure 27–5 Successive stages in the cleavage of a hen's egg. *A*, Cleavage is restricted to a small disc of cytoplasm on the upper surface of the egg yolk called the blastodermic disc. *B*, A subgerminal space appears beneath the blastodermic disc, separating it from the unsegmented yolk. *C*, The blastodermic disc cleaves into an upper epiblast and a lower hypoblast separated by the blastocoele.

ages separate an upper blastomere, which is a cell with a complete plasma membrane, separated from its neighbors and from the yolk, and a lower blastomere, which remains connected with the yolk. The blastomeres at the margin of the disc and the lower cells in contact with the yolk eventually lose the furrows that partially separated them and fuse into a continuous syncytium with many nuclei, termed the *periblast*, which does not participate in the formation of

the embryo but is believed to break down the yolk and make its nutrients available for the growing embryo.

In birds and some reptiles, the free blastomeres with complete plasma membranes become incorporated into two layers, an upper *epiblast*, and below that a thin layer of flat epithelial cells, the *hypoblast*. The hypoblast is separated from the epiblast by a cavity, the blastocoele, and is separated from the underly-

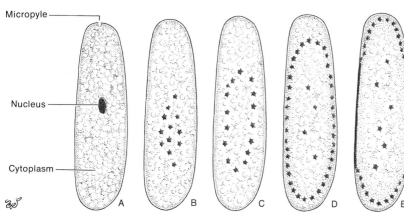

Figure 27–6 *A*, The centrolecithal egg of a beetle. *B–E*, Stages in the cleavage of the centrolecithal egg.

ing yolk by another cavity, the *subgerminal space.* The subgerminal space appears only under the central portion of the blastoderm. The area of the blastoderm over the subgerminal space is more transparent and is called the *area pellucida,* whereas the more opaque part of the blastoderm that rests directly on the yolk is called the *area opaca.*

Cleavage in Centrolecithal Eggs. Cleavage in centrolecithal eggs, such as those of insects, begins with the division of the nucleus in the central island of cytoplasm (Fig. 27-6). After several nuclear divisions without cytoplasmic division, the nuclei migrate out from the center of the egg. Each nucleus is surrounded by a bit of the original central cytoplasm. When the nuclei reach the surface of the egg the cytoplasm surrounding them fuses with the superficial layer of cytoplasm, resulting in a syncytium covering the surface of the egg. The cytoplasm subsequently becomes subdivided by furrows that extend in from the surface. These blastomeres are connected to the yolk mass for a time but eventually become separated from it; the constituents of the yolk are gradually used up as nutrients for the developing embryo. This stage can be compared to the formation of the blastula even though there is no cavity comparable to the blastocoele. The blastoderm surrounds a mass of uncleaved yolk rather than a cavity.

Segregation of Genetic Potentialities in Cleavage. An important question is whether the daughter nuclei formed during cleavage divisions are exactly equivalent or whether some sort of parceling out of potentialities occurs. It is conceivable that the specific fates of the different blastomeres may be governed by differences in the genetic information present in the various cleavage cells. There is now a wealth of experimental evidence indicating that there is no such segregation of genetic potentialities during cleavage. Isolated blastomeres, one of the first four or eight, can develop into an entire embryo with all of its parts but simply reduced in size. The limiting factor is not the genetic potential of the nucleus but the quantity of cytoplasm needed for development.

Another type of experiment avoids this limitation by transplanting nuclei from blastomeres or from cells of later stages of development into an enucleated uncleaved egg. A ripe frog's egg is prepared by removing its nucleus. A cell from an advanced stage of embryonic development is then separated from its neighbors and sucked up into a micropipette (Fig. 27-7). The plasma membrane of the cell is ruptured by the process and the nucleus plus some cytoplasmic debris is injected deep into the enucleated egg, after which the pipette is carefully withdrawn. The operated eggs begin cleaving and some develop normally and undergo metamorphosis. When transplanted in this fashion into an uncleaved egg, nuclei obtained from late blastula or early gastrula stages, when there are as many as 16,000 cells, can lead to the development of a normal embryo. Even nuclei from later stages of development, from the neural plate or from ciliated cells in the digestive tract of a swimming tadpole, may lead to the development of normal embryos when transplanted into enucleated eggs. The nuclei of cells from malignant tissue (cancer cells), but not from normal adult cells, supported development when transplanted into enucleated eggs. It appears that the chromosomes of nuclei from cells in advanced

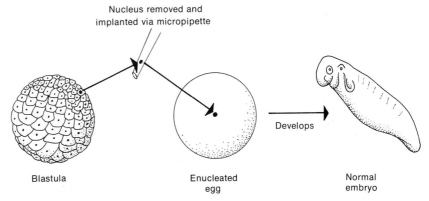

Figure 27-7 Diagram of an experiment in which a nucleus is transplanted from a cell of a gastrula to an enucleated egg. See text for discussion.

Nucleus removed and implanted via micropipette

Develops

Blastula

Enucleated egg

Normal embryo

stages of embryonic development or from adult cells are unable to divide rapidly enough to keep up with the rate of cytoplasmic division early in embryonic development. Chromosomal duplication occurs too slowly, and as a result the daughter cells receive incomplete sets of chromosomes and develop defectively.

Amphibian Gastrulation. Gastrulation in amphibia is a complex process that differs from that in *Amphioxus* because of the large, yolk-laden cells in the vegetal half of the blastula. A groove appears on one side of the blastula, and cells at the bottom of the groove stream into the interior of the embryo. The groove spreads transversely and its lateral ends extend along

the margin between the vegetal region and the upper part of the blastula until they meet on the ventral side of the embryo (Fig. 27–8). This groove, the **blastopore**, is produced by the invagination of presumptive endoderm and mesoderm into the interior of the embryo. The process involves the **invagination** of cells at the slit-shaped blastopore; the growth of cells of the roof of the blastocoele down over the lower, yolk-filled cells *(epiboly)*; and a rolling in of these cells when they reach the blastopore *(involution)*. By the time the blastopore has become ring-shaped, the yolk-filled cells of the vegetal pole remain as a *yolk plug* filling the space enclosed by the lips of the blasto-

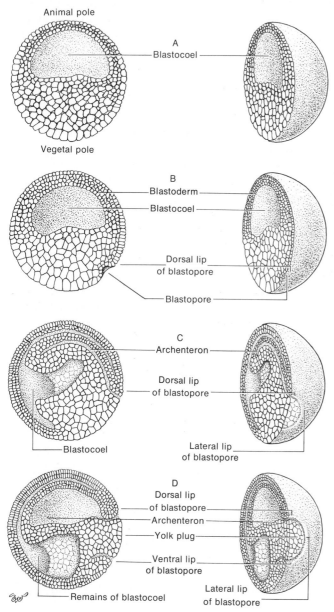

Figure 27–8 Stages in the development of a frog embryo. *A*, Late blastula. *B*, Early gastrula. *C*, Middle gastrula. *D*, Late gastrula.

pore. The rim of the blastopore continues to contract and eventually completely covers the yolk plug. A cavity leading from the groove on the surface of the embryo into the interior appears on the dorsal side, lined on all sides by cells that have invaginated from the surface. This cavity, the *archenteron,* is a narrow slit at first but gradually expands at the anterior end, encroaching on the blastocoele, which is eventually obliterated.

The cells of the animal region of the blastula increase their surface greatly during gastrulation. By the end of gastrulation, after the presumptive mesoderm and endoderm have gone through the blastopore into the interior of the embryo, they cover the entire embryo. The outer layer of cells includes the presumptive epidermis and the presumptive nervous system. The latter expands in a longitudinal direction but contracts in the transverse direction, and the presumptive nervous system becomes oval in shape, elongated in the anteroposterior axis. The presumptive notochord rolls over the dorsal lip of the blastopore into the interior of the embryo and becomes stretched longitudinally along the dorsal midline of the roof of the archenteron. The presumptive mesoderm rolls over the lateral and ventral lips of the blastopore, then moves anteriorly as a sheet of cells between the outer ectoderm and the inner endoderm, concentrated toward the dorsal side of the embryo. The mesodermal layer is thickest in the roof of the archenteron where it adjoins the notochord. More presumptive mesoderm rolls in over the lip of the blastopore, even after the yolk plug has disappeared from the surface. A portion of the presumptive endoderm lies at the equator of the blastula and is

taken into the interior in the original invagination of the blastopore. The remainder of the presumptive endoderm, the cells of the vegetal region, are taken into the interior more or less passively and form the floor of the archenteron.

Avian Gastrulation. Gastrulation in the bird is accomplished by a form of cell migration that differs from that in the frog; the cells from the epiblast appear to move downward singly. Initially a strip of epiblast extending forward in the midline of the embryo from the posterior edge of the area pellucida becomes thickened as the *primitive streak,* with a narrow furrow, the *primitive groove,* in its center. At the anterior end of the primitive streak is a thickened knot of cells, *Hensen's node.* The thickening of the epiblast in the primitive streak is brought about by the migration of cells in from the lateral portion of the epiblast. The cells of the epiblast invaginate at the primitive streak, move into the space between the epiblast and hypoblast, and reach the latter, forming a mass of moving cells (Fig. 27–9). The cells continue to migrate, moving laterally and anteriorly from the primitive streak area. The primitive streak is a dynamic structure and persists even though the cells composing it are constantly changing as they migrate in from the epiblast, sink down at the primitive streak, then move out laterally and anteriorly in the interior. The primitive streak, where invagination occurs, is considered to be homologous to the blastopore of the amphibian or amphioxus egg, but there is no cavity in the embryo of the chick that is homologous to the archenteron.

The presumptive notochord cells become concentrated in Hensen's node and then grow anteriorly as a narrow process from Hensen's

Figure 27–9 Gastrulation in the bird. The anterior half of the area pellucida of a chick embryo is cut transversely to show the migration of mesodermal and endodermal cells from the primitive streak.

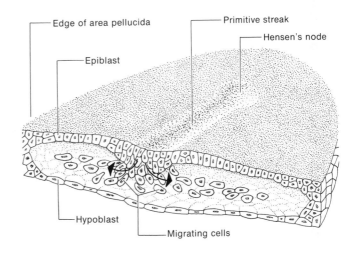

Edge of area pellucida

Primitive streak

Hensen's node

Epiblast

Hypoblast

Migrating cells

node just beneath the epiblast. Presumptive mesoderm grows laterally and anteriorly from the primitive streak between the epiblast, which becomes presumptive ectoderm, and the hypoblast. The original hypoblast cells plus other cells that migrate into the lower layer from the primitive streak form the presumptive endoderm, which forms the digestive tract and yolk sac.

Development of Mammalian Eggs. Cleavage of the mammalian egg typically results in a ball of cells, the *morula,* which becomes subdivided into an *inner cell mass* from which the embryo develops, and an enveloping hollow sphere of cells, the *trophoblast* (Fig. 26–16). The cells of the inner cell mass have a more basophilic cytoplasm. This reflects the rapid synthesis of RNA occurring within them. The cavity of the blastocyst may be compared to the blastocoele but the embryo is not a blastula, for its cells are differentiated into two types. The cells of the inner cell mass differentiate further into a thin layer of flat cells, the *hypoblast.* This, located on the interior surface of the mass, adjacent to the blastocoele, represents presumptive endoderm

(Fig. 27–10). The remaining cells of the inner cell mass become the epiblast. The cells of the hypoblast spread along the inner surface of the trophoblast and eventually surround the cavity of the blastocyst, forming a *yolk sac,* although the cavity is filled with fluid, not yolk.

As the hypoblast spreads out, the inner cell mass also spreads and becomes a disc-shaped plate of cells similar to the blastodisc of avian and reptilian eggs. The *blastodisc,* composed of an epiblast of thick columnar cells and a hypoblast of more irregular flat cells, becomes delimited from the rest of the embryo. Gastrulation begins with the formation of a primitive streak and Hensen's node in which cells migrate downward, laterally and anteriorly between the epiblast and hypoblast. Some of these migrating cells join the hypoblast and become presumptive endoderm; others migrate laterally between the epiblast and hypoblast to become presumptive mesoderm. The amniotic cavity appears as a crevice between the cells of the inner cell mass, which then enlarges. The cavity of the blastocyst becomes filled with mesenchymal cells, which are presumptive extraembryonic mesoderm. A

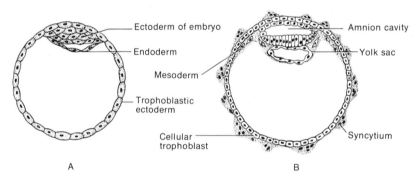

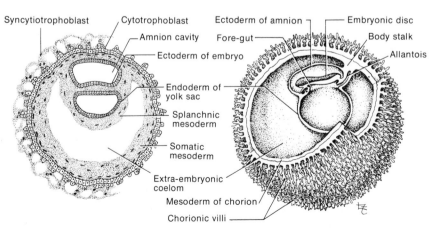

Figure 27–10 Diagrams of human embryos ten *(A)* to twenty *(D)* days old showing the formation of the amniotic and yolk sac cavities and the origin of the embryonic disc.

secondary, small, yolk sac cavity appears just beneath the hypoblast. The bilaminar embryonic disc comes to lie as a plate between the two cavities, connected to the trophoblast at the posterior end by a group of extra-embryonic mesoderm cells, the **body stalk** or **allantoic stalk.** The nonfunctional endodermal part of the allantois, which develops as a tube from the yolk sac, is rudimentary and never reaches the trophoblast Thus after two weeks of development the human embryo is a flat, two-layered disc of cells about 250 microns across, connected by a stalk to the trophoblast.

27–4 MORPHOGENETIC MOVEMENTS

Gastrulation in each of the types of embryos just discussed involves the movement or migration of cells, which occurs in specific ways and leads to specific arrangements of cells. These **morphogenetic movements** involve considerable parts of the embryo, which stretch, fold, contract or expand. The movements are not, apparently, analogous to the contraction of muscles or to some sort of amoeboid movement of the whole embryo. The forces driving these movements and the factors controlling their direction are unknown. If removed from the embryo and grown in organ culture, a section of presumptive ectoderm from the early gastrula of the frog will expand actively, just as it does *in situ* in contributing to the movements of gastrulation. A bit of the blastopore lip transplanted from one embryo into another embryo will invaginate and form an archenteric cavity independent of the archenteron of the host embryo. Invagination involves not only the movement of the cells but also changes in their shape — contractions at certain ends of the cells and expansions at others.

The movements of the cells, and the positions they take up, are guided, at least in part, by *selective affinities* of certain cells, which can be demonstrated if the cells of an early embryo are disaggregated and then incubated in various combinations. Epidermal cells become concentrated on the exterior of the cell mass, and mesodermal cells take up a position between the epidermis and the endoderm. Neural plate cells form hollow vesicles resembling a neural tube or brain vesicle and mesodermal cells tend to arrange themselves around coelomic cavities.

This sorting out of mixtures of cells is believed to result from their selective affinities. When cells touch as a result of random movement, they may remain in contact if held together or they may move apart if not bound strongly. You could postulate that different kinds of cells have qualitatively different specific affinities, or you could account for many of the experimental results by postulating quantitative differences in the degree of adhesiveness of different kinds of cells.

The continued mitotic divisions during gastrulation lead to more cells and more nuclear material but little or no change in the total volume or mass of the embryo. The rate of metabolism, as measured by the rate of oxygen consumption, increases two- or threefold over the rate during cleavage, presumably to supply biologically useful energy for the morphogenetic movements and for the sharply increased rate of synthesis of messenger RNA and of proteins.

27–5 DIFFERENTIATION AND ORGANOGENESIS

In all multicellular animals, except sponges and cnidarians, a third layer of cells, the mesoderm, develops between ectoderm and endoderm. In annelids, mollusks and certain other invertebrates, the mesoderm develops from special cells differentiated early in cleavage (Fig. 27–11). These migrate to the interior and come to lie between the ectoderm and endoderm. They multiply to form two longitudinal cords of cells, which develop into sheets of mesoderm. The coelomic cavity originates by the splitting of the sheets to form pockets and hence is called a **schizocoele.**

In primitive chordates (*Amphioxus*) the mesoderm arises as a series of bilateral pouches from the archenteron (Fig. 27–12). These lose their connection with the gut and fuse one with another to form a connected layer. The cavity of the original pouches remains as the coelom, termed an **enterocoele** because it is derived indirectly from the archenteron. The mesoderm is formed in amphibia by the invagination of cells over the lateral and ventral lips of the blastopore; the cells then move forward as a sheet between ectoderm and endoderm. The presumptive meso-

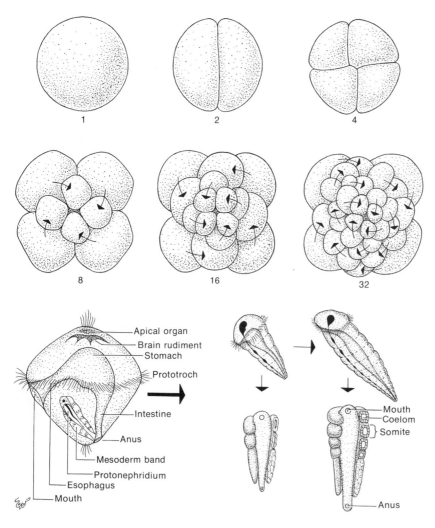

Figure 27–11 Development in annelids. *Upper half,* The successive cleavage divisions occur in a spiral pattern as indicated. *Lower left,* A typical trochophore. The upper half of the trochophore develops into the extreme anterior end of the adult worm and all the rest of the adult body develops from the lower half. A series of cavities appears within each mesodermal somite *(lower right),* which coalesce to form the coelom.

derm of the chick migrates down through the primitive streak area and then laterally and anteriorly between the presumptive ectoderm and endoderm.

However the mesoderm may originate in the various kinds of chordates, it splits into two sheets of cells which grow laterally and anteriorly between the ectoderm and endoderm. The cavity between the two sheets is the coelom, and when the digestive tract becomes separated as a tube from the yolk sac cavity the inner mesoderm grows around it, forming the muscles of the digestive tract. The endoderm forms only the inner lining of the digestive system.

The *notochord,* a flexible, unsegmented, skeletal rod that extends longitudinally along the dorsal midline of all chordate embryos, is formed at the same time as the mesoderm. Its exact mode of formation, like that of the mesoderm, differs from species to species. In all ver-

tebrates the notochord is a short-lived structure that is eventually replaced by the vertebral column, but in some lower vertebrates remnants of it can be found between the vertebrae, even in adults.

27–6 DEVELOPMENT OF THE NERVOUS SYSTEM

Although the two-week-old human embryo is a simple flat disc, the two-month-old embryo has nearly all its structures, at least in rudimentary form. The brain and spinal cord are among the first organs to appear. At about the third week the ectoderm just over the notochord in front of the primitive streak develops a thickened plate of cells called the *neural plate.* The center of this becomes depressed, and is known as the *neural groove* (Fig. 27–13), while

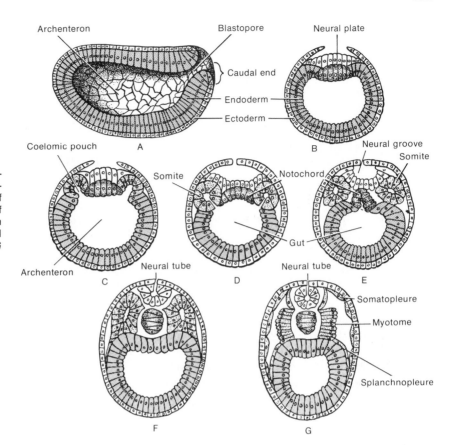

Figure 27–12 Stages in the development of an *Amphioxus* embryo, showing the formation of the mesoderm by the budding of pouches from the archenteron and the formation of the neural tube. *A* is a sagittal section; *B–G* are cross sections.

the outer edges of the plate rise in two longitudinal **neural folds** that meet at the anterior end and appear, when viewed from above, like a horseshoe. These folds gradually come together at the top, forming a hollow neural tube. The cavity of the anterior part of this neural tube becomes the ventricles of the brain, while the cavity of the posterior part becomes the neural canal, extending the length of the spinal cord. The brain region is the first

to form, and the long spinal cord develops slightly later (Fig. 27–14).

The anterior part of the neural tube, which gives rise to the brain, is much larger than the posterior part and continues to grow so rapidly that the head region comes to bend down at the anterior end of the embryonic disc. All the regions of the brain—forebrain, midbrain and hindbrain—are established by the fifth week of development, and a week or two later the

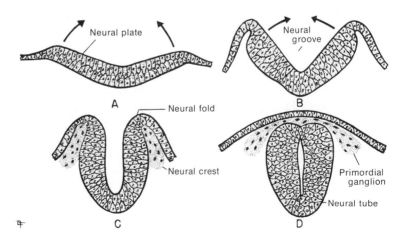

Figure 27–13 Cross sections of the ectoderm of human embryos at successively later stages illustrating the origin of the neural tube and the neural crest, which forms the dorsal root ganglia and the sympathetic nerve ganglia.

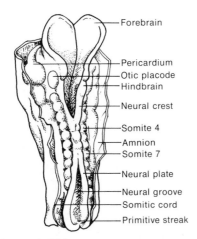

Forebrain
Pericardium
Otic placode
Hindbrain
Neural crest
Somite 4
Amnion
Somite 7
Neural plate
Neural groove
Somitic cord
Primitive streak

Figure 27–14 A dorsal view of a human embryo showing the closure of the neural tube between somites 4 and 7 and the open neural folds anterior and posterior to this.

outgrowths that will form the large cerebral hemispheres begin to grow.

The various motor nerves grow out of the brain or spinal cord but the sensory nerves have a separate origin. When the neural folds fuse to form the neural tube, bits of nervous tissue, known as the *neural crest*, are left over on each side of the tube (Fig. 27–13). These migrate downward from their original position and form the dorsal root ganglia of the spinal nerves and the postganglionic sympathetic neurons. From sensory cells in the dorsal root ganglia, dendrites grow out to the sense organs and axons grow in to the spinal cord. Other neural crest cells migrate and form the medullary cells of the adrenal glands, the neurilemma sheath cells of the peripheral neurons, and certain other structures.

A pair of saclike protrusions, the *optic vesicles*, appear on the lateral walls of the forebrain and grow laterally. The base of the vesicle becomes constricted as the optic nerve. The optic vesicle comes in contact with the inner surface of the overlying epidermis, then flattens out and invaginates to form a double-walled *optic cup.* The inner, much thicker layer of the cup becomes the sensory *retina* of the eye, and the outer, thin layer of the cup becomes the pigment layer of the retina. When the optic cup touches the overlying epidermis it stimulates the latter to develop into a *lens* rudiment. In birds and mammals the epidermis thickens and folds in to produce a pocket. This pinches off and forms a *lens vesicle* that lies in the opening of the optic cup and is surrounded

by the *iris,* formed from the rim of the optic cup. The cells on the inner side of the lens vesicle become columnar and then are transformed into long fibers. Their nuclei degenerate and the cytoplasm becomes hard, transparent and able to refract light.

27–7   DEVELOPMENT OF BODY FORM

The conversion of the two-week-old flat disc into a roughly cylindrical embryo is accomplished by three processes: (1) the growth of the embryonic disc, which is more rapid than the growth of the surrounding tissue; (2) the underfolding of the embryonic disc, especially at the front and rear ends; and (3) the constriction of the ventral body wall to form the future umbilical cord and to separate the embryo proper from the extra-embryonic parts. In addition, the body begins to separate into head and trunk, and the pectoral and pelvic appendages appear.

Growth is rapid at the anterior end of the embryonic disc, and soon the head region bulges forward from the original embryonic area. The tail, which even human embryos have at this stage, bulges to a lesser extent over the posterior end. The sides of the disc grow downward, eventually to form the sides of the body. The embryo becomes elongated because growth is more rapid at the head and tail ends than laterally. The enlarging of the embryo has been compared to the increase in size of a soap bubble blown from a pipe, which, as it grows, swells out in all directions above the mouth of the pipe (the yolk sac). What is to become the mouth and heart originally lies in front of the embryonic disc, and as the disc grows and bulges over the tissues in front, the mouth and heart swing underneath to the ventral side. A similar underfolding occurs at the posterior end. By such growth and underfolding the lateral and eventually the ventral walls of the body are formed, and the embryo becomes more or less cylindrical in shape.

While the embryo is still a simple disc, its entire undersurface is open to the yolk cavity. As the body wall folds, a *foregut* and a *hindgut* (which form, respectively, the anterior and posterior parts of the digestive tract) are cut off from the yolk sac but remain connected to it by the yolk stalk. As the embryo grows and folds,

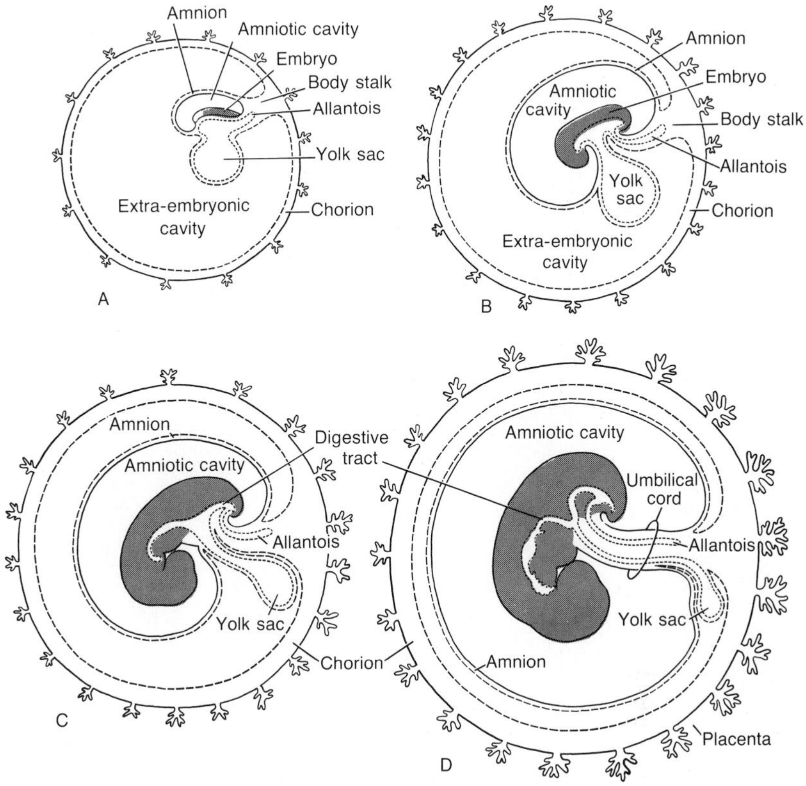

Figure 27–15 *A–D,* Successive stages in the development of the umbilical cord and body form in the human embryo. The solid lines represent layers of ectoderm; the dashed lines, mesoderm; and the dotted lines, endoderm.

the amnion also grows to enclose it, finally constricting the yolk stalk and the body stalk (with its allantois and blood vessels) together into a single cylindrical tube, the **umbilical cord.** This takes place about four weeks after development has commenced and allows the embryo to float free in the liquid-filled **amniotic cavity,** connected to the chorion and placenta only by the umbilical cord (Fig. 27–15).

The month-old embryo, about 5 mm. long, is now recognizable as a vertebrate of some kind. It has become cylindrical, with a relatively large head region and with prominent gills and a tail. Meanwhile, blocks of muscle, known as *somites,* are forming rapidly in the mesoderm on either side of the notochord and the beating heart is present as a large bulge on

the ventral surface behind the gills. The arms and legs are still mere buds on the sides of the body.

By the end of six weeks the embryo is about 12 mm. long; the head begins to be differentiated; the arms and legs have grown out, but the tail and gills are still present.

At the end of two months of growth, when the embryo is 25 mm. long, it begins to look definitely human. The face has begun to develop, showing the rudiments of eye, ear and nose. The arms and legs have developed, at first resembling tiny paddles, but by this stage the beginnings of fingers and toes are evident (Fig. 27–16). The tail, which was prominent during the fifth week of development, has begun to shorten and to be concealed by the

Figure 27–16 Stages in the development of the human arm *(upper row)* and leg *(lower row)* between the fifth and eighth weeks.

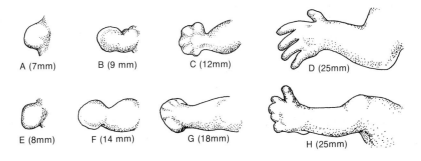

A (7mm)   B (9 mm)   C (12mm)   D (25mm)

E (8mm)   F (14 mm)   G (18mm)   H (25mm)

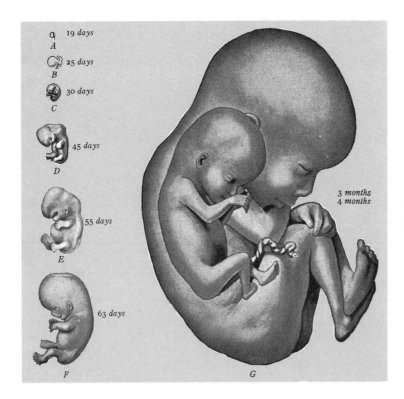

Figure 27–17 A graded series of human embryos. Note the characteristic position of the arms and legs in the four month fetus. (From Arey, L. B.: *Developmental Anatomy,* rev. 7th Ed. Philadelphia, W. B. Saunders Co., 1974.)

growing buttocks. As the heart moves posteriorly on the ventral side and the gill pouches become less conspicuous, a neck region appears. Now most of the internal organs are well laid out so that development in the remaining seven months consists mostly of an increase in size and the completion of some of the minor details of organ formation (Fig. 27–17).

The embryo is about 75 mm. long after three months of development, 250 mm. long after five months, and 50 cm. long after nine months. During the third month the nails begin forming and the sex of the fetus can be distinguished; by four months the face looks quite human; and by five months, hair appears on the body and head. During the sixth month, eyebrows and eyelashes appear. After seven months the fetus resembles an old person with red and wrinkled skin. During the eighth and ninth months, fat is deposited under the skin, causing the wrinkles partially to smooth out; the limbs become rounded, the nails project at the finger tips, the original coat of hair is shed, and the fetus is "at full term," ready to be born (Fig. 27–18). The total *gestation period,* or time of development, for human beings is about 280 days from the beginning of the last menstrual period before conception until the time of birth.

27–8 FORMATION OF THE HEART

Many organs develop in the embryo without having to function at the same time, but the heart and circulatory system must function while undergoing development. The heart forms first as a simple tube from the fusion of two thin-walled tubes beneath the developing head. In this early condition it is essentially like a fish heart, consisting of four chambers arranged in a series: the *sinus venosus,* which receives blood from the veins; the single *atrium;* the single *ventricle;* and the *arterial cone,* which leads to the aortic arches.

Initially the heart is a fairly straight tube, with the atrium lying posterior to the ventricle; but since the tube grows faster than the points to which its front and rear ends are attached, it bulges out to one side (Fig. 27–19). The ventricle then twists in an **S**-shaped curve down and in front of the atrium, coming to lie posterior and ventral to it as it does in the adult. The sinus venosus gradually becomes incorporated into the atrium as the latter grows around it, and most of the arterial cone is merged with the wall of the ventricle.

When it first appears, the embryonic heart is a single structure with only one of each chamber, whereas the adult heart is a double pump, with separate right and left atria and

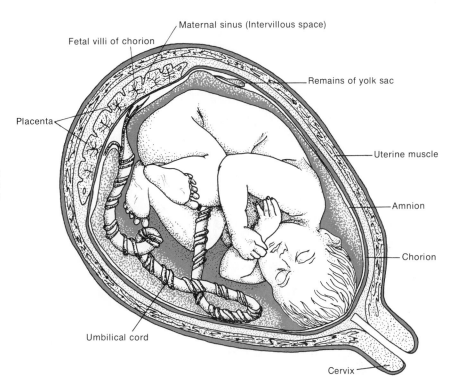

Figure 27–18 A diagrammatic section through the uterus showing the placenta and the fetus shortly before birth.

Maternal sinus (Intervillous space)

Fetal villi of chorion

Remains of yolk sac

Placenta

Uterine muscle

Amnion

Chorion

Umbilical cord

Cervix

ventricles. This separation prevents the mixing of aerated blood from the lungs with non-aerated blood from the rest of the body. The lungs are nonfunctional in the embryo, and not much blood passes through them. The heart begins separating into four chambers at an early stage. The two ventricles are completely separated by the end of the second month. The atria are partly separated, but complete separation does not occur until after birth, when the *oval window* between them finally closes (Fig. 27–20). Before birth this must be kept open to permit blood to get into the left side of the heart, for in the fetus only a small amount of blood passes through the lungs to the left atrium. Without this "window" the left side of the heart would be nearly empty, and most of the blood would pass through the right side only, entering the aorta through the ductus arteriosus (arterial duct).

27–9 DEVELOPMENT OF THE DIGESTIVE TRACT

The digestive tract is first formed as a separate foregut and hindgut by the growth and folding of the body wall, which cuts them off as two simple tubes from the original yolk sac (Fig. 27–15). These tubes grow as the rest of

the embryo grows, becoming greatly elongated. The lungs, liver and pancreas originate as hollow, tubular outgrowths from the original foregut and hence are composed of endoderm, but these outgrowths always are associated with some mesodermal tissue, which

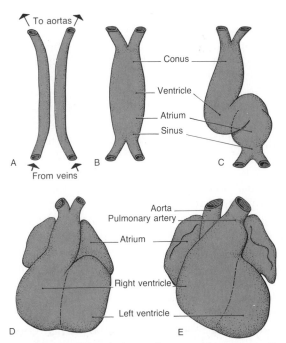

To aortas

Conus

Ventricle

Atrium

Sinus

A   B   C

From veins

Aorta

Pulmonary artery

Atrium

Right ventricle

Left ventricle

D   E

Figure 27–19 Ventral views of successive stages in the development of the heart. See text for discussion.

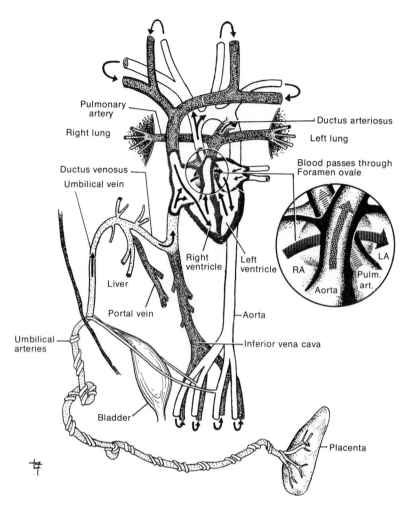

Pulmonary artery

Right lung

Ductus arteriosus

Left lung

Ductus venosus

Umbilical vein

Blood passes through
Foramen ovale

Right
ventricle

Left
ventricle

RA

LA

Pulm.
art.

Aorta

Liver

Portal vein

Aorta

Umbilical
arteries

Inferior vena cava

Bladder

Placenta

Figure 27–20 The human circulatory system before birth. The structures peculiar to the fetal circulatory system are the umbilical arteries and veins to the placenta; the arterial duct (ductus arteriosus), which connects the pulmonary artery and aorta; and the venous duct (ductus venosus), which connects the umbilical vein and the inferior vena cava. The oval window (foramen ovale) connects the right and left atria.

forms the blood and lymph vessels, connective tissue and muscles of these organs. The endoderm forms only the internal epithelium of the digestive tract and lungs, and the actual secretory cells of the pancreas and liver. Both the enzyme-secreting (acinar) cells and those of the islets of Langerhans in the pancreas are derived from tubular outgrowths from the foregut. The lung first develops as a single median outgrowth from the ventral side of the foregut. This single tube, the forerunner of the trachea, soon branches into two tubes, the rudiments of the bronchi, which in turn divide repeatedly and eventually give rise to the complex structure that is the adult lung.

The most anterior part of the foregut flattens out to become, in cross section, a flattened oval, rather than a circle, and develops into the *pharynx.* In the pharyngeal region a series of five paired pouches, the *gill pouches,* bud out laterally from the endoderm and meet a corresponding set of inpocketings from the overlying ectoderm. In lower vertebrates such as

fishes, the two sets of pockets fuse to make a continuous passage from the pharynx to the outside—the *gill slits,* which function as respiratory organs. In the human and other higher vertebrates this normally does not occur: the pouches exist but are nonfunctional vestiges that give rise to other structures or disappear. For example, the first pair of pouches becomes the cavities of the middle ear and their connection with the pharynx, the *eustachian tubes.* The second pair of pouches becomes a pair of tonsils, while parts of the third and fourth pairs of pouches become the thymus gland, and other parts of them become the parathyroids. The fifth pair of pouches becomes the ultimobranchial bodies, which in the human are incorporated into the thyroid as the parafollicular cells secreting calcitonin. The thyroid develops from a separate outgrowth on the floor of the pharynx.

The mouth cavity arises as a shallow pocket of ectoderm that grows in to meet the anterior end of the foregut; the membrane be-

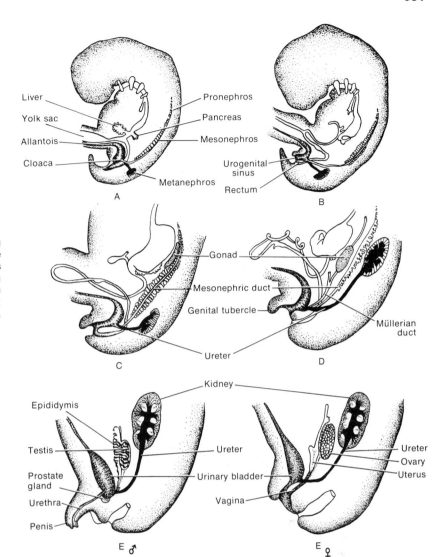

Figure 27-21 Diagrams showing stages in the development of the urinary and reproductive systems in the human. *A,* Early in the fifth week of development; *B,* early in the sixth week; *C,* seventh week; *D,* eighth week. *E* ♂, Three months, male; *E* ♀, three months, female.

tween the two ruptures and disappears during the fifth week of development. Similarly, the anus is formed from an ectodermal pocket which grows in to meet the hindgut; the membrane separating these two disappears early in the third month of development.

27-10 THE DEVELOPMENT OF THE KIDNEY

The development of the kidney provides one of the finest and most clear-cut examples of the principle of recapitulation, i.e., that "ontogeny recapitulates phylogeny." Within the subphylum of vertebrates are three different types of kidney. The earliest, or *pronephros,* is the adult kidney of certain primitive fishes. The second, or *mesonephros,* is the adult kid-

ney of amphibia and the higher fishes. The third, the *metanephros,* is the adult kidney of reptiles, birds and mammals. But in development, each of the higher animals repeats the evolutionary sequence of this organ. Thus frog embryos first develop a pronephros, which functions during early embryonic life, before the permanent mesonephros develops; and the human develops first a nonfunctional pronephros, then a mesonephros, which may be functional during fetal life, and finally the permanent metanephros (Fig. 27-21). The three kidneys develop one after another in both time and space, each new kidney lying posterior to the previous one.

The pronephros, which in the human embryo consists of about seven pairs of rudimentary kidney tubules, develops in the mesoderm and degenerates during the fourth week of embryonic life. From the tubules a pair of wolffian ducts grows back to the hindgut and connects with it.

The tubules of the mesonephros originate during the fourth week, reach their height at the end of the seventh week, and degenerate by the sixteenth week. These tubules connect with the ducts left by the degenerated pronephros and empty into them. In the female the mesonephros and its ducts degenerate completely except for a few nonfunctional remnants, but in the male some of the tubules remain and are converted into the epididymides, while the ducts become the vasa deferentia.

The metanephros of reptiles, birds and mammals develops as a pair of buds from the ducts of the mesonephros. The ureter and collecting tubules of the kidney develop from these buds, while the Bowman's capsules and convoluted tubules develop from the same sort of mesoderm that formed the tubules of the pronephros and mesonephros at a more anterior point. Later, these two portions unite to form the kidney tubules of the adult. The metanephros begins forming during the fifth week and is practically complete by the sixteenth week.

27–11 DEVELOPMENT OF THE GONADS AND REPRODUCTIVE PASSAGES

In many vertebrates a number of accessory structures have developed that facilitate the transfer of sperm from the male to the female reproductive tract and provide a place for the fertilized egg to develop. These structures have evolved either from, or in close association with, the urinary system and the two are frequently referred to as the *urogenital system.*

The sex of an embryo is determined at the time of fertilization by the XX-XY mechanism, but early in development the embryo has the potential of differentiating into either a male or female, for the primordia of both male and female duct systems are present (Fig. 27–22).

The paired paramesonephric or *mullerian ducts* develop just lateral to the wolffian ducts (Fig. 27–23) and are the precursors of the fallopian tubes, uterus and part of the vagina. The gonads develop as thickenings, *germinal ridges,* on the mesodermal epithelium lining the coelom. The ridge becomes composed of

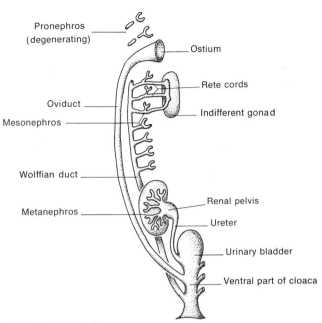

Figure 27–22 A ventral view of the urogenital organs of the sexually indifferent stage of an embryo.

several layers of cells and protrudes into the coelomic cavity, suspended from the peritoneal wall by a double layer of peritoneum, the *mesorchium* or *mesovarium.* The cells in the germinal ridge are of two types—one is very similar to other cells of the peritoneal epithelium and the second is a very different type, the *primordial germ cells* (Fig. 27–24). These are larger cells, with large, vesicular nuclei and clear cytoplasm that stains deeply with an alkaline phosphatase reaction. These primordial germ cells appear in the endodermal epithelium of the yolk sac in the vicinity of the allantoic stalk (p. 623). They migrate by amoeboid motion into the mesenchyme and eventually reach the germinal ridge. There are less than 100 primordial germ cells when they are first evident in the yolk sac but they multiply during their migration and some 5000 or more reach the germinal ridge. The mesenchyme of the germinal ridge is gradually replaced by fingerlike accumulations of cells, the *sex cords,* which migrate in from the mesonephric cord or the epithelium of the germinal ridge. The epithelium of the germinal ridge forms the cortex of the gonad and the sex cords compose its medulla. Development to this stage, that of the *indifferent gonad,* is similar in ovary and testis.

In male embryos the primordial germ cells migrate from the cortex into the primitive sex cords of the medulla; these become organized

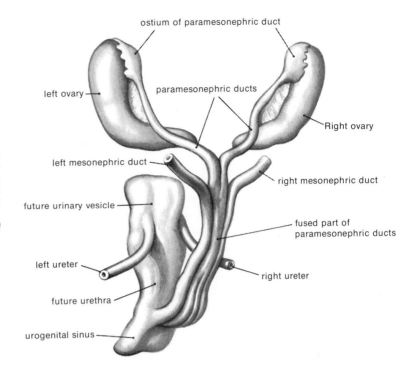

Figure 27-23 Connections of the genital and urinary ducts in a female human embryo of 29 mm. (beginning of third month) in which the genital ducts are still in the indifferent stage. (After Balinsky, B. I.: *An Introduction to Embryology*, 4th Ed. Philadelphia, W. B. Saunders Co., 1975.)

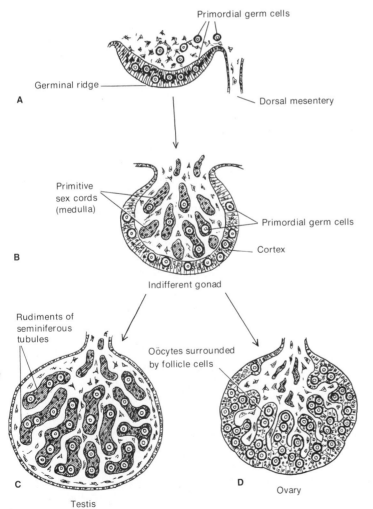

Figure 27-24 Diagram showing development of gonads in higher vertebrates. *A,* Germinal ridge stage; primordial germ cells partly embedded in epithelium of the ridge and located partly in the adjacent mesenchyme. *B,* Indifferent gonad, germ cells in the cortex and in primary sex cords. *C,* Gonad differentiating as testis; cortex reduced; germ cells in sex cords (future seminiferous tubules). *D,* Gonad differentiating as ovary; primary sex cords reduced; proliferating cortex contains the germ cells. (From Balinsky, B. I.: *An Introduction to Embryology.* 4th Ed. Philadelphia, W. B. Saunders Co., 1975.)

into the *seminiferous tubules.* The primordial germ cells become the spermatogonia and the *Sertoli cells* develop from the sex cords. The medulla of the testis becomes its functional part, while the cortex is reduced and converted into a thin epithelial layer covering the coelomic surface of the testis.

In female embryos the medulla is reduced, the primitive sex cords are resorbed, and the central part of the gonad becomes filled with loose mesenchyme permeated by blood vessels. The primordial germ cells remain in the cortex, which increases in thickness. Masses of cortical cells form primary follicles surrounding one or several primordial germ cells; these swell and become oögonia. These oögonia undergo rapid mitotic divisions and after 20 weeks of development in the human fetus, some 7,000,000 oögonia are present. Thereafter, mitosis ceases and no further ova are produced. The number is reduced by *atresia* (cell degeneration) to about 1,000,000 at the time of birth and to about 400,000 at the time of puberty. Only about 450 of these are destined to undergo ovulation during a woman's reproductive period; all the rest undergo atresia.

In male embryos the seminiferous tubules become connected to the *rete testis,* a system of thin tubules that develop from the dorsal part of the gonad and form connections with the adjoining mesonephric tubules (which become the *epididymis*) and through them to the wolffian duct, which becomes the *vas deferens.* This development proceeds under the stimulation of testosterone secreted by the interstitial cells of the developing testis. The testis appears to secrete a second hormone, possibly a peptide, that inhibits the development of the mullerian ducts so that no oviduct or uterus appears.

In female embryos the wolffian ducts degenerate when the mesonephros stops functioning as an excretory organ. The mullerian ducts open at the anterior end by a funnel-shaped *ostium* into the coelomic cavity and grow posteriorly and connect with the cloaca. The cloaca is subsequently divided by a *cloacal septum* into a ventral *urogenital sinus* and a dorsal *rectum.* The latter opens to the exterior by the anus. The mullerian ducts are lateral to the wolffian ducts initially, but the posterior ends of the mullerian ducts pass over the wolffian ducts. They fuse to form the uterus and cervix and combine with part of the urogenital sinus to form the vagina. The ventral part of the urogenital sinus into which the ureters open, together with the adjoining part of the allantoic stalk, expands to form the *urinary bladder.* The caudal part of the ventral urogenital sinus becomes a narrow tube, the *urethra.*

The external opening of the urogenital sinus is flanked on both sides by elongated thickenings, the *genital folds,* which meet anteriorly in front of the sinus and form a median outgrowth, the *genital tubercle,* the rudiment of the *phallus* (Fig. 27–25). From this sexually indifferent condition, under the stimulation of the sex hormones, development occurs toward the male or female condition. In the male the rudiment of the phallus grows and becomes the penis. The genital folds fuse on the posterior surface of the penis, forming a tube, the penile urethra, continuous with the urethra leading from the urinary bladder. In the female the phallus grows slightly and becomes the *clitoris* and the genital folds do not fuse but remain as the *labia minora.* The outer folds become the *labia majora,* comparable to the skin of the scrotum.

27–12 MORPHOGENETIC MOVEMENTS OF CELLS IN ORGANOGENESIS

The various rudiments of the organs arise by morphogenetic movements of sheets of epithelial cells or loose mesenchymal cells that are similar to those involved in gastrulation. Epithelial cells may undergo thickening in localized areas to form elongated cells; this occurs in the thickening of the ectoderm to form the neural plate. Epithelial layers may undergo folding to form grooves or pockets that, by continued folding, yield tubes or vesicles. Crevices may appear between epithelial layers or between masses of cells, splitting a single layer apart into two separate layers. Vesicles and tubes may be formed by the thickening of an epithelial layer followed by the formation of a cavity in the middle of the solid thickening, rather than by the folding of an epithelium. Masses of cells that were previously separated may fuse to form a new organ rudiment; for example, the edges of the neural plate fuse after the neural plate has been folded into a tube. Epithelial layers may break up to form mesenchyme, and mesenchyme

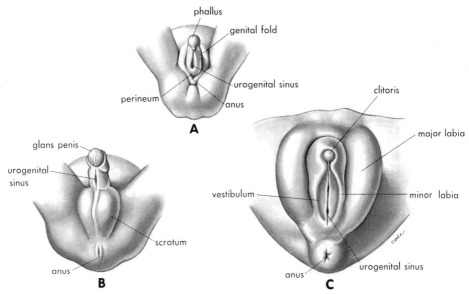

Figure 27–25 Development of external genitalia in human embryos. *A,* Indifferent stage in an embryo of 29 mm (beginning third month). *B,* Male embryo of 145 mm (16 weeks). *C,* Female embryo of 150 mm (16 weeks). (After Balinsky, B. I.: *An Introduction to Embryology,* 4th Ed. Philadelphia, W. B. Saunders Co., 1975.)

cells may secondarily be rearranged into an epithelium. Mesenchyme cells may aggregate in a mass without forming an epithelial sheet and later differentiate as cartilage, bone or muscle tissue. Mesenchyme cells are often accumulated near the surface of an epithelium or around a structure such as a vesicle or tube. They subsequently differentiate into cartilage or bone and produce a skeletal capsule around the organ developing from the vesicle or tube.

A sort of programmed cell death may be a morphogenetic force. The death of localized cells in part of a rudiment while the adjacent cells remain healthy and continue to proliferate may lead to a change in the shape of an organ. For example, the death of cells in a certain pattern during the development of the hand and foot separates the digits from one another.

The cellular migrations that are the basis of the morphogenetic movements of organogenesis can be readily observed when isolated embryonic cells are kept in culture medium. The factors that direct cell migration along certain channels and to certain destinations in the embryo are not at all clear but probably involve specific properties of the cell surface plus, in certain instances, the properties of the substrate along which the cells are migrating. The intercellular spaces of the mesenchymal areas of the embryo are filled with a colloidal,

partially gelated solution, which results in the stretching of molecular fibers in various directions through the space. These serve as a substratum for migrating mesenchyme cells and determine in part their direction of migration.

27–13 MALFORMATIONS

In view of the extreme complexity of the developmental process it is indeed remarkable that it occurs so regularly and that so few malformations occur. The development of a human arm involves the formation of 29 bones, each of a specific size and shape and each forming a joint with the next in a very specific way. It involves the formation of some 40 or more muscles, each with the proper origin, insertion and size. It involves the development of a large number of motor and sensory nerves, each with the proper synaptic connections on the motor end plates of muscles or in the sensory receptors in the skin, tendons and joints. And it involves the development of a large number of arteries and veins arranged in a fairly specific pattern to supply each of the parts of the arm.

About one child in 100 is born with some major defect, such as a cleft palate, a club foot or spina bifida. Some of these are inherited; others result from environmental factors. Experiments with fruit flies, frogs and mice have shown that x-rays, ultraviolet rays, temperature changes and a variety of chemical substances will induce alterations in development. The kind of defect produced depends on the *time* in development when the environmental agent is applied, and to a much lesser extent on the kind of agent used. For example, x-rays, the administration of cortisone and the lack of oxygen will all produce similar defects in mice—harelip and cleft palate—if applied at comparable times in development. There are certain critical periods in development during which particular organs are differentiating and growing most rapidly and are most susceptible to interference.

According to old beliefs, such developmental abnormalities are caused by fright or injury to the mother. It was even supposed that injury to a particular part of the mother's body resulted in malformation of that part of the fetus. However, the injuries commonly believed to cause malformations usually occur much later in pregnancy, long after the organs have completed differentiation.

27–14 TWINNING

In humans, apes and monkeys, as well as in many other species of mammals, offspring are usually produced singly, although in other animals more than one (up to 25 in the pig) are produced in a single litter. About once in every 88 human births, two individuals are delivered at the same time. More rarely, three, four, five, and even six children are born simultaneously. About three fourths of the twins (and triplets, quadruplets, quintuplets and so on) are the result of the simultaneous release of two eggs, one from each ovary, both of which are fertilized and develop. Such *fraternal twins* may be of the same or different sex and have only the same degree of family resemblance that brothers and sisters born at different times have. They are entirely independent individuals with different hereditary characteristics and result from the fertilization of two or more eggs ovulated simultaneously.

Purified human FSH (follicle-stimulating hormone) is used to induce ovulation in women whose infertility appears to be due to deficient production or release of gonadotropin. In the initial attempts to use this therapy many women apparently received too much

FSH, had multiple ovulations and released from two to ten or more ova, all of which were fertilized and began to develop. Most of the women with multiple fetuses have been unable to carry them to term but others with one or two fetuses have had normal births.

In contrast, *identical twins* (or triplets, and so on) are formed from a single fertilized egg that at some early stage of development divides into two (or more) independent parts, each of which develops into a separate fetus. Such twins, of course, are the same sex, have identical hereditary traits and are so similar that it is difficult to tell them apart.

Occasionally identical twins develop without separating completely and are born joined together *(Siamese twins)*. All grades of union have been known to occur, from almost complete separation to fusion throughout most of the body so that only the head or the legs are double. Sometimes the two twins are of different sizes and degrees of development, one being quite normal, while the second is only a partially formed parasite on the first. Such errors of development usually die during or shortly after birth.

27–15 CHANGES AT BIRTH

Great changes take place within a short time after a child is born. Hitherto it received both food and oxygen from the placenta; now its own digestive and respiratory systems must function. Correlated with these changes are major changes in the circulatory system; the umbilical arteries and veins are cut off, and the blood flow through the heart and lungs is altered.

It is believed that the first breath of the newborn infant is initiated by the accumulation of carbon dioxide in the blood after the umbilical cord is cut; this stimulates the respiratory center in the medulla. The resulting expansion of the lungs enlarges its blood vessels, which previously were partially collapsed, and blood from the right ventricle flows in ever-increasing amounts through the vessels of the lung, instead of through the arterial duct which connected the pulmonary artery and aorta during fetal life. The resulting increase in blood returning from the lungs to the left atrium results in the closing of the valvelike opening of the oval window (Fig. 27–26).

Figure 27-26 Changes in the human circulatory system at birth. *Left,* The circulatory system of the fetus; compare with Figure 27-20. *Right,* The circulatory system of the newborn child. Aerated blood is shown in white, nonaerated blood in dark stippling, and a mixture of the two in lighter stippling. In the embryo most of the blood entering the right atrium reaches the aorta either via the oval window (foramen ovale) or via the arterial duct (ductus arteriosus) *(inset).* The changes at birth are: (1) the loss of the placenta, (2) the expansion of the lungs, (3) the closing of the foramen ovale and (4) the closing and degeneration of the arterial duct.

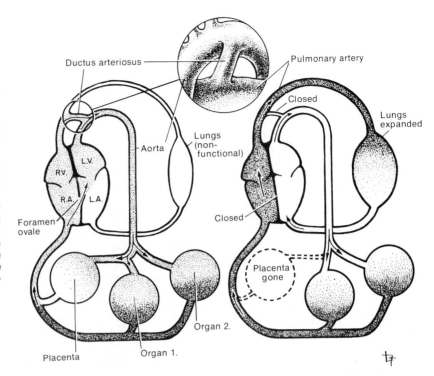

These changes take place within a short period after birth, and eventually the flap of the oval window grows into place and the arterial duct degenerates, so that the adult pattern of circulation is established.

Occasionally the oval window fails to close or the arterial duct fails to degenerate and there is a mixing of oxygenated and nonoxygenated blood. This results in a "blue baby," a child whose skin has a purplish hue because of the inadequate oxygenation of its blood. Delicate surgical procedures, developed after years of practice operations on dogs, now make it possible to operate on the heart itself and cure this condition.

27-16 POSTNATAL DEVELOPMENT

Development does not, of course, cease at birth. At birth the teeth and genital organs of the human infant are only partly formed and the body proportions are quite different from those of the adult. The head is proportionately much larger early in development than in the adult (Fig. 27-27). The head comprises about half the length of the two-month-old fetus but its growth terminates early in childhood so that the head of the adult is proportionately smaller than that of the newborn. The arms attain their proportionate size shortly after birth but the

legs attain theirs only after some 10 years of growth. The last human organs to mature are the genitals, which do not begin to grow rapidly until 12 to 14 years after the infant is born.

The degree of maturity and self-sufficiency of the newly hatched bird or newly born mammal varies widely from one species to another. Baby chicks and ducks can run around and eat solid food just after hatching, but baby robins are blind, have very few feathers and cannot stand. The newborn guinea pig has fur and teeth and can eat solid food. Newborn rats, mice and humans are quite helpless and require a lot of parental care to survive. The developmental processes that occur after birth involve some multiplication and differentiation of cells, but in large part they involve the growth of cells formed earlier. The weight or size of an animal or plant as a function of time usually yields an **S**-shaped growth curve, one remarkably like the growth curve of a population of individuals (p. 827). Although biologists can identify some of the factors ("environmental resistance") that stabilize the size of a population at a certain level (p. 829), very little is known of the factors that lead to the cessation of cell multiplication and growth. Some plants and animals do not, in fact, stop growing but continue to grow, though perhaps at a slower pace, all through their life.

The human egg is about 100 micrometers

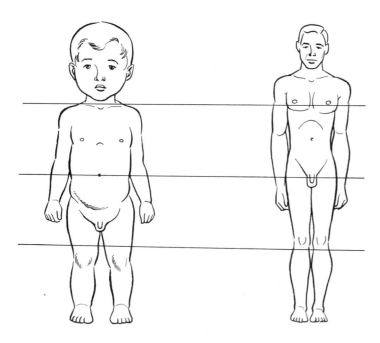

Figure 27-27 Relative sizes of the head, arms and legs in a young child *(left)* and in an adult *(right)*.

in diameter, just barely visible to the naked eye. A baby at birth is about 50 cm. long, roughly 5000 times as long as the egg. In developing from an infant to an adult, the height increases only an additional three and one half times to about 175 cm. The maximal rate of linear growth occurs before birth, in the fourth month of fetal life. There is a final growth spurt at the time of adolescence, which reaches its peak at about age 12 in girls and at about age 14 in boys.

Each structure and organ has its characteristic rate of growth. The growth rates of the various organs can be assigned to one of four types (Fig. 27-28). The growth curve of the skeleton follows that of the body as a whole. The brain and spinal cord grow relatively rapidly early in childhood and nearly reach their adult size by the age of 9. Lymphoid tissue, including the thymus, has a third type of growth curve, reaching a maximum at age 12 that exceeds the adult value. It then involutes until about age 20, when it attains adult values. The fourth pattern of growth is shown by the reproductive system, which grows very slowly until age 12 or thereabouts and then undergoes rapid growth at puberty.

27-17 THE AGING PROCESS

Since development in its broadest sense includes any biological change with time, it also includes those changes that result in the decreased functional capacities of the mature organism, the changes commonly called *aging.* The declining capacities of the various systems in the human body, though most apparent in the elderly, may begin much earlier in life, during childhood or even during prenatal life. The newborn female has only 400,000 oöcytes remaining of the 4,000,000 she had three months earlier in fetal life!

The various systems of the body may begin their decline at quite different times; the aging process is far from uniform in the various parts of the body. A 75-year-old man, for example, has lost 64 per cent of the taste buds, 44 per cent of the renal glomeruli and 37 per cent of the axons in his spinal nerves that he had at age 30. His nerve impulses are propagated at a rate 10 per cent slower, the blood supply to his brain is 20 per cent less, his glomerular filtration rate has decreased 31 per cent and the vital capacity of his lungs has declined 44 per cent. Relatively little is known about the aging process itself, but this is now an active field of scientific investigation. Although the marked improvements in medicine and public health have led to a larger fraction of the total human population surviving to an advanced age, there has been no concomitant increase in the maximum life expectancy for men or women.

A remarkable model of the aging process in man is provided by a rare, inherited type of abnormal development called *progeria.* Indi-

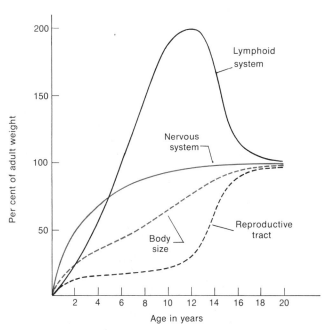

Figure 27-28 Diagram showing the relative rates of growth of the several different organ systems during human development.

viduals with this condition develop more or less normally until they are about one year old, then undergo changes that are considered typical of aging—they lose their hair, stop growing and develop the appearance of wizened old men or women. The collagen in the connective tissues of their skin becomes highly cross-linked, as is seen in old age. They usually die at age 10 to 15 of coronary artery disease secondary to extensive atherosclerosis.

Cells that differentiate and stop dividing appear to be more subject to the changes of aging than are those that continue to divide throughout life. Nerve and muscle cells, which lose the capacity for cell division at an earlier age, show a decline in their respective functional capacities at an earlier age than do tissues such as liver and spleen, which retain the capacity to undergo cell division.

Several theories have been advanced regarding the nature of the aging process—that it involves hormonal changes; that it involves the development of *autoimmune reactions* (allergies against certain components of the organism's own body that result in destruction of those components by antibodies); that it involves the accumulation of specific waste products within the cell (the "clinker" theory); that it involves changes in the molecular structure of macromolecules such as collagen (an increased cross-linkage between the helical chains); that there is a decrease in the elastic properties of connective tissues owing to an ac-

cumulation of calcium, which results in stiffening of the joints and hardening of the arteries; that it involves the peroxidation of certain lipids by free radicals; or that cells are destroyed by hydrolases released by the breaking of lysosomes. Other current theories suggest that aging involves the accumulation of somatic mutations caused by continued exposure to cosmic radiation and x-radiation, mutations that decrease the ability of the cell to carry out its normal functions at the normal rate. In all likelihood aging is a part of and due to the same kinds of developmental processes that bring about the increasing functional capacities of the various systems of the body during earlier development. They may be part of the program of timed development built into the genome. Like other developmental processes, aging may be accelerated by certain environmental influences and may occur at different rates in different individuals because of inherited differences. The best guarantee of a long life is to have long-lived parents and grandparents. There is some experimental evidence that aging, at least in rats, can be delayed by dietary means, by caloric restriction. The thin rats, by and large, live longer than the fat rats!

27-18 REGENERATION

Animals have to varying degrees the ability to repair the damage caused by a wound, by

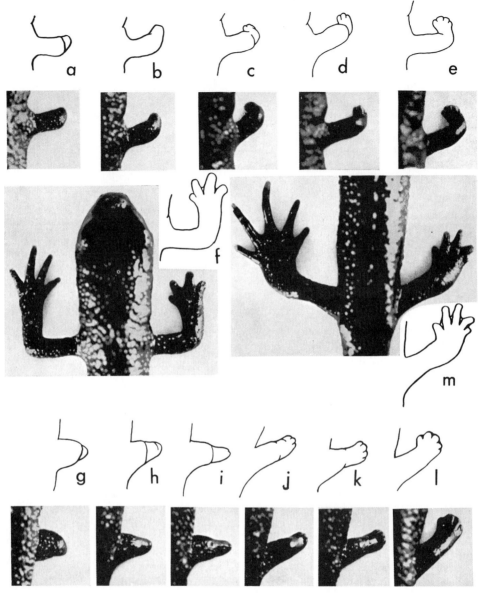

Figure 27–29 Stages in the regeneration of fore limbs *(a–f)* and hind limbs *(g–m)* in salamanders after the original limbs have been amputated. (From Balinsky, B. I.: *An Introduction to Embryology,* 4th Ed. Philadelphia, W. B. Saunders Co., 1975.)

the loss of an organ or by the severing of an appendage. The repair process involves a reinitiation of morphogenetic processes halted when development was completed. Hydras, planaria and earthworms have a high degree of ability to regenerate. If a planarian is cut in half, for example, each half will regenerate the missing part and two planaria are formed. Earthworms cut in half have a similar ability to regenerate the missing end. Echinoderms, such as the starfish, and crustaceans, such as the crab or lobster, can readily regenerate a lost arm or leg. Among vertebrates the salamanders

and newts are outstanding in their regenerative capacity and can regenerate a severed arm or leg. Certain lizards can regenerate a lost tail; indeed some lizards, when threatened, can sever their tail near the base, leaving the tail to distract their predator while they escape and later regenerate a new tail. The regenerated lizard tail differs from the original in the shape of the vertebrae and in the kind of scales covering it. These regenerative phenomena, especially the regeneration of limbs in salamanders and the regeneration of parts in planaria, have been studied extensively to learn about the regenerative process and about wound healing, and as a possible source of clues about normal differentiation.

Both adult and larval salamanders can regenerate limbs to a remarkable degree. After the limb has been amputated, the surrounding tissues close the wound and a mass of cells, a **blastema**, accumulates under the skin. The blastema grows rapidly by active proliferation of cells. At first it has a conical shape and then flattens dorsoventrally at the tip (Fig. 27–29). Rudiments of the digits appear and cell masses segregate in the proper pattern in the interior to become the rudiments of the bones and muscles. Organogenesis is followed by histological differentiation, and the entire regenerating limb continues to grow until it attains the size of a normal limb. The regenerated limb can move and eventually becomes indistinguishable from a normal limb. The cessation of growth when the normal size is reached is just as remarkable and unexplained as the initiation of growth when the limb is lost.

The presence of a nerve supply is required for regeneration to proceed normally. If the nerves supplying the arm or leg of the salamander are destroyed when the limb is removed the blastema stops growing and the development of the regenerating limb is halted (Fig. 27–30). However, if the regenerating limb has begun to differentiate, the nerve can be cut without interfering with the regenerative process. The nerve supply is required for the *initiation* of the regenerative process but not for its completion. If a nerve to a limb is cut across and the proximal end of the nerve is placed in a cut in the skin a short distance from the limb, a blastema will form as the wound heals. The nerve will initiate the development of a limb rudiment or even of a complete new limb in an area that does not normally form a limb. It appears that the regenerative stimulus supplied by the nerve is not the ordinary nerve impulse but perhaps some sort of growth-promoting neurosecretory product.

Regeneration can be inhibited by irradiating the blastema with x-rays in an appropriate dose, about 7000 r. The irradiated blastema regresses, becomes filled with connective tissue cells and may be completely resorbed. If a normal leg of an adult newt is irradiated with 7000 r, no effects are visible—neither the appearance nor the function of the leg is affected; however, at any subsequent time, even months later, if the irradiated leg is amputated, no regeneration will occur. X-rays have a similar inhibitory effect on all regenerative systems tested and thus cause some marked change in a fundamental property of cells.

27–19 WHAT REGULATES DEVELOPMENTAL PROCESSES?

One of the important unsolved problems of modern biology is the nature of the mechanisms that regulate developmental processes so that each organ appears at the proper time and in the proper spatial relations to the other organs. How can a single cell, a single fertilized egg, give rise to the many different types of cells that make up the adult organism, cells

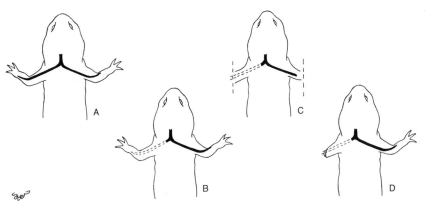

Figure 27–30 Diagram illustrating the role of an intact nerve supply in bringing about regeneration of amputated limbs in salamanders. *A*, Dorsal view of anterior end of a salamander showing the nerve supply to the forelimbs. *B*, The nerve supply to one forelimb is removed. *C*, Both forelimbs are amputated just above the elbow. *D*, The limb with a nerve supply regenerates; the other limb does not.

which differ so widely in their structure, functions and chemical properties? The advances in biochemical genetics in the past 25 years have permitted new intellectual and experimental approaches to this problem.

We now have a detailed working hypothesis as to how biological information is transferred from one generation of cells to the next and how this information may be transcribed and translated in each cell so that specific enzymes and other proteins are synthesized (p. 702). The operation of this system would produce a multicellular organism in which each cell would have the same assortment of enzymes as every other cell. Additional hypotheses are needed to account for: (1) the means by which the amount of any given enzyme produced in a cell is regulated; (2) the control of the time in the course of development when each kind of enzyme appears; and (3) the mechanism by which unique patterns of proteins are established in each of the several kinds of cells in a multicellular organism despite the fact that they all contain identical quotas of genetic information.

Any satisfactory model of the developmental process must be able to explain how genetic and nongenetic factors can interact to control the differentiation of cells and tissues. Why does the epicotyl of the developing pea seed become negatively geotropic and grow upward (p. 228), whereas the closely adjacent hypocotyl, composed of cells with the same genetic information as those of the epicotyl, becomes positively geotropic and grows downward? Why do the cells on the outer margin of the cambium differentiate into phloem (p. 263), while genetically identical cells on the inner margin of the cambium differentiate into xylem?

Biologists have speculated that differentiation might occur: (1) by some sort of segregation of genetic material during mitosis, (2) by the establishment of chemical gradients within the developing embryo, (3) by somatic mutations, (4) by the action of chemical organizers, or (5) by the induction of specific enzymes.

The Preformation Theory and Epigenesis. A theory widely held by early embryologists was that the egg or sperm contained a completely formed but minute germ that simply grew and expanded to form the adult. This *preformation theory* was gradually displaced by the contrasting theory of *epigenesis*, which stated that the unfertilized egg is structureless, not organized, and that development proceeds by the progressive differentiation of parts. However, development is not *simply* epigenetic. Certain potentialities, though not structures, may be localized in certain regions of the egg and early embryo; this restricts the developmental possibilities of that part. When the embryos of echinoderms or chordates are separated experimentally at the two- or four-cell stage, each of the separated cells will form an embryo complete in all details although smaller than normal. However, when embryos of annelids or mollusks are separated at the two-cell stage, neither cell can develop into a whole embryo. Each cell develops only into those structures it would have formed normally—half an embryo, perhaps, or some part of one. This localization of potentialities eventually occurs in the development of all eggs; it simply occurs earlier in some than in others.

Some sort of chemical or physiological differentiation must be present before any structural differentiation is visible, but the basic problem of how the chemical differentiation arises remains unsolved. By appropriate experiments it has been possible to map out the location of these potentialities in the frog, chick and other embryos (Fig. 27–31).

Differential Nuclear Division. Cellular differentiation might be explained if genetic material were parceled out differentially at cell division and the daughter cells received different kinds of genetic information. Although there are a very few clear instances of differential nuclear division in animals such as *Ascaris* and *Sciara*, this does not appear to be a general mechanism of differentiation. By and large, genes are neither lost nor gained during developmental processes. The generalization that the mitotic process ensures the exact distribution of genes to each cell of the organism is a valid one. Thus the differences in the kinds of enzymes and other proteins found in different cells of the same organism must arise by differences in the *activity* of the same set of genes in different cells. Experiments with nuclear transplants (p. 619) showed that even the nucleus from a differentiated cell taken from an advanced stage of embryonic development can, when placed in an enucleated egg, lead to the development of a normal embryo. Thus it

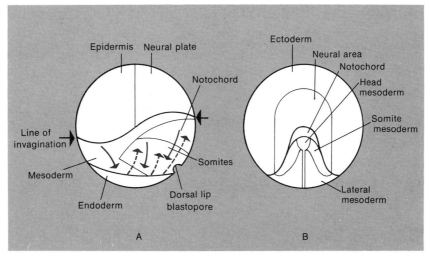

Figure 27–31 Embryo maps. *A*, Lateral view of a frog gastrula showing the presumptive fates of its several regions. The solid arrows indicate cell movements on the surface and the dashed arrows indicate cell movements in the interior. *B*, Top view of a chick embryo showing the location in the primitive streak stage of the cells that will form particular structures in the adult.

clearly had retained a full set of genetic information.

Differential Gene Activity. Some striking evidence regarding the differential activity of genes comes from cytologic studies of insect tissues. In certain insect tissues the chromosomes undergo repeated duplication, and the daughter strands line up exactly in register, locus by locus, so that characteristic bands appear along the length of the giant chromosome. When these bands are examined carefully, either in the same tissue at different times or in different tissues at the same time, certain differences in appearance become evident. A particular section of a chromosome may have the appearance of a diffuse puff (Fig. 27–32). Histochemical tests and autoradiographic evidence from experiments using tritium-labeled

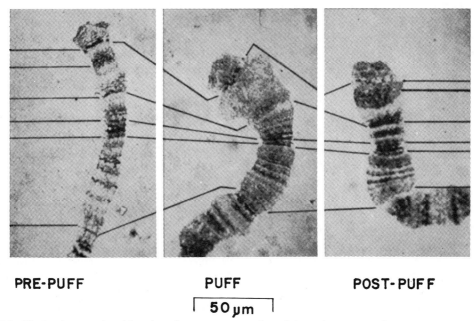

PRE-PUFF   PUFF   POST-PUFF

⌐ **50 µm** ⌐

Figure 27–32 Photomicrographs of the changing appearance of a polytene chromosome in a salivary gland of *Chironomus tentans* as a puff gradually appears on the chromosome. The corresponding bands in the three photomicrographs are indicated by connecting lines. The material that makes up the chromosomal puff has been shown histochemically and autoradiographically to be largely ribonucleic acid. (Courtesy of G. Rudkin.)

precursors have shown that the puff consists of RNA. It has been inferred that genes show this puffing phenomenon when they become active and that the puff represents the messenger RNA (mRNA) produced by the active gene in that band. It has been possible to correlate the appearance of puffs at certain regions in the chromosome with specific cellular events, such as the initiation of molting and pupation.

The turning on and off of the synthesis of specific proteins—differentiation at the molecular level—could occur by some process involving the genic DNA, the transcription of the DNA to form messenger RNA, the processing of the messenger RNA and its transport to the cytoplasm, the combination of the messenger RNA with the ribosome during protein synthesis, or even some transformation of the ultimate protein product. In view of the tremendous number of kinds of DNA that are represented by the genic complement of a cell in a multicellular plant or animal, we might ask what prevents that cell from producing continuously all the tremendous variety of messenger RNAs and their corresponding protein products that are possible. What determines which molecules of DNA are to be transcribed at any given moment in a given cell?

A mechanism that controls the transcription of DNA to regulate the production of messenger RNA would probably be the most economical one biologically, for it would clearly be to the cell's advantage not to have its ribosomes encumbered with nonfunctional molecules of messenger RNA. A cell optimally should produce only those kinds of messenger RNA that will code for the specific proteins required at that moment.

The continued synthesis of any protein requires the continued synthesis of its corresponding messenger RNA. Each kind of messenger RNA has a half-life ranging from a few minutes in certain microorganisms to 12 or 24 hours or longer in the human and other mammals. Although each molecule of RNA template can serve to direct the synthesis of many molecules of its protein, the RNA is eventually degraded and must be replaced. This provides a mechanism by which a cell can alter the kind of protein being synthesized as new types of messenger RNA replace the previous ones. Thus the cell can respond to exogenous stimuli with the production of new types of enzymes.

Enzyme Induction. The induction of enzymes by environmental stimuli has been cited as a model for embryonic differentiation. Bacteria, and to some extent animal cells, respond to the presence of certain substrate molecules by forming enzymes to metabolize them. Jacques Monod has suggested that extracellular or intracellular influences may initiate or suppress the synthesis of specific enzymes, thus affecting the chemical constitution of the cell and leading to differentiation. As the embryo develops, the gradients established as a result of growth and cell multiplication could result in quantitative and even qualitative differences in enzymes. The induction or inhibition of one enzyme could lead to the accumulation of another chemical product that would induce the synthesis of a new enzyme and confer a new functional activity on these cells.

Morphogenesis appears to be too complex a phenomenon to be explained in terms of any single process such as enzyme induction. Enzymes can be induced in an embryo by the injection of an appropriate substance. Adenosine deaminase, for example, has been induced in the chick embryo by the injection of adenosine; however, no enzyme has been induced that is not normally present to some extent in the embryo. Adult tissues such as the liver may show marked differences in their enzymatic activities, differences that might be the result of adaptations comparable to those seen in bacteria. Adaptive changes in enzymes, however, are temporary and reversible, whereas differentiation is a permanent, essentially irreversible process.

The DNA of the genes not being transcribed at any given moment may be bound to a histone or to an acidic nuclear protein that makes the DNA unavailable for the transcription system. The hypothesis that in the nucleus some genes are free and can be transcribed, whereas others are bound and not transcribable, is supported by a variety of experimental evidence. The protein pea seed globulin is synthesized in the seed of the pea plant but not in any other part of the plant. James Bonner has provided evidence that the DNA that codes for the synthesis of the globulin is bound to histone in the cells elsewhere in the plant. In the seed, the histone is removed; that particular segment of the DNA becomes free and can be transcribed, forming messenger RNA that codes for the synthesis of the globulin. The question of what controls the binding and release of a specific segment of DNA remains to be answered.

Other biologists, such as Albert Tyler, have postulated that DNA may be transcribed to form messenger RNA, but the mRNA is "masked" and inactive as a template for protein synthesis until it is subsequently unmasked by a separate process. This form of control may operate especially during early embryonic development. The mRNA synthesized in the oöcyte from the maternal genome may remain masked and inactive until activated by the fertilization process and freed to undergo translation.

Studies of the synthesis of RNA in the nucleus have shown that a large fraction of it, perhaps 80 per cent of the total RNA synthesized, is destroyed without leaving the nucleus; only about 20 per cent of the nuclear RNA is identical with RNA present in the cytoplasm. Some have suggested that this "heterogeneous" nuclear RNA serves in some way to regulate genic action. There appears to be far more DNA in the cells of multicellular organisms than is necessary to serve as template for the mRNA used in directing the synthesis of protein. Is the rest of the DNA part of some enormous, complex regulatory system that controls the activity of the DNA that does produce mRNA? Others interpret the rapid turnover of nuclear RNA as indicating that all genic DNA is transcribed into RNA all the time, but much of the RNA produced is rapidly destroyed before leaving the nucleus. Only the RNA that is processed by the addition of a polyadenylic acid tail survives and passes to the cytoplasm. This second interpretation implies that gene action is regulated not during transcription but subsequently, between transcription and translation, by some process that selectively stabilizes certain kinds of RNA.

Isozymes. The kinds of enzymes present in different cells and tissues of the same organism may show marked qualitative and quantitative differences. Mammalian liver cells, for example, have a glucose-6-phosphatase and can convert glycogen and other precursors to free glucose, whereas skeletal muscle cells lack this enzyme. Enzymes that catalyze the same reaction in different tissues may differ in molecular size, in their amino acid sequences, in their immunologic properties and in their responses to hormones and other control mechanisms.

Even within a single tissue or a single cell, multiple molecular forms of an enzyme, termed *isozymes,* may be found. All these proteins catalyze the same general reaction but have distinct chemical and physical properties. The different molecular forms may bear a different net charge and thus can be separated by electrophoresis.

The lactic dehydrogenase isozymes have been extensively studied. They are composed of subunits, four of which combine to form the active enzyme. There are two major kinds of subunits, A and B, each of which is a polypeptide chain with a specific gene-determined sequence of amino acids. The entire molecule is analogous to a hemoglobin molecule which is composed of two α and two β polypeptides; however, in the lactic dehydrogenase molecule, any combination of the two types of subunits is permissible. The combinations of two kinds of subunits taken four at a time (A_4, A_3B, A_2B_2, AB_3, B_4) add up to the five kinds of lactic dehydrogenases that are typically observed when the enzyme is extracted from a tissue and subjected to electrophoresis.

It appears that there are two genes, one for each of the subunits. Different types of tissues have characteristic ratios of the different chains in their tetramers, presumably reflecting differences in the relative activities of the two genes and differences in the rates of synthesis of the two subunits. What controls the relative activities of the two genes remains unknown. A very curious observation, one that requires explanation, is that the lactic dehydrogenase in the breast muscle of the chick changes during embryonic development from a pure B_4 isozyme through a series of intermediates to a pure A_4 type in the adult.

Organizers. When a piece of the dorsal lip of the blastopore of a frog gastrula is excised and implanted beneath the ectoderm of a second gastrula, the tissue heals in place and causes the development of a second brain, spinal cord and other parts at that site, so that a double embryo, or closely joined Siamese twins, result (Fig. 27–33). Many tissues show similar abilities to organize the development of an adjoining structure. The eye cup will initiate the formation of a lens from overlying ec-

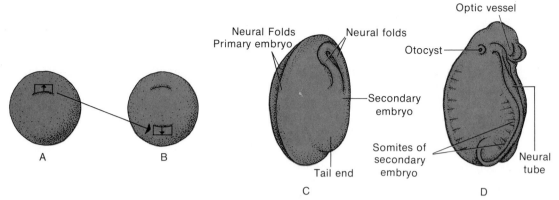

Figure 27-33 The induction of a second frog embryo by the implantation of a dorsal lip of a blastopore from embryo *A* onto the belly region of embryo *B*. Embryo *B* then develops through stage *C* to a double embryo *D*.

toderm even if it is transplanted to the belly region where the cells would normally form belly epidermis. Such experiments indicate that development involves a coordinated series of stimuli and responses, each step determining the succeeding one. The term organizer is applied to the region of the embryo with this property and also to the chemical substance released by that region which passes to the adjoining tissue and directs its development.

It had been widely accepted that organizers can transmit their inductive stimuli only when in direct contact with the reactive cells, but Niu and Twitty showed that induction may be mediated by diffusible substances that can operate without direct physical contact of the two tissues. Niu and Twitty grew small clusters of frog ectoderm, mesoderm and endoderm cells in tissue culture and found that ectoderm alone would never differentiate into nerve tissue. Ectoderm cells placed in a medium in which mesoderm cells had been grown for the previous week did differentiate into chromatophores and nerve fibers. No comparable differentiation occurred when ectoderm cells were placed in cultures that had contained endoderm. It appears that inductive tissues such as notochord and mesoderm cells contain and release diffusible substances, nucleoproteins, that can operate at a distance and induce the differentiation of ectoderm.

Morphogenetic Substances. Further evidence of diffusible morphogenetic substances is provided by Grobstein's experimental analy-sis of the differentiation of kidney tubules. Nephrogenic mesenchyme is normally induced to form renal tubules by the tip of the ureter (p. 632). If the two rudiments are separated by treatment with trypsin and the dissociated cells are grown in tissue culture, they will reaggregate and form tubules. Certain other tissues, the ventral portion of the spinal cord, for example, also proved to be efficient inducers of renal tubules. Grobstein then separated the inducing and reacting tissues by cellophane membranes of varying thicknesses and porosities. The inducing substance can pass through membranes up to 60 micrometers thick and with pores 0.4 micrometer in diameter. Electron micrographs revealed that induction occurred even when there was no cellular contact through the fine pores; the inducing principle is diffusible. It appears to be a large molecule and is at least in part protein, for it is inactivated by trypsin.

Other experiments of Grobstein have re-emphasized that extrinsic factors as well as nuclear factors may play a role in the process of differentiation. For example, embryonic pancreatic epithelium will continue to differentiate in organ culture only in the presence of mesenchyme cells. This requirement can be met not only by pancreatic mesenchyme from the mouse but also by mesenchyme from a variety of other sources, even embryonic chick mesenchyme. The mesenchyme can be replaced by a chick embryo juice and the active principle of the embryo juice appears to be a protein, for it is inactivated by trypsin but not by ribonuclease or deoxyribonuclease. It can be sedimented by high speed centrifugation and appears to be a large protein. This factor is only weakly effective in causing the differentiation of salivary gland epithelium in culture and is ineffective in inducing the formation of kidney tubules or of cartilage from their respective mesenchymes. Grobstein's experiments indicate that there is a spectrum of protein factors, each of which is more or less specific for the differentiation of one kind of cell.

The possible morphogenetic role of steroids has been explored by Dorothy Price of the University of Chicago. When the entire reproductive tract of a fetal male rat is dissected out and explanted, it grows normally in culture if the testes are present and in their normal spatial relationship to the tract. If both testes are removed from the explanted reproductive tract, its development is inhibited and the accessory organs, the vas deferens, seminal vesicles and prostate, do not differentiate. If the removed testes are replaced by pellets of testosterone implanted in their place, development of the explanted tract proceeds normally (Fig. 27–34). Thus testosterone can diffuse, at least over short distances, and induce the development of the male reproductive tract.

Suspensions of dissociated individual healthy cells can be prepared by treating a tissue briefly with a dilute solution of trypsin. When the suspension of cells is placed in tissue culture medium the cells may reaggregate and continue to differentiate in conformity with their previous pattern. The cells in tissue culture reaggregate not in a chaotic mass but in an ordered fashion, forming recognizable morphogenetic units. The cells appear to have specific affinities, for epidermal cells join with each other to form a sheet, disaggregated kidney cells join to form kidney tubules, and so on. The question of how one cell recognizes another cell and joins with it to form a tubule, sheet or other structural unit is indeed a fascinating one.

Disaggregated cells growing in culture can be subjected to a variety of experimental procedures to analyze the factors controlling their differentiation. Dissociated epidermal cells grown in the usual culture medium will reaggregate and differentiate as a stratified squamous epithelium. But if vitamin A is added to the culture medium or if the dissociated cells are exposed to vitamin A for as short a time as 15 minutes, the cells reaggregate and differentiate into a columnar epithelium with mucussecreting goblet cells. Vitamin A under these circumstances has a strong morphogenetic effect.

Embryonic tissues growing *in vivo* have differential sensitivities to changes in nutrients, to the presence of inhibitors and antimetabolites and to various environmental agents. Any of these factors, applied during the

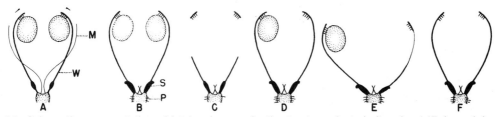

Figure 27–34 Schematic representation of fetal male reproductive tracts explanted after about 17 days of development. *A,* The tract at 17 days, at the time of explantation. *B–F,* Tracts cultured for 4 days, and hence at 21 days of development. *B,* Tract cultured with both testes present; the seminal vesicles and prostate glands developed. *C,* Tract cultured without testes showed regression of the wolffian ducts; no seminal vesicles and only a few prostate buds developed. *D,* Tract cultured with one testis in place developed normally. *E,* Tract cultured with one testis but spread apart showed some regression of the wolffian duct of the side without the testis, and the seminal vesicle on that side was smaller or absent. *F,* Tracts cultured with no testes but with testosterone micropellets present developed with normal wolffian ducts, seminal vesicles and prostatic buds. Müllerian duct *(M);* wolffian duct *(W);* seminal vesicles *(S);* prostate gland *(P).* (Reproduced by permission from Price and Pannabecker. *In* Wolstenholme, G. E. W., and Millar, S. C. P. (eds.): *Ciba Foundation Colloquia on Aging in Transient Tissues.* London, J. and A. Churchill, 1956.)

appropriate critical period in development, may change the course of development and differentiation and mimic the phenotype of a mutant gene, producing what is termed a *phenocopy.*

A striking demonstration that the same set of genes operating in dissimilar environments may have different morphological effects was provided by experiments with three races of frogs found in Florida, Pennsylvania and Vermont. Each of these races normally develops at a speed that is adapted to the length of the usual spring and summer season in its locale. Southern frogs develop slowly and northern frogs develop more rapidly. When northern frogs are raised under southern environmental conditions their development is overaccelerated, whereas when southern frogs are raised under northern conditions their development is over-retarded.

When an egg is fertilized with a sperm from a different race and the original egg nucleus is removed before the sperm nucleus can unite with it, it is possible to establish a cell with "northern" genes operating in "southern" cytoplasm or the reverse. Northern genes in southern cytoplasm resulted in poorly regulated development; the animal's head grew more rapidly than the posterior region and became disproportionately large. Southern genes introduced into northern cytoplasm led again to poorly regulated development but the head rather than the posterior region was retarded and disproportionately small. Genes from the Pennsylvania race of frog acted as "northern" with Florida cytoplasm but as "southern" with Vermont cytoplasm. The same set of genes had diverse morphological effects when they operated in different cytoplasmic environments.

Cellular differentiation may involve the differential activation of specific genes in different tissues; it may involve mechanisms affecting protein synthesis but operating at the ribosomal level or operating at the cell surface where the transport of substances into or out of the cells is regulated.

Differentiation may be controlled at least in part by influences originating outside the cell—by "organizers" from neighboring cells in early differentiation, by the mesenchymal proteins studied by Grobstein, or by hormones from distant cells. Such systemic influences participate in the integration of the differentiation of individual cells into the larger pattern of differentiation of the tissues of the whole organism. Eventually it should be possible for developmental biologists to bridge the gap between studies of development at the level of the whole organism and studies at the molecular level and trace in detail the sequence of events from the initial action of the gene to the final expression of the phenotype.

Our descriptive knowledge of the phenomena of fertilization, cleavage, gastrulation, morphogenesis and organogenesis is quite extensive, but our understanding of the fundamental molecular mechanisms involved in each of these processes is indeed rudimentary. This is a fertile field for future investigation.

QUESTIONS

1. Describe the basic differences between isolecithal and telolecithal eggs.
2. What parts of the adult body develop from the ectoderm? the mesoderm? the endoderm?
3. What is the primitive streak? What happens to it during development?
4. What becomes of the notochord in the later development of the vertebrates?
5. How is the spinal cord formed?
6. What becomes of the gill pouches in the human embryo?
7. What are "organizers"? How may they be involved in the control of development?
8. Distinguish clearly between fraternal and identical twins. What are Siamese twins?
9. What aberrations in development lead to the birth of a "blue baby"?
10. Describe the sequence of events from fertilization to the completion of neurulation in the frog.
11. Trace the development of the kidney in the human embryo.

SUPPLEMENTARY READING

Excellent general texts on embryology that include the details of organogenesis in each of the vertebrate types plus a synthesis of the experimental analysis of cellular differentiation are *An Introduction to Embryology* by B. I. Balinsky and *Modern Embryology* by C. W. Bodemer. A classic text, which describes the development of humans and other mammals, is *Developmental Anatomy* by L. B. Arey. A popularly written book, with superb illustrations, is *From Conception to Birth: The Drama of*

Life's Beginnings by Roberts Rugh and Landrum B. Shettles. Eugene Bell's *Molecular and Cellular Aspects of Development* presents reprints of many original papers dealing with the various facets of experimental embryology. A useful reference work for students interested in all aspects of human development is Frank Faulkner's *Human Development. Current Topics in Developmental Biology,* edited by A. Monroy and A. A. Moscona, is a series of volumes containing summaries of research in developmental biology.

There are a number of excellent books that discuss the biological aspects of the phenomenon of aging. These include A. Comfort's *Aging: The Biology of Senescence,* B. Strehler's *Time, Cells and Aging, Human Aging and Behavior* edited by G. A. Talland, L. Hayflick's *Human Cells and Aging,* and *Theoretical Aspects of Aging* edited by M. Rockstein, M. Sussman and J. Chesley. An excellent source book containing a collection of papers by many investigators in the field is *Organogenesis* by R. L. DeHaan and H. Ursprung.

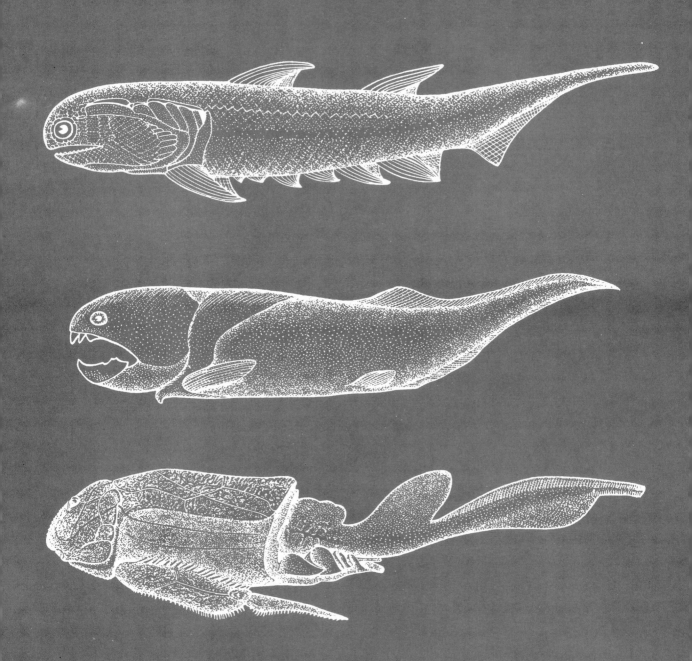

PART 7

HEREDITY AND EVOLUTION

INTRODUCTION

Within the brief span of a decade the foundations of our present concepts of heredity and evolution were laid. Charles Darwin's *The Origin of Species* (1859) presented evidence and argument supporting the concept of organic evolution and his theory of natural selection. Gregor Mendel discovered the basic laws of heredity in 1866, and Friedrich Miescher discovered nucleic acids in 1869. Although Darwin's views were widely discussed and accepted by most biologists within a few years, Mendel's and Miescher's contributions were unappreciated. Our present understanding of heredity really began in 1900 with the rediscovery of Mendel's laws and the subsequent development of the complex science of genetics. From his experiments on inheritance in peas Mendel made two key inferences: (1) The units of heredity, the genes, exist in pairs in individuals, but gametes have only one of each kind of gene. (2) During the formation of eggs and sperm each pair of genes separates independently of the members of other pairs of genes; the members of the pairs of genes are assorted at random in the gametes. The truth of these inferences has been demonstrated repeatedly.

Interest in the nucleic acids lagged until nearly 1950, when chemical and crystallographic data provided the basis for James Watson and Francis Crick's suggestion that the DNA molecule is a double helix. This was followed by Crick's deduction (1961) that three adjacent nucleotides in a DNA chain compose a triplet code, specifying a particular amino acid, and by Marshall Nirenberg's "cracking" of the genetic code. Evidence accumulated since then indicates that the genetic code is universal, that the same nucleotide triplet specifies a given amino acid in both bacteria and human beings.

Genetics and evolution have continued to develop together, and our increasing understanding of the genetics of populations and of differential reproduction has modernized Darwin's concept of natural selection. Molecular biologists, comparing the details of the molecular structure of complex proteins in different species, are rediscovering and revalidating the evolutionary relationships postulated decades ago on the basis of gross morphologic similarities and differences.

There is no reasonable alternative to the principle that all the species of plants and animals now living have arisen from pre-existing species by descent with modification, and essentially all biologists subscribe to this proposition. The patterns of basic macroscopic, microscopic and molecular resemblances among living things provide strong support for the theory of evolution. During the past century there has been an explosive growth of information relating to biology, stimulated in large part by the theory of evolution. All of this information is consistent with, and sustains further, the general concept of evolution and specific evolutionary relationships.

CHAPTER 28

TRANSMISSION GENETICS: THE CHROMOSOME THEORY OF HEREDITY

The essence of the reproductive process is the production of a new generation of offspring that resembles the parental generation, a process which clearly involves the transfer of biological information to the new organism via the egg and sperm. Human beings have been aware for many centuries that "like begets like" and that one of the prime characteristics of living things is their ability to reproduce their kind. This tendency of individuals to resemble their progenitors is called *heredity.* Resemblances between parents and offspring are close but are usually not exact. The offspring of a particular set of parents differ from each other and from their parents in many respects and to different degrees. Such differences, termed *variations,* are also characteristic of living things. Some variations are inherited, caused by the segregation of hereditary factors among the offspring. Other variations are not inherited but are due to the effects of temperature, moisture, food, light or other environmental factors on the development of the organism. The expression of inherited characteristics may be strongly influenced by the environment in which the individual develops. The branch of biology concerned with these phenomena of heredity and variation and the study of the laws governing similarities and differences among individuals related by descent is called *genetics.* Since its inception at the beginning of the present century, the science of genetics has advanced rapidly and is developing at an accelerated pace at present.

28-1 THE DEVELOPMENT OF GENETICS

In the eighteenth and nineteenth centuries several attempts were made to discover how specific characteristics are transmitted from one generation to the next. An important discovery was made in 1760 by the German botanist Kölreuter when he crossed two species of tobacco by placing pollen from one species on the stigmas of the other. The plants grown from the resulting seeds had characteristics intermediate between those of the two parents. Kölreuter made the logical inference that parental characteristics are transmitted through both the pollen (sperm) and the ovule (egg). However, he and his contemporary plant and animal breeders were unable to discover the nature of the hereditary mechanism, in part because the cytologic basis was unknown, but primarily because they attempted to study the inheritance of *all* the characteristics of the plant or animal at one time.

Gregor Mendel, an Austrian abbot who bred pea plants in the garden of his monastery at Brno, succeeded in discovering the basic laws of genetics where previous hydridizers had failed. He studied the inheritance of single contrasting characteristics; he counted and recorded the parents and offspring of each of his crosses. His knowledge of the principles of mathematics enabled him to interpret his data and led him to the hypothesis that each trait is determined by two genetic factors.

Mendel had several types of pea plants in his garden and kept records of the inheritance of seven clearly contrasting pairs of traits, such as yellow versus green seeds, round versus wrinkled seeds, green versus yellow pods, and so on. By crossbreeding and counting the types of offspring, Mendel was able to detect regularities in the pattern of inheritance that had escaped earlier breeders. Whenever he crossed plants with two different characteristics, such as yellow and green seeds, the plants in the next generation, the F_1 *generation*, were like one of the two parents. The second, or F_2 *generation*, included individuals of both parental types. When he counted these, he found that the two types of individuals were present in the F_2 generation in a ratio of approximately 3:1 (Table 28-1). For example, (Fig. 28-1) when he crossed tall plants with short plants, all the members of the F_1 generation were tall. When two of these first generation tall plants were crossed, the F_2 generation included some tall and some short plants—787 tall and 277 short. Clearly, in the first generation the genetic factor (gene) for shortness was hidden or overcome by the gene for tallness. Mendel termed the gene for tallness "dominant" and the gene for shortness "recessive."

Having discovered that the crossing of two

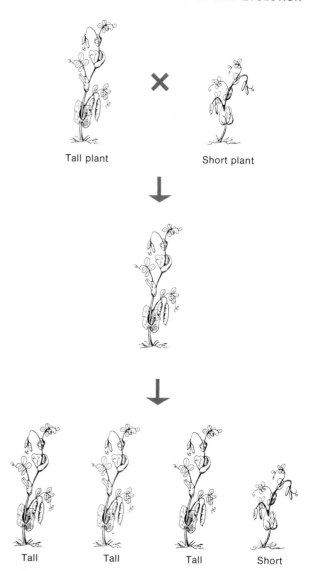

Figure 28-1 A diagram illustrating one of the crosses carried out by Gregor Mendel. Crossing a tall pea plant with a short pea plant yielded offspring all of which were tall. However, when these offspring were self-pollinated, the next generation included tall and short plants in a ratio of about 3:1.

first generation plants led to offspring in the second generation in a ratio of three with the dominant characteristic to one with the recessive characteristic, it occurred to Mendel that each plant must have two genetic factors, whereas each egg and sperm has only one. The first generation tall plants also had two genetic factors—one for tallness and one for shortness—but the tall gene was dominant and these plants were tall. However, when these F_1 plants formed eggs or sperm, the gene for tallness separated from the gene for shortness so that half of the eggs and half of the sperm contained a "tall" gene and half a "short" gene. (The genes are not tall or short, but cause the plants to grow to different heights.) The random fertilization of eggs by sperm led to four possible combinations of genes: one with two talls, *TT;* one with two shorts, *tt;* and two with one tall and one short, *Tt* and *tT.* The tall gene *(T)* is dominant to short *(t),* and, therefore three of the four kinds of offspring were tall plants and only one was short. It is now the convention to use capital letters for dominant genes and lower case letters for recessive genes, e.g., *T* for the gene for tall plants, and *t* for the gene for short plants.

Mendel's mathematical abilities enabled him to recognize that a 3:1 ratio would be expected among the offspring if each plant had two factors for any given characteristic rather than a single one. This brilliant piece of reasoning was supported when chromosomes were seen and the details of mitosis, meiosis, and fertilization became known.

Mendel reported his findings at a meeting of the Brno Society for the Study of Natural Science and published his results in the transactions of that society. The importance of his findings was not appreciated by the other biologists of the time, and they were neglected for nearly 35 years.

In 1900, Hugo DeVries in Holland, Karl Correns in Germany and Erich von Tschermak in Austria independently rediscovered the laws of inheritance that had been described by Mendel. On finding Mendel's paper in which these laws had been clearly stated 35 years before, they gave him credit for his discoveries by naming two of the fundamental laws of inheritance after him.

In the first decade of the twentieth century experiments with a wide variety of plants and animals, together with observations of human inheritance, showed that these same basic principles govern inheritance in all of these different organisms. W. S. Sutton in the United States and Theodore Boveri in Germany showed that the genes that Mendel described are located in the chromosomes within the nucleus. Some investigators studied inheritance in mice, rabbits, cattle and chickens, but the favorite subject for genetic studies became the fruit fly, *Drosophila.* These small insects have a short life cycle of 10 to 14 days, are eas-

ily raised in the laboratory, and have only four pairs of chromosomes. In certain of their tissues the chromosomes become very large and the details of their structure can be studied with the microscope. Hundreds of inherited variations involving eye colors, wing shapes and bristle patterns were detected and studied. It was eventually possible to map the specific location of each gene on a specific chromosome. T. H. Morgan and his associates carried out extensive experiments that revealed the genetic basis for the determination of sex and provided the explanation for certain unusual patterns of inheritance in which a trait is linked with the sex of the individual, the so-called sex-linked traits.

A further great advance came in 1927, when H. J. Muller showed that genes could be changed, i.e., could undergo mutations, when fruit flies and other organisms were exposed to x-rays. This provided many new mutant genes with which to study heredity. The nature of the mutations gave clues as to the nature and structure of the genes themselves. These studies were followed in the 1940's by experiments to examine the relationship of genes and enzymes. Investigators turned to the bread mold, *Neurospora,* in which a number of biochemical mutants, which lacked some specific enzyme, could be produced artificially and studied. In the past two decades the most popular organisms for genetic studies have been the intestinal bacterium *Escherichia coli* and some of the bacterial viruses or bacteriophages that infect this bacterium. Since the beginning of this century there has been a sustained interest in determining the inheritance of specific human traits and in determining the inheritance of desirable and undesirable traits in domestic animals and plants. Armed with the growing knowledge of genetic principles, geneticists have been able to breed almost to order cattle that can survive in hot climates, cows that produce large amounts of milk with a high content of butterfat, chickens that lay large eggs with thin shells, corn and wheat plants that are highly resistant to specific diseases, and so on.

28–2 CHROMOSOMES AND GENES

When a dividing cell is examined under the phase microscope or is fixed, stained, and examined under a regular microscope, elongate dark-staining bodies called *chromosomes* are visible within the nucleus (Fig. 28-2). Each chromosome consists of a central thread, a *chromonema,* along which lie a series of bead-like structures, the *chromomeres.* Each chromosome has, at a fixed point along its length, a small, clear, circular zone called a *centromere* that controls the movement of the chromosome during cell division. As the chromosome becomes shorter and thicker just before cell division occurs, the centromere region becomes accentuated and appears as a constriction. Chromosomes are distinctly visible by light microscopy only at the time of cell division. At other times they appear as long, thin, fine, dark-staining threads called *chromatin.* Even though in most organisms they are not visible by light microscopy, chromosomes are present as highly extended but distinct structural and functional entities between successive cell divisions.

Each cell of every organism of a given species contains a characteristic number of chromosomes. Each cell in the body of every human being has exactly 46 chromosomes (Fig. 28-3); many other species of animals and plants also have 46. It is not the *number* of chromosomes that differentiates the various species of animals but rather the nature of the hereditary factors in the chromosomes. A certain species of roundworm has only two chromosomes in each cell and some crabs have as many as 200 per cell. The highest chromosome number reported so far is about 1600, found in a radiolarian, a marine protist. Most species of animals and plants have chromosome numbers between 10 and 50. Numbers above and below this are comparatively rare.

Chromosomes always exist in pairs; there are invariably two of each kind in the somatic cells of higher plants and animals. Thus the 46 chromosomes in human cells consist of two of each of 23 different kinds. They differ in length and shape and in the presence of knobs or constrictions along their length. In most species,

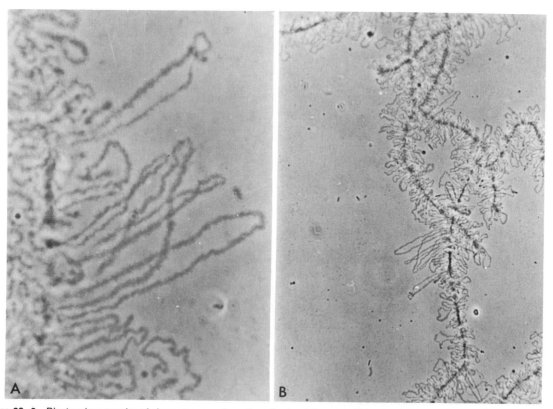

Figure 28-2 Photomicrographs of chromosomes from the oöcyte of the newt *Triturus viridescens* showing *(A)* the chromomeres (dark dots along the central thread) and *(B)* the loops radiating from the central thread. The presence of these loops gave the structures their name, "lampbrush" chromosomes. *A,* ×1600; *B,* ×1100. (Courtesy of Dr. Dennis Gould.)

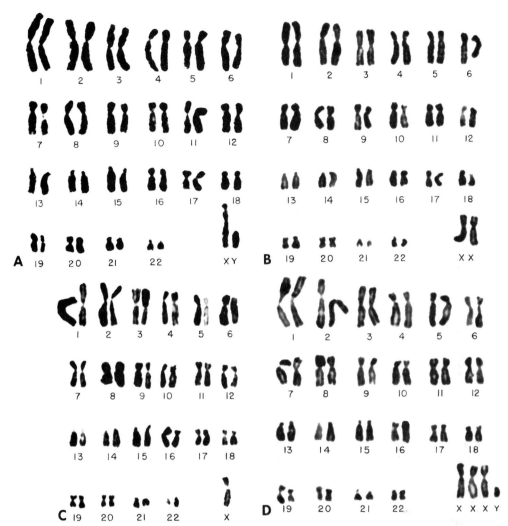

Figure 28-3 Human chromosomes. *A*, Normal male. *B*, Normal female. *C*, XO condition—gonadal dysgenesis. *D*, XXXY— an unusual example of Klinefelter's syndrome; the typical individual with Klinefelter's syndrome has an XXY pattern of chromosomes. (Courtesy of Dr. Melvin Grumbach.)

the chromosomes vary enough in these morphologic features so that cytologists can distinguish the different pairs.

28-3 GENES AND ALLELES

The laws of heredity follow directly from the behavior of the chromosomes in mitosis, meiosis and fertilization. Within each chromosome are many hereditary factors, or **genes,** each different from the others and each controlling the inheritance of one or more characteristics. The great regularity of the mitotic process (p. 57) ensures that each daughter cell will have two of each kind of chromosome and therefore two of each kind of gene. As the

chromosomes separate during meiosis (p. 589) and recombine in fertilization (p. 594), so, of course, must the paired genes separate and recombine. All of the phenomena of Mendelian genetics depend on these simple facts. Each chromosome behaves genetically as though it were composed of a string of genes arranged in a linear order. The members of a homologous pair of chromosomes have genes arranged in similar order. The gene for each trait occurs at a particular point in the chromosome called a *locus.* When the chromosomes undergo synapsis during meiosis, the homologous chromosomes become attached point by point and presumably gene by gene.

The inheritance of any trait can be studied only when there are two contrasting conditions, such as Mendel's yellow and green peas, brown versus blue eye color in humans, or brown versus black coat color in guinea pigs. An individual may have one or the other but not both of such contrasting conditions, originally termed allelomorphic traits, or *alleles.* More recently the terms gene and allele have been used interchangeably. The gene *B* for brown eyes is said to be an allele of gene *b* for blue eyes. This usage of the term allele emphasizes that there are two or more alternative kinds of genes located at a specific point or locus in the chromosome.

There may be a simple 1:1 relationship between a given gene and the trait that it controls, or one gene may participate in the control of several or many traits throughout the body, or many genes may cooperate to regulate the appearance of a single trait. As you will learn in the next chapter, each gene is a molecule of DNA in which biological information is stored as a triplet code in the sequence of nucleotides that compose the double helix of the DNA molecule. The information in each gene is "read out" and a specific protein is synthesized. The presence of the specific protein, an enzyme for example, provides the chemical basis for the trait.

28–4 A MONOHYBRID CROSS

The usage of genetic terms and some of the basic principles of genetics can be illustrated by considering a simple monohybrid cross, that is, a cross between two individuals that differ in a single pair of genes. The mating of a "pure" brown male guinea pig with a "pure" black female guinea pig is illustrated in Figure 28–4. During meiosis in the testes of the male the two *bb* genes separate so that each sperm has only one *b* gene. In the formation of ova in the female the *BB* genes separate so that each ovum has only one *B* gene. The fertilization of this egg by a *b* sperm results in an animal with the genetic formula *Bb*. These guinea pigs contain one gene for brown coat and one for black coat. What color would you expect them to be—dark brown, gray or perhaps spotted? In this instance they are just as black as the mother. The gene for black coat color is said to be *dominant* to the gene for

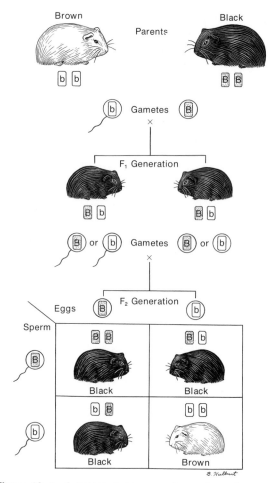

Figure 28–4 A monohybrid cross, the mating of one of a true-breeding strain of brown guinea pigs with a true-breeding strain of black guinea pigs. The F₁ generation includes only black individuals. The mating of two of these yields offspring in the ratio of 3 black:1 brown.

brown coat color. It will produce black coat color even when only one dose of the black gene is present *(Bb).* The brown gene is said to be *recessive* to the black one; it will produce brown coat color only when present in double dose *(bb).* The phenomenon of dominance supplies part of the explanation as to why an individual may resemble one of his parents more than the other despite the fact that both make equal contributions to his genetic constitution. In one species of animal black coat color may be dominant to brown, while in another species brown might be dominant to black. The particular genetic relations that obtain in any given species must be determined by experiment.

28-5 HOMOZYGOUS AND HETEROZYGOUS ORGANISMS

An animal or plant with two genes exactly alike, two black *(BB)* or two brown *(bb)*, is said to be *homozygous* for the trait. An organism with one dominant and one recessive gene *(Bb)* is said to be "hybrid," or *heterozygous.* Using these terms we can now formulate better definitions for dominant and recessive genes: A recessive gene is one that will produce its effect only when homozygous; a dominant gene is one that will produce its effect whether it is homozygous or heterozygous.

During meiosis in the gonads of the heterozygous black guinea pigs, the chromosome containing the *B* gene first synapses with, and then separates from, the chromosome containing the *b* gene, so that each sperm or egg has either a *B* gene or a *b* gene but never both. It follows that sperm (or eggs) containing *B* genes and those with *b* genes are formed in equal numbers by heterozygous *Bb* individuals. Since there are two types of eggs and two types of sperm, four combinations are possible in fertilization. There is no special attraction or repulsion between an egg and sperm containing the same kind of gene; hence these four possible combinations are equally probable. The combinations of eggs and sperm can be determined by algebraic multiplication:

$$(\tfrac{1}{2}\,B + \tfrac{1}{2}\,b) \text{ eggs} \times (\tfrac{1}{2}\,B + \tfrac{1}{2}\,b) \text{ sperm.}$$

The possible combinations of eggs and sperm may also be represented in a "checkerboard" or Punnett square (Fig. 28-4). The types of eggs are represented along the top, the types of sperm are indicated along the left side, and the squares are filled in with the resulting zygote combinations. Three fourths of all offspring will be *BB* or *Bb* and have black coat color, and one fourth will be *bb* and have brown coat color. The genetic mechanism responsible for the 3:1 ratios obtained by Mendel in his pea breeding experiments (Table 28-1) is now evident. The generation with which a particular experiment is begun is called the P_1 or *parental generation.* Offspring of this generation are called the F_1 or *first filial generation.* Those resulting when two F_1 individuals are bred constitute the F_2 or *second filial generation,* the grandchildren. Those resulting from the mating of two F_2 individuals make up the F_3 generation and so on.

28-6 PHENOTYPE AND GENOTYPE

The appearance of an individual with respect to a certain inherited trait is known as its *phenotype.* The organism's genetic constitution, usually expressed in symbols, is called its *genotype.* In the cross we have been considering, the phenotypic ratio of the F_2 generation is 3 black-coated guinea pigs: 1 brown-coated guinea pig and the genotypic ratio is 1 *BB*:2 *Bb*:1 *bb.* The phenotype may be a morphologic characteristic—shape, size, color—or a physiologic characteristic such as the presence or absence of a specific enzyme required for the metabolism of a specific substrate. Guinea pigs with the genotypes *BB* and *Bb* are alike phenotypically; they both have black coats. One third of the black guinea pigs in the F_2 generation of the mating of black × brown are homozygous, *BB*, and the other two thirds are heterozygous, *Bb.* Since animals cannot, in general, be self-fertilized, how do you think the geneticist can distinguish the homozygous *(BB)* and heterozygous *(Bb)* black-coated guinea pigs? He does this by a *test cross,* by mating each black guinea pig with a homozygous brown *(bb)* guinea pig (Fig. 28-5). If all of the offspring are black, what inference would you make about the genotype of the black parent? If any of the offspring are brown, what conclusion would you draw regarding the genotype of the black parent? Can you be more certain about one of these two inferences than the other?

Mendel did just such experiments and bred heterozygous tall pea plants with homozygous short ones. He predicted that the heterozygous parent would produce equal numbers of *T* and *t* gametes, whereas the homozygous short parent would produce only gametes containing *t,* and that this should lead to equal numbers of tall *(Tt)* and short *(tt)* individuals among the progeny. Thus, as a good hypothesis should, Mendel's hypothesis not only explained the known facts but enabled him to predict the results of other experiments.

This sort of testing is of great importance in the commercial breeding of animals or plants when the breeder is trying to establish a strain that will breed true for a certain charac-

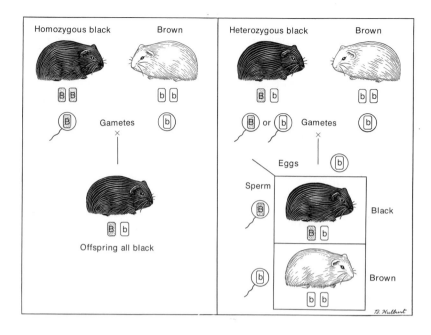

Figure 28-5 Diagram of test crosses, whereby a heterozygous black guinea pig may be distinguished from a homozygous black one. If a homozygous black is mated with a brown one *(left)*, all its offspring are black. If a heterozygous black is mated with a brown one *(right)*, half the offspring are black, and half brown.

teristic. When the individuals in a breeding stock are selected on the basis of their own phenotypes, the breeding program is not maximally effective because it does not differentiate between homozygous and heterozygous individuals. The method of *progeny selection* is one in which a breeder tests the genotypes of his breeding stock by making test matings and observing the offspring. If the offspring are superior with respect to the desired trait, the parents are thereafter used regularly for breeding. Two bulls, for example, may look equally healthy and vigorous, yet one will have daughters with qualities of milk production that are distinctly superior to those of the daughters of the other bull.

28-7 CALCULATING THE PROBABILITY OF GENETIC EVENTS

All genetic ratios are properly expressed in terms of probabilities. In the examples just dis-

cussed, we stated that the offspring of the mating of two individuals heterozygous for the same gene pair would appear in the ratio of three with the dominant trait and one with the recessive trait. If the number of offspring is large enough this ratio will be very closely approximated, as Mendel's experiments demonstrated (Table 28-1). However, if the number of offspring is small, the ratio of the two types may be quite different from the expected 3:1. Why should this be? If there are only four offspring, any distribution from all four with the dominant trait to all four with the recessive trait might be found, although the latter would occur only very rarely. A better statement is that there are three chances in four ($3/4$) that any particular offspring of two heterozygous individuals will show the dominant trait and one chance in four ($1/4$) that it will show the recessive trait.

The Product Law. One of the basic laws of probability, the *Product Law*, states that the

Table 28-1   AN ABSTRACT OF THE DATA OBTAINED BY MENDEL FROM HIS BREEDING EXPERIMENTS WITH GARDEN PEAS

Parental Characters	First Generation	Second Generation	Ratios
Yellow seeds × green seeds	all yellow	6022 yellow : 2001 green	3.01 : 1
Round seeds × wrinkled seeds	all round	5474 round : 1850 wrinkled	2.96 : 1
Green pods × yellow pods	all green	428 green : 152 yellow	2.82 : 1
Long stems × short stems	all long	787 long : 277 short	2.84 : 1
Axial flowers × terminal flowers	all axial	651 axial : 207 terminal	3.14 : 1
Inflated pods × constricted pods	all inflated	882 inflated : 299 constricted	2.95 : 1
Red flowers × white flowers	all red	705 red : 224 white	3.15 : 1

probability of two independent events occurring together is the product of the probabilities of each occurring separately. Using this principle we can calculate how frequently in a large series of matings, if each mating results in exactly four offspring, all four would show the recessive trait. How often in the mating of two heterozygous black guinea pigs, **Bb,** would one expect to get a litter of four brown guinea pigs? The probability that the first one will be brown is $\frac{1}{4}$. The probability that the second one will be brown is also $\frac{1}{4}$. The fertilization of each egg by a sperm is an independent event and we use the Product Law to calculate the combined probability of two (or more) events occurring together. Thus the probability that all four of the offspring of any given mating will be brown is $\frac{1}{4} \times \frac{1}{4} \times \frac{1}{4} \times \frac{1}{4}$, or $\frac{1}{256}$. In other words, there is one chance in 256 that all four guinea pigs will have brown coat color!

The probability that any given offspring will show the dominant trait, black coat color, is $\frac{3}{4}$. We can, in a similar fashion, calculate the probability that all four guinea pigs in the litter will be black. This is $\frac{3}{4} \times \frac{3}{4} \times \frac{3}{4} \times \frac{3}{4}$, or 81 chances in 256 that all four guinea pigs will have black coat color. You must not assume that if you actually mated 256 pairs of heterozygous guinea pigs and each had four offspring that you would be *guaranteed* of getting one litter of four brown-coated offspring. A great many people have lost vast sums of money in gambling by making a similar mistake! If you have made 255 such matings without getting a set of four brown offspring, what is the probability of getting four brown guinea pigs on the 256th try? You might be misled into thinking it is bound to happen, but in fact there is still only one chance in 256 that it will occur, since each of these matings is an independent event.

How many different ways are there of getting three black and one brown guinea pig in a given mating? A look at Figure 28–6 shows that there are four ways: The first to be born could be brown and the next three would be black. Or, the second one born could be brown and the first, third and fourth black. The other two possibilities are that the brown one would be the third or the fourth to be born. To calculate the probability that three of the four offspring will be black and one brown we must multiply the number of possible combinations by the probability of each type. There are $4 \times \frac{3}{4} \times \frac{3}{4} \times \frac{3}{4} \times \frac{1}{4}$, or $\frac{108}{256}$ chances that there will be three black guinea pigs and one brown guinea pig in a litter of four. It may be surprising at first that the 3 : 1 ratio is actually obtained in less than half of the total number of litters of four. However, when we add up the total numbers of black and brown offspring from a large number of matings, the 3 : 1 ratio is more and more closely approximated.

How many different ways are there of getting two black and two brown guinea pigs in a litter of four? There are six ways, as shown in Figure 28–6. From this you can calculate that the probability that two of the four offspring will be black and two brown is $6 \times \frac{3}{4} \times \frac{3}{4} \times \frac{1}{4} \times \frac{1}{4}$, or $\frac{54}{256}$. You can see that there are four different combinations that give one black and three brown. The first born may be black and the rest brown, or the second will be black and all others brown, or the third or the fourth will be black. The probability of getting one black guinea pig and three brown guinea pigs is $4 \times \frac{3}{4} \times \frac{1}{4} \times \frac{1}{4} \times \frac{1}{4}$, or $\frac{12}{256}$.

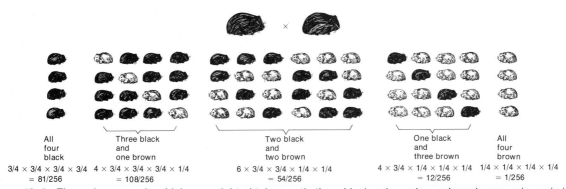

All four black	Three black and one brown	Two black and two brown	One black and three brown	All four brown
$3/4 \times 3/4 \times 3/4 \times 3/4$ $= 81/256$	$4 \times 3/4 \times 3/4 \times 3/4 \times 1/4$ $= 108/256$	$6 \times 3/4 \times 3/4 \times 1/4 \times 1/4$ $= 54/256$	$4 \times 3/4 \times 1/4 \times 1/4 \times 1/4$ $= 12/256$	$1/4 \times 1/4 \times 1/4 \times 1/4$ $= 1/256$

Figure 28–6 The various ways in which one might obtain exactly three black guinea pigs and one brown guinea pig in a litter of four.

All probabilities are expressed as fractions ranging from zero, expressing an impossibility, to one, expressing a certainty. Probabilities can be multiplied or added just like any other fractions. In this example there are no other possibilities; the four guinea pigs must be either four black, or three black and one brown, or two black and two brown, or one black and three brown, or four brown. The sum of the five probabilities will add up to one— $81/256 + 108/256 + 54/256 + 12/256 + 1/256 = 256/256 = 1$.

The color of the iris of the human eye is inherited by several pairs of genes, but one pair is the primary factor differentiating brown eye color from blue. The gene for brown eye color, *B*, is dominant to the gene for blue, *b*. If two heterozygous brown-eyed people marry, what is the probability that they will have a blue-eyed child? Clearly, there is one chance in four that any child of theirs will have blue eyes. Each mating is a separate, independent event and its result is not affected by the results of any previous matings. If these two brown-eyed parents have had three brown-eyed children and are expecting their fourth child, what is the probability that it will have blue eyes? Again, the unwary might guess that this one *must* have blue eyes, but in fact there is still only one chance in four of its having blue eyes and three chances in four that it will have brown eyes.

All genetic events are governed by the laws of probability. The prediction of any single event—e.g., predicting the characteristics of any single child—is highly uncertain; however, if the number of events is large enough, the laws of probability provide a reasonable prediction of the fraction of these events that will be of one type or the other.

28–8 INCOMPLETE DOMINANCE

From studies of the inheritance of many traits in a wide variety of organisms it is clear that one member of a pair of genes may not be completely dominant to the other. Indeed it may be improper to use the words dominant and recessive in such instances. For example, red and white are common flower colors in Japanese four-o'clocks. Each color breeds true when these plants are self-pollinated. What flower color might you expect in the offspring

of a cross between a red flowering plant and one that bears white flowers? Without knowing which is dominant you might predict that all would have red flowers or all would have white flowers. This cross was first made by the German botanist Karl Correns, who found that all the F_1 offspring have pink flowers! How can we explain that? Does this in any way prove that Mendel's assumptions about inheritance are wrong? Quite the contrary, for when two of these pink-flowered plants were crossed, offspring appeared in the ratio of one red-flowered to two pink-flowered to one white-flowered plant. In this instance, as in other aspects of science, the finding of results that differ from those predicted simply prompts the scientist to reexamine and modify his assumptions to account for the new exceptional results. The pink-flowered plants are clearly the heterozygous individuals and neither the red gene nor the white gene is completely dominant. When the heterozygote has a phenotype that is intermediate between those of its two parents, the genes are said to show the incomplete dominance, or to be *codominant.* In these crosses the genotypic and phenotypic ratios are identical.

Incomplete dominance is not unique to Japanese four-o'clocks. Red- and white-flowered sweet pea plants also produce pink-flowered plants when crossed. In both cattle and horses, reddish coat color is incompletely dominant to white coat color. The heterozygous individuals have roan-colored coats. If you saw a white mare nursing a roan-colored colt, what would you guess was the coat color of the colt's father? Is there more than one possible answer?

28–9 CARRIERS OF GENETIC DISEASES

Careful investigation of the phenotypes of many gene pairs has revealed slight differences between the homozygous dominant and the heterozygous individual. In the human many inherited diseases are transmitted by recessive genes, and it is important to be able to distinguish the homozygous normal individual from the heterozygous individual who may superficially appear to be normal but is a *carrier* for the trait. The mating of two carriers, two heterozygous individuals, would provide one chance in four for the appearance of a homozygous recessive individual showing the inherited disease. The human trait *sickle cell anemia* is inherited by a single pair of genes, but an individual must have two such genes to show the anemia. Thus the gene for sickle cell anemia may be said to be recessive to the normal gene. The red cells of a person with sickle cell anemia are shaped like a sickle or a half moon, whereas normal red cells are biconcave discs. The sickle cells contain hemoglobin molecules with a slightly different molecular structure from that found in normal red cells. The hemoglobin molecules in an individual with sickle cell anemia have the amino acid valine instead of glutamic acid at position 6, the sixth amino acid from the amino terminal end, in the β chain. The substitution of an amino acid with an uncharged side chain (valine) for one with a charged side chain (glutamate) makes the hemoglobin less soluble, and it tends to form crystals that break the red cell. The red cells of sickle cell anemics are more fragile than those of normal individuals. Individuals heterozygous for the sickling gene have a mixture of normal and abnormal hemoglobins in their red cells; about 45 per cent of the total hemoglobin has the abnormal chemical constitution. Their red cells do not usually undergo sickling but can be made to do so when the amount of oxygen in the blood is reduced. They are said to have sickle cell "trait." This provides a very simple test, using only a drop of blood, by which the heterozygote can be distinguished from the homozygous normal.

The human disease *phenylketonuria* is also inherited by a single pair of genes. The homozygous recessive individual lacks the enzyme needed to convert one amino acid, phenylalanine, to another, tyrosine. The phenylalanine accumulates in the tissues and causes retarded mental development. The heterozygote appears perfectly normal, but when given a standard amount of phenylalanine in his diet, he may accumulate more in his blood and excrete more in his urine. Unfortunately the results of such tests are not always clear-cut in distinguishing between homozygous normal and heterozygous carrier individuals; the values for the two groups overlap considerably. However, a very simple test, adding ferric chloride to the urine in the diapers of the newborn, distinguishes the homozygous phenylketonuric, who can then be kept on a diet containing a minimal amount of phenylalanine. This is effective in minimizing the amount of mental retardation as the infant develops.

28–10 DEDUCING GENOTYPES

The science of genetics resembles mathematics in that it consists of a few basic principles, which, once grasped, enable the student to solve a wide variety of problems. The genotypes of the parents can be deduced from the phenotypes of their offspring. In chickens, for example, the gene for rose comb *(R)* is dominant to the gene for single comb *(r)*. Suppose that a cock is mated to three different hens (Fig. 28–7). The cock and hens A and C have rose comb and hen B has a single comb. Breeding the cock with hen A produces a rose-combed chick, with hen B a single-combed chick, and with hen C a single-combed chick. What type of offspring can be expected from further matings of the cock with these hens?

Since the gene for single comb is recessive, all of the hens and chicks that are phenotypically single-combed must be homozygous *rr.* We can deduce then that hen B and the offspring of hens B and C are genotypically *rr.* All of those individuals that are phenotypically rose-combed must have at least one *R* gene, and the cock and hens A and C are therefore *R?.* The fact that the offspring of the cock and hen B was single-combed proves that the cock is heterozygous *Rr.* The single-combed chick received one *r* gene from its mother but must have received the other from its father. The fact that the offspring of the cock and hen C had a single comb proves that hen C also is heterozygous *Rr.* It is impossible to decide from the data given whether hen A is homozygous, *RR,* or heterozygous, *Rr.* Further breeding would be necessary to determine this. Additional matings of the cock with hen B would result in one half rose-combed and one half single-combed individuals; additional matings of the cock with hen C would produce three fourths rose-combed and one fourth single-combed chicks.

In working genetics problems it is well to use the following procedure to avoid errors: (1) Write down the symbols you are using for each

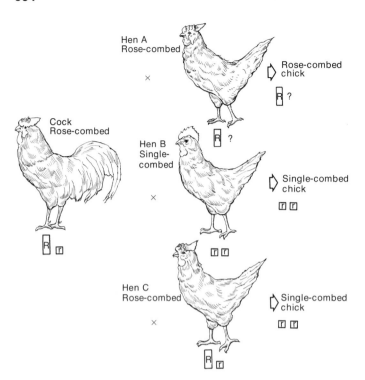

Figure 28-7 Deducing the parental genotypes from the phenotypes of the offspring. See text for discussion.

gene. (2) Determine the genotypes of the parents, deducing them from the phenotypes of the offspring if this is necessary. (3) Indicate the possible kinds of gametes formed by each of the parents. (4) Set up a "checkerboard," putting the possible types of sperm along its side and the possible types of eggs across its top. (5) Fill in the checkerboard and read off the genotypic and phenotypic ratios of the offspring.

28-11 MENDEL'S LAWS OF SEGREGATION AND INDEPENDENT ASSORTMENT

Frequently a geneticist must analyze the inheritance of two or more traits in the same group of individuals. A mating that involves individuals differing in two traits is called a *dihybrid cross.* The principles involved and the procedure of solving problems are exactly the same in monohybrid and in dihybrid or trihybrid crosses. In the latter the number of types of gametes is greater and the number of types of zygotes is correspondingly larger.

When two pairs of genes are located in different (nonhomologous) chromosomes, each pair is inherited independently of the other; that is, each pair separates during meiosis in-

dependently of the other. When a black, short-haired guinea pig (*BBSS*, since short hair is dominant to long hair) and a brown, long-haired guinea pig *(bbss)* are mated, the *BBSS* individual produces gametes that are all *BS*. The *bbss* guinea pig produces only *bs* gametes. Each gamete contains one and only one of each kind of gene. The union of *BS* gametes and *bs* gametes yields only individuals with the genotype *BbSs*. All of the offspring are heterozygous for hair color and for hair length and all are phenotypically black and short-haired.

When two of the F_1 individuals are mated, each produces four kinds of gametes in equal numbers—*BS, Bs, bS, bs;* thus 16 combinations are possible among the zygotes (Fig. 28-8). There are 9 chances in 16 of obtaining a black, short-haired individual; 3 chances in 16 of obtaining a black, long-haired individual; 3 chances in 16 of obtaining a brown, short-haired individual; and 1 chance in 16 of obtaining a brown long-haired individual.

Mendel's First Law, sometimes called the Law of Purity of Gametes or the Law of Segregation, is illustrated by the mating of the black and brown guinea pigs described previously. Mendel's First Law may be stated as follows: Genes exist in individuals in pairs, and in the formation of gametes each gene separates or segregates from the other member of

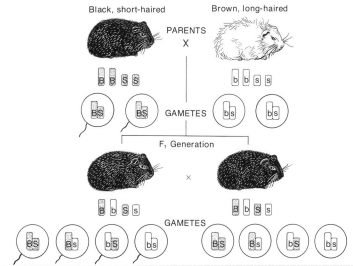

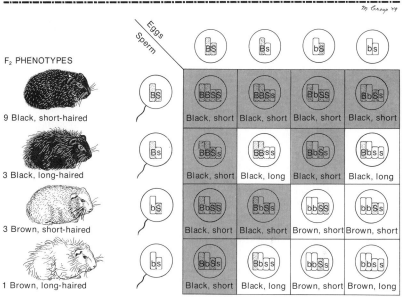

Figure 28-8 Diagram of a cross between a black, short-haired guinea pig and a brown, long-haired one, illustrating independent assortment.

the pair and passes into a different gamete, so that each gamete has one and only one of each kind of gene. Mendel's Second Law, the Law of Independent Segregation, is illustrated by this second mating. This law states that the members of one pair of genes separate or segregate from each other in meiosis independently of the members of other pairs of genes and come to be assorted at random in the resulting gamete. The segregation of the **B-b** genes is independent of the segregation of the **S-s** genes. This law does not apply if the two pairs of genes are located in the same pair of chromosomes.

In a similar fashion, problems involving three pairs of genes may be solved. An individual heterozygous for three pairs of genes lo-

cated in different pairs of chromosomes will yield eight types of gametes in equal numbers. The union of these eight types of eggs and eight types of sperm will yield 64 possible zygotes in the F₂ generation. In peas, as Mendel demonstrated, yellow seed color *(Y)* is dominant to green *(y)*, smooth seeds *(S)* are dominant to wrinkled *(s)*, and tall plants *(T)* are dominant to dwarf *(t)*. The mating of a homozygous yellow, smooth, tall plant *(YYSSTT)* with a homozygous green, wrinkled dwarf plant *(yysstt)* will produce offspring that are all yellow, smooth and tall *(YySsTt)*. When two of these F₁ plants are mated, F₂ offspring are produced in the ratio of 27 yellow, smooth, tall: 9 yellow, smooth, dwarf: 9 yellow, wrinkled, tall: 9 green, smooth, tall: 3 yellow, wrinkled, dwarf: 3 green, wrinkled, tall: 3 green, smooth, dwarf: 1 green, wrinkled, dwarf. Draw up a checkerboard to verify these numbers.

28-12 GENIC INTERACTIONS

In the examples discussed so far, the relationship between a gene and its phenotype has been direct, precise and exact. Each gene controls the appearance of a single trait. However, the relationship of gene to characteristic may be quite complex. Several pairs of genes may interact to affect a single trait, or one pair may inhibit or reverse the effect of another pair of genes; or a given gene may produce different effects when the environment is changed in some way. In all instances, the genes are inherited as units but they may interact in some complex fashion to produce the trait. The biochemical relations between gene and trait will be discussed further in the next chapter.

One of the simpler types of genetic interaction is illustrated by the inheritance of combs in poultry. The gene for rose comb, **R**, is dominant to that for single comb, **r**. Another pair of genes governs the inheritance of pea comb, **P**, versus single comb, **p**. A single-combed fowl must have the genotype *pprr*; a pea-combed fowl is either *PPrr* or *Pprr*; and a rose-combed fowl is either *ppRR* or *ppRr* (Fig. 28–9). When a homozygous pea-combed fowl is mated to a homozygous rose-combed one, the offspring have neither pea nor rose comb but a completely different type called walnut comb. The phenotype of walnut comb is produced whenever a fowl has one or two **R** genes plus one or two **P** genes. What would you predict about the types of combs among the offspring of the mating of two heterozygous walnut-combed fowls, *PpRr?*

Complementary Genes. Two pairs of genes that are inherited independently may interact in such a way that neither dominant can produce its effect unless the other is present. Such pairs of genes have been termed *complementary genes;* the action of each one "complements" the action of the other in the production of the phenotype. The presence of both dominants produces one trait; the alternate trait is produced by the absence of either one or both. In the course of experiments with sweet peas Bateson and Punnett found that

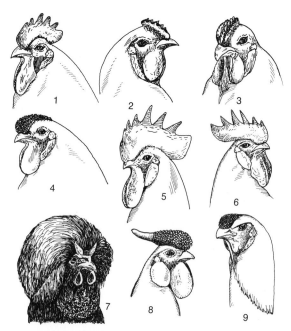

Figure 28-9 Heads of roosters, showing the different types of genetically determined combs. *1, 5,* and *6,* Single combs; *2* and *3,* pea combs; *4* and *8,* rose combs; *7,* V-shaped; *9,* strawberry. (Courtesy of the United States Department of Agriculture.)

crossing two white-flowered races of sweet peas gave offspring all of which had purple flowers! When two of the F_1 purple plants were crossed, the F_2 generation had offspring in the ratio of 9 purple-flowered plants to 7 white-flowered plants (Fig. 28–10). Two pairs of genes proved to be involved: gene **C** regulates an enzyme involved in the production of a white raw material from which purple pigment can be made by an enzyme produced by the second gene *(E)*. The homozygous recessive *cc* is unable to synthesize the raw material and the homozygous recessive *ee* lacks the enzyme to convert the raw material into a purple pigment. A race of sweet peas breeding true for purple flowers could be established by selecting two **CCEE** plants and mating them.

Supplementary Genes. Two independent pairs of genes that interact in such a way that one dominant will produce its effect whether or not the other is present but the second will produce its effect only in the presence of the first are termed **supplementary genes.** In guinea pigs, in addition to the **B** genes for black coat and **b** gene for brown coat described previously, a gene **C** produces an enzyme that converts a colorless precursor into the pigment melanin and hence is required for the produc-

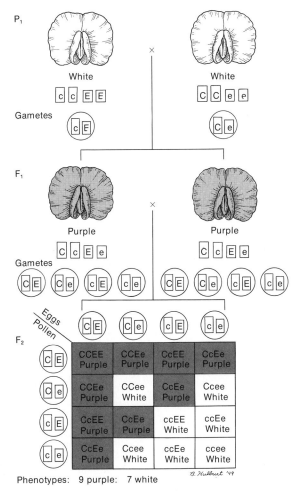

Phenotypes: 9 purple: 7 white

B. Hulburt '49

Figure 28–10 Diagram of a cross illustrating the action of complementary genes, the two pairs of genes which regulate flower color in sweet peas. Note that for a colored flower, the plant must have at least one C and one E gene.

tion of any kind of pigment. The homozygous recessive *cc* lacks the enzyme, produces no melanin, and the animal is a white-coated, pink-eyed *albino*, no matter what combination of *B* and *b* genes may be present. Thus when an albino with the genotype *ccBB* is mated to a brown guinea pig, *CCbb*, the F_1 generation will all be black-coated, *CcBb*. When two such animals are mated their offspring will be in the ratio of 9 black-coated: 3 brown-coated: 4 albino. Make a checkerboard to verify this.

The problems described in this text are obviously among the simplest of those dealt with by geneticists. Many traits are governed by a multitude of genes that interact with each other and with the environment to produce the final phenotype. The unscrambling of these in-

teractions provides the geneticist with many difficult problems. More than 12 pairs of genes interact in various ways to produce the coat color of rabbits, and more than 100 genes are concerned with the color and shape of the eyes in fruit flies.

28–13 POLYGENIC INHERITANCE

Many human characteristics—height, body form, intelligence and skin color—and many commercially important characteristics, such as milk production in cows, egg production in hens, the size of fruits and so on, cannot be separated into distinct alternate classes and are not inherited by a single pair of genes. Several, perhaps many, different pairs of genes affect each characteristic. The term *polygenic inheritance* is applied when two or more independent pairs of genes have similar and additive effects on the same characteristic. The inheritance of skin color in humans was studied by Davenport in Jamaica. He found that two major pairs of genes are involved, which he designated *A-a* and *B-b*. The capital letters represent genes producing dark skin—the more capital letters the darker the skin. The genes affect the character in an additive fashion. A full Negro has the genotype *AABB*, and a Caucasian has the genotype *aabb*. The F_1 offspring of a mating of *aabb* with *AABB* are all *AaBb* and have an intermediate skin color termed *mulatto.* The mating of two such mulattoes produces offspring with skin colors ranging from dark brown to white (Table 28–2).

Polygenic inheritance is characterized by an F_1 generation that is intermediate between

Table 28–2 POLYGENIC INHERITANCE OF SKIN COLOR IN HUMANS

Parents	AaBb (Mulatto)	×	AaBb (Mulatto)
Gametes	AB Ab aB ab		AB Ab aB ab

Offspring:
1 with 4 dominants—AABB—phenotypically Negro
4 with 3 dominants—2 AaBB and 2 AABb—phenotypically "dark"
6 with 2 dominants—4 AaBb, 1 AAbb, 1 aaBB—phenotypically mulatto
4 with 1 dominant—2 Aabb, 2 aaBb—phenotypically "light"
1 with no dominants—aabb—phenotypically Caucasian

the two parents and shows little variation, and an F_2 generation that shows a wide variation between the two parental types. Most of the F_2 generation have some intermediate phenotype and only a few show the traits of either grandparent. Of the 16 possible zygote combinations from a mating of *AaBb* with *AaBb,* only one, *AABB,* will be as dark as the Negro grandparent and only one, *aabb,* will be as light as the white grandparent. The genes *A* and *B* produce about the same amount of darkening of the skin; hence, the genotypes *AaBb, AAbb* and *aaBB* all produce similar phenotypes, mulatto skin color.

Skin color in humans is a rather simple example of polygenic inheritance because only two major pairs of genes are involved. The inheritance of height in humans is a more complex phenomenon involving perhaps ten or more pairs of genes. Tallness is recessive to shortness, thus the more capital letters in the genotype, the shorter the individual. Because of the many pairs of genes involved, and because height is modified by a variety of environmental conditions, the heights of adults range from perhaps 135 cm. to 215 cm. If we measured the heights of a thousand adult American men taken at random, we would find that only a few are as tall as 215 cm. or as short as 135 cm. The height of most would cluster around the mean, about 170 cm. When the number of people of each height is plotted against height in centimeters and the points connected, the result is a bell shaped curve called a *curve of normal distribution* (Fig. 28–11). If we measured the heights of one thousand adult American women and plotted the measurements in the same way, the data would generate a curve with a similar shape but the mean would be less; women, on the average, are shorter than men. If we measured the heights of one thousand men and women, what sort of curve would be generated by the data?

All living things show comparable variations in certain of their traits and these variations are usually distributed in a normal curve. If you measured the length of a thousand sea shells of the same species, or counted the number of kernels per ear in a thousand ears of corn, or the number of pigs per litter in a thousand litters, or weighed one thousand hen's eggs, you would find a normal curve of dis-

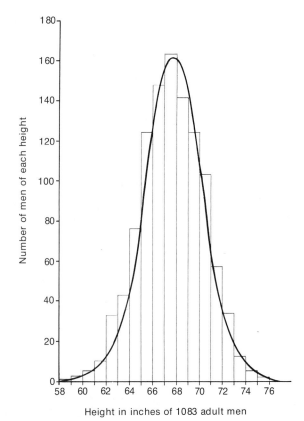

Height in inches of 1083 adult men

Figure 28–11 An example of a "normal curve," or curve of normal distribution, the heights of 1083 adult white males. The blocks indicate the actual number of men whose heights were within the unit range; for example, there were 163 men who were between 67 and 68 inches in height. The smooth curve is a normal curve based on the mean and standard deviation of the data.

tribution in each case. The variation resulting in this distribution may be caused by hereditary differences, by environmental differences, or by a combination of the two. How do geneticists go about establishing a breed of cow that will give more milk or a strain of hens that will lay bigger eggs, or a strain of corn with more kernels per ear? By selecting the organisms that approach the phenotype they are seeking and using them in further matings they gradually produce true breeding strains with the commercially desirable trait; that is, they select a strain homozygous for all of the dominant (or recessive) polygenes involved. It is clear that there is a limit to the effectiveness of breeding by selection. When a strain becomes homozygous for all the polygenes involved, further selective breeding cannot increase the desired quality.

28-14 MULTIPLE ALLELES

In the examples so far we have dealt with situations in which at any locus, at any given position on the chromosome, there is one of only two alternative kinds of genes, the dominant and recessive genes. At many, if not most, loci there may be additional possibilities, genes producing phenotypes different from both the dominant and the recessive. The term *multiple alleles* is applied to three or more genes that can occupy a single locus, that can fill the corresponding positions on a pair of homologous chromosomes. Each of the alleles produces a distinctive phenotype. Among the members of a population, any given individual may have any two of the genes but never more than two, and any gamete, of course, can have only one of them. But in the population as a whole there will be distributed three or more different alleles.

In rabbits, for example, a *C* gene causes fully colored coats. The homozygous recessive *(cc)* causes albino coat color. There are two other genes located at the same locus, c^h and c^{ch}. The gene c^h, when homozygous, causes the "Himalayan pattern" in which the body color is white but the tips of the ears, nose, tail and legs are colored. The c^{ch} gene, when homozygous, produces the "chinchilla pattern," in which the entire body has a light gray color. These genes can be arranged in a series—*C*, c^{ch}, c^h and *c*—in which each gene is dominant to the genes following it and recessive to those preceding it. In other series of multiple alleles, genes may be incompletely dominant so that the heterozygotes have a phenotype intermediate between those of their parents.

The human blood types O, A, B and AB, are inherited by multiple alleles. Gene I^A provides the code for the synthesis of a specific protein, *agglutinogen A*, in the red cell. Gene I^B leads to the production of a different protein, agglutinogen B. Gene *i* produces no agglutinogen. Gene *i* is recessive to the other two, but neither gene I^A nor I^B is dominant to the other. The symbols I^A, I^B and *i* are used to emphasize that all three are alleles at the same locus. Individuals with genotypes I^AI^A and I^Ai make up blood group A. Those with genotypes I^BI^B or I^Bi comprise blood group B. Individuals with blood group O have the genotype *ii*. When both the I^A and I^B genes are present in the same individual, both agglutinogens A and B are produced in the red cells and the individual belongs to blood group AB. These blood types are genetically determined and do not change during a person's lifetime. Determining the blood types of the individuals involved may be helpful in settling cases of disputed parentage (Table 28-3). Such blood tests can never prove that a certain man *is* the father of a certain child, but only whether or not he *could be* its father. They may definitely prove that he could *not* be the father of a certain child. Could an AB man be the father of an O child? Could an O man be the father of an AB child? Could a type B child with a type A mother have a type A father? A type O father?

Nearly a dozen other sets of blood types, including the MN groups and a series of Rh alleles are inherited by other genes, independently of the ABO blood types. Determining all of these types in a given person may be useful in establishing relationships that could not be made certain by the ABO blood type alone.

Table 28-3 EXCLUSION OF PATERNITY BASED ON ANALYSES OF BLOOD GROUPS

Child	Mother	Father Must Be of Type	Father Cannot Be of Type
O	O	O, A, B	AB
O	A	O, A, B	AB
O	B	O, A, B	AB
A	O	A, AB	O, B
A	A	A, B, AB, O	—
B	B	A, B, AB, O	—
A	B	A, AB	O, B
B	A	B, AB	O, A
B	O	B, AB	O, A
AB	A	B, AB	O, A
AB	B	A, AB	O, B
AB	AB	A, B, AB	O

28–15 LINKAGE AND CROSSING OVER

Each species of animal or plant has many more pairs of genes than it has pairs of chromosomes. Obviously there must be many genes per chromosome. Human beings have 23 pairs of chromosomes, some large and some smaller (Fig. 28–3), but thousands of pairs of genes. The chromosomes are inherited as units—they pair and separate during meiosis as units; thus, all the genes in any given chromosome tend to be inherited together. If the chromosomal units never changed, the traits would always be inherited together and linkage would be absolute. However, during meiosis when the chromosomes are pairing and undergoing synapsis, homologous chromosomes may exchange entire segments of chromosomal material, a process called *crossing over* (Fig. 28–12). This exchanging of segments occurs at random along the length of the chromosome. Several exchanges may occur at different points along the same chromosome at a single meiotic division. It follows that the greater the

distance between any two genes in the chromosome, the greater will be the likelihood that an exchange of segments between them will occur.

In fruit flies the pair of genes **V** for normal wings and **v** for vestigial wings and the pair of genes **B** for gray body color and **b** for black body color are located in the same pair of chromosomes. They tend to be inherited together and are said to be linked. What characteristics would you predict the offspring will have from a cross of a homozygous **VVBB** fly with a homozygous **vvbb** fly? They will all have gray bodies and normal wings and the genotype **VvBb**. When one of these F_1 heterozygotes is crossed with a homozygous **vvbb** fly (Fig. 28–13), the offspring appear in a ratio that differs from that of the ordinary test cross for a dihybrid. If the two pairs of genes were not linked but were in different chromosomes, the offspring would appear in the ratio of ¼ gray-bodied, normal-winged : ¼ black-bodied, normal-winged : ¼ gray-bodied, vestigial-winged : ¼ black-bodied, vestigial-winged flies. If

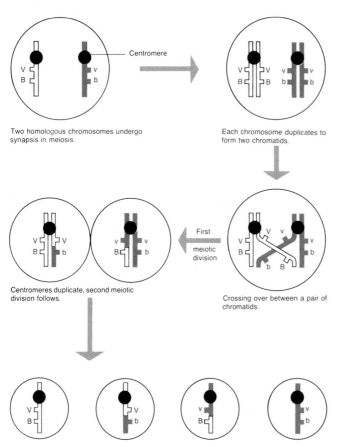

Two homologous chromosomes undergo synapsis in meiosis.

Each chromosome duplicates to form two chromatids.

Centromeres duplicate, second meiotic division follows.

First meiotic division

Crossing over between a pair of chromatids.

Four haploid gametes produced; here two crossover and two noncrossover gametes.

Figure 28–12 Diagram illustrating crossing over, the exchange of segments between chromatids of homologous chromosomes. Crossing over permits recombination of genes (e.g., *vB* and *Vb*); the farther apart genes are located on a chromosome, the greater is the probability that crossing over between them will occur.

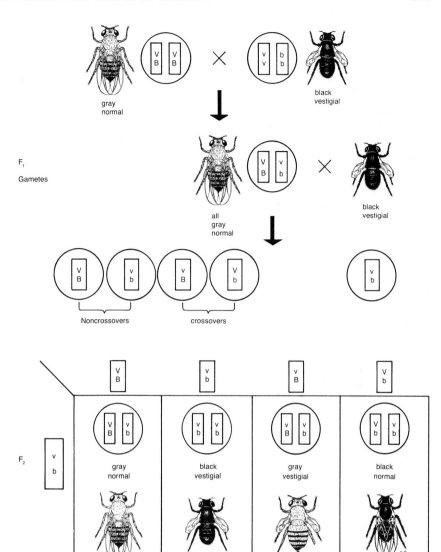

Figure 28-13 Diagram of a cross involving linkage and crossing over. The genes for vestigial versus normal wings and black versus gray body in fruit flies are linked; they are located in the same chromosome.

the genes were completely linked and no exchange of chromosomal segments occurred, then only the parental types—flies with gray bodies and normal wings and flies with black bodies and vestigial wings—would appear among the offspring, and these would be present in equal numbers (Fig. 28–13). However, there is an exchange of segments between the locus of gene **V** and the locus of gene **B**. Because of this crossing over of part of the chromosomes, some gray-bodied, vestigial-winged flies and some black-bodied, normal-winged flies (the crossover types) appear among the offspring (Fig. 28–13). Most of the offspring resemble the parents and are either gray-normal or black-vestigial. In this particular instance, crossing over occurs between

these two points in this chromosome in about one cell in every five or in 20 per cent of the total undergoing meiosis. In such crosses, about 40 per cent of the offspring are gray flies with normal wings. Another 40 per cent are black flies with vestigial wings. Ten per cent are gray flies with vestigial wings, and 10 per cent are black flies with normal wings. The distance between two genes in a chromosome is measured in "cross-over units," or "centimorgans" (named in honor of T. H. Morgan, the American geneticist who studied crossing over in the fruit fly), which represent the percentage of crossing over that occurs between them. Thus, **V** and **B** are said to be 20 centimorgans apart.

In a number of species the frequency of

Figure 28–14 Diagram illustrating the means of determining whether gene *C* lies between or to the right of genes *A* and *B* from the percentage of crossing over between each of the possible pairs.

crossing over between specific genes has been measured. All of the experimental results are consistent with the hypothesis that genes are present in a linear order in the chromosomes. Thus, if the three genes A, B and C occur in a single chromosome, the amount of crossing over between A and C is either the sum of, or the difference between, the amounts of crossing over between A and B and B and C. For example, if the crossing over between A and B is five centimorgans and between B and C is three centimorgans, the crossing over between A and C will be found to be either eight centimorgans (if C lies to the right of B) or two centimorgans (if C lies between A and B) (Fig. 28–14). By putting together the results of a great many such crosses, detailed maps of the location of specific genes on specific chromosomes have been made (Fig. 28–15).

Crossing over occurs at random, and more than one crossover between two loci in a single chromosome may occur at a given time. You can observe among the offspring only the frequency of *recombinations* and not the frequency of crossovers. The frequency of crossing over will be slightly larger than the observed frequency of recombination, because the simultaneous occurrence of two crossovers between two particular genes will lead to the reconstitution of the original combination of genes in a particular chromosome.

All the genes in a particular chromosome tend to be inherited together and compose a *linkage group.* The number of linkage groups

determined by genetic tests is always equal to the number of pairs of chromosomes. The most detailed chromosome maps are those for the bacterium *Escherichia coli*, which has one circular chromosome, and for fruit flies, which have four pairs of chromosomes. The chromosomes of corn, mice, *Neurospora* and certain other species of bacteria and viruses have been mapped in considerable detail.

Linkage provides an explanation for the common observation that certain traits in humans and other organisms tend to be inherited together. Such traits are determined by genes that are located rather close together in a given chromosome. Crossing over provides another means by which genetic recombinations may arise. It plays a role in evolution by making possible new combinations of genetic units in the offspring.

28–16 THE GENETIC DETERMINATION OF SEX

An exception to the general rule that all homologous pairs of chromosomes are identical in size and shape is provided by the *sex chromosomes.* The cells of the females of most species contain two identical sex chromosomes, called X chromosomes, but in males there is only one X chromosome and a smaller Y chromosome with which it undergoes synapsis during meiosis. Men have 22 pairs of ordinary chromosomes, or *autosomes,* plus one X and one Y chromosome; women have 22 pairs of autosomes plus two X chromosomes.

In some of the lower animals, for example in the fruit fly *Drosophila*, sex is determined by the ratio of the number of X chromosomes to the number of sets of autosomes. Males have one X to two haploid sets of autosomes, a ratio of 1:2 or 0.5. Females have two X chromosomes and two sets of autosomes, a ratio of 1.0. It is

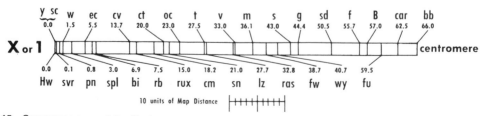

Figure 28–15 Crossover map of the X chromosome of the fruit fly *D. melanogaster.* (From Herskowitz, I. H.: *Genetics.* 2nd Ed. Boston, Little, Brown and Company, 1965.)

possible to set up abnormal flies with one X chromosome and three sets of autosomes, with a ratio of 0.33. These flies have all of their male characteristics exaggerated and are termed "supermales." It is also possible to set up individuals with three X chromosomes and two sets of autosomes (ratio = 1.5), which have all the female characteristics exaggerated and are termed "superfemales." The animals that have two X chromosomes and three sets of autosomes (ratio of 0.67) have characters intermediate between male and female flies and are termed "intersexes." These abnormal flies — supermales, superfemales and intersexes — are all sterile.

In humans and perhaps in other mammals, maleness is determined in large part by the presence of the Y chromosome. An individual with the XXY constitution is a nearly normal male in his external appearance, though with underdeveloped gonads (Klinefelter's syndrome, p. 725). An individual with one X but no Y chromosome has the appearance of an immature female (Turner's syndrome, p. 725).

In humans and in other species in which the normal male has one X and one Y chromosome, two kinds of sperm are produced: half contain an X chromosome and half contain a Y chromosome. All eggs contain one X chromosome. Fertilization of an X-bearing egg by an X-bearing sperm results in an XX, female, zygote, and the fertilization of an X-bearing egg by a Y-bearing sperm results in an XY, male, zygote. Since there are equal numbers of X- and Y-bearing sperm, about equal numbers of each sex are born. Some 106 boys are born to every 100 girls and the ratio of males to females at conception is even higher. One possible explanation of this numerical difference is that the Y chromosome is somewhat smaller than the X chromosome, and a sperm containing a Y chromosome might be lighter and able to swim a little faster than an X-bearing sperm. Consequently it might be able to win the race to the egg slightly more than half the time.

This XY mechanism of sex determination is believed to operate in all species of animals and plants with separate sexes. In birds and butterflies (Lepidoptera) the mechanism is reversed; males are XX and females are XY. Sex chromosomes have been detected in some plants, notably strawberries, and probably exist in other plants with separate sexes. The

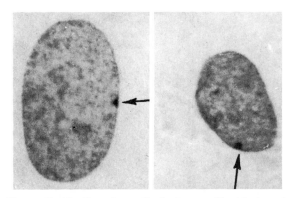

Figure 28–16 Sex chromatin in human fibroblasts cultured from skin of a female. The chromatin spot at the periphery of each nucleus is indicated by the arrow. (Feulgen, ×2200.) (Courtesy of Dr. Ursula Mittwoch, Galton Laboratory, University College, London.)

members of many species have the organs of both sexes present in each individual. In such organisms (termed hermaphroditic if animals and monoecious if plants) sex chromosomes have not been found.

In 1949, M. L. Barr found that certain cells show a "chromatin spot" at the edge of the nucleus (Fig. 28–16). These were evident in cells of human skin or from the mucosal lining of the mouth. Further investigation revealed that cells showing this spot had come from female individuals whereas cells that did not show the spot came from males. Using this characteristic it is possible to carry out "nuclear sexing" of individuals and determine whether an individual is genetically female or male.

This spot has been found to represent one of the two X chromosomes, which becomes dense and dark-staining. The other X chromosome resembles the autosomes and, during the interphase, is a fully extended thread not evident by light microscopy. From this and other evidence Mary Lyon has suggested that only one of the two X chromosomes in the female is active; the other is inactive. Which of the two becomes inactive in any given cell is a matter of chance, and the cells of a woman's body are of two kinds in which one or the other X chromosome is inactive. Since the two X chromosomes may have different genetic complements, the cells in a woman's body may differ in the effective genes present. In mice and cats, which have several sex-linked genes for certain coat colors, the female heterozygous for such genes may show patches of one coat color in the midst of areas of the other color. This phenomenon, termed **variegation,** is evident in tortoise-shell cats. This inactivation of one X chromosome apparently occurs early in embryonic development, and thereafter all the progeny of that cell have the same inactive X chromosome. Although one X chromosome appears to be inactive, there are marked abnormalities of development when one X chromosome is completely missing from the chromosomal complement of the cell (e.g., the XO condition of Turner's syndrome).

28–17 SEX-LINKED AND SEX-INFLUENCED TRAITS

The human X chromosome contains many genes, whereas the Y chromosome contains only a few, principally the genes for maleness. Traits controlled by genes located in the X chromosome are called *sex-linked* because their inheritance is linked with that of sex. A male offspring receives from his mother a single X chromosome and therefore all his genes for sex-linked characters. A female receives one X from her mother and one from her father. Males, having but one X chromosome, have only one of each kind of gene located in the X chromosome.

The normal eye color of the fruit fly is a dark red but white-eyed strains exist. The genes for red vs. white eye color are located in the X chromsome and hence are sex-linked. The Y chromosome contains no gene for eye color. The male, having only one gene for any sex-linked trait, cannot be either homozygous or heterozygous but is **hemizygous** for any gene located in the X chromosome. To avoid confusion the male genotype is written with the Y present. Red eye color *(R)* is dominant to white *(r).* When a homozygous red-eyed female is mated with a white-eyed male *(RR × rY)*, the offspring all have red eyes. The female offspring are **Rr** and the male offspring are **RY**. Crossing a white-eyed female with a red-eyed male *(rr × RY)* produces red-eyed female offspring, **Rr**, and white-eyed males, **rY**.

The human genes for hemophilia and color blindness are located in the X chromosome and the inheritance of these traits is sex-linked.

Hemophilia is a disease in which there is a deficiency in the formation of thromboplastin due to a deficiency of the so-called antihemo-

philic globulin. Individuals suffering from hemophilia have blood that does not clot properly; they will bleed profusely from even a small cut. If the sex-linked gene is recessive and relatively rare (that is, present in the population in low frequency), the trait will appear much more frequently in males than in females. Color blindness, for example, affects about 4 per cent of all human males but less than 1 per cent of human females. Hemophilia is a very rare trait in human males and was unknown in human females until a single instance of it was found in 1951. Queen Victoria of England was heterozygous for the gene for hemophilia and passed this to several of her sons and grandsons. This fact had a marked effect on the course of history, especially in Russia and Spain.

Not all the characteristics that are different in the two sexes are sex-linked. Some traits, known as "sex-influenced," are inherited by genes located in autosomes but have their appearance altered or influenced by the sex of the animal. Males and females with identical genotypes may have different phenotypes. In sheep, for example, a single pair of genes determines the presence or absence of horns. The gene **H** for the presence of horns is dominant in males but recessive in females, and its allele **h** for hornlessness is recessive in males but dominant in females. The genotype **HH** produces a horned animal regardless of sex. **Hh** produces the horned phenotype if the animal is male and a hornless phenotype if the animal is female, and **hh** produces a hornless animal whether it is a ram or a ewe.

In humans the gene for pattern baldness is sex-influenced, its expression being altered by the amount of male sex hormone present. There are many more bald men than women because only one gene for baldness will cause a man to lose his hair whereas two such genes are needed to produce a bald woman. Not all types of baldness are hereditary, of course; some types are caused by disease or other factors.

28–18 INBREEDING, OUTBREEDING AND HYBRID VIGOR

It is commonly believed that *inbreeding* — the mating of two closely related individuals such as brother and sister — is harmful and

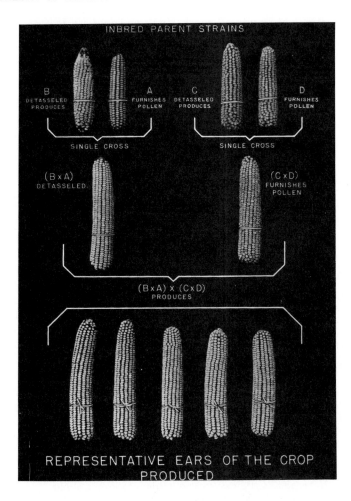

Figure 28–17 The mating of inbred strains of corn to produce the commercial variety with strong hybrid vigor. (Courtesy of the United States Department of Agriculture.)

leads to the production of idiots and monsters. Even the marriage of first cousins is forbidden by law in some states. There is nothing harmful, however, in inbreeding itself. Inbreeding procedures are used widely by geneticists to improve strains of cattle, corn or cantaloupes. It would not necessarily be a bad practice in human beings; it simply increases greatly the probability for recessive genes to become homozygous and thus to be expressed phenotypically. All organisms are heterozygous for many characteristics. Some of the recessive genes so hidden are for desirable characteristics, while others are for undesirable ones. A stock heterozygous for many recessive desirable traits may be improved by inbreeding. If the stock has many recessive undesirable traits, inbreeding will enable them to appear phenotypically. Human inbreeding increases the frequency of defects present at birth, termed *congenital anomalies.*

The mating of individuals of totally unrelated strains, termed *outbreeding,* frequently leads to offspring that are much better adapted for survival than either parent, a phenomenon termed *hybrid vigor.* A mule, the hybrid resulting from the mating of a horse and a donkey, is a strong, sturdy beast, better suited for many tasks than either parent. A large part of the corn grown in the United States is a special hybrid strain developed by the United States Department of Agriculture from the mating of four different strains (Fig. 28–17). Each year the seed to grow this uniformly fine hybrid corn must be obtained by mating the original strains. The hybrid is quite heterozygous and gives rise when mated to a wide variety of forms, none of which is as good as the original hybrid.

Hybrid vigor may be explained as follows: each of the parental strains of corn is homozygous for certain recessive undesirable genes but any two strains are homozygous for *different* undesirable genes. Each strain contains dominant genes to mask the recessive undesirable genes of the other strain. One strain then might have the genotype **AAbbCCdd** and another strain, the genotype **aaBBccDD**. The capital letters represent dominant genes for desirable traits, and the lower case letters represent recessive genes for undesirable traits. The hybrid offspring, with the genotype **AaBbCcDd**, would combine all the desirable and none of the undesirable traits of the two parental strains. The actual genetic situation in hybrid corn is, of course, much more complicated than this.

QUESTIONS

1. Show by diagrams how genes located in different pairs of chromosomes segregate independently in meiosis.
2. If a particular character in a certain species of animal were always transmitted from the mother to the offspring but never from the father to the offspring, what would you conclude about its mode of inheritance?
3. What do the following genetic symbols mean: **A, a, AA, aa, Aa**?
4. Define: gene, locus, allelomorph, dominant, recessive, homozygous, heterozygous, phenotype, genotype, chromatin spot.
5. In peas, yellow color is dominant to green. What will be the colors of the offspring of homozygous yellow × green? Heterozygous yellow × green? Heterozygous yellow × homozygous yellow? Heterozygous yellow × heterozygous yellow?
6. Could two blue-eyed parents have a brown-eyed child? Could two brown-eyed parents have a blue-eyed child?
7. If two animals heterozygous for a single pair of genes are mated and have 200 offspring, about how many will have the dominant phenotype?
8. Two long-winged flies were mated and the offspring included 77 with long wings and 24 with short wings. Is the short-winged condition dominant or recessive? What are the genotypes of the parents?
9. A blue-eyed man, both of whose parents were brown-eyed, marries a brown-eyed woman whose father was blue-eyed and whose mother was brown-eyed. This man and woman have a blue-eyed child. What are the genotypes of all the individuals mentioned?
10. Outline a breeding procedure whereby a true breeding strain of red cattle could be established from a roan bull and a white cow.
11. Suppose you learned that "shmoos" can have long, oval, or round bodies and that matings of shmoos gave the following:

long × oval gave 52 long : 48 oval
long × round gave 99 oval
oval × round gave 51 oval : 50 round
oval × oval gave 24 long : 53 oval : 27 round
What hypothesis about the inheritance of shmoo shape would be consistent with these results?

12. In rabbits, spotted coat **(S)** is dominant to solid color **(s)**, and black **(B)** is dominant to brown **(b)**. A brown spotted rabbit is mated to a solid black one and all the offspring are black spotted. What are the genotypes of the parents? What would be the appearance of the F_2 if two of these F_1 black spotted rabbits were mated?
13. The long hair of Persian cats is recessive to the short hair of Siamese cats, but the black coat color of Persians is dominant to the black-and-tan coat of Siamese. If a pure black, long-haired Persian is mated to a pure black-and-tan, short-haired Siamese, what will be the appearance of the F_1? If two of these F_1 cats are mated, what is the chance of obtaining in the F_2 a long-haired, black-and-tan cat?
14. In peas, tall plants **(T)** are dominant to dwarf **(t)**, yellow color **(Y)** is dominant to green **(y)** and smooth seed **(S)** is dominant to wrinkled seed **(s)**. What would be the phenotypes of the offspring of the following matings?
   a. **TtYySs × ttyyss**
   b. **TtyySs × ttYySs**
15. Distinguish between: complementary genes and supplementary genes; sex-linked character and sex-influenced character.
16. A walnut-combed rooster is mated to three hens. Hen A, which is walnut-combed, has offspring in the ratio of 3 walnut : 1 rose. Hen B, which is pea-combed, has offspring in the ratio of 3 walnut : 3 pea : 1 rose : 1 single. Hen C, which is walnut-combed, has only walnut-combed offspring. What are the genotypes of the rooster and the three hens?
17. What conditions result in the following phenotypic ratios?
   a. 3 : 1   e. 1 : 4 : 6 : 4 : 1
   b. 1 : 2 : 1   f. 2 : 1
   c. 9 : 3 : 3 : 1   g. 47 : 47 : 3 : 3
   d. 9 : 7

18. The weight of the fruit in one variety of squash is determined by three pairs of genes, **AABBCC**, producing 6-pound squashes, and **aabbcc**, producing 3-pound squashes. Each dominant gene adds ½ pound to the weight. When a 6-pound squash is crossed with a 3-pound squash, all the offspring weigh 4½ pounds. What would be the weights of the F_2 fruits if two of these F_1 plants were crossed?
19. Mrs. Doe and Mrs. Roe had babies at the same hospital at the same time. Mrs. Doe took home a girl and named her Nancy. Mrs. Roe took home a boy and named him Richard. However, she was sure she had had a girl and brought suit against the hospital. Blood tests showed that Mr. Roe was type O, Mrs. Roe was type AB, and Mr. and

Mrs. Doe were both type B. Nancy was type A and Richard type O. Had an exchange occurred?

20. Explain the mechanism of the genetic determination of sex in humans.

21. One pair of genes for coat color in cats is sex-linked. The gene *B* produces yellow coat, *b* produces black coat, and the heterozygote *Bb* produces tortoise-shell color. What kind of offspring result from the mating of a black male and a tortoise-shell female?

22. The barred pattern of chicken feathers is inherited by a pair of sex-linked genes, *B* for barred and *b* for no bars. If a barred female is mated to a non-barred male, what will be the appearance of the progeny? What commercial usefulness does this have?

23. What is meant by linkage? By crossing over?

24. What are the advantages and disadvantages of inbreeding?

25. What is meant by hybrid vigor? What is the genetic explanation of this phenomenon?

SUPPLEMENTARY READING

A discussion of genetics in terms of its hypotheses and how they were tested is presented in *Towards an Understanding of the Mechanism of Heredity* by H. L. Whitehouse. Alfred H. Sturtevant's *A History of Genetics* is a fine book for the informed layman, giving an authentic account of the rise of classic genetics, in which the author played a prominent role. Papers by many of the scientists responsible for important developments in genetics have been assembled and reprinted in *Classic Papers in Genetics*, edited by James A. Peters. A discussion of the sex chromatin and our increasing understanding of its significance is summarized in K. L. Moore's *The Sex Chromatin*. Elof Carlson's *The Gene: A Critical History* traces the development of genetic concepts. *Genetics* by M. W. Strickburger, *Genetics* by I. H. Herskowitz and *Heredity and Development* by J. A. Moore are standard college-level texts.

CHAPTER 29

THE STRUCTURE AND FUNCTION OF GENES

Our modern concepts of genetics originated with the rediscovery of Mendel's Laws in 1900. Since that time, geneticists have been attempting to determine the physical structure and chemical composition of the hereditary units — the genes — and to discover the mechanisms by which they transfer biological information from one cell to another and control the development and maintenance of the organism. During the first half of this century, much was learned about the complexity of the molecular structure of proteins. As a result, nearly all biochemists assumed that any complex biological unit with such marked specificity as the gene must also be a protein. There was great difficulty, however, in explaining how protein molecules could be duplicated precisely, as genes must be with each cell division. The theory that genetic information is transferred by nucleic acid molecules rather than by protein molecules developed gradually but is now firmly established.

29–1 THE ORIGIN OF THE "CENTRAL DOGMA"

More than a century ago (1869), Friedrich Miescher isolated from the nuclei of pus cells a new class of chemicals, which he called "nuclein." These substances, later called *nucleic acids,* were acidic in nature, were uniquely rich in phosphorus, and contained carbon, oxygen, hydrogen and nitrogen. Subsequent analyses revealed that there are two types of nucleic acids — *deoxyribonucleic acid,* or *DNA,* present in the nucleus; and *ribonucleic acid,* or *RNA,* present in the nucleus and cytoplasm. DNA was shown by P. A. Levene to be composed of four nitrogenous bases — two purines *(adenine* and *guanine)* and two pyrimidines *(cytosine* and *thymine);* a five-carbon sugar, *deoxyribose;* and phosphate groups. Levene showed that the purine or pyrimidine base is attached to the sugar by a glycosidic linkage and the sugar is attached to the phosphate by an ester bond (Figs. 3–5 and 3–12). The combination of base-sugar-phosphate constitutes the basic unit, termed a *nucleotide,* of nucleic acid. Four kinds of deoxynucleotides are found in

DNA; each contains one of the four kinds of nitrogenous bases—adenine, guanine, cytosine and thymine. Levene incorrectly concluded that all DNAs, of whatever source, are composed of *equal* amounts of these four nucleotides. Such simple molecules could not provide the basis for the biological specificity of the gene.

Our present belief that DNA and RNA are the primary agents for the transfer of biological information arose gradually, culminating in 1953 in the proposal by James Watson and Francis Crick of a model of the DNA molecule that explained how it could transfer information and undergo replication. This proposal stimulated an enormous flood of research, and has led to the present *"central dogma"* of biology: Genes are composed of DNA and are located within the chromosomes. Each gene contains information coded in the form of a specific sequence of purine and pyrimidine nucleotides within its DNA molecule. The unit of genetic information, called a **codon,** is a group of three adjacent nucleotides that specify a single amino acid in a polypeptide chain. Thus the genetic code is a *"triplet"* code. The DNA molecule consists of two complementary chains of polynucleotides twisted about each other in a regular helix and joined by hydrogen bonds between specific pairs of purine and pyrimidine bases. The DNA molecule is replicated when the two strands of the helix separate, and each acts as a template for the formation of a new complementary strand. Each pair of strands, one old and one new, then twist together to form two daughter helices.

29–2 GENETIC INFORMATION IS TRANSMITTED BY DNA

The DNA of each gene has a sequence of nucleotide triplets that differs from that of every other gene. This information is transcribed from the DNA of the gene to a kind of RNA termed *"messenger"* RNA, a nucleotide sequence complementary to the genic DNA. Messenger RNA is synthesized in the nucleus and passes to the ribosomes in the endoplasmic reticulum. It combines with five to ten ribosomes to form **polyribosomes** and serves as a template for the synthesis of an enzyme or some other specific protein.

For protein synthesis to occur, amino acids must be "activated" with ATP and then joined to specific adaptor molecules termed *"transfer"* RNA. Each kind of transfer RNA has a triplet code *(anticodon)* at some specific part of the molecule. The amino acid–transfer RNA complexes are arranged on the messenger RNA in an order dictated by the complementary nature of the nucleotide triplets in the messenger RNA codons and the transfer RNA anticodons. The information initially coded as a specific sequence of nucleotides in the DNA is **transcribed** as a specific sequence of nucleotides in messenger RNA, and is eventually **translated** into the specific order of the amino acids in the protein molecule.

Experimental Basis of the Gene Concept. The first direct evidence that DNA can transmit genetic information came from the experiments of Avery and his coworkers with the *"transforming agent"* isolated from pneumococci and certain other bacteria. These experiments in turn were based upon those of Fred Griffiths, an English bacteriologist. In 1928 Griffiths was studying two different strains of pneumococci, a virulent "smooth" one with a polysaccharide capsule, and a nonvirulent "rough" one without the capsule. When he injected live "rough" bacteria into mice, the mice would survive. If he injected mice with live "smooth" bacteria (the ones with a capsule), the mice died (Fig. 29–1). However, mice would survive if injected with heat-killed smooth bacteria. In a crucial experiment Griffiths injected mice with a mixture of live rough bacteria and heat-killed smooth bacteria. Although neither of these alone was harmful, the mixture of the two caused the death of the mice, and Griffiths could recover live smooth bacteria from the dead mice. From these and other experiments, Griffiths concluded that the live rough bacteria had been transformed into live smooth bacteria by some material from the dead smooth cells. These bacteria, when grown in culture, reproduced smooth bacteria. Thus it appeared that some sort of material from the dead bacteria had entered the live rough bacteria and changed them into smooth ones. There are several types of pneumococci, and in further experiments Griffiths found that injecting live Type II rough pneumococci into mice, together with heat-killed smooth Type III pneumococci, killed the mice. He could

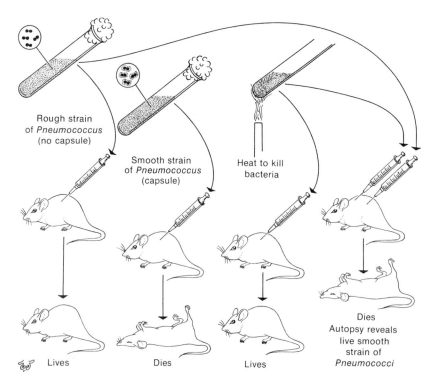

Figure 29-1 The experiments of Fred Griffiths, which demonstrated the transfer of genetic information from dead, heat-killed bacteria to living bacteria of a different strain. Although neither the rough strain of *Pneumococcus* nor heat-killed smooth strain pneumococci would kill a mouse, a combination of the two did. Autopsy of the dead mouse showed the presence of living, smooth strain pneumococci.

isolate live, virulent, "smooth" or encapsulated pneumococci from the dead mice, but these were of Type III and not of Type II! Again the conclusion was clear—some sort of genetic material had passed from the dead Type III cells to the living Type II cells, and this material changed them into Type III cells.

Methods were developed by which this bacterial transformation could be carried out in culture media rather than inside the bodies of mice. For example, live rough cells in a test tube could be transformed into smooth or encapsulated virulent bacteria by the fluid in which dead virulent cells had been dissolved. Finally in 1944, Avery and his colleagues at

the Rockefeller Institute showed unequivocally that the transforming substance is DNA. The transforming agent lost activity when treated with deoxyribonuclease (which hydrolyzes DNA) but not when treated with a proteolytic enzyme such as trypsin or chymotrypsin. Avery isolated and purified enough of the material to show that it was DNA with a high molecular weight.

During the 1940's, A. E. Mirsky and Hans Ris, working at the Rockefeller Institute, and André Boivin and Roger Vendrely, working at the University of Strasbourg, independently showed that the amount of DNA per nucleus is constant in all the body cells of a given orga-

Table 29-1   AMOUNT OF DEOXYRIBONUCLEIC ACID (DNA) PER NUCLEUS IN ANIMAL TISSUES, EXPRESSED AS MG. $\times 10^{-9}$

Species	Sperm	Red Cell	Liver	Heart	Kidney	Pancreas	Spleen
Shad	0.91	1.97	2.01				
Carp	1.64	3.49	3.33				
Brown trout	2.67	5.79					
Toad	3.70	7.33					
Frog		15.0	15.7				
Chicken	1.26	2.49	2.66	2.45	2.20	2.61	2.55
Dog			5.5		5.3		
Rat			9.47*	6.50	6.74	7.33	6.55
Ox			7.05		6.63	7.15	7.26
Human	3.25	7.30	10.36*		8.6		

*Cells such as those of the liver, which tend to become tetraploid, will have unusually high DNA contents.

nism. By making cell counts and chemical analyses, Mirsky and Vendrely showed that there is some 6×10^{-9} mg. of DNA per nucleus in somatic cells, but only 3×10^{-9} mg. of DNA per nucleus in egg cells or sperm cells (Table 29–1).

In tissues known to be *polyploid* (ones having more than two sets of chromosomes per nucleus), the amount of DNA per nucleus was found to be a corresponding multiple of the usual amount. For example, cells with four sets of chromosomes, termed *tetraploids,* were found to have 12×10^{-9} mg. of DNA per nucleus. From the amount of DNA per cell, one can estimate the number of nucleotide pairs per cell and thus the amount of genetic information present in each kind of cell (Table 29–2).

Only the amount of DNA and the amount of certain basic, positively charged proteins, termed *histones,* are relatively constant from one cell to the next. The amounts of other kinds of protein and of RNA vary considerably from cell to cell. Thus the fact that the amount of DNA, like the number of genes, is constant in all the cells of the body and the fact that the amount of DNA in germ cells is only half the amount in somatic cells is strong evidence that DNA is an essential part of the gene. However, because of the strong belief that only protein molecules have enough complexity to account for genetic specificity, this evidence was not

accepted by most biochemists. Biologists assumed that the genetic material in the chromosomes must be protein, even though it was shown that the amount of structural protein in the chromosomes is not constant, but varies with the activity of the cell.

Further evidence that DNA is the carrier of genetic information came from studies with bacteria and with the viruses that infect bacteria, called *bacteriophages* (p. 153). When bacterial cells infected with phage particles are broken open and examined with the electron microscope, there is no trace at all of the bacteriophage during the period of infection; there is no trace of any of the Ping-Pong paddlelike particles. Then complete phage particles begin to appear, and in addition, bits of incomplete phage particles are found mixed in with them (Fig. 29–2). During the period of infection, the number of phage particles increases in a regular fashion, but the rate is linear rather than geometrical. Bacteria, for example, increase geometrically, i.e., 1, 2, 4, 8, 16, 32, 64 and so on. In contrast, viruses appear in a linear progression: 1, 2, 3, 4, 5, 6, 7 and so on, as though they were being assembled in a factory on an assembly line.

Studies by A. D. Hershey and Martha Chase at Cold Spring Harbor, New York, showed that only the DNA and not the protein of the virus enters the bacterial cell. Hershey and Chase carried out an experiment to determine whether the phage, as it infects the bacterium, injects DNA, proteins, or both into the bacterial cell (Fig. 29–3). They took advantage of the fact that DNA contains phosphorus, whereas protein does not, and that protein contains some sulfur atoms, whereas DNA does not. By culturing bacteriophage on bacteria grown in a medium containing ^{32}P and ^{35}S, they were able to grow phage that contained ^{32}P in its DNA and ^{35}S in its protein. The radioactive phage particles were recovered and purified, and nonradioactive bacteria were infected with them. After the infection had begun, the bacterial cells were agitated in a blender to remove the extra virus, then broken apart and analyzed. The remains of the virus contained ^{35}S, but the bacterial cells contained radioactive phosphorus and very little, if any, radioactive sulfur. This is evidence that the DNA of the phage entered the cell, whereas the protein coat of the phage remained outside, attached to

Table 29–2 THE AMOUNT OF DNA PER CELL IN ANIMAL AND PLANT CELLS AND IN VIRUS PARTICLES

Source of Cells	DNA mg. $\times 10^{-9}$ per Cell	Nucleotide Pairs per Cell
Mammals	6	5.5×10^9
Birds	2	2×10^9
Reptiles	5	4.5×10^9
Amphibia	7	6.5×10^9
Fish	2	2×10^9
Insects	0.17–12	0.16×10^9
Crustacea	3	2.8×10^9
Mollusks	1.2	1.1×10^9
Echinoderms	1.8	1.7×10^9
Sponges	0.1	0.1×10^9
Higher plants	2.5–40	2.3×10^9
Fungi	0.02–0.17	0.02×10^9
Algae	3	2.8×10^9
Bacteria	0.002–0.06	2×10^6
T_2 bacteriophage	0.00024	2.2×10^5
λ bacteriophage	0.00008	7×10^4
Papilloma virus	–	6×10^3

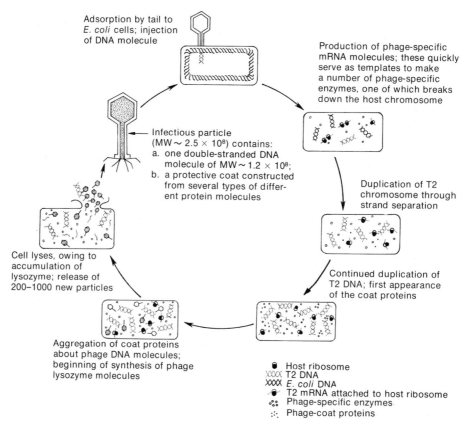

Figure 29–2 Diagram of the sequence of events following the infection of a bacterium by a T2 bacteriophage particle. Only the DNA from the phage enters the bacterial cell, yet this provides the information for the synthesis of new DNA and of new viral protein.

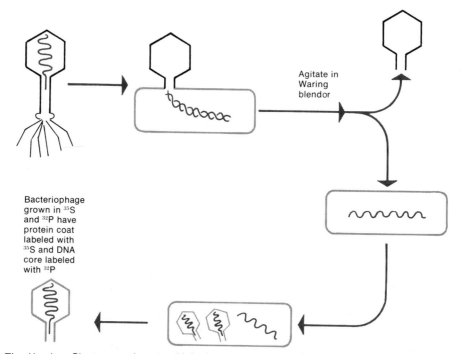

Figure 29–3 The Hershey-Chase experiment, which demonstrated that only DNA from bacteriophage is injected into bacteria, while the protein coat of the bacteriophage remains outside. All the genetic information needed for the synthesis of both new protein coat and new viral DNA is provided by the viral DNA.

the surface. This viral DNA, injected into the bacterial cell by the phage, in some way commandeers the machinery of the bacterial cell that ordinarily makes new bacteria and programs it to make new bacteriophage material instead. If the bacteriophage was allowed to multiply within the bacteria and then escape, the new generation of bacteriophage contained ^{32}P but no ^{35}S.

Bacterial Recombination. Yet another type of evidence that DNA is the genetic material came from experiments with different strains of bacteria begun by Lederberg and Tatum in 1946 (Fig. 29–4). One strain of bacteria contained a number of mutants that resulted in the loss of certain enzymes required for the synthesis of specific materials. This strain could survive if it were supplied with the products that it could not synthesize itself. Another strain of bacteria, with different mutants, required different nutritional materials to survive. In their key experiment, Lederberg and Tatum mixed these two strains together and grew them on a medium containing all the nutrients required by both strains. After they had grown and reproduced, samples of the progeny were transferred to a simple culture medium containing no special nutrients. Most of the resulting bacteria could not survive and died, but a few colonies did form. Subsequent analysis showed that these bacteria, like the so-called "wild type" bacteria, could grow and reproduce indefinitely without requiring special nutrients. Lederberg and Tatum suggested

that genetic material, DNA, had been transferred from one strain of bacteria to the other so that the genetic material for the enzymes missing in one strain was transferred from the other. The process was probably analogous to that by which transforming material is passed from one kind of pneumococcus to another. This was a novel concept at the time, because it was believed that bacteria could not undergo genetic recombination; however, subsequently Lederberg and Tatum's hypothesis has been found to be correct. When the process was studied by electron microscopy, it was possible to show a bridge of cytoplasm forming between two bacterial cells (Fig. 29–5). Usually only a portion of the donor's DNA enters the recipient cell before the two mating cells separate. Samples of bacteria are placed in a blender at various times after the cytoplasmic bridge has formed; varying amounts of DNA and genetic material are transferred across the bridge. The proportionality between the amount of DNA and the lengths of the genetic map transferred indicates that genetic information is contained in DNA.

Transduction. Bacterial genes may be transferred passively from one bacterium to another by a bacteriophage particle, a process that has been termed *transduction*. As a virus particle forms within the host cell, it may enclose and come to contain a small segment of the bacterial genetic material along with the phage DNA. When the phage is subsequently released, it becomes attached to a new bac-

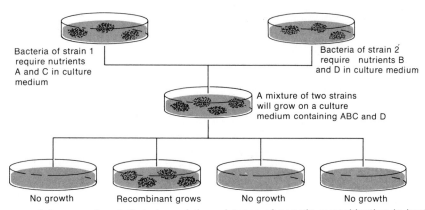

Figure 29–4 Diagram of the experiment that showed the existence of genetic recombination in bacteria. Bacteria of strain 1 require nutrients A and C, and bacteria of strain 2 require nutrients B and D in the culture medium. A mixture of the two strains will grow on a culture medium containing all four nutrients. Some of their offspring will grow on minimal culture media; these are bacteria in which genetic recombination has joined the wild type alleles of the mutants in one strain with the wild type alleles of the mutants in the other. The recombinant, like the original wild type, can grow without any added nutrients.

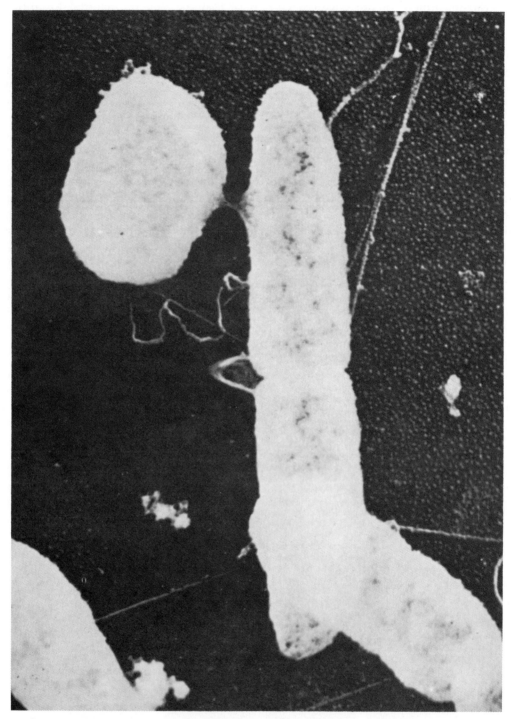

Figure 29–5 Conjugating bacteria conduct a transfer of genetic material. Long cell *(right)* is an Hfr "supermale" colon bacillus, which is attached by a short temporary bridge to a female colon bacillus. This electron micrograph, shown at a magnification of 100,000 diameters, was made by Thomas F. Anderson of the Institute for Cancer Research, Philadelphia, Pa.

terium and injects the segment of bacterial chromosome from the previous host into the new host, along with its own DNA. The segment of DNA may undergo "crossing over" with the new host's chromosome and thus incorporate genes from the previous host strain. Since only DNA is transferred in this way, this is further evidence confirming the hypothesis that genes are DNA.

Other Evidence. It had been known for a long time that nucleic acids absorb ultraviolet light very strongly, with the maximum absorption at a wavelength of 260 nm. It was also known that mutations could be produced by irradiating organisms with ultraviolet light. The wavelength of ultraviolet light most effective in producing mutations is also at 260 nm. When you compare the number of mutations produced per unit of energy delivered and the wavelengths at which that energy is delivered, you obtain an *action spectrum* for the production of mutations. The very close correlation between this action spectrum for the production of mutations and the absorption spectrum for nucleic acids suggests that genes are composed of nucleic acids. Mutations are produced when the nucleic acids absorb energy, and the absorbed energy in some way changes the nucleic acid molecule to produce a new mutant gene.

Using very gentle procedures, Stanley and Fraenkel-Conrat extracted from one plant virus both its nucleic acid and its proteins, which retained their biological properties; i.e., when protein and nucleic acid were mixed together they recombined and viral activity reappeared. When the reconstituted virus was applied to the leaf of the plant, the specific plant disease caused by the virus appeared. These investigators next extracted nucleic acid from one strain of virus and mixed it with protein extracted from another strain to produce a hybrid virus. This had the serological properties of the viral strain from which the protein was derived, but had the viral activity of the strain from which the nucleic acid came. The reconstituted virus produced the specific disease characteristic of the strain supplying the nucleic acid. In further experiments Fraenkel-Conrat found that isolated nucleic acid without any protein has some viral activity, although somewhat less than that of nucleic acid stabilized by the viral protein.

29–3 THE CONSTITUENTS OF DNA

The analyses of P. A. Levene had suggested that DNA, from whatever source, was composed of four nucleotides in equivalent amounts. However, during the 1940's Erwin Chargaff and his colleagues at Columbia University analyzed purified DNA from a variety of sources and showed clearly that the different nitrogenous bases do not occur in equal proportions. Chargaff found that although the proportion of these bases is the same in the DNA from all the cells of a given species, the DNAs from different species may differ markedly in the ratios of the constituent nucleotides. This suggested that the variations in the ratios of nitrogenous bases might represent a language. Although the ratios of purine and pyrimidine bases differed considerably in different samples of DNA, a pattern became apparent when these analyses were compared (Table 29–3). In all the samples the total amount of purines equaled the total amount of pyrimidines (A + G = T + C); the amount of adenine equaled the amount of thymine (A = T); and the amount of guanine equaled the amount of cytosine (G = C). DNA isolated from mammalian cells was in general rich in adenine and thymine and relatively poor in guanine and cytosine, whereas DNA isolated from bacterial sources was generally rich in guanine and cytosine and relatively poor in adenine and thymine. These findings constituted one of the important experimental bases on which the Watson-Crick model of DNA was eventually erected.

A further important clue about DNA structure came from Linus Pauling's studies of pro-

Table 29–3 RELATIVE AMOUNTS OF PURINES AND PYRIMIDINES IN SAMPLES OF DNA

Source	Adenine	Guanine	Cytosine	Thymine
Beef thymus	29.0	21.2	21.2	28.5
Beef liver	28.8	21.0	21.1	29.0
Beef sperm	28.7	22.2	22.0	27.2
Human thymus	30.9	19.9	19.8	29.4
Human liver	30.3	19.5	19.9	30.3
Human sperm	30.9	19.1	18.4	31.6
Hen red cells	28.8	20.5	21.5	29.2
Herring sperm	27.8	22.2	22.6	27.5
Wheat germ	26.5	23.5	23.0	27.0
Yeast	31.7	18.3	17.4	32.6
Vaccinia virus	29.5	20.6	20.0	29.9
T₂ bacteriophage	32.5	18.2	16.7	32.6

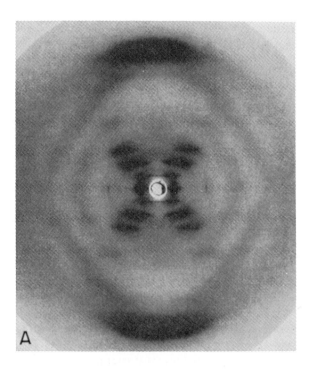

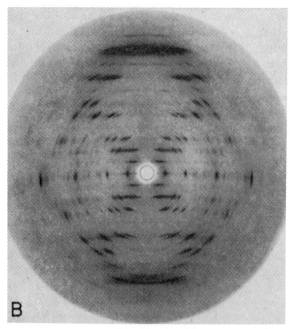

Figure 29-6 X-ray diffraction photographs of suitably hydrated fibers of DNA, showing the so-called B configuration. *A*, Pattern obtained using the sodium salt of DNA. *B*, Pattern obtained using the lithium salt of DNA. This pattern permits a most thorough analysis of DNA. The diagonal pattern of spots (reflections) stretching from 11 o'clock to 5 o'clock and from 1 o'clock to 7 o'clock provides evidence for the helical structure of DNA. The elongated horizontal reflections at the top and bottom of the photographs provide evidence that the purine and pyrimidine bases are stacked 0.34 nm. apart and are perpendicular to the axis of the DNA molecule. (Courtesy of Biophysics Research Unit, Medical Research Council, King's College, London.)

tein structure. Pauling had shown that there are several possible ways in which the amino acid chains of a protein may be held together. One of the favorite molecular structures, termed an *α-helix,* can be visualized as a peptide chain wound around a cylinder. This permits the formation of hydrogen bonds between the amino acids on successive turns of the screw. Pauling had described this α-helix form of protein molecules in 1950 and had suggested at that time that the structure of DNA might also prove to be some sort of helix held together by hydrogen bonds.

The primary clues about the structure of the DNA molecule came from studies using x-ray diffraction, carried out by Rosalind Franklin in the laboratory of M. H. F. Wilkins. When a pure crystal of DNA is bombarded with x-rays, the x-rays are diffracted or bent in specific directions as they pass through the substance. The amount and nature of the bending of the x-rays depends on the structure of the molecule itself. The pattern of x-ray diffraction (Fig. 29-6), although incomprehensible to the novice, provides to the experienced eye a number of clues about the structure of the molecule. From such x-ray diffraction pictures, Franklin and Wilkins inferred that the nucleotide bases (which are flat molecules) are stacked one on top of the other like a group of saucers. These x-ray diffraction patterns showed three major periodicities in crystalline DNA, one of 0.34 nm., one of 2.0 nm., and one of 3.4 nm.

29-4 THE WATSON-CRICK MODEL OF DNA

On the basis of Chargaff's analytical results and Franklin and Wilkins' x-ray diffraction patterns, Watson and Crick proposed in 1953 a model of the DNA molecule (Fig. 29-7) that has been very useful in providing a chemical explanation for many of its biological properties. The studies of Pauling and other chemists had provided a great deal of information about the exact distance between the atoms that are bonded together in a molecule, the angles between the bonds of a given atom, and the sizes of the atoms. Using this information, Watson and Crick began to build scale models of the component parts of DNA and then fit them together to agree with the various experimental

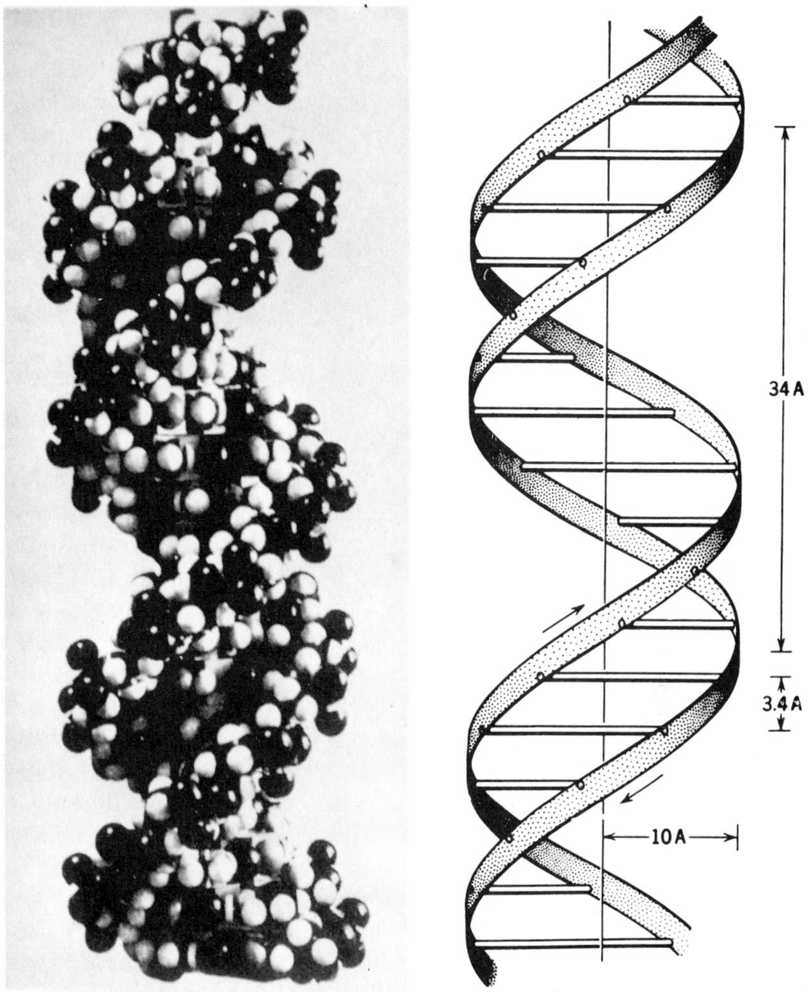

Figure 29-7 Photograph of a molecular model of deoxyribonucleic acid. (Courtesy of Dr. M. H. F. Wilkins). A schematic drawing of this two-stranded structure is shown on the right, together with certain of its dimensions in Ångstrom units. The arrows indicate that the two strands extend in opposite directions. (From Anfinsen, C. B.: *Molecular Basis of Evolution.* New York, John Wiley & Sons, 1959.)

data. It had been known that the adjacent nucleotides in DNA are joined in a chain by phosphodiester bridges, which link the 5′ carbon of the deoxyribose of one nucleotide with the 3′ carbon of the deoxyribose of the next nucleotide. It seemed clear to Watson and Crick that the 0.34 nm. periodicity found by Wilkins corresponded to the distance between successive nucleotides in the DNA chain. Further, it was a reasonable guess that the 2.0 nm. periodicity corresponded to the width of the chain. To explain the 3.4 nm. periodicity, they postulated, as Pauling had, that the chain was coiled in a helix. A *helix* is formed by winding a chain around a cylinder; in contrast, a *spiral* is formed by winding a chain around a cone. This 3.4 nm. periodicity corresponded to the distance between successive turns of the helix. A chain can be wound around a cylinder either loosely or tightly; this corresponds to the steepness of the pitch of the screw in the helix. Since 3.4 is just 10 times the 0.34

nm. distance between the successive nucleotides, it was clear that each full turn of the helix contained 10 nucleotides. From these data Watson and Crick could calculate the density of a chain of nucleotides coiled in a helix 2 nm. wide, with turns that were 3.4 nm. long. Such a chain would have a density only half as great as the known density of DNA. Consequently, they postulated that there were *two* chains—a *double helix* of nucleotides—that made up a DNA molecule.

The next problem, of course, was to determine the spatial relationships between the two chains that make up the double helix. Having tried a number of arrangements with their scale model, they found that the best fit with all the data was given by one in which the two nucleotide helices were wound in opposite directions (Fig. 29–8), with the sugar phosphate chains on the outside of the helix and the purines and pyrimidines on the inside, held together by **hydrogen bonds** between bases on

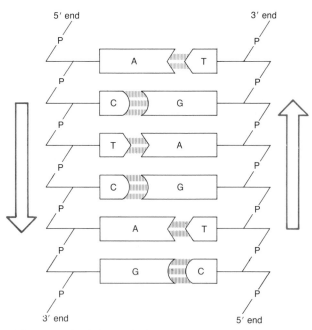

Figure 29–8 Schematic diagram of a portion of a DNA molecule showing the two polynucleotide chains joined by hydrogen bonds. The chains are not flat as represented here but are coiled around each other in helices (see Figure 29–7). The two strands extend in opposite directions as indicated by the arrows.

the opposite chains. These hydrogen bonds hold the chains together and maintain the helix. A double helix can be visualized by imagining the form that would be obtained by taking a ladder and twisting it into a helical shape, keeping the rungs of the ladder perpendicular. The sugar and phosphate molecules of the nucleotide chains make up the railings of the ladder, and the rungs are formed by the nitrogenous bases held together by hydrogen bonds.

Further study of the possible models made it clear to Watson and Crick that each crossrung must contain one purine and one pyrimidine. The space available with the 2.0 nm. periodicity would accommodate one purine and one pyrimidine, but not two purines, which would be too large, and not two pyrimidines, which would not come close enough together to form proper hydrogen bonds. Further examination of the detailed model showed that although a combination of adenine and cytosine was the proper size to fit as a rung on the ladder, they could not be arranged in such a way that they would form proper hydrogen bonds. A similar consider-

ation ruled out the pairing of guanine and thymine; however, adenine and thymine would form hydrogen bonds, and guanine and cytosine could form hydrogen bonds. The nature of the hydrogen bonds requires that adenine pair with thymine and that guanine pair with cytosine. This concept of *specific base pairing* provided a basis for Chargaff's rule that the amounts of adenine and thymine in any DNA molecule are always equal, and the amounts of guanine and cytosine are always equal. Two hydrogen bonds can form between adenine and thymine, and three hydrogen bonds between guanine and cytosine (Fig. 29–9). The specificity of the kind of hydrogen bond that can be formed assures that for every adenine in one chain, there will be a thymine in the other chain. Similarly for every guanine in the first chain there will be a cytosine in the second chain. Thus the two chains are complementary to each other; i.e., the sequence of nucleotides in one chain dictates the sequence of nucleotides in the other. The two strands are *antiparallel;* they extend in opposite directions and have their terminal phosphate groups at opposite ends of the double helix.

The most distinctive properties of the genetic material are that it carries information and undergoes replication. The Watson-Crick model explains how DNA molecules may carry out these two functions. When a DNA molecule undergoes replication, the two chains separate and each one brings about the formation of a new chain, which is complementary to it; thus two new chains are established (Fig. 29–10). The nucleotides in the new chain are assembled in a specific order, because each purine or pyrimidine in the original chain forms hydrogen bonds with the complementary pyrimidine or purine nucleotide triphosphate from the surrounding medium and lines them up in a complementary order. Phosphate ester bonds are formed by the reaction catalyzed by *DNA polymerase* to join the adjacent nucleotides in the chain, and a new polynucleotide chain results (Fig. 29–11). The new and the original chains then wind around each other, and two new DNA molecules are formed. Each chain, in other words, serves as a template or a mold against which a new partner chain is synthesized. The end result is two complete double chain molecules, each identical to the original double-chain molecule.

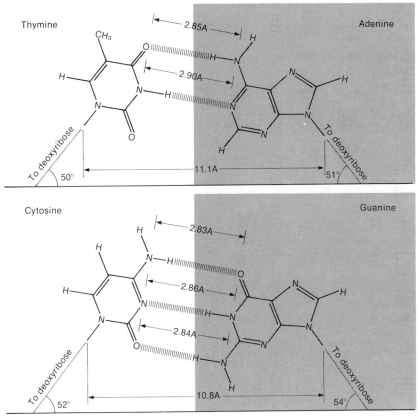

Figure 29-9 Diagram of the hydrogen bonding between the base pairs adenine and thymine *(above)* and guanine and cytosine *(below)* in DNA. The A-T pair has two hydrogen bonds and the G-C pair has three.

A second prime function of DNA, in addition to its role in replication, is that the information contained in its specific sequence of nucleotides must be transcribed some time between cell division. The product of the transcription process, messenger RNA, then combines with ribosomes to carry out the synthesis of enzymes and other specific proteins. Thus we can visualize how each gene can lead to the production of a specific enzyme. We shall return to this subject in Section 29-9.

29-5 THE SYNTHESIS OF DNA: REPLICATION

Within a few years after its formulation, the Watson-Crick model received strong experimental support from several sources. In 1957, Arthur Kornberg and his colleagues isolated the enzyme DNA polymerase from bacteria. This catalyzes the synthesis of DNA and requires as substrates the triphosphates of all four deoxyribonucleosides (abbreviated dATP,

dGTP, dCTP and dTTP). The reaction system further requires magnesium ions (Mg^{++}) and a small amount of high molecular weight DNA polymer to serve as primer or template for the reaction. The product of the reaction is more DNA polymer and one molecule of pyrophosphate (PP_i) for each molecule of deoxyribonucleotide incorporated.

$$\left.\begin{array}{l} \text{dATP} \\ \text{dGTP} \\ \text{dCTP} \\ \text{dTTP} \end{array}\right\}_n \quad \begin{array}{l} \text{DNA} \\ Mg^{++} \\ \xrightarrow{\hspace{1cm}} \text{DNA} + nPP_i \\ \text{DNA polymerase} \end{array}$$

The nucleoside triphosphate attacks the free 3'-hydroxyl of the last deoxyribose in the chain and forms an ester bond, freeing a molecule of pyrophosphate (Fig. 29–11). When the deoxyribonucleotide triphosphates were labeled with ^{14}C, the DNA polymer produced contained ^{14}C, permitting the inference that the labeled nucleotides had been incorporated into

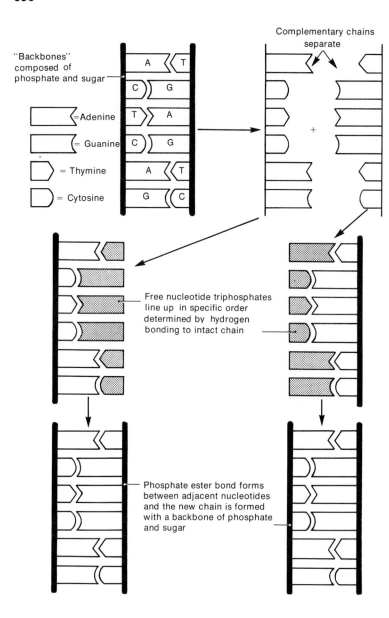

"Backbones" composed of phosphate and sugar

☐⟨ =Adenine

☐) = Guanine

☐ = Thymine

☐ = Cytosine

Complementary chains separate

Free nucleotide triphosphates line up in specific order determined by hydrogen bonding to intact chain

Phosphate ester bond forms between adjacent nucleotides and the new chain is formed with a backbone of phosphate and sugar

Figure 29-10 Diagrammatic scheme of how DNA molecules may undergo replication in the reaction catalyzed by the enzyme DNA polymerase.

the DNA chain. By appropriate experiments with ¹⁴C-labeled nucleotides, Kornberg could show that the ratios of A : T and of G : C in the DNA synthesized were the same as the corresponding ratios in the DNA used as primer. This suggested that the DNA produced is a copy of the primer DNA, as predicted by the Watson-Crick model.

The DNA polymerase from *Escherichia coli* will use template DNA isolated from any of a wide variety of sources—bacteria, viruses, mammalian cells and plant cells—and will produce DNA with a nucleotide ratio comparable to that of the template used. Thus the sequence of nucleotides in the product is dictated by the sequence in the primer DNA, and

not by the properties of the polymerase nor by the ratio of the substrate molecules present in the reaction mixture. Using a more highly purified enzyme, Kornberg was able, in 1968, to synthesize biologically active viral DNA, using viral DNA as primer. The DNA produced would infect bacteria just like "live" viruses.

Khorana and his colleagues synthesized a number of deoxyribonucleotide polymers containing adenylic and cytidylic acids, and other polymers containing alternating thymidylic and guanylic acids. Neither of these polymers alone was able to serve as primer or template in the DNA polymerase system; however, a mixture of the two, which form a synthetic double-stranded helix with conventional

Watson-Crick pairing of the bases, can serve as a template.

The DNA template appears to have two functions in the DNA polymerase system. First, it provides 3'-OH groups which are free to serve as the growing end of the DNA polymer. Next, the DNA template provides coded information. A double-stranded molecule is required because each strand of the pair serves as a *template* for the extension of the complementary strand. It also serves as a

primer for its own extension. The DNA-like polymer that is produced by the action of the DNA polymerase in the presence of a double-stranded template is also double-stranded. It has the same base composition as the template DNA. The ratios of the bases in the product are those predicted by the Watson-Crick model.

The synthesis of DNA occurs in the cells of higher organisms only during the interphase when chromosomes are in their extended form and are not readily visible. Thus, if an enzyme

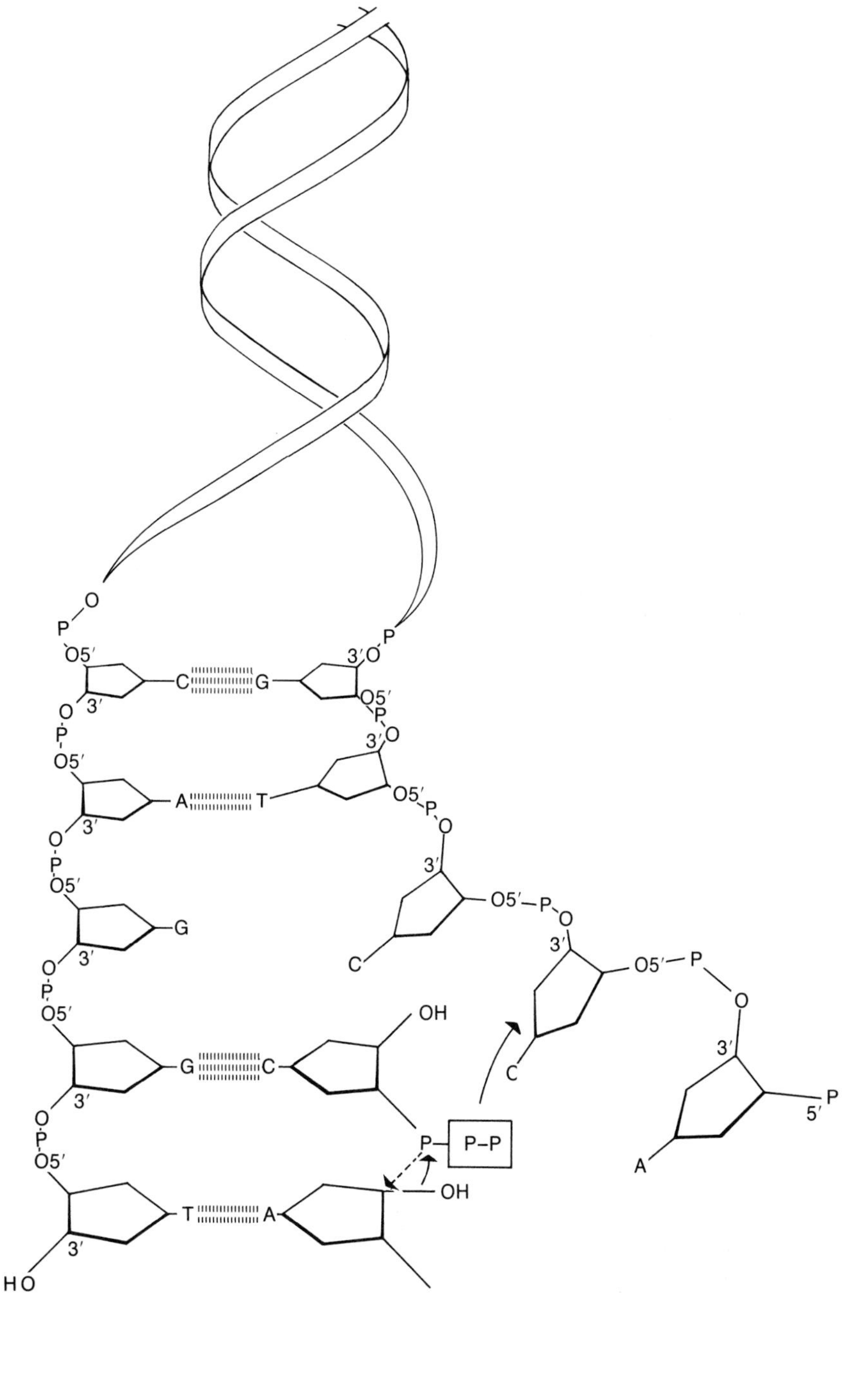

Figure 29–11 Mechanism of the replication of DNA by DNA polymerase. The two strands of the DNA double helix are shown separating. The left one, which runs from the 5' phosphate to 3' OH, is being copied, starting from the bottom. The newly synthesized chain begins with a 5' phosphate. In the new chain, as in the old, adenine forms base pairs with thymine, and cytidine forms base pairs with guanine. The new phosphodiester bond is made between adjacent bases in the forming chain by an attack by the 3' OH group of the deoxyribose on the bond between the inner phosphate and the outer two phosphates of the adjacent trinucleotide. The two outer phosphates are split off as inorganic pyrophosphate.

system similar to the Kornberg DNA polymerase catalyzes the synthesis of DNA *in vivo*, there must be some sort of biological signal which will initiate DNA synthesis at this time and will turn it off at other times. It appears that both the enzyme, the DNA polymerase, and the substrates dATP, dGTP, dCTP and dTTP are present all the time. The most likely explanation at present is that some sort of change in the DNA template initiates the synthesis of DNA at the appropriate time in the cell cycle and then turns it off.

Semiconservative Replication. During the replication of DNA, two new strands are formed, each of which is complementary to one of the existing DNA strands in the double-stranded helix. The double-stranded helix unwinds; one strand provides a template for one new strand, and the other original strand provides a template for the second new strand. This is called a "semiconservative" replication: the two original strands of DNA are retained or conserved in the product, one in each of the two daughter helices (Fig. 29–12).

Strong experimental evidence supports the concept that each eukaryotic chromosome is a single DNA molecule. The DNA molecules of the fruit fly *Drosophila* have been shown to be as long as 2.1 cm., just the length of the longest chromosome. A question that immediately comes to mind is how a eukaryotic cell manages to replicate such an enormous molecule in the allotted time during the cell cycle. The

answer is that replication does not simply begin at one end of the molecule and proceed to the other. Instead, replication begins at many sites along the chromosome (some two micrometers apart on the average in the rapidly dividing fertilized egg) and proceeds in *both* directions from the origin at about the same rate, one micrometer per minute.

The classic experiment of Meselson and Stahl used heavy nitrogen, ^{15}N, to distinguish "old" and "new" molecules of DNA and provided evidence that DNA replication is indeed carried out by a semiconservative process, at least in bacteria (Fig. 29–13). Bacteria grown for several generations in a medium containing heavy nitrogen had DNA (and RNA and protein) that was labeled with ^{15}N. When a sample of the DNA was isolated and centrifuged in a tube containing a cesium chloride density gradient, the DNA collected at a level in the tube that reflected its increased density owing to the presence of the heavy nitrogen atoms.

The bacteria were then transferred from the ^{15}N medium to a medium containing ordinary nitrogen, ^{14}N, and were allowed to divide once in this medium. When the DNA from this generation of bacteria was isolated and centrifuged, all the DNA was lighter, with the density expected if it had just half as many ^{15}N atoms as the DNA of the parental generation. If the Watson-Crick theory is correct and replication is semiconservative, this result would be expected, because one strand of the

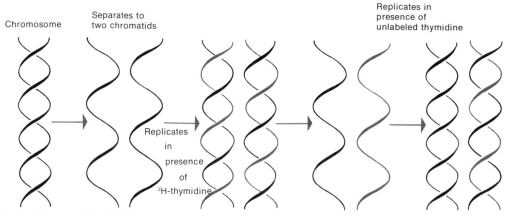

Figure 29–12 When rapidly dividing cells were grown in the presence of ^{3}H-thymidine and fixed for autoradiographic study, all the chromosomes were labeled (gray) and radioactivity was equally distributed between the two chromatids of each chromosome (center). If these cells were not fixed but allowed to divide once more in the presence of unlabeled thymidine and then fixed for study, only one of the two chromosomes would be labeled *(right)*. Thus each chromatid acts as a template for the formation of a new chromatid. The replication of chromosomes appears to be formally similar to the replication of the DNA double helix, but the relation between the cytologic and molecular events is not yet clear.

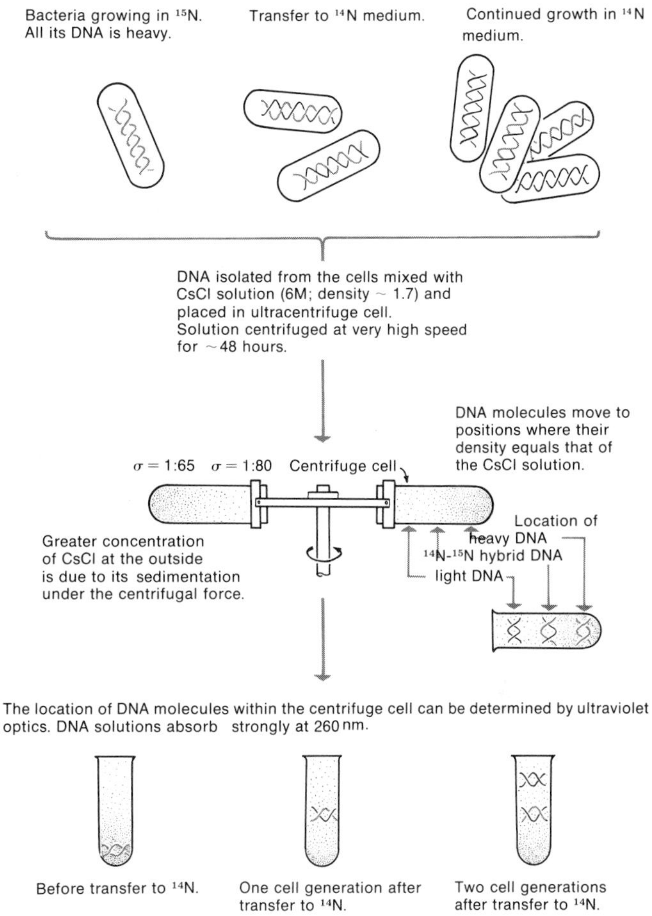

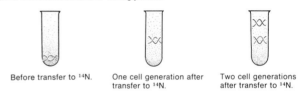

Bacteria growing in ¹⁵N. All its DNA is heavy.   Transfer to ¹⁴N medium.   Continued growth in ¹⁴N medium.

DNA isolated from the cells mixed with CsCl solution (6M; density ~ 1.7) and placed in ultracentrifuge cell. Solution centrifuged at very high speed for ~48 hours.

Figure 29-13 Diagram of the experiment of Meselson and Stahl, which indicated that DNA is replicated by a semiconservative mechanism: the two original strands of DNA are retained in the product, one in each daughter helix.

DNA molecules move to positions where their density equals that of the CsCl solution.

$\sigma = 1:65$ $\sigma = 1:80$ Centrifuge cell

Location of heavy DNA
¹⁴N-¹⁵N hybrid DNA
light DNA

Greater concentration of CsCl at the outside is due to its sedimentation under the centrifugal force.

The location of DNA molecules within the centrifuge cell can be determined by ultraviolet optics. DNA solutions absorb strongly at 260 nm.

Before transfer to ¹⁴N.   One cell generation after transfer to ¹⁴N.   Two cell generations after transfer to ¹⁴N.

double-stranded DNA in each organism would be labeled with ¹⁵N and the other would contain only ¹⁴N.

When these bacteria were allowed to divide a second time in the ¹⁴N medium, each molecule of DNA in the progeny again received one parental strand and one new strand containing only ¹⁴N. Some double-stranded DNA containing only ¹⁴N was formed and appeared as a light DNA on centrifugation. The parental strands containing ¹⁵N made complementary strands containing ¹⁴N, which were sedimented on centrifugation with a density characteristic of the half ¹⁵N, half ¹⁴N double-stranded state. Thus the original parental strands of DNA are not dispersed or split apart during the replication process but are conserved and passed to the next generation of cells. Each strand of the parental double helix is conserved in a different daughter cell, hence the process is termed semiconservative.

If the replication of the strands begins as the strands begin to untwist, **Y**-shaped mole-

cules of DNA should be evident during the replication process. Such **Y**-shaped regions have been found by autoradiography of chromosomes of *Escherichia coli* labeled with ³H-thymidine (Fig. 29–14).

29-6 THE GENETIC CODE

The Watson-Crick model of the DNA molecule implied that genetic information is transmitted by some specific sequence of its constituent nucleotides. In 1954, George Gamow, an imaginative physicist, was one of the first to suggest that the minimum coding relation between nucleotides and amino acids would be three nucleotides per amino acid. Four nucleotides taken two at a time provide for only 16 combinations ($4^2 = 16$), whereas four nucleotides taken three at a time provide for 64 combinations ($4^3 = 64$). At first glance, this would seem to provide many more code symbols than are needed, since there are only 20 different

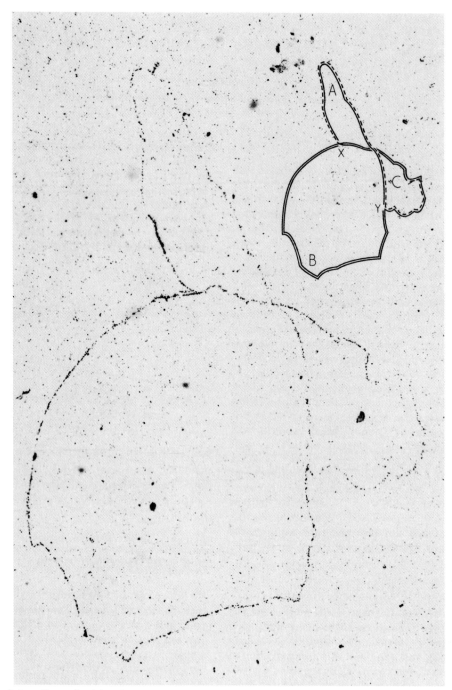

Figure 29–14 Autoradiograph of the chromosome of *E. coli* (strain K12 Hfr) labeled with tritiated thymidine for two generations and extracted by treatment of the cells with lysozyme. The inset shows the same chromosome in diagram. A predominantly half "hot" chromosome that has completed two-thirds of its second round of duplication is shown. Part of the still-unduplicated section is half marked with tracer (*Y* to *C*) and part is doubly marked with tracer (*C* to *X*). (From Cairns, J.: Cold Spring Harbor Symp. Quant. Biol. *28*:44, 1963.)

Table 29-4 THE GENETIC CODE: THE SEQUENCE OF NUCLEOTIDES IN THE TRIPLET CODONS OF MESSENGER RNA THAT SPECIFY A GIVEN AMINO ACID

First Position (5′ end)	Second Position	Third Position (3′ end)			
		U	C	A	G
U	U	Phe	Phe	Leu	Leu
	C	Ser	Ser	Ser	Ser
	A	Tyr	Tyr	Terminator	Terminator
	G	Cys	Cys	Terminator	Trp
C	U	Leu	Leu	Leu	Leu
	C	Pro	Pro	Pro	Pro
	A	His	His	Glu·NH$_2$	Glu·NH$_2$
	G	Arg	Arg	Arg	Arg
A	U	Ileu	Ileu	Ileu	Met
	C	Thr	Thr	Thr	Thr
	A	Asp·NH$_2$	Asp·NH$_2$	Lys	Lys
	G	Ser	Ser	Arg	Arg
G	U	Val	Val	Val	Val
	C	Ala	Ala	Ala	Ala
	A	Asp	Asp	Glu	Glu
	G	Gly	Gly	Gly	Gly

amino acids. It was believed at one time that some of these 64 combinations were simply "nonsense" codes that did not specify any amino acid. However, there is now strong evidence that all but three of the 64 combinations do, in fact, code for one or another amino acid, and that as many as six different nucleotide triplets may specify the same amino acid.

The fundamental characteristics of the genetic code are now well established: it is a triplet code with three adjacent nucleotide bases, termed a *codon,* specifying each amino acid (Table 29–4). For some years the question of whether the code was overlapping or not remained unsolved. For example, is the sequence CAG, AUC, GAC read only as CAG, AUC, GAC or can it also be read CAG, AGA, GAU, AUC, UCG, CGA, GAC? Is each nucleotide part of one codon or three? The amino acid sequences of each of the several mutant forms of the hemoglobin molecule have been analyzed. In each, only a single amino acid in the peptide chain is substituted. In contrast, if the code were overlapping and a given nucleotide were part of three adjacent codons, we would expect three adjacent amino acids to be changed. An overlapping code would restrict the possible orders of amino acids in a peptide. Thus the amino acid specified by CAG could be followed only by AG(X), by one of two amino acids. It is clear from analyses of amino

acid sequences in peptides that there are no such restrictions on the sequences possible. Finally, experiments with synthetic polynucleotides having known base sequences have shown conclusively that the code is not overlapping.

The code is commaless; no "punctuation" is necessary since the code is read out beginning at a fixed point and the entire strand is read, three nucleotides at a time, until the read-out mechanism comes to a specific "termination" code, which signals the end of the message. The nature of the signal, "begin reading here," at least in bacteria, also appears to be specified by a specific sequence of bases.

Experimental Evidence for a Triplet Code. From a mathematical analysis of the coding problem, Crick concluded early in 1961 that three consecutive nucleotides in a strand of messenger RNA provide the code that determines the position of a single amino acid in a polypeptide chain. Experimental evidence to support this was quickly forthcoming from the laboratory of Nirenberg and Matthaei. Using purified enzyme systems, they studied the incorporation of specific labeled amino acids into protein under the direction of artificial messenger RNAs of known composition. Nirenberg prepared a synthetic *polyuridylic acid* (UUUUU), using the enzyme polynucleotide phosphorylase. When this artificial messenger

RNA was added to a system of purified enzymes for the synthesis of proteins, *phenylalanine,* and no other amino acid, was incorporated into protein; the polypeptide that resulted contained only phenylalanine. The inference that UUU is the code for phenylalanine was inescapable. Further similar experiments by Nirenberg and by Severo Ochoa showed that polyadenylic acid provided the code for lysine and polycytidylic acid coded for proline. Making mixed nucleotide polymers (such as poly AC) and using them as artificial messengers made possible the assignment of many other nucleotide combinations to specific amino acids.

These experiments did not reveal the order of the nucleotides within the triplets, but this has been inferred from other kinds of experiments. Nirenberg and Leder discovered that even when no protein synthesis is occurring, specific amino acyl transfer RNA molecules will be attached to ribosomes when messenger RNA is present. Fortunately this effect does not require a long molecule of mRNA, which would be difficult to synthesize. Indeed, synthetic messenger RNA molecules as short as trinucleotides will suffice to promote the binding of specific amino acyl transfer RNAs to the ribosomes. It is possible to synthesize trinucleotides of known sequence; using these the coding assignment of all 64 possible triplets has been determined. For example, GUU, but not UGU nor UUG, induces the binding of valine transfer RNA to ribosomes. UUG induces the binding of leucine transfer RNA. Since GUU and UUG code for different amino acids, it follows that the reading of the code in the messenger RNA strand makes sense only in one direction.

Codon-Amino Acid Specificity. Careful examination of the coding relationships (Table 29–4) shows that there is a pattern to the "degeneracy" in the code. Bernfield and Nirenberg found, in 1965, that the binding of phenylalanine transfer RNA is induced by both UUU and UUC. The binding of serine transfer RNA is induced by both UCC and UCU, and proline transfer RNA is bound by CCC or CCU. In all these the two alternative triplets are identical except for the substitution of one pyrimidine for the other (C or U) at the 3′ end. In other instances, adenine and guanine can be interchanged in the 3′ position of the triplet.

For example, the code for lysine may be either AAG or AAA. For a number of amino acids, the first two nucleotides of the codon are specific but any of the four nucleotides may be present in the 3′ position. All of the four possible combinations will code for the same amino acid. In these, although the code may be read three nucleotides at a time, only the first two nucleotides appear to contain specific information. Only methionine and tryptophan have single triplet codes; all the other amino acids are specified by from two to as many as six different nucleotide triplets.

That messenger RNA is read three nucleotides at a time was firmly established by experiments carried out by Khorana and his colleagues. A poly UC messenger containing the regularly alternating base sequence UCUCUCUC was synthesized and used in a protein-synthesizing system. The resulting polypeptide contained a regular alternation of serine and leucine. Mathematical analysis of this result shows that the coding unit must contain an odd number of bases. Khorana then synthesized the nucleotide sequence AAGAAGAAGAAG. When this nucleotide polymer was used as template in a protein-synthesizing system, the result was either polylysine (AAG), polyglutamate (GAA) or polyarginine (AGA). The type of peptide synthesized depends on which nucleotide in the polynucleotide chain happens to be read first (Fig. 29–15). This *"frame shift" effect* (as in a movie camera) can be accounted for only if the chain is read in sequence, three nucleotides at a time, beginning from a fixed point. There are a few instances in which a specific nucleotide at the 5′ end of the codon can be changed and still provide a code for the same amino acid. For example, both UUG and CUG provide a code for leucine. Arginine is coded by six nucleotide triplets, two of which differ in containing A instead of C at the 5′ end. It appears that the middle nucleotide is the most informative one in the triplet, that the 5′ nucleotide is the next most informative, and the 3′ nucleotide is least specific.

Coding Degeneracy. The biological significance of the finding that the code is "degenerate," that more than one codon may specify a given amino acid and lead to its insertion into a growing peptide chain, is not fully understood. We must keep in mind that what has been

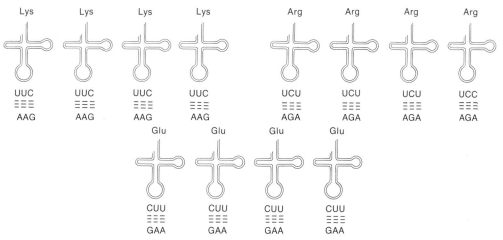

Figure 29–15 Diagram illustrating the "frame-shift" effect in the translation of the synthetic polynucleotide AAGAAGAA-GAAG. This led to the synthesis of polylysine, polyglutamate or polyarginine, depending on which nucleotide was read first. When read as AAG, polylysine was formed; when read as AGA, polyarginine was formed; and when read as GAA, polyglutamate was formed. The diagram emphasizes the three different codon-anticodon relationships possible.

termed a code is in fact a complex physical chemical system that depends for its operation on the specificity of several reactions catalyzed by enzymes. The sequences of nucleotides in DNA will have no meaning except in relation to a specific set of amino acyl transfer RNAs, each of which is synthesized by a specific enzyme. The protein is then synthesized on the ribosome by yet another enzyme system operating on the enzyme-messenger RNA-ribosome-transfer RNA complex. The specificity of the system as a whole depends on the ability of the enzyme to recognize specific amino acyl transfer RNAs when these are present at the ribosomal binding site.

Although it is known that there may be more than one transfer RNA for a given amino acid, it is not yet clear whether the number of amino acyl transfer RNA molecules equals the number of codons for that specific amino acid. Do the six different codons for serine correspond to six different serine-transfer RNA molecules, or is there only a single serine tRNA that responds to any of the six mRNA codons? Crick's *"wobble" hypothesis* (Fig. 29–16), states that the first two bases of a transfer RNA anticodon undergo hydrogen bonding specifically with the first two bases of the messenger RNA codon in antiparallel. However, the third base can undergo unusual base pairing—i.e., it can "wobble." His theory predicts that the inosine residue of alanine transfer RNA may pair either with U, C or A in a messenger codon. This could account for three of the four codons for alanine and would require only one additional transfer RNA to recognize the GCG code. The organisms that have been studied so far appear to have a set of at least 40 different

Figure 29–16 Diagram illustrating Crick's "wobble" hypothesis. The first two bases of the transfer RNA anticodon (CG) form the expected specific hydrogen bonds with the first two bases of the messenger RNA codon in antiparallel; however, the third base of the anticodon (I) can base pair with U, C or A in a messenger codon. This theory accounts for three of the four known codons for alanine.

transfer RNAs. Messenger RNA is read from the 5′ end toward the 3′ end; the codon at the 5′ end corresponds to the N-terminal amino acid of the polypeptide. Three of the 64 codons, UAA, UAG and UGA, do not specify any amino acid. Brenner suggested that these are "terminator" triplets, which signal the end of the polypeptide chain and cause the protein chain to become detached from the ribosome.

Universality of the Code. There is good reason to believe that the genetic code is universal; that is, a given codon specifies the same amino acid in the protein-synthesizing systems of all organisms, from viruses to the human. Synthetic poly U, for example, causes synthesis of polyphenylalanine in cell-free systems derived from bacteria, wheat germ, or rabbit reticulocytes. Perhaps even more convincing is the experimental finding that RNA isolated from tobacco mosaic virus and used as a template in the protein-synthesizing system derived from *E. coli* synthesizes proteins that are similar to native tobacco mosaic virus protein. RNA from the chick oviduct used as the template in a protein-synthesizing system of ribosomes and transfer RNA from *E. coli* resulted in the synthesis of *ovalbumin,* the characteristic protein of egg white. More indirect but quite persuasive evidence of universality can be obtained by looking carefully at the replacements of one amino acid for another found in the more than 60 mutant forms of human hemoglobin that have been analyzed. All of these replacements are ones that would be expected if there was a *single* base transformation between the triplets assigned to these amino acids on the basis of their behavior in the *E. coli* cell-free, protein-synthesizing system. The inference is that the code is the same in human beings and in bacteria.

One apparent exception to this generalization proved ultimately to be a striking confirmation of it. The amino acid sequence of a mutant form of hemoglobin, hemoglobin I, was initially reported as involving a substitution of aspartate for lysine. This would require a change in two of the three nucleotides in the codon, a "double mutation." This would be possible theoretically, but quite improbable statistically. However, a reinvestigation of the amino acid composition showed that the changed amino acid was glutamate, and a substitution of glutamate for lysine requires a change in only one of the three nucleotides in the codon.

Cytochrome c, a protein constituent of the electron transmitter system, has been isolated from several organisms and the amino acid sequence of each peptide has been determined. The differences from one species to another are very small, and the amino acid substitutions are those expected if a single nucleotide were substituted for another in that codon. Similar analyses of a number of other proteins have given similar results, and this indirect evidence for the universality of the genetic code is very strong.

Colinearity. DNA is a linear polynucleotide chain and a protein is a linear polypeptide chain. The sequence of amino acids in the peptide chain is dictated by the order of the corresponding nucleotide bases in codons in the messenger RNA, and this in turn is determined by the sequence of nucleotides in one of the two polynucleotide chains of the DNA molecule. Changing the sequence of nucleotides in the DNA molecule produces a corresponding change in the sequence of amino acids in peptides. The DNA molecule and the resulting polypeptide chain are said to be *colinear.*

This concept of colinearity was implicit in the original Watson-Crick model of the DNA molecule. Direct evidence of colinearity came from analyses by Charles Yanofsky of the genetic control of the enzyme tryptophan synthetase in bacteria.

Tryptophan synthetase is composed of four subunits, two A chains and two B chains. The A chain, studied by Yanofsky, is a polypeptide containing 267 amino acids. Yanofsky carefully mapped the genetic locations of each of a large number of mutants with altered A chains. He then collected A-chain polypeptides from each of the mutant strains and analyzed them to determine which amino acid had been changed. His analyses showed that the relative position of the changed nucleotide within a gene, as determined by genetic analysis, corresponds to the relative position of the altered amino acid in the peptide chain of the enzyme molecule, as determined by direct chemical analysis of the peptide. Each amino acid substitution could be accounted for by a change in a single nucleotide in a codon. Glycine (GGA), for example, was replaced by glutamate (GAA) in the tryptophan synthetase of one mutant and by arginine (AGA) in another. Other examples of colinearity have been derived from analyses of the genetic control of hemoglobin synthesis in man.

29–7 TRANSCRIPTION: THE SYNTHESIS OF RNA

RNA differs from DNA in containing ribose instead of deoxyribose, and uracil instead of thymine. Unlike DNA, the purine and pyrimidine nucleotides are usually not present in RNA in complementary ratios. This indicates that RNA is not a double helix like DNA but rather is single-stranded. The molecules of RNA are unbranched and contain the four kinds of ribonucleotides, A, G, C and U, linked by 3′, 5′ phosphodiester bonds.

The synthesis of proteins requires three kinds of RNA molecules: (1) *messenger RNA* transmits genetic information from the DNA molecule in the nucleus to the cytoplasm; (2) *ribosomal RNA* makes up a large portion of the cytoplasmic particles called *ribosomes* on which protein synthesis occurs; and (3) *transfer RNA* acts as an *adaptor* to bring the proper amino acid into line in the appropriate place in the growing polypeptide chain. Messenger RNA is synthesized by a *DNA-dependent RNA polymerase* first found in the nuclei of rat liver and identified subsequently as an important constituent of plant, bacterial and animal cells. The enzyme requires DNA as the template and uses as substrate the triphosphates of the four ribonucleosides commonly found in RNA. The products are RNA and inorganic pyrophosphate. The reaction system can use single-stranded or native DNA or a synthetic deoxyribopolynucleotide as template.

Transfer RNA and ribosomal RNA are also produced in the nucleus by DNA-dependent RNA-synthesizing systems and are transcribed from complementary deoxynucleotide sequences in DNA.

The RNA that is produced when carefully defined DNA templates are used is exactly that predicted by the kinds of base pairing permitted by the Watson-Crick model. The DNA template is double-stranded and contains two different but complementary template sequences that would have quite different genetic information. Only one DNA strand is apparently selected for transcription, and only

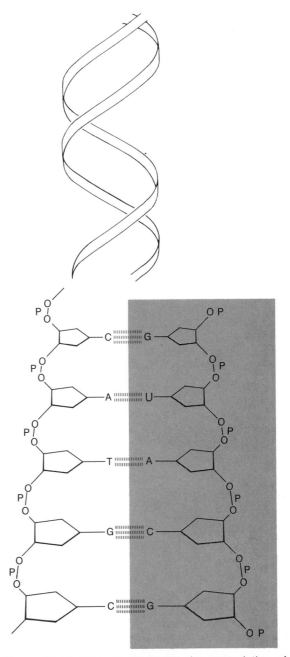

Figure 29–17 Diagram illustrating the transcription of DNA to form RNA, a process catalyzed by DNA-dependent RNA polymerase. The two strands of DNA are shown separating and the left one, which runs from the 5′ phosphate to the 3′ OH is being copied. The ribonucleoside triphosphates undergo complementary base pairing with the DNA nucleotides, then the enzyme joins the phosphodiester bonds. The growing RNA polynucleotide chain is shown in tint.

one kind of messenger RNA is produced (Fig. 29–17). The molecular basis for the distinction between the two complementary DNA strands is unknown. Biological information thus flows from DNA to RNA and then to protein: the specific sequence of nucleotides in DNA determines the sequence of nucleotides in RNA, and this, as we shall see, determines the sequence of amino acids in protein.

Electron microscopy has established that most cells contain an extensive system of tubules with thin membranes, termed *endoplasmic reticulum.* Associated with the endoplasmic reticulum, or floating freely in the cytoplasm, are small particles termed *ribosomes,* composed of RNA and protein.

Molecules of transfer RNA are considerably smaller than the molecules of messenger or ribosomal RNA. Each functions as a specific adaptor in protein synthesis, binding to and identifying one specific amino acid; i.e., each of the 20 amino acids is attached to one (or more) specific kinds of transfer RNA. One portion of the nucleotide sequence in transfer RNA represents an *anticodon,* a nucleotide triplet complementary to the codon in messenger RNA that specifies that amino acid. Transfer RNA is unusual in containing not only the four usual ribonucleotides, adenylic, guanylic, cytidylic and uridylic, but also small amounts of unusual nucleotides such as *6-methylamino adenylic acid, dimethylguanylic acid* and *thymine ribotide.*

Transfer RNAs are polynucleotide chains of some 70 nucleotides. Each of the several kinds of transfer RNA has an identical sequence of nucleotides, CCA, at the 3′ end to which the amino acid is attached. Each in addition has guanylic acid at the opposite, 5′, end of the nucleotide chain. The chain is doubled back on itself, forming three or more loops of unpaired nucleotides; the folding is stabilized by hydrogen bonds between complementary bases in the intervening portions of the chain to form a double helix (Fig. 29–18). The loop nearest the amino acid acceptor (CCA) has seven nucleotides, with cytidine, pseudouridine and thymidine at positions 21, 22 and 23 from the CCA end. The triplet that is complementary to the codon is located in a seven-membered middle loop and is preceded by uridine and followed by adenosine or a modified adenosine in the loop. Another unusual base, dimethylguanosine, is found eight positions before the anticodon, at the base of the larger (eight to twelve nucleotides) loop near the 5′ end. The folding results in a constant distance between the anticodon and the amino

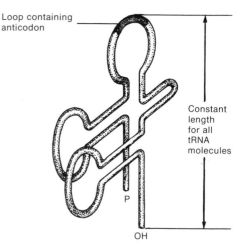

Figure 29–18 A diagram of the three-dimensional clover leaf structure of transfer RNA. One loop contains the triplet anticodon which forms specific base pairs with the mRNA codon. The amino acid is attached to the terminal ribose at the 3′ OH end, which has the sequence of CCA of nucleotides. Each transfer RNA also has guanylic acid, G, at the 5′ end (P). The pattern of folding permits a constant distance between anticodon and amino acid in all transfer RNAs examined.

acid in all of the tRNAs examined so far. The triplet anticodons found in these tRNAs agree very well (Table 29–5) with the complements predicted to the known mRNA codons for those amino acids.

The first transfer RNA to be completely analyzed was the transfer RNA for alanine derived from yeast cells. This dramatic achievement of Robert Holley and his colleagues was recognized by the awarding of a Nobel Prize to Holley in 1968. The alanine tRNA has 77 nucleotides arranged in a unique sequence. Nine of these are unusual bases, with one or more methyl groups that are added enzymatically after the nucleotides have been linked by phosphodiester bonds. Certain of these unusual bases cannot form conventional base pairs and may serve to disrupt the base pairing in other parts of the transfer RNA. This could expose specific chemical groups on the tRNA which form secondary bonds to messenger RNA or to the ribosome, or perhaps to the enzyme needed to attach the specific amino acid to its specific transfer RNA molecule. The exact sequence of nucleotides in several other transfer RNAs is now known. This implies that the sequence of the nucleotides in the genes that specify each of these particular kinds of transfer RNA is also known. The sequence in

Table 29–5 AGREEMENT BETWEEN THE ANTICODONS FOUND IN ISOLATED tRNAs AND THE ANTICODONS PREDICTED FROM THE GENETIC CODE

Amino Acid	mRNA Codons →	Complements ←	Observed Anticodons* ←
Alanine	GpCpA GpCpG GpCpC GpCpU	UpGpC CpGpC GpGpC ApGpC	IpGpC
Phenylalanine	UpUpC UpUpU	GpApA ApApA	OMGpApA
Tyrosine	UpApU UpApC	ApUpA GpUpA	GpψpA
Serine	ApGpC ApGpU UpCpC UpCpU UpCpA UpCpG	GpCpU ApCpU GpGpA ApGpA UpGpA CpGpA	IpGpA
Valine	GpUpU GpUpC GpUpA GpUpG	ApApC GpApC UpApC CpApC	IpApC

*The anticodons contain some "unusual" bases, including I (inosine), OMG (2′-O-methylguanosine) and ψ (pseudo-uridine).

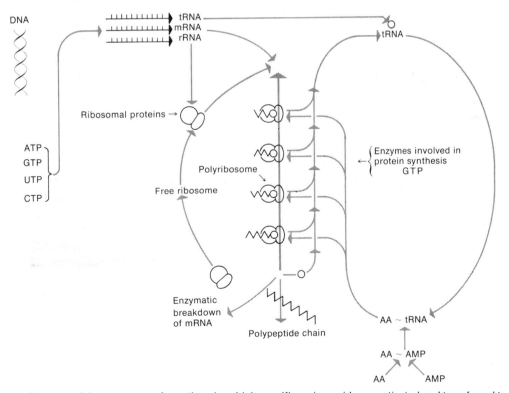

Figure 29–19 Diagram of the sequence of reactions by which specific amino acids are activated and transferred to specific transfer RNAs (aa-tRNA) and then, on the ribosome, transferred to a specific spot in the growing polypeptide chain. (After Watson, J. D.: *The Molecular Biology of the Gene.* New York, W. A. Benjamin, 1965.)

the genic DNA can be inferred from the Watson-Crick rules of specific base pairing. The gene specifying the transfer RNA for alanine would also be 77 nucleotides long. A complete, double-stranded DNA molecule with this sequence of 77 nucleotides was synthetized in 1970 by Khorana and was found to be transcribed to give alanine tRNA. This was the first gene to be synthesized.

29–8 REVERSE TRANSCRIPTION

The central dogma states that biological information flows from DNA to RNA to protein. An important exception to this rule was discovered by Howard Temin in 1964 when he found that infection with certain RNA tumor viruses, such as the Rous sarcoma virus, is blocked by inhibitors of DNA synthesis and by inhibitors of DNA transcription. This suggested that DNA synthesis and transcription are required for the multiplication of RNA tumor viruses and that information flows in the reverse direction; that is, from RNA to DNA. Temin proposed that a *DNA provirus* is formed as an intermediate in the replication of these RNA tumor viruses and in their cancer-producing effect. Temin's hypothesis required a new kind of enzyme — one that would synthesize DNA using RNA as a template. Just such an enzyme was discovered by Temin and by David Baltimore in 1970, discoveries for which they received the Nobel Prize in 1975. This RNA-directed DNA polymerase (also called reverse transcriptase) has been found to be present in all RNA tumor viruses.

An infecting RNA virus binds to and enters the host cell. The viral RNA is uncoated and enters the cell nucleus. There the viral reverse transcriptase forms a (−) DNA strand complementary to the (+) RNA, and subsequently a (+) DNA strand is synthesized from the (−) DNA strand. The complementary (±) DNA strands, the provirus, are integrated into the host cell's genome. After cell division the virus is transcribed and the viral proteins are formed in the cytoplasm as this RNA is translated. Viral (+) RNA molecules are formed and incorporated into mature virus particles with a protein coat.

The question of whether viruses cause human cancer is being investigated intensively in many laboratories at present. Suggestive evidence that viruses probably do play a role in certain human cancers is now at hand. First, human leukemias, sarcomas, lymphomas and breast adenocarcinomas have been shown to contain large RNA molecules similar to those of tumor viruses that cause cancer in mice. Second, these human cancer cells contain particles with reverse transcriptase activity. Third, the DNAs of some human cancer cells have viruslike sequences of nucleotides in their DNA, sequences not found in the DNA of comparable normal cells.

29–9 THE SYNTHESIS OF A SPECIFIC POLYPEPTIDE CHAIN

Activation of Amino Acids. Before the subunits, the amino acids, can be assembled into a peptide chain, each amino acid must be activated by an enzyme-mediated reaction with ATP (Fig. 29–19). There is a separate specific activating enzyme for each amino acid; all these enzymes are present in the cytoplasm and are not associated with any of the structural elements of the cell. The enzymes catalyze the reaction of the amino acid with ATP to form the amino acid-adenylic acid compound (aa-AMP) and release inorganic pyrophosphate. The same enzyme next catalyzes the transfer of the amino acid from the adenylic acid (AMP) to the specific transfer RNA for that amino acid.

$$aa - AMP + tRNA \longrightarrow aa\text{-}tRNA + AMP$$

At the end of the transfer RNA that contains the CCA nucleotides, the amino acid is attached to the ribose of the terminal adenylic acid. If these three nucleotides are removed the transfer RNA is unable to function.

The Role of Transfer RNA. Francis Crick had predicted on theoretical grounds that some sort of nucleic acid molecule must serve as an adaptor in the course of protein synthesis. He argued that since there is no simple correspondence between the molecular structures of a polynucleotide chain and a polypeptide chain that would enable the nucleotide sequence to specify the amino acid sequence directly, the amino acids might be lined up in appropriate register by means of small RNA adaptor molecules. These adaptor molecules could assemble at specific places on the nucleic acid template by their complementary sequences of nucleotides. This hypothesis has now been validated by a wide variety of experiments. For example, Chapeville prepared a complex of cysteine with its specific transfer RNA (cysteinyl-tRNA^{cys}) and then converted the cysteine to alanine while it was still bound to the transfer RNA (alanyl-tRNA^{cys}). When this was added to a protein-synthesizing system, a polypeptide was made that contained *alanine* at the sites in the peptide chain where cysteine should have been. Experiments such as this provided direct proof that the ordering of specific amino acids into their appropriate place is dictated by the specific transfer RNA and not by the amino acid that is bound to it.

Ribosomal Functions. The amino acid bound to its specific transfer RNA is transferred to the ribosomes. The role of the ribosome is to provide the proper orientation of the amino acid-transfer RNA precursor, the messenger or template RNA, and the growing polypeptide chain, so that the genetic code on the template or messenger RNA can be read accurately. There are some 15,000 ribosomes in a rapidly growing cell of *E. coli,* each with a molecular weight of nearly 3,000,000. These ribosomes account for nearly one third of the total mass of the cell.

The template for the synthesis of a specific protein is supplied by the messenger RNA formed on one strand of the double helix of DNA. The messenger RNA undergoes processing in the nucleus, in the course of which a long tail of polyadenylic acid containing about 100 molecules of adenylic acid is added. The addition of adenylic acid involves an enzymatic reaction utilizing ATP as the donor of adenylate. The poly A-rich RNA then passes out of the nucleus and becomes associated with the ribosomes. The poly A tail may play some role in transport through the nuclear membrane or in protecting the messenger RNA from destruction by ribonuclease.

Ribosomes from different kinds of cells may differ somewhat in their mass, in the composition of their RNA and in the ratio of RNA to protein, but there is a general similarity in their structures. Ribosomes can be separated into two subunits if placed in a solution with a low concentration of magnesium ion. The subunit structure is apparent in electron micro-

graphs; the smaller subunit seems to sit like a cap on the flat surface of the larger subunit. There are 21 proteins and 1 RNA in the smaller subunit of bacterial ribosomes, and 35 proteins and 2 RNA molecules in the larger subunit. Each subunit can be carefully separated into its constituent RNA and proteins. The 21 proteins and the RNA that compose the smaller subunit of bacterial ribosomes will undergo spontaneous reassembly and form a fully functional subunit. Similarly, the 35 proteins and the 2 RNA molecules of the larger subunit will reassemble spontaneously to form a fully functional subunit. These findings demonstrate that all the information needed for the correct assembly of the ribosome from its parts is contained in the structure of those parts; no factor from outside the ribosome is required. This indicates that the formation of the ribosome within the living cells is also a self-assembly process.

The ribosomes of higher organisms tend to be somewhat larger than the bacterial ones and are composed of two or four subunits. Each ribosome contains several dozen kinds of proteins bound to the RNA. Ribosomes have GTP-ase activity, which appears to play an important but undetermined role in the transfer of amino acids from transfer RNA to the forming peptide chain. Protein-synthesizing particles somewhat similar to ribosomes are also found in the nucleus, in chloroplasts and in mitochondria.

Protein synthesis has been studied intensively in preparations of rabbit reticulocytes, which are engaged primarily in making just one protein—hemoglobin. Alexander Rich and his coworkers showed that the ribosomes most active in protein synthesis are those that interact in clusters of five or more (Fig. 29–20). These clusters, termed *polysomes,* are held together by the strand of messenger RNA. Peptide chains are formed by the sequential addition of amino acids beginning at the N-terminal end, the end having a free amino group. Electron micrographs suggest that individual ribosomes become attached to one end of a polyribosome cluster and gradually move along the messenger RNA strand as the polypeptide chain attached to it increases in length by the sequential addition of amino acids (Figs. 29–21 and 29–22). Thus each ribosome appears to ride along the extended messenger RNA molecule, "reading the message" as it goes and in

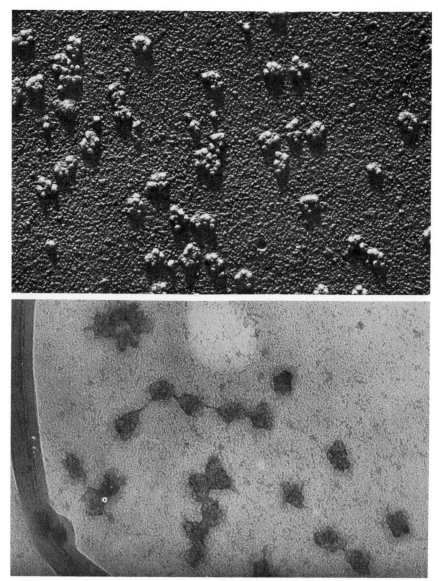

Figure 29–20 Electron micrographs of polyribosomes isolated from the reticulocytes of a rabbit. The upper preparation was shadowed with gold and the lower one was stained with uranyl nitrate. The electron micrographs show that polyribosomes tend to occur in clusters of four, five or six and the clusters are connected by a thin strand of mRNA. (Courtesy of Dr. Alexander Rich.)

some way bringing the transfer RNA molecule charged with its specific amino acid into line at the right position.

The incorporation of the first amino acid of a new polypeptide chain requires a special mechanism, since there is no preexisting polypeptide-transfer RNA complex to accept it at the combining site. The N-terminal amino acids of proteins are often substituted at their alpha amino end with an acetyl group or a formyl group. There is, in fact, a special enzyme that adds a formyl group to methionine after

methionine has been linked to its transfer RNA. Such N-formyl methionine molecules are incorporated preferentially at the N-terminal end of peptide chains, whereas methionines without the formyl group are incorporated in intermediate positions in the chain. Two different types of methionyl-transfer RNA may be involved, only one of which can receive a formyl group.

Three stages, initiation, elongation and termination, can be distinguished in the process of protein synthesis. The overall process

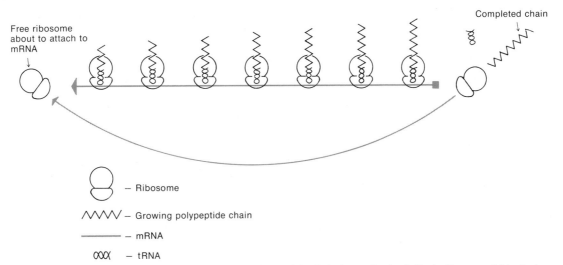

Figure 29–21 The postulated mechanism by which a polypeptide chain is synthesized. Each ribosome "rides" along the messenger RNA, reading and translating the genetic message. Amino acids are added as dictated by the specific base-pairing of the mRNA codon and the tRNA anticodon. (After Watson, J. D.: *The Molecular Biology of the Gene*. New York, W. A. Benjamin, 1965.)

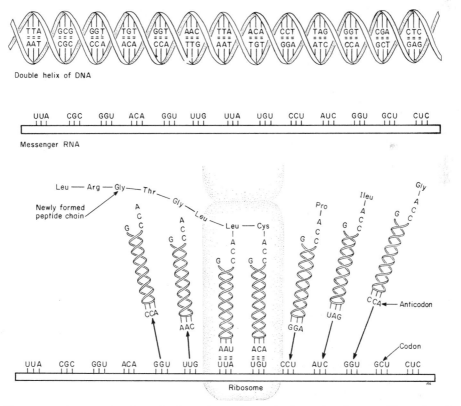

Figure 29–22 Diagram of the postulated mechanism of protein synthesis on the ribosome illustrating the relationship between the triplet code of the DNA helix, the complementary triplet code of messenger RNA and the complementary triplet code (anticodon) of transfer RNA. Molecules of tRNA charged with specific amino acids are depicted coming from the right, assuming their proper place on mRNA at the ribosome, transferring the amino acid to the growing peptide chain, and then *(left)* leaving the ribosome to be recharged with amino acids for further reactions. The growing polypeptide chain remains attached to its original ribosome.

requires the coordinated action of more than 100 different macromolecules, including mRNA, all the specific tRNAs, activation enzymes, initiator factors, elongation factors, and termination factors, in addition to the ribosomes themselves, each made of subunits and many kinds of protein and RNA. In *Escherichia coli,* in which the process has been studied in greatest detail, messenger RNA, formylmethionyl tRNA and the smaller ribosomal subunit come together to form an "initiation complex." The formylmethionyl tRNA recognizes the special initiation sequence on the messenger RNA (which is not necessarily at the 5′ end of the molecule). The larger ribosomal subunit then joins the complex to form a complete ribosome, and this is ready for the next phase. The elongation process begins with the binding to the complex of the next succeeding amino acyl tRNA, which is recognized by the specific codon next in line. The elongation cycle continues with the formation of a peptide bond, the elimination of the uncharged tRNA, and the transfer of the tRNA containing the peptide from one site within the ribosome to the other. This leaves the second site ready to accept the next amino acyl tRNA molecule. Both the binding of the amino acyl tRNA and the translocation from one site to the other require GTP. A specific protein *elongation factor* is required for the binding of the amino acyl tRNA to the ribosome. The synthesis of the peptide chain is terminated by "release factors" that recognize the terminator codons UAA, UGA and UAG. This leads to the hydrolysis of the bond between the polypeptide and the transfer RNA. When the peptide chain is completed and released from the ribosome, the ribosome then dissociates into its two subunits, the larger and smaller subunits.

The several steps involved in protein synthesis can be specifically inhibited by certain toxins and antibiotics. Streptomycin, for example, inhibits the initiation process, tetracyclines inhibit the binding of amino acyl tRNAs to the smaller subunit, and erythromycin binds to the larger subunit and inhibits the transfer of the growing peptide chain and its tRNA from one site within the ribosome to the next.

All the processes of gene replication, gene transcription and protein synthesis depend upon the formation of specific, though relatively weak, hydrogen bonds between specific pairs of purine and pyrimidine bases. The specificity of these bonds ensures the remarkable accuracy of the process; mistakes in base pairing occur less than one time in a thousand.

To determine in which direction synthesis of the peptide chain proceeds, tritium-labeled leucine was added to the hemoglobin-synthesizing system of rabbit reticulocytes in pulses of different duration. With a very long pulse, the labeled leucine was evenly distributed in all parts of the polypeptide; however, with short pulses, the labeled leucine was found only in those parts of the molecule synthesized last and these are at the C-terminal end. Thus the chain is synthesized beginning at the amino-terminal end and proceeding in sequence to the carboxyl-terminal end. In the cell-free system the synthesis of the complete α chain of hemoglobin, which is 141 amino acids long, requires 1.5 minutes; i.e., about two amino acids are added each second.

Two different enzymes are required to carry out the transfer of the amino acids from transfer RNA to the peptide linkage in the peptide chain. One enzyme binds the amino acid transfer RNA and the other is a synthetase that forms the peptide bond. The reactions that require GTP are the binding of the tRNA to the ribosome and the translocation of the amino acid-tRNA complex from one site in the ribosome to the other, rather than the synthesis of the peptide bond. GTP is used and the products are GDP and inorganic orthophosphate. In cell-free systems some 50 molecules of GTP are hydrolyzed for each amino acid transferred to a growing peptide. Presumably the protein-synthesizing system within the intact cell does not make such profligate use of its ~ P!

An overview of the several steps involved in the synthesis of a specific polypeptide chain is provided in Figure 29–23. Clearly much is yet to be learned about the biosynthesis of proteins. Even the best cell-free protein-synthesizing system operates at a rate only 0.01 as rapid as that within an intact, living cell.

29–10 CHANGES IN GENES: MUTATIONS

Although genes are remarkably stable and are transmitted to succeeding generations with great fidelity, they do from time to time undergo changes called *mutations.* After a gene

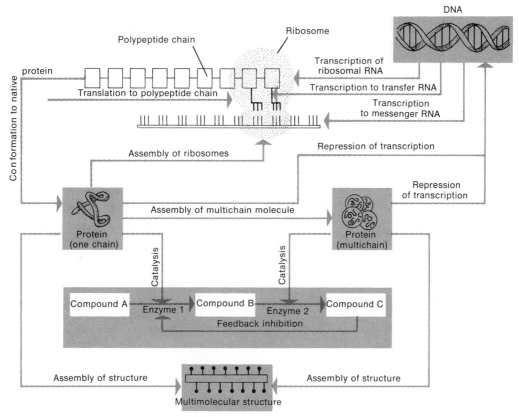

Figure 29–23 Overview of the process by which biological information is transferred from DNA via RNA to specific polypeptides. The peptide subunits are then assembled into multichain proteins.

has mutated to a new form, this new form is stable and usually has no greater tendency than the original gene to mutate again. A mutation can be defined as any inherited change not due to segregation or to the normal recombination of unchanged genetic material. Mutations provide the diversity of genetic material that makes possible the study of the process of inheritance. Investigations of the mechanisms of the mutation process have provided important clues to the nature of the genetic material itself.

Chromosomal mutations are accompanied by a visible change in the structure of the chromosome. A small segment of the chromosome may be missing (a *deletion*) or may be represented twice in the chromosome (a *duplication*) (Fig. 29–24). A segment of one chromosome may be transferred to a new position on a new chromosome (a *translocation*), or a segment may be turned end for end and attached to its usual chromosome (an *inversion*). *Point mutations* or gene mutations involve small changes in molecular structure that are

not evident under the microscope. These gene mutations involve some change in the sequence of nucleotides within a particular section of the DNA molecule, usually the substitution of one nucleotide for another in a given codon.

From your knowledge of the DNA molecule, you might predict that replacing one of the purine or pyrimidine nucleotides by an analogue such as azaguanine or bromouracil would result in mutation. In several experiments in which such analogues were incorpo-

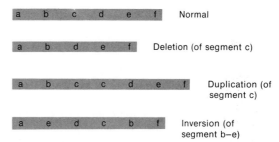

Figure 29–24 Diagram illustrating the types of mutations that involve changes in the structure of the chromosome.

rated into bacteriophage DNA, no mutations were evident. Because of the degeneracy of the genetic code (see Table 29–4), a number of changes in base pairs could occur without changing the amino acid specified. In other experiments, the incorporation of bromouracil into DNA did lead to an increased rate of mutation. Other chemicals known to be mutagenic include nitrogen mustards, epoxides, nitrous acid and alkylating agents. These are all chemicals that can react with specific nucleotide bases in the DNA and change their nature. When an analogue is incorporated into DNA it

may lead to mistakes in the pairing of nucleotides during subsequent replication processes. For example, when bromouracil is incorporated into DNA in place of thymine it will pair with guanine rather than with adenine, the normal pairing partner of thymine. This would lead to the substitution of a GC pair of nucleotides at the point in the double helix previously occupied by an AT pair of nucleotides (Fig. 29–25).

The presence of mutagenic materials may increase the frequency of mistakes in nucleotide base pairing, which would result in the

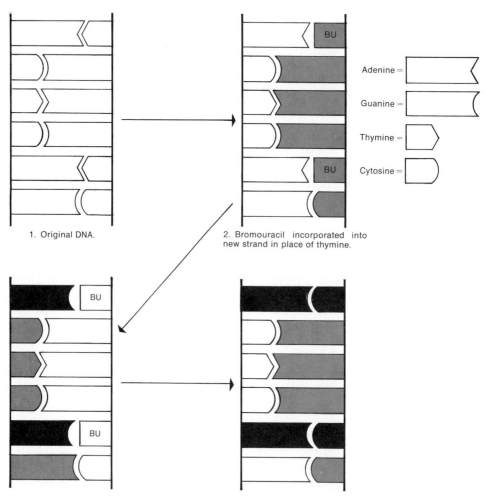

1. Original DNA.

2. Bromouracil incorporated into new strand in place of thymine.

Adenine =

Guanine =

Thymine =

Cytosine =

3. Strand with bromouracil leads to production of new strand with guanine paired to the bromouracil.

4. New, mutant DNA which contains no analogue bases, but has nucleotide sequence different from original, with GC pairs in place of AT.

Figure 29–25 Diagrammatic scheme of how an analogue of a purine or pyrimidine might interfere with the replication process and cause a mutation, an altered sequence of nucleotides in the DNA, indicated in black. The nucleotides of the new chain at each replication are indicated by the gray blocks. In this instance, two new GC pairs are indicated. A single substitution of a GC pair for an AT pair would be sufficient to cause a mutation if it occurred in the triplet code at a point that changed the kind of amino acid specified, i.e., in the first or second base of the triplet, or in the third member of certain triplets (see Table 29–4).

production of DNA molecules containing only natural bases but with an altered base order. The change in normal base order is significant, because during subsequent replications the altered base sequence will be reproduced by the normal process of DNA synthesis. The definition of mutation includes the restriction that the change that has been introduced into the DNA molecule must be propagated subsequently for an indefinite number of times.

Gene mutations generally result from errors in base pairing during the replication process; i.e., an AT base pair normally present may be replaced by GC, CG or TA pairs. The altered DNA will be transcribed to give an altered messenger RNA, and this will be translated into a peptide chain with one amino acid different from the normal kind of peptide. If the amino acid substitution occurs at or near the active site of the enzyme, the altered protein may have markedly decreased or altered enzymatic properties. However, if the amino acid substitution occurs elsewhere in the enzyme molecule, it may have little or no effect on the properties of the enzyme and may be undetected. The true number of gene mutations may indeed be much greater than the number observed.

If a single nucleotide pair were inserted into or deleted from the DNA molecule, it would shift the reading of the genetic message, alter all of the codons lying to the right of the substitution, and change completely the nature of the resulting peptide chain. Thus if the normal sequence is CAGTTCATG (read CAG, TTC, ATG), the insertion of a G between the two Ts results in CAGTGTCATG (read CAG, TGT, CAT, G . . .).

Gene mutations may be induced by x-rays, gamma rays, cosmic rays, ultraviolet rays and other types of radiation. How radiation may lead to changes in base pairs is not clear, but the radiant energy may react with water molecules to release short-lived, highly reactive free radicals that attack and react with specific bases. Mutations occur spontaneously at low but measurable rates that are characteristic of the species and of the gene. Some genes ("hot spots") are much more prone to undergo mutation than others. Spontaneous mutations may be caused by natural radiation such as cosmic rays or by errors in base pairing during replication. The rates of spontaneous mutations of different human genes range from 10^{-3} to 10^{-5} mutations per gene per generation. Since humans have a total of some 25,000 genes, this means that the total mutation rate is on the order of one mutation per person per generation. Each of us, in other words, has some mutant gene that was not present in either of our parents.

29–11 GENE-ENZYME RELATIONS

If each gene leads to the production of a specific enzyme by the method outlined in Section 29–9, we may next enquire how the presence or absence of a specific enzyme may affect the development of a specific trait. The expression of any structural or functional trait is the result of a number, perhaps a large number, of chemical reactions in series, with the product of each serving as the substrate for the next: A→B→C→D. The dark color of most mammalian skin or hair is due to the pigment *melanin* (D), produced from dihydroxyphenylalanine (C), derived in turn from tyrosine (B) and phenylalanine (A). Each reaction is controlled by an enzyme. The conversion of dihydroxyphenylalanine to melanin is mediated by *tyrosinase*. *Albinism*, characterized by the absence of melanin, results from the absence of tyrosinase. The gene for albinism (*a*) does not produce the enzyme tyrosinase, but its normal allele (*A*) does.

The earliest attempts to connect the action of a specific gene with a specific enzymatic reaction were studies of the inheritance of flower colors in which the specific flower pigments were extracted and analyzed. From studies of the inheritance of coat color in mammals and of eye color in insects, researchers were also able to relate specific genes with specific enzymic reactions in the synthesis of these pigments. A major advance in this field was made in 1941, when George Beadle and Edward Tatum, using the bread mold *Neurospora*, looked for mutations that interfere with reactions by which chemicals essential for its growth are produced. The wild type *Neurospora* requires as nutrients only sugar, salt, inorganic nitrogen and the vitamin biotin. A mixture of these makes up the so-called *minimal medium* for the growth of wild type *Neurospora*. Exposure of the conidia (haploid asexual spores) to x-rays or ultraviolet rays will produce mutations (Fig.

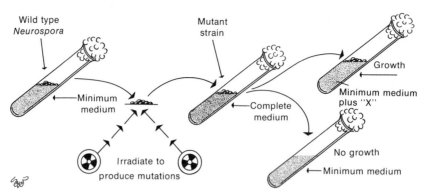

Figure 29–26 Production of mutant strains of *Neurospora* by x-radiation. The mutant strains produced can grow on complete medium but not on minimal medium; however, they can grow on minimal medium plus a single nutrient, X.

29–26). After irradiation the mold is transferred to a **complete medium,** an extract of yeast that contains all the known amino acids, vitamins, purines, pyrimidines and so on. Any nutritional mutant produced by the irradiation is able to survive and reproduce when grown on this complete medium. It can subsequently be tested for its ability to grow on minimal medium. If the irradiated mold is unable to grow on minimal medium, it is concluded that the mutant is unable to produce some compound essential for growth. By trial and error, by adding substances to the minimal medium in groups or singly, the required substance may be identified. Genetic tests show that the mutant strain produced by irradiation differs from the normal wild type by a single gene, and chemical tests show that the addition of a single chemical substance to the minimal medium will enable the mutant strain to grow normally.

Beadle and Tatum made the inference that each normal gene produces a single enzyme that regulates a single step in the biosynthesis of that particular chemical. The mutant gene does not produce the enzyme, and therefore that strain of the organism must be supplied with the product of the reaction that is impaired. In certain instances it has been possible to extract the particular enzyme from the cells of normal *Neurospora* but not from cells of the mutant strain. The biosynthesis of each compound involves a number of different steps, each mediated in turn by a separate gene-controlled enzyme. Indeed, biologists estimate the minimal number of steps involved in the synthesis of a given substance from the number of different mutants that will interfere with its production.

Similar one-to-one relationships of gene, enzyme and biochemical reaction in man were described by the English physician A. E. Garrod in 1908. **Alkaptonuria** is an inherited condition in which a substance in the patient's urine turns black when exposed to air. **Homogentisic acid,** a normal intermediate in the metabolism of phenylalanine and tyrosine, is excreted in the urine of alkaptonurics. The tissues of normal individuals have an enzyme that oxidizes homogentisic acid so that it is ultimately excreted as carbon dioxide and water (Fig. 29–27). Patients with alkaptonouria lack this enzyme because they lack the gene that controls its production. In these patients homogentisic acid accumulates in the tissues and blood and is excreted in the urine. Garrod coined the term **inborn errors of metabolism** to describe alkaptonuria and comparable conditions such as **phenylketonuria** and albinism.

29–12 THE OPERON CONCEPT: CONTROL OF PROTEIN SYNTHESIS

It appears from many kinds of experimental evidence that genes normally do not operate to produce the maximum number of enzymes all the time. Each gene appears to be "repressed" to a greater or lesser extent under normal conditions. Then, in response to some sort of environmental demand for that particular enzyme, the gene becomes "derepressed" and this leads to an increased production of the enzyme. When a single gene is fully derepressed it can cause the synthesis of fantastically large amounts of an enzyme. One enzyme may comprise 5 to 8 or more per cent of

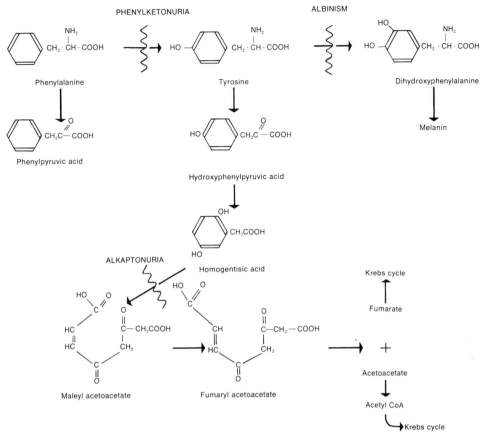

Figure 29–27 Pathway by which phenylalanine and tyrosine are metabolized. Mutants that interfere with the production of the enzymes that catalyze each of these steps may result in "inborn errors of metabolism," such as phenylketonuria, albinism and alkaptonuria.

the total protein of the cell! If all enzymes were produced at a similar very high rate, metabolic chaos would ensue. Thus the phenomena of gene repression and derepression appear to be necessary to provide a means of increasing or decreasing the rate of synthesis of some particular enzyme in response to variations in the environmental requirement for that enzyme.

In unicellular organisms and in the cells of multicellular organisms there may be wide variations in the number of enzymes per cell and in the amount of a given enzyme per cell. Thus there must be some mechanism that controls how much of a particular enzyme is synthesized in a given cell at a given moment. The rate of synthesis of a protein may be controlled in part by the genetic apparatus and in part by factors from the external environment. Most of the data relating to the control of protein synthesis in cells has come from studies of micro-

bial systems, especially in the colon bacillus *Escherichia coli.*

From the length of the chromosome of *E. coli* and the estimate that the average gene contains about 1500 nucleotide pairs (and codes for a polypeptide chain of 500 amino acids), it has been calculated that the genes of *E. coli* code for 2000 to 4000 different polypeptides. Estimates place the number of different enzymes required by an *E. coli* growing on glucose at about 800. Some of these must be present in large amounts; others are required in smaller quantities.

Inducers. *E. coli* cells growing on glucose contain very little of the enzyme β-*galactosidase.* When grown on lactose as the sole carbon source, the cells require β-galactosidase to cleave lactose to glucose and galactose. Under these conditions β-galactosidase makes up some 3 per cent of the total protein of the cell; there are perhaps 3000 molecules of enzyme in each cell of *E. coli,* which is at least a thousandfold increase over the amount present in cells growing on glucose. Two other enzymes, a *galactoside permease* and a *galactoside transacetylase,* respond equally dramatically to the presence of lactose in the incubation medium. A substance such as lactose that elicits an increased amount of an enzyme is called an *inducer,* and enzymes that respond to inducers are termed *inducible enzymes.*

Repressors. Cells of *E. coli* grown in a medium without any amino acids contain the the whole spectrum of enzymes required to synthesize all of the 20 or so amino acids needed for the assembly of protein molecules. The introduction of one or more amino acids to the incubation medium greatly decreases the amount of the biosynthetic enzymes required for the production of that amino acid. Biosynthetic enzymes that are reduced in amount by the presence of the end product of a biosynthetic sequence (e.g., the amino acid) are called *repressible enzymes,* and the small molecule (the amino acid) that brings about the repression is called a *corepressor.*

Both induction and repression of enzymes are adaptive phenomena that appear to be of survival value to bacteria—it would be a waste of energy and material to synthesize a battery of enzymes not immediately required. The responses to inducers and repressors are not "all or none" like the responses of nerve conduction or muscle contraction but instead allow for the formation of intermediate amounts of enzyme in response to intermediate conditions in the environment.

Regulator Genes. Differences in the amount of a specific protein in bacteria are generally due to variations in the rate of synthesis of that enzyme and are not due to altered stability or to a change in the rate of degradation. The rate of synthesis is controlled in turn by the amount of mRNA present per cell. Jacob and Monod have postulated that the amount of effective mRNA template for a given enzyme is controlled by a special kind of molecule, a protein called a *repressor,* which blocks the synthesis of messenger RNA. They have further postulated that these repressor proteins are coded for by special genes termed *regulator genes.* Gilbert and his colleagues at Harvard reported the isolation and characterization of the repressor for β-galactosidase from *E. coli.*

This is a protein with a molecular weight of about 150,000. From their data they calculated that there are about 10 molecules of β-galactosidase repressor per cell.

Repressors have been postulated to block the synthesis of specific proteins by combining with a specific site on the DNA and blocking its transcription to form messenger RNA. Alternatively, the repressor might combine with the molecule of messenger RNA, prevent its being attached to the ribosome and thereby increase the probability of its enzymatic degradation.

If regulator genes made repressors all the time, the synthesis of messenger RNA would always be inhibited. Thus it is necessary to postulate further that repressors may exist in either active or inactive forms, depending on whether the repressors are combined with specific small molecules—the inducers or corepressors. The attachment of an inducer inactivates the repressor; thus, the combination of lactose with the repressor blocks the repressor and permits the synthesis of the enzyme.

The attachment of the corepressor, e.g., the amino acid, to the repressor increases the number of active repressor molecules. These bind to the DNA, decrease the transcription of the DNA and the number of messenger RNA molecules that are produced, and code for the enzyme. The repressor is thought to be bound to its specific inducer or corepressor by weak bonds such as hydrogen bonds (Fig. 29–28).

There is experimental evidence that two or more enzymes may vary in amount in a coordinate fashion, suggesting that they are under the control of the same repressor system. It has been inferred that a single repressor may control the formation of the mRNA for these "coordinately repressed" enzymes. Frequently, but not always, genetic mapping shows that the genes for these coordinately repressed enzymes are closely linked on the chromosome and that a single mRNA molecule may carry the message for the synthesis of two or more enzymes.

The genes whose codes are transcribed on a single mRNA molecule and that are under the control of a single repressor have been termed an *operon* by Jacob and Monod. Originally it was believed that all the genes controlled by the same repressor must lie closely adjacent in a chromosome, but more recently examples have been found of widely separated

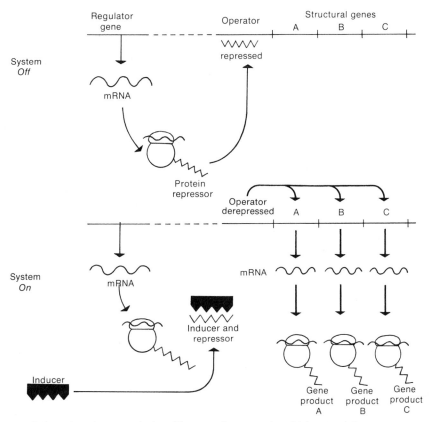

Figure 29-28 The regulation of genic transcription. Diagram of a means by which a regulator gene may produce a messenger RNA that codes for a protein repressor. The repressor in turn inhibits an operator and thus prevents the transcription of the structural genes *(above). Below,* The messenger RNA produced by the regulator gene again produces a protein repressor but this is combined with an inducer. The operator gene is thereby derepressed, permitting the transcription of structural genes A, B and C and the formation of the protein gene products A, B and C.

genes that appear to be coordinately repressed by the same repressor.

A further entity has been postulated to account for the control of the operon: the *operator site.* Operators are believed to lie adjacent to the genes in the operon and are thought to be the sites on the DNA to which active repressor molecules are bound, thereby inhibiting the synthesis of mRNA by the genes in the adjacent operon. In the absence of active repressors, the genes in the operon are free to be transcribed, mRNA is formed and enzymes are produced on the ribosome. If the operator site is absent, the repressor cannot inhibit the synthesis of the specific mRNA and the corresponding enzyme is produced all the time. It is possible that the operator site in some way provides a code that directs the RNA polymerase to begin transcribing the DNA at that specific point at the end of the operon.

If there are such things as regulator genes, it has been pointed out that the transcription of the regulator gene must also be under regulation so that the synthesis of repressors is controlled. One cannot postulate an infinite sequence of repressors, each of which represses the synthesis of the next. Either a repressor can repress its own synthesis or repressors are synthesized without such control.

In summary, in postulating this overall control mechanism, based on their studies of β-galactosidase in *E. coli,* Jacob and Monod suggested that in addition to the structural genes that provide the code for the synthesis of specific proteins there are "regulator genes" that code for the synthesis of repressors; that the repressors may be active or inactive, depending on whether they are bound to small molecules, the inducers or corepressors; and that active repressors bind to operator sites in the DNA and turn off the transcription of the adjacent structural genes. The structural genes under the control of one repressor are termed an operon.

There is clear evidence that the synthesis of many proteins in *E. coli* and in other organisms is not influenced by substances such as corepressors and inducers in the external environment; either their operator sites are turned on all the time or their genes are not associated with an operator site.

29–13   INFORMATION TRANSFER OUTSIDE THE NUCLEUS

It has been known since early in this century that a few traits appear to be inherited not by the usual nuclear genes but by some mechanism restricted to the cytoplasm. Such traits are inherited exclusively from the maternal parent; the egg, but not the sperm, supplies cytoplasm to the zygote.

Not all biological information is transferred by nuclear DNA; both mitochondria and plastids such as chloroplasts contain DNA (Fig. 29–29). About 2 per cent of the DNA of a liver cell is located in its mitochondria. Mitochondrial DNA may undergo mutations, leading to changes in the sequence of amino acids in the structural proteins of the mitochondria. Nass and Nass found fibrous structures in chick mitochondria that are removed by treatment with deoxyribonuclease. Mitochondria contain a DNA polymerase that carries out the replication of mitochondrial DNA; they also have a DNA-dependent RNA polymerase and can synthesize RNA.

The amount of DNA in the mitochondria from beef heart corresponds to a helix of molecular weight 3×10^7; however, the DNA in mitochondria exists as several pieces, each a closed circle of molecular weight 1×10^7. The replication of mitochondrial DNA is completely independent of the nuclear DNA; it is both synthesized and degraded more rapidly than nuclear DNA. The biological information in the mitochondrial DNA specifies some, but not all, of the mitochondrial proteins; the remainder are controlled by nuclear genes.

Plastids such as chloroplasts also contain DNA and RNA and, like the mitochondria, have the capacity for independent growth and division and for the synthesis of specific proteins.

The protist *Euglena* normally has chloroplasts but can survive without them if supplied with appropriate nutrients. Euglenas that are deprived of their chloroplasts never develop new ones; however, if they are again supplied with chloroplasts these bodies will undergo division and appear in all daughter cells. If the *Euglena* is supplied with structurally different chloroplasts from a mutant strain, these unusual chloroplasts will undergo division and appear in the daughter cells. This is cogent ev-

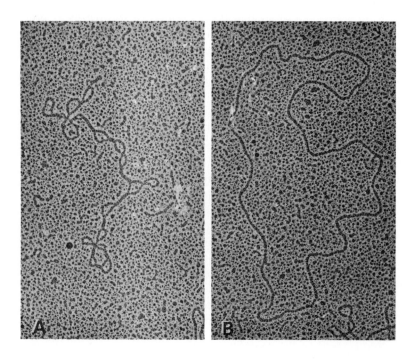

Figure 29–29 DNA extracted from rat liver mitochondria and observed by the spreading technique. *A,* A DNA molecule in a twisted circle configuration. *B,* A DNA molecule in the open circle configuration. (From De Robertis, E. D., Saez, F. A., and De Robertis, E. M. F., Jr.: *Cell Biology,* 6th Ed. Philadelphia, W. B. Saunders Co., 1975; courtesy of B. Stephens.)

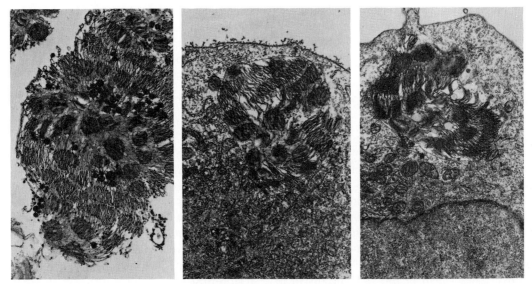

Figure 29–30 Electron micrographs of spinach chloroplasts that have undergone phagocytosis by mouse cells grown in cell culture but have retained their structural integrity. Magnification ×12,000. (Courtesy of Dr. Margit Nass.)

idence that the control of chloroplast structure is at least in part under the control of genetic material (DNA) in the plastids themselves.

Mammalian cells grown in cell culture were shown by Margit Nass in 1969 to be able to take up by phagocytosis chloroplasts prepared from spinach leaves. The chloroplasts survived for at least six cell generations (six days) and retained their structural integrity (Fig. 29–30). At the appropriate time they divided and were assorted at random to the daughter cells. The chloroplasts taken up could be reisolated from the mouse cells several days later and retained their ability to carry out at least some of the reactions of photosynthesis — the fixation of carbon dioxide and the photochemical reduction of dichlorophenolindophenol. The DNA from the reisolated chloroplasts was still in macromolecular form.

29–14 LETHAL GENES

Certain genes produce such a tremendous deviation from the normal development of an organism that it is unable to survive. The presence of these lethal genes can be detected by certain upsets in the expected genetic ratios. For example, a certain strain of mice had individuals with yellow coat color but experimenters found it impossible to establish a true breeding strain with yellow coat. Instead, when two yellow mice were bred, offspring were produced in the ratio of 2 yellow : 1 nonyellow.

A yellow mouse bred to a black mouse gave half yellow mice and half black mice among the offspring. Then investigators noticed that the litters of yellow × yellow matings were only about three quarters as large as other litters. They reasoned that one quarter of the embryos, those homozygous for yellow, did not develop. When the uterus of the mother was opened early in pregnancy the abnormal embryos, those homozygous for the yellow coat gene, were found. Embryos homozygous for yellow begin development, then cease developing, die, and are resorbed.

Many, perhaps most, of the lethal genes appear to have no effect when heterozygous but cause death when homozygous, and these are called recessive lethal genes. They can be detected only by special genetic techniques. Genetic analyses of wild populations of fruit flies and other organisms revealed the presence of many recessive lethals.

Several genes are suspected of acting as lethals in human embryos. Mohr and Wriedt reported a case in which two people with brachyphalangy (short fingers) married and produced a child with no fingers or toes who died shortly after birth. In the light of our present theory about the relations between genes and development, we can suppose that a lethal gene is a mutant that causes the absence of some enzyme of primary importance in intermediary metabolism. The absence of this enzyme prevents the development of the organism.

29–15 PENETRANCE AND EXPRESSIVITY

Each recessive gene described so far produces its trait when it is homozygous, and each dominant gene produces its effect when it is homozygous or heterozygous. Geneticists, however, have found that many genes do not always produce their phenotypes when they should. Genes that always produce the expected phenotype are said to have complete *penetrance.*

If only 70 per cent of the individuals of a stock homozygous for a certain recessive gene show the character phenotypically, the gene is said to have 70 per cent penetrance. The term penetrance refers to the statistical regularity with which a gene produces its effect when present in the requisite homozygous (or heterozygous) state. The percentage of penetrance of a given gene may be altered by changing the conditions of temperature, moisture, nutrition, and so forth under which the organism develops.

Some stocks that are homozygous for a recessive gene may show wide variations in the appearance of the character. Fruit flies homozygous for a recessive gene producing shortening and scalloping of the wings exhibit variations in the *degree* of shortening and scalloping. Such differences are known as variations in the *expressivity,* or expression, of the gene.

The expressivity of the gene may also be altered by changing the environmental conditions during the organism's development. Many of the human morphologic traits show similar variations in the expression of the trait among different members of a family or even on the right and left sides of one person. In view of the long and sometimes tenuous connection between the gene and the final production of the trait, it is easy to understand why the expression of the trait might vary or why the mutant trait might be completely absent.

In summary, the primary genetic material in all living systems is nucleic acid. The genetic material is DNA in all organisms except a few viruses in which only RNA is present. Genetic information is transferred from one generation to the next in the form of specific sequences of nucleotides in the nucleic acid chain. The coding unit is a sequence of three nucleotides (a codon) that specifies the position of a single amino acid in the peptide chain. A gene is a localized sequence of nucleotides in DNA that carries the information for the synthesis of a peptide chain with a specific sequence of amino acids. The model of DNA proposed by Watson and Crick provides for gene specificity, gene replication, gene transcription and gene mutation. A gene mutation is believed to be an alteration of the specific sequence of nucleotides in the nucleic acid molecule, which leads to the formation of a different type of protein or perhaps prevents the formation of that protein. Each gene is believed to control the production of a single specific protein.

QUESTIONS

1. Discuss the evidence that deoxyribonucleic acid is an integral part of the gene.
2. What is the nature of a "transforming agent"? What importance may this phenomenon have for our understanding of the chemical basis of inheritance?
3. What evidence regarding the nature of the genetic material has been obtained from experiments with bacteriophages?
4. Discuss the chemical composition of DNA.
5. How does the Watson-Crick model of DNA account for its observed properties? How does the Watson-Crick model explain the process of gene replication?
6. What are the two prime functions of DNA and how are these believed to be carried out?
7. Discuss the mechanism by which the enormously long DNA molecule of the eukaryotic chromosome is replicated rapidly.
8. How can one estimate the total number of genes in any given organism?
9. Define the term mutation. What types of mutations can be distinguished and how may each be produced?
10. How is DNA synthesized enzymically? What substances are needed to carry out the process?
11. Discuss the problem of the "coding" of information in the gene.
12. Discuss the current theory regarding the transfer of specific information from the gene to the site of protein synthesis in the cell.
13. Distinguish between messenger RNA, ribosomal RNA and transfer RNA. What is the role of each in the synthesis of proteins?

14. What is meant by an "inborn error of metabolism"? What are some examples of inborn errors of metabolism in human beings?
15. What is meant by a gene repressor? A derepressor?
16. Distinguish between "structural" genes, "regulatory" genes and "operator" genes.
17. What is a reverse transcriptase and what is its function? How might it be used in biochemical research?
18. Distinguish between penetrance and expressivity.

SUPPLEMENTARY READING

Molecular Biology of the Gene, by Nobel laureate James D. Watson, provides a masterful summary of biochemical genetics. Most of the examples in the book are drawn from studies of inheritance in bacteria and viruses. An excellent dissertation of the genetic factors regulating the synthesis of specific peptide chains is *The Biosynthesis of Macromolecules* by Vernon Ingram. A highly readable and popular account of the chemistry of the gene is *The Language of Life* by George and Muriel Beadle. W.

Hayes' *The Genetics of Bacteria and their Viruses* is a concise and well written survey of one of the most exciting frontiers of biology in the twentieth century. *Episomes* by A. Campbell is a modern account of the occurrence of DNA molecules in the cytoplasm, how they may be incorporated into chromosomes, their biological significance and their relationship to viruses. *The Double Helix* by James D. Watson is an interesting autobiographical account of the role the author played in the discovery of the model of the DNA molecule. There have been many popular articles describing various aspects of biochemical genetics published in *Scientific American;* many of these are available as offprints and are included in Donald Kennedy's *The Living Cell: Readings from the Scientific American.* H. L. Whitehouse's *Towards an Understanding of the Mechanism of Heredity* discusses the field of genetics in terms of the several hypotheses that have been put forward and how they have been tested. Gunther Stent's *Molecular Genetics: An Introductory Narrative* is an excellent source of facts and theories about biochemical genetics by one of the major investigators in this field. *DNA Synthesis* by Arthur Kornberg is a clearly written summary of the experiments by the author and others that have led to our present understanding of this important biological process.

CHAPTER 30

HUMAN INHERITANCE: POPULATION GENETICS

For the study of inheritance in any species, geneticists would prefer (1) to have *isogenic strains*, standard stocks of genetically identical individuals, (2) to mate members of different isogenic strains, and (3) to raise the offspring under carefully controlled conditions. The organisms favored in genetic studies are bacteria, molds and fruit flies, which produce many offspring and have only a short time between successive generations.

Judged by these criteria, man is not a very favorable subject for studies of inheritance, for members of the human race are genetically diverse—they are heterozygous for many genes—and there are wide variations in their physical, biologic and social environments. Genetic considerations rarely influence the choice of a mate; even large human families are small by genetic standards; and 20 to 30 years elapse between generations.

Despite these difficulties a great deal has been learned about human inheritance and the field is progressing rapidly. Some of the phenomena in human inheritance, originally quite puzzling, have been clarified by the solutions of analogous problems in the inheritance of bacteria, molds, flies or mice. In the field of chemical genetics many important discoveries have been made first with human material and then have been confirmed by experiments using bacteria or molds. The first evidence that genes determine the sequence of amino acids in proteins came from studies of the different types of human hemoglobin.

Geneticists studying human inheritance were forced to devise methods of measuring the relative frequency of contrasting traits in an entire population of individuals and of calculating the frequency of specific alleles. Each pair of genes, such as *A* and *a*, is distributed in the population in such a way that any member may have the genotype *AA*, the genotype *Aa* or the genotype *aa*. If there is no selective advantage for any of these three genotypes, the frequency of these genes in successive generations of individuals will remain unchanged. As long as individuals with each of these three genotypes are just as likely to mate and have offspring as individuals with the other genotypes, the three genotypes will be present in succeeding generations in the same proportion as in the initial generation. This concept of genetic equilibrium is fundamental to all studies of population genetics.

30-1 THE LAWS OF PROBABILITY

The early studies of human heredity usually dealt with readily identified single traits and their distribution among the members of a family, as illustrated by the pedigree shown in Figure 30-1. The discovery of methods of making inferences about the mode of inheritance of a trait from its distribution in an entire population, based on the laws of probability, has been of great usefulness in human genetics.

Genetic events are governed by the laws of probability. The prediction of any single event is highly uncertain, of course, but in a large number of events the laws of probability provide a reasonable prediction of the fraction of these events that will be of one type or the other.

Three types of probabilities can be distinguished. *A priori probabilities* are those which can be specified in advance from the nature of the event. For example, in flipping a coin the probability of obtaining heads is one in two and in casting dice the probability of obtaining a two is one in six (there are six sides to a die); these probabilities are independent of whether or not the event actually occurs. In contrast, *empiric probabilities* are obtained by counting the number of times a given event occurs in a certain number of trials. For example, if a surgeon performs a certain type of operation on 500 people and 40 of them die, then the probability of death in this type of operation is 40 out of 500 or 0.08. Such empiric probabilities have to be used in many fields of research where there is no theoretical basis, no *a priori* basis, for predicting the outcome. This type of probability is used in setting up the "risk tables" used widely by insurance firms. If a scientist collects data about the numbers of individuals with certain traits in a given population and wants to know the probability that these numbers agree with the ratio expected on the basis of some genetic theory (1:1 or 3:1, and so on) he uses the methods of *sampling probability.*

If two events are independent the probability of their coinciding is the *product* of their individual probabilities. For example, the probability of obtaining heads on the first toss of a coin is $\frac{1}{2}$ and the probability of obtaining heads on the second toss of a coin (an independent event) is $\frac{1}{2}$. The probability of obtaining heads twice on successive tosses of the coin is the product of their probabilities, $\frac{1}{2} \times \frac{1}{2}$, or $\frac{1}{4}$: there is one chance in four of obtaining heads twice on two successive tosses of a coin. This "product rule" of probability also holds for three or more independent events. For example, the probability of choosing at random an individual who is male, has blood group A and was born in June is $0.5 \times 0.4 \times 0.084 = 0.0168$.

The probability that one or another of two mutually exclusive events will occur is the *sum* of their separate probabilities. For example, in rolling dice the probability that the die will come up *either* two or five is $\frac{1}{6} + \frac{1}{6} = \frac{1}{3}$.

The application of these considerations to genetics is illustrated in Figure 30-2. Let us consider the probability that a son will inherit from his father the particular allele that the father inherited in turn from his father, the child's grandfather. The father has two alleles $A^P A^M$ obtained from the grandfather and grandmother respectively. He will pass on to his son one of these two. The "favorable" event is the transmission of the A^P gene and the total number of events is two; thus, the probability that the son will receive from his father the same allele that the father received from the grandfather is one in two. The probability that the child will receive from his father the allele that the father obtained from the

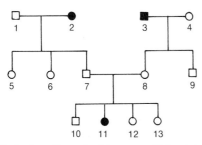

Figure 30-1 A pedigree for blue eyes. Males are indicated by squares and females by circles. Individuals showing the trait under study are indicated by black symbols and those not showing the trait by white symbols. Relationships are indicated by connecting lines; all members of the same generation are placed in the same row. Thus, 11 is a blue-eyed girl whose sisters, 12 and 13, and brother, 10, are brown-eyed. Her father, 7, and mother, 8, as well as her aunts, 5 and 6, uncle, 9, paternal grandfather, 1, and maternal grandmother, 4, are brown-eyed. Her paternal grandmother, 2, and maternal grandfather, 3, are blue-eyed.

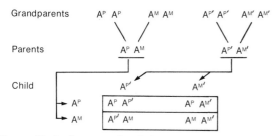

Figure 30-2 An example of the application of the laws of probability to genetics, illustrating both the "product law" of independent events and the "sum law" of mutually exclusive events. See text for discussion.

grandfather (A^P) and that he will also obtain from his mother the allele that she received from the grandmother ($A^{M'}$) is the product of their independent occurrences, $1/2 \times 1/2 = 1/4$. The probability that the child will obtain either the two alleles from the two grandfathers $A^P\,A^{P'}$ or the two alleles from the two grandmothers $A^M\,A^{M'}$ is the sum of their independent probabilities, $1/4 + 1/4 = 1/2$.

30-2 THE MATHEMATICAL BASIS OF POPULATION GENETICS

The question that sometimes puzzles beginning geneticists is why, if brown eye genes are dominant to blue eye genes, haven't all the blue eye genes disappeared? The answer lies partly in the fact that a recessive gene, such as the one for blue eyes, is not changed by having existed for a generation next to a brown eye gene in a heterozygous individual, **Bb**. The remainder of the explanation lies in the fact that as long as there is no selection for either eye color, that is, as long as peo-

ple with blue eyes are just as likely to marry and have as many children as people with brown eyes, successive generations will have the same proportions of blue- and brown-eyed people as the initial one.

The reason for this may not be immediately obvious, but a brief exercise in mathematics will show why it is true. If we consider the distribution of a single pair of genes, **A** and **a**, any member of the population will have two of these genes and hence have the genotype **AA**, **Aa** or **aa**. No other possibilities exist. Now, let us suppose that these genotypes are present in the population in a ratio of $1/4$ **AA**: $1/2$ **Aa**: $1/4$ **aa**. If all the members of the population select their mates at random, without regard for whether they are **AA**, **Aa** or **aa**, and if all the types of pairs produce, on the average, comparable numbers of offspring, the succeeding generation will also have genotypes in the ratio $1/4$ **AA**: $1/2$ **Aa**: $1/4$ **aa**. This can be demonstrated (Table 30-1) by listing the possible types of matings, the frequency of their random occurrence, and the kinds and proportions of offspring produced by each type of mating. When the types of offspring are summed, we find that the next generation also has genotypes in the ratio $1/4$ **AA**: $1/2$ **Aa**: $1/4$ **aa**.

30-3 THE HARDY-WEINBERG PRINCIPLE

Within a decade of the rediscovery of Mendel's Laws, indeed, a year before Johannson in 1909 proposed the word "gene" for Mendel's "genetic factors," G. H. Hardy, an English mathematician, and G. Weinberg, a German physician, independently observed that the frequencies of the members of a pair of

Table 30-1   THE OFFSPRING OF THE RANDOM MATING OF A POPULATION COMPOSED OF 1/4 **AA**, 1/2 **Aa** AND 1/4 **aa** INDIVIDUALS

Mating		Frequency	Offspring
MALE	FEMALE		
AA	× AA	1/4 × 1/4	1/16 AA
AA	× Aa	1/4 × 1/2	1/16 AA + 1/16 Aa
AA	× aa	1/4 × 1/4	1/16 Aa
Aa	× AA	1/2 × 1/4	1/16 AA + 1/16 Aa
Aa	× Aa	1/2 × 1/2	1/16 AA + 1/8 Aa + 1/16 aa
Aa	× aa	1/2 × 1/4	1/16 Aa + 1/16 aa
aa	× AA	1/4 × 1/4	1/16 Aa
aa	× Aa	1/4 × 1/2	1/16 Aa + 1/16 aa
aa	× aa	1/4 × 1/4	1/16 aa
			Sum: 4/16 AA + 8/16 Aa + 4/16 aa

allelic genes in a population are described by the expansion of a binomial equation: $(pA + qa)^2 = p^2AA + 2pq\ Aa + q^2aa$. In this expression, p represents the frequency of the A gene, and q the frequency of the a gene in the population. Since the gene must be either A or a, $p + q = 1$. Thus, if we know the value of either p or q we can calculate the value of the other.

When we consider all the matings in any given generation, a p number of A-containing eggs and a q number of a-containing eggs are fertilized by a p number of A-containing sperm and a q number of a-containing sperm: $(pA + qa) \times (pA + qa)$. The proportion of the types of offspring of all of these matings is described by the algebraic product, $p^2\ AA + 2pq\ Aa + q^2\ aa$. If p, the frequency of the A gene, equals $\frac{1}{2}$, then q, the frequency of the a gene, equals $1 - p$, or $1 - \frac{1}{2}$, or $\frac{1}{2}$. From the formula, the frequency of genotype AA, p^2, equals $(\frac{1}{2})^2$ or $\frac{1}{4}$; the frequency of Aa, $2\ pq$, equals $2 \times \frac{1}{2} \times \frac{1}{2}$ or $\frac{1}{2}$ and the frequency of aa, q^2, equals $(\frac{1}{2})^2$ or $\frac{1}{4}$. Any population in which the distribution of alleles A and a conforms to the relation $p^2AA + 2\ pqAa + q^2aa$ is in **genetic equilibrium**. The proportions of these alleles in successive generations will be the same (unless altered by selection or mutation).

30-4 GENE POOLS AND GENOTYPES

The genetic constitution of a population of a given organism is termed the **gene pool**. Stated differently, all the genes of all the individuals in a population make up the gene pool. This may be contrasted to the **genotype**, which is the genetic constitution of a single *individual*. Any individual may have only two alleles of any given gene. In contrast, the gene pool of the population may contain any number of different alleles of a specific gene. The ABO blood groups (p. 359) are inherited by three alleles, I^A, I^B and i. In the population there are three alleles, but any given individual can have no more than two of the three.

The frequencies of the O, A, B and AB blood groups have been measured in several different populations (Table 30–2), and from these we can calculate the gene frequencies for I^A, I^B and i. Calculating the gene frequencies for a three-allele system is a little more complex mathematically than for a two-allele system, and we shall not go into it here. Expansion of the equation $(pI^A + qI^B + ri^2$, in which p, q and r represent the frequencies of genes I^A, I^B and i, respectively, gives $p^2\ I^AI^A + 2pq\ I^AI^B + 2pr\ I^Ai + q^2\ I^B + 2qr\ I^Bi + r^2ii$. The frequency of the I^B allele varies most strikingly (Table 30–2), with a 20-fold difference between the frequency of the allele in the African blacks and the American Indians.

The gene pools of different populations may differ in the ratios or proportions of the specific alleles. One population may have the alleles A and a in a ratio of 0.5 to 0.5. Another population of the same species may have the two alleles in the ratio 0.7A:0.3a. The next generation, and each succeeding generation, will contain an identical gene pool, 0.7A and 0.3a. The three kinds of genotypes will be present in the ratio 0.49AA:0.42Aa:0.09aa in succeeding generations provided that: (1) there are no mutations for A or a; (2) the three kinds of genotypic individuals have equal probabilities of surviving, mating and producing offspring, and there is no selection of mates according to these genotypes; and (3) the population of individuals is large enough so that chance cannot play a role in determining gene frequencies.

This principle that a population is genetically stable in succeeding generations is termed the **Hardy-Weinberg Principle**. (The mathematical expression of this principle was presented in Section 30–3.) Since the principle was established, J. B. S. Haldane, R. A.

Table 30-2 THE FREQUENCY OF THE ABO BLOOD GROUPS IN DIFFERENT POPULATIONS

Population	Frequency of Blood Groups				Gene Frequencies		
	O	A	B	AB	I^A	I^B	i
Northern Europeans	.40	.45	.10	.05	.29	.08	.63
African blacks	.42	.24	.28	.06	.13	.23	.64
American Indians	.67	.29	.03	.01	.17	.01	.82

*The gene frequencies are calculated from the distribution of the blood groups using the formula $(pI^A + qI^B + ri)^2$, just as the frequencies of **A** and **a** were calculated from the formula $(p\mathbf{A} + q\mathbf{a})^2$.

Fisher and Sewall Wright have developed mathematical methods for analyzing the inheritance of a given trait in a population. Subsequently it has become clear that the process of evolution, stated in the simplest terms, represents departure from the Hardy-Weinberg principle of genetic stability. Evolution involves changes in the gene pool of a population that result from mutations and selection. Thus a knowledge of the Hardy-Weinberg Principle is needed to understand the mechanism of evolutionary change.

30–5 ESTIMATING THE FREQUENCY OF GENETIC "CARRIERS"

Neither the value of p nor of q, which are *gene frequencies*, can be measured directly. However, since the recessive phenotype can be distinguished you can determine q^2, the frequency of genotype *aa*. From this you can calculate the gene frequencies q (which is the square root of q^2) and p (which is $1 - q$). Finally you can calculate the frequencies of the other genotypes, p^2AA and $2 pqAa$. To calculate the number of individuals in a population who are genetic carriers for a given trait (i.e., are heterozygotes, *Aa*) you need to know only that it is inherited by a single pair of genes and the frequency with which the homozygous recessive individuals appear in the population. Consideration of some specific examples of the genetics of populations should help in understanding these principles.

Albinism. Albinos are individuals with no pigment at all in their skin or hair. *Albinism* is an inherited trait in which the individual lacks a specific enzyme, *tyrosinase*. Tyrosinase catalyzes one of the reactions involved in the production of the dark pigment *melanin*. Albinism is inherited by a single pair of genes, and albinos, the homozygous recessive individuals, occur about once in 20,000 births. From this fact you can calculate that the frequency of *aa* individuals (q^2) is $\frac{1}{20,000}$. The value of q can be calculated by taking the square root of q^2. The square root of $\frac{1}{20,000}$ is about $\frac{1}{141}$. Since $p = 1 - q$ or $1 - \frac{1}{141}$, $p = \frac{140}{141}$. From these values for p

and q, you can calculate the value of $2 pq$, which represents the frequency of the genetic "carrier" *Aa* individuals: $2 \times \frac{140}{141} \times \frac{1}{141} = \frac{1}{70}$. Surprising as it may seem, one person in 70 is a *carrier* of albinism, although only one person in 20,000 is homozygous and displays the trait. At first glance it may seem odd that there are so many carriers in a population that contains so few homozygous recessives, yet reflecting on the mathematical relations involved should lead you to realize that this must be true. When q is small (such as $\frac{1}{141}$), then q^2 will be very small, but $2 pq$ will be much larger.

Inheritance of Taste. As another example of the usefulness of the methods of *population genetics*, let us consider the inheritance of another human trait. Human beings differ in their ability to taste **phenylthiocarbamide** (PTC) and related compounds with the thiocarbamide group $(-NCS)$, which contains one atom each of nitrogen, carbon and sulfur. Some people find that PTC has a bitter taste, others report it to be completely tasteless. From the distribution of "tasters" and "nontasters" in specific families, L. H. Snyder assumed that the ability to taste PTC is inherited by a single pair of genes and that tasting (*T*) is dominant to nontasting (*t*). Snyder's genetic survey revealed that 70.2 per cent of a white population of the United States were tasters and 29.8 per cent were nontasters. From these values he calculated (Table 30–3) that 12.4 per cent of the children from marriages of tasters with tasters would be nontasters and, further, that in marriages of tasters with nontasters, 35.4 per cent of the children would be nontasters. The percentages Snyder actually found were 12.3 per cent and 33.6 per cent respectively. Thus the observed percentages agreed closely with those predicted by the Hardy-Weinberg Principle. From this Snyder concluded that his original assumption, that the tasting-nontasting trait is inherited by a single pair of genes, was correct. Nontasters, although constituting some 30 per cent of the white population, are very rare in populations of blacks, Eskimos and American Indians. Thus the *T-t* gene pools are quite different in these different populations.

Inheritance of MN Blood Groups. The MN and ABO blood groups are inherited independently. Each of us is either A, B, O or AB and also either M, N or MN. The distribution

Table 30-3 THE INHERITANCE OF THE ABILITY TO TASTE PHENYLTHIOCARBAMIDE

Marriage	Number of Families	Offspring		Proportion Nontasters	
		TASTERS	NONTASTERS	OBSERVED	CALCULATED*
Taster × taster	425	929	130	0.123	0.124
Taster × nontaster	289	483	278	0.336	0.354
Nontaster × nontaster	86	5	218	0.979	1.0

*From the Hardy-Weinberg Principle it follows that among marriages of parents with unlike traits (i.e., taster × nontaster), the fraction of the offspring with the recessive trait is $q/1 + q$. Among marriages of parents with like traits (i.e., taster × taster), the fraction of offspring with the recessive trait, nontaster, is $(q/1 + q)^2$. Since 29.8 per cent of the population were nontasters, $q^2 = 0.298$, $q = 0.545$, $q/1 + q = 0.354$ and $(q/1 + q)^2 = 0.124$.

of these blood groups in the white population of the United States has been found to be 29.16 per cent type M, 49.58 per cent type MN and 21.2 per cent type N. Let us test the theory that these traits are inherited by a single pair of codominant genes—i.e., MN represents the heterozygous individual. To do this we determine whether the distribution of phenotypes found would be expected on the basis of the Hardy-Weinberg Principle of population genetics: $p^2\text{MM} + 2\ pq\text{MN} + q^2\text{NN}$.

We begin with the fraction of the population that is phenotypically N and assume that $q^2 = 0.2126$. From this we can calculate that $q = \sqrt{0.2126}$, or 0.46. The frequency, p, of the M gene is $1 - 0.46$, or 0.54. The square of this is $0.54^2 = 0.29$, which agrees well with the observed frequency of blood group M—29.16 per cent. Furthermore, $2\ pq = 2 \times 0.54 \times 0.46 = 0.49$, which agrees with the 49.58 per cent observed to have blood type MN. The excellent agreement between the observed and predicted values supports the hypothesis that M and N blood types are inherited by a single pair of codominant genes. The heterozygous individual has both M and N antigens in his blood cells and is phenotypically MN. To test this hypothesis differently, repeat the calculations but begin with $p^2 = 0.2916$ and calculate p, then $q = 1 - p$, and so on.

Many of the diseases of humans that are known to be inherited are determined by recessive genes. The undesirable trait—the disease condition—represents the homozygous recessive individual. When a man and a woman ask a geneticist whether they should have children, one of his primary concerns is whether both may be heterozygous for the same unwanted recessive trait. If they are, then there is one chance in four that any of their offspring will show this inherited disease.

Phenylketonuria (PKU). *Phenylketonuria* is an inherited disease in which there is a deficiency of the enzyme that in normal individuals converts phenylalanine to tyrosine. Phenylalanine and phenylpyruvic acid accumulate in the blood stream and tissues and are excreted in the urine. The accumulation of phenylalanine in the nervous tissues interferes with their function; hence individuals with phenylketonuria have a mental deficiency associated with their disease. The gene frequency q for this trait in the United States population is about 0.005 and the gene frequency p of the normal allele is 0.995. The frequency of heterozygotes is about one in every 100 persons ($2 \times 0.995 \times 0.005 = 0.01$). The heterozygous individual has a somewhat lower than normal ability to metabolize phenylalanine; however, the results of phenylalanine metabolism tests given to heterozygotes overlap those given to normal people, so that it is not possible to distinguish between the two with confidence. If a combination of family history and the phenylalanine metabolism tests makes it likely that both husband and wife are heterozygous carriers for phenylketonuria (or for any other recessive trait) it becomes possible to predict the probability (one in four) that they will have a child showing the abnormality. It is also possible to predict the possibility that they will have one, two or three normal children. To do this one simply makes the appropriate expansion of the binomial and calculates the value of the appropriate term in the expansion.

To calculate, for example, the probability that a man and his wife, both heterozygous carriers for phenylketonuria, will have three normal children out of three, you multiply $(p + q)^3$. In this, p represents the probability of a normal child, i.e., $\frac{3}{4}$, and q represents the probability of the phenylketonuric child, i.e., $\frac{1}{4}$, in a mating of two heterozygous individuals. Since we are calculating the probability of *three* normals in a total of three offspring we multiply $(p + q) \times (p + q) \times (p + q)$ or $(p + q)^3$. The algebraic product is $p^3 + 3\,p^2q + 3\,pq^2 + q^3$. The part of the product that represents the probability of the combination of three normal children is p^3. The next term, $3\,p^2q$, represents the probability of two normal children and one phenylketonuric child in a total of three children. To calculate the probability we take $\left(\frac{3}{4}\right)^3 = \frac{27}{64}$ from which we may conclude that there are 27 chances out of 64 that a pair of heterozygous individuals could have three normal children. By similar reasoning and calculation you can see that q^3 is $\left(\frac{1}{4}\right)^3$ or $\frac{1}{64}$. In other words there is one chance in 64 that if the man and his wife have three children, all three will be phenylketonurics.

30–6 HUMAN CYTOGENETICS

Many of the basic principles of genetics were discovered by experiments with lower organisms in which it was possible to relate genetic data with cytologic events. This involved making smears of cells and examining them under the microscope to see the number and the structure of the chromosomes present. Some of the organisms used in genetics, such as the fruit fly *Drosophila*, have few chromosomes (four pairs), and in the salivary glands and certain other tissues the chromosomes are quite large and the details of their structure are readily evident.

The normal human karyotype for males and females is shown in Figure 28–3. Cells from the bone marrow, blood or skin are incubated until they begin to undergo mitosis and then are treated with colchicine to stop the process in the metaphase. The cells are placed in hypotonic solution, causing them to swell and enabling the chromosomes to spread out so that they can be visualized more readily. The preparation is fixed and stained and the chromosomes are photographed. Each chromosome is cut out of the photographic print and aligned so that the homologous pairs are placed together. Chromosomes are identified by their length, by the position of the centromere, and by the presence of knobs or satellites. The

gradation in size between several of the pairs is quite fine, and although the largest chromosome is some five times as long as the smallest chromosome, there are only slight differences between some of the intermediate-sized ones. Sometimes chromosome pairs are simply allotted to one of seven different groups.

Although the normal human chromosome number is 46, some rare instances of abnormal chromosome numbers have been reported. An individual may be polyploid, having one or more complete extra sets of chromosomes, or he may have one or two extra chromosomes and hence a total chromosome number of 47 or 48. A severely defective male child was found to be a **triploid** individual with a total of 69 chromosomes. He had 66 autosomes, two X chromosomes and one Y chromosome. It seems likely that this zygote was formed from a normal haploid egg that was fertilized by an unusual diploid sperm or from an exceptional diploid egg fertilized by a normal haploid sperm.

Nondisjunction refers to the failure of a pair of homologous chromosomes to separate normally during the reduction division. Two X chromosomes, for example, might fail to separate and both might enter the egg nucleus while the polar bodies have no X chromosome. Alternatively the two joined X chromosomes might go into the polar body, leaving the female pronucleus with no X chromosomes. Nondisjunction of the XY chromosomes in the male might lead to the formation of sperm that have both an X and a Y chromosome or to sperm with neither an X nor a Y chromosome. Chromosomal nondisjunction may occur during either the first or second meiotic division; it may also occur during mitotic divisions and lead to the establishing of a clone of abnormal cells in an otherwise normal individual.

Cytogenetic studies have clarified the origin of one of the more distressing abnormal conditions in humans, that of **Down's syndrome**, or "mongolism." Individuals suffering from this syndrome have abnormalities of the face, eyelids, tongue and other parts of the body and are greatly retarded in both their physical and mental development. The term mongolism was originally applied to this condition because affected individuals often show a fold of the eyelid similar to that typical of members of the Mongolian race. Down's syndrome is a relatively common congenital malformation, occurring in 0.15 per cent of all births. Down's syndrome is a hundredfold more likely in the offspring of women 45 years or older than in the offspring of mothers under 19. The occurrence of Down's syndrome, however, is independent of the age of the father, and it is also independent of the number of preceding pregnancies in the woman. Cytogenetic studies revealed that individuals with Down's syndrome have one extra small chromosome 21, making a total of 47. The presence of this extra small chromosome is believed to arise by nondisjunction in the maternal oöcyte.

It is not clear why the DNA transcribing system does not simply ignore the redundant bit of genetic information and produce cells identical to those of the normal individual, but the presence of this extra chromosome leads to the complex physical and mental abnormalities that characterize Down's syndrome. Whether the extra genes in the third chromosome 21 lead to the production of an extra amount of certain enzymes and whether this is the basis for the abnormal physical and mental development is not known. When a certain chromosome or part of a chromosome has been added or deleted ("genetic imbalance"), comparable defects will be observed in all types of organisms.

Down's syndrome should be inherited as though it were a dominant gene since an af-flicted individual would form gametes half of which would have the normal complement of 23 chromosomes and half of which would have 24 chromosomes. In the rare cases in which persons with Down's syndrome have had offspring they have produced normal and afflicted children in about equal proportions.

In another condition caused by an altered chromosome number, individuals are outwardly nearly normal males but have small testes and produce few or no sperms. Their seminiferous tubules are very aberrant in appearance, and the individuals may have gynecomastia (a tendency for formation of femalelike breasts). This condition, called *Klinefelter's syndrome,* usually becomes apparent only after puberty, when the small testes and gynecomastia may bring the individual to the attention of his physician. The cells of these individuals show a chromatin spot and at one time they were thought to be XX individuals, i.e., genetic females. However, when their chromosomes were examined cytologically and counted, it was found that they have 47 chromosomes; their cells have two X and one Y chromosome. The fact that they are nearly normal males in their external appearance emphasizes the male-determining effect of the Y chromosome in man.

In *Turner's syndrome* the external genitalia, though feminine, are those of an immature female. The internal reproductive tract is

Table 30–4 SOME HUMAN CHROMOSOMAL ABNORMALITIES

Abnormality	Genetic Features	Clinical Aspects
Turner's syndrome (gonadal dysgenesis)	XO	Short stature, streak ovary, juvenile female genitalia, poorly developed breasts
Klinefelter's syndrome	XXY	Gynecomastia, small testes
Triple X females	XXX	Two "Barr bodies" present, fairly normal females but secondary sex characteristics may be poorly developed
Down's syndrome	Trisomy 21	Epicanthal folds, protruding tongue, hypotonia, mental retardation
Trisomy 18	Trisomy 18	Mental retardation, multiple congenital malformations
D trisomy	Trisomy 15	Mental retardation, severe multiple anomalies, cleft palate, polydactyly, central nervous system defects, eye defects
Translocation mongolism	15/21, 21/22, or 21/21 translocation	Mongolism, clinically similar to trisomy 21
Philadelphia chromosome	Deletion of one arm of chromosome 21	Chronic granulocytic leukemia
Orofaciodigital syndrome	Translocation of part of chromosome 6 to 1	Defects of upper lip, palate, and mouth, stubby toes with short nails
Cri du chat syndrome	Deletion of short arm of chromosome 5	Mental retardation, facial anomalies

*From Page, E. W., Villee, C. A., and Villee, D. B.: *Human Reproduction,* 2nd Ed. Philadelphia, W. B. Saunders Co., 1976.

present and resembles that of an immature but perfectly formed female. The uterus is present but small, and the gonads may be absent. The cells of these individuals are "chromatin negative," which suggests that they are males. However, they have only 45 chromosomes; they have one X chromosome but no Y chromosome. This type of disorder again emphasizes the importance of the Y chromosome in determining the male characteristics.

An individual with an extra chromosome, with three of one kind, is said to be *trisomic*, and an individual lacking one of a pair is said to be *monosomic*. Thus individuals with Down's syndrome are trisomic for chromosome 21 and individuals with Turner's syndrome are mono-

somic for the X chromosome. The features of certain human chromosomal abnormalities are summarized in Table 30–4.

30–7 THE INHERITANCE OF PHYSICAL TRAITS IN HUMANS

The development of each organ of the body is regulated by a large number of genes. The age at which a particular gene expresses itself phenotypically may vary widely. Most characteristics develop long before birth, but some, such as hair and eye color, may not appear until shortly after birth. Some, such as amaurotic idiocy, become evident in early

Table 30–5   SOME HUMAN TRAITS KNOWN TO BE INHERITED

Dominant	Recessive
Hair, Skin, Nails, Teeth	
Dark hair	Blond hair
Nonred hair	Red hair
Curly hair	Straight hair
Abundant body hair	Little body hair
Early baldness (dominant in male)	Normal
White forelock	Self-color
Piebald (skin and hair spotted with white)	Self-color
Pigmented skin, hair, eyes	Albinism
Black skin (2 pairs of genes, dominance incomplete)	White skin
Ichthyosis (scaly skin)	Normal
Epidermis bullosa (hypersensitivity to slight abrasions)	Normal
Absence of enamel of teeth	Normal
Normal	Absence of sweat glands
Eyes	
Brown	Blue or gray
Hazel or green	Blue or gray
"Mongolian fold"	No fold
Congenital cataract	Normal
Nearsightedness	Normal vision
Farsightedness	Normal vision
Astigmatism	Normal vision
Glaucoma	Normal
Aniridia (absence of iris)	Normal
Congenital displacement of lens	Normal
Normal	Optic atrophy (sex-linked)
Normal	Microphthalmus
Features	
Free ear lobes	Attached ear lobes
Broad lips	Thin lips
Large eyes	Small eyes
Long eyelashes	Short eyelashes
Broad nostrils	Narrow nostrils
High, narrow bridge of nose	Low, broad bridge
"Roman" nose	Straight nose
Skeleton and Muscles	
Short stature (many genes)	Tall stature
Achondroplasia (dwarfism)	Normal
Ateliosis (midget)	Normal

childhood and still others, such as glaucoma and Huntington's chorea, develop only after the individual has reached maturity. Some of the human traits whose inheritance has been studied are listed in Table 30–5. Some of these conditions may be due to two or more different genes, so that they are inherited by dominant genes in one pedigree and by recessive genes in another.

A distinction is made between traits that are inherited and those that are congenital. A *congenital trait* is simply one that is present at birth; it may or may not be inherited, that is, transmitted from one generation to the next. Some congenital abnormalities are inherited while others are produced by some accident in the developmental process. For example, if a woman contracts German measles during the first three months of pregnancy there is a greatly increased probability that her offspring will show some sort of congenital malformation. Conversely, not all inherited conditions are congenital, that is, they are not evident at birth; some appear later in life.

30–8 THE INHERITANCE OF MENTAL ABILITIES IN HUMANS

The inheritance of mental ability or intelligence is one of the most important, yet one of the most difficult, problems of human genetics.

Table 30–5 SOME HUMAN TRAITS KNOWN TO BE INHERITED *(Continued)*

Dominant	Recessive
Polydactyly (more than 5 digits on hands or feet)	Normal
Syndactyly (webbing of 2 or more fingers or toes)	Normal
Brachydactyly (short digits)	Normal
Cartilaginous growths on bones	Normal
Progressive muscular atrophy	Normal
Circulatory and Respiratory Systems	
Hereditary edema (Milroy's disease)	Normal
Blood groups A, B, and AB	Blood group O
Hypertension (high blood pressure)	Normal
Normal	Hemophilia (sex-linked)
Normal	Sickle cell anemia
Excretory System	
Polycystic kidney	Normal
Endocrine System	
Normal	Diabetes mellitus
Digestive System	
Enlarged colon (Hirschsprung's disease)	Normal
Nervous System	
Tasters (of phenylthiocarbamide)	Nontasters
Normal	Congenital deafness
Normal	Spinal ataxia
Huntington's chorea	Normal
Normal	Amaurotic idiocy
Migraine (sick headache)	Normal
Normal	Dementia praecox (several pairs of genes involved)
Normal	Phenylketonuria
Paralysis agitans	Normal
Cancers	
Normal	Xeroderma pigmentosum
Recklinghausen's disease	Normal
Normal	Retinal glioma

The interpretation of genetic data can easily become colored by political, sociologic, psychologic or educational theories. It is difficult to set up a test that measures innate intellectual capacity and is not influenced by the previous training of the individual. The older psychologic tests, such as the Stanford-Binet, are prepared by setting tasks for children and then determining what capacity can normally be expected of children of each age. In taking the test the child is given progressively more difficult problems until he is finally unable to solve them.

When such tests are given, a wide range of mental ability, from complete incompetence to excellent comprehension, has been found for each age group. A child of six who can solve problems ordinarily solved by children eight years old obviously is superior to one of six who can do only those normally done by six year olds. The "mental age" as determined by this test is divided by the actual chronologic age and the quotient is multiplied by 100 to give the **intelligence quotient,** or I.Q. When the intelligence quotients of a large number of people are measured they form a curve of normal distribution from zero to over 140, with the largest number of scores in the normal class and progressively fewer scores in the classes farther from normal.

That the mental capacities of people form a continuous series from idiot to genius, with the distribution of I.Q.'s conforming to a normal curve, suggests that intelligence is inherited by a system of polygenes, and other evidence substantiates this hypothesis. Mental retardation may be caused by diseases such as syphilis or meningitis, by injuries sustained during birth or by other environmental factors, but the majority of cases are due to inheritance.

More recent intelligence tests have provided measures of primary abilities such as the ability to reason inductively, the ability to memorize, and the ability to visualize objects in three dimensions. Special abilities—musical, artistic, mechanical and mathematical—have a hereditary basis and their inheritance is separate from that of general intelligence. Since musical ability is a complex function of pitch discrimination, tone memory, and a sense of rhythm, melody and harmony, it is not surprising that its inheritance should be complex.

30–9 GENETIC COUNSELING

Parents who have had one abnormal child or who have a parent or some other member of the family affected with a hereditary disease and hence are concerned about the risk of abnormality in a first or subsequent child may seek genetic advice. Genetic clinics are available in most metropolitan centers to provide such advice.

Advice, of course, can be given only in terms of the *probability* that any given offspring will have a particular condition. The geneticist needs a carefully taken family history of each parent and may use tests for the detection of heterozygous carriers of certain conditions. Some diseases are inherited by a single pair of genes, and the probabilities are then easily calculated. For example, if one parent is affected with a trait that is inherited as an autosomal dominant, such as perineal muscular atrophy or Huntington's chorea, the probability that any given child will have the trait is 0.5. The birth of one child with a trait such as albinism or phenylketonuria establishes that both parents are heterozygous carriers, and the probability that a subsequent child will be affected is 0.25. For a disease inherited by a recessive gene on the X chromosome, such as hemophilia, the probability depends on whether the father or mother has the disease or is a carrier for it. A normal woman and an affected man will have daughters who are carriers and sons who are normal. The probability that a carrier woman and a normal man will have an affected son is 0.5, and the probability that they will have a carrier daughter is 0.5.

In conditions in which the method of inheritance is unknown or is doubtful, an estimate of the probability of appearance of a given trait can be obtained from a table of empiric risk. It is difficult to give a precise probability, for the trait may be inherited in different ways in different pedigrees. For example, retinitis pigmentosa appears to be inherited as a sex-linked trait in some pedigrees and as an autosomal trait in others.

Mental deficiency, epilepsy, deafness, congenital heart disease, anencephaly, harelip, spina bifida and hydrocephalus are conditions about which inquiries are commonly made. With such conditions the possibility must be considered that some environmental factor may have played a role in the appearance of the abnormality in the previous child. Did the mother have some sort of infectious disease during pregnancy (e.g., German measles)? Was she on some sort of drug therapy? Was she subjected to any sort of radiation? By dissecting out possible environmental contributions, the geneticist can make a better estimate of the probability that genetic factors are involved and a better estimate of the probability of recurrence of the particular trait in some subsequent offspring.

30–10 HEREDITY AND ENVIRONMENT: TWIN STUDIES

In many of the lower organisms it is possible to make a direct test of the relative effects of heredity and environment on the expression of some given trait because it is possible to establish and use in testing a strain of organisms that are genetically identical. Variations in the phenotypes of such organisms can be attributed to environmental factors. No such *isogenic strain* of human beings exists. With lower organisms it is possible to control the environmental variations very strictly, whereas it is not possible to control at will the environment in which human development is occurring. The experimental design that most closely approaches these ideal conditions is a comparison of identical and fraternal twins.

The frequency of twinning varies considerably in different human races; the rate in the United States among the white population is one twin birth in every 88.6 births. *Identical twins* originate from a single fertilized egg that begins development and subsequently splits into two embryos. The two members of the pair are genetically identical and are as similar as the two sides of the body of a single individual. Nonidentical, or fraternal, twins result from the fertilization of two eggs by two separate sperm and are no more similar genetically than are brothers and sisters born at different times.

An estimate of the number of twins that are identical (*monozygotic*) and fraternal (*dizygotic*) can be obtained from the sex ratios of the twin pairs. Dizygotic twins should occur in the ratio of $1/4$ male-male, $1/4$ female-female and

$1/2$ male-female. Monozygotic twins will be either male-male or female-female but will not include any male-female pairs. The number of dizygotic, like-sexed pairs will be equal to the number of dizygotic, unlike-sexed pairs. Thus, multiplying the number of unlike-sexed pairs of twins by two gives an estimate of the total number of dizygotic twins. Subtracting this from the total gives an estimate of the number of monozygotic pairs of twins. Applying this method to the number of twins in the white population of the United States yields an estimate that 34.2 per cent of all twin births are monozygotic twins. In other words, about 1 of 3 pairs of twins are identical. In the Japanese population, where the over-all frequency of twinning is lower (0.7 per cent instead of 1.129 per cent as in the United States), the fraction of monozygotic twins is considerably higher. The frequency of monozygotic twinning in the American and Japanese populations is nearly the same but the Japanese population has fewer dizygotic twins.

The diagnosis of whether a particular pair of twins are monozygotic or dizygotic cannot always be made from the nature of the amnion, chorion and placenta. Although twins with two separate amnions, two separate chorions and two separate placentas are likely to be dizygotic, monozygotic twins can also show this condition. A pair of twins with one amnion, one chorion and one placenta are indeed monozygotic.

The most decisive means of identifying monozygotic twins takes advantage of the **homograft reaction.** When a piece of skin is removed from one person and grafted to another, it will remain healthy and continue to grow in its new place only if both donor and host are identical twins. If they are dizygotic twins, rather than monozygotic twins, the homograft reaction between host and graft will occur and the graft will be rejected and sloughed off in a period of about four to six weeks.

It is now abundantly clear that both physical and mental traits are the result of the interplay of both genetic and environmental factors. A few genes, such as those that determine the blood groups, produce their effect regardless of the environment. The expression of other genes may be markedly affected by altered environment. Studies from the field of biochemical genetics suggest that the greater the number of biochemical steps that intervene between a given gene and the final appearance of its trait, the greater is the opportunity for environmental influences to come into play.

When one fraternal twin is mentally retarded, the other is also in about 25 per cent of pairs. In identical twins, however, when one is mentally retarded the other has the trait in nearly every case, evidence of the very important role of heredity in producing the trait. Identical twins are much more similar in their intelligence than are fraternal twins; indeed, identical twins reared apart are more similar in intelligence than are fraternal twins reared together.

Children reared together in an orphanage, where the environment is fairly constant, show just as much variability in intelligence as do children reared separately in their own homes. Even when children are adopted early in infancy there is a greater correlation between the intelligence of the child and its true parents than between the child and its foster parents. It may be concluded from these studies that the upper limit of a person's mental ability is determined genetically, but how fully these inherited abilities are developed is determined by environmental influences, by training and by experience.

It is easy to understand why the offspring of intelligent parents are sometimes less intelligent than either parent. Since the coordinate action of many pairs of genes is involved in intelligence, the fortuitous combination of those that produced the intelligent parents may be broken up by genic segregation. Conversely, the chance recombination of favorable genes may produce a brilliant child from average parents (but geniuses are never produced by feeble-minded parents).

30–11 GENETIC EQUILIBRIUM AND GENETIC DRIFT

The essence of the Hardy-Weinberg Principle is that under specific conditions the frequencies of the several alleles of a given gene remain constant from one generation to the next. The first condition for genetic equilibrium is either that there must be no mutations or that the rates of forward and reverse

mutations must be in equilibrium. That is, either there is no change of gene A into gene a, or else the rate at which gene A mutates to gene a (forward mutation) is equal to the rate at which gene a mutates to gene A (reverse mutation). Genes undergo mutations continually; indeed, there is no way to prevent mutations from occurring. Further, the rates of forward and reverse mutations will rarely be equal. Thus there is usually a tendency, termed *mutation pressure*, for one of the alleles to increase in frequency and another allele to decrease in frequency. This mutation pressure may be countered by some other factor, such as selection. Even though mutations occur constantly, they occur at random. Mutations are seldom the major factor in producing changes in gene frequencies in a population. They increase genetic variability and ultimately provide the raw material of evolution, but mutations alone are unlikely to determine the nature or direction of evolutionary change.

To maintain its genetic equilibrium, a population must be large enough so that chance events are unlikely to change gene frequencies. The Hardy-Weinberg Principle is based on statistical concepts. Its operation requires that the sample size be large enough to minimize the possibility of chance deviations. In small (less than 100 members) isolated breeding populations of a given species, there is a relatively high probability that one allele or the other will be lost from the population by chance even though it may determine a trait that is of adaptive value. In such small populations there is a strong tendency for the population to become homozygous for one allele or the other. In contrast, large breeding populations have a tendency to remain more variable, and to include many heterozygous individuals. The production of random evolutionary changes by chance in small breeding populations is termed *genetic drift*. Genetic drift results in changes in the gene pool of a population and thus produces evolutionary change; however, such evolutionary changes are aimless, at random, and not adaptive. Genetic drift may explain the common observation that closely related species in different parts of the world frequently differ in curious, even bizarre, ways that have no adaptive value.

The maintenance of genetic equilibrium also requires that the population not lose genes from its gene pool by the outward *migration* of certain members of the population. The population must also not receive new genes from immigrants from other populations. The populations of some species in nature are essentially isolated and do not undergo gene migration. Other populations do interbreed to some extent with neighboring populations and a considerable amount of gene migration occurs. New genes are introduced into the populations, genetic variability is increased, and this increased genetic variability may play a role in the evolution of that species or population.

30–12 FACTORS CHANGING GENE FREQUENCIES: DIFFERENTIAL REPRODUCTION

To maintain genetic equilibrium, the members of the population must mate completely at random. The male must not select his mate because she has the same phenotype he has. Indeed, more than simply the selection of mates is involved, for *random reproduction* implies that all the many factors that contribute to success in reproduction, to success in producing viable offspring, are also operating at random and independently of the genotypes of the individuals involved. The selection of a mate, the fertility of the pair, the fraction of the resulting zygotes that complete development to birth, the survival of the young to reproductive age and their fertility are all factors that may influence the relative effectiveness of certain types or strains of organisms to perpetuate their kind. If individuals with certain genotypes are better able to raise large numbers of offspring to the age when they in turn reproduce, genetic equilibrium will not be maintained. Instead, the frequency of certain genes in the population will increase as a result of *differential reproduction.* This change in gene frequencies in the gene pool of a population caused by differential reproduction is the modern way of thinking about Darwin's concept of natural selection, the basis of evolution. If differential reproduction does not occur, the gene pool does not change and the Hardy-Weinberg concept of genetic equilibrium applies. If differential reproduction does occur (and it usually does) the gene pool changes. Evolutionary changes resulting from differential re-

production are characteristic of nearly all populations of organisms, including human.

30–13 EVOLUTION: THE FAILURE TO MAINTAIN GENETIC EQUILIBRIUM

Evolution by natural selection, stated in its simplest terms, means that individuals with certain genotypes and certain traits have more surviving offspring in the next generation than other individuals do. They thus contribute a proportionately greater percentage of genes to the gene pool of the next generation than do organisms with contrasting traits. The evolutionary changes resulting from differential reproduction tend to improve the average ability of members of the population to produce successive generations with genotypes like their parents. As we shall see, natural selection does not operate upon the phenotypes of single genes but upon the total phenotypic effect of the entire array of genes present. One group of organisms may survive despite some clearly disadvantageous character. Another group may be eliminated despite certain traits that are highly advantageous for survival. In evolution you don't necessarily get counted out because of a single bad trait, nor do you get a gold star for a single good adaptive trait. The organisms that ultimately survive and serve as parents of the next generation are ones whose total spectrum of qualities renders them a little better able to survive and reproduce their kind.

Thus, mutation is of importance in supplying the raw materials of evolution. Gene migration and genetic drift play generally minor roles in evolution; the major factor in evolution is differential reproduction. In the absence of mutation and selection, the gene frequencies in a population tend to remain constant in successive generations and the Hardy-Weinberg Principle of genetic equilibrium applies.

QUESTIONS

1. In what ways are human beings not favorable for studies of inheritance? What means have been devised for overcoming the difficulties of studying human inheritance?
2. Compare and contrast a priori probability, empiric probability and sampling probability.

3. Discuss the "product rule" of probability.

4. What is the probability of drawing a heart from a full deck of cards? If the first card drawn is a heart, what is the probability that the second card will also be a heart? What is the probability of drawing an ace from a full deck of cards? What is the probability of drawing the ace of hearts?

5. What is the probability of rolling a seven with a pair of dice?

6. What are the implications of the Hardy-Weinberg Principle? How is this applied to studies of human genetics?

7. How does one determine whether or not a given population is in genetic equilibrium?

8. Compare the metabolic blocks of alkaptonuria, phenylketonuria and albinism.

9. What is meant by "nondisjunction"? What human abnormalities appear to be the result of nondisjunction?

10. Distinguish between congenital and hereditary traits.

11. Could two dark-haired parents have a red-haired child? What genotypes must the two parents have to permit this?

12. In what ways are studies of twins useful in supplying information about the relative importance of inheritance and environment in determining a given trait?

13. What is the evidence that intelligence is inherited? How might a child be more intelligent than either of its parents?

14. In each of the pedigrees shown below, determine the method of inheritance of the trait and, as far as possible, fill in the genotypes of each individual. Assume that an individual who marries into the family and does not exhibit a trait does not carry any recessive gene for it.

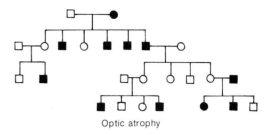

Optic atrophy

15. Distinguish between monozygotic and dizygotic twins. What physical characteristics can be used to identify monozygotic twins?

16. Contrast the meanings of the terms gene pool and genotype.

17. What is meant by differential reproduction?

18. What are the effects on genetic equilibrium of mutation pressure? of genetic drift? of gene migration?

SUPPLEMENTARY READING

Curt Stern's *Principles of Human Genetics* is an excellent, well written text on general genetics with special emphasis on human inheritance. Two newer, briefer texts with excellent discussions of human genetics and cytogenetics are *Genetics in Medicine* by Thompson and Thompson and V. A. McKusick's *Human Genetics. You and Heredity* by A. Scheinfeld contains a popular account of the inheritance of human traits. Descriptions of some of the inherited abnormalities are found in *An Introduction to Medical Genetics* by J. A. Fraser-Roberts and in *Outline of Human Genetics* by L. S. Penrose. *The Biochemistry of Human Genetics*, edited by J. E. Wolstenholme and C. M. O'Connor, includes papers by many notable research workers presented at a symposium on the subject. *The Metabolic Basis of Inherited Disease*, edited by J. B. Stanbury, J. B. Wyngaarden and D. S. Fredrickson, gives detailed presentations of the inborn errors of metabolism found in humans. *Radiation and Human Mutation* by H. J. Muller describes some of the dangers to the human germ plasm from the radiations resulting from x rays and nuclear energy. *Human Afflictions and Chromosomal Aberrations* by R. Turpin and J. Lejeune is an excellent source book for information on human genetics.

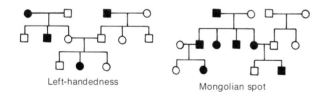

Left-handedness   Mongolian spot

CHAPTER 31

PRINCIPLES AND THEORIES OF EVOLUTION

In previous chapters we have had brief glimpses of some of the immense variety of living organisms that inhabit every nook and cranny of the sea and land. The concept of **evolution** is based on detailed comparisons of the structures of living and fossil forms, on the sequences of appearance and extinction of species in past ages, on the physiologic and biochemical similarities and differences between species, and on analyses of the genetic constitutions of present-day plants and animals. The concept of evolution has been implicitly and explicitly involved in many of the subjects discussed previously and is not a new topic at this point. Our understanding of the concept of evolution can emerge logically and naturally from our understanding of genetics. This is not, however, how the concept of evolution arose historically. It came as a result of a large number of observations of similarities and differences in structures and functions of the various kinds of animals and plants in different parts of the world. It came from Charles Darwin's profound insight into the arrangement of the pieces of the puzzle of how these similarities and differences may have arisen.

Evolution means an unfolding or unrolling—a gradual, orderly change from one condition to another. The planets and stars, the earth's topography, the chemical compounds of the universe, and even the chemical elements and their subatomic particles have undergone gradual, orderly changes, sometimes termed *inorganic evolution.* The principle of *organic evolution* states that all the various plants and animals existing at the present time have descended from other, usually simpler, organisms by gradual modifications which have accumulated in successive generations. One of the major trends in the evolution of most plants and animals has been toward increased adaptation to some particular environment, and this has frequently involved increased specializations and complexity of structures and their functions.

The process of evolution has not ceased; on the contrary, it is occurring more rapidly today than in many of the past ages. In the last few hundred thousand years, hundreds of species of animals and plants have become extinct and other hundreds have arisen. Although the process is usually too gradual to be observed, there are notable examples of evolutionary change within the time of recorded history. For example, early in the fifteenth century a litter of rabbits was released on Porto Santo, a small island near Madeira. There were no other rabbits and no carnivorous enemies on the island and the rabbits multiplied with amazing speed. By the nineteenth century they were strikingly different from the ancestral European stock, only half as large as their European relatives, with a different color pattern, and a more nocturnal way of life. More importantly, they could not produce offspring when bred with members of the European species. Within four hundred years, then, a new species of rabbit had developed.

31–1 HISTORICAL DEVELOPMENT OF THE CONCEPT OF EVOLUTION

The idea that the present forms of life arose from earlier, simpler ones was not new when Darwin's *The Origin of Species* was published in 1859. Elements of the theory of organic evolution are to be found in the writings of certain early Greek philosophers — Thales (640?–546 B.C.), Anaximander (611–547 B.C.), Empedocles (495?–435? B.C.) and Epicurus 342?–270 B.C.). The spirit of this age of Greek philosophy was somewhat like our own, for simple, natural explanations were sought for all phenomena. Little biology was known and the Greeks' ideas about evolution were extremely vague; they can scarcely be said to foreshadow our present theory of organic evolution.

Aristotle (384–322 B.C.), a great biologist as well as philosopher, worked out an elaborate theory of gradually evolving life forms termed the "ladder of nature." He held the metaphysical belief that nature strives to change from the simple and imperfect to the more complex and perfect. The Roman poet Lucretius (96?–55 B.C.) also gave an evolutionary explanation of the origin of plants and animals in his poem *De Rerum Natura*. With the Renaissance, interest in the natural sciences quickened, and from the fourteenth century on, an increasingly large number of people found the concept of organic evolution reasonable. In *The Origin of Species,* Darwin lists about twenty thinkers who had considered the theory seriously, among them his own grandfather, Erasmus Darwin (1731–1802), and the French scientist Lamarck (1744–1829).

Long before Darwin, odd fragments resembling bones, teeth and shells had been discovered buried in the ground. Some of these corresponded to parts of familiar, living animals, but others were strangely unlike any known form. Many of the objects found in rocks high in the mountains resembled parts of marine animals. Leonardo da Vinci, a true universal genius, correctly interpreted these curious finds in the fifteenth century, and gradually others accepted his explanation that they were the remains of animals that had existed in previous ages but that had become extinct. Such evidence of former life suggested the theory of **catastrophism,** the idea that a succession of fires and floods have periodically destroyed all living things and necessitated the repopulation of the world by successive acts of special creation.

Three Englishmen in the eighteenth and early nineteenth centuries laid the foundations of modern geology. In 1785, James Hutton developed the concept (termed **uniformitarianism)** that the geologic forces at work in the past were the same as those of the present. After a careful study of the erosion of valleys by rivers and the formation of sedimentary deposits at the river mouths, he concluded that the processes of erosion, sedimentation, disruption and uplift, over long periods of time, could account for the formation of the fossil-bearing rock strata. In 1802, John Playfair's *Illustrations of the Huttonian Theory of the Earth* was published, in which he gave further explanation and examples of the idea of uniformitarianism in geologic processes.

Sir Charles Lyell, one of the most influential geologists of his time, did much in his *Principles of Geology* (1832) to establish the principle of uniformity. By demonstrating the validity of geologic evolution he proved irrefutably that the earth is much older than a few thousand years, old enough for the process of organic evolution to have occurred. He was a close personal friend of Darwin's and had great influence on his thinking, and through his work he paved the way for and made possible the ideas presented in *The Origin of Species*.

Jean Baptiste de Lamarck. The earliest theory of organic evolution to be logically developed was that of Jean Baptiste de Lamarck, the great French zoologist whose *Philosophie Zoologique* was published in 1809. Lamarck, like most biologists of his time, believed that organisms are guided through their lives by innate and mysterious forces that enable them to adapt and overcome adverse environmental forces. He believed that, once made, these adaptations are transmitted from generation to generation—that is, that acquired characteristics are inherited. In developing this notion, Lamarck went on to state that new organs arise in response to demands of the environment and that their size is proportional to their "use or disuse." Such changes in size were believed to be inherited by succeeding generations. Lamarck explained the evolution of the giraffe's long neck by suggesting that an ancestor took to browsing on the leaves of trees, instead of on grass, and in reaching up, stretched and elongated its neck; its offspring then supposedly inherited the longer neck.

The Lamarckian theory of the inheritance of acquired characteristics is an attractive one. It would explain the complete adaptation of many plants and animals to the environment, but it is unacceptable because overwhelming genetic evidence indicates that *acquired characteristics cannot be inherited.* Many experiments have been performed in attempts to demonstrate the inheritance of acquired traits, but all have ended in failure. From what we now know about the mechanism of heredity, it is obvious that acquired traits cannot be inherited, for such characteristics are in the *body* cells only, whereas an inherited trait is transmitted by the *gametes*—the eggs and sperm.

Charles Darwin and Alfred Russel Wallace. Darwin's contribution to the body of scientific knowledge was twofold: he presented a mass of detailed evidence and cogent argument to prove that organic evolution has occurred, and he devised a theory—that of *natural selection*—to explain how it operates.

Although his university training was in theology, Darwin was extremely interested in both biology and geology and while at Cambridge became acquainted with Professor Henslow the naturalist. Through his help Darwin, just out of college and only 22 years old, was appointed to the position of naturalist on the *H. M. S. Beagle*, which was to make a five-year cruise around the world to gather data for oceonographic charts for the British navy. Darwin studied the animals, plants and geologic formations of the east and west coasts of South America, making extensive collections and notes. The *Beagle* then went to the Galápagos Islands, west of Ecuador, where Darwin was fascinated by the diversity of the giant tortoises and the finches that lived on each of the islands. As Darwin mused over these observations, he was led to reject the theory of special creation and seek an alternative explanation for his observations.

According to his journal, the idea of natural selection occurred to Darwin shortly after his return to England in 1836, but he spent the next 20 years or so accumulating the vast body of facts that eventually became *The Origin of Species*. In 1858 he received a manuscript from Alfred Russel Wallace, a young naturalist who was studying the distribution of plants and animals in the East Indies and the Malay Peninsula. In this paper Wallace set forth his concept of natural selection, which he had reached independently, stimulated, as Darwin had been, by Malthus' book on population growth and pressure and the struggle for existence. By mutual agreement, Darwin and Wallace presented a joint paper on their theory at the meeting of the Linnean Society in London in 1858, and Darwin's monumental work was published the next year.

31–2 THE DARWIN-WALLACE THEORY OF NATURAL SELECTION

The explanation advanced by Darwin and Wallace of the way evolution occurs may be summarized as follows:

1. Variation is characteristic of every group of animals and plants. Darwin and Wallace assumed that variation was one of the innate properties of living things. Inherited and noninherited variations can now be distinguished. Only inherited variations, produced by mutations, are important in evolution.

2. More of each kind of organism begin to grow than can possibly obtain food, survive and reproduce. Yet, since the number of members of each species remains fairly constant under natural conditions, it must be assumed that most of the offspring in each generation perish. If all the offspring of any species remained alive and reproduced, they would soon crowd all other species from the earth.

3. Since a larger number of individuals are born than can survive, there is a struggle for survival, a competition for food and space. This may be an active kill-or-be-killed contest or one less immediately apparent but no less real, such as the struggle of plants or animals to survive drought, cold or other unfavorable environmental conditions.

4. Those organisms with variations that better equip them to survive in a given environment will be favored over other organisms that are less well adapted. The ideas of the "struggle for survival" and "survival of the fittest" are the core of Darwin's and Wallace's theory of natural selection.

5. The surviving individuals will give rise to the next generation, and in this way the "successful" variations are transmitted to the next generation and the next and so on.

This process would tend to provide successive generations of organisms with better adaptations to their environment. Indeed, as the environment changed, further adaptations would follow. The operation of natural selection over many years could lead ultimately to the development of descendants that are quite different from their ancestors — different enough to be recognized as a separate kind of animal or plant. Certain members of the population with one group of variations might become adapted to environmental changes in one way, while other members with a different set of variations might become adapted in a different way. Thus two or more different kinds of organisms may arise from a single ancestral group. Darwin and Wallace recognized that animals and plants may exhibit variations that are neither a help nor a hindrance to them in their survival in a given environment. These variations will not be affected directly by natural selection, and the transmission of such neutral variations to succeeding generations will be governed by chance.

Darwin's theory of natural selection was so reasonable and so well supported by his arguments that most biologists soon accepted it. Some objected that the theory could not explain the presence of many apparently useless structures in an organism. Many of the differences between species may not be important for survival but are simply incidental effects of genes that have physiological effects of survival value. Other "useless" or nonadaptive differences may be controlled by genes that are closely linked in the chromosome to other genes controlling characteristics that are important for survival. Still other nonadaptive characteristics may become fixed in a population by chance, by the phenomenon termed genetic drift.

The concepts of the struggle for survival and the survival of the fittest were key points in the Darwin-Wallace theory of natural selection, but biologists have come to realize that actual physical struggle between animals for survival, or competition between plants for space, sun or water, is probably less important as an evolutionary force than Darwin had imagined.

31–3 MODERNIZATION OF THE DARWIN-WALLACE HYPOTHESIS

Populations. The evolution of any kind of organism occurs over many generations during which individuals are born and die but the population continues. Thus the unit in evolution is not the individual but rather a *population* of individuals. A population of similar individuals living within a defined area and interbreeding is termed a **deme,** or **genetic population.** The territorial limits of any given deme may be vague and difficult to define and the number of individuals in the deme may fluctuate widely from time to time. A deme commonly overlaps with one or more adjacent demes to some extent. The next larger unit of population in nature is the **species,** which is composed of a series of intergrading demes.

The relative frequencies of the genes in a

population will remain constant from one generation to the next if (1) the population is large, (2) there is no selection for or against any specific gene or allele, i.e., if mating occurs at random, (3) no mutations occur, and (4) there is no migration of individuals into or out of the population. The operation of the Hardy-Weinberg Principle will result in maintaining a given gene frequency in a population. The essential feature of the process of evolution is a gradual *change* in the gene frequencies of a population when this Hardy-Weinberg equilibrium is upset either because mutations occur, because reproduction is nonrandom, or because the population is small, so that the gene frequencies in successive generations will be determined by chance events.

The demes and species found in nature tend to remain unchanged for many generations. This fact implies that there has been no change in the genetic make-up of the demes and no change in the environmental factors that determine survival. When the characteristics of the population do change, this reflects either changes in the genetic factors brought about by mutations or changes in the environmental factors that lead to the selective survival of one or another phenotype.

Gene Pools. One of the basic concepts of population genetics, one of prime importance in understanding evolution, is that each population is characterized by a certain *gene pool.* Each *individual* in the population is genetically unique and has a specific genotype. However, if we count all of the alleles of a given gene ($A_1 A_2 A_3 \ldots$), either in the entire population or in some statistically valid sample of the population, we can then calculate the fraction of the total pool represented by allele A_1, by allele A_2 and so on. A population with a gene pool that is constant from one generation to the next is said to be in *genetic equilibrium;* i.e., the frequency of each allele in the population remains unchanged in successive generations.

In contrast, a population undergoing evolution is one in which the gene pool is changing from generation to generation. The gene pool of a population may be changed by mutations, by the introduction of genes from some outside population into this population, or by natural selection. Recombinations brought about by crossing over or by the assortment of chromosomes in meiosis may lead to new combinations of genes. The new phenotypes resulting from such recombinations may have some specific advantage or disadvantage for survival that would ultimately be reflected in a change in the gene pool.

Most, perhaps all, genes have many different effects on the phenotype (they are said to be *pleiotropic*). Some of the effects of a given gene may be advantageous for survival (termed *positive selection pressure*) and others may be disadvantageous (termed *negative selection pressure*). Whether the frequency of a given allele increases or decreases will depend on whether the sum of the positive selection pressures due to its advantageous effects is greater or lesser than the sum of the negative selection pressures due to its harmful effects.

In the reproduction of small populations chance alone may play a considerable role in determining the composition of the succeeding generation. The equilibrium of the genetic pool of the population can be changed by chance processes rather than by natural selection. This role of chance in the evolution of small breeding populations has been described by Sewall Wright as *genetic drift.* Within small interbreeding populations, heterozygous gene pairs tend to become homozygous for one allele or the other by chance rather than by selection. This may lead to the accumulation of certain disadvantageous characters and the subsequent elimination of the group possessing those characters.

The role that genetic drift actually plays in the evolution of organisms in nature has been a subject for debate among biologists, but there seems little doubt that it does play at least a minor role. Certainly many animal and plant populations in nature are divided into subgroups small enough to be affected by the chance events underlying genetic drift. Genetic drift represents an exception to the Hardy-Weinberg Principle governing the tendency for a population to keep its gene pool constant, to maintain its proportion of homozygous and heterozygous individuals. The Hardy-Weinberg Principle is based on statistical events and, like all statistical laws, holds only when the number of individuals involved is large enough. Genetic drift may explain the common observation that closely related species in different parts of the world frequently differ in curious, even bizarre, ways that appear to have no particular adaptive value.

The role of chance in evolution is particularly evident when a species moves into a new area, for the number of individuals moving into that area is usually small. These first colonizers from which the entire new population develops rarely constitute a representative sample of the gene pool of the original population, but instead differ from the parent population in the frequencies of specific genes. These differences may be quite marked but the new colonizing generation tends to differ from the parent population in ways that are random rather than selective.

This effect is most apparent on islands and other areas of geographic isolation and helps to account for the differences evident in island populations as compared to their mainland relatives. When a species is expanding continuously, the populations at the edge of the range, invading new areas, are likely to be small and differ genetically from the main body of the population. In all of these situations, when the breeding population is small, chance rather than selection may play a large role in determining the evolution of a particular group.

31–4 DIFFERENTIAL REPRODUCTION

Evolution by natural selection implies that individuals with certain traits have more surviving offspring in the next generation. In this way they contribute a proportionately greater percentage of genes to the gene pool of the next generation than do organisms with other traits. New inherited variations arise primarily by mutation. If organisms with the new mutation survive and have, on the average, more offspring that survive than do the organisms without that particular mutant allele, the gene pool of the population will gradually change. Thus the number of these mutant alleles in the population will increase in succeeding generations.

This process, termed *differential reproduction* or nonrandom reproduction, implies that the conditions of the Hardy-Weinberg equilibrium do not apply in the population at that time. The individuals that produce more surviving offspring in the next generation are usually, but not necessarily, those that are best adapted to survive in the given environment. Well adapted individuals may be healthier, better able to obtain food and mates, and better

able to care for their offspring; however, of primary importance in evolution is how many of their offspring survive to be parents of the next succeeding generation.

The ultimate raw material of evolution is a *mutation* which establishes an alternative allele at a given locus and makes possible an alternative phenotype. Evolutionary changes are possible only when there are alternative phenotypes that may survive or perish. However, it is important to keep in mind that the process of selection does not operate gene by gene, but rather individual by individual, and on the basis of the effects of the individual's entire genetic system. The forces of natural selection operate on the entire individual and not on single traits.

When a mutation first appears, only one or a very few organisms in the population will bear the mutant gene, and these will breed with other members of the population from which the mutant arose. The change in the gene pool so that the mutant gene appears with greater and greater frequency in the population is a gradual process which may occur over many generations. In evolutionary changes in a large population, the success or lack of success of some new mutant gene will depend largely on its ability to confer on its possessors the capacity to leave a larger number of surviving individuals in the next generation.

Populations become diverse by the action of the evolutionary forces of mutation, genetic drift, and the transfer of genes from one population to another through migration and hybridization. Although these processes generally operate at random, a key feature of evolutionary change is the tendency of organisms to become adapted to survive and reproduce in a given environment. The evolutionary process itself is not random with respect to establishing adaptive features of the organism undergoing evolution.

The process of differential reproduction may not be random with respect to (1) union of male and female gametes, (2) the production of viable zygotes, or (3) the development and survival of the zygotes until they become adults and produce their own offspring. Nonrandom reproduction tends to produce nonrandom, directional changes in the genetic pool, which lead to nonrandom, directional evolution. In a strict sense, random mating implies that any male is as likely to mate with any one female as with any other, or more generally, that the gametes of any two individuals in a population will be equally likely to unite. However, in nature, mating is usually not completely at random.

There are well established behavior patterns of courtship and mating in many species, which lead to the acceptance or refusal of one individual by another in mating. Such behavior patterns compose one of the forces that may direct differential reproduction through *nonrandom mating.* For example, in many fishes or birds, some brightly colored part on the male (Fig. 31–1) serves as a stimulus to the female, which is necessary before copulation can be begun. Mutations that lead to the formation of bigger, brighter spots tend to make those males more attractive to females and may confer selective advantages on their possessors. Conversely, mutations that would lead to the formation of smaller, duller spots would have a negative selection pressure. Darwin recognized this kind of evolutionary force and termed it *sexual selection,* but it is simply one kind of natural selection, one factor that may result in differential reproduction.

Both the number of gametes produced by an individual and the proportion of the gametes that will unite with others to form zygotes may be under genetic control. Both of these factors affect differential reproduction by what might be termed *differential fecundity,* differences in the number of viable zygotes that are produced in a given mating. Organisms in which the probability of survival of any given individual is low will usually have high fecundity to ensure survival of the species. This may be an evolutionary advantage in one species, but if the probability of survival of any given individual is high, an extremely high fecundity may actually reduce the chance of survival of the offspring by reducing the opportunity for parental care and feeding. Newborn mammals, and especially newborn primates which require a great deal of parental care and feeding, would have greatly reduced chances of survival if the size of the litter were increased. Each member of a set of twins, triplets, and so on tends to have a lower birthweight than a singleton newborn, and this also decreases the probability of its survival.

After a viable zygote has been formed by the union of two gametes, it must pass through a long period of development and growth before it becomes sexually mature and able to contribute offspring to the next generation. The differential success of organisms in a population in surviving to the reproductive age and contributing to the next generation's gene pool has been responsible for the more obvious features of organic evolution.

To survive to sexual maturity and to reproduce, each organism must be able to withstand

Figure 31–1 The male *(above)* and the female *(below)* of the red-winged blackbird *(Agelaius phoeniceus).* Note the white spot on the male's wing. (From Orr, R. T.: *Vertebrate Biology,* 4th Ed. Philadelphia. W. B. Saunders Co., 1976.)

a variety of physical factors in its environment, such as the amount of sunlight, moisture, temperature, gravity, light and darkness. In addition, each organism must live amidst other organisms, which leads to competition to eat and avoid being eaten. Plant species must compete for room in the soil and for sunlight, as well as for water and inorganic salts. Each plant is constantly threatened by animals that may eat it before it has had the opportunity to reach sexual maturity and release the spores or seeds that provide for the next generation. Animals are under similar pressure to avoid being eaten or killed and to find food for themselves. Any adaptation that improves an organism's ability to find food and avoid being eaten may play an important role in its differential reproduction. There are many ways in which the process of differential reproduction can be facilitated, and a correspondingly large number of ways in which natural selection may operate.

It is important to keep in mind that natural selection generally does not operate on the phenotypes of single genes but rather on the phenotypic effect of the entire genetic system, or *genome.* One group of organisms may survive despite some clearly disadvantageous characteristic, while another group of organisms may be eliminated despite certain traits that appear to be highly advantageous for survival. The plants and animals that ultimately survive and are the parents of the next generation are those with qualities whose sum total renders them a little better able than their competitors to survive and reproduce their kind. Since the environment may change from time to time, the characteristics that are of adaptive value at one time may be useless or even deleterious at another.

31-5 MUTATIONS—THE RAW MATERIALS OF EVOLUTION

The term mutation was coined by the Dutch botanist Hugo de Vries, one of the rediscoverers of Mendel's laws. De Vries carried out genetic experiments with the evening primrose and other plants that grew wild in Holland. When he transplanted these into his garden and crossed them, some of the resulting plants were unusual and differed markedly from the original wild plant. These unusual

forms bred true in subsequent generations. For such sudden changes in the character of an organism, de Vries used the word mutation. Darwin had described such sudden changes (called sports by earlier breeders) but believed that they occurred too rarely to be of importance in evolution. The vast number of genetic experiments with plants and animals carried on since 1900 have shown that mutations do occur constantly and that the changes in the phenotype produced by such mutations may be of adaptive value and contribute to the survival of the organism. As the gene theory has developed, this word mutation has come to refer to a sudden, random, discontinuous change in the gene (or chromosome), although it is still used, to some extent, to refer to the new type of plant or animal. Hundreds of different mutations have been observed in the plants and animals most widely used in genetic experiments—corn and fruit flies. Among the mutations that have been observed in the fruit fly (Fig. 31–2) are changes in body color from yellow to brown, gray and black; eye colors ranging from red, brown and purple to white; wings that are curled, crumpled, shortened or completely absent; oddly shaped legs and bristles; and such remarkable changes as the development of a pair of legs on the forehead in place of the antennae. The six-toed cats of Cape Cod and the short-legged breed of Ancon sheep are examples of mutations among domestic animals.

The adaptive importance of one or a few mutations is quite evident in bacteria. Mutations that change the nutritional requirements of bacteria (p. 710) can have life-or-death consequences for their possessors. Mutations that increase bacterial resistance to antibiotics, e.g., penicillin-resistant strains of staphylococci, greatly increase the bacteria's ability to survive, and coping with these antibiotic-resistant forms has been a severe problem in many hospitals.

Some mutations produce barely distinguishable changes in the structure or function of the organism in which they occur. Other mutations produce a major change early in development and lead to multiple marked changes in the resulting body form or function. Usually when such a major change occurs in the control of some early stage of development, the result is a nonviable monster that dies almost immedi-

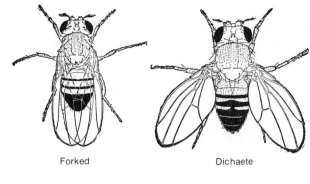

Forked   Dichaete

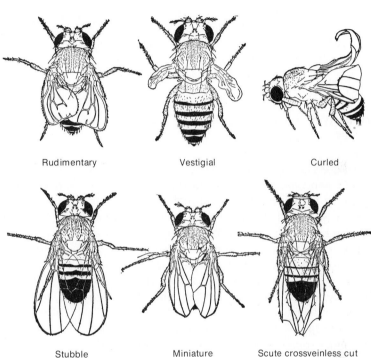

Figure 31-2 Some wing and bristle mutants in the fruit fly *Drosophila melanogaster.* (Drawn by E. M. Wallace. From Sturtevant and Beadle: *An Introduction to Genetics.* Philadelphia, W. B. Saunders Co., 1939.)

Rudimentary   Vestigial   Curled

Stubble   Miniature   Scute crossveinless cut

ately. A few such major changes may give rise to forms that are enabled by their mutation to occupy some new environment. These constitute what Richard Goldschmidt of the University of California called *"hopeful monsters."* He suggested, for example, that the ancestral type of bird, *Archaeopteryx,* evolved into the modern bird by a mutation that in a single step changed the shape of its tail. *Archaeopteryx* had a long, reptilelike tail that was covered with feathers. If a single mutation caused a shortening of the tail, than a "hopeful monster" with the fanshaped arrangement of feathers characteristic of modern birds might have resulted (Fig. 31–3). This fan-shaped tail, better suited for flying than the former, long, reptile-like tail, would give its possessors a selective advantage during evolution. Such skeletal

changes are known to occur as the result of a single mutation. The Manx cat, for example, owes its stubby tail to a mutation that caused the shortening and fusing of most the tail vertebrae. A similar mutation in an ancestral *Archaeopteryx* might have led to the present fan-shaped arrangement of the tail in modern birds.

Types of Mutations. Some mutations, chromosomal mutations (p. 707), are accompanied by a visible change in the structure of the chromosome or by a change in the total number of chromosomes per cell. A single chromosome may be added to, or deleted from, the usual diploid set, or the entire set of chromosomes may be doubled or tripled, yielding organisms called *polyploids.* Polyploid plants and animals are usually larger and more robust

Figure 31–3 A comparison of the structure of the tail of the primitive bird Archaeopteryx *(right)* and the tail of a modern bird *(left)*.

than their diploid parents (Fig. 31–4). Changes in chromosome number are observed more frequently in plants than in animals because the nature of the reproductive process in plants permits these altered chromosome numbers to be passed from one generation to the next. Some of the cultivated varieties of tomatoes, corn, wheat, and other plants owe their vigor and large fruit size to the fact that they are polyploids.

The physical basis of gene mutations, involving some alteration in the nucleotide sequence in genic DNA, and the nature of certain other chromosomal mutations were discussed previously in Chapter 29, Section 29–10.

31–6 BALANCED POLYMORPHISMS

Geneticists use the term *polymorphism* ("many shapes") to refer to two or more types of individuals that differ discontinuously—that is, without intergrading intermediate forms—in some genetically determined characteristic.

Figure 31–4 Photographs of diploid *(left)* and tetraploid *(right)* Easter lilies illustrating the differences in size typically seen in polyploid plants. (Courtesy of S. L. Emsweller, Bureau of Plant Industry, U. S. Department of Agriculture. Science in Farming, Yearbook of Agriculture, 1943–47.)

The human blood groups O, A, B and AB (p. 359) provide a classic example of a polymorphism. Within a given population you might expect the selection process to yield a population completely homozygous for whichever member of a pair of alleles determines the trait with the greatest adaptive value. This does indeed happen in some evolutionary lines, but it is not the only possibility in differential reproduction. For example, the individuals heterozygous for the gene for sickle cell anemia *(Ss)* are somewhat more resistant to malaria than are homozygous normal individuals *(SS)*. In Central Africa where malaria is endemic, there is strong positive selection pressure for the heterozygous individuals *(Ss)* because of their resistance to malaria, a strong negative selection pressure against the homozygous sickle cell anemics *(ss)* and a slight negative selection pressure against the homozygous normal *(SS)* individuals because of their lack of resistance to malaria. These separate and opposing forces act to maintain an equilibrium, a *balanced polymorphism*, of homozygotes and heterozygotes.

The existence of variation itself may be of adaptive value for a population, because a completely homozygous population would have no genetic substratum on which natural selection could act. A population that has a good prognosis for survival in the future is one that has maintained enough variation to permit further adaptive changes. Observations on wild populations of fruit flies and other organisms have shown that their genetic pools do change adaptively, even in response to such changes in the environment as the alternations of the seasons.

The heterozygous individual may have greater fitness for reproducing and surviving than either of the corresponding homozygous individuals. Obviously, however, the heterozygous state cannot be maintained in a population unless a certain number of the somewhat less fit homozygous individuals are also produced. The relative selective values of the heterozygous and homozygous states will determine the particular ratio of alleles in the genetic pool that will result in the optimal proportion of heterozygotes and homozygotes.

31–7 ADAPTIVE RADIATION

Because of the constant competition for food and living space, each group of organisms

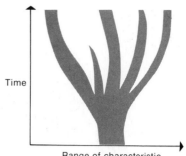

Figure 31–5 Diagram illustrating the increasing range of a characteristic as evolution occurs with time. Four different stocks are indicated as emerging by evolution from a single relatively homogenous stock.

tends to spread out and occupy as many different habitats as possible. This process of evolution from a single ancestral species of a variety of forms that occupy somewhat different habitats is termed *adaptive radiation* (Fig. 31–5). It is clearly advantageous in evolution in enabling the organisms to tap new sources of food or to escape from some of their enemies.

One of the classic examples of adaptive radiation is the evolution of placental mammals (Fig. 31–6). From a primitive, insect-eating, five-toed, short-legged creature that walked with the soles of its feet flat on the ground have evolved all the present-day types of placental mammals. These include dogs and deer adapted for terrestrial life in which running rapidly is important for survival; squirrels and primates adapted for life in the trees; bats equipped for flying; beavers and seals, which maintain an amphibious existence; the completely aquatic whales, porpoises and sea cows; and the burrowing animals—moles, gophers and shrews. In each of these, the number and shape of the teeth, the length and number of leg bones, the number and attachment sites of muscles, the thickness and color of fur, the length and shape of the tail and so on have undergone changes that increase the adaptation of the animal to its particular environment.

A comparable adaptive radiation of the marsupials in Australia has resulted in species that are adapted for many modes of life. These include carnivorous marsupials such as the Tasmanian wolf, ant-eating types, burrowing molelike marsupials, semiarboreal phalangers and koala bears, plains-dwelling herbivorous kangaroos, and rabbitlike bandicoots. Yet anoth-

Figure 31-6 Adaptive radiation. All the various mammals shown have evolved from the common ancestor depicted in the center. Each of the descendants has become adapted to a different type of habitat.

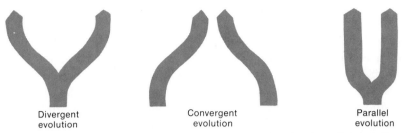

Divergent evolution

Convergent evolution

Parallel evolution

Figure 31-7 Diagram illustrating the difference between divergent, convergent and parallel evolution. A single stock may branch to give two diverging stocks which become more and more different as evolution proceeds. In convergent evolution two stocks originally quite different come to resemble each other more and more as time passes, probably because they occupy a comparable habitat and become adapted to similar conditions. In parallel evolution a single stock branches into two, which then evolve in parallel fashion for a long period of time as each stock independently responds to similar environmental influences.

er classic example of adaptive radiation is that of the reptiles during the Mesozoic Era (p. 764).

Adaptive radiation may take place on a very small scale, as represented by the variety of ground finches today on the Galápagos Islands west of Ecuador. Some of these birds live on the ground and feed on seeds, others feed mainly on cactus, and still others have taken to living in trees and eating insects. These variations in feeding have been accompanied by evolutionary changes in the size and structure of the beak. The essence of adaptive radiation, then, is the evolution from a single ancestral form of a variety of different forms, each of which is adapted and specialized in some unique way to survive in a particular habitat.

Adaptive radiation that gives rise to several different types of descendants, adapted in different ways to different environments, may be termed divergent evolution. The opposite phenomenon, *convergent evolution,* also occurs fairly frequently; i.e., two or more quite unrelated groups may, in becoming adapted to a similar environment, develop characteristics that are more or less similar (Fig. 31-7). For example, wings have evolved not only in birds but also in mammals (bats), in reptiles (pterosaurs) and in insects. A very similar streamlined shape, dorsal fins, tail fins, and flipperlike fore and hind limbs have evolved in dolphins and porpoises (which are mammals), the extinct ichthyosaurs (which were reptiles) and in both bony and cartilaginous fishes (Fig. 31-8). Moles and gophers have adapted to a burrowing life and have evolved similar fore and hind leg structures adapted for digging. The mole is an insectivore and the gopher is a rodent. The eye of the squid and the eye of a vertebrate

such as the fish are also very similar in structure although very different in their embryonic origin.

31-8 SPECIATION

The unit of classification for both plants and animals is the species. It is difficult to give a definition of this term that can be applied uniformly throughout all five kingdoms, but a *species* may be defined as a population of individuals with similar structural and functional characteristics that have a common ancestry and

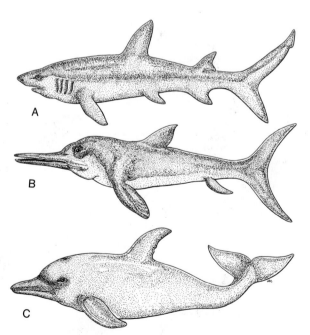

Figure 31-8 Convergent evolution. *A,* Shark; *B,* ichthyosaur (a fossil reptile); and *C,* dolphin (a mammal), all of which have a marked superficial similarity because of their adaptation to similar environments.

in nature breed only with each other. A species is a collection of **demes**, or populations within which interbreeding may occur—a group of populations with a common gene pool. It is implicit in this definition that there is no free flow of genes between two different species.

In providing an explanation for the origin of a new species, we must describe how the summation of unit evolutionary changes in a population may culminate eventually in the establishment of new species (and of new genera, families and orders). This requires that **reproductive barriers** arise between the incipient species as they are becoming established. When interbreeding between subgroups of a population becomes progressively less frequent and the resulting hybrids become progressively less fertile, the several groups eventually become different species. Any factor that decreases the amount of interbreeding between groups of organisms is termed an **isolating mechanism.**

A very common type of isolation is **geographic**, the separation of groups of related organisms by some physical barrier, such as a river, desert, glacier, mountain or ocean (Fig. 31–9). In a mountainous region the individual mountain ranges afford effective barriers between the valleys. Valleys only a short distance

apart but separated by ridges always covered with snow typically have species of plants and animals that are peculiar to those valleys. Thus there are usually more different species in a given area in a mountainous region than on open plains. For example, in the mountains of the western United States there are 23 species and subspecies of rabbits, whereas in the much larger plains area of the Midwest and the East there are only 8 species of rabbits.

The Isthmus of Panama provides another striking example of geographic isolation. On either side of the isthmus, the phyla and classes of marine invertebrates are made up of different but closely related species. For some 16,000,000 years during the Tertiary Period, there was no connection between North and South America, and marine animals could migrate freely between what is now the Gulf of Mexico and the Pacific Ocean. When the Isthmus of Panama reemerged, the closely related groups of animals were isolated, and the differences between the fauna in the two regions represent the subsequent accumulation of hereditary differences. The digging of the Panama Canal provided a means for some of the more mobile forms to migrate from one ocean to the other. However, much of the canal is filled with fresh water and is 28 meters

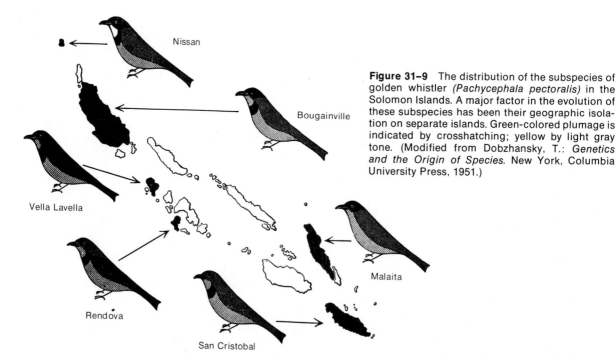

Figure 31–9 The distribution of the subspecies of golden whistler *(Pachycephala pectoralis)* in the Solomon Islands. A major factor in the evolution of these subspecies has been their geographic isolation on separate islands. Green-colored plumage is indicated by crosshatching; yellow by light gray tone. (Modified from Dobzhansky, T.: *Genetics and the Origin of Species.* New York, Columbia University Press, 1951.)

above sea level. This is a formidable barrier to the migration of most invertebrates!

Geographic isolation is usually not permanent, and the two previously isolated groups may come into contact again and resume interbreeding unless *genetic isolation* or interspecific sterility has arisen in the meantime. Genetic isolation results from mutations that occur independently of mutations for structural or functional features. Genetic isolation may appear only after a long period of geographic isolation has produced striking differences between two groups of organisms, or it may originate within a single, otherwise homogeneous group of organisms. For example, in a species of fruit fly, *Drosophila pseudoobscura*, a mutation for genetic isolation has produced two groups of flies that, though externally indistinguishable, are completely sterile when cross-mated. These two groups are isolated genetically as effectively as if they lived on different continents. As generations pass and different mutations accumulate in each group by chance and by selection, the two groups will undoubtedly become visibly different. Two groups of organisms living in the same geographic area may be *ecologically isolated* if they occupy different habitats. Marine animals living in the intertidal zone are effectively isolated from other organisms living only a few feet away below the low-tide mark. Ecologic isolation might result from the simple fact that two groups of organisms breed at different times of the year.

31–9 THE ORIGIN OF SPECIES BY HYBRIDIZATION

Although members of different species are usually not interfertile, occasionally members of two different but closely related species may interbreed to produce yet a third species by *hybridization.* Such phenomena make it difficult to establish a hard and fast definition of species. By hybridization, the best characters of each of the original species may be combined into a single descendant, thereby creating a new type better able to survive than either of its parents. If the new form combined the worst characters of both parents it would obviously be at a serious disadvantage and would be unlikely to survive.

When different species with different chromosome numbers are crossed, the offspring are usually sterile. The unlike chromosomes cannot pair in meiosis, and the resulting eggs and sperm do not receive the proper assortment of chromosomes—one of each kind. When within such *interspecific hybrids* the number of chromosomes is doubled, meiosis can take place in a normal fashion, and normal, fertile eggs and sperm will be produced. Thereafter the hybrid species will breed true, and indeed will not produce fertile offspring when bred with either of the parental species. Many related species of higher plants have chromosome numbers that are multiples of some basic number. The species of wheat include ones with 14, 28 and 42 chromosomes; there are species of roses with 14, 28, 42 and 56 chromosomes; and species of violets with every multiple of 6 from 12 to 54.

That such natural series arise by hybridization and doubling of the chromosomes is supported by laboratory experiments yielding similar series. One of the more famous of these experimental crosses was made by Karpechenko, who crossed the radish with a cabbage, hoping perhaps to get a plant with a cabbage top and a radish root. Radishes and cabbages belong to different genera, but both have 18 chromosomes. The resulting hybrid also had 18 chromosomes, 9 from its radish parent and 9 from its cabbage parent. Since the radish and cabbage chromosomes were unlike, they could not pair during meiosis, and the hybrid was almost completely sterile. By chance, however, a few of the eggs and pollen formed contained all 18 chromosomes, and a mating between two of these resulted in a plant with 36 chromosomes. This plant was fertile, for during meiosis the pairs of radish chromosomes underwent synapsis as did the pairs of cabbage chromosomes. The hybrid exhibited some of the characteristics of each parent, and bred true for these characteristics. Unfortunately it had a radishlike top and a cabbagelike root. Since it could not be crossed readily with either of its parent species, it was, in effect, a new species produced by hybridization, followed by the doubling of the number of chromosomes.

A similar occurrence in nature has been documented in the marsh grasses. One of these, *Spartina townsendi,* first appeared more than 100 years ago in the harbor of Southampton, England, in company with two other species, *Spartina stricta* and *Spartina alterniflora*. The new species, *Spartina townsendi*, was much more vigorous than either of the parents and was soon widespread. It was especially valuable in collecting and holding soil and was transplanted to the Holland dykes and to other parts of the world. Because in many characteristics it was intermediate between the two species with which it was first found, it was believed to have originated as a hybrid. When it was possible to examine the chromosome numbers, the hypothesis was confirmed, for *Spartina townsendi* was found to have 126 chromosomes, *Spartina stricta* 56 and *Spartina alterniflora* 70. Thus there is no doubt that the new species arose by hybridization and doubling of the chromosomes.

31-10 PHYLOGENY

The evolutionary history of any group of organisms is termed its **phylogeny.** It is basic to many aspects of biological research to know which organisms are most closely related—i.e., which ones have common ancestors in the recent past, and which ones have common ancestors only in the more distant past. To establish the phylogenetic relationships of a group of organisms, each investigator must examine as many characteristics of each type as possible, looking for patterns of similarities and dissimilarities that may provide clues. Phylogeneticists originally were restricted largely to comparing morphologic characters—patterns of bones, muscles and nerves—but now a host of physiologic, biochemical, immunologic and cytologic characters can be examined and used to test the validity of the relationships inferred on the basis of morphologic characters. It is reassuring to find that the evolutionary relationships inferred on the basis of the newest, most sophisticated biochemical analyses of the types of proteins found in different species agree remarkably well with the evolutionary relationships described a century ago on the basis of gross morphologic similarities.

31-11 THE ORIGIN OF LIFE

The current theories of mutation, natural selection and population dynamics provide a satisfactory explanation of how the animals and plants living today evolved from earlier forms by descent with modification. The question of the ultimate origin of living things on this planet has been given serious consideration by a number of biologists. Some have theorized that some kind of spore or germ may have been carried through space from some other planet to this one. This is unsatisfactory, not only because it begs the question of the ultimate source of the spores, but also because it seems most unlikely that any sort of living thing could survive the extreme cold and intense irradiation encountered in interplanetary travel. Evidence for life in other parts of the cosmos came from the discovery in 1961 of what were identified as fossils of microscopic organisms, somewhat like algae, in meteorites, but this of course is no proof that *living* organisms could be transported through space.

It is considered extremely improbable that any type of life is arising from spontaneous generation at the present time. Francesco Redi's experiments, made about 1680, showed that maggots do not arise *de novo* from decaying meat and thus laid to rest the old superstition that animals could appear by spontaneous generation. Some 200 years later Louis Pasteur showed conclusively that microorganisms such as bacteria do not arise by spontaneous generation but come only from previously existing bacteria (Fig. 2–3). Other investigators have shown since then that even the smallest organisms, the filtrable viruses, do not come from nonviral material by spontaneous generation. The multiplication of viruses requires the presence of previously existing viruses. Although the spontaneous generation of life at present is unlikely, it is most probable that billions of years ago, when chemical and physical conditions on the earth's surface were quite different from those at present, the first living things did arise from nonliving material. What has changed so radically to account for the fact that spontaneous generation no longer occurs?

This concept (that the first living things did evolve from nonliving things) and suggestions as to what the sequence of events may have been were put forward by Pflüger, J. B. S. Haldane, R. Beutner, and particularly by the Russian biochemist A. I. Oparin in his book, *The Origin of Life* (1938). The earth originated some five billion years ago, either as a part broken off from the sun or by the gradual condensation of interstellar dust. Most authorities now agree that the earth was very hot and molten when it was first formed and that conditions consistent with life appeared on the earth only perhaps three billion years ago. Twenty-two different amino acids were isolated recently from Precambrian rocks from South Africa that are at least 3.1 billion years old. At that time the earth's atmosphere contained essentially no free oxygen—all the oxygen atoms were combined as water or as oxides. The primitive atmosphere was strongly reducing, composed of methane, ammonia and water originating by "out-gassing" from the earth's interior.

Reactions by which organic substances can be synthesized from inorganic ones are well known. Originally the carbon atoms in the earth's crust were present mainly as metallic carbides. These could react with water to form acetylene, which would subsequently polymerize to form compounds with long chains of carbon atoms. High-energy radiation such as cosmic rays can catalyze the synthesis of organic compounds. This was shown by Melvin Calvin's experiments in which solutions of carbon dioxide and water were irradiated in a cyclotron, and formic, oxalic and succinic acids, which contain one, two and four carbons respectively, were obtained. These acids, as you know, are intermediates in certain metabolic pathways of living organisms. Irradiating solutions of inorganic compounds with ultraviolet light or passing electric charges through the solutions to simulate lightning also produces organic compounds. Harold Urey and Stanley Miller in 1953 exposed a mixture of water vapor, methane, ammonia and hydrogen gases to electric discharges for a week and demonstrated the production of amino acids such as glycine and alanine, together with other complex organic compounds. The earth's atmosphere in prebiotic times probably contained water vapor, methane, ammonia and hydrogen gas, from which irradiation could produce a tremendous variety of organic materials. Amino acids and other compounds could be produced in nature at the present time by lightning discharges or ultraviolet radiations; however, any organic compound produced in this way might undergo spontaneous oxidation or it would be taken up and degraded by molds, bacteria and other organisms.

The details of the chemical reactions that could give rise, without the intervention of living things, to carbohydrates, fats and amino acids have been worked out by Oparin and extended by Calvin and others. Most, if not all, of the reactions by which the more complex organic substances were formed probably occurred in the sea, in which the inorganic precursors and the organic products of the reaction were dissolved and mixed. The sea became a sort of dilute broth in which these molecules collided, reacted and aggregated to form new molecules of increasing size and complexity (this might be called the "chicken soup" theory of evolution). As more has been learned of the role of specific hydrogen bonding and other weak intermolecular forces in the pairing of specific nucleotide bases and the effectiveness of these processes in the transfer of biological information, it has become clear that similar forces could have operated early in evolution before "living" organisms appeared.

The forces of intermolecular attraction and the tendency for certain molecules to form liquid crystals provide us with means by which large complex specific molecules can be formed spontaneously. Oparin suggested that a kind of natural selection could have operated in the evolution of these complex molecules before anything recognizable as life was present. As the molecules came together to form colloidal aggregates, these aggregates began to compete with one another for raw materials. Some of the aggregates that had some particularly favorable internal arrangement would acquire new molecules more rapidly than others and would eventually become the dominant types.

Once some protein molecules had been formed and had achieved the ability to catalyze reactions, the rate of formation of additional molecules would be greatly speeded up. When combined with nucleic acids, these complex protein molecules should eventually acquire the ability to catalyze the synthesis of molecules like themselves. These hypothetical *autocatalytic* particles made of nucleic acids and proteins would have some of the properties of a virus or perhaps of a free gene.

A major step in the evolution of these early prebiotic aggregates would have been the development of a protein-lipid *membrane* surrounding the aggregate that would permit the accumulation of some molecules and the exclusion of others. All the viruses known at present are parasites that can live only within the cells of higher animals and plants; however, if there were free-living viruses—ones that do not produce a disease—they would be very difficult to detect.

The first living organisms, having arisen in a sea of organic molecules and in contact with an atmosphere lacking oxygen, presumably obtained energy by the fermentation of certain of these organic substances. The first organisms were almost certainly *heterotrophs*, and they could survive only as long as the supply of organic molecules that had been accumulated in the sea broth in the past lasted. Before the supply was exhausted, however, some of the heterotrophs evolved further and became autotrophs, able to make their own organic molecules by chemosynthesis or photosynthesis. One of the by-products of photosynthesis is gaseous oxygen; indeed, all the oxygen in the atmosphere has been produced and is still produced by photosynthesis.

An explanation of how an autotroph may have evolved from one of these primitive, fermenting heterotrophs was presented by N. H. Horowitz in 1945. Horowitz postulated that an organism would acquire, by successive gene mutations, the enzymes needed to synthesize complex substances from simple substances, but these enzymes would be acquired in the *reverse* order of the sequence in which they are ultimately used in normal metabolism. For example, let us suppose that our first primitive heterotroph required an organic compound, Z, for its growth. This substance, Z, and a vast variety of other organic compounds, Y, X, W, V, U and so forth, were present in the organic sea broth that was the environment of this heterotroph. They had been synthesized previously by the action of nonliving factors of the environment. The heterotroph would be able to survive as long as the supply of compound Z lasted. If a mutation occurred for a new enzyme enabling the heterotroph to synthesize Z from substance Y, the strain of heterotroph with this mutation would be able to survive when the supply of substance Z was exhausted. A second mutation that established an enzyme catalyzing a reaction by which substance Y could be made from substance X would again have survival value when the supply of Y was exhausted. Similar mutations, setting up enzymes enabling the organism to use successively simpler substances, W, V, U . . . ,

and eventually some inorganic substance, A, would result in an organism able to make substance Z, which it needs for growth, out of substance A. When by other series of mutations the organism could synthesize all its requirements from simple inorganic compounds, as the green plants can, it would have become an autotroph. Once the first simple autotrophs had evolved, the way was clear for the further evolution of the enormous variety of plants, bacteria, molds and animals that now inhabit the world.

From arguments such as these we are drawn to the conclusion that the origin of life as an orderly, natural process on this planet was not only possible, it was almost inevitable. Furthermore, with the vast number of planets in all the known galaxies of the universe, there must be many that have conditions that permit the origin of life. It is probable, then, that there are other planets—perhaps many other planets—on which life as we know it exists. Wherever the physical environment will support life, living things should, if given enough time, appear and ramify into a wide variety of types. Some of these may be quite unlike the ones on this planet, but others might be quite similar to those found here. Some might indeed be like ourselves. Living things on other planets might have a completely different kind of genetic code or might be made up of elements other than carbon, hydrogen, oxygen and nitrogen.

It seems unlikely that we will ever know how life originated, whether it happened only once or many times, or whether it might happen again. The theory (1) that organic substances were formed from inorganic substances by the action of physical factors in the environment; (2) that they interacted to form more and more complex substances, finally enzymes, and then self-reproducing systems (free genes); (3) that these free genes diversified and united to form primitive and perhaps viruslike heterotrophs; (4) that lipid-protein membranes evolved to separate these prebiotic aggregates from the surrounding environment; and (5) that autotrophs then evolved from the primitive heterotrophs, has the virtue of being quite plausible. Many of the components of this theory have been subjected to experimental verification.

31–12 PRINCIPLES OF EVOLUTION

There are differences of opinion among investigators as to the precise molecular events that underlie mutations, the kinds of mutations involved in evolution, and the degree to which such factors as natural selection, isolation, genetic recombination, hybridization and the size of the breeding population affect the evolution of some particular organism. However, there are several fundamental facts about which they are agreed: Changes in the chromosomes and genes are the raw materials of evolution; some sort of isolation is necessary to establish a new species; and natural selection by differential reproduction is involved in the survival of some, but not all, of the mutations that occur. In addition, there are five principles of evolution to which nearly all biologists can subscribe:

1. Evolution occurs more rapidly at some times than at others. At the present time it is occurring rapidly, with many new forms appearing and many old ones becoming extinct. What factors are responsible for this?

2. Evolution does not proceed at the same rate among different types of organisms. At one extreme are the lamp shells, or brachiopods, some species of which have remained unchanged for the last 500,000,000 years at least, for fossil shells found in rocks deposited at that time are identical with those of animals living today. In contrast, several species of humans have appeared and become extinct in the past few hundred thousand years. In general, evolution occurs rapidly when a new species first appears, and then gradually slows down as the group becomes established.

3. New species do not evolve from the most advanced and specialized forms already living, but from relatively simple, unspecialized forms. Thus the mammals did not evolve from the large, specialized dinosaurs but from a group of rather small and unspecialized reptiles.

4. Evolution is not always from the simple to the complex. There are many examples of "regressive" evolution, in which a complex form has given rise to simpler ones. Most parasites have evolved from free-living ancestors that were more complex than the present forms; wingless birds, such as the cassowary,

have descended from birds that could fly; many wingless insects have evolved from winged ones; the legless snakes came from reptiles with appendages; whales, with no hind legs, evolved from mammals that had two pairs of legs. These are all reflections of the fact that mutations occur at random, and the resulting changes in phenotype do not necessarily increase the complexity or the "perfection" of the organism. If there is some advantage to a species in having a simpler structure, or in doing without some structure altogether, any mutations that happen to occur for such conditions will tend to accumulate by natural selection.

5. Evolution occurs by populations, not by individuals, by the processes of mutation, differential reproduction, natural selection and genetic drift.

QUESTIONS

1. What is implied in the theory of uniformitarianism?
2. Explain briefly the concept of organic evolution.
3. In what ways does Lamarck's theory of adaptation not agree with present evidence?
4. What contributions did Darwin make to the theory of evolution?
5. Describe briefly the Darwin-Wallace theory of natural selection. What is meant by "survival of the fittest"? Do you think this applies to human populations?
6. Why is it that only inherited changes are important in the evolutionary process?
7. What is meant by convergent evolution? balanced polymorphism? gene pool? differential reproduction?
8. After a mutation has occurred in a population, what events must take place if the mutant trait is to become established in the population?
9. What is meant by the term genetic drift? What role may this play in evolution?
10. Define the term adaptive radiation and give examples other than those cited in the text.
11. Discuss the current theory of the steps involved in the establishment of a new species of plant or animal.
12. From what you have learned of the processes of heredity and evolution, how do you think new species arise—by the accumulation of small mutations or by a few mutations with large phenotypic effects? Give the reasons for your answer.
13. Differentiate the several types of mutations. How are mutations produced?
14. Why do nearly all of the mutations occurring at

the present time have a detrimental effect on the organism in which they occur?

15. Contrast hybridization with other ways in which new species may be produced.

16. What contributions to the principles of evolution were made by Erasmus Darwin, Alfred Russel Wallace, Thomas Huxley, Thomas Malthus, Jean Baptiste de Lamarck and Hugo de Vries?

17. Discuss the role of isolation in the origin of species.

18. Describe the steps by which simple inorganic substances may have undergone chemical evolution to yield the complex system of organic chemicals we recognize as a living thing. Which of these steps have been duplicated experimentally?

SUPPLEMENTARY READING

Charles Darwin's classic, *The Origin of Species,* is available in a number of modern editions and is well worth sampling for its clear and logical argument and its wealth of examples. His chronicle of the expedition during which he made the observations that led him eventually to his theory of evolution was published as *The Voyage of the Beagle,* which provides an excellent introduction to the man and his work. Darwin's life and his voyage are engagingly described in *Darwin and the Beagle* by A. Moorhead, *The Process of Evolution* by P. R. Ehrlich and R. W. Holm and *Processes of Organic Evolution* by G. L. Stebbins, Jr., are clear, literate presentations of the theories of evolution, updating the theory of natural selection. The impact of the theory of evolution on Victorian England and a vivid portrayal of Thomas Huxley's championing of Darwin's theory is presented in *Apes, Angels and Victorians* by William Ervine. *From the Greeks to Darwin* by Henry Fairfield Osborn in an interesting history of the early ideas on evolution. Nontechnical discussions of some present views regarding evolution are found in *The Meaning of Evolution* by George Gaylord Simpson, in *Evolution* by J. M. Savage, in *A Synthesis of Evolutionary Theory* by H. H. Ross, and in *Process and Patterns in Evolution* by T. H. Hamilton. More technical books detailing special phases of evolution are *Animal Evolution: A Study of Recent Views on Its Causes* by G. S. Carter, *Variation and Evolution in Plants* by G. L. Stebbins, Jr., *Animal Species and Evolution* by E. Mayr, and *The Major Features of Evolution* by G. G. Simpson. This last book discusses the relationship between paleontology and genetics. *Genetics and the Origin of Species* by Theodosius Dobzhansky presents the neo-Darwinian viewpoint of the importance of natural selection. *The Material Basis of Evolution* by Richard Goldschmidt gives the detailed arguments for the importance in evolution of large macromutations.

Theories of the origin of life are discussed in *The Origin of Life* by A. I. Oparin and in other books with the same title by J. Keosian and by C. Ponnamperuma. *Chemical Evolution* by M. E. Calvin is an excellent source book for an understanding of the processes by which life may have originated. The biography of Alfred Russel Wallace, entitled *Darwin's Moon,* by A. William-Ellis gives a vivid portrayal of Wallace's life and times and reemphasizes his solid contributions to advancing the theory of evolution. A more general history of the development of evolutionary concepts is given in *Darwin's Century: Evolution and Men Who Discovered It* by L. C. Eiseley.

CHAPTER 32

THE EVIDENCE FOR EVOLUTION

The evidence that organic evolution has occurred is so overwhelming that no one who is acquainted with it has any doubt that new species are derived from previous ones by descent with modification. The extensive fossil record provides direct evidence of organic evolution and gives the details of the evolutionary relationships of many lines of descent.

Even if the remarkably detailed fossil record did not exist, studies of the anatomy, physiology and biochemistry of modern plants and animals, their development and cytogenetics, and the manner in which they are distributed over the earth's surface have provided incontrovertible proof that organic evolution has occurred.

32–1 THE FOSSIL RECORD

The science of *paleontology* deals with the finding, cataloguing and interpretation of the abundant and diverse evidence of life in former times. The term *fossil* (Latin, *fossilium*, something dug up) refers not only to the bones, shells, teeth and other hard parts of a plant or animal body that have been preserved, but also to any impression or trace left by some previously existing organism (Fig. 32–1). In view of the large number of fossils of plants and animals that have been found, it is sobering to realize that only a small fraction of all the organisms that ever lived have been preserved as fossils and only a small fraction of these fossils have been dug up and studied to date.

Footprints or trails made in soft mud,

which subsequently hardened, are a common type of fossil. From such remains one can infer something of the structure and body proportions of the animals that made them. In 1948, tracks were discovered near Pittsburgh of an amphibian from the Pennsylvanian period, some 250,000,000 years ago. That the animal moved by hopping, rather than by walking, was evident from the fact that the footprints lay opposite each other, in pairs.

Most of the vertebrate fossils are skeletal parts, from which it is possible to deduce the animal's posture and style of walking. From the bone scars, indicating muscle attachments, paleontologists can deduce the general position and size of the muscles and, from this, the contours of the body. On the basis of such considerations, reconstructions are made of the

Figure 32-1 One of the more famous examples of a fossil, the remains of *Archaeopteryx*, a tailed, toothed bird from the Jurassic period. (Courtesy of the American Museum of Natural History, New York.)

animal as it looked in life. Such qualities as the texture and color of the fur or scales can, of course, only be guessed at.

In one interesting and striking type of fossil, the original hard parts and, more rarely, soft tissues have been replaced by minerals — a process known as **petrifaction.** The minerals that replace the tissues may be iron pyrites, silica, calcium carbonate or other substances. The petrified muscles from a shark more than 300,000,000 years old were so well preserved by this process that not only individual muscle fibers but their cross striations could be observed in thin sections under the microscope. The Petrified Forest in Arizona is a famous example of the process of petrifaction.

Molds and casts are superficially similar to petrified fossils but are produced differently. *Molds* were formed by the hardening of the material surrounding the buried organism, followed by the decay and removal of the organism by seepage of the ground water. Sometimes the molds were subsequently filled with minerals, which in turn hardened to form *casts* — replicas of the original structures.

Occasionally, paleontologists are fortunate enough to find organisms frozen in the soil or ice of the far North, usually in Siberia and Alaska. The remains of woolly mammoths more than 25,000 years old have been found so well preserved that their flesh was eaten by dogs. Other forms — plants, insects and spiders — have

been preserved in amber, a fossil resin from pine trees. Originally the resin was a sap soft

Figure 32-2 Two termites imbedded in amber. These insects, dating from the Middle Tertiary (about 38,000,000 years ago), have been preserved almost perfectly. (From Buchsbaum, R.: *Animals Without Backbones.* Rev. Ed. Chicago, University of Chicago Press, 1948. Photograph by P. S. Tice.)

enough to engulf the fragile insect and penetrate every part; then it gradually hardened, preserving the animal intact (Fig. 32–2).

The formation and preservation of a fossil require that some structure be buried. This may take place at the bottom of a body of water, or on land by the accumulation of windblown sand, soil or volcanic ash. The people and animals in Pompeii were preserved almost perfectly by the volcanic ash from the eruption of Vesuvius. Sometimes animals were trapped and entombed in a bog, quicksand or an asphalt pit. The famous La Brea tar pits in Los Angeles have provided superb fossils of Pleistocene animals.

32–2 THE GEOLOGIC TIME TABLE

The earth's crust consists of five major rock strata, each subdivided into minor strata, lying one on top of the other. These sheets of rock were formed by the accumulation of mud or sand at the bottom of oceans, seas and lakes, and each contains certain characteristic fossils that serve to identify deposits made at the same time in different parts of the world. Geologic time has been divided into *eras, periods* and *epochs* (Table 32–1) according to the succession of these rock strata one on the other. The duration of each period or epoch can be estimated from the relative thickness of the sedimentary deposits, although, obviously, the rate of deposition varied at different times and in different places.

The layers of sedimentary rock should occur in the sequence of their deposition, with the newer, later strata on top of the older, earlier ones, but subsequent geologic events may have changed the relationship of the layers. Not all the strata occur in any one region, for some lands were exposed when others were submerged. In some regions the strata formed previously have subsequently emerged and eroded away, so that relatively recent strata were then deposited directly upon very ancient ones. Moreover, certain sections of the earth's crust have undergone tremendous foldings and splittings, so that older layers may now rest on top of newer ones. Sometimes the age of a rock stratum can be determined by a study of its fossil content, for some kinds of fossils were deposited in only one era or period.

Certain radioactive elements are transformed into other elements at rates that are slow and essentially unaffected by the temperatures and pressures to which the rock has been subjected. Half of a given sample of uranium will be converted into lead in 4.5 billion years, and by measuring the proportion of uranium and lead in a piece of crystalline rock, an accurate estimate of the absolute age of the rock can be made. By this method the oldest rocks of the earliest geologic period are calculated to be about 3,500,000,000 years old, and the latest Cambrian rocks to be 500,000,000 years old. These dates have been confirmed by newer methods in which the radioactive decay of rubidium-87 (half-life of 47 billion years!) and potassium-40 have been utilized to measure the ages of micas and feldspars. Events in more recent times can be dated by the decay of radioactive carbon-14, which has a half-life of 5568 years.

Relatively short periods of geologic time can be determined by counting the yearly deposits of clay on the bottom of lakes and ponds. Advances in isotopic techniques have made possible some astonishing conclusions in the field of geology. For example, the proportion of the various oxygen isotopes in the calcium carbonate secreted by living organisms depends upon the temperature; consequently, by analyzing the oxygen isotopes in the calcium carbonate of fossil shells, it is possible to estimate the temperature of the sea in which those animals lived, hundreds of millions of years ago.

Between the major eras there were widespread geologic disturbances, called *revolutions*, which raised or lowered vast regions of the earth's surface and created or eliminated shallow inland seas. These revolutions changed the distribution of sea and land organisms and wiped out many of the previous life forms. The era known as the Paleozoic ended with the revolution that raised the Appalachian mountains and, it is thought, killed all but 3 per cent of the then existing forms of life. Similarly, the Rocky Mountain Revolution, which raised the Andes, Alps and Himalayas, as well as the Rockies, resulted in the annihilation of most of the reptiles of the Mesozoic era.

The raising and lowering of portions of the earth's crust result from the slow movements of the enormous tektonic plates that compose the

Table 32–1 GEOLOGIC TIME TABLE

Era	Period	Epoch	Duration in Millions of Years	Time from Beginning of Period to Present (Millions of Years)	Geologic Conditions	Plant Life	Animal Life
Cenozoic (Age of Mammals)	Quaternary	Recent	0.011	0.011	End of last ice age; climate warmer	Decline of woody plants; rise of herbaceous ones	Age of humans
		Pleistocene	1.9	1.9	Repeated glaciation; 4 ice ages	Great extinction of species	Extinction of great mammals; first human social life
	Tertiary	Pliocene	4	6	Continued rise of mountains of western North America; volcanic activity	Decline of forests; spread of grasslands; flowering plants, monocotyledons developed	Man evolved from humanlike apes; elephants, horses, camels almost like modern species
		Miocene	19	25	Sierra and Cascade mountains formed; volcanic activity in northwest U.S.; climate cooler		Mammals at height of evolution; first humanlike apes
		Oligocene	13	38	Lands lower; climate warmer	Maximum spread of forests; rise of monocotyledons, flowering plants	Archaic mammals extinct; rise of anthropoids; forerunners of most living genera of mammals
		Eocene	16	54	Mountains eroded; no continental seas; climate warmer		Placental mammals diversified and specialized; hoofed mammals and carnivores established
		Paleocene	11	65			Spread of archaic mammals

Rocky Mountain Revolution (Little Destruction of Fossils)

Era	Period	Epoch	Duration in Millions of Years	Time from Beginning of Period to Present (Millions of Years)	Geologic Conditions	Plant Life	Animal Life
Mesozoic (Age of Reptiles)	Cretaceous		70	135	Andes, Alps, Himalayas, Rockies formed late; earlier, inland seas and swamps; chalk, shale deposited	First monocotyledons; first oak and maple forests; gymnosperms declined	Dinosaurs reached peak, became extinct; toothed birds became extinct; first modern birds; archaic mammals common
	Jurassic		46	181	Continents fairly high; shallow seas over some of Europe and western U.S.	Increase of dicotyledons; cycads and conifers common	First toothed birds; dinosaurs larger and specialized; insectivorous marsupials
	Triassic		49	230	Continents exposed; widespread desert conditions; many land deposits	Gymnosperms dominant, declining toward end; extinction of seed ferns	First dinosaurs, pterosaurs and egg-laying mammals; extinction of primitive amphibians

Appalachian Revolution (Some Loss of Fossils)

Era	Period	Epoch	Duration in Millions of Years	Time from Beginning of Period to Present (Millions of Years)	Geologic Conditions	Plant Life	Animal Life
Paleozoic (Age of Ancient Life)	Permian		50	280	Continents rose; Appalachians formed; increasing glaciation and aridity	Decline of lycopods and horsetails	Many ancient animals died out; mammal-like reptiles, modern insects arose
	Pennsylvanian		40	320	Lands at first low; great coal swamps	Great forests of seed ferns and gymnosperms	First reptiles; insects common; spread of ancient amphibians
	Mississippian		25	345	Climate warm and humid at first, cooler later as land rose	Lycopods and horsetails dominant; gymnosperms increasingly widespread	Sea lilies at height; spread of ancient sharks
	Devonian		60	405	Smaller inland seas; land higher, more arid; glaciation	First forests; land plants well established; first gymnosperms	First amphibians; lungfishes, sharks abundant
	Silurian		20	425	Extensive continental seas; lowlands increasingly arid as land rose	First definite evidence of land plants; algae dominant	Marine arachnids dominant; first (wingless) insects; rise of fishes
	Ordovician		75	500	Great submergence of land; warm climates even in Arctic	Land plants probably first appeared; marine algae abundant	First fishes, probably fresh water; corals, trilobites abundant; diversified mollusks
	Cambrian		100	600	Lands low, climate mild; earliest rocks with abundant fossils	Marine algae	Trilobites, brachiopods dominant; most modern phyla established

Second Great Revolution (Considerable Loss of Fossils)

Era	Period	Epoch	Duration in Millions of Years	Time from Beginning of Period to Present (Millions of Years)	Geologic Conditions	Plant Life	Animal Life
Proterozoic			1000	1600	Great sedimentation; volcanic activity later; extensive erosion, repeated glaciations	Primitive aquatic algae and fungi	Various marine protozoa; towards end, mollusks, worms, other marine invertebrates

First Great Revolution (Considerable Loss of Fossils)

Era	Period	Epoch	Duration in Millions of Years	Time from Beginning of Period to Present (Millions of Years)	Geologic Conditions	Plant Life	Animal Life
Archeozoic			2000	3600	Great volcanic activity; some sedimentary deposition; extensive erosion	No recognizable fossils; indirect evidence of living things from deposits of organic material in rock	

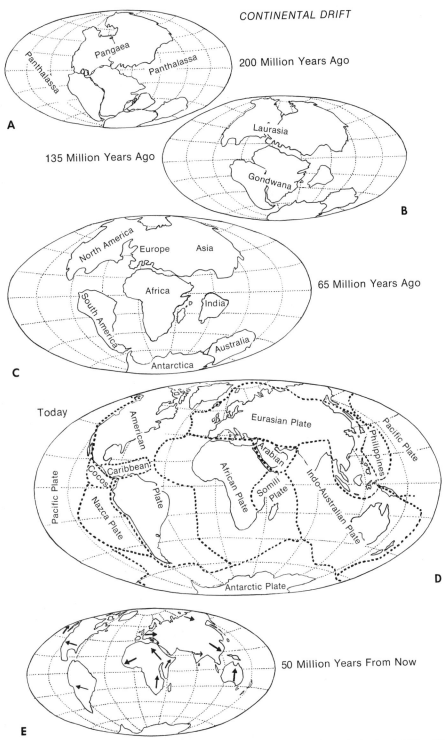

Figure 32–3 Continental drift. *A,* The supercontinent Pangaea of the Triassic period, about 200 million years ago. *B,* Break-up of Pangaea into Laurasia (Northern Hemisphere) and Gondwana (Southern Hemisphere) 135 million years ago in the Cretaceous period. *C,* Further separation of land masses, which occurred in the Tertiary period, 65 million years ago. Note that Europe and North America are still joined and that India is a separate land mass. *D,* The continents today. *E,* Projected positions of the continents in 50 million years. (From Norstog, K., and Long, R. W.: *Plant Biology.* Philadelphia, W. B. Saunders Co., 1976.)

crust and float on the underlying molten core. These movements are continuing and build up tensions in the crust that, when finally released, result in earthquakes. The movements of these plates have produced tremendous changes in the earth's geography (Fig. 32–3).

32–3 PRECAMBRIAN LIFE

The Archeozoic Era. The rocks of the oldest geologic era are very deeply buried in most parts of the world, but are exposed at the bottom of the Grand Canyon and along the shores of Lake Superior. The oldest era does not begin with the origin of the earth but with the formation of the earth's crust, when rocks and mountains were already in existence and the processes of erosion and sedimentation had begun. The two-billion year long Archeozoic was characterized by catastrophic and widespread volcanic activity and deep-seated upheavals that climaxed in the raising of mountains. The heat, pressure and churning associated with these movements probably destroyed most of the fossils, but some evidence of life still remains. Scattered throughout the Archeozoic rocks are traces of graphite or pure carbon, probably the transformed remains of plant and animal bodies. From the amount of graphite in these rocks we can infer that there was an abundance of life in the Archeozoic seas.

The Proterozoic Era. The second era, some one billion years in length, was characterized by the deposition of large quantities of sediment and by at least one great period of glaciation, during which ice sheets stretched to within 20 degrees of the equator. The fossils found in Proterozoic rocks show not only that life was present but also that evolution had proceeded quite far before the end of the era. Plants and animals were differentiated; multicellular organisms had evolved; and representatives of some of the major groups of plants and animals had appeared. Since 1947, several rich deposits of Precambrian fossils have been found in South Australia. These include jellyfish, corals, segmented worms and two animals with no resemblance to any known fossil or living form. The bodies of these animals were soft and were strengthened only by spicules of calcium carbonate.

32–4 THE PALEOZOIC ERA

Between the strata of the late Proterozoic and the earliest layers of the third major era, the Paleozoic, is a considerable gap, caused by a geologic revolution. During the 370,000,000 years of the Paleozoic, members of every phylum and class of animals appeared except the birds and mammals. Since some of these animals existed for a relatively short time, their fossils enable geologists to correlate rocks of the same era found in different localities.

The Cambrian Period. The earliest subdivision of the Paleozoic era, the *Cambrian period*, is represented by rocks rich in fossils, so that the reconstructions of what the world was like in those days are probably quite accurate. The forms living in this period were so varied and complex that they must have evolved from ancestors dating back to the Proterozoic era, at the latest, and possibly to the Archeozoic. All the present-day animal phyla, except the chordates, were represented, and all plants and animals lived in the sea. (The land must have been a weird, lifeless waste until the late Ordovician or Silurian, when plants became established on land.) There were primitive, shrimplike crustaceans and arachnidlike forms, some of whose descendants exist almost unchanged today (i.e., the horseshoe crab). The sea floor was covered with simple sponges, corals, echinoderms growing on stalks, snails, pelecypods, primitive cephalopods, brachiopods and trilobites. *Brachiopods*, sessile, bivalved plankton feeders, flourished in the Cambrian and the rest of the Paleozoic. The *trilobites* (Fig. 14–9) were primitive arthropods with flattened, elongated bodies covered dorsally by a hard shell. The shell had two longitudinal grooves that divided the body into three lobes. There were a pair of legs on each somite but the last, and each leg had an outer gill branch and an inner walking or swimming branch. Most trilobites were 5 to 8 cm. long, but a few were as large as 60 cm. There were both unicellular and multicellular algae. One of the best-preserved collections of Cambrian fossils was found in the mountains of British Columbia; it includes annelids, crustacea, and a connecting link between annelids and arthropods that is similar to the living peripatus.

Evolution since the Cambrian has been

marked not by the establishment of entirely new body patterns but rather by the ramification of the lines already present and by the replacement of original, primitive forms with better-adapted ones. The fact that no new phyla have originated since the early Paleozoic does not necessarily mean that no other patterns of animal organization are possible or that mutations for new patterns did not occur. It probably indicates only that by that time the existing forms had reached a degree of adaptation to the environment that gave them a marked advantage over any new, unadapted types.

The Ordovician Period. During the Cambrian period the continents gradually had begun to be covered with water, and in the Ordovician period this submergence reached its maximum, so that much of what is now land was covered by shallow seas. Inhabiting the seas were giant cephalopods—squid or nautiluslike animals with straight shells 5 to 7 meters long and 30 cm. in diameter. The first traces of the vertebrates are found in Ordovician rocks; these small *ostracoderms* were jawless, armored, bottom-dwelling fishes without fins (Fig. 32–4). Their armor consisted of a heavy, bony covering over the head and thick scales over the trunk and tail; otherwise they were similar to the jawless lamprey eels of

today. They lived in fresh water and their armor plate was a defense against *eurypterids*, the carnivorous, giant water scorpions—sometimes 3 meters long—that also lived in fresh water.

The Silurian Period. Two events of great biologic importance occurred in the Silurian period: the land plants evolved, and air-breathing animals appeared. The first land plants resembled ferns rather than mosses, and ferns were the dominant plants of the Devonian and Mississippian periods that followed. The first air-breathing land animals were arachnids, resembling to some extent modern scorpions. The continental areas, which had been low during the Cambrian and Ordovician, rose—especially in Scotland and northeastern North America—and the climate became much colder.

The Devonian Period. During the Devonian, the original ostracoderms evolved into a great variety of fishes, and the period is frequently called the Age of Fishes. Some descendants of the ostracoderms evolved jaws and paired appendages and changed from filter-feeding bottom dwellers to active predators. The earliest jawed fishes, the spiny-skinned sharks, are placed in the subclass Acanthodii. Another group of primitive jawed fishes living in the Devonian were the placoderms, small, armored fresh water fishes with a variable number (as many as seven) of paired fins (Fig. 32–5). One of the best known placoderms is *Dunkleosteus*, a monster that attained a length of 3 meters. The head and anterior part of the trunk had a bony armor, but the remainder of the body was naked. The evolution of jaws from a portion of the gill arch skeleton enabled the placoderms and their descendants to become adapted to new modes of life. The success of the jawed vertebrates undoubtedly contributed to the extinction of the ostracoderms.

True sharks appeared in fresh water during the Devonian but tended to migrate to the ocean and to lose their cumbersome armor plate. The ancestors of the bony fishes also appeared in Devonian fresh water streams and had evolved by the middle of the Devonian into three main types: lungfishes, lobe-finned fishes and ray-finned fishes. All had lungs and an armor of bony scales. A few lungfishes have survived to the present, and the ray-finned fishes, after undergoing a slow evolution in the

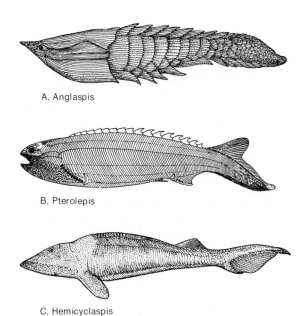

A, Anglaspis

B, Pterolepis

C, Hemicyclaspis

Figure 32–4 Three fossil ostracoderms—primitive, jawless, limbless fishes. (From Romer, A. S.: *The Vertebrate Body*, 4th Ed. Philadelphia, W. B. Saunders Co., 1970.)

remaining Paleozoic and early Mesozoic eras, ramified greatly in the latter part of the Mesozoic to give rise to the modern bony fishes, or *teleosts.* The lobe-finned fishes, which were the ancestors of the land vertebrates, were almost extinct by the end of the Paleozoic, and it was once believed that they had vanished with the end of the Mesozoic. But since 1939, several specimens of living coelacanths nearly 2 meters long have been caught in the deep waters off the east coast of South Africa (Figs. 15–7 and 32–6).

The latter part of the Devonian was marked by the appearance of the first land vertebrates, called *labyrinthodonts,* clumsy, salamanderlike, ancient amphibians with short necks and heavy muscular tails (Fig. 32–7). These creatures, whose skulls were encased in bony armor, were similar in most respects to the lobe-fins, but had evolved limbs strong enough to support the weight of the body on land. These earliest arms and legs were five-fingered, a pattern that has generally been retained by the higher vertebrates. The Devonian was the first period characterized by true forests; ferns, club mosses, horsetails and primitive gymnosperms—the "seed ferns"—all flourished. In-

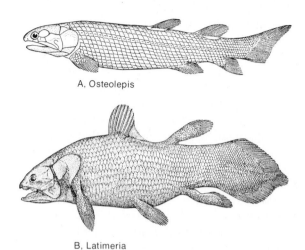

A, Osteolepis

B, Latimeria

Figure 32–6 Crossopterygians. *A,* Typical fossil form from the Devonian. *B,* The living coelacanth, found in the Indian Ocean off Africa in 1938. (See also Fig. 15–7.) (From Romer, A. S.: *The Vertebrate Body,* 4th Ed. Philadelphia, W. B. Saunders Co., 1970.)

sects and millipedes are believed to have originated in the late Devonian.

The Carboniferous Period. The Mississippian and Pennsylvanian periods are frequently grouped together as the Carboniferous, for during this time there flourished the great swamp forests whose remains gave rise to the major coal deposits of the world. The land was covered with low swamps, filled with horsetails, ferns, seed ferns and large-leaved evergreens (Fig. 32–8). The first reptiles, the

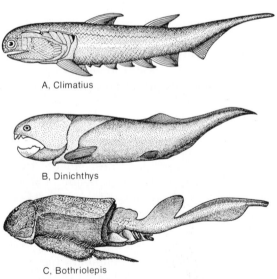

A, Climatius

B, Dinichthys

C, Bothriolepis

Figure 32–5 Acanthodians and placoderms from the Devonian period. *A, Climatius,* a "spiny-skinned shark," with large fin spines and five pairs of accessory fins between the pectoral and pelvic pairs. *B, Dinichthys,* a giant arthrodire that grew to a length of 10 meters. Its head and thorax were covered by bony armor, but the rest of the body and tail were naked. *C, Bothriolepis,* a placoderm with a single pair of jointed flippers projecting from the body. (From Romer, A. S.: *The Vertebrate Body,* 4th Ed. Philadelphia, W. B. Saunders Co., 1970.)

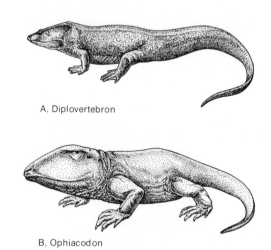

A, Diplovertebron

B, Ophiacodon

Figure 32–7 *A, Diplovertebron,* a primitive Paleozoic amphibian (labyrinthodont). *B, Ophiacodon,* an early Permian pelycosaur. Although the pelycosaurs were primitive reptiles, they had certain characteristics indicating that they represent a first stage in the evolution of the mammals. (From Romer, A. S.: *The Vertebrate Body,* 4th Ed. Philadelphia, W. B. Saunders Co., 1970.)

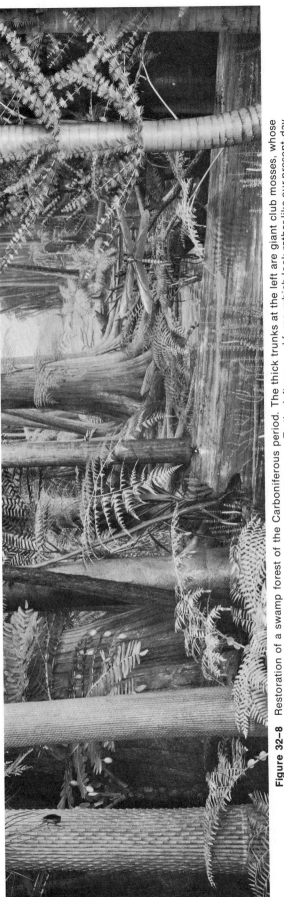

Figure 32-8 Restoration of a swamp forest of the Carboniferous period. The thick trunks at the left are giant club mosses, whose grasslike leaves and large cones can be seen in the upper left corner. To the left are seed ferns, which look rather like our present-day ferns but have seeds. The tree to the right is a calamite, a giant horsetail, with leaves arranged in whorls. A large insect, an ancestor of the dragonflies, is visible just to the left of the calamite. (Copyright, Chicago Natural History Museum.)

Figure 32-9 Texas in the Permian period, about 230,000,000 years ago. Various pelycosaurs are shown. Some of them had large fins, others were essentially like lizards. (Copyright, Chicago Natural History Museum; from the painting by Charles R. Knight.)

cotylosaurs, or stem reptiles, similar to their antecedent amphibians, appeared in the Pennsylvanian period, flourished in the final Paleozoic period—the Permian—and became extinct early in the Mesozoic era. Whether the most primitive reptile known (called *Seymouria* for the town in Texas near which its fossil remains were found) should be considered an amphibian about to become a reptile, or a reptile just over the border from an amphibian is debatable. One of the main differences between reptiles and amphibians is the type of egg laid: amphibians lay jelly-coated eggs in the water, and reptiles lay shell-covered eggs on land. Since no seymourian eggs have survived, it may never be possible to decide to which class this animal belonged. *Seymouria* was a large, sluggish beast resembling a lizard. Its short, stubby legs extended laterally from the body as do a salamander's, instead of being closer together and extending directly down to form pillarlike supports for the body.

Two important groups of winged insects evolved during the Carboniferous—the ancestors of the cockroaches, which reached a length of 10 cm., and ancestral dragonflies, some of which had a wingspread of 75 cm.

The Permian Period. The final period of the Paleozoic was characterized by great changes in climate and topography. The level of the continents rose all over the world, so that the shallow seas that covered the region from Nebraska to Texas at the beginning of the period drained off, leaving the land a salt desert. At the end of the Permian a general folding of the earth's crust, called the *Appalachian Revolution,* raised the great mountain chain from Nova Scotia to Alabama. These mountains originally were higher than the present Rockies. Other ranges were brought into existence in Europe at this time. A great glaciation, spreading from the Antarctic, covered most of the southern hemisphere, extending almost to the equator in Brazil and Africa; North America was one of the few parts of the world to escape glaciation at that time, but even its climate became much colder and drier than it had been during most of the Paleozoic. Many of the Paleozoic forms of life, apparently unable to adapt to the climatic changes, became extinct during the Appalachian Revolution. Even many of the marine forms became extinct, owing to the cooling of

Figure 32–10 A mammal-like reptile *(Lycaenops)* from the late Permian period in South Africa. (From Romer, A. S.: *The Vertebrate Body,* 4th Ed. Philadelphia, W. B. Saunders Co., 1970.)

the water and the decrease in the amount of space available caused by the diminishing of the shallow seas.

From the primitive cotylosaurs there evolved during the late Carboniferous and early Permian the group of reptiles believed to be in the direct line giving rise to the mammals. These were the *pelycosaurs* (Fig. 32–9), carnivorous reptiles that were more slender and lizardlike than the stem reptiles. In the latter part of the Permian there evolved, probably from the pelycosaurs, another group of reptiles with a few more mammalian characteristics—the *therapsids* (Fig. 32–10). One of these, *Cynognathus,* the "dog-jawed" reptile, was a slender, lightly built animal about 150 cm. long, with a skull intermediate between that of a reptile and that of a mammal. The teeth, instead of being conical and all alike, as reptilian teeth are, were differentiated into incisors, canines and molars. In the absence of information about the animal's soft parts, whether it had scales or hair, whether or not it was warm blooded, and whether it suckled its young, it is called a reptile, but if more evidence were available, it might be classified as a very early mammal. The therapsids were widespread in the late Permian, but were crowded out early in the Mesozoic by the great variety of other reptiles.

32–5 THE MESOZOIC ERA

The Mesozoic era began about 230,000,000 years ago and lasted some 167,000,000 years. The outstanding feature of the Mesozoic era was the origin, differentiation and final extinction of a great variety of reptiles. For this reason the Mesozoic is commonly called the Age of Reptiles.

Figure 32-11 During much of the Cretaceous period, a shallow inland sea covered the western half of the Mississippi Valley. Three reptiles characteristic of this time and place are shown. In the center is a large mosasaur about 10 meters long; to the right is a giant marine turtle, 2.5 meters long; and flying at the left are a number of reptiles of the genus *Pteranodon*, forms with short tails and a long crest extending back from the skull. (Copyright, Chicago Natural History Museum; from the painting by Charles R. Knight.)

There were six major evolutionary lines of reptiles; the most primitive line included the ancient cotylosaurs and the turtles. The turtles originated in the Permian and have evolved the most complicated armor of any land animal, consisting of scales derived from the epidermis fused to the underlying ribs and breastbone. With this protection, both marine and land forms have survived with few structural changes since before the time of the dinosaurs. Their legs extend laterally, making locomotion difficult and slow, and their skulls are unpierced behind the eye sockets, a feature essentially unchanged from the ancient cotylosaurs.

A second group of reptiles to survive with relatively few changes from the ancestral stem reptiles are the *lizards*—the most abundant of living reptiles—and the snakes. For the most part the lizards have kept the primitive type of locomotion with the legs extended laterally, although many can run rapidly. Most of them are small, but the monitor lizard of the East Indies attains a length of 4 meters, and some fossil ones were 8 meters long. The *mosasaurs*, marine lizards of the Cretaceous (Fig. 32–11), attained a length of 13 meters, and had a long tail useful in swimming. During the Cretaceous, snakes evolved from lizard ancestors. The important difference between snakes and lizards is not the loss of legs (some lizards are legless), but certain changes in the skull and jaws of the snake enabling it to open its mouth wide enough to swallow an animal larger than itself. A representative of an ancient line that has managed somehow to survive in New Zealand is the lizardlike *tuatara*. It shares several traits with the ancestral cotylosaurs, one of which is the presence of a third eye on the top of its head.

The main group of Mesozoic reptiles were the *archosaurs*, or "ruling reptiles." The only living members of this group are the alligators and crocodiles. At an early point in their evolution from stem reptiles the ruling reptiles, then about 1 meter long, became adapted to two-legged locomotion—their front legs became short, while the hind legs became long, stout and considerably modified. These animals rested or walked on all fours, but in emergencies they reared up and ran on the two hind legs, balanced by their fairly long tail. From the early archosaurs developed many different,

specialized forms; some continued to use two-legged locomotion but others reverted to walking on all fours. These descendants include the *phytosaurs*—aquatic, alligatorlike reptiles, common during the Triassic; the *crocodiles*, which evolved during the Jurassic and replaced the phytosaurs as aquatic forms; and the *pterosaurs*, or flying reptiles. The pterosaurs included animals the size of a robin, as well as the largest animal ever to fly, a pterosaur with a wingspread of 15.5 meters, discovered in 1975 in western Texas. There were two types of flying reptiles, one with a long tail that had a steering rudder at the end, the other with a short tail. Both these types apparently were fisheaters, and they probably flew long distances over the water in search of food. Their legs were not adapted for standing, and it is believed that, like bats, they rested by clinging to some support and hanging suspended.

Of all the reptilian branches, the most famous are the *dinosaurs* (meaning "terrible reptiles"). These were divided into two main types: one with a birdlike pelvis, the other with a reptilian pelvis.

The *Saurischia* (reptile pelvis) first evolved in the Triassic, and remained in existence until the Cretaceous. The early ones were fast, carnivorous, two-legged forms the size of a rooster, which probably preyed upon lizards and the archaic mammals then in existence. Throughout the Jurassic and Cretaceous, this group showed a tendency to grow larger, culminating in the gigantic carnivore of the Cretaceous, *Tyrannosaurus* (Fig. 32–12). Beginning in the late Triassic, other Saurischia changed to a plant diet, reverted to a four-legged gait and, during the Jurassic and Cretaceous, evolved into tremendous amphibious forms. Among these—the largest four-footed animals that ever lived—were *Brontosaurus*, with a length of 21 meters; *Diplodocus*, which attained a length of 29 meters; and *Brachiosaurus*, the biggest of them all, with an estimated weight of 50 tons.

The other group of dinosaurs, the *Ornithischia* (birdlike pelvis), were herbivores, probably from the beginning of their evolution. Although some of them walked upright, the majority had a four-legged gait. Having lost their front teeth, they developed a stout, horny, birdlike beak. In some forms this was broad and ducklike (hence the name "duck-billed"

(*Text continued on page 768.*)

Figure 32–12 Giant dinosaurs from the Cretaceous period of western North America. The largest flesh-eating dinosaur known was *Tyrannosaurus*, two of which are shown attacking the herbivorous, horned dinosaur *Triceratops*. *Tyrannosaurus* reached a length of 15 meters and a height of 6 meters. Its head was as much as 2 meters long and was equipped with many sharp teeth. The front legs were small and completely useless; it walked on its powerful hind legs and balanced with its long tail. *Triceratops* was armed with a horn on the nose and a pair of horns over the eyes and was protected by a bony ruff covering the neck and shoulders. The rest of the body was covered with a leathery hide, so that it was vulnerable except when facing its enemy. (Copyright, Chicago Natural History Museum; from the painting by Charles R. Knight.)

Figure 32-13 Western Canada in the Cretaceous period, about 110,000,000 years ago. The land was low, well-watered and covered with numerous swamps. Most of the dinosaurs were harmless, plant-eating forms of the Ornithischia group of reptiles, characterized by birdlike pelvic bones. Two types of duck-billed dinosaurs can be seen—three large, uncrested ones to the right, and two types of crested ones in the left background. In the middle foreground is a heavily armored, four-footed dinosaur covered with bony plates and spines. In the center background are two ostrich dinosaurs—tall, slender animals, with the general proportions of an ostrich, but with short forelegs and a long, slender tail. (Copyright, Chicago Natural History Museum; from the painting by Charles R. Knight.)

Figure 32-14 Scene off the coast of North America in Jurassic times, about 155,000,000 years ago. Two types of marine reptiles are shown: plesiosaurs, with long necks, broad, flat bodies, and sturdy, paddle-shaped limbs; and ichthyosaurs, with fishlike fins and tails. Both types were fisheaters. (Copyright, Chicago Natural History Museum; from the painting by Charles R. Knight.)

dinosaurs). Webbed feet were characteristic of this type; other species developed great armor plates as protection against the carnivorous saurischians. *Ankylosaurus* (Fig. 32–13), dubbed "the reptilian tank," had a broad, flat body, covered with an armor plate and large, laterally projecting spines. Still other ornithischians of the Cretaceous period developed bony plates around the head and neck. One of these, *Triceratops*, had two horns over its eyes and another over its nose—each 1 meter long.

Two other groups of Mesozoic reptiles, separate from each other and from the dinosaurs, were the marine **plesiosaurs** and **ichthyosaurs** (Fig. 32–14). The extremely long neck of the plesiosaurs took up over half of their total length of 15 meters. The trunk was broad, flat and rather turtlelike, the tail was small, and the animal paddled along by means of finlike arms and legs. The porpoiselike ichthyosaurs (fish reptiles) had a body form superficially like that of a fish or a whale, with a short neck, a large dorsal fin, and a sharklike tail. They swam by wiggling their tails, and used their feet only for steering. The ichthyosaur young were apparently born alive, after having hatched from eggs within the mother, for the adults were too specialized to come out on land, and a reptilian egg will drown in water. The presence of skeletons of the young within the body cavity of adult fossils has strengthened this theory.

At the end of the Cretaceous a great many reptiles became extinct; they were apparently unable to adapt to the marked changes brought about by the Rocky Mountain Revolution. As the climate became colder and drier many of the plants that served as food for the herbivorous reptiles disappeared. Some of the herbivorous reptiles were too large to walk about on land when the swamps dried up. The smaller, warm-blooded mammals appeared and were better able to compete for food; many of them ate reptilian eggs. The demise of the many kinds of reptiles was probably the result of a combination of a whole host of factors, rather than any single one.

Although the reptiles were the dominant animals of the Mesozoic, many other important organisms evolved during that time: snails and bivalves increased in number and kind; sea urchins reached their peak; mammals originated in the Triassic; and teleost fishes and birds arose during the Jurassic. Most of the modern orders of insects appeared early in the Mesozoic. During the early Triassic the most abundant plants were seed ferns, cycads and conifers; but by the Cretaceous, many others, resembling present-day species, had appeared—sycamores, magnolias, palms, maples and oaks.

Excellent fossils, some even showing the outlines of feathers in the earliest species of bird, have been preserved from the Jurassic (Fig. 32–1). *Archaeopteryx* was about the size of a crow, had rather feeble wings, jawbones armed with teeth, and a long reptilian tail covered with feathers (Fig. 32–15 A). Cretaceous

Figure 32–15 *A*, A restoration of *Archaeopteryx*, the earliest known bird. *B*, A restoration of *Hesperornis*, a large diving bird of the Cretaceous. (Courtesy of the American Museum of Natural History, New York.)

rocks have yielded fossils of two other primitive birds—*Hesperornis* (Fig. 32–15 *B*), an aquatic diving bird that had lost the ability to fly, and *Ichthyornis*, a powerful flying bird, about the size of a pigeon, with reptilian teeth. Birds did not evolve from the flying reptiles, the pterosaurs, but from a group of primitive dinosaurs called thecodonts. Modern toothless birds evolved early in the following era.

32–6 THE CENOZOIC ERA

With equal justice, the Cenozoic Era could be called the Age of Mammals, the Age of Birds, the Age of Insects or the Age of Flowering Plants, for it is marked by the evolution of all these forms. It extends from the Rocky Mountain Revolution, some 63,000,000 years ago, to the present, and is subdivided into two periods. The earlier *Tertiary* period lasted some 62,000,000 years, and the *Quaternary* period includes the last million or million and a half years.

The Tertiary Period. The Tertiary is sub-divided into five epochs, named, from earliest to latest, *Paleocene, Eocene, Oligocene, Miocene* and *Pliocene*. The Rockies, formed at the beginning of the Tertiary, were considerably eroded by the time of the Oligocene, giving the North American continent a gently rolling topography. In the Miocene another series of uplifts raised the Sierra Nevadas and a new set of Rockies, and resulted in the formation of the western deserts. The climate was milder in the Oligocene than it is at present, and palms grew as far north as Wyoming. The uplift begun in the Miocene continued in the Pliocene and, coupled with the ice ages of the Pleistocene, killed many of the mammals and other forms that had evolved. The final elevation of the Colorado Plateau, which also caused the cutting of Grand Canyon, occurred almost entirely in the short Pleistocene and Recent epochs.

The earliest fossils of true mammals were deposited late in the Triassic, but by the Jurassic there were four orders of mammals, all about the size of a rat or small dog. The earliest mammals (*monotremes*) were egg-laying animals, and their only living survivors are the

Figure 32–16 Two living examples of monotremes—mammals that lay eggs. *A*, The duck-billed platypus. Note the short fur, webbed feet, horny, duck-shaped beak and unusual tail. *B*, The spiny anteater, about 45 cm. long, is covered with strong, pointed spines, yellow with black tips. Its narrow black snout is cylindrical, and it captures its food (mostly ants) with the long, protrusible tongue, which is covered with sticky saliva. (Courtesy of the Australian News and Information Bureau.)

duck-billed platypus and the spiny anteater of Australia (Fig. 32–16). Both have fur and suckle their young but lay eggs like turtles. The ancestral, egg-laying mammals certainly must have differed from the specialized platypus and anteater, but the fossil records of those early forms are incomplete. The present-day monotremes have been able to survive this long only because they lived in Australia, which, until recently, had no placental mammals to offer competition.

By the Jurassic and Cretaceous, most mammals were advanced enough to bring forth their young alive, although the most primitive of them—the *marsupials*—gave birth to underdeveloped young that had to remain for several months in a pouch of the mother's abdomen containing the nipples. The Australian marsupials, freed like the monotremes of competition from the better-adapted placental mammals that were responsible for the extinction of their cousins on other continents, evolved into a wide variety of types that superficially resemble some of the placentals. There are marsupial mice, shrews, cats, moles, bears, and one species of wolf, as well as a number of forms with no placental counterparts, such as the kangaroo, wombat and wallaby (Fig. 32–17). During the Pleistocene there were giant kangaroos and wombats the size of a rhinoceros in Australia. The opossum more closely resembles the primitive, ancestral marsupial type than do any of these more specialized forms; it is the only marsupial found outside Australia and South America.

The advanced, modern placental mammals, including the human, are distinctive in bringing forth their young alive and ready to live an independent existence. They all evolved from a tiny, shrewlike, insect-eating, tree-dwelling ancestor known from fossils in Cretaceous rocks. Some of these ancestral mammals remained in the trees and gave rise, through a series of intermediate forms, to the primates—monkeys, apes and humans. Others lived on or under the ground and, during the Paleocene, evolved into all the other mammals living today. The archaic mammals of the Paleocene had conical, reptilian teeth, five digits on each foot, and a small brain; they also walked on the soles of their feet instead of on their toes. During the Tertiary the evolution of grasses, which served as food, and forests,

Figure 32–17 Kangaroo mother with her young in her pouch. (Courtesy of the Australian News and Information Bureau.)

which afforded protection, were important factors in leading to changes in the mammalian body pattern. Concomitant with a tendency toward increased size, the mammals all displayed tendencies toward an increase in the relative size of the brain and toward changes in the teeth and feet. As modern forms better equipped for survival arose, the archaic mammals became extinct.

Although fossils of both marsupial and placental animals have been found in Cretaceous rocks, it is rather surprising to find the remains of highly developed mammals in strata of the early Tertiary. Whether they actually arose at that time or had existed before in the highlands and had not been preserved is unknown.

In the Paleocene and Eocene epochs the first carnivores, called *creodonts,* arose from the primitive insect-eating placental mammals (Fig. 32–18). They were replaced in the Eocene and Oligocene by more modern forms ancestral to the present-day carnivores, such as cats, dogs, bears and weasels, as well as to the web-footed, marine carnivores—the seals and walruses. One of the most famous fossil carnivores, the saber-toothed tiger (Fig. 32–19),

Figure 32–18 Recreation of an archaic meat-eating mammal, a creodont from the Eocene period, eating a tiny ancestral horse, *Eohippus*. (Copyright, American Museum of Natural History, New York.)

became extinct only recently in the Pleistocene. These animals had tremendously elongated, knifelike upper canine teeth and a lower jaw that could be swung down and out of the way, allowing the teeth to be used as sabers for stabbing the prey.

The larger herbivorous mammals, most of which have hooves, are sometimes referred to as the **ungulates.** They do not form a single, natural group, but consist of several independent lines. Although both horses and cows have hooves, they are no more closely related than either one is to a tiger. The molar teeth of ungulates are flattened and enlarged to facilitate the chewing of leaves and grass. Their legs have become elongated and adapted for the rapid movement necessary to escape predators. The earliest ungulates, the **condylarths,** appeared in the Paleocene; they had long bodies and tails, flat, grinding molars, and short legs ending in five toes, each of which bore a hoof. Corresponding to the archaic carnivores, or creodonts, were the archaic ungulates called **uintatheres.** During the Paleocene and Eocene

Figure 32–19 Recreation of a scene at the Rancho La Brea tar pits (now a part of Los Angeles, California) in the Pleistocene. Many well-preserved specimens of animals now extinct have been found imbedded in the asphalt. In the left foreground are two saber-toothed tigers; in the right foreground, three large ground sloths; the giant vultures, now extinct, had a wingspread of 3 meters. In the background are mastodons and dire wolves. (Copyright, American Museum of Natural History, New York, from a painting by Charles R. Knight.)

some of these were as large as elephants, and some had three large horns projecting from the top of the head.

The fossil records of several ungulate lines—the horse, the camel and the elephant—are complete, and it is possible to trace the evolution of these animals from small, primitive, five-toed creatures. The chief evolutionary tendencies in the ungulates have been toward an increase in the over-all size of the body and a decrease in the number of toes. The ungulates were early divided into two groups, one characterized by an even number of toes, and including the cow, sheep, camel, deer, giraffe, pig and hippopotamus; the other characterized by an odd number of toes, and including the horse, zebra, tapir and rhinoceros. The elephants and their recently extinct relatives, the mammoths and mastodons, can be traced back to a trunkless Eocene ancestor the size of a hog. This primitive form, called *Moeritherium*, was close to the stem that also gave rise to such dissimilar creatures as the coney (a small, woodchucklike animal found in Africa and Asia) and the sea cow.

The whales and porpoises descended from whalelike forms of the Eocene, called **zeuglodonts**, which in turn are believed to have evolved from the creodonts. The evolutionary history of the bats can be traced to ancestral, winged types of the Eocene, descendants of the primitive insectivores. The evolutionary history of some of the other mammals—the rodents, rabbits and edentates (anteaters, sloths and armadillos)—is less well known.

The Quaternary Period. The Quaternary period includes the final million or million and a half years of the earth's history and is divided into two epochs, the *Pleistocene* and the *Recent*, which began some 11,000 years ago with the recession of the last ice sheet. The Pleistocene was marked by four periods of glaciation, between which the sheets of ice retreated. At their greatest extent these ice sheets covered nearly 4,000,000 square miles of North America, extending south as far as the Ohio and Missouri rivers. The Great Lakes were carved out by the advancing glaciers and changed their outlines radically a number of times. From time to time they emptied into the Mississippi. It is estimated that in the past, when the Mississippi drained lakes as far west

as Duluth and as far east as Buffalo, its volume was more than 60 times as great as at present. During the Pleistocene glaciations, enough water was removed from the sea and locked in the ice to lower the sea level 65 to 100 meters. This created land connections, highways for the dispersal of many land forms, between Siberia and Alaska at Bering Strait and between England and the continent of Europe.

The plants and animals of the Pleistocene were similar to those alive today. It is sometimes difficult to distinguish between Pleistocene and Pliocene deposits, because the organisms were similar and nearly modern in form. A considerable number of mammals, including the saber-toothed tiger, mammoth and giant ground sloth, became extinct during the Pleistocene, after the appearance of primitive man. The Pleistocene was marked by the extinction of many species of plants, especially woody ones, and the appearance of numerous herbaceous forms.

The paleontologic record makes it impossible to doubt that the present species arose from previously existing, different ones. The fossil record is not equally clear for all lines of evolution. Most plant tissues are too soft to leave good fossil remains and the connecting links between the animal phyla were apparently soft-bodied forms that left no fossil traces. For many lines of evolution, especially the vertebrates, the successive steps are well known.

32–7 OTHER EVIDENCE FOR EVOLUTION

Evidence from Taxonomy. The characteristics of living things are such that they can be fitted into a hierarchical scheme of categories—species, genera, families, orders, classes and phyla. The most reasonable explanation for this is that the hierarchical scheme indicates evolutionary relationships. If the kinds of plants and animals were not related by evolutionary descent, their characteristics would be present in a confused, random pattern and no such hierarchy of forms could be established.

The classification of modern-day organisms into well defined groups is possible only because most of the intermediate forms have become extinct. If every animal and plant that ever lived were still living today, it would be a difficult matter to divide the living world into neat taxonomic categories, for there would be a continuous series of forms grading from lowest to highest. The species now living have been called "islands in a sea of death" and have been likened also to the terminal twigs of a tree of which the trunk and main branches have disappeared. The problem of the taxonomist is to reconstruct the missing branches and put each twig on the proper branch.

Evidence from Morphology. Comparisons of the structure of groups of animals and plants show that organ systems have a fundamentally similar pattern that is varied to some extent among the members of a given phylum. The skeletal, circulatory and excretory systems of vertebrates provide particularly clear illustrations of this.

Only similarities based on homologous organs are valid in attributing evolutionary relationships. *Homologous organs,* you will recall, are basically similar in their structure, in their relationships to adjacent structures, in their embryonic development and in their nerve and blood supply. A seal's front flipper, a bat's wing, a cat's paw, a horse's front leg and the human hand and arm (Fig. 13–1), though superficially dissimilar and adapted for quite different functions, are homologous organs. Each consists of almost the same number of bones, muscles, nerves and blood vessels, arranged in the same pattern, and with very similar modes of development. The existence of such homologous organs is a strong argument for a common evolutionary origin.

VESTIGIAL ORGANS. Most plants and animals contain organs or parts of organs that are useless and degenerate, often undersized or lacking some essential part, as compared to homologous structures in related organisms. In the human body there are more than 100 such *vestigial organs,* including the appendix, the coccyx (the fused tail vertebrae), the wisdom teeth, the nictitating membrane of the eye, the body hair and the muscles that move the ears and nose (Fig. 32–20). Vestigial organs are the remnants of ones which were functional in some ancestral animal. Because of a change in the environment or mode of life of the species, the organ became unnecessary for survival and, gradually, nonfunctional. Is this consistent with Lamarck's concept of the role of "use and disuse" in evolution? Mutations are constantly occurring that decrease the size and function of various organs. If the organs are necessary for survival, the organisms undergoing such mutations will be eliminated. If the organs are not necessary for survival, they will become reduced in size and vestigial and eventually will be eliminated.

Whales and pythons have vestigial hind leg bones embedded in the flesh of the abdomen; the wingless birds have vestigial wing bones; many blind, burrowing or cave-dwelling animals have vestigial eyes; and so on.

Evidence from Comparative Biochemistry.

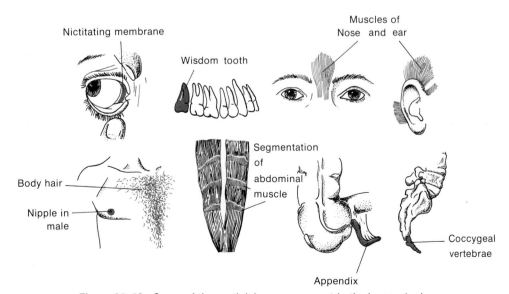

Figure 32–20 Some of the vestigial organs present in the human body.

Evidence for evolutionary relationships can be obtained from functional and chemical similarities and differences as well as from morphologic similarities and differences. For example, the degree of similarity between the plasma proteins of various animals may be tested by the antigen-antibody technique. An animal, such as a rabbit, is given repeated injections of the protein to be tested, such as human serum, which contains proteins foreign to the rabbit's blood. The plasma cells of the rabbit respond by producing antibodies specific for human blood-protein antigens. The antibodies are then obtained by withdrawing blood from the rabbit and allowing it to clot (the antibodies are in the serum). Even a dilute sample of the serum, when mixed with human blood, results in a visible precipitation caused by the combination of antigens and antibodies. The strength of the reaction can be measured by making successive dilutions of the human serum, mixing each dilution with a fresh sample of the antibody (rabbit serum) and observing the dilution at which precipitation no longer occurs. A precipitation will be formed with nonhuman blood only if the serum is very much more concentrated, and even then the precipitation may be greatly delayed. By using a series of rabbits, each injected with the blood of different species, it has been possible to obtain a series of antibodies, each specific for the blood proteins of a particular species of animal.

Thousands of tests involving different animals have revealed a basic similarity between the blood proteins of all the mammals, the degree of relationship being indicated by how

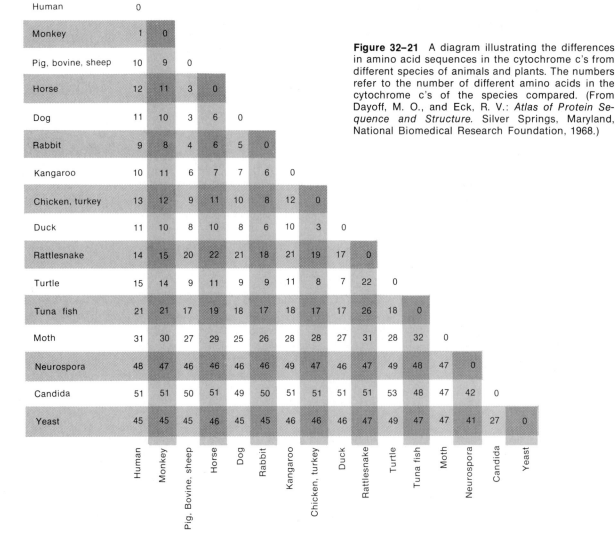

Figure 32–21 A diagram illustrating the differences in amino acid sequences in the cytochrome c's from different species of animals and plants. The numbers refer to the number of different amino acids in the cytochrome c's of the species compared. (From Dayoff, M. O., and Eck, R. V.: *Atlas of Protein Sequence and Structure*. Silver Springs, Maryland, National Biomedical Research Foundation, 1968.)

much the antigen and antibody solutions can be diluted and still result in visible precipitation. The closest "blood relations" of humans, as determined in this way, are, in descending order, the great apes, the Old World monkeys, the New World prehensile-tailed monkeys and the tarsioids. Of all the types of primate blood, the lemur's results in the least precipitation when combined with antibodies specific for human serum. The biochemical relationships of a variety of animals and plants tested in this way correlate well with the relationships determined by other means.

Investigations of the sequence of amino acids in the α and β chains of hemoglobins from different species have revealed great similarities, of course, and specific differences, the pattern of which demonstrates the order in which the underlying mutations, the changes in nucleotide base pairs, must have occurred in evolution. The evolutionary relationships inferred from these studies agree completely with those based on anatomic studies. Analyses of the sequence of amino acids in the protein portion of the cytochrome enzymes provide further corroborating evidence of evolutionary relationships (Fig. 32–21). Thus, evidence of evolutionary relations can be obtained by studies of similarities and differences in *molecular* structure as well as by studies of gross structure.

Comparisons of the properties of specific enzymes from different organisms have revealed similarities and differences from which evolutionary relationships can be deduced. The rates of reaction of lactic dehydrogenase and certain other enzymes isolated from different species with the normal pyridine nucleotide coenzyme (NAD) relative to the rates with analogues of NAD can be used to demonstrate evolutionary relationships. Enzymes from animals deemed to be closely related on the basis of anatomic or other evidence show very similar patterns of their rates of reaction, whereas enzymes from animals deemed to be only distantly related show very different patterns.

It might seem unlikely that an analysis of the urinary wastes of different species would provide evidence of evolutionary relationship, yet this is true. The kind of waste excreted depends upon the particular kinds of enzymes present, and the enzymes are determined by genes that have been selected in the course of evolution. The waste products of the metabolism of adenine and guanine are excreted by humans and other primates as uric acid, by other mammals as allantoin, by amphibians and most fishes as urea, and by most invertebrates as ammonia. Vertebrate evolution has been marked by the successive loss of enzymes required for the stepwise breakdown of uric acid. Joseph Needham made the interesting observation that the chick embryo in the early stages of development excretes ammonia, then urea, and finally uric acid. The enzyme uricase, which catalyzes the first step in the degradation of uric acid, is present in the early chick embryo but disappears later. The adult frog excretes urea but the tadpole excretes ammonia. These are biochemical examples of recapitulation.

Evidence from Embryology. The importance of embryologic evidence for evolution was stressed by Darwin and brought into even greater prominence by Ernst Haeckel in 1866, when he developed his theory that embryos, in the course of development, repeat the evolutionary history of their ancestors in some abbreviated form, a phenomenon known as *recapitulation*. This idea, succinctly stated as "Ontogeny recapitulates phylogeny," stimulated research in embryology and focused attention on the general resemblance between embryonic development and the evolutionary process. It is now clear that the embryos of the higher animals resemble the *embryos* of lower forms, not the adults as Haeckel had believed. The early stages of all vertebrate embryos are remarkably similar (Fig. 32–22) and it is not easy to differentiate a human embryo from the embryo of a pig, chick, frog or fish.

In recapitulating its evolutionary history in a few days, weeks or months, the embryo eliminates some steps and alters and distorts others. Early mammalian embryos have many characteristics in common with those of fish, amphibia and reptiles but have in addition other structures enabling them to survive and develop within the mother's uterus rather than within an eggshell. These secondary traits may alter the original characteristics common to all vertebrates, so that the basic resemblances are blurred. The concept of recapitulation must be used with due caution, but it can be helpful in understanding such curious and complex patterns of development as those of the vertebrate

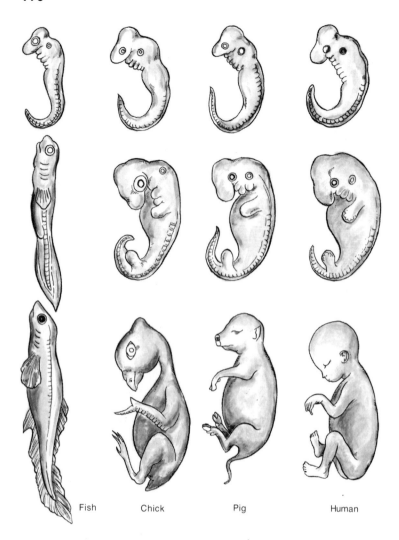

Fish   Chick   Pig   Human

Figure 32–22 Successive stages *(top to bottom)* in the embryonic development of the fish, chick, pig and human. Note that the earlier stages of development *(top row)* are remarkably similar and that differences become more marked as development proceeds *(bottom row)*.

circulatory and excretory systems. It is also useful, when not taken too literally, in getting a broad picture of the whole of development. The fertilized egg may be compared to the single-celled flagellate ancestor of all animals, and the blastula to a colonial protozoan or some spherical multicellular form that may have been the ancestor of all the Metazoa. Haeckel believed that the ancestor of the cnidarians and all the higher animals was a gastrula-like animal, *Gastrea,* with two layers of cells and a central cavity connected by a blastopore to the outside.

After the gastrulation stage, development follows one of two main lines: in one (the echinoderms and chordates) the blastopore—the opening from the gastrular cavity—becomes the anus or comes to lie near the anus; in the other (the annelids, mollusks,

arthropods and others) the blastopore becomes the mouth or comes to lie near the mouth.

In both lines, a third layer of cells—the mesoderm—develops between the ectoderm and endoderm. In the chordate-echinoderm line this develops, at least in part, as pouches or evaginations from the primitive digestive tract, whereas in the annelid line the mesoderm originates from special cells differentiated early in development.

Shortly after the appearance of the mesoderm, all chordate embryos develop a dorsal, hollow nerve cord; a notochord (the internal supporting rod for the body); and perforations in the pharynx (the gill slits). The early human embryo resembles a fish embryo, with gill slits; pairs of aortic arches, or blood vessels traversing the gill bars; a fishlike heart with a single atrium and ventricle; a primitive pronephros or fish kidney; and a tail, complete with muscles for wagging it. Later the human embryo resembles a reptilian embryo: its gill slits close; the bones that make up each vertebra and that had been separate, as in fish embryos, fuse; a new kidney—the mesonephros—forms, and the pronephros disappears; and the atrium becomes divided into right and left chambers. Still later the human embryo develops a mammalian, four-chambered heart, and a third, completely new kidney (the metanephros), while the notochord regresses, and so on. During the seventh month of intrauterine development the human embryo resembles—in being completely covered with hair and in the relative size of body and limbs—a baby ape more than it does an adult human.

Evidence from Genetics. The selection and breeding of domesticated animals and cultivated plants for the past several thousand years provide us with models of how some of the evolutionary forces operate. All the varieties of present-day dogs are descended from one or a few related species of wild dog or wolf, and yet they vary tremendously in many characteristics. Compare, for example, the size of the chihuahua and the St. Bernard or Great Dane; the head shape of the bulldog and collie; the body proportions of the cocker, dachshund and Russian wolfhound. If these varieties were found in the wild, they would undoubtedly be assigned to different species, and perhaps even to different genera. But since all are known to come from common ancestors, and since all are interfertile, they are regarded as varieties or races of a single species.

The plant breeders who developed the present varieties of cultivated plants have similarly produced, by selection and interbreeding, a tremendous variety of plants from one or a few forms. For example, the cliff cabbage, which still grows wild in Europe, is the ancestor not only of our cultivated cabbage but also of such dissimilar plants as cauliflower, kohlrabi, Brussels sprouts, broccoli and kale. Many varieties of wheat, too, have been produced by selection, each adapted for certain growing conditions. Thus there are winter wheats, spring wheats, wheats resistant to drought, to rusts, and to other pests. The cultivated species of tobacco has been traced back to a cross between two species of wild tobacco; corn has been traced to teosinte (a grasslike

plant growing wild in the Andes and Mexico). Breeding experiments and observations indicate that species are not, as Linnaeus believed, unchangeable biologic entities, each of which was created separately, but groups of organisms that have arisen from other species and that can give rise to still others.

The number and the detailed structure of the chromosomes of related species can be compared by cytologic methods. Such studies have provided useful evidence concerning the evolutionary history of fruit flies, jimson weeds, primroses and many other plants and animals.

32–8 BIOGEOGRAPHY

Not all plants and animals are found in all parts of the world; they are not even found everywhere that they could survive, as one would expect if climate and topography were the only factors determining distribution. Central Africa, for example, has elephants, gorillas, chimpanzees, lions and antelopes, while Brazil, with a similar climate and other environmental conditions, has none of these, but does have prehensile-tailed monkeys, sloths and tapirs. The present distribution of organisms is understandable only on the basis of the evolutionary history of each species.

The *range* of a given species—that is, the portion of the earth in which it is found—may be only a few square miles or, as with humans, almost the entire world. In general, closely related species do not have identical ranges, nor are their ranges far apart; they are usually adjacent, but separated by a barrier of some sort, such as a mountain or a desert. This generalization, formulated by David Starr Jordan and known as *Jordan's Rule*, follows from the role of isolation in the formation of species.

As one would expect, regions such as Australia and New Zealand, which have been separated from the rest of the world for a long time, have a flora and fauna peculiar to them. Australia has a population of monotremes and marsupials found nowhere else. During the Mesozoic, Australia was isolated from the rest of the world, so that its primitive mammals never had any competition from the better-adapted placental mammals, which eliminated the monotremes and most of the marsupials ev-

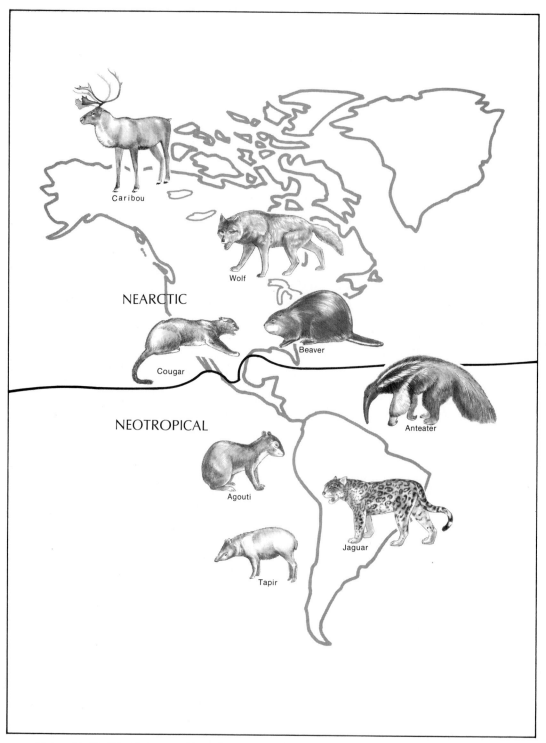

Figure 32–23   The biogeographic realms of the world, with some of their characteristic animals.

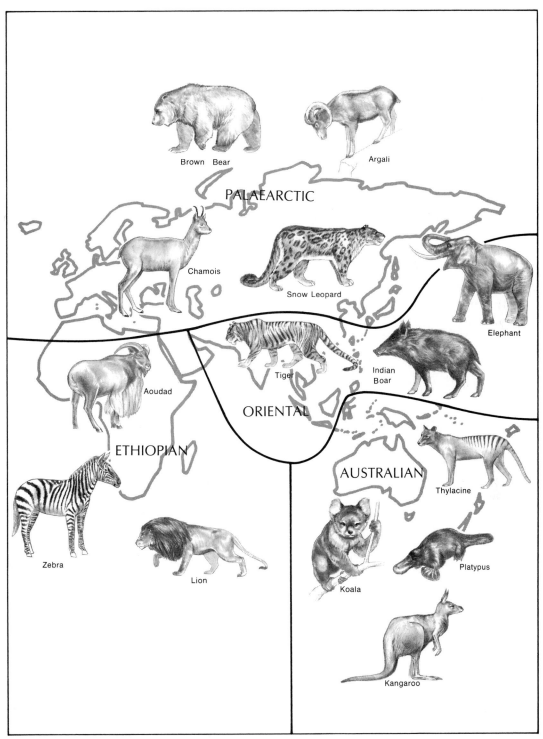

Figure 32-23 *Continued.*

erywhere else. The primitive mammals gave rise to a variety of forms that were able to take advantage of the different habitats available.

The kinds of animals and plants found on oceanic islands resemble, in general, those of the nearest mainland yet include some species found nowhere else. Darwin studied the flora and fauna of the Cape Verde Islands, some 400 miles west of Dakar, and of the Galápagos Islands, a comparable distance west of Ecuador. On each archipelago the plants and terrestrial animals were indigenous, but those of the Cape Verdes resembled African species and those of the Galápagos resembled South American ones. Organisms from the neighboring continent migrated or were carried to the island and subsequently evolved into new species. The animals and plants found on oceanic islands are only those that could survive the trip there. There are no frogs or toads on the Galápagos, even though there are woodland spots ideally suited for such creatures, because neither the animals nor their eggs can survive exposure to sea water. There are no terrestrial mammals, either, although there are many bats, as well as land and sea birds. The occurrence of these particular forms—closely related to, yet not identical with, those of the Ecuador coast—suggests strongly that after the first animals and plants arrived on the islands, mutations took place that changed the species slightly, and these changes were retained because of isolation. These forms offer a good example of the evolutionary process.

Alligators are found only in the rivers of the southeastern United States and in the Yangtze River of China, and sassafras, tulip trees and magnolias grow only in the eastern United States, Japan and eastern China. Early in the Cenozoic the northern hemisphere was much flatter than it is now and the North American continent was connected with eastern Asia by a land bridge at Bering Strait and possibly with Greenland. The climate of this region was much warmer than at present, and fossil evidence shows that alligators, magnolia trees and sassafras were distributed over the entire region. Later in the Cenozoic, as the Rockies increased in height, the western part of North America became colder and dry, causing the plants adapted to a warm, humid climate to become extinct. Then, with the Pleistocene glaciations, the ice sheets moving from the north met the desert and mountain regions in western North America, eliminating any surviving temperate plants; and in Europe the polar glaciations nearly met the glacier spreading from the Alps, so that many of the temperate plants there became extinct. In southeastern United States and eastern China there were regions untouched by the glaciation in which the magnolia trees and alligators survived. Because the alligators and magnolia trees of the two regions have been separated for several million years, they have followed separate evolutionary pathways and are slightly different, but they are still closely related species of the same genera.

The facts about the distribution of plants and animals constitute the science of *biogeography,* one of the basic tenets of which is that each species of animal and plant originated only once. The particular place where this occurred is known as the species' *center of origin.* The center of origin is not a single point but the range of the population when the new species was formed. From its headquarters each species spreads out until halted by a barrier of some kind—physical, such as an ocean or mountain; environmental, such as an unfavorable climate; or biologic, such as the absence of food or the presence of organisms that prey upon it or compete with it for food or shelter.

32-9 THE BIOGEOGRAPHIC REALMS

Careful studies of the distribution of plants and animals over the earth have revealed the existence of six major biogeographic realms, each characterized by certain unique organisms (Fig. 32–23). Although the biogeographic realms were originally defined on the basis of mammalian distribution, they have since been found valid for many other classes of plants and animals. The areas of any one division are often widely separated, with great variations of climate and topography, but it has been possible, during most geologic ages, for organisms to pass more or less freely from one part of the realm to another. In contrast, the realms have been separated from each other by major physical barriers.

The *Palearctic* realm comprises Europe, Africa north of the Sahara Desert, and Asia north of the Himalaya and Nan-Ling mountains, plus Japan, Iceland, the Azores and the Cape Verde Islands. Some of the indigenous animals are moles, deer, oxen, sheep, goats, robins and magpies. A few species of some of these forms are also found in the Nearctic realm.

The *Nearctic* realm includes only Greenland and North America as far south as the northern plateau of Mexico. Besides many of the same forms characteristic of the Palearctic, it supports species of mountain goats, prairie dogs, opossums, skunks, raccoons, bluejays and turkey buzzards found nowhere else. The land bridge connecting North America and Asia at Bering Strait in former geologic times was used by many migrating animals and plants. The fauna and flora of the Palearctic and Nearctic realms are similar in many respects and the two are sometimes combined as the *Holarctic* realm.

The *Neotropical* realm consists of South America, Central America, southern Mexico and the islands of the West Indies. Its distinctive fauna includes alpacas, llamas, prehensile-tailed monkeys, bloodsucking bats, sloths, tapirs, capybaras, anteaters and a host of bird species—toucans, puff birds, tinamous and others—found nowhere else in the world.

The part of Africa south of the Sahara, plus the island of Madagascar makes up the *Ethiopian* realm. The gorilla, chimpanzee, zebra, hippopotamus, giraffe, aardvark and many birds, reptiles and fishes live only in this region.

The *Oriental* realm includes India, Sri Lanka (formerly Ceylon), Indochina, southern China, the Malay peninsula and some of the islands of the East Indies. Outstanding among the mammals peculiar to it are the orang-utan, black panther, tiger, water buffalo, Indian elephant, gibbon and tarsier.

The sixth and last realm, called the *Australian*, includes Australia, New Zealand, New Guinea and other islands of the East Indies. The imaginary dividing line between the Oriental and Australian realms, known as *Wallace's Line*, separates Bali and Lombok, goes through the Straits of Macassar between Borneo and Celebes, and then passes east of the Philippines. Although the islands of Bali and Lombok are separated by a channel only 20 miles wide, their respective animals and plants are more dissimilar than are those of England and Japan, almost half a world apart. Native to the Australian realm are the duck-billed platypus, kangaroo, wombat, koala bear and other marsupials, and among its strange assortment of birds are the emu and cassowary (both flightless birds), the lyrebird and the cockatoo.

Why certain animals and plants are present in one region but are excluded from another in which they are well adapted to survive (and in which they flourish when introduced by humans) can be explained only by their evolutionary history.

QUESTIONS

1. List the various kinds of paleontologic evidence. Why is it that fossils are frequently difficult to find?
2. What are some of the factors that interfere with our obtaining a complete and unbiased picture of life in the past from a study of the fossil record?
3. Name some of the forms of life known to have existed in the Proterozoic era.
4. What were the most common animals of the Cambrian period? Where did they live and how were they adapted to survive?
5. Explain how an estimate of the age of a rock is made on the basis of the radioactive elements present.
6. When did vertebrates first appear? What did they look like and where did they live?
7. Compare and contrast the ostracoderms and the placoderms as to structure and mode of life.
8. What were the two major events of the Silurian period? What is the significance of the name Carboniferous?
9. What was the Appalachian Revolution? When did it occur and what were its effects?
10. What is meant by the term stem reptile? What relation do turtles bear to these animals?
11. What factors may have contributed to the extinction of the dinosaurs?
12. Compare the structure of the wings of a bat, a bird, and a pterosaur.
13. When did mammals originate? Compare monotremes and marsupials.
14. Briefly trace the evolution of the various phyla from the beginning of life to the present.
15. Give, briefly, the various types of evidence from living organisms which bear on the question of evolution.
16. What is the theory of recapitulation? How has it been modified and what is its significance in its present form?
17. Discuss the genetic basis for the theory of recapitulation.

18. Why are marsupials widespread in Australia and almost nonexistent elsewhere?
19. What is the explanation for the observation that the animals and plants of England and Japan are very similar despite the fact that they lie on nearly the opposite sides of the world?
20. Name some vestigial organs found in the human body. What functional organs are they the remains of?
21. Define the terms "range" and "center of origin."
22. What is Jordan's Rule?
23. Describe the methods used to determine evolutionary relationships from the nature of serum proteins. From the nature of tissue enzymes.
24. Discuss the proposition that the hierarchical scheme of animal and plant classification is evidence for organic evolution.
25. List the biogeographic realms. For each, list several unique animals and plants.

SUPPLEMENTARY READING

A. S. Romer's *Vertebrate Paleontology* is a standard source book and his *The Vertebrate Story* is a well written, nontechnical account of the evolution of vertebrates. *The Life of Vertebrates* by J. Z. Young is a fascinating account of the evolution and adaptations of vertebrates. The text is a skillful blend of morphologic and physiologic perspectives. *Evolution Emerging* by W. K. Gregory is a superbly illustrated source book on vertebrate evolution. E. E. Olson's *Vertebrate Paleozoology*, E. H. Colbert's *Evolution of the Vertebrates* and N. J. Berrill's *The Origin of Vertebrates* are well written elementary texts for the general student. Many interesting observations regarding fossil plants are contained in H. N. Andrews' *Ancient Plants and the World They Lived In.* A fascinating account of evolutionary studies of plants is given in *Plants, Life and Man* by Edgar Anderson. Sherwin Carlquist's *Island Life, A Natural History of the Islands of the World* is an informative and readable book about some of the curious evolutionary pathways followed by plants and animals on isolated islands. Two excellent books on the evolution of the invertebrates are R. R. Schrock and W. H. Twenhofel's *Principles of Invertebrate Paleontology* and *Invertebrate Fossils* by R. C. Moore, C. G. Lalicker and A. G. Fischer. The newer theories about the movements of the continents over the surface of the earth are discussed in N. Calder's *The Restless Earth.*

An Introduction to Comparative Biochemistry by E. B. Baldwin is a classic exposition of some of the biochemical similarities among animals that aid in defining evolutionary relationships. A brief essay on the same subject is *Of Molecules and Men* by F. Crick. Comparisons of the molecular structures of hemoglobins, cytochromes and certain enzymes in different species of organisms can be useful in tracing evolutionary relationships; the comparisons parallel those made previously of the gross morphology of animals and plants. Such comparisons are detailed in G. H. Juke's *Molecules and Evolution,* H. F. Blum's *Time's Arrow and Evolution,* M. E. Calvin's *Chemical Evolution* and in *Evolving Genes and Proteins* edited by V. Bryson and H. J. Vogel. A detailed discussion of some of the biochemical facts bearing on evolutionary theory is presented in *Biochemical Evolution* edited by Marcel Florkin. E. Florey's *Introduction to General and Comparative Animal Physiology* is an excellent text of general physiology that emphasizes the evolutionary aspects of the subject.

CHAPTER 33

THE PRIMATES AND HUMAN EVOLUTION

The line of evolution leading from the primitive, jawless fishes, the ostracoderms, to the mammals was traced in Chapter 32. The fossil records of horses, elephants, camels and many other mammals are quite complete, but those of the primates are still somewhat fragmentary. For the most part our primate ancestors lived in tropical forests, where animal remains are likely to undergo rapid decay before they can be fossilized. In the past decade many important new primate fossils have been found in North, East, and South Africa, in Europe, in India and in Southeast Asia. We can get some idea of what our ancestral primates might have looked like from the relatively primitive primates that have survived to the present.

The earliest placental mammals were small, tree-dwelling, insect-eating animals that appeared in the middle of the Mesozoic era, during the Jurassic period. Deposits in Montana from the end of the Cretaceous period have yielded remnants of what may be the earliest primate fossil, *Purgatorius,* together with bones of dinosaurs. The genus *Purgatorius,* a group of hedgehoglike animals, survived into the Paleocene epoch.

33–1 THE PRIMATES

The line of evolution leading to the primates appears to have begun with the tree shrews, squirrel-like animals with characteristics intermediate between those of the insectivores and the higher primates. Fossil tree shrews have been found in Paleocene deposits, and a few tree shrews, such as *Tupaia* (Fig. 33–1), still survive in the forests of Malaya and the Philippines. Tree shrews have a long snout and tail and opposable first toes. Their toes are tipped with claws instead of flat nails as those of the higher primates are. Primate evolution in general has been characterized by adaptation for arboreal life, and only in some of the larger apes and the human has this tendency been reversed in part.

Primates are rather unspecialized mammals; their adaptations for arboreal life include

Figure 33–1 *Tupaia*, a primitive tree shrew found in Malaya and the Philippines. (Courtesy of William Montagna, Oregon Regional Primate Research Center.)

prehensile hands and feet with opposable thumbs and great toes; digits tipped with flattened nails; long, flexible, mobile arms and legs; upright posture; well developed brains; and binocular vision. The eyes of primates are equipped with cones as well as rods and hence primates have color vision. Primates depend more on vision and less on olfaction to keep informed about their environment. The two suborders of primates are the Prosimii and the Anthropoidea.

The earliest primates, the prosimians, were abundant in the early Paleocene over much of the world, with the exception of Australia and South America. At that time North America was joined directly to Europe and many genera of mammals, including two or three of primates, were present in both the Old and New Worlds. The prosimians were distributed throughout the northern part of the Northern Hemisphere, and their evolution and adaptive radiation occurred during a time when the climate in that area was much milder than it is at present. Fossils of the early prosimian *Plesiadapis* have been found in Paleocene deposits in North America and France, and several million years later *Adapis* lived in Europe and the related genera *Smilodectes* and *Notharctus* lived in North America. Their skeletons show arboreal adaptations and similarities

to certain features of living lemurs and lorises. During the Eocene, North America and Europe became separated and the equator migrated southward; the primates in these two regions subsequently evolved along separate pathways. The lemurs, bush babies and tarsiers (Figs. 33–2 and 33–3) living today in Africa and Southeast Asia are prosimians that have descended from these early forms with relatively little change.

Lemurs (Fig. 33–2) live in the tropics of Africa and Asia and are especially abundant on the island of Madagascar, where they have managed to survive because no other primates were present to compete with them. These small, nocturnal, tree-dwelling animals, superficially rather like squirrels, have a fairly long muzzle, eyes directed more to the side than forward, and a well developed tail. Although some of the toes end in claws, the thumbs and big toes are covered by nails and are widely separated from the other digits.

Tarsius (Fig. 33–3 A) is about the size of a rat but has long hind legs and moves by hopping. It is characterized by enormous eyes directed forward, providing for stereoscopic and night vision, and by a relatively large brain. The muzzle is short, so that *Tarsius* has a face similar to that of a higher primate. The

Figure 33–2 Ring-tailed lemur, one of the most primitive of living primates. (Courtesy of William Montagna, Oregon Regional Primate Research Center.)

Figure 33–3 A group of prosimian primates. *A,* The tarsier, *Tarsius,* lives in the East Indies. Note the large, forward-directed eyes adapted for binocular night vision and the adhesive pads on the tips of the digits, which assist the animal in clinging to the branches of trees. *B,* The bush baby, *Galago,* is widely distributed in the forests and savannahs of Africa south of the Sahara. *C,* The slow loris, *Nycticebus,* is found in Sri Lanka and southern India south of the Tatti River. *D,* The potto, *Perodicticus,* lives in tropical forests of Africa from the Guinea coast and Congo River to the Rift Valley. (Courtesy of William Montagna, Oregon Regional Primate Research Center.)

upper lip is not attached to the gum as in the lower primates but is free as in monkeys, apes and humans. Its toes are long, slender and supplied with adhesive pads.

From the prosimians, three major groups evolved in the Eocene: the **Ceboidea,** or New World monkeys; the **Cercopithicoidea,** or Old World monkeys; and the **Hominoidea,** or great apes and humans. The ceboids, found in South and Central America, have widely separated nostrils that are directed forward and sideward, giving the nose a broad, flat appearance. The New World monkeys have one more bicuspid tooth on each side of both upper and lower jaws than do the Old World monkeys, apes and

humans. The ceboids represent a group of primates isolated in South America during the Tertiary that underwent an evolution independent of the other monkeys. Among the living ceboids are the howler monkeys; the marmoset, or squirrel monkey; the capuchin monkey—the organ grinder's companion; and the spider monkey (Fig. 33–4). All these are found in tropical forest environments and show a variety of adaptations to the different environments, some of which parallel those of the Old World monkeys. Most of the ceboids have a well developed prehensile tail that they use as a fifth hand for grasping objects and hanging from trees, and the group is sometimes called

Figure 33-4 Spider monkey, a New World monkey with a strong prehensile tail, used in swinging from tree to tree. (Courtesy of the San Diego Zoo.)

the prehensile-tailed monkeys. The prehensile tail actually has skin ridge patterns on the tip, like our fingerprints, to improve the grasp.

The large group of Old World monkeys includes the macaque (Fig. 33–5), guenon, mandrill, mangabey, baboon, langur, proboscis monkey and many others. These cercopithicoids are characterized by much narrower noses, with the nostrils closer together and directed downward. All of them tend to sit upright and all have buttocks with bare, hardened sitting pads, called *ischial callosities*,

which are frequently a brilliant red or blue. The mandrills and baboons have taken to living on the ground instead of in trees, but they walk on all fours and have an elongated snout and large canine teeth. Baboons are intelligent animals that travel in troops and cooperate in obtaining food and protecting their females and young. The cercopithicoid tail is not prehensile and is used in balancing.

In addition to human beings, the hominoids include a variety of fossil apes and "ape men" plus four genera of living apes—the gibbon, orang-utan, chimpanzee and gorilla, all members of the family Pongidae. The great apes have rudimentary tails or no tail at all, arms that are longer than their legs, a semierect posture, opposable thumbs and great toes, and chests that are broad like that of the human rather than thin and deep like those of most mammals. Their brains are larger than those of the lower primates and more like that of the human in pattern and relative size of the parts. They range in size from the gibbon, which is 90 cm. tall, to the gorilla, which is 180 cm. tall. Some large gorillas weigh as much as 280 kg.

Gibbons (Fig. 33–6), found in Southeast Asia, are the smallest of the apes, with slim bodies and long spidery arms that reach the ground when the animal walks erect. The gibbon is able to walk on its hind legs and does so when on the ground, holding its long arms up and out for balance. Its usual mode of locomotion, swinging through the trees from limb to limb (brachiating), allows the animal to clear as much as 13 meters at a time. The spectacular aerial acrobatics of the gibbon (*Hylobates*) require great ability, coordination, good eyesight and the ability to make rapid judgments of distance and possible landing sites. Gibbons are generally herbivorous, but occasionally eat insects, a bird's egg, or even a bird. A caged

Figure 33–6 An anthropoid, the white-banded gibbon. These apes use their long arms to swing from tree to tree with great agility. (Courtesy of the San Diego Zoo.)

gibbon has been seen to swing from one arm and pluck a bird out of the air in midflight! Gibbons do not live in troops like baboons but in pairs, each couple living with its young offspring and defending its territory.

The orang-utans (Fig. 33–7), genus *Pongo*, natives of the rain forests in Borneo and Sumatra, are bulky, powerful animals covered with long, reddish-brown hair. Although short-legged and scarcely 150 cm. tall, orangs may weigh as much as 175 kg. Orangs have enormously long arms, with a span of 2.5 meters, and long, slender hands and feet. They are successful arboreal animals, but because of their weight they must move more deliberately in the trees than the gibbons do. Orangs are shy, solitary animals that eat fruit and leaves and build nests in trees. The orang-utan is now an endangered species, in risk of becoming extinct.

Chimpanzees (*Pan*) (Fig. 33–8) and gorillas (*Gorilla*) (Fig. 33–9) both live in tropical Africa and have many characteristics in common. An adult chimpanzee weighs about 50 kg. and is between 150 and 165 cm. tall. Like the orang-utans, chimpanzees build nests in trees for

Figure 33–5 A female Japanese macaque *(Macaca fuscata)*, one of the Old World monkeys. (Courtesy of William Montagna, Oregon Regional Primate Research Center.)

sleeping at night and for noonday siestas. Although primarily brachiating tree-dwellers, they are quite at home on the ground. They have longer legs and shorter arms than the orang. Both chimpanzees and gorillas are "knuckle-walkers," walking with the hand doubled to support the weight on the middle joint of the fingers. The lips of the chimpanzee are free, not bound to the gum, and they can produce a wide range of facial expressions by moving the fine muscles of the face. Although they can make a variety of vocal sounds they cannot be taught to talk. Psychological studies of chimpanzees and gorillas have shown that they are curious, perceptive, able to reason and have strong emotions and social instincts. Gorillas are less docile, imitative and suggestible than chimpanzees.

Gorillas (Fig. 33–9) are not only the largest of the primates; they are, pound for pound, the strongest—several times stronger than the human. The gorilla's legs are relatively short, the arms are more human in their proportions

Figure 33–8 Pigmy chimpanzee. (Courtesy of William Montagna. Oregon Regional Primate Research Center.)

Figure 33–7 Adult male orang-utan. (Courtesy of William Montagna. Oregon Regional Primate Research Center.)

than those of the other apes, and the hands are relatively short and wide—quite like human hands. The massive head has large bony crests on top of the skull to which are attached the neck and jaw muscles. Gorillas are ground-dwellers, although they occasionally build sleeping nests in low trees, and have feet adapted for walking rather than for swinging through the trees. Like humans, gorillas walk on the soles of the feet with the toes extended rather than curled under in the fashion of the other apes. Gorillas normally walk on all fours but may rear up on their hind legs when attacking. They are ordinarily herbivorous, prefering foods such as bananas, carrots and nuts, but they will eat meat and insects.

The human is more similar to the chimpanzee and gorilla than to any other primate but differs in enough characteristics to warrant being placed in a separate family, the Hominidae. The differences between the other great apes and ourselves are rather small differences in the proportion of parts, and these correlate

Figure 33–9 A gorilla, the largest of the anthropoid apes. (Courtesy of William Montagna, Oregon Regional Primate Research Center.)

with our adaptation for a terrestrial rather than an arboreal life. Almost every bone, muscle, internal organ and blood vessel of the ape is repeated in the human. The close relationship of humans to chimpanzees and gorillas is supported by the close similarities of their karyotypes, but there are some respects in which the great apes are more specialized than humans and in which the human more closely resembles the gibbon or Old World monkeys. Some of the characters that distinguish the human are: (1) Our brain is larger, being two and one-half to three times as large as the gorilla's. (2) The human nose has a prominent bridge and peculiar elongated tip. (3) Our upper lip has a median furrow and our lips are rolled outward, revealing the mucous membranes. (4) We have a jutting chin and apes have none. (5) Our great toe is not opposable but is in line with the others. (6) The human foot is adapted for bearing weight by being arched, both lengthwise and crosswise. (7) We are relatively hairless.

(8) Our canine teeth project very little, if at all, beyond the line of the other teeth. (9) We have an erect posture. (10) Our legs are longer than our arms.

Antigen-antibody tests of similarities in serum proteins show that, of all the apes and monkeys, gorillas and chimpanzees have serum proteins most nearly like those of the human. The amino acid sequence in the chimpanzee's hemoglobin is identical to that of the human; that of the gorilla and rhesus monkey differ from the human's in 2 and 15 amino acids, respectively.

No single ape resembles man in all traits more than do the rest. The gorilla, for instance, has hands, feet and a pelvis more like the human's than are those of any other ape (Fig. 33–10), but the chimpanzee's skull and skin color are most like the human being's. The orang-utan is the only ape to have exactly the same number of ribs that we have and it also has our high forehead. The gibbon most closely resembles us in the relative length of its legs, its posture and its gait on the ground, but it is covered with dense hair. With respect to any structure or proportion of parts, the difference between man, gorilla and chimpanzee is less than the difference between any of these and monkeys.

33–2 FOSSIL PRIMATES

During the late Cretaceous there were tree-living insectivores that resembled present-day tree shrews and that are believed to have given rise to the primates. The earliest fossils that are clearly primates are lemuroids and tarsioids found in Paleocene rocks in both North America and Europe. Since that time the lemurs have apparently evolved separately and have not given rise to other forms. The New and Old World monkeys evolved separately from the prosimians. There is some debate as to whether New World monkeys, which are found principally in South America, came by raft from Africa when Africa and South America were separated by only a rather wide strait, or whether they migrated from North America via a series of islands that separated the two continents. The New World monkeys have had a separate evolutionary history since the Eocene, when Europe and North America parted com-

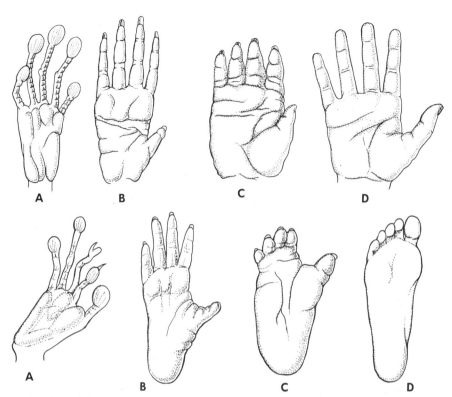

Figure 33–10 The hands and feet of tarsier *(A)*, orang-utan *(B)*, gorilla *(C)* and human *(D)*. Each of these shows special adaptations for some function. The hands and feet of the tarsier, with enlarged skin pads at the tip, are adapted for grasping branches. The fingers of the orang-utan are very long, adapted for brachiating, and the thumb is reduced in size. The fingers of the gorilla are shortened, adapted for walking. The human's fingers are less specialized and, with a fully opposable thumb, are more generally useful in handling objects. The feet of tarsier, orang-utan and gorilla are handlike, with opposable first toes, but the human foot shows marked adaptation for walking. The first toe is not opposable and all the toes are short. All the joints at the base of the toes are in line and form the ball of the foot. The presence of this ball permits humans to balance on their feet and to walk erect. (From Curtis, Helena: *Biology*. New York, Worth Publishers, 1968, p. 797.)

pany, and represent side branches rather than segments of the main evolutionary trunk leading to man.

The oldest fossils of Old World monkeys are *Parapithecus* and several related genera that have been found in lower Oligocene rocks in an area of the Egyptian desert known as the Fayum, 60 miles southwest of Cairo. *Parapithecus* was smaller than any of the modern monkeys or apes and was near the stem leading to humans (it had the same dental formula). *Parapithecus* may represent a common ancestor of today's Old World monkeys, anthropoid apes and man. From that point the evolution of the modern Old World monkeys diverged from that of the hominoids. Another Oligocene fossil from the Fayum, *Aelopithecus,* is believed to foreshadow the later fossil *Pliopithecus* and the present-day gibbon. The fossil *Aelopithecus,* with its deep

gibbonlike mandibular symphysis, supports the hypothesis that the gibbons become distinct from the other great apes (orangs, chimpanzees and gorillas) long before they evolved separately from the humans.

Another important find from the Fayum is the nearly complete skull of *Aegyptopithecus,* which dates from about 29 million years ago and is very much like *Dryopithecus africanus* found in East Africa in the Miocene. *Aegyptopithecus* may be the most likely of the known fossil forms to be a common ancestor of the great apes and humans. With the discovery of more Oligocene fossils it is now clear that there was considerable variety among the forms living at that time and that different ones seemed to be leading toward different later fossil or present-day forms. However, whether the four separate lines leading to the human, the African great apes, the gibbons, and the Old

World monkeys were distinct in Oligocene times is still under debate.

The line leading to gibbons may have become distinct from that of monkeys by the Miocene, for fossils of *Pliopithecus* with gibbonlike features have been found in France, Czechoslovakia and East Africa. Most other fossils of higher primates from the Tertiary are grouped into a single genus, *Dryopithecus* (Fig. 33–11), first known from teeth and apelike bits of jawbones from Miocene and Pliocene deposits in Europe and the Sewalik Hills in India. *Dryopithecus* is now known to have been a wide-ranging genus extending from western Europe to China. Several East African Miocene fossils formerly called proconsul are now included in this same genus. The teeth range from ones smaller than those of a chimpanzee to ones as large as those of a gorilla. Several fine specimens of *Dryopithecus*, including a nearly complete skull, have been discovered by Mary and Louis Leakey in Kenya, East Africa. There were many species of *Dryopithecus*, one of which, *Dryopithecus indicus*, is thought to be in the ancestry of *Gigantopithecus* (p. 794). *Dryopithecus afri-*

canus is thought to be ancestral to the chimpanzee and *Dryopithecus major* may be ancestral to the gorilla. Another species, *Dryopithecus sivalensis*, is believed to have resembled the present-day orang-utan. Thus by the middle of the Miocene the evolutionary lines leading to the various types of modern anthropoids were distinct.

Early Pliocene deposits from northern Italian bogs have yielded fossil remains of *Oreopithecus*, an apelike animal with humanlike teeth and jaws. Other fossil apes (*Ramapithecus*) with humanlike teeth and jaws have been found in early Pliocene deposits in northwestern India and Kenya. The face of *Ramapithecus*, reconstructed from bits of jawbone, appears to be similar to that of *Australopithecus*, but *Ramapithecus* lived about 10 million years earlier. Because of the very short, deep face of *Ramapithecus* anthropologists have inferred that the animal could not have relied on its teeth as weapons as much as *Dryopithecus* and most other primates other than humans do and, therefore, must have used its hands for hunting and defense. This has the further implication that the animal

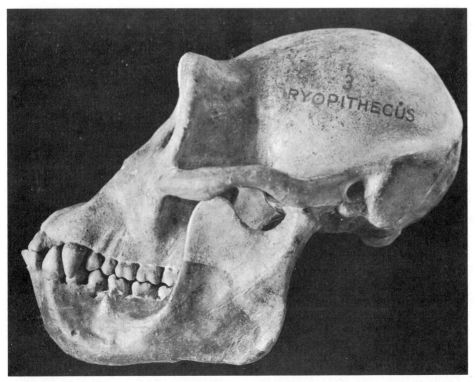

Figure 33–11   Restored skull of *Dryopithecus,* the fossil "oak ape," possibly the ancestor common to gorillas, chimpanzees and humans. (Courtesy of the American Museum of Natural History, New York.)

was specialized for bipedal locomotion. Hands used for manipulation rather than walking are one of the most characteristic human anatomical adaptations and one that probably was a prerequisite for the development and use of tools. From this line of argument it has been postulated that *Ramapithecus* is the actual ancestor of *Homo sapiens* that lived in the late Miocene. Similar specimens from about 14 million years ago have been found by Leakey in Kenya at Fort Turner. He originally named these *Kenyapithecus*, but these specimens are now assigned to *Ramapithecus*. Other *Ramapithecus* fossils have been found in Hungary. Thus, by the beginning of the Pliocene there was a widespread primate genus with small canine teeth, as one would expect to find in our ancestors. Other anthropologists do not agree that this evidence is sufficient to identify *Ramapithecus* as a hominid, and the question cannot be settled until limb bones are discovered and described so that evidence of locomotion can be added to that provided by the jawbone.

33–3   THE AUSTRALOPITHECINES

A series of fossil anthropoids, named *Australopithecus* by their discoverer, R. A. Dart, have been found in South and East Africa in Taung cave deposits nearly 2 million years old. The first of these fossils, the skull of a baby, was found in 1925 and the specimen was named *Australopithecus*. Subsequently, Dart and his associate Broom found adult skulls and parts of skeletons. Although these were originally given separate names they are now believed to be animals very closely related to, if not identical with, the original *Australopithecus*. Dart noted in these specimens resemblances to human skulls not shared by chimpanzees or gorillas and believed that *Australopithecus* marks the achievement of hominid status in the line of evolution leading to humans. Subsequently Dart and Broom found similar specimens in four other places in South Africa, including Sterkfontein, Kromdraai and Swartkrans. The cave at Swartkrans has produced more than 500 fossil remains representing some 60 individuals, together with some stones that clearly were tools used by these animals.

From the various sites in South Africa nearly 1500 hominid specimens have been obtained. The specimens from Taung and Makapan are smaller and smoother than those from Swartkrans, and although some anthropologists believe that there are two or more species, a "robust" species and a "gracile" species, others question whether these are truly distinct species or simply represent differences between local populations at different times or between the sexes. Far to the north in Tanzania in a dry ravine called Olduvai Gorge, Louis and Mary Leakey found a nearly complete skull of *Australopithecus* in a layer that contained a number of stone tools. This specimen was originally termed *Zinjanthropus*, but it is now recognized as one of the robust kinds of *Australopithecus*. Other specimens have been found in Olduvai Gorge and in other adjacent regions in Kenya and southern Ethiopia. The Leakey's son found other fossils nearby, near the shore of Lake Rudolf (recently renamed Lake Turkana). These fossils are also assigned to the robust form of *Australopithecus*. In addition, many other bones have been found, so that the reconstruction can be quite complete. At this same site, volcanic material has been found that has been dated as from as long as 2.6 million years ago.

Australopithecus apparently came down out of the trees and lived in caves. It hunted animals, used wooden clubs and crude stone tools (called Oldowan choppers, these appeared about 2.6 million years ago), and may have learned to use fire. The structure of the pelvis and leg bones and the location of the foramen magnum in the skull suggest that these creatures had a fairly erect posture, and certain features of the skeleton, particularly the pelvis and the curve of the lumbar vertebrae, are similar to those of humans and typical of animals with bipedal locomotion. The face of *Australopithecus* is short but large with essentially no forehead, the front teeth are small and the skull capacity of 650 ml. is small by human standards but large for an ape. The cheekbone and jaw hinge are very similar to those of humans; the molars and the small canine and front teeth are also humanlike and suggest that the animal was a herbivore. Thus, it appears that the first fossil members of the Hominidae appeared some 5 million years ago in the Pliocene and early Pleistocene and that at least the early members of the genus *Australopithecus* were ancestors of our species.

33-4 FOSSIL HOMINIDS

The hominid stock appears to have diverged from the line of evolution leading to the great apes sometime after the Miocene. The remains of a number of creatures with characteristics intermediate between those of the fossil ape and the living human have been found in Pliocene and Pleistocene deposits in widely scattered parts of Africa, Asia and Europe. Some half dozen specimens were found in 1960 at Olduvai Gorge in East Africa by Louis Leakey. He named them *Homo habilis,* for he believed that they were more advanced than the australopithecines. His reconstruction of *Homo habilis* suggested a creature about 125 cm. tall with small humanlike teeth and feet that were more human- than apelike.

The fossil remains of the hominids are mostly pieces of skull, jawbone and teeth, with a few other skeletal parts such as femurs, pelvis, and ribs. It is understandable that the fossil record of man's immediate ancestors is incomplete, for these creatures were too intelligent to be caught in quicksands or tar pits, and being primarily forest dwellers their dead bodies must have been quickly and completely devoured by other animals. The characteristics distinguishing humans from the apes did not appear simultaneously in a single form; these ape men were essentially human in some respects but apelike in others. Whether they were apes or humans is a matter of semantics, but they were large-brained anthropoids that walked erect, had well formed hands and made tools of stone and bone. We have a fairly clear idea of what these individuals looked like from their fossil remains and we also know quite a bit about how they lived from the tools, weapons, ornaments and other cultural remains that have been found.

Other remains found in 1953 at Swartkrans were identified as truly human and given the name of *Telanthropus.* Subsequently, *Telanthropus* has been reclassified as *Homo erectus,* and most anthropologists consider *Homo habilis* also to be a member of *Homo erectus.*

One of the earliest hominids was the Java man, whose remains were found in 1891 on the banks of the Solo River in eastern Java in Pleistocene deposits some 830,000 years old. Their discoverer, Eugene Dubois, originally named these remains *Pithecanthropus erectus,* but the name has been changed to *Homo erectus* now that many more hominid fossils have been found and the degree of difference from our species, *Homo sapiens,* is such that they are better considered as two species of the same genus rather than two different genera. The skull capacity of these specimens ranges from 700 to 1100 ml., well above that usually found in *Australopithecus* (about 650 ml.) but less than modern man's 1200 to 1500 ml.

Reconstructions from the skeletal parts that have been found indicate that an adult Java man was about 170 cm. tall, weighed perhaps 70 kg., and walked erect with a stride similar to modern man's. His face was projecting and chinless and his jaws were massive and equipped with a set of huge teeth, although the canine teeth were not enlarged tusks as in the apes. He had a broad, low-bridged nose and a heavy bony ridge over the eyes. He apparently learned to use and make tools, for the rock strata bearing his remains also contain primitive stone implements. These individuals probably traveled in small family groups, living in caves and hunting in the forest. Since 1891, excavations have turned up several more fossil *Homo erectus* specimens in Java, as well as the remains of a larger and apparently earlier type called *Meganthropus. Meganthropus,* like *Telanthropus,* seems to be intermediate in some respects between *Australopithecus* and *Homo erectus.*

Investigations during the early 1920's of limestone caves in Choukoutien near Peking revealed many animal fossils and among them two teeth that had belonged to a primitive ape man of the Middle Pleistocene, some 300,000 to 700,000 years ago. Their discoverer, Davidson Black, named them *Sinanthropus pekinensis,* but the discovery of subsequent specimens showed that this was only a variant of *Homo erectus,* like the specimens discovered in Java. Further excavations of the cave revealed parts of some 40 individuals, male and female, young and old, of this same species. Peking man (Fig. 33–12) had a skull much like that of Java man but with a somewhat greater brain capacity, averaging 1075 ml. The skull

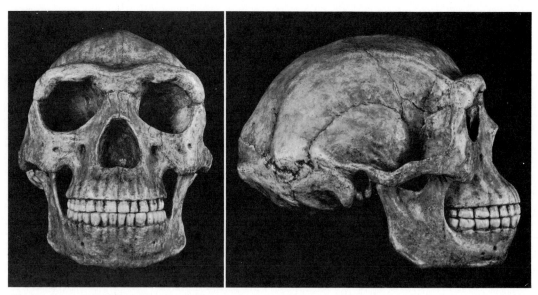

Figure 33–12 Front and side views of a reconstructed skull of Peking man, *Homo erectus*. Note the massive bony ridges over the eyes, the low, retreating forehead, the protruding jaws and the absence of a chin. (Courtesy of the American Museum of Natural History, New York.)

had heavy bony ridges over the eyes, a low, slanting forehead, a massive, chinless, rather apelike jaw, and a nose that was broad and flat. The difference in size between males and females may have been greater in Peking man than in modern man, for the remains fall into two rather distinct groups, one much larger than the other. Franz Weidenreich, the anthropologist who has studied Peking man most intensively, found that Java and Peking man are identical in 57 out of 74 features of the skull and that there are clear differences in only 4 characteristics, one of which is the aforementioned difference in size of males and females. This variety of *Homo erectus* was clearly using stone tools more than half a million years ago.

A curious anthropologic story is that of *Gigantopithecus*, the "Hong Kong drugstore giant." In the late 1930's, Von Koenigswald bought a number of fossil teeth in a Chinese apothecary shop, some of which he could identify as those of an orang-utan. Three, however, were larger than those of any known primate. Their discoverer believed them to be the remains of a giant ape man, which he named *Gigantopithecus*, perhaps related to *Meganthropus* and Java and Peking man. In addition to the specimens of *Homo erectus* from China, Java and Africa, some specimens found in Hungary, France and Greece may be assigned to *Homo erectus* or perhaps may be

transitional forms between *Homo erectus* and *Homo sapiens*.

33–5 THE NEANDERTHALOIDS

The very first human fossil was found in 1856 in a cave in the Neander Valley near Düsseldorf and was given the name *Homo neanderthalensis*. The skull excited a lively controversy at first—some scientists correctly guessing it to be the remains of a primitive man and others guessing it to be the skull of a congenital idiot. One surmised it was the skull of a Russian soldier killed in the Napoleonic Wars. Since that time many similar skulls have turned up in widely separated parts of Europe, Asia Minor, North Africa, Siberia and the islands of the Mediterranean. These remains are always associated with a particular Stone Age culture known as Mousterian (named after the Moustier Cave on the bank of the Vézère River in France). Neanderthals lived in Europe for thousands of years during and after the third and final interglacial period ranging from 35,000 to 70,000 years ago. Neanderthal man was between 150 and 165 cm. tall and powerfully built. He walked upright. His skull was large and massive (Fig. 33–13), with a thick, bony ridge over the eyes and a receding forehead. His nose was broad and short and he had

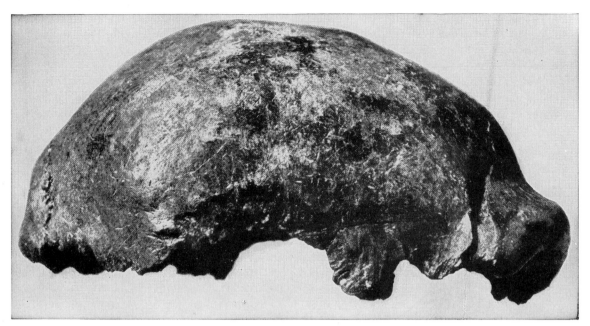

Figure 33-13 Skull cap of Neanderthal man. Note the heavy eyebrow ridges to the right and the extreme thickness of the bone. (Courtesy of the American Museum of Natural History, New York.)

almost no chin at all. Neanderthal man's brain was as large or larger than that of modern man, with some specimens having skull capacities as great as 1750 ml. Other Neanderthal cranial capacities ranged down to about 1200 ml. The proportions of the parts of the brain estimated from casts indicate that he was probably quite similar to modern man in general intelligence.

The Neanderthals were a society of small family groups who lived primarily in caves, used fire, made flint weapons, hunted a variety of game ranging in size up to the mammoth and rhinoceros, and buried their dead reverently with food and ornaments. As more of the weapons, tools, utensils and ornaments made by Neanderthal man have been found, we are coming to realize that he was very capable and intelligent and could, in all probability, handle complex abstract ideas.

Skeletal remains have been found in many parts of the world of humanlike individuals with many characteristics in common with the classic Neanderthals. These, termed Neanderthaloids, include individuals from Mount Carmel in Israel, the Shanidar Valley of northern Iraq, Morocco, Libya, Rhodesia, and South Africa. Most of these finds have been dated at about 40,000 years ago. As yet there is no consensus among anthropologists as to whether these should be considered a separate species,

Homo neanderthalensis; a separate subspecies, *Homo sapiens neanderthalensis;* or simply variants of one species of *Homo sapiens.* An almost complete skeleton of a Neanderthal was found at La Chapelle-Aux-Saints, France.

Reconstructions of the Neanderthals usually depict them as standing stooped, with the knees bent forward. This supposition is based on the position of the foramen magnum, the angle at the base of the skull and the rugged limb bones with enlarged joints and bent shafts. The bowed shaft of the bones in the thigh and leg and the direction of the joints in the leg and feet have suggested to some anthropologists that Neanderthals had a somewhat apelike, bent-knee gait. Other anthropologists suggest that bowing of the leg bones may have been caused by rickets. A reexamination of the leg bones of the individual from La Chapelle-Aux-Saints led to the conclusion that this individual suffered from a severe disease of the joints. Six fossil thigh bones from Java and limb bones from other Neanderthaloid specimens are essentially modern in type. Although several features of the Neanderthal skeleton are quite distinctive, none contradicts the hypothesis that Neanderthal had a typically upright human posture. Many skulls of Neanderthal man, as well as those from the caves in China and Java, have been found without other parts of the skeleton and with the bottom of the skull broken open. This suggests that the brains were removed for cannibalism, perhaps in some sort of ritual.

Another primitive skull, to which the name Rhodesian man was given, was found in 1921 in a quarry at Broken Hill, Rhodesia. The skull, in almost perfect condition, has an extremely large eyebrow ridge and a receding forehead like a gorilla's but a cranial capacity of about 1300 ml. The teeth are large and set in a large jaw, but they are definitely human rather than apelike. They are badly decayed, a condition found rarely in nonhuman primates and other animals. This skull has some characteristics of *Homo erectus*, and in other respects it is Neanderthaloid. Other skeletal remains from South Africa and the Transvaal have similar characteristics and have been dated as about 55,000 years old.

On the banks of the Solo River in Java, near the site where *Pithecanthropus* was found, have been found fossils of a second human type which used both stone and bone implements. Eleven skulls of Solo man have been discovered, all with the bases smashed in, suggesting that Solo man, like other Neanderthaloids and Peking man, considered brains a delicacy. Solo man resembles the Neanderthals in having a heavy ridge over his eyes and a receding forehead with a brain capacity of about 1300 ml. These individuals are not modern in type, but are intermediate between *Homo erectus* and the Neanderthals in many respects.

33–6 THE ORIGIN OF *HOMO SAPIENS*

The species *Homo sapiens* includes not only all the living races of man, but also some extinct ones such as the Cro-Magnon. Remains of individuals that were clearly similar to modern man, *Homo sapiens sapiens*, go back at least 30,000 years. There are some bits of evidence that suggest to some anthropologists that modern man has been on the earth for a much longer time and that he was a contemporary of, and coexisted with, Neanderthal man. In this view Neanderthal man was a side branch that became extinct without contributing to the later evolution of modern man. This argument is based on a relatively few fossil skulls.

The Galley Hill fossil, found in 1888 in the Thames valley below London, is a nearly complete human skeleton, and the gravel pit containing it is of the middle Pleistocene, about half a million years ago. Galley Hill man was short, 158 cm. in height, and stocky with no feature more apelike than some that can be found in living human races. His skull capacity was about 1400 ml. and the cast of his brain shows an essentially modern development of the various brain regions. The skull bones are quite thick, but the eyebrow ridges are not excessively large. The age of these remains is in dispute because the skeleton was removed before a qualified geologist could attest to the antiquity of the deposit.

At Swanscombe, England, in 1935, several pieces of skull were found deep in Middle Pleistocene deposits. Associated with the bones were hand axes and flakes of the Middle Acheulian type, together with bones of animals that lived during the long second interglacial period. Fluorine analysis of the human bones indicates a degree of fossilization comparable to that of the middle Pleistocene animal bones from the same deposit. These then constitute claims for an antiquity greater than that of any of the Neanderthaloid specimens known. Unfortunately the Swanscombe specimen consists only of a small fragment of the skull vault, pieces of parietal bone and a part of the occiput. The skull is distinguished by its unusual thickness and by a large area for the insertion of neck muscles. One expert has pointed out similarities between this and the Broken Hill skull of Rhodesian man, who was Neanderthaloid. An intensive search of the area some years later resulted in the finding of further bits of the skull that match those found earlier like pieces of a jigsaw puzzle. The opinion of Sir Arthur Keats and Ernest A. Hooton that the Swanscombe and Galley Hill fossils are members of the species *Homo sapiens* that lived in the Middle Pleistocene is not shared by all anthropologists.

Parts of two skulls were excavated at Fontéchevade in France in 1947, together with some stone flake tools of the Lower Paleolithic and animal fossils of forms associated with a warm climate consistent with an interglacial period. The pieces of skull are thick and broad, resembling the Swanscombe skull.

What caused the relatively sudden disappearance of the Neanderthals some 30,000 years ago and their replacement by modern *Homo sapiens sapiens* is still an unanswered question. One of the more widely held theories suggests that modern man brought with him some disease to which he had developed immunity but the Neanderthals had not. Some of these fossils show features that are intermediate between Neanderthal and modern skeletons, and others show signs of disease.

One of the first fossils discovered came from the Cro-Magnon rock shelters on the banks of the Vézère River in the Dordogne Valley of south central France. The remains of at least five individuals were found in 1868. The best preserved skull was that of an old man who had lost his teeth but had the facial features of a European of today. The various remains were at first thought to be representative of a single race, and they do share characteristics such as a long, massive skull without eyebrow ridges, a prominent chin and a high forehead, together with a rather large brain capacity (as much as 1800 ml.) (Fig. 33–14). However, further studies have shown considerable heterogeneity and suggest that the fossils did not belong to a single race. They may have been contemporaries of the Neanderthals and may have hastened their extinction.

Many other similar specimens have been found, together with characteristic tools made of bone and pressure-flaked stone of the Mag-

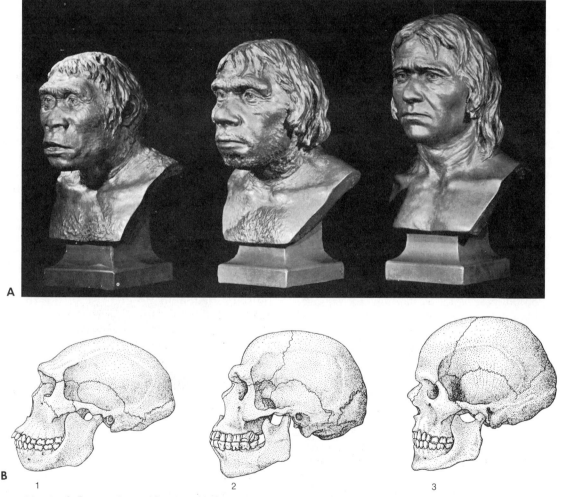

Figure 33–14 *A*, Restorations of Dr. J. H. McGregor of what prehistoric men probably looked like. From left to right, the Java man *(Homo erectus)*, Neanderthal man *(Homo sapiens neanderthalensis)* and Cro-Magnon man *(Homo sapiens sapiens)*. (Courtesy of Dr. J. H. McGregor and the American Museum of Natural History, New York.) *B*, Lateral views of the skull and lower jaw of *(1)* Java man, *(2)* Neanderthal man and *(3)* Cro-Magnon man. (Reprinted from *Evolution* by Theodore H. Eaton, Jr. By permission of W. W. Norton & Company, Inc., Copyright © 1970 by W. W. Norton & Company, Inc.)

dalenian culture period extending from 15,000 to 10,000 B.C. The men and women of the so-called Upper Paleolithic culture period in Europe were all *Homo sapiens sapiens* and not Neanderthals. Those most like the old Cro-Magnon man are usually grouped together as Cro-Magnon man, and those that diverge in some respects are given other designations. The finding of many more specimens has shown that there was not a sharp break between Neanderthals and Cro-Magnons in Europe but rather a transition.

The skulls of all these individuals, Upper Paleolithic men and women from diverse regions, lack marked racial features but have a common set of characters—a big jaw, narrow head and large but divided brow ridges. It is of interest that the physical characteristics of the people living in any part of the world today cannot be matched exactly with those of any group of their predecessors who lived in the same region 10,000 or more years ago, adding to the ample evidence that large migrations took place in prehistoric times as well as in historic times. The newcomers might be peaceful immigrants or conquering warriors, but they almost always mingled with and bred with the indigenous people rather than simply killing them off and superseding them. Furthermore, evolutionary changes within groups have continued in the last 10,000 years and continue today.

The North American continent was first peopled with humans some 30,000 years ago. The earliest authenticated date for human skeletal material in the New World is a bit of skull bone found in California and shown to be more than 26,000 years old by carbon 14 radioactivity. Skeletal remains some 12,000 years old have been found in Peru, others 11,000 years old in Mexico, and another specimen about 10,000 years old was found in Arlington Springs, California. Only the shallow Bering Strait separates Alaska from Siberia and the level of the sea was lowered during glacial periods, so that America and Asia were repeatedly connected by land bridges. From radiocarbon dating of specimens that indicate the height of the sea at various times in the past there is evidence that the last land bridge from Asia to America existed from about 25,000 years ago until about 11,000 years ago. The human skeletal remains found in North America most probably represent individuals whose ancestors arrived here by that land bridge.

Some of the skeletons of the Upper Pleistocene age associated with Mousterian stone implements found at Mount Carmel show a curious mixture of Neanderthal characteristics and others like those of Cro-Magnons. These and other remains suggest to many anthropologists that Neanderthals did not become extinct in the strict sense of the term but were absorbed by interbreeding with the various races of *Homo sapiens.*

33-7 HUMAN CHARACTERISTICS

In the course of his evolution from ape men (Fig. 33-15), modern man has not increased greatly in height and his frame has become less massive. We differ from contemporary apes in being well adapted to a bipedal gait and using our hands not for locomotion but for the manufacture and manipulation of tools. We have an omnivorous diet that includes both vegetable and animal materials. We stand completely erect and our heads are balanced on a relatively slender neck instead of jutting forward from the shoulder and being held in place by a set of massive neck muscles as in the apes. Many of our features are correlated with our upright stance. We have a lumbar curve in the vertebral column that places our center of gravity over the pelvis and hind legs. Our ilium (one of the bones of the pelvis) is broad and flaring and provides a large surface for the attachment of the gluteal and other muscles that hold us erect (Fig. 33-16). Our legs are longer and stronger than our arms. The distal end of our femur is brought close to the midline, which gives us a knock-kneed appearance but places our foot under the projection of the body's center of gravity and enables us to balance on one foot when the other is off the ground (Fig. 33-17). Our foot has lost the grasping ability of our more primitive ancestors, for all our toes are short and parallel with each other (Fig. 33-10). We have a large heel-bone and our carpals and metacarpals form strong supporting arches. The first toe, which is closest to the line of support, is enlarged as the great toe. Our head is balanced on the top of the vertebral column and the foramen mag-

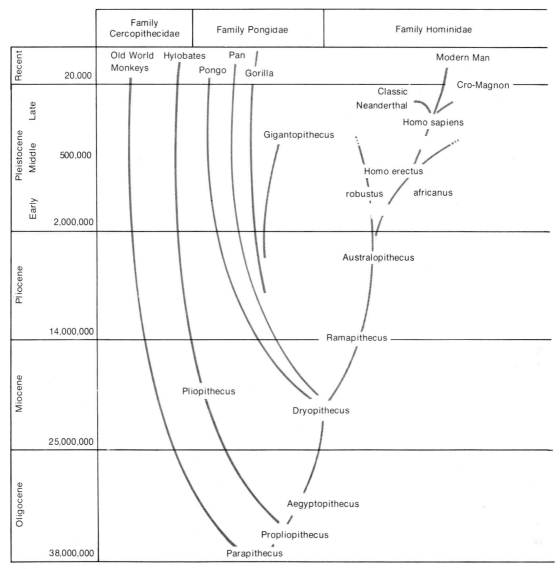

Figure 33-15 A phylogeny of catarrhine primates. The families refer to recent species. (From Villee, C. A., Walker, W. F., Jr., and Barnes, R. D.: *General Zoology*, 4th Ed. Philadelphia, W. B. Saunders Co., 1973.)

num is far under the skull. The nuchal area on the back of the skull to which neck muscles are attached has been much reduced and the mastoid process is enlarged. Some investigators have postulated that the loss of body hair was an adaptation to promote heat loss in our primitive ancestors who were chasing game in the savannah.

Since our ancestors no longer used their hands for locomotion they could be used in other ways. Our ancestors retained and improved upon the grasping ability of the primitive primate hand as they began to use bones and stones as clubs and to make stone tools. Our thumb is longer and the metacarpal portion of our hand is shorter than in apes (Fig. 33-10).

A careful search of the homesites of our primitive ancestors and a catalogue of the bones present indicate that they ate rodents, lizards and possibly insects, as well as plant material. Later our ancestors used fire and softened the food by cooking it. The human teeth and jaws are less massive than those of the apes and our face does not protrude as much. Our tooth row is more rounded and the canines

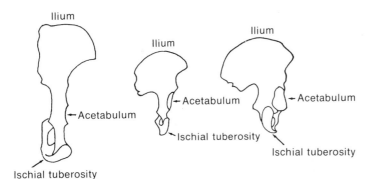

Ilium

Ilium

Ilium

← Acetabulum

← Acetabulum

← Acetabulum

↙ Ischial tuberosity

Ischial tuberosity ↘

Ischial tuberosity

CHIMPANZEE   AUSTRALOPITHECUS   MODERN MAN

Figure 33–16 Lateral views of the pelvis of a chimpanzee, *Australopithecus,* and modern man. (From Coon, C. S.: *The Origin of Races.* Copyright © 1962 by Carleton S. Coon. Reprinted by permission of Alfred A. Knopf, Inc.)

are small (Fig. 33–18). Our human ancestors defended themselves with tools and weapons rather than with their teeth. Correlated with the reduction in jaw size has been a reduction in the size and complexity of the teeth, and there is now a strong tendency for the third molars (wisdom teeth) to be vestigial.

The brain of early man was not much larger than an ape's, but that of modern man is

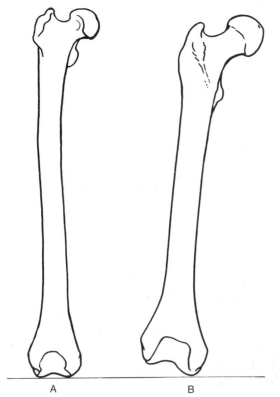

A   B

Figure 33–17 Femora of *(A)* an extinct ape, *Dryopithecus,* and *(B)* a modern man. When the distal articular surface is in the horizontal plane, the distal end of the human femur inclines medially. (Modified from Campbell, B.: *Human Evolution,* 2nd Ed. Chicago, Aldine Publishing Co., 1974. Copyright © 1974 by Aldine Publishing Co.)

considerably larger. The large brain may have evolved quickly, probably during the period when man began to hunt antelope and other larger animals, a practice that required more sophisticated weapons, a cooperative social structure and effective means of communicating and sharing ideas. The human forehead has become more vertical, the bony ridges over the eyes have diminished and the face, particularly the jaws, has become smaller in relation to the rest of the skull (Fig. 33–19). The evolutionary trend toward greater intelligence has made man less dependent upon sheer physical strength for getting food and fighting enemies. Speech, tools and weapons were developed, and humans began to live in clans and tribes, completing the transition from the ancestral solitary arboreal primates to ground-dwelling civilized animals.

33–8 CULTURAL EVOLUTION

Although most of the evidence of the path of human evolution comes from the fossils that have been discovered and described, some corroborative evidence comes from cultural implements, such as tools, weapons, cooking utensils, ornaments and art objects, that our ancestors left behind. The science of *archeology,* concerned with the cultural significance of such objects, is complex and fascinating and we can do no more here than indicate its importance.

The objects made and used by man, called *artifacts,* were deposited like fossils at widely separated times and so are found in different layers of the soil, the later ones usually lying above the earlier ones. When fossils and ar-

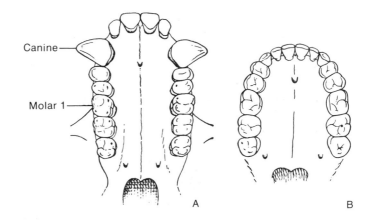

Figure 33–18 Palate and upper teeth of *A,* gorilla; *B, Australopithecus; C,* modern human. (From Clark, W. E.: *The Antecedents of Man.* New York, Harper & Row, 1952.)

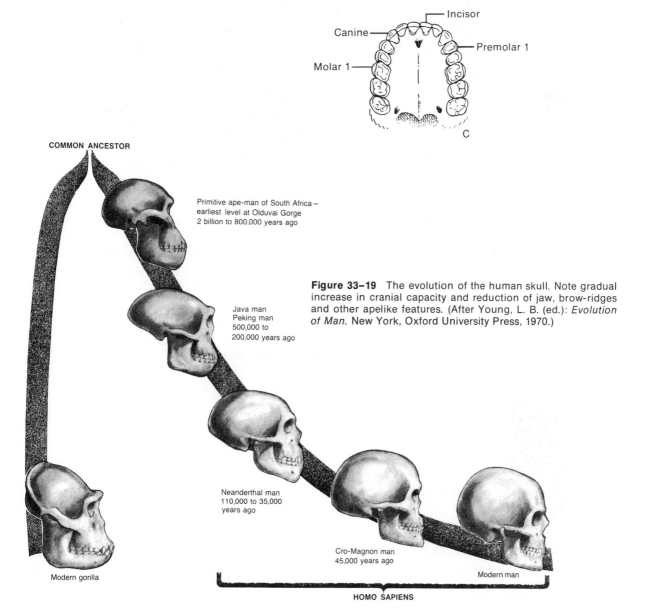

Figure 33–19 The evolution of the human skull. Note gradual increase in cranial capacity and reduction of jaw, brow-ridges and other apelike features. (After Young, L. B. (ed.): *Evolution of Man.* New York, Oxford University Press, 1970.)

tifacts are found together and the date of the culture associated with the artifact is known, the anthropologist is able to determine the age of the fossils.

Australopithecus, like their anthropoid contemporaries and the modern primates, ate a varied diet of fruit, buds, shoots, insects, eggs and an occasional bird or small mammal, but their diet included more meat than that of other anthropoids. The evidence for this is the finding of broken bones of small animals and of the young of larger animals such as antelopes at their campsites. It is presumed that australopithecines ate more meat than other contemporary primates because they were more efficient at catching animals, perhaps because they used simple weapons in their hunting and formed bands to hunt.

Figure 33-20 Lower and Middle Paleolithic industries. *A,* A flake scraper from the site and period of Swanscombe man. Hgt., 7.6 cm. *B,* A chopping tool of fossil wood, collected in Burma by Hallam L. Movius, Jr. Hgt., 7.6 cm. *C,* A Middle Paleolithic core with striking platform and flake scars. Width, 14 cm. *D,* A double-bladed knife from Bisitun, Iran. Hgt., 6.4 cm. *E,* A Neanderthal spear point from the same site. Hgt., 6.4 cm. (From Coon, C. S.: *The Story of Man.* New York, Alfred A. Knopf, Inc., 1954.)

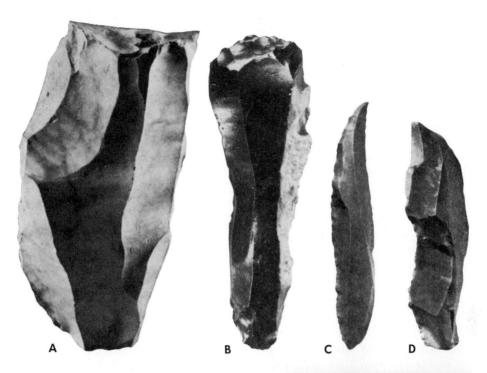

Figure 33–20 *Continued* Upper Paleolithic industries. *A,* A blade core, Aurignacian, France. Hgt., 11.4 cm. *B,* An end scraper on a blade, Turaif, Saudi Arabia. Hgt., 8.25 cm. *C,* A backed blade, France. Hgt., 5.7 cm. *D,* A burin, Turaif, Saudi Arabia. The working edge is up. Hgt., 5 cm. *E,* Bone needles made with burins, France. Hgts., 3.8 cm., 4.7 cm., 4.4 cm. (From Coon, C. S.: *The Story of Man.* New York, Alfred A. Knopf, Inc., 1954.)

During the Miocene, the species *Homo erectus* hunted not only the smaller animals but also some of the very large animals that have subsequently become extinct. His campsites are associated with the bones of large animals such as the mammoth, elephant, rhinoceros, bear and hippopotamus. The stone implements found at these sites are not the simple chopping stones used by *Australopithecus,* for *Homo erectus* had acquired a more sophisticated Acheulian culture. He was chipping flint and other fine-grain stones on all surfaces to fashion a hand axe (Fig. 33–20) and produced large stone flakes to use in a variety of cutting tools. *Homo erectus* apparently had learned to use fire, for the bones in his campsites appear to be charred. The ability to use fire and to cure hides to make clothing would have been essential for him to live in central Europe and Asia at a time when the continental glaciers were advancing. From the remains it is possible to infer that *Homo erectus* was an intelligent human who lived in groups and had the ability to communicate and teach the young to make tools and to hunt and to transmit knowledge of the seasons and habits of game.

The Neanderthals often built their fires on hearths on cave floors, and the Mousterian culture of the Neanderthals included a large kit of tools such as stone axes, scrapers, borers, knives, spear points and saw-edged and notched tools, probably used in making spear handles and other wooden implements. The stones show a secondary chipping that was done to refine the shapes and sharpen the edges. The Mousterian culture is characterized by implements made by chipping flakes from a piece of flint and then sharpening the edges by removing more flakes with a bone tool. The most common weapon seems to have been a triangular piece of stone, the forerunner of the spear and the arrowhead.

Later in the Upper Paleolithic culture associated with Cro-Magnon man, an improved method of tool-making was discovered. Flakes were removed from the pieces of flint by steady and carefully applied pressure rather than by blows. This enabled Cro-Magnon man to produce long, slender, knifelike blades, many of which, elaborately and skillfully carved, are great works of art. Upper Paleolithic human beings, the Cro-Magnons and others, were painters as well as skilled craftsmen. Their cave paintings found in France and Spain show a remarkable grasp of the principles of design (Figs. 33–21 and 33–22). Caves discovered in the Valley of the Dordogne in 1948 and 1956 have a wealth of beautifully preserved paintings of contemporary animals. The tools of the Upper Paleolithic Cro-Magnon man included chisels and awls—tools to make tools. Using these, Cro-Magnon man made from bone or ivory a variety of points for spears, harpoons and fishing hooks. He invented the needle with an eye.

Figure 33–21 Restoration by Charles R. Knight of a group of Upper Paleolithic artists drawing animals on the wall of a cave. (Courtesy of the American Museum of Natural History, New York.)

Figure 33–22 The art of Upper Paleolithic humans: paintings from the wall of the cavern of Lascaux, Dordogne, France. (Photo by Windels Montignac.)

Humans of the Mesolithic or Middle Stone Age culture were still hunters and food gatherers but had domesticated the dog. They had invented the sled and snowshoes for survival in the cold glacial periods and had developed tailored fur clothing. They may have attempted to write, for pebbles of this age have been found marked with red ochre dots, bars and crosses. They lived in small, isolated breeding groups, which would favor the occurrence of genetic drift and lead to the formation of divergent groups.

The Neolithic or New Stone Age culture originated somewhere between Egypt and India. It is characterized by implements bearing the marks of careful grinding and polishing and by the beginnings of agriculture and animal husbandry. The earliest animals to be domesticated after the dog were the pig, sheep, goat and cow; the horse was not domesticated until much later. Our human ancestors gradually changed from wandering hunters and food gatherers to more settled food producers, raising grain, making pottery and cloth and living in villages. The increase in food supply led to an increase in the size of the population; breeding groups became larger and interbred with neighboring ones and the tendency toward genetic drift was greatly decreased. The remains of bowls, pitchers, and other utensils have been helpful to archeologists, for each cultural group since the time of the Neolithic has used certain distinctive methods for making and decorating their pottery. Other inventions of the ingenious Neolithic people were the dugout canoe and the wheel. With the Neolithic age we come to historical times, for the oldest cultures of Egypt and Mesopotamia were Neolithic. The use of metals—first copper and then bronze (a copper-tin alloy)—for making vessels, tools and weapons began about 4000 B.C. At about 1400 B.C. men of the Near East initiated the Iron Age by mastering the technique of deriving iron from its ores.

33–9 THE PRESENT RACES OF HUMANS

A race is a subdivision of a species that consists of a population with a characteristic combination of gene frequencies, one that differs from that of other populations of the species. The several populations of human beings around the world are characterized by such distinctive gene frequencies. Some examples of local races are the Ainus of northern Japan, the Nordics of northern Europe, the Eskimos of Arctic America and the American Indians. Races are populations that differ from one another not in any single feature, but in having different frequencies of many genes and therefore characteristics that affect body proportions, skull shape, degree of skin pigmentation, texture of head hair, abundance of body hair, form of eyelids, thickness of lips, frequencies of various blood groups, tasting abilities and many other anatomical and physiological traits. While some of these differences, such as degree of skin pigmentation, are probably adaptive, the significance of others is unknown. Although it is possible to recognize many individuals as belonging to one or another of the broad racial groups there are always some ambiguities. Since different populations of the same species can and do interbreed, the offspring may be difficult to classify and constitute one of the sources of ambiguity. Secondly, there is a great deal of genetic variation *within* each human population as well as differences *between* populations, so that the phenotypes of certain members of a population may be quite unusual for that racial group. Thus one can provide a fairly straightforward definition for the general concept of race and can observe racial differences between different populations and individuals. The scientific study of specific races is, however, a much more difficult matter, and attempts to define behavioral differences among the races have been unsuccessful. The statements that "All peoples have the same mentality" or that "Different races have different mentalities" are equally without scientific foundation. Race is a term that has been abused by persons who would like to emphasize human differences for the sake of establishing or maintaining a superior economic or social position.

Anthropologists believe that the various populations living today represent localized differences in larger geographic groups. The anthropologist Carlton Coon recognizes five major racial groups of human beings (Fig. 33–23): (1) the **Caucasoids**, which include the Nordic, Alpine and Mediterranean races of Europe, the Armenoids and Dinarics of eastern Europe, the Near East and North Africa, and the Hindus of India; (2) the **Mongoloids**, which include the Chinese, Japanese, Ainus, Eskimos and American Indians; (3) the **Congoids**, or Negroes and Pygmies; (4) the **Capoids**, or Bushmen and Hottentots of Africa; and (5) the **Australoids**, which include the Australian aborigines, Negritos, Tasmanians and Papuomelanesians. Anthropologists are not in agreement as to how these groups should be classified, and any individual within a group may be phenotypically so unlike his parents as to be placed in a separate group. Each quality,

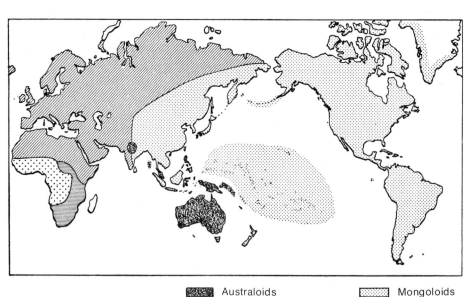

Figure 33–23 Probable distribution of the five major racial groups of modern man shortly after the ending of the Pleistocene. (From Coon, C. S.: *The Origin of Races.* Copyright © 1962 by Carleton S. Coon. Reprinted by permission of Alfred A. Knopf, Inc.)

Australoids   Mongoloids
Caucasoids   Capoids
Congoids

such as skin color, may vary tremendously within each race so that a member of the Caucasoid race may have skin as dark as a typical Negro's and a Chinese may have skin as white as a Caucasian's.

Skin color, the color of the hair and eyes, the waviness and texture of the hair, the shape of the head and its features, the arrangement of the whorls and loops on the skin of the finger tips and the proportions of the various parts of the body are some of the characteristics that differentiate human populations. Anthropologists have paid particular attention to the ratio between the breadth and the length of the head. When a living person's head is measured this ratio is called the *cephalic index.* When the measurements are made of a skull the ratio is called the *cranial index.* By convention, a skull with a breadth less than 75 per cent (a cranial index of 75) of its length is termed long-headed, or *dolichocephalic.* One with a cranial index of 80 or more is said to be broad, round-headed or *brachycephalic;* and a skull with a cranial index between 75 and 80 is said to be *mesocephalic.* Similarly, noses are classified according to the ratio of breadth to length, the *nasal index,* and faces are classified on the basis of the *facial index,* the ratio of length to breadth.

Whenever any large, populous and diverse species of animal or plant has been studied intensively by many scientists, there have been widely divergent opinions as to how many subspecies or races can be recognized. This is certainly true of *Homo sapiens* and various authorities divide humankind into as few as 3 or as many as 30 different races.

It should be clear why it is difficult to make generalizations about the mental or physical characteristics of any single race of modern humans. No race is "pure"; the evolutionary history of human beings has been one of continuous intermixture of races as peoples migrated, invaded and conquered their neighbors or were conquered by them. Each of these populations has great potentialities and each has made important contributions to civilization.

QUESTIONS

1. What characteristics distinguish humans from the great apes?
2. Why is it incorrect to say that humans evolved from monkeys? What did we evolve from?
3. Describe the current theory of the course of evolution from primitive insectivores to present-day humans.
4. Distinguish between ceboid and cercopithicoid monkeys.
5. How have human physical features altered during our evolution from the ancestral lemuroid prosimians?
6. Compare the structures and functions of gibbons, orangs and gorillas. Which of these show the best adaptations to arboreal life?
7. Compare the appearance, characteristics and lifestyles of Neanderthal and Cro-Magnon humans. What became of each?
8. Define the term dolichocephalic. How does one measure the cephalic index of the human?
9. What is an archeological artifact? Of what use are artifacts in tracing human evolution?
10. In what ways do the Upper Paleolithic and Neolithic cultures differ?
11. Why is genetic drift less important in human evolution at present than it was 20,000 or more years ago?
12. Compare the method of locomotion of the gibbon, the gorilla and the present-day human.
13. In what respects can the following be considered "milestones" in the evolution of the primates: *Parapithecus, Aelopithecus, Aegyptopithecus, Dryopithecus, Ramapithecus* and *Australopithecus?*
14. How do you suppose anthropologists go about reconstructing an ape man such as *Gigantopithecus* on the basis of three fossil teeth?
15. Distinguish among the following types of early humans: Neanderthals, Cro-Magnons, Rhodesian man and Galley Hill man.

SUPPLEMENTARY READING

Excellent descriptions of our ancestral hominids and anthropoids are given in the *Ascent of Man* by David Pilbeam, *The Evolution of the Genus* Homo by William Howells and Joseph B. Birdsell's *Human Evolution.* An interesting account of the discovery of ape men by one of the major investigators is *Apes, Giants and Man* by Franz Weidenreich. *The Races of Europe* by Carleton S. Coon is a treatise describing in detail the many subdivisions of the Caucasian race. *The Distribution of Man* by W. W. Howells describes the origin and migrations of the races of humans. Standard texts of physical anthropology are Gabriel Lasker's *Physical Anthropology: A Perspective* and Frederick S. Hulse's *The Human Species.* *The Evolution of Man and Society* by C. D. Darlington and *The Nature of Human Nature* by Alexander Comfort are well written expositions of human biology, variability and evolution. *The Brain in Hominid Evolution* by P. V. Tobias provides a thorough review of the evolution of the brain and its relation to cultural development. Beginning in 1972, Time-Life Books has published a series of superbly illustrated books on human evolution, entitled *The Emergence of Man.*

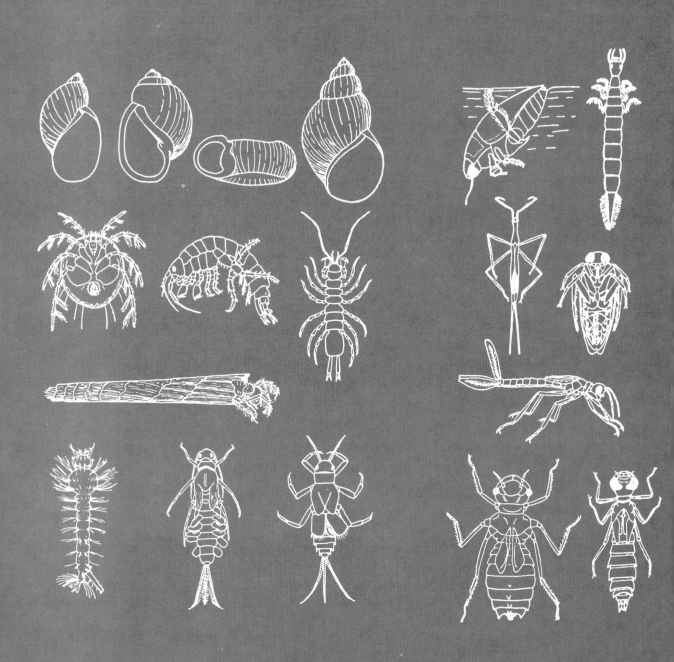

PART 8

ECOLOGY

INTRODUCTION

As more is learned about any kind of plant or animal, it becomes increasingly clear that each species has undergone adaptations to survive under a particular set of environmental circumstances. Each may show adaptations to wind, sun, moisture, temperature, salinity and other aspects of the physical environment as well as adaptations to specific plants and animals that live in the same region. The study of the interrelationships of organisms with their physical and biotic environments is termed *ecology*. This word is very much in the public consciousness today as human beings begin to be aware of some of their past and current ecological practices. It is important for everyone to know and appreciate the principles of this aspect of biology so that they can form intelligent opinions regarding such things as insecticides, detergents, mercury pollution, sewage disposal and power dams and their effects on us, on our civilization and on the world we live in.

The Greek *oikos* means "house" or "place to live," and ecology (*oikos logos*) is literally the study of organisms "at home," in their native environment. The term was proposed by the German biologist Ernst Haeckel in 1869, but many of the concepts of ecology antedated the term by a century or more. Ecology is concerned with the biology of groups of organisms and their relations to the environment. The term *autecology* refers to studies of individual organisms, or of populations of single species, and their relations to their environment. The contrasting term, *synecology*, refers to studies of groups of organisms associated to form a functional unit of the environment. Groups of organisms may be associated in three different levels of organization—populations, communities and ecosystems. In ecologic usage, a *population* is a group of individuals of any one kind of organism, a group of individuals of a single species. A community in the ecologic sense, a *biotic community*, includes all the populations occupying a given defined physical area. An *ecosystem* consists of the community plus the physical, nonliving environment. Thus synecology is concerned with the many relationships within communities and ecosystems. The ecologist deals with such questions as who lives in the shade of whom, who eats whom, who plays a role in the propagation and dispersion of whom, and how energy flows from one individual to the next in a food chain. The ecologist attempts to define and analyze those characteristics of populations that are distinct from the characteristics of individuals as well as factors responsible for the aggregation of populations into communities.

CHAPTER 34

PRINCIPLES
OF ECOLOGY

The fundamentals of *ecology*, the study of the interrelations between living things and their physical and biotic environment, were introduced in Chapter 6. Now that we have considered certain details of plant and animal structure and function and have gained some idea of how these forms arose in evolution, we are ready to reconsider the problems of ecology in more detail.

In earlier chapters we examined the reactions by which energy and materials flow through plant and animal cells and through individual organisms. We can now examine the concept that the flow of energy and materials through all the organisms living in a given area and composing a specific ecosystem is regulated by control systems analogous to those operating in the cells and tissues of a single organism.

34–1 THE CONCEPTS OF RANGES AND LIMITS

Probably no species of plant or animal is found everywhere in the world; some parts of the earth are too hot, too cold, too wet, too dry or too something else for the organism to survive there. Even if the environment does not kill the adult directly it can effectively keep the species from becoming established if it prevents its reproducing or kills off the egg, embryo or some other stage in the life cycle.

Most species of organisms are not even found in all the regions of the world where they could survive. The existence of barriers prevents their further dispersal and enables us to distinguish the major biogeographic realms (p. 780), characterized by certain assemblages of plants and animals.

Biologists early in the nineteenth century were aware that each species requires certain materials for growth and reproduction and can be restricted if the environment does not provide a certain minimal amount of each one of these materials. Justus Liebig stated in 1840 what is now known as his Law of the Minimum, that is, the rate of growth of each organism is limited by whatever essential nutrient is present in a minimal amount. Liebig, who studied the factors affecting the growth of plants, found that the yield of crops was often limited not by a nutrient required in large amounts, such as water or carbon dioxide, but by something needed only in trace amounts, such as boron or manganese. Liebig's law is strictly applicable only under steady-state conditions, when the inflow of energy and materials equals the outflow. In addition, there may be interactions between factors such that a very high concentration of one nutrient may alter the rate of utilization of another (the rate-limiting one) and hence alter the effective minimal amount required. Certain plants, for example, require less zinc when growing in the shade than when growing in the sunlight.

V. E. Shelford pointed out in 1913 that *too much* of a certain factor would act as a limiting factor just as well as too little of it and that the distribution of each species is determined by its *range of tolerance* to variations in each environmental factor. Much work has been done to define the limits of tolerance, the limits within which species can exist, and this concept, sometimes called Shelford's Law of Tolerance, has been helpful in understanding the distribution of organisms.

It has usually been found that certain stages in the life cycle are critical in limiting organisms—seedlings and larvae are usually more sensitive than adult plants and animals. Adult blue crabs, for example, can survive in water with a low salt content and can migrate for some distance upriver from the sea, but their larvae cannot, and so the species cannot become permanently established there.

Some organisms have very narrow ranges of tolerance to environmental factors; others can survive within much broader limits. Any given organism may have narrow limits of tolerance for one factor and wide limits for another. Ecologists use the prefixes *steno-* and *eury-* to refer to organisms with narrow and wide, respectively, ranges of tolerance to a given factor. A stenothermic organism is one that will tolerate only narrow variations in temperature. The housefly is a eurythermic organism, for it can tolerate temperatures ranging from 5 to 45°C. The adaptation to cold of the antarctic fish *Trematomus bernacchi* is remarkable. It is extremely stenothermic and will tolerate temperatures only between −2°C and +2°C. At 1.9°C this fish is immobile from heat prostration!

Temperature is an important limiting factor, as the relative sparseness of life in the desert and arctic demonstrates. Most of the animals that do live in the desert have adapted to the rigors of the environment by living in burrows during the day and coming out to forage only at night. Many animals escape the bitter cold of the northern winter not by migrating south but by burrowing beneath the snow. Measurements made in Alaska show that when the surface temperature is −55°C, the temperature 60 cm. under the snow, at the surface of the soil, is −7°C.

Although the ringnecked pheasant has been introduced into the southern states a number of times and the adults survive well, the developing eggs are apparently killed by the high daily temperatures and are unable to complete development.

Light.   The amount of light is an important factor in determining the distribution and behavior of both plants and animals. Light is, of course, the ultimate source of energy for life on this planet, yet prolonged exposure of cells to light of high intensity may be fatal. Both plants and animals have evolved mechanisms and responses to protect them against too much (or too little) light.

The amount of daylight per day, known as the *photoperiod*, has a marked influence on the time of flowering of plants, the time of migration of birds, the time of spawning of fish and the seasonal changes of color of certain birds and mammals. The effects of the photoperiod on vertebrates appear to be mediated by some neurohormonal mechanism involving the hypothalamus, the pituitary and the pineal. Knowledge of photoperiod phenomena has proved to be of considerable economic importance. Chicken farmers have found that using artificial illumination in the hen house, and thereby extending the photoperiod, stimulates the hens to lay more eggs.

Water.   Water is a physiologic necessity for all living things but is also a limiting factor, primarily for land organisms. The total amount of rainfall, its seasonal distribution, the humidity and the supply of ground water are some of the factors limiting distribution of plants and animals. Some lakes and streams, particularly in the western and southwestern United States, periodically become dry or almost dry, and the fish and other aquatic animals are killed. During periods of low water the water temperature may rise enough to kill off the aquatic forms.

Many of the protozoa survive the drying of the puddles in which they normally live by forming thick-walled protective cysts. As mentioned before, some animals have adapted to desert conditions by digging and living in burrows in which the temperature is lower and the humidity is higher than at the surface. Measurements have shown that the temperature in the burrow of a kangaroo rat 60 cm. underground may be only 16°C when the surface temperature is over 38°C. The desert plants must stay on the surface and have had to evolve structures to prevent water loss and to resist high temperatures.

An excess of water is fatal to certain animals. Earthworms may be driven from their burrows by heavy rainfall. Oxygen is only sparingly soluble in water, and the earthworm cannot get enough oxygen when immersed. Knowledge of the limits of water tolerance can be used by man to regulate insect pests. For example, wire worms, pests attacking West Coast crops, were found to have rather narrow limits of tolerance to water and to be most sensitive as larvae and pupae. They can be destroyed by exceeding the maximum limit of tolerance—by flooding irrigated fields—or by planting alfalfa or wheat, which dry out the soil below the limit of tolerance of the larvae.

Other Environmental Factors. Atmospheric gases are usually not limiting for land organisms except for forms living deep in the soil, on mountain heights or within the bodies of other animals. In aquatic environments the amount of dissolved oxygen present may vary considerably and be the limiting factor for certain forms.

The oxygen tension in stagnant ponds or in streams fouled by sewage and industrial wastes may become so low that it is incompatible with many forms of life. Some parasites have adapted to the low oxygen tension within the host's gut by evolving metabolic pathways by which biologically useful energy can be released from foodstuffs without the utilization of free oxygen.

The trace elements necessary for plant and animal life may be present in too small amounts and be limiting; deficiencies of cobalt and copper produce severe deficiency diseases in plants and grazing animals—certain regions of Australia are unsuitable for raising cattle or sheep because of this. Other trace elements that may be limiting factors are manganese, zinc, iron, sulfur, selenium and boron.

The amount of carbon dioxide in the air is remarkably constant, but the amount dissolved in water varies widely. An excess of carbon dioxide may be a limiting factor for fish and insect larvae. The hydrogen ion concentration, pH, of water is related physicochemically to the carbon dioxide concentration, and it, too, may be an important limiting factor in aquatic environments.

Water currents are factors limiting certain aquatic plants and animals—there are marked differences in the flora and fauna of a still pond and a swiftly flowing creek.

The type of soil, the amount of topsoil and its pH, porosity, slope, water-retaining properties and so on are limiting factors for many plants. The ability of many animals to survive in a given region depends on the presence of certain plants to provide shelter, cover and food. Grasses, shrubs and trees each provide shelter for certain kinds of land animals, and seaweeds and fresh water aquatic plants play a similar role for aquatic animals. Even fire may be a factor of ecologic importance. The continued existence of the fine forests of long-leaf pines in the southeastern states is due to their superior resistance to fire. In the absence of occasional small ground fires, these pines are gradually replaced by small hardwoods, much less valuable as timber and much more readily killed by fire.

In summary, whether or not an animal can become established in a given region is the result of a complex interplay of physical factors, such as temperature, light, water, winds and salts, and biotic factors, such as the plants and other animals in that region that may serve as food, compete for food or space or act as predators or parasites.

34–2 STRUCTURAL ADAPTATIONS

In the course of evolution, organisms have undergone successive structural adaptations and readaptations as the environment changed or as they migrated to a new environment. As a result, many organisms today have structures or physiologic mechanisms that are useless or even deleterious but that were advantageous at an earlier time when the organism was adapted to a different environment.

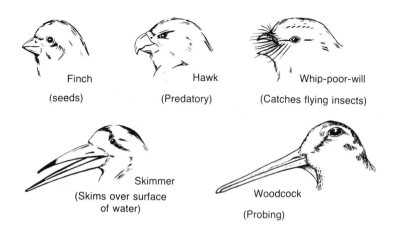

Finch
(seeds)

Hawk
(Predatory)

Whip-poor-will
(Catches flying insects)

Skimmer
(Skims over surface
of water)

Woodcock
(Probing)

Figure 34–1 Diagrams of the bills of various birds, illustrating their adaptation to the type of food eaten. (From Villee, C. A., Walker, F. W., Jr., and Barnes, R. D.: *General Zoology,* 4th Ed. Philadelphia, W. B. Saunders Co., 1973.)

The adaptations of the mouthparts of certain animals to the kind of food eaten are among the most striking that can be cited. The mouthparts of certain insects are adapted for sucking nectar from certain species of plants; others are specialized for sucking blood, for biting or for chewing vegetation. The bills of various kinds of birds (Fig. 34–1) and the teeth of various kinds of mammals may be highly adapted for particular kinds of food.

In many animals, the specialized adaptation to a certain way of life is simply the latest stage in a series of adaptations. For example, both the human and the baboon, whose immediate ancestors were tree-dwellers, have returned to the ground and have become readapted to walking.

Readaptation may be a very complicated process. The present-day Australian tree-climbing kangaroos are the descendants of an original ground-dwelling marsupial. From the ground-dwellers evolved forms that, in adaptive radiation, took to the trees and developed limbs adapted to tree climbing (or perhaps the sequence of events was the reverse—first the evolution of specialized limbs, then the adoption of a tree-dwelling mode of life). Some of these tree-dwellers eventually left the trees and became readapted to ground life, and the hind legs became lengthened, strengthened and adapted for leaping. But finally some of these kangaroos went back to the trees, although their legs were by then so highly specialized for leaping that they could not be used for grasping a tree trunk. Consequently the present-day tree kangaroos must climb like bears, by bracing their feet against the tree trunk. A comparison of the feet of existing Australian marsupials reveals all the stages in

this complicated, shifting process of adaptation.

34–3 PHYSIOLOGIC ADAPTATIONS

One of the major struggles among organisms stems from the competition for food; hence a mutation enabling an animal to use a new type of food is extremely advantageous. This may be accomplished in a number of ways—by the evolution of a new digestive or energy-liberating enzyme system, for example. A mutation resulting in a new energy-liberating enzyme enabled the sulfur bacteria to obtain energy from hydrogen sulfide, a substance that is poisonous to almost all other organisms. The evolution of a special enzyme for reducing disulfide bridges (—S—S—) gave the clothes moth its unique ability to digest wool; the molecules of wool protein are cross-linked with many disulfide bridges.

Another type of favorable mutation is one that decreases the growing season of a plant or the total length of time required for an insect to develop. Such mutations enable the organism to survive farther from the equator and open up new areas of living space and new sources of food.

Any mutation that increases a species' limits of temperature tolerance—i.e., makes it more eurythermic—may enable it to live in a new region, at a higher latitude or a higher altitude. Other organisms have solved the problem of living in arctic regions by becoming dormant during the cold season or by developing behavior patterns of migration. Many birds, but only a few mammals, migrate south to avoid the cold northern winter.

A number of kinds of mammals—monotremes, shrews, rodents and bats—hibernate over the winter. It is doubtful whether carnivores such as bears and skunks actually hibernate; they simply sleep for long periods at a time.

In true *hibernation* the body temperature falls to just a degree or two higher than the surrounding air temperature, metabolism decreases greatly, and the heart rate and rate of respiration become very slow. No food is eaten and the animal uses up its stores of body fat, awakening in the spring in an emaciated condition (and probably "as hungry as a bear"). What induces hibernation and what wakes the animal in the spring are not completely clear; changes in the environmental temperature are, of course, important, but changes within the animal's body are also involved.

Birds and mammals are unique in possessing mechanisms for controlling body temperature, keeping it constant despite wide fluctuations in the environmental temperature. These thermostated animals are said to be *homoiothermic* ("warm-blooded" is not quite correct; they are really "constant-temperature-blooded"). Fish, amphibians, reptiles and all the invertebrates are *poikilothermic*—their body temperature fluctuates with that of the environment (again, "cold-blooded" is not exactly descriptive; their body temperature is simply determined by the temperature of the environment).

Marine fish are usually adapted to survive within a certain range of pressures and hence at a particular depth. Surface animals are crushed by the terrific pressures of the deep, and deep-sea animals usually burst when brought to the surface. The whale, however, is able to withstand great changes in pressure, and can dive to depths of 800 meters without injury. Presumably its lung alveoli collapse when the pressure on the body reaches a certain point and gases are no longer absorbed into the blood.

Humans can survive pressures as high as six atmospheres if the pressure is increased and then decreased slowly. The increase in pressure increases the amount of gases dissolved in the blood and body fluids. If the pressure is decreased suddenly, these gases come out of solution and form bubbles throughout the body. Those in the blood impede circulation and bring about the symptoms of diver's disease, or "the bends." The pilot of a jet plane may climb so rapidly that the atmospheric pressure is reduced fast enough to bring bubbles of gas out of solution in his blood and cause a type of the bends.

34-4 COLOR ADAPTATIONS

Adaptations for survival are evident in the color and pattern of plants and animals as well as in their structure and physiologic processes. We can distinguish three types of color adaptation: concealing or *protective coloration,* which enables the organism to blend with its background and be less visible to predators; *warning coloration,* which consists of bright, conspicuous colors and is assumed by poisonous or unpalatable animals to warn potential predators not to eat them; and *mimicry,* in which the organism resembles some other living or nonliving object—a twig, leaf, stone, or perhaps some other animal which, being poisonous, has warning coloration.

Concealing coloration may serve to hide an animal that wants to escape the notice of potential predators or it may be assumed by a predator in order to be unnoticed by his potential prey. Examples of concealing coloration are legion—the white coats of arctic mammals and the stripes and spots of tigers, leopards, zebras and giraffes, which, though conspicuous in a zoo, blend imperceptibly with the moving pattern of light and dark typical of their native savannah. Some animals—frogs, flounders, chameleons, crabs and others—have a remarkable ability to change their color and pattern as they move from a dark to a light background or from one that is uniform to one that is mottled (Fig. 34-2).

To demonstrate experimentally that concealing coloration does have survival value, F. B. Isely fastened grasshoppers with different body colors to plots of different colored soils—light, dark, grassy, sandy and so on. After these plots had been exposed to the predatory activities of wild birds or chickens for a given length of time, the survivors were tabulated. It was found that there was a significantly higher percentage of survivors among those grasshoppers that matched their background.

Figure 34-2 The remarkable ability of the flounder to change its color and pattern to conform with its background. *Left,* A flounder on a uniform, light background; *right,* the same fish placed on a spotted, darker background.

H. B. D. Kettlewell of Oxford University made careful field observations of the frequency of capture of light-colored and dark-colored mutants of the moth *Biston betularia* by birds. In the woods of Dorsetshire, where there is little pollution and the tree trunks are light-colored, the light-colored moths were significantly less likely to be caught and eaten by birds. In woods in the industrial Midlands, where leaves and tree trunks are blackened with soot, the dark-colored moths were significantly less likely to be caught and eaten.

When an animal is equipped with poison fangs, a stinging mechanism or some chemical that gives it a noxious taste, it is probably to its advantage to have the fact widely advertised, and, in fact, many animals so equipped do have warning colors.

An interesting example is a species of European toad with a bright scarlet belly. Certain chemicals secreted by its skin glands make the toad extremely unpalatable, and whenever a potential predator, such as a stork, swoops over a congregation of toads, they flop on their backs, exposing their scarlet bellies as a warning. The storks and other birds apparently become conditioned by the association of the red color and the bad taste and avoid the toads assiduously.

Other animals survive by mimicking one of these protectively colored animals; for instance, some harmless, defenseless and palatable animals are identical in shape and color with a poisonous or noxious animal of quite a different family or order, and, being mistaken for it by predators, are left alone. Many tropical insects have evolved this type of protection, termed **Batesian mimicry.** If the animal is noxious enough, there will be significant protection afforded to its mimics even if the mimics outnumber their noxious models. The evolution of markedly similar shapes and color patterns by two species, both of which are noxious, is termed **Müllerian mimicry.** It is believed that the similar appearance of the two

species increases the probability that predators will learn to avoid that particular shape and color pattern.

The reality of the selective advantage of color adaptations has been debated. It has been argued that animal vision may be so different from human vision (certain animals may be colorblind or may be able to see ultraviolet or infrared light) that an animal that appears to be protectively colored to a human may be quite evident to its natural predators. However, many experimental studies, such as the experiments with grasshoppers and moths, have shown that protective coloration does have survival value.

Color and patterns may serve to attract other organisms when that is necessary for survival. The red and blue ischial callosities of monkeys and the gay, extravagant plumage of various birds apparently have an attraction for members of the opposite sex that plays an important role as a prelude to mating. Vividly colored flowers seem to attract the birds and insects whose activities are needed to ensure the pollination of the plant or the dispersal of its seed.

34-5 ADAPTATIONS OF SPECIES TO SPECIES

The evolution and adaptation of each species have not occurred in a biologic vacuum, independent of other forms. On the contrary, the adaptation of each species has been influenced markedly by the concurrent adaptations of other species. As a result of this, many types of cross dependency between species have arisen. Some of the clearest and best understood of these involve insects.

Insects are necessary for the pollination of a great many plants; some plants are so dependent on certain insects that they are unable to survive in a given region unless those particular insects are present. For example, the

Smyrna fig could not be grown in California, even though all climatic conditions were favorable, until the fig insect, which pollinates the plant, was introduced.

Birds, bats and even snails serve as pollen transporters for some plants, but insects are the prime pollinators. Flowering plants have evolved bright colors and fragrances, presumably to attract insects and birds and ensure pollination. There has been some doubt as to whether insects can detect different odors and colors. The experiments of Karl von Frisch show that honey bees, at least, are able to differentiate colors and scents and that they are guided in their visits to flowers by these stimuli.

Some of the species-to-species adaptations are so exact that neither form can exist in a region without the other. The yucca plant and the yucca moth, like the fig and fig insect, have evolved to a point of complete interdependence (Fig. 34–3 A). The yucca moth, by a

Figure 34–3 The yucca plant *(A)* is pollinated only by the yucca moth, one of which is shown in the open flower at the right. (From Weatherwax, P.: *Botany,* 3rd Ed. Philadelphia, W. B. Saunders Co., 1956.) *B,* A larva of the butterfly family Lycaenidae eating a lupine seed pod while being tended by ants. The species to which the larva belongs has adapted to overcome toxic substances in the food plant, and the ants earn the sticky substance given off by the larva when they fend off parasites or predators. It is interesting to note that the ants eat the plant in the absence of caterpillars and aphids.

A

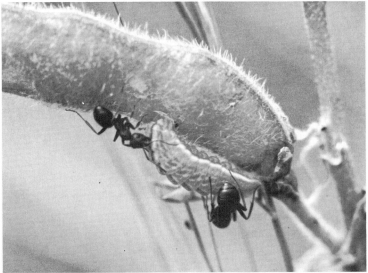

B

series of unlearned acts, goes to a yucca flower, collects some pollen, and takes it to a second flower. There it pushes its ovipositor (egg-laying organ) through the wall of the ovary of the flower and lays an egg. It then carefully places some pollen on the stigma. The yucca plant in this way is sure to be fertilized and produce seeds; the larva of the yucca moth feeds on these yucca seeds. The yucca produces a large number of seeds and is not injured by the loss of the few eaten by the moth larva.

Other examples of species-to-species adaptations are the host-parasite, prey-predator, commensals and mutualistic interdependences discussed in Chapter 6 (see also Fig. 34–3 B).

34–6 HABITAT AND ECOLOGIC NICHE

The concepts of habitat and ecologic niche were introduced in Chapter 6. The term *habitat* is widely used and simply means the place where an organism lives. The *ecologic niche* is a more inclusive term that includes not only the physical space occupied by an organism but also its functional role as a member of the community—that is, its trophic position and its position in the gradients of temperature, moisture, *p*H and other conditions of the environment. The ecologic niche of an organism depends not only on where it lives, but also on what it does—that is, how it transforms energy, how it behaves in response to and modifies its physical and biotic environment, and how it is acted upon by other species. A common analogy is that the habitat is the organism's "address" and the ecologic niche is the organism's "profession," biologically speaking. To describe the complete ecologic niche of any species would require detailed knowledge of a large number of biologic characteristics and physical properties of the organism and its environment. Since this is very difficult to obtain, the concept of ecologic niche is used most

often to describe differences between species with regard to one or a few of their major features.

Charles Elton, in 1927, was one of the first to use the term ecologic niche in the sense of the functional status of an organism in its community. The term has somewhat different meanings to different ecologists. To some it is the ultimate distributional unit within which each species is held by its structural and behavioral limitations. To others it is the functional status of an organism in its community, and these ecologists emphasize the energy relations of the species. To others, such as G. E. Hutchinson, the niche is a multidimensional space within which the environment permits an individual or species to survive indefinitely. The ecologic niche is an abstraction that includes all the physical, chemical, physiologic and biotic factors that an organism needs in order to survive. To describe the ecologic niche of any given species we must know what it eats and what eats it, what its activities and range of movements are, and what effects it has on other organisms and on the nonliving parts of the surroundings. Two different species of aquatic insects (Fig. 34–4) may live in the same habitat, such as the waters of a small, shallow, vegetation-choked pond, but occupy quite different ecologic niches. The backswimmer, *Notonecta,* is a predator that swims about catching and eating other animals. The water boatman, *Corixa,* looks very much like the backswimmer but plays a very different role in the community since it feeds largely on decaying vegetation.

Two species of organisms that occupy the same or similar ecologic niches in different geographical locations are termed *ecological equivalents.* The array of species present in a given type of community in different biogeographic regions may differ widely. However, similar ecosystems tend to develop wherever there are similar physical habitats; the equiva-

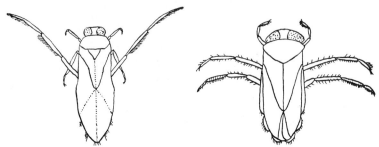

Figure 34–4 *Notonecta,* the backswimmer *(left),* and *Corixa,* the water boatman *(right),* are two aquatic bugs occupying the same habitat—the shallow, vegetation-choked edges of ponds and lakes—but having different ecologic niches.

Table 34-1 ECOLOGICAL EQUIVALENTS: SPECIES OCCUPYING COMPARABLE ECOLOGIC NICHES ON NORTH AMERICAN COASTS

	Grazers on Intertidal Rocks	Fish Feeding on Plankton	Bottom-Dwelling Carnivores
Northeast coast	*Littorina littorea* (periwinkle)	Alewife, Atlantic herring	*Homarus* (lobster)
Gulf coast	*Littorina irrorata*	Menhaden, threadfin	*Menippe* (stone crab)
Northwest coast	*Littorina danaxis, Littorina scutelata*	Sardine, Pacific herring	*Paralithodes* (king crab)
Tropical coast	*Littorina ziczac*	Anchovy	*Panulirus* (spiny lobster)

Adapted from Odum, E. P.: *Fundamentals of Ecology*, 3rd Ed. Philadelphia, W. B. Saunders Co., 1971.

lent functional niches are occupied by whatever biological groups happen to be present in the region. Thus, a savannah biome tends to develop wherever the climate permits the development of extensive grasslands, but the species of grass and the species of animals eating the grass may be quite different in different parts of the world. On each of the four continents there at one time were grasslands with large grazing herbivores present. These herbivores were all ecological equivalents. However, in North America the grazing herbivores were bison and prong-horn antelope; in Eurasia, the saga antelope and wild horses; in Africa, other species of antelope and zebra; and in Australia, the large kangaroos. In all four regions these native herbivores have been replaced to a greater or lesser extent by man's domesticated sheep and cattle. As examples of ecological equivalents, the species occupying three marine ecologic niches in four different regions of the coast are shown in Table 34-1. The same kinds of ecologic niches are usually present in similar habitats in different parts of the world. Comparisons of such habitats and analyses of the similarities and differences in the species that are ecological equivalents in these different habitats have been helpful in clarifying the interrelations of these different ecologic niches in any given habitat.

34-7 THE PHYSICAL ENVIRONMENT

The numbers and distribution of plants are influenced by climate *(climatic factors)* and by soil *(edaphic factors).* The kinds of plants, together with climatic and edaphic factors, influence the number and distribution of the various kinds of animals, and this, in turn, may influence plants. The soil on the surface of the earth is quite nonuniform. The original crust of the earth, the igneous rock, was relatively uniform, but weathering, erosion and sedimentation have produced marked geochemical differentiation of the earth's surface. The distribution of organisms is greatly affected by the kind of soil present. The converse relationship is also true—namely, that the organisms present make a major contribution to the kind of soil. Organisms are the sources of the great deposits of fossil fuels, such as peat, coal and oil.

The weathering of the earth's crust has led to the deposition on the surface of the earth of a coat, the characteristics of which depend on the kind of parent rock that was weathered, the kind of weathering processes to which it has been subjected, and its overall age. The term *soil* is applied to this mixture of weathered rock plus organic debris. Most soils remain where they were formed from the parent rock, but some have been carried from their place of origin to another location by the wind (sand dunes or loess), by water (alluvial deposits at the deltas of rivers) or by glaciers. Most of the soils of Canada and the northeastern United States were deposited by the action of glaciers.

After its formation, a soil undergoes development, controlled both directly and indirectly by the climate. The temperature and rainfall determine the rate at which materials in solution and suspension are transported by percolating water out of the soil to places where they accumulate. The nature of the climate determines the kind of vegetation present, and this, in turn, determines what kind of organic materials will be available to be incorporated into the soil. In cold, humid regions where rainfall is greater than evaporation, the vegetation produces an acid humus, giving an ash-grey soil termed *podzol.* In tropical regions with high temperatures and heavy rainfall there is little acidity from the decay of tropical vegetation and this results in red *lateritic* soil with a high iron content. In regions with low rainfall, unevenly distributed over the year, and with high rates of evaporation, the soil tends to be calcified, rich in calcium carbonate. The different characteristics of the various types of soil determine the kinds of plants that can grow in the region. Soils supply anchorage for the plants, water, mineral nutrients and aeration of the roots. Some of the important characteristics of a soil are its texture—whether it is gravel, sand, silt or clay, its organic content, the amount of soil water present, the amount of air entrapped in the soil and its acidity and salinity.

As soils develop they tend to become stratified. The uppermost layer is one from which nutrients have been removed by water percolating through it. Below this is a layer of accumulated materials derived from the layer above. The bottom layer is composed of unweathered parent material.

34–8 SOLAR RADIATION

Perhaps the most outstanding feature of the earth is the nonuniformity of its physical conditions, which range from Arctic tundra to tropical rain forests. Even the oceans are very patchy, nonuniform places. The earth derives nearly all its energy from the sun, but even the sun's energy is not uniformly distributed over the face of the globe. The solar radiation reaching the surface of the earth varies with the length of the path that the sun's rays take through the atmosphere and whether it is vertical or at an angle; the area of horizontal surface over which is spread a "bundle" of the sun's rays of a given cross-sectional area; the

distance of the earth from the sun (which changes seasonally because of the elliptical orbit of the earth around the sun); the amount of water vapor, dust and pollutants in the atmosphere; and the total length of the day (the photoperiod). At higher latitudes the angle of incidence of the sun's rays is less than at middle latitudes, and the energy is spread more thinly. The rays must pass through a thicker layer of atmosphere (Fig. 34–5), and consequently the polar regions receive less radiant energy in the course of a year than do equatorial regions.

The major variations in the amount of incoming solar energy are related to the movements the earth makes with respect to the sun (Fig. 34–6). One complete annual orbit of the earth around the sun requires 365¼ days. The axis of the earth is tilted 23½ degrees relative to its plane of orbit, and therefore the distribution of energy varies throughout the year. The northern hemisphere receives more radiant energy during the period between March 21 and September 21 than in the other half of the year, not only because there are more hours of daylight but also because the angle of incidence of sunlight is more nearly vertical during that period.

The rotation of the earth on its own axis every 24 hours produces day and night and the energy changes associated with these periods. The changes in temperature lag behind the

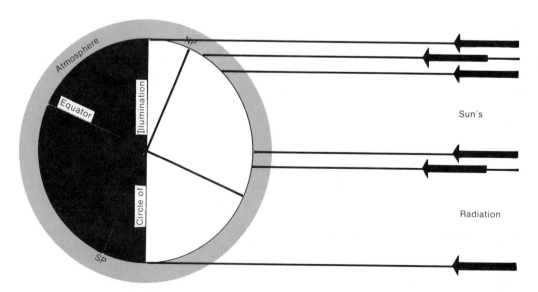

Figure 34–5 Circle of illumination, areas of daylight and darkness, angles of sun's rays at different latitudes and differences in areas affected and thickness of atmosphere penetrated at time of summer solstice. (After Ward, H. B., and Powers, W. E.: *Introduction to Weather and Climate.* Evanston, Ill., Northwestern University Press, 1942.)

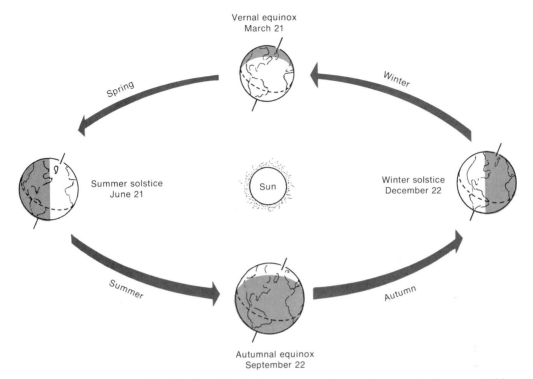

Figure 34-6 The sunlit portions of the Northern Hemisphere are seen to vary from greater than one half in summer to less than one half in winter. The proportion of any latitude that is sunlit is also the proportion of the 24-hour day between sunrise and sunset. (From MacArthur, R. H., and Connell, J. H.: *The Biology of Populations.* New York, John Wiley & Sons, Inc., 1966.)

changes in the amount of light energy received. The maximum temperature during the day is usually in midafternoon, and the lowest temperature is just before sunrise. The temperature of the soil tends to lag even more than this. The atmosphere changes the distribution of the energy of different wave lengths so that the nature of sunlight actually reaching the earth is different from the sunlight some 100 miles above the atmosphere. Radiant energy of short wave lengths is not absorbed by water or water vapor; it thus passes through the atmosphere with little diminution. Some of the energy of sunlight is absorbed by the earth, some is reflected back as longer wave lengths or heat. Snow, water and light-colored soils reflect heat, whereas bare ground with dark soils absorb it. This situation has been altered further by man's activities in paving large areas of the earth and in building, plowing, removing trees and causing air pollution. It is estimated that as much as 40 per cent of the heat of the atmosphere is derived from the condensation of water vapor derived from the evaporation of water from the surface of the ocean. The moisture-laden air rises, moves to higher lati-

tudes where it is cooled, and gives up its moisture as clouds or rain. The heat is then absorbed by the wet atmosphere. The atmosphere, heated from below and radiating heat back to the surface of the earth, serves as a heat trap, as does the roof of a greenhouse whose glass substitutes for clouds and water vapor.

34-9 ENERGY FLOW

Most of the sun's energy that reaches the earth is eventually lost as heat. A small proportion of the energy of sunlight is absorbed by plants, and a small portion of this is transformed into the potential energy of stored food products. The rest of the energy leaves the plant and becomes part of the earth's general heat loss. All living things except green plants obtain their energy by taking in the products of photosynthesis, carried out by green plants, or the products of chemosynthesis, carried out by microorganisms. Each organism is in a dynamic state and its constituents are constantly being degraded and rebuilt. Thus each organism can be considered as a sort of durable framework through which energy and matter flow in this dynamic state. When carbon or nitrogen atoms enter an organism, they are synthesized into the compounds characteristic for that organism, then are subsequently returned to the environment. The entire earth has a finite amount of carbon, nitrogen and other atoms, and these must constantly be recycled (p. 131). The degradation of organic molecules by decomposer organisms is of great importance in preventing the catastrophe that would occur if all the carbon or nitrogen atoms became bound in some life form and could no longer be used.

The various kinds of organisms in nature are balanced with their environment, but for many organisms this balance is a precarious one. Members of the human race have upset the initial balance of nature in an alarming number of instances, and many of these, once undone, are very difficult, if not impossible, to reestablish. The reasons for this will become more apparent when we discuss community successions in the next chapter. In tropical Africa and South America the natives clear portions of the rain forest for fields to grow crops. Without fertilization the cleared area may produce a crop for only a few years. When it becomes unproductive the area is abandoned and more forest is felled to provide new crop land. The abandoned area can probably never again be covered with a mature rain forest. The thin tropical soils have very meager supplies of mineral nutrients; these are leached out of the soil by the heavy rains. The mature rain forest is in precarious balance with the soil and can maintain itself only as long as the balance is not disturbed. Once the balance has been upset, the forest is irretrievably lost. Human beings in the more technologically advanced countries have done similar things, much more efficiently and with equally devastating results. Many tragic examples can be given of our blunders in upsetting critical ecological balances. The organisms living in nature are parts of complex interacting communities of many species and are not single, isolated species. The evolution of organisms is determined not simply by the limitations and peculiarities of the temperature, soil, pH, salinity and other factors of the abiotic environment, but also by their relationships with other organisms living

in that region. It is true that organisms make the community, but it is equally true that the community makes the organisms.

The carbon, nitrogen and various other cycles in nature operate to conserve the limited amount of usable matter on the earth. In contrast, the amount of energy available is very great and is constantly being renewed in sunlight. The flow of energy is not cyclic but is one-way. By measuring the amount of energy taken up and given off by each kind of organism, the ecologist can determine the functional structure of the organisms living together in a community. From this can be calculated how much life can be supported in a given area and how many individuals of each species the area can support. As the potential energy of sunlight is transferred from plants and other primary producers through herbivores and their carnivorous predators and parasites, and ultimately, following their deaths, through the decomposer microorganisms, a large portion of the energy is lost at each step as heat. Because of this progressive loss of energy as heat, the total energy flow at each succeeding level is less and less. When an animal eats food, less than 20 per cent of the foodstuff is eventually converted into the flesh of the animal that is doing the eating. The domestic pig is one of the more efficient converters: under the best feeding practices, a pig will convert about 20 per cent of the mass of the food that it eats into pork chops and bacon.

The transfer of energy through a biological community begins when the energy of sunlight is fixed in a green plant by photosynthesis. It is estimated that only 8 per cent of the energy of the sun reaching the planet strikes green plants and that only 2 per cent of this is utilized in photosynthesis. Part of this energy is used by the plant itself to drive the many processes required for maintenance. The amount left over that is stored and expressed as growth represents the *net primary production.*

The net primary production of a field of sugar cane in Hawaii was 190 Kcal. per m.^{2} per day, and the average insolation was about 4000 Kcal. per m.^{2} per day. From this we can calculate that the net efficiency of the sugar cane is about 4.8 per cent. Such values can be achieved only by crops under intensive cultivation during a favorable growing season. On an overall, annual basis sugar cane fields have

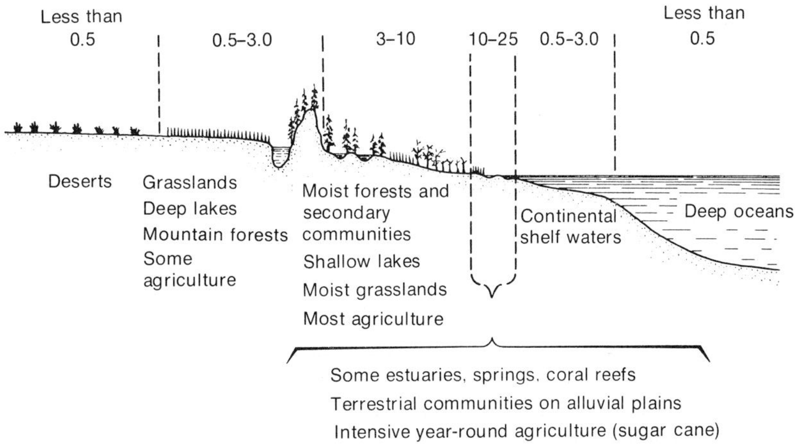

Figure 34-7 The world distribution of primary production, in grams of dry matter per square meter day, as indicated by average daily rates of gross production in major ecosystems. (From Odum, E. P.: *Fundamentals of Ecology,* 3rd Ed. Philadelphia, W. B. Saunders Co., 1971.)

an efficiency of about 1.9 per cent and tropical forests have an efficiency of about 2 per cent. The stored energy accumulates as living material or biomass. Part is recycled each season by death and decomposition of the organisms; the part that remains alive is called the *standing crop biomass.* This, of course, can vary greatly with the season. In grasslands there is an annual turnover of the biomass, but in forests much of the energy is tied up in wood. The most productive ecosystems on an energy basis are coral reefs and the estuaries of rivers (Fig. 34-7). The least productive are deserts and the open ocean. By and large the production of plant material in each given area of the earth has reached an optimum level that is limited only by soil and by the climate. The various kinds of consumers depend upon the production of green plants. Of the net production that is available to herbivores, not all is assimilated. For example, a grasshopper assimilates only about 30 per cent of its food, but some mice assimilate nearly 90 per cent. Most of this goes into the maintenance of the organism and is eventually lost as heat in the process of respiration. A small residue is stored in the form of new tissue and new individuals; it is this stored energy in the herbivore that is available to the next trophic layer, the carnivores. This enormous decrease in biomass at each stage is the basis for the concept of food chains and the pyramidal nature of the successive levels in the food chain.

34-10 FOOD CHAINS AND PYRAMIDS

The number of organisms of each species — or more precisely their total mass — is de-

termined by the rate of flow of energy through the biological part of the ecosystem that includes them.

The transfer of energy from its ultimate source in plants, through a series of organisms each of which eats the preceding and is eaten by the following, is known as a *food chain.* The number of steps in the series is limited to perhaps four or five because of the great decrease in available energy at each step. The percentage of food energy consumed that is converted to new cellular material, and thus is available as food energy for the next animal in the food chain, is known as the percentage efficiency of energy transfer.

The flow of energy in ecosystems, from sunlight through photosynthesis in autotrophic producers, through the tissues of herbivorous primary consumers and the tissues of carnivorous secondary consumers, determines the number and total weight (*biomass*) of organisms at each level in the ecosystem. The flow of energy is greatly reduced at each successive level of nutrition because of the heat losses at each transformation of energy and this decreases the biomass in each level.

Some animals eat but one kind of food and, therefore, are members of a single food chain. Other animals eat many different kinds of food and not only are members of different food chains but also may occupy different positions in different food chains. An animal may be a primary consumer in one chain, eating green plants, but a secondary or tertiary consumer in other chains, eating herbivorous animals or other carnivores (Fig. 34-8).

Humans are at the end of a number of food chains; for example, man eats a fish such as a black bass, which ate little fish, which ate

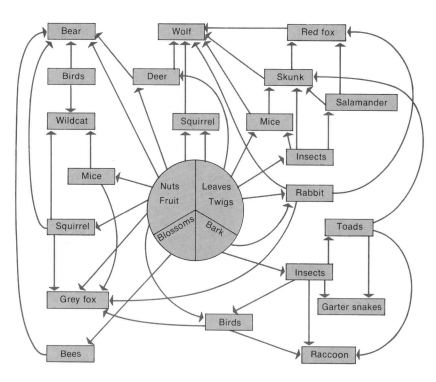

Figure 34–8 Diagram of the interrelationships in a food chain in a deciduous forest in Illinois. (After Shelford, V. E.: Ecological monographs 21:183–214, 1951.)

small invertebrates, which ate algae. The ultimate size of the human population (or the population of any animal) is limited by the length of the food chain, the per cent efficiency of energy transfer at each step in the chain and the amount of light energy falling on the earth.

Since humans can do nothing about increasing the amount of incident light energy and very little about the per cent efficiency of energy transfer, they can increase their food energy only by shortening their food chain, i.e., by eating the primary producers, plants, rather than animals. In overcrowded countries such as India and China, people are largely vegetarians because this food chain is shortest and a given area of land can in this way support the greatest number of people. Steak is a luxury in both ecologic and economic terms, but hamburger is just as much an ecologic luxury as steak is.

In addition to predator food chains, such as the man–black bass–minnow–crustacean one, there are parasite food chains. For example, mammals and birds are parasitized by fleas; in the fleas live protozoa, which are in turn the hosts of bacteria. Since the bacteria might be

parasitized by viruses, there could be a five-step parasite food chain.

A third type of food chain is one in which plant material is converted into dead organic matter, *detritus*, before being eaten by animals such as millipedes and earthworms on land, by marine worms and mollusks, or by bacteria and fungi. In a community of organisms in the shallow sea, about 30 per cent of the total energy flows via detritus chains, but in a forest community, with a large biomass of plants and a relatively small biomass of animals, as much as 90 per cent of energy flow may be via detritus pathways. In an intertidal salt marsh, where most of the animals—shellfish, snails and crabs—are detritus eaters, 90 per cent or more of the energy flow is via detritus chains.

Since in any food chain there is a loss of energy at each step, it follows that there is a smaller biomass in each successive step. H. T. Odum has calculated that 8100 kg. of alfalfa plants are required to provide the food for 1000 kg. of calves, which provide enough food to keep one 12-year-old, 48-kg. boy alive for 1 year. Although boys eat many things other than veal, and calves other things besides alfalfa, these numbers illustrate the principle of a food chain. A food chain may be visualized as a *pyramid;* each step in the pyramid is much smaller than the one on which it feeds. Since the predators are usually larger than the ones on which they prey, the pyramid of numbers of individuals in each step of the chain is even more striking than the pyramid of the mass of individuals in successive steps: one boy requires 4.5 calves, which require 20,000,000 alfalfa plants.

QUESTIONS

1. How would you define ecology? Differentiate between autecology and synecology.
2. Discuss the factors that may prevent a given species of animal or plant from becoming established in a given region.
3. What limits the number of steps in a food chain?
4. Define the term biomass. What factors determine it?
5. Contrast the predator type of food chain with one involving detritus as a step in the chain.
6. What kinds of structural adaptations are evident in a mole? a deer? a tiger? a beaver?
7. What is meant by hibernation? How can you determine whether an animal is hibernating or simply asleep?
8. Why is poikilothermic a better term for a lizard than "cold-blooded"?
9. Give examples from your experience of concealing coloration and warning coloration.
10. What experiments could you devise to determine whether color adaptations have a selective advantage for a population of animals or plants?

SUPPLEMENTARY READING

There are several excellent general texts of ecology, such as *Concepts of Ecology* by E. J. Kormondy and *Ecology: The Experimental Analysis of Distribution and Abundance* by C. J. Krebs. E. P. Odum's *Fundamentals of Ecology* provides broad coverage of the field, emphasizing especially the energy relationships in ecology. R. L. Smith's *Ecology and Field Biology* emphasizes the natural history aspects of the science. An eloquent essay on the relevance of the ecosystem to man is Aldo Leopold's "The Land Ethic" in his book *A Sand County Almanac. The Principles of Microbial Ecology* by T. D. Brock emphasizes a number of aspects of the role of microorganisms in ecological relationships that are frequently passed over lightly in general texts.

A wonderfully illustrated account of animal camouflage is found in H. B. Cott's *Adaptive Coloration of Animals. Mimicry in Plants and Animals* by W. Wickler provides many examples of camouflage in plants as well as in animals. A classic and complete text of soil science is *The Nature and Properties of Soil* by N. C. Brady and H. O. Buckman. A good summary of the principles regulating the energy environment in which we live is to be found in D. M. Gates' *Energy Exchange in the Biosphere.* The principles of ecological regulation are presented in D. Lack's *The Natural Regulation of Animal Numbers* and in L. B. Slobodkin's *Growth and Regulation of Animal Populations.* The aspects of ecology concerned primarily with animals are reviewed in T. O. Browning's *Animal Populations,* in S. C. Kendeigh's *Animal Ecology* and in C. Elton's *The Ecology of Animals.* General treatments of the ecology of plants are presented in H. J. Oosting's *The Study of Plant Communities* and in M. Treshow's *Environment and Plant Response.*

CHAPTER 35

SYNECOLOGY: COMMUNITIES, BIOMES AND LIFE ZONES

Each region of the earth—sea, lake, forest, prairie, tundra, desert—is inhabited by a characteristic assemblage of animals and plants. These are interrelated in many and diverse ways as competitors, commensals, predators and so on. The members of each assemblage are not determined by chance but by the sum of the many interacting physical and biotic factors of the environment. The ecologist refers to the organisms living in any given area as a **biotic community;** this is composed of smaller groups, or **populations,** groups of individuals of any one kind of organism.

35-1 POPULATIONS AND THEIR CHARACTERISTICS

A population may be defined as a group of organisms of the same species that occupy a given area. It has characteristics that are a function of the whole group and not of the individual members; these are *population density, birth rate, death rate, age distribution, biotic potential, rate of dispersion* and *growth form.* Although individuals are born and die, individuals do not have birth rates and death rates; these are characteristics of the population as a whole. Modern ecology deals especially with communities and populations; the study of community organization is a particularly active field at present. Population and community relationships are often more important in determining the occurrence and survival of organisms in nature than are the direct effects of physical factors in the environment.

One important attribute of a population is its *density*—the number of individuals per unit area or volume, e.g., human inhabitants per square mile, trees per acre in a forest, millions of diatoms per cubic meter of sea water. This is a measure of the population's success in a given region. Frequently in ecologic studies it is important to know not only the population density but also whether it is changing and, if so, what the rate of change is.

Population density is often difficult to measure in terms of individuals, but measures such as the number of insects caught per hour in a standard trap, the number of sea urchins caught in a standard "sea mop," or the number of birds seen or heard per hour are usable substitutes. A method that will give good results when used with the proper precautions is that of capturing, let us say, 100 animals, tagging them in some way and then releasing them. On some subsequent day, another 100 animals are trapped and the proportion of tagged animals is determined. This assumes that animals caught once are neither more likely nor less likely to be caught again and that both sets of trapped animals are random samples of the population. If the 100 animals caught on the second day include 20 tagged ones, the total population of tagged and untagged animals in the area of the traps is 500; x/100 = 100/20, hence x = 500.

For many types of ecologic investigations, an estimate of the number of individuals per total area or volume, known as the crude density, is not sufficiently precise. Only a fraction of that total area may be a suitable habitat for the population, and the size of the individual members of a population may vary tremendously. Ecologists, therefore, calculate an *ecologic density*, defined as the number, or more exactly as the mass, of individuals per area or volume of habitable space. Trapping and tagging experiments might give an estimate of 500 rabbits per square mile, but if only half of that square mile actually consists of areas suitable for rabbits to live in, then the ecologic density will be 1000 rabbits per square mile of rabbit habitat. With species whose members vary greatly in size, such as fish, live weight or some other estimate of the total mass of living fish is a much more satisfactory estimate of density than simply the number of individuals present.

A graph in which the number of organisms or the logarithm of that number is plotted against time is a *population growth curve* (Fig. 35–1). Such curves are characteristic of populations rather than of a single species and are amazingly similar for populations of almost all organisms from bacteria to human beings.

Population growth curves have a characteristic shape. When a few individuals enter a previously unoccupied area, growth is slow at first (the positive acceleration phase), then becomes rapid and increases exponentially (the logarithmic phase). The growth rate eventually slows down as environmental resistance gradually increases (the negative acceleration phase) and finally reaches an equilibrium, or saturation level. The upper asymptote of the sigmoid curve is termed the *carrying capacity* of the given environment.

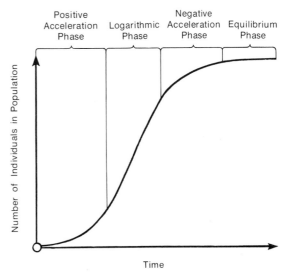

Figure 35–1 A typical sigmoid (S-shaped) growth curve of a population, one in which the total number of individuals is plotted against the time. The absolute units of time and the total number in the population would vary from one species to another, but the shape of the growth curve would be similar for all populations.

The birth rate, or natality, of a population is simply the number of new individuals produced per unit time. The *maximum birth rate* is the largest number of organisms that could be produced per unit time under ideal conditions, when there are no limiting factors. This is a constant for a species, determined by physiologic factors such as the number of eggs produced per female per unit time, the proportion of females in the species and so on. The actual birth rate is usually considerably less than this, for not all the eggs laid are able to hatch, not all the larvae or young survive and so on. The size and composition of the population and a variety of environmental conditions affect the actual birth rate. It is difficult to determine the maximum natality, for it is difficult to be sure that all limiting factors have been removed. However, under experimental conditions, one can get an estimate of this value, which is useful in predicting the rate of increase of the population and in providing a yardstick for comparison with the actual birth rate.

The mortality rate of a population refers to the number of individuals dying per unit time. There is a theoretical *minimum mortality*, somewhat analogous to the maximum birth rate, which is the number of deaths that would occur under ideal conditions, deaths due simply to the physiologic changes of old age. This minimum mortality rate is also a constant for a given population. The actual mortality rate will vary, depending upon physical factors and on the size, composition and density of the population.

By plotting the number of survivors in a population against time, one gets a *survival curve* (Fig. 35–2). If the units of the time axis are expressed as the percentage of total life span, the survival curves for organisms with very different total life spans can be compared. Modern medical and public health practices have greatly improved the average life expectancy in developed countries, and the curve for human survival approaches the curve for minimal mortality. From such curves one can determine at what stage in the life cycle a particular species is most vulnerable. Reducing or increasing mortality in this vulnerable period will have the greatest effect on the future size of the population. Since the death rate is more variable and affected to a greater extent by en-

vironmental factors than the birth rate, it has a primary role in controlling the size of a population.

It is obvious that populations that differ in the relative numbers of young and old will have different characteristics, different birth and death rates and different prospects. Death rates typically vary with age, and birth rates are usually proportional to the number of individuals able to reproduce. Three ages can be distinguished in a population in this respect: prereproductive, reproductive and postreproductive. A. J. Lotka has shown from theoretical considerations that a population will tend to become stable and have a constant proportion of individuals of these three ages.

Censuses of the ages of plant or animal populations are of value in predicting population trends. Rapidly growing populations have a high proportion of young forms. The age of fishes can be determined from the growth rings on their scales, and studies of the age ratios of commercial fish catches are of great use in predicting future catches and in preventing overfishing of a region.

The term *biotic potential*, or reproductive potential, refers to the inherent power of a population to increase in numbers when the age ratio is stable and all environmental conditions are optimal. The biotic potential is defined mathematically as the slope of the population growth curve during the logarithmic phase of growth. When environmental conditions are less than optimal, the rate of population growth is less. The difference between the potential ability of a population to increase and the actual change in the size of the population is a measure of environmental resistance.

Even when a population is growing rapidly in number, each *individual* organism of the reproductive age carries on reproduction at the same rate as at any other time; the increase in numbers is due to increased survival. At a conservative estimate, one man and one woman, with the cooperation of their children and grandchildren, could give rise to 200,000 progeny within a century, and a pair of fruit flies could multiply to give 3368×10^{52} offspring in a year. Since optimal conditions are not maintained, such biologic catastrophes do not occur, but the situations in India, Africa and elsewhere indicate the tragedy implicit in the tendency toward human overpopulation.

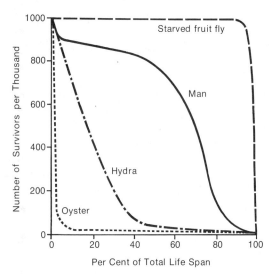

Figure 35–2 Survival curves of four different animals, plotted as number of survivors left at each fraction of the total life span of the species. The total life span for humans is about 100 years; the solid curve indicates that about 10 per cent of the babies born die during the first few years of life. Only a small fraction of the human population dies between ages 5 and 45, but after 45 the number of survivors decreases rapidly. Starved fruit flies live only about five days, but almost the entire population lives the same length of time and dies at once. The vast majority of oyster larvae die but the few that become attached to the proper sort of rock or to an old oyster shell survive. The survival curve of hydras is one typical of most animals and plants, in which a relatively constant fraction of the population dies off in each successive time period.

The sum of the physical and biologic factors that prevent a species from reproducing at its maximum rate is termed the *environmental resistance.* Environmental resistance is often low when a species is first introduced into a new territory, so that the species increases in number at a fantastic rate, as when the rabbit was introduced into Australia and the English sparrow and Japanese beetle were brought into the United States. But as a species increases in number the environmental resistance to it also increases, in the form of both the organisms that prey upon it or parasitize it and the competition between the members of the species for food and living space.

In an essay in 1798 the Englishman Robert Malthus pointed out this tendency for populations to increase in size until checked by the environment. He realized that these same principles control human populations and suggested that wars, famines and pestilences are inevitable and necessary as brakes on population growth. Since Malthus' time, productive capacity of human beings has increased tremendously as has the total human population. But Malthus' basic principle, that there are physical limits to the amount of food that can be produced for any species, remains true. The earth has a finite carrying capacity for human beings just as it does for any other animal. As environmental resistance increases, the rate of increase of the human population will eventually have to decrease. An equilibrium will be reached either by decreasing the birth rate or by increasing the mortality rate.

35–2 POPULATION CYCLES

Once a population becomes established in a certain region and has reached the equilibrium level, the numbers will vary up and down from year to year, depending on variations in environmental resistance or on factors intrinsic to the population. Some of these population variations are completely irregular, but others are regular and cyclic.

One of the best known of these is the regular nine to ten year cycle of abundance and scarcity of the snowshoe hare and the lynx in Canada, which can be traced from the records of the number of pelts received by the Hudson's Bay Company. The peak of the hare population comes about a year before the peak of the lynx population (Fig. 35–3). Since the lynx feeds on the hare, it is obvious that the lynx cycle is related to the hare cycle.

Lemmings and voles are small mouselike animals living in the northern tundra region. Every three or four years there is a great increase in the number of lemmings; they eat all the available food in the tundra and then migrate in vast numbers looking for food. They may invade villages in hordes, and finally many reach the sea and drown. The numbers of arctic foxes and snowy owls, which feed on lemmings, increase similarly. When the lemming population decreases, the foxes starve and the owls migrate south—thus there is an invasion of snowy owls in the United States every three or four years.

Although some cycles recur with great reg-

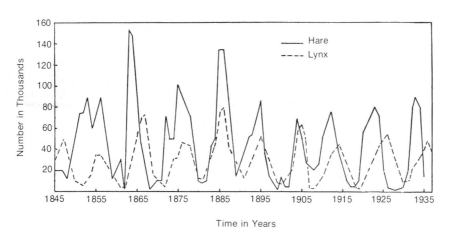

Figure 35–3 Changes in the abundance of the lynx and the snowshoe hare, as indicated by the number of pelts received by the Hudson's Bay Company. This is a classic case of cyclic oscillation in population density. (Redrawn from MacLulich, D. A.: Fluctuations in the numbers of the varying hare (*Lepus americanus*). Univ. Toronto Studies, Biol. series, no. 43, 1937.)

ularity, others do not. For example, in the carefully managed forests of Germany the numbers of four species of moths whose caterpillars feed on pine needles were estimated from censuses made each year from 1880 to 1940. The numbers varied from less than one to more than 10,000 per thousand square meters. The cycles of maxima and minima of the four species were quite independent and were irregular in their frequency and duration. Ecologically speaking, each species was marching to its own tune.

Attempts to explain these vast oscillations in numbers on the basis of climatic changes have been unsuccessful. At one time it was believed that these were caused by sunspots, and the sunspot and lynx cycles do appear to have corresponded during the early part of the nineteenth century. However, the cycles are of slightly different lengths and by 1920 were completely out of phase, sunspot maxima corresponding to lynx minima. Attempts to correlate these cycles with other periodic weather changes and with cycles of disease organisms have been unsuccessful.

The snowshoe hares die off cyclically even in the absence of predators and known disease organisms. The animals apparently die of "shock," characterized by low blood sugar, exhaustion, convulsions and finally death, symptoms that resemble the "alarm response" induced in laboratory animals subjected to physiologic stress.

This similarity led J. J. Christian to propose in 1950 that their death, like the alarm response, is the result of some upset in the adrenal-pituitary system. As population density increases, there is increasing physiologic stress on individual hares, owing to crowding and competition for food. Some individuals are forced into poorer habitats where food is less abundant and predators more abundant. The physiologic stresses stimulate the adrenal medulla to secrete epinephrine, which stimulates the pituitary via the hypothalamus to secrete more ACTH. This, in turn, stimulates the adrenal cortex to produce corticosteroids (p. 566), an excess or imbalance of which produces the alarm response or physiologic shock. In the latter part of the winter of a year of peak abundance, with the stress of cold weather, lack of food and the onset of the new reproductive season putting additional demands on the pitu-

itary to secrete gonadotropins, the adrenal-pituitary system fails, becomes unable to maintain its normal control of carbohydrate metabolism, and low blood sugar, convulsions and death ensue. This is a reasonable hypothesis, but the appropriate experiments and observations in the wild needed to test it have not yet been made.

35–3 POPULATION DISPERSION AND TERRITORIALITY

Populations have a tendency to disperse, or spread out in all directions, until some barrier is reached. Within the area, the members of the population may occur at random (this is rarely found), they may be distributed more or less uniformly throughout the area (this occurs when there is competition or antagonism to keep them apart) or, most commonly, they may occur in small groups or clumps.

Aggregation in clumps may increase the competition between the members of the group for food or space, but this is more than counterbalanced by the greater survival power of the group during unfavorable periods. It can be shown experimentally that a group of animals has much greater resistance than a single individual to adverse conditions such as desiccation, heat, cold or poisons. The combined effect of the protective mechanisms of the group is effective in countering the adverse environment, whereas that of a single individual is not.

Aggregation may be caused by local habitat differences, weather changes, reproductive urges or social attractions. Such aggregations of individuals may have a definite organization involving social hierarchies of dominant and subordinate individuals arranged in a "peck order."

Other species of animals are regularly found spaced apart; each member tends to occupy a certain area or **territory**, which he defends against intrusion by other members of the same species and sex. Usually a male establishes a territory of his own (perhaps by fighting with other males) and then, by making himself conspicuous, tries to entice a female to share the territory with him.

It has been suggested that territoriality may have survival value for a species in ensuring an adequate amount of food, nesting materials and cover for the young, in protecting the female and young against other males, and in limiting the population to a density that can be supported by the environment. Many species of birds and some mammals, fish, crabs and insects establish such territories, either as regions for gathering food or as nesting areas.

35-4 BIOTIC COMMUNITIES

A biotic community is an assemblage of populations living in a defined area or habitat; it can be either large or small. The interactions of the various kinds of organisms maintain the structure and function of the community and provide the basis for the ecologic regulation of community succession. The concept that animals and plants live together in an orderly manner, not strewn haphazardly over the surface of the earth, is one of the important principles of ecology.

Sometimes adjacent communities are sharply defined and separated from each other; more frequently they blend imperceptibly together. Why certain plants and animals constitute a given community, how they affect each other, and how humans can control them to their advantage are some of the major problems of ecologic research.

In trying to control some particular species, it has frequently been found more effective to modify the community rather than to attempt direct control of the species itself. For example, the most effective way to increase the quail population is not to raise and release birds, nor even to kill off predators, but to maintain the particular biotic community in which quail are most successful.

Although each community may contain hundreds or thousands of species of plants and animals, most of these are relatively unimportant, and only a few exert a major control on the community owing to their size, numbers or activities. In land communities these major species are usually plants, for they both produce food and provide shelter for many other species. Many land communities are named for their dominant plants—sagebrush, oak-hickory, pine and so on. Aquatic communities, containing no conspicuous large plants, are usually named for some physical characteristic—stream rapids community, mud flat community or sandy beach community.

In ecologic investigations it is unnecessary (and indeed usually impossible) to consider all the species present in a community. Usually a study of the major plants that control the community, the larger populations of animals and the fundamental energy relations (the food chains) of the system will define the ecologic relations within the community. For example, in studying a lake one would first investigate the kinds, distribution and abundance of the important producer plants and the physical and chemical factors of the environment that might be limiting. Then the reproductive rates, mortality rates, age distributions and other population characteristics of the important game fish would be determined. A study of the kinds, distribution and abundance of the primary and secondary consumers of the lake, which constitute the food of the game fish, and the nature of other organisms that compete with these fish for food would elucidate the basic food chains in the lake. Quantitative studies of these would reveal the basic energy relationships of the system and show how efficiently the incident light energy is being converted into the desired end product, the flesh of game fish. On the basis of this knowledge the lake could intelligently be managed to increase the production of game fish.

Detailed studies of simpler biotic communities, such as those of the Arctic or the desert where there are fewer organisms and their interrelations are more evident, have provided a basis for studying and understanding the much more varied and complex forest communities.

A thorough ecologic investigation of a particular region requires that the region be studied at regular intervals throughout the year for a period of several years. The physical, chemical, climatic and other factors of the region are carefully evaluated, and an intensive study is made of a number of carefully delimited areas that are large enough to be representative of the region but small enough to be studied quantitatively. The number and kinds of plants and animals in these study areas are estimated by suitable sampling techniques. Estimates are made periodically throughout the year to determine not only the components of the community at any one time but also their seasonal and annual variations. The biologic and physical data are correlated, the major and minor communities of the region are identified, and the food chains and other important ecologic relationships of the members of the community are analyzed. The particular adaptations of the animals and plants for their respective roles in the community can then be appreciated.

35–5 COMMUNITY SUCCESSION

Any given area tends to have an orderly sequence of communities that change together with the physical conditions and lead finally to a stable mature community, or *climax community.* The entire sequence of communities characteristic of a given region is termed a *sere,* and the individual transitional communities are called *seral stages,* or seral communities. In successive stages there is not only a change in the species of organisms present but also an increase in the number of species and in the total biomass.

These series are so regular in many parts of the world that an ecologist, recognizing the particular seral community present in a given area, can predict the sequence of future changes. The ultimate causes of these successions are not clear. Climate and other physical factors play some role, but the succession is directed in part by the nature of the community itself, for the action of each seral community is to make the area less favorable for itself

Figure 35–4 Dune succession. Diagram illustrating stages in succession. Note the beach grasses on the fore dunes, shrubs on older dunes, and ancient dunes with an established forest in the background. (From Norstog, K., and Long, R. W.: *Plant Biology.* Philadelphia, W. B. Saunders Co., 1976.)

Sand, present initially.   Sand, added by waves and wind.   Humus, added by plants and animals.

and more favorable for other species until the stable climax community is reached. Physical factors such as the nature of the soil, the topography and the amount of water may cause the succession of communities to stop short of the expected climax community in what is called an *edaphic climax.*

Occasionally the organisms that man wants to encourage for his own ends—timber, game birds, fresh water game fish—are members of a seral stage in community succession rather than of the climax community. Then the ecologist has the difficult problem of trying to manipulate the community to halt the succession and maintain the desired seral community.

One of the classic studies of ecologic succession was made on the shores of Lake Michigan (Fig. 35–4). As the lake has become smaller it has left successively younger sand dunes, and one can study the stages in ecologic succession as one goes away from the lake. The youngest dunes, nearest the lake, have only grasses and insects; the next older ones have shrubs such as cottonwooods, then there are evergreens and finally there is a beech-maple climax community, with deep rich soil full of earthworms and snails.

As the lake retreated it also left a series of ponds. The youngest of these contain little rooted vegetation and lots of bass and bluegills. Later the ponds become choked with vegetation and smaller in size as the basins fill. Finally the ponds become marshes and then dry ground, invaded by shrubs and ending in the beech-maple climax forest. Man-made ponds, such as those behind dams, similarly tend to become filled up.

Another dramatic example of community succession began on August 7, 1883, when a volcanic explosion occurred on the Indonesian island Krakatoa, causing part of the island to disappear. The remainder was covered with hot volcanic debris to a depth of 60 meters, and all life was obliterated. A year later some grass and a single spider were found. By 1908, 202 species of animals had taken up residence on the island. This increased to 621 species by 1919, and to 880 species by 1934, when there was a young forest on one part of the island.

Ecologic succession can be demonstrated in the laboratory. If a few pieces of dry grass are placed in a beaker of pond water, a population of bacteria will appear in a few days. Next,

flagellates appear and eat the bacteria, then ciliated protozoa such as paramecia appear and eat the flagellates. Finally predator protozoa such as *Didinium* will appear and eat the paramecia. The protozoa, present as spores or cysts attached to the grass, emerge in a definite succession of protozoan communities.

Biotic communities show marked *vertical stratification,* determined in large part by vertical differences in physical factors such as temperature, light and oxygen. The operation of such physical factors in determining vertical stratification in lakes and in the ocean is quite evident. In a forest there is a vertical stratification of plant life, from mosses and herbs on the ground to shrubs, low trees and tall trees. Each of these strata has a distinctive animal population. Even such highly motile animals as birds are restricted, more or less, to certain layers. Some species of birds are found only in shrubs, others only in the tops of tall trees. There are daily and seasonal changes in the populations found in each stratum, and some animals are found first in one, then in another layer as they pass through their life histories. These strata are interrelated in many diverse ways, and most ecologists consider them to be subdivisions of one large community rather than separate communities. Vertical stratification, by increasing the number of ecologic niches in a given surface area, reduces competition between species and enables more species to coexist in a given area.

35–6 THE CONCEPT OF THE ECOSYSTEM

The biotic community comprises all the living organisms that inhabit a certain area. A larger unit, termed the *ecosystem,* includes the organisms in a given area and the encompassing physical environment. In the ecosystem a flow of energy, derived from organism-environment interactions, leads to a clearly defined trophic structure with biotic diversity and to the cyclic exchange of materials between the living and nonliving parts of the system. From the trophic (nourishment) standpoint, an ecosystem has two components: an *autotrophic* part, in which light energy is captured or "fixed" and used to build simple inorganic substances into complex organic substances; and a *heterotrophic* part, in which the complex molecules undergo rearrangement, utilization and decomposition. In describing an ecosystem it is convenient to recognize and tabulate the following components: There are (1) the inorganic substances, such as carbon dioxide, water, nitrogen and phosphate, that are involved in material cycles; (2) the organic compounds such as proteins, carbohydrates and lipids that are synthesized in the biotic phase; (3) the climate, temperature and other physical factors; (4) the producers, autotrophic organisms (mostly green plants) that can manufacture complex organic materials from simple inorganic substances; (5) the macroconsumers, or *phagotrophs*, heterotrophic organisms (mostly animals) that ingest other organisms or chunks of organic matter; and (6) the microconsumers, or *saprotrophs* (Fig. 35–5), heterotrophic organisms (mostly fungi and bacteria) that break down the complex compounds of dead organisms, absorb some of the decomposition products and release inorganic nutrients that are made available to the producers to complete the various cycles of elements.

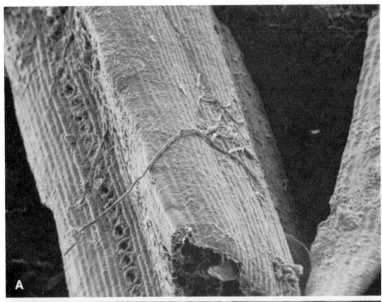

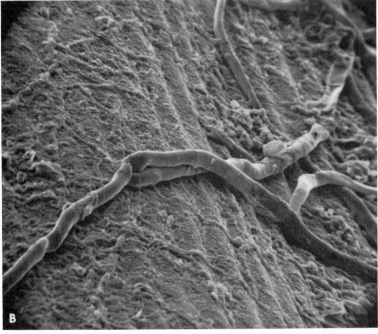

Figure 35–5 Scanning electron micrograph of a branching fungal mycelium on a pine needle undergoing decomposition in the litter on the forest floor. *A,* A section of pine needle, magnified 100×, showing the mycelium on its surface. *B,* An enlarged view of the mycelium, magnified 500×. (From Odum, E. P.: *Fundamentals of Ecology,* 3rd Ed. Philadelphia, W. B. Saunders Co., 1971. Courtesy of Dr. Robert Todd, Institute of Ecology, University of Georgia.)

The producers, phagotrophs and saprotrophs make up the biomass of the ecosystem, the living weight. In analyzing an ecosystem, the investigator studies the energy circuits present, the food chains, the patterns of biologic diversity in time and space, the nutrient cycles, the development and evolution of the ecosystem and the factors that control the composition of the ecosystem. It is important to realize that the ecosystem is the basic functional unit in ecology and includes both the biotic communities and the abiotic environment in a given region, each of which influences the properties of the other and both of which are needed to maintain life on the earth.

The term *ecosystem* was proposed by the British ecologist A. G. Tansley in 1935, but the concept of the unity of organisms and environment can be traced back to very early biological literature. The concept of the ecosystem is a very broad one and emphasizes the obligatory relationships, the causal relationships and the interdependence of biotic and abiotic components. The ecosystem is the level of biological organization particularly suitable for the application of the techniques of systems analysis. An entity may be considered an ecosystem if its major components are present and operate together to achieve some sort of functional stability, even though this may persist for only a short time. Thus, a pond, a lake, a tract of forest and an aquarium in the laboratory are examples of ecosystems. Even a temporary pond is a definite ecosystem. It contains characteristic organisms and undergoes characteristic processes even though its existence may be limited to a relatively brief period of time.

The interaction of autotrophic and heterotrophic components is a universal feature of all ecosystems, whether they are located on land, in fresh water or in the ocean. Frequently the autotrophic and heterotrophic components are partially separated spatially. The greatest amount of autotrophic metabolism occurs in a "green belt" stratum in which light energy is available. Below this lies a "brown belt" in which the most intense heterotrophic metabolism takes place. In the brown belt organic matter tends to accumulate both in soils and in sediments. The two functions may also be partially separated in time, for there may be a considerable delay in the heterotrophic utilization of the products of autotrophic organisms. In a forest, for example, the products of photosynthesis tend to accumulate in the form of leaves, wood and the food stored in seeds and roots. A relatively long time may elapse before these materials become litter and soil and available to the heterotrophic system.

The two major energy circuits in any ecosystem are the *grazing circuit,* in which animals eat living plants or parts of plants, and the contrasting *organic detritus circuit,* in which dead materials accumulate and are decomposed by bacteria and fungi. From an operational standpoint the living and nonliving parts of ecosystems are tightly interwoven and difficult to separate. Both inorganic compounds and organic compounds not only are found within and without living organisms, but also are in a constant state of flux between living and nonliving conditions. A few substances, such as ATP, are found uniquely inside living cells. In contrast, humic substances, the resistant end products of decomposition, are never found inside living cells, yet are a major and characteristic component of all ecosystems. DNA and chlorophyll may occur both inside and outside organisms, but are nonfunctional when outside the cell. Ecologists can measure the amount of ATP, humus and chlorophyll in a given area or volume to provide an index of the biomass, the decomposition and the production, respectively, in that ecosystem.

The three living components of an ecosystem, the producers, phagotrophs and saprotrophs, are roughly equivalent to plants, animals and bacteria plus fungi, respectively. However, it should be realized that these are functional classifications and that some species of organisms occupy intermediate positions, while others shift their mode of nutrition according to the circumstances of the environment. The heterotrophs can be separated into large and small consumers in an arbitrary fashion.

An excellent way to study ecology is to investigate a small pond, a meadow or an old field. Any area exposed to light—a lawn, a flower-box in the window or a laboratory aquarium—can be used in an initial study of an ecosystem as long as the physical dimensions and biotic diversity of the area are not so great that observation of the whole is difficult.

35–7 THE HABITAT APPROACH

The subject of ecology can be approached through discussions of the principles and concepts of the science as they apply to different levels of organization, the individual, population, community and ecosystem. Another general approach, the *habitat approach*, describes the distinctive features of the major habitats and their subdivisions, how they are organized, the organisms present in each and the ecologic role of these organisms in that region (i.e., the identity of the major producers, consumers and decomposers).

Four major habitats can be distinguished: *marine, estuarine, fresh water* and *terrestrial.* No plant or animal is found in all four major habitats, and indeed no animal or plant is found everywhere within any one of these. Every species of animal and plant tends to produce more offspring than can survive within the normal range of the organism. There is strong *population pressure* tending to force the individuals of each species to spread out and become established in new territories. Competing species, predators, lack of food, adverse climate and the unsuitability of the adjacent regions, perhaps owing to the lack of some requisite physical or chemical factor, all act to counterbalance the population pressure and to prevent the spread of the species. Since all these factors are subject to change, the range of a species tends to be dynamic rather than static and may change quite suddenly. The spread of a species is prevented by geographic *barriers*, such as oceans, mountains, deserts and large rivers, and is facilitated by *highways*, such as land connections between continents. The present distribution of plants and animals is determined by the barriers and highways that exist now and those that have existed in the geologic past.

The biogeographic realms, discussed on page 780, are regions made up of whole continents, or of large parts of a continent, separated by major geographic barriers and characterized by the presence of certain unique animals and plants. Within these biogeographic realms, and established by a complex interaction of climate, other physical factors and biotic factors, are large, distinct, easily differentiated community units called *biomes.* A biome is a large community unit characterized by the kinds of plants and animals present. This may be contrasted to the ecosystem, which is a natural unit of living and nonliving components that interact to form a stable system in which the exchange of materials follows a circular path. Thus an ecosystem might be a small pond (Fig. 6–5) or a large area coextensive with a biome but including the physical environment as well as the populations of animals, plants and microorganisms.

In each biome the *kind* of climax vegetation is uniform—grasses, conifers, deciduous trees—but the particular *species* of plant may vary in different parts of the biome. The kind of climax vegetation depends upon the physical environment, and the two together determine the kind of animals present. The definition of biome includes not only the actual climax community of a region but also the several intermediate communities that precede the climax community.

There is usually no sharp line of demarcation between adjacent biomes; instead each blends with the next through a fairly broad transition region termed an *ecotone.* There is, for example, an extensive region in northern Canada where the tundra and coniferous forests blend in the tundra-coniferous forest ecotone. The ecotonal community typically consists of some organisms from each of the adjacent biomes plus some that are characteristic of, and perhaps restricted to, the ecotone. There is a tendency (called the *edge effect*) for the ecotone to contain both a greater number of species and a higher population density than either adjacent biome.

Some of the biomes recognized by ecologists are *tundra, coniferous forest, deciduous forest, broad-leaved evergreen subtropical forest, grassland, desert, chaparral* and *tropical rain forest.* These biomes are distributed, though somewhat irregularly, as belts around the world (Fig. 35–6), and as one travels from the equator to the pole he may traverse tropical rain forests, grassland, desert, deciduous forest, coniferous forest and finally reach the tundra in northern Canada, Alaska or Siberia.

Since climatic conditions at higher altitudes are in many ways similar to those at higher latitudes, there is a similar succession of biomes on the slopes of high mountains (Fig. 35–7). For example, as one goes from the San

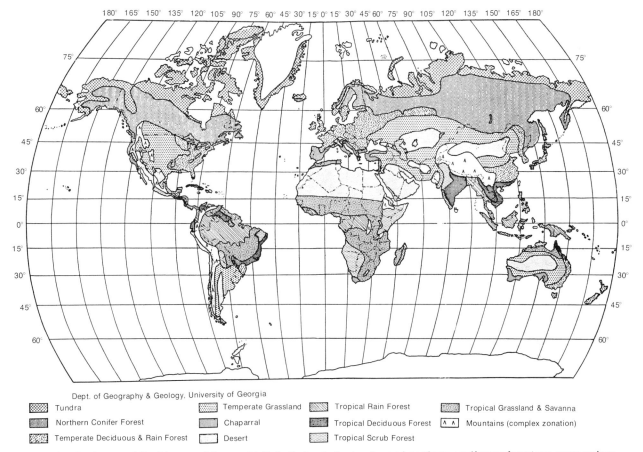

Dept. of Geography & Geology, University of Georgia

Tundra	Temperate Grassland	Tropical Rain Forest	Tropical Grassland & Savanna
Northern Conifer Forest	Chaparral	Tropical Deciduous Forest	Mountains (complex zonation)
Temperate Deciduous & Rain Forest	Desert	Tropical Scrub Forest	

Figure 35–6 A map of the biomes of the world. Note that only the tundra and northern coniferous forest are more or less continuous bands around the world. Other biomes are generally isolated in different biogeographic realms and may be expected to have ecologically equivalent but taxonomically unrelated species. (From Odum, E. P.: *Fundamentals of Ecology*, 3rd Ed. Philadelphia, W. B. Saunders Co., 1971.)

Joaquin Valley of California into the Sierras, one passes from desert through grassland and chaparral to deciduous forest and coniferous forest and then, above timberline, to a region resembling the tundra of the Arctic.

35–8 THE TUNDRA BIOME

Between the Arctic Ocean and polar ice-caps and the forests to the south lies a band of treeless, wet, arctic grassland called the *tundra* (Fig. 35–8). Some five million acres of tundra stretch across northern North America, northern Europe and Siberia. The primary characteristics of this region are the low temperatures and short growing season. The amount of precipitation is rather small but water is usually not a limiting factor because the rate of evaporation is also very low.

The ground usually remains frozen except for the uppermost 10 or 20 cm., which thaw during the brief summer season. The permanently frozen deeper soil layer is called the *permafrost.* The rather thin carpet of vegetation includes lichens, mosses, grasses, sedges and a few low shrubs. The animals that have adapted to survive in the tundra are caribou or reindeer, the arctic hare, arctic fox, polar bear, wolf, lemming, snowy owl, ptarmigan and, during the summer, swarms of flies, mosquitoes and a host of migratory birds.

The caribou and reindeer are highly migratory because there is not enough vegetation produced in any one local area to support them. Although casual inspection might suggest that tundras are rather barren areas, a surprisingly large number of organisms have become adapted to survive the cold. During the long daylight hours of the brief summer,

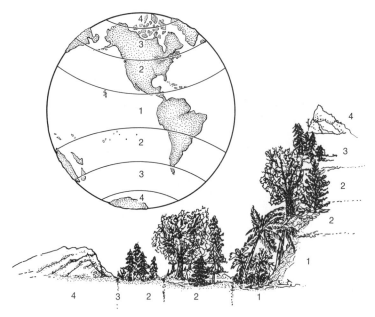

Figure 35–7 Diagram showing correspondence of life zones at successively higher altitudes at the same latitude (*1* to *4, right*) and at successively higher latitudes at the same altitude (*1* to *4, left,* and *inset*).

1. TROPICAL ZONE
Tropical forests

2. TEMPERATE ZONE
Deciduous and coniferous forests

3. ALPINE ZONE
Low herbaceous vegetation, mosses and lichens

4. POLAR ZONE
Snow and ice

the rate of primary production is quite high. The production from the vegetation on the land, from the plants in the many shallow ponds that dot the landscape and from the phytoplankton in the adjacent Arctic Ocean provides enough food to support a variety of resident mammals and many kinds of migratory birds and insects.

35–9 THE FOREST BIOMES

Several different types of forest biomes can be distinguished, and these are arranged generally on a gradient from north to south or from high altitude to lower altitude. Adjacent to the tundra region either at high latitude or high altitude is the *northern coniferous forest* (Fig.

Figure 35–8 The tundra biome. View of the tundra vegetation in Iceland showing "lumpy" nature of the low tundra. (From Norstog, K., and Long, R. W.: *Plant Biology.* Philadelphia, W. B. Saunders Co., 1976.)

Figure 35–9 The coniferous forest biome covers parts of Canada, northern Europe, and Siberia and extends southward at higher altitudes on the larger mountain ranges. (From Orr, R. T.: *Vertebrate Biology,* 4th Ed. Philadelphia, W. B. Saunders Co., 1976.)

35–9), which stretches across both North America and Eurasia just south of the tundra. This is characterized by spruce, fir and pine trees and by animals such as the snowshoe hare, the lynx and the wolf.

The evergreen conifers provide dense shade throughout the year; this tends to inhibit development of shrubs and a herbaceous undergrowth. The continuous presence of green leaves permits photosynthesis to occur throughout the year despite the low temperature during the winter and results in a fairly high annual rate of primary production.

These coniferous forests are the major source of commercial lumber around the world. After they have fallen, the needles decay very slowly and the soil develops a characteristic condition with relatively little humus. The northern coniferous forest, like the tundra, shows a very marked seasonal periodicity, and the populations of animals tend to show marked peaks and depressions in numbers.

A distinctive subdivision of the northern coniferous forest biome, perhaps distinctive enough to be considered a separate biome, is

Figure 35–10 The piñon-juniper biome in Arizona. The small piñon pines and cedars each grow some distance from the neighboring trees, giving an open, parklike appearance to the woodland. (U.S. Forest Service photo.)

the pigmy conifer or *piñon-juniper biome* found in west central California and in the Great Basin and Colorado River regions of Nevada, Utah, Colorado, New Mexico and Arizona. This occupies a belt between the desert or grasslands at lower altitudes and the true northern coniferous forest found at higher altitudes where there is more rainfall. In this region the annual rainfall of 25 to 50 cm. is irregularly distributed throughout the year. The small piñon pines and cedars tend to be widely spaced, and the biome has an open, parklike appearance (Fig. 35–10). The pine nuts and cedar berries are eaten by the piñon jay, the gray titmouse and the bush tit.

Along the west coast of North America from Alaska south to central California is a region termed the *moist coniferous forest biome*, characterized by a much greater humidity, somewhat higher temperatures, and smaller seasonal ranges than those of the classic coniferous forests farther north. There is high rainfall, from 75 to 375 cm. per year, and, in addition, a great deal of moisture is contributed by the frequent fogs. There are forests of

Sitka spruce in the northern section, western hemlock, arbor vitae, and Douglas fir in the Puget Sound area, and the coastal redwood, *Sequoia sempervirens*, in California. These forests typically have a luxuriant ground cover of ferns and other herbaceous plants. The potential production of this region is very great, and with careful foresting and replanting, the annual crop of lumber is very high.

The *temperate deciduous forest biome* (Fig. 35–11) is found in areas with abundant, evenly distributed rainfall (75 to 150 cm. annually) and moderate temperatures with distinct summers and winters. Temperate deciduous forest biomes originally covered eastern North America, all of Europe, parts of Japan and Australia, and the southern portion of South America.

The trees present—beech, maple, oak, hickory and chestnut—lose their leaves during half the year; thus, the contrast between winter and summer is very marked. The undergrowth of shrubs and herbs is generally well developed. The animals originally present in the forest were deer, bears, squirrels, gray foxes,

Figure 35–11 A beech-maple forest in New England. (From Norstog, K., and Long, R. W.: *Plant Biology*. Philadelphia, W. B. Saunders Co., 1976.)

bobcats, wild turkeys and woodpeckers. Much of this forest region has now been replaced by cultivated fields and cities.

In regions of fairly high rainfall but where temperature differences between winter and summer are less marked, as in Florida, is the *broad-leaved evergreen subtropical forest biome.* The vegetation includes live oaks, magnolias, tamarinds and palm trees, with many vines and epiphytes such as orchids and Spanish moss.

The variety of life reaches its maximum in the *tropical rain forests* (Fig. 35–12), which occupy low-lying areas near the equator with annual rainfalls of 200 cm. or more. The thick rain forests, with a tremendous variety of plants and animals, are found in the valleys of the Amazon, Orinoco, Congo and Zambesi Rivers and in parts of Central America, Malaya, Borneo and New Guinea.

The extremely dense vegetation makes it difficult to study or even photograph the rain forest biome. The vegetation is vertically stratified, with tall trees often covered with vines, creepers, lianas and epiphytes. Under the tall trees is a continuous evergreen carpet, the canopy layer, some 25 to 35 meters tall. The lowest layer is an understory that becomes dense where there is a break in the canopy.

No single species of animal or plant is present in large enough numbers to be dominant. The diversity of species is remarkable; there may be more species of plants and insects in a few acres of tropical rain forest than in all of Europe. The trees of the tropical rain forest are usually evergreen and rather tall. Their roots are often shallow and have swollen bases or flying buttresses.

The tropical rain forest is the ultimate of jungles, although the low light intensity at the ground level may result in sparse herbaceous vegetation and actual bare spots in certain areas. Many of the animals live in the upper layers of the vegetation. Among the characteristic animals are monkeys, sloths, termites, ants, anteaters, many reptiles and many brilliantly colored birds—parakeets, toucans and birds of paradise.

35–10 THE GRASSLAND BIOME

The *grassland biome* (Fig. 35–13) is found where rainfall is about 25 to 75 cm. per year, not enough to support a forest, yet more than that of a true desert. Grasslands typically occur in the interiors of continents—the prairies of the western United States and those of Argentina, Australia, southern Russia and Siberia. Grasslands provide natural pasture for grazing animals, and our principal agricultural food

Figure 35–12 The rain forest biome: border of a clearing in the Ituri Forest of Nala, the Congo. (Photograph by Herbert Lang. Courtesy of The American Museum of Natural History, New York.)

Figure 35–13 A region of shortgrass grassland with a herd of bison, originally one of the major grazing animals in the grassland biome of western United States and Canada. The bison in the center is wallowing. (From Odum, E. P.: *Fundamentals of Ecology*, 3rd Ed. Philadelphia, W. B. Saunders Co., 1971.)

plants have been developed by artificial selection from the grasses.

The mammals of the grassland biome are either grazing or burrowing forms—bison, antelope, zebras, wild horses and asses, rabbits, ground squirrels, prairie dogs and gophers. These characteristically aggregate into herds or colonies, which probably provides some protection against predators. The birds characteristic of grasslands are prairie chickens, meadowlarks and rodent hawks.

The species of grasses present in any given grassland may range from tall species, 150 to 250 cm. in height, to short species of grass that do not exceed 15 cm. in height. Some species of grass grow in clumps or bunches and others spread out and form sods with underground rhizomes. The roots of the several species of grass found in grasslands all penetrate deeply into the soil and the weight of the roots of a healthy plant will be several times the weight of the shoot.

Trees and shrubs may occur in grasslands either as scattered individuals or in belts along the streams and rivers. The soil of grasslands is very rich in humus because of the rapid growth and decay of the individual plants. The grassland soils are well suited for growing cultivated food plants such as corn and wheat, which are species of cultivated grasses. The grasslands are also well adapted to serve as natural pastures for cattle, sheep and goats. However, when grasslands are subjected to consistent overgrazing and overplowing, they can be turned into man-made deserts.

There is a broad belt of tropical grassland, or **savannah**, in Africa lying between the Sahara desert and the tropical rain forest of the Congo basin. Other savannahs are found in South America and Australia. Although the annual rainfall is high, as much as 125 cm., a distinct, prolonged dry season prevents the development of a forest. During the dry season there may be extensive fires, which play an important role in the ecology of the region. In this region (Fig. 35–14) are great numbers and great varieties of grazing animals and predators such as lions.

How to make best use of these African grasslands is a problem now facing the new nations of Africa as they work to raise the level of nutrition in their human populations. Many ecologists are of the opinion that it would be better to harvest the native herbivores—antelope, hippopotamuses and wildebeests—on a sustained yield basis rather than try to exterminate them completely and substitute cattle. The diversity of the natural population would mean broader use of all the resources of primary production, and the native species are immune to the many tropical parasites and diseases that plague the cattle that have been introduced.

35–11 THE CHAPARRAL BIOME

In mild, temperate regions of the world with relatively abundant rain in the winter but very dry summers the climax community in-

Figure 35–14 The savannah biome; characteristic animals of the African grasslands—zebra and wildebeest—Kruger National Park, Transvaal, Republic of South Africa. (From Odum, E. P.: *Fundamentals of Ecology,* 2nd Ed. Philadelphia, W. B. Saunders Co., 1959. Photograph by Herbert Lang.)

cludes trees and shrubs with hard, thick evergreen leaves (Fig. 35–15). This type of vegetation is called chaparral in California and Mexico, macchie around the Mediterranean and mellee scrub on Australia's south coast.

The trees and shrubs common in California's chaparral are chamiso and manzanita. Eucalyptus trees introduced from Australia's south coast into California's chaparral region have prospered mightily and have replaced to a considerable extent the native woody vegetation in areas near cities.

Mule deer and many kinds of birds live in the chaparral during the rainy season but move north or to higher altitudes to escape the hot, dry summer. Brush rabbits, wood rats, chipmunks, lizards, wren-tits and brown towhees are characteristic animals of the chaparral biome. During the hot, dry season, there is an ever present danger of fire, which may sweep rapidly over the chaparral slopes. Following a fire, the shrubs sprout vigorously after the first rains and may reach maximum size within twenty years.

35–12 THE DESERT BIOME

In regions with less than 25 cm. of rain per year, or in certain hot regions where there may be more rainfall but with an uneven distribution in the annual cycle, vegetation is sparse and consists of greasewood, sagebrush or cactus. The individual plants in the desert are typically widely spaced, with large bare areas separating them. In the brief rainy season, the California desert becomes carpeted with an amazing variety of wildflowers and grasses, most of which complete their life cycle from seed to seed in a few weeks. The animals present in the desert are reptiles, insects and burrowing rodents such as the kangaroo rat and pocket mouse, both of which are able to live

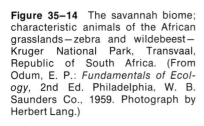

Figure 35–15 An example of the chaparral biome in California. The shrubs in the picture are *Eriodictyon tomentosum.* (U.S. Forest Service photo.)

without drinking water by extracting the water from the seeds and succulent cactus they eat.

The small amount of rainfall may be due to continued high barometric pressure, as in the Sahara and Australian deserts; a geographical position in the rain shadow of a mountain, as in the western North American deserts; or to high altitude, as in the deserts in Tibet and Bolivia. The only absolute deserts, where little or no rain ever falls, are those of northern Chile and the central Sahara.

Careful measurements of the amount of dry matter produced for a given area in the course of a year show a clear linear relationship with the amount of rainfall, at least up to 60 cm. per year. This illustrates clearly the primary role of moisture as a limiting factor in the productivity of the desert. Where the soil is favorable, an irrigated desert can be extremely productive because of the large amount of sunlight.

Two types of deserts can be distinguished on the basis of their average temperatures: "hot" deserts, such as that found in Arizona, characterized by the giant saguaro cactus, palo verde trees and the creosote brush; and "cool" deserts, such as that present in Idaho, dominated by sagebrush (Fig. 35–16).

Certain reptiles and insects are well adapted for survival in deserts because of their thick, impervious integuments and the fact that they excrete dry waste matter. A few species of mammals have become secondarily adapted to the desert by excreting very concentrated urine and avoiding the sun by remaining in their burrows during the day. The camel and the desert birds must have an occasional drink of water but can go for long periods of time using the water stored in the body.

When deserts are irrigated, the large volume of water passing through the irrigation system may lead to the accumulation of salts in the soil as some of the water is evaporated, and this will eventually limit the area's productivity. The water supply itself can fail if the watershed from which it is obtained is not cared for properly. The ruins of old irrigation systems and of the civilizations they supported in the deserts of North Africa and the Near East remind us that the irrigated desert will retain its productivity only when the entire system is kept in appropriate balance.

Figure 35–16 Two types of desert in western North America. *Above,* A "cool" desert in Idaho, dominated by sagebrush. (U. S. Forest Service photo.) *Below,* A rather "hot" desert in Arizona, with giant cactus (Sagauro) and palo verde trees, in addition to creosote bushes and other desert shrubs. In extensive areas of desert country, the desert shrubs alone dot the landscape. (U. S. Soil Conservation Service photo.)

35–13 THE EDGE OF THE SEA: MARSHES AND ESTUARIES

Where the sea meets the land there may be one of several kinds of ecosystems with distinctive characteristics: a rocky shore, a sandy beach, an intertidal mud flat or a tidal estuary containing salt marshes. The word estuary refers to the mouth of a river or a coastal bay where the salinity is less than in the open ocean, intermediate between sea water and fresh water. Most estuaries, particularly those in temperate and arctic regions, undergo marked variations in temperature, salinity and other physical properties in the course of a year, and to survive there, estuarine organisms must have a wide range of tolerance to these changes (they must be euryhaline, eurythermic, and so on).

The waters of estuaries are among the most naturally fertile in the world, frequently having a much greater productivity than the adjacent sea or the fresh water up the river. This high productivity is brought about by the action of the tides, which promote a rapid circulation of nutrients and aid in the removal of waste products, and by the presence of many kinds of plants, which provide for an extensive photosynthetic carpet. These include the phytoplankton, the algae living in and on the mud, sand, rocks or other hard surfaces, and the large attached plants, "seaweeds," eel grasses and marsh grasses.

Some of the marsh grass is eaten by insects and other terrestrial herbivores, but most of it is converted to detritus and is consumed by the clams, crabs and other marine detritus eaters. Estuaries may have a high productivity of fish, oysters, shrimp and other seafood, which can be tapped by "mariculture," as is done in the oyster farms of Japan where oysters are grown suspended on rafts hanging from floats. This is an excellent way of obtaining protein foods as a harvest of the natural productivity of the estuaries. The farms must be spaced well apart and protected from pollution.

Estuaries and marshes are among the ecologic regions of the world that are most seriously threatened by human activities. They were long considered to be worthless regions in which waste materials could be dumped. Many have been irretrievably lost by being drained, filled and converted to housing developments or industrial sites. We are just beginning to appreciate that the best interests of all are served by maintaining estuaries in their natural state and protecting them from waste material and thermal and oil pollution.

35–14 MARINE LIFE ZONES

There has recently been a great upsurge of interest in oceanography in general and in marine ecology in particular as we have begun to appreciate that we have much to learn about the mysterious sea. The oceans, which cover 70 per cent of the earth's surface, constitute one of the great reservoirs of living things and of the essential nutrients needed by both land and marine organisms. It is clear that the total weight of living things (the biomass) in the ocean far exceeds that of all living things on land and in fresh water.

The seas are continuous one with another, and marine organisms are restrained from spreading to all parts of the ocean only by factors such as temperature, salinity and depth. The salinity of the open ocean is about 35 parts per thousand. In the western Baltic it is 12 parts per thousand and 0.6 parts per thousand in the Gulf of Finland owing to the inflow of fresh water. In the Red Sea, with no source of fresh water and a high rate of evaporation, it reaches 46.5 parts per thousand. The temperature of the oceans ranges from about $-2°C$ in the polar seas to $32°C$ or more in the tropics, but the annual range of variation in any given region is usually no more than $6°C$.

The waters of the seas are continually moving in vast currents, such as the Gulf Stream, the North Pacific Current and the Humboldt Current, which circle in a clockwise fashion in the northern hemisphere and counterclockwise in the southern hemisphere. These currents not only influence the distribution of marine forms but also have marked effects on the climates of the adjacent land masses. In addition, there are very slow currents of cold, dense water flowing at great depths from the polar regions toward the equator.

Where the wind consistently moves surface water away from steep coastal slopes, water from the deep is brought to the surface by a process termed *upwelling*. This water is cold and rich in nutrients that have accumulated in the depths. Regions of upwelling typically occur on the western coasts of continents, as in California, Peru and Portugal, and are the most productive of all marine areas. The upwelling produced by the Peru Current has created one of the richest fisheries in the world and supports large populations of seabirds that deposit nitrate- and phosphate-rich *guano* on the headlands and adjacent coastal islands. These upwellings are very important in returning elements such as phosphorus to the surface for recycling.

Although the saltiness of the open ocean is relatively uniform, the concentrations of phosphates, nitrates and other nutrients vary widely in different parts of the sea and at different times of the year and are usually the major factors limiting the biologic productivity of the seas in a given region.

Like the land, the ocean consists of regions characterized by different physical conditions, and consequently inhabited by specific kinds of plants and animals. All of the animal phyla except Onychophora and all of the classes except amphibians, centipedes, millipedes and insects are well represented in the oceans. Ctenophores, brachiopods, echinoderms, chaetognaths and a few lesser phyla are found only in the oceans.

A gently sloping continental shelf usually extends some distance offshore; beyond this the ocean floor (the *continental slope*) drops steeply to the abyssal region. The region of shallow water over the continental shelf is called the *neritic zone;* it can be subdivided into *supratidal* (above the high tide mark), *intertidal* (between the high and low tide lines, a region also known as the littoral) and *subtidal* regions (Fig. 35–17).

The open sea beyond the edge of the continental shelf is the *oceanic zone.* The upper part of the ocean, into which enough light can penetrate to be effective in photosynthesis, is known as the *euphotic zone.* The average lower limit of this is at about 100 meters, but in

a few regions of clear tropical water this may extend to twice that depth. The regions of the ocean beneath the euphotic zone are called the *bathyal zone* over the continental slope, to a depth of perhaps 2000 meters; the depths of the ocean beyond that constitute the *abyssal zone.*

The floor of the ocean is not flat, but is thrown into gigantic ridges and valleys. Some of the ridges rise nearly to the surface (or above it where there are oceanic islands) and some of the valleys lie 10,000 meters below the surface of the sea. Huge underwater avalanches occur from time to time as parts of the ridges tumble into the valleys.

Some organisms are bottom dwellers, called *benthos,* and creep or crawl over the bottom or are *sessile* (attached to it). Others are *pelagic,* living in the open water, and are either active swimmers, *nekton,* or organisms that float with the current, *plankton.* The plankton includes algae, protozoa, small larval forms of a variety of animals, and a few worms. The nekton includes jellyfish, squid, fishes, turtles, seals and whales. Some of the benthic animals—crabs, snails, starfish, certain worms—crawl over the substrate. Clams and worms burrow into the sand, mud or rock of the sea bottom. Sponges, sea anemones, corals, bryozoa, crinoids, oysters, barnacles and tunicates are attached to the substratum.

The intertidal zone is one of the most favorable of all the habitats in the world, and many biologists believe that life may have originated here. The abundance of water, light,

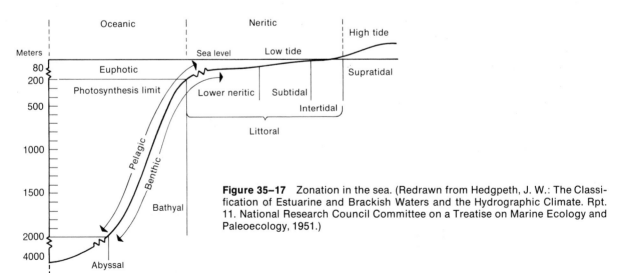

Figure 35–17 Zonation in the sea. (Redrawn from Hedgpeth, J. W.: The Classification of Estuarine and Brackish Waters and the Hydrographic Climate. Rpt. 11. National Research Council Committee on a Treatise on Marine Ecology and Paleoecology, 1951.)

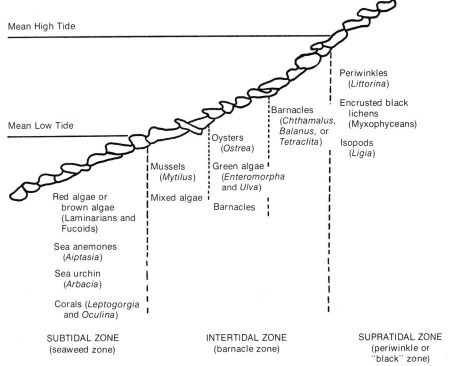

Figure 35–18   Distribution of plants and animals on a rocky shore at Beaufort, North Carolina. (From Odum, E. P.: *Fundamentals of Ecology,* 3rd Ed. Philadelphia, W. B. Saunders Co., 1971.)

oxygen, carbon dioxide and minerals makes it extremely salutary for plants, and the dense growth of plants, providing food and shelter, makes it an excellent habitat for animals. The plants of the region are primarily a wide variety of algae plus a few grasses. Many of the animals are sessile and are more or less permanently fixed to the sea bottom, though they may be pelagic at some stage of their life cycle. These sessile animals are usually restricted to certain depths of the intertidal zone (Fig. 35–18). There is keen competition among the plants for space and among the animals for space and food, so the forms living here have had to evolve special adaptations to survive.

The gravitational pulls of sun and moon each cause two bulges of the water on opposite sides of the earth, the high tides. The tides advance westward since the earth is rotating eastward on its axis. Since the earth rotates once a day on its axis, there are two high tides and two low tides per day. The periodicity of the tides is about 12.5 hours; hence the tides are about 50 minutes later on each succeeding day. Twice a month, at full and new moon, the earth, sun and moon are in line; the pulls of the sun and moon on the waters of the earth

are additive; and the difference between high and low tide is greater than normal (the spring tides). At the quarter moons, when sun and moon are pulling at right angles, the difference between high and low tide is less than usual (neap tides). The difference in tidal range between high and low tides ranges from about 30 cm. in the open ocean to 15 meters or more in the Bay of Fundy and the English Channel.

Since the intertidal zone is exposed to air twice daily, its inhabitants have had to develop some sort of protection against desiccation. Some animals avoid this by burrowing into the damp sand or rocks until the tide returns; others have developed shells which can be closed, retaining a supply of water inside. Many plants contain jellylike substances such as agar, which absorb large quantities of water and retain it while the tide is out.

One of the outstanding characteristics of this region is the ever present action of the waves, and the organisms living on a sandy or rocky beach have had to evolve ways of resisting wave action. The many seaweeds have tough pliable bodies, able to bend with the waves without breaking, while the animals either are encased in hard calcareous shells, such as those of mollusks, bryozoa, starfish, barnacles and crabs, or are covered by a strong leathery skin that can bend without breaking, like that of the sea anemone and octopus.

The successive zones of the intertidal region can be seen clearly on most rocky shores (Fig. 35–19). At the uppermost end is a zone of bare rock marking the transition between land and sea. Next is a spray zone with dark patches of algae on which the periwinkles *(Littorina)* graze. Below this is the zone regularly covered by the high tide; rocks in this zone are encrusted with barnacles, limpets and mussels. On the layer of rocks below this there is a cover of rock kelp and Irish moss containing small crabs and snails. Rocky coasts typically have tide pools which contain characteristic assemblages of animals and plants.

The sandy shore may be an even more harsh environment than the rocky shore. It is subject to all the extremes of the latter plus the inconvenience of a constantly shifting substratum. The last makes life on the surface almost impossible, and so life has retreated below the surface. Zonation on a sandy beach is illustrated in Figure 35–20. It does not, however, conform to a universal pattern as does that of rocky shores. In the example depicted, the supralittoral zone is inhabited by ghost crabs and beach hoppers. These animals spend most of the daytime hidden in damp burrows, and forage at night. Ghost crabs nightly go to the water to dampen their gill chambers. The intertidal zone is not as rich here as on the rocky shore, but it is the home of ghost shrimps, clams and bristle worms. Lower down on the beach are lugworms, trumpet worms and other species of clams. Two interesting inhabitants of this zone are the mole crab and the coquina clam. As waves roll up the beach these two small creatures emerge from the sand, ride the waves up the beach, and as the velocity of the water decreases, burrow quickly into the sand as the waves retreat. Once settled the crab extends its antennae and the clam its siphon to extract particulate food from the receding waves.

The subtidal zone is also thickly populated, for it has plenty of light and the nutrients required by plants. The absence of the periodic

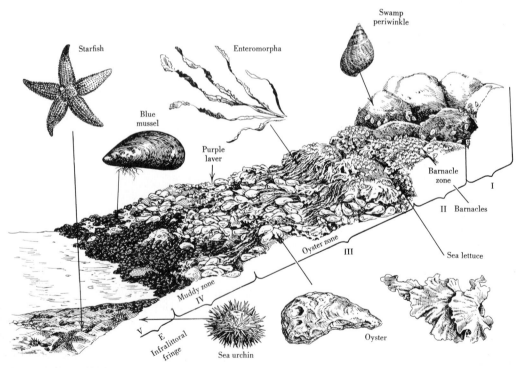

Figure 35–19 Zonation along a rocky shore—mid-Atlantic coastline. (*I*) Bare rock with some black algae and swamp periwinkle; (*II*) barnacle zone; (*III*) oyster zone: oysters, *Enteromorpha,* sea lettuce, and purple laver; (*IV*) muddy zone: mussel beds; (*V*) infralittoral fringe: starfish and so on. Note absence of kelps. (Zonation drawing based on Stephenson, 1952; sketches done from life or specimens. From Smith, R. L.: *Ecology and Field Biology.* New York, Harper & Row, Publishers, 1966.)

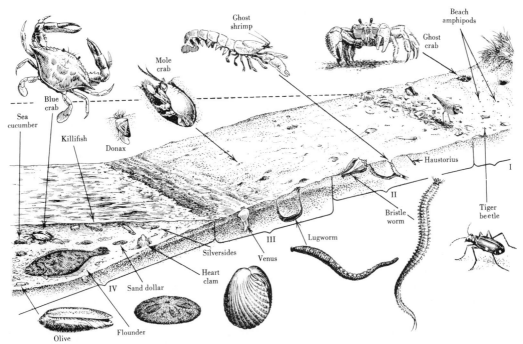

Figure 35–20 Life on a sandy ocean beach along the Atlantic Coast. Although strong zonation is absent, organisms still change on a gradient from land to sea. (*I*) Supratidal zone: ghost crabs and sand fleas; (*II*) flat beach zone: ghost shrimp, bristle worms, clams; (*III*) intratidal zone: clams, lugworms, mole crabs; (*IV*) subtidal zone: the dashed line indicates high tide. (From Smith, R. L.: *Ecology and Field Biology.* New York, Harper & Row, Publishers, 1966.)

exposure to air and the diminished wave action permit many plants and animals to live here that could not survive in the tidal zone. Here live many species of fish and many single-celled algae; the larger seaweeds, which require a substratum for attachment, are found only in the shallower parts of the region.

The subtidal zone is the home of cockles, razor clams, moon shells and other mollusks. Still farther out are starfish, sand dollars, sea cucumbers, killifish, silversides and flounders.

In this area of the marine environment, as in all others, the *primary producer organisms*, equivalent to the flowering plants on land, are the *phytoplankton* (Fig. 35–21), consisting principally of diatoms and dinoflagellates. It is difficult to appreciate their importance because they are so small. An absolutely minimal estimate would place their density at 12,500,000 individuals per cubic foot. In temperate regions the phytoplankton undergoes two seasonal population explosions or "blooms," one in the spring, the other in late summer or fall. The mechanism is similar to that responsible for "blooms" in lakes. In the wintertime low temperatures and reduced light restrict photo-

synthesis to a low level; however, these factors do not prevent bacteria and other microorganisms from generating high concentrations of nitrogen, phosphorus and other nutrient elements. When spring brings higher temperatures and more light, photosynthesis accelerates, and there is an ample supply of nutrients. The nutrient supply is ample because the winter mixing of surface and deep water brings up nutrients that have fallen to and accumulated at the bottom. Within a fortnight, the diatoms multiply 10,000-fold. This prodigious growth accounts for the spring bloom. Soon, however, the nutrients are exhausted. Replacement from lower layers no longer occurs because warming of the surface water keeps it on top and prevents mixing. Nutrients are now locked in the bodies of animals that have eaten the phytoplankton or are slowly falling to the bottom in dead bodies. Whereas temperature and light were the limiting factors during the winter, nutrient level is the limiting factor during the summer, especially since existing phytoplankton is now being consumed by animals. Now nutrients begin to accumulate again in lower layers. As fall approaches the upper

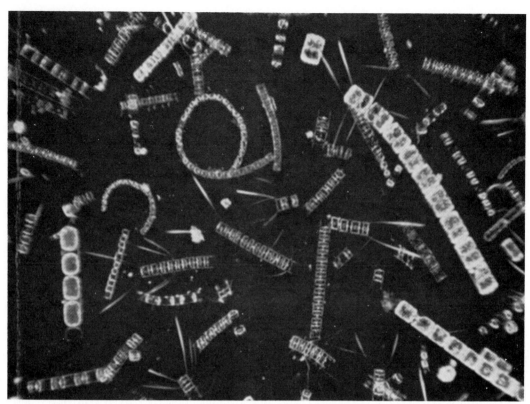

Figure 35–21   Living plants of the plankton (phytoplankton). ×110. Chains of cells of several species of *Chaetoceros* (those with spines), a chain of *Thalassiosira condensata* (at and pointing to bottom right corner) and a chain of *Lauderia borealis* (above the last named). By electronic flash. (Copyright by Douglis P. Wilson.)

layers of water begin to cool again. The accompanying density change, together with the autumn equinoctial gales, begins mixing the water again. Water rich in phosphates and nitrates is brought up from below. Other forms of phytoplankton, especially nitrogen-fixing blue-green algae, now bloom until reduced nutrients or temperature again intercedes (Fig. 35–22).

The other important population in the ocean is the *zooplankton* (Fig. 35–23). The zooplankton comprises all the animals carried passively by moving water. Every major phylum of the animal kingdom is represented, if not as adults, then at least as eggs or larval stages. All the organisms are small, but not so small as not to be visible through a 6X hand lens. The beauty and almost limitless variety of forms of these animals beggar description. A sample haul near the Isle of Man gave an average of 4500 animals per cubic meter. The same hauls gave about 727,000 planktonic plants per cubic meter.

The active surface dwellers of the *neritic*

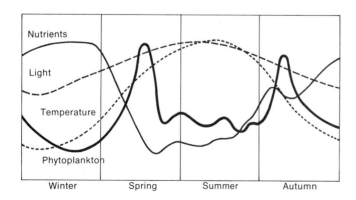

Figure 35–22   The probable mechanisms for phytoplankton "blooms." See text for explanation. (From Odum, E. P.: *Fundamentals of Ecology*, 3rd Ed. Philadelphia, W. B. Saunders Co., 1971.)

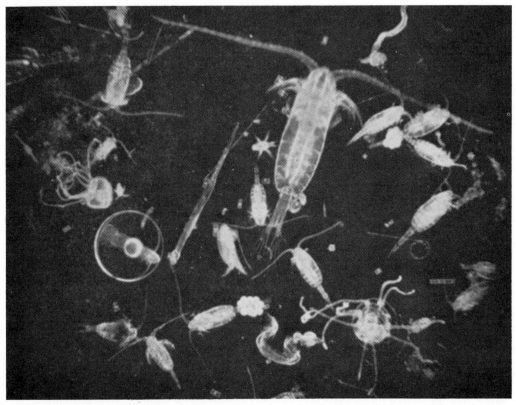

Figure 35–23 Living animals of the plankton (zooplankton), ×16. The copepods *Calanus finmarchicus* (the largest animal) and *Pseudocalanus elongatus* (similar in shape, but much smaller than *Calanus,* and the one with the cluster of eggs); two small anthomedusae with long tentacles; a fish egg (the circular object); a young arrow-worm *Sagitta* (to the right of the fish egg); small nauplius (larval stage) of copepod (close to left side of *Calanus*) and the planktonic tunicate *Oikopleura* (curly objects at top right and middle bottom). By electronic flash, but partially narcotized. (Copyright by Douglis P. Wilson.)

zone are many bony fishes, large crustacea, turtles, seals, whales and a host of sea birds. All these *consumers* are restricted in their distribution by temperature, salinity and nutrients. There is horizontal zonation and vertical stratification. The greatest numbers of fishes (individuals, not species) are found in northern waters and where there are cold upwellings. Only a few species make up the bulk of commercial fisheries. Three fourths of the world's catch consists of herring, cod, haddock, pollock, salmon, flounder, sole, plaice, halibut, mackerel, tuna and bonito groups. The cod and flounder groups are bottom fish. Upon these and other marine organisms feed a great number of birds. Sandpipers, plovers, herons, curlews and others search the supra- and intertidal zones; cormorants, sea ducks and pelicans, the subtidal zones; and petrels and shearwaters, the lower neritic zone farther out to sea.

The *oceanic region* is less rich in species and numbers than the coastal areas, but it has its characteristic species. Many of these are transparent or bluish and since the sediment-free water of the open sea is marvelously transparent, these animals are nearly invisible. Animals that are too thick to be transparent frequently have smooth shiny and silvery bodies that make them invisible by mirroring the water in which they swim. Among the most characteristic animals of the open ocean are the baleen and toothed whales, the only mammals that are truly and absolutely marine. The baleen, or whalebone, whales live on phytoplankton, which they strain from the water. The toothed whales live on nekton, and giant squids compose part of their diet. The distribution of these whales is correlated directly or indirectly with the distribution of phytoplankton. The flora and fauna of arctic seas differ from those of tropical seas. The boundaries between the two are not sharp and shift with the seasons. The major tropical oceans, the Atlantic and the Indo-Pacific, are separated from each other by regions of cold water. Each contains a large number of different species of animals and plants.

Above the surface of the open sea are the oceanic birds—the petrels, albatrosses and frigate birds. They are not truly marine because they must come to land to breed. Apart from that they may spend all their lives at sea out of sight of land. Like the whales their distribution may be worldwide but not uniform. They are restricted by the distribution of fishes, which in turn are dependent on the phytoplankton.

Eighty-eight per cent of the ocean is more than 1 mile deep. The ocean is continuous throughout the world except for the deep water of the Arctic Ocean, which is cut off from the rest by a narrow submerged mountain range connecting Greenland, Iceland and Europe. This is the area of great pressure and of perpetual night. Since no photosynthesis is possible, the only source of energy is the constant rain of organic debris, the bodies and waste products of organisms in the surface layers, that falls toward the bottom. The other prerequisite for life, oxygen, gets to the bottom by means of the oceanic circulation discussed previously.

Little is known of life in the deep because of the enormous difficulty of observation. The *pelagic* animals are strong swimmers, not easily caught in nets. The scant knowledge available has been gleaned from studies of net hauls and by observation from special undersea craft or via underwater television. In general, the animals of this world are either filter-feeders, which sieve out particles before they reach the bottom, grubbers, which ingest sediment, or predators on the filter-feeders and grubbers.

Most of the fish of the abyssal region are rather small and peculiarly shaped; many are equipped with luminescent organs, which may serve as lures for the forms preyed upon. The majority of the deep-sea creatures are related to shallow-sea forms, and they must have migrated to their present habitat recently (geologically speaking), for none is older than the Mesozoic.

Since the number of members of any one species in these vast depths is small, reproduc-

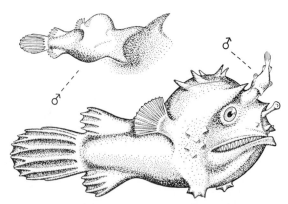

Figure 35-24 Sexual parasitism in the deep-sea angler fish. *Photocorynus spiniceps.* The small male is permanently attached to the female; he has no independent existence but is nourished by the blood of the female. (From Allee, W. C., et al.: *Principles of Animal Ecology.* Philadelphia, W. B. Saunders Co., 1949.)

tion is more of a problem than in any other region, and some fish have evolved a curious adaptation to ensure that reproduction will occur. At an early age the male becomes attached to and fuses with the head of the female, where he continues to live as a small (2.5 cm.) parasite (Fig. 35-24). In due course he becomes mature and when the female lays her eggs, he releases his sperm into the water to fertilize them.

The bottom of the sea is a soft ooze, made of the organic remains and shells of foraminifera, radiolaria, and other animals and plants. Many invertebrates live on the ocean floor even at great depths (Fig. 35-25). These are usually characterized by thin, almost transparent shells, whereas the related shallow-water forms, exposed to wave action, have thick, hard shells. Apparently even the greatest "deeps" are inhabited, for tube-dwelling worms have been dredged from depths of 8000 meters, and sea urchins, starfish, bryozoa and brachiopods have been found at depths of 6000 meters. Animals living on these bottom oozes typically have long thin appendages and possess spines or stalks.

35-15 FRESH WATER LIFE ZONES

Fresh water habitats can be divided into *standing water*—lakes, ponds and swamps—and *running water*—rivers, creeks and springs —each of which can be further subdivided.

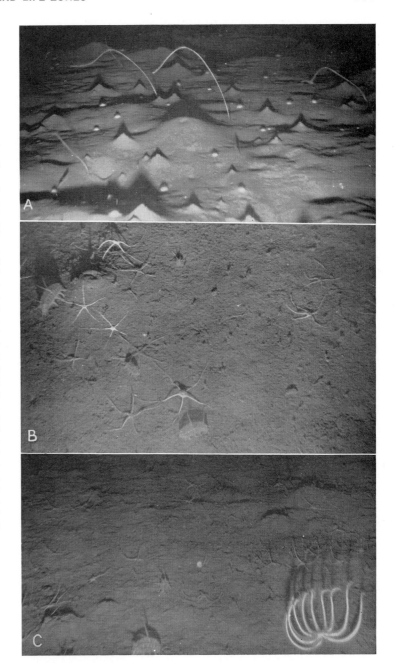

Figure 35–25 Photographs (by means of a benthograph, a special underwater camera) of the ocean bottom at three different depths off southern California (San Diego Trough). *A,* At 95 meters. Note abundant sea urchins (probably *Lytechinus*) appearing as globular, light-colored bodies, and the long, curved sea whips (probably *Acanthoptilum*). Burrowing worms have built the conical piles of sediment at the mouth of their burrows. *B,* At 1200 meters. Vertical photograph of about 4 square meters of bottom composed of green silty mud having a high organic content. Note the numerous brittle stars (Ophiuroidea) and several large sea cucumbers (Holothuroidea). The latter have not been identified as to species as they have never been dredged from the sea and have only been seen in bottom photographs! *C,* At 1400 meters. Note in the right foreground the ten arms of what is probably a comatulid crinoid (a relative of the starfish that is attached to the bottom by a stalklike part). Small worm tubes and brittle stars litter the surface, and two sea cucumbers may be seen in the left foreground. Continual activity of burrowing animals keeps the sea bottom "bumpy." The bottom edge of the picture represents a distance of about 2 meters. (*A* and *C* from Emery, K. O.: Submarine photography with the benthograph. Scient. Monthly 75:3–11, 1952. *B,* official Navy photo, courtesy of the U. S. Navy Electronics Laboratory, San Diego, Calif.)

The biologic communities of fresh water habitats are in general more familiar than the salt water ones, and many of the animals used as specimens in biology classes are from fresh water—amoebas, hydras, planarias, crayfish and frogs.

Standing water, such as a lake, can be divided (much as the zones of the ocean were distinguished) into the shallow water near the shore (the *littoral zone*), the surface waters away from the shore (the *limnetic zone*) and the deep waters under the limnetic zone (the *profundal zone*).

Aquatic life is probably most prolific in the littoral zone. Within this zone the plant communities form concentric rings around the pond or lake as the depth increases (Fig. 35–26). At the shore proper are the cattails, bulrushes, arrowheads and pickerelweeds—the emergent, firmly rooted vegetation linking water and land environments. Out slightly deeper are the rooted plants with floating

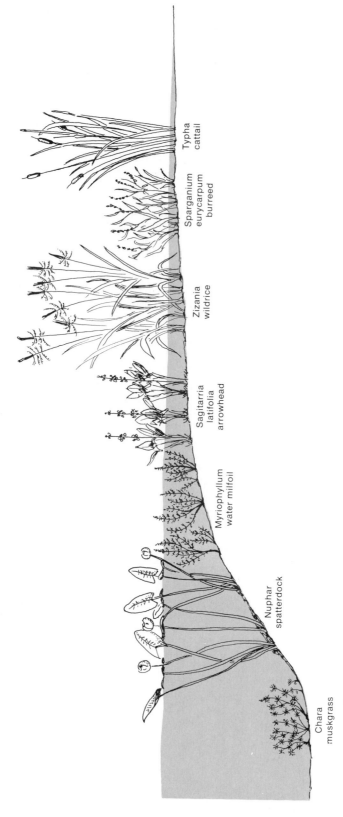

Figure 35–26 Zonation of vegetation about ponds and along river banks. Note the changes in vegetation with water depth. (After Dansereau, P.: *Biogeography: An Ecological Perspective.* New York, The Ronald Press Company, 1959; from Smith, R. L.: *Ecology and Field Biology.* New York, Harper & Row, Publishers, 1966.)

leaves such as the water lilies. Still deeper are the fragile thin-stemmed water weeds, rooted but totally submerged. Here also are found diatoms, blue-green algae and green algae. Common green pond scum is one of the latter.

The littoral zone is also the scene of the greatest concentration of animals (Fig. 35–27), distributed in recognizable communities. In or on the bottom are various dragonfly nymphs, crayfish, isopods, worms, snails and clams. Other animals live in or on plants and other objects projecting up from the bottom. These include the climbing dragonfly and damselfly nymphs, rotifers, flatworms, bryozoa, hydra, snails and others. The zooplankton consists of water fleas such as *Daphnia*, rotifers and ostracods. The larger freely swimming fauna *(nekton)* includes diving beetles and bugs, dipterous larvae (e.g., mosquitoes) and large numbers of many other insects. Among the ver-

tebrates are frogs, salamanders, snakes and turtles. Floating members of the community *(neuston)* include whirligig beetles, water striders and numerous protozoa. Many pond fish (sunfish, top minnows, bass, pike and gar) spend much of their time in the littoral zone.

The limnetic, or open-water, zone is occupied by many microscopic plants (dinoflagellates, *Euglena, Volvox*), many small crustaceans (copepods, cladocera and so on) and many fish.

Deep (profundal) life consists of bacteria, fungi, clams, bloodworms (larvae of midges), annelids and other small animals capable of surviving in a region of little light and low oxygen.

As compared to ponds where the littoral zone is large, the water usually shallow and temperature stratification usually absent, lakes have large limnetic and profundal zones, a

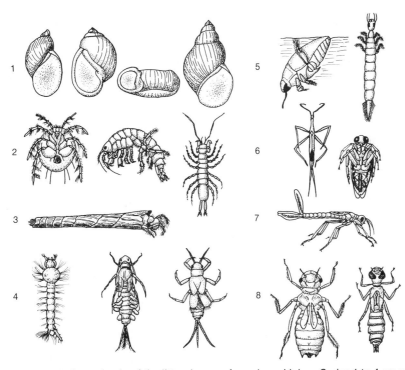

Figure 35–27 Some representative animals of the littoral zone of ponds and lakes. Series 1 to 4 are primarily herbivorous forms (primary consumers); series 5 to 8 are predators (secondary consumers). *1,* Pond snails *(left to right): Lymnaea* (pseudosuccinea) *columella; Physa gyrina; Helisoma trivolvis; Campeloma decisum. 2,* Small arthropods living on or near the bottom or associated with plants or detritus *(left to right):* a water mite, or Hydracarina *(Mideopsis);* an amphipod *(Gammarus);* an isopod *(Asellus). 3,* A pond caddis fly larva *(Triaenodes),* with its thin, light portable case. *4, (Left to right)* A mosquito larva *(Culex pipiens);* a clinging or periphytic mayfly nymph *(Cloeon);* a benthic mayfly nymph *(Caenis)*—note gill covers which protect gills from silt. *5,* A predatory diving beetle, *Dytiscus,* adult and *(right)* larva. *6,* Two predaceous Hemipterans, a water scorpion, *Ranatra* (Nepidae), and *(right)* a backswimmer, *Notonecta. 7,* A damsel fly nymph, *Lestes* (Odonata-Zygoptera); note three caudal gills. *8,* Two dragonfly nymphs (Odonata-Anisoptera), *Helocordulia,* a long-legged sprawling type (benthos), and *(right) Aeschina,* a slender climbing type (periphyton). (After Pennak, R. W.: *Fresh-water Invertebrates of the United States.* © 1953, The Ronald Press Company, New York.)

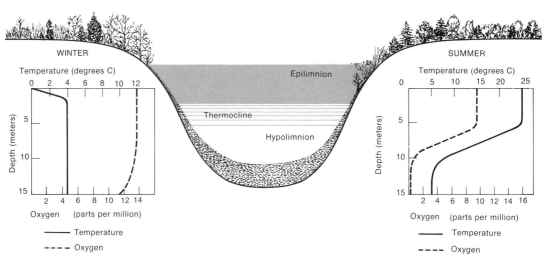

Figure 35–28 Thermal stratification in a north temperate lake (Linsley Pond, Conn.). Summer conditions are shown on the right, winter conditions on the left. Note that in summer the oxygen-rich circulating layer of water, the epilimnion, is separated from the cold, oxygen-poor hypolimnion waters by a broad zone, called the thermocline, which is characterized by a rapid change in temperature and oxygen with increasing depth. (From *Life in the Depths of a Pond* by E. S. Deevey, Jr. Copyright © by Scientific American, Inc. All rights reserved.)

marked **thermal stratification** and a seasonal cycle of heat and oxygen distribution. In the summertime, the surface water *(epilimnion)* of lakes becomes heated while that below *(hypolimnion)* remains cold. There is no circulatory exchange between upper and lower layers, with the result that the lower layers frequently become deprived of oxygen. Between the two is a region of steep temperature decline *(thermocline)*. As the cooler weather of fall approaches, the surface water cools, the temperature is equal at all levels, the water of the whole lake begins to circulate and the deep is again oxygenated. This is the *fall overturn.* In winter, as the surface temperature drops to 4°C, the water at the surface becomes less dense, remains at the surface and impedes circulation. The bottom is now warmer than the top. Because bacterial decomposition and respiration are less at low temperatures and cold water holds more oxygen, there is usually no great winter stagnation. The formation of ice may, however, cause oxygen depletion and result in a heavy winterkill of fish. The *spring overturn* occurs when the ice melts and the heavier surface water sinks to the bottom (Fig. 35–28).

Moving waters differ in three major aspects from lakes and ponds: current is a controlling and limiting factor; land-water interchange is great because of the small size and depth of moving water systems as compared

with lakes; oxygen is almost always in abundant supply except when there is pollution. The extremes of temperature tend to be greater than in standing water. Plants and animals living in streams are usually attached to surfaces or, in the case of animals, are exceptionally strong swimmers. Characteristic stream organisms are: caddis fly larvae, blackfly larvae, attached green algae, encrusting diatoms and aquatic mosses.

Fresh water habitats change much more rapidly than other life zones; ponds become swamps, swamps become filled in and converted to dry land, and streams erode their banks and change their course. The kinds of plants and animals present may change markedly and show ecologic successions similar to those on land. The large lakes, such as the Great Lakes, are relatively stable habitats and have more stable populations of plants and animals. Lake Baikal in the Soviet Union is the oldest and deepest lake in the world, formed during the Mesozoic era, and it contains many species of fish and other animals found nowhere else.

35–16 THE DYNAMIC BALANCE OF NATURE

The concept of the dynamic state of the cellular constituents was discussed in Chapter

5 and we learned that the protein, fat, carbohydrate and other constituents of the cells are constantly being broken down and resynthesized. A biotic community undergoes an analogous constant reshuffling of its constituent parts; the concept of the dynamic state of biotic communities is an important ecologic principle. Plant and animal populations are constantly subject to changes in their physical and biotic environment and must adapt or die. In addition, communities undergo a number of rhythmic changes — daily, tidal, lunar, seasonal and annual — in the activities or movements of their constituent organisms. These result in periodic changes in the composition of the community as a whole. A population may vary in size, but if it outruns its food supply, like the Kaibab deer or the lemmings, equilibrium is quickly restored. Communities of organisms are comparable in many ways to a many-celled organism, and exhibit growth, specialization and interdependence of parts, characteristic form, and even development from immaturity to maturity, old age and death.

QUESTIONS

1. What is meant by a biotic community? Give examples from your own experience.
2. What characteristics are peculiar to a population as a whole and not to its individual members?
3. What is meant by a survival curve? Discuss the importance of such curves to a life insurance company.
4. Which varies more in different populations, birth rates or death rates? Why?
5. Define the terms biotic potential and environmental resistance. Draw a population growth curve and indicate graphically the relation of these terms to the curve.
6. Explain why there is a tendency toward an orderly sequence of communities leading to a climax community. What is the climax community in your region?
7. Discuss the measures that could be taken to increase the number of beavers in Pennsylvania and the number of quails in Virginia.
8. Discuss the factors that tend to keep the size of a population of animals in the wild relatively constant. What factors tend to cause cyclic variations in the size of a population of animals in the wild?
9. What is a biome? How does it differ from a biotic community?
10. Why are similar biomes found at high latitudes and high altitudes? Would you expect to find exactly the same species of plants and animals in the tundra region of Alaska and in the tundra region of the Andes? Why? What is meant by the term ecological equivalents?
11. What are the chief differences between the intertidal zone and the shallow sea?
12. Compare the adaptations made by barnacles, snails, and starfish that enable them to survive in the intertidal zone.
13. Discuss the implications of the phrase "the dynamic state of communities."
14. Differentiate between plankton and nekton. Give examples of each.
15. Describe the subdivisions of the marine habitat and give examples of animals typically found in each.
16. Compare the flora and fauna of a swiftly flowing stream and a quiet lake.
17. Define the term ecotone. Give an example of an ecotone.

SUPPLEMENTARY READING

Texts of ecology that emphasize the population aspects of the subject are *The Biology of Populations* by R. H. MacArthur and J. H. Connell, *An Introduction to Ecology and Populations* by T. C. Emmel and *Readings in Population and Community Ecology* by W. E. Hagen. *Insect Societies* and *Sociobiology,* both by E. O. Wilson, provide masterful coverage of these fascinating subjects.

A general discussion of each of the major biomes and life zones is presented in E. P. Odum's *Fundamentals of Ecology.* More extensive discussions of the specific biomes can be found in Marston Bates' *The Forest and the Sea,* A. Hardy's *The Open Sea,* two of Rachel Carson's books, *The Sea Around Us* and *The Edge of the Sea,* D. P. Wilson's *Life of the Shore and Shallow Sea,* R. C. Coker's *Lakes, Streams and Ponds,* George Reid's *Ecology of Inland Waters and Estuaries, The Tropics* by E. Aubert de la Rue, F. Bourliere and J. P. Harroy, E. C. Jaeger's *The North American Deserts,* Willy Ley's *The Poles,* A. S. Leopold's *The Desert,* Peter Farb's *The Forest,* and P. W. Richard's *The Tropical Rain Forest.*

The September, 1969, issue of *Scientific American* was devoted to the oceans. An excellent reference source for estuarine ecology is *Estuaries,* edited by George Lauff. The biogeography of island communities is outlined in *The Theory of Island Biogeography* by R. MacArthur and E. O. Wilson. An excellent source of material relating to a variety of ecosystems is the collection edited by G. Van Dyne, *The Ecosystem Concept in Natural Resource Management.* The vegetation of the deciduous forest communities is summarized in E. Lucy Braun's *Deciduous Forests of Eastern North America,* and the ecology of the grassland biome is discussed in James Malin's *The Grasslands of North America.*

CHAPTER 36

HUMAN ECOLOGY

In recent years there has been a dramatic upsurge of interest in ecology coupled with a shift in emphasis by applied ecologists from populations to ecosystems. As ecologic principles are implemented in the control of the environment the human population is becoming aware of the "big picture" of water cycles, nutrient cycles, productivity, food chains, global pollution and related topics. Ecologists are turning their attention to the control and management of the human race as well as to the management of fish, wildlife, forests and other natural resources. The continued explosive growth of the human population has forced humans to consider the ultimate limitations of the earth rather than simply local limitations. In his famous "Essay on Population" (1798), Thomas Malthus predicted that human populations would increase faster than their food supply. It is now becoming clear that this planet is simply a big "space ship" with definite limits to both its productivity and its ability to cope with pollutants. A key question in ecology is whether we are in greater danger of running out of resources or of being overcome by polluted air and water.

A great deal of rhetoric regarding ecology and the future of mankind has been delivered. Predictions have ranged from statements that it is already too late—that the human race is doomed—to ones that deny that there is any ecological problem at all. It is becoming clear that technology alone cannot solve the problems of population and pollution; in addition, moral, economic and legal constraints are required, arising from full and complete public understanding that human beings and the environment are a unit. The technological advances to solve many aspects of the population and pollution problems have been made; what is urgently needed is a general and worldwide appreciation of the nature of the problem by informed laymen and a willingness on the part of everyone to apply the principles of ecology for the benefit of all. In the immediate future this probably will depend not so much on further advances in the environmental sciences, but rather on the use of principles and techniques from economics, law, politics, urban planning and other areas of humanistic science to solve ecologic problems.

Among the many ways in which the knowledge of the principles of ecology can be used to further human society one of the more important is the rational conservation of our natural resources. Conservation does not mean simply hoarding—not using the resources at all—nor does it imply a simple rationing of our supplies so that some will be left for the future. True conservation implies taking full advantage of our knowledge of ecology and managing the world's ecosystems so as to establish a balance of harvest and renewal, thus ensuring a continuous yield of useful plants, animals and other material. This should at the same time guarantee the preservation of an environment of high quality that offers aesthetic and recreational uses as well as material products. The record of human squandering of natural resources is indeed a dark one. The slaughter of the bison that once roamed the western plains, the decimation of the whales, the depletion of our supplies of many kinds of fresh water and marine fishes, the extinction of the passenger pigeon and other birds, the razing of thousands of square miles of forest (and the burning of even more by careless use of fire!), the pollution of streams with sewage, industrial and farm wastes, the careless cultivation of land that has resulted in the complete ruin of many square miles of valuable farmland and the silting of streams—these are some of the more flagrant examples of natural resources wasted beyond hope of rapid reclamation. Although countermeasures have been taken by state and federal departments of conservation, the chief task at present is to make the population at large realize the urgency of the problem and the magnitude of the job to be done and to obtain general support for the measures that must be taken. For many apsects of the conservation problem other additional basic research is needed to determine the possible effects of particular conservation measures on the ecology of the entire region.

36-1 AGRICULTURE

At the end of the Pleistocene, some 10,000 years ago, our human ancestors gradually changed from nomadic hunters and fishers into farmers. The earliest farmers raised wheat and barley and domesticated sheep and goats. Agricultural societies first appeared in the Near East in what is now Turkey, Iran and Iraq; they then spread across Europe, and by about 6000 years ago had reached Britain. Agricultural communities apparently originated separately in Central and South America and in China and India. The consequences of the change from a hunter society to an agricultural society were truly profound. The populations were no longer wanderers; they could store food, accumulate possessions and own land. The change to an agricultural economy led to an increase in the population. Estimates place

the number of human beings living some 10,000 years ago at about five million. In the next 4000 years the population increased to about 85 million and by the time of Christ there were some 133 million people scattered over the globe. Thus, the human population increased more than 25-fold in an 8000 year period.

The more intensive cultivation of soil needed to feed this ever-increasing population has taken its toll of the land available for farming. Wind and water have caused soil erosion through all geologic ages, but unwise farming and forestry practices have greatly increased the rate of erosion in certain parts of the globe. A century or more of destructive exploitation of farm lands by planting one crop such as corn or cotton year after year has caused severe damage to certain areas and resulted in erosion and depletion of soil nutrients. The Soil Conservation Program sponsored jointly by federal and local agencies and based on sound ecologic principles has been effective in countering erosion. The rotation of crops, contour farming, the establishment of windbreaks to prevent soil erosion by winds, and the use of proper fertilizers to renew the soil are all measures that have proved effective in maintaining a balanced ecosystem. Successful farming must, of course, follow the principles of good land use. It is not conservation to reclaim marginal land for agricultural purposes or to build expensive dams and canals to irrigate land unless the land can produce crops that can make the reclamation and irrigation worthwhile.

If the grasslands of regions with slight rainfall are plowed and planted with wheat a dust bowl will inevitably develop. If the land is kept as grassland and grazed in moderation the soil will be kept in place and no dust bowl will develop; the land can be used economically year after year. By destroying the grass covering the soil, overgrazing can lead to destructive erosion just as improper plowing does. Overgrazing also leads to the invasion of the grassland by undesirable weeds and desert shrubs that are subsequently difficult to eradicate so the grass can grow again. Poor land use affects not only the unwise farmer but also the whole population, which is eventually taxed to pay for the rehabilitation of the land.

The ecologists specializing in the management of land have classified land into eight categories, on the basis of its slope, kind of soil and natural biotic communities. These categories range from type I, which is excellent for farming and can be cultivated continuously, through three classes that can be used for farming only with special care, and another three classes that are suitable only for permanent pasture or forest, to type VIII, suitable only to be left as it is for game, for recreational and scenic purposes or for watershed protection (Fig. 36–1).

The control of insect pests by chemical pesticides must be carried out cautiously, with possible ecologic upsets in mind. Spraying orchards, forests and marshes may destroy not only the pests but also useful insects such as honey bees, which pollinate many kinds of fruit trees and crops, and useful insect parasites.

DDT and related chemicals may kill other animals in addition to insects; amphibians and reptiles are the most vulnerable vertebrates. The vertebrates are less sensitive than insects, and DDT applied at a level of about 1 pound per acre is effective in controlling insects without endangering the vertebrates. However, when applied at a level of 5 to 10 pounds per acre, some of the useful larger animals are killed along with the insects. Insect pests may actually increase after the use of DDT; the chemical may kill greater numbers of insect enemies of the pest than of the pests themselves. A number of strains of insects resistant to DDT have developed. However, the World Health Organization warns that banning DDT

Figure 36–1 Classification of land according to its usefulness. Types I and II may be cultivated continuously; types III and IV are subject to erosion and must be cultivated with great care; types VI and VII are suitable for pasture or forests but not for cultivation; type VIII is productive only as a habitat for game. (U.S. Soil Conservation Service photo.)

before cheap, safe and effective substitutes are found would be a disaster to world health. The agricultural production of the United States might decrease about 30 per cent if pesticides were banned, and the poorer nations of the world would suffer even more from decreased food supplies and increased incidences of insect-borne diseases.

The great increase in agricultural productivity, the "green revolution," has been accomplished with the use of fertilizers, irrigation, pest control, and genetic selection of specific strains of high-yield crop plants such as wheat and rice. This industrialization of agriculture has required the input of considerable amounts of fuel energy. Doubling the crop yield, for example, is achieved only by a tenfold increase in use of fertilizers, pesticides and fuel energy. As a result, industrialized agriculture is a major cause of both air and water pollution.

36-2 FORESTRY

The management of our forests is an important aspect of applied ecology. Until very recently most of the timber cut came from the accumulated growth of trees over the centuries. The lumber industry must adjust itself to bring the amount harvested into balance with the annual growth of the forests. Careful forest management has been carried on in Europe for many decades but is only beginning in this country. Proper timber management in our national and state forests has been important in demonstrating to the owners of private forests the results that can be obtained in this way. Since, in some regions, the desirable timber trees are members of the climax community, the ecologic problem is simply to find the best way to speed the return of the climax community after the trees have been cut. In other regions, the desirable trees are earlier seral stages of the ecologic succession, and forest management involves establishing means of preventing the succession to the climax community. This is also true of many kinds of animals; most game birds and many of the most valuable game fish are members of, and thrive best in, an early seral stage of their community.

36-3 WILDLIFE MANAGEMENT

The management of our fish and wildlife resources is a field of applied ecology sup-

ported by wide public interest, especially by sportsmen's clubs and associations. "Wildlife" used in this connection usually means game and fur-bearing animals. Since the various types of wildlife are adapted to different stages of ecologic succession, their management requires a knowledge of and the proper use of these stages. As the Middle West became more and more intensively farmed, and the original forests and prairies were reduced to small patches, the prairie chickens and ruffed grouse that were adapted to these habitats were greatly decreased in numbers. However, this region has been partially restocked with game birds by introducing ring-necked pheasants and Hungarian partridges, which had become adapted to living in the intensively farmed regions of Europe.

Of the three general methods used to increase the population of game animals—laws restricting the number killed, artificial stocking, and the improvement of the habitat—the latter is the most effective. If the game habitats are destroyed or drastically altered, protective laws and artificial stocking of the region are useless.

Protective laws must operate to prevent a population from getting too large as well as too small. Deer populations, in the absence of natural predators but subject to a constant, moderate amount of hunting, may increase to a point where they actually ruin the vegetation of the forest. Hunting should be restricted, of course, when populations are small and increased when they are larger. This requires accurate annual estimates of the population density of the game species.

Stocking a region artificially with game animals is effective only if they are being introduced into a new region or into one from which they have been killed off. Beavers, for example, were trapped to extermination in Pennsylvania, but restocking with Canadian beavers has been very successful, and it is estimated that there are some fifteen to twenty thousand beavers busy building dams in Pennsylvania. These beaver dams are now an important factor in flood control in that region. The principles of population growth make it clear that if game animals of a certain species are already present, artificially stocking that region with additional members of the species will be futile. Stocking a region with a completely new species must be done cautiously, or the species may succeed so well as to become a pest and upset the biotic community, as has happened with rabbits in Australia and the English sparrow in the United States.

The management of the fish in a pond may be directed toward providing sport for hook and line fishermen or toward raising a crop of food fish and draining the pond at regular intervals to harvest the crop. To provide the best sport fishing a lake or pond should be stocked with a combination of the sport fish and its natural prey; stocking a pond with large-mouth bass plus bluegills gives seven to ten times more bass in three years than does stocking with bass alone. Stocking with fish must be done with care, for if a lake that already has about as many fish occupying a certain ecologic niche as possible is stocked with more of the same kind, there will be a decrease in the rate of growth and the average size of the fish. It has been found that sport fishing with hook and line is not likely to overfish a lake; the lake is more likely to be underfished, and the resulting crowding leads to a decrease in the average size of the fish in the population.

36–4 AQUACULTURE

The building of dams raises intricate ecologic problems, for dams may be intended for power, for flood control, for the prevention of soil erosion, for irrigation or for the creation of recreational areas. Since no one dam can satisfactorily accomplish all these objectives, the primary objective must be clearly delineated and the secondary results must be understood. A contrast of two proposals for dealing with the same watershed (Table 36–1) shows that the multiple dam plan costs less, destroys a smaller area of productive farmland, impounds more water and is more effective in controlling floods and soil erosion. Since the productivity of a lake is inversely proportional to its depth, the series of small reservoirs would have greater productivity of fish.

To be most effective, the building of a dam must be accompanied by measures to decrease soil erosion upstream, or the storage reservoir will fill with silt in a few years.

The management of the fish population in the lakes created by large dams is more difficult than the management in a pond. Sport fishing is usually very good when a dam has first been built, but gradually the silting up of the reservoir and the decrease in productivity

Table 36–1 A COMPARISON OF A SINGLE MAIN RIVER RESERVOIR PLAN WITH A PLAN FOR MULTIPLE SMALLER HEADWATERS RESERVOIRS

	Main Stream Reservoir	Multiple Headwaters Reservoirs
Number of reservoirs	1	34
Drainage area, square miles	195	190
Flood storage, acre feet	52,000	59,100
Surface water area for recreation, acres	1,950	2,100
Flood pool, acres	3,650	5,100
Bottom farmland inundated, acres	1,850	1,600
Bottom farmland protected, acres	3,371	8,080
Total cost	$6,000,000	$1,983,000

From Odum, E. P.: *Fundamentals of Ecology*, 3rd Ed. Philadelphia, W. B. Saunders Co., 1971. The costs in 1977 dollars would, of course, be greatly inflated!

change the nature of the fish community from game fish to less desirable catfish and shiners.

The term ecologic backlash has come into use to describe unforeseen, detrimental consequences of a project that either cancel out the anticipated gains of a particular modification of the environment or create more problems than were solved. A dam built primarily for hydroelectric power on the Zambesi River in Africa has had many undesirable side effects. The large lake shore created by the dam has greatly increased the habitat for the tsetse fly and has caused a severe outbreak of disease in cattle. Similar dams in other parts of Africa have increased the incidence of sleeping sickness among humans. Because the temperature of the water in the lake is high, the lake does not undergo spring and fall overturns, which mix the waters of lakes in the temperate zone. This greatly decreases the lake's productivity, and the fish catch has not compensated for the lost productivity of the grazing and agricultural land covered by the waters of the lake. The displacement of people from the rich river bank lands covered by the dammed waters to less suitable land has led to increased soil erosion. Some of the displaced people have moved to cities, causing further social upheaval. The regulated flow of water downstream of the dam has proved to be more damaging than the natural flooding that previously inundated the bottom lands annually. Maintenance of the fertility of these lands will require the input of expensive chemical fertilizers. There is evidence that, similarly, the Aswan dam on the Nile is not an unmixed blessing.

Lake Baikal in Siberia is a very deep and ancient lake that was formed by earth move-

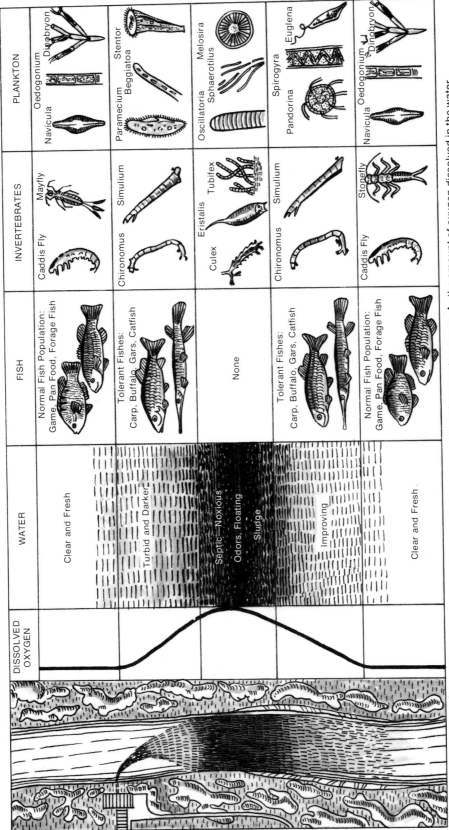

Figure 36–2 Pollution of a stream with untreated sewage and its subsequent recovery. As the amount of oxygen dissolved in the water decreases (left), fishes disappear and only organisms able to obtain oxygen from the surface or those that tolerate low oxygen tensions are able to survive. When the sewage has been reduced by bacteria, the population of animals and plants in the stream returns to normal. (Redrawn from Odum, E. P.: Fundamentals of Ecology, 3rd Ed. Philadelphia, W. B. Saunders Co., 1971.)

ment during the Mesozoic era, some 150,000,000 years ago. It contains some 20 to 25 per cent of the world's fresh water supply. It is often called the "Australia" of fresh water lakes because of its long isolation and its unique collection of plants and animals. Three hundred seventy-six of its 384 species of arthropods and 29 of its 36 species of fish are found nowhere else in the world. The ecology of Lake Baikal is now threatened by Soviet industry. Its water is polluted by wastes discharged from an enormous paper factory and the trees of its watershed are being felled for conversion to paper. The soil in the watershed is very thin and soil erosion occurs rapidly after the trees are felled. The soil is washed into the lake, making its originally crystal clear waters turbid and further altering the ecosystem. The region south and east of the lake is desert and with this change in the watershed the desert may spread to the shores of the lake.

The three chief sources of stream pollution are industrial materials, which either are directly toxic themselves or reduce the oxygen supply in the water; sewage and other materials that decrease the oxygen content of the water and introduce bacteria and other septic organisms (Fig. 36–2); and turbidity caused by soil erosion in the watershed. As the silt settles out downstream it may cover up the spawning grounds of fish and have other direct deleterious effects. Erosion can be prevented by proper soil management, industrial wastes can be minimized by suitable design of the manufacturing process, and properly treated sewage can be emptied into a stream without deranging its ecologic relations. The proper design and operation of "waste management parks" can utilize heated water from power plants (thermal pollution) and certain types of treated and diluted domestic and industrial wastes as energy subsidies for adapted species of fish and other animals. These can ultimately be harvested to provide food and other useful products.

36–5   MARINE FISHERIES AND ESTUARINE MARICULTURE

The primary productivity of the sea, as measured by the pounds of organic carbon produced per year per acre of surface, is very high. The productivity of the western Atlantic off the coast of North America is 2.5 to 3.5 tons of organic carbon per acre and that of Long Island Sound is 2.5 to 4.5 tons per acre. The productivity of the average forest is about 1 ton per acre; most cultivated land fixes only about $3/4$ ton of organic carbon per acre; and only the rich, intensively cultivated cornfields of Ohio produce as much as 4 tons per acre. Despite this high productivity, man's actual harvest from the ocean, in terms of pounds of fish caught per acre of surface, is very low. Only the rich fishing grounds of the North Sea produce as much as 15 pounds of fish per acre. The ecologic reasons for this are clear: the fish are secondary or tertiary consumers and are on top of a vast pyramid of producers. There are many organisms competing for the food energy fixed by the algae in addition to the edible fish and crustacea harvested from the sea.

The annual harvest of marine fish over the entire globe was 18 million tons in 1938, reached 70 million tons in 1970, but has been declining since then. Almost 80 per cent comes from the North Atlantic, the northern and western Pacific and off the west coasts of Peru and Ecuador in South America. About half the total harvest is used as human food; the remainder is used as food for pets, poultry and livestock.

Humans could undoubtedly recover for their use much more of the biologic productivity of the sea. Although they might be reluctant to eat marine algae themselves, the algae might be filtered from sea water and processed so as to be suitable as feed for cattle or some other gastronomically acceptable animal.

Careful studies by the United States Fish and Wildlife Service of the fish population of George's Bank and other commercially fished areas have led to recommendations about the rate of fishing and the size of nets used to ensure that the fish are harvested at an optimal size for greatest yield at present and in the future. These areas, which had been fished so extensively that some of the most desirable species were reduced greatly in numbers, are now beginning to revive under careful management.

The shellfish—oysters, clams, shrimps and lobsters—present somewhat different and more difficult problems, for their habitat is more limited than that of commercial fish and they are more affected by adverse environmental changes. Oysters, whose food consists of algae or detritus of a certain size filtered from the sea water by their gills, are unable to use algae of a different size. Oysters were unable to survive in certain bays of Long Island Sound when ducks were raised commercially in large numbers on the adjacent shore. The wastes from the duck farms were washed into the bays, and the addition of this organic matter changed the community ecology in such a way that the normal food of the oysters, diatoms, was replaced by other algae that could not be used by the oysters. Once an oyster bed has been seriously depleted it may fail to recover even if seeded with oyster larvae, because the larvae require a favorable surface for attachment, and the most favorable is the shell of an old oyster. In commercial oyster farms the larvae are provided with artificial sites for attachment. Once they have become attached they may be moved to other waters, even from one ocean to another, to complete their growth in waters that are favorable for feeding although not favorable for the reproduction of the species. The intensive cultivation (mariculture) of oysters, clams and other shellfish in estuaries and other marine environments has provided an important source of protein-rich food in Japan, Indonesia and Australia. It does require a considerable input of energy, at present largely in the form of human hand-labor. By using the raft method of oyster culture, rather than simply harvesting the wild population, the production of oyster meat in a given area can be increased five- to tenfold.

36–6 MINERAL RESOURCES

Because it was assumed that the supply of mineral resources would last for centuries to come and that there was no way of conserving them, little attention has been paid to this aspect of conservation. However, both of these assumptions are quite wrong. The relationship between the resources available and their utilization is expressed by the demographic quotient, Q:

$$Q = \frac{\text{total resources available}}{\text{population density} \times \text{per capita consumption}}$$

The quality of modern life goes down as the quotient goes down, and it is decreasing rapidly at present. Even if the total resources available could be kept constant by recycling, the value of Q decreases as the population density increases and as the per capita consumption increases. The utilization of natural resources in the United States is increasing at a rate of about 10 per cent per year. The demands for iron, copper, lead, tin and even the relatively abundant metal aluminum are such that the country is no longer self-sufficient either in mineral resources or in the fossil fuels needed to extract the new mineral resources. At the moment biotic and mineral resources are more critical and will be exhausted sooner than energy sources. Natural gas may soon be depleted and oil will follow a few decades later; but coal will last for a considerably longer time, hopefully until the complete safety of methods for producing large amounts of electricity in nuclear reactors can be assured, or until commercially feasible methods of utilizing solar energy have been achieved. The limiting factor in the utilization of industrial energy will be pollution rather than supply. In the report of the Committee on Pollution of the National Academy of Sciences, *Resources and Man* (1969), a total of 26 recommendations were made. These can be summarized in the statement that control of the size of human populations and improved management and recycling of our resources are urgently needed, beginning immediately!

36–7 PUBLIC HEALTH

Many aspects of the field of public health require the application of ecologic principles; the prevention of the spread of diseases carried by animals is an ecologic as well as a medical problem. The most effective way of eliminating malaria, for example, is to eliminate the particular species of mosquito that is the vector of the malaria parasite, yet this must be done without destroying the useful insects of the region. The mosquitoes that transmit malaria in different parts of the world have quite different ecologic niches, and therefore measures that may be effective in mosquito control in one region may be quite ineffective in another. The malaria of the southeastern United States is transmitted by mosquitoes living in marshes;

Italian malarial mosquitoes live in cool running water in the uplands; and Puerto Rican malarial mosquitoes live in brackish (slightly salty) water. Careful ecologic surveys of each region are necessary to formulate the proper measures to control the insects.

The size of the populations of rats, mice and many insect pests increases with the size of cities and the correlated tendency toward the development of slums in the older parts of the town. A survey in England in 1953 revealed that only 0.1 per cent of the houses in towns with less than 25,000 houses were infested with bedbugs, but more than 1.0 per cent were infested in towns with more than 100,000 houses! Careful ecologic studies in Baltimore showed that although professional crews of rat trappers might catch as much as half of the rat population, the number of rats quickly returned to its former level. Cats proved to be greatly overrated as rat predators and were not effective in controlling the rat population. However, by changing the essential elements of the rats' habitat, by improving sanitation and thus decreasing the garbage on which the rats fed and the wastes in which they hid, the rat population was reduced to about 10 per cent of its former size. It remained at this lower level because that was the total number of rats that could survive in the altered environment.

36–8 POLLUTION

The ultimate cause of pollution is people, and as the number of people increases there is a corresponding increase in the amount of pollution. As the number of people has increased and the amount of energy used by each person has also increased, the total energy demand has increased at a very high rate. This is reflected in the air pollutants from the exhaust stacks of industrial plants and electric generating plants burning fossil fuels and from the exhaust gases of automobiles transporting people to and from their work and play.

The National Academy of Sciences, in the report by the Committee on Pollution, defined pollution as follows:

Pollution is an undesirable change in the physical, chemical or biological characteristics of our air, land and water that may or will harmfully affect human life or that of desirable species; industrial processes, living conditions and cultural assets; or that may or will waste or deteriorate our raw material resources. Pollutants are residues of the things we make, use and throw away. Pollution increases not only because as people multiply the space available to each person becomes smaller, but also because the demands per person are continually increasing so that each throws away more year by year. As the earth becomes more crowded there is no longer an "away." One person's trash basket is another's living space.*

Pollution is undoubtedly the most important limiting factor for humans. It is difficult to put a price tag on the cost of pollution, but it is clear that pollution places a burden on human society from (1) the loss of resources owing to unnecessarily wasteful exploitation, (2) the cost of abating and controlling pollution and (3) the cost in human health.

From an ecological viewpoint two types of pollution can be recognized—pollution involving biodegradable pollutants and pollution involving nondegradable pollutants. Biodegradable pollutants such as domestic sewage can be decomposed rapidly by natural processes or by carefully engineered systems, such as a community's sewage treatment plant. Problems arise when the input of degradable pollutants into the environment exceeds the environment's capacity to decompose or disperse them. The problems with disposable sewage result in general from the fact that urban populations have grown much faster than their sewage treatment facilities. Biodegradable pollutants can be dealt with by a combination of mechanical and biological treatments; but there are limits to the total amount of organic matter that can be decomposed in a given area, and there is an overall limit to the amount of carbon dioxide (one of the decomposition products) that can be released into the air.

The nondegradable pollutants include metals such as mercury, trace metals, steel and aluminum cans and organic chemicals such as DDT that are degraded only very slowly. Some of these nondegradable pollutants accumulate as they proceed along the food chains. Dealing with these pollutants is a much more difficult and expensive problem. Some of these pollu-

*From *Resources and Man* (1969), the National Research Council, National Academy of Sciences, Washington, D.C.

tants, such as aluminum cans, are actually a resource if they are recycled instead of thrown away. Others, such as DDT, can be replaced by more degradable substitutes once they are found.

The design and operation of an efficient waste treatment plant is a very complex subject, beyond the scope of this book, but it is obvious that such operations must be carried out with the best possible application of ecological principles. It is perfectly feasible to produce drinkable water from sewage waste, but this is a very expensive procedure. (However, it is less expensive than producing drinking water from sea water.)

The pollution of the air in industrialized countries such as the United States is a problem of first magnitude. Although the pollution of the air over the entire country is serious enough, the most striking aspect of air pollution is the local concentration of pollutants that may occur over cities such as Los Angeles and New York during a temperature inversion. This occurs when air is trapped under a warm upper layer that prevents a vertical rise of the pollutants. The kinds of pollutants and their relative proportions vary from place to place but usually include carbon monoxide, oxides of nitrogen and sulfur, hydrocarbons, particulate matter and lead. Certain combinations of pollutants may react in the environment to produce additional pollution. For example, certain components in automobile exhaust can combine in the presence of sunlight to produce even more toxic substances called *photochemical smog.* One of the components of photochemical smog blocks a key reaction in photosynthesis and kills plants by inhibiting their production of food.

The subject of pesticides and herbicides has been an extremely controversial one in the past decade. Early in this century, when farms were small and diversified and there was plenty of farm labor, the increase in the numbers of insect pests was blocked by simple inorganic salts. The industrialization of agriculture that has occurred within the last few decades required, and was made possible by, broad spectrum poisons such as the organochlorides and organophosphates. Unfortunately these poisons were used much too enthusiastically, and since many are persistent and are degraded very slowly, these broad spectrum poisons have accumulated to such an extent in many parts of the environment that they can no longer be used.

Two highly undesirable effects of the saturation of the environment with DDT and other organochlorides are the development of strains of insects that are resistant to the DDT and the accumulation of these toxic materials in the food chain. Many of the problems can be traced to the fact that new insecticides have been tested, usually superficially, at the level of the organism against which they are directed and then have been used at the ecosystem level without controlled tests of their effects on the entire ecosystem.

It is clear that if man is to raise enough food crops to supply his burgeoning population he must have some way of dealing with the insect pests that are competing for this same supply of food. To protect ourselves against the diseases transmitted by insects, such as malaria, yellow fever and dengue (mosquitoes), sleeping sickness (tse-tse flies), plague (fleas), typhus (fleas, ticks, lice or mites) and typhoid (frequently transmitted by flies) we must continue to use effective insecticides. Chemical insecticides must be used only in appropriate places and amounts and with adequate controls. Certain kinds of insects can be removed by introducing appropriate predators or parasites, by supplying hormonal factors such as juvenile hormone that will prevent the insect from completing its life cycle, by using sex lures and other pheromones that will confuse the reproductive efforts of the insects, or by sterilizing large numbers of male insect pests by chemicals or radiation and then releasing the sterile males into the natural environment. For other insects, however, the only effective countermeasures are chemical insecticides.

As previously mentioned, if all chemical pesticides were banned, the agricultural production of the United States would decrease by 30 per cent or more, according to the United States Department of Agriculture, and food prices would soar. Some partisans have claimed that 50 million Americans would be in danger of starving to death if chemical pesticides were withheld from agriculture. The poor and undeveloped nations of the world would probably suffer even more from disease and hunger than the developed nations if DDT were banned. In pest control programs on many kinds of crops the organochloride pesticides are being replaced by organophosphorus and carbamate compounds, which are less resistant and more biodegradable. However, larger amounts of these pesticides are usually required to control pests in a given area and they may be more toxic to humans than DDT. Maintaining agriculture production with the aid of chemical pesticides and herbicides without damaging the ecosystem irretrievably will be a difficult task.

Much interest has centered on the discovery of substantial amounts of mercury in both fresh water and marine fish. The ocean contains a great many metals, including gold, silver and platinum as well as mercury. These metals have been washed into the ocean over the course of billions of years and are not primarily the result of industrial pollution. In contrast, the mercury in certain fresh water ecosystems such as the Great Lakes is primarily the result of industrial pollution, and measures can and should be taken to eliminate further input. Mercury from sea water undoubtedly accumulates in the flesh of all kinds of fish. It is also present in flesh of lobsters, shrimp, clams, oysters and other shellfish. It should be noted that one commonly prescribed medicine, calomel, contains mercury.

Other pollutants in the environment are the various radioactive materials such as strontium-90. Strontium-90 is a beta emitter, having a half-life of 28 years. Strontium is very similar to calcium in its chemical and physical properties and moves with it in its natural cycle. Calcium and strontium are washed out of rocks and move down rivers to the ocean. In the Far North radioactive fall-out of strontium-90 has been absorbed by lichens, and the caribou and reindeer that eat the lichens have the strontium concentrated in their flesh. Humans who eat the flesh of the animals or drink milk from them may then accumulate strontium in their bones. In other parts of the world strontium-90 has accumulated in the bones of both adults and children who obtain it in cow's milk; the cows obtained it by eating vegetation polluted with strontium-90.

Little strontium-90 has been added to the environment since 1963. If the ban on nuclear testing can be continued, strontium-90, and the related cesium-137, will gradually disappear from the environment. Both are by-products of fission reactions and enter the environment through fall-out or through faulty waste disposal; they are especially hazardous because they accumulate in successive living organisms along the food chain. Studies in Great Britain showed that strontium-90 was 21 times more concentrated in grass than in soil and some 700 times more concentrated in sheep than in the grass.

Most human beings now live in urban areas, and urban areas not only pollute the air but also modify the climate. Because of the absorption of solar radiation by vertical surfaces and the production of heat by the machines in a city, the air temperature in a city will be 1 to 3° F higher and the humidity will be about 6 per cent lower than in the surrounding countryside. Because of particulate air pollution, cloudiness will be 10 per cent greater and there will be 30 to 100 per cent more fog, the higher value occurring in the winter. Precipitation will be 10 per cent greater, there will be 15 per cent less sunshine and 10 to 30 per cent less ultraviolet radiation in the city than in the countryside.

In 1970, the average population density in the United States was one person to every 10 acres of ice-free land area. The average population density of the world was at the same level. Even if the anticipated reduction in birth rate occurs, the population of the United States will double in the next 30 years. By the year 2000 there will be only 5 acres of land for every man, woman and child. Although as little as $1/3$ acre can produce enough calories to sustain a person, the kind of optimal quality diet that includes meats, fruits, and vegetables requires at least $1\frac{1}{2}$ acres per person. Another acre per person is needed to produce the paper, wood and cotton needed by each person, and an additional $1/2$ acre is needed for the roads, airports, buildings and other areas not involved in food production. Only about 2 acres per person remain for all the many other uses that humans have for land. An affluent, developed nation actually requires more space and resources per person than an undeveloped nation. From this it follows that the key in determining the optimum human population density should be the adequate and usable *pollution-free living space* and not simply the food needed. E. P. Odum, in his book *Fundamentals of Ecology* (1971), has said, in essence, that this earth can feed more warm bodies sustained as so many domestic animals in a polluted feed lot than it can support quality human beings who have a right to a pollution-free environment, a reasonable chance for personal liberty, and a variety of options for the pursuit of happiness. To obtain this latter goal, at least one-third of all land should remain protected for use as national, state or municipal parks, greenbelts, refuges, wilderness areas and so on. If the land is under private ownership, it should be protected by zoning or other legal means.

36–9 HUMAN ECOLOGY

No great amount of thought is required to realize that ecologic principles apply to human populations as well as to populations of animals and plants. Human ecology deals not only with the dynamics of human populations but also with the relationship of human beings to the many physical and biotic factors which impinge upon them. By appreciating that human populations are a part of larger units—of biotic communities and ecosystems—we can deal with our own special problems more intelligently. Human beings have wittingly and unwittingly exercised a great deal of control over their environment and have modified the communities and ecosystems of which we are a part. However, this control is far from complete, and humans must, like other animals and plants, adapt to those situations that we cannot change. By understanding and cooperating with the various cycles of nature, we will have a better chance of surviving in the future than if we blindly attempt to change and control them.

The human population is clearly in danger of multiplying beyond the ability of the earth to support it. In the past several centuries, the population of the world has increased tremendously as new territories have been opened for exploitation and as methods of producing food have become more efficient. The population explosion is not due to an increased birth rate but to a decreased death rate. This has resulted from improved public health measures, such as pure drinking water, pasteurized milk, and improved sewage disposal, and from improvements in medical practice and drugs, such as antibiotics. At present the rate of population increase is greater in some nations than in others and generally is greater in the less developed nations. Strong measures are being taken by nations such as India, which has raised the minimum age for marriage for men and women and is considering requiring sterilization after a family has had two children.

Most biologists and social scientists believe that the danger of human overpopulation is both great and imminent. It has been amply shown that the Malthusian Principle, that populations have an inherent ability to grow exponentially, is true for organisms generally, and the growth of the human population in the past 300 years has followed an exponential curve. The productivity and carrying capacity of the earth for human beings can be maintained and perhaps increased somewhat, but eventually the human biomass must be brought into equilibrium with the space and food available. The human population is probably in greater danger of running out of drinkable water and breathable air than it is of exhausting its food supply! Some limitation of human reproduction is clearly inevitable. It remains to be seen whether the human species will do this voluntarily or involuntarily.

QUESTIONS

1. What is meant by conservation? What conservation measures are being taken in your state?
2. What methods may be taken to increase the number of game fish in a large lake? The number of game birds in a forest?
3. What ecologic problems may be raised by the damming of a river in the temperate zone? How may the problems differ when a river in the tropics is dammed? What ecologic problems are caused by mining operations? By the building of a large factory for the manufacture of chemicals?
4. Discuss the ecologic principles involved in the operation of an oyster farm.
5. Outline a program for dealing with insect pests in an agricultural community.
6. What measures can be taken (1) immediately and (2) eventually to reduce and eliminate pollution in a river such as the Cuyahoga (Ohio) or the Schuylkill (Pennsylvania)?
7. What are the major sources of air pollution in cities such as Los Angeles, New York and Tokyo? What practicable measures can be taken to reduce these sources?
8. What sources of energy can be tapped to supply the increasing needs of the ever-growing human population with the least harmful effects on the environment?

9. What measures can be taken to control the population of rats, mice and bugs in cities?
10. In what ways are ecology and public health related?
11. What is meant by human ecology? How is it related to sociology?

SUPPLEMENTARY READING

A thorough ecologic analysis of man's impact on the productivity of the biosphere is presented in *Population Resources and Environment: Issues in Human Ecology,* by P. R. Ehrlich, and A. H. Ehrlich, and in M. A. Benarde's *Our Precarious Habitat, An Integrated Approach to Understanding Man's Effect on His Environment.* An alternative view of the ecologic crisis is presented in Barry Commoner's *The Closing Circle.* The anthology edited by George W. Cox, *Readings in Conservation Ecology,* is an excellent source book for the subject, as is *Biology and the Future of Man,* edited by Philip Handler. Ian McHarg outlines his ideas for improving the design of human communities in his *Design With Nature.*

Population, Evolution and Birth Control, edited by Garrett Hardin, is a thoughtful analysis of the present human predicament. A recent overview of the conservation movement is Alexander Adams' *Eleventh Hour: A Hard Look at Conservation.* Some outrageous examples of the devastation of a continent by poor ecologic planning are presented in G. Marine's *America The Raped.*

One of the books that touched off popular concern about damage to the environment was Rachel Carson's beautifully written *Silent Spring.* A follow-up to this is found in *Since Silent Spring* by Frank Graham, Jr. A well written book on this subject is Berton Roueche's *What's Left.* Some good books on the pesticide problem are R. L. Doutt and W. W. Kilgore's *Pest Control, Biological Effects of Pesticides in Mammalian Systems,* edited by H. F. Kraybill, and *Chemical Fallout,* edited by M. W. Miller and G. G. Berg, a collection of 26 papers discussing pesticides in nature, their mechanism of action and their efforts on animal and human populations. Excellent background reading on the problems of the supply of food for the burgeoning human population is provided by *Seed to Civilization: The Story of Man's Food,* by C. B. Heiser, Jr.

APPENDIX I

Physical and Chemical Concepts

MATTER AND ENERGY

The universe consists of *matter* and *energy*, which are related by the Einstein equation $E = mc^2$, where E = energy, m = mass and c = the velocity of light, a constant. This equation provides the theoretical basis for the conversion of matter to energy that occurs within the sun or in an atomic bomb or nuclear reactor. In the familiar, everyday world, however, matter and energy are separate and distinguishable. Matter occupies space and has mass, and energy is the ability to produce a change or motion in matter—the ability to do work. Energy may take the form of heat, light, electricity, motion or chemical energy. The kinds of energy transformations important in biological phenomena are discussed in Chapter 5.

Matter may exist in the solid, liquid or gaseous state, depending on the strength of the forces holding the particles together and on the temperature. With increasing temperature the motion of the particles increases and the forces binding the particles together decrease. Ice melts, becomes liquid water, when the temperature is raised above 0°C, and water becomes a gas, water vapor, at a temperature of 100°C.

ATOMIC STRUCTURE

Regardless of the form—gaseous, liquid or solid—that matter may assume, it is composed of units called atoms. In nature there are 92 chemically different kinds of atoms, ranging from the lightest, hydrogen, to the heaviest, uranium. In addition, 13 atoms larger than uranium (the "transuranium elements") have been identified as products of man-made nuclear reactions. All atoms, natural and synthetic alike, are much smaller than the tiniest particle visible in the usual electron microscope. By special scanning electron microscopy with magnifications of 5,500,000 times, Albert Crewe has been able to photograph certain larger atoms such as uranium and thorium. The structure and properties of atoms have been inferred from experiments made with many types of elaborate apparatus.

The concept that matter is composed of indivisible particles can be traced to the Greek philosopher Democritus, who suggested in the fourth century B.C. that substances such as wood were composed of small particles, which he termed atoms. In his view, a change in the substance, such as the conversion of wood to charcoal, reflected a re-arrangement of the atoms. John Dalton theorized in 1805 that each chemical element is composed of identical atoms and that the atoms of each kind of element are different in weight from the atoms of all other elements. In 1900 Thompson and Millikan found that there are nega-tively charged subatomic particles, termed *electrons,* associated with the atom. Thompson postulated that the atom was a sphere of uniform density and uniform positive charge, with electrons imbedded on the surface. Experiments of Rutherford and Geiger in 1910 suggested a new model of the atom, in which all of the positive charge and most of the mass were concentrated in a small central portion of the atom, the nucleus. Rutherford believed that the electrons orbit the nucleus at relatively great distances from it. In 1913 Niels Bohr showed that the orbits of electrons are not fixed. An electron may undergo an increase in energy state and move from one orbit to another farther from the nucleus.

Further discoveries of positively charged *protons* and electrically neutral particles, *neutrons,* in the nucleus, led to further changes in the concept of the atom. The atomic nucleus is composed of protons and neutrons; protons have a unit positive charge and neutrons are elec-trically neutral. Neutrons and protons have nearly identical masses and the electron has a unit negative charge, but only 1/1800 the mass of a proton.

The *atomic weight* of an element is essentially equal to the sum of the masses of the protons and neutrons in the nucleus; the mass of the electrons can be disregarded. Both neutrons and protons are assigned a mass of 1; hence a carbon atom with 6 neutrons and 6 protons in its nucleus has an atomic weight of 6 + 6, or 12. The electrical neutrality of the atom is maintained by the presence of a number of negatively charged electrons just equal to the number of positively charged protons in the nucleus.

The key to our present understanding of atomic structure is based on experiments to determine the frequency and amount of light emitted or absorbed by hydrogen atoms. Light is emitted from a hydrogen dis-charge tube in certain sharply defined frequencies, between which no light is emitted. The frequencies are spaced systematically, some in the visible and some in the ultraviolet region of the spectrum, with a regular decrease in the space between the lines as the frequency increases. Analyses of the hydrogen spectrum led to the concept that the dif-ferent lines represent different energy levels of the hydrogen atom which are related in an integral fashion. This concept of the behavior of matter, called quantum mechanics, was developed by several scien-tists in the late 1920's. An important feature of quantum mechanical theory is that the motion of an electron is described by the quantum numbers and orbitals. *Quantum numbers* are integers, 1, 2, 3 and so on, which identify the energy state of an atom, and the term *orbital* describes the motion of an electron in space corresponding to a particular energy level.

Quantum mechanics describes how the electron moves around the nucleus; it defines the probability that an electron will be found a certain distance out from the nucleus. However, it does not tell us the *exact* path along which the electron moves — it does not describe the trajectory of the electron from one point to the next; it can simply predict the probability of finding an electron at any given point in space. When this probability is considered over a period of time it can give us an averaged picture of how an electron behaves. An orbital is such an averaged picture of the motion of an electron. Quantum mechanics does provide a prediction of how the electron orbitals will change as the principal quantum number, n, increases.

In the 1920's, Louis de Broglie extended Einstein's concept of the wave-particle duality of the photon; i.e., whatever their mass, all moving objects have a wavelength associated with them. Schrödinger applied a wave equation to analyze the behavior of an electron in the neighborhood of the positively charged nucleus. The solutions of this equation are wave functions (ψ), which give the amplitude of the electron wave as a function of position. The square of this wave function indicates how the probability of finding the electron varies with position. The Schrödinger equation has been solved exactly only for the simplest atom, hydrogen, but approximate wave functions have been calculated for more complex atoms.

The orbital occupied by an electron is determined by the amount of energy the electron possesses, and electrons can move from one orbital to another if they gain or lose energy. Furthermore, it is possible to show from the Schrödinger equation that there can be no more than two electrons in any one orbital. Electron orbitals have different sizes and shapes. The first orbital, nearest the atomic nucleus and with the lowest energy level, is spherical; others are spherical or elliptical or are represented by more complex three-dimensional coordinates. The elliptical orbitals have various orientations in space and are assigned to one of three axes (Cartesian coordinates), each at right angles to the other two. The orbitals are arranged in a series of increasing energy levels; the innermost orbital (with the least energy) is $1s$ (spherical, one orbital). At the second energy level there also is a spherical orbital, $2s$ (larger than $1s$ since the more energetic electrons can wander through a larger space), plus three additional, elliptical orbitals, each oriented at right angles to the other two. These are called $2p_x$, $2p_y$ and $2p_z$; each of these, like every other orbital, can hold two electrons, so there can be a total of six $2p$ electrons. In the third energy level there is one $3s$ spherical orbital, three $3p$ elliptical orbitals and five more complex orbitals, termed $3d$. In addition to the d orbitals, orbitals termed f, with even more complex configurations, are found in the very large atoms. The letters were derived from earlier designations of certain lines in atomic spectra — sharp (s), principal (p), diffuse (d) and fine (f).

The imposition of boundary conditions on the wave functions in the Schrödinger equation restricts the values of the variables of the wave functions to a set of integers. The variables and their integral values are termed quantum numbers: the principal quantum number, n; the azimuthal quantum number, l; the magnetic quantum number, m; and the electron spin quantum number, s. The number in the designation of orbitals, e.g., $2s$, refers to the principal quantum number, n, and the letter refers to the value of the azimuthal quantum number, l : s ($l = 0$), p ($l = 1$), d ($l = 2$) and f ($l = 3$). It is helpful to remember that the total number of orbitals in each series is the *square* of the principal quantum number, *n*. Thus there is one orbital in $1s$ $(1^2 = 1)$, 4 in $2s$ plus $2p$ $(2^2 = 4)$, 9 in $3s$, $3p$ and $3d$ $(3^2 = 9)$, 16 in $4s$, $4p$, $4d$ and $4f$ $(4^2 = 16)$ and so forth.

The electrons in the several orbital series can be diagrammed with a horizontal line representing the orbital and a vertical arrow representing each electron (remember there can be no more than two electrons in any given orbital). Thus the $1s$ orbital, or the lowest energy state (the "ground state") of hydrogen, the simplest atom with a single electron, is represented:

$$\underline{+}$$

The six electrons of the carbon atom are assigned:

$$
\begin{array}{lllll}
\uparrow & 2p & \underline{+} & \underline{+} & \underline{+} \\
\text{increasing energy} & 2s & \underline{+} & & \\
& 1s & \underline{+\!\!\!+} & &
\end{array}
$$

Note that when an orbital series has fewer electrons (e.g., 6) than the number required to fill all the orbitals (in this case, 10), the electrons do not pair but occupy separate orbitals. The orbital series of three other atoms of great importance in biological phenomena, nitrogen, oxygen and phosphorus, are diagrammed in the adjacent figure.

The atoms of each element have a characteristic number of electrons. The distribution and behavior of these electrons, especially those in the outermost shell, determine the chemical properties of the atom. The simplest atom, hydrogen, has a single electron around the nucleus. In its most stable form (its ground state), the hydrogen atom has one electron moving around the nucleus in a spherical region; by suitable mathematical manipulations, one can calculate the most probable distance of the electron from the nucleus. The electron does not circle the nucleus in a single "orbit," as the earth circles the sun; instead it whirls around the nucleus, now close to it, then further away.

At least a dozen kinds of subatomic particles other than protons, neutrons and electrons are now known, but we need not consider them. The different kinds of matter are produced by differences in the number and arrangement of these basic particles. Living systems are composed of exactly the same kinds of atoms, with the same kind of atomic structure, as nonliving systems.

ELEMENTS

An *element* is a substance composed of atoms, all of which have the same number of protons in the nucleus and therefore the same number of electrons circling in orbitals. Most elements have a tendency to unite with other elements to form compounds, but a few, such as the "noble" gases, helium, neon, argon, krypton and xenon, occur in nature uncombined with other elements.

The unique "aliveness" of living things does not reflect the presence of some rare or unique element. On the contrary, four elements, carbon, oxygen, hydrogen and nitrogen, make up some 96 per cent of the human body. Another four, calcium, phosphorus, potassium and sulfur, constitute another 3 per cent of the body weight. Minute amounts of iodine, iron, sodium, chlorine, magnesium, copper, manganese, cobalt, zinc and perhaps a few other elements complete the list. All these elements, especially the first four, are abundant in the atmosphere, the earth's crust and the sea. Life depends upon the complexity of the interrelations of these common and abundant elements.

The assignment of electrons to orbital series in the atoms of nitrogen, oxygen and phosphorus. Each horizontal line represents an orbital, and each vertical arrow an electron.

ISOTOPES

Most elements are composed of two or more kinds of atoms that differ in the number of neutrons in their nuclei. There are 3 kinds of hydrogen, 5 kinds of carbon and 16 kinds of lead. The different types of atoms of an element are called *isotopes* (iso, equal or same + *topos*, Greek for place) because they occupy the same place in the periodic table of the elements. All the isotopes of any given element have the same number of protons but different numbers of neutrons in the nucleus. Although the isotopes of a given element have the same chemical properties, they can be differentiated physically. Some are radioactive and can be detected and measured by the kind and amount of radiation they emit. Others can be differentiated by the slight difference in the mass of the atoms caused by an extra neutron in the nucleus. Substances containing ^{15}N, heavy nitrogen, instead of ^{14}N, the usual isotope, or ^2H, heavy hydrogen (deuterium), instead of ^1H, will have a greater mass, which can be detected with a mass spectrometer.

A tremendous insight into the details of the metabolic activities of cells has been gained by preparing substances such as sugar labeled with radioactive carbon, ^{11}C or ^{14}C, or heavy carbon, ^{13}C, in place of ordinary carbon, ^{12}C. The labeled substance is administered to a plant or animal, or cells are incubated in a solution containing it, and the labeled products resulting from the cell's or organism's normal metabolic processes are isolated and identified. By such experiments, it has been possible to trace step by step the sequence of reactions undergone by a given compound and to determine the form in which the labeled atoms finally leave the cell or organism. The rate of formation of bone, for example, and the effect of vitamin D and parathyroid hormone on this process can be studied with the aid of radioactive calcium, ^{45}Ca. Many biological problems that could not be attacked in any other way can be solved by this method.

IONS

The number of electrons in the outermost energy (valence) shell varies from zero to eight in different atoms. If there are zero or eight electrons in the outer shell, the element is chemically inactive and will not readily combine with other elements (e.g., the "noble" gases). When there are fewer than eight electrons, the atom tends to gain or lose some in order to achieve an outer shell of eight. Since the number of protons in the atomic nucleus remains the same, the gain or loss of electrons results in an electrically charged atom called an *ion.*

Atoms with one, two or three electrons in the outer energy shell tend to lose them to other atoms and become positively charged *cations* because of the excess protons in the nucleus. When a hydrogen atom, for example, loses its sole electron it becomes hydrogen ion, H^+. Sodium also has only a single electron in its outer energy shell and becomes sodium ion, Na^+, when it loses that electron. Atoms with five, six or seven electrons in the outer shell tend to gain electrons from other atoms and thus become negatively charged *anions* because of the excess electrons. Chlorine has seven electrons in its outer energy shell and has a strong tendency to gain an electron (to fill the outer energy shell of eight electrons) and become chloride ion, Cl^-. Atoms with four electrons in the outer shell (such as carbon) tend to share them with neighboring atoms. Both positively and negatively charged atoms are called *ions.* Because particles with opposite electrical charges attract each other, positive and negative ions tend to unite, forming an ionic, or electrostatic, bond.

COMPOUNDS

A chemical compound is a substance composed of two or more different kinds of atoms or ions joined together; it thus can be decomposed or split apart into two or more simpler substances. The amounts of the elements in a given compound are always present in a definite proportion by weight. This reflects the fact that atoms are attached to one another by chemical bonds in a precise way to form the compound. An assembly of atoms held together by chemical bonds is called a *molecule.* A molecule is the smallest particle of a compound that has the composition and properties of a larger part of it. A molecule is made of two or more atoms, which may be the same, as in a molecule of oxygen or nitrogen, or from different elements. A molecule made of two or more different kinds of atoms is a chemical compound. The properties of a chemical compound are usually quite different from the properties of its component elements. Each water molecule, for example, contains two atoms of hydrogen and one atom of oxygen, and the chemical properties of water are quite different from those of either hydrogen or oxygen. Like all chemical formulas, that of water, H_2O, states the kinds of atoms that are present in the molecule and their relative proportions.

MIXTURES

In contrast to a chemical compound such as water, a mixture is composed of two or more kinds of atoms or molecules that may be combined in varying proportions. Water and alcohol may be mixed in any ratio, and air is a mixture of varying amounts of oxygen and nitrogen plus small amounts of water vapor, carbon dioxide, argon and other gases. Although a pure compound will exhibit certain fixed and unchanging chemical and physical properties by means of which it can be identified, a mixture will have properties that vary with the relative abundance of its constituent parts.

MOLECULAR WEIGHTS

The weight of any single atom or molecule is much too small to be expressed conveniently in terms of grams or micrograms. Instead these weights are expressed in terms of the atomic weight unit (dalton), approximately the weight of a proton or neutron. On this scale the lightest element, hydrogen, has an atomic weight of approximately 1, carbon an atomic weight of approximately 12, and oxygen an atomic weight of approximately 16. The molecular weight is the sum of the atomic weights of the atoms in the molecule; thus the molecular weight of water is 18, or $(2 \times 1) + 16$. A molecule of the single sugar glucose, composed of 6 carbon atoms, 12 hydrogen atoms, and 6 oxygen atoms, has the formula $C_6H_{12}O_6$ and a molecular weight of $(6 \times 12) + (12 \times 1) + (6 \times 16)$, or 180 daltons.

BONDS

The constituent atoms of a molecule are joined together by forces called chemical bonds. Ionic, covalent and hydrogen bonds are of importance in the molecules present in biological materials. Ionic and hydrogen bonds are relatively weak and easily broken. Covalent bonds are strong, and their formation is endergonic, that is, energy must be supplied to form them. Both the formation and the cleavage of covalent bonds are carried out by enzymatic reactions within the cell.

Ionic bonds result from the attraction of particles with unlike charges, i.e., a sodium ion, which is a positively charged atom, and a chloride ion, which is a negatively charged atom. These unite to form sodium chloride, or table salt. Ionic bonds may also form between positively and negatively charged molecules.

Hydrogen bonds are formed when a hydrogen atom is shared between two atoms, one of which is usually oxygen. They tend to form between any hydrogen atom covalently bonded to oxygen or nitrogen and any strongly electronegative atom, usually oxygen or nitrogen in another molecule or in another part of the same molecule. Hydrogen bonds are weak and are readily formed and broken. They have a specific length and a specific direction, which is of great importance in their role in determining the structure of macromolecules such as proteins and nucleic acids. Hydrogen bonds are geometrically quite precise. The water molecules in liquid water are held together largely by hydrogen bonds. This results from the fact that the oxygen atom and two hydrogen atoms in a water molecule form a triangle. The electrons used to bond hydrogen to oxygen are more strongly attracted to the oxygen nucleus than to the hydrogen nuclei and tend to be located nearer the oxygen atom. Because of this, the two hydrogen atoms have a small local positive charge, and the oxygen atom has a small local negative charge, although the water molecule as a whole is electrically neutral. Molecules that are positive at one end and negative at the other are said to be "polar." Such molecules are usually soluble in water since the electrostatic attraction of the negative and positive charges tends to align the water molecules around them. When the positively charged hydrogen atom of one water molecule is next to an atom carrying an electronegative charge, such as the oxygen atom in another water molecule, the attraction between them forms a hydrogen bond.

In a covalent bond, the electrons associated with two different atoms are shared and each is subject to the attractive forces of both atomic nuclei. This is clearly seen in the covalent bond of the hydrogen molecule H_2; the bond holding the two atoms together results from the fact that each of the two electrons is attracted simultaneously to two protons. The two atoms are also separated by repulsive forces—the two protons repel each other and the two electrons do the same. There is a stable bond length in the hydrogen molecule that is determined by the balance between the forces of attraction and those of repulsion. Similar factors determine the length of other covalent bonds, such as —C—C—, —C—N— or —C=O. Helium gas is monatomic, He, not He_2. How would you explain the lack of bonding between helium atoms?

Two of the carbon atom's six electrons are in the $1s$ orbital, and there is one in each of the others, $2s$, $2p_x$, $2p_y$ and $2p_z$. This arrangement is most stable because it keeps the electrons as far apart as possible. The four electrons in the $2s$ and $2p$ orbitals are "valence" electrons available for covalent bonding. A simple way of showing this is the "electron dot" method, in which each electron in the outer shell is shown as a dot around the letter representing the atom, for example, $\cdot\overset{\cdot}{C}\cdot$. When one carbon and four hydrogens share electron pairs, a molecule of methane (CH_4) is formed: $H\!:\!\overset{\displaystyle H}{\underset{\displaystyle H}{\overset{\cdot\cdot}{C}}}\!:\!H$. The electrons whirl around both the carbon nucleus and the four hydrogen nuclei; each atom shares its outer shell electrons with the other, thereby completing the $1s$ orbital of each hydrogen and the $2s$ and $2p$ orbitals of the carbon. Such a bond is termed a nonpolar covalent bond.

The nitrogen atom has seven electrons, two in $1s$ and five in the $2s$ and $2p$ orbitals. The most stable state of the nitrogen atom, in which the electrons are farthest apart, is the one in which there are two electrons in $2s$ and one each in the $2p$ orbitals. The five electrons in the outer shell are available for electron sharing to form covalent bonds, as shown by $\cdot\overset{\cdot}{N}\cdot$. When a nitrogen atom shares electrons with three hydrogen atoms, a molecule of ammonia (NH_3), as shown by $H\!:\!\overset{\displaystyle\cdot\cdot}{\underset{\displaystyle H}{N}}\!:\!H$, is formed.

The oxygen atom has eight electrons, two in the $1s$ orbital and six occupying the $2s$ and $2p$ orbitals. The state of the oxygen atom in which the electrons are farthest apart is that in which there are two electrons in $2s$, two electrons in one $2p$ orbital and one in each of the other $2p$ orbitals. These six in the outer shell are available for electron sharing to form covalent bonds $:\!\overset{\cdot\cdot}{O}\cdot$. When oxygen shares electrons with two hydrogen atoms, H_2O is formed, $H\!:\!\overset{\cdot\cdot}{\underset{\cdot\cdot}{O}}\!:\!H$. In this covalent bond the shared electrons tend to be pulled more strongly to one element than to the other. Such a bond is called a polar covalent bond. In a water molecule the electrons tend to lie closer to the oxygen nucleus than to the hydrogen nucleus, giving the oxygen atom a partial negative charge and the hydrogen atoms a partial positive charge. Covalent bonds may have all degrees of polarity, from ones in which the electrons are exactly shared, as in the hydrogen molecule, to ones in which the electrons are much closer to one atom than to the other and the bond is therefore quite polar. In a sense, an ionic bond is simply one extreme of this, in which the electrons are pulled completely from one atom to the other.

Two other types of weak bonds, Van der Waals bonds and hydrophobic bonds, are of particular importance in the structure of protein molecules.

The covalent bonds of great importance in joining molecules together are formed by removing an OH group from one molecule and an H from the other. In biosynthetic reactions, the bond is usually formed not by actually removing a molecule of water but by substituting a phosphate group or some other group for an OH group on one molecule and then removing the phosphate and an H from the second molecule to liberate inorganic phosphate and form the bond.

The bonds joining sugar molecules together are *glycosidic bonds*, formed by removing an H from an alcohol group of one sugar and an OH from an aldehyde group of the other. The bonds joining the amino acids in proteins are *peptide bonds*, formed by removing an OH from the carboxyl group of one amino acid and an H from the amino group of another. The fatty acids and glycerol of neutral fats are joined by *ester bonds*, formed by removing an OH from the carboxyl group of a fatty acid and an H from an alcohol group of glycerol. Other ester bonds of great biological importance are *phosphate esters*, formed by removing an H from phosphoric acid and an OH from a sugar, and *thioesters*, which involve the removal of an OH from the carboxyl group of an acid and an H from an SH group rather than an OH group. Nucleotides have a glycosidic bond between the sugar and the purine or pyrimidine and a phosphate ester bond joining the phosphate to the sugar. Coenzyme A, for example, forms thioesters (sulfur esters) with a variety of substances. Acetyl coenzyme A is the thioester of acetic acid and coenzyme A.

There is a chemical tradition to refer to molecules other than carbonates, which contain the element carbon, as organic compounds, and to refer to all other ones as inorganic compounds. However, inorganic compounds do play important roles in the physiology of living organisms.

The outer orbit of the carbon atom contains four electrons (two in $2s$ and two in $2p$) which can be shared in a number of different ways with adjacent atoms. Carbon can form covalent bonds with a variety of elements and thus is a constituent of a wider variety of compounds than any other element. It was believed at one time that organic compounds were in some way uniquely different from others and could be produced only by living matter. This hypothesis was disproved in 1828 when the German chemist Wöhler succeeded in synthesizing *urea*, an organic product found in the urine of many animals, from the inorganic compounds ammonium sulfate and potassium cyanate. Since that time many thousands of organic compounds have been prepared by chemical synthesis. Some of these compounds are complex molecules of great biologic importance, such as vitamins, hormones and antibiotics and other drugs.

ACIDS, BASES AND SALTS

Among the compounds present in living systems are water, carbon dioxide, acids, bases and salts. An *acid* is a compound that releases hydrogen ions, H^+, when dissolved in water. Acids turn blue litmus paper to red and have a sour taste. Hydrochloric (HCl) and sulfuric (H_2SO_4) are inorganic acids; lactic (from sour milk) and acetic (from vinegar) are two common organic acids. A *base* is a compound that releases hydroxyl ions (OH^-) when dissolved in water. It may be defined alternatively as a compound that accepts hydrogen ions from another compound. Bases turn red litmus paper blue. Sodium hydroxide (NaOH) and ammonium hydroxide (NH_4OH) are common inorganic bases, and purines and pyrimidines (constituents of nucleic acids) are common organic bases.

For convenience, the degree of acidity or alkalinity of a fluid, its hydrogen ion concentration, may be expressed in terms of pH, the logarithm of the reciprocal of the hydrogen ion concentration, $\log 1/[H^+]$. Most animal and plant cells are neither strongly acid nor alkaline, but contain an essentially neutral mixture of acidic and basic substances. The hydrogen ion concentration of such a solution (or of pure water) is about 10^{-7} molar, and thus its pH is 7.0. At pH 7.0 the concentrations of free H^+ ions and of free OH^- ions are exactly equal. Any considerable change in the pH of a cell is inconsistent with life. Since the scale is a logarithmic one, a solution with a pH of 6 has a hydrogen ion concentration 10 times greater than a solution with a pH of 7, and is much more acidic.

When an acid and a base are mixed, the hydrogen ion of the acid unites with the hydroxyl ion of the base to form a molecule of water (H_2O). The remainder of the acid (anion) combines with the rest of the base (cation) to form a salt. Hydrochloric acid, for example, reacts with sodium hydroxide to form water and sodium chloride, common table salt:

$$H^+Cl^- + Na^+OH^- \longrightarrow H_2O + Na^+Cl^-$$

A salt may be defined as a compound in which the hydrogen atom of an acid is replaced by some metal.

When a salt, an acid or a base is dissolved in water it dissociates into its constituent ions. These charged particles can conduct an electric current; hence these substances are known as *electrolytes.* Sugars, alcohols and the many other substances that do not separate into charged particles when dissolved, and therefore do not conduct an electric current, are called nonelectrolytes.

Cells and extracellular fluids contain a variety of mineral salts, of which sodium, potassium, calcium and magnesium are the chief cations (positively charged ions) and chloride, bicarbonate, phosphate and sulfate are the important anions (negatively charged ions). Although the body fluids of terrestrial animals differ considerably from sea water in their total salt content, they resemble it in general in the kinds of salts present and in their relative concentrations. The total concentration of salts in the body fluids of most marine animals is equal to that in sea water, about 3.4 per cent. Vertebrates, whether terrestrial, fresh water or marine, have less than 1 per cent of salts in their body fluids. The body fluids of fresh water and terrestrial invertebrates contain 0.3 to 0.7 per cent salts. Life processes require the presence of certain salts in relative concentrations that lie within certain limits.

APPENDIX II

The Classification of Living Things

The system of cataloguing organisms by phylum, class, order, family, genus and species was described in Chapter 6. In this synoptic survey the phyla are arranged according to the "Five Kingdom" system advanced by R. H. Whittaker (Science, *163*:150–160, 1969). The advantages and disadvantages of this system are discussed in Whittaker's article and on page 128 of this book. As biologists have discovered more of the lower organisms and have learned more about their structure, function and mode of reproduction, the traditional separation of the living world into "plant" and "animal" has become more difficult to maintain. The separation of the prokaryotic organisms as the kingdom Monera and of the primarily unicellular eukaryotic organisms as the kingdom Protista had been advanced previously by several authorities. Whittaker suggests that the fungi, multinucleate eukaryotic organisms that are typically syncytial, lack photosynthetic pigments and obtain their nutrients by absorbing them from the environment, constitute a distinct kingdom. This leaves the plant kingdom as a more homogenous group of multicellular, eukaryotic, photosynthetic, nonmotile organisms that exhibit differentiation of cells into tissues. The animal kingdom then includes only multicellular, eukaryotic organisms without photosynthetic pigments that obtain their nutrition primarily by ingesting other organisms and digesting them in an internal cavity. The animals show tissue differentiation to a marked degree. Thus, in broad view, the five kingdoms are the prokaryotes (kingdom Monera), the unicellular eukaryotes (kingdom Protista) and three kingdoms of multicellular eukaryotes distinguished by their mode of nutrition—photosynthetic (the plants), absorptive (the fungi) and ingestive (animals). In the following sections, the numbers in parentheses are estimates of the known species in the group; for many of the groups there are many additional species not yet described.

KINGDOM MONERA

These prokaryotic organisms lack nuclear membranes, plastids, mitochondria and advanced (9 + 2–strand) flagella. They are typically solitary unicellular or colonial unicellular organisms (but one group is mycelial). The predominant mode of nutrition is absorptive, but some groups are photosynthetic or chemosynthetic. Reproduction is primarily asexual by fission or budding; protosexual phenomena also occur. The organisms are nonmotile or move by the beating of simple flagella or by gliding.

BRANCH MYXOMONERA. Without flagella, motility (if present) by gliding.

Phylum Cyanophyta. Blue-green algae with no distinct nuclei or chloroplasts; unicellular or filamentous. (2500)

Phylum Myxobacteriae. Unicellular or filamentous gliding bacteria.

BRANCH MASTIGOMONERA. Motile by simple flagella (and related nonmotile forms).

Phylum Schizophyta. True bacteria. (3000)

Phylum Actinomycota. Branching, filamentous bacteria that form a mycelial structure.

Phylum Spirochaetae. Spirochetes, move by bending of unique axial filament.

KINGDOM PROTISTA

Primarily solitary unicellular or colonial unicellular eukaryotic organisms that do not form tissues. Simple multinucleate organisms or stages of life cycles occur in a number of groups. The organisms possess nuclear membranes and mitochondria; in many forms, plastids, (9 + 2–strand) flagella and other organelles are present. The nutritive modes of these organisms include photosynthesis, absorption, ingestion and combinations of these. Their reproductive cycles typically include both asexual divisions of haploid forms and true sexual processes with karyogamy and meiosis. The organisms move by advanced flagella or by other means or are nonmotile.

Phylum Euglenophyta. Euglenoid organisms; with nuclei and chloroplasts; lack an outer cellulose wall; pigmented eye spot present. (450)

Phylum Xanthophyta. The yellow-green algae. (400)

Phylum Chrysophyta. The golden brown algae and the diatoms. (10,000)

Phylum Pyrrophyta. Dinoflagellates and cryptomonads. Unicellular, flagellate; brown, red or blue algae. (1100)

Phylum Hyphochytridiomycota. Hyphochytrids.

Phylum Plasmodiophoromycota. Plasmodiophores.

Phylum Sporozoa. Sporozoans. Parasitic protists that reproduce by spores and have no method of locomotion.

Phylum Cnidosporidia. Cnidosporidians.

Phylum Zoomastigina. Animal flagellates. Protozoa that move by whiplike cytoplasmic protrusions, flagella.

Phylum Sarcodina. Rhizopods, protozoa that move by pseudo-podia.

Phylum Ciliophora. Ciliates and suctorians; protozoa that move by beating of cilia; adult suctorians attach to substrate by a stalk.

KINGDOM PLANTAE

Multicellular organisms with walled and frequently vacuolate eukaryotic cells and with photosynthetic pigments in plastids. Some closely related organisms lack the pigments or are unicellular or syn-cytial. Principal nutritive mode is photosynthesis, but a number of lines have become absorptive. Primarily nonmotile, living anchored to a sub-strate. Structural differentiation leading toward organs of photosyn-thesis, anchorage and support, and in higher forms toward specialized photosynthetic, vascular and covering tissues. Reproduction is pri-marily sexual, with cycles of alternating haploid and diploid genera-tions; the haploid generation is progressively reduced in the higher members of the kingdom.

Division Rhodophyta. Red algae; multicellular, usually marine; some have bodies impregnated with calcium carbonate. Chlorophyll a and (in some) d, with r-phycocyanin and r-phycoerythrin also present; food storage as floridean starch; flagella lacking. (4000)

Division Phaeophyta. Brown algae; multicellular, often large bodies—the large seaweeds and kelps. Chlorophyll a and c, with fucoxanthin also present; food storage as laminarin and mannitol; zoospores with two lateral flagella, one of whiplash and one of tinsel type. (1500)

SUBKINGDOM EUCHLOROPHYTA

Chlorophyll a and b; food storage as starch within plastids; ances-tral flagellation included two or more anterior whiplash flagella.
BRANCH CHLOROPHYCOPHYTA. Primarily aquatic, without marked somatic cell differentiation.

Division Chlorophyta. Green algae. (10,000)

Division Charophyta. Stoneworts.
BRANCH METAPHYTA. Primarily terrestrial, with somatic cell and tissue differentiation.

Division Bryophyta. Embryophyte plants without conducting tissues. Multicellular plants, usually terrestrial, with a marked alterna-tion of sexual and asexual generations. The prominent plant is the gametophyte (sexual generation), on which the sporophyte is dependent. (25,000)

Class Bryopsida. Mosses. The gametophyte plant has an erect stem, and its leaves are arranged in a spiral. (15,000)

Class Marchantiopsida. The liverworts. Usually sim-ple, flat plants living in moist, shady places. (9000)

Class Anthocerotopsida. Hornworts. (300)

Division Rhyniophyta. Leafless, rootless, homosporous, vascular plants. All are extinct.

Division Lycopodiophyta. The clubmosses, with simple conducting systems and small green leaves. (900)

Class Lycopodiopsida. The homosporous clubmosses.

Class Isoetopsida. The heterosporous clubmosses and quillworts.

Division Equisetophyta. The scouring rushes. Horsetails with simple conducting systems, jointed stems and reduced, scalelike leaves.

Division Polypodiophyta. The true ferns. (9000) Ferns are generally homosporous; however, water ferns are heterosporous. Gametophyte is usually free-living and photosynthetic.

Division Pinophyta. The gymnosperms. Conifers, cycads and most other evergreen trees and shrubs. No true flowers or ovules are present; the seeds are borne naked on the surface of the cone scales. (640)

Class Lyginopteridopsida. The seed ferns—the most primitive of the seed plants—known only from fossils (late Paleozoic era).

Class Bennettitopsida. The cycadeoids, all extinct.

Class Cycadopsida. The cycads, the most primitive living seed plants, found in tropical and subtropical regions. (120)

Class Ginkgoopsida. The ginkgo, or maidenhair tree, is the only living member of this group.

Class Pinopsida. The conifers, the common evergreen trees and shrubs, with needle-shaped leaves, plus some extinct, large-leaved evergreen trees, the *Cordaites*. Fossil remains of these have been found in deposits from the Devonian to the Permian. (575)

Class Gnetopsida. Climbing shrubs or small trees found in tropical or semitropical regions, with many characteristics in common with the angiosperms. (75)

Division Magnoliophyta. The angiosperms. Flowering plants, with seeds enclosed in an ovary. (250,000)

Class Magnoliopsida. The Dicotyledonae, or dicots. Most flowering plants. Embryos with two cotyledons or seed-leaves; vascular bundles in a ring in the stem; leaves with netlike venation; flower parts (sepals, petals, stamens, and carpels) in fives, fours or twos. (175,000)

Class Liliopsida. The Monocotyledonae, or monocots. The grasses, lilies and orchids. Leaves with parallel veins, stems in which the vascular bundles are scattered, and flower parts in threes or sixes. The embryo has only one seed-leaf. (75,000)

KINGDOM FUNGI

Primarily (excepting subkingdom Gymnomycota) multinucleate organisms with eukaryotic nuclei dispersed in a walled and often septate mycelial syncytium; plastids and photosynthetic pigments lacking. Nutrition absorptive. Limited or absent differentiation of somatic tissue; differentiation of reproductive tissue and elaboration of life cycle marked in higher forms. Primarily nonmotile (but with protoplasmic flow in the mycelium), living imbedded in a medium or food supply. Reproductive cycles typically including both sexual and asexual processes; mycelia mostly haploid in lower forms but dikaryotic in many higher forms.

SUBKINGDOM GYMNOMYCOTA

Organisms with life cycles including separate cells, aggregations of cells, and sporulation stages.

Phylum Myxomycota. Syncytial or plasmodial slime molds. (450)
Phylum Acrasiomycota. Cellular or pseudoplasmodial slime molds. (25)
Phylum Labyrinthulomycota. Cell-net slime molds. (15)

SUBKINGDOM DIMASTIGOMYCOTA

Biflagellate (heterokont) zoospores present, chytrid to simple mycelial organization, cellulose walls.

Phylum Oömycota. Oösphere fungi, the water molds. (200)

SUBKINGDOM EUMYCOTA

Predominantly mycelial organization; zoospores uniflagellate if present; chitin walls; other characters as stated for kingdom.
BRANCH OPISTHOMASTIGOMYCOTA. Uniflagellate (opisthokont) zoospores present; chytrid to simple mycelial organization; mainly aquatic.

Phylum Chytridiomycota. True chytrids and related fungi.
BRANCH AMASTIGOMYCOTA. Flagellated zoospores absent; simple to advanced mycelial organization (but secondarily unicellular in yeasts); mainly terrestrial.

Phylum Zygomycota. Conjugation fungi, the black molds. (300)
Phylum Ascomycota. Sac fungi: mildews, morels, truffles, yeasts. (30,000)
Phylum Basidiomycota. Club fungi, mushrooms, toadstools, rusts and smuts. (15,000)
Fungi Imperfecti. A heterogeneous collection of fungi, principally Ascomycota, that lack sexual stages and are not easily assigned to one of the other groups.

KINGDOM ANIMALIA

Multicellular organisms with eukaryotic cells lacking stiff cell walls, plastids and photosynthetic pigments. Nutrition is primarily ingestive with digestion in an internal cavity, but some forms are absorptive and a number of groups lack an internal digestive cavity. Level of organization and tissue differentiation in higher forms far exceeding that of other kingdoms, with evolution of sensory-neuro-motor systems and motility of the organism (or in sessile forms, of its parts) based on contractile fibrils. Reproduction predominantly sexual; haploid stages other than the gametes almost lacking above the lowest phyla.

SUBKINGDOM AGNOTOZOA

Nutrition absorptive and ingestive by surface cells; internal digestive cavity and tissue differentiation lacking. Minute, motile by cilia.
Phylum Mesozoa. Mesozoans.

SUBKINGDOM PARAZOA

Nutrition primarily ingestive by individual cells lining internal water canals. Cell differentiation present but tissue differentiation lacking or very limited; cells with some motility but the organism is nonmotile.

Phylum Porifera. The sponges (sessile, aquatic animals), both fresh water and marine. The simplest of the many-celled animals, resembling in many respects a protozoan colony. The body is perforated with many pores to admit water, from which food is strained. Three classes: Calcarea (calcareous spicules), Hexactinellida (siliceous spicules) and Demospongiae (protein spicules; bath sponges). (4000)

Phylum Archaeocyatha. Extinct.

SUBKINGDOM EUMETAZOA

Advanced multicellular organization with tissue differentiation, other characteristics of the kingdom.

BRANCH RADIATA. Animals with radial or biradial symmetry.

Phylum Cnidaria. Radially symmetrical, aquatic animals with a central gastrovascular cavity. The body wall consists of two layers of cells, in the outer of which are stinging cells, nematocysts. (10,000)

Class Hydrozoa. Hydralike animals, either single or colonial. There is usually an alternation of a hydralike (asexual) generation with a jellyfish (sexual) generation.

Class Scyphozoa. True jellyfishes.

Class Anthozoa. The corals and sea anemones, which have no alternation of generations. The digestive cavities of these animals are divided by mesenteries to increase the effective surface

Phylum Ctenophora. The comb jellies or sea walnuts. These animals lack the stinging capsules of cnidarians and move by means of eight comblike bands of cilia. (100)

BRANCH BILATERIATA. Animals of bilateral symmetry.

Phylum Platyhelminthes. The flatworms, having flat and either oval or elongated, bilaterally symmetrical bodies and three cell layers. The excretory organs are flame cells, protonephridia. There is a true central nervous system. (14,000)

Class Turbellaria. Nonparasitic flatworms with a ciliated epidermis.

Class Trematoda. The flukes, parasitic, nonciliated flatworms with one or more suckers. Many are internal parasites with complicated life cycles.

Class Cestoda. The tapeworms, parasitic flatworms with no digestive tract; the body consists of a head and a chain of "segments" or individuals which bud from the head.

Phylum Nemertea. The proboscis worms. Nonparasitic, usually marine animals with a complete digestive tract and a protrusible proboscis armed with a hook for capturing prey. The simplest animal with a vascular system carrying blood. (550)

Phylum Nematoda. The roundworms. An extremely numerous phylum. Characterized by elongated, cylindrical, bilaterally symmetrical bodies; they live as parasites in plants and animals or are free-living in the soil or water. (8000)

Phylum Acanthocephala. The hook-headed worms. Parasitic worms with no digestive tract and a head armed with many recurved hooks. (100)

Phylum Nematomorpha. The horsehair worms. Extremely thin, brown or black worms about 15 cm. long, resembling a horsehair. The adults are free-living, but the larvae are parasitic in insects. (200)

Phylum Rotifera. Small, wormlike animals, commonly called "wheel animalcules," with a complete digestive tract, flame cells and a circle of cilia on the head, the beating of which suggests a wheel. (1200)

Phylum Gastrotricha. Microscopic, wormlike animals resembling the rotifers but lacking the crownlike circle of cilia. (100)

Phylum Entoprocta. Bryozoa or "moss" animals, with mouth and anus within the U-shaped circle of ciliated tentacles termed the lophophore. Stalked, sessile, microscopic animals, mostly marine forms. (60)

Phylum Ectoprocta. Sessile colonial bryozoa with a lophophore of ciliate tentacles and an anus that opens outside the lophophore. Coelom present. (5000)

Phylum Brachiopoda. The lamp shells. Marine animals with two hard shells (one dorsal and one ventral), superficially like a clam. They obtain food by means of lophophore. (200 at present; 3000 extinct)

Phylum Phoronida. Wormlike marine forms that secrete and live in a leathery tube; they have a U-shaped digestive tract and a lophophore. (10)

Phylum Annelida. The segmented worms. There is a distinct head, digestive tract, coelom, closed circulatory system and, in some, non-jointed appendages. The digestive system is divided into specialized regions. (10,000)

Class Polychaeta. Mostly marine worms. Each segment of their bodies has a pair of paddlelike structures (parapodia) for swimming and many bristles (chaetae). Some burrow in sand and mudflats; some live in calcareous tubes which they secrete; others swim freely in the ocean.

Class Oligochaeta. Fresh water or terrestrial worms, with no parapodia and few bristles per segment.

Class Archiannelida. Primitive annelids without bristles or external segmentation.

Class Hirudinea. The leeches—flattened annelids lacking bristles and parapodia, but with suckers at anterior and posterior ends.

Phylum Onychophora. Rare, tropical animals, structurally intermediate between annelids and arthropods, with an annelidlike excretory system and an insectlike respiratory system. Body segmented, with a hemocoel and one pair of unjointed appendages per segment. Only a few species known.

Phylum Arthropoda. Segmented animals with jointed appendages and a hard, chitinous skin; body divided into head, thorax and abdomen. Hemocoel present. (800,000)

SUBPHYLUM TRILOBITA. Trilobites, primitive marine arthropods that originated in the Cambrian period and became extinct in the Permian. Segmented body divided by two longitudinal furrows into three lobes. All segments except last had a pair of biramous appendages.

SUBPHYLUM CHELICERATA. Chelicerae on third segment, no antennae.

Class Xiphosura. Horseshoe crabs. Book gills on opisthosoma.

Class Eurypterida. Opisthosoma divided into mesosoma and metasoma. Mesosomal appendages were gill-like.

Class Pycnogonida. Sea spiders. Body greatly reduced in size.

Class Arachnida. Spiders, scorpions, ticks and mites. Adults have no antennae; the first pair of appendages ends in pincers, the second pair is used as jaws and the last four pairs are used for walking.

SUBPHYLUM CRUSTACEA. Lobsters, crabs, barnacles, water fleas and sowbugs. Animals that are usually aquatic, have two pairs of antennae and respire by means of gills.

SUBPHYLUM LABIATA. Antennae on second segment, nothing on third, mandibles on fourth. Second maxillae form lower lip.

Class Chilopoda. The centipedes. Each body segment, except the head and tail, has a pair of legs.

Class Diplopoda. The millipedes. Each external segment (really two segments fused) bears two pairs of legs.

Class Insecta. The largest group of animals, mostly terrestrial. The body is divided into a distinct head, with four pairs of appendages; the thorax, with three pairs of legs and usually two pairs of wings; and the abdomen, which has no appendages. Respiration by means of tracheae. There are about 24 different orders of insects, of which the following are common:

> *Order Orthoptera.* Grasshoppers, crickets and praying mantids.
> *Order Isoptera.* Termites.
> *Order Odonata.* Dragonflies and damselflies.
> *Order Anoplura.* Lice.
> *Order Hemiptera.* Water boatmen, bedbugs and backswimmers.
> *Order Homoptera.* Cicadas, aphids and scale insects.
> *Order Coleoptera.* Beetles, weevils and fireflies.
> *Order Lepidoptera.* Butterflies and moths.
> *Order Diptera.* Flies, mosquitoes and gnats.
> *Order Hymenoptera.* Ants, wasps, bees and gallflies.

Phylum Mollusca. Unsegmented, soft-bodied animals, usually covered by a shell and with a ventral, muscular foot. Respiration is by means of gills, protected by a fold of the body wall — the mantle. (80,000)

Class Amphineura. Chitons, marine forms with a shell composed of eight plates.

Class Scaphopoda. Tooth shells, marine forms living in sand or mud; tubular shells open at both ends.

Class Gastropoda. Snails, slugs, whelks, abalones; asymmetrical animals with a single spiral shell or no shell.

Class Pelecypoda. Clams, mussels, oysters, scallops. These lack a head and have a hatchet-shaped foot for burrowing. The shell consists of two plates or valves (the animals are called bivalves), one on each side of the body.

Class Cephalopoda. Squids, cuttlefish, octopuses. Marine animals having a well developed "head-foot," with eight or ten tentacles, and well developed eyes and nervous system.

Phylum Brachiata or Pogonophora. Beard worms.

Phylum Chaetognatha. Arrow worms, free-swimming marine worms with a complete digestive tract and a body cavity (coelom) that develops from pouches of the digestive tract as in the echinoderms and lower chordates. (30)

Phylum Echinodermata. Marine animals that are radially symmetrical as adults, bilaterally symmetrical as larvae. The skin contains calcareous, spine-bearing plates. The animals have a unique water vascular system of canals, and tube feet for locomotion. Respiration is by skin gills or by out-pocketings of the digestive tract. (6000)

Class Asteroidea. The starfishes. The body is a central disc with broad arms (usually five) not sharply marked off from the disc.

Class Ophiuroidea. The brittle stars and serpent stars. The animals have long, narrow arms sharply differentiated from the central, disc-shaped body.

Class Echinoidea. The sea urchins and sand dollars. Spherical or flattened oval animals with many long spines.

Class Holothuroidea. Sea cucumbers. Long, ovoid, soft-bodied armless echinoderms, usually with a ring of tentacles around the mouth.

Class Crinoidea. Sea lilies and feather stars. The body is cup-shaped and attached by a stalk to the substrate. Most of these are known only as fossils; only a few species survive.

Phylum Hemichordata. The acorn worms. Marine animals with an anterior muscular proboscis, connected by a collar region to a long wormlike body. The larval form resembles an echinoderm larva. (100)

Phylum Chordata. Bilaterally symmetrical animals with a notochord, gill clefts in the pharynx and a dorsal, hollow neural tube. (70,000)

SUBPHYLUM UROCHORDATA. The tunicates or sea squirts. The adults are saclike, attached, filter-feeding animals with a tunic of cellulose; they often form colonies, whereas the larval forms are free-swimming and have a notochord in the tail region.

SUBPHYLUM CEPHALOCHORDATA. Amphioxus. Marine animals with a segmented, elongated fishlike body. They burrow in the sand and take in food by the beating of cilia on the anterior end. They have a notochord extending from the tip of the head to the tip of the tail.

SUBPHYLUM VERTEBRATA. Animals having a definite head, a backbone of vertebrae, a well developed brain and, usually, two pairs of limbs. They have a ventrally located heart, a pair of well developed eyes and other sense organs.

SUPERCLASS PISCES. Aquatic vertebrates, fishes.

Class Agnatha. Lampreys, hagfishes and fossil ostracoderms. Vertebrates without jaws or paired fins.

Class Placodermi. Primitive jawed fishes of the Paleozoic era. Bony head shield movably articulated with the trunk, known only from fossils.

Class Chondrichthyes. Sharks, rays, skates and chimaeras. Fishes with a cartilaginous skeleton and scales of dentin and enamel imbedded in the skin.

Class Osteichthyes. The bony fishes. Lungs or swim bladder usually present and body usually covered with body scales. The sturgeon, bowfin, salmon and lungfishes.

SUPERCLASS TETRAPODA. Four-legged land vertebrates.

Class Amphibia. Frogs, toads, salamanders and the extinct forms, labyrinthodonts. As larvae these forms breathe by gills; as adults they breathe by lungs. There are two pairs of five-toed limbs; the skin is usually scaleless, moist and slimy.

Class Reptilia. Lizards, snakes, turtles, crocodiles, the extinct dinosaurs and other forms. The body is covered with horny scales derived from the epidermis of the skin. The animals breathe by means of lungs and have a three-chambered heart.

Class Aves. The birds. Warm-blooded, typically winged animals whose skin is covered with feathers. Present-day birds are toothless, but the primitive ones had reptilian teeth. The forelimbs are modified as wings.

Class Mammalia. Warm-blooded animals whose skin is covered with hair. The females have mammary glands, which secrete milk for the nourishment of the young.

> *Subclass Prototheria.* The monotremes, primitive forms that retain many reptilian features, such as egg-laying and cloacae. Most of them are extinct; only two species survive — the duck-billed platypus and the spiny anteater.
>
> *Subclass Metatheria.* The pouched mammals or marsupials. The young are born alive but in a very undeveloped state. They complete development attached to teats located in a pouch on the mother's abdomen. The subclass includes opossums and a variety of forms found only in Australia — kangaroos, wallabies, koala bears, wombats and so on.
>
> *Subclass Eutheria.* The placental mammals. The young develop within the uterus of the mother, obtaining nourishment via the placenta.
>
> *Order Insectivora.* Primitive, insect-eating mammals; moles and shrews.
>
> *Order Chiroptera.* Bats.
>
> *Order Carnivora.* Dogs, cats, bears, sea lions and seals.
>
> *Order Rodentia.* Rats, squirrels, beavers and porcupines.
>
> *Order Lagomorpha.* Rabbits and hares.
>
> *Order Primates.* Monkeys, apes and humans.
>
> *Order Artiodactyla.* Even-toed ungulates; cattle, deer, camels and hippopotamuses.
>
> *Order Perissodactyla.* Odd-toed ungulates; horses, zebras and rhinoceroses.
>
> *Order Edentata.* Armadillos, sloths and ant-eaters.
>
> *Order Proboscidea.* Elephants.
>
> *Order Cetacea.* Whales, dolphins and porpoises.
>
> *Order Sirenia.* Sea cows — large, plant-eating aquatic mammals.

GLOSSARY

abscission layer A special layer of thin-walled cells, loosely joined together, extending across the base of the petiole, thus weakening the base of the leaf and finally permitting the leaf to fall.

absorption (ab-sorp'shun) [L. *ab* away + *sorbere* to suck in]. The taking up of a substance, as by the skin, mucous surfaces or lining of the digestive tract.

absorption spectrum A measure of the amount of energy at specific wavelengths that has been absorbed as light passes through a substance. Each type of molecule has a characteristic absorption spectrum.

acclimatization (ă-klī'ma-ti-za'shun). Gradual physiological changes in an organism in response to slow, relatively long-lasting changes in the environment.

acetylcholine (as"ĕ-til-ko'lēn). The acetic acid ester of the organic base choline, normally secreted at the ends of many neurons; responsible for the transmission of a nerve impulse across a synapse.

achondroplasia (ah-kon"dro-pla'ze-ah) [Gr. *a*- not + *chondros* cartilage + *plassein* to form + *-ia*]. A hereditary disturbance of growth and maturation of the bones which results in inadequate bone formation and a characteristic type of dwarfism.

acid (as'id) [L. *acidus*, from *acere* to be sour]. A substance whose molecules or ions release hydrogen ions (protons) in water. Acids have a sour taste, turn blue litmus paper red and unite with bases to form salts.

acidosis (as"ĭ-do'sis). A pathologic condition resulting from the accumulation of acid or the loss of base in the body; characterized by an increased hydrogen ion concentration (decreased pH).

acromegaly (ak'ro-meg'ah-le) [Gr. *akron* extremity + *megalē* great]. A condition characterized by overgrowth of the extremities of the skeleton, the nose, jaws, fingers and toes. This may be produced by excessive secretion of growth hormone from the pituitary.

acrosome (ak'ro-sōm) [Gr. *akron* extremity + *soma* body]. A cap-like structure covering the head of the spermatozoon.

actin (ak'tin). A protein found in muscle that together with myosin is responsible for the contraction and relaxation of muscle.

action current A slight current that can be detected with appropriate sensitive devices when any tissue becomes active— when a muscle contracts, a gland secretes or a nerve conducts an impulse.

active transport   The transfer of a substance into or out of a cell across the cell membrane against a concentration gradient by a process which requires the expenditure of energy.

adaptation   The fitness of an organism for its environment; the process by which it becomes fit; a characteristic that enables the organism to survive in its environment.

adaptive radiation   The evolution from a single ancestral species of a variety of species that occupy different habitats.

adenine (ad′ĕ-nīn) [Gr. *aden* a gland].   A purine (nitrogenous base) that is a component of nucleic acids and of nucleotides important in energy transfer—adenosine triphosphate (ATP), adenosine diphosphate (ADP) and adenylic acid (AMP).

adenosine triphosphate (ah-den′o-sin).   An organic compound containing adenine, ribose and three phosphate groups; of prime importance for energy transfers in biological systems.

adipose (ad′ĭ-pōs) [L. *adiposus* fatty].   Referring to the tissue in which fat is stored or to the fat itself.

adventitious root (ad″ven-tish′us) [L. *ad* to + *venire* to come].   A root that originates from an unusual place, such as a stem.

aeciospores (ē′sĭ-o-spōr″) [Gr. *aikia* injury + *sporas* seed].   Thin-walled binucleate spores of the wheat rust produced in the spring on the leaves of barberry plants.

aerobic (a-er-o′bik) [Gr. *aero* air].   Growing or metabolizing only in the presence of molecular oxygen.

agglutination (ah-gloo″tĭ-na′shun) [L. *agglutinare* to glue to a thing].   The collection into clumps of cells or particles distributed in a fluid.

Agnatha (ăg′na-tha) [*a-* not + Gr. *gnathos* jaw].   The jawless fishes. A class of vertebrates including lampreys, hagfishes and many extinct forms.

algae (al′je) [L. (pl.) seaweeds].   Any of a large group of plants that contain chlorophyll but that do not form embryos during development and lack vascular tissues.

allantois [ah-lan′to-is) [Gr. *allas* sausage + *eidos* form].   One of the extraembryonic membranes of reptiles, birds and mammals; a pouch growing out of the posterior part of the digestive system and serving as an embryonic urinary bladder or as a source of blood vessels to and from the chorion or placenta.

allele (ah-lēl′) [Gr. *allēlōn* of one another].   One of a group of alternative forms of a gene that may occur at a given site (locus) on a chromosome.

allergy   A hypersensitivity to some substance in the environment, manifested as hay fever, skin rash, asthma, food allergies, etc.

alveolus (al-ve′o-lus) [L. (dim.) *alveus* hollow].   A small saclike dilatation or cavity.

amoeboid motion (ah-me′boid) [Gr. *amoibē* change + *eidos* form].   The movement of a cell by means of the slow oozing of the cellular contents.

amensalism (a-men′sal-izm).   A relationship between two species whereby one is adversely affected by the second, but the second species is unaffected by the presence of the first.

amino acid (am′ĭ-no).   An organic compound containing an amino group (—NH₂) and a carboxyl group (—COOH); amino acids may be linked together to form the peptide chains of protein molecules.

amnion (am'ne-on) [Gr. *amnion* lamb].   One of the extraembryonic membranes of reptiles, birds and mammals: a fluid-filled sac around the embryo.

anabolism (ah-nab'o-lizm) [Gr. *anabole* a throwing up].   Chemical reaction in which simpler substances are combined to form more complex substances, resulting in the storage of energy, the production of new cellular materials and growth.

anaerobic (an"a-er-o'bik) [Gr. *an-* not + *aero* air + *bios* life].   Growing or metabolizing only in the absence of molecular oxygen.

analogous (ah-nal'o-gus) [Gr. *analogos* according to a due ratio, conformable, proportionate].   Similar in function or appearance but not in origin or development.

anaphase (an'ah-fāz) [Gr. *ana* up, back, again + *phasis* phase].   Stage in mitosis or meiosis, following the metaphase, in which the chromosomes move apart toward the poles of the spindle.

anaphylaxis (an"ah-fi-lak'sis) [Gr. *ana* up, back, again + *phylaxis* protection].   An unusual or exaggerated reaction of the organism to a foreign protein or other substance.

androgen (an'dro-jen) [Gr. *andros* man + *gennan* to produce].   Any substance that possesses masculinizing activities, such as testosterone or one of the other male sex hormones.

angiosperm (an'je-o-sperm"). [Gr. *angeion* vessel + *sperma* seed].   The traditional name for plants having flowers and fruits and seeds in an enclosed ovary.

anion (an'i-on) [Gr. *ana* up + *iōn* going].   An ion carrying a negative charge.

anisogamy (ăn"ī-sŏg'a-me).   Reproductive process involving motile gametes of similar form but dissimilar size.

anther (an'ther) [Gr. *anthēros* blooming].   "Male" parts of the flower in a flowering plant; the portion of the stamen that contains the pollen sacs (microsporangia) in which haploid microspores or pollen grains are formed.

antheridium (an"ther-id'e-im) [L. *anthera* medicine made from flowers + Gr. *idion*, a diminutive ending].   Male organ of a cryptogamic plant in which sperm are produced.

antherozoid (an'ther-o-zoid").   The motile fertilizing cell of fungi.

anthocyanins (an"tho-si'ah-nin) [Gr. *antho* flower + *kyanos* blue].   A class of pigments of blue, red and violet flowers. They are glycosides, yielding anthocyanidin and a sugar on hydrolysis.

antibiotic (an"ti-bi-ot'ik) [Gr. *anti* against + *bios* life].   Substance produced by microorganisms that has the capacity, in dilute solutions, to inhibit the growth of or to destroy bacteria and other microorganisms; used largely in the treatment of infectious diseases of man, animals and plants.

antibody (an'ti-bod"e).   A protein produced in response to the presence of some foreign substance in the blood or tissues.

anticodon (ăn'ti-kōdŏn).   A sequence of three nucleotides in transfer RNA that is complementary to, and combines with, the three nucleotide codon on messenger RNA, thereby binding the amino acid-transfer RNA combination to the mRNA.

antidiuretic hormone (an"tĭ-di"u-ret'ik hor'mōn) [Gr. *anti* against + *diouretikos* promoting urine]. A hormone secreted by the posterior lobe of the pituitary which controls the rate at which water is reabsorbed by the kidney tubules.

antigen (an'tĭ-jen) [Gr. *anti* against + *gennan* to produce]. A foreign substance, usually protein or protein-polysaccharide complex in nature, that elicits the formation of specific antibodies within an organism.

antimetabolite (an″tĭ-mĕ-tab'o-līt) [Gr. *anti* against + *metaballein* to turn about, change, alter]. Substance bearing a close structural resemblance to one required for normal physiological functioning; it exerts its effect by replacing or interfering with the utilization of the essential metabolite.

antitoxin (an″ti-tok'sin) [Gr. *anti* against + *toxicon* poison]. An antibody produced in response to the presence of a toxin (usually protein) released by a bacterium.

aphasia (ah-fa'ze-ah) [Gr. *a-* not + *phasis* speech]. Inability to recognize certain kinds of symbols (such as writing or speech) due to injury or disease of the brain centers.

apical meristem (ap'e-kal mer'ĭ-stem) [L. *apex* tip, summit + Gr. *merizein* to divide]. Undifferentiated embryonic tissue of plants located in tips of stems or roots.

apoenzyme (ap″o-en'zīm) [Gr. *apo* from + *en* in + *zymē* leaven]. Protein portion of an enzyme; requires the presence of a specific coenzyme to become a complete functional enzyme.

archegonium (ar″ke-go'ne-um) [Gr. *archē* beginning + *gonos* offspring]. The female organ of a cryptogamic plant in which eggs are produced.

archenteron (ar-ken'ter-on) [Gr. *archē* beginning + *enteron* intestine]. The central cavity of the gastrula, lined with endoderm, which forms the rudiment of the digestive system.

arteriole (ar-te're-ol) [Gr. *arteria* artery]. A minute arterial branch, especially one just proximal to a capillary.

artery A vessel through which the blood passes away from the heart to the various parts of the body; typically has thick, elastic walls.

arthropod (ar'thro-pod) [Gr. *arthron* joint + *pous* foot]. An invertebrate, such as an insect or a crustacean, that has jointed legs.

ascospores (as'ko-spor) [Gr. *askos* bag + *sporos* seed]. Set of spores, usually eight, contained in a special spore case.

ateliosis (ah-te″le-o'sis) [Gr. *ateleia* incompleteness]. Hypophyseal infantilism, as seen in a midget, for example.

atom The smallest quantity of an element that can retain the chemical properties of the element, composed of an atomic nucleus containing protons and neutrons together with electrons that circle the nucleus in specific orbits.

atomic orbital Distribution of an electron around the atomic nucleus.

atresia (ah-tre'ze-ah) [Gr. *a-* not + *trēsis* a hole + *-ia*]. Absence or closure of a normal body orifice, passage or cavity.

atrium (a'tre-um) [Gr. *atrion* hall]. A chamber affording entrance to another structure or organ; a chamber of the heart receiving blood from a vein and pumping it into a ventricle.

autosome (aw′to-sōm) [Gr. *autos* self + *sōma* body]. Any ordinary paired chromosome, as distinguished from a sex chromosome.

autotrophy (aw-tot′ro-fe) [Gr. *autos* self + *trophis* to nourish]. State of being self-nourishing, manufacturing organic nutrients from inorganic raw materials.

auxins (awk′sin) [Gr. *auxē* increase]. Hormonelike substances in plants that promote growth by elongation.

avicularia (a-vik″u-la′ri-a) [L. *avicula* bird]. Specialized members of a colony of ectoproct bryozoa that resemble the head of a bird.

axon (ak′son) [Gr. *axōn* axle]. Nerve fiber that conducts nerve impulses away from the cell body.

bacillus (bah-sil′us) [L. *bacillum* little staff]. A rod-shaped bacterium.

bacteriophage (bak-te′re-o-fāj″) [L. *bacterion* little rod + Gr. *phagein* to eat]. Virus that infects and may kill bacteria.

bacterium (bak-te′re-um) [L. *bactērion* little rod]. Small, typically one-celled microorganisms characterized by the absence of a formed nucleus.

balanced polymorphism (pol″e-mor′fizm) [Gr. *poly* many + *morphē* form]. An equilibrium mixture of homozygotes and heterozygotes maintained by separate and opposing forces of natural selection.

basal metabolic rate The amount of energy expended by the body just to keep alive, when no food is being digested and no muscular work is being done.

base A compound that releases hydroxyl ions (OH^-) when dissolved in water; turns red litmus paper blue.

basidium (bah-sid′e-um) [Gr. *basis* base]. The clublike spore-producing organ of certain of the higher fungi.

benthos (ben′thos) [Gr. *benthos* bottom of the sea]. The flora and fauna of the bottom of oceans or lakes.

bicuspid (bi-kus′pid) [L. *bi* two + *cuspis* point]. Having two cusps, flaps or points.

binomial nomenclature System of naming organisms by the combination of the names of genus and species.

bioassay (bi″o-as-sā) [Gr. *bios* life + assay]. Determination of the effectiveness of a biologically active substance by noting its effect on a living organism.

biogenesis (bi″o-jen′e-sĭs). The generalization that all living things come only from preexisting living things.

biologic clocks Means by which activities of plants or animals are adapted to the regularly recurring changes in the external physical conditions, and perhaps to changes in internal milieu as well.

biological oxidation Process in which electrons removed from an atom or molecule are transferred through the electron transmitter system of the mitochondrion.

bioluminescence (bi″o-loo″mĭ-nes′ens) [Gr. *bios* life + L. *lumen* light]. Emission of light by living cells or by enzyme systems prepared from living cells.

biomass The total weight of all the organisms in a particular habitat.

biome (bi′ōm) [Gr. *bios* life + *ome* mass]. Large, easily differentiated community unit arising as a result of complex interactions of climate, other physical factors and biotic factors.

biosphere The entire zone of air, land and water at the surface of the earth that is occupied by living things.

biotic potential Inherent power of a population to increase in numbers when the age ratio is stable and all environmental conditions are optimal.

birefringence (bi″re-frin′jens) [L. *bi* two + *refringere* to break up]. Property of a substance in solution to refract light differently in different planes.

blastocoele (blas′to-sēl) [Gr. *blastos* germ + *koilos* hollow]. The fluid-filled cavity of the blastula, the mass of cells produced by cleavage of a fertilized ovum.

blastula (blas′tu-lah) [Gr. *blastos* germ]. Usually spherical structure produced by cleavage of a fertilized ovum, consisting of a single layer of cells surrounding a fluid-filled cavity.

Bowman's capsule [*Sir William Bowman,* nineteenth century British physician]. Double-walled, hollow sac of cells that surrounds the glomerulus at the end of each kidney tubule.

Brachiopoda (bra″ki-op′o-da) [L. *brachium* arm]. A phylum of marine organisms that possess a pair of shells and, internally, a pair of coiled arms that bear ciliated tentacles.

brachycephalic (brak″e-se-fal′ik) [Gr. *brachys* short + *kephale* head]. Having a skull that is broad; roundheaded; with a breadth more than 80 per cent of its length.

brachydactyly (brak″e-dak′ti-li) [Gr. *brachys* short + *daktylos* finger]. Abnormal shortness of the fingers and toes.

brachyphalangy (brak″e-fah-lan′je) [Gr. *brachy* short + *phalanx* log]. Abnormal shortness of one or more of the phalanges of a finger or toe.

branchial (brang′ke-al) [Gr. *branchion* a gill]. Pertaining to gills or the gill region.

Brownian movement [*Robert Brown,* nineteenth century Scottish botanist]. Motion of small particles in solution or suspension resulting from their being bumped by water molecules.

Bruce effect The blocking of pregnancy in a newly impregnated mouse by a pheromone, the odor of a foreign male.

brush border The many fine hairlike processes extending from the free surface of certain epithelial cells, e.g., the cells of the proximal convoluted tubules of the mammalian kidney.

Bryophyta (bri″o-fīt′a) [Gr. *bryo* moss + *phyton* plant]. A division of the plant kingdom comprising mosses, liverworts and hornworts.

bryozoa (bri″o-zo′a) [Gr. *bryo* moss + *zoe* life]. Moss animals; the colonies of some species, delicately branched and beautiful, are sometimes mistaken for seaweed; other species form colonies that appear as thin, lacy encrustations on rocks.

budding Asexual reproduction in which a small part of the parent's body separates from the rest and develops into a new individual, eventually either taking up an independent existence or becoming a more or less independent member of the colony.

buffers Substances in a solution that tend to lessen the change in hydrogen ion concentration (*p*H) that otherwise would be produced by adding acids or bases.

calcitonin (kal″sĭ-to′nin). A polypeptide hormone composed of 32 amino acids in a single chain, secreted by parafollicular cells in the thyroid; counters the effect of parathyroid hormone and causes the deposition of calcium and phosphate in bones.

calorie The amount of heat required to raise 1 gram of water 1 degree centigrade (strictly, from 14.5 to 15.5°C). A kilocalorie is a unit 1000 times larger, the amount of heat required to raise 1 kilogram of water 1 degree centigrade.

calyx (ka′liks) [Gr. *kalyx* a bud, a cup]. A cup-shaped organ or cavity; the outermost circle of leaves (sepals) in a complete flower.

cambium (kam′be-um) [L. *cambialis* change]. A layer of meristematic cells in stems and roots of many tracheophytes that divide to produce secondary xylem and secondary phloem.

capillaries (kap′i-lar″e) [L. *capillaris* hairlike]. Microscopic thin-walled vessels located in the tissues, which connect arteries and veins and through the walls of which substances pass to the tissue fluid.

carbohydrate (kar″bo-hi′drāt). Compounds containing carbon, hydrogen and oxygen, in the ratio of 1C:2H:1O; e.g., sugars, starches and cellulose.

carnivore (kar′ni-vōr) [L. *carno* flesh + *vorare* to devour]. An animal that eats flesh.

carotene (kar′o-tēn) [L. *carota* carrot]. Yellow to orange-red pigments found in carrots, sweet potatoes, leafy vegetables, etc., which can be converted in the animal body to vitamin A.

Casparian strip A band of suberized material around the radial walls of the cells of the endodermis.

catabolism (kah-tab′o-lizm) [Gr. *katabolē* a throwing down]. Chemical reactions by which complex substances are converted, within living cells, into simpler compounds with the release of energy.

catalyst (kat′ah-list) [Gr. *katalysis* dissolution]. A substance that regulates the speed at which a chemical reaction occurs without affecting the end point of the reaction and without being used up as a result of the reaction.

cation (kat′i-on) [Gʀ. *kata* down + *ion* going]. An ion bearing a positive charge.

ceboids New World prehensile-tailed monkeys.

cecum (se′kum) [L. *caecum* blind]. A blind pouch into which open the ileum, the colon and the vermiform appendix.

cell constant animals An extreme example of mosaic development which results in all individuals in a species having exactly the same number of cells in comparable tissues performing similar functions.

cell theory The generalization that all living things are composed of cells and cell products, that new cells are formed by the division of preexisting cells, that there are fundamental similarities in the chemical constituents and metabolic activities of all cells, and that the activity of an organism as a whole is the sum of the activities and interactions of its independent cell units.

cells   The microscopic units of structure and function that compose the bodies of plants and animals.

centriole (sen'trĭ-ōl) [L. *centrum* center].   A small dark-staining organelle lying near the nucleus in the cytoplasm of animal cells that forms the spindle during mitosis and meiosis.

centromere (sen'tro-mer) [Gr. *kentro* center + *meros* part].   The point on a chromosome to which the spindle fiber is attached; during mitosis or meiosis it is the first part of the chromosome to pass toward the pole.

cercaria (ser-ka're-ah) [Gr. *kerkos* tail].   The final free-swimming larval stage of a trematode parasite, which encysts in a fish.

cercopithecoid   An Old World monkey; has a tail but does not use it as a limb.

cerebellum (ser″e-bel'lum) [L. (dim.) *cerebrum* brain].   The part of the vertebrate brain that controls muscular coordination.

cerebrum (ser'e-brum) [L. *cerebrum* brain].   The main portion of the vertebrate brain, occupying the upper part of the cranium; the two cerebral hemispheres, united by the corpus callosum, form the largest part of the central nervous system in the human.

chelicera (kē-lis'er-a) [Gr. *chele* claw + *keras* horn].   A pair of pincerlike head appendages found in spiders, scorpions and other arachnids.

chemoreceptor (ke″mo-re-sep'tor).   A sense organ or sensory cell that responds to chemical stimuli.

chemotropism (ke-mot'ro-pizm) [Gr. *chemeia* chemistry + *tropos* a turning].   A growth response to a chemical stimulus.

chimaera (ki-me'rah) [Gr. *chimaira* a mythological fire-spouting monster with a lion's head, goat's body and serpent's tail].   An individual organism whose body contains cell populations derived from different zygotes of the same or of different species; occurring spontaneously, as in twins, or produced artificially, as an organism that develops from combined portions of different embryos, or one in which tissues or cells of another organism have been introduced.

chitin (ki'tin) [Gr. *chiton* tunic].   An insoluble, horny protein-polysaccharide that forms the exoskeleton of arthropods and the cell walls of many fungi.

chlorenchyma (klo-ren'kĭ-mah) [Gr. *chloros* green + *chymos* juice].   The chlorophyll-bearing tissue of plants.

chlorophyll (klo'ro-fil) [Gr. *chloros* green + *phyllon* leaf].   Pigments that give plants their green color and are of paramount importance in transforming radiant energy to chemical energy in the process of photosynthesis.

chloroplast (klo'ro-plast) [Gr. *chloros* green + *plastos* formed].   A chlorophyll-bearing intracellular organelle of plant cells; site of photosynthesis.

choanocyte (ko'ă-no″sīt) [Gr. *choane* funnel + *kytos* hollow vessel].   A unique cell type having a flagellum surrounded by a thin cytoplasmic collar; characteristic of sponges and one group of protozoa.

chorion (ko're-on).   An extraembryonic membrane in reptiles,

birds and mammals that forms an outer cover around the embryo
and in mammals contributes to the formation of the placenta.

chromatin (kro'mah-tin) [Gr. *chroma* color].   The readily stainable
portion of the cell nucleus, forming a network of fibrils within the
nucleus; composed of DNA and proteins.

chromatin spot   An aggregation of chromatin at the periphery of
the nucleus, evident in cells of human skin or from the mucosal
lining of the mouth; makes possible the determination of the
"nuclear sex" of an individual. Most of the cells of a female and
none of the cells of a male have a chromatin spot.

chromatophore (kro'mah-to-fōr") [Gr. *chroma* color + *pherein* to
bear].   Any pigmentary cell or color-producing plastid, such as
those of the deep layers of the epidermis; a chlorophyll-containing
granule in certain bacteria.

chromomere (kro'mo-mēr) [Gr. *chroma* color + *meros* part].   One
of a linear series of beadlike structures composing a chromosome.

chromosomes (kro'mo-sōm) [Gr. *chroma* color + *soma* body].
Filamentous or rod-shaped bodies in the cell nucleus that contain
the hereditary units, the genes.

Chrysophyta (kri-sof'i-ta) [Gr. *chrysos* yellow + *phyton* plant].
Golden-brown algae; division name.

cilia (sil'e-ah) [L. *cilium* eyelid].   Small, bristlelike, cytoplasmic
projections on the free surface of cells; they beat in coordinated
fashion to move the cell or its environment.

circadian rhythms (ser"kah-de'an) [L. *circa* about + *dies* a day].
Repeated sequences of events that occur at about 24-hour intervals.

cleidoic egg (kli-do'ik) [Gr. *kleidouchos* holding the keys]. The
eggs of reptiles, birds and primitive mammals which are self-
sufficient and in which the embryo develops directly into a miniature adult without passing through a larval stage.

climax community   The final, stable and mature community in
a series that appears in succession (termed seral stages). The
climax community is in equilibrium with the environmental conditions and is composed of a definite group of plant and animal
species.

cline (klīn) [Gr. *klinein* to slope]. Continuous series of differences
in structure or function exhibited by the members of a species
along a line extending from one part of their range to another.

clitoris (kli'to-ris) [Gr. *kleitoris* small hill].   A small, erectile body
at the anterior part of the vulva, which is homologous to the male
penis.

cloaca (klo-a'kah) [L. *cloaca* a sewer].   A common chamber receiving the discharge of the digestive, excretory and reproductive
systems in most of the lower vertebrates.

clone   A population of cells descended by mitotic division from a
single ancestral cell.

cobalamin (ko-bal'ah-min).   Vitamin B_{12}; substance essential to
the manufacture of red cells.

cocci (kok'si) [L.; Gr. *kokkos* berry].   Spherical bacterial cells,
usually less than 1 μm in diameter.

cochlea (kok'le-ah) [Gr. *kochlias* snail].   Part of the inner ear;

a spirally coiled tube of two and a half turns, resembling a snail's shell.

codon A sequence of three adjacent nucleotides that code for a single amino acid.

coelom (se′lom) [Gr. *koilia* cavity]. Body cavity of triploblastic animals lying within the mesoderm and lined by it.

coenzyme (ko-en′zīm) [L. *cum* with + Gr. *en* in + *zymē* leaven]. A substance that is required for some particular enzymatic reaction to occur; participates in the reaction by donating or accepting some reactant; loosely bound to enzyme.

coleoptile (ko″le-op′til) [Gr. *koleo* a sheath + *ptile* a feather]. Hollow, sheathlike cylindrical structure that envelops the unexpanded leaves of a monocot shoot.

colinearity The correspondence between the linear sequence of the nucleotide codons, the RNA, and the linear sequence of amino acids in the polypeptide coded for by that sequence.

collagen (kol′ah-jen) [Gr. *kolla* glue + *gennan* to produce]. Protein in connective tissue fibers which is converted to gelatin by boiling.

collenchyma (ko-leng′kĭ-ma) [Gr. *kolla* glue + *en* in + *chymos* juice]. Tissue occurring just beneath the epidermis of stems and leaf stalks which provides the plant with support; composed of cells with walls thickened in the corner.

colloid (kol′oid) [Gr. *kollodes* glutinous]. A two-phase system in which particles of one phase, ranging in size from 1 to 100 mμ, are dispersed in the second phase; a gelatinous material secreted by cuboidal epithelial cells, arranged in hollow spheres one cell thick, in the thyroid.

commensalism (kŏ-men′sal-izm″) [L. *cum* together + *mensa* table]. A relationship between two species in which one is benefited and the second is neither harmed nor benefited by existing together.

community An assemblage of populations that live in a defined area or habitat, which can either be very large or quite small. The organisms constituting the community interact in various ways with one another.

cone (kōn) [L. *conus* cone]. In zoology, the conical photoreceptive cell of the retina which is particularly sensitive to bright light and which, by distinguishing light of various wave lengths, mediates color vision. In botany, the reproductive structure of gymnosperms.

conifers (ko′ni-fer) [L. *conus* cone + *ferre* to bear]. Gymnosperms bearing needlelike leaves that are well adapted to withstand heat and cold.

conjugation (kon″ju-ga′shun) [L. *conjugatio* a blending]. The act of joining together; form of sexual reproduction in which nuclear material is exchanged during the temporary union of two cells; occurs in many ciliate protozoa and in bacteria.

Conservation of Energy, Law of A fundamental law of physics that states that in any given system the amount of energy is constant; energy is neither created nor destroyed, but only transformed from one form to another.

Conservation of Matter, Law of A fundamental law of physics that states that in any chemical reaction atoms are neither created nor destroyed but simply change partners.

consumer organisms Those elements of an ecosystem, plants or animals, that eat other plants or animals.

contraception (kon-trah-sep'shun) [L. *contra* against + *conceptus* conceiving]. Methods of birth control which involve the use of hormones to prevent ovulation or of mechanical or chemical agents to prevent the sperm from reaching and fertilizing the egg.

convergent evolution (kon-ver'jent) [L. *cum* together + *vergere* to incline]. The independent evolution of similar structures, which carry on similar functions, in two or more organisms of widely different, unrelated ancestry.

copulation (kop"u-la'shun) [L. *copulare* to join together]. Sexual union; act of physical joining of two animals during which sperm cells are transferred from one to the other.

corpus allatum (kor'pus al-la'tum) [L. *corpus* body + *adlatus* added]. An endocrine gland located in the head of insects just behind the brain; it secretes juvenile hormone.

corpus callosum (kor'pus kah-lo'sum) [L. *corpus* body + *callosus* hard]. A large commissure of fibers interconnecting the two cerebral hemispheres in mammals.

corpus luteum (kor'pus lu-tē'um) [L. *corpus* body + *luteus* yellow]. A yellow glandular mass in the ovary formed by the cells of an ovarian follicle that has matured and discharged its ovum.

corpus striatum (kor'pus stri-a'tum) [L. *corpus* body + *striatum* striped]. A large subcortical mass of neuron cell bodies and fibers in the base of each cerebral hemisphere.

cortex (kor'teks) [L. *cortex* bark]. The outer layer of an organ; in plants, the tissue beneath the epidermis.

cotyledon (kot"ĭ-le'don) [Gr. *kotyledon* a cup-shaped hollow]. The seed leaf of the embryo of a plant.

covalent bond Chemical bond involving one or more shared pairs of electrons.

cretinism (kre'tin-izm). A chronic condition in the young due to congenital lack of thyroid secretion; retarded physical and mental development.

crossing over Process during meiosis in which the homologous chromosomes undergo synapsis and exchange segments.

ctenophores (tēn'o-fors) [Gr. *ktenos* comb]. Marine animals ("comb jellies") whose bodies consist of two layers of cells enclosing a mass of jelly; the outer surface is covered with eight rows of cilia, resembling combs, by which the animal moves through the water.

cutin (ku'tin) [L. *cutis* skin]. A waxy, waterproof material that prevents the loss of water from a leaf surface.

cycads (si'kad). Members of one of the classes of woody seed plants (the Cycadopsida) which live mainly in tropical and semitropical regions and have either short, tuberous, underground stems or erect, cylindrical stems above the ground.

cyclosis (si-klo'sis) [Gr. *kyklōsis* a surrounding, enclosing]. Circular movement of the cytoplasm, found typically in the cells of the leaves of plants.

cytochromes (si'to-krom) [Gr. *kytos* hollow vessel + *chroma* color]. The iron-containing heme proteins of the electron transmitter system that are alternately oxidized and reduced in biological oxidation.

cytokinesis (si"to-ki-ne'sis) [Gr. *kytōs* hollow vessel + *kinesis* motion]. The division of the cytoplasm during mitosis or meiosis.

dalton [*John Dalton*, eighteenth century British physicist]. The unit of molecular weight; the weight of one hydrogen atom.

deamination (de-am"i-na'shun). Removal of an amino group ($-NH_2$) from an amino acid or other organic compound.

decarboxylation (de"kar-bok"sĭ-la'shun). Removal of a carboxyl group ($-COOH$) from an organic compound.

deciduous (de-sid'u-us) [L. *decidere* to fall off]. Not permanent; to fall off at maturity.

dehydrogenation (de-hi"dro-jen-a'shun) [L. *de* apart + Gr. *hydōr* water]. A form of oxidation in which hydrogen atoms are removed from a molecule.

delamination (de"lam-i-na'shun) [L. *de* apart + *lamina* plate]. Separation of the blastoderm into an upper ectoderm and a lower endoderm during embryonic development.

deme (dēm). A population of very similar organisms interbreeding in nature and occupying a circumscribed area.

denaturation (de-na-tūr-a'shun). Alteration of physical properties and three dimensional structure of a protein, nucleic acid or other macromolecule by mild treatment that does not break the primary structure.

dendrite (den'drīt) [Gr. *dendron* tree]. Nerve fiber, typically branched, that conducts a nerve impulse toward the cell body.

denitrify (de-nī'tri-fi). To convert ammonia to atmospheric nitrogen (e.g., by enzymes of certain bacteria).

deoxyribose (de-ok"se-ri'bōs). A five-carbon sugar with one less oxygen atom than the parent sugar, ribose; constituent of DNA.

dermis (der'mis) [Gr. *derma* skin]. The deeper layer of the skin of vertebrates.

desmosomes Discontinuous buttonlike plaques present on the two opposing cell surfaces and separated by the intercellular space; they apparently serve to hold the cells together.

detoxification (de-tok"si-fi-ka'shun). Enzymatic processes that reduce the toxicity of a substance.

deuterostome (du'ter-o-stōm) [Gr. *deuteros* second + *stoma* mouth]. An animal in which the site of the blastopore is posterior (far from the mouth), which forms anew at the anterior end.

diapause (di'a-poz) [Gr. *dia* through + *pausis* a stopping]. Inactive state of an insect during the pupal stage.

diastole (di-as'to-le) [Gr. *diastolē* a drawing asunder; expansion]. Relaxation of the heart muscle, especially that of the ventricle, during which the lumen becomes filled with blood.

diatoms (dī'ă-tŏm"). Unicellular, microscopic algae with a regularly shaped, siliceous cell wall.

dicotyledon (di-kot"i-le'dun) [Gr. *dis* double + *kotyledon* a cup-shaped hollow]. A flowering plant with embryos having two seed leaves, or cotyledons.

differentiation Development toward a more mature state; a process changing a relatively unspecialized cell to a more specialized cell.

diffusion The movement of molecules from a region of high concentration to one of lower concentration, brought about by their kinetic energy.

digitigrade (dij'i-ti-grād") [L. *digitus* finger or toe + *gradus* a step]. Locomotion by walking on the toes; applied to animals—dogs and cats—in which only the digits touch the ground.

dioecious (di-e'shus) [Gr. *dis* double + *oikos* house]. Plant species possessing staminate and pistillate flowers on separate individuals.

diploid (dip'loid) [Gr. *diploos* twofold]. A chromosome number twice that found in gametes; containing two sets of chromosomes.

disaccharides (di-sak'ah-rid). Sugars that yield two monosaccharides on hydrolysis; e.g., sucrose, lactose and maltose.

distal (dis'tal) [L. *distans* distant]. Remote; farther from the point of reference.

DNA Deoxyribose nucleic acid; present in chromosomes and contains genetic information coded in specific sequences of its constituent nucleotides.

dolichocephalic (dol"i-ko-se-fal'ik) [Gr. *dolichos* long + *kephalē* head]. Longheaded; a skull with a breadth less than 75 per cent of its length.

Down's syndrome A congenital malformation in which individuals have abnormalities of the face, eyelids, tongue and other parts of the body and are greatly retarded in both their physical and mental development; results from a trisomy of chromosome 21 or 18.

DPN (See *NAD*.)

drupe (droop) [L. *drupa* an overripe olive]. Stone fruits in which the outer part of the ovary wall forms a skin, the middle part becomes fleshy and juicy, and the inner part forms a hard pit or stone around the seed; e.g., peaches, plums, apricots.

ecdysone (ek-di'son) [Gr. *ekdysis* a getting out]. The hormone that induces molting (ecdysis) in arthropods.

Echinodermata (e-kin"o-der-mat'a) [L. *echinus* hedgehog, sea urchin + Gr. *derma* skin]. Phylum of spiny-skinned marine animals (e.g., starfish, sea urchins, sea cucumbers).

ecologic niche The status of an organism within a community or ecosystem; depends on the organism's structural adaptations, physiologic responses and behavior.

ecology (e-kol'o-je) [Gr. *oikos* house + *logos* word, discourse]. The study of the interrelations between living things and their environment, both physical and biotic.

ecosystem (ek"o-sis'tem). A natural unit of living and nonliving parts that interact to produce a stable system in which the exchange of materials between living and nonliving parts follows a circular path.

ecotone   A fairly broad transition region between adjacent biomes; contains some organisms from each of the adjacent biomes plus some that are characteristic of, and perhaps restricted to, the ecotone.

ectoderm (ek′to-derm) [Gr. *ektos* without + *derma* skin].   The outer of the two germ layers of the gastrula; gives rise to the skin and nervous system.

edaphic factors (ē-dăf′ik).   Factors in the soil that influence the distribution and numbers of plants and animals.

effector (ef-fek′tor).   Structures of the body by which an organism acts; means by which it reacts to stimuli; e.g., muscles and glands.

electrolyte (e-lek′tro-līt) [Gr. *elektron* amber + *lytos* soluble].   Substance that dissociates in solution into charged particles, ions, and thus permits the conduction of an electric current through the solution.

electron transmitter system   System of enzymes localized within the mitochondria that transfers electrons from foodstuff molecules to oxygen.

element (el′ĕ-ment).   One of the hundred or so types of matter, natural or man-made, composed of atoms, all of which have the same number of protons in the atomic nucleus and the same number of electrons circling in the orbits.

embolus (em′bo-lus) [Gr. *embolos* plug].   A thrombus or any other particle carried by the blood stream that blocks a blood vessel.

embryo (em′bre-o) [Gr. *en* in + *bryein* to swell].   The early stage of development of an organism; the developing product of fertilization of an egg.

emulsion (e-mul′shun) [L. *emulsum* to milk out].   A colloid in which one liquid phase is dispersed in another liquid phase.

endergonic (end″er-gon′ik) [Gr. *endon* within + *ergon* work].   A reaction characterized by the absorption of energy; requires energy to occur.

endocrine (en′do-krin) [Gr. *endon* within + *krinein* to separate].   Secreting internally; applied to organs whose function is to secrete into the blood or lymph a substance that has a specific effect on another organ or part.

endoderm (en′do-derm) [Gr. *endon* within + *derma* skin].   The inner germ layer of the gastrula, lining the archenteron; becomes the digestive tract and its outgrowths—the liver, lungs and pancreas.

endoskeleton (en″do-skel′ĕ-ton) [Gr. *endon* within + *skeleton* a dried body].   Bony and cartilaginous supporting structures within the body; provide support from within.

endosperm (en′do-sperm) [Gr. *endon* within + *sperma* seed].   Nutritive tissue from the female gametophyte that surrounds and nourishes the developing embryo of seed plants. It is haploid in gymnosperms and triploid in angiosperms.

engram (en′gram) [Gr. *en* in + *gramma* mark].   The term applied to the presumed change that occurs in the brain as a consequence of learning; a memory trace.

entropy   A randomized state of energy that is unavailable to do work.

environmental resistance   The sum of the physical and biologic factors that prevent a species from reproducing at its maximum rate.

enzyme (en′zīm) [Gr. *en* in + *zyme* leaven]. A protein catalyst produced within a living organism that accelerates specific chemical reactions.

epiboly (e-pib′o-le) [Gr. *epibolē* cover]. A method of gastrulation by which the smaller blastomeres at the animal pole of the embryo grow over and enclose the cells of the vegetal hemisphere.

epicotyl (ep″i-kot′il). The part of the axis of a plant embryo or seedling above the point of attachment of the cotyledons.

epidermis (ep″i-der′mis) [Gr. *epi* on + *derma* skin]. The outermost layer of cells of an organism.

epididymis (ep″i-did′i-mis) [Gr. *epi* on + *didymos* testis]. Complexly coiled tube adjacent to the testis where sperm are stored.

epigenesis (ep″i-jen′e-sis) [Gr. *epi* on + *genesis* to be born]. The theory that development proceeds from a structureless cell by the successive formation and addition of new parts that do not preexist in the fertilized egg.

epiglottis (ep″i-glot′is) [Gr. *epi* on + *glottis* the tongue]. The lidlike structure that covers the glottis, the entrance to the larynx.

epilimnion (ĕp-i-lĭm′nĭ-ŏn). The uppermost layer, or surface water, of a lake or pond.

epiphyte (ep″i-fīt) [Gr. *epi* on + *phyton* plant]. A plant that grows upon another plant, for position and support only.

epithelium (ep″i-the′le-um) [Gr. *epi* on + *thēlē* nipple]. The layer of tissue covering the internal and external surfaces of the body, including the lining of vessels and other small cavities; consists of cells joined by small amounts of cementing substances.

equilibrium (e″kwĭ-lib′re-um) [L. *aequus* equal + *libra* balance]. A state of balance; a condition in which opposing forces exactly counteract each other.

estrogen (es′tro-jen). One of the female sex hormones, produced by the ovarian follicle, which promotes the development of the secondary sex characteristics.

estrus (es′trus) [Gr. *oistros* anything that drives mad, any vehement desire]. The recurrent, restricted period of sexual receptivity in female mammals, marked by intense sexual urge.

ethology [Gr. *ethos* custom + *logos* study of]. The study of the whole range of animal behavior under natural conditions.

etiolation (e″ti-o-la′shun) [Gr. *aitio* a cause]. A blanching of color in a plant due to lack of chlorophyll when grown in the dark; the plants have small leaves and long, weak stems.

eukaryotic [Gr. *eu* good + *karyon* kernel]. Applied to organisms that have nuclei surrounded by membranes, Golgi apparatus and mitochondria.

eustachian tube (u-sta′ke-an) [*Bartolommeo Eustachio*, sixteenth century Italian anatomist]. The auditory tube passing between the middle ear cavity and pharynx of most terrestrial vertebrates; it permits the equalization of pressure on the tympanic membrane.

eutherian (u-thēr′ĭ-an) [Gr. *eu* good + *therion* beast]. One of the placental mammals in which a well formed placenta is present and the young are born at a relatively advanced stage of development; includes all living mammals except monotremes and marsupials.

excretion (eks-kre'shun) [L. *ex* out + *cernere* to shift, separate]. Removal of metabolic wastes by an organism.

exergonic (ek"ser-gon'ik) [L. *ex* out + Gr. *ergon* work]. A reaction characterized by the release of energy.

exotoxin (ek"so-tok'sin) [L. *exo* outside + *toxicum* poison]. An extremely potent poison secreted by the bacterial cell to the outside environment.

expressivity (eks"pres-siv'i-te) [L. *expressus*]. The extent to which a heritable trait is manifested by an individual carrying the principal gene conditioning it.

extensor (eks-ten'sor) [L. *extendere* to stretch]. A muscle that serves to extend or straighten a limb.

facilitation (fah-sil"i-ta'shun) [L. *facilis* easy]. The promotion or hastening of any natural process; the reverse of inhibition.

feedback control System in which the accumulation of the product of a reaction leads to a decrease in its rate of production or a deficiency of the product leads to an increase in its rate of production.

fermentation (fer"men-ta'shun) [L. *fermentum* leaven]. Anaerobic decomposition of an organic compound by an enzyme system; energy is made available to the cell for other processes.

ferredoxin (fer"ĕ-dok'sin). An iron-containing protein, a component of the electron transmitter system in the chloroplast, that undergoes cyclic oxidation and reduction during photosynthesis.

fertilization (fer"tĭ-lĭ-za'shun) [L. *fertilis* to bear, produce]. The fusion of a spermatozoon with an ovum to initiate development of the resulting zygote.

fetus (fe'tus) [L. *fetus* fruitful]. The unborn offspring after it has largely completed its embryonic development; from the third month of pregnancy to birth in humans.

fission (fish'un) [L. *fissio* to cleave]. Process of asexual reproduction in which an organism divides into two approximately equal parts.

flagellates (flaj'ĕ-lāt) [L. *flagellum* whip]. Microorganisms furnished with one or more slender, whiplike processes termed flagella.

flavin (flā'vin). A yellowish compound, a component of certain coenzymes and prosthetic groups of certain oxidation-reduction enzyme systems.

flexor (flek'sor) [L. *flectere* to bend]. A muscle that serves to bend a limb.

fluorescence (floo"o-res'ens). The emission of light by a substance that has absorbed radiation of a different wavelength; results when an excited singlet state decays to the ground state, an extremely rapid process that is independent of temperature.

follicle (fol'lĭ-k'l) [L. *folliculus* small bag]. A small sac of cells in the mammalian ovary that contains a maturing egg.

food chain A sequence of organisms through which energy is transferred from its ultimate source in a plant; each organism eats the preceding and is eaten by the following member of the sequence.

foramen ovale (fo-ra'men) [L. *forare* to bore, pierce]. The oval window between the right and left atria, present in the fetus and by means of which blood entering the right atrium may enter the aorta without passing through the lung.

foraminifera (fo-ram"ĭ-nif'er-ah) [L. *forare* to *bore* + *ferre* to bear]. Amoeboid protozoa that secrete chalky, many-chambered shells with pores through which the animal extends its pseudopods.

fossils (fos'ils) [L. *fossilis* to dig]. Any remains of an organism that have been preserved in the earth's crust.

fovea (fo've-ah) [L. *fovea* a small pit]. A small pit in the surface of a structure or organ; specifically, a pit in the center of the retina that contains only cones and provides for keenest vision.

fruit (froot) [L. *fructus* fruit]. The ripened ovary of a plant including the seed and its envelopes.

fucoxanthin (fu"ko-zan'thin) [L. *fucus* rock lichen + Gr. *xanthos* yellow]. The brown pigment found in diatoms, brown algae and dinoflagellates.

fundus (fun'dus) [L. *fundus* bottom]. The bottom or base of an organ; the part of a hollow organ farthest from its opening.

gamete (gam'ēt) [Gr. *gametē* wife]. A reproductive cell; an egg or sperm whose union, in sexual reproduction, initiates the development of a new individual.

gametophyte (gam'ĕ-to-fīt) [Gr. *gametē* wife + *phyton* plant]. The haploid or sexual (gamete-producing) stage in the life cycle of a plant.

ganglion (gang'gle-on) [Gr. *gangli* knot]. A knotlike mass of the cell bodies of neurons located outside the central nervous system.

gastrodermis (gas"tro-der'mis) [Gr. *gastēr* stomach + *derma* skin]. The tissue lining the gut cavity that is responsible for digestion and absorption.

gastrula (gas'troo-lah) [Gr. *gastēr* stomach]. Early embryonic stage that follows the blastula; consists initially of two layers, the ectoderm and the endoderm, and of two cavities, the blastocoele between ectoderm and endoderm and the archenteron, formed by invagination, which lies within the endoderm and opens to the exterior through the blastopore.

gastrulation (gas"troo-la'shun) [Gr. *gastēr* stomach]. The process by which the young embryo becomes a gastrula and acquires first two and then three layers of cells.

gemma cups (jem'ah) [L. eye, or bud, of a plant]. Cup-shaped vegetative buds in bryophytes which develop asexually into new entire plants.

gene (jēn) [Gr. *gennan* to produce]. The biologic unit of genetic information, self-reproducing and located in a definite position (locus) on a particular chromosome.

genetic drift The tendency, within small interbreeding populations, for heterozygous gene pairs to become homozygous for one allele or the other by *chance* rather than by selection.

genetic equilibrium The situation in which the distribution of alleles in a population is constant in successive generations (unless altered by section or mutation).

genome (je'nōm) [Gr. *gennan* to produce + *ōma* mass, abstract entity]. A complete set of hereditary factors, contained in the haploid assortment of chromosomes.

genotype (jen'o-tīp) [Gr. *gennan* to produce + *typos* type]. The fundamental hereditary constitution, assortment of genes, of any given organism.

genus (je'nus) [L. birth, race, kind, sort]. A rank in taxonomic classification in which closely related species are grouped together.

geotropism (je-ot'ro-pizm) [Gr. *ge* earth + *tropos* a turning]. A growth response toward or away from the earth; the influence of gravity on growth.

gibberellins (jib-er-ĕl'ins). One group of chemically defined substances that occur naturally and have functions in the control of growth and development in flowering plants; they promote the elongation of shoots in young plants of certain species.

gill The respiratory organ of aquatic animals, usually a thin-walled projection from the body surface or from some part of the digestive tract.

gizzard (giz'erd) [L. *gigeria* cooked entrails of poultry]. A portion of the digestive tract specialized for mechanical digestion.

globulin (glob'u-lin) [L. *globulus* globule]. One of a class of proteins in blood plasma, some of which (gamma-globulins) function as antibodies.

glomerulus (glo-mer'u-lūs) [L. *glomus* ball]. A tuft of minute blood vessels or nerve fibers; specifically, the knot of capillaries at the proximal end of a kidney tubule.

glycolysis (gli-kol'ĭ-sis) [Gr. *glykys* sweet + *lysis* solution]. The metabolic conversion of sugars into simpler compounds.

goiter (goi'ter). An enlargement of the thyroid gland, causing a swelling in the front part of the neck; may result from overactivity of the thyroid or from deficiency of iodine.

Golgi bodies [*Camillo Golgi*, nineteenth century Italian histologist]. A type of cell organelle found in the cytoplasm of all cells except mature sperm and red blood cells; believed to play a role in the secretion of cell products.

gonad (gon'ad) [Gr. *gonē* seed]. A gamete-producing gland; an ovary or testis.

grana [L. *grana* grain]. Small bodies within chloroplasts that contain alternate layers of chlorophyll, protein and lipid and that are the functional units of photosynthesis.

guttation The appearance of water droplets on leaves, forced out through leaf pores by root pressure.

gymnosperm (jim'no-sperm) [Gr. *gymnos* naked + *sperma* seed]. A seed plant in which the seeds are not enclosed in an ovary; the conifers and a number of extinct plants.

habitat (hab'ĭ-tat) [L. *habitus,* from *habere* to hold]. The natural abode of an animal or plant species; the physical area in which it may be found.

habituation   Applied to the process by which organisms become accustomed to a stimulus and cease to respond to it.

haploid (hap′loid) [Gr. *haploos* simple, single].   Having a single set of chromosomes, as normally present in a mature gamete.

Hardy-Weinberg law   The relative frequencies of the members of a pair of allelic genes in a population are described by the expansion of the binomial equation $a^2 + 2ab + b^2$.

haversian canals (ha-ver′shan) [*Clopton Havers*, seventeenth century English physician].   Channels extending through the matrix of bone and containing blood vessels and nerves.

hemoglobin (he″mo-glo′bin) [Gr. *haima* blood].   The red, iron-containing, protein pigment of the erythrocytes that transports oxygen and carbon dioxide and aids in regulation of pH.

hemophilia (he″mo-fil′e-ah) [Gr. *haima* blood + *philein* to love].   Hereditary disease in which the formation of thromboplastin is impaired owing to a deficiency of the so-called antihemophilic globulin; blood does not clot properly; "bleeder's disease."

hepatic (he-pat′ik) [Gr. *hēpatikos* liver].   Pertaining to the liver.

herbaceous (her-ba′shus) [L. *herba* herb].   Pertaining to, or having the characteristics of, an herb; nonwoody.

herbivore (her′bĭ-vōr) [L. *herba* herb + *vorare* to devour].   A plant-eating animal.

hermaphroditism (her-maf′ro-dit-izm) [Gr. *Hermes* + *Aphrodite*, a mythological god and goddess, whence *hermaphroditos,* a person having the attributes of both sexes].   A state characterized by the presence of both male and female sex organs in the same organism.

heterogamy (het″er-og′ah-me) [Gr. *heteros* other + *gamōs* marriage].   Reproduction involving the union of two gametes that differ in size and structure; e.g., egg and sperm.

heterografts (het′er-o-grafts) [Gr. *heteros* other].   Grafts of tissue obtained from the body of an animal of a species other than that of the recipient.

heterosis [Gr. *heteros* other].   Hybrid vigor; the offspring from the mating of individuals of totally unrelated strains frequently are much better adapted for survival than either parent.

heterotrophs (het′er-o-trofs) [Gr. *heteros* other + *trophos* feeder].   Organisms that cannot synthesize their own food from inorganic materials and therefore must live either at the expense of autotrophs or upon decaying matter.

heterozygous (het″er-o-zi′gus) [Gr. *heteros* other + *zygos* yoke].   Possessing two different alleles for a given character at the corresponding loci of homologous chromosomes.

hibernation (hi″ber-na′shun) [L. *hiberna* winter].   The dormant state of decreased metabolism in which certain animals pass the winter.

homeostasis (ho″me-o-sta′sis) [Gr. *homois* unchanging + *stasis* standing].   The tendency to maintain uniformity or stability in the internal environment of the organism.

hominid [L. *homo* man].   Pertaining to the family of humans; a living or extinct human or humanlike type.

homograft reaction [Gr. *homos* same + graft].   The rejection by the host organism of a graft of tissue from an organism of the same species but a different genotype.

homoiothermic (ho-moi′o-ther″mik) [Gr. *homois* unchanging + *thermē* heat]. Constant-temperature animals; e.g., birds and mammals that maintain a constant body temperature despite variations in environmental temperature.

homologous structures (ho-mol′o-gus) [Gr. *homologos* agreeing, corresponding]. Those structures of various animals that arise from common rudiments and are similar in basic plan and development.

homozygous (ho″mo-zi′gus) [Gr. *homos* same + *zygos* yoke]. Possessing an identical pair of alleles at the corresponding loci of homologous chromosomes for a given character or for all characters.

hormones (hor″mōns) [Gr. *hormaein* to set in motion, spur on]. Substances produced in cells in one part of the body that diffuse or are transported by the blood stream to cells in other parts of the body where they regulate and coordinate their activities.

humus (hu′mus) [L. *humus* earth, ground]. Organic matter in the soil; a dark mold of decayed vegetable tissue that gives soil a brown or black color.

hybrid vigor (hī′brid) [L. *hybrida* mongrel]. The mating of genetically dissimilar individuals of totally unrelated strains, which may yield offspring that are better adapted to survive than either parent strain.

hydrogen bond A weak bond between two molecules formed when a hydrogen atom is shared between two atoms, one of which is usually oxygen; of primary importance in the structure of nucleic acids and proteins.

hydrolysis (hi-drol′ĭ-sis) [Gr. *hydōr* water + *lysis* dissolution]. The splitting of a compound into parts by the addition of water between certain of its bonds, the hydroxyl group being incorporated in one fragment, and the hydrogen atom in the other.

hydrophytes (hi′dro-fīts) [Gr. *hydor* water + *phyton* plant]. Plants that grow in a very wet environment, either completely aquatic or rooted in water or mud but with stems and leaves above the water.

hydroponics (hi″dro-pon′iks) [Gr. *hydōr* water + L. *ponere* place]. Soil-less culture of plants; roots are immersed in a nutrient-rich aqueous medium.

hypersensitivity (hi″per-sen′sĭ-tiv′ĭ-te) [Gr. *hyper* above + sensitivity]. A state of altered reactivity; abnormally increased sensitivity; ability to react with characteristic symptoms to the presence of certain substances (allergens) in amounts innocuous to normal individuals.

hypertonic (hi″per-ton′ik) [Gr. *hyper* above + *tonos* tone]. Having a greater concentration of solute molecules and a lower concentration of solvent (water) molecules and hence an osmotic pressure greater than that of the solution with which it is compared.

hypha (hi′fah) [Gr. *hyphe* a web]. One of the filaments composing the mycelium of a fungus.

hypocotyl (hi″po-ko-tl) [Gr. *hypo* under + *kotyle* hollow]. The part of the axis of a plant embryo or seedling below the point of attachment of the cotyledons.

hypothalamus (hi″po-thal′ah-mus) [Gr. *hypo* under + *thalamos* inner chamber].   A region of the forebrain, the floor of the third ventricle, which contains various centers controlling visceral activities, water balance, temperature, sleep, etc.

hypothesis (hi-poth′ĕ-sis) [Gr. *hypo* under + *thesis* setting down].   A supposition assumed as a basis of reasoning which can then be tested by further controlled experiments.

hypotonic (hi″po-ton′ik) [Gr. *hypo* under + *tonos* tone].   Having a lower concentration of solute molecules and a higher concentration of solvent (water) molecules and hence an osmotic pressure lower than that of the solution with which it is compared.

immune reaction (ĭ-mūn′) [L. *immunis* safe].   The production of antibodies in response to antigens.

immunologic tolerance (i-mu″ne-loj′ik).   The ability of an organism to accept cells transplanted from a genetically distinct organism; results from the exposure of the organism to an antigen before it has developed the capacity to react to it, after which development of capacity to react may be delayed or postponed indefinitely.

implantation (im″plan-ta′shun) [L. *in* into + *plantare* to set].   The insertion of a part or tissue in a new site in the body; the attachment of the developing embryo to the epithelial lining (endometrium) of the uterus.

imprinting   A form of rapid learning by which a young bird or mammal forms a strong social attachment to an object within a few hours after hatching or birth.

induction (in-duk′shun) [L. *inducere* to lead in].   The production of a specific morphogenic effect in one tissue of a developing embryo through the influence of an organizer or another tissue.

inflammation (in″flah-ma′shun) [L. *inflammare* to set on fire].   The reactions of tissues to injury: pain, increased temperature, redness and accumulation of leukocytes.

ingestion (in-jes′chun) [L. *in* into + *gerere* to carry].   The act of taking food into the body by mouth.

insight learning   The appearance of a new response in an organism as a result of its evaluation of previous experience.

instinct   A genetically determined pattern of behavior or responses that is not based on the individual's previous experience.

integument (in-teg′u-ment) [L. *integumentum*, from *in* on + *tegere* to cover].   Skin; the covering of the body.

interferon (in″ter-fēr′on).   A protein formed during the interaction of animal cells with viruses, which is capable of conferring on fresh animal cells of the same species resistance to infection with a wide range of viruses.

internode (in′ter-nōd) [L. *inter* between + *nodus* knot].   The section of a stem between two nodes.

invagination (in-vaj″ĭ-na′shun) [L. *in* within + *vagina* sheath].   The infolding of one part within another, specifically a process of gastrulation in which one region folds in to form a double-layered cup.

inversion, chromosomal   Turning a segment of a chromosome end for end and attaching it to the same chromosome.

ion (i'on) [Gr. *iōn* going]. An atom or a group of atoms bearing an electric charge, either positive (cation) or negative (anion).

isogamy (i-sog'ah-me) [Gr. *isos* equal + *gamos* marriage]. Reproduction resulting from the union of two gametes that are identical in size and structure.

isomer (i'so-mer) [Gr. *isos* equal + *meros* part]. A molecule with the same chemical formula as another but a different structural formula; e.g., glucose and fructose.

isotonic or **isosmotic** (i-so-ton'ik, i-sos-mot'ik). Having identical concentrations of solute and solvent molecules and hence the same osmotic pressure as the solution with which it is compared.

isotopes (i'so-tōps) [Gr. *isos* equal + *topos* place]. Alternate forms of a chemical element having the same atomic number (that is, the same number of nuclear protons and orbital electrons) but possessing different atomic masses (that is, different numbers of neutrons).

isozymes (i'so-zīms) [Gr. *isos* equal + *zyme* leaven]. Different molecular forms of proteins with the same enzymatic activity.

juvenile hormone An arthropod hormone that preserves juvenile morphology during a molt. Without it, metamorphosis toward the adult form takes place.

karyokinesis (kar''e-o-ki-ne'sis) [Gr. *karyon* nucleus or nut + *kinesis* motion]. The phenomena involved in division of the nucleus in mitosis.

karyotype [Gr. *karyon* nucleus or nut + *typos* type]. A characterization of a set of chromosomes of an individual with regard to their number, size and shape.

keratin (ker'ah-tin) [Gr. *keratos* horn]. A horny, water-insoluble protein found in the epidermis of vertebrates and in nails, feathers, hair, horn and the like.

ketone bodies Incompletely oxidized fatty acids which are toxic in high concentrations; excreted in the urine, causing an acidosis.

kinesis (ki-ne''sis) [Gr. *kinēsis* movement]. The activity of an organism in response to a stimulus; the direction of the response is not controlled by the direction of the stimulus (in contrast to a taxis).

kinesthesis (kin'es-the'sis) [Gr. *kinē* movement + *aisthesis* perception]. Sense that gives us our awareness of the position and movement of the various parts of the body.

kinins (ki'nins). Group of polypeptides produced in blood and tissues that act on blood vessels, smooth muscles and certain nerve endings; e.g., bradykinin or kallidin; one of a group of compounds containing adenine that stimulates cell division and growth of plant cells in tissue culture.

labyrinthodont (lab''ĭ-rin'tho-dont) [Gr. *labyrinthos* labyrinth + *odontos* tooth]. A member of a subclass of extinct amphibians in

entiated embryonic tissue of plants, capable of producing additional cells by mitotic divisions.

merozoite (mer″o-zo′it) [Gr. *meros* part + *zōon* animal]. One of the young forms derived from the splitting up of the schizont in the human cycle of the malarial parasite, *Plasmodium;* it is released into the circulating blood and attacks new erythrocytes.

mesenchyme (mes″eng-kīm) [Gr. *mesos* middle + *enchyme* an infusion]. A meshwork of loosely associated, often stellate cells; found in the embryos of vertebrates and the adults of some invertebrates.

mesoderm (mes′o-derm) [Gr. *mesos* middle + *derma* skin]. The middle layer of the three primary germ layers of the embryo, lying between the ectoderm and the endoderm.

mesoglea (mes″o-gle′ah) [Gr. *mesos* middle + *gloia* glue]. A gelatinous matrix located between the ectoderm and endoderm of cnidarians.

mesonephros (mes″o-nef′ros) [Gr. *mesos* middle + *nephros* kidney]. An embryonic vertebrate kidney that succeeds the pronephros; its tubules develop adjacent to the middle portion of the coelom and drain into the archinephric duct.

mesophyll (mes′o-fil) [Gr. *mesos* middle + *phyllon* leaf]. Thin-walled cells in the interior of a leaf; rich in chloroplasts.

mesophytes (mes′o-fīts) [Gr. *mesos* middle + *phytos* plant]. Common land plants that live in a climate with an average amount of moisture.

messenger RNA A particular kind of ribonucleic acid that is synthesized in the nucleus and passes to the ribosomes in the cytoplasm; combines with RNA in the ribosomes and provides a template for the synthesis of an enzyme or some other specific protein.

metabolism (mě-tab′o-lizm) [Gr. *metaballein* to turn about, change, alter]. The sum of all the physical and chemical processes by which living organized substance is produced and maintained; the transformations by which energy and matter are made available for the uses of the organism.

metamerism (met-am′er-izm) [Gr. *meta* with + *meros* part]. The state of being made up of serial segments, as in annelids and chordates.

metamorphosis (met″ah-mor′fo-sis) [Gr. *meta* after, beyond, over + *morphōsis* a shaping, bringing into shape]. An abrupt transition from one developmental stage to another, e.g., from a larva to an adult.

metanephros (met″ah-nef′ros) [Gr. *meta* after, beyond, over + *nephros* kidney]. The adult kidney of reptiles, birds and mammals.

metaphase (mět′ah-fāz) [Gr. *meta* after, beyond, over + *phasis* to make to appear]. The middle stage of mitosis during which the chromosomes line up in the equatorial plate and separate lengthwise.

Metazoa (met″ah-zo′ah) [Gr. *meta* after, beyond, over + *zōon* animal]. Division of the animal kingdom that embraces all multicellular animals whose cells become differentiated to form tissues; all animals except the Protozoa.

microsporangia (mī″kro-spo-ran′je-a) [Gr. *mikros* small + *sporos* seed + *angeion* vessel]. Small sacs (pollen sacs) that contain microspore mother cells, which divide by meiosis to form microspores.

microspores (mī″kro-spors′). [Gr. *mikros* small + *sporos* seed]. Small asexual haploid spores that germinate to produce male gametophytes.

microtubule A cytoplasmic organelle, an elongate slender tube; contains a specific protein, tubulin.

mimicry (mim′ĭk-re′) [Gr. *minos* to imitate]. An adaptation for survival in which an organism resembles some other living or nonliving object.

miracidium (mir″ah-sid′ĭ-um) [Gr. *meirakidion* youthful person]. The first larval stage of parasitic flukes.

mitochondria (mīt″o-kon′dre-ah) [Gr. *mitos* thread + *chondrion* granule]. Spherical or elongate intracellular organelles that contain the electron transmitter system and certain other enzymes; site of oxidative phosphorylation.

mitosis (mī-to′sis) [Gr. *mitos* thread + *osis* state or condition]. A form of cell or nuclear division by means of which each of the two daughter nuclei receives exactly the same complement of chromosomes as the parent nucleus had.

mixture A combination of two or more kinds of atoms or molecules that may be combined in varying proportions.

mole (mōl) [L. *moles* a shapeless mass]. The amount of a chemical compound whose mass in grams is equivalent to its molecular weight, the sum of the atomic weights of its constituent atoms.

molecule (mol′e-kūl) [L. *molecula* little mass]. The smallest particle of a covalently bonded element or compound, having the composition and properties of a larger part of the substance.

molting [L. *mutare* to change]. The shedding and replacement of an outer covering such as hair, feathers or exoskeleton.

Monera (mō-ne′rah) [Gr. *moneres* single]. A kingdom of organisms that includes the simplest prokaryotic microorganisms, the bacteria and blue-green algae, forms lacking true nuclei or plastids and in which sexual reproduction is very rare or absent.

mongolism (mŏn′go-lism). See *Down's syndrome*.

monocotyledon (mŏn′o-kot″ĭ-le′don) [Gr. *monos* single + *kotyledon* a cup-shaped hollow]. A flowering plant with one seed leaf, or cotyledon; one of the two large groups of angiosperms, e.g., grasses, lilies, orchids.

monoecious (mŏn-e′shus) [Gr. *monos* single + *oikos* house]. Bearing both male and female cones or flowers (microsporophylls and megasporophylls) on the same plant.

monomer (mŏn′ō-měr) [Gr. *monos* single + *meros* part]. A simple molecule of a compound of relatively low molecular weight which can be linked with others to form a polymer.

morphogenesis (mor″fo-jen′e-sis) [Gr. *morphē* form + *gennan* to produce]. The development of form, size and other features of a particular organ or part of the body.

motor unit All the skeletal muscle fibers that are stimulated by a single motor neuron.

mucosa (mū-ko′sah).   Mucous membrane; e.g., the lining of the digestive tract.

multiple alleles   Three or more alternate conditions of a single locus that produce different phenotypes.

mutation   A stable, inherited change in a gene.

mutualism (mū′tū-al-izm).   An association whereby two organisms of different species each gain from being together and are unable to survive separately.

mycelium (mī-sē′lē-um) [Gr. *mykes* fungus + *hēlos* nail].   The entire mass of branching filaments (hyphae) that constitute one fungus.

myelin (mī′ĕ-lin) [Gr. *myelos* marrow].   The fatty material that forms a sheath around the axons of nerve cells in the central nervous system and in certain peripheral nerves.

myofibrils (mī″ō-fī′brĭls) [Gr. *mys* muscle + L. *fibrilla* small fiber].   Microscopic, extended contractile fibers composed of the proteins myosin and actin.

myopia (mī-ō′pē-ah) [Gr. *myein* to shut + *ōps* eye].   Nearsightedness; the eyeball is too long and the retina too far from the lens; light rays converge at a point in front of the retina and are again diverging when they reach it, resulting in a blurred image.

myosin (mī′ō-sĭn) [Gr. *mys* muscle].   A soluble protein found in muscle; in combination with actin, functions in the contraction and relaxation of muscle fibers.

myxedema (mĭk″sĕ-dē′mah) [Gr. *myxa* mucus + *oidēma* swelling].   A condition that results from a deficiency of thyroxine secretion in an adult; characterized by a low metabolic rate and decreased heat production.

NAD   Abbreviation of nicotinamide adenine dinucleotide, a coenzyme that functions as a hydrogen acceptor in biological oxidations (also called DPN).

NADP   Abbreviation of nicotinamide adenine dinucleotide phosphate, a coenzyme that functions as a hydrogen acceptor in biological oxidations (also called TPN).

nares (na′rēz) [L. *nares* nostrils].   The openings of the nasal cavities. External nares open to the body surface; internal nares, to the pharynx.

nastic movement (nă′stĭk).   A response of a plant to external stimuli that is independent of the direction from which the stimuli come.

nekton (nĕk′tŏn) [Gr. *nēktōs* swimming].   Collective term for the organisms that are active swimmers.

nematocyst (nĕm′ah-tō-sĭst) [Gr. *nēma* thread + *kystis* bladder].   A minute stinging structure found on cnidarians and used for anchorage, for defense and for the capture of prey.

nephridium (nĕ-frĭd′ē-um) [Gr. *nephros* kidney].   The excretory organ of the earthworm and other annelids which consists of a ciliated funnel, opening into the next anterior coelomic cavity and connected by a tube to the outside of the body.

nephron (nĕf′fron) [Gr. *nephros* kidney]. The anatomical and functional unit of the vertebrate kidney.

neuroglia (nū-rŏg′lĭ-ă). Connecting and supporting cells in the central nervous system surrounding the neurons.

neurohumor (nū″rō-hū′mŏr) [Gr. *neuron* nerve + L. *humor* a liquid]. A substance secreted by the tip of a neuron that is able to activate a neighboring neuron or muscle.

neuron (nū′rŏn) [Gr. *neuron* nerve]. A nerve cell with its processes, collaterals and terminations; the structural unit of the nervous system.

neurosecretion (nū″rō-sē-krē′shun) [Gr. *neuron* nerve + L. *secretio*, from *secernere* to secrete]. The production of hormones by nerve cells.

neuroses (nū-rō′sēs) [Gr. *neuron* nerve + *osis* state or condition]. Comparatively mild and common psychic disorders with a great variety of symptoms, including anxiety, fear, shyness and oversensitiveness.

neurula (nu′roo-lah) [Gr. *neuron* nerve]. The early embryonic stage during which the primitive nervous system forms.

neutrons (nū′trons). Electrically uncharged particles of matter existing along with protons in the atomic nucleus of all elements except the mass 1 isotope of hydrogen.

node (nōd) [L. *nodus* knot]. The point on a stem where a leaf or bud develops; a swelling or protuberance.

nondisjunction The failure of a pair of homologous chromosomes to separate normally during the reduction division at meiosis; both members of the pair are carried to the same daughter nucleus, and the other daughter cell is lacking in that particular chromosome.

notochord (no′to-kord) [Gr. *nōton* back + *chordē* cord]. The rod-shaped body in the anteroposterior axis that serves as an internal skeleton in the embryos of all chordates and in the adults of some; replaced by a vertebral column in most adult chordates.

notum (no′tum) [Gr. *nōton* back]. The dorsal part of the body. In arthropods, the dorsal element of each segment.

nucleolus (nū-klē′ō-lus) [L. (dim.) *nux* nut]. A spherical body found within the cell nucleus rich in ribonucleic acid and believed to be the site of synthesis of ribosomes.

nucleotide (nū′klē-ō-tīd). A molecule composed of a phosphate group, a 5-carbon sugar—ribose or deoxyribose—and a nitrogenous base—a purine or a pyrimidine; one of the subunits into which nucleic acids are split by the action of nucleases.

nutrient (nū′trē-ĕnt) [L. *nutriri* to nourish]. A general term for any substance that can be used in the metabolic processes of the body.

nymph (nimf) [L. *nympha* young woman]. A juvenile insect that often resembles the adult and that will become an adult without an intervening pupal stage.

ocellus (o-sel′us) [L. (dim) *oculus* eye]. A simple light receptor found in many different types of invertebrate animals.

olfaction (ol-fak′shun) [L. *olfacere* to smell]. The act of smelling.

ommatidium (om″ah-tid′ĭ-um) [Gr. (dim.) *omma* eye]. One of the elements of a compound eye, itself complete with lens and retina.

ontogeny (ŏn-tŏj′ĕ-nē) [Gr. *ōn* existing + *gennan* to produce]. The complete developmental history of the individual organism.

Onychophora (ŏn-ĭ-kof′o-rah) [Gr. *onyx* nail + *phoros* bearing]. Rare, tropical, caterpillarlike animals, structurally intermediate between annelids and arthropods, possessing an annelidlike excretory system, an insectlike respiratory system, and claw-tipped short legs.

oögamy (ō-ŏg′ă-mē). The fertilization of a large nonmotile female gamete by a small, motile male gamete.

oögenesis (ō″ō-jĕn′ĕ-sĭs) [Gr. *ōon* egg + *genesis* production]. The origin and development of the ovum.

oögonium (ō″ō-gō′nē-um) [Gr. *ōon* egg + *gonē* generation]. The primordial cell from which the ovarian egg arises; undergoes growth to become a primary oöcyte.

operator site An entity postulated to account for the control of the operon. The operator site is adjacent to the structural genes in the operon and is believed to be the site on the DNA to which repressor molecules are bound, thereby inhibiting the synthesis of mRNA by the genes in the adjacent operon.

operon The genes whose codes are transcribed on a single mRNA molecule and are under the control of a single repressor.

orbital (or′bĭ-tal) [L. *orbitalis* mark of a wheel]. The distribution of an electron around the atomic nucleus.

organelle (or″gan-el′) [Gr. *organon* bodily organ]. One of the specialized structures within a cell, e.g., the mitochondria, Golgi complex, ribosomes, contractile vacuole, etc.

organizer A part of an embryo that influences some other part and directs its histological and morphological differentiation.

orthogenesis (ŏr″thō-jĕn′ĕ-sis) [Gr. *orthos* straight + *genesis* production]. Evolution progressing in a given direction; straight-line evolution.

osmosis (ŏs-mō′sĭs) [Gr. *ōsmos* impulsion]. The passage of solvent molecules from the lesser to the greater concentration of solute when two solutions are separated by a membrane that selectively prevents the passage of solute molecules but is permeable to the solvent.

outbreeding The mating of individuals of unrelated strains.

ovulation (ŏv″u-la′shun) [L. *ovulum* little egg + *atus* process, product]. The discharge of a mature ovum from the graafian follicle of the ovary.

ovule (o′vūl) [L. *ovulum* little egg]. A megasporangium within the ovary of a seed plant enclosed within one or more integuments.

ovum (ō′vum) [L. *ovum* egg]. The female reproductive cell, which after fertilization by a sperm develops into a new member of the same species.

oxidation The process in which electrons are removed from an atom or molecule.

oxidative phosphorylation (ok″sĭ-da′tiv fos″fōr-ĭ-la′-shun). The conversion of inorganic phosphate to the energy-rich phosphate of ATP by reactions coupled to the transfer of electrons in the electron transmitter system of the mitochondria.

oxygen debt The accumulation of lactic acid in muscles during violent exercise.

pacemaker The part whose rate of reaction sets the pace for a series of interrelated reactions; e.g., the sinoatrial node initiates the heart beat and regulates the rate of contraction of the heart.

palisade cells A compact layer of cylindrical cells located in the mesophyll layer near the upper epidermis of a leaf.

paramylum Carbohydrate storage compound present in the euglenoids, chemically distinct from both starch and glycogen.

parapodia (par″ah-po′de-ah) [L. *para* beyond + Gr. *podion* little foot]. Paired, thickly bristled paddles extending laterally from each segment of polychaete worms.

parasitism (par″ah-sīt″izm) [Gr. *parasitos* one who eats at the table of another + *ismos* condition]. A type of heterotrophic nutrition found among both plants and animals; an organism that lives in or on the living body of a plant or animal (host) and obtains its nourishment from it.

parasympathetic (par′ah-sim″pah-thet′ik) [Gr. *para* beyond + *sym* with + *pathos* feeling]. A segment of the autonomic nervous system; fibers originate in the brain and the pelvic region of the spinal cord and innervate primarily the internal organs.

parathyroids (par″ah-thi′roids) [Gr. *para* beyond + *thyreoeidēs* shieldlike]. Small, pea-sized glands situated in the substance of the thyroid gland; their secretion is concerned chiefly with regulating the metabolism of calcium and phosphorus by the body.

parenchyma (par-eng′kĭ-mah) [Gr. *para* beside + *en* in + *khein* to pour]. Plant cells that are relatively unspecialized, are thin-walled, contain chlorophyll and are typically rather loosely packed; function in photosynthesis and in the storage of nutrients.

parthenogenesis (par″thĕ-no-jen′e-sis) [Gr. *parthenos* virgin + *genesis* production]. The development of an unfertilized egg into an adult organism; common among honey bees, wasps and certain other arthropods.

parturition (par′tu-rish′un) [L. *parturito* childbirth]. The process of giving birth to a child.

pelagic (pe-laj′ik) [Gr. *pelagiōs* living in the sea]. An organism that inhabits open water, as in mid-ocean.

pepsin (pep′sin) [Gr. *pepsis* digestion]. A proteolytic enzyme secreted by the cells lining the stomach; functions only in a very acid medium and works optimally at pH 2.

perennial (per-en′ĭ-al) [L. *per* through + *annus* year]. A plant that lives throughout the year and survives for several years.

pericycle (per″e-si′kl) [Gr. *peri* around + *kyklos* circle]. A single layer of parenchymal cells capable of being transformed into meristem to give rise to the root cambium and cork cambium and to branch roots.

peripheral resistance The state of constriction or relaxation of the blood vessels; plays an important role in determining blood pressure.

peristalis (per″ĭ-stal′sis) [Gr. *peri* around + *stalsis* contraction]. Powerful, rhythmic waves of muscular contraction and relaxation in the walls of hollow tubular organs, such as the ureter or the parts of the digestive tract, that serve to move the contents through the tube.

permeability (per″me-ah-bil′ĭ-te) [L. *per* through + *meare* to pass]. The property of a membrane that permits the passage of a given substance.

petals (pĕt′l) [Gr. *petalon* a leaf]. The whorl of modified leaves that constitute part of a flower; lie inside the whorl of sepals and outside the whorl of stamens; typically have bright colors or attractive scents to attract insects or birds and ensure pollination.

petiole (pet′e-ōl) [L. *petiolus* a little foot, a fruit stalk]. The stalk by which a leaf is attached to a stem.

*p***H** [p(otential of) H(ydrogen)]. The negative logarithm of the hydrogen ion concentration, by which the degree of acidity or alkalinity of a fluid may be expressed.

phagocytosis (fag″o-si-to′sis) [Gr. *phagein* to eat + *kytos* hollow vessel + *osis* state or condition]. The engulfing of microorganisms, other cells and foreign particles by a cell such as a white blood cell.

pharynx (far′inks) [Gr. *pharunx* throat]. That part of the digestive tract from which the gill pouches or slits develop; in higher vertebrates it is bounded anteriorly by the mouth and nasal cavities and posteriorly by the esophagus and larynx.

phenocopy (fe′no-kop″e) [Gr. *phainein* to show + L. *copia* abundance, number]. The appearance in an individual of a trait characteristic of another genotype; results from physical or chemical influences in the environment that change the course of development and produce a trait that mimics that of an individual with a different genotype.

phenotype (fe′no-tīp) [Gr. *phainein* to show + *typos* type]. The outward, visible expression of the hereditary constitution of an organism.

pheromone (fēr′o-mōn) [Gr. *phorein* to carry]. A substance secreted by one organism to the external environment that influences the development or behavior of other members of the same species.

phloem (flo′em) [Gr. *phloios* bark]. One type of vascular tissue in plants, characterized by the presence of sieve tubes; transports organic nutrients both up and down the stem or root.

phosphorescence (fos″fo-res′ens) [Gr. *phōs* light + *phorein* to carry]. The emission of light without appreciable heat, caused by the decay of a molecule in the triplet state to the ground state.

phosphorylation (fos″fōr-ĭ-la′shun) [Gr. *phōs* light + *phorein* to carry]. The introduction of a phosphate group into an organic molecule.

photolysis (fo-tol′ĭ-sis) [Gr. *phōs* light + *lysis* loosing]. The splitting of a molecule under the action of light; e.g., the cleavage of water in photosynthesis by the radiant energy absorbed by chlorophyll.

photon (fō′tŏn) [Gr. *phōs* light + *ton* slice]. A particle of electromagnetic radiation, one quantum of radiant energy.

photoperiodism (fo″to-pe′re-od-izm) [Gr. *phōs* light + *peri* around + *hodos* way + *ismos* state]. The physiologic response of animals and plants to variations of light and darkness.

photosynthesis (fo″to-sin′the-sis) [Gr. *phōs* light + *synthesis* putting together]. The process of synthesizing carbohydrates from cardon dioxide and water, utilizing the radiant energy of light captured by the chlorophyll in plant cells.

phototropism (fo-tot′ro-pizm) [Gr. *phos* light + *tropos* a turning]. The growth response of an organism to light.

phycocyanin (fi″ko-si′an-in) [Gr. *phykos* seaweed + *kyanos* blue]. A blue chromoprotein found in blue-green algae.

phycoerythrin (fi″ko-er′ĭ-thrin) [Gr. *phykos* seaweed + *erythros* red]. A red chromoprotein found in red algae.

phylogeny (fi-loj′e-ne) [Gr. *phylon* tribe + *genesis* generation]. The complete evolutionary history of a group of organisms.

phylum (fi′lum) [Gr. *phylon* race]. A primary, large, main division of the animal kingdom, including organisms that are assumed to have a common ancestry.

phytoplankton (fi″to-plank′ton) [Gr. *phyton* plant + *planktos* wandering]. Microscopic floating plants, most of which are algae, that are distributed throughout the ocean or a lake.

pi electrons Mobile electrons located in a system of conjugated single and double bonds that are associated not with a single atom or bond but with the conjugated system as a whole.

pinocytosis (pi″no-si-to′sis) [Gr. *pinein* to drink + *kytos* cell + *osis* state or condition]. "Cell drinking"; the engulfing and absorption of droplets of liquids by cells.

pistil (pis′til) [L. *pistillus* a pestle]. The organ of a flower that consists of an ovary, style and stigma and produces megaspores.

pituitary (pĭ-tu′ĭ-tār″e) [L. *pituitarius* secreting phlegm]. A small gland that lies just below the hypothalamus of the brain, to which it is attached by a narrow stalk; the anterior lobe forms in the embryo as an outgrowth of the roof of the mouth and the posterior lobe grows down from the floor of the brain.

placenta (plah-sen′tah) [L. *placenta* a flat cake]. A structure formed in part from tissues derived from the embryo and in part from maternal tissues—the lining of the uterus—by means of which the embryo receives nutrients and oxygen and eliminates wastes.

Placodermi (plak′o-der″mi) [Gr. *plakos* a tablet, a flat plate + *derma* skin]. The earliest of the jawed fishes, known only from fossils; believed to be ancestral to both bony and cartilaginous fishes.

plankton (plank′ton) [Gr. *planktōs* wandering]. Minute, free-floating organisms, both plants and animals, that live in practically all natural waters.

plantigrade (plan′tĭ-grād) [L. *planta* sole + *gradus* step]. A type of locomotion adapted for a comparatively slow gait, characterized by walking on the full sole of the foot.

plasma membrane (plaz′mah mem′brān) [Gr. *plasma* anything formed or molded + L. *membrana* skin covering]. A living, functional part of the cell through which all nutrients entering the cell and all waste products or secretions leaving it must pass.

plasmodium (plaz-mo′de-um) [Gr. *plasma* anything formed +

odēs like].   Multinucleate, amoeboid mass of living matter that comprises the diploid phase of slime molds; a single-celled organism that reproduces by spore formation and causes malaria.

plasmolysis (plaz-mol'ĭ-sis) [Gr. *plasma* anything formed + *lysis* dissolution].   Contraction of the cytoplasm of a cell due to the loss of water by osmotic action.

plastid (plas'tid) [Gr. *plastos* formed + *idion* diminutive suffix]. A specialized organelle of the cell; e.g., chloroplast or amyloplast.

pleiotropic gene (plī″ō-trō'pik) [Gr. *pleiōn* more + *tropos* turn]. A gene that affects a number of different characteristics in a given individual.

plexus (plek'sus) [L. *plexus* a braid].   A network of interconnecting structures, such as nerves (e.g., the brachial plexus of nerves supplying the arm).

ploidy (ploi'de) [Gr. *ploos* fold + *odēs* like, resembling].   Relating to the number of sets of chromosomes in a cell.

poikilothermic (poi'kĭ-lo-ther'mik) [Gr. *poikilos* varied + *thermē* heat].   Having a body temperature that fluctuates with that of the environment; "cold-blooded."

polar body   Small cell that consists of practically nothing but a nucleus; formed during oögenesis, maturation of the egg, and appears as a speck at the animal pole of the egg.

pollen (pol'en) [L. *pollen* fine dust].   The mass of microspores (male fertilizing elements) of seed plants.

polygenes (pol″e-jēns′) [Gr. *poly* many + *gennan* to produce]. Two or more pairs of genes that affect the same trait in an additive fashion.

polymorphism (pol″e-mor'fizm) [Gr. *poly* + *morphē* form].   Differences in form among the members of a species; occurrence of several distinct phenotypes in a population.

polyploids (pol'e-ploid″) [Gr. *poly* many + *ploos* folds].   Organisms that have more than two full sets of homologous chromosomes.

polyps (pol'ip) ,Gr. *polypous* morbid excrescences].   Hydralike animals; the sessile stage of the life cycle of certain cnidarians; protruding growths from a mucous membrane.

population   The group of individuals of a given species inhabiting a specified geographic area.

Porifera (po-rif'ĕ-rah) [L. *porus* pore + *ferre* to bear].   The phylum of sponges; the body is perforated with many pores to admit water, from which food is strained.

portal system (por'tal) [L. *porta* a gate].   A group of veins that drain one region and lead to a capillary bed in another organ rather than directly to the heart (e.g., the renal and hepatic portal systems).

precursor (pre-kur'sor) [L. *praecurrere* to run before].   A substance that precedes another substance in a metabolic pathway; a substance from which another substance is synthesized.

predation (pre-da'shun) [L. *praedatio* plunder].   Relationship in which one species adversely affects the second but cannot live without it; the first species kills and devours the second.

primitive streak (prim'i-tive) [L. *primitivus* first (in point of time)]. A longitudinal groove that develops on the embryonic disc of the eggs of fishes, reptiles, birds and mammals as a consequence of the movement of cells and formation of mesoderm; it is homologous to the lips of the blastopore and marks the future longitudinal axis of the embryo.

primordium (pri-mor'de-um) [L. *primordium* the beginning]. The earliest discernible indication during embryonic development of an organ or part.

proboscis (pro-bos'is) [Gr. *pro* before + *boskein* to feed, graze]. Any tubular process of the head or snout of an animal, usually used in feeding.

progeny selection (proj'e-ne se-lek'shun) [L. *progignere* to bring forth]. A breeding program in which the genotype is determined by making test matings and observing the offspring.

progeria (prō-jĕ'rī-a). A rare inherited type of abnormal human development, characterized by the appearance in a young child of changes typical of the aging process.

progesterone (pro-jes'ter-ōn) [L. *pro* before + *gestus* to bear, carry, conduct]. The hormone produced in the corpus luteum of the ovary and in the placenta; acts with estradiol to regulate estrous and menstrual cycles and to maintain pregnancy.

proglottid (pro-glot'id) [L. *pro* before, in front of + *glottis* the tongue]. The body sections of a tapeworm.

prokaryotic [L. *pro* before + Gr. *karyon* nucleus or nut]. Applied to organisms that lack membrane-bound nuclei, plastids and Golgi apparatus; the bacteria and blue-green algae.

prophase (pro'fāz) [L. *pro* before + Gr. *phasis* an appearance]. The first stage in mitosis, during which the chromatin threads condense, distinct chromosomes become evident and a spindle forms.

proprioceptor [L. *proprius* one's own]. Internal sense cells that give information to the brain regarding movements, position of body and muscle stretch.

prosimian (pro"sim"ĭ-an) [L. *pro* before, in front of + *simia* an ape]. A primitive living primate or an early ancestral primate.

prostate (pros'tāt) [Gr. *prostates* one who stands before]. The largest accessory sex gland of male mammals; it surrounds the urethra at the point where the vasa deferentia join it, and it secretes a large portion of the seminal fluid.

prosthetic group (pros-thet'ik) [Gr. *prostithenai* to add]. A cofactor tightly bound to an enzyme.

protean behavior (pro'te-un) [Gr. *Proteus* the sea god of Greek mythology who changed shape unpredictably when seized]. An irregular, unpredictable sequence of movements by prey when pursued by predators.

proteins (prō'tēn) [Gr. *prōtos* first]. Macromolecules containing carbon, hydrogen, oxygen, nitrogen and usually sulfur and phosphorus; composed of chains of amino acids bound in peptide bonds; one of the principal types of compounds present in all cells.

prothallus (prō-thăl'us). The independent gametophyte generation of ferns and related lower vascular plants.

Protista (pro-tis'tah) [Gr. *protista* the very first, from *protos* first]. Kingdom of living organisms, including the protozoa, flagellates and certain algae.

protocooperation (pro"to-co-op"er-a'shun) [Gr. *prōtos* first + L. *cooperatio* to work]. Relationship in which each of two populations benefits by the presence of the other but can survive in its absence.

proton (pro'ton) [Gr. *prōtos* first]. A basic physical particle present in the nuclei of all atoms that has a positive electric charge and a mass similar to that of a neutron; a hydrogen ion.

protonema (pro"to-nē'mah) [Gr. *prōtos* first + *nematos* a thread]. A filamentous green body of mosses from which grows an erect stem, to which is attached a spiral whorl of one-cell thick leaves.

protonephridium (pro"to-nef-rid'e-um) [Gr. *prōtos* first + *nephridios* kidneys]. The flame-cell excretory organs of lower invertebrates and of some larval higher animals.

protozoa (pro"to-zo'ah) [Gr. *protos* first + *zōon* animal]. The single-celled, animal-like amoebas, ciliates, flagellates and sporozoa.

pseudocoelom (su"do-se'lom) [Gr. *pseudēs* false + *koilia* cavity]. A body cavity between the mesoderm and endoderm; a persistent blastocoele.

pseudopod (su'do-pod) [Gr. *pseudēs* false + *pous* foot]. A temporary cytoplasmic protrusion of an amoeba or amoeboid cell; functions in locomotion and feeding.

pulvinus (pul-vi'nus) [L. *pulvinus* cushion]. A cushionlike enlargement of a petiole at its point of insertion in the stem.

pupa (pu'pah) [L. a doll]. A stage in the development of an insect, between the larva and the imago (adult); a form that neither moves nor feeds.

purines (pu'rēns) [blend of *pure* and *urine*]. Organic bases with carbon and nitrogen atoms in two interlocking rings; components of nucleic acids, ATP, NAD and other biologically active substances.

putrefaction (pu"trĕ-făk'shun) [L. *putrefactio* decaying]. The enzymatic anaerobic degradation of proteins and amino acids.

pyrenoid (pi'rĕ-noid) [Gr. *pyrēn* fruit stone + *eidos* form]. Starch-containing granular bodies seen in the chromatophores of certain protozoa.

pyrimidines (pi-rim'ĭ-din). Nitrogenous bases composed of a single ring of carbon and nitrogen atoms; components of nucleic acids.

quantasome (kwon'tah-sōm). Unit structures containing about 230 molecules of chlorophyll present within the lamellar structure of the grana within the chloroplasts.

quantum (kwon'tum) [L. *quantum* as much as]. A unit of radiant energy; has no electric charge and very little mass; the energy of a quantum is an inverse function of the wavelength of the radiation.

quillworts (kwil'worts). A group of heterosporous lycopodiophyta that have slender, quill-like leaves resembling those of a bunch of garlic.

race A division of a species; a population that differs from other populations with respect to the frequency of one or more genes; a subgroup of a species distinguished by a certain combination of morphologic and physiologic traits.

radicle (rad'ĭ-k'l) [L. *radiculus* root]. Root portion of the hypocotyl of seed plants.

radula (raj'oo-la) [L. *radula* a scraper]. A rasplike structure in the alimentary tract of chitons, snails, squids and certain other mollusks.

range The portion of the earth in which a given species is found.

rassenkreis [*Ger. Rassenkreis* race-circle]. Series of geographic subspecies in a population that is spread over a wide territory; each subspecies differs in some respects from its neighboring ones but interbreeds with them, and consequently the groups at the two ends of the series may be quite different and have markedly reduced interfertility.

reabsorption The selective removal of certain substances from the glomerular filtrate by the cells of the convoluted tubules of the kidney and their secretion into the bloodstream.

recapitulation The tendency for embryos in the course of development to repeat, perhaps in an abbreviated fashion, the sequence of stages in the embryonic development of their evolutionary ancestors.

receptor A sensory nerve ending that responds to a given type of stimulus; a compound (usually a protein) present in the target cell of a hormone that specifically takes up and binds that hormone.

recessive genes Genes that do not express their phenotype unless carried by both members of a set of homologous chromosomes; i.e., genes that produce their effect only when homozygous, when present in "double dose."

redia (re'dĭ-ah) [*Francesco Redi*, seventeenth century Italian naturalist]. The second stage in the life cycle of flukes. It reproduces asexually in snails.

reduction The addition of electrons to an atom or molecule; opposite of oxidation.

reflex (re'fleks) [L. *reflexus* bent back]. An inborn, automatic, involuntary response to a given stimulus which is determined by the anatomic relations of the involved neurons; the functional units of the nervous system.

reflex arc A sequence of sensory, connector and motor neurons that conduct the nerve impulses for a given reflex.

refractory period The period of time that must elapse after the response of a neuron or muscle fiber to one impulse before it can respond again.

regeneration Regrowth of a lost or injured tissue or part of an organism.

regulator genes Special genes that provide codes for the synthesis of repressor proteins.

releasing hormones Short peptides synthesized in the hypothalamus and secreted into the hypothalamo-hypophysial portal system and carried to the pituitary, where they initiate the synthesis and release of specific pituitary hormones.

renal (rē'nal) [L. *renalis* kidney]. Pertaining to the kidney.

renal corpuscle The complex formed by a glomerulus and the surrounding Bowman's capsule of a kidney tubule; filtration, the first step in urine formation, occurs here.

rennin (ren'in). Enzyme secreted by the gastric mucosa that converts the milk protein, casein, from a soluble to an insoluble substance, thereby curdling the milk.

repressor The protein substance produced by a regulator gene that represses protein synthesis in a specific gene.

resonating system A system of atoms bonded together in such a way that the external electrons can be arranged in many different ways without moving any of the constituent atoms.

respiration (res"pĭ-ra'shun) [L. *respirare* to breathe]. Process by which animal and plant cells utilize oxygen, produce carbon dioxide and conserve the energy of foodstuff molecules in biologically useful forms such as ATP; the act or function of breathing.

reticulum (re-tik'u-lum) [L. (dim.) *rete* net]. A network of fibrils or filaments, either within a cell or in the intercellular matrix.

retina (rĕt'ĭ-nah) [L. *rete* net]. The innermost of the three tunics of the eyeball, which surrounds the vitreous body and is continuous posteriorly with the optic nerve; contains the light-sensitive receptor cells, rods and cones.

rhizoids (ri'zoids) [Gr. *rhiza* root + *eides* form]. Colorless, hairlike absorptive filaments that extend from the base of the stem of bryophytes, fern prothallia and certain fungi and lichens; serve as roots.

rhizome (ri'zōm) [Gr. *rhizōma* root stem]. An underground stem, present in ferns and grasses, that gives rise to above-ground leaves.

Rhodophyta (rō-dō-fī'tah) [Gr. *rhodon* rose + *phykos* seaweed]. The division of red algae, which are found almost entirely in the oceans.

rhodopsin (rō-dop'sin) [Gr. *rhodon* rose + *opsis* sight]. A substance in the retina of the eye (visual purple) made up of retinal, a derivative of vitamin A, and a protein, opsin; undergoes a chemical reaction triggered by light that stimulates the receptor cell to send an impulse to the brain, resulting in the sensation of sight.

ribosomes (ri-bo-sōms). Minute granules, composed of protein and ribonucleic acid, either free in the cytoplasm or attached to the membranes of the endoplasmic reticulum of a cell; the site of protein synthesis.

rickettsia (rik-et'se-ah). [*Howard T. Ricketts*, nineteenth century American pathologist]. A type of disease organism intermediate in size and complexity between a virus and a bacterium; parasitic within cells of insects and ticks, transmitted to humans by the bite of the infected insect or tick.

RNA Ribonucleic acid, a nucleic acid containing the sugar ribose; present in both nucleus and cytoplasm and of prime importance in the synthesis of proteins.

rod (rŏd) [Anglo-Saxon *rodd*]. In zoology, the rod-shaped photoreceptive cells of the retina, which are particularly sensitive to dim light and mediate black and white vision.

root pressure The positive pressure of the sap in the roots of plants, generated by the hypertonicity of the sap with respect to the water in the surrounding soil.

saccule (sak'ūl) [L. *sacculus* a little sac]. Small, hollow sac in the inner ear lined with sensitive hair cells and containing small stones made of calcium carbonate; contains receptors for the sense of static balance.

saprobic nutrition (sap″ro-bĭk nu-trish'un) [Gr. *sapros* rotten + L. *nutritio* nourishing]. A type of heterotrophic nutrition in which organisms absorb their required nutrients through the cell membrane following the extracellular digestion of nonliving organic material.

schizocoel (skiz'o-cēl) [Gr. *schizein* to divide + *koilia* cavity]. A body cavity formed by the splitting of embryonic mesoderm into two layers.

sclerenchyma (skle-reng'kĭ-mah) [Gr. *skleros* hard + *enchyma* infusion]. Supporting tissue of plants consisting of cells with greatly thickened walls impregnated with lignin.

secondary response A rapid production of antibodies induced by a second injection of antigen several days, weeks or even months after the primary injection.

secretion (sē-krē'shun) [L. *secretio,* from *secernere* to secrete]. The production and release by a cell of some substance that is used elsewhere in the body in some process.

segmentation (seg″men-ta'shun) [L. *segmentum* a piece cut off]. Division of a body or structure into more or less similar parts.

sepals (se'pals) [Gr. *skepē* covering]. The outermost parts of a flower, usually green and most like ordinary leaves.

sere (sēr). A sequence of communities that replace one another in succession in a given area; the transitory communities are called seral stages. The series ends with a climax community typical of the climate in that part of the world.

serum (se'rum) [L. *serum* whey]. The clear portion of a biological fluid separated from its particulate elements; light yellow liquid left after clotting of blood has occurred.

sinoatrial node (si″no-a'tre-al) [L. *sino* a hollow + *atrium* hall + *nodus* knot]. A small mass of nodal tissue located at the point where the superior vena cava empties into the right atrium; initiates the heart beat and regulates the rate of contraction.

solute (so'lūt) [L. *solvere* to dissolve]. A substance dissolved in a true solution; a solution consists of a solute and a solvent.

solvent (sol'vent) [L. *solvere* to dissolve]. The fluid medium in which the solute molecules are dissolved in a true solution; a liquid that dissolves or that is capable of dissolving.

somites (so'mīts) [Gr. *soma* body]. Paired, blocklike masses of mesoderm, arranged in a longitudinal series alongside the neural tube of the embryo, forming the vertebral column and dorsal muscles.

species (spe'shēz) [L. *species* sort, kind]. A unit of taxonomic classification for both plants and animals; a population of similar individuals, alike in their structural and functional characteristics, which in nature breed only with each other and which have a common ancestry.

sphincter (sfingk'ter) [Gr. *sphinkter* to bind tight].   A group of circularly arranged muscle fibers the contractions of which close an opening (e.g., the pyloric sphincter at the end of the stomach).

spiracle (spir'ah-k'l) [L. *spirare* to breathe].   A breathing opening, such as the opening on the body surface of a trachea in insects or a modified gill opening in cartilaginous fishes through which some water enters the pharynx.

sporangium (spo-ran'je-um) [Gr. *sporos* seed + *angeion* vessel].   A structure within which asexual spores or sporelike bodies are produced.

spore (spōr) [Gr. *sporos* seed].   An asexual reproductive element, usually unicellular, or an organism, such as a protozoan or a cryptogamic plant, that can develop directly into an adult.

sporophyte (spo″ro-fīt) [Gr. *sporos* seed + *phyton* plant].   The diploid, asexual, spore-producing stage in the alternation of generations in the life cycles of plants.

Sporozoa (spo″ro-zo'ah) [Gr. *sporos* seed + *zōon* animal].   Subphylum of protozoa (spore formers); these organisms have no special method of locomotion and are parasitic; one kind is the human parasite causing malaria.

stamen (stā'men) [L. *stamen* warp, thread, fiber].   The structure of a flower that produces microspores, or pollen; consists of a slender filament tipped by an anther containing microsporangia in which the haploid microspores are produced.

stapes (sta'pēz) [L. *stapes* stirrup].   The innermost of the small bones in the middle ear cavity, shaped somewhat like a stirrup.

statocyst (stat'o-sist) [Gr. *statos* standing + *kystis* sac].   A cellular cyst containing one or more granules; is used in a variety of animals to sense the direction of gravity.

steatopygia (stē″ah-to-pij'ē-ah) [Gr. *steatos* fat + *pygē* buttock].   An excessive accumulation of fat on the buttocks and thighs.

stele (stēl) [Gr. *stechelo* stem].   Vascular cylinder of stem and root; term for the pericycle and the tissues within it – xylem, phloem and parenchyma.

steroids (ste'roids) [Gr. *stereos* solid + *eides* like].   Complex molecules containing carbon atoms arranged in four interlocking rings, three of which contain six carbon atoms each and the fourth of which contains five; the male and female sex hormones and the adrenal cortical hormones.

stigma (stig'mah) [Gr. *stigma* mark].   The uppermost part of a pistil; secretes a moist, sticky substance to trap and hold the pollen grains that reach it.

stimulus (stim'u-lus) [L. *stimulus* goad].   Any agent, act, or influence that produces functional or trophic reaction in a receptor or in an irritable tissue.

stipe (stīp) [L. *stipes* a stock, post, branch].   A short stalk or stemlike structure that is a part of the body of certain brown algae.

stoma (sto'mah) [Gr. *stoma* mouth].   In botany, a minute opening on the surface of a leaf, surrounded by a pair of guard cells that regulate the size of the opening.

stoneworts (stōn'wort).   Multicellular green algae found in fresh water ponds; resemble miniature trees, with structures that superficially look like, and serve the functions of, roots, stems, leaves and seeds, though they are not anatomically like their counterparts in higher plants.

strobilus (stro′bi-lus) [Gr. *strobilos* a round ball].   A cone formed at the tip of a stem by a group of sporophylls.

style (stīl) [Gr. *stilos* a pillar].   A long, slender part of a pistil that connects the ovary with the stigma.

suberin (soo′ber-in) [L. *suber* the cork tree].   An insoluble waxy material present in the walls of cork and endodermis cells that makes them waterproof.

subgerminal space (sub-jer′mĭ-nal) [L. *sub* under + *germinalis* germ].   The shallow cavity under the dividing cells of a hen's egg; not homologous to the blastocoele of the frog egg.

suspensor (sus-pen′sor) [L. *suspendere* to hang].   A filament of cells produced by cell divisions that the plant zygote undergoes following fertilization; the embryo of the plant forms from the end cell of this filament.

symbiosis (sim″bi-o′sis) [Gr. *symbiōsis* to live together].   The living together of two dissimilar organisms; association may form mutualism, commensalism, parasitism or amensalism.

synapse (sĭn′aps) [Gr. *synapsis* conjunction].   The junction between the axon of one neuron and the dendrite of the next.

synapsis (sĭ-nap′sis) [Gr. *synapsis* conjunction].   The pairing and union side by side of homologous chromosomes from the male and female pronuclei early in meiosis.

syncytium (sin-sit′-e-um) [gr. *syn* together + *kytos* a hollow vessel]. A multinucleate mass of cytoplasm produced by the merging of cells.

synergistic (sin″er-jis′tik) [Gr. *syn* with + *ergon* work].   Acting together; enhancing the effect of another force or agent.

syngamy (sin′gah-me) [Gr. *syn* with + *gamos* marriage].   Sexual reproduction; the union of the gametes in fertilization.

systole (sĭs′to-le) [Gr. *systolē* a drawing together].   The contraction of the heart; the interval between the first and second heart sounds during which blood is forced into the aorta and pulmonary arteries.

taiga (tī′ga) [Russian].   Northern coniferous forest biome found primarily in Canada, northern Europe and Siberia.

taxis (tak′sis) [Gr. *taxis* a drawing up in rank and file].   An orientation movement in response to a stimulus in a direction determined by the direction of the stimulus; found in animals, some lower plants and the male sex cells of mosses or ferns.

taxonomy (taks-on′o-me) [Gr. *taxis* a drawing up in rank and file + *nomos* law].   The science of naming, describing and classifying organisms.

tectorial membrane (tek-to′re-al) [L. *tectum* roof + *membrana* skin covering].   The roof membrane of the organ of Corti in the cochlea of the ear.

telophase (tĕl′o-fāz) [Gr. *telos* end + *phasis* phase].   The last of the four stages of mitosis, during which the two daughter nuclei appear and the cytoplasm usually divides.

template (tem′plāt) [L. *templum* a small timber]. A pattern or mold that guides the formation of a duplicate.

territoriality (ter″i-tor′ĭ-al′ĭte) [L. *territorium* the earth]. Behavior pattern in which one organism (usually a male) delineates a territory of his own and defends it against intrusion by other members of the same species and sex.

testis (tes′tis) [L. *testis* testicle]. The male gonad that produces spermatozoa; in humans and certain other mammals the testes are situated in the scrotal sac.

tetanus (tet′ah-nus) [Gr. *tetanos,* from *teinein* to stretch]. Sustained, steady maximal contraction of a muscle, without distinct twitching, resulting from a rapid succession of nerve impulses.

tetany (tet′ah-ne) [Gr. *tetanos* stretched]. A syndrome manifested by sharp flexion of the wrist and ankle joints, muscle twitchings, cramps and convulsions; occurs in parathyroid hypofunction.

tetrad (tĕt′răd) [Gr. *tetra* four]. A bundle of four homologous chromosomes produced at the end of the first meiotic prophase.

tetraploid (tĕt′rah-ploid″) [Gr. *tetra* four + *ploos* fold]. An individual or cell having four sets of chromosomes.

tetrapoda (tet-rap′o-dah) [Gr. *tetra* four + *podos* foot]. Four-limbed vertebrates; the amphibia, reptiles, birds and mammals.

thalamus (thal′ah-mus) [Gr. *thalamos* inner chamber]. The lateral walls of the diencephalon; the main relay center for sensory impulses going to the cerebrum; also interacts with the cerebrum in complex ways.

thallus (thal′us) [Gr. *thallos* green shoot]. A simple plant body not differentiated into root, stem and leaf.

theory (the′o-re) [Gr. *theōria* speculation as opposed to practice]. A formulated hypothesis supported by a large body of observations and experiments.

therapsids (ther-ap′sids). A group of mammallike reptiles of the Permian period from which mammals evolved.

Thermodynamics, First Law of (ther″mo-di-nam′-iks) [Gr. *thermē* heat + *dynamis* power]. Law that states that energy is neither created nor destroyed but only transformed from one kind to another.

thigmotropism (thig-mot′rō-pizm) [Gr. *thigma* touch + *trope* turn]. The orientation of an organism in response to the stimulus of contact or touch.

threshold (thresh′old). The value at which a stimulus just produces a sensation, is just appreciable, or comes just within the limits of perception.

thrombin (throm′bin) [Gr. *thrombos* lump, curd, clot]. The enzyme derived from prothrombin that converts fibrinogen to fibrin; participates in blood clotting.

thrombus (throm′bus) [Gr. *thrombos* lump, curd, clot]. A clot in a blood vessel or in one of the cavities of the heart, which remains at the point of its formation.

thylakoid (thī′lă-koid). An individual photosynthetic lamella, typically arranged in stacks to form grana.

tissue (tish′u) [L. *texere* to weave]. Group of similarly specialized cells that together perform certain special functions; e.g., muscle tissue, bone tissue, nerve tissue.

tonus (to'nus) [Gr. *tonos* strain, tone]. The continuous partial contraction of muscle.

tornaria (tor-na're-ah) [L. *tornare* to turn]. The free-swimming hemichordate larva that shows many similarities to echinoderm larvae.

toxin (tok'sin) [L. *toxicum* poison]. Poisonous substance produced by one organism that usually affects one particular organ or organ system, rather than the body as a whole, of another organism.

tracheids (trā'kē-id) [Gr. *tracheia* rough, rugged]. Elongate xylem cells, with pointed ends and a pattern of thickenings on the walls, that are the first to develop in the xylem of higher plants.

tracheophyte (trā'ke-ō-fīt") [L. *trāchēa* trachea + Gr. *phyton* plant]. A vascular plant — has xylem and phloem.

transducers (trans-du'sers) [L. *transducere* to lead across]. Devices receiving energy from one system in one form and supplying it to a second system in a different form; e.g., converting radiant energy to chemical energy.

transduction (trans-duk'shun) [L. *transducere* to lead across]. The transfer of a genetic fragment from one cell to another; e.g., from one bacterium to another by a virus.

transfer RNA A form of RNA composed of about 70 nucleotides which serve as adaptor molecules in the synthesis of proteins. An amino acid is bound to a specific kind of transfer RNA and then arranged in order by the complementary nature of the nucleotide triplet (codon) in template or messenger RNA and the triplet anti-codon of transfer RNA.

transforming agents Substances isolated from pneumococci and and certain other bacteria that bring about a permanent, inherited change when applied to another strain of that type of bacteria.

translocation (trans"lo-ka'shun) [L. *trans* through + *locus* place]. The process of transporting organic substances in phloem; the transfer of a fragment of one chromosome to a different, non-homologous chromosome.

transpiration (tran"spi-ra'shun) [L. *trans* through + *spiratio* exhalation]. Evaporation of water from the leaves of a plant, aids in drawing water up the stem.

transverse plane A section in a bilaterally symmetrical animal that includes a dorso-ventral axis and a left-right axis but is at right angles to the anteroposterior axis.

trichocyst (trik'o-sist) [Gr. *trichos* hair + *kystis* a sac or bladder]. A cellular organelle in the cytoplasm of ciliated protozoa such as *Paramecium* that can discharge a filament that may aid in trapping and holding prey.

trilobite (trī'lo-bīt) [L. *tres* three + *lobus* lobe]. Marine arthropods of the Paleozoic era characterized by two dorsal longitudinal furrows that separated the body into three lobes.

triplet code The sequences of three nucleotides that compose the codons, the units of genetic information in DNA that specify the order of amino acids in a peptide chain.

triplet state The state resulting when an electron is activated by absorbing a photon, moves to an outer orbital of higher energy and pairs with an electron of like spin.

triploid (trip'loid) [Gr. *triploos* triple + *eidēs* like]. An individual or cell having three sets of chromosomes.

trochophore (tro'ko-fōr) [Gr. *trochos* wheel + *phoros* bearing]. A larval form, similar to the larva of mollusks, that characterizes the development of polychaetes and archiannelids.

troop The social unit of many primate species, consisting of several males, three to many females, and their offspring.

trophallaxis (tro″fah-lak'sis) [Gr. *trephein* to nourish + *allaxis* exchange]. The mutual exchange of food and secretions among the members of an insect colony.

tropism (trō'pizm) [Gr. *tropē* a turning]. A growth response in a nonmotile organism, elicited by an external stimulus.

tubers (tu'bers) [L. *tuber* a hump, swelling, knob]. Thickened underground stems that are adapted for food storage; found in plants such as the potato.

tundra [Russian]. A treeless plain between the taiga in the south and the polar ice cap in the north; characterized by low temperatures, a short growing season and ground that is frozen most of the year.

turgor pressure (tur'gor presh'ur) [L. *turgor* swelling + *pressure* to squeeze]. An internal pressure generated as water enters a plant cell by osmosis and the cytoplasm is pressed against the cell wall; prevents the entrance of any additional water.

turnover number The number of molecules of substrate acted upon by one molecule of enzyme per minute.

ubiquinone Coenzyme Q, a component of the electron transmitter system; consists of a head, a six-membered carbon ring, which can take up and release electrons, and a long tail composed of a chain of carbon atoms.

umbilicus (um-bil'i-cus) [L. *umbilicus* the navel]. The navel; the scar marking the site of attachment of the umbilical cord in the fetus.

ungulates (ung″gu-lātes) [L. *ungula* hoof]. Four-legged mammals in which the digits may be more or less fused and their ends protected with a horny coating, or hoof.

unguligrade (ung-gwil'ĭ-grād) [L. *ungula* hoof + *gradus* a step]. A type of locomotion in which the animal runs upon the tips of one or two digits of each limb.

urea (u-re'ah) [Gr. *ouron* urine]. The diamide of carbonic acid, NH_2CONH_2; one of the water-soluble end products of protein metabolism.

ureter (u-re'ter) [Gr. *ourein* to urinate]. The fibromuscular tube that conveys urine from the kidney to the bladder.

urethra (u-re'thrah) [Gr. *ourein* to urinate]. The membranous canal conveying urine from the bladder to the exterior of the body.

uterus (u'ter-us) [L. *uterus* the womb]. The womb; the hollow, muscular organ of the female reproductive tract in which the fetus undergoes development.

utricle (u'tre-k'l) [L. *utriculus* a bag].   The larger of the two divisions of the membranous labyrinth; contains the receptors for dynamic body balance.

vaccine (vak'sēn) [L. *vaccinus* a cow].   The commercially produced antigen of a particular disease, strong enough to stimulate the body to make antibodies but not sufficiently strong to cause the disease's harmful effects.

vacuole (vak'u-ōl) [L. *vacuus* empty + -*ole* dim. ending].   Small space within a cell, filled with watery liquid and separated from the rest of the cytoplasm by a vacuolar membrane.

vagina (vah-ji'nah) [L. *vagina* a scabbard].   In many kinds of animals, the terminal portion of the female reproductive tract; receives the male copulatory organ.

valence (va'lens) [L. *valentia* strength].   An expression of the number of atoms of hydrogen (or its equivalent) that one atom of a chemical element can hold in combination, if negative, or displace in a reaction, if positive; the number of electrons gained, lost or shared by the atom in forming bonds with one or more other atoms.

ventricle (ven'trĭ-k'l) [L. (dim.) *venter* the stomach].   A cavity in an organ, such as one of the several cavities of the brain or one of the chambers of the heart that receives blood from the atria.

vesicle [L. *vesicula* a little bladder].   Any small sac or open space.

vestigial (ves-tij'e-al) [L. *vestigium* footprint, trace, sign].   Useless, incomplete or undersized; said of an organ present in one organism that is a remnant of a homologous organ that functioned in an ancestral organism.

villus (vil'lus) [L. *villus* tuft of hair].   A small, fingerlike vascular process of protrusion, especially a protrusion from the free surface of a membrane, such as the lining of the intestine.

virus (vi'rus) [L. *virus* slimy liquid, poison].   Minute infectious agent, composed of a nucleic acid core and a protein shell; may reproduce and mutate within a host cell.

vital capacity (vi'tal kah-pas'i-te) [L. *vita* life + *capacitas*, from *capere* to take].   The total amount of air displaced when one breathes in as deeply as possible and then breathes out as completely as possible.

vitamin (vi-tah-min) [L. *vita* life].   An organic substance necessary in small amounts for the normal metabolic functioning of a given organism; must be present in the diet because the organism cannot synthesize an adequate amount of it.

vitreous (vit're-us) [L. *vitreus* glassy].   Glasslike or hyaline; designates the vitreous body of the eye, which contains clear transparent jelly that fills the posterior part of the eyeball.

viviparous (vi-vip'ah-rus) [L. *vivus* alive + *parere* to bring forth, produce].   Bearing living young that develop from eggs within the body of the mother, deriving nutrition either from yolk or directly from the maternal organism through a special organ, the placenta, which is an outgrowth of the embryo.

warning coloration Adaptation for survival that consists of bright, conspicuous colors and is assumed by poisonous or unpalatable animals to warn potential predators not to eat them.

woody perennials Plants that live longer than two years and have a thick, tough, lignified stem, or trunk, covered with a layer of cork.

x organ Organ present in crustacea that produces hormones that regulate molting, metabolism, reproduction, the distribution of pigment in the compound eyes and the control of pigmentation of the body.

xanthophyll (zan′tho-fil) [Gr. *xanthos* yellow + *phyllon* leaf]. The yellow pigment of plants, occurring along with carotene in green leaves, grass and other vegetable matter.

xerophthalmia (ze″rof-thal′me-ah) [Gr. *xēros* dry + *ophthalmos* eye]. A type of blindness characterized by an abnormally dry, lusterless and horny layer of epithelium over the cornea; due to a deficiency of vitamin A.

xerophytes (ze′ro-fīts) [Gr. *xēros* dry + *phyton* plant]. Plants, such as yuccas and cactuses, that are adapted to live where soil water is scarce.

xylem (zi′lem) [Gr. *xylon* wood]. The tissue in tracheophytes that conducts water and dissolved salts; consists of tracheids and vessels; may also provide mechanical support.

yolk sac A pouchlike outgrowth of the digestive tract of certain vertebrate embryos that grows around the yolk, digests it, and makes it available to the rest of the organism.

zoospore (zo′o-spōr) [Gr. *zoon* animal + *sporos* seed]. A flagellated, motile spore produced asexually.

zygote (zi′got) [Gr. *zygotos* yoked together]. The cell formed by the union of two gametes; a fertilized egg.

BIBLIOGRAPHY

Ackroyd, J. F.: *Immunological Methods.* Philadelphia, F. A. Davis Co., 1964.

Adams, Alexander: *Eleventh Hour: A Hard Look at Conservation.* New York, G. P. Putnam's Sons, 1970.

Adrian, Edgar: *The Physical Background of Perception.* London, Oxford University Press, 1947.

Afzelius, B.: *Anatomy of the Cell.* Chicago, University of Chicago Press, 1966.

Agrios, G. N.: *Plant Pathology.* New York, Academic Press, 1969.

Ahmadjian, V.: *The Lichen Symbiosis.* Waltham, Mass., Blaisdell Publishing Co., 1967.

Aidley, D. J.: *The Physiology of Excitable Cells.* New York, Cambridge University Press, 1971.

Alexopoulos, C. J.: *Introductory Mycology,* 2nd Ed. New York, John Wiley & Sons, 1962.

Allee, W. C., Emerson, A. E., Park, O., Park, T., and Schmidt, K. P.: *Principles of Animal Ecology.* Philadelphia, W. B. Saunders Co., 1949.

Allen, A. A.: *The Book of Bird Life,* 2nd Ed. Princeton, D. Van Nostrand Co., 1961.

Allen, J. M. (ed.): *The Molecular Control of Cellular Activity.* New York, McGraw-Hill Book Co., 1962.

Alpern, M., Lawrence, M., and Wolsk, B.: *Sensory Processes.* Belmont, Calif., Brooks-Cole Publishing Co., 1967.

Amoore, J. E., Johnson, J. W., Jr., and Rubin, M.: *The Stereochemical Theory of Odor.* Sci. Amer. *210* (2), 81 (February, 1964).

Andersen, H. T.: *The Biology of Marine Mammals.* New York, Academic Press, 1969.

Anderson, Edgar: *Plants, Life and Man.* Berkeley, Calif., University of California Press, 1967.

Andrews, H. N.: *Ancient Plants and the World They Lived In.* Ithaca, N. Y., Comstock Publishing Associates, 1947.

Arey, L. B.: *Developmental Anatomy,* 8th Ed. Philadelphia, W. B. Saunders Co., 1974.

Armstrong, E. A.: *Bird Display and Behavior.* Gloucester, Mass., Peter Smith, 1965.

Asdell, S. A.: *Patterns of Mammalian Reproduction,* 2nd Ed. Ithaca, N. Y., Comstock Publishing Associates, 1964.

Aubert de la Rue, E., Bourliere, F., and Harroy, J. P.: *The Tropics.* New York, Alfred A. Knopf, 1957.

Austen, C. R.: *Fertilization.* Englewood Cliffs, N. J., Prentice-Hall, 1965.

Baker, J. J. W., and Allen, G. E.: *Matter, Energy and Life: An Introduction for Biology Students.* Reading, Mass., Addison-Wesley, 1965.

Baker, J. J. W., and Allen, G. E.: *Hypothesis, Prediction, and Implication in Biology.* Reading, Mass., Addison-Wesley, 1971.

Baldwin, Ernest B.: *An Introduction to Comparative Biochemistry,* 4th Ed. Cambridge, Cambridge University Press, 1964.

Baldwin, Ernest B.: *Dynamic Aspects of Biochemistry,* 4th Ed. Cambridge, Cambridge University Press, 1964.

Balinsky, B. I.: *An Introduction to Embryology,* 4th Ed. Philadelphia, W. B. Saunders Co., 1975.

Bardach, John: *Downstream.* New York, Grosset and Dunlap, 1964.

Bardach, John: *Harvest of the Sea.* New York, Harper & Row, Publishers, 1968.

Barnes, R. D.: *Invertebrate Zoology,* 3rd Ed. Philadelphia, W. B. Saunders Co., 1973.

Barrington, E. J. W.: *The Biology of the Hemichordates and Protochordata.* San Francisco, W. H. Freeman, 1965.

Bastock, M.: *Courtship: A Zoological Study.* London, Heinemann Educational, 1967.

Bates, M.: *The Forest and the Sea.* New York, Random House, 1960.

Beadle, G. W., and Beadle, M.: *The Language of Life.* Garden City, N. Y., Doubleday and Co., 1966.

Beaumont, William: *Experiments and Observations on the Gastric Juice and the Physiology of Digestion.* Cambridge, Harvard University Press, 1929.

Beermann, W., and Clever, U.: *Chromosome Puffs.* Sci. Amer., *210* (4), 50 (April, 1964).

Békésy, G. von: *The Ear.* Sci. Amer., *197* (2), 66 (August, 1957).

Bell, Eugene (ed.): *Molecular and Cellular Aspects of Development.* New York, Harper & Row Publishers, 1965.

Bell, P. R., and Woodstock, C. L. F.: *The Diversity of Green Plants.* Reading, Mass., Addison-Wesley, 1968.

Bellairs, A.: *The Life of Reptiles.* London, Weidenfeld and Nicolson, 1969.

Benarde, M. A.: *Our Precious Habitat: An Integrated Approach to Understanding Man's Effect on His Environment.* New York, W. W. Norton & Co., 1970.

Berrill, N. J.: *The Origin of Vertebrates.* London, Oxford University Press, 1955.

Beveridge, W. I. B.: *The Art of Scientific Investigation.* New York, W. W. Norton & Co., 1957.

Bierhorst, D. W.: *Morphology of Vascular Plants.* New York, The Macmillan Company, 1971.

Biggs, R., and MacFarlane, R. G.: *Human Blood Coagulation and Its Disorders,* 3rd Ed. Philadelphia, F. A. Davis Co., 1962.

Birdsell, J. B.: *Human Evolution,* 2nd Ed. New York, Random House, 1975.

Bloom, W., and Fawcett, D.: *Textbook of Histology,* 10th Ed. Philadelphia, W. B. Saunders Co., 1973.

Blum, Harold F.: *Time's Arrow and Evolution,* Rev. Ed. Gloucester, Mass., Peter Smith, 1969.

Bodemer, C. W.: *Modern Embryology.* New York, Holt, Rinehart and Winston, 1968.

Bold, H. C.: *Morphology of Plants,* 3rd Ed. New York, Harper & Row, Publishers, 1974.

Boney, A. D.: *A Biology of Marine Algae.* London, Hutchinson & Co., 1966.

Bonner, J. T.: *The Ideas of Biology.* New York, Harper & Row, Publishers, 1962.

Bonner, J. T.: *The Cellular Slime Molds,* 2nd Ed. Princeton, Princeton University Press, 1966.

Bonner, J., and Galston, A. W.: *Principles of Plant Physiology,* 2nd Ed. San Francisco, W. H. Freeman and Co., 1959.

Bonner, J., and Varner, J. E. (eds.): *Plant Biochemistry.* New York, Academic Press, 1965.

Boolootian, R. A. (ed.): *Physiology of Echinodermata.* New York, John Wiley & Sons, 1966.

Boyd, W. C.: *Introduction to Immunochemical Specificity.* New York, Interscience Publications, 1962.

Boyle, Robert: *The Hudson River.* New York, W. W. Norton & Co., 1969.

Boynton, Holmes: *The Beginnings of Modern Science.* New York, Classics Club, 1948.

Brady, N. C. (ed.): *Agriculture and the Quality of our Environment.* Washington, D. C., Amer. Assoc. Adv. Sci., 1967.

Brady, N. C., and Buckman, H. O.: *The Nature and Properties of Soil,* 6th Ed. New York, The Macmillan Company, 1960.

Braun, E. Lucy: *Deciduous Forests of Eastern North America.* Philadelphia, The Blakiston Company, 1950.

Braun, W.: *Bacterial Genetics,* 2nd Ed. Philadelphia, W. B. Saunders Co., 1965.

Brock, T. D.: *The Principles of Microbial Ecology.* Englewood Cliffs, N. J., Prentice-Hall, 1966.

Brock, T. D.: *Biology of Microorganisms.* Englewood Cliffs, N. J., Prentice-Hall, 1970.

Brown, Harrison: *The Challenge of Man's Future.* New York, Viking Press, 1954.

Brown, Harrison, Bonner, J., and Weir, J.: *The Next 100 Years.* New York, Viking Press, 1957.

Browning, T. O.: *Animal Populations.* New York, Harper & Row, Publishers, 1963.

Buettner-Janusch, John: *Physical Anthropology: A Perspective.* New York, John Wiley & Sons, 1973.

Bullock, T. H., and Horridge, G. A.: *Structure and Function in the Nervous Systems of Invertebrates.* 2 Vols. San Francisco, W. H. Freeman & Co., 1969.

Bullock, William: *The History of Bacteriology.* London, Oxford University Press, 1938.

Burnett, J. H.: *Fundamentals of Mycology.* New York, St. Martin's Press, 1968.

Calder, N.: *The Restless Earth.* New York, Viking Press, 1972.

Calvin, Melvin: *Chemical Evolution.* New York, Oxford University Press, 1969.

Campbell, A.: *Episomes.* New York, Harper & Row, Publishers, 1969.

Cannon, W. B.: *The Wisdom of the Body.* New York, W. W. Norton & Co., 1932.

Cannon, W. B.: *The Way of an Investigator.* New York, W. W. Norton & Co., 1945.

Carlquist, Sherwin: *Island Life, A Natural History of the Islands of the World.* Garden City, N. Y., Doubleday & Co., 1965.

Carlson, E. A.: *The Gene: A Critical History.* Philadelphia, W. B. Saunders Co., 1966.

Carpenter, P. L.: *Microbiology,* 3rd Ed. Philadelphia, W. B. Saunders Co., 1972.

Carpenter, P. L.: *Immunology and Serology,* 3rd Ed. Philadelphia, W. B. Saunders Co., 1975.

Carson, Rachel L.: *The Sea Around Us.* New York, Oxford University Press, 1951.

Carson, Rachel L.: *The Edge of the Sea.* Boston, Houghton Mifflin, 1955.

Carson, Rachel L.: *Silent Spring.* Boston, Houghton Mifflin, 1962.

Carter, G. S.: *Animal Evolution: A Study of Recent Views of Its Causes.* London, Sidgwick & Jackson, 1951.

Chapman, R. F.: *The Insects: Structure and Function.* New York, American Elsevier Publishing Co., 1969.

Christensen, H. N.: *Neutrality Control in the Living Organism.* Philadelphia, W. B. Saunders Co., 1971.

Clark, W. E. LeGros: *The History of the Primates.* London, British Museum, 1949.

Cloudsley-Thompson, J. L.: *Spiders, Scorpions, Centipedes and Mites.* New York, Pergamon Press, 1958.

Cobb, Stanley: *Foundations of Neuropsychiatry,* 5th Ed. Baltimore, Williams & Wilkins Co., 1952.

Cochrane, D. M.: *Living Amphibians of the World.* New York, Doubleday & Co., 1961.

Cohen, I. B.: *Science, Servant of Man.* Boston, Little, Brown & Co., 1948.

Coker, R. E.: *Lakes, Streams and Ponds.* Chapel Hill, University of North Carolina Press, 1954.

Colbert, E. H.: *Evolution of the Vertebrates,* 2nd Ed. New York, John Wiley & Sons, 1969.

Comfort, A.: *Aging: The Biology of Senescence.* New York, Holt, Rinehart and Winston, 1964.

Comfort, A.: *The Nature of Human Nature.* New York, Harper & Row, Publishers, 1967.

Commoner, Barry: *The Closing Circle.* New York, Alfred A. Knopf, 1971.

Conant, J. B.: *On Understanding Science.* New Haven, Yale University Press, 1947.

Conant, J. B.: *Science and Common Sense.* New Haven, Yale University Press, 1951.

Constantinides, P. C., and Carey, N.: *The Alarm Reaction.* Sci. Amer. *180* (3), 20 (March, 1949).

Corner, E. J. H.: *The Life of Plants.* New York, Mentor Books, New American Library, 1968.

Cott, H. B.: *Adaptive Coloration in Animals.* Oxford, Oxford University Press, 1940.

Cox, George W. (ed.): *Readings in Conservation Ecology.* New York, Appleton-Century-Crofts, 1969.

Crick, F. H. C.: *Of Molecules and Men.* Seattle, University of Washington Press, 1966.

Cronquist, A.: *The Evolution and Classification of Flowering Plants.* Boston, Houghton Mifflin Co., 1968.

Crow, James F.: *Ionizing Radiation and Evolution.* Sci. Amer. *201* (3), 138 (September, 1959).

Curtis, H.: *The Viruses.* New York, Natural History Press, 1966.

Dales, R. P.: *Annelids.* London, Hutchinson University Library, 1963.

Darlington, C. D.: *The Evolution of Man and Society.* New York, Simon and Schuster, 1969.

Darwin, Charles: *The Origin of Species.* 1859. Available in a number of recent reprint editions.

Daubenmire, R.: *Plant Communities.* New York, Harper & Row, Publishers, 1968.

Davson, H.: *A Textbook of General Physiology,* 3rd Ed. Boston, Little, Brown & Co., 1964.

Dawson, E. Y.: *Marine Botany: An Introduction.* New York, Holt, Rinehart and Winston, 1966.

DeHaan, R. L., and Ursprung, H. (eds.): *Organogenesis.* New York, Holt, Rinehart and Winston, 1965.

Delevoryas, T.: *Plant Diversification.* New York, Holt, Rinehart and Winston, 1966.

DeRobertis, E. D. P., Nowinski, W. W., and Saez, F. A.: *Cell Biology,* 6th Ed. Philadelphia, W. B. Saunders Co., 1975.

Dethier, V. G.: *The Physiology of Insect Senses.* New York, Barnes & Noble, 1963.

Dethier, V. G.: *The Hungry Fly.* Cambridge, Harvard University Press, 1976.

Dethier, V. G., and Stellar, E.: *Animal Behavior,* 3rd Ed. Englewood Cliffs, N. J., Prentice-Hall, 1970.

Deutsch, J. A.: *The Structural Basis of Behavior.* Cambridge, Cambridge University Press, 1960.

Diamond, M. (ed.): *Perspectives in Reproduction and Sexual Behavior.* Bloomington, University of Indiana Press, 1968.

Dixon, M., and Webb, E. C.: *The Enzymes,* 2nd Ed. New York, Academic Press, 1964.

Dobell, Clifford: *Antony van Leeuwenhoek and His "Little Animals."* London, John Bale Medical Pub., 1932.

Dobzhansky, Theodosius: *Genetics and the Origin of Species,* 3rd Ed. New York, Columbia University Press, 1951.

Dobzhansky, Theodosius: *Evolution, Genetics and Man.* New York, John Wiley & Sons, 1955.

Dodson, Edward O., and Dodson, Peter.: *Evolution: Process and Product,* 2nd Ed. New York, D. Van Nostrand Co., 1976.

Doutt, R. L., and Kilgore, W. W.: *Pest Control.* New York, Academic Press, 1967.

Doyle, W. T.: *Nonvascular Plants: Form and Function.* Belmont, Calif., Wadsworth Publishing Co., 1965.

Drill, V. A.: *The Oral Contraceptives.* New York, McGraw-Hill Book Co., 1966.

Dubos, René J.: *Louis Pasteur, Free Lance of Science.* Boston, Little, Brown & Co., 1950.

Du Praw, E. J.: *Cell and Molecular Biology.* New York, Academic Press, 1968.

Ehrlich, Paul: *The Population Bomb.* New York, Ballantine Books, 1968.

Ehrlich, Paul, and Ehrlich, Anne: *Population, Resources, Environment: Issues in Human Ecology,* 2nd Ed. San Francisco, W. H. Freeman & Co., 1972.

Ehrlich, P. R., and Holm, R. W.: *The Process of Evolution.* New York, McGraw-Hill Book Co., 1963.

Eiseley, L. C.: *Darwin's Century: Evolution and the Men Who Discovered It.* Garden City, N. Y., Doubleday and Co., 1958.

Elton, C.: *The Ecology of Animals,* 4th Ed. New York, Barnes and Noble, 1969.

Emmel, T. C.: *An Introduction to Ecology and Populations.* New York, W. W. Norton & Co., 1973.

Englemann, F.: *Physiology of Insect Reproduction.* New York, Pergamon Press, 1970.

Esau, Katherine: *Anatomy of Seed Plants.* New York, John Wiley & Sons, 1960.

Etkin, William: *Social Behavior and Organization Among Vertebrates.* Chicago, University of Chicago Press, 1964.

Evans, H. E.: *Life on a Little-Known Planet.* New York, E. P. Dutton Co., 1968.

Eyzaguirre, C.: *Physiology of the Nervous System.* Chicago, Year Book Medical Publishers, 1969.

Falkner, Frank (ed.): *Human Development.* Philadelphia, W. B. Saunders Co., 1966.

Farb, P.: *The Forest.* New York, Time-Life Nature Library, 1961.

Fawcett, D.: *The Cell.* Philadelphia, W. B. Saunders Co., 1966.

Feibleman, J. K.: *Testing Hypotheses by Experiment.* Perspectives in Biology and Medicine *4,* 91 (1960).

Florey, E.: *General and Comparative Animal Physiology.* Philadelphia, W. B. Saunders Co., 1966.

Fogg, G. E.: *Photosynthesis,* 2nd Ed., New York, American Elsevier Publishing Co., 1972.

Ford, Thomas R., and de Jong, Gordon F. (eds.): *Social Demography.* Englewood Cliffs, N. J., Prentice-Hall, 1970.

Foster, A. S., and Gifford, E. M., Jr.: *Comparative Morphology of Vascular Plants,* 2nd Ed. San Francisco, W. H. Freeman and Co., 1974.

Freedman, R., and Morris, J. E.: *The Brains of Animals and Man.* New York, Holiday House, 1972.

Frisch, L.: *The Genetic Code.* Cold Spring Harbor Symposia on Quantitative Biology, Vol. 31, 1967.

Frobisher, M.: *Fundamentals of Microbiology,* 9th Ed. Philadelphia, W. B. Saunders Co., 1974.

Fulton, John: *Selected Readings in the History of Physiology.* Springfield, Ill., Charles C Thomas, 1930.

Gabriel, M. L., and Fogel, S.: *Great Experiments in Biology,* New York, Prentice-Hall, 1955.

Gaebler, O. H. (ed.): *Enzymes; Units of Biological Structure and Function.* New York, Academic Press, 1956.

Gaito, J.: *Macromolecules and Behavior.* Amsterdam, North-Holland Publishing Co., 1966.

Galston, A. W.: *The Green Plant.* Englewood Cliffs, N. J., Prentice-Hall, 1968.

Galston, A. W., and Davies, P. J.: *Hormonal Regulation in Higher Plants.* Science *163,* 1288–1297, 1969.

Gamow, G.: *Mr. Thompkins Explores the Atom.* New York, Cambridge University Press, 1967.

Gardner, Martin: *Fads and Fallacies In the Name of Science.* New York, Dover Publications, 1957.

Garrett, H. E.: *Great Experiments in Psychology,* 3rd Ed. New York, Appleton-Century-Crofts, 1951.

Gates, D. M.: *Energy Exchange in the Biosphere.* New York, Harper & Row, Publishers, 1962.

Giese, Arthur C.: *Cell Physiology,* 4th Ed. Philadelphia, W. B. Saunders Co., 1974.

Glaessner, Martin: *Precambrian Animals.* Sci. Amer. *204* (3), 72 (March, 1961).

Goldschmidt, Richard B.: *The Material Basis of Evolution.* New Haven, Yale University Press, 1940.

Gordon, M. S.: *Animal Function: Principles and Adaptations.* New York, The Macmillan Company, 1968.

Graham, Frank, Jr.: *Since Silent Spring.* Boston, Houghton Mifflin, 1970.

Gray, W. D., and Alexopoulos, C. J.: *Biology of the Myxomycetes.* New York, Ronald Press, 1968.

Greenberg, D. M.: *Metabolic Pathways,* 3rd Ed. (3 Vols.). New York, Academic Press, 1967–1969.

Gregory, R. L.: *Eye and Brain: The Psychology of Vision,* 2nd Ed. New York, McGraw-Hill Book Co., 1973.

Gregory, W. K.: *Evolution Emerging; A Survey of Changing Patterns From Primeval Life to Man.* New York, The Macmillan Company, 1951.

Greulach, V. A.: *Plant Structure and Function.* New York, The Macmillan Company, 1973.

Grobstein, C.: *The Strategy of Life.* San Francisco, W. H. Freeman and Co., 1965.

Guthrie, Douglas: *A History of Medicine.* Philadelphia, J. B. Lippincott Co., 1946.

Guyton, A. C.: *Basic Human Physiology.* Philadelphia, W. B. Saunders Co., 1971.

Guyton, A. C.: *Function of the Human Body,* 4th Ed. Philadelphia, W. B. Saunders Co., 1974.

Hall, Thomas S.: *A Source Book in Animal Biology.* New York, McGraw-Hill Book Co., 1951.

Hamilton, T. H.: *Process and Patterns in Evolution.* New York, The Macmillan Company, 1967.

Handler, Philip (ed.): *Biology and the Future of Man.* New York, Oxford University Press, 1970.

Hardin, Garrett (ed.): *Population, Evolution and Birth Control.* San Francisco, W. H. Freeman and Co., 1969.

Hardy, A. C.: *The Open Sea: The World of Plankton.* New York, Houghton Mifflin, 1957.

Hardy, A. C.: *The Open Sea: Fish and Fisheries.* New York, Houghton Mifflin, 1969.

Harris, J. W.: *The Red Cell.* Cambridge, Harvard University Press, 1963.

Hartman, C. G.: *Possums.* Austin, University of Texas Press, 1953.

Harvey, E. N.: *Bioluminescence.* New York, Academic Press, 1952.

Harvey, William: *Anatomical Studies on the Motion of the Heart and Blood.* (Translated by C. D. Leake). Springfield, Ill., Charles C Thomas, 1931.

Hayes, W.: *The Genetics of Bacteria and Their Viruses,* 3rd Ed. New York, John Wiley & Sons, 1973.

Hayflick, L.: *Human Cells and Aging.* Sci. Amer. *218,* 32 (March, 1968).

Hazen, W. E.: *Readings in Population and Community Ecology,* 3rd Ed. Philadelphia, W. B. Saunders Co., 1975.

Heiser, C. B., Jr.: *Seed to Civilization, the Story of Man's Food.* San Francisco, W. H. Freeman and Co., 1973.

Henderson, L. J.: *The Fitness of the Environment.* New York, The Macmillan Company, 1913.

Herskowitz, I. H.: *Genetics.* Boston, Little, Brown & Co., 1965.

Hillman, W. S.: *The Physiology of Flowering.* New York, Holt, Rinehart and Winston, 1962.

Hinde, R. A.: *Animal Behavior,* 2nd Ed. New York, McGraw-Hill Book Co., 1970.

Hirsch, J. (ed): *Behavior-Genetic Analysis.* New York, McGraw-Hill Book Co., 1967.

Hoar, W. S.: *General and Comparative Physiology.* Englewood Cliffs, N. J., Prentice-Hall, 1966.

Hodgkin, A. L.: *The Conduction of the Nerve Impulse.* Springfield, Ill., Charles C Thomas, 1964.

Hokin, L. E. (ed.): *Metabolic Transport.* New York, Academic Press, 1972.

Holton, Gerald: *The Making of Modern Science — Biographical Sketches.* Boston, Houghton Mifflin, 1971.

Howells, W. W.: *Mankind in the Making,* Rev. Ed. Garden City, N.Y., Doubleday and Co., 1967.

Howells, William: *The Evolution of the Genus* Homo. Reading, Mass., Addison-Wesley, 1973.

Hulse, F. S.: *The Human Species,* 2nd Ed. New York, Random House, 1971.

Hyman, Libbie H.: *The Invertebrates,* Vols. 1–6. New York, McGraw-Hill Book Co., 1940–1967.

Ingram, V. M.: *The Biosynthesis of Macromolecules.* New York, W. A. Benjamin, 1965.

Irvine, William: *Apes, Angels, and Victorians.* New York, McGraw-Hill Book Co., 1955.

Isaacs, Alick: *Interferon.* Sci. Amer. *204* (5), 51 (May, 1961).

Jackson, D. F. (ed.): *Algae, Man and the Environment.* Syracuse, Syracuse University Press, 1968.

Jackson, R. M., and Raw, F.: *Life in the Soil.* New York, St. Martin's Press, 1966.

Jacobson, L. O., and Doyle, M. (eds.): *Erythropoiesis.* New York, Grune and Stratton, 1962.

Jaeger, E. C.: *The North American Deserts.* Palo Alto, Stanford University Press, 1957.

Jolly, Alison: *The Evolution of Primate Behavior.* New York, The Macmillan Company. 1972.

Jones, M., Netterville, J. T., Johnston, D. O., and Wood, J. L.: *Chemistry, Man and Society,* 2nd Ed. Philadelphia, W. B. Saunders Co., 1976.

Jukes, T. H.: *Molecules and Evolution.* New York, Columbia University Press, 1966.

Kaestner, A.: *Invertebrate Zoology.* (3 Vols.). New York, Wiley-Interscience, 1967–1969.

Katz, B.: *Nerve, Muscle and Synapse.* New York, McGraw-Hill Book Co., 1966.

Kennedy, D. (ed.): *The Living Cell: Readings from* Scientific American. San Francisco, W. H. Freeman and Co., 1965.

Keosian, J.: *The Origin of Life,* 2nd Ed. New York, Van Nostrand-Reinhold Publishing Co., 1968.

Kistner, R.: *The Pill.* New York, Delacorte Press, 1969.

Kormondy, E. J.: *Concepts of Ecology.* Englewood Cliffs, N. J., Prentice-Hall, Inc., 1969.

Kornberg, A.: *DNA Synthesis.* San Francisco, W. H. Freeman and Co., 1974.

Kraybill, H. F. (ed.): *Biological Effects of Pesticides in Mammalian Systems.* New York, N.Y., Acad. Sci., 1969.

Krebs, C. J.: *Ecology: The Experimental Analysis of Distribution and Abundance.* New York, Harper & Row, Publishers, 1972.

Kummer, Hans: *Primate Societies: Group Techniques of Ecological Adaptation.* Chicago, Aldine, Atherton, Inc., 1971.

Lack, D.: *Darwin's Finches.* Sci. Amer. *188* (4), 66 (April, 1953).

Lack, D.: *The Natural Regulation of Animal Numbers.* New York, Oxford University Press, 1954.

Lasker, G. W.: *Physical Anthropology,* 2nd Ed. New York, Holt, Rinehart and Winston, 1976.

Lauff, George (ed.): *Estuaries.* Washington, D.C., Amer. Assoc. Adv. Sci., Publ. 83, 1967.

Ledbetter, M. C., and Porter, K.: *Introduction to the Fine Structure of Plant Cells.* New York, Springer-Verlag, 1970.

Lehninger, A. L.: *The Mitochondrion.* New York, W. A. Benjamin, 1964.

Lehninger, A. L.: *Bioenergetics.* New York, W. A. Benjamin, 1965.

Lehninger, A. L.: *Biochemistry*, 2nd Ed. New York, Worth Publications, 1975.

Leopold, A. C.: *The Desert*. New York, Time-Life Nature Library, 1961.

Leopold, A. C.: *Plant Growth and Development*. New York, McGraw-Hill Book Co., 1964.

Leopold, Aldo: *A Sand County Almanac*. New York, Oxford University Press, 1949.

Lerner, A. B.: *Hormones and Skin Color*. Sci. Amer. *205* (1), 87 (July, 1961).

Levey, R. H.: *The Thymus Hormone*. Sci. Amer. *211* (1), 104 (July, 1964).

Ley, Willy: *The Poles*. New York, Time-Life Nature Library, 1962.

Lindauer, M.: *Communication Among Social Bees*. Cambridge, Harvard University Press, 1961.

Loewy, A. G., and Siekevitz, P.: *Cell Structure and Function*, 2nd Ed. New York, Holt, Rinehart and Winston, 1969.

Lorenz, Konrad: *King Solomon's Ring*. New York, Thomas Y. Crowell Company, 1952.

Lorenz, Konrad: *On Aggression*. New York, Harcourt, Brace & World, Inc., 1966.

Lull, R. S.: *Organic Evolution*, 2nd Ed. New York, The Macmillan Company, 1947.

Luria, S. E.: *Life: The Unfinished Experiment*. New York, Charles Scribner's Sons, 1973.

Luria, S. E., and Darnell, J. E.: *General Virology*, 2nd Ed. New York, John Wiley & Sons, 1967.

MacArthur, R. H., and Connell, J. H.: *The Biology of Populations*. New York, John Wiley & Sons, 1969.

MacArthur, R. H., and Wilson, E. O.: *The Theory of Island Biogeography*. Princeton, Princeton University Press, 1967.

MacGinitie, G. E.: *Natural History of Marine Animals*. New York, McGraw-Hill Book Co., 1949.

McCormack, Arthur: *The Population Problem*. New York, Thomas Y. Crowell Company, 1970.

McElroy, W. D., and Glass, B. (eds.): *Light and Life*. Baltimore, The Johns Hopkins Press, 1961.

McGaugh, J. L., Weinberger, N. M., and Whalen, R. E.: *Psychobiology*. San Francisco, W. H. Freeman and Co., 1967.

McGill, T. E.: *Readings in Animal Behavior*. New York, Holt, Rinehart and Winston, 1965.

McGilvery, R. W.: *Biochemistry*. Philadelphia, W. B. Saunders Co., 1970.

McGilvery, R. W.: *Biochemical Concepts*. Philadelphia, W. B. Saunders Co., 1975.

McHarg, Ian: *Design with Nature*. Garden City, N.Y., Natural History Press, 1969.

Malin, James: *The Grasslands of North America*. Lawrence, Kan., James Malin, 1956.

Manning, A.: *An Introduction to Animal Behavior*, 2nd Ed. London, Edward Arnold Publishers, 1972.

Marine, G.: *America The Raped*. New York, Simon and Schuster, 1969.

Markert, C. L. (ed.): *Isozymes*. New York, Academic Press, 1975.

Marler, P., and Hamilton, W. J.: *Mechanisms of Animal Behavior*. New York, John Wiley & Sons, 1966.

Marshall, N. B.: *The Life of Fishes*. Cleveland, World Publishing Co., 1966.

Masters, W. H., and Johnson, V. F.: *Human Sexual Response*. Boston, Little, Brown & Co., 1966.

Matthews, G. V.: *Bird Navigation*, 2nd Ed. Cambridge, Cambridge University Press, 1968.

Mayr, E.: *Animal Species and Evolution*. Cambridge, Harvard University Press, 1963.

Mazzeo, J.: *The Design of Life*. New York, Pantheon Books, 1967.

Meglitsch, P. A.: *Invertebrate Zoology*, 2nd Ed. New York, Oxford University Press, 1972.

Merrill, J. P.: *The Artificial Kidney*. Sci. Amer. *205* (1), 53 (July, 1961).

Metz, C. B., and Monroy, A. (eds.): *Fertilization*. New York, Academic Press, 1967.

Meyer, B. S., Anderson, D. B., and Böhning, R. H.: *Introduction to Plant Physiology.* New York, J. D. Van Nostrand Co., 1960.

Miller, M. W., and Berg, G. G. (eds.): *Chemical Fallout.* Springfield, Ill., Charles C Thomas, 1969.

Milne, L. J., and Milne, M. J.: *The Mating Instinct.* Boston, Little, Brown & Co., 1954.

Monroy, A., and Moscona, A. A.: *Current Topics in Developmental Biology.* New York, Academic Press, published annually.

Moore, J. A.: *Heredity and Development.* New York, Oxford University Press, 1972.

Moore, K. L. (ed.): *The Sex Chromatin.* Philadelphia, W. B. Saunders Co., 1966.

Moore, R. C., Lalicker, C. G., and Fischer, A. G.: *Invertebrate Fossils.* New York, McGraw-Hill Book Co., 1952.

Moorhead, A.: *Darwin and the Beagle.* Harmondsworth, England, Penguin Books, 1971.

Moran, J. M., Morgan, M. D., and Wiersma, J. H.: *An Introduction to Environmental Sciences.* Boston, Little, Brown & Co., 1973.

Morton, J. E.: *Molluscs,* 4th Ed. London, Hutchinson University Library, 1967.

Moscona, A. A.: *Tissues from Dissociated Cells.* Sci. Amer. *200* (5), 132 (May, 1959).

Neurath, H.: *Protein-Digesting Enzymes.* Sci. Amer. *211* (6), 58 (December, 1964).

Nichols, D.: *Echinoderms.* London, Hutchinson University Library, 1966.

Nirenberg, M. W.: *The Genetic Code II.* Sci. Amer. *208* (3), 57 (March, 1963).

Noble, E. R., and Noble, G. A.: *Parasitology,* 3rd Ed. Philadelphia, Lea and Febiger, 1971.

Nordenskiold, E.: *The History of Biology.* New York, Tudor Publishing Co., 1966.

Norstog, K., and Long, R. W.: *Plant Biology.* Philadelphia, W. B. Saunders Co., 1976.

Oatley, Keith: *Brain Mechanisms and Mind.* New York, E. P. Dutton Co., 1972.

Odum, E. P.: *Fundamentals of Ecology,* 3rd Ed. Philadelphia, W. B. Saunders Co., 1971.

Olds, J.: *Pleasure Centers in the Brain.* Sci. Amer. *195* (4), 105 (October, 1956).

Olson, E. E.: *Vertebrate Paleozoology.* New York, Wiley-Interscience, 1971.

Oosting, H. J.: *The Study of Plant Communities,* 3rd Ed. San Francisco, W. H. Freeman and Co., 1969.

Oparin, A. I.: *Life, Its Nature, Origin and Development.* New York, Academic Press, 1962.

Orr, R. T.: *Vertebrate Biology,* 4th Ed. Philadelphia, W. B. Saunders Co., 1976.

Osborn, Fairfield: *Our Plundered Planet.* Boston, Little, Brown & Co., 1948.

Osborn, Frederick: *Preface to Eugenics,* 2nd Ed. New York, Harper & Bros., 1951.

Osborn, Henry Fairfield: *From the Greeks to Darwin.* New York, The Macmillan Company, 1913.

Osborn, Henry Fairfield: *Men of the Old Stone Age,* 3rd Ed. New York, Charles Scribner's Sons, 1918.

Page, E. W., Villee, C. A., and Villee, D. B.: *Human Reproduction,* 2nd Ed. Philadelphia, W. B. Saunders Co., 1976.

Pavlov, I. P.: *Conditioned Reflexes.* New York, International Publishers, 1941.

Penrose, L. S.: *Outline of Human Genetics.* New York, John Wiley & Sons, 1959.

Peters, J. A. (ed.): *Classic Papers in Genetics.* Englewood Cliffs, N. J., Prentice-Hall, 1960.

Pfeiffer, J. E.: *The Emergence of Man,* 2nd Ed. New York, Harper & Row, Publishers, 1972.

Pilbeam, David: *The Ascent of Man.* New York, The Macmillan Company, 1972.

Pitts, R. F.: *The Physiology of the Kidney and Body Fluids,* 2nd Ed. Chicago, Year Book Medical Publishers, 1968.

Ponnamperuma, C.: *The Origins of Life.* New York, E. O. Dutton & Co., 1972.

Porter, K. R.: *Herpetology.* Philadelphia, W. B. Saunders Co., 1972.

Porter, K. R., and Bonneville, M. A.: *Fine Structure of Cells and Tissues,* 3rd Ed. Philadelphia, Lea & Febiger, 1968.

Prosser, C. L., and Brown, F. A.: *Comparative Animal Physiology,* 3rd Ed. (2 Vols.). Philadelphia, W. B. Saunders Co., 1973.

Pycraft, W. P.: *The Courtship of Animals,* 2nd Ed. London, Hutchinson & Co., 1933.

Racker, E.: *Mechanisms in Bioenergetics.* New York, Academic Press, 1965.

Ranson, S. W., and Clark, S. L.: *The Anatomy of the Nervous System,* 10th Ed. Philadelphia, W. B. Saunders Co., 1959.

Raven, P., and Curtis, H.: *Biology of Plants.* New York, Worth Publications, 1970.

Ray, Peter M.: *The Living Plant,* 2nd Ed. New York, Holt, Rinehart and Winston, 1971.

Raymond, P. E.: *Prehistoric Life.* Cambridge, Harvard University Press, 1939.

Rebuck, J. W. (ed.): *The Lymphocyte and Lymphocytic Tissue.* New York, Paul B. Hoeber, 1960.

Reed, S. C.: *Counseling in Medical Genetics,* 2nd Ed. Philadelphia, W. B. Saunders Co., 1963.

Reid, George: *Ecology of Inland Waters and Estuaries.* New York, Reinhold Publishing Corp., 1961.

Rich, A.: *Polyribosomes.* Sci. Amer. *209* (6), 68 (December, 1963).

Richard, P. W.: *The Tropical Rain Forest.* New York, Cambridge University Press, 1952.

Richardson, M.: *Translocation in Plants.* New York, St. Martin's Press, 1968.

Ricketts, E. F., and Calvin, J.: *Between Pacific Tides.* Stanford, Stanford University Press, 1939.

Rockstein, M., Sussman, M. L., and Chesky, J. (eds.): *Theoretical Aspects of Aging.* New York, Academic Press, 1974.

Romer, A. S.: *The Vertebrate Story.* Chicago, University of Chicago Press, 1959.

Romer, A. S.: *Vertebrate Paleontology,* 3rd Ed. Chicago, University of Chicago Press, 1966.

Romer, A. S.: *The Vertebrate Body,* 4th Ed. Philadelphia, W. B. Saunders Co., 1970.

Rosebury, Theodor: *Peace or Pestilence.* New York, Whittlesey House, 1949.

Rosenberg, J. L.: *Photosynthesis.* New York, Holt, Rinehart and Winston, 1965.

Ross, H. H.: *A Synthesis of Evolutionary Theory.* Englewood Cliffs, N. J., Prentice-Hall, 1964.

Rothfield, L. I. (ed.): *Structure and Function of Biological Membranes.* New York, Academic Press, 1971.

Roueché, Berton: *What's Left.* Boston, Little, Brown & Co., 1968.

Round, F. E.: *The Biology of the Algae.* London, Edward Arnold, 1965.

Ruch, T. C., and Patton, H. D.: *Physiology and Biophysics,* 20th Ed. (3 Vols.). Philadelphia, W. B. Saunders Co., 1973.

Ruch, T. C., Patton, H. D., Woodbury, J. W., and Towe, A. L.: *Neurophysiology,* 2nd Ed. Philadelphia, W. B. Saunders Co., 1965.

Rugh, Roberts, and Shettles, Landrum B.: *From Conception to Birth: The Drama of Life's Beginnings.* New York, Harper & Row, 1971.

Runcorn, S. K.: *Corals as Paleontologic Clocks.* Sci. Amer. *215* (4), 26 (October, 1966).

Rustad, Ronald C.: *Pinocytosis.* Sci. Amer. *204*(4), 120 (April, 1961).

Sagan, Carl: *The Cosmic Connection.* Garden City, N. Y., Doubleday and Co., 1973.

Salisbury, F. B., and Ross, C.: *Plant Physiology.* Belmont, Calif., Wadsworth Publishing Co., 1969.

Savage, J. M.: *Evolution.* New York, Holt, Rinehart and Winston, 1963.

Scharrer, E., and Scharrer, B.: *Neuroendocrinology.* New York, Columbia University Press, 1963.

Schmidt-Nielsen, Knut: *How Animals Work.* New York, Cambridge University Press, 1972.

Schmitt, W. L.: *Crustaceans.* Ann Arbor, University of Michigan Press, 1965.

Schneiderman, H. A., and Gilbert, L. I.: *Control of Growth and Development of Insects.* Science *143*, 373 (January 24, 1964).

Schoenheimer, Rudolph: *The Dynamic State of the Body Constituents.* Cambridge, Harvard University Press, 1949.

Schrier, A. M., Harlow, H. F., and Stollnitz, F. (eds.): *The Behavior of Nonhuman Primates.* New York, Academic Press, 1965.

Schroedinger, E.: *What is Life?* Cambridge, Cambridge University Press, 1944.

Schultz, L. P.: *The Ways of Fishes.* New York, D. Van Nostrand Co., 1948.

Sears, Paul B.: *Deserts on the March.* Norman, Okla., University of Oklahoma Press, 1935.

Sedgwick, W. T., Tyler, H. V., and Bigelow, R. P.: *A Short History of Science.* New York, The Macmillan Company, 1939.

Sherfey, Mary Jane: *The Nature and Evolution of Female Sexuality.* New York, Vintage Books, 1973.

Shrock, R. R., and Twenhofel, W. H.: *Principles of Invertebrate Paleontology.* New York, McGraw-Hill Book Co., 1953.

Singer, Charles: *A History of Biology,* Rev. Ed. New York, Abelard-Schumann, 1959.

Slobodkin, L. B.: *Growth and Regulation of Animal Populations.* New York, Henry Holt & Co., 1961.

Sluckin, W.: *Imprinting and Early Learning.* London, Methuen & Co., 1964.

Smith, Geeds: *Plague on Us.* New York, Commonwealth Fund, 1941.

Smith, Homer W.: *Principles of Renal Physiology.* New York, Oxford University Press, 1956.

Smith, Robert L.: *Ecology and Field Biology,* 2nd Ed. New York, Harper & Row, Publishers, 1974.

Smith, T. L., and Zopf, P. E., Jr.: *Demography: Principles and Practices.* Philadelphia, F. A. Davis Co., 1970.

Solomon, A. K.: *Pores in the Cell Membrane.* Sci. Amer. *203* (6), 146 (December, 1960).

Speirs, R. S.: *How Cells Attack Antigens.* Sci. Amer. *212* (2), 118 (February, 1964).

Stanbury, J. B., Wyngaarden, J. B., and Fredrickson, D. S. (eds.): *The Metabolic Basis of Inherited Disease,* 3rd Ed. New York, Blakiston Division, McGraw-Hill Book Co., 1972.

Stanier, R. Y.: *The Microbial World,* 3rd Ed. Englewood Cliffs, N. J., Prentice-Hall, 1970.

Stebbins, G. Ledyard, Jr.: *Variation and Evolution in Plants.* New York, Columbia University Press, 1950.

Stebbins, G. Ledyard, Jr.: *Processes of Organic Evolution,* 2nd Ed. Englewood Cliffs, N. J., Prentice-Hall, 1971.

Steeves, T. A., and Sussex, I. M.: *Patterns in Plant Development.* Englewood Cliffs, N. J., Prentice-Hall, 1972.

Stent, G.: *Molecular Genetics: An Introductory Narrative.* San Francisco, W. H. Freeman and Co., 1971.

Stern, Curt: *Principles of Human Genetics,* 3rd Ed. San Francisco, W. H. Freeman and Co., 1970.

Stevens, S. S., and Warshofsky, F.: *Sound and Hearing.* New York, Time-Life Life Science Library, 1965.

Steward, F. C.: *Plants at Work.* Reading, Mass., Addison-Wesley, 1964.

Steward, F. C.: *Growth and Organization in Plants.* Reading, Mass., Addison-Wesley, 1968.

Steward, W. D. P.: *Nitrogen Fixation in Plants.* London, Athone Press, 1966.

Strehler, B.: *Time, Cells and Aging.* New York, Academic Press, 1962.

Strickberger, M. W.: *Genetics,* 2nd Ed. New York, The Macmillan Company, 1976.

Stryer, L.: *Biochemistry.* San Francisco, W. H. Freeman and Co., 1975.

Sturtevant, A. H.: *A History of Genetics.* New York, Harper & Row, Publishers, 1965.

Sutcliffe, J.: *Plants and Water.* New York, St. Martin's Press, 1968.

Swanson, C. P.: *The Cell,* 3rd Ed. Englewood Cliffs, N. J., Prentice-Hall, Inc., 1969.

Talland, G. A. (ed.): *Human Aging and Behavior.* New York, Academic Press, 1968.

Tepperman, J.: *Metabolic and Endocrine Physiology,* 2nd Ed. Chicago, Year Book Medical Publishers, 1968.

Thomas, L.: *The Lives of a Cell: Notes of a Biology Watcher.* New York, Viking Press, 1974.

Thompson, D'Arcy W.: *On Growth and Form,* Rev. Ed. New York, The Macmillan Company, 1942.

Thompson, J. S., and Thompson, M. W.: *Genetics in Medicine,* 2nd Ed. Philadelphia, W. B. Saunders Co., 1973.

Thorpe, W. H.: *Learning and Instinct in Animals.* Cambridge, Harvard University Press, 1956.

Tinbergen, N.: *The Study of Instinct.* Oxford, Oxford University Press, 1952.

Tinbergen, N.: *Social Behavior in Animals.* New York, John Wiley & Sons, 1953.

Torrey, J. G.: *Development of Flowering Plants.* New York, The Macmillan Company, 1967.

Treshow, M.: *Environment and Plant Response.* New York, McGraw-Hill Book Co., 1970.

Turner, C. D., and Bagnara, J. T.: *General Endocrinology,* 6th Ed. Philadelphia, W. B. Saunders Co., 1976.

Turpin, R., and Lejeune, J.: *Human Afflictions and Chromosomal Aberrations.* London, Pergamon Press, 1969.

Tyler, Albert: *Fertilization and Antibodies.* Sci. Amer. *192* (5), 36 (May, 1955).

Vallery-Radot, R.: *Life of Pasteur.* New York, Doubleday, Doran & Co., 1928.

Vander, A. J., Sherman, J. H., and Luciano, D. S.: *Human Physiology.* New York, McGraw-Hill Book Co., 1970.

Van der Kloot, W. G.: *Behavior.* New York, Holt, Rinehart and Winston, 1968.

Van Dyne, G. (ed.): *The Ecosystem Concept in Natural Resource Management.* New York, Academic Press, 1969.

Van Lawick-Goodall, Jane: *In the Shadow of Man.* Boston, Houghton Mifflin, 1971.

Vaughan, T. A.: *Mammalogy.* Philadelphia, W. B. Saunders Co., 1972.

Villee, C. A. (ed.): *The Control of Ovulation.* London, Pergamon Press, 1961.

Villee, C. A., Walker, W. F., and Barnes, R. B.: *General Zoology,* 4th Ed. Philadelphia, W. B. Saunders Co., 1973.

Villee, D. B.: *Human Endocrinology.* Philadelphia, W. B. Saunders Co., 1975.

Vogt, William: *Road to Survival.* New York, Sloane, 1948.

von Frisch, Karl: *Bees, Their Vision, Chemical Senses and Language.* Ithaca, N. Y., Cornell University Press, 1950.

Waddington, C. H.: *Principles of Development and Differentiation.* New York, The Macmillan Company, 1966.

Wagner, R. H.: *Environment and Man.* New York, W. W. Norton & Co., 1971.

Wald, George: *Innovation in Biology.* Sci. Amer. *199* (3), 100 (September, 1958).

Wald, George: *Light and Life.* Sci. Amer. *201* (4), 92 (October, 1959).

Ward, Barbara: *Spaceship Earth.* New York, Columbia University Press, 1966.

Ward, M. A. (ed.): *Man and His Environment.* New York, Pergamon Press, 1970.

Wardlaw, C. W.: *Morphogenesis in Plants.* London, Methuen & Co., 1968.

Watson, E. V.: *The Structure and Life of Bryophytes,* 2nd Ed. London, Hutchinson & Co., Ltd., 1967.

Watson, J. D.: *The Double Helix.* New York, Atheneum, 1968.

Watson, J. D.: *Molecular Biology of the Gene,* 2nd Ed. New York, W. A. Benjamin, 1971.

Weaver, J. E., and Clement, F. E.: *Plant Ecology,* 3rd Ed. New York, McGraw-Hill Book Co., 1947.

Weber, G. (ed.): *Recent Advances in Enzyme Regulation.* New York, Pergamon Press. Published annually.

Weidenreich, Franz: *Apes, Giants, and Man.* Chicago, University of Chicago Press, 1947.

Wells, M. J.: *Brain and Behavior in Cephalopods.* London, Heinemann Educational, 1962.

Welty, J. C.: *The Life of Birds,* 2nd Ed. Philadelphia, W. B. Saunders Co., 1975.

White, E. H.: *Chemical Background for the Biological Sciences.* Englewood Cliffs, N. J., Prentice-Hall, 1964.

Whitehouse, H. L.: *Towards an Understanding of the Mechanism of Heredity.* New York, St. Martin's Press, 1965.

Wickler, W.: *Mimicry in Plants and Animals.* London, World University Press, 1968.

Wiggers, Carl J.: *The Heart.* Sci. Amer. *196* (5), 87 (May, 1957).

Wightman, W. P. D.: *The Growth of Scientific Ideas.* New Haven, Yale University Press, 1951.

Williams, R. H.: *Textbook of Endocrinology,* 5th Ed. Philadelphia, W. B. Saunders Co., 1974.

Williams-Ellis, A.: *Darwin's Moon.* London and Edinburgh, Blackie & Son, 1966.

Willier, B. H., Weiss, P., and Hamburger, V. (eds.): *Analysis of Development.* Philadelphia, W. B. Saunders Co., 1955.

Wilson, C. L., and Loomis, W. E.: *Botany.* New York, Holt, Rinehart & Winston, 1962.

Wilson, Douglas P.: *They Live in the Sea.* London, Wm. Collins Sons & Co., 1947.

Wilson, Douglas P.: *Life of the Shore and Shallow Sea.* London, Nicholson and Watson, 1951.

Wilson, E. Bright: *An Introduction to Scientific Research.* New York, McGraw-Hill Book Co., 1952.

Wilson, E. O.: *Pheromones.* Sci. Amer. *208* (5), 95 (May, 1963).

Wilson, E. O.: *The Insect Societies.* New York, Academic Press, 1971.

Wilson, E. O.: *Sociobiology.* Cambridge, Harvard University Press, 1975.

Wolfe, A. V., and Crowder, M. A.: *An Introduction to Body Fluid Metabolism.* Baltimore, Williams & Wilkins Co., 1964.

Wolfe, S. L.: *Biology of the Cell.* Belmont, Calif., Wadsworth Publishing Co., 1972.

Wolman, Abel: *Water, Health and Society.* Bloomington, Ind., Indiana University Press, 1969.

Wolstenholme, G. E. W., and O'Connor, C. M. (eds.): *The Biochemistry of Human Genetics.* Ciba Foundation Symposium. Boston, Little, Brown & Co., 1959.

Woodward, John D.: *Biotin.* Sci. Amer. *204* (6), 139 (June, 1961).

Yonge, C. M.: *A Year on the Great Barrier Reef.* New York, Putnam, 1930.

Yound, J. Z.: *The Life of Mammals.* New York, Oxford University Press, 1957.

Yound, J. Z.: *The Life of Vertebrates,* 2nd Ed. Oxford, Clarendon Press, 1963.

Young, W. C. (ed.): *Sex and Internal Secretions,* 3rd Ed. Baltimore, Williams & Wilkins Co., 1961.

Zilboorg, G.: *A History of Medical Psychology.* New York, W. W. Norton and Co., 1941.

Zim, H. S., and Ingle, L.: *Seashores.* Racine, Wis., Western Publishing Co., 1955.

Zimmermann, M. H.: *How Sap Moves in Trees.* Sci. Amer. *208* (3), 87 (March, 1963).

Zinsser, Hans: *Rats, Lice and History.* Boston, Little, Brown & Co., 1935.

Zweifach, Benjamin: *The Microcirculation of the Blood.* Sci. Amer. *200* (1), 54 (January, 1959).

INDEX

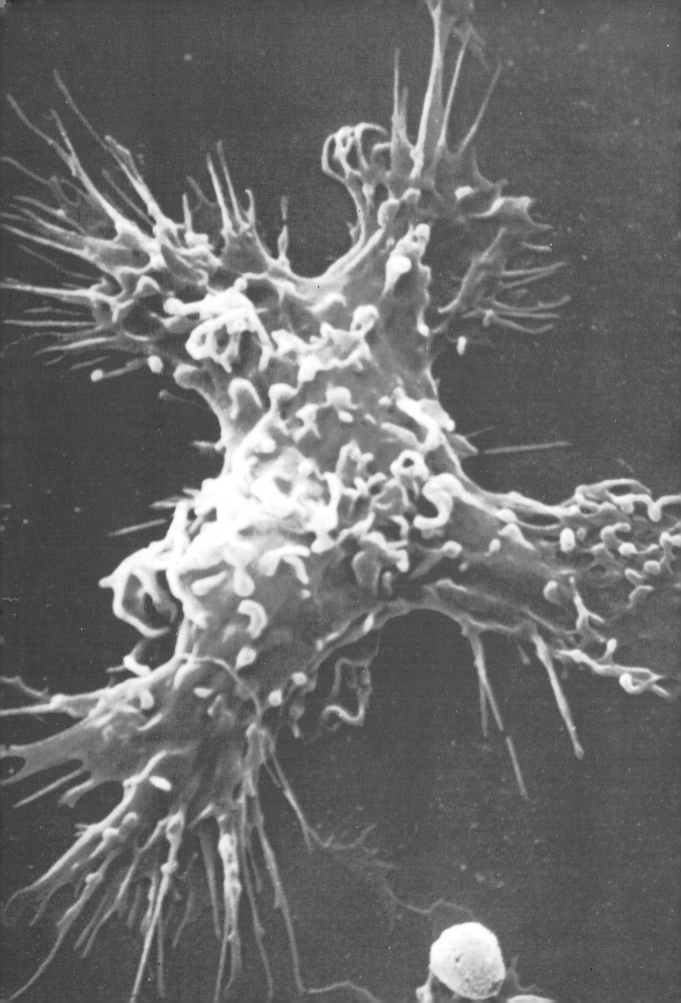